TARGET NEET

Fully Solved

Past Years Solved Papers + 10 Mock Papers

• **Corporate Office :** 45, 2nd Floor, Maharishi Dayanand Marg, Corner Market, Malviya Nagar, New Delhi-110017
Tel. : 011-49842349 / 49842350

By :
Preetima Bajpai
Sanjeev Kumar Jha
Sanju Priyadarshi

Typeset by Disha DTP Team

For further information about the books from DISHA,
Log on to **www.dishapublication.com** or email to **info@dishapublication.com**

Content

TEST SERIES

TREND ANALYSIS OF AIPMT/NEET PAPERS (2009-2019)

PHYSICS

Ch. No.	Chapter Name	Number of Question(s)in											
		2009	**2010**	**2011**	**2012**	**2013**	**2014**	**2015**	**2016 Ph-1**	**2016 Ph-2**	**2017**	**2018**	**2019**
1	Units and Measurements	1	1	1	1	2	1	1	0	1	1	1	2 (Q. 17, 38)
2	Motion in a Straight Line	2	3	2	2	2	0	1	1	1	1	1	1 (Q. 15)
3	Motion in a Plane	2	2	2	2	2	1	1	1	0	1	1	2 (Q. 36, 45)
4	Laws of Motion	1	1	2	0	2	3	2	2	2	2	2	2 (Q. 14, 20)
5	Work, Energy and Power	4	2	3	2	2	1	3	1	4	1	2	2 (Q. 30, 33)
6	System of Particles and Rotational Motion	3	2	2	3	2	2	3	4	3	3	3	3 (Q. 13, 21, 34)
7	Gravitation	1	2	3	3	2	3	2	2	2	2	2	2 (Q. 26, 31)
8	Mechanical Properties of Solids	0	0	0	0	1	1	1	0	0	1	1	1 (Q. 24)
9	Mechanical Properties of Fluids	0	0	0	0	1	1	1	1	2	1	1	2 (Q. 35, 40)
10	Thermal Properties of Matter	2	2	0	2	2	2	2	3	2	2	1	1 (Q. 22)
11	Kinetic Theory	1	0	0	0	1	1	2	1	2	1	1	–
12	Thermodynamics	2	1	2	0	2	2	2	3	1	2	3	2 (Q. 6, 10)
13	Oscillations	2	2	2	0	0	1	2	0	1	2	1	3 (Q. 11, 12, 18)
14	Waves	2	2	1	1	2	–	1	3	2	2	2	–
15	Electric Charges and Fields	1	2	1	2	1	3	1	1	1	1	2	3 (Q. 2, 28,43)
16	Electrostatic Potential and Capacitance	1	1	2	1	1	3	1	1	1	2	–	1 (Q. 37)
17	Current Electricity	5	4	3	4	2	4	3	2	2	2	3	3 (Q. 5, 8, 39)
18	Moving Charges and Magnetism	1	0	2	2	2	2	2	3	3	2	2	2 (Q. 7, 23)
19	Magnetism and Matter	2	2	1	1	1	1	0	1	1	1	1	1 (Q. 3)
20	Electromagnetic Induction	2	5	1	2	1	1	1	0	0	1	1	2 (Q. 16, 44)
21	Alternating Current	1	1	2	1	2	1	1	2	2	1	1	–
22	Electromagnetic Waves	–	–	1	–	1	1	1	1	1	1	1	1 (Q. 9)
23	Ray Optics and Optical Instruments	0	2	2	5	2	1	2	3	4	2	2	3 (Q. 27, 32, 42)
24	Wave Optics	2	1	–	1	2	2	2	2	1	3	3	1 (Q. 4)
25	Dual Nature of Radiation and Matter	3	3	8	3	2	2	2	2	2	2	2	1 (Q. 19)
26	Atoms	2	2	1	3	1	2	2	1	1	1	1	1 (Q. 41)
27	Nuclei	3	2	2	4	2	1	1	1	–	1	1	1 (Q. 29)
28	Semiconductor Electronics : Materials, Devices and Simple Circuits	4	5	4	5	2	2	2	3	3	3	3	2 (Q. 1, 25)
	Total Questions	**50**	**50**	**50**	**50**	**45**	**45**	**45**	**45**	**45**	**45**	**45**	**45**

Note : Number in brackets indicating serial number of questions asked in NEET 2019 Question Paper.

TREND ANALYSIS OF AIPMT/NEET PAPERS (2009-2019)

CHEMISTRY

Ch. No.	Chapter Name	Number of Question(s)in											
		2009	2010	2011	2012	2013	2014	2015	2016 Ph-1	2016 Ph-2	2017	2018	2019
1	Some Basic Concepts of Chemistry	1	1	0	0	1	2	1	0	1	1	2	1 (Q. 90)
2	Structure of Atom	3	1	3	2	3	2	0	1	2	1	1	2 (Q. 75, 89)
3	Classification of Elements and Periodicity in Properties	1	2	0	1	0	1	1	1	0	1	0	1 (Q. 46)
4	Chemical Bonding and Molecular structure	4	4	2	3	4	3	5	2	3	4	2	4 (Q. 52, 53, 65, 86)
5	States of Matter	1	0	4	1	2	1	0	1	0	0	2	1 (Q. 72)
6	Thermodynamics	2	2	2	3	0	2	0	1	1	2	1	2 (Q. 70, 84)
7	Equilibrium	3	6	2	3	2	4	3	2	2	3	3	3 (Q. 51, 54, 88)
8	Redox Reactions	0	1	0	2	0	1	0	1	1	0	1	1 (Q. 69)
9	Hydrogen	0	0	0	0	0	1	0	1	0	0	0	1 (Q. 47)
10	The s-Block Elements	1	2	2	1	0	0	2	1	1	3	3	2 (Q. 48, Q. 74)
11	The p-Block Elements (gp - 13 & 14)	2	1	1	1	3	0	0	0	1	1	2	2 (Q. 80, 87)
12	Organic Chemistry-Some Basic Principles and Techniques	4	7	4	2	4	2	4	0	2	3	2	2 (Q. 64, 85)
13	Hydrocarbons	2	2	1	1	1	1	4	3	4	3	2	2 (Q. 66, 82)
14	Environmental Chemistry	0	0	1	1	1	1	0	1	0	1	1	1 (Q. 57)
15	The Solid State	1	1	0	2	2	1	1	2	1	1	1	1 (Q. 73)
16	Solutions	1	3	3	1	1	0	3	2	2	1	0	2 (Q. 61, 77)
17	Electrochemistry	3	0	3	1	3	2	1	1	5	1	1	2 (Q. 56, 76)
18	Chemical Kinetics	4	3	2	2	2	0	2	2	1	2	2	2 (Q. 68, 83)
19	Surface Chemistry	0	0	1	2	0	1	1	1	2	1	1	1 (Q. 50)
20	General Principles and Processes of Isolation of Elements	0	0	2	3	0	0	1	1	0	1	1	1 (Q. 59)
21	The p-Block Elements (gp - 15, 16, 17 & 18)	3	1	0	2	3	1	1	5	3	2	2	4 (Q. 60, 61, 62, 81)
22	The d and f-Block Elements	2	2	2	1	3	5	5	2	2	2	2	0
23	Coordination Compounds	2	3	3	1	2	2	2	1	1	4	3	1 (Q. 79)
24	Haloalkanes and Haloarenes	1	0	0	0	0	1	1	2	2	0	0	0
25	Alcohols, Phenols and Ethers	3	0	0	1	1	3	1	2	0	3	4	1 (Q. 67)
26	Aldehydes, Ketones and Carboxylic Acids	1	3	1	3	2	1	3	1	1	2	1	0
27	Amines	1	2	2	0	2	2	1	3	3	2	2	2 (Q. 55, 63)
28	Biomolecules	2	1	1	3	0	2	0	3	3	1	2	1 (Q. 78)
29	Polymers	1	1	1	2	2	2	1	1	1	1	1	1 (Q. 71)
30	Chemistry in Everyday Life	1	1	2	0	1	1	1	1	0	1	0	1 (Q. 49)
	Total Questions	**50**	**50**	**50**	**50**	**45**	**45**	**45**	**45**	**45**	**45**	**45**	**45**

Note : Number in brackets indicating serial number of questions asked in NEET 2019 Question Paper.

TREND ANALYSIS OF AIPMT/NEET PAPERS (2009-2019)
BIOLOGY

Ch. No.	Chapter Name	Number of Question(s)in											
		2009	2010	2011	2012	2013	2014	2015	2016 Ph-1	2016 Ph-2	2017	2018	2019
1	The Living World	1	0	1	0	1	0	0	1	1	0	1	1 (Q.180)
2	Biological Classification	3	3	8	4	1	6	2	4	4	4	4	1 (Q.174)
3	Plant Kingdom	3	3	1	3	4	4	4	2	2	4	4	1 (Q.162)
4	Animal Kingdom	2	4	3	1	4	4	3	3	2	5	3	2 (Q.136,157)
5	Morphology of Flowering Plants	2	3	4	7	2	5	5	7	5	2	2	1 (Q.135)
6	Anatomy of Flowering Plants	5	3	4	5	3	2	4	0	3	4	4	3 (Q.133, 140, 173)
7	Structural Organisations in Animals	6	1	3	2	1	2	2	2	2	1	1	3 (Q.107, 163, 175)
8	Cell: The Unit of Life	4	4	4	5	3	4	4	3	3	1	4	3 (Q.155, 164, 165)
9	Biomolecules	0	1	2	3	4	2	1	5	3	2	1	2 (Q.120, 122)
10	Cell Cycle and Cell Division	2	2	1	2	2	3	2	4	3	2	1	2 (Q.104, 123)
11	Transport in Plants	1	0	0	0	1	0	1	1	0	2	2	0
12	Mineral Nutrition	2	3	4	3	1	1	2	1	2	0	1	0
13	Photosynthesis in Higher Plants	2	3	2	2	1	1	2	4	2	1	2	0
14	Respiration in Plants	1	1	0	0	2	1	0	1	1	2	2	1 (Q.91)
15	Plant Growth and Development	2	2	0	0	2	4	2	1	1	1	0	1 (Q.92)
16	Digestion and Absorption	4	2	2	1	1	2	2	2	1	4	1	3 (Q.96, 101, 103)
17	Breathing and Exchange of Gases	0	2	0	1	2	1	2	2	2	1	1	2 (Q.117, 167)
18	Body Fluids and Circulation	5	3	5	0	1	2	3	2	2	1	2	2 (Q.102,144)
19	Excretory Products and their Elimination	2	3	5	1	1	1	2	1	1	1	2	2 (Q.113, 169)
20	Locomotion and Movement	2	0	0	1	3	1	2	1	1	2	2	2 (Q.138, 150)
21	Neural Control and Coordination	2	1	2	3	2	3	2	2	1	3	2	2 (Q.115, 130)
22	Chemical Coordination and Regulation	1	4	2	3	3	2	1	3	4	3	3	2 (Q.108, 154)
23	Reproduction in Organisms	1	0	1	1	3	0	0	1	0	0	3	0

TREND ANALYSIS OF AIPMT/NEET PAPERS (2009-2019)
BIOLOGY

Ch. No.	Chapter Name	Number of Question(s)in											
		2009	2010	2011	2012	2013	2014	2015	2016 Ph-1	2016 Ph-2	2017	2018	2019
24	Sexual Reproduction in Flowering Plants	3	4	5	3	6	3	4	2	4	6	3	6 (Q.109, 112, 119, 128,141,153)
25	Human Reproduction	7	11	3	4	3	3	3	2	5	3	4	2 (Q.111, 172)
26	Reproductive Health	0	1	2	2	3	3	1	2	2	2	1	2 (Q.147,171)
27	Principles of Inheritance and Variation	5	5	4	2	4	4	3	6	3	5	5	6 (Q.99, 114, 126, 134,156,160)
28	Molecular basis of Inheritance	5	4	1	4	2	4	6	3	4	7	5	6 (Q.131, 132, 137, 139, 166, 168)
29	Evolution	1	1	1	4	4	2	1	3	3	0	3	4 (Q.152,159, 170,179)
30	Human Health and Disease	5	3	5	6	2	2	6	3	2	3	5	5 (Q.94, 95, 116, 148, 161)
31	Strategies for Enhancement in Food Production	2	2	3	2	1	1	2	1	3	2	1	3 (Q.93, 125, 176)
32	Microbes in Human Welfare	0	3	2	7	2	1	0	2	1	3	1	6 (Q.98, 118, 121, 124, 143, 158)
33	Biotechnology: Principles and Processes	1	5	2	5	2	3	0	3	4	4	2	2 (Q.97, 178)
34	Biotechnology and its Applications	3	3	3	1	1	2	3	0	1	1	2	1 (Q.106)
35	Organisms and Populations	2	2	4	0	2	4	2	3	4	3	4	2 (Q.105, 142)
36	Ecosystem	2	2	2	6	3	1	6	2	2	1	1	2 (Q.100, 110)
37	Biodiversity and its Conservation	2	1	3	2	1	4	2	2	3	3	2	1 (Q.127)
38	Environmental Issues	8	4	4	3	3	2	3	3	2	1	4	6 (Q.129, 145, 146, 149, 151, 177)
	Total Questions	**100**	**100**	**100**	**100**	**90**	**90**	**90**	**90**	**90**	**90**	**90**	**90**

Note : Number in brackets indicating serial number of questions asked in NEET 2019 Question Paper.

NEET 2019 Solved Paper

Held on : 5th May 2019

PHYSICS

1. 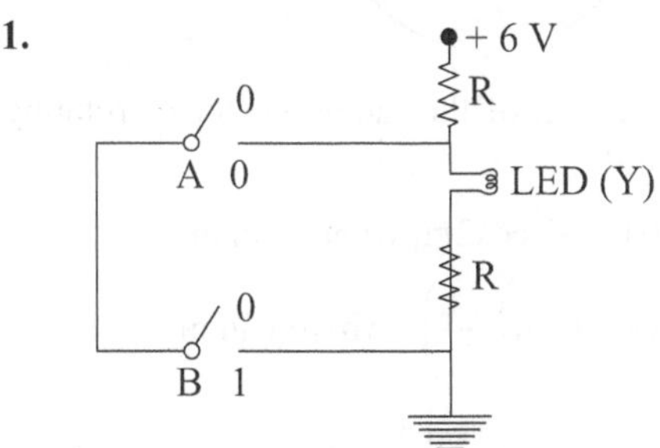

The **correct** Boolean operation represented by the circuit diagram drawn is :

(1) AND (2) OR
(3) NAND (4) NOR

2. A hollow metal sphere of radius R is uniformly charged. The electric field due to the sphere at a distance r from the centre :

(1) increases as r increases for $r < R$ and for $r > R$
(2) zero as r increases for $r < R$, decreases as r increases for $r > R$
(3) zero as r increases for $r < R$, increases as r increases for $r > R$
(4) decreases as r increases for $r < R$ and for $r > R$

3. At a point A on the earth's surface the angle of dip, $\delta = +25°$. At a point B on the earth's surface the angle of dip, $\delta = -25°$. We can interpret that:

(1) A and B are both located in the northern hemisphere.
(2) A is located in the southern hemisphere and B is located in the northern hemisphere.
(3) A is located in the northern hemisphere and B is located in the southern hemisphere.
(4) A and B are both located in the southern hemisphere.

4. In a double slit experiment, when light of wavelength 400 nm was used, the angular width of the first minima formed on a screen placed 1 m away, was found to be 0.2°. What will be the angular width of the first minima, if the entire experimental apparatus is immersed in water? ($\mu_{water} = 4/3$)

(1) 0.266° (2) 0.15°
(3) 0.05° (4) 0.1°

5. Six similar bulbs are connected as shown in the figure with a DC source of emf E, and zero internal resistance.

The ratio of power consumption by the bulbs when (i) all are glowing and (ii) in the situation when two from section A and one from section B are glowing, will be:

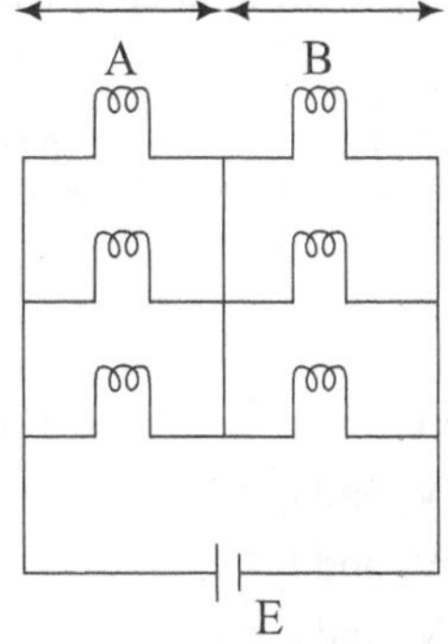

(1) 4 : 9 (2) 9 : 4
(3) 1 : 2 (4) 2 : 1

6. In which of the following processes, heat is neither absorbed nor released by a system ?

(1) isothermal (2) adiabatic
(3) isobaric (4) isochoric

7. A cylindrical conductor of radius R is carrying a constant current. The plot of the magnitude of the magnetic field, B with the distance, d, from the centre of the conductor, is **correctly** represented by the figure :

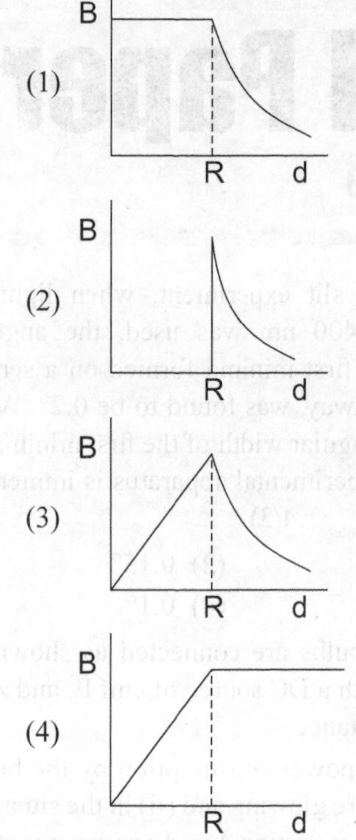

8. In the circuits shown below, the readings of the voltmeters and the ammeters will be :

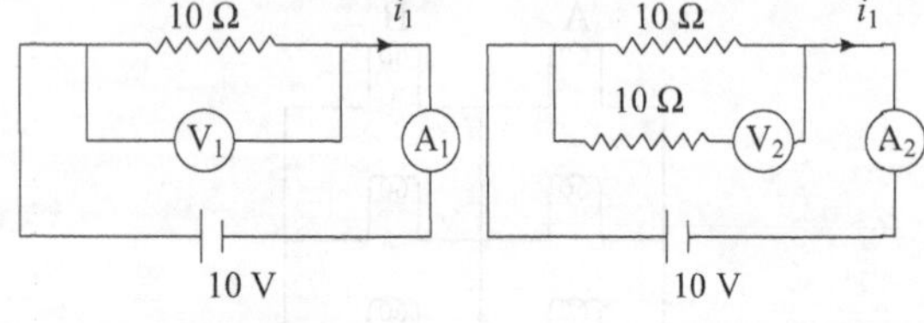

(1) $V_2 > V_1$ and $i_1 = i_2$
(2) $V_1 = V_2$ and $i_1 > i_2$
(3) $V_1 = V_2$ and $i_1 = i_2$
(4) $V_2 > V_1$ and $i_1 > i_2$

9. Which colour of the light has the longest wavelength ?
(1) red (2) blue
(3) green (4) violet

10. Increase in temperature of a gas filled in a container would lead to:
(1) increase in its mass
(2) increase in its kinetic energy
(3) decrease in its pressure
(4) decrease in intermolecular distance

11. The radius of circle, the period of revolution, initial position and sense of revolution are indicated in the fig.

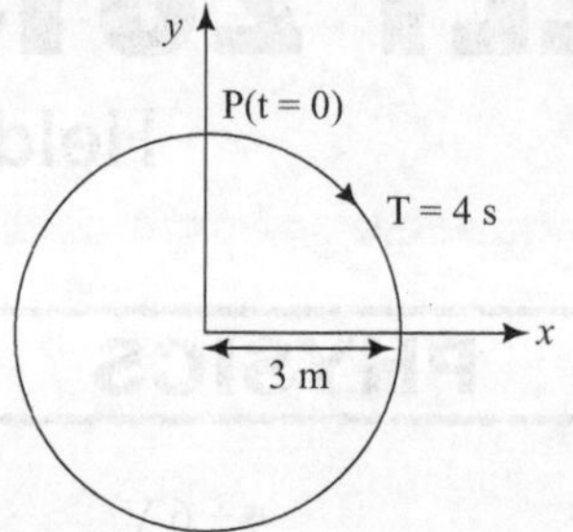

y - projection of the radius vector of rotating particle P is :
(1) $y(t) = -3 \cos 2\pi t$, where y in m
(2) $y(t) = 4 \sin\left(\frac{\pi t}{2}\right)$, where y in m
(3) $y(t) = 3 \cos\left(\frac{3\pi t}{2}\right)$, where y in m
(4) $y(t) = 3 \cos\left(\frac{\pi t}{2}\right)$, where y in m

12. Average velocity of a particle executing SHM in one complete vibration is:
(1) $\frac{A\omega}{2}$ (2) $A\omega$
(3) $\frac{A\omega^2}{2}$ (4) zero

13. A solid cylinder of mass 2 kg and radius 4 cm is rotating about its axis at the rate of 3 rpm. The torque required to stop after 2π revolutions is :
(1) 2×10^{-6} N m (2) 2×10^{-3} N m
(3) 12×10^{-4} N m (4) 2×10^{6} N m

14. A block of mass 10 kg is in contact against the inner wall of a hollow cylindrical drum of radius 1 m. The coefficient of friction between the block and the inner wall of the cylinder is 0.1. The minimum angular velocity needed for the cylinder to keep the block stationary when the cylinder is vertical and rotating about its axis, will be : (g = 10 m/s²)
(1) $\sqrt{10}$ rad/s (2) $\frac{10}{2\pi}$ rad/s
(3) 10 rad/s (4) 10π rad/s

15. The speed of a swimmer in still water is 20 m/s. The speed of river water is 10 m/s and is flowing due east. If he is standing on the south bank and wishes to cross the river along the shortest path, the angle at which he should make his strokes w.r.t. north is given by :
(1) 30°west (2) 0°
(3) 60° west (4) 45° west

16. A 800 turn coil of effective area 0.05 m^2 is kept perpendicular to a magnetic field 5×10^{-5} T. When the plane of the coil is rotated by 90° around any of its coplanar axis in 0.1 s, the emf induced in the coil will be :
(1) 2 V (2) 0.2 V
(3) 2×10^{-3} V (4) 0.02 V

17. In an experiment, the percentage of error occurred in the measurement of physical quantities A, B, C an D are 1%, 2%, 3% and 4% respectively. Then the maximum percentage of error in the measurement X, where $X = \frac{A^2 B^{1/2}}{C^{1/3} D^3}$ will be :
(1) $\left(\frac{3}{13}\right)\%$ (2) 16%
(3) −10% (4) 10%

18. The displacement of a particle executing simple harmonic motion is given by
$y = A_0 + A \sin\omega t + B \cos\omega t$.
Then the amplitude of its oscillation is given by :
(1) $A_0 + \sqrt{A^2 + B^2}$ (2) $\sqrt{A^2 + B^2}$
(3) $\sqrt{A_0^2 + (A + B)^2}$ (4) A+B

19. An electron is accelerated through a potential difference of 10,000 V. Its de Broglie wavelength is, (nearly): ($m_e = 9 \times 10^{-31}$ kg)
(1) 12.2×10^{-13} m (2) 12.2×10^{-12} m
(3) 12.2×10^{-14} m (4) 12.2 nm

20. A mass m is attached to a thin wire and whirled in a vertical circle. The wire is most likely to break when:
(1) the mass is at the highest point
(2) the wire is horizontal
(3) the mass is at the lowest point
(4) inclined at an angle of 60° from vertical

21. Two particles A and B are moving in uniform circular motion in concentric circles of radii r_A and r_B with speed υ_A and υ_B respectively. Their time period of rotation is the same. The ratio of angular speed of A to that of B will be :
(1) $r_A : r_B$ (2) $\upsilon_A : \upsilon_B$
(3) $r_B : r_A$ (4) 1 : 1

22. A copper rod of 88 cm and an aluminium rod of unknown length have their increase in length independent of increase in temperature. The length of aluminium rod is : ($\alpha_{Cu} = 1.7 \times 10^{-5} K^{-1}$ and $\alpha_{Al} = 2.2 \times 10^{-5} K^{-1}$)
(1) 6.8 cm (2) 113.9 cm
(3) 88 cm (4) 68 cm

23. Ionized hydrogen atoms and α-particles with same momenta enters perpendicular to a constant magnetic field, B. The ratio of their radii of their paths $r_H : r_\alpha$ will be :
(1) 2 : 1 (2) 1 : 2
(3) 4 : 1 (4) 1 : 4

24. When a block of mass M is suspended by a long wire of length L, the length of the wire becomes (L + *l*). The elastic potential energy stored in the extended wire is :
(1) *Mgl* (2) MgL
(3) $\frac{1}{2}Mgl$ (4) $\frac{1}{2}MgL$

25. For a p-type semiconductor, which of the following statements is **true** ?
(1) Electrons are the majority carriers and trivalent atoms are the dopants.
(2) Holes are the majority carriers and trivalent atoms are the dopants.
(3) Holes are the majority carriers and pentavalent atoms are the dopants.
(4) Electrons are the majority carriers and pentavalent atoms are the dopants.

26. The work done to raise a mass m from the surface of the earth to a height h, which is equal to the radius of the earth, is :
(1) mgR (2) 2 mgR
(3) $\frac{1}{2}$mgR (4) $\frac{3}{2}$mgR

27. In total internal reflection when the angle of incidence is equal to the critical angle for the pair of media in contact, what will be angle of refraction ?

(1) 180°
(2) 0°
(3) equal to angle of incidence
(4) 90°

28. Two point charges A and B, having charges +Q and –Q respectively, are placed at certain distance apart and force acting between them is F. If 25% charge of A is transferred to B, then force between the charges becomes :

(1) F (2) $\frac{9F}{16}$ (3) $\frac{16F}{9}$ (4) $\frac{4F}{3}$

29. α-particle consists of:

(1) 2 protons and 2 neutrons only
(2) 2 electrons, 2 protons and 2 neutrons
(3) 2 electrons and 4 protons only
(4) 2 protons only

30. Body A of mass 4m moving with speed u collides with another body B of mass 2m, at rest. The collision is head on and elastic in nature. After the collision the fraction of energy lost by the colliding body A is :

(1) $\frac{1}{9}$ (2) $\frac{8}{9}$ (3) – (4) $\frac{5}{9}$

31. A body weighs 200 N on the surface of the earth. How much will it weigh half way down to the centre of the earth ?

(1) 150 N (2) 200 N
(3) 250 N (4) 100 N

32. Pick the **wrong** answer in the context with rainbow.

(1) When the light rays undergo two internal reflections in a water drop, a secondary rainbow is formed.
(2) The order of colours is reversed in the secondary rainbow.
(3) An observer can see a rainbow when his front is towards the sun.
(4) Rainbow is a combined effect of dispersion, refraction and reflection of sunlight.

33. A force F = 20 + 10y acts on a particle in y-direction where F is in newton and y in meter. Work done by this force to move the particle from y = 0 to y = 1 m is :

(1) 30 J (2) 5 J (3) 25 J (4) 20 J

34. A disc of radius 2 m and mass 100 kg rolls on a horizontal floor. Its centre of mass has speed of 20 cm/s. How much work is needed to stop it ?

(1) 3 J (2) 30 kJ (3) 2 J (4) 1J

35. A small hole of area of cross-section 2 mm² is present near the bottom of a fully filled open tank of height 2 m. Taking g = 10 m/s², the rate of flow of water through the open hole would be nearly:

(1) 12.6×10^{-6} m³/s (2) 8.9×10^{-6}m³/s
(3) 2.23×10^{-6} m³/s (4) 6.4×10^{-6} m³/s

36. When an object is shot from the bottom of a long smooth inclined plane kept at an angle 60° with horizontal, it can travel a distance x_1 along the plane. But when the inclination is decreased to 30° and the same object is shot with the same velocity, it can travel x_2 distance. Then $x_1 : x_2$ will be :

(1) $1 : \sqrt{2}$ (2) $\sqrt{\ } : 1$
(3) $1 : \sqrt{3}$ (4) $1 : 2\sqrt{3}$

37. A parallel plate capacitor of capacitance 20 μF is being charged by a voltage source whose potential is changing at the rate of 3 V/s. The conduction current through the connecting wires, and the displacement current through the plates of the capacitor, would be, respectively:

(1) zero, 60 μA (2) 60 μA, 60 μA
(3) 60 μA, zero (4) zero, zero

38. The unit of thermal conductivity is :

(1) J m K^{-1} (2) J m^{-1} K^{-1}
(3) W m K^{-1} (4) W m^{-1} K^{-1}

39. Which of the following acts as a circuit protection device ?

(1) conductor (2) inductor
(3) switch (4) fuse

40. A soap bubble, having radius of 1 mm, is blown from a detergent solution having a surface tension of 2.5×10^{-2} N/m. The pressure inside the bubble equals at a point Z_0 below the free surface of water in a container. Taking g = 10 m/s², density of water = 10³ kg/m³, the value of Z_0 is :

(1) 100 cm (2) 10 cm
(3) 1 cm (4) 0.5 cm

41. The total energy of an electron in an atom in an orbit is –3.4 eV. Its kinetic and potential energies are, respectively:
(1) –3.4 eV, –3.4 eV
(2) –3.4 eV, –6.8 eV
(3) 3.4 eV, –6.8 eV
(4) 3.4 eV, 3.4 eV

42. Two similar thin equi-convex lenses, of focal length *f* each, are kept coaxially in contact with each other such that the focal length of the combination is F_1. When the space between the two lenses is filled with glycerin (which has the same refractive index ($\mu = 1.5$) as that of glass) then the equivalent focal length is F_2. The ratio $F_1 : F_2$ will be :
(1) 2 : 1 (2) 1 : 2
(3) 2 : 3 (4) 3 : 4

43. Two parallel infinite line charges with linear charge densities + λ C/m and –λ C/m are placed at a distance of 2R in free space. What is the electric field mid-way between the two line charges?
(1) zero (2) $\frac{2\lambda}{\pi \epsilon_0 R}$ N/C
(3) $\frac{\lambda}{\pi \epsilon_0 R}$ N/C (4) $\frac{\lambda}{2\pi\varepsilon_0 R}$ N/C

44. In which of the following devices, the eddy current effect is **not** used ?
(1) induction furnace
(2) magnetic braking in train
(3) electromagnet
(4) electric heater

45. A particle moving with velocity $\vec{V}$ is acted by three forces shown by the vector triangle PQR. The velocity of the particle will :

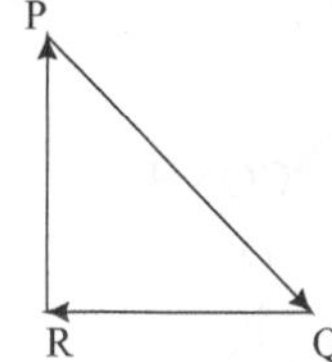

(1) increase
(2) decrease
(3) remain constant
(4) change according to the smallest force $\overrightarrow{QR}$

CHEMISTRY

46. For the second period elements the correct increasing order of first ionisation enthalpy is :
(1) Li < Be < B < C < N < O < F < Ne
(2) Li < B < Be < C < O < N < F < Ne
(3) Li < B < Be < C < N < O < F < Ne
(4) Li < Be < B < C < O < N < F < Ne

47. The method used to remove temporary hardness of water is:
(1) Calgon's method
(2) Clark's method
(3) Ion-exchange method
(4) Synthetic resins method

48. Which of the following is an amphoteric hydroxide ?
(1) $Sr(OH)_2$ (2) $Ca(OH)_2$
(3) $Mg(OH)_2$ (4) $Be(OH)_2$

49. Among the following, the narrow spectrum antibiotic is:
(1) penicillin G (2) ampicillin
(3) amoxycillin (4) chloramphenicol

50. Which mixture of the solutions will lead to the formation of negatively charged colloidal $[AgI]I^-$ sol. ?
(1) 50 mL of 1 M $AgNO_3$ + 50 mL of 1.5 M KI
(2) 50 mL of 1 M $AgNO_3$ + 50 mL of 2 M KI
(3) 50 mL of 2 M $AgNO_3$ + 50 mL of 1.5 M KI
(4) 50 mL of 0.1 M $AgNO_3$ + 50 mL of 0.1 M KI

51. Conjugate base for Brönsted acids H_2O and HF are :
(1) OH^- and H_2F^+, respectively
(2) H_3O^+ and F^-, respectively
(3) OH^- and F^-, respectively
(4) H_3O^+ and H_2F^+, respectively

52. The number of sigma (σ) and pi (π) bonds in pent-2-en-4-yne is:
(1) 10 σ bonds and 3 π bonds
(2) 8 σ bonds and 5 π bonds
(3) 11 σ bonds and 2 π bonds
(4) 13 σ bonds and no π bond

53. The correct structure of tribromooctaoxide is:

(1) 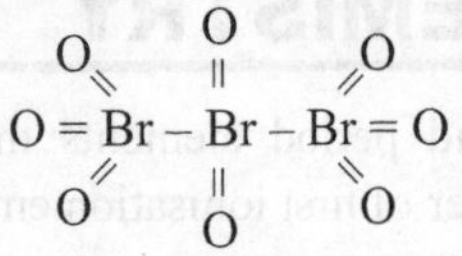
```
 O   O   O
  \\ ||  //
O = Br – Br – Br = O
  //  ||  \\
 O    O    O
```

(2) 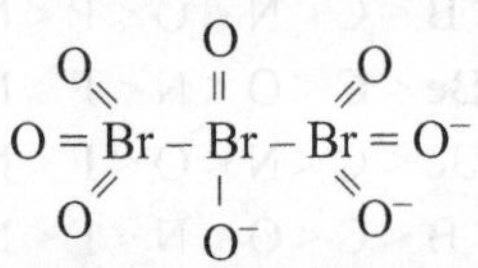
```
 O    O    O
  \\  ||  //
O = Br – Br – Br = O⁻
  //  |   \\
 O    O⁻   O⁻
```

(3) 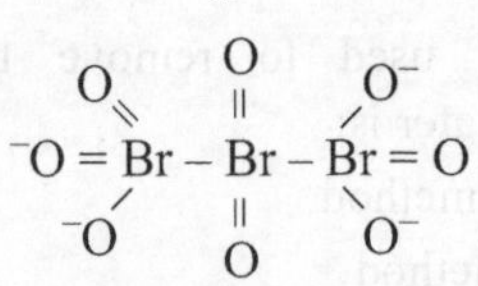
```
 O    O    O⁻
  \\  ||  /
⁻O = Br – Br – Br = O
  /   ||  \\
⁻O    O    O⁻
```

(4) 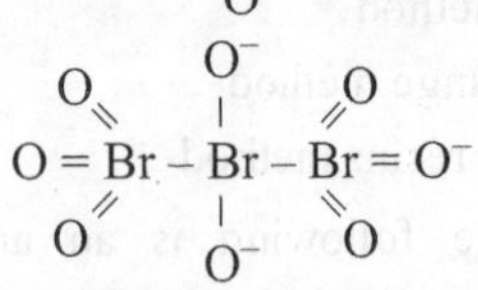
```
 O    O⁻   O
  \\  |   //
O = Br – Br – Br = O⁻
  //  |   \\
 O    O    O
```

54. pH of a saturated solution of $Ca(OH)_2$ is 9. The solubility product (K_{sp}) of $Ca(OH)_2$ is:

(1) 0.5×10^{-15} (2) 0.25×10^{-10}
(3) 0.125×10^{-15} (4) 0.5×10^{-10}

55. The correct order of the basic strength of methyl substituted amines in aqueous solution is :

(1) $(CH_3)_2NH > CH_3NH_2 > (CH_3)_3N$
(2) $(CH_3)_3N > CH_3NH_2 > (CH_3)_2NH$
(3) $(CH_3)_3N > (CH_3)_2NH > CH_3NH_2$
(4) $CH_3NH_2 > (CH_3)_2NH > (CH_3)_3N$

56. For a cell involving one electron $E^{\ominus}_{cell} = 0.59$ V at 298 K, the equilibrium constant for the cell reaction is :

$$\left[\text{Given that } \frac{2.303RT}{F} = 0.059 \text{ V at T} = 298 \text{ K}\right]$$

(1) 1.0×10^{2} (2) 1.0×10^{5}
(3) 1.0×10^{10} (4) 1.0×10^{30}

57. Among the following, the one that is **not** a green house gas is:

(1) nitrous oxide (2) methane
(3) ozone (4) sulphur dioxide

58. The mixture that forms maximum boiling azeotrope is:

(1) Water + Nitric acid
(2) Ethanol + Water
(3) Acetone + Carbon disulphide
(4) Heptane + Octane

59. Which one is malachite from the following ?

(1) $CuFeS_2$ (2) $Cu(OH)_2$
(3) Fe_3O_4 (4) $CuCO_3.Cu(OH)_2$

60. Match the following:

(a) Pure nitrogen	(i) Chlorine
(b) Haber process	(ii) Sulphuric acid
(c) Contact proces	(iii) Ammonia
(d) Deacon's process	(iv) Sodium azide or Barium azide

Which of the following is the **correct** option ?

	(a)	(b)	(c)	(d)
(1)	(i)	(ii)	(iii)	(iv)
(2)	(ii)	(iv)	(i)	(iii)
(3)	(iii)	(iv)	(ii)	(i)
(4)	(iv)	(iii)	(ii)	(i)

61. Which is the correct thermal stability order for H_2E (E = O, S, Se, Te and Po) ?

(1) $H_2S < H_2O < H_2Se < H_2Te < H_2Po$
(2) $H_2O < H_2S < H_2Se < H_2Te < H_2PO$
(3) $H_2Po < H_2Te < H_2Se < H_2S < H_2O$
(4) $H_2Se < H_2Te < H_2Po < H_2O < H_2S$

62. Identify the incorrect statement related to PCl_5 from the following:

(1) Three equatorial P – Cl bonds make an angle of 120° with each other
(2) Two axial P – Cl bonds make an angle of 180° with each other
(3) Axial P – Cl bonds are longer than equatorial P – Cl bonds
(4) $PC1_5$ molecule is non-reactive

63. The major product of the following reaction is :

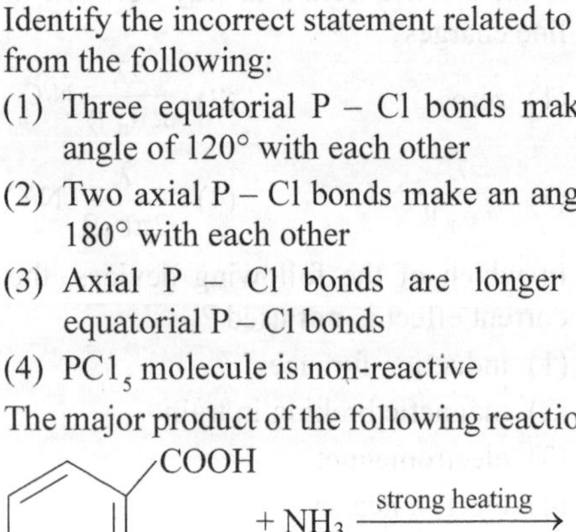

(Benzene ring with COOH and COOH on adjacent carbons) $+ NH_3 \xrightarrow{\text{strong heating}}$

(1) 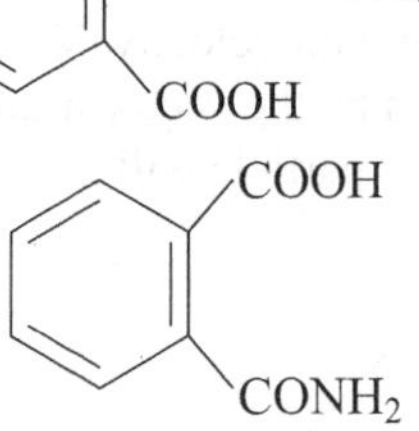 (Benzene ring with COOH and $CONH_2$ on adjacent carbons)

(2) (Benzene ring fused to five-membered ring with two C=O and NH)

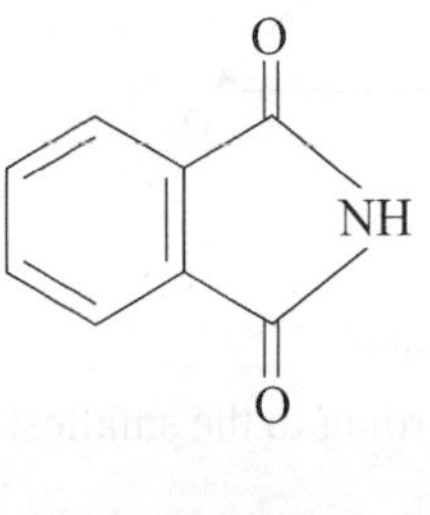

(3) [benzene ring with COOH and NH_2 ortho]

(4) [benzene ring with NH_2 and NH_2 ortho]

64. The compound that is most difficult to protonate is :

(1) $H-O-H$

(2) H_3C-O-H

(3) $H_3C-O-CH_3$

(4) $Ph-O-H$

65. The manganate and permanganate ions are tetrahedral, due to:

(1) The π-bonding involves overlap of *p*-orbitals of oxygen with *d*-orbitals of manganese

(2) There is no π-bonding

(3) The π-bonding involves overlap of *p*-orbital of oxygen with *p*-orbitals of manganese

(4) The π-bonding involves overlap of *d*-orbital of oxygen with *d*-orbitals of manganese

66. The most suitable reagent for the following conversion, is:

$$H_3C-C\equiv C-CH_3 \longrightarrow \text{(}H_3C, CH_3 \text{ cis; } H, H\text{)}$$

cis -2-butene

(1) Na/liquid NH_3 (2) H_2, Pd/C, quinoline

(3) Zn/HCl (4) Hg^{2+}/H^+, H_2O

67. The structure of intermediate A in the following reaction, is:

$$C_6H_5-CH(CH_3)_2 \xrightarrow{O_2} A \xrightarrow[H_2O]{H^+} C_6H_5OH + H_3C-CO-CH_3$$

(1) $C_6H_5-O-CH(CH_3)_2$

(2) $C_6H_5-C(CH_3)_2-O-O-H$

(3) $C_6H_5-O-O-CH(CH_3)_2$

(4) $C_6H_5-CH(CH_3)-CH_2-O-O-H$ (HC attached to ring, bearing CH_3 and $CH_2-O-O-H$)

68. If the rate constant for a first order reaction is k, the time (t) required for the completion of 99% of the reaction is given by :

(1) t = 0.693/k (2) t = 6.909/k

(3) t = 4.606/k (4) t = 2.303/k

69. Which of the following reactions are disproportionation reaction?

(a) $2Cu^+ \rightarrow Cu^{2+} + Cu^0$

(b) $3MnO_4^{2-} + H^+ \rightarrow 2MnO_4^- + MnO_2 + 2H_2O$

(c) $2KMnO_4 \xrightarrow{\Delta} K_2MnO_4 + MnO_2 + O_2$

(d) $2MnO_4^- + 3Mn^{2+} + 2H_2O \rightarrow 5MnO_2 + 4H^{\oplus}$

Select the **correct** option from the following:

(1) (a) and (b) only (2) (a), (b) and (c)

(3) (a), (c) and (d) (4) (a) and (d) only

70. In which case change in entropy is negative ?

(1) Evaporation of water

(2) Expansion of a gas at constant temperature

(3) Sublimation of solid to gas

(4) $2H(g) \rightarrow H_2(g)$

71. The biodegradable polymer is:
(1) nylon-6, 6 (2) nylon 2-nylon 6
(3) nylon-6 (4) Buna-S

72. A gas at 350 K and 15 bar has molar volume 20 percent smaller than that for an ideal gas under the same conditions. The **correct** option about the gas and its compressibility factor (Z) is:
(1) $Z > 1$ and attractive forces are dominant
(2) $Z > 1$ and repulsive forces are dominant
(3) $Z < 1$ and attractive forces are dominant
(4) $Z < 1$ and repulsive forces are dominant

73. A compound is formed by cation C and anion A. The anions form hexagonal close packed (hcp) lattice and the cations occupy 75% of octahedral voids. The formula of the compound is:
(1) C_2A_3 (2) C_3A_2
(3) C_3A_4 (4) C_4A_3

74. Enzymes that utilize ATP in phosphate transfer require an alkaline earth metal (M) as the cofactor. M is:
(1) Be (2) Mg (3) Ca (4) Sr

75. $4d$, $5p$, $5f$ and $6p$ orbitals are arranged in the order of decreasing energy. The **correct** option is:
(1) $5f > 6p > 5p > 4d$ (2) $6p > 5f > 5p > 4d$
(3) $6p > 5f > 4d > 5p$ (4) $5f > 6p > 4d > 5p$

76. For the cell reaction
$2Fe^{3+}(aq) + 2I^-(aq) \rightarrow 2Fe^{2+}(aq) + I_2(aq)$
E°_{cells} 0.24 V at 298 K. The standard Gibbs energy (Δ, G°) of the cell reaction is:
[Given that Faraday constant $F = 96500\ C\ mol^{-1}$]
(1) -46.32 kJ mol^{-1} (2) -23.16 kJ mol^{-1}
(3) 46.32 kJ mol^{-1} (4) 23.16 kJ mol^{-1}

77. For an ideal solution, the **correct** option is:
(1) $\Delta_{mix} S = 0$ at constant T and P
(2) $\Delta_{mix} V \neq 0$ at constant T and P
(3) $\Delta_{mix} H = 0$ at constant T and P
(4) $\Delta_{mix} G = 0$ at constant T and P

78. The non-essential amino acid among the following is:
(1) valine (2) leucine
(3) alanine (4) lysine

79. What is the **correct** electronic configuration of the central atom in $K_4[Fe(CN)_6]$ based on crystal field theory ?
(1) $t_{2g}^4\ e_g^2$ (2) $t_{2g}^6\ e_g^0$
(3) $e^3\ t_2^3$ (4) $e^4\ t_2^2$

80. Which of the following is **incorrect** statement?
(1) PbF_4 is covalent in nature
(2) $SiCl_4$ is easily hydrolysed
(3) GeX_4 (X = F, Cl, Br, I) is more stable than GeX_2
(4) SnF_4 is ionic in nature

81. Match the Xenon compounds in **Column-I** with its structure in **Column-II** and assign the correct code:

Column-I	**Column-II**
(a) XeF_4	(i) pyramidal
(b) XeF_6	(ii) square planar
(c) $XeOF_4$	(iii) distorted octahedral
(d) XeO_3	(iv) square pyramidal

Code:

	(a)	**(b)**	**(c)**	**(d)**
(1)	(i)	(ii)	(iii)	(iv)
(2)	(ii)	(iii)	(iv)	(i)
(3)	(ii)	(iii)	(i)	(iv)
(4)	(iii)	(iv)	(i)	(ii)

82. An alkene "A" on reaction with O_3 and $Zn - H_2O$ gives propanone and ethanal in equimolar ratio. Addition of HCl to alkene "A" gives "B" as the major product. The structure of product "B" is:

(1) $Cl-CH_2-CH_2-\underset{CH_3}{\overset{CH_3}{CH}}$

(2) $H_3C-CH_2-\overset{CH_2Cl}{CH}-CH_3$

(3) $H_3C-CH_2-\underset{Cl}{\overset{CH_3}{C}}-CH_3$

(4) $H_3C-\underset{Cl}{CH}-\underset{CH_3}{\overset{CH_3}{CH}}$

83. For the chemical reaction
$N_2(g) + 3H_2(g) \rightleftharpoons 2NH_3(g)$
the gas is **correct** option is:

(1) $-\frac{1}{3}\frac{d[H_2]}{dt} = -\frac{1}{2}\frac{d[NH_3]}{dt}$

(2) $-\frac{d[N_2]}{dt} = 2\frac{d[NH_3]}{dt}$

(3) $-\frac{d[N_2]}{dt} = \frac{1}{2}\frac{d[NH_3]}{dt}$

(4) $3\frac{d[H_2]}{dt} = 2\frac{d[NH_3]}{dt}$

84. Under isothermal condition, a gas at 300 K expands front 0.1 L to 0.25 L against a constant external pressure of 2 bar. The work done by the gas is [Given that 1 L bar = 100 J]

(1) –30 J (2) 5 kJ
(3) 25 J (4) 30 J

85. Among the following, the reaction that proceeds through an electrophilic substitution, is:

(1) $C_6H_5\overset{+}{N}_2Cl^- \xrightarrow{Cu_2Cl_2} C_6H_5Cl + N_2$

(2) $C_6H_6 + Cl_2 \xrightarrow{AlCl_3} C_6H_5Cl + HCl$

(3) $C_6H_6 + Cl_2 \xrightarrow{\text{UV light}} C_6H_6Cl_6$ (hexachlorocyclohexane)

(4) $C_6H_5CH_2OH + HCl \xrightarrow{\text{Heat}} C_6H_5CH_2Cl + H_2O$

86. Which of the following diatomic molecular species has only π bonds according to Molecular Orbital Theory ?

(1) O_2 (2) N_2
(3) C_2 (4) Be_2

87. Which of the following species is **not** stable ?

(1) $[SiF_6]^{2-}$ (2) $[GeCl_6]^{2-}$
(3) $[Sn(OH)_6]^{2-}$ (4) $[SiCl_6]^{2-}$

88. Which will make basic buffer ?

(1) 50 mL of 0.1 M NaOH + 25 mL of 0.1 M CH_3COOH
(2) 100 mL of 0.1 M CH_3COOH + 100 mL of 0.1 M NaOH
(3) 100 mL of 0.1 M HCl + 200 mL of 0.1 M NH_4OH
(4) 100 mL of 0.1 M HCl + 100 mL of 0.1 M NaOH

89. Which of the following series of transitions in the spectrum of hydrogen atom falls in visible region ?

(1) Lyman series (2) Balmer series
(3) Paschen series (4) Brackett series

90. The number of moles of hydrogen molecule required to produce 20 moles of ammonia through Haber' s process is:

(1) 10 (2) 20
(3) 30 (4) 40

BIOLOGY

91. Conversion of glucose to glucose-6-phosphate, the first irreversible reaction of glycolysis, is catalysed by:

(1) Aldolase
(2) Hexokinase
(3) Enolase
(4) Phosphofructokinasc

92. What is the site of perception of photoperiod necessary for induction of flowering in plants ?

(1) Lateral buds (2) Pulvinus
(3) Shoot apex (4) Leaves

93. Which of the following is **true** for Golden rice?

(1) It is vitamin A enriched, with a gene from daffodil.
(2) It is pest resistant, with a gene from *Bacillus thuringiensis*.
(3) It is drought tolerant, developed using *Agrobacterium* vector.
(4) It has yellow grains, because of a gene introduced from a primitive variety of rice.

94. Identify the **correct** pair representing the causative agent of typhoid fever and the confirmatory test for typhoid.

(1) *Plasmodium vivax* / UTI test
(2) *Streptococcus pneumoniae* / Widal test
(3) *Salmonella typhi* / Anthrone test
(4) *Salmonella typhi* / Widal test

95. Colostrum, the yellowish fluid, secreted by mother during the initial days of lactation is very essential to impart immunity to the newborn infants because it contains:

(1) Natural killer cells
(2) Monocytes
(3) Macrophages
(4) Immunoglobulin A

96. It takes very long time for pineapple plants to produce flowers. Which combination of hormones can be applied to artificially induce flowering in pineapple plants throughout the year to increase yield?

(1) Auxin and Ethylene
(2) Gibberellin and Cytokinin
(3) Gibberellin and Abscisic acid
(4) Cytokinin and Abscisic acid

97. DNA precipitation out of a mixture of biomolecules can be achieved by treatment with:

(1) Isopropanol
(2) Chilled ethanol
(3) Methanol at room temperature
(4) Chilled chloroform

98. What triggers activation of protoxin to active *Bt* toxin of *Bacillus thuringiensis* in boll worm?

(1) Body temperature
(2) Moist surface of midgut
(3) Alkaline pH of gut
(4) Acidic pH of stomach

99. The frequency of recombination between gene pairs on the same chromosome as a measure of the distance between genes was explained by:

(1) T.H. Morgan (2) Gregor J. Mendel
(3) Alfred Sturtevant (4) Sutton Boveri

100. Which of the following is the most important cause for animals and plants being driven to extinction?

(1) Habitat loss and fragmentation
(2) Drought and floods
(3) Economic exploitation
(4) Alien species invasion

101. Identify the cells whose secretion protects the lining of gastro-intestinal tract from various enzymes.

(1) Chief cells (2) Goblet cells
(3) Oxyntic cells (4) Duodenal cells

102. Match the Column-I with Column-II.

Column-I	Column-II
(a) P-wave	(i) Depolarisation of ventricles
(b) QRS complex	(ii) Repolarisation of ventricles
(c) T - wave	(iii) Coronary ischemia
(d) Reduction in the size of T-wave	(iv) Depolarisation of atria
	(v) Repolarisation of atria

Select the **correct** option.

	(a)	(b)	(c)	(d)
(1)	(iv)	(i)	(ii)	(iii)
(2)	(iv)	(i)	(ii)	(v)
(3)	(ii)	(i)	(v)	(iii)
(4)	(ii)	(iii)	(v)	(iv)

103. Match the following structures with their respective location in organs:

(a) Crypts of Lieberkuhn	(i) Pancreas
(b) Glisson's capsule	(ii) Duodenum
(c) Islets of Langerhans	(iii) Small intestine
(d) Brunner's glands	(iv) Liver

Select the **correct** option from the following:

	(a)	(b)	(c)	(d)
(1)	(iii)	(i)	(ii)	(iv)
(2)	(ii)	(iv)	(i)	(iii)
(3)	(iii)	(iv)	(i)	(ii)
(4)	(iii)	(ii)	(i)	(iv)

104. Cells in G_0 phase :
(1) exit the cell cycle
(2) enter the cell cycle
(3) suspend the cell cycle
(4) terminate the cell cycle

105. The Earth Summit held in Rio de Janeiro in 1992 was called:
(1) to reduce CO_2 emissions and global warming.
(2) for conservation of biodiversity and sustainable utilisation of its benefits.
(3) to assess threat posed to native species by invasive weed species.
(4) for immediate steps to discontinue use of CFCs that were damaging the ozone layer.

106. Which of the following glucose transporters is insulin-dependent?
(1) GLUT I (2) GLUT II
(3) GLUT III (4) GLUT IV

107. Which of the statements given below is **not** true about formation of annual rings in trees?
(1) Annual ring is a combination of spring wood and autumn wood produced in a year.
(2) Differential activity of cambium causes light and dark bands of tissue - early and late wood respectively.
(3) Activity of cambium depends upon variation in climate.
(4) Annual rings are not prominent in trees of temperate region.

108. Match the following hormones with the respective disease:

(a) Insulin	(i) Addison's disease
(b) Thyroxine	(ii) Diabetes insipidus
(c) Corticoids	(iii) Acromegaly
(d) Growth hormone	(iv) Goitre
	(v) Diabetes mellitus

Select the **correct** option.

	(a)	**(b)**	**(c)**	**(d)**
(1)	(v)	(i)	(ii)	(iii)
(2)	(ii)	(iv)	(iii)	(i)
(3)	(v)	(iv)	(i)	(iii)
(4)	(ii)	(iv)	(i)	(iii)

109. In some plants, the female gamete develops into embryo without fertilisation. This phenomenon is known as :
(1) Autogamy (2) Parthenocarpy
(3) Syngamy (4) Parthenogenesis

110. Which of the following ecological pyramids is generally inverted?
(1) Pyramid of numbers in grassland
(2) Pyramid of energy
(3) Pyramid of biomass in a forest
(4) Pyramid of biomass in a sea

111. Extrusion of second polar body from egg nucleus occurs:
(1) after entry of sperm but before fertilisation
(2) after fertilisation
(3) before entry of sperm into ovum
(4) simultaneously with first cleavage

112. Pinus seed **cannot** germinate and establish without fungal association. This is because:
(1) its embryo is immature.
(2) it has obligate association with mycorrhizae.
(3) it has very hard seed coat.
(4) its seeds contain inhibitors that prevent germination.

113. Which of the following factors is responsible for the formation of concentrated urine ?
(1) Low levels of antidiuretic hormone.
(2) Maintaining hyperosmolarity towards inner medullary interstitium in the kidneys.
(3) Secretion of erythropoietin by juxtaglomerular complex.
(4) Hydrostatic pressure during glomerular filtration.

114. In *Antirrhinum* (Snapdragon), a red flower was crossed with a white flower and in F_1 generation, pink flowers were obtained. When pink flowers were selfed, the F_2 generation showed white, red and pink flowers. Choose the **incorrect** statement from the following:
(1) This experiment does not follow the principle of dominance.
(2) Pink colour in F_1 is due to incomplete dominance.

(3) Ratio of F_2 is $\frac{1}{4}$ (Red): $\frac{2}{4}$ (Pink): $\frac{1}{4}$ (White)
(4) Law of Segregation does not apply in this experiment.

115. Which part of the brain is responsible for thermoregulation?
(1) Cerebrum
(2) Hypothalamus
(3) Corpus callosum
(4) Medulla oblongata

116. Which of the following sexually transmitted diseases is **not** completely curable?
(1) Gonorrhoea (2) Genital warts
(3) Genital herpes (4) Chlamydiasis

117. Respiratory Quotient (RQ) value of tripalmitin is:
(1) 0.9 (2) 0.7 (3) 0.07 (4) 0.09

118. Select the correct group of biocontrol agents.
(1) *Bacillus thuringiensis*, Tobacco mosaic virus, Aphids
(2) *Trichoderma, Baculovirus, Bacillus thuringiensis*
(3) *Oscillatoria, Rhizobium, Trichoderma*
(4) *Nostoc, Azospirillium, Nucleopolyhedrovirus*

119. Which one of the following statements regarding post-fertilisation development in flowering plants is **incorrect**?
(1) Ovary develops into fruit.
(2) Zygote develops into embryo.
(3) Central cell develops into endosperm.
(4) Ovules develop into embryo sac.

120. Concanavalin A is:
(1) an alkaloid (2) an essential oil
(3) a lectin (4) a pigment

121. Match the following organisms with the products they produce:

(a) *Lactobacillus*	(i) Cheese
(b) *Saccharomyces cerevisiae*	(ii) Curd
(c) *Aspergillus niger*	(iii) Citric acid
(d) *Acetobacter aceti*	(iv) Bread
	(v) Acetic acid

Select the **correct** option.

	(a)	(b)	(c)	(d)
(1)	(ii)	(iv)	(v)	(iii)
(2)	(ii)	(iv)	(iii)	(v)
(3)	(iii)	(iv)	(v)	(i)
(4)	(ii)	(i)	(iii)	(v)

122. Consider the following statements:
(A) Coenzyme or metal ion that is tightly bound to enzyme protein is called prosthetic group.
(B) A complete catalytic active enzyme with its bound prosthetic group is called apoenzyme.

Select the **correct** option.
(1) Both **(A)** and **(B)** are true.
(2) **(A)** is true but **(B)** is false.
(3) Both **(A)** and **(B)** are false.
(4) **(A)** is false but **(B)** is true.

123. The **correct** sequence of phases of cell cycle is:
(1) $M \rightarrow G_1 \rightarrow G_2 \rightarrow S$
(2) $G_1 \rightarrow G_2 \rightarrow S \rightarrow M$
(3) $S \rightarrow G_1 \rightarrow G_2 \rightarrow M$
(4) $G_1 \rightarrow S \rightarrow G_2 \rightarrow M$

124. *Thiobacillus* is a group of bacteria helpful in carrying out:
(1) Nitrogen fixation
(2) Chemoautotrophic fixation
(3) Nitrification
(4) Denitrification

125. Select the **incorrect** statement.
(1) Inbreeding increases homozygosity.
(2) Inbreeding is essential to evolve purelines in any animal.
(3) Inbreeding selects harmful recessive genes that reduce fertility and productivity.
(4) Inbreeding helps in accumulation of superior genes and elimination of undesirable genes.

126. What map unit (Centimorgan) is adopted in the construction of genetic maps?
(1) A unit of distance between two expressed genes, representing 10% cross over.
(2) A unit of distance between two expressed genes, representing 100% cross over.

(3) A unit of distance between genes on chromosomes, representing 1% cross over.

(4) A unit of distance between genes on chromosomes, representing 50% cross over

127. Which one of the following is **not** a method of *in situ* conservation of biodiversity?

(1) Biosphere Reserve

(2) Wildlife Sanctuary

(3) Botanical Garden

(4) Sacred Grove

128. Placentation, in which ovules develop on the inner wall of the ovary or in peripheral part, is:

(1) Basal (2) Axile

(3) Parietal (4) Free central

129. Due to increasing air-borne allergens and pollutants, many people in urban areas are suffering from respiratory disorder causing wheezing due to:

(1) benign growth on mucous lining of nasal cavity.

(2) inflammation of bronchi and bronchioles.

(3) proliferation of fibrous tissues and damage of the alveolar walls.

(4) reduction in the secretion of surfactants by pneumocytes.

130. Which of the following statements is **correct**?

(1) Cornea is an external, transparent and protective proteinacious covering of the eye-ball.

(2) Cornea consists of dense connective tissue of elastin and can repair itself.

(3) Cornea is convex, transparent layer which is highly vascularised.

(4) Cornea consists of dense matrix of collagen and is the most sensitive portion of the eye.

131. Purines found both in DNA and RNA are:

(1) Adenine and thymine

(2) Adenine and guanine

(3) Guanine and cytosine

(4) Cytosine and thymine

132. Expressed Sequence Tags (ESTs) refers to:

(1) Genes expressed as RNA

(2) Polypeptide expression

(3) DNA polymorphism

(4) Novel DNA sequences

133. Phloem in gymnosperms lacks:

(1) Albuminous cells and sieve cells

(2) Sieve tubes only

(3) Companion cells only

(4) Both sieve tubes and companion cells

134. What is the genetic disorder in which an individual has an overall masculine development gynaecomastia, and is sterile?

(1) Turner's syndrome

(2) Klinefelter's syndrome

(3) Edward syndrome

(4) Down's syndrome

135. Grass leaves curl inwards during very dry weather. Select the most appropriate reason from the following:

(1) Closure of stomata

(2) Flaccidity of bulliform cells

(3) Shrinkage of air spaces in spongy mesophyll

(4) Tyloses in vessels

136. Consider following features:

(a) Organ system level of organisation

(b) Bilateral symmetry

(c) True coelomates with segmentation of body

Select the correct option of animal groups which possess all the above characteristics.

(1) Annelida, Arthropoda and Chordata

(2) Annelida, Arthropoda and Mollusca

(3) Arthropoda, Mollusca and Chordata

(4) Annelida, Mollusca and Chordata

137. Under which of the following conditions will there be no change in the reading frame of following mRNA ?

5′ AACAGCGGUGCUAUU 3′

(1) Insertion of G at 5th position

(2) Deletion of G from 5th position

(3) Insertion of A and G at 4th and 5th positions respectively

(4) Deletion of GGU from 7th, 8th and 9th positions

138. Select the **correct** option:

(1) 8th, 9th and 10th pairs of ribs articulate directly with the sternum.

(2) 11th and 12th pairs of ribs are connected to the sternum with the help of hyaline cartilage.

(3) Each rib is a flat thin bone and all the ribs are connected dorsally to the thoracic vertebrae and ventrally to the sternum.

(4) There are seven pairs of vertebrosternal, three pairs of vertebrochondral and two pairs of vertebral ribs.

139. The shorter and longer arms of a submetacentric chromosome are referred to as:

(1) s-arm and l-arm respectively

(2) p-arm and q-arm respectively

(3) q-arm and p-arm respectively

(4) m-arm and n-arm respectively

140. Xylem translocates:

(1) Water only

(2) Water and mineral salts only

(3) Water, mineral salts and some organic nitrogen only

(4) Water, mineral salts, some organic nitrogen and hormones

141. Persistent nucellus in the seed is known as:

(1) Chalaza (2) Perisperm

(3) Hilum (4) Tegmen

142. Match Column-I with Column-II.

Column-I	Column-II
(a) Saprophyte	(i) Symbiotic association of fungi with plant roots
(b) Parasite	(ii) Decomposition of dead organic materials
(c) Lichens	(iii) Living on living plants or animals
(d) Mycorrhiza	(iv) Symbiotic association of algae and fungi

Choose the **correct** answer from the options given below:

	(a)	(b)	(c)	(d)
(1)	(i)	(ii)	(iii)	(iv)
(2)	(iii)	(ii)	(i)	(iv)
(3)	(ii)	(i)	(iii)	(iv)
(4)	(ii)	(iii)	(iv)	(i)

143. Which of the following can be used as a biocontrol agent in the treatment of plant disease?

(1) *Trichoderma* (2) *Chlorella*

(3) *Anabaena* (4) *Lactobacillus*

144. What would be the heart rate of a person if the cardiac output is 5 L, blood volume in the ventricles at the end of diastole is 100 mL and at the end of ventricular systole is 50 mL?

(1) 50 beats per minute

(2) 75 beats per minute

(3) 100 beats per minute

(4) 125 beats per minute

145. Which of the following Protocols did aim for reducing emission of chlorofluorocarbons into the atmosphere?

(1) Montreal Protocol

(2) Kyoto Protocol

(3) Gothenburg Protocol

(4) Geneva Protocol

146. Polyblend, a fine powder of recycled modified plastic, has proved to be a good material for:

(1) making plastic sacks

(2) use as a fertiliser

(3) construction of roads

(4) making tubes and pipes

147. Which of the following contraceptive methods do involve a role of hormone?

(1) Lactational amenorrhea, Pills, Emergency contraceptives

(2) Barrier method, Lactational amenorrhea, Pills

(3) CuT, Pills, Emergency contraceptives

(4) Pills, Emergency contraceptives, Barrier methods

148. Drug called 'Heroin' is synthesised by :

(1) methylation of morphine

(2) acetylation of morphine

(3) glycosylation of morphine

(4) nitration of morphine

149. Which of the following pairs of gases is mainly responsible for green house effect?

(1) Ozone and Ammonia

(2) Oxygen and Nitrogen

(3) Nitrogen and Sulphur dioxide

(4) Carbon dioxide and Methane

150. Which of the following muscular disorders is inherited?

(1) Tetany
(2) Muscular dystrophy
(3) Myasthenia gravis
(4) Botulism

151. Which one of the following equipments is essentially required for growing microbes on a large scale, for industrial production of enzymes?

(1) BOD incubator (2) Sludge digester
(3) Industrial oven (4) Bioreactor

152. The concept of *"Omnis cellula-e cellula"* regarding cell division was first proposed by:

(1) Rudolf Virchow
(2) Theodore Schwann
(3) Schleiden
(4) Aristotle

153. What is the fate of the male gametes discharged in the synergid?

(1) One fuses with the egg, other(s) degenerate(s) in the synergid.
(2) All fuse with the egg.
(3) One fuses with the egg, other(s) fuse(s) with synergid nucleus.
(4) One fuses with the egg and other fuses with central cell nuclei.

154. How does steroid hormone influence the cellular activities?

(1) Changing the permeability of the cell membrane.
(2) Binding to DNA and forming a gene-hormone complex.
(3) Activating cyclic AMP located on the cell membrane.
(4) Using aquaporin channels as second messenger.

155. Which of the following pair of organelles **does not** contain DNA?

(1) Mitochondria and Lysosomes
(2) Chloroplast and Vacuoles
(3) Lysosomes and Vacuoles
(4) Nuclear envelope and Mitochondria

156. A gene locus has two alleles A, a. If the frequency of dominant allele A is 0.4, then what will be the frequency of homozygous dominant, heterozygous and homozygous recessive individuals in the population ?

(1) 0.36 (AA); 0.48 (Aa); 0.16 (aa)
(2) 0.16 (AA); 0.24 (Aa); 0.36 (aa)
(3) 0.16 (AA); 0.48 (Aa); 0.36 (aa)
(4) 0.16 (AA); 0.36 (Aa); 0.48 (aa)

157. Match the following organisms with their respective characteristics:

(a) *Pila*	(i) Flame cells
(b) *Bombyx*	(ii) Comb plates
(c) *Pleurobrachia*	(iii) Radula
(d) *Taenia*	(iv) Malpighian tubules

Select the **correct** option from the following:

	(a)	(b)	(c)	(d)
(1)	(iii)	(ii)	(i)	(iv)
(2)	(iii)	(iv)	(ii)	(i)
(3)	(ii)	(iv)	(iii)	(i)
(4)	(iii)	(ii)	(iv)	(i)

158. Which of the following is a commercial blood cholesterol lowering agent?

(1) Cyclosporin A (2) Statin
(3) Streptokinase (4) Lipases

159. Variations caused by mutation, as proposed by Hugo de Vries, are:

(1) random and directional
(2) random and directionless
(3) small and directional
(4) small and directionless

160. Select the **incorrect** statement.

(1) Male fruit fly is heterogametic.
(2) In male grasshoppers, 50% of sperms have no sex-chromosome.
(3) In domesticated fowls, sex of progeny depends on the type of sperm rather than egg.
(4) Human males have one, of their sex-chromosome much shorter than the other.

161. Which of the following immune responses is responsible for rejection of kidney graft?

(1) Auto-immune response
(2) Humoral immune response
(3) Inflammatory immune response
(4) Cell-mediated immune response

162. From evolutionary point of view, retention of the female gametophyte with developing young embryo on the parent sporophyte for some time, is first observed in:

(1) Liverworts (2) Mosses
(3) Pteridophytes (4) Gymnosperms

163. Select the **correct** sequence of organs in the alimentary canal of cockroach starting from mouth:

(1) Pharynx → Oesophagus → Crop → Gizzard → Ileum → Colon → Rectum
(2) Pharynx → Oesophagus → Gizzard → Crop → Ileum → Colon → Rectum
(3) Pharynx → Oesophagus → Gizzard → Ileum → Crop → Colon → Rectum
(4) Pharynx → Oesophagus → Ileum → Crop → Gizzard → Colon → Rectum

164. Which of the following statements regarding mitochondria is **incorrect**?

(1) Outer membrane is permeable to monomers of carbohydrates, fats and proteins.
(2) Enzymes of electron transport are embedded in outer membrane.
(3) Inner membrane is convoluted with infoldings.
(4) Mitochondrial matrix contains single circular DNA molecule and ribosomes.

165. Which of the following statements is **not** correct?

(1) Lysosomes have numerous hydrolytic enzymes.
(2) The hydrolytic enzymes of lysosomes are active under acidic pH.
(3) Lysosomes are membrane bound structures.
(4) Lysosomes are formed by the process of packaging in the endoplasmic reticulum.

166. Which of the following features of genetic code does allow bacteria to produce human insulin by recombinant DNA technology?

(1) Genetic code is not ambiguous
(2) Genetic code is redundant
(3) Genetic code is nearly universal
(4) Genetic code is specific

167. Tidal volume and Expiratory reserve volume of an athlete is 500 mL and 1000 mL respectively. What will be his Expiratory capacity if the residual volume is 1200 mL?

(1) 1500 mL (2) 1700 mL
(3) 2200 mL (4) 2700 mL

168. Match the following genes of the Lac operon with their respective products:

(a) i gene	(i) β-galactosidase
(b) z gene	(ii) Permease
(c) a gene	(iii) Repressor
(d) y gene	(iv) Transacetylase

Select the **correct** option.

	(a)	(b)	(c)	(d)
(1)	(i)	(iii)	(ii)	(iv)
(2)	(iii)	(i)	(ii)	(iv)
(3)	(iii)	(i)	(iv)	(ii)
(4)	(iii)	(iv)	(i)	(ii)

169. Use of an artificial kidney during hemodialysis may result in:

(a) Nitrogenous waste build-up in the body.
(b) Non-elimination of excess potassium ions.
(c) Reduced absorption of calcium ions from gastro-intestinal tract.
(d) Reduced RBC production.

Which of the following options is the most **appropriate**?

(1) (a) and (b) are correct
(2) (b) and (c) are correct
(3) (c) and (d) are correct
(4) (a) and (d) are correct

170. Match the hominids with their **correct** brain size:

(a) *Homo habilis*	(i) 900 cc
(b) *Homo neanderthalensis*	(ii) 1350 cc
(c) *Homo erectus*	(iii) 650 - 800 cc
(d) *Homo sapiens*	(iv) 1400 cc

Select the **correct** option.

	(a)	(b)	(c)	(d)
(1)	(iii)	(i)	(iv)	(ii)
(2)	(iii)	(ii)	(i)	(iv)
(3)	(iii)	(iv)	(i)	(ii)
(4)	(iv)	(iii)	(i)	(ii)

171. Select the hormone-releasing Intra-Uterine Devices.
(1) Vaults, LNG-20
(2) Multiload 375, Progestasert
(3) Progestasert, LNG-20
(4) Lippes Loop, Multiload 375

172. Select the **correct** sequence for transport of sperm cells in male reproductive system
(1) Testis → Epididymis → Vasa efferentia → Rete testis → Inguinal canal → Urethra
(2) Seminiferous tubules → Rete testis → Vasa efferentia → Epididymis → Vas deferens → Ejaculatory duct → Urethra → Urethral meatus
(3) Seminiferous tubules → Vasa efferentia → Epididymis → Inguinal canal → Urethra
(4) Testis → Epididymis → Vasa efferentia → Vas deferens → Ejaculatory duct → Inguinal canal → Urethra → Urethral meatus

173. What is the direction of movement of sugars in phloem?
(1) Non-multidirectional
(2) Upward
(3) Downward
(4) Bi-directional

174. Which of the following statements is **incorrect**?
(1) Viroids lack a protein coat.
(2) Viruses are obligate parasites.
(3) Infective constituent in viruses is the protein coat.
(4) Prions consist of abnormally folded proteins.

175. The ciliated epithelial cells are required to move particles or mucus in a specific direction. In humans, these cells are mainly present in:
(1) Bile duct and bronchioles
(2) Fallopian tubes and pancreatic duct
(3) Eustachian tube and salivary duct
(4) Bronchioles and fallopian tubes

176. Which of the following statements is **incorrect**?
(1) Morels and truffles are edible delicacies.
(2) *Claviceps* is a source of many alkaloids and LSD.
(3) Conidia are produced exogenously and ascospores endogenously.
(4) Yeasts have filamentous bodies with long thread-like hyphae.

177. Which of these following methods is the most suitable for disposal of nuclear waste?
(1) Shoot the waste into space.
(2) Bury the waste under Antarctic ice-cover.
(3) Dump the waste within rocks under deep ocean.
(4) Bury the waste within rocks deep below the Earth's surface.

178. Following statements describe the characteristics of the enzyme Restriction Endonuclease. Identify the **incorrect** statement.
(1) The enzyme cuts DNA molecule at identified position within the DNA.
(2) The enzyme binds DNA at specific sites and cuts only one of the two strands.
(3) The enzyme cuts the sugar-phosphate backbone at specific sites on each strand.
(4) The enzyme recognizes a specific palindromic nucleotide sequence in the DNA.

179. In a species, the weight of newborn ranges from 2 to 5 kg. 97% of the newborn with an average weight between 3 to 3.3 kg survive whereas 99% of the infants born with weights from 2 to 2.5 kg or 4.5 to 5 kg die. Which type of selection process is taking place?
(1) Directional selection
(2) Stabilising selection
(3) Disruptive selection
(4) Cyclical selection

180. Select the **correctly** written scientific name of Mango which was first described by Carolus Linnaeus:
(1) *Mangifera indica* Car. Linn.
(2) *Mangifera indica* Linn.
(3) *Mangifera indica*
(4) *Mangifera Indica*

HINTS & SOLUTION

PHYSICS

1. (3) From the given logic circuit, LED will glow, when voltage across LED is high.

Truth Table

A	B	Y
0	0	1
0	1	1
1	0	1
1	1	0

Boolean expression $Y = \overline{A.B}$

i.e., circuit represents NAND gate.

2. (2) Charge Q will be distributed over the surface of hollow metal sphere.

(i) For r < R (inside)

[At a point inside the hollow sphere]

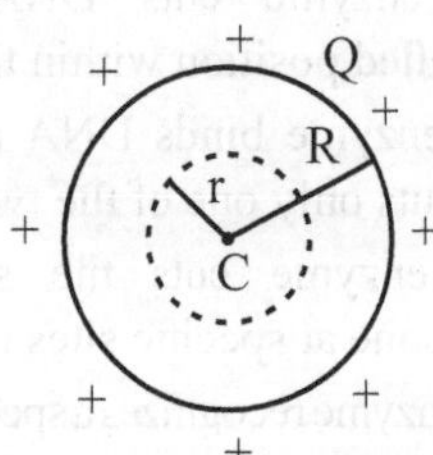

By Gauss's law, $\oint \vec{E}_{in} \cdot \overrightarrow{dS} = \frac{q_{en}}{\varepsilon_0} = 0$

As enclosed charge is = 0

So, $E_{in} = 0$ the electric field inside the hollow sphere is always zero.

(ii) For r > R (outside)

[At a point outside hollow sphere]

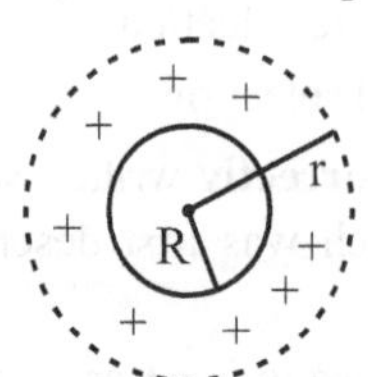

By Gauss's law, $\oint \vec{E}_0 \cdot \overrightarrow{dS} = \frac{q_{en}}{\varepsilon_0}$ ($\because q_{en} = Q$)

$\therefore E_0 4\pi r^2 = \frac{Q}{\varepsilon_0}$

$\therefore E_0 \propto \frac{1}{r^2}$

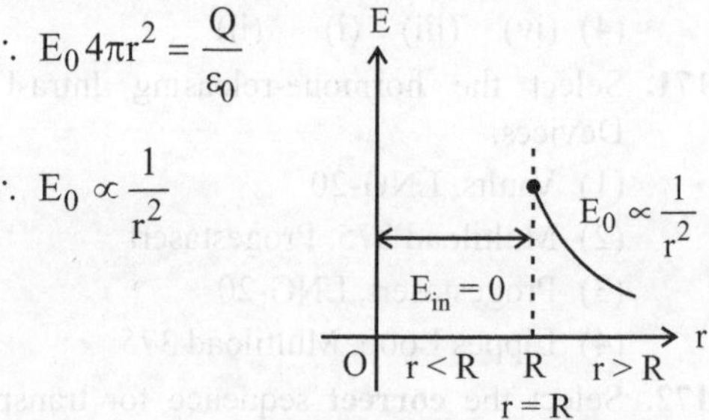

3. (3) As we know that the angle of dip is the angle between earth's resultant magnetic field from horizontal.

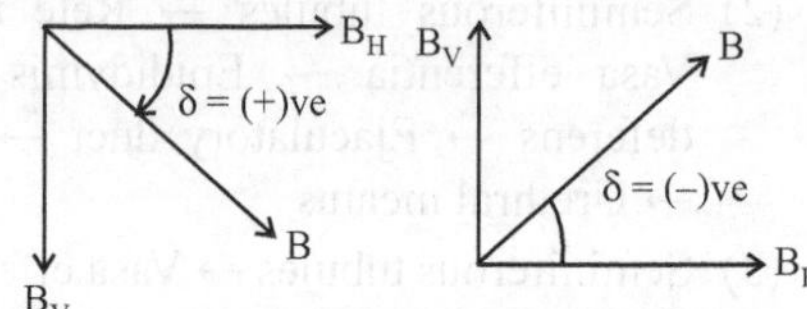

At equator, dip is zero. At Northern hemisphere, dip is positive. At southern hemisphere, dip is negative.

4. (2) For double slit experiment angular fringe width $\theta_0 = \frac{\beta}{D}$

Angular fringe width (in water)

$$\theta_w = \frac{\beta}{\mu D} = \frac{\theta_0}{\mu}$$

$$= \frac{0.2^\circ}{\left(\frac{4}{3}\right)} = 0.15^\circ$$

5. (2) When all bulbs are glowing

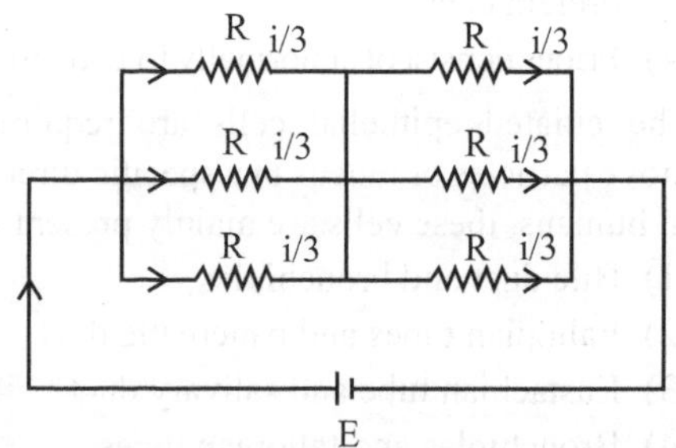

$$R_{eq} = \frac{R}{3} + \frac{R}{3} = \frac{2R}{3}$$

$$\text{Power } (P_i) = \frac{E^2}{R_{eq}} = \frac{3E^2}{2R} \quad \ldots(i)$$

When two from section A and one from section B are glowing, then

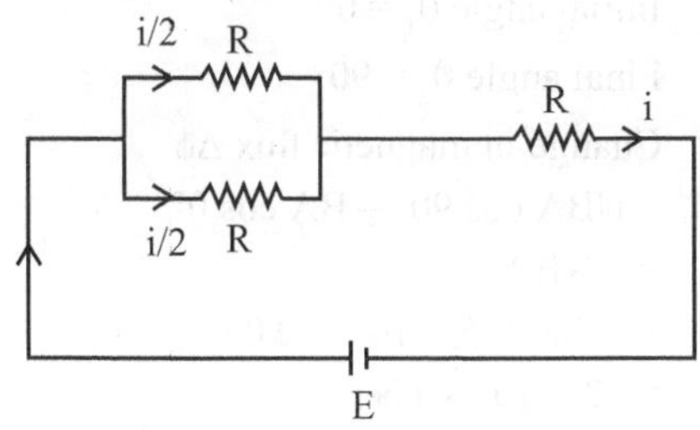

$$R_{eq} = \frac{R}{2} + R = \frac{3R}{2}$$

$$\text{Power } (P_f) = \frac{2E^2}{3R} \quad \ldots(ii)$$

Dividing equation (i) by (ii) we get

$$\frac{P_i}{P_f} = \frac{3E^2 3R}{2R\ 2E^2} = 9:4$$

6. **(2)** In adiabatic process, there is no exchange of heat, with the surroundings.

7. **(3)** Inside (d < R)

Magnetic field inside conductor

$B = Kd$ ..... (i)

This is straight line passing through origin

At surface (d = R)

$$B = \frac{\mu_0}{2\pi}\frac{1}{d}$$

Maximum at surface

Outside (d > R)

$$B = \frac{\mu_0}{2\pi}\frac{1}{d}$$

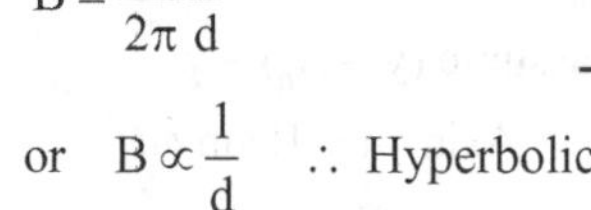

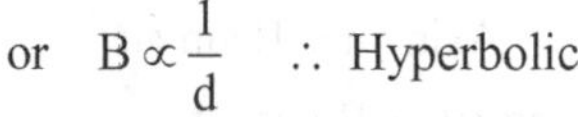

or $B \propto \frac{1}{d}$ $\therefore$ Hyperbolic

8. **(3)** Resistance for ideal voltmeter = ∞

Resistance for ideal ammeter = 0

For I[st] circuit,

$$V_1 = i_1 \times 10 = \frac{10}{10} \times 10 = 10 \text{ volt}$$

For IInd circuit,

$$V_2 = i_2 \times 10 = \frac{10}{10} \times 10 = 10 \text{ volt}$$

$$V_1 = V_2 \text{ and, } i_1 = i_2 = \frac{10V}{10\Omega} = 1A$$

9. **(1)** $\lambda_{Red} > \lambda_{Green} > \lambda_{Blue} > \lambda_{Violet.}$ Red has the longest wavelength.

10. **(2)** $U = \frac{F}{2}nRT$

As $U \propto T$

$\therefore$ Increase in temperature would lead to the increase in kinetic energy of gas.

11. **(4)** At t = 0, y = 3, which is maximum displacement so equation will be cosine function.

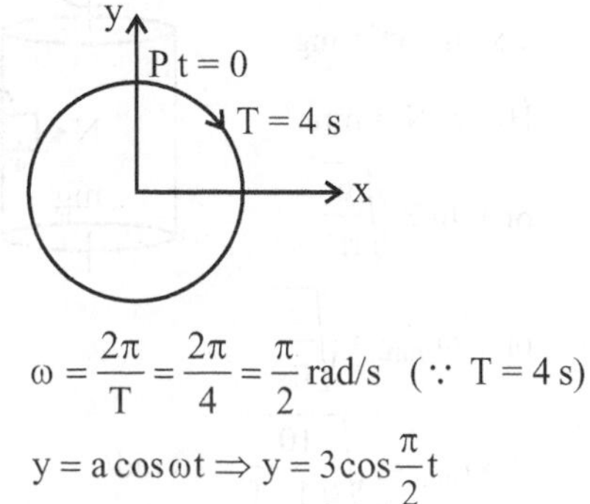

$$\omega = \frac{2\pi}{T} = \frac{2\pi}{4} = \frac{\pi}{2} \text{ rad/s} \quad (\because T = 4 \text{ s})$$

$$y = a\cos\omega t \Rightarrow y = 3\cos\frac{\pi}{2}t$$

12. **(4)** Displacement of the particle in one complete vibration is zero, so, average velocity in one complete vibration will be

$$= \frac{\text{Displacement}}{\text{Time interval}} = \frac{y_f - y_i}{T} = 0$$

13. **(1)** According to the Work-energy theorem,

$$W = \frac{1}{2}I\left(\omega_f^2 - \omega_i^2\right)$$

Given that,

$\theta = 2\pi$ revolution/minute

$\theta = 2\pi \times 2\pi = 4\pi^2$ rad

$$\omega_i = 3 \times \frac{2\pi}{60} \text{ rad/s}$$

$\omega_f = 0$ rad/s

Putting the values of ω_f and ω_i we get

$$\Rightarrow -\tau\theta = \frac{1}{2} \times \frac{1}{2}mr^2\left(0^2 - \omega_i^2\right)$$

$$\Rightarrow -\tau = \frac{\frac{1}{2}\times\frac{1}{2}\times 2\times\left(4\times10^{-2}\right)\left(-3\times\frac{2\pi}{60}\right)^2}{4\pi^2}$$

$\Rightarrow \tau = 2 \times 10^{-6}$ N-m

14. (3) Given mass of block, m = 10 kg; radius of cylindrical drum, r = 1m; coefficient of friction between the block and the inner wall of the cylinder $\mu = 0.1$;

Minimum angular velocity ω_{min}

For equilibrium of the block limiting friction

$f_L \geq mg$

$\Rightarrow \mu N \geq mg$

$\Rightarrow \mu r\omega^2 \geq mg$

Hrere, $N = mr\omega^2$

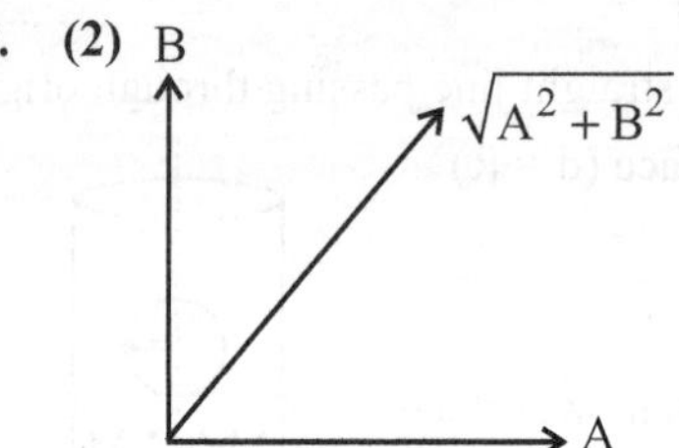

or, $m \geq \sqrt{\frac{g}{r\mu}}$

or, $\omega_{min} = \sqrt{\frac{g}{r\mu}}$

$\therefore \omega_{min} = \sqrt{\frac{10}{0.1\times1}} = 10$ rad/s

15. (1) Velocity of swimmer w.r.t. river $V_{SR} = 20$ m/s

Velocity of river w.r.t. ground $V_{RG} = 10$ m/s

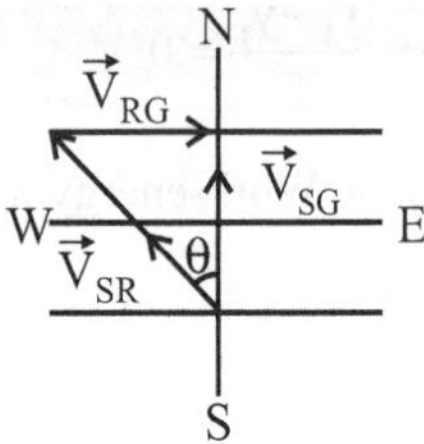

$\vec{V}_{SG} = \vec{V}_{SR} + \vec{V}_{RG}$

$\sin\theta = \left|\frac{\vec{V}_{RG}}{\vec{V}_{SR}}\right| \Rightarrow \sin\theta = \frac{10}{20}$

$\Rightarrow \sin\theta = \frac{1}{2} \quad \therefore \theta = 30°$ west

i.e.,to cross the river along the shortest path, swimmer should make his strokes 30° west.

16. (4) Given :

Magnetic field $B = 5 \times 10^{-5}$ T

Number of turns in coil N = 800

Area of coil A = 0.05 m^2

Time taken to rotate = $\Delta t = 0.1$ s

Initial angle $\theta_1 = 0°$

Final angle $\theta_2 = 90°$

Change in magnetic flux $\Delta\phi$

= NBA cos 90° – BA cos 0°

= – NBA

$= -800 \times 5 \times 10^{-5} \times 0.05$

$= -2 \times 10^{-2}$ weber

$$\theta = -\frac{\Delta\phi}{\Delta t} = \frac{-(-)2\times10^{-3}\text{ Wb}}{0.1\text{ s}} = 0.02 \text{ V}$$

[Note : This is average emf induced]

17. (2) Given, $x = \frac{A^2B^{1/2}}{C^{1/3}D^3}$

$$\% \text{ error}, \frac{\Delta x}{x}\times100 = 2\frac{\Delta A}{A}\times100 + \frac{1}{2}\frac{\Delta B}{B}\times 100 + \frac{1}{3}\frac{\Delta C}{C}\times100 + 3\frac{\Delta D}{D}\times100$$

$= 2\times1\% + \frac{1}{2}\times2\% + \frac{1}{3}\times3\% + 3\times4\%$

= 2% + 1% + 1% + 12% = 16%

18. (2)

Graph: B on vertical axis, A on horizontal axis, resultant $\sqrt{A^2+B^2}$.

Given equation

$y = A_0 + A\sin\omega t + B\sin\omega t$

Now assume $(y - A_0) = \gamma$

$y - A_0 = A\sin\omega t + B\sin\omega t$

$\gamma = A\sin\omega t + B\cos\omega t$

$= \sqrt{A^2+B^2}\ \sin(\omega t + \phi)$

which is S.H.M.

where $\cos\phi = \frac{A}{\sqrt{A^2+B^2}}$

and $\sin\phi = \frac{B}{\sqrt{A^2+B^2}}$

so, resultant amplitude

$= \sqrt{A^2+B^2}$

19. **(2)** de-Broglie wavelength of electron

$$\lambda = \frac{12.27}{\sqrt{V}}\text{Å} = \frac{12.27\times10^{-10}}{\sqrt{10000}} = 12.27\times10^{-12}\ \text{m}$$

20. **(3)** $T - mg = \frac{mv^2}{r}$ [centripetal force $= \frac{mv^2}{r}$]

$$\Rightarrow \quad T = mg + \frac{mv^2}{r}$$

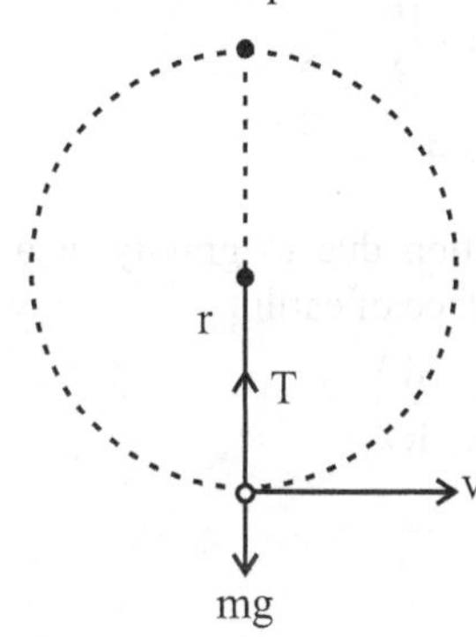

As the velocity is maximum at lowest point so tension is maximum at the lowest position of mass, so the chance of breaking is maximum.

21. **(4)** Let T_A and T_B are the time periods of particle A and B respectively. According to question,

$T_A = T_B = T$

If ω_A and ω_B are their angular speeds, then

$$\omega_A = \frac{2\pi}{T_A} \text{ and } \omega_B = \frac{2\pi}{T_B}$$

$$\therefore \quad \frac{\omega_A}{\omega_B} = \frac{T_B}{TA} = \frac{T}{T} = 1:1$$

22. **(4)** $l'_{Cu} = l_{Cu}(1 + \alpha_{Cu}\Delta T)$

$\Delta l_{Cu} = l_{Cu}\alpha_{Cu}\Delta T$

$l'_{Al} = l_{Al}(1 + \alpha_{Al}\Delta T)$

$\Delta l_{Al} = l_{Al}\alpha_{Al}\Delta T$

Since, change in length is independent of temperature

$\therefore \alpha_{Cu} l_{Cu} = \alpha_{Al} l_{Al}$

$\Rightarrow 1.7\times10^{-5}\times 88\text{ cm} = 2.2\times10^{-5}\times l_{Al}$

$\Rightarrow l_{Al} = \frac{1.7\times88}{2.2} = 68\text{ cm}$

23. **(1)** Radius of the path $= r = \frac{mv}{qB} = \frac{P}{qB}$

For H^+ion, $r_H = \frac{p_H}{eB}$

For α particle

$$r_\alpha = \frac{p_\alpha}{2eB}$$

$$\frac{r_H}{r_\alpha} = \frac{\frac{p}{eB}}{\frac{p}{2eB}} \quad \text{[as given } p_H = p_\alpha = p]$$

$$\Rightarrow \frac{r_H}{r_\alpha} = \frac{2}{1}$$

24. **(3)** Here,

$Kx_0 = Mg$

where K = force constant

$$\Delta E = \frac{1}{2}Kx_0^2$$

$$= \frac{1}{2}\frac{Mg}{x_0}\times x_0^2$$

$$= \frac{1}{2}Mgx_0$$

Stored elastic potential energy in extended wire, $= \frac{1}{2}Mg\ell$ [here $x_0 = \ell$]

25. **(2)** In p-type semiconductor, trivalent impurities are added to intrinsic semiconductor, which creates holes which are majority charge carriers.

26. **(3)** Mass to be raised = *m*

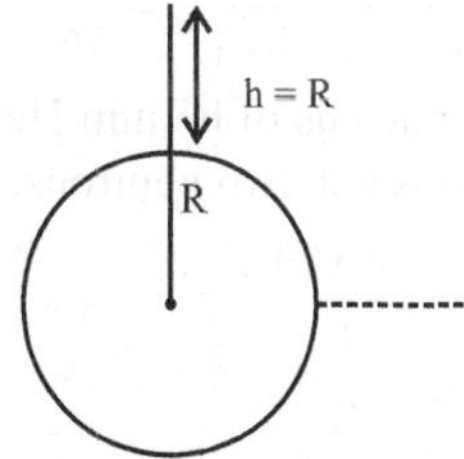

Potential energy at the surface of the earth

$$U_{surface} = \frac{-GMm}{R}$$

Potential energy at a height from the surface of the earth h = R

$$U_{height} = \frac{-GMm}{2R}$$

According to work-energy theorem, work done = change in PE

$$\therefore \; W = U_{height} - U_{surface}$$

$$\Rightarrow \frac{-GMm}{2R} - \left(-\frac{GMm}{R}\right)$$

$$= \frac{GMm}{2R} = \frac{gR^2m}{2R} = \frac{mgR}{2} \quad \left(\because GM = gR^2\right)$$

27. (4) For total internal reflection when $i = i_c$, then

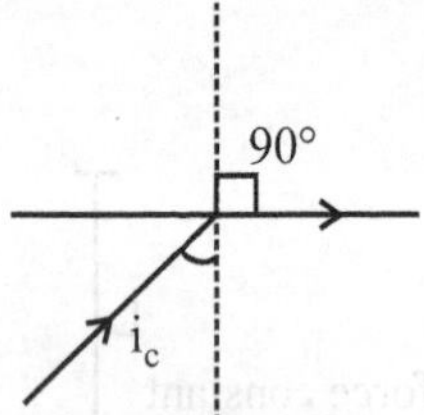

refracted ray grazes with the surface. That means the angle of refraction r = 90°.

28. (2) We know that,

$$F = \frac{kQ^2}{r^2} \qquad +Q \; A \longrightarrow r \longleftarrow B \; -Q$$

If 25% of charges of A transfer to B then

$$q_A = Q - \frac{Q}{4} = \frac{3Q}{4} \text{ and } q_B = -Q + \frac{Q}{4} = \frac{-3Q}{4}$$

$$q_A \bullet \text{———} r \text{———} \bullet q_B$$

$$F_1 = \frac{kq_Aq_B}{r^2}$$

$$\Rightarrow F_1 = \frac{k\left(\frac{3Q}{4}\right)^2}{r^2} = \frac{9}{16}\frac{kQ}{r^2} = \frac{9F}{16}$$

29. (1) α-particle is nucleus of helium He^{++} which has two protons and two neutrons.

30. (2) $m_1u_1 + m_2u_2 = m_1v_1 + m_2v_2$

Also, $\frac{1}{2}m_1u_1^2 + \frac{1}{2}m_2u_2^2$

$$= \frac{1}{2}m_1v_1^2 + \frac{1}{2}m_2v_2^2$$

For elastic collision

$$v_2 = \frac{2m_1u_1}{m_1 + m_2}$$

as $u_2 = 0$; $u_1 = u$

$m_1 = 4m$ and $m_2 = 2m$

so, $v_2 = \frac{4}{3}u$

$$\frac{\Delta KE}{KE} = \frac{\frac{1}{2}\times 2m\times\left(\frac{4}{3}u\right)^2}{\frac{1}{2}\times 4m\times u^2}$$

$$= \frac{\frac{1}{2}\times 2\times\frac{16}{9}}{\frac{1}{2}\times 4} = \frac{8}{9}$$

31. (4) Acceleration due to gravity at a depth d from surface of earth

$$g' = g\left(1 - \frac{d}{R}\right) \qquad \text{...(i)}$$

Here, —

$$\therefore \; g' = \frac{g}{2}$$

Multiplying by mass 'm' on both sides of equation (i) we get

$$W' = W\left(1 - \frac{d}{R}\right) \qquad (\because W = mg)$$

$$= 200\left(1 - \frac{R}{2R}\right) = \frac{200}{2} = 100 \text{ N}$$

32. (3) Rainbow will be observed only when the sun is at the back side of observer.

33. (3) Work done by variable force

Work done, $W = \int_{y_i}^{y_f} F dy \Rightarrow \int_{y=0}^{y_f=1} F.dy$

where, F = 20 + 10 y

$$\therefore \; W = \int_0^1 (20 + 10y)\,dy$$

$$= \left[20y + \frac{10y^2}{2}\right]_0^1 = 25 \text{ J}$$

34. (1) Work done to stop the disc = change in total kinetic energy of disc

Final KE = 0

Initial KE = Translational K.E. + Rotational K.E.

$$= \frac{1}{2}mv^2 + \frac{1}{2}I\omega^2$$

$$= \frac{1}{2}mv^2 + \frac{1}{2} \times \frac{mR^2}{2} \times \left(\frac{v}{R}\right)^2$$

$$= \frac{1}{2}mv^2 + \frac{1}{4}mv^2 = \frac{3}{4}mv^2$$

$$= \frac{3}{4} \times 100 \times (20 \times 10^{-2})^2 = 3J$$

$|\Delta KE| = 3J$

35. (1) Volumetric flow rate of liquid

$Q = au = a\sqrt{2gh}$

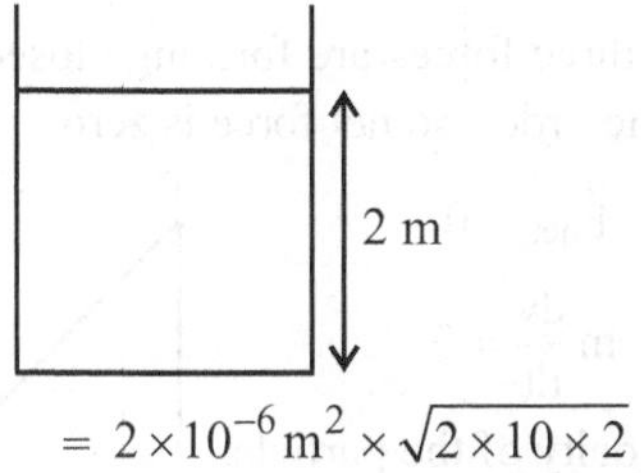

$$= 2 \times 10^{-6} m^2 \times \sqrt{2 \times 10 \times 2}\ m/s$$

$= 2 \times 2 \times 3.14 \times 10^{-6}\ m^3/s$

$= 12.56 \times 10^{-6}\ m^3/s$

$= 12.6 \times 10^{-6}\ m^3/s$

36. (3) When object shot from inclined plane kept at 60° and velocity say u, then distance travel

$$x_1 = \frac{u^2}{2g\sin 60°}$$

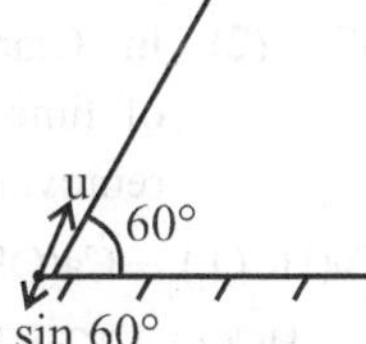

When object shot from inclined plane kept at 30° and same velocity u, then distance travel

$$x_2 = \frac{u^2}{2g\sin 30°}$$

u

30°

g sin 30°

$$\therefore \frac{x_1}{x_2} = \frac{\frac{u^2}{2g\sin 60°}}{\frac{u^2}{2g\sin 30°}} = \frac{\sin 30°}{\sin 60°} = \frac{1 \times 2}{2 \times \sqrt{3}} = 1 : \sqrt{3}$$

37. (2) Given, capacitance of capacitor C = 20 μF $= 20 \times 10^{-6}$ F

Rate of change of potential $\left(\frac{dV}{dt}\right) = 3$ v/s

$\because q = CV$

or, $\frac{dq}{dt} = C\frac{dV}{dt}$

or, $i_C = 20 \times 10^{-6} \times 3 = 60 \times 10^{-6}\ A = 60\ \mu A$

As we know that $i_D = i_C = 60\ \mu A$

[Alternate solution : we know that $i_d = i_c$ and the displacement current has finite value when p.d across capacitor is changing. So, (1), (3) and (4) are eliminated. Hence (2)]

38. (4) In steady state, the amount of heat flowing from one face to the other face in time dt is given by

$$H = \frac{kA(T_1 - T_2)dt}{\ell}$$

$$\Rightarrow \frac{dH}{dt} = \frac{kA}{\ell}\Delta T \text{ (k = coefficient of}$$

thermal conductivity)

$$\therefore \quad = \frac{dH}{A\,dt \quad T}$$

Unit of k = $Wm^{-1}\ K^{-1}$

39. (4) Fuse is an safety device that operates to provide over current protection of an electrical circuit. A fuse is mainly a metal wire that melts when too much current flows through it due to low melting point and protects electric appliances.

40. (3) Pressure inside the soap bubble

$$= P_0 + \frac{4T}{R}$$

And pressure at a point below the surface of water = $\rho g Z_0 + P_0$

[where P_0 = atmospheric pressure]

By equating these pressure we get,

$$P_0 + \frac{4T}{R} = P_0 + \rho g Z_0$$

$$Z_0 = \frac{4T}{R \times \rho g}$$

$$Z_0 = \frac{4 \times 2.5 \times 10^{-2}}{10^{-3} \times 1000 \times 10} m$$

$Z_0 = 1$ cm

41. **(3)** According to Bohr's model of H-atom, the relation between kinetic energy, potential energy and total energy

$$K.E. = |TE| = \frac{|P.E.|}{2}$$

so, K.E. = – (T.E.)

$\therefore$ K.E. = 3.4 eV

and P.E. = 2 (T.E)

$\Rightarrow$ P.E. = – 6.8 eV

42. **(2)**

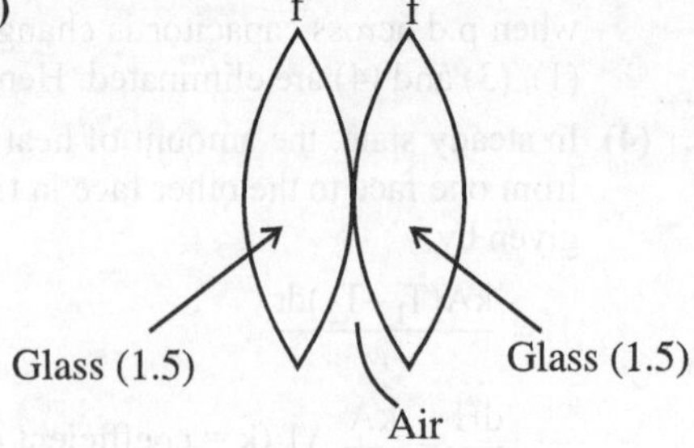

Equivalent focal length in air

$$\frac{1}{F_1} = \frac{1}{f} + \frac{1}{f} = \frac{2}{f}$$

$$\Rightarrow F_1 = \frac{f}{2} \quad ...(i)$$

When glycerin is filled inside, it behaves like a concave lens of focal length (–f)

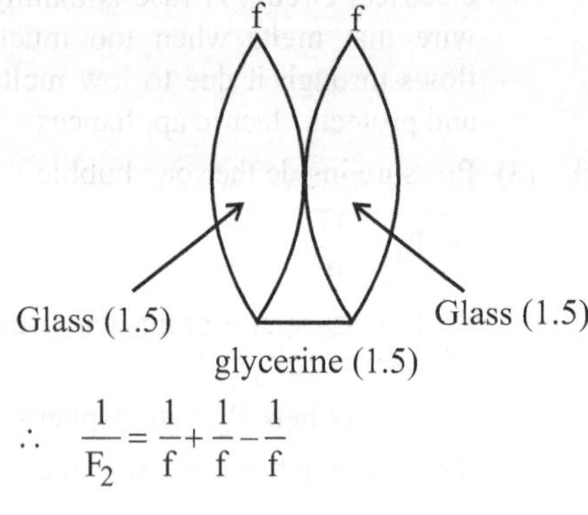

$$\therefore \quad \frac{1}{F_2} = \frac{1}{f} + \frac{1}{f} - \frac{1}{f}$$

$$\Rightarrow F_2 = f \quad ...(ii)$$

Dividing (i) by (ii), we get

$$\frac{F_1}{F_2} = \frac{1}{2}$$

43. **(3)** Electric field due to line charge (1)

$$\vec{E}_1 = \frac{\lambda}{2\pi\varepsilon_0 R}\hat{i}\,N/C$$

Electric field due to line charge (2)

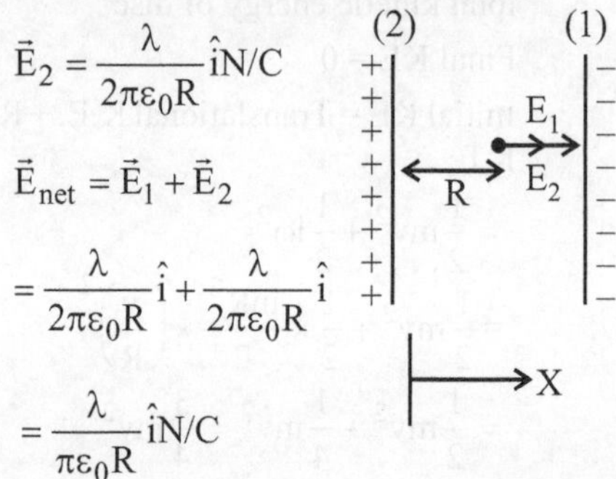

$$\vec{E}_2 = \frac{\lambda}{2\pi\varepsilon_0 R}\hat{i}\,N/C$$

$$\vec{E}_{net} = \vec{E}_1 + \vec{E}_2$$

$$= \frac{\lambda}{2\pi\varepsilon_0 R}\hat{i} + \frac{\lambda}{2\pi\varepsilon_0 R}\hat{i}$$

$$= \frac{\lambda}{\pi\varepsilon_0 R}\hat{i}\,N/C$$

[Note : Both $\vec{E}_1$ and $\vec{E}_2$ are in the same direction.]

44. **(4)** An electric heater works on the principle of Joule's heating effect and it does not involve Eddy currents.

45. **(3)** As three forces are forming closed loop in same order, so net force is zero.

i.e., $\vec{F}_{net} = 0$

or $m\frac{d\vec{v}}{dt} = 0$

$\therefore$ Velocity of the particle,

$\vec{v}$ = constant

P

Q

R

CHEMISTRY

46. **(2)** Consider the stability of electronic configuration after loss of one electron.

47. **(2)** In Clark's method calculated amount of lime is added to hard water for the removal of temporary hardness.

$$Ca(HCO_3)_2 + Ca(OH)_2 \longrightarrow 2CaCO_3\downarrow + 2H_2O$$

$$Mg(HCO_3)_2 + 2Ca(OH)_2 \longrightarrow 2aCO_3 + Mg(OH)_2\downarrow + 2H_2O$$

48. **(4)** Amphoteric hydroxide means it can react with both acid and base.

$$Be(OH)_2 + 2HCl \longrightarrow BeCl_2 + 2H_2O$$

$$Be(OH)_2 + 2NaOH \longrightarrow Na_2[Be(OH)_4]$$

49. **(1)** Narrow spectrum antibiotics are used for the specific infection when the causative organism is known. Penicillin G is narrow spectrum antibiotic used to treat infections caused by susceptible bacteria.

50. **(2)** $AgNO_3 + KI \longrightarrow \underset{\text{Negatively charged colloid}}{AgI} + KNO_3$

A solution of $AgNO_3$ and KI will form a negatively charged colloidal sol, $[AgI]I^-$, only when KI is present in excess (i.e., KI behaves as a solvent).

Millimole of KI is maximum in option (2) $(50 \times 2 = 100)$

51. **(3)** When a proton is removed from an acid, we obtain its conjugate base.

$H_2O \rightleftharpoons H^+ + OH^-$

$HF \rightleftharpoons H^+ + F^-$

52. **(1)** One double bond and one triple bond means 3π bonds $\therefore$ option (1)

$$H-\overset{H}{\underset{H}{\overset{|_1}{C}}}-\overset{H}{\overset{|_2}{C}}=\underset{H}{\overset{3}{C}}-\overset{4}{C}\equiv\overset{5}{C}-H$$

Pent-2-en-4-yne

53. **(1)**

$$O_2\overset{+6}{Br}-\overset{+4}{Br}(=O)_2-\overset{+6}{Br}O_2 \quad (\text{each } O=Br)$$

Br_3O_8

$3x - (2 \times 8) = 0$

$3x - 16 = 0$

$3x = 16$

$x = \dfrac{16}{3}$

Fractional oxidation state is the average oxidation state of the element.

$\dfrac{6+4+6}{3} \quad \dfrac{16}{3}$

54. **(1)** $Ca(OH)_2 \rightleftharpoons Ca^{2+} + 2OH^-$

$pH = 9, pOH = 14 - 9 = 5$

$[OH^-] = 10^{-5}$

$\left[Ca^{2+}\right] = \dfrac{10^{-5}}{2}$

$K_{sp} = [Ca^{2+}]\,[OH^-]^2$

$= \left(\dfrac{10^{-5}}{2}\right) \times (10^{-5})^2$

$= 0.5 \times 10^{-15}$

55. **(1)** Account for the inductive effect, solvation effect (H-bonding.) and steric hinderance for basic character in aqueous solutions

$(CH_3)_2NH > CH_3NH_2 > (CH_3)_3N$

56. **(3)** $E^\circ_{cell} = \dfrac{2.303RT}{nF}\log K$

Given : $E^\circ_{cell} = 0.59\,V$,

$n = 1$

$0.59 = \dfrac{0.059}{1}\log K$

$\dfrac{0.59}{0.059} = \log K$

$10 = \log K$

$K = 10^{10}$

57. **(4)** Green house gases present in the Earth's atmosphere are water vapours, carbondioxide, methane, nitrous oxide and ozone. Sulphurdioxide is not a green house gas.

58. **(1)** The solutions which show a large negative deviation from Raoult's law form maximum boiling azeotrope. Nitric acid and water forms a maximum boiling azeotrope with a boiling point of 393.5 K

59. **(4)** Malachite is a copper carbonate hydroxide mineral with the formula $CuCO_3.Cu(OH)_2$.

60. **(4)**

Pure nitrogen	Sodium azide or Barium azide
Haber process	Ammonia
Contact process	Sulphuric acid
Deacon's process	Chlorine

61. **(3)** On going down the group, bond dissociation enthalpy of the hydrides of oxygen family decreases. Therefore, thermal stability also decreases.

62. **(4)** PCl_5 is very reactive due to the presence of weak axial bonds. It is used in the synthesis of various organic compounds.

63. (2) Phthalic acid (benzene-1,2-dicarboxylic acid, $C_6H_4(COOH)_2$) $+ NH_3 \longrightarrow$

$C_6H_4(COO^{\oplus}NH_4)_2 \xrightarrow[-2H_2O]{\Delta} C_6H_4(CONH_2)_2 \xrightarrow[-NH_3]{\text{Strong Heating}}$ phthalimide ($C_6H_4(CO)_2NH$)

64. (4) In phenol ($C_6H_5\ddot{O}H$) lone pair of e^-s present on oxygen atom is involved in delocalisation. Therefore, it is most difficult to protonate the phenol amongst the given compounds, due to less availability of e^-s.

65. (1) Manganate ion: Mn bonded to O (double bond), O (double bond), O^-, O^-

Permanganate ion: Mn bonded to O^-, O (double bond), O (double bond), O (double bond)

66. (2) Alkynes can be reduced to cis-alkenes with the use of Lindlar's catalyst

67. (2) Cumene ($C_6H_5CH(CH_3)_2$) $\xrightarrow{O_2}$ Cumene hydroperoxide ($C_6H_5C(CH_3)_2{-}O{-}O{-}H$)

$\xrightarrow[H_2O]{H^+}$ Phenol (C_6H_5OH) $+ H_3C{-}CO{-}CH_3$

68. (3) For a first order reaction,

$$t = \frac{2.303}{k} \log \frac{[A]_o}{[A]}$$

for 99% completion of the reaction,

$$t = \frac{2.303}{k} \log \frac{100}{100-99}$$

$$t = \frac{2.303}{k} \log 10^2$$

$$t = \frac{4.606}{k}$$

69. (1) In a disproportionation reaction, one species undergoes both oxidation and reduction.

$$2\overset{+}{Cu} \longrightarrow \overset{2+}{Cu} + Cu^{\circ}$$

(Reduction / Oxidation)

$$3\overset{+6}{Mn}O_4^{2-} + 4H^+ \longrightarrow 2\overset{+7}{Mn}O_4^- + \overset{+4}{Mn}O_2 + 2H_2O$$

(Reduction / Oxidation)

70. (4) In $2H(g) \longrightarrow H_2(g)$, no. of species decreases, therefore entropy decreases.

71. (2) Nylon 2-nylon 6 is a biodegradable polymer.

72. (3) Compressibility factor, $Z = \frac{PV}{nRT}$

Given: At 350 K and 15 bar,

molar volume < volume of ideal gas

$\therefore Z < 1$

Therefore, attractive forces are dominant and the gas can be compressed easily.

73. (3)

	Oh void (C)	:	HCP (A)
No of ions ⇒	$6 \times \frac{75}{100}$	:	6
	$\frac{3}{4}$	:	1
	3	:	4 C_3A_4

74. (2) Enzyme that utilise ATP in phosphate transfer require an alkaline earth metal (M) Mg as the cofactor.

75. (1)

	$n + \ell$
5*f*	5 + 3 = 8
6*p*	6 + 1 = 7
5*p*	5 + 1 = 6
4*d*	4 + 2 = 6

5*f* > 6*p* > 5*p* > 4*d*

76. **(1)** $\Delta G = -nFE^\circ$
$= -2 \times 96500 \times 0.24 = -46320$ J/mol
$= -46.32$ kJ/mol

77. **(3)** ΔH_{mix} is zero at constant T and P.

78. **(3)** Alanine is non-essential amino acid.

79. **(2)** $K_4\left[Fe(CN)_6\right]^{4-} \longrightarrow 4K^+ + [Fe(CN)_6]^{4-}$
$x - 6 = -4$
$x = +2$
Fe: $4s^2\ 3d^6$
Fe^{2+} : $3d^6$

⇅	↑	↑	↑	↑

In spherical field

e_g: [ ] [ ]
t_{2g}: [⇅] [⇅] [⇅]

In the presence of $6CN^-$ (strong field ligand)

80. **(1)** PbF_4 is ionic in nature

81. **(2)**

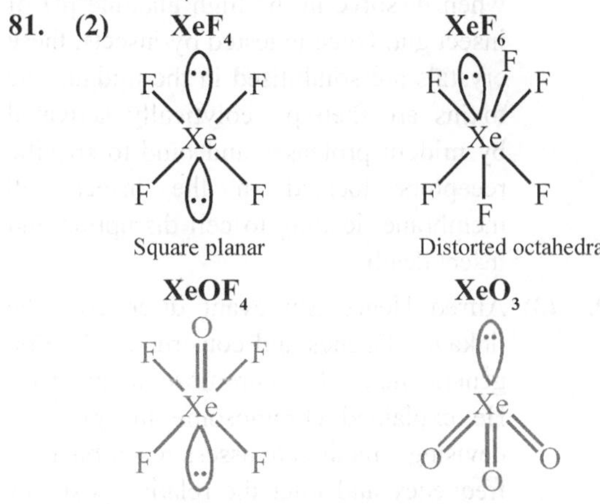

Square planar Distorted octahedral

Square pyramidal Pyramidal

82. **(3)**

$$CH_3-\underset{CH_3}{\overset{|}{C}}=CH-CH_3$$

$\xrightarrow{O_3/Zn/H_2O}$ $CH_3-\overset{O}{\overset{||}{C}}-CH_3$ + CH_3CHO

$\xrightarrow{HCl}$ $CH_3-\underset{Cl}{\overset{CH_3}{C}}-CH_2-CH_3$

83. **(3)** $N_2(g) + 3H_2(g) \rightleftharpoons 2NH_3(g)$

$$\frac{-d[N_2]}{dt} = \frac{-1}{3}\frac{d[H_2]}{dt} = \frac{1}{2}\frac{d[NH_3]}{dt}$$

84. **(1)** $W = -P_{ext}(V_2 - V_1)$ (Irreversible isothermal expansion)
$= -2\,(0.25 - 0.1)$
$= -2\,(0.15)$
$= -0.3$ L bar
$= -0.3 \times 100$ J
$= -30$ J

85. **(2)** (i) $AlCl_3 + C\ell - C\ell \rightarrow [AlCl_4]^- + Cl^+$
Electrophile

(ii) Benzene $+ Cl^+ \longrightarrow$ Chlorobenzene (C₆H₅Cl)

86. **(3)** Only π bond is present in C_2 molecule.
$\sigma 1s^2 \sigma^* 1s^2 \sigma 2s^2 \sigma^* 2s^2 \pi 2p_x^2 = \pi 2p_y^2$

87. **(1)** $[SiCl_6]^{2-}$ does not exist because six large chloride ions cannot be accommodated around Si^{4+} due its small size.

88. **(3)** Basic Buffer is made up of weak base & its salt.

	HCl +	$NH_4OH \longrightarrow$	$NH_4Cl + H_2O$
Initial	10 m mol	20 m mol	0
Final	0	10 m mol	10 m mol

89. **(2)** Balmer series

90. **(3)** $N_2 + 3H_2 \longrightarrow 2NH_3$

$1 \text{ Mol } NH_3 = \frac{3}{2} \text{ mol } H_2$

$20 \text{ mol } NH_3 = \frac{3}{2} \times 20 \text{ mol } H_2 = 30 \text{ mol } H_2$

$\therefore$ 30 moles of H_2 are required.

BIOLOGY

91. **(2)** Glycolysis is the anaerobic enzymatic conversion of glucose to lactate or pyruvate, resulting in energy stored in the form of ATP, as occurs in muscle.

The two irreversible steps in Glycolysis are:

(i) Conversion of glucose to glucose 6-phosphate, catalysed by hexokinase. It is the first step of activation phase of glycolysis.

(ii) Conversion of fructose 6-phosphate to fructose 1,6 biphosphate catalysed by phosphofructokinase-1.

92. (3) During flowering, photoperiodic stimulus is perceived by leaves of plants.

93. (1) Golden rice is a form of rice with biosynthesis of beta-carotene (a form of vitamin A) which gives its golden or yellow colouring. It is considered a type of genetically modified rice (GMO) and a fortified food that is produced to combat a vitamin A deficiency in areas where this vitamin is scarce.

It was in the late 1990s that two European scientists developed Golden Rice after inserting beta carotene-enriched rice gene from daffodil to rice.

94. (4) Typhoid is a bacterial infection that can lead to a high fever, diarrhoea, and vomiting. It is caused by the bacteria *Salmonella typhimurium (S. typhi).* Typhoid is diagnosed by Widal test. This test was developed by Georges Ferdinand Widal in 1896 and helps to detect presence of salmonella antibodies in a patient's serum.

95. (4) Colostrum, as a thin yellowish milk, is the first secretion for a few days after childbirth. Colostrum serves as an immune system enhancer to the new born infants. It contains immune cells, high protein and protective antibody (such as IgA) content. IgA help protect the mucous membranes in the throat, lungs, and intestines of the infant.

96. (1) A pineapple plant flowers only once, and produces one pineapple. Then it dies. But before it dies it also produces offspring.

Auxin and ethylene can be applied to artificially induce flowering in pine apple to increase yield throughout the year. Applied auxin mimic the action of ethylene by stimulating ethylene formation, and that ethylene, not auxin, causes pineapples to flower. Ethylene initiates flowering and synchronising fruit-set in pineapples.

97. (2) During the isolation of desired gene, chilled ethanol is used for the precipitation of DNA. Ethanol is used in DNA extraction to force the DNA to precipitate in a solution. In order to collect a DNA sample, cells are broken down through agitation, and then mixed with water, salt and ethanol to create an aqueous solution. Ethanol along with salt work to prevent the DNA from dissolving into the water, instead causing it to precipitate out so it can be separated and extracted using a centrifuge.

98. (3) *Bacillus thuringiensis* (Bt) is a Gram positive, spore-forming bacterium that synthesises parasporal crystalline inclusions containing Cry and Cyt proteins, some of which are toxic against a wide range of insect orders (for example, moth larva that attacks the fruiting bodies of certain crop), nematodes and human-cancer cells.

These toxins are produced in inactive protoxins form but become activated when dissolve in the high alkaline pH of insect gut. Once ingested by insects, these crystals are solubilised in the midgut, the toxins are then proteolytically activated by midgut proteases and bind to specific receptors located in the insect cell membrane, leading to cell disruption and insect death.

99. (3) Alfred Henry Sturtevant discovered the linkage of genes and constructed the first genetic map of a chromosome in 1913. He explained chromosome mapping by devising a method to assess recombination frequency and infer the relative positions and distances of linked genes on a chromosome based on the average number of crossovers between them during meiosis.

100. (1) Habitat loss and fragmentation are the most important cause for animals and plants being driven to extinction. Habitats of various organisms are altered or destroyed by uncontrolled and unsustainable human activities (such as deforestation, slash and burn agriculture, mining, and urbanisation). This will lead to breaking up of the habitat into small pieces, which effect the movement of migratory animals and also, decrease the genetic exchange between populations leading to a declination of species.

101. (2) Goblet cells secretes mucus. It is secreted along entire digestive tract where it lubricates food and lining, coats lining and protects it from mechanical digestion from acid and digestive enzymes.

102. (1) Electrocardiogram is a diagnostic tool, used to assess the electrical and muscular functions of the heart. It consists of waveform components which indicate electrical events during one heart beat. These waveforms are labelled P, Q, R, S, T and U.

- P wave indicates atrial depolarisation *i.e.* that the atria are contracting, pumping blood into the ventricles.
- The QRS complex represents ventricular depolarisation and contraction.
- T wave is upwards waveform representing ventricular repolarisation. The normal T wave is slightly asymmetric with a steeper downward slope. Reduction in the size of T wave represents insufficient supply of oxygen *i.e.* coronary ischemia. Coronary ischemia is an intermediate condition in coronary artery disease during which the heart tissue is slowly or suddenly starved of oxygen and other nutrients.

103. (3) Crypts of Lieberkuhn are tubular glands that lie between the finger-like projections of the inner surface of the small intestine. Glisson's capsule is a layer of connective tissue present in liver. Islets of langerhans constitute the endocrine portion of pancreas and are responsible for the production and release of hormones that regulate glucose levels. Brunner's glands are found in submucosa of duodenum and protect the intestinal walls from gastric juices.

104. (1) G_0 or resting phase is a non-growing state distinct from interphase. It is a period in the cell cycle in which cells exist in a quiescent state. It is viewed as either an extended G_1 phase, where the cell is neither dividing nor preparing to divide, or a distinct quiescent stage that occurs outside of the cell cycle.

105. (2) Earth Summit (Rio Summit)-1992, called upon all nations to take appropriate measures for conservation of biodiversity and sustainable utilisation of its benefits.

106. (4) GLUT1 and GLUT4 are glucose transport proteins which facilitate glucose transport into insulin-sensitive cells. GLUT1 is insulin-independent and is widely distributed in different tissues. GLUT4 is insulin-dependent and is responsible for the majority of glucose transport into muscle and adipose cells in anabolic conditions.

107. (4) Growth rings are formed by the seasonal activity of cambium. In plants of temperate regions, cambium is more active in spring (less lignin deposition in xylem) and less active in autumn (more lignin deposition in xylem) seasons. In temperate regions climatic conditions are not uniform throughout the year. However in tropics climatic conditions are uniform throughout the year.

108. (3) Insulin is a pancreatic hormone, secreted from alpha cells of islets of Langerhans. Its deficiency leads to diabetes mellitus (characterised by high blood sugar levels over a prolonged period).

Thyroxine is an iodine containing hormone, secreted by the thyroid gland. It plays an important role in regulating the body's metabolic rate, heart and digestive function, muscle control, brain development and bone maintenance. Hyper secretion or hyposecretion of thyroxine can be associated with enlargement of thyroid gland called goitre.

Deficiency of corticoids (cortisol and aldosterone) secreted from adrenal gland (cortex region) leads to Addison's disease. Cortisol regulates the body's reaction to stressful situations. Aldosterone helps with sodium and potassium regulation. The adrenal cortex also produces sex hormones (androgens).

Acromegaly is an abnormal growth of the hands, feet, and face, caused by overproduction of growth hormone by the pituitary gland.

109. (4) The phenomenon in which female gamete develops into embryo without getting fused with male gamete (fertilisation) is called parthenogenesis.

Autogamy is a form of self-fertilisation in which fission of the cell nucleus occurs without division of the cell, the two pronuclei so formed reuniting to form the synkaryon. Parthenocarpy is the production of fruit without fertilisation of an egg in the ovary. Synkaryon is a nucleus formed by the fusion of two pre-existing nuclei.

110. (4) Ecological pyramid is a graphical representation that illustrates the flow of energy through different forms of life in an ecosystem. It is generally inverted in case of pyramid of biomass in a sea because the biomass of the fishes is more than the phytoplanktons (producers).

Features of pyramid of biomass in a sea:

- Food chain in ocean - Phytoplankton → zooplankton → predatory fish
- The biomass of consumers (zooplankton) is greater than phytoplankton (the producer) because of their small size and low weight. The predatory fish is much larger than zooplankton.
- The producers of ocean or aquatic ecosystem are phytoplankton and they have short life span and turn over and have lower mass than zooplankton.
- The biomass of tropic level is depends on the longevity of member.

111. (1) Extrusion of second polar body from egg nucleus occurs after entry of sperm into the secondary oocyte but before fertilisation. The entry of sperm into the ovum induces completion of the meiotic division of the secondary oocyte. Entry of sperm causes breakdown of metaphase promoting factor (MPF) and turns on anaphase promoting complex (APC).

112. (2) Fungus associated with roots of *Pinus* increases minerals & water absorption for the plant by increasing surface area and in turn fungus gets food from plant. Therefore, mycorrhizal association is obligatory for *Pinus* seed germination.

113. (2) Loop of Henle along with vasa rectae plays an important role in the concentration of urine. The proximity between loop of Henle and vasa recta as well as counter current in them help in maintaining an increasing osmolarity towards the inner medullary interstitium. This mechanism helps to maintain a concentration gradient in medullary interstitium so human urine is nearly four times concentrated than initial filtrate formed. The urine concentrating mechanism plays a fundamental role in regulating water and sodium excretion.

114. (4) Genes for flower colour in snapdragon shows incomplete dominance which is an exception of Mendel's first principle (*i.e.* Law of dominance), whereas Law of segregation is universally applicable. Incomplete dominance is a process when a dominant allele, or form of a gene, does not completely mask the effects of a recessive allele, and the organism's resulting physical appearance shows a blending of both alleles.

115. (2) Hypothalamus (a region of the forebrain) is the thermoregulatory centre of brain. It coordinates both the autonomic nervous system and the activity of the pituitary, controlling body temperature, thirst, hunger, and other homeostatic systems, and involved in sleep and emotional activity.

116. (3) Genital herpes is caused by two types of viruses- herpes simplex virus type 1 (HSV-1) and herpes simplex virus type 2 (HSV-2). At present there is no cure for type-II-herpes simplex virus. Other non curable STIs are hepatitis-B and HIV.

117. (2) Respiratory quotient may be defined as a ratio indicating the relation of the volume of carbon dioxide given off in respiration to that of the oxygen consumed.

Respiratory Quotient (RQ)

$$= \frac{\text{Amount of } CO_2 \text{released}}{\text{Amount of } CO_2 \text{consumed}}$$

$$2(C_{51}H_{98}O_6) + 145O_2 \rightarrow 102CO_2 + 98H_2O$$
Tripalmitin + Energy

$$RQ = \frac{102CO_2}{145O_2} = 0.7$$

118. (2) Biocontrol agents are those natural organisms like parasitism, predation and other mechanisms which play an important role in controlling the plant pests like nematodes weeds, insects, and mites and helps in maintaining and balancing the plant species along with their natural enemies. Fungus *Trichoderma*, Baculoviruses (NPV) and *Bacillus thuringiensis* are used as biocontrol agents.

Rhizobium, Nostoc, Azospirillum and *Oscillatoria* are used as biofertilisers, whereas TMV is a pathogen and aphids are pests that harm crop plants.

119. (4) Post-fertilisation is a series of events that takes place after fertilisation. Following are the post-fertilisation changes in flowering plants:

- Ovule → Seed
- Ovary → Fruit
- Zygote → Embryo
- Central cell → Endosperm

120. (3) Concanavalin A is a secondary metabolite, e.g. lectin. Lectin is a mannose/glucose-binding protein isolated from Jack beans (*Canavalia ensiformis*). They are not antibodies and do not originate in an immune system but bind specifically to carbohydrate-containing receptors on cell surfaces (as of red blood cells).

121. (2) Microbes are used in production of several household and industrial products –

- *Lactobacillus* – Production of curd
- *Saccharomyces cerevisiae* – Bread making
- *Aspergillus niger* – Citric acid production
- *Acetobacter aceti* – Acetic acid

122. (2) Coenzyme or metal ion that is tightly bound to enzyme protein is called prosthetic group. A complete catalytic active enzyme with its bound prosthetic group is called holoenzyme

A protein that forms an active enzyme system by combination with a coenzyme and determines the specificity of this system for a substrate is called apoenzyme.

123. (4) Cell cycle is the complete series of events from one cell division to the next. The correct sequence of phases of cell cycle is $G_1 \rightarrow S \rightarrow G_2 \rightarrow M$.

124. (4) *Thiobacillus denitrificans* cause denitrification *i.e.*, conversion of oxides of nitrogen to free N_2.

125. (3) Inbreeding exposes harmful recessive genes that are eliminated by selection. It also helps in accumulation of superior genes and elimination of less desirable genes. Therefore this is selection at each step & which increase the productivity of inbred population. Close and continued inbreeding usually reduces fertility and even productivity.

126. (3) 1 map unit represent 1% cross over. Map unit is used to measure genetic distance. This genetic distance is based on average number of cross over frequency.

127. (3) Botanical garden is an *ex - situ* conservation (offsite conservation) method to conserve biodiversity.

128. (3) In parietal placentation the ovules develop on the inner wall of ovary or in peripheral part and number of placentae corresponds to the number of carpels. It occurs in bicarpellary or multicarpellary but unilocular ovary, e.g. mustard, *Argemone* etc.

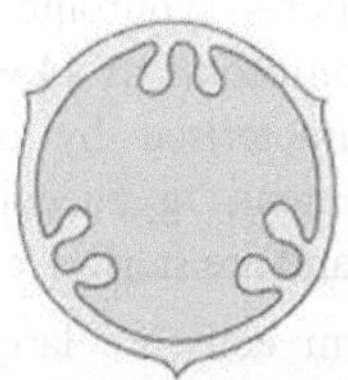

Parietal placentation

129. (2) Asthma (an allergic condition) is a difficulty in breathing causing wheezing due to inflammation of bronchi and bronchioles. It can be due to increasing air born allergens and pollutants. Many people in urban areas are suffering from this respiratory disorder.

130. (4) Cornea is the transparent part of the eye that covers the front portion of the eye. It covers the pupil (the opening at the centre of the eye), iris (the coloured part of the eye), and anterior chamber (the fluid-filled inside of the eye). Its main function is to refract, or bend, light. Cornea consists of dense matrix of collagen and corneal epithelium. It is the most sensitive part of eye.

131. (2) Purines, a heterocyclic compound, such as adenine and guanine are found both in DNA and RNA.

132. (1) Expressed Sequence Tags (ESTs) are DNA sequences (genes) that are expressed as mRNA for protein synthesis. These are used in human Genome Project.

133. (4) Phloem, a complex tissue, is found in highly organised plants ("higher plants") and characterised by the possession of certain specialised cells, the sieve elements, and which functions as the major channel of rapid conduction of sugars over fairly long distances in the plant body. In addition to sieve elements, it always contains parenchyma cells, usually of more than one type, and frequently includes sclerenchyma cells.

Phloem in gymnosperms lacks both sieve tube and companion cells.

134. (2) Klinefelter's syndrome is a condition in which one or more extra X chromosomes are present in a male. Individuals with Klinefelter's syndrome have trisomy of sex chromosome as 44 + XXY (47). They are characterised by overall masculine development, gynaecomastia (enlarged breast) and are sterile.

135. (2) Bulliform cells are large, bubble-shaped epidermal cells that occur in groups on the upper surface of the leaves of many grasses. These cells become flaccid due to water loss. Loss of turgor pressure in these cells causes leaves to "roll up" during water stress.

136. (1) Organ system of organisation, bilateral symmetry and true coelomates with segmented body are found in annelid, arthropoda and chordates.

In mollusca , the body is unsegmented.

137. (4) There will be no change in reading frame of m-RNA by deletion of GGU from 7th, 8th, and 9th position.

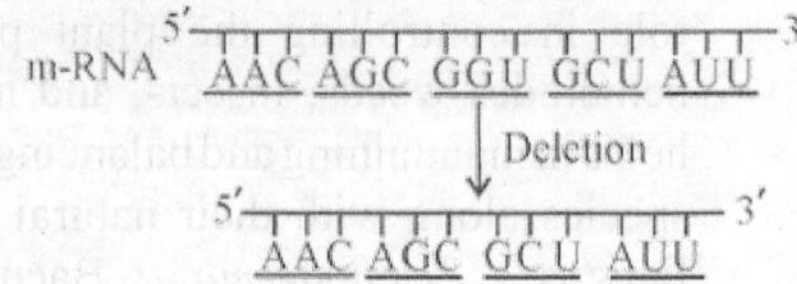

138. (4) There are 12 pairs of ribs in human body. First seven pairs are attached to the sternum ventrally with the help of hyaline carlilage, also called as true ribs. These are called vertebrosternal ribs. The next three pairs (8th, 9th and 10th) do not articulate directly with the sternum, also called vertebrochondral (false) ribs . The last two pairs (11th and 12th) are the free floating ribs because they are not connected ventrally.

139. (2) Each chromosome has two arms, labeled *p* (the shorter of the two) and *q* (the longer). In submetacentric chromosome, the centromere slightly away from the middle of the chromosome resulting into one shorter arm and one longer arm.

140. (4) Xylem is a type of complex tissue . it translocates water, mineral salts, organic nitrogen and hormones.

141. (2) In some seeds like black pepper and beet, the ramnents of nucellus are persistant, this persistant nucellus is called perisperm. It is present in the seeds in the form of a layer of nutritive tissue .

142. (4) Saprophytes are organisms which live on dead organic matter and help in their decomposition.

Parasites is an organisms that live in or on other living plants and animals and dependent on them for their food.

Lichens represent a type of symbiotic association of algae and fungi, in which, both of them dependent on each other for their food and shelter.

Mycorrhiza is also a type of symbiotic association of fungi and plant roots, e.g., *Cycas* coralloid root.

143. (1) *Trichoderma* is a very effective biological mean for plant disease management especially the soil born. It is a free-living fungus which is common in soil and root ecosystems. It reduces growth, survival or infections caused by pathogens by different mechanisms like competition, antibiosis, mycoparasitism, hyphal interactions, and enzyme secretion.

144. (3) Cardiac output is the product of the heart rate (HR), or the number of heart beats per minute (bpm),and the stroke volume (SV), which is the volume of blood pumped from the ventricle per beat; thus,

Cardiac output = stroke volume × Heart rate

As per the given information,

Cardiac output = 5L or 5000 ml

Blood volume in ventricles at the end of diastole = 100 ml

Blood volume in ventricles at the end of systole = 50 ml

∴ Stroke volume = 100 – 50

= 50 ml.

So,

5000 ml = 50 ml × Heart rate

So,

Heart rate = 100 beats per minute.

145. (1) The Montreal Protocol on substances that deplete the Ozone Layer (the Montreal Protocol) is an international agreement made in 1987. It was designed to stop the production and import of ozone depleting substances (chlorofluorocarbons and reduce their concentration in the atmosphere to help protect the earth's ozone layer.

146. (3) Mr. Ahmed Khan, Managing Director of KK Plastic Waste Management has discovered a solution to utilise the waste plastic for the construction of roads. His company developed a technology which turns the plastic waste into a polyblend which can be used in laying of roads. Polyblend is developed and mixed with bitumen.

When bitumen is mixed with Plastic/ Polymer coated aggregate a portion of bitumen diffuses through the polymer layer & binds with aggregate. During this process 3-D internal cross linked network structure are formed which results into stronger bonding.

Therefore this blend is much effctive for laying roads. The cost for construction of the road is slightly high but it would be compensated by the durability and echo friendly properties.

147. (1) In lactational amenorrhea, pills and emergency contraceptives methods, there is the involvement of hormones.

Lactational amenorrhea (absence of menstruation) is based on the fact that ovulation and therefore the cycle do not occur during the period of intense lactation following parturition. Prolactin is the major hormone responsible for milk production and is present in sufficient quantities in almost all women to allow the establishment of normal lactation.

Emergency contraception methods include emergency contraception pills (ECP), intrauterine device, e.g., LNG-20 (Levonor- gestrel) and ulipristal acetate.

CuT and barrier method do not involve any hormonal role.

148. (2) Heroin is synthesised by acetylation of morphine.

Heroin, commonly called smack and is chemically diacetylmorphine is made from the resin of poppy (*Papaver somniferum*) plants. Milky, sap-like opium is first removed from the pod of the poppy flower. This opium is refined to make morphine, then further refined into different forms of heroin. The synthesis of heroin is a simple one step acetylation reaction, and typically is performed by the addition of a large excess of acetic anhydride directly to morphine followed by heating the resulting solution to, or near, boiling. Generally the final product is isolated by treating the cooled reaction mixture with sodium carbonate and collecting the heroin base by filtration.

149. (4) Gases, responsible for Green house effect are carbon dioxide and methane.

A green house gas is a gas that absorbs and emits radiant energy within the thermal infrared range. Greenhouse gases cause the greenhouse effect. The primary greenhouse gases in Earth's atmosphere are water vapour, carbon dioxide, methane, nitrous oxide and ozone. Without greenhouse gases, the average temperature of Earth's surface would be about −18 °C (0 °F), rather than the present average of 15 °C (59 °F).

150. (2) Muscular dystrophy is a inheritable disease that gradually cause the muscles to weaken, leading to an increasing level of disability.

Tetany, Myasthenia gravis and Botulism are not inheritable diseases.

151. (4) To produce enzyme in large quantity, equipment required are bioreactors. Large scale production involves use of bioreactors.

152. (1) Concept of "*Omnis cellula-e cellula*" regarding cell division was proposed by Rudolph Virchow.

153. (4) In flowering plants, out of the two male gametes discharged in synergids, one fuses with the egg and other fuses with the secondary or definitive nucleus present in central cell.

Egg (n) + 1st male gamete (n) → Zygote (2n)

Secondary nucleus (2n) + 2nd male gamete (n) → Endosperm Nucleus(3n)

(central cell nuclei)

154. (2) Steroid hormones pass through the plasma membrane of a target cell and adhere to intracellular receptors residing in the cytoplasm or in the nucleus. In the nucleus, the hormone-receptor complex binds to a DNA sequence, called a hormone response element , which triggers gene transcription and translation.

155. (3) Lysosomes and vacuoles do not contain DNA.

156. (3) Frequency of dominant allele (p) = 0.4

Frequency of recessive allele (q) = 1 – 0.4 = 0.6

Frequency of homozygous dominant individuals (AA)

$= p^2 = (0.4)^2 = \mathbf{0.16(AA)}$

Frequency of heterozygous individuals (Aa)

$= 2pq = 2(0.4)(0.6) = \mathbf{0.48(Aa)}$

Frequency of homozygous recessive individuals (aa)

$= q^2 = (0.6)2 = \mathbf{0.36(aa)}$

157. (2) (a) *Pila* is a member of mollusc. The mouth contains a rasping organ for feeding called radula.

(b) *Bombyx* is an arthropod. In *Bombyx* excretion takes place through malpighian tubules.

(c) *Pleurobrachia* is a ctenophore. The body bears eight external rows of ciliated comb plates, which help in locomotion.

(d) *Taenia* is a platyhelminth. In it specialised cells, called flame cells are present which help in osmoregulation and excretion.

158. (2) Statins are drugs that can lower blood cholesterol.

It is obtained from a yeast (Fungi) called *Monascus purpureus.*

They work by blocking a substance, body needs to make cholesterol.

159. (2) According to Hugo de Vries, mutations are random and directionless.

Hugo deVries believed that mutation caused speciation and hence called it saltation (single step large mutation).

160. (3) In birds female heterogamety is found thus sex of progeny depends on the types of egg rather than the type of sperms of birds (fowls).

Birds (fowls) — ♂ ⟶ Sperm = A + Z type (100%)
Birds (fowls) — ♀ ⟶ eggs ⟨ A + Z (50%), A + W (50%)

161. (4) Transplant rejection occurs when transplanted tissue is rejected by the recipient's immune system, which destroys the transplanted tissue. The body is able to differentiate self and nonself and the cell-mediated response, which is responsible for graft rejection.

162. (3) In Pteridophyte, megaspore is retained for some time in female gametophyte, however the permanent retention is required for seed formation in Gymnosperms. That's why Pteridophytes exhibit precursor to seed habit only.

163. (1) The correct sequence of organs in the alimentary canal of cockroach starting from mouth is :

Pharynx → Oesophagus → Crop → Gizzard → Ileum → Colon → Rectum

164. (2) In mitochondria, enzymes for electron transport are present in the inner membrane.

165. (4) Lysosomes are membrane bound structures, contained numerous hydrolytic enzymes which are active under acidic pH.

Lysosomes are formed by budding from the Golgi complex. Lysosome enzymes are made by proteins from the endoplasmic reticulum.

166. (3) In recombinant DNA technology bacteria are able to produce human insulin, used to treat diabetes because genetic code is nearly universal.

167. (1) Expiratory capacity is the total volume of air that a person can expireafter a normal inspiration. this includes tidal volume and expirtory reserve volume (TV + ERV).

Tidal volume of an athelete = 500 ml

Expiratory reserve volume of an athelete = 1000 ml

Expiratory capacity = TV + ERV
= 500 + 1000
= 1500 ml

168. (3) Lac operon consist of one regulatory genes (*i* gene) and three structural genes (*z, y,* and *a*).

- *i* gene codes for Repressor.
- *z* gene codes for Beta-galactosidase.
- *y* gene codes for Permease.
- *a* gene codes for Transacetylase.

169. (3) Statements (c) and (d) are correct.

Dialysis is a procedure that is a substitute for many of the normal functions of the kidney.

(a) It remove waste, salt and extra water to prevent them from building up in the body.

(b) It keep a safe level of certain chemicals in your blood, such as potassium, sodium and bicarbonate helping to control blood pressure.

(c) There will be reduced absorption of calcium ions from gastrointestinal tract.

(d) RBC production will be reduced, due to reduced erythropoietin hormone.

170. (3) The correct match of hominids and their brain sizes are:

Homo habilis – 650-800 cc

Homo neanderthalensis –1400 cc

Homo erectus –900 cc

Homo sapiens – 1350 cc

171. (3) Progestasert and LNG-20 are hormone releasing IUDs (Intra Uterine Devices). These are the devices which are used in birth control by making uterus unsuitable for implantation and the cervix hostile to sperms.

172. (2) The correct sequence for transport of sperm cells in male reproductive system is:

Seminiferous tubules → Rete testis → Vasa efferentia→ Epididymis →Vas deferens →Ejaculatory duct → Urethra → Urethral meatus

173. (4) The direction of movement of sugar in phloem is bi-directional.

The Pressure Flow Hypothesis, also known as the Mass Flow Hypothesis, is the best-supported theory to explain the movement of sap through the phloem. It was proposed by Ernst Munch, a German plant physiologist in 1930. A high concentration of organic substances, particularly sugar, inside cells of the phloem at a source, such as a leaf, creates a diffusion gradient (osmotic gradient) that draws water into the cells from the adjacent xylem. This creates turgor pressure, also known as hydrostatic pressure, in the phloem. Movement of phloem sap occurs by bulk flow (mass flow) from sugar sources to sugar sinks. The movement in phloem is bidirectional, whereas, in xylem cells, it is unidirectional (upward). Because of this multi-directional flow, coupled with the fact that sap cannot move with ease between adjacent sieve-tubes, it is not unusual for sap in adjacent sieve-tubes to be flowing in opposite directions.

174. (3) Infective constituent in viruses is either DNA or RNA, not protein. The simplest form consist of two basic components: nucleic acid (single- or double-stranded RNA or DNA) and a protein coat, the capsid, which functions as a shell to protect the viral genome from nucleases and which during infection attaches the virion to specific receptors exposed on the prospective host cell.

175. (4) In humans, ciliated columnar epithelium is present in bronchioles and fallopian tubes. Their function is to move particles or mucus in a specific direction over the epithelium.

176. (4) Yeast is an unicellular sac fungi. It lacks filamentous structure or hyphae.

177. (4) It has been recommended that storage of nuclear waste after sufficient pre treatment, should be done in suitably shielded containers and buried within the rocks deep below the earth's surface (500 m deep).

178. (2) A restriction enzyme or restriction endonuclease is an enzyme that cleaves DNA into fragments at or near specific recognition sites within molecules known as restriction sites. Restriction enzymes are one class of the broader endonuclease group of enzymes. Each restriction endonuclease functions by inspecting the length of a DNA sequence. Once it finds its specific recognition sequence, it will bind to the DNA and cut each of the two strands of the double helix at specific points in their sugar phosphate backbone. These enzymes predictably cut both strands because the sequences they recognize are palindromic. That is the recognition sequences of identical bases on both DNA strands.

179. (2) Stabilising selection is a type of natural selection in which the population mean stabilises on a particular non-extreme trait value. This is thought to be the most common mechanism of action for natural selection because most traits do not appear to change drastically over time. The result of stabilising is the over-representation in a specific trait. For example, human birth weight, the number of eggs a bird lays, and the density of cactus spines.

The given data shows stabilising selection as most of the newborn having average weight between 3 to 3.3 kg survive and babies with less and more weight have low survival rate.

180. (2) According to rules of binomial nomenclature, correctly written scientific name of mango is *Mangifera indica* Linn.

This system of nomenclature was given by Carl Linnaeus. The scientific name of mango is given as *Mangifera indica Linn. Mangifera* indicates the 'genus' while *indica* represents a particular species or 'specific epithet' and Linn indicates the Biologist Linnaeus who first described the species of mango.

NEET 2018

PHYSICS

1. A tuning fork is used to produce resonance in a glass tube. The length of the air column in this tube can be adjusted by a variable piston. At room temperature of 27°C two successive resonances are produced at 20 cm and 73 cm of column length. If the frequency of the tuning fork is 320 Hz, the velocity of sound in air at 27°C is
   (1) 330 m/s (2) 339 m/s
   (3) 300 m/s (4) 350 m/s
2. An electron falls from rest through a vertical distance h in a uniform and vertically upward directed electric field E. The direction of electric field is now reversed, keeping its magnitude the same. A proton is allowed to fall from rest in it through the same vertical distance h. The time of fall of the electron, in comparison to the time of fall of the proton is
   (1) smaller (2) 5 times greater
   (3) equal (4) 10 times greater
3. A pendulum is hung from the roof of a sufficiently high building and is moving freely to and fro like a simple harmonic oscillator. The acceleration of the bob of the pendulum is 20 m/s^2 at a distance of 5 m from the mean position. The time period of oscillation is
   (1) 2π s (2) π s
   (3) 1 s (4) 2 s
4. The electrostatic force between the metal plates of an isolated parallel plate capacitor C having a charge Q and area A, is
   (1) independent of the distance between the plates
   (2) linearly proportional to the distance between the plates
   (3) inversely proportional to the distance between the plates
   (4) proportional to the square root of the distance between the plates
5. Current sensitivity of a moving coil galvanometer is 5 div/mA and its voltage sensitivity (angular deflection per unit voltage applied) is 20 div/V. The resistance of the galvanometer is
   (1) 40 Ω (2) 25 Ω
   (3) 500 Ω (4) 250 Ω
6. A thin diamagnetic rod is placed vertically between the poles of an electromagnet. When the current in the electromagnet is switched on, then the diamagnetic rod is pushed up, out of the horizontal magnetic field. Hence the rod gains gravitational potential energy. The work required to do this comes from
   (1) the current source
   (2) the magnetic field
   (3) the induced electric field due to the changing magnetic field
   (4) the lattice structure of the material of the rod
7. An inductor 20 mH, a capacitor 100 μF and a resistor 50Ω are connected in series across a source of emf, V = 10 sin 314 t. The power loss in the circuit is
   (1) 0.79 W (2) 0.43 W
   (3) 1.13 W (4) 2.74 W
8. A metallic rod of mass per unit length 0.5 kg m^{-1} is lying horizontally on a smooth inclined plane which makes an angle of 30° with the horizontal. The rod is not allowed to slide down by flowing a current through it when a magnetic field of induction 0.25 T is acting on it in the vertical direction. The current flowing in the rod to keep it stationary is
   (1) 7.14 A (2) 5.98 A
   (3) 11.32 A (4) 14.76 A
9. A carbon resistor of (47 ± 4.7) kΩ is to be marked with rings of different colours for its identification. The colour code sequence will be
   (1) Violet – Yellow – Orange – Silver
   (2) Yellow – Violet – Orange – Silver
   (3) Green – Orange – Violet – Gold
   (4) Yellow – Green – Violet – Gold

10. A set of 'n' equal resistors, of value 'R' each,are connected in series to a battery of emf 'E' and internal resistance 'R'. The current drawn is I. Now, the 'n' resistors are connected in parallel to the same battery. Then the current drawn from battery becomes 10 I. The value of'n' is
(1) 10 (2) 11
(3) 9 (4) 20

11. A battery consists of a variable number 'n' of identical cells (having internal resistance 'r' each) which are connected in series. The terminals of the battery are short-circuited and the current I is measured. Which of the graphs shows the correct relationship between I and n?

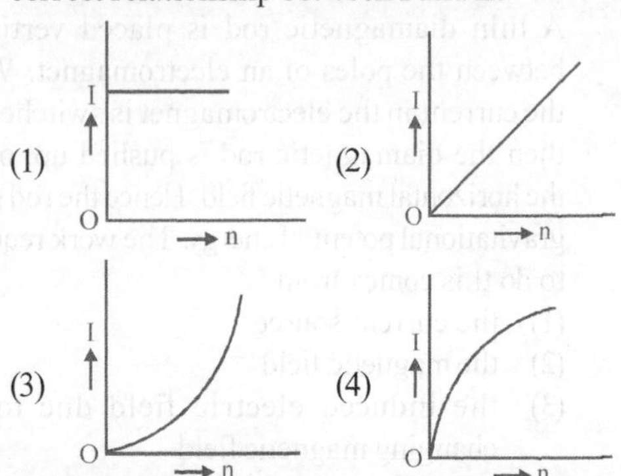

12. In Young's double slit experiment the separation d between the slits is 2 mm, the wavelength λ of the light used is 5896 Å and distance D between the screen and slits is100 cm. It is found that the angular width of the fringes is 0.20°. To increase the fringe angular width to 0.21° (with same λ and D) the separation between the slits needs to be changed to
(1) 1.8mm (2) 1.9mm
(3) 1.7mm (4) 2.1mm

13. An astronomical refracting telescope will have large angular magnification and high angular resolution, when it has an objective lens of
(1) small focal length and large diameter
(2) large focal length and small diameter
(3) small focal length and small diameter
(4) large focal length and large diameter

14. Unpolarised light is incident from air on a plane surface of a material of refractive index 'μ'. At a particular angle of incidence 'i', it is found that the reflected and refracted rays are perpendicular to each other. Which of the following options is correct for this situation?
(1) Reflected light is polarised with its electric vector parallel to the plane of incidence
(2) Reflected light is polarised with its electric vector perpendicular to the plane of incidence
(3) $i = \tan^{-1}\left(\frac{1}{\mu}\right)$
(4) $i = \sin^{-1}\left(\frac{1}{\mu}\right)$

15. An em wave is propagating in a medium with a velocity $\vec{V} = V\hat{i}$. The instantaneous oscillating electric field of this em wave is along +y axis. Then the direction of oscillating magnetic field of the em wave will be along
(1) –z direction (2) +z direction
(3) –x direction (4) –y direction

16. The refractive index of the material of a prism is $\sqrt{2}$ and the angle of the prism is 30°.One of the two refracting surfaces of the prism is made a mirror inwards, by silver coating. A beam of monochromatic light entering the prism from the other face will retrace its path (after reflection from the silvered surface) if its angle of incidence on the prism is
(1) 60° (2) 45°
(3) Zero (4) 30°

17. An object is placed at a distance of 40 cm from a concave mirror of focal length 15 cm.If the object is displaced through a distance of 20 cm towards the mirror, the displacement of the image will be
(1) 30 cm away from the mirror
(2) 36 cm away from the mirror
(3) 36 cm towards the mirror
(4) 30 cm towards the mirror

18. The magnetic potential energy stored in a certain inductor is 25 mJ, when the current in the inductor is 60 mA. This inductor is of inductance
(1) 0.138 H (2) 138.88 H
(3) 13.89 H (4) 1.389 H

19. For a radioactive material, half-life is 10 minutes. If initially there are 600 number of nuclei, the time taken (in minutes) for the disintegration of 450 nuclei is
(1) 20 (2) 10
(3) 15 (4) 30

20. The ratio of kinetic energy to the total energy of an electron in a Bohr orbit of the hydrogen atom, is
(1) 1 : 1 (2) 1 : –1
(3) 1 : –2 (4) 2 : –1

21. An electron of mass m with an initial velocity $\vec{V} = V_0\hat{i}$ ($V_0 > 0$) enters an electric field $\vec{E} = -E_0\hat{i}$ (E_0 = constant > 0) at t = 0. If λ_0 is its de-Broglie wavelength initially, then its de-Broglie wavelength at time t is

(1) $\dfrac{\lambda_0}{\left(1+\dfrac{eE_0}{mV_0}t\right)}$ (2) $\lambda_0\left(1+\dfrac{eE_0}{mV_0}t\right)$

(3) λ_0 (4) $\lambda_0 t$

22. When the light of frequency $2\nu_0$ (where ν_0 is threshold frequency), is incident on a metalplate, the maximum velocity of electronsemitted is v_1. When the frequency of the incident radiation is increased to $5\nu_0$, the maximum velocity of electrons emitted from the same plate is v_2. The ratio of v_1 to v_2 is

(1) 1 : 2 (2) 1 : 4

(3) 2 : 1 (4) 4 : 1

23. In the combination of the following gates the output Y can be written in terms of inputs A and B as

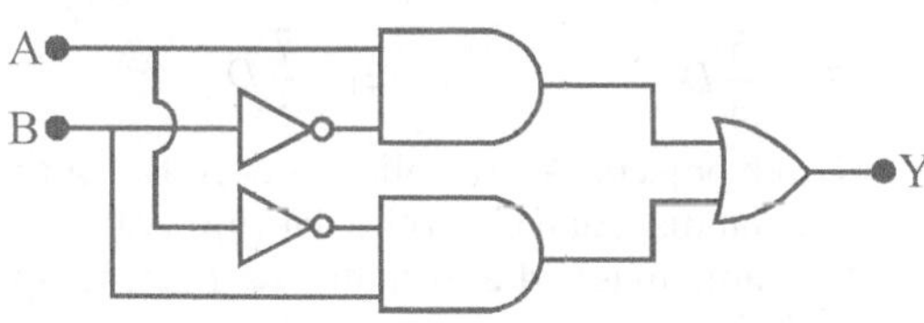

(1) $\overline{A \cdot B}$ (2) $A\cdot\overline{B}+\overline{A}\cdot B$

(3) $\overline{A+B}$ (4) $\overline{A\cdot B+A\cdot B}$

24. In the circuit shown in the figure, the input voltage V_i is 20 V, $V_{BE} = 0$ and $V_{CE} = 0$. The values of I_B, I_C and β are given by

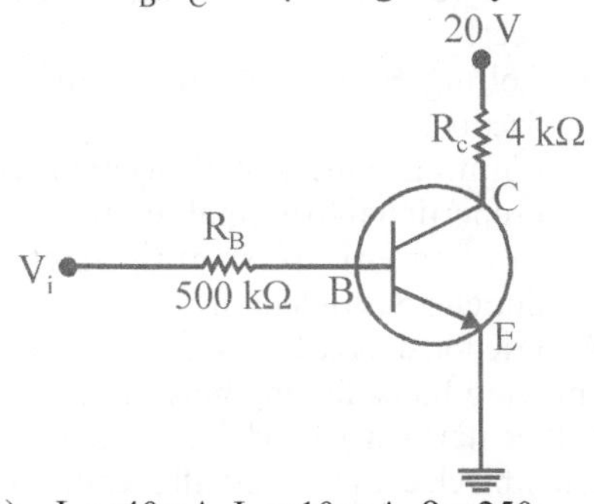

(1) $I_B = 40\ \mu A, I_C = 10\ mA, \beta = 250$

(2) $I_B = 25\ \mu A, I_C = 5\ mA, \beta = 200$

(3) $I_B = 40\ \mu A, I_C = 5\ mA, \beta = 125$

(4) $I_B = 20\ \mu A, I_C = 5\ mA, \beta = 250$

25. In a p-n junction diode, change in temperature due to heating

(1) Affects only reverse resistance

(2) Affects only forward resistance

(3) Affects the overall V-I characteristics of p-n junction

(4) Does not affect resistance of p-n junction

26. A solid sphere is rotating freely about its symmetry axis in free space. The radius of the sphere is increased keeping its mass same. Which of the following physical quantities would remain constant for the sphere?

(1) Angular velocity

(2) Moment of inertia

(3) Angular momentum

(4) Rotational kinetic energy

27. The kinetic energies of a planet in an elliptical orbit about the Sun, at positions A, B and C are K_A, K_B and K_C, respectively. AC is the major axis and SB is perpendicular to AC at the position of the Sun S as shown in the figure. Then

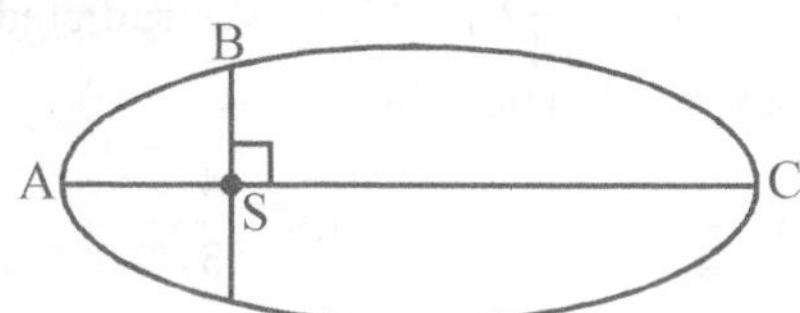

(1) $K_A < K_B < K_C$ (2) $K_A > K_B > K_C$

(3) $K_B > K_A > K_C$ (4) $K_B < K_A < K_C$

28. If the mass of the Sun were ten times smaller and the universal gravitational constant were ten times larger in magnitude, which of the following is not correct?

(1) Raindrops will fall faster

(2) Walking on the ground would become more difficult

(3) 'g' on the Earth will not change

(4) Time period of a simple pendulum on the Earth would decrease

29. A solid sphere is in rolling motion. In rolling motion a body possesses translational kinetic energy (K_t) as well as rotational kinetic energy (K_r) simultaneously. The ratio $K_t : (K_t + K_r)$ for the sphere is

(1) 7 : 10 (2) 5 : 7

(3) 2 : 5 (4) 10 : 7

30. A small sphere of radius 'r' falls from rest in a viscous liquid. As a result, heat is produced due to viscous force. The rate of production of heat when the sphere attains its terminal velocity, is proportional to

(1) r^3 (2) r^2

(3) r^4 (4) r^5

31. A sample of 0.1 g of water at 100°C and normal pressure (1.013×10^5 Nm^{-2}) requires 54 cal of heat energy to convert to steam at 100°C. If the volume of the steam produced is 167.1 cc, the change in internal energy of the sample, is
(1) 104.3 J (2) 208.7 J
(3) 84.5 J (4) 42.2 J

32. Two wires are made of the same material and have the same volume. The first wire has cross-sectional area A and the second wire has cross-sectional area 3A. If the length of the first wire is increased by $\Delta \ell$ on applying a force F, how much force is needed to stretch the second wire by the same amount?
(1) 9 F (2) 6 F
(3) F (4) 4 F

33. The power radiated by a black body is P and it radiates maximum energy at wavelength, λ_0. If the temperature of the black body is now changed so that it radiates maximum energy at wavelength $\frac{3}{4}\lambda_0$, the power radiated by it becomes nP. The value of n is
(1) $\frac{3}{4}$ (2) $\frac{4}{3}$
(3) $\frac{81}{256}$ (4) $\frac{256}{81}$

34. At what temperature will the rms speed of oxygen molecules become just sufficient for escaping from the Earth's atmosphere?
(Given :
Mass of oxygen molecule (m) = 2.76×10^{-26} kg
Boltzmann's constant $k_B = 1.38 \times 10^{-23}$ JK^{-1})
(1) 2.508×10^4 K (2) 8.360×10^4 K
(3) 1.254×10^4 K (4) 5.016×10^4 K

35. The volume (V) of a monatomic gas varies with its temperature (T), as shown in the graph. The ratio of work done by the gas, to the heat absorbed by it, when it undergoes a change from state A to state B, is

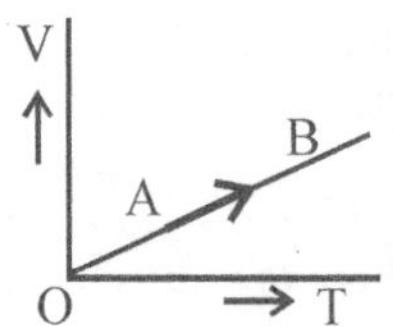

(1) $\frac{2}{5}$ (2) $\frac{2}{3}$
(3) $\frac{2}{7}$ (4) $\frac{1}{3}$

36. The fundamental frequency in an open organ pipe is equal to the third harmonic of a closed organ pipe. If the length of the closed organ pipe is 20 cm, the length of the open organ pipe is
(1) 13.2 cm (2) 8 cm
(3) 16 cm (4) 12.5 cm

37. The efficiency of an ideal heat engine working between the freezing point and boiling point of water, is
(1) 26.8% (2) 20%
(3) 12.5% (4) 6.25%

38. A body initially at rest and sliding along a frictionless track from a height h (as shown in the figure) just completes a vertical circle of diameter AB = D. The height h is equal to

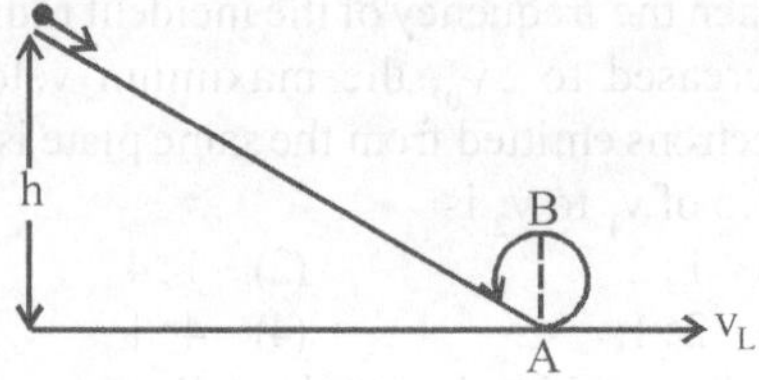

(1) $\frac{3}{2}D$ (2) D
(3) $\frac{5}{4}D$ (4) $\frac{7}{5}D$

39. Three objects, A : (a solid sphere), B : (a thin circular disk) and C : (a circular ring), each have the same mass M and radius R. They all spin with the same angular speed ω about their own symmetry axes. The amounts of work (W) required to bring them to rest, would satisfy the relation
(1) $W_C > W_B > W_A$ (2) $W_A > W_B > W_C$
(3) $W_A > W_C > W_B$ (4) $W_B > W_A > W_C$

40. Which one of the following statements is incorrect?
(1) Rolling friction is smaller than sliding friction.
(2) Limiting value of static friction is directly proportional to normal reaction.
(3) Coefficient of sliding friction has dimensions of length.
(4) Frictional force opposes the relative motion.

41. A moving block having mass m, collides with another stationary block having mass 4m. The lighter block comes to rest after collision. When the initial velocity of the lighter block is v, then the value of coefficient of restitution (e) will be
(1) 0.5 (2) 0.25
(3) 0.4 (4) 0.8

42. A block of mass m is placed on a smooth inclined wedge ABC of inclination θ as shown in the figure. The wedge is given an acceleration 'a' towards the right. The relation between a and θ for the block to remain stationary on the wedge is

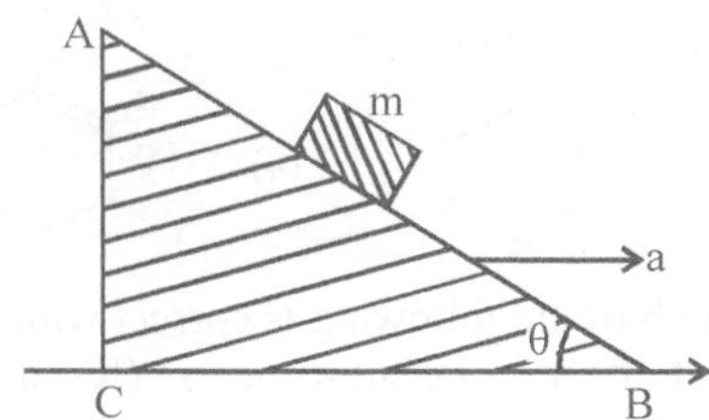

(1) $a = \frac{g}{\text{cosec}\,\theta}$ (2) $a = \frac{g}{\sin\theta}$
(3) $a = g\tan\theta$ (4) $a = g\cos\theta$

43. A toy car with charge q moves on a frictionless horizontal plane surface under the influence of a uniform electric field $\vec{E}$. Due to the force q $\vec{E}$, its velocity increases from 0 to 6 m/s in one second duration. At that instant the direction of the field is reversed. The car continues to move for two more seconds under the influence of this field. The average velocity and the average speed of the toy car between 0 to 3 seconds are respectively
(1) 2 m/s, 4 m/s (2) 1 m/s, 3 m/s
(3) 1.5 m/s, 3 m/s (4) 1 m/s, 3.5 m/s

44. The moment of the force, $\vec{F} = 4\hat{i} + 5\hat{j} - 6\hat{k}$ at (2, 0, –3), about the point (2, –2, –2), is given by
(1) $-8\hat{i} - 4\hat{j} - 7\hat{k}$ (2) $-4\hat{i} - \hat{j} - 8\hat{k}$
(3) $-7\hat{i} - 4\hat{j} - 8\hat{k}$ (4) $-7\hat{i} - 8\hat{j} - 4\hat{k}$

45. A student measured the diameter of a small steel ball using a screw gauge of least count 0.001 cm. The main scale reading is 5 mm and zero of circular scale division coincides with 25 divisions above the reference level. If screw gauge has a zero error of –0.004 cm, the correct diameter of the ball is
(1) 0.521 cm (2) 0.525 cm
(3) 0.529 cm (4) 0.053 cm

CHEMISTRY

46. The correct order of N-compounds in its decreasing order of oxidation states is
(1) HNO_3, NO, N_2, NH_4Cl
(2) HNO_3, NO, NH_4Cl, N_2
(3) NH_4Cl, N_2, NO, HNO_3
(4) HNO_3, NH_4Cl, NO, N_2

47. The correct order of atomic radii in group 13 elements is
(1) B < Al < In < Ga < Tl
(2) B < Al < Ga < In < Tl
(3) B < Ga < Al < In < Tl
(4) B < Ga < Al < Tl < In

48. Considering Ellingham diagram, which of the following metals can be used to reduce alumina?
(1) Fe (2) Zn
(3) Cu (4) Mg

49. Which one of the following elements is unable to form MF_6^{3-} ion?
(1) Ga (2) Al
(3) In (4) B

50. Which of the following statements is not true for halogens?
(1) All form monobasic oxyacids
(2) All are oxidizing agents
(3) Chlorine has the highest electron-gain enthalpy
(4) All but fluorine shows positive oxidation states

51. In the structure of ClF_3, the number of lone pair of electrons on central atom 'Cl' is
(1) One (2) Two
(3) Three (4) Four

52. The difference between amylose and amylopectin is
(1) Amylopectin has 1 → 4 α-linkage and 1 → 6 α-linkage
(2) Amylose has 1 → 4 α-linkage and 1 → 6 β-linkage
(3) Amylose is made up of glucose and galactose
(4) Amylopectin has 1 → 4 α-linkage and 1 → 6 β-linkage

53. Regarding cross-linked or network polymers, which of the following statements is incorrect?
(1) They contain covalent bonds between various linear polymer chains.
(2) They are formed from bi- and tri-functional monomers.
(3) They contain strong covalents bonds in their polymer chains.
(4) Examples are bakelite and melamine.

54. A mixture of 2.3 g formic acid and 4.5 g oxalic acid is treated with conc. H_2SO_4. The evolved gaseous mixture is passed through KOH pellets. Weight (in g) of the remaining product at STP will be
(1) 1.4 (2) 3.0
(3) 4.4 (4) 2.8

55. Which of the following oxides is most acidic in nature?
(1) MgO (2) BeO
(3) CaO (4) BaO

56. Nitration of aniline in strong acidic medium also gives *m*-nitroaniline because
(1) Inspite of substituents nitro group always goes to only *m*-position.
(2) In electrophilic substitution reactions, amino group is meta directive.
(3) In acidic (strong) medium aniline is present as anilinium ion.
(4) In absence of substituents, nitro group always goes to *m*-position.

57. The compound A on treatment with Na gives B, and with PCl_5 gives C. B and C react together to give diethyl ether. A, B and C are in the order
(1) $C_2H_5OH, C_2H_6, C_2H_5Cl$
(2) $C_2H_5OH, C_2H_5Cl, C_2H_5ONa$
(3) $C_2H_5OH, C_2H_5ONa, C_2H_5Cl$
(4) $C_2H_5Cl, C_2H_6, C_2H_5OH$

58. Hydrocarbon (A) reacts with bromine by substitution to form an alkyl bromide which by Wurtz reaction is converted to gaseous hydrocarbon containing less than four carbon atoms. (A) is
(1) $CH\equiv CH$ (2) $CH_2=CH_2$
(3) CH_4 (4) CH_3-CH_3

59. The compound C_7H_8 undergoes the following reactions:

$$C_7H_8 \xrightarrow{3Cl_2/\Delta} A \xrightarrow{Br_2/Fe} B \xrightarrow{Zn/HCl} C$$

The product 'C' is
(1) *m*-bromotoluene
(2) *o*-bromotoluene
(3) *p*-bromotoluene
(4) 3-bromo-2,4,6-trichlorotoluene

60. Which oxide of nitrogen is not a common pollutant introduced into the atmosphere both due to natural and human activity?
(1) N_2O_5 (2) NO_2
(3) NO (4) N_2O

61. Which of the following molecules represents the order of hybridisation sp^2, sp^2, sp, sp from left to right atoms?
(1) $HC\equiv C-C\equiv CH$
(2) $CH_2=CH-C\equiv CH$
(3) $CH_3-CH=CH-CH_3$
(4) $CH_2=CH-CH=CH_2$

62. Which of the following carbocations is expected to be most stable?

(1) NO_2, $\oplus$, Y, H (2) NO_2, $\oplus$, Y, H

(3) NO_2, H, Y, $\oplus$ (4) NO_2, H, Y, $\oplus$

63. Which of the following is correct with respect to – I effect of the substituents? (R = alkyl)
(1) $-NH_2 < -OR < -F$
(2) $-NR_2 < -OR < -F$
(3) $-NR_2 > -OR > -F$
(4) $-NH_2 > -OR > -F$

64. In the reaction

OH (phenol) $+ CHCl_3 + NaOH \longrightarrow$ O^-Na^+, CHO

The electrophile involved is
(1) Dichloromethyl cation $(\overset{\oplus}{C}HCl_2)$
(2) Formyl cation $(\overset{\oplus}{C}HO)$
(3) Dichlorocarbene $(:CCl_2)$
(4) Dichloromethyl anion $(\overset{\ominus}{C}HCl_2)$

65. Carboxylic acids have higher boiling points than aldehydes, ketones and even alcohols of comparable molecular mass. It is due to their
(1) Formation of intramolecular H-bonding
(2) Formation of carboxylate ion
(3) Formation of intermolecular H-bonding
(4) More extensive association of carboxylic acid via van der Waals force of attraction

66. Compound A, $C_8H_{10}O$, is found to react with NaOI (produced by reacting Y with NaOH) and yields a yellow precipitate with characteristic smell.
A and Y are respectively

(1) $H_3C-C_6H_4-CH_2-OH$ and I_2

(2) $C_6H_5-CH_2-CH_2-OH$ and I_2

(3) CH_3–(benzene ring with CH_3)–OH and I_2

(4) (benzene ring)–$\underset{\large OH}{CH}-CH_3$ and I_2

67. Identify the major products P, Q and R in the following sequence of reactions:

$$C_6H_6 + CH_3CH_2CH_2Cl \xrightarrow{\text{Anhydrous } AlCl_3}$$

$$P \xrightarrow[\text{(ii) } H_3O^+/\Delta]{\text{(i) } O_2} Q + R$$

	P	Q	R
(1)	$C_6H_5CH_2CH_2CH_3$	C_6H_5CHO	CH_3CH_2-OH
(2)	$C_6H_5CH_2CH_2CH_3$	C_6H_5CHO	C_6H_5COOH
(3)	$C_6H_5CH(CH_3)_2$	C_6H_5OH	$CH_3-CO-CH_3$
(4)	$C_6H_5CH(CH_3)_2$	C_6H_5OH	$CH_3CH(OH)CH_3$

68. Which of the following compounds can form a zwitterion?

(1) Aniline (2) Acetanilide

(3) Glycine (4) Benzoic acid

69. For the redox reaction

$$MnO_4^- + C_2O_4^{2-} + H^+ \longrightarrow Mn^{2+} + CO_2 + H_2O$$

The correct coefficients of the reactants for the balanced equation are

	MnO_4^-	$C_2O_4^{2-}$	H^+
(1)	16	5	2
(2)	2	5	16
(3)	5	16	2
(4)	2	16	5

70. Which one of the following conditions will favour maximum formation of the product in the reaction, $A_2(g) + B_2(g) \rightleftharpoons X_2(g)\ \Delta_r H = -X$ kJ :

(1) Low temperature and high pressure

(2) Low temperature and low pressure

(3) High temperature and low pressure

(4) High temperature and high pressure

71. When initial concentration of the reactant is doubled, the half-life period of a zero order reaction

(1) is halved

(2) is doubled

(3) remains unchanged

(4) is tripled

72. The correction factor '*a*' to the ideal gas equation corresponds to

(1) Density of the gas molecules

(2) Volume of the gas molecules

(3) Forces of attraction between the gas molecules

(4) Electric field present between the gas molecules

73. The bond dissociation energies of X_2, Y_2 and XY are in the ratio of 1 : 0.5 : 1. ΔH for the formation of XY is –200 kJ mol^{-1}. The bond dissociation energy of X_2 will be

(1) 200 kJ mol^{-1} (2) 100 kJ mol^{-1}

(3) 400 kJ mol^{-1} (4) 800 kJ mol^{-1}

74. Magnesium reacts with an element (X) to form an ionic compound. If the ground state electronic configuration of (X) is $1s^2 2s^2 2p^3$, the simplest formula for this compound is

(1) Mg_2X_3 (2) MgX_2

(3) Mg_3X_2 (4) Mg_2X

75. Iron exhibits bcc structure at room temperature. Above 900°C, it transforms to fcc structure. The ratio of density of iron at room temperature to that at 900°C (assuming molar mass and atomic radii of iron remains constant with temperature) is

(1) $\frac{\sqrt{3}}{\sqrt{2}}$ (2) $\frac{4\sqrt{3}}{3\sqrt{2}}$

(3) $\frac{1}{2}$ (4) $\frac{3\sqrt{3}}{4\sqrt{2}}$

76. Consider the following species :

CN^+, CN^-, NO and CN

Which one of these will have the highest bond order?

(1) NO (2) CN^-

(3) CN (4) CN^+

77. Which one is a wrong statement?
(1) Total orbital angular momentum of electron in 's' orbital is equal to zero
(2) An orbital is designated by three quantum numbers while an electron in an atom is designated by four quantum numbers
(3) The value of m for d_z2 is zero
(4) The electronic configuration of N atom is

$1s^2$ $2s^2$ $2p_x^1$ $2p_y^1$ $2p_z^1$
[↑↓] [↑↓] [↑ | ↑ | ↓]

78. The correct difference between first and second order reactions is that
(1) The rate of a first-order reaction does not depend on reactant concentrations, the rate of a second-order reaction does depend on reactant concentrations
(2) The half-life of a first-order reaction does not depend on $[A]_0$, the half-life of a second-order reaction does depend on $[A]_0$
(3) The rate of a first-order reaction does depend on reactant concentrations, the rate of a second-order reaction does not depend on reactant concentrations
(4) A first-order reaction can be catalyzed, a second-order reaction cannot be catalyzed

79. In which case is number of molecules of water maximum?
(1) 18 mL of water (2) 0.18 g of water
(3) 10^{-3} mol of water
(4) 0.00224 L of water vapours at 1 atm and 273 K

80. Among CaH_2, BeH_2, BaH_2, the order of ionic character is
(1) $BeH_2 < CaH_2 < BaH_2$
(2) $CaH_2 < BeH_2 < BaH_2$
(3) $BaH_2 < BeH_2 < CaH_2$
(4) $BeH_2 < BaH_2 < CaH_2$

81. Consider the change in oxidation state of bromine corresponding to different emf values as shown in the diagram below :

$$BrO_4^- \xrightarrow{1.82\ V} BrO_3^- \xrightarrow{1.5\ V} HBrO \xrightarrow{1.595\ V} Br_2 \xrightarrow{1.0652\ V} Br^-$$

Then the species undergoing disproportionation is
(1) BrO_3^- (2) BrO_4^-
(3) HBrO (4) Br_2

82. The solubility of $BaSO_4$ in water is 2.42×10^{-3} gL^{-1} at 298 K. The value of its solubility product (K_{sp}) will be
(Given molar mass of $BaSO_4$ = 233 g mol^{-1})
(1) $1.08 \times 10^{-10} mol^2L^{-2}$
(2) $1.08 \times 10^{-12} mol^2L^{-2}$
(3) $1.08 \times 10^{-8} mol^2L^{-2}$
(4) $1.08 \times 10^{-14} mol^2L^{-2}$

83. Following solutions were prepared by mixing different volumes of NaOH and HCl of different concentrations :
a. $60\ mL\ \frac{M}{10}\ HCl + 40\ mL\ \frac{M}{10}\ NaOH$
b. $55\ mL\ \frac{M}{10}\ HCl + 45\ mL\ \frac{M}{10}\ NaOH$
c. $75\ mL\ \frac{M}{5}\ HCl + 25\ mL\ \frac{M}{5}\ NaOH$
d. $100\ mL\ \frac{M}{10}\ HCl + 100\ mL\ \frac{M}{10}\ NaOH$
pH of which one of them will be equal to 1?
(1) b (2) a
(3) c (4) d

84. On which of the following properties does the coagulating power of an ion depend?
(1) The magnitude of the charge on the ion alone
(2) Size of the ion alone
(3) The sign of charge on the ion alone
(4) Both magnitude and sign of the charge on the ion

85. Given van der Waals constants for NH_3, H_2, O_2 and CO_2 are respectively 4.17, 0.244, 1.36 and 3.59, which one of the following gases is most easily liquefied?
(1) NH_3 (2) H_2
(3) CO_2 (4) O_2

86. Iron carbonyl, $Fe(CO)_5$ is
(1) Tetranuclear (2) Mononuclear
(3) Dinuclear (4) Trinuclear

87. The type of isomerism shown by the complex $[CoCl_2(en)_2]$ is
(1) Geometrical isomerism
(2) Coordination isomerism
(3) Linkage isomerism
(4) Ionization isomerism

88. Which one of the following ions exhibits *d-d* transition and paramagnetism as well?
(1) CrO_4^{2-} (2) $Cr_2O_7^{2-}$
(3) MnO_4^{2-} (4) MnO_4^-

89. The geometry and magnetic behaviour of the complex $[Ni(CO)_4]$ are
(1) Square planar geometry and diamagnetic

(2) Tetrahedral geometry and diamagnetic
(3) Tetrahedral geometry and paramagnetic
(4) Square planar geometry and paramagnetic

90. Match the metal ions given in Column I with the spin magnetic moments of the ions given in Column II and assign the correct code :

	Column I		Column II
a.	Co^{3+}	i.	$\sqrt{8}$ BM
b.	Cr^{3+}	ii.	$\sqrt{35}$ BM
c.	Fe^{3+}	iii.	$\sqrt{3}$ BM
d.	Ni^{2+}	iv.	$\sqrt{24}$ BM
		v.	$\sqrt{15}$ BM

	a	b	c	d
(1)	iv	v	ii	i
(2)	i	ii	iii	iv
(3)	iii	v	i	ii
(4)	iv	i	ii	iii

BIOLOGY

91. The difference between spermiogenesis and spermiation is
(1) In spermiogenesis spermatids are formed, while in spermiation spermatozoa are formed.
(2) In spermiogenesis spermatozoa are formed, while in spermiation spermatids are formed.
(3) In spermiogenesis spermatozoa are formed, while in spermiation spermatozoa are released from sertoli cells into the cavity of seminiferous tubules.
(4) In spermiogenesis spermatozoa from sertoli cells are released into the cavity of seminiferous tubules, while in spermiation spermatozoa are formed.

92. The amnion of mammalian embryo is derived from
(1) ectoderm and mesoderm
(2) endoderm and mesoderm
(3) ectoderm and endoderm
(4) mesoderm and trophoblast

93. The contraceptive 'SAHELI'
(1) blocks estrogen receptors in the uterus, preventing eggs from getting implanted.
(2) increases the concentration of estrogen and prevents ovulation in females.
(3) is a post-coital contraceptive.
(4) is an IUD.

94. Hormones secreted by the placenta to maintain pregnancy are
(1) hCG, hPL, progestogens, prolactin
(2) hCG, hPL, estrogens, relaxin, oxytocin
(3) hCG, progestogens, estrogens, glucocorticoids
(4) hCG, hPL, progestogens, estrogens

95. Match the items given in Column I with those in Column II and select the correct option given below :

Column I		Column II
A. Proliferative Phase	i.	Breakdown of endometrial lining
B. Secretory Phase	ii.	Follicular Phase
C. Menstruation	iii.	Luteal Phase

	A	B	C
(1)	iii	ii	i
(2)	i	iii	ii
(3)	iii	i	ii
(4)	ii	iii	i

96. All of the following are part of an operon except
(1) an operator (2) structural genes
(3) a promoter (4) an enhancer

97. A woman has an X-linked condition on one of her X chromosomes. This chromosome can be inherited by
(1) Only daughters
(2) Only sons
(3) Both sons and daughters
(4) Only grandchildren

98. According to Hugo de Vries, the mechanism of evolution is
(1) Multiple step mutations
(2) Saltation
(3) Minor mutations
(4) Phenotypic variations

99. AGGTATCGCAT is a sequence from the coding strand of a gene. What will be the corresponding sequence of the transcribed mRNA?
(1) AGGUAUCGCAU
(2) UGGTUTCGCAT
(3) UCCAUAGCGUA
(4) ACCUAUGCGAU

100. Among the following sets of examples for divergent evolution, select the incorrect option :
(1) Forelimbs of man, bat and cheetah
(2) Heart of bat, man and cheetah
(3) Eye of octopus, bat and man
(4) Brain of bat, man and cheetah

101. Conversion of milk to curd improves its nutritional value by increasing the amount of
(1) Vitamin D (2) Vitamin A
(3) Vitamin E (4) Vitamin B_{12}

102. Which of the following is not an autoimmune disease?
(1) Psoriasis
(2) Rheumatoid arthritis
(3) Vitiligo
(4) Alzheimer's disease

103. The similarity of bone structure in the forelimbs of many vertebrates is an example of
(1) Homology (2) Analogy
(3) Adaptive radiation
(4) Convergent evolution

104. Which of the following characteristics represent 'Inheritance of blood groups' in humans?
A. Dominance B. Co-dominance
C. Multiple allele
D. Incomplete dominance
E. Polygenic inheritance
(1) B, C and E (2) A, B and C
(3) A, C and E (4) B, D and E

105. In which disease does mosquito transmitted pathogen cause chronic inflammation of lymphatic vessels?
(1) Elephantiasis (2) Ascariasis
(3) Amoebiasis (4) Ringworm disease

106. All of the following are included in 'ex-situ conservation' except
(1) Wildlife safari parks (2) Sacred groves
(3) Seed banks (4) Botanical gardens

107. Which part of poppy plant is used to obtain the drug "Smack"?
(1) Flowers (2) Latex
(3) Leaves (4) Roots

108. In a growing population of a country,
(1) pre-reproductive individuals are more than the reproductive individuals.
(2) reproductive individuals are less than the post-reproductive individuals.
(3) pre-reproductive individuals are less than the reproductive individuals.
(4) reproductive and pre-reproductive individuals are equal in number.

109. Which one of the following population interactions is widely used in medical science for the production of antibiotics?
(1) Commensalism (2) Mutualism
(3) Amensalism (4) Parasitism

110. Match the items given in Column I with those in Column II and select the correct option given below :

	Column-I		Column-II
A.	Eutrophication	i.	UV-B radiation
B.	Sanitary landfill	ii.	Deforestation
C.	Snow blindness	iii.	Nutrient enrichment
D.	Jhum cultivation	iv.	Waste disposal

	A	B	C	D
(1)	ii	i	iii	iv
(2)	i	iii	iv	ii
(3)	i	ii	iv	iii
(4)	iii	iv	i	ii

111. Which of the following options correctly represents the lung conditions in asthma and emphysema, respectively?
(1) Inflammation of bronchioles; Decreased respiratory surface
(2) Increased number of bronchioles; Increased respiratory surface
(3) Decreased respiratory surface; Inflammation of bronchioles
(4) Increased respiratory surface; Inflammation of bronchioles

112. Match the items given in Column I with those in Column II and select the correct option given below :

	Column I		Column II
A.	Tricuspid valve	i.	Between left atrium and left ventricle
B.	Bicuspid valve	ii.	Between right ventricle and pulmonary artery
C.	Semilunar valve	iii.	Between right atrium and right ventricle

	A	B	C
(1)	iii	i	ii
(2)	i	iii	ii
(3)	ii	i	iii
(4)	i	ii	iii

113. Match the items given in Column I with those in Column II and select the correct option given below:

	Column I		Column II
A.	Tidal volume	i.	2500 – 3000 mL
B.	Inspiratory Reserve volume	ii.	1100 – 1200 mL
C.	Expiratory Reserve volume	iii.	500 – 550 mL
D.	Residual volume	iv.	1000 – 1100 mL

	A	B	C	D
(1)	iii	ii	i	iv
(2)	iii	i	iv	ii
(3)	iv	iii	ii	i
(4)	i	iv	ii	iii

114. Which of the following is an amino acid derived hormone?
(1) Epinephrine (2) Ecdysone
(3) Estriol (4) Estradiol

115. Which of the following structures or regions is incorrectly paired with its functions?
(1) Medulla oblongata : controls respiration and cardiovascular reflexes.
(2) Limbic system : consists of fibre tracts that interconnect different regions of brain; controls movement.
(3) Corpus callosum : band of fibers connecting left and right cerebral hemispheres.
(4) Hypothalamus : production of releasing hormones and regulation of temperature, hunger and thirst.

116. The transparent lens in the human eye is held in its place by
(1) ligaments attached to the ciliary body
(2) ligaments attached to the iris
(3) smooth muscles attached to the ciliary body
(4) smooth muscles attached to the iris

117. Which of the following hormones can play a significant role in osteoporosis?
(1) Aldosterone and Prolactin
(2) Progesterone and Aldosterone
(3) Parathyroid hormone and Prolactin
(4) Estrogen and Parathyroid hormone

118. Which of the following gastric cells indirectly help in erythropoiesis?
(1) Chief cells (2) Mucous cells
(3) Parietal cells (4) Goblet cells

119. Match the items given in Column I with those in Column II and select the correct option given below :

	Column I		Column II
A.	Fibrinogen	(i)	Osmotic balance
B.	Globulin	(ii)	Blood clotting
C.	Albumin	(iii)	Defence mechanism

	A	B	C
(1)	(iii)	(ii)	(i)
(2)	(i)	(ii)	(iii)
(3)	(ii)	(iii)	(i)
(4)	(i)	(iii)	(ii)

120. Which of the following is an occupational respiratory disorder?
(1) Anthracis (2) Silicosis
(3) Emphysema (4) Botulism

121. Calcium is important in skeletal muscle contraction because it
(1) Binds to troponin to remove the masking of active sites on actin for myosin.
(2) Activates the myosin ATPase by binding to it.
(3) Prevents the formation of bonds between the myosin cross bridges and the actin filament.
(4) Detaches the myosin head from the actin filament.

122. Select the incorrect match :
(1) Lampbrush chromosomes – Diplotene bivalents
(2) Allosomes – Sex chromosomes
(3) Polytene chromosomes – Oocytes of amphibians
(4) Submetacentric chromosomes – L-shaped chromosomes

123. Nissl bodies are mainly composed of
(1) Proteins and lipids
(2) DNA and RNA
(3) Free ribosomes and RER
(4) Nucleic acids and SER

124. Which of these statements is incorrect?
(1) Enzymes of TCA cycle are present in mitochondrial matrix
(2) Glycolysis occurs in cytosol
(3) Oxidative phosphorylation takes place in outer mitochondrial membrane
(4) Glycolysis operates as long as it is supplied with NAD that can pick up hydrogen atoms

125. Which of the following events does not occur in rough endoplasmic reticulum?
(1) Protein folding
(2) Protein glycosylation
(3) Phospholipid synthesis
(4) Cleavage of signal peptide

126. Many ribosomes may associate with a single mRNA to form multiple copies of a polypeptide simultaneously. Such strings of ribosomes are termed as
(1) Polysome (2) Polyhedral bodies
(3) Nucleosome (4) Plastidome

127. Which of the following terms describe human dentition?
(1) Thecodont, Diphyodont, Homodont
(2) Thecodont, Diphyodont, Heterodont
(3) Pleurodont, Diphyodont, Heterodont
(4) Pleurodont, Monophyodont, Homodont

128. Identify the vertebrate group of animals characterized by crop and gizzard in its digestive system
(1) Amphibia (2) Reptilia
(3) Osteichthyes (4) Aves

129. Which one of these animals is not a homeotherm?
(1) *Macropus* (2) *Chelone*
(3) *Psittacula* (4) *Camelus*

130. Which of the following features is used to identify a male cockroach from a female cockroach?
(1) Presence of a boat shaped sternum on the 9th abdominal segment
(2) Presence of caudal styles
(3) Presence of anal cerci
(4) Forewings with darker tegmina

131. Which of the following organisms are known as chief producers in the oceans?
(1) Dinoflagellates (2) Diatoms
(3) Euglenoids (4) Cyanobacteria

132. Ciliates differ from all other protozoans in
(1) using flagella for locomotion
(2) having a contractile vacuole for removing excess water
(3) having two types of nuclei
(4) using pseudopodia for capturing prey

133. Which of the following animals does not undergo metamorphosis?
(1) Earthworm (2) Tunicate
(3) Starfish (4) Moth

134. Match the items given in Column I with those in Column II and select the correct option given below:

	Column I (Function)		Column II (Part of Excretory system)
A.	Ultrafiltration	i.	Henle's loop
B.	Concentration of urine	ii.	Ureter
C.	Transport of urine	iii.	Urinary bladder
D.	Storage of urine	iv.	Malpighian corpuscle
		v.	Proximal convoluted tubule

	A	B	C	D
(1)	iv	v	ii	iii
(2)	iv	i	ii	iii
(3)	v	iv	i	iii
(4)	v	iv	i	ii

135. Match the items given in Column I with those in Column II and select the correct option given below :

	Column I		Column II
A.	Glycosuria	i.	Accumulation of uric acid in joints
B.	Gout	ii.	Mass of crystallised salts within the kidney
C.	Renal calculi	iii.	Inflammation in glomeruli
D.	Glomerular nephritis	iv.	Presence of in glucose urine

	A	B	C	D
(1)	iii	ii	iv	i
(2)	i	ii	iii	iv
(3)	iv	i	ii	iii
(4)	ii	iii	i	iv

136. What is the role of NAD^+ in cellular respiration?
(1) It functions as an enzyme.
(2) It functions as an electron carrier.
(3) It is the final electron acceptor for anaerobic respiration.
(4) It is a nucleotide source for ATP synthesis.

137. Which one of the following plants shows a very close relationship with a species of moth, where none of the two can complete its life cycle without the other?
(1) *Hydrilla* (2) *Yucca*
(3) *Viola* (4) Banana

138. Oxygen is not produced during photosynthesis by
(1) Green sulphur bacteria
(2) *Nostoc*
(3) *Chara* (4) *Cycas*

139. In which of the following forms is iron absorbed by plants?
(1) Ferric (2) Ferrous
(3) Both ferric and ferrous
(4) Free element

140. Double fertilization is
(1) Fusion of two male gametes of a pollen tube with two different eggs
(2) Fusion of one male gamete with two polar nuclei
(3) Syngamy and triple fusion
(4) Fusion of two male gametes with one egg

141. Which of the following elements is responsible for maintaining turgor in cells?
(1) Magnesium (2) Sodium
(3) Calcium (4) Potassium

142. Pollen grains can be stored for several years in liquid nitrogen having a temperature of
(1) –120°C (2) –80°C
(3) –160°C (4) –196°C

143. Which among the following is not a prokaryote?
(1) *Saccharomyces* (2) Mycobacterium
(3) *Oscillatoria* (4) *Nostoc*

144. The two functional groups characteristic of sugars are
(1) Hydroxyl and methyl
(2) Carbonyl and methyl
(3) Carbonyl and hydroxyl
(4) Carbonyl and phosphate

145. Which of the following is not a product of light reaction of photosynthesis?
(1) ATP (2) NADH
(3) Oxygen (4) NADPH

146. Stomatal movement is not affected by
(1) Temperature (2) Light
(3) CO_2 concentration (4) O_2 concentration

147. The Golgi complex participates in
(1) Fatty acid breakdown
(2) Formation of secretory vesicles
(3) Activation of amino acid
(4) Respiration in bacteria

148. Which of the following is true for nucleolus?
(1) Larger nucleoli are present in dividing cells
(2) It is a membrane-bound structure
(3) It is a site for active ribosomal RNA synthesis
(4) It takes part in spindle formation

149. Stomata in grass leaf are
(1) Dumb-bell shaped (2) Kidney shaped
(3) Barrel shaped (4) Rectangular

150. The stage during which separation of the paired homologous chromosomes begins is
(1) Pachytene (2) Diplotene
(3) Zygotene (4) Diakinesis

151. Which of the following is commonly used as a vector for introducing a DNA fragment in human lymphocytes?
(1) Retrovirus (2) Ti plasmid
(3) pBR 322 (4) λ phage

152. Use of bioresources by multinational companies and organisations without authorisation from the concerned country and its people is called
(1) Bio-infringement (2) Biopiracy
(3) Bioexploitation (4) Biodegradation

153. In India, the organisation responsible for assessing the safety of introducing genetically modified organisms for public use is
(1) Indian Council of Medical Research (ICMR)
(2) Council for Scientific and Industrial Research (CSIR)
(3) Genetic Engineering Appraisal Committee (GEAC)
(4) Research Committee on Genetic Manipulation (RCGM)

154. The correct order of steps in Polymerase Chain Reaction (PCR) is
(1) Extension, Denaturation, Annealing
(2) Annealing, Extension, Denaturation
(3) Denaturation, Annealing, Extension
(4) Denaturation, Extension, Annealing

155. Select the correct match
(1) Ribozyme - Nucleic acid
(2) F_2 × Recessive parent - Dihybrid cross
(3) G. Mendel - Transformation
(4) T.H. Morgan - Transduction

156. A 'new' variety of rice was patented by a foreign company, though such varieties have been present in India for a long time. This is related to
(1) Co-667 (2) Sharbati Sonora
(3) Basmati (4) Lerma Rojo

157. Select the correct match
(1) Alec Jeffreys - *Streptococcus pneumoniae*
(2) Alfred Hershey and Martha Chase - TMV
(3) Francois Jacob and Jacques Monod - Lac operon
(4) Matthew Meselson and F. Stahl - *Pisum sativum*

158. Which of the following has proved helpful in preserving pollen as fossils?
(1) Pollenkitt (2) Cellulosic intine
(3) Sporopollenin (4) Oil content

159. The experimental proof for semiconservative replication of DNA was first shown in a
(1) Fungus (2) Bacterium
(3) Virus (4) Plant

160. Which of the following pairs is wrongly matched?
(1) Starch synthesis in pea : Multiple alleles
(2) ABO blood grouping : Co-dominance
(3) T.H. Morgan : Linkage
(4) XO type sex determination : Grasshopper

161. Offsets are produced by
(1) Meiotic divisions (2) Mitotic divisions
(3) Parthenogenesis (4) Parthenocarpy

162. Select the correct statement
(1) Franklin Stahl coined the term "linkage"
(2) Punnett square was developed by a British scientist
(3) Transduction was discovered by S. Altman
(4) Spliceosomes take part in translation

163. Which of the following flowers only once in its life-time?
(1) Bamboo species (2) Jackfruit
(3) Papaya (4) Mango

164. Niche is
(1) all the biological factors in the organism's environment

(2) the physical space where an organism lives
(3) the functional role played by the organism where it lives
(4) the range of temperature that the organism needs to live

165. In stratosphere, which of the following elements acts as a catalyst in degradation of ozone and release of molecular oxygen?
(1) Carbon (2) Cl
(3) Oxygen (4) Fe

166. What type of ecological pyramid would be obtained with the following data?
Secondary consumer : 120 g
Primary consumer : 60 g
Primary producer : 10 g
(1) Inverted pyramid of biomass
(2) Pyramid of energy
(3) Upright pyramid of biomass
(4) Upright pyramid of numbers

167. Which of the following is a secondary pollutant?
(1) CO (2) CO_2
(3) O_3 (4) SO_2

168. World Ozone Day is celebrated on
(1) 5th June (2) 21st April
(3) 22nd April (4) 16th September

169. Natality refers to
(1) Death rate (2) Birth rate
(3) Number of individuals entering a habitat
(4) Number of individuals leaving the habitat

170. Match the items given in Column I with those in Column II and select the correct option given below:

	Column I		Column II
A.	Herbarium	(i)	It is a place having a collection of preserved plants and animals
B.	Key	(ii)	A list that enumerates methodically all the species found in an area with brief description aiding identification
C.	Museum	(iii)	Is a place where dried and pressed plant specimens mounted on sheets are kept
D.	Catalogue	(iv)	A booklet containing a list of characters and their alternates which are helpful in identification of various taxa.

	A	B	C	D
(1)	(i)	(iv)	(iii)	(ii)
(2)	(iii)	(ii)	(i)	(iv)
(3)	(iii)	(iv)	(i)	(ii)
(4)	(ii)	(iv)	(iii)	(i)

171. Which one is wrongly matched?
(1) Uniflagellate gametes – *Polysiphonia*
(2) Biflagellate zoospores – *Brown algae*
(3) Unicellular organism – *Chlorella*
(4) Gemma cups – *Marchantia*

172. After karyogamy followed by meiosis, spores are produced exogenously in
(1) *Neurospora* (2) *Alternaria*
(3) *Saccharomyces* (4) *Agaricus*

173. Winged pollen grains are present in
(1) Mustard (2) *Cycas*
(3) *Pinus* (4) Mango

174. Pneumatophores occur in
(1) Halophytes
(2) Free-floating hydrophytes
(3) Submerged hydrophytes
(4) Carnivorous plants

175. Plants having little or no secondary growth are
(1) Grasses
(2) Deciduous angiosperms
(3) Cycads (4) Conifers

176. Casparian strips occur in
(1) Epidermis (2) Pericycle
(3) Endodermis (4) Cortex

177. Secondary xylem and phloem in dicot stem are produced by
(1) Apical meristems (2) Vascular cambium
(3) Axillary meristems (4) Phellogen

178. Select the wrong statement :
(1) Cell wall is present in members of Fungi and Plantae
(2) Mushrooms belong to Basidiomycetes
(3) Mitochondria are the powerhouse of the cell in all kingdoms except Monera
(4) Pseudopodia are locomotory and feeding structures in Sporozoans

179. Which of the following statements is correct?
(1) Ovules are not enclosed by ovary wall in gymnosperms
(2) *Selaginella* is heterosporous, while *Salvinia* is homosporous
(3) Stems are usually unbranched in both Cycas and *Cedrus*
(4) Horsetails are gymnosperms

180. Sweet potato is a modified
(1) Stem (2) Adventitious root
(3) Rhizome (4) Tap root

HINTS & SOLUTIONS

PHYSICS

1. **(2)** Two successive resonance are produced at 20 cm and 73 cm of column length

$\therefore \quad \frac{\lambda}{2} = (73-20) \times 10^{-2}$ m

$\Rightarrow \quad \lambda = 2 \times (73-20) \times 10^{-2}$

Velocity of sound, $v = n\lambda$

$= 2 \times 320\,[73-20] \times 10^{-2}$

$= 339.2\ ms^{-1}$

2. **(1)** As we know, $F = qE = ma$

$\Rightarrow \quad a = \frac{qE}{m}$

$h = \frac{1qE}{2m}t^2$

$\therefore \quad t = \sqrt{\frac{2hm}{qE}}$

i.e., time $t \propto \sqrt{m}$ as 'q' is same for electron and proton.

Since, electron has smaller mass so it will take smaller time.

3. **(2)** From question, acceleration, $a = 20\ m/s^2$, and displacement, $y = 5m$

$|a| = \omega^2 y$

$\Rightarrow \quad 20 = \omega^2(5)$

$\Rightarrow \quad \omega - 2$ rad/s

Time period of pendulum,

$T = \frac{2\pi}{\omega} = \frac{2\pi}{2} = \pi$ s

4. **(1)** Electrostatic force between the metal plates

$F_{plate} = \frac{Q^2}{2A_{\varepsilon_0}}$

For isolated capacitor Q = constant

Clearly, F is independent of the distance between plates.

5. **(4)** Current sensitivity of moving coil galvanometer

$I_s = \frac{NBA}{C} \qquad$ ...(i)

Voltage sensitivity of moving coil galvanometer,

$V_s = \frac{NBA}{CR_G} \qquad$ ...(ii)

Dividing eqn. (i) by (ii)

Resistance of galvanometer

$R_G = \frac{I_s}{V_s} = \frac{5 \times 1}{20 \times 10^{-3}} = \frac{5000}{20} = 250\ \Omega$

6. **(1)** Rod gains gravitational potential energy which comes from energy of current source.

7. **(1)** Power dissipated in an LCR series circuit connected to an a.c. source of emf E

$P = E_{rms} i_{rms} \cos\phi = \frac{E_{rms}^2 R}{Z^2} = \frac{E_{rms}^2 R}{R^2 + \left(\omega L - \frac{1}{C\omega}\right)^2}$

$= \frac{\left(\frac{10}{\sqrt{2}}\right)^2 \times 50}{(50)^2 + \left(314 \times 20 \times 10^{-3} - \frac{1}{314 \times 100 \times 10^{-6}}\right)^2}$

Solving we get, P = 0.79 W

8. **(3)** From figure, for equilibrium,

mg sin 30° = I/B cos 30°

$\Rightarrow \quad I = \frac{mg}{\ell B} \tan 30°$

$= \frac{0.5 \times 9.8}{0.25 \times \sqrt{3}} = 11.32$ A

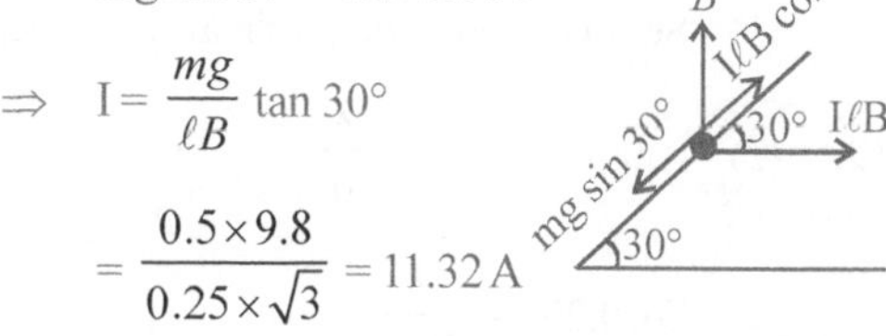

9. **(2)** Colour code for carbon resistor

0	Black		
1	Brown	Tolerance :	± 5% Gold
2	Red		± 10% Silver
3	Orange		± 20% No colour
4	Yellow		
5	Green		
6	Blue		
7	Violet		
8	Grey		
9	White		

$(47 \pm 4.7)\ k\Omega = 47 \times 10^3 \pm 10\%$

$\therefore$ Yellow - Violet - Orange - Silver

10. (1) In series grouping equivalent resistance
$R_{series} = nR$
In parallel grouping equivalent resistance

$R_{parallel} = \frac{R}{n}$

$I = \frac{E}{nR + R}$ ...(i)

$10\,I = \frac{E}{\frac{R}{n} + R}$ ...(ii)

Dividing eq. (ii) by (i), $10 = \frac{(n+1)R}{\left(\frac{1}{n}+1\right)R}$

Solving we get, n = 10

11. (1) Short circuited current,

$I = \frac{n\varepsilon}{nr} = \frac{\varepsilon}{r}$

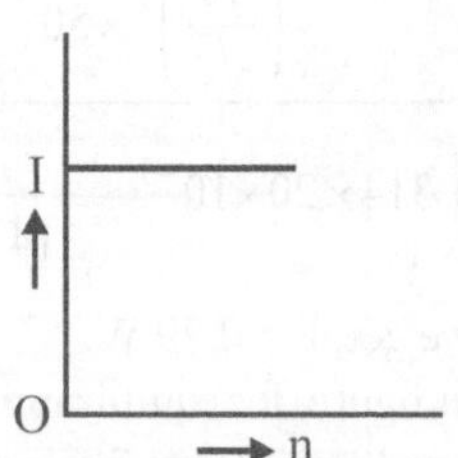

So, I is independent of n and I is constant.

12. (2) Angular width $= \frac{\lambda}{d}$

So, $0.20° = \frac{\lambda}{2\text{mm}}$

$\Rightarrow \lambda = 0.20° \times 2$

Again, $0.21° = \frac{\lambda}{d}$

Now putting the value of λ

$d = \frac{0.20° \times 2\text{mm}}{0.21°}$

$\therefore d = 1.9\,\text{mm}$

13. (4) For telescope, angular magnification $= \frac{f_0}{f_E}$

So, focal length of objective lens should be large.

Angular resolution $= \frac{D}{1.22\lambda}$ So, D should be large.

So, objective lens of refracting telescope should have large focal length (f_0) and large diameter D for larger angular magnification.

14. (2) When reflected light rays and refracted rays are perpendicular, reflected light is polarised with electric field vector perpendicular to the plane of incidence.

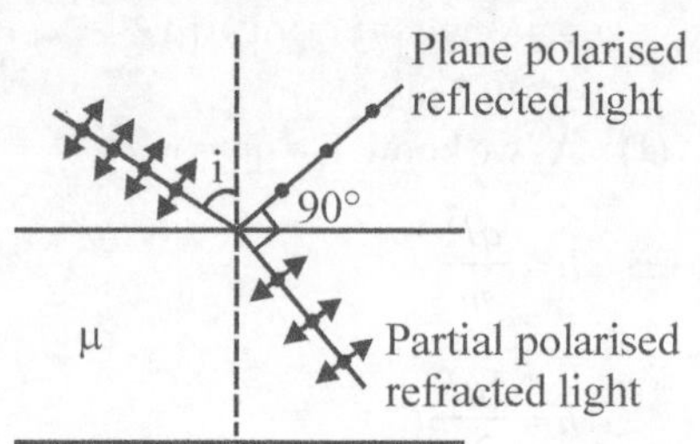

Also, tan i = μ (i = Brewester angle)

15. (2) As we know,

$\vec{E} \times \vec{B} = \vec{V}$

$(E\hat{j}) \times (\vec{B}) = V\hat{i}$

($\because$ Electric field vector is along +y axis)

So, $\vec{B} = B\hat{k}$

i.e., direction of magnetic field vector is along +z direction.

16. (2) For retracing the path, light ray should be normally incident on silvered face.

A = r + O ⇒ r = 30°

Applying Snell's law at point M,

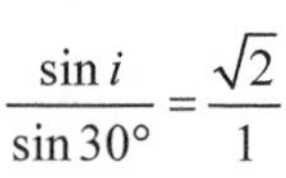

$\frac{\sin i}{\sin 30°} = \frac{\sqrt{2}}{1}$

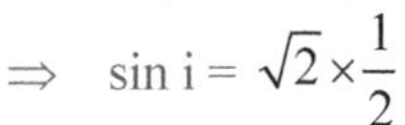

$\Rightarrow \sin i = \sqrt{2} \times \frac{1}{2}$

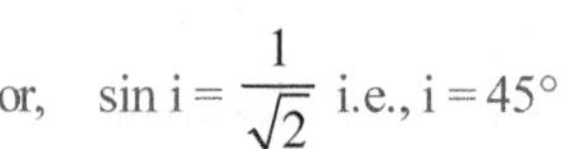

or, $\sin i = \frac{1}{\sqrt{2}}$ i.e., i = 45°

17. (2)

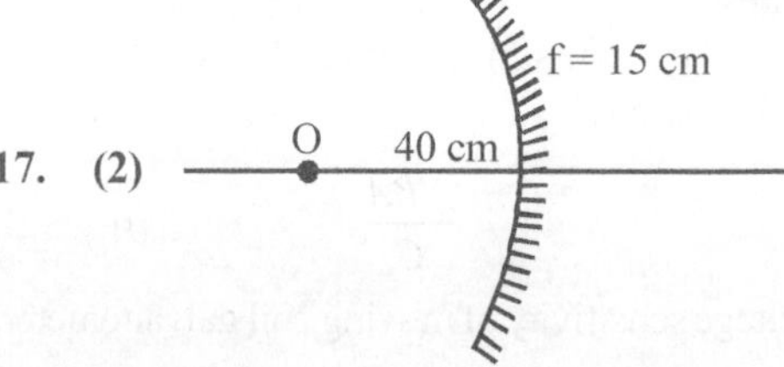

using mirror formula, $\frac{1}{f}=\frac{1}{v_1}+\frac{1}{u}$

$$-\frac{1}{15}=\frac{1}{v_1}+\frac{1}{u} \Rightarrow \frac{1}{v_1}=\frac{1}{-15}+\frac{1}{40}$$

$\therefore \quad v_1=-24$ cm

When object is displaced by 20 cm towards mirror

Now, $\quad u_2=-20$

So, $\frac{1}{f}=\frac{1}{v_2}+\frac{1}{u_2}$

$$\frac{1}{-15}=\frac{1}{v_2}-\frac{1}{20} \Rightarrow \frac{1}{v_2}=\frac{1}{20}-\frac{1}{15}$$

$\therefore \quad v_2=-60$ cm

Therefore image shifts away from mirror by $=60-24=36$ cm

18. **(3)** From question energy stored in inductor, $U=25\times10^{-3}$ J

Current, I = 60 mA

Energy stored in inductor $U=\frac{1}{2}LI^2$

$$25\times10^{-3}=\frac{1}{2}\times L\times(60\times10^{-3})^2$$

$$L=\frac{25\times2\times10^6\times10^{-3}}{3600}=13.89\text{ H}$$

19. **(1)** Number of nuclei remaining, N = 600 – 450 = 150 after time 't'

$$\frac{N}{N_0}=\frac{150}{600}=\frac{1}{4}$$

$$N=N_0e^{-\lambda t} \Rightarrow \ell n\frac{N_0}{N}=\lambda t$$

$$\Rightarrow t=\frac{1}{\lambda}\ell n\frac{N_0}{N}$$

$$\Rightarrow t=\frac{2.303\times T_{\frac{1}{2}}}{0.693}\log_{10}\frac{N_0}{N}$$

$$=\frac{2.303\times10}{0.693}\log_{10}4$$

20. **(2)** In a Bohr orbit of the hydrogen atom Kinetic energy,

$$k=\frac{kze^2}{2r_n}$$

Total energy, $E=\frac{-kze^2}{2r_n}$

So, Kinetic energy : total energy = 1 : –1

21. **(1)** Initial de-Brogile wavelength

$$\lambda_0=\frac{h}{mV_0} \quad ...(i)$$

$\vec{E}_0$ ← ; F ← ⊖ → V_0

Acceleration of electron

$$a=\frac{eE_0}{m} \quad (\because F=ma=eE_0)$$

Velocity after time 't'

$$V=\left(V_0+\frac{eE_0}{m}t\right)$$

$$\text{So, } \lambda=\frac{h}{mV}=\frac{h}{m\left(V_0+\frac{eE_0}{m}t\right)}$$

$$=\frac{h}{mV_0\left[1+\frac{eE_0}{mV_0}t\right]}=\frac{\lambda_0}{\left[1+\frac{eE_0}{mV_0}t\right]} \quad ...(ii)$$

Dividing eqs. (ii) by (i),

de-Broglie wavelength $\lambda=\frac{\lambda_0}{\left[1+\frac{eE_0}{mV_0}t\right]}$

22. **(1)** Using Einstein's photoelectric equation, $E=W_0+k_{max}$

When light of frequency, $2v_0$ is incident on a metal plate,

$$h(2v_0)=hv_0+\frac{1}{2}mv_1^2$$

$$hv_0=\frac{1}{2}mv_1^2 \quad ...(i)$$

when light of frequency, $5v_0$ is incident on a metal plate

$$h(5v_0)=hv_0+\frac{1}{2}mv_2^2$$

$$4h\nu_0 = \frac{1}{2}mv_2^2 \quad ...(ii)$$

Dividing eqn. (i) by (ii),

$$\frac{1}{4} = \frac{v_1^2}{v_2^2}$$

$$\therefore \quad \frac{v_1}{v_2} = \frac{1}{2}$$

23. **(2)**

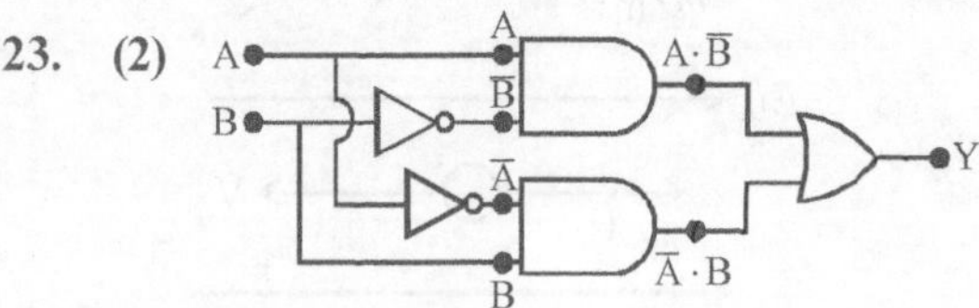

$Y = (A \cdot \overline{B} + \overline{A} \cdot B)$

24. **(3)** From question, $V_{BE} = = 0$, $V_i = 20$ V
$V_{CE} = 0$
$V_b = 0$ (earthed)
$I_B = ?$, $I_C = ?$, $\beta = ?$

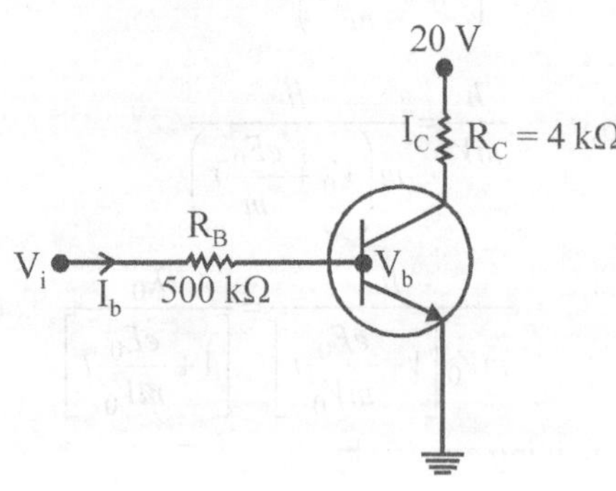

$$I_C = \frac{(20-0)}{4\times10^3} = 5\times10^{-3} = 5 \text{ mA}$$

$V_i = V_{BE} + I_B R_B$

or, $V_i = 0 + I_B R_B$

$\Rightarrow$ $20 = I_B \times 500 \times 10^3$

$$\Rightarrow \quad I_B = \frac{20}{500\times10^3} = 40\ \mu A$$

$$\beta = \frac{I_C}{I_b} = \frac{25\times10^{-3}}{40\times10^{-6}} = 125$$

25. **(3)** On heating, number of electron-hole pairs increases, so overall resistance of diode will change.
Hence forward biasing and reversed biasing both are changed.

26. **(3)** Angular momentum, remains conserved until the torque acting on sphere remains zero.
$\tau_{ex} = 0$

So, $\frac{dL}{dt} = 0$

i.e., angular momentum L = constant.

27. **(2)** Speed of the planet will be maximum when its distance from the sun is minimum as mvr = constant.

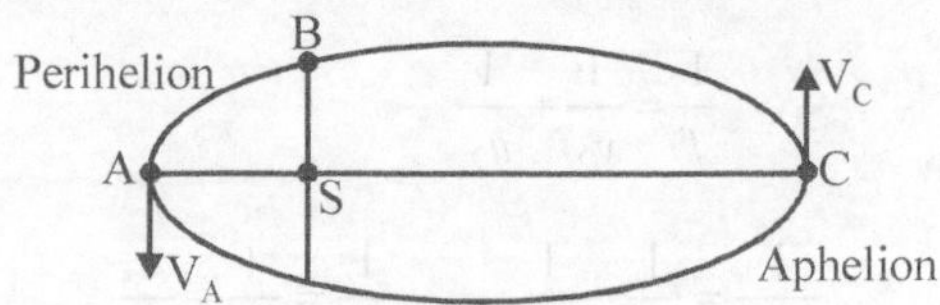

Point A is perihelion and C is aphelion.
Clearly, $V_A > V_B > V_C$
So, $K_A > K_B > K_C$

28. **(3)** If universal gravitational constant becomes ten times, then $G' = 10$ G

Acceleration due to gravity, $g = \frac{GM}{R^2}$

So, acceleration due to gravity increases.

29. **(2)** In rolling motion, rotational kinetic energy.

$$K_t = \frac{1}{2}mv^2$$

$$\text{And, } K_t + K_r = \frac{1}{2}mv^2 + \frac{1}{2}I\omega^2$$

$$= \frac{1}{2}mv^2 + \frac{1}{2}\left(\frac{2}{5}mr^2\right)\left(\frac{v}{r}\right)^2 = \frac{7}{10}mv^2$$

$$\therefore \quad \frac{K_t}{K_t + K_r} = \frac{\frac{1}{2}mv^2}{\frac{7}{10}mv^2} = \frac{5}{7}$$

30. **(4)** Power = rate of production of heat = F.V
$= 6\pi\eta r V_T \cdot V_T = 6\pi\eta r V_T^2$
($\because F = 6\pi\eta V_T r$ stoke's formula)

$V_T \propto r^2$

$$\because \quad V_T = \frac{2}{9}\frac{r^2(\rho-\sigma)}{\eta}g$$

$\therefore$ Power $\propto r^5$

31. **(2)** Using first law of thermodynamics equation,

$\Delta Q = \Delta U + \Delta W$

$\Rightarrow$ 54×4.18

$= \Delta U + 1.013 \times 10^5(167.1 \times 10^{-6} - 0)$

$(\because \Delta W = P\Delta V)$

$\Rightarrow$ $\Delta U = 208.7$ J

32. **(1)** Wire 1: → F A, 3ℓ

$$\Delta\ell = \left(\frac{F}{AY}\right)3\ell \quad ...(i)$$

Wire 2: → F′ 3A, ℓ

$$\Delta\ell = \left(\frac{F'}{3AY}\right)\ell \quad ...(ii)$$

From equation (i) & (ii),

$$\Delta\ell = \left(\frac{F}{AY}\right)3\ell = \left(\frac{F'}{3AY}\right)\ell \text{ or, } F' = 9F$$

33. **(4)** From Wien's law

$\lambda_{max} T$ = constant

i.e., $\lambda_{max_1} T_1 = \lambda_{max_2} T_2$

$$\Rightarrow \lambda_0 T = \frac{3\lambda_0}{4} T'$$

$$\Rightarrow T' = \frac{4}{3} T$$

Power radiated $P \propto T^4$

$$\text{So, } \frac{P_2}{P_1} = n = \left(\frac{T'}{T}\right)^4 = \left(\frac{4}{3}\right)^4 = \frac{256}{81}$$

34. **(2)** Let at temperature T rms speed of oxygen molecules become just sufficient for escaping from the Earth's atmosphere

$V_{escape} = 11200$ m/s

$$\text{Also, } V_{rms} = V_{escape} = \sqrt{\frac{3k_B T}{m_{O_2}}} = 11200 \text{ m/s}$$

Putting value of K_B and m_{O_2} we get,

$T = 8.360 \times 10^4$ K

35. **(1)** Gas is monatomic, so $C_p = \frac{5}{2}R$

Given process is isobaric

$\therefore$ $dQ = n\,C_p\,dT$

$$\Rightarrow dQ = n\left(\frac{5}{2}R\right)dT$$

$dW = P\,dV = n\,RdT$

$$\therefore \text{Required ratio} = \frac{dW}{dQ} = \frac{nRdT}{n\left(\frac{5}{2}R\right)dT} = \frac{2}{5}$$

36. **(1)** For closed organ pipe, third harmonic

$$n = \frac{(2N-1)V}{4\ell} = \frac{3V}{4\ell} \quad (\because N = 2)$$

For open organ pipe, fundamental frequenty

$$n = \frac{NV}{2\ell} = \frac{V}{2\ell'} \quad (\because N = 1)$$

$$\text{According to question, } \frac{3V}{4\ell} = \frac{V}{2\ell'}$$

$$\Rightarrow \ell' = \frac{4\ell}{3 \times 2} = \frac{2\ell}{3} = \frac{2 \times 20}{3} = 13.33 \text{ cm}$$

37. **(1)** Efficiency of ideal heat engine,

$$\eta = \left(1 - \frac{T_2}{T_1}\right)$$

Sink temperature, $T_2 = 100°C = 100 + 273 = 373$ K

Source temperature, $T_1 = 0°C = 0 + 273 = 273$ K

$$\text{Percentage efficiency, } \%\eta = \left(1 - \frac{T_2}{T_1}\right) \times 100$$

$$= \left(1 - \frac{273}{373}\right) \times 100 = \left(\frac{100}{373}\right) \times 100 = 26.8\%$$

38. **(3)** As track is frictionless, so total mechanical energy will remain constant

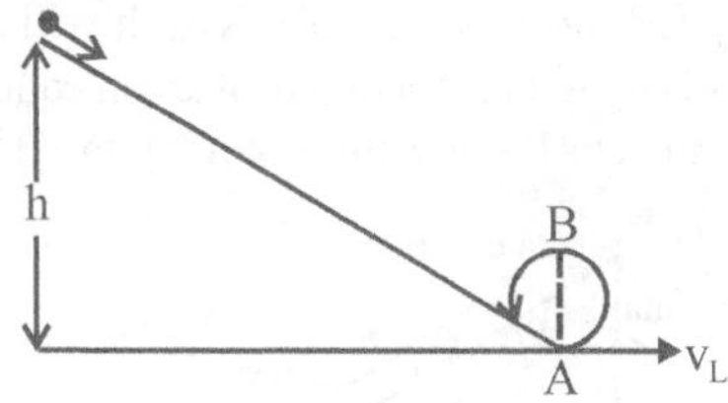

$$\text{i.e., } 0 + mgh = \frac{1}{2} m v_L^2 + 0$$

$$\text{Using } v^2 - u^2 = 2gh, \; h = \frac{v_L^2}{2g} \quad (\because u = 0)$$

For completing the vertical circle, $v_L \geq \sqrt{5gR}$

$$\text{or, } h = \frac{5gR}{2g} = \frac{5}{2}R = \frac{5}{4}D$$

39. **(1)** Work done required to bring them rest

$\Delta W = \Delta KE$ (work-energy theorem)

$$\Delta W = \frac{1}{2}I\omega^2 \left(\Delta kE_{rot} = \frac{1}{2}I\omega^2 \right)$$

or, $\Delta W \propto I$ (for same ω)

$$I_{\text{solid sphere}} = \frac{2}{5}MR^2, \ I_{\text{Disk}} = \frac{1}{2}MR^2$$

$$I_{\text{Ring}} = MR^2$$

$\therefore \quad W_C > W_B > W_A$

40. **(3)** Coefficient of friction or sliding friction has no dimension.

$$f = \mu_s N \Rightarrow \mu_s = \frac{f}{N} = [M^0L^0T^0]$$

41. **(2)**

m →v, 4m v = 0 — Before Collision

m v = 0, 4m →v′ — After Collision

According to law of conservation of linear momentum,

$$mv + 4m \times 0 = 4mv' + 0 \Rightarrow v' = \frac{v}{4}$$

Coefficient of restitution,

$$e = \frac{\text{Relative velocity of separation}}{\text{Relative velocity of approach}} = \frac{\frac{v}{4}}{v}$$

or, $$e = \frac{1}{4} = 0.25$$

42. **(3)** Let the mass of block is m. It will remains stationary if forces acting on it are in equilibrium. i.e., $ma \cos\theta = mg \sin\theta \Rightarrow a = g \tan\theta$

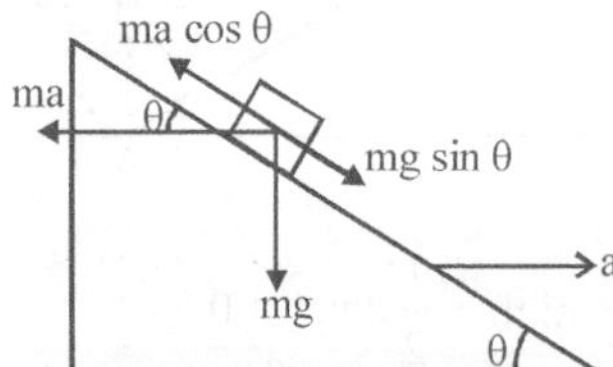

Here ma = Pseudo force on block, mg = weight.

43. **(2)**

A: t = 0, v = 0 —a→ t = 1, v = 6 ms^{-1} ←a— t = 2, B, v = 0; C: t = 3, v = –6 ms^{-1} ←a

Acceleration, $a = \frac{v-u}{t} = \frac{6-0}{1}$

$= 6 \text{ ms}^{-2}$

For t = 0 to t = 1 s,

$$S_1 = \frac{1}{2} \times 6(1)^2 = 3 \text{ m} \quad \text{...(i)}$$

For t = 1 s = to t = 2 s,

$$S_2 = 6.1 - \frac{1}{2} \times 6(1)^2 = 3\text{m} \quad \text{...(ii)}$$

For t = 2 s to t = 3 s,

$$S_3 = 0 - \frac{1}{2} \times 6(1)^2 = -3 \text{ m} \quad \text{(iii)}$$

Total displacement $S = S_1 + S_2 + S_3 = 3$ m

Average velocity $= \frac{3}{3} = 1 \text{ ms}^{-1}$

Total distance travelled = 9 m

Average speed $= \frac{9}{3} = 3 \text{ ms}^{-1}$

44. **(3)** Moment of force, $\vec{\tau} = \vec{r} \times \vec{F}$

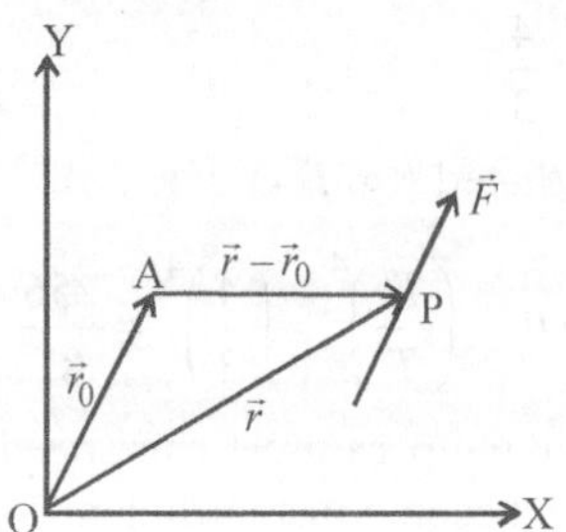

$$\vec{\tau} = (\vec{r} - \vec{r}_0) \times \vec{F}$$

$$\vec{r} - \vec{r}_0 = (2\hat{i} + 0\hat{j} - 3\hat{k}) - (2\hat{i} - 2\hat{j} - 2\hat{k})$$

$$= 0\hat{i} + 2\hat{j} - \hat{k}$$

$$\vec{\tau} = (0\hat{i} + 2\hat{j} - \hat{k})(4\hat{i} + 5\hat{j} - 6\hat{k})$$

$$\vec{\tau} = \begin{vmatrix} \hat{i} & \hat{j} & \hat{k} \\ 0 & 2 & -1 \\ 4 & 5 & -6 \end{vmatrix} = -7\hat{i} - 4\hat{j} - 8\hat{k}$$

45. **(3)** Diameter of the ball

= MSR + CSR × (ℓeast count) – zero error

= 0.5 cm + 25 × 0.001 – (–0.004)

= 0.5 + 0.025 + 0.004 = 0.529 cm

CHEMISTRY

46. (1) $\overset{+5}{HNO_3}, \overset{+2}{NO}, \overset{0}{N_2}, \overset{-3}{NH_4Cl}$

47. (3)

48. (4) Mg has more $-\Delta G°$ value than alumina. So it will be in the lower part of Ellingham diagram. Metals which have more $-\Delta G°$ value can reduce those metal oxides which have less $-\Delta G°$ value.

49. (4) MF_6^{3-}
Boron belongs to 2nd period and it does not have vacant *d*-orbital.

50. (4) Due to high electronegativity and small size, F forms only one oxoacid, HOF known as fluoric (I) acid. Oxidation number of F is +1 in HOF.

51. (2) The structure of ClF_3 is

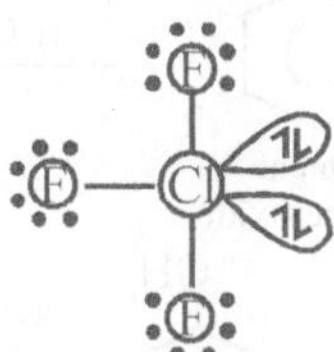

The number of lone pair of electrons on central Cl is 2.

52. (1) Amylose and amylopectin are polymers of α-D-glucose, so β-link is not possible. Amylose is linear with $1 \to 4$ α-linkage whereas amylopectin is branched and has both $1 \to 4$ and $1 \to 6$ α-linkages.
So option (1) should be the correct option.

53. (3) Cross-linked or network polymers are usually formed from bi-functional and tri-functional monomers and contain strong covalent bonds between various linear polymer chains like melamine, bakelite etc.

54. (4) $HCOOH \xrightarrow[\text{Dehydrating agent}]{H_2SO_4} CO + H_2O$ [H_2O absorbed by H_2SO_4]

At start (moles) $= \frac{2.3}{46} = \frac{1}{20}$ 0   0

Final moles   0 $\frac{1}{20}$ $\frac{1}{20}$

$H_2C_2O_4 \xrightarrow{H_2SO_4} CO + CO_2 + H_2O$ [H_2O absorbed by H_2SO_4]

At start (moles) $= \frac{4.5}{90} = \frac{1}{20}$ 0   0   0

Final moles   0 $\frac{1}{20}$ $\frac{1}{20}$ $\frac{1}{20}$

CO_2 is absorbed by KOH.
So the remaning product is only CO.
Moles of CO formed from both reactions

$$= \frac{1}{20} + \frac{1}{20} = \frac{1}{10}$$

Left mass of CO = moles × molar mass

$$= \frac{1}{10} \times 28 = 2.8 \text{ g}$$

55. (2) In metals moving down the group metallic character increases, so basic nature increases hence most acidic will be BeO.

$$\xrightarrow[\text{increasing basic character}]{BeO < MgO < CaO < BaO}$$

56. (3)

$$C_6H_5NH_2 \xrightarrow[\text{nitrating mixture}]{H^+} C_6H_5\overset{+}{N}H_3 \xrightarrow{\overset{+}{N}O_2} m\text{-}O_2N\text{-}C_6H_4\overset{+}{N}H_3$$

In acidic medium aniline is protonated to form anilinium ion which is meta-directing.

57. (3) $\underset{(A)}{C_2H_5OH} \xrightarrow{Na} \underset{(B)}{C_2H_5O^-Na^+}$

$\underset{(A)}{C_2H_5OH} \xrightarrow{PCl_5} \underset{(C)}{C_2H_5Cl}$

$\underset{(B)}{C_2H_5O^-Na^+} + \underset{(C)}{C_2H_5Cl} \xrightarrow{S_N2} C_2H_5OC_2H_5$

So the correct option is (3)

58. (3) $CH_4 \xrightarrow[h\nu]{Br_2} CH_3-Br \xrightarrow[\text{ether}]{Na} CH_3-CH_3$ (less than four 'C')

59. (1) $C_6H_5CH_3 \xrightarrow[\Delta]{3Cl_2} C_6H_5CCl_3 \xrightarrow[Fe]{Br_2} m\text{-}Br\text{-}C_6H_4CCl_3 \xrightarrow[HCl]{Zn} m\text{-}Br\text{-}C_6H_4CH_3$

m-Bromotoluene

60. (1) Nitrous oxide (N_2O) occurs naturally in environment.

In automobile engine, when fuel is burnt dinitrogen and dioxygen combine to yield NO and NO_2.

61. (2) $\overset{sp^2}{CH_2} = \overset{sp^2}{CH} - \overset{sp}{C} \equiv \overset{sp}{CH}$

62. (4) $-NO_2$ group is meta-directing group

(Less stable due to more e^- withdrawing effect of $-NO_2$)

(More stable due to less e^- withdrawing effect of $-NO_2$) greater no. of resonating structures.

63. (1) – I effect increases on increasing electronegativity of atom. So, correct order of – I effect is

$-NH_2 < -OR < -F.$

*Most appropriate answer is option (1), however option (2) may also be correct answer.

64. (3) $CHCl_3 + NaOH \longrightarrow \bar{C}Cl_3 + H_2O$

$\downarrow -Cl$ (α-elimination)

$:CCl_2$ dichlorocarbene (electrophile)

65. (3) Carboxylic acids have higher boiling points than aldehydes, ketones and even alcohols of comparable molecular mass.

This is due to more extensive association through intermolecular H-bonding.

66. (4) Haloform reaction is shown by compound having

$CH_3 - \underset{\underset{O}{\|}}{C} -$ or $CH_3 - \underset{\underset{OH}{|}}{CH} -$ Group

$C_6H_5 - \underset{\underset{OH}{|}}{CH} - CH_3 \xrightarrow[NaOH + I_2]{NaOI \text{ or}} C_6H_5 - \underset{\underset{O}{\|}}{C} - \overset{-\ +}{ONa} + \underset{\text{Yellow ppt.}}{CHI_3}$

67. (3) Mechanism :

$CH_3 - CH_2 - CH_2 - Cl \xrightarrow{AlCl_3} CH_3 - CH_2 - \overset{+}{CH_2} + \overset{-}{AlCl_4}$ (1° Carbocation)

$CH_3 - \overset{+}{CH} - CH_3$ (2° Carbocation, via H shift)

$CH_3 - CH - CH_3$ (with phenyl; ESR with benzene) (P)

$\underset{\text{Cumene (P)}}{C_6H_5CH(CH_3)_2} \xrightarrow[\Delta]{O_2} \underset{\text{Cumene hydroperoxide}}{C_6H_5C(CH_3)_2-O-O-H} \xrightarrow[\Delta]{H_3O^+} \underset{\text{Phenol (Q)}}{C_6H_5OH} + \underset{\text{Acetone (R)}}{CH_3-\underset{\underset{O}{\|}}{C}-CH_3}$

68. (3) $\underset{\text{Glycine}}{HOOC - CH_2 - NH_2} \rightleftharpoons \underset{\text{Zwitter ion}}{\bar{O}OC - CH_2 - \overset{+}{N}H_3}$

69. (2) $\overset{+7}{Mn}O_4^- \longrightarrow \overset{+2}{Mn^{2+}}$; $5e^-$ gain ...(i)

$\overset{+3}{C_2}O_4^{2-} \longrightarrow \overset{+4}{C}O_2$; $2e^-$ loss ...(ii)

Multiplying (i) by 2 and (ii) by 5 to balance e^-

$2\,MnO_4^- + 5\,C_2O_4^{2-} \longrightarrow 2\,Mn^{2+} + 10\,CO_2$

On balancing charge;

$2\,MnO_4^- + 5\,C_2O_4^{2-} + 16\,H^+ \longrightarrow 2\,Mn^{2+} + 10\,CO_2 + 8\,H_2O$

70. (1) $A_2(g) + B_2(g) \rightleftharpoons X_2(g); \Delta H = -x$ kJ

On increasing pressure equilibrium shifts in a direction where number of moles decreases i.e. forward direction.

On decreasing temperature, equilibrium shifts in exothermic direction i.e., forward direction.

So, high pressure and low temperature favours maximum formation of product.

71. **(2)** $(t_{1/2})_{zero} = \frac{[A]_o}{2k}$

$\therefore$ If $[A]_o$ = doubled, $t_{1/2}$ = doubled

72. **(3)** In real gas equation,
van der waal constant $(a) \propto$ forces of attraction.

73. **(4)** Let B.E of x_2, y_2 and xy are x kJ mol^{-1}, 0.5 x kJ mol^{-1} and x kJ mol^{-1} respectively

$$\frac{1}{2}x_2 + \frac{1}{2}y_2 \rightarrow xy; \ \Delta H = -200 \text{ kJ mol}^{-1}$$

$$\Delta H = -200 = \Sigma\,(B.E)_{Reactants} - \Sigma(B.E)_{Product}$$

$$= \left[\frac{1}{2}\times(x) + \frac{1}{2}\times(0.5x)\right] - [1\times(x)]$$

On solving, x = 800 kJ mol^{-1}

74. **(3)** Element (X) electronic configuration
$1s^2\,2s^2\,2p^3$
So, valency of X will be 3.
Valency of Mg is 2.
Formula of compound formed by Mg and X will be Mg_3X_2.

75. **(4)** For bcc lattice : Z = 2, $a = \frac{4r}{\sqrt{3}}$

For fcc lattice : Z = 4, $a = 2\sqrt{2}r$

$$\therefore \frac{d_{25\,°C}}{d_{900\,°C}} = \frac{\left(\frac{ZM}{N_A a^3}\right)_{bcc}}{\left(\frac{ZM}{N_A a^3}\right)_{fcc}}$$

$$= \frac{2}{4}\left(\frac{2\sqrt{2}r}{\frac{4r}{\sqrt{3}}}\right)^3 = \frac{3\sqrt{3}}{4\sqrt{2}}$$

76. **(2)** NO : $(\sigma 1s)^2, (\sigma^*1s)^2, (\sigma 2s)^2, (\sigma^*2s)^2, (\sigma 2p_z)^2, (\pi 2p_x)^2 = (\pi 2p_y)^2, (\pi^*2p_x)^1 = (\pi^*2p_y)^0$

$$B.O. = \frac{10-5}{2} = 2.5$$

CN^- : $(\sigma 1s)^2, (\sigma^*1s)^2, (\sigma 2s)^2, (\sigma^*2s)^2, (\pi 2p_x)^2 = (\pi 2p_y)^2, (\sigma 2p_z)^2$

$$B.O. = \frac{10-4}{2} = 3$$

CN : $(\sigma 1s)^2, (\sigma^*1s)^2, (\sigma 2s)^2, (\sigma^*2s)^2, (\pi 2p_x)^2 = (\pi 2p_y)^2, (\sigma 2p_z)^1$

$$B.O. = \frac{9-4}{2} = 2.5$$

CN^+ : $(\sigma 1s)^2, (\sigma^*1s)^2, (\sigma 2s)^2, (\sigma^*2s)^2, (\pi 2p_x)^2 = (\pi 2p_y)^2$

$$B.O. = \frac{8-4}{2} = 2$$

Hence, option (2) should be the right answer.

77. **(4)** The correct configuration of 'N' is

$1s^2$	$2s^2$	$2p_x^1$	$2p_y^1$	$2p_z^1$
⇅	⇅	1	1	1

78. **(2)** $(t_{1/2})$ 1st order = Independent of concentration

$(t_{1/2})$ 2nd order $\propto \frac{1}{[A]_o}$

79. **(1)**
(1) Mass of water = 18 × 1 = 18 g

Molecules of water = mole × $N_A = \frac{18}{18}N_A = N_A$

(2) Molecules of water = mole × N_A

$= \frac{0.18}{18}N_A = 10^{-2}N_A$

(3) Molecules of water = mole × $N_A = 10^{-3}N_A$

(4) Moles of water = $\frac{0.00224}{22.4} = 10^{-4}$

Molecules of water = mole × $N_A = 10^{-4}N_A$

80. **(1)** $BeH_2 < CaH_2 < BaH_2$
Smaller the size of cation, more will be its polarising power. Hence BeH_2 will be least ionic.

81. **(3)** Calculate $E°_{cell}$ corresponding to each compound undergoing disproportionation reaction. The reaction for which $E°_{cell}$ comes out + ve is spontaneous.

$HBrO \longrightarrow Br_2$ E° = 1.595 V, SRP (cathode)

$HBrO \longrightarrow BrO_3^-$ E° = –1.5 V, SOP (anode)

$2HBrO \longrightarrow Br_2 + BrO_3^-$

$E°_{cell}$ = SRP (cathode) – SRP (anode)
= 1.595 – 1.5
= 0.095 V

$E°_{cell} > 0 \Rightarrow \Delta G° < 0$ [spontaneous]

82. **(1)** Solubility of $BaSO_4 = 2.42 \times 10^{-3}$ gL^{-1}

$$\therefore s = \frac{2.42\times10^{-3}}{233} = 1.038\times10^{-5} \text{ mol L}^{-1}$$

$K_{sp} = s^2 = (1.038 \times 10^{-5})^2 = 1.08 \times 10^{-10}$ mol^2 L^{-2}

83. (3) Meq. of HCl $= 75\times\frac{1}{5}\times 1 = 15$

Meq. of NaOH $= 25\times\frac{1}{5}\times 1 = 5$

Meq. of HCl in resulting solution = 10

Molarity of $[H^+]$ in resulting mixture

$$=\frac{10}{100}=\frac{1}{10}$$

$$pH = -\log[H^+] = -\log\left[\frac{1}{10}\right] = 1.0$$

84. (4) According to Hardy Schulze rule, coagulating power of an ion depends on both magnitude and sign of the charge on the ion.

85. (1) van der waal constant '*a*', signifies intermolecular forces of attraction.
Higher is the value of '*a*', easier will be the liquefaction of gas.

86. (2) $Fe(CO)_5$

$$EAN = Z - O.N. + 2(C.N.) = 26 - 0 + 2(5) = 26 + 10 = 36$$

Only one central metal atom/ion is present and it follows EAN rule, so it is mononuclear.

87. (1) In the given complex, the CN of Co is 6, and the complex has octahedral geometry.

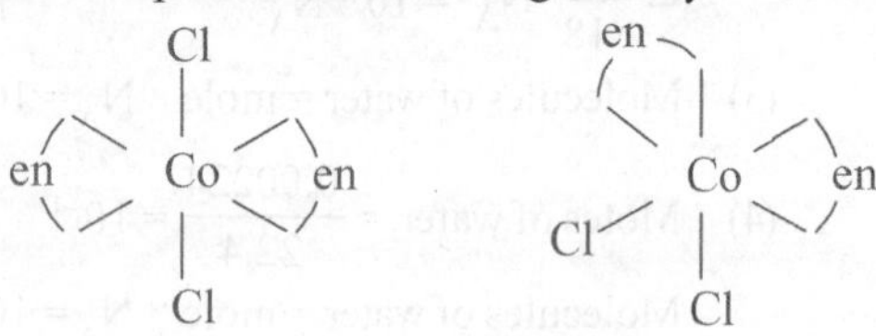

trans (optically inactive) *cis* (optically active)

88. (3)

CrO_4^{2-}	Cr^{6+} diamagnetic
$Cr_2O_7^{2-}$	Cr^{6+} diamagnetic
MnO_4^-	Mn^{7+} diamagnetic
MnO_4^{2-}	Mn^{6+} paramagnetic

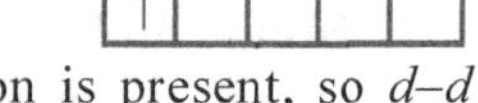

Thus unpaired electron is present, so *d–d* transition is possible.

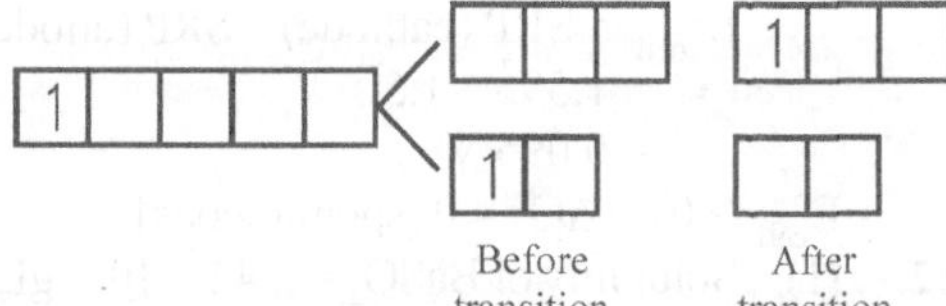

89. (2) Ni(28) : $[Ar]3d^8 4s^2$

$\because$ CO is a strong field ligand, so unpaired electrons get paired. Hence, configuration would be:

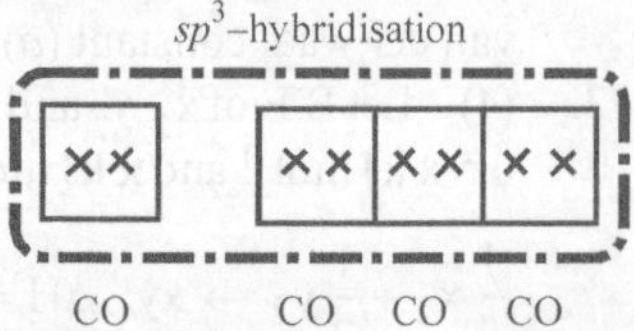

For, four 'CO' ligands hybridisation would be sp^3 and thus the complex would be diamagnetic and of tetrahedral geometry.

90. (1) $Co^{3+} = [Ar]\,3d^6$, unpaired $e^-(n) = 4$

Spin magnetic moment

$$= \sqrt{4(4+2)} = \sqrt{24}\text{ B.M.}$$

$Cr^{3+} = [Ar]\,3d^3$, unpaired $e^-(n) = 3$

Spin magnetic moment

$$= \sqrt{3(3+2)} = \sqrt{15}\text{ B.M.}$$

$Fe^{3+} = [Ar]\,3d^5$, unpaired $e^-(n) = 5$

Spin magnetic moment $= \sqrt{5(5+2)} = \sqrt{35}$ B.M.

$Ni^{2+} = [Ar]\,3d^8$, unpaired $e^-(n) = 2$

Spin magnetic moment $= \sqrt{2(2+2)} = \sqrt{8}$ B.M.

BIOLOGY

91. (3) Spermiogenesis is conversion of spermatids into spermatozoa whereas spermiation is the release of the sperms from sertoli cells into the cavity of seminiferous tubule.

92. (1) The extraembryonic membranes are amnion, chorion, allantois and yolk sac. Amnion is derived from mesoderm on the outer side and ectoderm on the inner side. Chorion is formed from trophoectoderm and mesoderm, whereas allantois and yolk sac membrane have mesoderm on outerside and endoderm in inner side.

93. (1) Saheli is world's first and only oral non steroidal contraceptive pill, dosage-once a week pill. Its functioning is based upon selective estrogen receptor modulation and prevents egg from getting implanted.

94. (4) The placenta is an endocrine gland that is only present during pregnancy. It relases hCG, hPL, progestogens, estrogens, etc. Human chorionic gonadotropic hormone (hCG) stimulates the corpus luteum during pregnancy to release estrogen and progesterone. Human placental lactogen (hPL) is involved in growth of body of mother and breast. Progesterone maintains pregnancy.

95. **(4)** In proliferative phase, the follicles start developing, called follicular phase. Secretory phase is also called as luteal phase mainly controlled by progesterone secreted by corpus luteum. Menstruation involves breakdown of overgrown endometrial lining.

96. **(4)** Unlike eukaryotes in prokaryotes the genes are organised into operon. Operon is a co-ordinated group of genes which are all transcribed together & regulate a metabolic pathway as a unit.

97. **(3)** Woman acts as a carrier. Both son & daughter inherit X-chromosome. Although only son would be the diseased one.

$$X^cX \times XY$$
$$\downarrow$$
$$X^cX \quad X^cY \quad XX \quad XY$$

98. **(2)** As per mutation theory which was given by Hugo de Vries state that the evolution is a discontinuous phenomenon or saltatory phenomenon/saltation.

99. **(1)** Coding strand and mRNA have the similar nucleotide sequence except, Thymine (T) is replaced by Uracil (U) in mRNA.

100. **(3)** Divergent evolution demonstrates how species can have common (homologous) anatomical structures, whereas eye of octopus, bat and man are examples of analogous organs showing convergent evolution.

101. **(4)** Curd has enriched presence of vitamins specially Vit-B_{12}, which improves its nutritional value than milk.

102. **(4)** Alzheimer's disease is a neurodegenerative disorder due to deficiency of neurotransmitter acetylcholine. Rheumatoid arthritis is an autoimmune disorder in which antibodies are produced against the synovial membrane and cartilage. Vitiligo causes white patches on skin also characterized as autoimmune disorder. Psoriasis is a skin disease that causes itchy or sore patches of thick red skin and is also autoimmune.

103. **(1)** Bird and bat wings are analogous, as forelimb is homologous. In different vertebrates, bones of forelimbs are similar but their forelimbs are adapted in different way as per their adaptation, show homology.

104. **(2)** I^AI^O, I^BI^O - Dominant-recessive relationship
I^AI^B - Codominance
I^A, I^B & I^O - Three different allelic forms of a gene (multiple allelism)

105. **(1)** Lymphatic filariasis, also known as elephantiasis, is a human disease caused by parasitic worms known as filarial worms. It is caused by roundworm, *Wuchereria bancrofti* and it is transmitted by culex mosquito.

106. **(2)** In-situ conservation is the on-site conservation or the conservation of genetic resources in natural populations of plant or animal species. E.g. Sacred groves.

107. **(2)** 'Smack' also called as brown sugar/Heroin is formed by acetylation of morphine. It is obtained from the latex of unripe capsule of Poppy plant (*Papaver somniferum*).

108. **(1)** Whenever the pre-reproductive individuals or the younger population size is larger than the reproductive group, the population will be an increasing population.

109. **(3)** Amensalism/antibiosis, association between organisms of two different species in which one is inhibited or destroyed and the other is unaffected. These are chemicals secreted by one microbial group (eg : *Penicillium*) which harm other microbes (*eg : Staphylococcus*). It has no effect on *Penicillium* or the organism which produces it.

110. **(4)**

111. **(1)** Asthma is a common long-term inflammatory disease of the airways of the lungs. It is a difficulty in breathing causing wheezing due to inflammation of bronchi and bronchioles. Emphysema is a chronic disorder in which alveolar walls are damaged due to which respiratory surface is decreased, and is mainly caused by smoking.

112. **(1)** Tricuspid valves are AV valve present between right atrium and right ventricle in heart. Bicuspid valves (mitral valves) are AV valve present between left atrium and left ventricle. Semilunar valves are present at the openings of aortic and pulmonary aorta.

113. **(2)** Tidal volume (TV) is volume of air inspired or expired during normal respiration. It is approximately 500 mL. Inspiratory reserve volume (IRV) is additional volume of air a person can inspire by a forceful inspiration. It is around 2500 - 3000 mL. Expiratory reserve volume (ERV) is additional volume of air a person can be expired by a forceful expiration. This averages 1000 - 1100 mL. Residual volume (RV) is volume of air remaining in lungs even after forceful expiration. This averages 1100 - 1200 mL.

114. (1) Epinephrine is derived from tyrosine by the removal of carboxyl group. It is a catecholamine.

115. (2) The limbic system (emotional motor system) is responsible for the experience and expression of emotion but not movement. It is located in the core of the brain and includes the amygdala, hippocampus and hypothalamus.

116. (1) The muscles that move the eyeball are attached to the sclera. Suspensory ligament of lens - a series of fibers that connect the ciliary body of the eye with the lens, holding it in place.

117. (4) Osteoporosis has 3 causes: excess parathyroid hormone, advanced age, and lack of estrogen in older females. Estrogen promotes the activity of osteoblast and inhibits osteoclast. Parathormone promotes mobilisation of calcium from bone into blood. Excessive activity of parathormone causes demineralisation leading to osteoporosis.

118. (3) Parietal or oxyntic cell is a source of HCl and intrinsic factor. HCl converts iron present in diet from ferric to ferrous form so that it can be absorbed easily and used during erythropoiesis. Intrinsic factor is essential for the absorption of vitamin B_{12} and its deficiency causes pernicious anaemia.

119. (3) Fibrinogen forms fibrin strands during coagulation. These strands forms a network and the meshes of which are occupied by blood cells, this structure finally forms a clot. Antibodies are derived from Gamma-Globulin fraction of plasma proteins which means globulins are involved in defence mechanisms. Albumin is a plasma protein mainly responsible for Blood Colloidal Osmotic Pressure (BCOP).

120. (2) Silicosis is an occupational disease caused due to excess inhalation of silica dust in the workers involved grinding or stone breaking industries.

121. (1) The sarcoplasmic reticulum to release calcium ions into the muscle interior where they bind to troponin, thus causing tropomyosin to shift from the face of the actin filament to which myosin heads need to bind to produce contraction.

122. (3) Polytene chromosomes are found in salivary glands of insects of order Diptera.

123. (3) Nissl bodies are present in the soma or cell body of a neuron. When observed under electron microscope, they appear to be composed of rough endoplasmic reticulum (RER) and free ribosomes hence help in protein synthesis.

124. (3) Oxidative phosphorylation takes place in inner mitochondrial membrane.

125. (3) Phospholipid synthesis does not take place in rough endoplasmic reticulum (RER). Smooth endoplasmic reticulum (SER) is involved in lipid synthesis.

126. (1) A polyribosome (or polysome) is a complex of an mRNA molecule and two or more ribosomes that act to translate mRNA instructions into polypeptides.

127. (2) In thecodont teeth are present in the sockets of the jaw bone called alveoli. In diphyodont teeth erupts twice, temporary milk or deciduous teeth are replaced by a set of permanent or adult teeth. In heterodont dentition it consists of different types of teeth namely incisors, canine, premolars and molars.

128. (4) The digestive tract of aves has additional chambers in their digestive system as crop and gizzard. Crop is concerned with storage of food grains, whereas gizzard is a masticatory organ in birds used to crush food grain.

129. (2) The two extremes in the animal kingdom are endothermic homeotherms and ectothermic poikilotherms. Most mammals, including humans, as well as most birds are endothermic homeotherms, while most fish, invertebrates, reptiles, and amphibians are ectothermic poikilotherms. *Chelone* (Turtle) belongs to class reptilia which is poikilotherm or cold blooded.

130. (2) Males bear a pair of short, thread like anal styles which are absent in females. Anal/caudal styles arise from 9th abdominal segment in male cockroach.

131. (2) Diatoms are the chief producers in some oceans and in some seasons as they are the primary producers and the food chain in marine ecosystem depends on it.

132. (3) Ciliates differs from other protozoans in having two types of nuclei. E.g., *Paramoecium* have two types of nuclei i.e. macronucleus & micronucleus.

133. (1) Metamorphosis refers to transformation of larva into adult. Animal that perform metamorphosis are said to have indirect development. Metamorphosis includes, in insects, the transformation of a maggot into an adult fly and a caterpillar into a butterfly and, in amphibians, the changing of a tadpole into a frog. In earthworm development is direct which means no larval stage are there and hence no metamorphosis.

134. (2) In renal physiology, ultrafiltration occurs at the barrier between the blood and the filtrate in the glomerular capsule (Bowman's capsule) in the kidneys. Concentration of urine refers to water absorption from glomerular filtrate created by counter-current mechanism in Henle's loop. Urine is carried from kidney to bladder through ureter. Urinary bladder is for storage of urine.

135. (3) Glycosuria denotes presence of glucose in the urine. Gout is due to deposition of uric acid crystals in the joint. Renal calculi are precipitates of calcium phosphate produced in the pelvis of the kidney. Glomerular nephritis is the inflammatory condition of glomerulus characterised by proteinuria and haematuria.

136. (2) In cellular respiration, NAD^+ act as an electron carrier.

137. (2) In obligate mutualism, one organism cannot survive without the other. *Yucca* have an obligate mutualism with a species of moth i.e. *Pronuba.*

138. (1) Green sulphur bacteria do not use H_2O as source of proton, therefore they do not evolve O_2.

139. (1) Iron is absorbed by plants in the form of ferric ions.

140. (3) Double fertilization is a unique phenomenon that occurs in angiosperms only. Double fertilization = Syngamy + Triple fusion

141. (4) Plants require potassium ions (K^+) for protein synthesis and for the opening and closing of stomata, which is regulated by proton pumps to make surrounding guard cells either turgid or flaccid.

142. (4) Pollen grains can be stored for several years in liquid nitrogen at -196°C (Cryopreservation).

143. (1) *Saccharomyces* i.e. yeast is an eukaryote (unicellular fungi). Mycobacterium is a bacterium. *Oscillatoria* and *Nostoc* are cyanobacteria.

144. (3) Sugar is a carbohydrate. These are polyhydroxy aldehyde, ketone or their derivatives, which means they have carbonyl and hydroxyl groups in its structure.

145. (2) ATP, NADPH and oxygen are products of light reaction, while NADH is a product of respiration process.

146. (2) Light, temperature and concentration of CO_2 affect opening and closing of stomata. They are not affected by O_2 concentration.

147. (2) Golgi complex after processing, packages them in vesicles, and either stores them for later use or sends them out of the cell. It is also the organelle that builds lysosomes (cell digestion machines).

148. (3) The nucleolus (plural nucleoli) is a large, distinct, spheroidal subcompartment of the nucleus of eukaryote cells that is the site of ribosomal RNA (rRNA) synthesis and assembly of ribosomal subunits.

149. (1) Grass being a monocot, has Dumb-bell shaped stomata in their leaves.

150. (2) The guard cells in dicot plants are kidney shaped and dumbbell shaped in monocots.

151. (1) Retrovirus is commonly used as vector for introducing a DNA fragment in human lymphocyte.

152. (2) Biopiracy term used for or refers to the use of bioresources by multinational companies and other organisation without proper authorisation from the countries and people concerned with compensatory payment.

153. (3) Indian Government has setup organisation such as GEAC (Genetic Engineering Appraisal Committee) which will make decisions regarding the validity of GM research and safety of introducing GM-organism for public services.

154. (3) PCR is based on three simple steps required for any DNA synthesis reaction: (i) denaturation of the template into single strands; (ii) annealing of primers to each original strand for new strand synthesis; and (iii) extension of the new DNA strands from the primers.

155. (1) Ribozyme is a catalytic RNA, which is nucleic acid.

156. (3) In 1997, an American company got patent rights on Basmati rice through the US patent and trademark office that was actually been derived from Indian farmer's varieties. The diversity of rice in India is one of the richest in the world, 27 documented varieties of Basmati are grown in India. Indian basmati was crossed with semi-dwarf varieties and claimed as an invention or a novelty. Sharbati sonora and Lerma Rojo are varieties of wheat.

157. (3) Francois Jacob and Jacque Monod proposed model of gene regulation known as operon model/lac operon. Alec Jeffreys gave DNA fingerprinting technique. Matthew Meselson and F. Stahl gave semi-conservative DNA replication in E.coli. Alfred Hershey and Martha Chase proved DNA as genetic material not protein.

158. (3) Sporopollenin cannot be degraded by enzyme; strong acids and alkali, therefore it is helpful in preserving pollen as fossil.

159. (2) Semi-conservative DNA replication was first shown in bacterium *escherichia coli* by Matthew Meselson and Franklin Stahl.

160. (1) Starch synthesis in pea is controlled by

pleiotropic gene. Pleiotropy occurs when one gene influences two or more seemingly unrelated phenotypic traits.

161. (2) An offset is a small, virtually complete daughter plant that has been naturally and asexually produced on the mother plant. They are clones, meaning that they are genetically identical to the mother plant. Offset is a vegetative part of a plant, formed by mitosis.

162. (2) Punnett (British scientists) devised the "Punnett Square" to depict the number and variety of genetic combinations, and had a role in shaping the Hardy-Weinberg law. Franklin Stahl proved semi-conservative mode of replication. Transduction was discovered by Zinder and Lederberg. Spliceosome formation is part of post-transcriptional change in eukaryotes.

163. (1) A monocarpic plant flowers and produces seeds only once before dying, e.g. bamboo. A polycarpic plant reproduces sexually more than once in its lifetime, e.g. jackfruit, papaya, mango, etc.

164. (3) Joseph Grinnell in 1917 coined the term niche, which he used as largely equivalent to a species habitat. It refers the functional role played by the organism where it lives.

165. (2) Chlorofluorocarbons (CFCs) and other halogenated ozone depleting substances (ODS) are mainly responsible for man-made chemical ozone depletion. CFCs rise into the stratosphere where they are eventually broken down by ultraviolet (UV) rays from the Sun. This causes them to release free chlorine. The chlorine reacts with oxygen which leads to the chemical process of destroying ozone molecules.

166. (1) The given data depicts the inverted pyramid of biomass, usually present in aquatic ecosystem. Upright pyramid of biomass and numbers are not possible, as the data depicts primary producer is less than primary consumer and this is less than secondary consumers. Pyramid of energy is always upright.

167. (3) A primary pollutant is an air pollutant emitted directly from a source. A secondary pollutant is not directly emitted as such, but forms when other pollutants (primary pollutants) react in the atmosphere. O_3 (ozone) is a secondary pollutant.

168. (4) World Ozone day is celebrated on 16th September. 5th June is World Environment Day. 21st April is National Yellow Bat Day. 22nd April is National Earth Day.

169. (2) Natality is the birth rate within a population. When compared with the death or mortality rate, the growth or decrease in a population can be determined.

170. (3) Herbarium - Dried and pressed plant specimen.
- Key - Identification of various taxa.
- Museum - Plant and animal specimen are preserved.
- Catalogue - Alphabetical listing of species.

171. (1) *Polysiphonia* is a genus of red algae, where asexual spores and gametes are non-motile or non-flagellated.

172. (4) In *Agaricus* (a genus of basidiomycetes), basidiospores or meiospores are produced exogenously. *Neurospora* (a genus of ascomycetes) produces ascospores as meiospores but endogenously inside the ascus). *Alternaria* (a genus of deuteromycetes) does not produce sexual spores. *Saccharomyces* (Unicellular ascomycetes) produces ascospores, endogenously.

173. (3) Winged pollen grains are present in *Pinus*. Each pollen grain has two wing-like structures which enables it to float in air, as an adaptation for dispersal by the wind.

174. (1) Some lateral roots of mangroves become specialized as pneumatophores in saline mud flats. These are lateral roots that grow upward (negative geotropism) for varying distances and function as the site of oxygen intake for the submerged primary root system.

175. (1) Grasses are monocots and monocots usually do not have secondary growth. Palm like monocots have anomalous secondary growth.

176. (3) Casparian strip is a band of cell wall material deposited in the radial and transverse walls of the endodermis. Casparian strip is made of suberin and sometimes lignin.

177. (2) Secondary tissues are generated from the growth of a cambium. Vascular cambium gives rise to secondary xylem to the inside, and to secondary phloem to the outside.

178. (4) Pseudopodia are locomotory structures in sarcodines (Amoeboid).

179. (1) The gymnosperms are a group of seed-producing plants. The name is based on the unenclosed condition of their seeds (called ovules in their unfertilized state).

180. (2) Sweet potato is a modified adventitious root for storage of food. Rhizomes are underground modified stem. Tap root is primary root directly elongated from the radicle.

NEET 2017

PHYSICS

1. A spring of force constant k is cut into lengths of ratio 1 : 2 : 3. They are connected in series and the new force constant is k'. Then they are connected in parallel and force constant is k″. Then k' : k″ is : -

(1) 1 : 9 (2) 1 : 11 (3) 1 : 14 (4) 1 : 6

2. The ratio of resolving powers of an optical microscope for two wavelengths $\lambda_1 = 4000$ Å and $\lambda_2 = 6000$ Å is :-

(1) 9 : 4 (2) 3 : 2 (3) 16 : 81 (4) 8 : 27

3. The two nearest harmonics of a tube closed at one end and open at other end are 220 Hz and 260 Hz. What is the fundamental frequency of the system?

(1) 20 Hz (2) 30 Hz (3) 40 Hz (4) 10 Hz

4. Consider a drop of rain water having mass 1 g falling from a height of 1 km. It hits the ground with a speed of 50 m/s. Take 'g' constant with a value 10 m/s^2.

The work done by the (i) gravitational force and the (ii) resistive force of air is :-

(1) (i) 1.25 J (ii) –8.25 J
(2) (i) 100 J (ii) 8.75 J
(3) (i) 10 J (ii) – 8.75 J
(4) (i) – 10 J (ii) –8.25 J

5. A physical quantity of the dimensions of length that can be formed out of c, G and $\frac{e^2}{4\pi\varepsilon_0}$ is [c is velocity of light, G is universal constant of gravitation and e is charge] :-

(1) $c^2\left[G\frac{e^2}{4\pi\varepsilon_0}\right]^{1/2}$ (2) $\frac{1}{c^2}\left[\frac{e^2}{G4\pi\varepsilon_0}\right]^{1/2}$

(3) $\frac{1}{c}G\frac{e^2}{4\pi\varepsilon_0}$ (4) $\frac{1}{c^2}\left[G\frac{e^2}{4\pi\varepsilon_0}\right]^{1/2}$

6. Two rods A and B of different materials are welded together as shown in figure. Their thermal conductivities are K_1 and K_2. The thermal conductivity of the composite rod will be :-

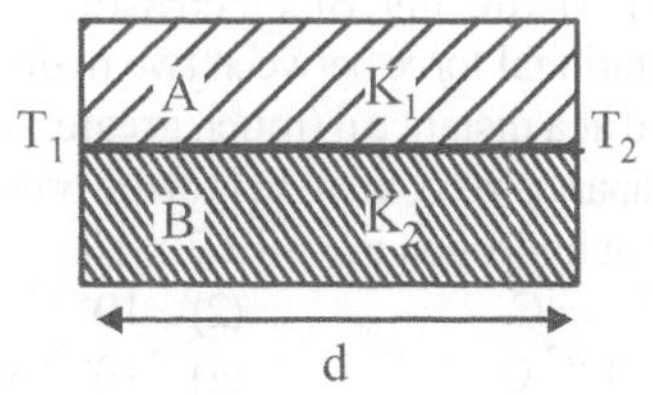

(1) $\frac{3(K_1+K_2)}{2}$ (2) K_1+K_2

(3) $2(K_1+K_2)$ (4) $\frac{K_1+K_2}{2}$

7. A capacitor is charged by a battery. The battery is removed and another identical uncharged capacitor is connected in parallel. The total electrostatic energy of resulting system :-

(1) decreases by a factor of 2
(2) remains the same
(3) increases by a factor of 2
(4) increases by a factor of 4

8. In a common emitter transistor amplifier the audio signal voltage across the collector is 3V. The resistance of collector is 3 kΩ. If current gain is 100 and the base resistance is 2 kΩ, the voltage and power gain of the amplifier is :-

(1) 15 and 200 (2) 150 and 15000
(3) 20 and 2000 (4) 200 and 1000

9. Thermodynamic processes are indicated in the following diagram :

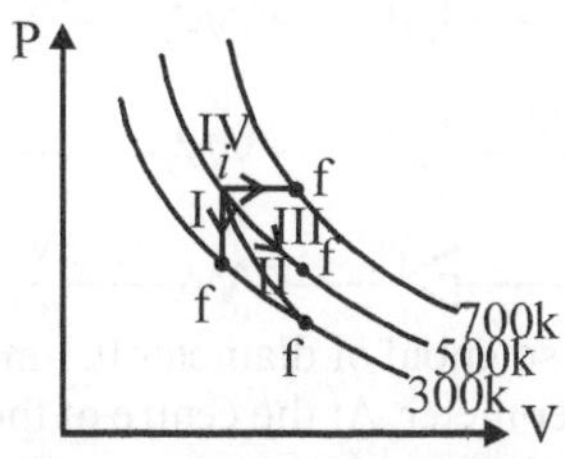

Match the following

	Column-1		Column-2
P.	Process I	a.	Adiabatic
Q.	Process II	b.	Isobaric
R.	Process III	c.	Isochoric
S.	Process IV	d.	Isothermal

(1) $P \rightarrow c, Q \rightarrow a, R \rightarrow d, S \rightarrow b$
(2) $P \rightarrow c, Q \rightarrow d, R \rightarrow b, S \rightarrow a$
(3) $P \rightarrow d, Q \rightarrow b, R \rightarrow a, S \rightarrow c$
(4) $P \rightarrow a, Q \rightarrow c, R \rightarrow d, S \rightarrow b$

10. Suppose the charge of a proton and an electron differ slightly. One of them is – e, the other is (e + Δe). If the net of electrostatic force and gravitational force between two hydrogen atoms placed at a distance d (much greater than atomic size) apart is zero, then Δe is of the order of [Given mass of hydrogen $m_h = 1.67 \times 10^{-27}$ kg]
(1) 10^{-23}C (2) 10^{-37} C
(3) 10^{-47} C (4) 10^{-20} C

11. The resistance of a wire is 'R' ohm. If it is melted and stretched to 'n' times its original length, its new resistance will be :-
(1) $\frac{R}{n}$ (2) n^2R (3) $\frac{R}{n^2}$ (4) nR

12. The given electrical network is equivalent to :

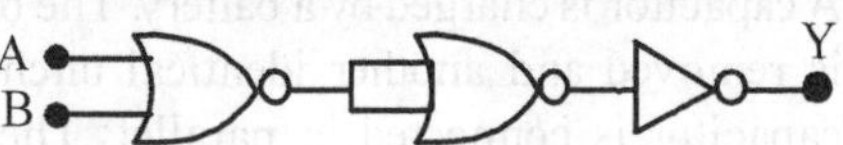

(1) OR gate (2) NOR gate
(3) NOT gate (4) AND gate

13. The de-Broglie wavelength of a neutron in thermal equilibrium with heavy water at a temperature T (Kelvin) and mass m, is :-
(1) $\frac{h}{\sqrt{3mkT}}$ (2) $\frac{2h}{\sqrt{3mkT}}$
(3) $\frac{2h}{\sqrt{mkT}}$ (4) $\frac{h}{\sqrt{mkT}}$

14. Which one of the following represents forward bias diode ?
(1) –4V —▷|— R —3V
(2) –2V —▷|— R +2V
(3) 3V —▷|— R 5V
(4) 0V —▷|— R –2V

15. A long solenoid of diameter 0.1 m has 2×10^4 turns per meter. At the centre of the solenoid, a coil of 100 turns and radius 0.01 m is placed with its axis coinciding with the solenoid axis. The current in the solenoid reduces at a constant rate to 0A from 4 A in 0.05 s. If the resistance of the coil is $10\pi^2\Omega$. the total charge flowing through the coil during this time is :-
(1) 16 μC (2) 32 μC
(3) 16 π μC (4) 32 π μC

16. Preeti reached the metro station and found that the escalator was not working. She walked up the stationary escalator in time t_1. On other days, if she remains stationary on the moving escalator, then the escalator takes her up in time t_2. The time taken by her to walk up on the moving escalator will be:
(1) $\frac{t_1t_2}{t_2 - t_1}$ (2) $\frac{t_1t_2}{t_2 + t_1}$
(3) $t_1 - t_2$ (4) $\frac{t_1 + t_2}{2}$

17. Young's double slit experment is first performed in air and then in a medium other than air. It is found that 8th bright fringe in the medium lies where 5th dark fringe lies in air. The refractive index of the medium is nearly :-
(1) 1.59 (2) 1.69 (3) 1.78 (4) 1.25

18. A beam of light from a source L is incident normally on a plane mirror fixed at a certain distance x from the source. The beam is reflected back as a spot on a scale placed just above the source I. When the mirror is rotated through a small angle θ, the spot of the light is found to move through a distance y on the scale. The angle θ is given by :-
(1) $\frac{y}{x}$ (2) $\frac{x}{2y}$ (3) $\frac{x}{y}$ (4) $\frac{y}{2x}$

19. If θ_1 and θ_2 be the apparent angles of dip observed in two vertical planes at right angles to each other, then the true angle of dip θ is given by :-
(1) $\tan^2\theta = \tan^2\theta_1 + \tan^2\theta_2$
(2) $\cot^2\theta = \cot^2\theta_1 - \cot^2\theta_2$
(3) $\tan^2\theta = \tan^2\theta_1 - \tan^2\theta_2$
(4) $\cot^2\theta = \cot^2\theta_1 + \cot^2\theta_2$

20. Two cars moving in opposite directions approach each other with speed of 22 m/s and 16.5 m/s respectively. The driver of the first car blows a horn having a frequency 400 Hz. The frequency heard by the driver of the second car is [velocity of sound 340 m/s] :-
(1) 361 Hz (2) 411 Hz (3) 448 Hz (4) 350 Hz

21. Two blocks A and B of masses 3 m and m respectively are connected by a massless and inextensible string. The whole system is suspended by a massless spring as shown in figure. The magnitudes of acceleration of A and B immediately after the string is cut, are respectively :-

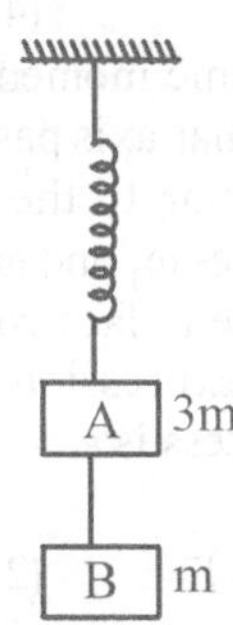

(1) $\frac{g}{3}, g$ (2) g, g (3) $\frac{g}{3}, \frac{g}{3}$ (4) $g, \frac{g}{3}$

22. A thin prism having refracting angle 10° is made of glass of refractive index 1.42. This prism is combined with another thin prism of glass of refractive index 1.7. This combination produces dispersion without deviation. The refracting angle of second prism should be :-

(1) 6° (2) 8° (3) 10° (4) 4°

23. The acceleration due to gravity at a height 1 km above the earth is the same as at a depth d below the surface of earth. Then :-

(1) d = 1 km (2) $d = \frac{3}{2}$ km

(3) d = 2 km (4) $d = \frac{1}{2}$ km

24. A potentiometer is an accurate and versatile device to make electrical measurements of E.M.F. because the method involves :-

(1) Potential gradients
(2) A condition of no current flow through the galvanometer
(3) A combination of cells, galvanometer and resistances
(4) Cells

25. A spherical black body with a radius of 12 cm radiates 450 watt power at 500 K. If the radius were halved and the temperature doubled, the power radiated in watt would be :-

(1) 450 (2) 1000 (3) 1800 (4) 225

26. Figure shows a circuit that contains three identical resistors with resistance R = 9.0 Ω each, two identical inductors with inductance L = 2.0 mH each, and an ideal battery with emf ε = 18 V. The current 'i' through the battery just after the switch closed is,...... :-

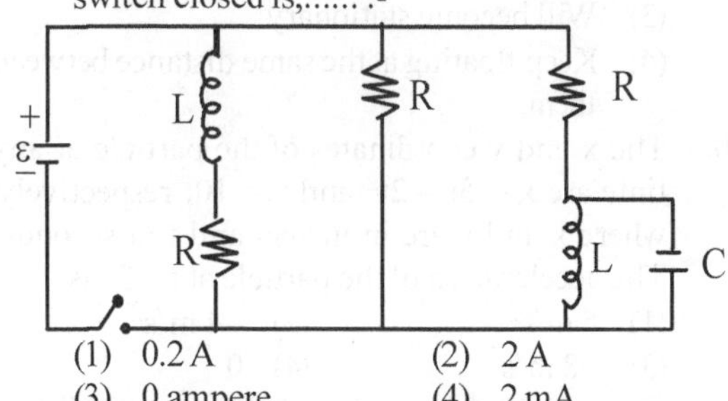

(1) 0.2 A (2) 2 A
(3) 0 ampere (4) 2 mA

27. Radioactive material 'A' has decay constant '8 λ' and material 'B' has decay constant 'λ'. Initially they have same number of nuclei. After what time, the ratio of number of nuclei of material 'B' to that 'A' will be $\frac{1}{e}$?

(1) $\frac{1}{7\lambda}$ (2) $\frac{1}{8\lambda}$ (3) $\frac{1}{9\lambda}$ (4) $\frac{1}{\lambda}$

28. The diagrams below show regions of equipotentials.

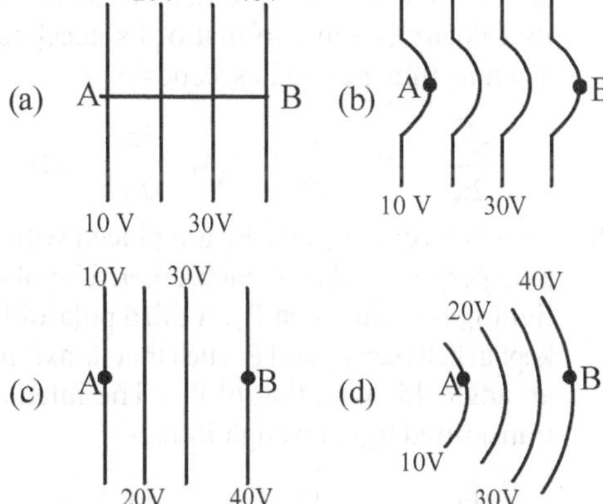

A positive charge is moved from A to B in each diagram.

(1) In all the four cases the work done is the same
(2) Minimum work is required to move q in figure (a)
(3) Maximum work is required to move q in figure (b)
(4) Maximum work is required to move q in figure (c)

29. Two astronauts are floating in gravitational free space after having lost contact with their spaceship. The two will :-

(1) Move towards each other.
(2) Move away from each other.
(3) Will become stationary
(4) Keep floating at the same distance between them.

30. The x and y coordinates of the particle at any time are $x = 5t - 2t^2$ and $y = 10t$ respectively, where x and y are in meters and t in seconds. The acceleration of the particle at t = 2s is

(1) 5 m/s^2 (2) -4 m/s^2
(3) -8 m/s^2 (4) 0

31. One end of string of length l is connected to a particle of mass 'm' and the other end is connected to a small peg on a smooth horizontal table. If the particle moves in circle with speed 'v' the net force on the particle (directed towards centre) will be (T represents the tension in the string) :-

(1) $T+\frac{mv^2}{l}$ (2) $T-\frac{mv^2}{l}$
(3) Zero (4) T

32. A particle executes linear simple harmonic motion with an amplitude of 3 cm. When the particle is at 2 cm from the mean position, the magnitude of its velocity is equal to that of its acceleration. Then its time period in seconds is :-

(1) $\frac{\sqrt{5}}{2\pi}$ (2) $\frac{4\pi}{\sqrt{5}}$ (3) $\frac{2\pi}{\sqrt{3}}$ (4) $\frac{\sqrt{5}}{\pi}$

33. Two Polaroids P_1 and P_2 are placed with their axis perpendicular to each other. Unpolarised light I_0 is incident on P_1. A third polaroid P_3 is kept in between P_1 and P_2 such that its axis makes an angle 45° with that of P_1. The intensity of transmitted light through P_2 is :-

(1) $\frac{I_0}{4}$ (2) $\frac{I_0}{8}$ (3) $\frac{I_0}{16}$ (4) $\frac{I_0}{2}$

34. The bulk modulus of a spherical object is 'B'. If it is subjected to uniform pressure 'p', the fractional decrease in radius is :-

(1) $\frac{B}{3p}$ (2) $\frac{3p}{B}$ (3) $\frac{p}{3B}$ (4) $\frac{p}{B}$

35. In an electromagnetic wave in free space the root mean square value of the electric field is E_{rms} = 6V/m. The peak value of the magnetic field is :-

(1) 2.83×10^{-8} T (2) 0.70×10^{-8} T
(3) 4.23×10^{-8} T (4) 1.41×10^{-8} T

36. A rope is wound around a hollow cylinder of mass 3 kg and radius 40 cm. What is the angular acceleration of the cylinder if the rope is pulled with a force of 30 N ?

(1) 0.25 rad/s^2 (2) 25 rad/s^2
(3) 5 m/s^2 (4) 25 m/s^2

37. Two discs of same moment of inertia rotating about their regular axis passing through centre and perpendicular to the plane of disc with angular velocities ω_1 and ω_2. They are brought into contact face to face coinciding the axis of rotation. The expression for loss of energy during this process is:-

(1) $\frac{1}{4}I(\omega_1-\omega_2)^2$ (2) $I(\omega_1-\omega_2)^2$
(3) $\frac{1}{8}(\omega_1-\omega_2)^2$ (4) $\frac{1}{2}I(\omega_1+\omega_2)^2$

38. The photoelectric threshold wavelength of silver is 3250×10^{-10}m. The velocity of the electron ejected from a silver surface by ultraviolet light of wavelength 2536×10^{-10} m is (Given $h = 4.14 \times 10^{-15}$ eVs and $c = 3 \times 10^8 \text{ ms}^{-1}$)

(1) $\approx 0.6 \times 10^6 \text{ ms}^{-1}$ (2) $\approx 61 \times 10^3 \text{ ms}^{-1}$
(3) $\approx 0.3 \times 10^6 \text{ ms}^{-1}$ (4) $\approx 6 \times 10^5 \text{ ms}^{-1}$

39. A 250-Turn rectangular coil of length 2.1 cm and width 1.25 cm carries a current of 85 μA and subjected to magnetic field of strength 0.85 T. Work done for rotating the coil by 180° against the torque is:-

(1) 4.55 μJ (2) 2.3 μJ
(3) 1.15 μJ (4) 9.1 μJ

40. The ratio of wavelengths of the last line of Balmer series and the last line of Lyman series is :-

(1) 1 (2) 4 π (3) 0.5 (4) 2

41. A carnot engine having an efficiency of $\frac{1}{10}$ as heat engine, is used as a refrigerator. If the work done on the system is 10 J, the amount of energy absorbed from the reservoir at lower temperature is :-

(1) 90 J (2) 99 J (3) 100 J (4) 1 J

42. A gas mixture consists of 2 moles of O_2 and 4 moles of Ar at temperature T. Neglecting all vibrational modes, the total internal energy of the system is :-

(1) 15 RT (2) 9 RT (3) 11 RT (4) 4 RT

43. An arrangement of three parallel straight wires placed perpendicular to plane of paper carrying same current 'I along the same direction is shown in fig. Magnitude of force per unit length on the middle wire 'B' is given by :-

(1) 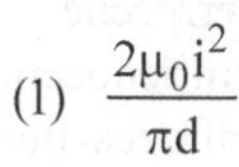
$\frac{2\mu_0 i^2}{\pi d}$

(2) 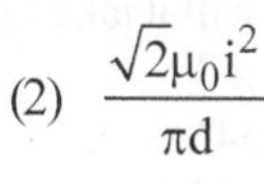
$\frac{\sqrt{2}\mu_0 i^2}{\pi d}$

(3) 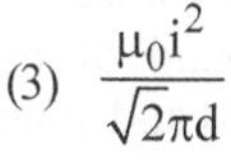
$\frac{\mu_0 i^2}{\sqrt{2}\pi d}$

(4) $\frac{\mu_0 i^2}{2\pi d}$

44. A U tube with both ends open to the atmosphere, is partially filled with water. Oil, which is immiscible with water, is poured into one side until it stands at a distance of 10 mm above the water level on the other side. Meanwhile the water rises by 65 mm from its original level (see diagram). The density of the oil is :-

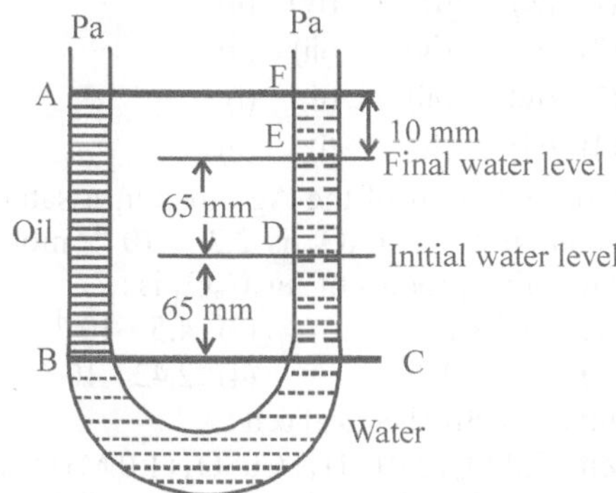

(1) 425 kg m^{-3} (2) 800 kg m^{-3}
(3) 928 kg m^{-3} (4) 650 kg m^{-3}

45. Which of the following statements are **correct** ?
(a) Centre of mass of a body always coincides with the centre of gravity of the body
(b) Centre of mass of a body is the point at which the total gravitational torque on the body is zero
(c) A couple on a body produce both translational and rotation motion in a body
(d) Mechanical advantage greater than one means that small effort can be used to lift a large load

(1) (a) and (b) (2) (b) and (c)
(3) (c) and (d) (4) (b) and (d)

CHEMISTRY

46. Name the gas that can readily decolourise acidified $KMnO_4$ solution :
(1) SO_2 (2) NO_2 (3) P_2O_5 (4) CO_2

47. Mechanism of a hypothetical reaction $X_2 + Y_2 \rightarrow 2XY$ is given below :
(i) $X_2 \rightarrow X + X$ (fast)
(ii) $X + Y_2 \rightleftharpoons XY + Y$ (slow)
(iii) $X + Y \rightarrow XY$ (fast)
The overall order of the reaction will be :
(1) 2 (2) 0 (3) 1.5 (4) 1

48. The element Z = 114 has been discovered recently.
It will belong to which of the following family/group and electronic configuration ?
(1) Carbon family, [Rn] $5f^{14}\, 6d^{10}\, 7s^2\, 7p^2$
(2) Oxygen family, [Rn] $5f^{14}\, 6d^{10}\, 7s^2\, 7p^4$
(3) Nitrogen family, [Rn] $5f^{14}\, 6d^{10}\, 7s^2\, 7p^6$
(4) Halogen family, [Rn] $5f^{14}\, 6d^{10}\, 7s^2\, 7p^5$

49. The heating of phenyl-methyl ethers with HI produces
(1) Iodobenzene (2) Phenol
(3) Benzene (4) Ethyl chlorides

50. Which one is the correct order of acidity ?
(1) $CH \equiv CH > CH_3 - C \equiv CH > CH_2 = CH_2 > CH_3 - CH_3$
(2) $CH \equiv CH > CH_2 = CH_2 > CH_3 - C \equiv CH > CH_3 - CH_3$
(3) $CH_3 - CH_3 > CH_2 = CH_2 > CH_3 - C \equiv CH > CH \equiv CH$
(4) $CH_2 = CH_2 > CH_3 - CH = CH_2 > CH_3 - C \equiv CH > CH \equiv CH$

51. Predict the correct intermediate and product in the following reaction :

$$H_3C - C \equiv CH \xrightarrow[HgSO_4]{H_2O,\, H_2SO_2} \underset{(A)}{\text{Intermediate}} \rightarrow \underset{(B)}{\text{product}}$$

(1) A : $H_3C - C(OH) = CH$ B : $H_3C - C(SO_4) = CH_2$
(2) A : $H_3C - C(=O) = CH_3$ B : $H_3C - C \equiv CH$
(3) A : $H_3C - C(OH) = CH_2$ B : $H_3C - C(=O) = CH_3$
(4) A : $H_3C - C(SO_4) = CH_2$ B : $H_3C - C(=O) - CH_3$

52. The equilibrium constant of the following are :

$N_2 + 3H_2 \rightleftharpoons 2NH_3$ K_1

$N_2 + O_2 \rightleftharpoons 2NO$ K_2

$H_2 + \frac{1}{2}O_2 \rightarrow H_2O$ K_3

The equilibrium constant (K) of the reaction :

$$2NH_3 + \frac{5}{2}O_2 \overset{K}{\rightleftharpoons} 2NO + 3H_2O, \text{ will be;}$$

(1) $K_2K_3^3/K_1$ (2) K_2K_3/K_1

(3) $K_2^3K_3/K_1$ (4) $K_1K_3^3/K_2$

53. Which one is the most acidic compound ?

(1) OH (ring)

(2) OH (ring), NO_2

(3) OH, O_2N, NO_2, NO_2 (ring)

(4) OH, CH_3 (ring)

54. The correct increasing order of basic strength for the following compounds is :

(I) NH_2 (benzene ring) (II) NH_2, NO_2 (benzene ring) (III) NH_2, CH_3 (benzene ring)

(1) III < I < II (2) III < II < I

(3) II < I < III (4) II < III < I

55. Ionic mobility of which of the following alkali metal ions is lowest when aqueous solution of their salts are put under an electric field ?

(1) K (2) Rb (3) Li (4) Na

56. The most suitable method of separation of 1 : 1 mixture of ortho and para-nitrophenols is :

(1) Chromatography (2) Crystallisation

(3) Steam distillation (4) Sublimation

57. $HgCl_2$ and I_2 both when dissolved in water containing I ions the pair of species formed is :

(1) HgI_2, I^- (2) HgI_4^{2-}, I_3^-

(3) Hg_2I_2, I^- (4) HgI_2, I_3^-

58. Mixture of chloroxylenol and terpineol acts as :

(1) antiseptic (2) antipyretic

(3) antibiotic (4) analgesic

59. An example of a sigma bonded organometallic compound is :

(1) Grignard's reagent (2) Ferrocene

(3) Cobaltocene (4) Ruthenocene

60. A first order reaction has a specific reaction rate of 10^{-2} sec^{-1}. How much time will it take for 20g of the reactant to reduce to 5 g ?

(1) 138.6 sec (2) 346.5 sec

(3) 693.0 sec (4) 238.6 sec

61. Match the interhalogen compounds of column-I with the geometry in column II and assign the correct code :

Column-I	Column-II
(A) XX'	(i) T-shape
(B) XX'_3	(ii) Pentagonal bipyramidal
(C) XX'_5	(iii) Linear
(D) XX'_7	(iv) Square-pyramidal
	(v) Tetrahedral

Code :

	(A)	(B)	(C)	(D)
(1)	(iii)	(i)	(iv)	(ii)
(2)	(v)	(iv)	(iii)	(ii)
(3)	(iv)	(iii)	(ii)	(i)
(4)	(iii)	(iv)	(i)	(ii)

62. Concentration of the Ag^+ ions in a saturated solution of $Ag_2C_2O_4$ is 2.2×10^{-4} mol L^{-1}. Solubility product of $Ag_2C_2O_4$ is :

(1) 2.66×10^{-12} (2) 4.5×10^{-11}

(3) 5.3×10^{-12} (4) 2.42×10^{-8}

63. In the electrochemical cell :

$Zn \mid ZnSO_4\,(0.01M) \mid\mid CuSO_4\,(1.0\,M) \mid Cu$, the emf of this Daniel cell is E_1. When the concentration of $ZnSO_4$ is changed to 1.0M and that of $CuSO_4$ changed to 0.01M, the emf changes to E_2. From the followings, which one is the relationship between E_1 and E_4? (Given, $\frac{RT}{F} = 0.059$)

(1) $E_1 < E_2$ (2) $E_1 > E_2$

(3) $E_2 = 0 \neq E_1$ (4) $E_1 = E_2$

64. Which of the following pairs of compounds is isoelectronic and isostructural ?

(1) TeI_2, XeF_2 (2) IBr_2^-, XeF_2

(3) IF_3, XeF_2 (4) $BeCl_2, XeF_2$

65. The IUPAC name of the compound

O O
‖ ‖
H – C

is :

(1) 5-formylhex-2-en-3-one
(2) 5-methyl-4-oxohex-2-en-5-al
(3) 3-keto-2-methylhex-5-enal
(4) 3-keto-2-methylhex-4-enal

66. Which one is the wrong statement ?
(1) The uncertainty principle is $\Delta E \times \Delta t \geq h / 4\pi$
(2) Half filled and fully filled orbitals have greater stability due to greater exchange energy, greater symmetry and more balanced arrangement.
(3) The energy of 2s orbital is less than the energy of 2p orbital in case of Hydrogen like atoms
(4) de-Broglies's wavelength is given by $\lambda = \frac{h}{mv}$, where m = mass of the particle, v = group velocity of the particle

67. Which is the incorrect statement ?
(1) Density decreases in case of crystals with Schottky's defect
(2) NaCl(s) is insulator, silicon is semiconductor, silver is conductor, quartz is piezo electric crystal
(3) Frenkel defect is favoured in those ionic compounds in which sizes of cation and anions are almost equal
(4) $FeO_{0.98}$ has non stoichiometric metal deficiency defect

68. The species, having bond angles of 120° is :
(1) ClF_3 (2) NCl_3 (3) BCl_3 (4) PH_3

69. For a given reaction, $\Delta H = 35.5$ kJ mol^{-1} and $\Delta S = 83.6$ JK^{-1} mol^{-1}. The reaction is spontaneous at : (Assume that ΔH and ΔS do not vary with temperature)
(1) T > 425 K (2) All temperatures
(3) T > 298 K (4) T < 425 K

70. Which of the following is a sink for CO ?
(1) Micro organism present in the soil
(2) Oceans
(3) Plants
(4) Haemoglobin

71. If molality of the dilute solutions is doubled, the value of molal depression constant (K_f) will be :
(1) halved (2) tripled
(3) unchanged (4) doubled

72. Which of the following is dependent on temperature?
(1) Molarity (2) Mole fraction
(3) Weight percentage (4) Molality

73. Which one of the following statements is not correct?
(1) The value of equilibrium constant is changed in the presence of a catalyst in the reaction at equilibrium
(2) Enzymes catalyse mainly bio-chemical reactions
(3) Coenzymes increase the catalytic activity of enzyme
(4) Catalyst does not initiate any reaction

74. Identify A and predict the type of reaction

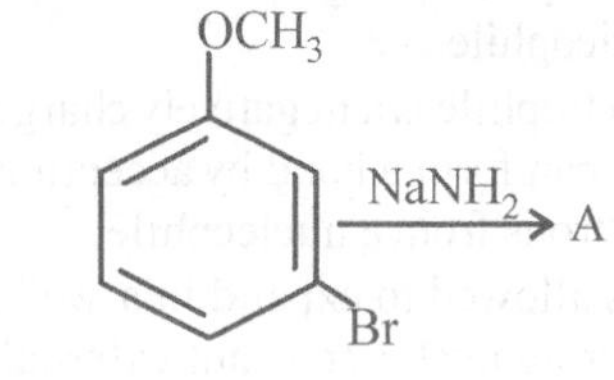

(1) OCH_3, NH_2 and elimination addition reaction

(2) OCH_3, Br and cine substitution reaction

(3) OCH_3 and cine substitution reaction

(4) 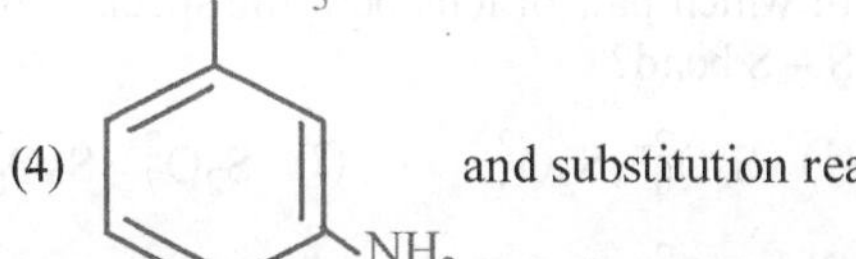

and substitution reaction

75. The correct order of the stoichiometries of AgCl formed when $AgNO_3$ in excess is treated with the complexes : $CoCl_3.6NH_3$, $CoCl_3.5NH_3$, $CoCl_3.4NH_3$ respectively is :
(1) 3 AgCl, 1 AgCl, 2 AgCl
(2) 3 AgCl, 2 AgCl, 1 AgCl
(3) 2 AgCl, 3 AgCl, 1 AgCl
(4) 1 AgCl, 3 AgCl, 2 AgCl

76. The correct statement regarding electrophile is -
(1) Electrophile is a negatively charged species and can form a bond by accepting a pair of electrons from another electrophile
(2) Electrophiles are generally neutral species and can form a bond by accepting a pair of electrons from a nucleophile
(3) Electrophile can be either neutral or positively charged species and can form a bond by accepting a pair of electrons from a nucleophile
(4) Electrophile is a negatively charged species and can form a bond by accepting a pair of electrons from a nucleophile

77. A gas is allowed to expand in a well insulated container against a constant external pressure of 2.5atm from an initial volume of 2.50 L to a final volume of 4.50L. The change in internal energy ΔU of the gas in joules will be :
(1) –500J (2) –505J
(3) +505J (4) 1136.25J

78. Which of the following reactions is appropriate for converting acetamide to methanamine ?
(1) Hoffmann hypobromamide reaction
(2) Stephens reaction
(3) Gabriels phthalimide synthesis
(4) Carbylamine reaction

79. With respect to the conformers of ethane, which of the following statements is true ?
(1) Bond angle changes but bond length remains same
(2) Both bond angle and bond length change
(3) Both bond angles and bond length remains same
(4) Bond angle remains same but bond length changes

80. In which pair of ions both the species contain S – S bond?
(1) $S_4O_6^{2-}, S_2O_3^{2-}$ (2) $S_2O_7^{2-}, S_2O_8^{2-}$
(3) $S_4O_6^{2-}, S_2O_7^{2-}$ (4) $S_2O_7^{2-}, S_2O_3^{2-}$

81. It is because of inability of ns^2 electrons of the valence shell to participate in bonding that :
(1) Sn^{2+} is oxidising while Pb^{4+} is reducing
(2) Sn^{2+} and Pb^{2+} are both oxidising and reducing
(3) Sn^{4+} is reducing while Pb^{4+} is oxidising
(4) Sn^{2+} is reducing while Pb^{4+} is oxidising

82. Correct increasing order for the wavelengths of absorption in the visible region the complexes of Co^{3+} is :
(1) $[Co(H_2O)_6]^{3+}, [Co(en)_3]^{3+}, [Co(NH_3)_6]^{3+}$
(2) $[Co(H_2O)_6]^{3+}, [Co(NH_3)_6]^{3+}, [Co(en)_3]^{3+}$
(3) $[Co(NH_3)_6]^{3+}, [Co(en)_3]^{3+}, [Co(H_2O)_6]^{3+}$
(4) $[Co(en)_3]^{3+}, [Co(NH_3)_6]^{3+}, [Co(H_2O)_6]^{3+}$

83. Consider the reactions :

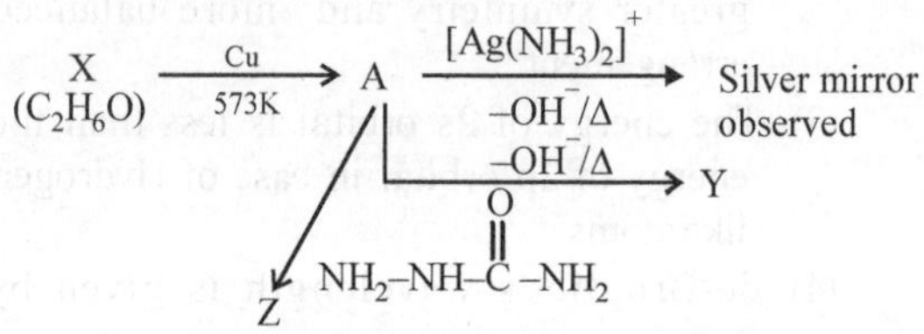

X (C_2H_6O) $\xrightarrow[573K]{Cu}$ A $\xrightarrow[-\bar{O}H/\Delta]{[Ag(NH_3)_2]^+}$ Silver mirror observed

A $\xrightarrow{-\bar{O}H/\Delta}$ Y

A $\xrightarrow{NH_2-NH-\overset{O}{\overset{\|}{C}}-NH_2}$ Z

Identify A, X, Y and Z
(1) A-Methoxymethane, X-Ethanol, Y-Ethanoic acid, Z-Semicarbazide.
(2) A - Ethanal, X-Ethanol, Y - But - 2-enal, Z-Semicarbazone
(3) A-Ethanol, X-Acetaldehyde, Y - Butanone, Z-Hydrazone
(4) A-Methoxymethane, X-Ethanoic acid, Y-Acetate ion, Z-hydrazine.

84. Of the following, which is the product formed when cyclohexanone undergoes aldol condensation followed by heating ?

(1) 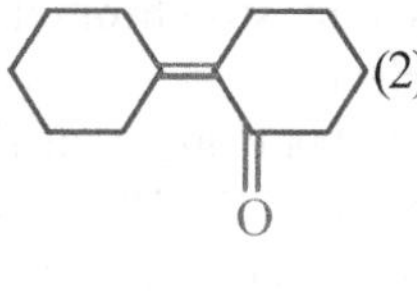 [structure: O] (2) 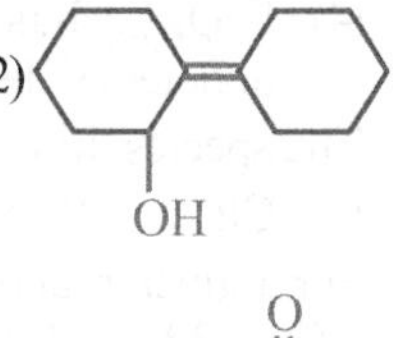 [structure: OH]

(3) 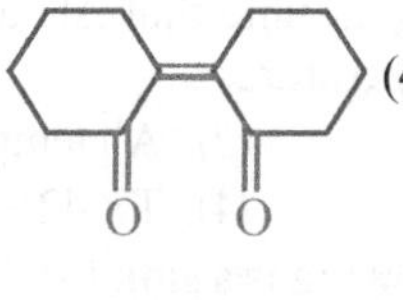 [structure: O O] (4) 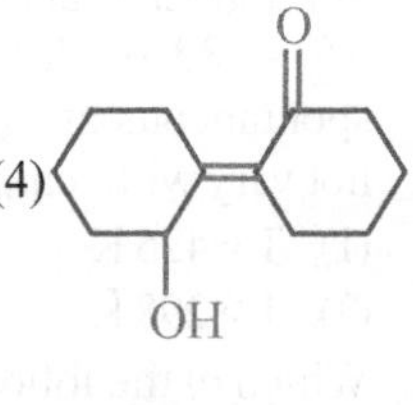 [structure: O, OH]

85. Which of the following pairs of species have the same bond order ?
(1) O_2, NO^+ (2) CN^-, CO
(3) N_2, O_2^- (4) CO, NO

86. Extraction of gold and silver involves leaching with CN^- ion. Silver is later recovered by :
(1) distillation
(2) zone refining
(3) displacement with Zn
(4) liquation

87. A 20 litre container at 400 K contains CO_2(g) at pressure 0.4 atm and an excess of SrO (neglect the volume of solid SrO). The volume of the container is now decreased by moving the movable piston fitted in the container. The maximum volume of the container, when pressure of CO_2 attains its maximum value, will be :
(Given that : $SrCO_3(s) \rightleftharpoons SrO(s) + CO_2(g)$, $K_p = 1.6$ atm)
(1) 10 litre (2) 4 litre (3) 2 litre (4) 5 litre

88. Pick out the correct statement with respect to $[Mn(CN)_6]^{3-}$
(1) It is sp^3d^2 hybridised and tetrahedral
(2) It is d^2sp^3 hybridised and octahedral
(3) It is dsp^2 hybridised and square planar
(4) It is sp^3d^2 hybridised and octahedral

89. The reason for greater range of oxidation states in actinoids is attributed to :
(1) actinoid contraction
(2) 5*f*, 6*d* and 7*s* levels having comparable energies
(3) 4*f* and 5*d* levels being close in energies
(4) the redioactive nature of actinoids

90. Which of the following statements is not correct?
(1) Ovalbumin is a simple food reserve in egg-white
(2) Blood proteins thrombin and fibrinogen are involved in blood clotting
(3) Denaturation makes the proteins more active
(4) Insulin maintains sugar level in the blood of a human body

BIOLOGY

91. Which one of the following statements is correct, with reference to enzymes ?
(1) Holoenzyme = Apoenzyme + Coenzyme
(2) Coenzyme = Apoenzyme + Holoenzyme
(3) Holoenzyme = Coenzyme + Co-factor
(4) Apoenzyme = Holoenzyme + Coenzyme

92. A decrease in blood pressure / volume will not cause the release of :
(1) Atrial natriuretic factor
(2) Aldosterone
(3) ADH
(4) Renin

93. Which cells of "Crypts of Lieberkuhn" secrete antibacterial lysozyme ?
(1) Paneth cells (2) Zymogen cells
(3) Kupffer cells (4) Argentaffin cells

94. Which of the following are not polymeric ?
(1) Proteins (2) Polysaccharides
(3) Lipids (4) Nucleic acids

95. Functional megaspore in an angiosperm develops into ?
(1) Endosperm (2) Embryo sac
(3) Embryo (4) Ovule

96. Myelin sheath is produced by :
(1) Astrocytes and Schwann cells
(2) Oligodendrocytes and Osteoclasts
(3) Osteoclasts and Astrocytes
(4) Schwann cells and Oligodendrocytes

97. Attractants and rewards are required for :
(1) Entomophily (2) Hydrophily
(3) Cleistogamy (4) Anemophily

98. Receptor sites for neurotransmitters are present on :
(1) Pre-synaptic membrane
(2) Tips of axons
(3) Post-synaptic membrane
(4) Membrane of synaptic vesicles

99. Coconut fruit is a :
(1) Berry (2) Nut
(3) Capsule (4) Drupe

100. Adult human RBCs are enucleated. Which of the following statement(s) is/are **most appropriate** explanation for this feature ?
(a) They do not need to reproduce
(b) They are somatic cells
(c) They do not metabolize
(d) All their internal space is available for oxygen transport
(1) only (a) (2) (a), (c) and (d)
(3) (b) and (c) (4) only (d)

101. Capacitation occurs in :
(1) Epididymis
(2) Vas deferens
(3) Female reproductive tract
(4) Rete testis

102. Which of the following are found in extreme saline conditions ?
(1) Eubacteria (2) Cyanobacteria
(3) Mycobacteria (4) Archaebacteria

103. Asymptote in a logistic growth curve is obtained when :
(1) K = N (2) K > N
(3) K < N (4) The value of 'r' approaches zero

104. Artificial selection to obtain cows yielding higher milk output represents :
(1) Directional as it pushes the mean of the character in one direction
(2) Disruptive as it splits the population into two, one yielding higher output and the other lower output
(3) Stabilizing followed by disruptive as it stabilizes the population to produce higher yielding cows
(4) Stabilizing selection as it stabilizes this character in the population

105. Select the mismatch :
(1) *Rhodospirillum* - Mycorrhiza
(2) *Anabaena* - Nitrogen fixer
(3) *Rhizobium* - Alfalfa
(4) *Frankia* - Alnus

106. Good vision depends on adequate intake of carotene rich food.
Select the best option from the following statements :
(i) Vitamin A derivatives are formed from carotene
(ii) The photopigments are embedded in the membrane discs of the inner segment
(iii) Retinal is a derivative of Vitamin A
(iv) Retinal is a light absorbing part of all the visual photopigments
Options :
(1) (i), (iii) and (iv) (2) (i) and (iii)
(3) (ii), (iii) and (iv) (4) (i) and (iv)

107. The DNA fragments separated on an agarose gel can be visualised after staining with :
(1) Acetocarmine
(2) Aniline blue
(3) Ethidium bromide
(4) Bromophenol blue

108. The hepatic portal vein drains blood to liver from:
(1) Stomach (2) Kidneys
(3) Intestine (4) Heart

109. The vascular cambium normally gives rise to :
(1) Primary phloem (2) Secondary xylem
(3) Periderm (4) Phelloderm

110. Thalassemia and sickle cell anemia are caused due to a problem in globin molecule synthesis. Select the correct statement :
(1) Both are due to a quantitative defect in globin chain synthesis
(2) Thalassemia is due to less synthesis of globin molecules
(3) Sickel cell anemia is due to a quantitative problem of globin molecules
(4) Both are due to a qualitative defect in globin chain synthesis

111. The genotypes of a husband and Wife are I^AI^B and I^Ai.
Among the blood types of their children, how many different genotypes and phenotypes are possible?
(1) 3 genotypes ; 4 phenotypes
(2) 4 genotypes ; 3 phenotypes
(3) 4 genotypes ; 4 phenotypes
(4) 3 genotypes ; 3 phenotypes

112. Which of the following facilitates opening of stomatal aperture ?
(1) Decrease in turgidity of guard cells
(2) Radial orientation of cellulose microfibrils in the cell wall of guard cells
(3) Longitudinal orientation of cellulose microfibrils in the cell wall of guard cells
(4) Contraction of outer wall of guard cells

113. In Bougainvillea thorns are the modifications of:
(1) Adventitious root (2) Stem
(3) Leaf (4) Stipules

114. Which one of the following is related to Ex-situ conservation of threatened animals and plants ?
(1) Biodiversity hot spots
(2) Amazon rainforest
(3) Himalayan region
(4) Wildlife safari parks

115. Root hairs develop from the region of :
(1) Elongation
(2) root cap
(3) Meristematic activity
(4) Maturation

116. A disease caused by an autosomal primary non-disjunction is :
(1) Klinefelter's Syndrome
(2) Turner's Syndrome
(3) Sickel Cell Anemia
(4) Down's Syndrome

117. The water potential of pure water is :
(1) Less than zero
(2) More than zero but less than one
(3) More than one
(4) Zero

118. Which of the following options gives the correct sequence of events during mitosis ?
(1) Condensation → nuclear membrane disassembly → arrangement at equator → centromere division → segregation → telophase
(2) Condensation → crossing over → nuclear membrane disassembly → segregation → telophase
(3) Condensation → arrangement at equator → centromere division → segregation → telophase
(4) Condensation → nuclear membrane disassembly → crossing over → segregation → telophase

119. The process of separation and purification of expressed protein before marketing is called :
(1) Downstream processing
(2) Bioprocessing
(3) Postproduction processing
(4) Upstream processing

120. A temporary endocrine gland in the human body is :
(1) Corpus cardiacum (2) Corpus luteum
(3) Corpus allatum (4) Pineal gland

121. Which of the following is made up of dead cells?
(1) Collenchyma
(2) Phellem
(3) Phloem
(4) Xylem parenchyma

122. An example of colonial alga is :
(1) *Volvox* (2) *Ulothrix*
(3) *Spirogyra* (4) *Chlorella*

123. Match the following sexually transmitted diseases (Column-I) with their causative agent (Column-II) and select the correct option :

Column-I	Column-II
(a) Gonorrhea	(i) HIV
(b) Syphilis	(ii) *Neisseria*
(c) Genital Warts	(iii) *Treponema*
(d) AIDS	(iv) Human papilloma-Virus

	(a)	(b)	(c)	(d)
(1)	(iii)	(iv)	(i)	(ii)
(2)	(iv)	(ii)	(iii)	(i)
(3)	(iv)	(iii)	(ii)	(i)
(4)	(ii)	(iii)	(iv)	(i)

124. The function of copper ions in copper releasing IUD's is :
(1) They inhibit gametogenesis
(2) They make uterus unsuitable for implantation
(3) They inhibit ovulation
(4) They suppress sperm motility and fertilising capacity of sperms

125. Which of the following in sewage treatment removes suspended solids ?
(1) Secondary treatment
(2) Primary treatment
(3) Sludge treatment
(4) Tertiary treatment

126. An important characteristic that Hemichordates share with Chordates is :
(1) Ventral tubular nerve cord
(2) Pharynx with gill slits
(3) Pharynx without gill slits
(4) Absence of notochord

127. The final proof for DNA as the genetic material came from the experiments of :
(1) Hershey and Chase
(2) Avery, Mcleod and McCarty
(3) Hargobind Khorana
(4) Griffith

128. Among the following characters, which one was not considered by Mendel in his experiments on pea ?
(1) Trichomes – Glandular or non-glandular
(2) Seed – Green or Yellow
(3) Pod – Inflated or Constricted
(4) Stem – Tall or Dwarf

129. Plants which produce characteristic pneumatophores and show vivipary belong to :
(1) Halophytes (2) Psammophytes
(3) Hydrophytes (4) Mesophytes

130. The pivot joint between atlas and axis is a type of :
(1) Cartilaginous joint (2) Synovial joint
(3) Saddle joint (4) Fibrous joint

131. With reference to factors affecting the rate of photosynthesis, which of the following statements is not correct ?
(1) Increasing atmospheric CO_2 concentration up to 0.05% can enhance CO_2 fixation rate
(2) C_3 plants respond to higher temperatures with enhanced photosynthesis while C_4 plants have much lower temperature optimum
(3) Tomato is a greenhouse crop which can be grown in CO_2 enriched atmosphere for higher yield
(4) Light saturation for CO_2 fixation occurs at 10% of full sunlight

132. DNA fragments are:
(1) Negatively charged
(2) Neutral
(3) Either positively or negatively charged depending on their size
(4) Positively charged

133. Which of the following components provides sticky character to the bacterial cell ?
(1) Nuclear membrane (2) Plasma membrane
(3) Glycocalyx (4) Cell wall

134. Which of the following options best represents the enzyme composition of pancreatic juice ?
(1) amylase, pepsin, trypsinogen, maltase
(2) peptidase, amylase, pepsin, rennin
(3) lipase, amylase, trypsinogen, procarboxypeptidase
(4) amylase, peptidase, trypsinogen, rennin

135. Which among these is the correct combination of aquatic mammals ?
(1) Dolphins, Seals, Trygon
(2) Whales, Dolphins, Seals
(3) Trygon, Whales, Seals
(4) Seals, Dolphins, Sharks

136. Fruit and leaf drop at early stages can be prevented by the application of:
(1) Ethylene (2) Auxins
(3) Gibberellic acid (4) Cytokinins

137. Select the correct route for the passage of sperms in male frogs:
(1) Testes → Vasa efferentia → Kidney → Seminal Vesicle → Urinogenital duct → Cloaca
(2) Testes → Vasa efferentia → Bidder's canal → Ureter → Cloaca
(3) Testes → Vasa efferentia → Kidney → Bidder's canal → Urinogenital duct → Cloaca
(4) Testes → Bidder's canal → Kidney → Vasa efferentia → Urinogenital duct → Cloaca

138. In case of a couple where the male is having a very low sperm count, which technique will be suitable for fertilisation ?
(1) Gamete intracytoplasmic fallopian transfer
(2) Artificial Insemination
(3) Intracytoplasmic sperm injection
(4) Intrauterine transfer

139. Which ecosystem has the maximum biomass ?
(1) Grassland ecosystem
(2) Pond ecosystem
(3) Lake ecosystem
(4) Forest ecosystem

140. Lungs are made up of air-filled sacs, the alveoli. They do not collapse even after forceful expiration, because of:
(1) Inspiratory Reserve Volume
(2) Tidal Volume
(3) Expiratory Reserve Volume
(4) Residual Volume

141. Presence of plants arranged into well defined vertical layers depending on their height can be seen best in:
(1) Tropical Rain Forest
(2) Grassland
(3) Temperate Forest
(4) Tropical Savannah

142. Which of the following statements is correct ?
(1) The descending limb of loop of Henle is impermeable to water.
(2) The ascending limb of loop of Henle is permeable to water.
(3) The descending limb of loop of Henle is permeable to electrolytes.
(4) The ascending limb of loop of Henle is impermeable to water.

143. Alexander Von Humboldt described for the first time:
(1) Laws of limiting factor
(2) Species area relationships
(3) Population Growth equation
(4) Ecological Biodiversity

144. Zygotic meiosis is characteristic of;
(1) *Fucus* (2) *Funaria*
(3) *Chlamydomonas* (4) *Marchantia*

145. If there are 999 bases in an RNA that codes for a protein with 333 amino acids, and the base at position 901 is deleted such that the length of the RNA becomes 998 bases, how many codons will be altered ?
(1) 11 (2) 33 (3) 333 (4) 1

146. Flowers which have single ovule in the ovary and are packed into inflorescence are usually pollinated by:
(1) Bee (2) Wind (3) Bat (4) Water

147. Transplantation of tissues/organs fails often due to non-acceptance by the patient's body. Which type of immune-response is responsible for such rejections ?
(1) Cell - mediated immune response
(2) Hormonal immune response
(3) Physiological immune response
(4) Autoimmune response

148. Life cycle of *Ectocarpus* and *Fucus* respectively are:
(1) Diplontic, Haplodiplontic
(2) Haplodiplontic, Diplontic
(3) Haplodiplontic, Haplontic
(4) Haplontic, Diplontic

149. A gene whose expression helps to identify transformed cell is known as :
(1) Vector (2) Plasmid
(3) Structural gene (4) Selectable marker

150. A dioecious flowering plant prevents both :
(1) Autogamy and geitonogamy
(2) Geitonogamy and xenogamy
(3) Cleistogamy and xenogamy
(4) Autogamy and xenogamy

151. Which statement is wrong for Krebs' cycle ?
(1) There is one point in the cycle where FAD+ is reduced to $FADH_2$
(2) During conversion of succinyl CoA to succinic acid, a molecule of GTP is synthesised
(3) The cycle starts with condensation of acetyl group (acetyl CoA) with pyruvic acid to yield citric acid
(4) There are three points in the cycle where NAD^+ is reduced to $NADH + H^+$

152. Phosphoenol pyruvate (PEP) is the primary CO_2 acceptor in:
(1) C_4 plants (2) C_2 plants
(3) C_3 and C_4 plants (4) C_3 plants

153. During DNA replication, Okazaki fragments are used to elongate:
(1) The lagging strand towards replication fork.
(2) The leading strand away from replication fork.
(3) The lagging strand away from the replication fork.
(4) The leading strand towards replication fork.

154. Which of the following RNAs should be most abundant in animal cell ?
(1) t-RNA (2) m-RNA
(3) mi-RNA (4) r-RNA

155. GnRH, a hypothalamic hormone, needed in reproduction, acts on:
(1) anterior pituitary gland and stimulates secretion of LH and FSH.
(2) posterior pituitary gland and stimulates secretion of oxytocin and FSH.
(3) posterior pituitary gland and stimulates secretion of LH and relaxin.
(4) anterior pituitary gland and stimulates secretion of LH and oxytocin.

156. What is the criterion for DNA fragments movement on agarose gel during gel electrophoresis ?
(1) The smaller the fragment size, the farther it moves
(2) Positively charged fragments move to farther end
(3) Negatively charged fragments do not move
(4) The larger the fragment size, the farther it moves

157. Hypersecretion of Growth Hormone in adults does not cause further increase in height, because:
(1) Epiphyseal plates close after adolescence.
(2) Bones loose their sensitivity to Growth Hormone in adults.
(3) Muscle fibres do not grow in size after birth.
(4) Growth Hormone becomes inactive in adults.

158. DNA replication in bacteria occurs:
(1) Within nucleolus
(2) Prior to fission
(3) Just before transcription
(4) During S phase

159. Which one from those given below is the period for Mendel's hybridization experiments ?
(1) 1840 - 1850 (2) 1857 - 1869
(3) 1870 - 1877 (4) 1856 - 1863

160. Viroids differ from viruses in having;
(1) DNA molecules without protein coat
(2) RNA molecules with protein coat
(3) RNA molecules without protein coat
(4) DNA molecules with protein coat

161. MALT constitutes about _______ percent of the lymphoid tissue in human body.
(1) 20% (2) 70% (3) 10% (4) 50%

162. Which of the following is correctly matched for the product produced by them ?
(1) *Methanobacterium* : Lactic acid
(2) *Penicillium notatum* : Acetic acid
(3) *Sacchromyces cerevisiae* : Ethanol
(4) *Acetobacter aceti* : Antibiotics

163. Which among the following are the smallest living cells, known without a definite cell wall, pathogenic to plants as well as animals and can survive without oxygen ?
(1) *Pseudomonas* (2) *Mycoplasma*
(3) *Nostoc* (4) *Bacillus*

164. Which of the following represents order of Horse' ?
(1) Perissodactyla (2) Caballus
(3) Ferus (4) Equidae

165. Frog's heart when taken out of the body continues to beat for sometime.
Select the best option from the following statements.
(a) Frog is a poikilotherm.
(b) Frog does not have any coronary circulation.
(c) Heart is "myogenic" in nature.
(d) Heart is autoexcitable
Options:
(1) Only (d) (2) (a) and (b)
(3) (c) and (d) (4) Only (c)

166. Homozygous purelines in cattle can be obtained by :
(1) mating of unrelated individuals of same breed.
(2) mating of individuals of different breed.
(3) mating of individuals of different species.
(4) mating of related individuals of same breed.

167. Identify the wrong statement in context of heartwood:
(1) It is highly durable
(2) It conducts water and minerals efficiently
(3) It comprises dead elements with highly lignified walls
(4) Organic compounds are deposited in it

168. Anaphase Promoting Complex (APC) is a protein degradation machinery necessary for proper mitosis of animal cells. If APC is defective in a human cell, which of the following is expected to occur ?
(1) Chromosomes will be fragmented
(2) Chromosomes will not segregate
(3) Recombination of chromosome arms will occur
(4) Chromosomes will not condense

169. Which of the following cell organelles is responsible for extracting energy from carbohydrates to form ATP ?
(1) Ribosome (2) Chloroplast
(3) Mitochondrion (4) Lysosome

170. Mycorrhizae are the example of:
(1) Amensalism (2) Antibiosis
(3) Mutualism (4) Fungistasis

171. Out of 'X' pairs of ribs in humans only 'Y' pairs are true ribs. Select the option that correctly represents values of X and Y and provides their explanation:
(1) X = 12, Y = 5 True ribs are attached dorsally to vertebral column and sternum on the two ends.
(2) X = 24, Y = 7 True ribs are dorsally attached to vertebral column but are free on ventral side.
(3) X = 24, Y = 12 True ribs are dorsally attached to vertebral column but are free on ventral side.
(4) X = 12, Y = 7 True ribs are attached dorsally to vertebral column and ventrally to the sternum.

172. In case of poriferans, the spongocoel is lined with flagellated cells called:
(1) oscula
(2) choanocytes
(3) mesenchymal cells
(4) ostia

173. Which one of the following statements is not valid for aerosols ?
(1) They alter rainfall and monsoon patterns
(2) They cause increased agricultural productivity
(3) They have negative impact on agricultural land
(4) They are harmful to human health

174. A baby boy aged two years is admitted to play school and passes through a dental check - up. The dentist observed that the boy had twenty teeth. Which teeth were absent?
(1) Canines (2) Pre-molars
(3) Molars (4) Incisors

175. Select the mismatch
(1) *Cycas* – Dioecious
(2) *Salvinia* – Heterosporous
(3) *Equisetum* – Homosporous
(4) *Pinus* – Dioecious

176. The morphological nature of the edible part of coconut is:
(1) Cotyledon (2) Endosperm
(3) Pericarp (4) Perisperm

177. Double fertilization is exhibited by :
(1) Algae (2) Fungi
(3) Angiosperms (4) Gymnosperms

178. Spliceosomes are not found in cells of;
(1) Fungi (2) Animals
(3) Bacteria (4) Plants

179. The association of histone H1 with a nucleosome indicates:
(1) DNA replication is occurring.
(2) The DNA is condensed into a Chromatin Fibre.
(3) The DNA double helix is exposed.
(4) Transcription is occurring.

180. The region of Biosphere Reserve which is legally protected and where no human activity is allowed is known as:
(1) Buffer zone (2) Transition zone
(3) Restoration zone (4) Core zone

HINTS & SOLUTIONS

PHYSICS

1. **(2)** Let ℓ be the complete length of the spring.
Lengths when cut in ratio, 1 : 2 : 3 are

$\frac{\ell}{6}, \frac{\ell}{3}$ and $\frac{\ell}{2}$

Spring constant (k) $\propto \frac{1}{\text{length}(\ell)}$

Spring constant for given segments
$k_1 = 6k, k_2 = 3k$ and $k_3 = 2k$
When they are connected **in series**

$$\frac{1}{k'} = \frac{1}{6k} + \frac{1}{3k} + \frac{1}{2k} \Rightarrow \frac{1}{k'} = \frac{6}{6k}$$

$\therefore$ Force constant $k' = k$
And when they are connected **in parallel**
$k'' = 6k + 3k + 2k \Rightarrow k'' = 11k$
Then the ratio's

$\frac{k'}{k''} = \frac{1}{11}$ i.e., $k' : k'' = 1 : 11$

2. **(2)** Resolving power of a microscope =

$$\frac{2\mu \sin\theta}{\lambda}$$

i.e., $R \propto \frac{1}{\lambda}$ or, $\frac{R_1}{R_2} = \frac{\lambda_2}{\lambda_1}$

Given that the two wavelengths,

$\lambda_1 = 4000 Å$, $\lambda_2 = 6000 Å$

$$\therefore \frac{R_1}{R_2} = \frac{6000 Å}{4000 Å} = \frac{3}{2}$$

3. **(1)** Difference in two successive frequencies of closed pipe

$$\frac{2v}{4l} = 260 - 220 = 40 \text{ Hz}$$

or $\frac{2v}{4l} = 40 \text{ Hz} \Rightarrow \frac{v}{4\ell} = 20\text{Hz}$

Which is the fundamental frequency of system (or closed organ pipe)

4. **(3)** From work energy theorem,
$W_g + W_a = \Delta K.E$

or, $mgh + W_a = \frac{1}{2}mv^2 - 0$

$$10^{-3} \times 10 \times 10^3 + W_a = \frac{1}{2} \times 10^{-3} \times (50)^2$$

$\Rightarrow W_a = -8.75$ J
which is the work done due to air resistance
Work done due to gravity = mgh = $10^{-3} \times 10 \times 10^3 = 10$ J

5. **(4)** Dimensions of length is related as,

$$L = [C]^x [G]^y \left[\frac{e^2}{4\pi\varepsilon_0}\right]^z \Rightarrow \frac{e^2}{4\pi\varepsilon_0} = ML^3T^{-2}$$

$L = [LT^{-1}]^x [M^{-1}L^3T^{-2}]^y [ML^3T^{-2}]^z$
$[L] = [L^{x+3y+3z} M^{-y+z} T^{-x-2y-2z}]$
Comparing both sides
$-y + z = 0 \Rightarrow y = z$ ...(i)
$x + 3y + 3z = 1$ ...(ii)
$-x - 4z = 0$ ...(iii)
From (i), (ii) & (iii)

$z = y = \frac{1}{2}$, $x = -2$

Hence, $L = C^{-2}\left[G \cdot \frac{e^2}{4\pi\varepsilon_0}\right]^{1/2}$

6. **(4)** Heat current $H = H_1 + H_2$

$$= \frac{K_1A(T_1 - T_2)}{d} + \frac{K_2A(T_1 - T_2)}{d}$$

$$\frac{K_{EQ}2A(T_1 - T_2)}{d} = \frac{A(T_1 - T_2)}{d}[K_1 + K_2]$$

Hence equivalent thermal conductivities for two rods of equal area is given by

$$K_{EQ} = \frac{k_1 + k_2}{2}$$

7. **(1)** When battery is replaced by another uncharged capacitor

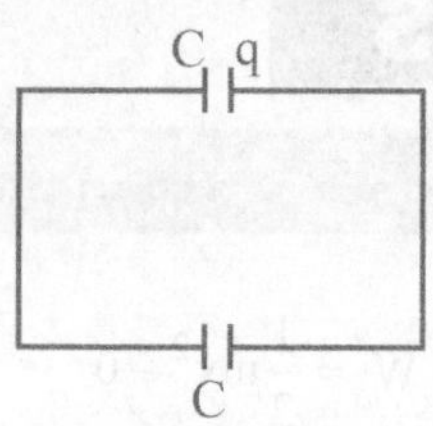

As uncharged capacitor is connected parallel

So, $C' = 2C$

and $V_c = \frac{q_1 + q_2}{C_1 + C_2}$

$V_c = \frac{q + 0}{C + C} \Rightarrow V_c = \frac{V}{2}$

Initial Energy of system, $U_i = \frac{1}{2}CV^2$ ... (i)

Final energy of system, $U_f = \frac{1}{2}(2C)\left(\frac{V}{2}\right)^2$

$= \frac{1}{2}CV^2\left(\frac{1}{2}\right)$ ...(ii)

From equation (i) and (ii)

$U_f = \frac{1}{2}U_i$

i.e., Total electrostatic energy of resulting system decreases by a factor of 2

8. **(2)** Given, current gain $\beta = 100, R_c = 3k\Omega, R_b = 2k\Omega$

Voltage gain $(A_v) = \beta\frac{R_c}{R_b} = 100\left(\frac{3}{2}\right) = 150$

Power gain $= A_v\,\beta$

$= 150(100) = 15000$

9. **(1)** Process I volume is constant. hence, it is isochoric

In process IV, pressure is constant hence, it is isobaric

10. **(2)** According to question, the net electrostatic force (F_E) = gravitational force (F_G)

$F_E = F_G$

or $\frac{1}{4\pi\varepsilon_0}\frac{\Delta e^2}{d^2} = \frac{Gm^2}{d^2}$

$\Rightarrow \Delta e = m\sqrt{\frac{G}{K}}$ $\left(\frac{1}{4\pi\varepsilon_0} = k = 9\times10^9\right)$

$= 1.67 \times 10^{-27}\sqrt{\frac{6.67\times10^{-11}}{9\times10^9}}$

$\Delta e \approx 1.436 \times 10^{-37}\,C$

11. (2) We know that, $R = \frac{\rho\ell}{A}$

or $R = \frac{\rho\ell^2}{\text{Volume}} \Rightarrow R \propto \ell^2$

According to question $\ell_2 = n\ell_1$

$\frac{R_2}{R_1} = \frac{n^2 l_1^2}{l_1^2}$ or, $\frac{R_2}{R_1} = n^2$

$\Rightarrow R_2 = n^2 R_1$

12. **(2)**

$y_1 = \overline{A + B}$

$y_2 = \overline{y_1 + y_1} = \overline{y_1} = A + B$

$y = y_2 = \overline{A + B}$ i.e NOR Gate

13. **(1)** We know that,

de-Broglie wavelength λ

$= \frac{\hbar}{P} = \frac{h}{\sqrt{2m(KE)}}$

K.E. of thermal neutron $= \frac{3}{2}kT$

$= \frac{h}{\sqrt{2m\left(\frac{3}{2}kT\right)}} \Rightarrow \lambda = \frac{h}{\sqrt{3mkT}}$

14. **(4)**

In forward bias, $V_1 > V_2$ i.e., in figure (iv) p-type semiconductor is at higher potential w.r.t. n-type semiconductor.

15. **(2)** Given, no. of turns $n = 100$

radius, $r = 0.01\,m$

resistance, $R = 10\pi^2\,\Omega$

As we know,

$\varepsilon = -N\frac{d\phi}{dt}$

$$\frac{\varepsilon}{R} = -\frac{N}{R}\frac{d\phi}{dt}, \quad \Delta I = -\frac{N}{R}\frac{d\phi}{dt}$$

$$\frac{\Delta q}{\Delta t} = -\frac{N}{R}\frac{\Delta \phi}{\Delta t} \Rightarrow \Delta q = -\left[\frac{N}{R}\left(\frac{\Delta\phi}{\Delta t}\right)\right]\Delta t$$

'–' ve sign shows that induced emf opposes the change of flux.

$$\Delta q = \frac{\mu_0 n i \pi r^2}{R}$$

$$\Delta q = \frac{4\pi \times 10^{-7} \times 100 \times 4 \times \pi \times (0.01)^2}{10\pi^2}$$

$\Delta q = 32\mu C$

16. (2) Velocity of preeti w.r.t. elevator $= \frac{d}{t_1} = v_1$

Velocity of elevator w.r.t. ground

$v_2 = \frac{d}{t_2}$ then velocity of preeti w.r.t. ground

$v = v_1 + v_2$

$$\frac{d}{t} = \frac{d}{t_1} + \frac{d}{t_2} \Rightarrow \frac{1}{t} = \frac{1}{t_1} + \frac{1}{t_2}$$

$t = \frac{t_1 t_2}{(t_1 + t_2)}$ (time taken by preeti to walk up on the moving escalator)

17. (3) According to question

8th bright fringe in medium = 5th dark fring in air

$$Y_{8th\ bright} = 8\frac{\lambda D}{\mu d}$$

$$Y_{5th\ dark} = (2 \times 5 - 1)\frac{\lambda D}{2d} = \frac{9}{2}\frac{\lambda D}{d}$$

$$\Rightarrow \frac{9}{2}\frac{\lambda D}{d} = 8\frac{\lambda D}{\mu d}$$

or, refractive index $\mu = \frac{16}{9} = 1.78$

18. (4) When mirror is rotated by angle θ reflected ray will be rotated by 2θ.

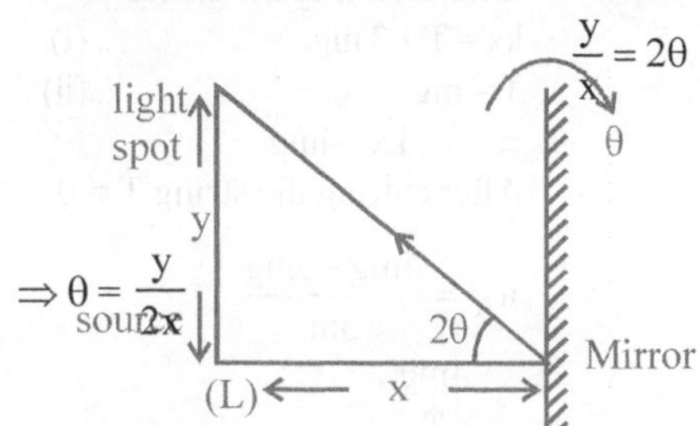

$$\Rightarrow \theta = \frac{y}{2x}$$

19. (4) If θ_1 and θ_2 are Opparent angles of dip Let α be the angle which one of the plane make with the magnetic meridian.

$\tan\theta_1 = \frac{v}{H\cos\alpha}$ i.e., $\cos\alpha = \frac{v}{H\tan\theta_1}$ ...(i)

$\tan\theta_2 = \frac{v}{H\sin\alpha}$, i.e., $\sin\alpha = \frac{v}{H\tan\theta_2}$ ...(ii)

Squaring and adding (i) and (ii), we get

$$\cos^2\alpha + \sin^2\alpha = \left(\frac{V}{H}\right)^2\left(\frac{1}{\tan^2\theta_1} + \frac{1}{\tan^2\theta_2}\right)$$

i.e., $1 = \frac{V^2}{H^2}\left[\cot^2\theta_1 + \cot^2\theta_2\right]$

or $\frac{H^2}{V^2} = \cot^2\theta_1 + \cot^2\theta_2$

i.e., $\cot^2\theta = \cot^2\theta_1 + \cot^2\theta_2$

20. (3) As we known from Doppler's Effect

$$f_{apprent} = f_0\left[\frac{v + v_0}{v - v_s}\right] = 400\left[\frac{340 + 16.5}{340 - 22}\right]$$

$f_{appprent} = 448$ Hz

A $v_s = 22$ m/s → ← $v_0 = 16.5$ m/s B

$f_0 = 400$ Hz

21. (1)

kx
3m
T
3mg
m
mg

Before cutting the string

$kx = T + 3\,mg$ ...(i)

$T = mg$ ...(ii)

$\Rightarrow \quad kx = 4mg$

After cutting the string $T = 0$

$$a_A = \frac{4mg - 3mg}{3m}$$

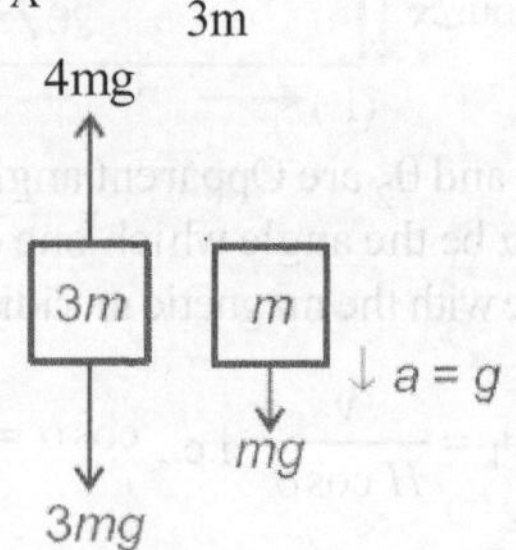

$$a_A = \frac{g}{3} \uparrow$$

and $a_B = \frac{mg}{m} = g \downarrow$

22. **(1)** For dispersion without deviation

$(\mu - 1)A_1 + (\mu' - 1)A_2 = 0$

$|(\mu - 1)A_1| = |(\mu' - 1)A_2|$

$(1.42–1) \times 10° = (1.7–1)A_2$

$4.2 = 0.7A_2 \Rightarrow A_2 = 6°$

23. **(3)** Above earth surface Below earth surface

$$g_h = g\left(1 - \frac{2h}{R_e}\right) \qquad g_d = g\left(1 - \frac{d}{R_e}\right)$$

According to question, $g_h = g_d$

$$g\left(1 - \frac{2h}{R_e}\right) = g\left(1 - \frac{d}{R_e}\right)$$

Clearly,

$d = 2h = 2\text{ km}$

24. **(2)** Reading of potentiometer is accurate because during taking reading it does not draw any current from the circuit.

25. **(3)** Given $r_1 = 12\text{ cm}, r_2 = 6\text{ cm}$

$T_1 = 500\text{ K}$ and $T_2 = 2 \times 500 = 1000\text{ K}$

$P_1 = 450$ watt

Rate of power loss $P \propto r^2T^4$

$$\frac{P_1}{P_2} = \frac{r_1^2T_1^4}{r_2^2T_2^4} \Rightarrow P_2 = P_1\frac{r_2^2T_2^4}{r_1^2T_1^4}$$

Solving we get, $P_2 = 1800$ watt

26. **(2)**

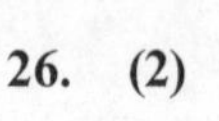

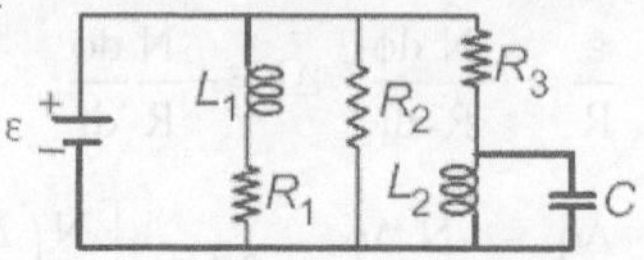

At $t = 0$, no current flows through R_1 and R_3

$\therefore$

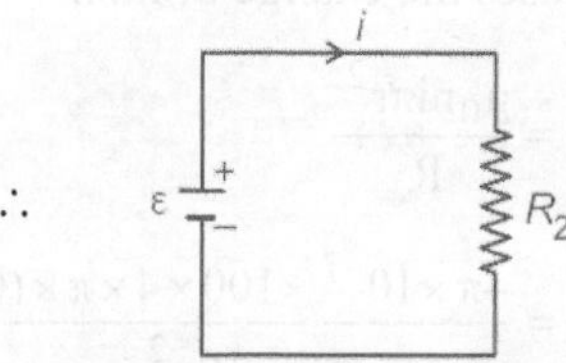

Current through battery just after the switch closed is

$$i = \frac{\varepsilon}{R_2} = \frac{18}{9} = 2A$$

27. **(1)** Given, $\lambda_A = 8\lambda, \lambda_B = \lambda$

$$N_B = \frac{N_A}{e}$$

$$\Rightarrow \quad N_o e^{-\lambda_0 B^t} = N_o\frac{e^{-\lambda_A t}}{e}$$

$e^{-\lambda t} = e^{-8\lambda t}e^{-1}, \; e^{-\lambda t} = e^{-8\lambda t - 1}$

Comparing both side powers

$-\lambda t = -8\lambda t - 1$

$-1 = 7\lambda t \Rightarrow t = -\frac{1}{7\lambda}$

The best possible answer is $t = -\frac{1}{7\lambda}$

28. **(1)** As the regions are of equipotentials, so Work done $W = q\Delta V$

ΔV is same in all the cases hence work-done will also be same in all the cases.

29. **(1)** Both the astronauts are in the condition of weightness. Gravitational force between them pulls towards each other. Hence Astronauts move towards each other under mutual gravitional force.

30. **(2)** Given:

$x = 5t - 2t^2 \qquad y = 10t$

$$v_x = \frac{dx}{dt} = 5 - 4t \qquad v_y = \frac{dy}{dt} = 10$$

$a_x = \frac{dv_x}{dt} = -4 \qquad a_y = \frac{dv_y}{dt} = 0$

$\vec{a} = a_x i + a_y j \qquad \vec{a} = -4i \text{ m/s}^2$

Hence, acceleration of particle at t = 2 s = –4m/s²

31. **(4)** Net force on particle in uniform circular motion is centripetal force $\left(\frac{mv^2}{\ell}\right)$ which provided by tension in string so the net force will be equal to tension i.e., T.

32. **(2)** Given, Amplitude A = 3 cm
Which particle is at x = 2 cm
According to question, velocity = acceleration

$\omega\sqrt{A^2 - x^2} = x\omega^2$

$\sqrt{(3)^2 - (2)^2} = 2\left(\frac{2\pi}{T}\right)$

$\sqrt{5} = \frac{4\pi}{T} \Rightarrow T = \frac{4\pi}{\sqrt{5}}$

33. **(2)**

According to malus law, $I = I_0 \cos^2\theta$

$I_1 = \frac{I_0}{2}$

$I_2 = \frac{I_0}{2}\cos^2 45° = \frac{I_0}{2} \times \frac{1}{2} = \frac{I_0}{4}$

$I_3 = \frac{I_0}{4}\cos^2 45° \Rightarrow I_3 = \frac{I_0}{8}$

34. **(3)** Bulk modulus is given by

$B = \frac{P}{\left(\frac{\Delta V}{V}\right)} \Rightarrow \frac{\Delta V}{V} = \frac{P}{B}$

$3\frac{\Delta R}{R} = \frac{P}{B}$ (here, $\frac{\Delta R}{R}$ = fractional decreases in radius)

$\Rightarrow \frac{\Delta R}{R} = \frac{P}{3B}$

35. **(1)** Given, $E_{rms} = 6$ V/m

$\frac{E_{rms}}{B_{rms}} = c \Rightarrow B_{rms} = \frac{E_{rms}}{c}$ ...(i)

$B_{rms} = \frac{B_0}{\sqrt{2}} \Rightarrow B_0 = \sqrt{2}\,B_{rms}$

$B_0 = \sqrt{2} \times \frac{E_{rms}}{C}$ From equation (i)

$= \frac{\sqrt{2} \times 6}{3 \times 10^8} = 2.83 \times 10^{-8}$ T

36. **(2)** Given, mass of cylinder m = 3kg
R = 40 cm = 0.4 m
F = 30 N ; α = ?
As we know, torque τ = Iα
$F \times R = MR^2\alpha$

$\alpha = \frac{F \times R}{MR^2}$

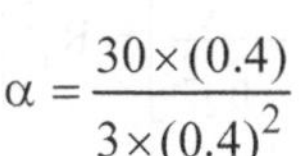

$\alpha = \frac{30 \times (0.4)}{3 \times (0.4)^2}$

or, $\alpha = 25 \text{rad/s}^2$

37. **(1)** Here, $I\omega_1 + I\omega_2 = 2I\omega$

$\Rightarrow \omega = \frac{\omega_1 + \omega_2}{2}$

$(K.E.)_i = \frac{1}{2}I\omega_1^2 + \frac{1}{2}I\omega_2^2$

$(K.E.)_f = \frac{1}{2} \times 2I\omega^2 = I\left(\frac{\omega_1 + \omega_2}{2}\right)^2$

Loss in K.E. = $(K.E)_f - (K.E)_i$

$= \frac{1}{4} I(\omega_1 - \omega_2)^2$

38. (1, 4) Both answers are correct
Given,
$\lambda_0 = 3250 \times 10^{-10}$ m, $\lambda = 2536 \times 10^{-10}$ m

$\phi = \frac{hc}{\lambda_0} = \frac{4.14 \times 10^{-15} \times 3 \times 10^8}{3250 \times 10^{-10}} = 3.82\text{eV}$

$hv = \frac{hc}{\lambda} = \frac{4.14 \times 10^{-15} \times 3 \times 10^8}{2536 \times 10^{-10}} = 4.89\text{eV}$

According to Einstein's photoelectric equation,
$K_{max} = hv - \phi$
$KE_{max} = (4.89-3.82)\text{eV} = 1.077\text{ eV}$

$\frac{1}{2}mv^2 = 1.077 \times 1.6 \times 10^{-19}$

$\Rightarrow v = \sqrt{\frac{2 \times 1.077 \times 1.6 \times 10^{-19}}{9.1 \times 10^{-31}}}$

or, $v = 0.6 \times 10^6$ m/s or 6×10^5 m/s

39. (4) Work done, $W = MB(\cos\theta_1 - \cos\theta_2)$
When it is rotated by angle 180° then
$W = MB(\cos 0° - \cos 180°) = MB(1 + 1)$
$W = 2MB \Rightarrow W = 2(NIA)B$
$= 2 \times 250 \times 85 \times 10^{-6}[1.25 \times 2.1 \times 10^{-4}] \times 85 \times 10^{-2}$
$= 9.1\ \mu J$

40. (2) For last line of Balmer series : $n_1 = 2$ and $n_2 = \infty$

$\frac{1}{\lambda_B} = RZ^2\left[\frac{1}{n_1^2} - \frac{1}{n_2^2}\right] = R_1^2\left[\frac{1}{2^2} - \frac{1}{\infty^2}\right]$

$y_B = \frac{4}{R}$ ...(1)

For last line of Lyman series : $n_1 = 1$ and $n_2 = \infty$

$\frac{1}{\lambda_L} = RZ^2\left[\frac{1}{n_1^2} - \frac{1}{n_2^2}\right] = RZ^2\left[\frac{1}{1^2} - \frac{1}{\infty^2}\right]$

$\lambda_L = \frac{1}{R}$ ...(ii)

Dividing equation (i) by (ii)

$\frac{\lambda_B}{\lambda_L} = \frac{\frac{4}{R}}{\frac{1}{R}}$

Ratio of wavelengths is $\frac{\lambda_B}{\lambda_L} = 4$

41. (1) Given, efficiency of engine, $n = \frac{1}{10}$
work done on system W = 10J
Coefficient of performance of refrigerator

$\beta = \frac{Q_2}{W} = \frac{1-\eta}{\eta} = \frac{1 - \frac{1}{10}}{\frac{1}{10}} = \frac{\frac{9}{10}}{\frac{1}{10}} = 9$

Energy absorbed from reservoir
$Q_2 = \beta w$, $Q_2 = 9 \times 10 = 90$ J

42. (3) Internal energy of the system is given by

$U = \frac{f}{2}nRT$

Degree of freedom
$F_{diatomic} = 5$, $f_{monoatomic} = 3$
and, number of moles
$n(O_2) = 2$, $n(Ar) = 4$

$U_{total} = \frac{5}{2}(2)RT + \frac{3}{2}(4)RT = 11RT$

43. (3) Force per unit length between two parallel current carrying conductors,

$F = \frac{\mu_0 i_1 i_2}{2\pi d}$

Since same current flowing through both the wires
$i_1 = i_2 = i$

so $F_1 = \frac{\mu_0 i^2}{2\pi d} = F_2$

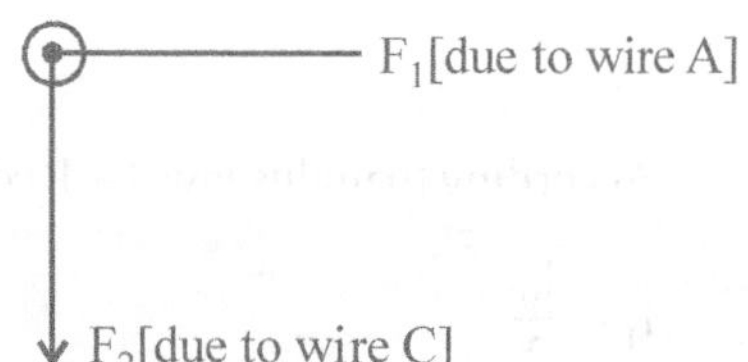

∴ Magnitude of force per unit length on the middle wire 'B'

$F_{net} = \sqrt{F_1^2 + F_2^2} = \frac{\mu_0 i^2}{\sqrt{2}\pi d}$

44. **(3)** Here, $h_{oil} \times \rho_{oil} \times g = h_{water} \times \rho_{water} \times g$

$\rho_0 g \times 140 \times 10^{-3} = \rho_w g \times 130 \times 10^{-3}$

$\rho_{oil} = \frac{130}{140} \times 10^3 \approx 928 kg/m^3$ [$\because \rho_w = 1$ kgm^{-3}]

45. **(4)** Centre of mass may or may not coincide with centre of gravity. Net torque of gravitational pull is zero about centre of mass.

$\tau_g = \Sigma\tau_i = \Sigma r_i \times m_i g = 0$

Mechanical advantage, $M.A. = \frac{Load}{Effort}$

If $M.A. > 1 \Rightarrow Load > Effort$

CHEMISTY

46. **(1)** Potassium permanganate has a purple colour. When sulphur dioxide reacts with potassium permanganate the solution decolourizes.

$5SO_2 + 2KMnO_4 + 2H_2O \rightarrow 2H_2SO_4 + 2MnSO_4 + K_2SO_4$

47. **(3)** The overall reaction rate depends on the rate of the slowest step.

i.e., Overall rate = Rate of slowest step (ii)

$= k[X][Y_2]$ ... (1)

k = rate constant

Assuming step (i) to be reversible, its equilibrium constant,

$k_{eq} = \frac{[X]^2}{[X_2]} \Rightarrow [X]^2$

$= k_{eq}[X_2]; [X] = k_{eq}^{\frac{1}{2}}[X_2]^{\frac{1}{2}}$

From eq (1) and (2)

$Rate = k k_{eq}^{\frac{1}{2}}[X_2]^{\frac{1}{2}}[Y_2]$

$Overall\ order = \frac{1}{2} + 1 = \frac{3}{2} = 1.5$

48. **(1)** Z = 114 belong to Group 14, carbon family

Electronic configuration

$= [Rn]5f^{14}6d^{10}7s^27p^2$

49. **(2)** When Ar – O – R ethers are reacted with HI, they are cleaved at weaker O – R bond to give phenol and alkyl iodide.

$C_6H_5-O-CH_3 \xrightarrow{HI} C_6H_5-OH + CH_3I$

50. **(1)** Correct order is

$H-C\equiv C-H > H_3C-C\equiv C-H > H_2C=CH_2 > CH_3-CH_3$

(Two acidic hydrogens) (One acidic hydrogen)

51. **(3)** Hydration of alkynes give ketones.

$H_3C-C\equiv CH \longrightarrow H_3C-\underset{(A)}{C(OH)}=CH_2 \xrightarrow{Tautomerism} H_3C-\underset{(B)}{C(=O)}-CH_3$

52. **(1)** (i)

$N_2 + 3H_2 \rightleftharpoons 2NH_3; K_1 = \frac{[NH_3]^2}{[N_2][H_2]^3}$

(ii) $N_2 + O_2 \rightleftharpoons 2NO; K_2 = \frac{[NO]^2}{[N_2][O_2]}$

(iii) $H_2 + \frac{1}{2}O_2 \longrightarrow H_2O; K_3 = \frac{[H_2O]}{[H_2][O_2]^{1/2}}$

Applying (II + 3 × III – I) we will get

$2NH_3 + \frac{5}{2}O_2 \overset{K}{\rightleftharpoons} 2NO + 3H_2O;$

$K = \frac{[NO]^2}{[N_2][O_2]} \times \frac{[H_2O]^3}{[H_2]^3 \times [O_2]^{3/2}} \Big/ \frac{[NH_3]^2}{[N_2][H_2]^3}$

$\therefore K = K_2 \times K_3^3 / K_1$

53. **(3)** Electron withdrawing $-NO_2$ group has very strong –I and –R effects so, compound 3 will be most acidic.

54. **(3)** $-NO_2$ group has strong – R effect and $-CH_3$ shows +R effect.

$\therefore$ Order of basic strength is

p-nitroaniline (NH_2, NO_2) < aniline (NH_2) < p-toluidine (NH_2, CH_3)

55. **(3)** Li^+ being smallest, has maximum charge density.

$\therefore$ Li^+ is most heavily hydrated among all alkali metal ions. Effective size of Li^+ in aqueous solution is therefore, largest. So, moves slowest under electric field.

56. **(3)** Steam distillation is the most suitable method of separation of 1 : 1 mixture of ortho and para nitrophenols as there is intramolecular hydrogen bonding in o-nitrophenol.

57. (2) In a solution containing $HgCl_2$, I_2 and I^-, both $HgCl_2$ and I_2 compete for I^-.
Since formation constant of $[HgI_4]^{2-}$ is very large (1.9×10^{30}) as compared with I_3 ($K_f = 700$)
$\therefore$ I^- will preferentially combine with $HgCl_2$.

$$HgCl_2 + 2I^- \rightarrow \underset{\text{Red ppt}}{HgI_2} \downarrow + 2Cl^-$$

$$HgI_2 + 2I^- \rightarrow \underset{\text{soluble}}{[HgI_4]^{2-}}$$

58. (1) Dettol is a mixture of chloroxylenol and terpineol which is very commonly known antiseptic.

59. (1) Grignard's reagent (RMgX) is a σ-bonded organometallic compound.

60. (1) Half life for a first order reaction,

$$t_{1/2} = \frac{0.693}{K} \quad \text{So, } t_{1/2} = \frac{0.693}{10^{-2}} \text{sec}.$$

Also, for the reduction of 20 g or reactant to 5 g, two half lives will be required.
$\therefore$ For 20 g of the reactant to reduce to 5g, time taken,

$$t = 2 \times \frac{0.693}{10^{-2}} \text{sec} = 138.6 \text{ sec.}$$

61. (1) $XX' \rightarrow$ Linear (e.g. ClF, BrF)
$XX_3' \rightarrow$ T-Shape (e.g. ClF_3, BrF_3)
$XX_5' \rightarrow$ Square pyramidal (e.g. $BrF_5 IF_5$)
$XX_7' \rightarrow$ Pentagonal bipyramidal (e.g. IF_7)

62. (3)

$$Ag_2C_2O_4(s) \rightleftharpoons \underset{2s}{2Ag^+(aq)} + \underset{s}{C_2O_4^{2-}(aq)}$$

$K_{SP} = [Ag^+]^2 [C_2O_4^{2-}]$
$[Ag^+] = 2.2 \times 10^{-4}$ M
Given that:
$\therefore$ Concentration of $C_2O_4^{2-}$ ions,

$$[C_2O_4^{2-}] = \frac{2.2 \times 10^{-4}}{2} M = 1.1 \times 10^{-4} M$$

$\therefore$ $K_{SP} = (2.2 \times 10^{-4})^2 (1.1 \times 10^{-4})$
$= 5.324 \times 10^{-12}$

63. (2) For cell,
$Zn|ZnSO_4(0.01\ M) || CuSO_4(1.0\ M)|Cu$

$$E_{cell} = E°_{cell} - \frac{2.303RT}{nF} \frac{\log[Zn^{2+}]}{[Cu^{2+}]}$$

$$\therefore E_1 = E°_{cell} - \frac{2.303RT}{2 \times F} \times \log \frac{(0.01)}{1}$$

When concentrations are changed for $ZnSO_4$ and $CuSO_4$, we can write

$$E_2 = E°_{cell} - \frac{2.303RT}{2F} \times \log \frac{1}{0.01} \therefore E_1 > E_2$$

64. (2) IBr_2^-, XeF_2
Total number of valence electrons are equal in both the species and both the species exhibit linear shape.

65. (4) $H-\overset{O}{\overset{\|}{C}}_1-\overset{O}{\overset{\|}{C}}_2 \ldots$ (structure with carbons numbered 1–6)

3-keto-2-methylhex-4-enal
Aldehydes get higher priority over ketone and alkene in numbering of principal carbon chain.

66. (3) For hydrogen like atoms energy of 2s-orbital and 2p-orbital is equal.

67. (3, 4) Those ionic compounds in which size of cation and anion is largely different show Frenkel defect.
Non-stoichiometric ferrous oxide is $Fe_{0.93-0.96}O_{1.00}$ and it is due to metal deficiency defect.

68. (3) BCl_3 is trigonal planar and hence the bond angle is 120°.

Cl, B, Cl, Cl, 120°

69. (1) Given ΔH 35.5 kJ mol^{-1}
$\Delta S = 83.6\ JK^{-1}\ mol^{-1}$ $\because \Delta G = \Delta H - T\Delta S$
For a reaction to be spontaneous, ΔG = –ve
i.e., $\Delta H < T\Delta S$

$$\therefore T > \frac{\Delta H}{\Delta S} = \frac{35.5 \times 10^3\ Jmol^{-1}}{83.6\ JK^{-1}}$$

So, the given reaction will be spontaneous at T > 425 K

70. (1) Microorganisms present in the soil is a sink for CO.

71. (3) K_f (molal depression constant) only depends on the nature of the solvent and is independent of the concentration of the solution.

72. (1) Molarity depends on the volume of a solution which can be changed with change in temperature.

73. (1) A catalyst speeds up both forward and backward reaction with the same rate.
So, equilibrium constant is not affected by the presence of a catalyst at any given temperature.

74. (4)

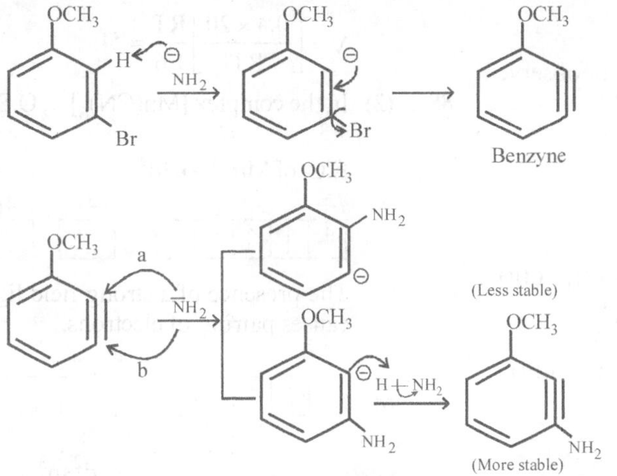

$$\text{(o-bromoanisole)} \xrightarrow{NH_2^-} \longrightarrow \text{Benzyne}$$

Benzyne

(Less stable)

(More stable)

More stable as –ve charge is close to electron withdrawing group.

Also, incoming nucleophile gets attached on same 'C' on which 'Br' (Leaving group) was present.

$\therefore$ not a cine substitution reaction

75. (2)

$$\left[Co(NH_3)_6\right]Cl_3 \xrightarrow{AgNO_3} 3\text{mol of AgCl}$$

$$\left[Co(NH_3)_5 Cl\right]Cl_2 \xrightarrow{AgNO_3} 2\text{mol of AgCl}$$

$$\left[Co(NH_3)_4 Cl_2\right]Cl \xrightarrow{AgNO_3} 1\text{mol of AgCl}$$

Complexes are respectively $[Co(NH_3)_6Cl_3$, $[Co(NH_3)_5Cl]Cl_2$ and $[Co(NH_3)_4Cl_2]Cl$

76. (3)

77. (2) The system is in isolated state.

$\because$ For an adiabatic process, q = 0

$\Delta U = q + w$

$\therefore \Delta U = w = -p\Delta V = -2.5 \text{ atm} \times (4.5 - 2.5) \text{ L}$

$= -2.5 \times 2 \text{ L-atm} = -5 \times 101.3 \text{ J} = -506.5 \text{ J} \approx -505 \text{J}$

78. (1) $\underset{\text{acetamide}}{CH_3 - \overset{O}{\overset{\|}{C}} - NH_2} + Br_2 + 4NaOH \xrightarrow{\Delta}$

$\underset{\text{methanamine}}{CH_3 - NH_2} + 2NaBr + Na_2CO_3 + 3H_2O$

It is called Hoffmann Bromamide reaction.

79. (3) There is no change in bond angles and bond lengths in the conformations of ethane.

80. (1) $:\ddot{O}^- - \overset{O}{\overset{\|}{\underset{O}{\underset{\|}{S}}}} - S - S - \overset{O}{\overset{\|}{\underset{O}{\underset{\|}{S}}}} - \ddot{O}:^-$ $S_4O_6^{2-}$ $S_2O_3^{2-}$

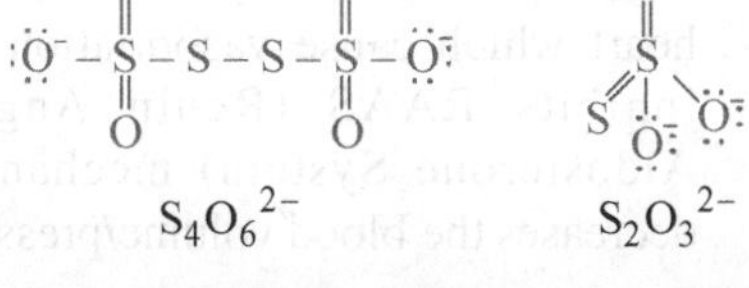

81. (4) Inertness of ns^2 electrons of the valence shell to participate in bonding on moving down the group in heavier p-block elements is called **inert pair effect**.

As a result, Pb(II) is more stable than Pb(IV)

Sn(IV) is more stable than Sn(II)

$\therefore$ Pb(IV) is easily reduced to Pb(II) and can acts as an oxidising agent whereas Sn(II) is easily oxidised to Sn(IV) and can acts as a reducing agent.

82. (4) The order of the ligand in the spectrochemical series

$H_2O < NH_3 < en$

Hence, the wavelength of the light observed will be in the order

$[Co(H_2O)_6]^{3+} < [Co(NH_3)_6]^{3+} < [Co(en)_3]^{3+}$

Thus, wavelength absorbed will be in the opposite order

i.e., $[Co(en)_3]^{3+}$, $[Co(NH_3)_6]^{3+}$, $[Co(H_2O)_6]^{3+}$

83. (2) Since 'A' gives positive silver mirror test therefore, it must be an aldehyde of α-Hydroxyketone. Also, reaction with OH^- i.e., aldol condensation (by assuming alkali to be dilute) indicates that A is aldehyde as aldol reaction of ketones is reversible and carried out in special apparatus. It indicates that A is an aldehyde

$$\underset{(X)}{CH_3 - CH_2OH} \xrightarrow[573K]{Cu} CH_3 - CHO$$

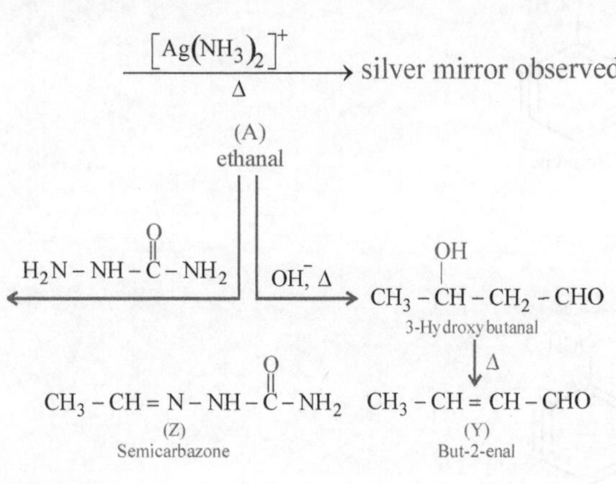

84. (1)

$$\text{cyclohexanone} + \text{cyclohexanone} \xrightarrow[\text{(ii) }\Delta]{\text{(i) OH}} \text{2-cyclohexylidenecyclohexanone}$$

85. (2) CN^- and CO have same no. of electrons and have same bond order equal to 3.

86. (3) Zn being more reactive than Ag and Au, displaces them.

$$4Ag + 8NaCN + 2H_2O + O_2 \xrightarrow{\text{Leaching}} 4Na[Ag(CN)_2] + 4NaOH$$

Soluble Sodium dicyanoargentate (I)

Soluble cyanide compound can be treated with Zn to give metal by displacement.

$$2Na[Ag(CN)_2] + Zn \xrightarrow{\text{Displacement}} Na_2[Zn(CN)_4] + 2Ag\downarrow$$

87. (4) Max. pressure of CO_2 = Pressure of CO_2 at equilibrium

For reaction,

$$SrCO_3(s) \rightleftharpoons SrO(s) + CO_2$$

$K_p = P_{CO_2} = 1.6\,\text{atm}$ = maximum pressure of CO_2

volume of container at this stage.

$$V = \frac{nRT}{P} \quad \ldots\text{(i)}$$

Since container is sealed and reaction was not earlier at equilibrium.

$\therefore$ n = constant.

$$n = \frac{PV}{RT} = \frac{0.4 \times 20}{RT} \quad \ldots\text{(ii)}$$

Put equation (ii) in equation (i)

$$V = \left[\frac{0.4 \times 20}{RT}\right]\frac{RT}{1.6} = 5L$$

88. (2) In the complex $[Mn(CN)_6]^{3-}$, O.S. of Mn is +3

E.C. of $Mn^{+3} \rightarrow 3d^4$

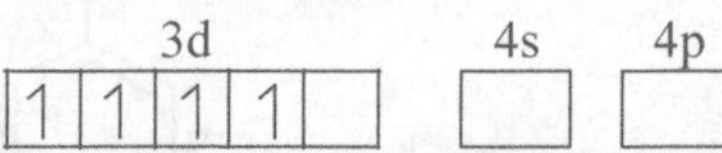

The presence of a strong field ligand CN^- causes pairing of electrons.

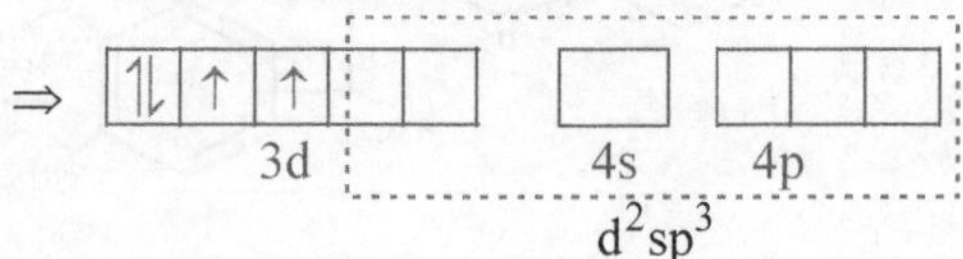

As, coordination number of Mn = 6, so it will form an octahedral complex.

$\therefore$ $[Mn(CN)_6]^{3-}$

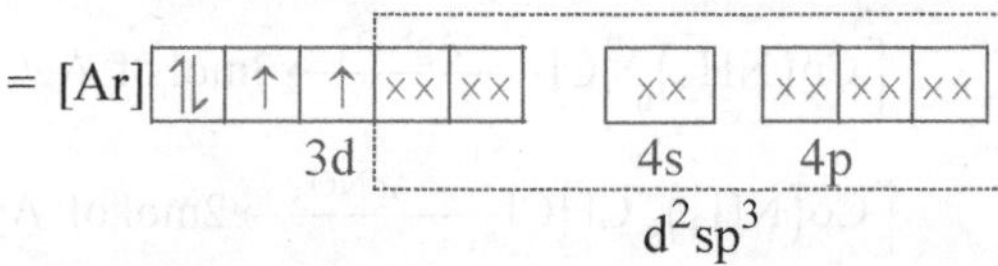

89. (2) Minimum or comparable energy gap between 5f, 6d and 7s subshell makes electron excitation easier, hence there is a greater range of oxidation states in actinoids.

90. (3) Due to denaturation of proteins, helix get uncoiled and protein loses its biological activity.

BIOLOGY

91. (1) Holoenzyme is conjugated enzyme which consists protein part called apoenzyme and a non-protein called cofactor. Coenzyme are also organic compounds but their association with apoenzyme is only transient and and as cofactors.

92. (1) A decrease in blood pressure / volume stimulates the release of renin, aldosterone, and ADH while increase in blood pressure / volume stimulates the release of Atrial Natriuretic Factor (ANF) secreted by atria of heart which cause vasodilation and also inhibits RAAS (Renin Angiotensin Aldosterone System) mechanism that decreases the blood volume/pressure.

93. (1) – Kupffer-cells are phagocytic cells present in liver.
– Zymogen cells are enzyme producing cells which are secreted by pancreas.
– Paneth cell secretes lysozyme which acts as anti-bacterial agent, into the humen of intestine.
– Argentaffin cells are hormone producing cells.

94. (3) – Nucleic acids are polymers of nucleotides
– Proteins are polymers of amino acids
–Polysaccharides are polymers of monosaccharides
–Lipids are the tri-esters of fatty acids with glycerol.

95. (2) The first cell of female gametophytic generation in angiosperm is megaspore. It undergoes three successive mitotic division to form 8-nucleated and 7-celled embryo sac.

96. (4) Myelin sheath wrapped around the nerve axon. Oligodendrocytes are neuroglial cells which produce myelin sheath in central nervous system while Schwann cell produces myelin sheath in peripheral nervous system.

97. (1) Insect pollinated plants provide rewards as edible pollen grain and nectar as usual rewards. In order to materialize and maximize pollination flowers have developed a set of attributes which are aimed at attracting the pollinators called attractants.

98. (3) Pre-synaptic membrane is involved in the release of neurotransmitter in the chemical synapse. The receptors sites for neurotransmitters are present on post-synaptic membrane of neuron.

99. (4) Coconut fruit is a drupe. A drupe is a fleshy fruit with thin skin and central stone containing the seed.

100. (4) In Human RBCs, nucleus get degenerates during maturation and it provide more space for oxygen carrying pigment (Haemoglobin). It lacks many cell organelles including mitochondria so respires anaerobically.

101. (3) Capacitation is increase in fertilising capacity of sperms which occurs in female reproductive tract. It is required to render sperm to fertilize an oocyte.

102. (4) Archaebacteria are able to survive in harsh conditions due to the presence of branched lipid chain in cell membrane that reduces fluidity of cell membrane.
It include halophiles which are exclusively found in saline habitats.

103. (1) In logistic growth curve, the curve has an upper a symptote known as carrying capacity (K) is obtained when the maximum population size is at $\frac{dN}{dt} = 0$. A population growing in a habitat with limited resources shows logistic growth curve.

For logistic growth $\frac{dN}{dt} = rN\left(\frac{K-N}{K}\right)$

If $K = N$ then $\frac{K-N}{K} = 0$

$\therefore$ the $\frac{dN}{dt} = 0$.

the population reaches asymptote.

104. (1) Artificial selection to obtain cow yielding higher milk output will shift the peak to one direction, so this represent an example of Directional selection. In stabilizing selection, the organisms with the mean value of the trait are selected. In disruptive selection, both extremes get selected.

105. (1) *Rhodospirillum* is facultative anaerobe and free living nitrogen fixer. Mycorrhiza show symbiotic relationship between fungi and roots of higher plants.

106. (1) Carotene is the source of retinal which is involved in formation of rhodopsin of rod cells. Retinal, a derivative of vitamin A, is the light-absorbing part of all visual photopigments. Photopigments are occur entirely on the surface of membrane disc.

107. (3) Ethidium bromide (Et Br) is used to stain the DNA fragments and will appear as orange coloured bands when kept under UV light.

108. (3) In hepatic portal system, hepatic portal vein drains blood to liver from intestine.

109. (2) During secondary growth in plants vascular cambium gives rise to secondary xylem and secondary phloem. Phelloderm is formed by cork cambium. Periderm is the corky outer layer formed in secondary thickening.

110. (2) Thalassemia is a quantitative problem of synthesising very few globin molecules while sickle cell anaemia is a qualitative problem of synthesising an incorrectly functioning globin.

111. (2) Husband × Wife

I^AI^B I^Ai

♀ \ ♂	I^A	I^B
I^A	I^AI^A	I^AI^B
i	I^Ai	I^Bi

Number of genotypes = 4
Number of phenotypes = 3
I^AI^A and I^Ai = A
I^AI^B = AB
I^Bi = B

112. (2) Cellulose microfibrils are arranged radially rather than longitudinally which makes easy for the stomata to open.

113. (2) Thorns are hard, pointed straight structures for protection against grazing animal. These are modified stem.

114. (4) Ex-situ conservation is offsite strategy for conservation of biological diversity in zoological park and botanical gardens respectively.

115. (4) In roots, the root hairs develops from zone of maturation. This zone is differentiated zone thus bearing root hairs.

116. (4) Down's syndrome is caused by non-disjunction of 21st chromosome i.e. Trisomy.

117. (4) By convention, the water potential of pure water at standard temperature, which is not under any pressure, is taken to be zero. $(\Psi_w = 0)$

118. (1) The correct sequence of events occur during mitosis would be as follows

(i) DNA condensation occurs so that chromosomes become visible during early to mid-prophase.

(ii) Disassembly of nuclear membrane begins at late prophase or transition to metaphase.

(iii) Chromosomes align at equator during metaphase.

(iv) Centromere division occurs during anaphase forming daughter chromosomes.

(v) During anaphase segregation also occurs in which daughter chromosomes separate and move to opposite poles.

(vi) Telophase finally leads to formation of two daughter nuclei.

119. (1) The various stages of processing that occur after the completion of fermentation or biosynthetic stage which include separation and purification of product called downstream processing.

120. (2) Corpus luteum is the temporary endocrine gland formed in the ovary after ovulation. It release hormones like progesterone, oestrogen etc.

121. (2) Cork cambium undergoes periclinal division and cuts off thick walled suberised dead cells towards outside i.e. phellem (cork) and it cuts off thin walled living cells i.e., phelloderm on inner side.

122. (1) *Volvox* is motile colonial fresh water green alga. It forms spherical colonies.

123. (4) Gonorrhoea – *Neisseria* (Bacteria)
Syphilis – *Treponema* (Bacteria)
Genital Warts – *Human papilloma virus*
AIDS – HIV (Virus)

124. (4) Cu^{2+} interfere in the sperm movement, which suppress the sperm motility and fertilising capacity of sperms.

125. (2) Primary treatment is a physical process which involves two process, i.e. filtration and sedimentation of big solid waste.

126. (2) Pharyngeal gill slits are present in hemichordates and in chordates. Notochord is present in chordates only. Ventral tubular nerve cord is present in non-chordates.

127. (1) Hershey and Chase proved that DNA as genetic material. They used bacteriophage for their experiment.

128. (1) During his experiments Mendel have taken seven characters in a pea plant. In which nature of trichromes i.e., gladular or non-glandular was not considered by Mendel.

129. (1) Halophytes growing in saline soils show vivipary for seed germination and have pneumatophores for gaseous exchange.

130. (2) Pivot joint is a type of synovial joint which provide freely movement between atlas and axis vertebrae of vertebral column.

131. (2) In C_3 plants photosynthesis decreases at higher temperature due to increased photorespiration.
C_4 plants have higher temperature optimum because of the presence of enzyme called pyruvate phosphate dikinase, which is sensitive to low temperature.

132. (1) DNA fragments are negatively charged because of presence of phosphate group.

133. (3) Sticky character of the bacterial wall is due to glycocalyx which is rich in glycoproteins.

134. (3) Rennin and Pepsin enzymes are present in the gastric juice whereas Maltase is present in the intestinal juice.

135. (2) Sharks and Trygon (sting ray) are the members of cartilaginous fish while Whale, Dolphin and Seals are aquatic mammals belong to class mammalia.

136. (2) Auxins helps to prevent premature leaf and fruit fall.

137. (3) In male frog the sperms passage is: Testes → Vasa efferentia → Kidney → Bidder's canal → Urinogenital duct → Cloaca.

138. (2) Infertility cases due to inability of the male partner to inseminate the female or due to very low sperm count in the ejaculates, could be corrected by using artificial insemination (AI) technique. In this procedure semen is injected directly into the vagina or uterus.

139. (4) Forest ecosystem has the maximum biomass. Some very high productive ecosystem are
- Tropical rain forest
- Coral reef
- Estuaries
- Sugarcane fields

140. (4) Volume of air present in lungs after forceful expiration as residual volume prevents the collapsing of alveoli.

141. (1) Thc tropical rain forest have five vertical strata on the basis of plants height i.e., ground vegetation, shrubs, short canopy trees, tall canopy trees and tall emergent trees.

142. (4) Descending limb of loop of Henle is permeable to water but impermeable to electrolytes whereas ascending limb is impermeable to water but permeable to electrolytes.

143. (2) Alexander Von Humboldt noticed that within a region species richness increases with the increase in area.

144. (3) Chlamydomonas has haplontic life cycle hence shows zygotic meiosis.

145. (2) If deletion happen at 901st position than the remaining 98 bases specifying for 33 codons of amino acids will be altered.

146. (2) Wind pollination or anemophily occurs in flowers which are having a single ovule in each ovary, and numerous flowers packed in an inflorescence. It is a non-directional pollination.

147. (1) Cell mediated immune response causes non-acceptance or rejection of graft or transplanted tissues/organs.

148. (2) *Ectocarpus* exhibits haplodiplontic life cycle while *Fucus* has diplontic life cycle.

149. (4) Selectable markers in recombinant DNA technology, helps in identification and elimination of non-transformants and selectively permits the growth of the transformants.

150. (1) Autogamy occurs in bisexual flowers. Geitonogamous flowers are unisexual but present in the same plant. Dioecious condition is observed when unisexual male and female flowers are present on different plants and it prevents both autogamy and geitonogamy.

151. (3) Krebs cycle begins with condensation of acetyl CoA (2C) with oxaloacetic acid (4C) to form citric acid (6C).

152. (1) In the mesophyll cell cytoplasm of C_4 plants like sugarcane, maize, sorghum etc. PEP is 3C compound which serves as primary CO_2 acceptor

153. (3) Two DNA polymerase molecules simultaneously work at the DNA fork, one on the leading strand and the other on the lagging strand.
DNA polymerase synthesizes each Okazaki fragment at lagging strand in 5′-3′ direction. As the replication fork opens further, new okazaki fragments appear. The first Okazaki fragment appears away from the replication fork and thus the direction of elongation would be away from replication fork.

154. (4) Ribosomal RNA (rRNA) is most abundant in animal cell. It constitutes 80% of total RNA of the cell.

155. (1) GnRH is secreted by hypothalamus which stimulates anterior pituitary gland for the secretion of gonadotropins (FSH and LH).

156. (1) DNA fragments during gel electrophoresis, separate (resolve) according to their size due to sieving effect provided by agarose gel.

157. (1) Epiphyseal plate which is responsible for

bone growth close after adolescence so hypersecretion of growth hormone in adults does not cause further increase in height. The epiphyseal plate is a hyaline cartilage plate in the metahphysis at each end of a long bone.

158. (2) In bacteria DNA replication occurs in cytoplasm prior to fission. Prokaryotes due to their primitive nature do not show well marked S-phase

159. (4) Mendel conducted hybridization experiments for 7 years on Pea plant between 1856 to 1863 and his data was published in 1865.

160. (3) Viroids in nature are sub-viral agents as infectious RNA particles, without protein coat.

161. (4) MALT or Mucosa Associated Lymphoid Tissue constitutes about 50 percent of the lymphoid tissue in human body. It is scattered along mucosal lining in the human body.

162. (3) *Saccharomyces cerevisiae* commonly know as Brewer's yeast, causes fermentation of carbohydrates and produces ethanol.

163. (2) *Mycoplasmas* are smallest, prokaryotes lacking cell wall and are pleomorphic in nature. These are pathogenic to both plants and animals.

164. (1) Horse belongs to
Orders - Perissodactyla
Family - Equidae
Genus - *Equus*
Species - *ferus*
Subspecies - *caballus*

165. (3) The vertebrates process myogenic heart which is self contractile system or autoexcitable; it will thus keep working outside the body for some time.

166. (4) Inbreeding increases homozygosity. So, mating of the related individuals of same breed will give homozygous purelines.

167. (2) Heartwood is inactive physiologically due to deposition of organic compounds and formation of tyloses so it will not conduct water and minerals.

168. (2) During anaphase, Anaphase Promoting Complex (APC) is a protein necessary for separation of daughter chromosomes. A defective APC will cause the chromosomes fail to segregate during anaphase.

169. (3) The site of aerobic oxidation of carbohydrates in cells to generate ATP are mitochondria.

170. (3) Mycorrhizae is a symbiotic association between fungi and roots of higher plants.

171. (4) In human, 12 pairs of ribs are present out of which 7 pairs of ribs (1st to 7th pair) are dorsally attached to vertebral column and ventrally to the sternum.

172. (2) In poriferans (sponges) choanocytes (collar cells) form lining of spongocoel . Flagella present in collar cells provide circulation to water in water canal system.

173. (2) Aerosols through its direct or indirect effects on plants can cause various problems in agriculture. However, continuous increase in air pollution may represent a threat to agriculture in the future that is persistent and largely irreversible.

174. (2) Total number of teeth in a human child is 20. In primary dentition premolars are absent.

175. (4) Pinus is monoecious plant comprising of both male and female cones on same plant.

176. (2) The edible part of coconut is its endosperm. Coconut has double endosperm, liquid endosperm and cellular.

177. (3) Double fertilization is a unique feature exhibited only by angiosperms. It involves both syngamy and triple fusion.

178. (3) In eukaryotes spliceosomes are used in removal of introns during post-transcriptional processing of hnRNA. They are absent in prokaryotes.

179. (2) The association of H1 protein indicates the complete formation of nucleosome which requires DNA condensation. The DNA is therefore in condensed form.

180. (4) Biosphere reserve is a protected area for wildlife with multipurpose activities.
It has three zones
(a) Core zone – does not have any human interference
(b) Buffer zone – has limited human activity
(c) Transition zone – allows human settlement, grazing cultivation, etc.

NEET/ AIPMT 2016

(Held On : May 1, 2016)

PHYSICS

1. From a disc of radius R and mass M, a circular hole of diameter R, whose rim passes through the centre is cut. What is the moment of inertia of the remaining part of the disc about a perpendicular axis, passing through the centre ?
(1) $15\,MR^2/32$ (2) $13\,MR^2/32$
(3) $11\,MR^2/32$ (4) $9\,MR^2/32$

2. A square loop ABCD carrying a current i, is placed near and coplanar with a long straight conductor XY carrying a current I, the net force on the loop will be :

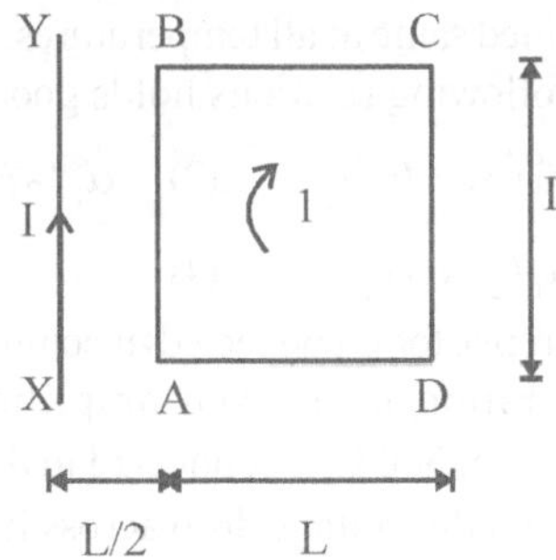

(1) $\frac{2\mu_0 Ii}{3\pi}$ (2) $\frac{\mu_0 Ii}{2\pi}$
(3) $\frac{2\mu_0 IiL}{3\pi}$ (4) $\frac{\mu_0 IiL}{2\pi}$

3. The magnetic susceptibility is negative for :
(1) diamagnetic material only
(2) paramagnetic material only
(3) ferromagnetic material only
(4) paramagnetic and ferromagnetic materials

4. A siren emitting a sound of frequency 800 Hz moves away from an observer towards a cliff at a speed of 15ms^{-1}. Then, the frequency of sound that the observer hears in the echo reflected from the cliff is :
(Take velocity of sound in air = 330 ms^{-1})
(1) 765 Hz (2) 800 Hz
(3) 838 Hz (4) 885 Hz

5.

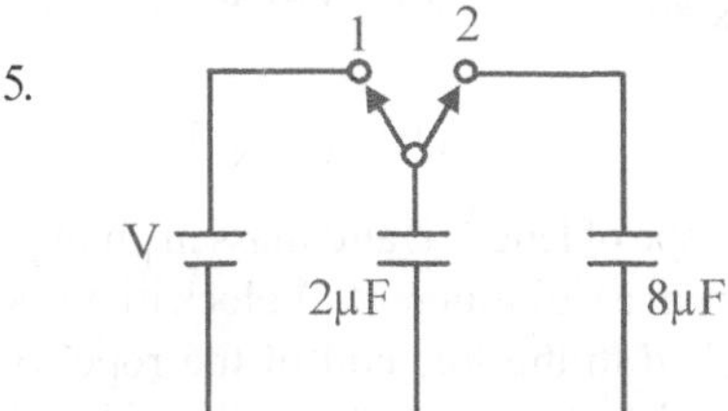

A capacitor of 2μF is charged as shown in the diagram. When the switch S is turned to position 2, the percentage fo its stored energy dissipated is :
(1) 0% (2) 20%
(3) 75% (4) 80%

6. In a diffraction pattern due to a single slit of width 'a', the first minimum is observed at an angle 30° when light of wavelength 5000 Å is incident on the slit. The first secondary maximum is observed at an angle of :
(1) $\sin^{-1}\left(\frac{1}{4}\right)$ (2) $\sin^{-1}\left(\frac{2}{3}\right)$
(3) $\sin^{-1}\left(\frac{1}{2}\right)$ (4) $\sin^{-1}\left(\frac{3}{4}\right)$

7. At what height from the surface of earth the gravitation potential and the value of g are $-5.4 \times 10^7\text{ J kg}^{-2}$ and 6.0 ms^{-2} respectively ?
Take the radius of earth as 6400 km :
(1) 2600 km (2) 1600 km
(3) 1400 km (4) 2000 km

8. Out of the following options which one can be used to produce a propagating electromagnetic wave ?
(1) A charge moving at constant velocity
(2) A stationary charge
(3) A chargeless particle
(4) An accelerating charge

9. Two identical charged spheres suspended from a common point by two massless strings of lengths l, are initially at a distance d (d << l) apart because of their mutual repulsion. The charges begin to leak from both the spheres at a constant rate. As a result, the spheres approach each other with a velocity v. Then v varies as a function of the distance x between the spheres, as :

(1) $v \propto x^{\frac{1}{2}}$ (2) $v \propto x$

(3) $v \propto x^{-\frac{1}{2}}$ (4) $v \propto x^{-1}$

10. A uniform rope of length L and mass m_1 hangs vertically from a rigid support. A block of mass m_2 is attached to the free end of the rope. A transverse pulse of wavelength λ_1 is produced at the lower end of the rope. The wavelength of the pulse when it reaches the top of the rope is λ_2 the ratio λ_2/λ_1 is

(1) $\sqrt{\frac{m_1}{m_2}}$ (2) $\sqrt{\frac{m_1+m_2}{m_2}}$

(3) $\sqrt{\frac{m_2}{m_1}}$ (4) $\sqrt{\frac{m_1+m_2}{m_1}}$

11. A refrigerator works between 4°C and 30°C. It is required to remove 600 calories of heat every second in order to keep the temperature of the refrigerated space constant. The power required is:

(Take 1 cal = 4.2 joules)

(1) 2.365 W (2) 23.65 W

(3) 236.5 W (4) 2365 W

12. An air column, closed at one end and open at the other, resonates with a tuning fork when the smallest length of the column is 50 cm. The next larger length of the column resonating with the same tuning fork is :

(1) 66.7 cm (2) 100 cm

(3) 150 cm (4) 200 cm

13. Consider the junction diode as ideal. The value of current flowing through AB is :

A 1kΩ B

+4V –6V

(1) 0 A (2) 10^{-2} A

(3) 10^{-1} A (4) 10^{-3} A

14. The charge flowing through a resistance R varies with time t as $Q = at - bt^2$, where a and b are positive constants. The total heat produced in R is:

(1) $\frac{a^3R}{6b}$ (2) $\frac{a^3R}{3b}$

(3) $\frac{a^3R}{2b}$ (4) $\frac{a^3R}{b}$

15. A black body is at a temperature of 5760 K. The energy of radiation emitted by the body at wavelength 250 nm is U_1, at wavelength 500 nm is U_2 and that at 1000 nm is U_3. Wien's constant, $b = 2.88 \times 10^6$ nmK. Which of the following is correct ?

(1) $U_1 = 0$ (2) $U_3 = 0$

(3) $U_1 > U_2$ (4) $U_2 > U_1$

16. Coefficient of linear expansion of brass and steel rods are α_1 and α_2. Lengths of brass and steel rods are ℓ_1 and ℓ_2 respectively. If $(\ell_2 - \ell_1)$ is maintained same at all temperatures, which one of the following relations holds good ?

(1) $\alpha_1 \ell_{2_2} = \alpha_2 \ell_{1_2}$ (2) $\alpha_1 \ell_2 = \alpha_2 \ell_1$

(3) $\alpha_1 \ell_2 = \alpha_2 \ell_1$ (4) $\alpha_1 \ell_1 = \alpha_2 \ell_2$

17. A npn transistor is connected in common emitter configuration in a given amplifier. A load resistance of 800 Ω is connected in the collector circuit and the voltage drop across it is 0.8 V. If the current amplification factor is 0.96 and the input resistance of the circuit is 192Ω, the voltage gain and the power gain of the amplifier will respectively be :

(1) 4, 3.84 (2) 3.69, 3.84

(3) 4, 4 (4) 4, 3.69

18. The intensity at the maximum in a Young's double slit experiment is I_0. Distance between two slits is $d = 5\lambda$, where λ is the wavelength of light used in the experiment. What will be the intensity in front of one of the slits on the screen placed at a distance D = 10 d ?

(1) I_0 (2) $\frac{I_0}{4}$

(3) $\frac{3}{4}I_0$ (4) $\frac{I_0}{2}$

19. A uniform circular disc of radius 50 cm at rest is free to turn about an axis which is perpendicular to its plane and passes through its centre. It is subjected to a torque which produces a constant angular acceleration of 2.0 rad s^{-2}. its net acceleration in ms^{-2} at the end of 2.0 s is approximately:
(1) 8.0 (2) 7.0
(3) 6.0 (4) 3.0

20. An electron of mass m and a photon have same energy E. The ratio of de-Broglie wavelengths associated with them is :
(1) $\frac{1}{2}\left(\frac{E}{2m}\right)^{\frac{1}{2}}$ (2) $\left(\frac{E}{2m}\right)^{\frac{1}{2}}$
(3) $c(2mE)^{\frac{1}{2}}$ (4) $\frac{1}{xc}\left(\frac{2m}{E}\right)^{\frac{1}{2}}$

21. A disk and a sphere of same radius but different masses roll off on two inclined planes of the same altitude and length. Which one of the two objects gets to the bottom of the plane first ?
(1) Disk
(2) Sphere
(3) Both reach at the same time
(4) Depends on their masses

22. The angle of incidence for a ray of light at a refracting surface of a prism is 45°. The angle of prism is 60°. If the ray suffers minimum deviation through the prism, the angle of minimum deviation and refractive index of the material of the prism respectively, are :
(1) $45°, \frac{1}{\sqrt{2}}$ (2) $30°, \sqrt{2}$
(3) $45°, \sqrt{2}$ (4) $30°, \frac{1}{\sqrt{2}}$

23. When an α-particle of mass 'm' moving with velocity 'v' bombards on a heavy nucleus of charge 'Ze', its distance of closest approach from the nucleus depends on m as :
(1) $\frac{1}{m}$ (2) $\frac{1}{\sqrt{m}}$
(3) $\frac{1}{m^2}$ (4) m

24. A particle of mass 10 g moves along a circle of radius 6.4 cm with a constant tangential acceleration. What is the magnitude of this acceleration if the kinetic energy of the particle becomes equal to 8×10^{-4} J by the end of the second revolution after the beginning of the motion ?
(1) 0.1 m/s^2 (2) 0.15 m/s^2
(3) 0.18 m/s^2 (4) 0.2 m/s^2

25. The molecules of a given mass of a gas have r.m.s. velocity of 200 ms^{-1} at 27°C and 1.0×10^5 Nm^{-2} pressure. When the temperature and pressure of the gas are respectively, 127°C and 0.05×10^5 Nm^{-2}, the r.m.s. velocity of its molecules in ms^{-1} is :
(1) $100\sqrt{2}$ (2) $\frac{400}{\sqrt{3}}$
(3) $\frac{100\sqrt{2}}{3}$ (4) $\frac{100}{3}$

26. A long straight wire of radius a carries a steady current I. The current is uniformly distributed over its cross-section. The ratio of the magnetic fields B and B', at radial distances $\frac{a}{2}$ and 2a respectively, from the axis of the wire is :
(1) $\frac{1}{4}$ (2) $\frac{1}{2}$
(3) 1 (4) 4

27. A particle moves so that its position vector is given by $\vec{r} = \cos\omega t\hat{x} + \sin\omega t\hat{y}$. Where ω is a constant. Which of the following is true ?
(1) Velocity and acceleration both are perpendicular to $\vec{r}$
(2) Velocity and acceleration both are parallel to $\vec{r}$
(3) Velocity is perpendicular to $\vec{r}$ and acceleration is directed towards the origin
(4) Velocity is perpendicular to $\vec{r}$ and acceleration is directed away from the origin

28. What is the minimum velocity with which a body of mass m must enter a vertical loop of radius R so that it can complete the loop ?
(1) $\sqrt{gR}$ (2) $\sqrt{2gR}$
(3) $\sqrt{3gR}$ (4) $\sqrt{5gR}$

29. When a metallic surface is illuminated with radiation of wavelength λ, the stopping potential is V. If the same surface is illuminated with radiation of wavelength 2λ, the stopping potential is $\frac{V}{4}$. The threshold wavelength for the metallic surface is :
(1) 4λ (2) 5λ
(3) $\frac{5}{2}\lambda$ (4) 3λ

30. A gas is compressed isothermally to half its initial volume. The same gas is compressed separately through an adiabatic process until its volume is again reduced to half. Then :-
(1) Compressing the gas isothermally will require more work to be done.
(2) Compressing the gas through adiabatic process will require more work to be done.
(3) Compressing the gas isothermally or adiabatically will require the same amount of work.
(4) Which of the case (whether compression through isothermal or through adiabatic process) requires more work will depend upon the atomicity of the gas.

31. A potentiometer wire is 100 cm long and a constant potential difference is maintained across it. Two cells are connected in series first to support one another and then in opposite direction. The balance points are obtained at 50 cm and 10 cm from the positive end of the wire in the two cases. The ratio of emf's is :-
(1) 5 : 1 (2) 5 : 4
(3) 3 : 4 (4) 3 : 2

32. A astronomical telescope has objective and eyepiece of focal lengths 40 cm and 4 cm respectively. To view an object 200 cm away from the objective, the lenses must be separated by a distance :-
(1) 37.3 cm (2) 46.0 cm
(3) 50.0 cm (4) 54.0 cm

33. Two non-mixing liquids of densities ρ and $n\rho$ $(n > 1)$ are put in a container. The height of each liquid is h. A solid cylinder of length L and density d is put in this container. The cylinder floats with its axis vertical and length $pL (p < 1)$ in the denser liquid. The density d is equal to :-
(1) $\{1+(n+1)p\}\rho$ (2) $\{2+(n+1)p\}\rho$
(3) $\{2+(n-1)p\}\rho$ (4) $\{1+(n-1)p\}\rho$

34. To get output 1 for the following circuit, the correct choice for the input is

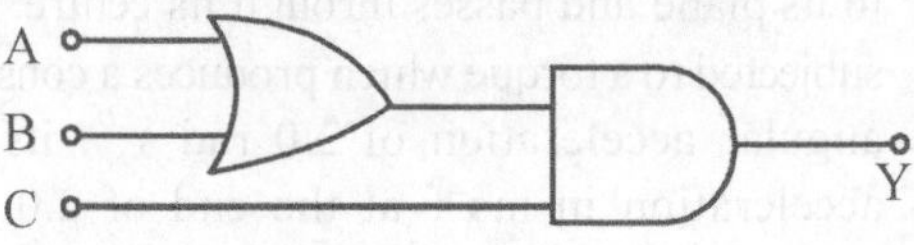

(1) A = 0, B = 1, C = 0
(2) A = 1, B = 0, C = 0
(3) A = 1, B = 1, C = 0
(4) A = 1, B = 0, C = 1

35. A piece of ice falls from a height h so that it melts completely. Only one-quarter of the heat produced is absorbed by the ice and all energy of ice gets converted into heat during its fall. The value of h is :
[Latent heat of ice is 3.4×10^5 J/kg and g = 10 N/kg]
(1) 34 km (2) 544 km
(3) 136 km (4) 68 km

36. The ratio of escape velocity at earth (v_e) to the escape velocity at a planet (v_p) whose radius and mean density are twice as that of earth is :-
(1) 1 : 2 (2) $1 : 2\sqrt{2}$
(3) 1 : 4 (4) 1 : 2

37. If the magnitude of sum of two vectors is equal to the magnitude of difference of the two vectors, the angle between these vectors is :-
(1) 0° (2) 90°
(3) 45° (4) 180°

38. Given the value of Rydberg constant is $10^7 m^{-1}$, the wave number of the last line of the Balmer series in hydrogen spectrum will be :-
(1) $0.025 \times 10^4 m^{-1}$ (2) $0.5 \times 10^7 m^{-1}$
(3) $0.25 \times 10^7 m^{-1}$ (4) $2.5 \times 10^7 m^{-1}$

39. A body of mass 1 kg begins to move under the action of a time dependent force $\vec{F}=(2t\hat{i}+3t^2\hat{j})$ N, where $\hat{i}$ and $\hat{j}$ are unit vectors alogn x and y axis. What power will be developed by the force at the time t?
(1) $(2t^2 + 3t^3)$W (2) $(2t^2 + 4t^4)$W
(3) $(2t^3 + 3t^4)$ W (4) $(2t^3 + 3t^5)$W

40. An inductor 20 mH, a capacitor 50 μF and a resistor 40Ω are connected in series across a source of emf V = 10 sin 340 t. The power loss in A.C. circuit is :-
(1) 0.51 W (2) 0.67 W
(3) 0.76 W (4) 0.89 W

41. If the velocity of a particle is $v = At + Bt^2$, where A and B are constants, then the distance travelled by it between 1s and 2s is :-
(1) $\frac{3}{2}A + 4B$ (2) $3A + 7B$
(3) $\frac{3}{2}A + \frac{7}{3}B$ (4) $\frac{A}{2} + \frac{B}{3}$

42. A long solenoid has 1000 turns. When a current of 4A flows through it, the magnetic flux linked with each turn of the solenoid is 4×10^{-3} Wb. The self inductance of the solenoid is :
(1) 4H (2) 3H
(3) 2H (4) 1H

43. A small signal voltage $V(t) = V_0 \sin \omega t$ is applied across an ideal capacitor C :-
(1) Current I (t), lags voltage V(t) by 90°.
(2) Over a full cycle the capacitor C does not consume any energy from the voltage source.
(3) Current I (t) is in phase with voltage V(t).
(4) Current I (t) leads voltage V(t) by 180°.

44. Match the corresponding entries of column-1 with column-2 (Where m is the magnification produced by the mirror):

	Column-1		Column-2
(A)	$m = -2$	(a)	Convex mirror
(B)	$m = -\frac{1}{2}$	(b)	Concave mirror
(C)	$m = +2$	(c)	Real image
(D)	$m = +\frac{1}{2}$	(d)	Virtual image

(1) A → b and c, B → b and c, C → b and d, D → a and d.
(2) A → a and c, B → a and d, C → a and b, D → c and d
(3) A → a and d, B → b and c, C → b and d, D → b and c
(4) A → c and d, B → b and d, C → b and c, D → a and d

45. A car is negotiating a curved road of radius R. The road is banked at an angle θ. the coefficient of friction between the tyres of the car and the road is μ_s. The maximum safe velocity on this road is :
(1) $\sqrt{gR^2 \frac{\mu_s + \tan\theta}{1 - \mu_s \tan\theta}}$
(2) $\sqrt{gR \frac{\mu_s + \tan\theta}{1 - \mu_s \tan\theta}}$
(3) $\sqrt{\frac{g}{R} \frac{\mu_s + \tan\theta}{1 - \mu_2 \tan\theta}}$
(4) $\sqrt{\frac{g}{R^2} \frac{\mu_s + \tan\theta}{1 - \mu_s \tan\theta}}$

CHEMISTRY

46. Consider the molecules CH_4, NH_3 and H_2O. Which of the given statements is false ?
(1) The H–C–H bond angle in CH_4, the H–N–H bond angle in NH_3, and the H–O–H bond angle in H_2O are all greater than 90°
(2) The H–O–H bond angle in H_2O is larger than the H–C–H bond angle in CH_4.
(3) The H–O–H bond angle in H_2O is smaller than the H–N–H bond angle in NH_3.
(4) The H–C–H bond angle in CH_4 is larger than the H–N–H bond angle in NH_3.

47. In the reaction

$$H\text{–}C \equiv CH \xrightarrow[(2)CH_3CH_2Br]{(1)NaNH_2 / liq.NH_3} X \xrightarrow[(2)CH_3CH_2Br]{(1)NaNH_2 / liq.NH_3} Y$$

X and Y are
(1) X = 1-Butyne ; Y = 3-Hexyne
(2) X = 2-Butyne ; Y = 3-Hexyne
(3) X = 2-Butyne ; Y = 2-Hexyne
(4) X = 1-Butyne ; Y = 2-Hexyne

48. Among the following, the correct order of acidity is
(1) $HClO_3 < HClO_4 < HClO_2 < HClO$
(2) $HClO < HClO_2 < HClO_3 < HClO_4$
(3) $HClO_2 < HClO < HClO_3 < HClO_4$
(4) $HClO_4 < HClO_2 < HClO < HClO_3$

49. The rate of a first -order reaction is 0.04 mol $\ell^{-1}s^{-1}$ at 10 seconds and 0.03 mol $\ell^{-1}s^{-1}$ at 20 seconds after initiation of the reaction. The half-life period of the reaction is
(1) 24.1 s (2) 34.1 s
(3) 44.1 s (4) 54.1 s

50. Which one of the following characteristics is associated with adsorption ?
(1) ΔG is negative but ΔH and ΔS are positive
(2) ΔG, ΔH and ΔS all are negative
(3) ΔG and ΔH are negative but ΔS is positive
(4) ΔG and ΔS are negative but ΔH is positive

51. In which of the following options the order of arrangement does not agree with the variation of property indicated against it ?
(1) $Al^{3+} < Mg^{2+} < Na^+ < F^-$ (increasing ionic size)
(2) B < C < N < O (increasing first ionisation enthalpy)
(3) I < Br < Cl < F (increasing electron gain enthalpy)
(4) Li < Na < K < Rb (increasing metallic radius)

52. Which of the following statements is false ?
(1) Mg^{2+} ions form a complex with ATP
(2) Ca^{2+} ions are important in blood clotting
(3) Ca^{2+} ions are not important in maintaining the regular beating of the heart.
(4) Mg^{2+} ions are important in the green parts of plants.

53. Which of the following statements about hydrogen is incorrect ?
(1) hydrogen has three isotopes of which tritium is the most common.
(2) Hydrogen never acts as cation in ionic salts
(3) Hydronium ion, H_3O^+ exists freely in solution
(4) Dihydrogen does not act as a reducing agent

54. The correct statement regarding a carbonyl compound with a hydrogen atom on its alphacarbon, is :
(1) a carbonyl compound with a hydrogen atom on its alpha-carbon never equilibrates with its corresponding enol.
(2) a carbonyl compound with a hydrgen atom on its alpha-carbon rapidly equilibrates with its corresponding enol and this process is known as aldehyde-ketone equilibration.
(3) a carbonyl compound with a hydrogen atom on its alpha-carbon rapidly equilibrates with its corresponding enol and this process is known as carbonylation.
(4) a carbonyl compound with a hydrogen atom on its alpha-carbon rapidly equilibrates with its corresponding enol and this process is known as keto-enol tautomerism.

55. MY and NY_3, two nearly insoluble salts, have the same K_{sp} values of 6.2×10^{-13} at room temperature. Which statement would be true in regard to MY and NY_3 ?
(1) The molar solubilities of MY and NY_3 in water are identical.
(2) The molar solubility of MY in water is less than that of NY_3
(3) The salts MY and NY_3 are more soluble in 0.5 M KY than in pure water.
(4) The addition of the salt of KY to solution of MY and NY_3 will have no effect on their solubilities.

56. In a protein molecule various amino acids are linked together by
(1) α-glycosidic bond
(2) β-glycosidic bond
(3) peptide bond
(4) dative bond

57. Natural rubber has
(1) all cis-configuration
(2) all trans-configuration
(3) alternate cis-and trans-configuration
(4) random cis-and trans-configuration

58. Match items of Column I with the items of Column II and asign the correct code :

	Column-I		Column-II
(a)	Cyanide process	(i)	Ultrapure Ge
(b)	Froth floatation process	(ii)	Dressing of ZnS
(c)	Electrolytic reduction	(iii)	Extraction of Al
(d)	Zone refining	(iv)	Extraction of Au
		(v)	Purification of Ni

Code :

	(a)	(b)	(c)	(d)
(1)	(iv)	(ii)	(iii)	(i)
(2)	(ii)	(iii)	(i)	(v)
(3)	(i)	(ii)	(iii)	(iv)
(4)	(iii)	(iv)	(v)	(i)

59. Which one of the following statements is correct when SO_2 is passed through acidified $K_2Cr_2O_7$ solution ?
(1) The solution turns blue
(2) The solution is decolourized
(3) SO_2 is reduced
(4) Green $Cr_2(SO_4)_3$ is formed

60. The electronic configurations of Eu(Atomic No. 63), Gd(Atomic No. 64) and Tb (Atomic No. 65) are
(1) $[Xe]4f^76s^2$, $[Xe]4f^8 6s^2$ and $[Xe]4f^85d^16s^2$
(2) $[Xe]4f^75d^16s^2$, $[Xe]4f^7 5d^1 6s^2$ and $[Xe]4f^96s^2$
(3) $[Xe]4f^65d^16s^2$, $[Xe]4f^75d^16s^2$ and $[Xe]4f^85d^16s^2$
(4) $[Xe]4f^76s^2$, $[Xe]4f^75d^16s^2$ and $[Xe]4f^96s^2$

61. Two electrons occupying the same orbital are distinguished by
(1) Principal quantum number
(2) Magnetic quantum number
(3) Azimuthal quantum number
(4) Spin quantum number

62. Which copper is heated with conc. HNO_3 it produces
(1) $Cu(NO_3)_2$ and NO_2
(2) $Cu(NO_3)_2$ and NO
(3) $Cu(NO_3)_2$, NO and NO_2
(4) $Cu(NO_3)_2$ and N_2O

63. Which of the following reagents would distinguish cis-cyclopenta-1,2-diol from the trans-isomer?
(1) Acetone
(2) Ozone
(3) MnO_2
(4) Aluminium isopropxide

64. The correct thermodynamic conditions for the spontaneous reaction at all temperatures is
(1) $\Delta H < 0$ and $\Delta S = 0$
(2) $\Delta H > 0$ and $\Delta S < 0$
(3) $\Delta H < 0$ and $\Delta S > 0$
(4) $\Delta H < 0$ and $\Delta S < 0$

65. Lithium has a bcc structure. Its density is 530 kg m^{-3} and its atomic mass is 6.94 g mol^{-1}. Calculate the edge length of a unit cell of Lithium metal. ($N_A = 6.02 \times 10^{23}\ mol^{-1}$)
(1) 154 pm (2) 352 pm
(3) 527 pm (4) 264 pm

66. Which one of the following orders is correct for the bond dissociation enthalpy of halogen molecules?
(1) $I_2 > Br_2 > Cl_2 > F_2$
(2) $Cl_2 > Br_2 > F_2 > I_2$
(3) $Br_2 > I_2 > F_2 > Cl_2$
(4) $F_2 > Cl_2 > Br_2 > I_2$

67. Which of the following is an analgesic ?
(1) Novalgin (2) Penicillin
(3) Streptomycin (4) Chloromycetin

68. Equal moles of hydrogen and oxygen gases are placed in a container with a pin-hole through which both can escape. What fraction of the oxygen escapes in the time required for one-half of the hydrogen to escape ?
(1) 1/8 (2) 1/4
(3) 3/8 (4) 1/2

69. Consider the nitration of benzene using mixed conc of H_2SO_4 and HNO_3. If a large amount of $KHSO_4$ is added to the mixture, the rate of nitration will be
(1) faster (2) slower
(3) unchanged (4) doubled

70. Predict the correct order among the following :-
(1) lone pair- lone pair > lone pair - bond pair > bond pair - bond pair
(2) lone pair - lone pair > bond pair - bond pair > lone pair - bond pair
(3) bond pair - bond pair > lone pair - bond pair > lone pair - lone pair
(4) lone pair - bond pair > bond pair - bond pair > lone pair - lone pair

71. The product obtained as a result of a reaction of nitrogen with CaC_2 is
(1) $Ca(CN)_2$ (2) CaCN
(3) $CaCN_3$ (4) Ca_2CN

72. Consider the following liquid - vapour equilibrium.

Liquid $\rightleftharpoons$ Vapour

Which of the following relations is correct ?

(1) $\frac{d\ell nG}{dT^2} = \frac{\Delta H_v}{RT^2}$ (2) $\frac{d\ell nP}{dT} = \frac{\Delta H_v}{RT}$

(3) $\frac{d\ell nP}{dT^2} = \frac{-\Delta H_v}{T^2}$ (4) $\frac{d\ell nP}{dT} = \frac{-\Delta H_v}{RT^2}$

73. Match the compounds given in column I with the hybridisation and shape given in column II and mark the correct option.

	Column-I		Column-II
(a)	XeF_6	(i)	Distorted octahedral
(b)	XeO_3	(ii)	Square planar
(c)	$XeOF_4$	(iii)	Pyramidal
(d)	XeF_4	(iv)	Square pyramidal

Code :

	(a)	(b)	(c)	(d)
(1)	(i)	(iii)	(iv)	(ii)
(2)	(i)	(ii)	(iv)	(iii)
(3)	(iv)	(iii)	(i)	(ii)
(4)	(iv)	(i)	(ii)	(iii)

74. Which of the following has longest C–O bond length? (Free C–O bond length in CO is 1.128Å)

(1) $Ni(CO)_4$ (2) $[Co(CO)_4]^-$

(3) $[Fe(CO)_4]^{2-}$ (4) $[Mn(CO)_6]^+$

75. The pressure of H_2 required to make the potential of H_2-electrode zero in pure water at 298 K is :

(1) 10^{-14} atm (2) 10^{-12} atm

(3) 10^{-10} atm (4) 10^{-4} atm

76. The addition of a catalyst during a chemical reaction alters which of the following quantities?

(1) Entropy

(2) Internal energy

(3) Enthalpy

(4) Activation energy

77. The ionic radii of A^+ and B^- ions are 0.98×10^{-10} m and 1.81×10^{-10} m. The coordination number of each ion in AB is

(1) 6 (2) 4

(3) 8 (4) 2

78. Which is the correct statement for the given acids?

(1) Phosphinic acid is a diprotic acid while phosphonic acid is a monoprotic acid

(2) Phosphinic acid is a monoprotic acid while phosphonic acid is a diprotic acid

(3) Both are triprotic acids

(4) Both are diprotic acids

79. Fog is colloidal solution of

(1) Liquid in gas (2) Gas in liquid

(3) Solid in gas (4) Gas in gas

80. Which of the following statement about the composition of the vapour over an ideal 1 : 1 molar mixture of benzene and toluene is correct? Assume that the temperature is constant at 25°C. (**Given :** Vapour Pressure Data at 25°C, benzene = 12.8 kPa, toluene = 3.85 kPa)

(1) The vapour will contain a higher percentage of benzene

(2) The vapour will contain a higher percentage of toluene

(3) The vapour will contain equal amounts of benezene and toluene

(4) Not enough information is given to make a predication

81. The correct statement regarding the comparison of staggered and eclipsed conformation of ethane, is

(1) The staggered conformation of ethane is less stable than eclipsed conformation, because staggered conformation has torsional strain

(2) The eclipsed conformation of ethane is more stable than staggered conformation, because eclipsed conformation has no torsional strain

(3) The eclipsed conformation of ethane is more stable than staggered conformation even though the eclipsed conformation has torsional strain

(4) The staggered conformation of ethane is more stable than eclipsed conformation, because staggered conformation has no torsional strain.

82. The reaction

Cyclopentyl–OH $\xrightarrow{NaOH}$ Cyclopentyl–$O^-\overset{\oplus}{Na}$ $\xrightarrow{Me-I}$ Cyclopentyl–O–Me

Can be classified as :-

(1) Williamson ether synthesis reaction

(2) Alcohol formation reaction

(3) Dehydration reaction

(4) Williamson alcohol synthesis reaction

83. The product formed by the reaction of an aldehyde with a primary amine is

(1) Schiff base (2) Ketone

(3) Carboxylic acid (4) Aromatic acid

84. Which of the following biphenyls is optically active ?

(1) O_2N I

(2) Br Br I I

(3) I I

(4) CH_3 CH_3

85. For the following reactions :-

(a) $CH_3CH_2CH_2Br + KOH \rightarrow CH_3CH{=}CH_2 + KBr + H_2O$

(b) H_3C CH_3 Br + KOH $\rightarrow$ H_3C CH_3 OH + KBr

(c) + Br_2 $\rightarrow$ Br Br

Which of the following statements is correct ?

(1) (a) and (b) are elimination reaction and (c) is addition reaction

(2) (a) is elimination, (b) is substitution and (c) is addition reaction

(3) (a) is elimination, (b) and (c) are substitution reactions

(4) (a) is substitution, (b) and (c) are addition reaction

86. At 100°C the vapour pressure of a solution of 6.5g of a solute in 100 g water is 732 mm. If K_b = 0.52, the boiling point of this solution will be

(1) 101°C (2) 100°C

(3) 102°C (4) 103°C

87. The correct statement regarding RNA and DNA, respectively is

(1) The sugar component in RNA is arabinose and the sugar component in DNA is 2'-deoxyribose.

(2) The sugar component in RNA is ribose and the sugar component in DNA is 2'-deoxyribose.

(3) The sugar component in RNA is arabinose

(4) The sugar component in RNA is 2'-deoxyribose and the sugar component in DNA is arabinose

88. The correct statement regarding the basicity of arylamines is

(1) Arylamines are generally less basic than alkylamines because the nitrogen lone-pair electrons are delocalized by interaction with the aromatic ring π electron system.

(2) Arylamines are generally more basic than alkylamines because the nitrogen lone-pair electrons are not delocalized by interaction with the aromatic ring π electron system.

(3) Arylamines are generally more basic than alkylamines because of aryl group.

(4) Arylamines are generally more basic than alkylamines, because the nitrongen atom in arylamines is sp-hybridized.

89. Which one given below is a non-reducing sugar?

(1) Maltose (2) Lactose

(3) Glucose (4) Sucrose

90. The pair of electron in the given carbanion, $CH_3C \equiv C^-$, is present in which of the following orbitals ?

(1) 2p (2) sp^3

(3) sp^2 (4) sp

BIOLOGY

91. Gause's principle of competitive exclusion states that :
    (1) more abundant species will exclude the less abundant species through competition.
    (2) competition for the same resources excludes species having different food preferences.
    (3) no two species can occupy the same niche indefinitely for the same limiting resources.
    (4) larger organisms exclude smaller ones through competition.
92. The two polypeptides of human insulin are linked together by
    (1) hydrogen bonds
    (2) phosphodiester bond
    (3) covalent bond
    (4) disulphide bridges
93. The coconut water from tender coconut represents
    (1) endocarp
    (2) fleshy mesocarp
    (3) free nuclear proembryo
    (4) free nuclear endosperm
94. Which of the following statements is wrong for viroids?
    (1) They lack a protein coat
    (2) They are smaller than viruses
    (3) They cause infections
    (4) Their RNA is of high molecular weight
95. Which of the following featrues is not present in the Phylum - Arthropoda ?
    (1) Chitinous exoskeleton
    (2) Metameric segmentation
    (3) Parapodia
    (4) Jointed appendages
96. Which of the following most appropriately describes haemophilia ?
    (1) Recessive gene disorder
    (2) X - linked recessive gene disorder
    (3) Chromosomal disorder
    (4) Dominant gene disorder
97. Emerson's enhancement effect and Red drop have been instrumental in the discovery of
    (1) photophosphorylation and non-cyclic electron transport
    (2) two photosystems operating simultaneously
    (3) photophosphorylation and cyclic electron transport
    (4) oxidative phosphorylation
98. In which of the following, all three are macronutrients?
    (1) Boron, zinc, manganese
    (2) Iron, copper, molybdenum
    (3) Molybdenum, magnesium, manganese
    (4) Nitrogen, nickel, phosphorus
99. Name the chronic respiratory disorder caused mainly by cigarette smoking
    (1) emphysema
    (2) asthma
    (3) respiratory acidosis
    (4) respiratory alkalosis
100. A system of rotating crops with legume or grass pasture to improve soil structure and fertility is called
    (1) Ley farming
    (2) Contour farming
    (3) Strip farming
    (4) Shifting agriculture
101. Mitochondria and chloroplast are
    (a) semi-autonomous organelles
    (b) formed by division of pre-existing organelles and they contain DNA but lack protein synthesizing machinery

    Which one of the following options is correct ?
    (1) Both (a) and (b) are correct
    (2) (b) is true but (a) is false
    (3) (a) is true but (b) is false
    (4) Both (a) and (b) are false
102. In context of Amniocentesis, which of the following statement is incorrect ?
    (1) It is usually done when a woman is between 14-16 weeks pregnant.
    (2) It is used for prenatal sex determination
    (3) It can be used for detection of Down syndrome
    (4) It can be used for detection of Cleft palate
103. In a chloroplast the highest number of protons are found in
    (1) stroma
    (2) lumen of thylakoids
    (3) inter membrane space
    (4) antennae complex
104. Photosensitive compound in human eye is made up of
    (1) Guanosine and Retinol
    (2) Opsin and Retinal
    (3) Opsin and Retinol
    (4) Transducin and Retinene

105. Spindle fibres attach on to
(1) Telomere of the chromosome
(2) Kinetochore of the chromosome
(3) Centromere of the chromosome
(4) Kinetosome of the chromosome

106. Which is the National Aquatic Animal of India ?
(1) Gangetic shark
(2) River dolphin
(3) Blue whale
(4) Sea-horse

107. Which of the following is required as inducer(s) for the expression of Lac operon ?
(1) Glucose
(2) Galactose
(3) Lactose
(4) Lactose and galactose

108. Which of the following pairs of hormones are not antagonistic (having opposite effects) to each other?
(1) Parathormone – Calcitonin
(2) Insulin – Glucagon
(3) Aldosterone – Atrial Natriuretic Factor
(4) Relaxin – Inhibin

109. Microtubules are the constituents of
(1) Cilia, Flagella and Peroxisomes
(2) Spindle fibres, Centrioles and Cilia
(3) Centrioles, Spindle fibres and Chromatin
(4) Centrosome, Nucleosome and Centrioles

110. A complex of ribosomes attached to a single strand of RNA is known as
(1) Polysome
(2) Polymer
(3) Polypeptide
(4) Okazaki fragment

111. Fertilization in humans is practically feasible only if
(1) the sperms are transported into vagina just after the release of ovum in fallopian tube
(2) the ovum and sperms are transported simultaneously to ampullary isthmic junction of the fallopian tube
(3) the ovum and sperms are transported simultaneously to ampullary - isthmic junction of the cervix
(4) the sperms are transported into cervix within 48 hrs of release of ovum in uterus

112. Asthma may be attributed to
(1) bacterial infection of the lungs
(2) allergic reaction of the mast cells in the lungs
(3) inflammation of the trachea
(4) accumulation of fluid in the lungs

113. The *Avena* curvature is used for bioassay of :
(1) ABA (2) GA_3
(3) IAA (4) Ethylene

114. The standard petal of a papilionaceous corolla is also called
(1) Carina (2) Pappus
(3) Vexillum (4) Corona

115. Tricarpellary syncarpous gynoecium is found in flowers of
(1) Liliaceae (2) Solanaceae
(3) Fabaceae (4) Poaceae

116. One of the major components of cell wall of most fungi is
(1) Chitin (2) Peptidoglycan
(3) Cellulose (4) Hemicellulose

117. Select the incorrect statement :
(1) FSH stimulates the sertoli cells which help in spermiogenesis
(2) LH triggers ovulation in ovary
(3) LH and FSH decrease gradually during the follicular phase
(4) LH triggers secretion of androgens from the Leydig cells

118. In meiosis crossing over is initiated at
(1) Pachytene (2) Leptotene
(3) Zygotene (4) Diplotene

119. A tall true breeding garden pea plant is crossed with a dwarf true breeding garden pea plant. When the F_1 plants were selfed the resulting genotypes were in the ratio of
(1) 1 : 2 : 1 : : Tall homozygous : Tall heterozygous : Dwarf
(2) 1 : 2 : 1 : : Tall heterozygous : Tall homozygous : Dwarf
(3) 3 : 1 : : Tall : Dwarf
(4) 3 : 1 : : Dwarf : Tall

120. Which of the following is the most important cause of animals and plants being driven to extinction ?
(1) Over - exploitation
(2) Alien species invasion
(3) Habitat loss and fragmentation
(4) Co-extinctions

121. Which of the following is a characteristic feature of cropland ecosystem ?
(1) Absence of soil organisms
(2) Least genetic diversity
(3) Absence of weeds
(4) Ecological succession

122. Changes in GnRH pulse frequency in females is controlled by circulating levels of
(1) estrogen and progesterone
(2) estrogen and inhibin
(3) progesterone only
(4) progesterone and inhibin

123. Which of the following is not a feature of the plasmids?
(1) Independent replication
(2) Circular structure
(3) Transferable
(4) Single - stranded

124. Which of the following features is not present in *Periplaneta americana* ?
(1) Schizocoelom as body cavity
(2) Indeterminate and radial cleavage during embryonic development
(3) Exoskeleton composed of N-acetylglucosamine
(4) Metamerically segmented body

125. In higher vertebrates, the immune system can distinguish self-cells and non-self. If this property is lost due to genetic abnormality and it attacks self-cells, then it leads to
(1) allergic response
(2) graft rejection
(3) auto-immune disease
(4) active immunity

126. Match the terms in Column-I with their description in Column-II and choose the correct option :

	Column-I		Column-II
(a)	Dominance	(i)	Many genes govern a single character
(b)	Codominance	(ii)	In a heterozygous organism only one allele expresses itself
(c)	Pleiotropy	(iii)	In a heterozygous organism both alleles express themselves fully
(d)	Polygenic inheritance	(iv)	A single gene influences many characters

	(a)	(b)	(c)	(d)
(1)	(ii)	(i)	(iv)	(iii)
(2)	(ii)	(iii)	(iv)	(i)
(3)	(iv)	(i)	(ii)	(iii)
(4)	(iv)	(iii)	(i)	(ii)

127. Joint Forest Management Concept was introduced in India during
(1) 1960 s (2) 1970 s
(3) 1980 s (4) 1990 s

128. Pick out the correct statements :
(a) Haemophilia is a sex-linked recessive disease
(b) Down's syndrome is due to aneuploidy
(c) Phenylketonuria is an autosomal recessive gene disorder.
(d) Sickle cell anaemia is a X-linked recessive gene disorder
(1) (a) and (d) are correct
(2) (b) and (d) are correct
(3) (a), (c) and (d) are correct
(4) (a), (b) and (c) are correct

129. Which one of the following statements is wrong?
(1) Cyanobacteria are also called blue-green algae
(2) Golden algae are also called desmids
(3) Eubacteria are also called false bacteria
(4) Phycomycetes are also called algal fungi

130. Proximal end of the filament of stamen is attached to the
(1) Anther
(2) Connective
(3) Placenta
(4) Thalamus or petal

131. Which of the following approaches does not give the defined action of contraceptive ?

(1)	Barrier methods	prevent fertilization
(2)	Intra uterine devices	Increase phagocytosis of sperms, suppress sperm motility and fertilizing capacity of sperms
(3)	Hormonal contraceptives	Prevent/retard entry of sperms, prevent ovulation and fertilization
(4)	Vasectomy	Prevents spermatogenesis

132. The taq polymerase enzyme is obtained from
(1) *Thermus aquaticus*
(2) *Thiobacillus ferroxidans*
(3) *Bacillus subtilis*
(4) *Pseudomonas putida*

133. Identify the correct statement on 'inhibin'
(1) Inhibits the secretion of LH, FSH and Prolactin.
(2) Is produced by granulose cells in ovary and inhibits the secretion of FSH.
(3) Is produced by granulose cells in ovary and inhibits the secretion of LH.
(4) Is produced by nurse cells in testes and inhibits the secretion of LH.

134. Which part of the tobacco plant is infected by *Meloidogyne incognita* ?
(1) Flower (2) Leaf
(3) Stem (4) Root

135. Antivenom injection contains preformed antibodies while polio drops that are administered into the body contain
(1) Activated pathogens
(2) Harvested antibodies
(3) Gamma globulin
(4) Attenuated pathogens

136. Which one of the following cell organelles is enclosed by a single membrane ?
(1) Mitochondria (2) Chloroplasts
(3) Lysosomes (4) Nuclei

137. Lack of relaxation between successive stimuli in sustained muscle contraction is known as
(1) Spasm (2) Fatigue
(3) Tetanus (4) Tonus

138. Which of the following is not a stem modification?
(1) Pitcher of *Nepenthes*
(2) Thorns of citrus
(3) Tendrils of cucumber
(4) Flattened structures of Opuntia

139. Water soluble pigments found in plant cell vacuoles are
(1) Xanthophylls (2) Chlorophylls
(3) Carotenoids (4) Anthocyanins

140. Select the correct statement :-
(1) Gymnosperms are both homosporous and heterosporous
(2) *Salvinia*, *Ginkgo* and *Pinus* all are gymnosperms
(3) *Sequoia* is one of the tallest trees
(4) The leaves of gymnosperms are not well adapted to extremes of climate

141. Which of the following is not required for any of the techniques of DNA fingerprinting available at present ?
(1) Polymerase chain reaction
(2) Zinc finger analysis
(3) Restriction enzymes
(4) DNA-DNA hybridization

142. Which type of tissue correctly matches with its location ?

	Tissue	Location
(1)	Smooth muscle	Wall of intestine
(2)	Areolar tissue	Tendons
(3)	Transitional epithelium	Tip nose
(4)	Cuboidal epithelium	Lining of stomach

143. A plant in your garden avoids photorespiratory losses, has improved water use efficiency shows high rates of photosynthesis at high temperatures and has improved efficiency of nitrogen utilisation. In which of the following physiological groups would you assign this plant?
(1) C_3 (2) C_4
(3) CAM (4) Nitrogen fixer

144. Which of the following structures is homologus to the wing of a bird ?
(1) Dorsal fin of a Shark
(2) Wing of a Moth
(3) Hind limb of Rabbit
(4) Flipper of Whale

145. Which of the following characteristic features always holds true for the corresponding group of animals?

(1)	Cartilaginous endoskeleton	Chondrichthyes
(2)	Viviparous	Mammalia
(3)	Possess a mouth with an upper and a lower jaw	Chordata
(4)	3 - chambered heart with one incompletely divided ventricle	Reptilia

146. Which of the following statements is not true for cancer cells in relation to mutations ?
(1) Mutations in proto-oncogenes accelerate the cell cycle.
(2) Mutations destroy telomerase inhibitor.
(3) Mutations inactive the cell control.
(4) Mutations inhibit production of telomerase.

147. The amino acid Tryptophan is the precursor for the synthesis of
(1) Melatonin and Serotonin
(2) Thyroxine and Triiodothyronine
(3) Estrogen and Progesterone
(4) Cortisol and Cortisone

148. Following are the two statements regarding the origin of life
(a) The earliest organisms that appeared on the earth were non-green and presumably anaerobes.
(b) The first autotrophic organisms were the chemoautotrophs that never released oxygen.

Of the above statements which one of the following options is correct ?

(1) (a) is correct but (b) is false.
(2) (b) is correct but (a) is false.
(3) Both (a) and (b) are correct.
(4) Both (a) and (b) are false.

149. Reduction in pH of blood will
(1) reduce the rate of heart beat.
(2) reduce the blood supply to the brain.
(3) decrease the affinity of hemoglobin with oxygen.
(4) release bicarbonate ions by the liver.

150. Analogous structures are a result of
(1) Divergent evolution
(2) Convergent evolution
(3) Shared ancestry
(4) Stabilizing selection

151. Which of the following is a restriction endonuclease?
(1) Hind II (2) Protease
(3) DNase I (4) RNase

152. The term ecosystem was coined by
(1) E.P. Odum (2) A.G. Tansley
(3) E. Haeckel (4) E. Warming

153. Which one of the following statements is wrong?
(1) Sucrose is a disaccharide.
(2) Cellulose is a polysaccharide.
(3) Uracil is a pyrimidine.
(4) Glycine is a sulphur containing amino acid.

154. In bryophytes and pteridophytes, transport of male gametes requires
(1) Wind (2) Insects
(3) Birds (4) Water

155. When does the growth rate of a population following the logistic model equal zero ? The logistic model is given as $dN/dt = rN(1–N/K)$
(1) when N/K is exactly one.
(2) when N nears the carrying capacity of the habitat.
(3) when N/K equals zero.
(4) when death rate is greater than birth rate.

156. Which one of the following statements is not true?
(1) Tapetum helps in the dehiscence of anther
(2) Exine of pollen grains is made up of sporopollenin
(3) Pollen grains of many species cause severe allergies
(4) Stored pollen in liquid nitrogen can be used in the crop breeding programmes

157. Which of the following would appear as the pioneer organisms on bare rocks?
(1) Lichens (2) Liverworts
(3) Mosses (4) Green algae

158. Which one of the following is the starter codon?
(1) AUG (2) UGA
(3) UAA (4) UAG

159. Which one of the following characteristics is not shared by birds and mammals ?
(1) Ossified endoskeleton
(2) Breathing using lungs
(3) Viviparity
(4) Warm blooded nature

160. Nomenclature is governed by certain universal rules. Which one of the following is contrary to the rules of nomenclature?
(1) Biological names can be written in any language
(2) The first word in a biological name represents the genus name, and the second is a specific epithet
(3) The names are written in Latin and are italicised
(4) When written by hand, the names are to be underlined

161. Blood pressure in the pulmonary artery is
(1) same as that in the aorta.
(2) more than that in the carotid.
(3) more than that in the pulmonary vein.
(4) less than that in the venae cavae.

162. Cotyledon of maize grain is called
(1) plumule (2) coleorhiza
(3) coleoptile (4) scutellum

163. In the stomach, gastric acid is secreted by the
(1) gastrin secreting cells
(2) parietal cells
(3) peptic cells
(4) acidic cells

164. Depletion of which gas in the atmosphere can lead to an increased incidence of skin cancers
(1) Nitrous oxide (2) Ozone
(3) Ammonia (4) Methane

165. Chrysophytes, Euglenoids, Dinoflagellates and Slime moulds are included in the kingdom
(1) Monera (2) Protista
(3) Fungi (4) Animalia

166. Water vapour comes out from the plant leaf through the stomatal opening. Through the same stomatal opening carbon dioxide diffuses into the plant during photosynthesis. Reason out the above statements using one of following options
(1) Both processes cannot happen simultaneously.
(2) Both processes can happen together because the diffusion coefficient of water and CO_2 is different.
(3) The above processes happen only during night time.
(4) One process occurs during day time, and the other at night.

167. In mammals, which blood vessel would normally carry largest amount of urea ?
(1) Renal Vein (2) Dorsal Aorta
(3) Hepatic Vein (4) Hepatic Portal Vein

168. Seed formation without fertilization in flowering plants involves the process of
(1) Sporulation
(2) Budding
(3) Somatic hybridization
(4) Apomixis

169. Which of the following is wrongly matched in the given table ?

	Microbe	Product	Application
(1)	Trichoderma polysporum	Cyclosporin A	immunosuppressive drug
(2)	Monascus purpureus	Statins	lowering of blood cholesterol
(3)	Streptococcus	Streptokinase	removal of clot from blood vessel
(4)	Clostridium butylicum	Lipase	removal of oil stains

170. In a testcross involving F_1 dihybrid flies, more parental-type offspring were produced than the recombinant-type offspring. This indicates
(1) the two genes are located on two different chromosomes.
(2) chromosomes failed to separate during meiosis.
(3) the two genes are linked and present on the same chromosome.
(4) both of the characters are controlled by more than one gene.

171. It is much easier for a small animal to run uphill than for a large animal, because
(1) it is easier to carry a small body weight.
(2) smaller animals have a higher metabolic rate.
(3) small animals have a lower O_2 requirement.
(4) the efficiency of muscles in large animals is less than in the small animals.

172. Which of the following is not a characteristic feature during mitosis in somatic cells ?
(1) Spindle fibres
(2) Disappearance of nucleolus
(3) Chromosome movement
(4) Synapsis

173. Which of the following statements is not correct?
(1) Pollen grains of many species can germinate on the stigma of a flower, but only one pollen tube of the same species grows into the style.
(2) Insects that consume pollen or nectar without bringing about pollination are called pollen/nectar robbers.
(3) Pollen germination and pollen tube growth are regulated by chemical components of pollen interacting with those of the pistil.
(4) Some reptiles have also been reported as pollinators in some plant species.

174. Specialised epidermal cells surrounding the guard cells are called
(1) Complementary cells
(2) Subsidiary cells
(3) Bulliform cells
(4) Lenticels

175. Which of the following guards the opening of hepatopancreatic duct into the duodenum ?
(1) Semilunar valve (2) Ileocaecal valve
(3) Pyloric sphincter (4) Sphincter of Oddi

176. Stems modified into flat green organs performing the functions of leaves are known as
(1) Cladodes (2) Phyllodes
(3) Phylloclades (4) Scales

177. The primitive prokaryotes responsible for the production of biogas from the dung of ruminant animals, include the
(1) Halophiles
(2) Thermoacidiophiles
(3) Methanogens
(4) Eubacteria

178. A river with an inflow of domestic sewage rich in organic waste may result in
(1) drying of the river very soon due to algal bloom.
(2) increased population of aquatic food web organisms.
(3) an increased production of fish due to biodegradable nutrients.
(4) death of fish due to lack of oxygen.

179. A cell at telophase stage is observed by a student in a plant brought from the field. He tells his teacher that this cell is not like other cells at telophase stage. There is no formation of cell plate and thus the cell is containing more number of chromosomes as compared to other dividing cells. This would result in
(1) Aneuploidy
(2) Polyploidy
(3) Somaclonal variation
(4) Polyteny

180. A typical fat molecule is made up of
(1) three glycerol molecules and one fatty acid molecule
(2) one glycerol and three fatty acid molecules
(3) one glycerol and one fatty acid molecule
(4) three glycerol and three fatty acid molecules

HINTS & SOLUTIONS

PHYSICS

1. (2) Moment of inertion of complete disc about point 'O'.

$$I_{\text{Total disc}} = \frac{MR^2}{2}$$

Mass of removed disc

$$M_{\text{Removed}} = \frac{M}{4} \text{ (Mass} \propto \text{area)}$$

Moment of inertia of removed disc about point 'O'.

I_{Removed} (about same perpendicular axis)

$= I_{an} + mx^2$

$$= \frac{M}{4}\frac{(R/2)^2}{2} + \frac{M}{4}\left(\frac{R}{2}\right)^2 = \frac{3MR^2}{32}$$

Therefore the moment of inertia of the remaining part of the disc about a perpendicular axis passing through the centre,

$$I_{\text{Remaing disc}} = I_{\text{Total}} - I_{\text{Removed}}$$

$$= \frac{MR^2}{2} - \frac{3}{32}MR^2 = \frac{13}{32}MR^2$$

2. (1) The direction of current in conductor

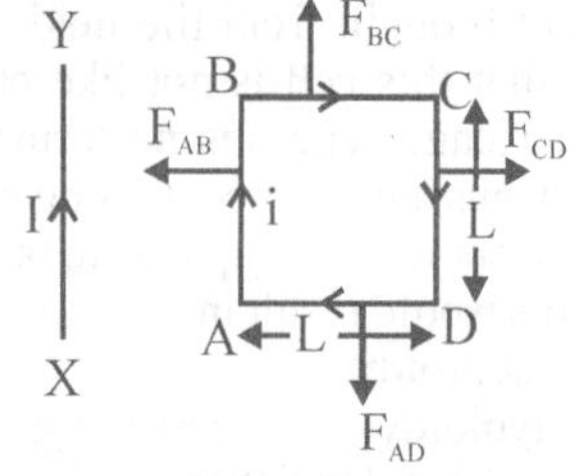

XY and AB is same

$\therefore \quad F_{AB} = i\ell B \qquad$ (attractive)

$$F_{AB} = i(L).\frac{\mu_0 I}{2\pi\left(\frac{L}{2}\right)}(\leftarrow) = \frac{\mu_0 iI}{\pi}(\leftarrow)$$

F_{BC} opposite to F_{AD}

$F_{BC}(\uparrow)$ and $F_{AD}(\downarrow)$

$\Rightarrow$ cancels each other

$F_{CD} = i\ell B$ (repulsive)

$$F_{CD} = i(L)\frac{\mu_0 I}{2\pi\left(\frac{3L}{2}\right)}(\rightarrow) = \frac{\mu_0 iI}{3\pi}(\rightarrow)$$

Therefore the net force on the loop

$$F_{net} = F_{AB} + F_{BC} + F_{CD} + F_{AD}$$

$$\Rightarrow F_{net} = \frac{\mu_0 iI}{\pi} - \frac{\mu_0 iI}{3\pi} = \frac{2\mu_0 iI}{3\pi}$$

3. (1) Magnetic susceptibility χ for dia-magnetic materials only is negative and low $|\chi| = 1$; for paramagnetic substances low but positive $|\chi| = 1$ and for ferromagnetic substances positive and high $|\chi| = 10^2$.

4. (3) According to Doppter's effect in sound, Apparent frequency,

Observer source 15 m/s

$$n' = \frac{v}{v - v_s} n_0$$

$$= \frac{330}{330-15}(800) = \frac{330 \times 800}{315} = 838 \text{ Hz}$$

The frequency of sound observer hears in the echo reflected from the cliff is 838 Hz.

5. (4) When S and 1 are connected

The 2μF capacitor gets charged. The potential difference across its plates will be V.

The potential energy stored in 2 μF capacitor

$$U_i = \frac{1}{2}CV^2 = \frac{1}{2} \times 2 \times V^2 = V^2$$

When S and 2 are connected

The 8μF capacitor also gets charged. During this charging process current flows in the wire and some amount of energy is dissipated as heat. The energy loss is

$$\Delta U = \frac{1}{2}\frac{C_1 C_2}{C_1 + C_2}(V_1 - V_2)^2$$

Here, $C_1 = 2\mu F, C_2\ 8\ \mu F, V_1 = V, V_2 = 0$

$$\therefore\ \Delta U = \frac{1}{2} \times \frac{2 \times 8}{2+8}(V-0)^2 = \frac{4}{5}V^2$$

The percentage of the energy dissipated

$$= \frac{\Delta U}{Ui} \times 100 = \frac{\frac{4}{5}V^2}{V^2} \times 100 = 80\%$$

6. (4) For the first minima,

$$\theta = \frac{\eta\lambda}{a} \Rightarrow \sin 30° = \frac{\lambda}{a} = \frac{1}{2}$$

First secondary maxima will be at

$$\sin\theta = \frac{3\lambda}{2a} = \frac{3}{2}\left(\frac{1}{2}\right) \Rightarrow \theta = \sin^{-1}\left(\frac{3}{4}\right)$$

7. (1) As we know, gravitational potential (v) and acceleration due to gravity (g) with height

$$V = \frac{-GM}{R+h} = -5.4 \times 10^7 \quad \ldots(1)$$

$$\text{and } g = \frac{GM}{(R+h)^2} = 6 \quad \ldots(2)$$

Dividing (1) and (2)

$$\frac{\frac{-GM}{R+h}}{\frac{GM}{(R+h)^2}} = \frac{-5.4 \times 10^7}{6}$$

$$\Rightarrow \frac{5.4 \times 10^7}{(R+h)} = 6$$

$\Rightarrow$ R + h = 9000 km so, h = 2600 km

8. (4) To generate electromagnetic waves we need accelerating charge particle.

9. (3) From figure $\tan\theta = \dfrac{F_e}{mg} \simeq \theta$

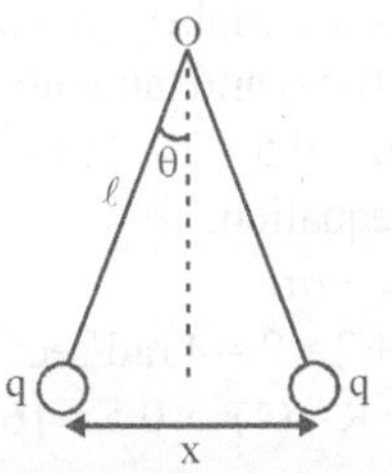

$$\frac{kq^2}{x^2 mg} = \frac{x}{2\ell}$$

or $\quad x^3 \propto q^2 \quad \ldots(1)$

or $\quad x^{3/2} \propto q \quad \ldots(2)$

Differentiating eq. (i) w.r.t. time

$$3x^2\frac{dx}{dt} \propto 2q\frac{dq}{dt} \text{ but } \frac{dq}{dt} \text{ is constant}$$

so $x^2(v) \propto q$ Replace q from eq. (2)

$x^2(v) \propto x^{3/2}$ or $v \propto x^{-1/2}$

10. (2) From figure, tension $T_1 = m_2 g$

$T_2 = (m_1 + m_2)g$

As we know

Velocity $\propto \sqrt{T}$ So,

$\lambda \propto \sqrt{T}$

$$\Rightarrow \frac{\lambda_1}{\lambda_2} = \frac{\sqrt{T_1}}{\sqrt{T_2}}$$

$$\Rightarrow \frac{\lambda_2}{\lambda_1} = \sqrt{\frac{m_1 + m_2}{m_2}}$$

11. (3) Coefficient of performance of a refrigerator,

$$\beta = \frac{Q_2}{W} = \frac{T_2}{T_1 - T_2}$$

(Where Q_2 is heat removed)

Given: $T_2 = 4°C = 4 + 273 = 277\,k$

$T_1 = 30°C = 30 + 273 = 303\,k$

$$\therefore\ \beta = \frac{600 \times 4.2}{W} = \frac{277}{303 - 277}$$

$\Rightarrow$ W = 236.5 joule

$$\text{Power } P = \frac{W}{t} = \frac{236.5 \text{ joule}}{1 \text{ sec}} = 236.5 \text{ watt.}$$

12. (3) For a closed organ pipe first minimum resonating length

$$L_1 = \frac{\lambda}{4} = 50 \text{ cm}$$

$\therefore$ Next or second resonating length, L_2

$$= \frac{3\lambda}{4} = 150 \text{ cm}$$

13. (2) Since diode is in forward bias, so the value of current flowing through AB.

$$i = \frac{\Delta V}{R} = \frac{4-(-6)}{1 \times 10^3} = \frac{10}{10^3} = 10^{-2} A$$

14. (1) Given: Charge $Q = at - bt^2$

$\therefore$ Current $i = \frac{\partial Q}{\partial t} = a - 2bt$

$\{\text{for } i = 0 \Rightarrow t = \frac{a}{2b}\}$

From joule's law of heating, heat produced
$dH = i^2 R dt$

$$H = \int_0^{a/2b} (a - 2bt)^2 R dt$$

$$H = \left.\frac{(a - 2bt)^3 R}{-3 \times 2b}\right|_0^{\frac{a}{2b}} = \frac{a^3 R}{6b}$$

15. (4) According to wein's displacement law, maximum amount of emitted radiation corresponding to $\lambda_m = \frac{b}{T}$

$$\lambda_m = \frac{2.88 \times 10^6 \text{ nmK}}{5760 \text{K}} = 500 \text{ nm}$$

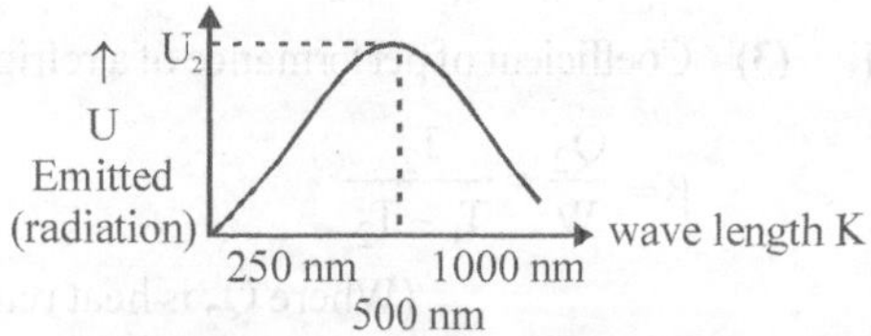

From the graph $U_1 < U_2 > U_3$

16. (4) From question, $(\ell_2 - \ell_1)$ is maintained same at all temperatures hence change in length for both rods should be same
i.e., $\Delta\ell_1 = \Delta\ell_2$
As we know, coefficient of linear expansion,

$$\alpha = \frac{\Delta\ell}{\ell_0 \Delta T}$$

$\ell_1\alpha_1\Delta T = \ell_2\alpha_2\Delta T$
$\ell_1\alpha_1 = \ell_2\alpha_2$

17. (1) Given: amplification factor $\alpha = 0.96$
load resistance, $R_L = 800\ \Omega$
input resistance, $R_i = 192\Omega$

So, $\beta = \frac{\alpha}{1-\alpha} = \frac{0.96}{0.04} \Rightarrow \beta = 24$

Voltage gain for common emitter configuration

$$A_v = \beta \cdot \frac{R_L}{R_i} = 24 \times \frac{800}{192} = 100$$

Power gain for common emitter configuration
$P_v = \beta A_v = 24 \times 100 = 2400$
Voltage gain for common base configuration

$$A_v = \alpha, \frac{R_L}{R_P} = 0.96 \times \frac{800}{192} = 4$$

Power gain for common base configuration
$P_v = A_v\alpha = 4 \times 0.96 = 3.84$

18. (4) Let P is a point infront of one slit at which intensity is to be calculated. From figure,

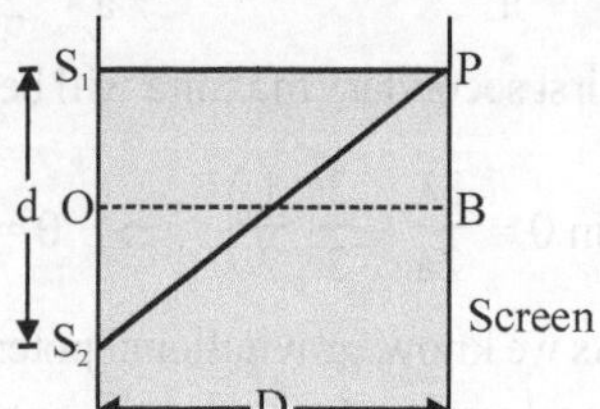

Path difference $= S_2P - S_1P$

$$= \sqrt{D^2 + d^2} - D$$

$$= D\left(1 + \frac{1}{2}\frac{d^2}{D^2}\right) - D$$

$$= D\left[1 + \frac{d^2}{2D^2} - 1\right] = \frac{d^2}{2D}$$

$$\Delta x = \frac{d^2}{2 \times 10d} = \frac{d}{20} = \frac{5\lambda}{20} = \frac{\lambda}{4}$$

Place difference,

$$\Delta\phi = \frac{2\pi}{\lambda}, \frac{\lambda}{4} = \frac{\pi}{2}$$

So, resultant a intensity at the desired point 'p' is

$$I = I_0 \cos^2\frac{\phi}{2} = I_0 \cos^2\frac{\pi}{4} = \frac{I_0}{2}$$

19. (1) Given: Radius of disc, R = 50 cm
angular acceleration $\alpha = 2.0 \text{ rads}^{-2}$;
time t = 2s
Particle at periphery (assume) will have both radial (one) and tangential acceleration
$a_t = R\alpha = 0.5 \times 2 = 1 \text{ m/s}^2$
From equation,
$\omega = \omega_0 + \alpha t$
$\omega = 0 + 2 \times 2 = 4 \text{ rad/sec}$
$a_c = \omega^2 R = (4)^2 \times 0.5 = 16 \times 0.5 = 8 \text{m/s}^2$

Net acceleration,

$$a_{total} = \sqrt{a_t^2 + a_c^2} = \sqrt{1^2 + 8^2} \approx 8 \text{ m/s}^2$$

20. **(1)** For electron De-Broglie wavelength,

$$\lambda_e = \frac{h}{\sqrt{2mE}}$$

For photon E = pc

$\Rightarrow$ De-Broglie wavelength, $\lambda_{Ph} = \frac{hc}{E}$

$$\therefore \frac{\lambda_e}{\lambda_{Ph}} = \frac{h}{\sqrt{2mE}} \times \frac{E}{hc} = \left(\frac{E}{2m}\right)^{1/2} \frac{1}{c}$$

21. **(2)** Time of descent $\propto \frac{K^2}{R^2}$

Order of value of $\frac{K^2}{R^2}$

for disc; $\frac{K^2}{R^2} = \frac{1}{2} = 0.5$

for sphere; $\frac{K^2}{R^2} = \frac{2}{5} = 0.4$

(sphere) < (disc)

$\therefore$ Sphere reaches first

22. **(2)** **Given :** Angle of incidence

$i = 45°$

Angle of prism, $A = 60°$

Angle of minimum deviation,

$\delta_m = 2i - A = 30°$

Refractive index of material of prism.

$$\mu = \frac{\sin\left(\frac{A+\delta_m}{2}\right)}{\sin A/2}$$

$$= \frac{\sin 45°}{\sin 30°} = \frac{1}{\sqrt{2}} \cdot \frac{2}{1} = \sqrt{2}$$

23. **(1)** At closest distance of approach, the kinetic energy of the particle will convert completely into electrostatic potential energy.

Kinetic energy K.E. $= \frac{1}{2}mv^2$

Potential energy P.E. $= \frac{KQq}{r}$

$$\frac{1}{2}mv^2 = \frac{KQq}{r} \Rightarrow r \propto \frac{1}{m}$$

24. **(1)** **Given:** Mass of particle, $M = 10g = \frac{10}{1000}$ kg

radius of circle R = 6.4 cm

Kinetic energy E of particle $= 8 \times 10^{-4}$J

acceleration $a_t = ?$

$$\frac{1}{2}mv^2 = E \Rightarrow \frac{1}{2}\left(\frac{10}{1000}\right)v^2$$

$= 8 \times 10^{-4}$

$\Rightarrow v^2 = 16 \times 10^{-2}$

$\Rightarrow v = 4 \times 10^{-1} = 0.4$ m/s

Now, using

$v^2 = u^2 + 2a_t s$ $(s = 4\pi R)$

$$(0.4)^2 = 0^2 + 2a_t\left(4 \times \frac{22}{7} \times \frac{6.4}{100}\right)$$

$$\Rightarrow a_t = (0.4)^2 \times \frac{7 \times 100}{8 \times 22 \times 6.4} = 0.1 \text{ m/s}^2$$

25. **(2)** Here $v_1 = 200$ m/s; temperature

$T_1 = 27°C$

$= 27 + 273 = 300$ k

temperature $T_2 = 127°$ C

$= 127 + 273 = 400$ k $v_2 = ?$

R.M.S. Velocity, $V \propto \sqrt{T}$

$$\Rightarrow \frac{v}{200} = \sqrt{\frac{400}{300}}$$

$$\Rightarrow v = \frac{200 \times 2}{\sqrt{3}} \text{ m/s}$$

$$\Rightarrow v = \frac{400}{\sqrt{3}} \text{ m/s}$$

26. **(3)** For points inside the wire i.e., $(r \le R)$

Magnetic field $B = \frac{\mu_0 Ir}{2\pi R^2}$

For points outside the wire $(r \ge R)$

Magnetic field, $B' = \frac{\mu_0 I}{2\pi R}$

$$\therefore \frac{B}{B'} = \frac{\frac{\mu_0 I(a/2)}{2\pi a^2}}{\frac{\mu_0 I}{2\pi(2a)}} = 1:1$$

27. (3) Given: Position vector

$\vec{r} = \cos wt\ \hat{x} + \sin \omega t\ \hat{y}$

$\therefore$ Velocity, $\vec{v} = -\omega \sin \omega t\ \hat{x} + \omega \cos \omega t\ \hat{y}$

and acceleration,

$\vec{a} = -\omega^2 \cos \omega t\ \hat{x} + \omega \sin \omega t\ \hat{y} = -\omega^2\ \vec{r}$

$\vec{r} \cdot \vec{v} = 0$ hence $\vec{r} \perp \vec{v}$ and

$\vec{a}$ is directed towards the origin.

28. (4) To complete the loop a body must enter a vertical loop of radius R with the minimum velocity $v = \sqrt{5gR}$.

29. (4) According to Einstein's photoelectric effect,

$$eV = \frac{hc}{\lambda} - \frac{hc}{\lambda_0} \quad \ldots(i)$$

$$eV/4 = \frac{hc}{2\lambda} - \frac{hc}{\lambda_0} \quad \ldots(ii)$$

Dividing equation (i) by (ii)

$$\Rightarrow \quad 4 = \frac{\frac{1}{\lambda} - \frac{1}{\lambda_0}}{\frac{1}{2\lambda} - \frac{1}{\lambda_0}} \text{ on solving we get,}$$

$\lambda_0 = 3\lambda$

30. (2) W_{ext} = negative of area with volume-axis W(adiabatic) > W(isothermal)

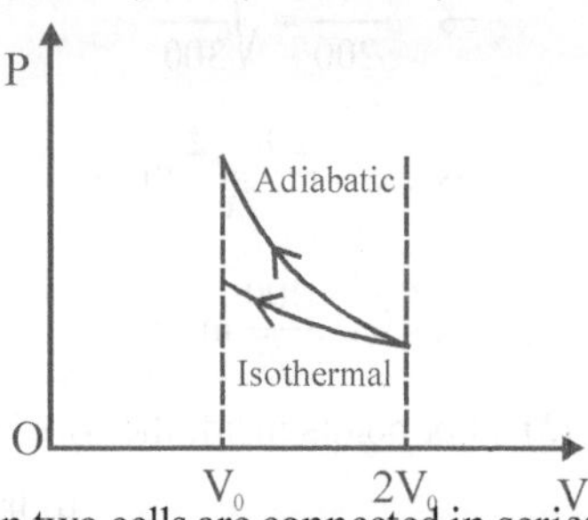

31. (4) When two cells are connected in series i.e., $E_1 + E_2$ the balance point is at 50 cm. And when two cells are connected in opposite direction i.e., $E_1 - E_2$ the balance point is at 10 cm. According to principle of potential

$$\frac{E_1 + E_2}{E_1 - E_2} = \frac{50}{10}$$

$$\Rightarrow \quad \frac{2E_1}{2E_2} = \frac{50+10}{50-10} \quad \Rightarrow \quad \frac{E_1}{E_2} = \frac{3}{2}$$

32. (d) **Given:** Focal length of objective, $f_0 = 40$cm

Focal length of eye – piece $f_e = 4$ cm

image distance, $v_0 = 200$ cm

Using lens formula for objective lens

$$\frac{1}{v_0} - \frac{1}{u_0} = \frac{1}{f_0} \quad \Rightarrow \quad \frac{1}{v_0} = \frac{1}{f_0} + \frac{1}{u_0}$$

$$\Rightarrow \quad \frac{1}{v_0} = \frac{1}{40} + \frac{1}{-200} = \frac{+5-1}{200}$$

$\Rightarrow \quad v_0 = 50$ cm

Tube length $\ell = |v_0| + f_e = 50 + 4 = 54$ cm.

33. (4) As we know,

Pressure P = Vdg

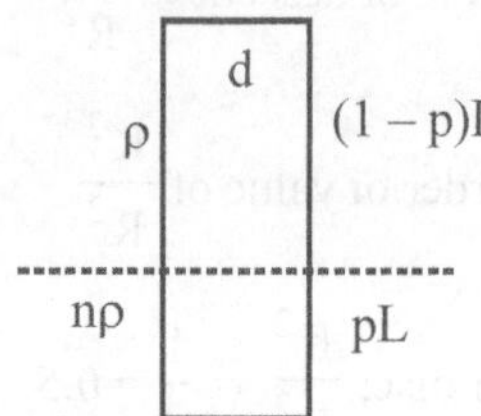

Here, $LAdg = (pL)A(n\rho)g + (1-p)LA\rho g$

$\Rightarrow \quad d = (1-p)\rho + pn\rho = [1 + (n-1)p]\rho$

34. (4) The Boolen expression for the given combination is output

$Y = (A + B) \cdot C$

Truthtable

A	B	C	$Y = (A+B) \cdot C$
0	0	0	0
1	0	0	0
0	1	0	0
0	0	1	0
1	1	0	0
0	1	1	1
1	0	1	1
1	1	1	1

Hence, A = 1, B = 0, C = 1

35. (3) According to question only one-quarter of the heat produced by falling piece of ice is absorbed in the melting of ice.

i.e., $\frac{mgh}{4} = mL$

$$\Rightarrow \quad h = \frac{4L}{g} = \frac{4 \times 3.4 \times 10^5}{10} = 136 \text{ km}.$$

36. **(2)** As we know, escape velocity,

$$V_e = \sqrt{\frac{2GM}{R}} = \sqrt{\frac{2G}{R}\cdot\left(\frac{4}{3}\pi R^3\rho\right)} \propto R\sqrt{\rho}$$

$$\therefore \quad \frac{V_e}{V_p} = \frac{R_e}{R_p}\sqrt{\frac{\rho_e}{\rho_p}}$$

$$\Rightarrow \quad \frac{V_e}{V_p} = \frac{R_e}{2R_e}\sqrt{\frac{\rho_e}{2\rho_e}}$$

$$\therefore \quad \text{Ratio } \frac{V_e}{V_p} = 1 : 2\sqrt{2}$$

37. **(2)** $|\vec{A}+\vec{B}| = |\vec{A}-\vec{B}|$

Squaring on both sides

$$|\vec{A}+\vec{B}|^2 = |\vec{A}-\vec{B}|^2$$

$$\Rightarrow \vec{A}\cdot\vec{A} + 2\vec{A}\cdot\vec{B} + \vec{B}\cdot\vec{B}$$

$$= \vec{A}\cdot\vec{A} - 2\vec{A}\cdot\vec{B} + \vec{B}\cdot\vec{B}$$

$$\Rightarrow 4\vec{A}\cdot\vec{B} = 0 \Rightarrow 4AB\cos\theta = 0$$

$$\Rightarrow \cos\theta = 0 \Rightarrow \theta = 90^\circ$$

38. **(3)** According to Bohr's theory, the wave number of the last line of the Balmer series in hydrogen spectrum,

For hydrogen atom z = 1

$$\frac{1}{\lambda} = RZ^2\left(\frac{1}{n_2^2} - \frac{1}{n_1^2}\right)$$

$$= 10^7 \times 1^2\left(\frac{1}{2^2} - \frac{1}{\infty^2}\right)$$

$$\Rightarrow \quad \text{wave number } \frac{1}{\lambda} = 0.25 \times 10^{-7}\ m^{-1}$$

39. **(4)** Given force $\vec{F} = 2t\hat{i} + 3t^2\hat{j}$

According to Newton's second law of motion,

$$m\frac{d\vec{v}}{dt} = 2t\hat{i} + 3t^2\hat{j} \quad (m = 1\ kg)$$

$$\Rightarrow \quad \int_0^{\vec{v}} d\vec{v} = \int_0^t \left(2t\hat{i} + 3t^2\hat{j}\right)dt$$

$$\Rightarrow \quad \vec{v} = t^2\hat{i} + t^3\hat{j}$$

Power $P = \vec{F}\cdot\vec{v}\ (2t\hat{i} + 3t^2\hat{j})\cdot(t^2\hat{i} + t^3\hat{j})$

$= (2t^3 + 3t^5)W$

40. **(1)** Given: L = 20 mH; C = 50 μF; R = 40 Ω

V = 10 sin 340 t

$$\therefore \quad V_{runs} = \frac{10}{\sqrt{2}}$$

$$X_C = \frac{1}{\omega C} = \frac{1}{340 \times 50 \times 10^{-6}} = 58.8\ \Omega$$

$$X_L = \omega L = 340 \times 20 \times 10^{-3} = 6.8\ \Omega$$

$$\text{Impedance, } Z = \sqrt{R^2 + (X_C - X_L)^2}$$

$$= \sqrt{40^2 + (58.8 - 6.8)^2} = \sqrt{4304}\ \Omega$$

Power loss in A.C. circuit,

$$P = i_{rms}^2 R = \left(\frac{V_{rms}}{Z}\right)^2 R$$

$$= \left(\frac{10/\sqrt{2}}{\sqrt{4304}}\right)^2 \times 40 = \frac{50 \times 40}{4304} \simeq 0.51\ W$$

41. **(3)** Given: velocity

$$V = At + Bt^2 \Rightarrow \frac{dx}{dt} = At + Bt^2$$

By integrating we get distance travelled

$$\Rightarrow \quad \int_0^x dx = \int_1^2 \left(At + Bt^2\right)dt$$

Distance travelled by the particle between 1s and 2s

$$x = \frac{A}{2}\left(2^2 - 1^2\right) + \frac{B}{3}\left(2^3 - 1^3\right) = \frac{3A}{2} + \frac{7B}{3}$$

42 **(4)** Here, number of turns n = 100; current through the solenoid i = 4A; flux linked with each turn = 4×10^{-3} Wb

∴ Total flux linked, and total

$= 1000[4 \times 10^{-3}] = 4$ Wb

$\phi_{total} = 4 \Rightarrow Li = 4$

$\Rightarrow L = 1$ H

43. **(2)** As we know, power $P = V_{rms} \cdot I_{rms} \cos\phi$

as $\cos\phi = 0$ $(\because \phi = 90^\circ)$

∴ Power consumed = 0

(in one complete cycle)

44. (1) Magnitude m = +ve $\Rightarrow$ virtual image
m = –ve $\Rightarrow$ real image
magnitude of magnification,
$|m| > 1 \Rightarrow$ magnified image
$|m| < 1 \Rightarrow$ diminished image

45. (2) On a banked road,

$$\frac{V_{max}}{Rg} = \left(\frac{\mu_s + \tan\theta}{1 - \mu_s \tan\theta}\right)$$

Maximum safe velocity of a car on the banked road

$$V_{max} = \sqrt{Rg\left[\frac{\mu_s + \tan\theta}{1 - \mu_s \tan\theta}\right]}$$

CHEMISTRY

46. (2) CH_4 NH_3 H_2O

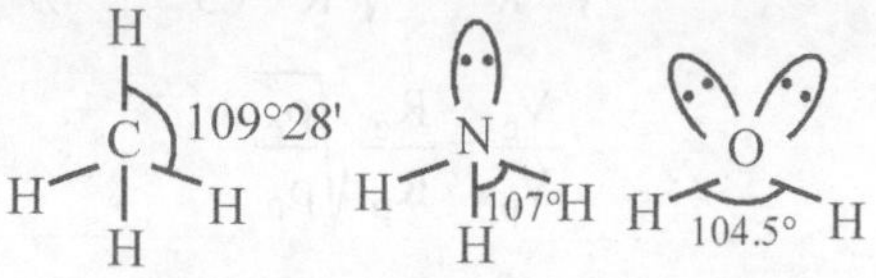

H–C–H with bond angle 109°28'; N with lone pair, H–N–H bond angle 107°; O with two lone pairs, H–O–H bond angle 104.5°

Tetrahedral; Trigonal pyramidal Bent

Note: The geometry of H_2O should have been tetrahedral if there are all bond pairs. But due to presence of two lone pairs the shape is distorted tetrahedral. Hence bond angle reduced to 104.5° from 109.5°.

47. (1)

$$HC\equiv CH \xrightarrow[\text{liq. } NH_3]{NaNH_2} HC\equiv \overset{\ominus}{C}\overset{\oplus}{Na} \xrightarrow{H_3C-CH_2\cdot Br} \underset{\substack{(X)\\ \text{1-Butyne}}}{HC\equiv C-CH_2-CH_3} \xrightarrow[\text{liq. } NH_3]{NaNH_2} H_3C-CH_2-C\equiv \overset{\ominus}{C}\overset{\oplus}{Na} \xrightarrow{H_3C-CH_2-Br} \underset{\text{3-Hexyne}}{H_3C-CH_2-C\equiv C-CH_2-CH_3}$$

48. (2) Acidic strength increases as the oxidation number of central atom increases.

$$\underset{+1}{HClO} < \underset{+3}{HClO_2} < \underset{+5}{HClO_3} < \underset{+7}{HClO_4}$$

49. (1) For a first order reaction

$$k = \frac{2.303}{(t_2 - t_1)}\log\frac{(a - x_1)}{(a - x_2)}$$

$$k = \frac{2.303}{(20 - 10)}\log\left(\frac{0.04}{0.03}\right)$$

$$k = \frac{2.303 \times 0.1249}{10}$$

$$\frac{0.6932}{t_{1/2}} = \frac{2.303 \times 0.1249}{10}$$

$$t_{1/2} = \frac{0.6932 \times 10}{2.303 \times 0.1249} = 24.1 \text{ sec}$$

50. (2) Adsorption is spontaneous process, therefore change in the free energy (ΔG) for the process is negative.
According to Gibbs's Helmholtz eqn.
$\Delta G = \Delta H - T\Delta S$

ΔS is negative because adhering of gas molecules to the surface lowers the randomness.
$\therefore$ ΔG can be –ve only when ΔH is –ve.

51. (2) The correct order is
B < C < O < N
Generaly ionisation energy increases across a period. But here first I.E. of O is less than the first I.E. of N. This is due to the half-filled 2p orbital in N($1s^2, 2s^2, 2p^3$) which is more stable than the 2p orbital in O ($1s^2, 2s^2, 2p^4$)

(3) The correct order of electron affinity is
I < Br < F < Cl
Halogens have high electron affinities which decreases on moving down the group. However, fluorine has lower value than chlorine which is due to its small size and repulsion between the electron added and electrons already present.

52. (3)

53. (1 & 4)
Among the three isotopes of hydrogen, Protium ($_1H^1$) is most common. It is an energetic reducing agent. It reduces oxides,

chlorides and sulphides of certain metals and produce free metals at ordinary temperature.

$$CuO + 2H \rightarrow Cu + H_2O$$

54. (4) Keto-enol tautomerism is possible only in those aldehydes and ketones which have atleast one α-hydrogen atom, which can convert the ketonic group to the enolic group. e.g.

$$CH_3-\overset{\overset{O}{\|}}{C}-CH_2-\overset{\overset{O}{\|}}{C}CH_3 \rightleftharpoons$$

Ketonic form

$$CH_3-\overset{\overset{OH}{|}}{C}=CH-\overset{\overset{O}{\|}}{C}-CH_3$$

Enolic form

55. (2) $MY \rightleftharpoons M^+ + Y^-$

$K_{SP} = s^2 = 6.2 \times 10^{-13}$

$s = \sqrt{6.2 \times 10^{-13}}$

$s = 7.87 \times 10^{-7}\ mol\ L^{-1}$

$NY_3 \rightleftharpoons N^{3+} + 3Y^-$

$K_{SP} = s \times (3s)^3 = 27s^4 = 6.2 \times 10^{-13}$

$$s = \left(\frac{6.2 \times 10^{-13}}{27}\right)^{1/4}$$

$s = 3.89 \times 10^{-4}\ mol\ L^{-1}$

∴ molar solubility of NY_3 is more than MY in water.

56. (3) Peptide bond

$$-\underset{\underset{O}{\|}}{C}-NH-$$

57. (1) Natural rubber is found to be a polymer of cis-isoprene i.e. it is cis-polyisoprene

$$H_2C=\underset{\underset{CH_3}{|}}{C}-CH=CH_2 \xrightarrow{\text{polymerisation}}$$

Isoprene

$$\left[-CH_2\ \ CH_2-\right] \quad (H_3C)C=C(H)$$

cis-polyisoprene

58. (1) Highly electropositive metals like Al, K, Na etc. are extracted by the electrolytic reduction.

- zone refining method is used for obtaining metals of high purity e.g. Ge
- Froth flotation process is suitable for sulphide ores
- Cyanide process is used for the extraction of gold.

59. (4) $K_2Cr_2O_7 + SO_2 + H_2SO_4 \rightarrow K_2SO_4 + \underset{\text{green colour}}{Cr_2(SO_4)_3} + H_2O$

60. (4) $Eu(63) = [Xe]\,4f^7\,6s^2$

$Gd(64) = [Xe]\,4f^7\,5d^1\,6s^2$

$Tb(65) = [Xe]\,4f^9\,6s^2$

61. (4) Two electrons occupying the same orbital should have opposite spins i.e. they differ in spin quantum number.

62. (1) Less active metals like Cu, Ag etc react with conc acid and form nitrate and nitrogen dioxide.

63. (1)

64. (3) $\Delta G = \Delta H - T{\cdot}\Delta S$

For a spontaneous reaction ΔG = –ve (always)

which is possible only if

$\Delta H < 0$ and $\Delta S > 0$

∴ spontaneous at all temperatures.

65. (2) density $(\rho) = \dfrac{Z \times M}{N_A \times a^3}$

For bcc structure

$Z = 2, \rho = 530\ kg\ m^{-3} = 0.530\ g\ cm^{-3}$

$$0.530 = \frac{2 \times 6.94}{6.02 \times 10^{23} \times a^3}$$

$a^3 = 4.348 \times 10^{-23}\ cm^3$

$a = 3.52 \times 10^{-8}\ cm$

$a = 352\ pm$

66. (2) Bond dissociation enthalpy decreases as the bond distance increases from F_2 to I_2. This is due to increase in the size of the atom, on moving from F to I.

F – F bond dissociation enthalpy is smaller then Cl – Cl and even smaller than Br – Br. This is because F atom is very small and hence the three lone pairs of electrons on each F atom repel the bond pair holding the F-atoms in F_2 molecules.

The increasing order of bond dissociation enthalphy is

$$I_2 < F_2 < Br_2 < Cl_2$$

67. (1) Novalgin is most widely used as analgesic. Analgesics are pain releiving

68. (1) Given, $n_{H_2} = n_{O_2}$ and $t_{H_2} = t_{O_2}$

According to Graham's law of diffusion for two different gases.

$$\frac{r_{H_2}}{r_{O_2}} = \frac{v_1/t_1}{v_2/t_2} \Rightarrow \sqrt{\frac{M_{O_2}}{M_{H_2}}} = \sqrt{\frac{32}{2}}$$

$$\frac{1/2}{1/x} = \sqrt{16} = 4$$

$$\frac{x}{2} = 4$$

$\therefore x = 8$

$\therefore$ Fraction of $O_2 = 1/8$

69. (2) The presence of large amount of $KHSO_4$ will decrease ionisation of H_2SO_4 that result in lesser ionisation of nitric acid and lesser formation of nitronium ion $[NO_2^+]$. Hence the rate of nitration will be slower.

70. (1) According to VSEPR theory order of repulsion in between lp – lp, lp – bp and bp – bp is as under

lp – lp > lp – bp > bp – bp

71. (Bonus)

$CaC_2 + N_2 \rightarrow CaCN_2 + C$

72. (4) Clausius – Clapeyron's equation

$$\frac{d \ln P}{dT} = \frac{\Delta H_v}{RT^2}$$

73. (1)

XeF$_6$ **XeO$_3$**

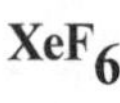

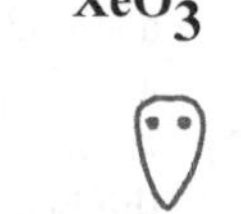

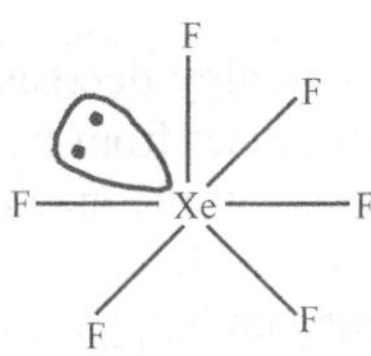

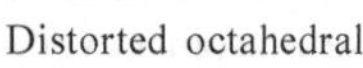

Distorted octahedral Pyramidal

XeOF$_4$: **XeF$_4$:**

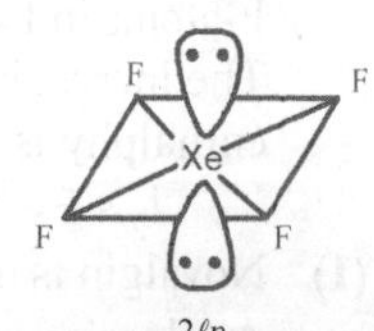

Square pyramidal Square planar

74. (3) $[Fe(CO)_4]^{2-}$

Since metal atom is carrying maximum –ve charge therefore it would show maximum synergic bonding as a resultant C—O bond length would be maximum.

75. (1) $2H^+(aq) + 2e^- \rightarrow H_2(g)$

$$\therefore \quad E = E^0 - \frac{0.0591}{2} \log \frac{P_{H_2}}{[H^+]^2}$$

$$0 = 0 - 0.0295 \log \frac{P_{H_2}}{(10^{-7})^2}$$

$$\frac{P_{H_2}}{(10^{-7})^2} = 1$$

$$P_{H_2} = 10^{-14} \text{ atm}$$

76. (4) A catalyst provides an alternative route for the reaction with a lower activation energy.

77. (1) radii ratio $= \frac{r_+}{r_-} = \frac{0.98 \times 10^{-10}}{1.81 \times 10^{-10}} = 0.54$

Since, the radii ratio is in between 0.414 to 0.732

so, coordination number is 6.

78. (2) Phosphinic acid as shown in structure below has one P—OH bend thus it is monobasic or monoprotic

O
||
P
H — | — OH (Monoprotic)
H

Phosphonic acid as shown in structure has two P–OH bonds thus it is dibasic or diprotic

O
||
P (Diprotic acid)
H — | — OH
OH

79. (1) Fog is a colloidal system having dispersed phase as liquid and dispersion medium as gas.

80. (1) Let us consider that A is benzene and B is toluene

1 : 1 molar mixture of A and B

$$\therefore \quad x_A = \frac{1}{2} \text{ and } x_B = \frac{1}{2}$$

Total pressure of solution (P) = $P_A^o x_A + P_B^o x_B$

$$P = 12.8 \times \frac{1}{2} + 3.85 \times \frac{1}{2} = 8.325 \text{ kPa}$$

$$Y_A = \frac{P_A^o x_A}{P} = \frac{12.8 \times \frac{1}{2}}{8.325} = 0.768$$

$\therefore$ $y_B = 1 - y_A = 1 - 0.768 = 0.232$

so, the vapour will contain higher percentage of benzene.

81. **(4)** In stggered conformation any two hydrogen atoms on adjacent carbon atoms are as far apart as possible thereby minimising repulsion between the electron clouds of σ-bonds of two non-bonded H-atomic (torsional strain)

Staggered form ⇌ Eclipsed form

No torsional strain

82. **(1)** This is an example of Williamson ether synthesis reaction in which sodium alkoxide reacts with alkyl halide and gives ether.

83. **(1)** Schiff base is formed when 1° amine reacts with aldehydes.

$$\text{R(H)C=O} + \text{R'}-\text{NH}_2 \xrightarrow{H^+} \text{R(H)C=N}-\text{R'}$$

Aldehyde primary amine Schiff base

84. **(2)** [biphenyl with Br, Br and I, I substituents] is optically active due to absence of plane of symmetry and center of symmetry.

85. **(2)** (a) $CH_3CH_2CH_2\text{–}Br + KOH$
$\rightarrow CH_3CH\text{=}CH_2 + KBr + H_2O$

This is dehydrohalogenation reaction which is an example of elimination reaction.

(b) $(H_3C)(CH_3)C\text{–}Br + KOH \rightarrow (H_3C)(CH_3)C\text{–}OH + KBr$

Replacement of Br^- by OH^- is substitution reaction thus it is a nucleophilic substitution reaction.

(c) Cyclohexene $+ Br_2 \rightarrow$ 1,2-dibromocyclohexane (Br, Br)

Above reaction involves addition of Br_2 across double bond. Thus it is called addition reaction.

86. **(1)** $\left(\frac{P^o - P_s}{P^o}\right) = \frac{n}{N} = \frac{W_1}{M_1} \times \frac{M_2}{W_2}$

Where, W_1 = wt of solute

W_2 = wt of solvent

M_1 = Mass of solute

M_2 = Mass of solvent

at 100°C, P^o = 760 mm

$$\frac{760 - 732}{760} = \frac{6.5 \times 18}{M_1 \times 100}$$

$M_1 = 31.75 \text{ g mol}^{-1}$

$$\Delta T_b = m \times K_b = \frac{W_1 \times 1000}{M_1 \times W_2} \times K_b$$

$$\Delta T_b = \frac{0.52 \times 6.5 \times 1000}{31.75 \times 100} = 1.06°C$$

$\therefore$ boiling point of solution
= 100°C + 1.06°C = 101°C

87. **(2)** Sugar in DNA is 2-deoxyribose whereas sugar in RNA is ribose.

88. **(1)** Arylamines are generally less basic than alkylamines due to following factors

(a) Due to resonance in aromatic amines.

(b) Lower stability of anilinium ion

89. **(4)** Sucrose is non-reducing disaccharide as the two monosaccharide units are linked through their respective carbonyl groups.

90. **(4)** $CH_3\text{—}C \equiv C^{\ominus}$

No.of σ bp – 1, ℓp – 1] 2 & hybridisation is sp

BIOLOGY

91. **(3)** Gause's principle of competitive exclusion can be restated to say that no two species can occupy the same niche indefinitely when resources are limiting. Certainly species can and do coexist while competing for some of the same resources. Nevertheless, Gause's theory predicts that when two species coexist on a long-term basis, either resources must not be limited or their niches will always differ in one or more features; otherwise, one species will outcompete the other and the extinction of the second species will inevitably result, a process referred to as competitive exclusion.

92. **(4)** Insulin is a hormone consisting of 2 polypeptide chains. Each chain is composed of a specific sequence of amino acid residues connected by peptide bonds. In humans, chain A has 21 amino acids, and chain B has 30. Post translational modifications result in the connection of these two chains by disulfide bridges. Cysteine residues on A7 and B7, as well as A20 to B19 are covalently connected by disulfide bridges.

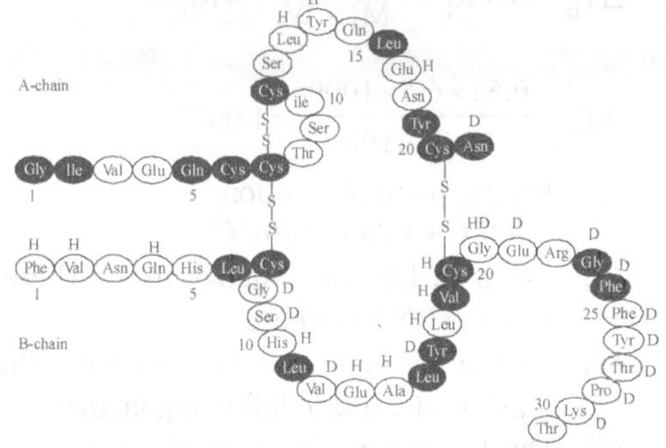

93. **(4)** Coconut water is the clear liquid inside young green coconuts (fruits of the coconut palm). In early development, it serves as a suspension for the endosperm of the coconut during the nuclear phase of development. As growth continues, the endosperm matures into its cellular phase and deposits into the rind of the coconut meat.

94. **(4)** Viroids, the smallest known pathogens, are naked, circular, single-stranded RNA molecules that do not encode protein yet replicate autonomously when introduced into host plants. Viroids only infect plants; some cause economically important diseases of crop plants, while others appear to be benign.

95. **(3)** All arthropods possess a stiff exoskeleton (external skeleton) composed primarily of chitin. Arthropod bodies are divided into segments. Parapodia are paired, lateral appendages extending from the body segments. Arthropod appendages may be either biramous (branched) or uniramous (unbranched). They do not possess jointed appendages.

96. **(2)** Hemophilia A and hemophilia B are inherited in an X-linked recessive pattern. The genes associated with these conditions are located on the X chromosome, which is one of the two sex chromosomes. In males (who have only one X chromosome), one altered copy of the gene in each cell is sufficient to cause the condition. In females (who have two X chromosomes), a mutation would have to occur in both copies of the gene to cause the disorder. Because it is unlikely that females will have two altered copies of this gene, it is very rare for females to have hemophilia. A characteristic of X-linked inheritance is that fathers cannot pass X-linked traits to their sons.

97. **(2)** Wavelengths beyond 700nm are apparently of insufficient energy to drive any part of photosynthesis. So a huge drop in efficiency has been noticed at 700nm. This phenomenon is called as "Red drop effect". In other words there is a sharp decrease in quantum yield at wavelengths greater than 680nm. The number of oxygen molecules released per light quanta absorbed is called as quantum yield of photosynthesis. This effect was first of all noticed by Robert Emerson. Later on Emerson and his group observed that if chlorella plants are given the inefficient far red light and red light of shorter wavelengths in alternate fashion, the quantum yields were greater than could be expected from adding the rates found when either color was provided alone. This synergistic effect or enhancement is known as EEE or "Emerson Enhancement Effect". This was the first good evidence that there are two photo systems; one absorbs far red light and other red light and both of them must operate to drive photosynthesis most effectively.

98. (Bonus) Nitrogen, phosphorus, potassium, calcium, sulfur and magnesium are the macronutrients. In the above given options no three element denotes macronutrients.

99. (1) Emphysema results when the delicate linings of the air sacs in the lungs become damaged beyond repair. Most commonly, the toxins in cigarette smoke create the damage. Emphysema is called smoker's disease.

100. (1) The growing of grass or legumes in rotation with grain or tilled crops as a soil conservation measure is called Ley farming.

101. (3) Mitochondria & chloroplast are semi-autonomous cell organelle which are formed by division of pre-existing organelle & contain DNA but they also contain protein synthesizing machinery, thus (a) is true & (b) is false.

102. (4) Amniocentesis (also referred to as amniotic fluid test or AFT) is a medical procedure used in prenatal diagnosis of chromosomal abnormalities and fetal infections, and also used for sex determination in which a small amount of amniotic fluid, which contains fetal tissues, is sampled from the amniotic sac. Cleft palate is a developmental abnormality which can only be detected by sonography.

103. (2) Proton concentration is higher in the lumen of thylakoid due to photolysis of water, H+ pumping and NADP reductase activity in stroma. During the light-dependent reaction, protons are pumped across the thylakoid membrane into the lumen making it acidic down to pH 4.

104. (2) Retinal is also known as retinaldehyde. It was originally called retinene, and renamed afterwards it was discovered to be vitamin A aldehyde. Retinal is one of the many forms of vitamin A (the number of which varies from species to species). Retinal is a polyene chromophore, bound to proteins called opsins, and is the chemical basis of animal vision.

105. (2) Attachment of microtubules to chromosomes is mediated by kinetochores, which actively monitor spindle formation and prevent premature anaphase onset during mitosis.

106. (2) River Dolphin is the National Aquatic Animal of India. This mammal is also said to represent the purity of the holy Ganga as it can only survive in pure and fresh water. Platanista gangetica has a long pointed snout and also have visible teeth in both the upper and lower jaws.

107. (3) Lac operon is an inducible operon. Lactose is the substrate for the enzyme beta-galactosidase and it also regulates switching on and off of the operon. Hence, it is termed as inducer. Inducers function by disabling repressors. The gene is expressed because an inducer binds to the repressor. The binding of the inducer to the repressor prevents the repressor from binding to the operator. RNA polymerase can then begin to transcribe operon genes.

108. (4) Calcitonin exerts its effects by interacting with target cells, primarily in bone and kidney. The actions of PTH and calcitonin are antagonistic on bone resorption but synergistic on decreasing the renal tubular reabsorption of phosphorus. Insulin and glucagon work synergistically to keep blood glucose concentrations normal. An elevated blood glucose concentration results in the secretion of insulin and glucose is transported into body cells. Aldosterone secretion may be inhibited by potassium depletion, inhibitors of the renin-angiotensin system, dopamine and atrial natriuretic factor. The latter appears to be an important physiological regulator of aldosterone secretion. ANF inhibits basal, ACTH, Angiotensin II and potassium-stimulated aldosterone production in vitro by a direct action on the adrenal gland. Inhibin is a hormone secreted by granulosa cells of the ovary which inhibits the secretion of FSH (Follicle Stimulating Hormone) whereas relaxin produced by ovary and placenta at the time of childbirth, softens the ligament in the pelvis and widens the cervix.

109. (2) Microtubules are long, hollow cylinders made up of polymerised α - and β -tubulin dimers. Microtubules are part of a structural network (the cytoskeleton) within the cell's cytoplasm. Roles of the microtubule cytoskeleton include mechanical support, organization of the cytoplasm, transport, motility and chromosome segregation so they are present in cilia and flagella for cellular movement, centrioles for chromosomal movement and spindle fibres for structural organization as well as chromosomal movement during nuclear division.

110. (1) A polysome or polyribosome is a complex of an mRNA molecule and two or more ribosomes, which is formed during the active translation process. They were initially named as ergosomes in 1963. However, further research by Jonathan Warner and Alex Rich characterized polysome.

111. (2) The word ampulla is derived from the Latin word 'flask'. Being the second portion of the fallopian tube, it is the intermediate dilated portion which immediately curves over the ovary. This is the common site of human fertilization as both the ovum and sperms are simultaneously transported here.

112. (2) A mast cell or a mastocyte is typically a white blood cell. It is a special kind of granulocyte, which is a part of the immune system and laden with histamine and heparin. Besides these, mast cells also secrete the prostaglandin (PG) D2, and leukotriene (LT) C4, which are capable of inducing bronchoconstriction and mucosal edema, both features of asthma.

113. (3) Auxin has been clearly demonstrated in the leaf sheath or coleoptile of oat plant (*Avena sativa*). This plant coleoptile has been used for the test of hormone Auxin (IAA) participating in the growth of the plant.

114. (3) Papilionaceous flowers are flowers with the characteristic irregular and butterfly-like corolla. A single, large upper petal is known as the banner or vexillum and the name has been derived from an ancient military standard.

115. (1) The Liliaceae are a huge variety, comprising about 280 genera and 4000 species of perennial herbs mostly. It includes starchy rhizomes, corms or bulbs, characterized by alternate leaves which are whorled. The flowers are often showy, mostly bisexual and actinomorphic. Also, they exhibit tricarpellary syncarpous gynoecium. Example of such a class is the *Erythronium montanum*, the alpine fawn lily.

116. (1) A cell wall is a rigid structural layer, which provides protection and structural support to the cells. The composition of cell walls varies from species to genera. In the case of fungi, the cell wall is composed of strong covalent linkages of chitin, glucans and glycoproteins. Alternatively, in case of land plants, the cell wall is composed of cellulose and hemicellulose and Archean cell walls have peptidoglycans.

117. (3) Considering the female reproductive endocrinology, ovulation is the process of the monthly release of the viable oocyte from the ovary between the time of menarche and menopause. During this time, there is a surge in the production of LH and FSH, termed as gonadotropins, thereby initiating estradiol and progesterone secretion from the ovary. Both these hormones are very important for the menstrual cycle.

118. (1) Meiosis is a process (discovered by German biologist, Oscar Hertwig) where a single cell divides twice to produce four cells containing half the original amount of genetic information. These cells are our sex cells - sperm in males, eggs in females. Crossing over is the process by which two chromosomes, exchange some distal part of their DNA. This process occurs in the Pachytene stage of Prophase I of meiosis.

119. (1)

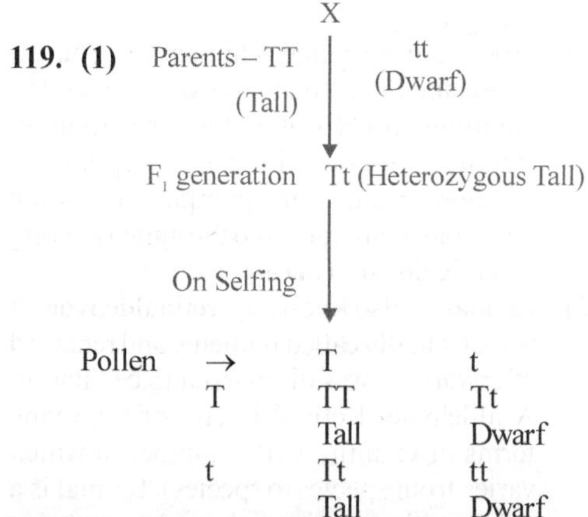

Phenotypic ratio : 3: 1 (Tall :Dwarf)

Genotypic Ratio: 1:2:1 (Homozygous Tall : Heterozygous Tall : Dwarf)

120. (3) The primary cause of human-induced extinction events is simply human overpopulation of planet Earth. The most important causal anthropogenic activities are habitat destruction and fragmentation.

121. (2) Cropland ecosystem is largest anthropogenic ecosystem characterized by less diversity and high productivity.

122. (1) At the pituitary, GnRH stimulates the synthesis and secretion of the gonadotropins, follicle-stimulating hormone (FSH), and luteinizing hormone (LH). These processes are controlled by the size and frequency of GnRH pulses, as well as by feedback from progesterone and estrogens. Low-frequency GnRH pulses are required for FSH release, whereas high-frequency GnRH pulses stimulate LH pulses in a one to one manner.

123. (4) Plasmid has an extra chromosomal, double stranded circular DNA.

124. (2) *Periplanata americana* shows spiral and determinate types of cleavage during embryonic development which is a feature of Protostomes.

125. (3) An autoimmune disease is a pathological state arising from an abnormal immune response of the body to substances and tissues that arc normally present in the body.

126. (2)

127. (3) Joint Forest Management originated in West Bengal accidentally at the Arabari Forest Range in West Midnapore, near Midnapore town in 1971. After the initial successes in West Bengal and Haryana, the JFM schemes received national importance in the legislation of 1988.

128. (4) Sickle cell disease is inherited in an autosomal recessive pattern.

129. (3) Eubacteria are the True bacteria.

130. (4) The proximal end is attached to the thalamus whereas the distal end bears anther.

131. (4) Vasectomy is a surgical procedure for male sterilization or permanent contraception. During the procedure, the male vas deferens are severed and then tied or sealed in a manner so as to prevent sperm from entering into the seminal stream (ejaculate) and thereby prevent fertilization.

132. (1) The Taq polymerase enzyme is obtained from *Thermus aquaticus* which lives in hot springs.

133. (2) In both females and males, inhibin inhibits FSH production. In females, FSH stimulates the secretion of inhibin from the granulosa cells of the ovarian follicles in the ovaries. In turn, inhibin suppresses FSH. In males, androgens stimulate inhibin production. It is secreted from the Sertoli cells, located in the seminiferous tubules inside the testes.

134. (4) *Meloidogyne incognita* is a nematode (roundworm) in the family Heteroderidae. It is commonly called the "southern root-knot nematode" or the "cotton root-knot nematode".

135. (4) Oral Polio Vaccine consists of a mixture of attenuated (weakened) poliovirus strains of all three poliovirus types.

136. (3) Double Membrane bound Organelles: Mitochondria, Chloroplasts, Endoplasmic Reticulum, Golgi Body, and Nucleus. Single Membrane bound Organelles: Lysosomes, Peroxisomes, and Vacuoles. Organelles lacking any membrane: Ribosomes, Centrioles, Nucleolus.

137. (3) Sustained contraction with no relaxation phase is called muscle tetanus.

138. (1) Pitcher of *Nepenthes* or pitcher plant is modification of leaf whereas thorns in citrus tendrils of cucumber and flattened structure of *Opuntia* are all stem modification.

139. (4) Many leaves produce water-soluble vacuolar pigments, which are stored within cell vacuoles (microscopic water sacs within each cell). Two major classes of leaf vacuolar pigments are anthocyanins and betalains.

140. (3) *Sequoia semepervirans* is one of tallest trees.

141. (2) Zinc-finger analysis is used for protein analysis. The zinc finger proteins are a super family of proteins involved in numerous activities of plant growth and development.

142. (1) Wall of intestine is made of smooth muscle. Tendons consist of dense regular connective tissue fascicles encased in dense irregular connective tissue sheaths. Tip of nose consists of squamous epithelium. Lining of epithelium is made of columnar epithelium.

143. (2) C4 plants are adapted to hot and dry climate and lack photorespiration due to Kranz anatomy and have Greater productivity of biomass.

144. (4) Wings of a bird and flippers of a whale are modified forelimbs.

145. (1) Chondrichthyes always have cartilaginous endoskeleton. Most mammals are viviparous, giving birth to live young. However, the five species of monotreme, the platypuses and the echidnas, lay eggs. Chordates have jawless animals (Agnatha) as well. Most reptiles have 3 chambered heart. Crocodilians have 4 chambered heart. Turtles have 3 chambered heart but with an incomplete wall in the single ventricle, so their heart are functionally 4 chambered.

146. (4) Telomerase production is increased in cancer. Telomerase has been examined in hundreds of studies as a potentially sensitive biomarker for screening, early cancer detection, prognosis or in monitoring as an indication of residual disease.

147. (1) Tryptophan is a precursor to neuro-transmitters serotonin and melatonin. Thyroxine (3,5,3',5'-tetraiodothyronine) is produced by follicular cells of the thyroid gland. It is produced as the precursor thyroglobulin. Estrogen is biosynthesized from progesterone (arrived at in two steps from cholesterol, via intermediate pregnenolone). Cortisone is one of several end-products of a process called steroidogenesis. Cortisol is produced in the adrenal cortex of kidney.

148. (3) Both statements are correct because primitive atmosphere was reducing and chlorophyll appeared later on. Chemoautotrophs were the first autotrophic organisms unable to perform photolysis of water and never released oxygen.

149. (3) Reduction of pH of blood will decrease the affinity of hemoglobin with oxygen which in turn causes Acidosis.

150. (2) Analogous structures are those that have the same function, but they are not derived from a common ancestor and have undergone different patterns of development ie., Convergent evolution which is natural selection that favors the same type of structure in different ancestors.

151. (1) A restriction enzyme or restriction endonuclease is an enzyme that cuts DNA at or near specific recognition nucleotide sequences known as restriction sites. Hind II among these is a type of restriction endonuclease.

152. (2) Sir Arthur George Tansley was an English botanist and a pioneer in the science of ecology who coined the term ecosystem.

153. (4) Glycine (abbreviated as Gly or G) is the smallest of the 20 amino acids commonly found in proteins, and indeed is the smallest possible (having a hydrogen substituent as its side-chain). The formula is NH_2CH_2COOH. Its codons are GGU, GGC, GGA, GGG of the genetic code.

154. (4) Bryophytes have neither pollen nor flowers and rely on water to carry the male gametes (the sperm) to the female gametes (the eggs). The antherozoids (male gametes of pteridophytes) are armed with hair-like or whip-like cilia or flagellae and are able to swim through water; they do not travel great distances and are only released when free water is available.

155. (1) In logistic growth model population growth equation is described as

$$\frac{dN}{dt} = rN\left(1 - \frac{N}{K}\right)$$

N = population density at time t;
r = Intrinsic rate of natural increase;
K = carrying capacity

When N/K = 1; $1 - \frac{N}{K} = 0$

Therefore, $\frac{dN}{dt} = 0$

156. (1) Tapetum is important for the nutrition and development of pollen grains, as well as a source of precursors for the pollen coat.

157. (1) Lichens produce small amounts of carbonic acids from their "roots" and these slowly dissolve the rock, releasing nutrients. Other nutrients are obtained from water-borne or air-borne particles of both organic and inorganic materials. The body of a lichen contains an alga; this is able to photosynthesise sugars, some of which are passed on to the fungal component of the lichen. In return the alga gets a tiny but significant amount of shelter within the tissues of the lichen.

158. (1) The start codon is the first codon of a messenger RNA (mRNA) transcript translated by a ribosome. The start codon always codes for methionine in eukaryotes and a modified Met (fMet) in prokaryotes. The most common start codon is AUG.

159. (3) Giving birth to living young that develop within the mother's body rather than hatching from eggs. All mammals except the monotremes are viviparous.

160. (1) Binomial nomenclature (also called binominal nomenclature or binary nomenclature) is a formal system of naming species of living things by giving each a name composed of two parts, both of which use Latin grammatical forms, although they can be based on words from other languages.

161. (3) Arteries have higher blood pressure than vein because blood is forced inside them from heart and also their lumen is narrow.

162. (4) The cotyledons are known as seed leaves, they are attached to the embryonic axis. Dicotyledons typically have two cotyledons and monocotyledons have only one cotyledon. The single shield-shaped cotyledon in grains known as scutellum.

163. (2) The main constituent of gastric acid is hydrochloric acid which is produced by parietal cells (also called oxyntic cells) in the gastric glands in the stomach.

164. (2) The ozone layer or ozone shield refers to a region of Earth's stratosphere that absorbs most of the Sun's ultraviolet (UV) radiation. It contains high concentrations of ozone (O3) relative to other parts of the atmosphere.

165. (2) All unicellular eukaryotic organism like diatoms, desmids (chrysophytes), euglenoids, dinoflagellates and slime mould are included in Protista.

166. (2) In actively growing plants, water is continuously evaporating from the surface of leaf cells through stomatal opening exposed to air. This is called transpiration. Through the same stomatal opening carbon dioxide diffuses into the plant during photosynthesis. Simultaneously as both are the process of simple diffusion occurs in order of diffusion pressure gradient or diffusion coefficient.

167. (3) Hepatic Vein, because the liver produces urea and other waste materials and then it pours it all in the right ventricle of the heart for oxygenation. The heart distributes the blood to various parts of the body from here. So the impure blood brought by the Hepatic Vein and other blood vessels get distributed through the aorta. This will automatically imply that the Renal Artery will contain lesser impure blood than the Hepatic Vein.

168. (4) Apomixis (asexual seed formation) is the result of a plant gaining the ability to bypass the most fundamental aspects of sexual reproduction: meiosis and fertilization. Without the need for male fertilization, the resulting seed germinates a plant that develops as a maternal clone.

169. (4) *Clostridium butylicum* is used for butyric acid production

170. (3) When two genes in a dihybrid cross are situated on the same chromosome, the proportion of parental gene combinations are much higher than the non-parental or recombinant type. This is also called incomplete linkage.

171. (2) Basal metabolic rate is inversely proportional to body size. So smaller animals have a higher metabolic rate. Hence more energy production.

172. (4) Synapsis is the pairing of two homologous chromosomes that occurs during meiosis. It allows matching-up of homologous pairs prior to their segregation, and possible chromosomal crossover between them. Synapsis takes place during prophase I, Zygotene of meiosis.

173. (1) Pollen grains of different species are incompatible, so they fail to germinate. If the pollen-stigma interaction is compatible, the pollen grain hydrates and germinates shortly following landing on the stigma.

174. (2) Specialized epidermal cells surrounding the guard cells are known as Subsidiary or accessory cell.

175. (4) The sphincter of Oddi (or hepatopancreatic sphincter) is a muscular valve that controls the flow of digestive juices (bile and pancreatic juice) through the ampulla of Vater into the second part of the duodenum.

176. (3) Phylloclade to refer a portion of a flattened leaf-like stem or branch with multiple nodes and internodes, and cladode for a single internode of a phylloclade. In some plants, the petioles are flattened and widened, to become phyllodes or phyllodia.

177. (3) Methanogens are microorganisms that produce methane as a metabolic byproduct in anoxic conditions. They are obligate anaerobic ancient and primitive bacteria. They are involved in methanogenesis.

178. (4) Domestic sewage rich in organic waste leads to increase in Biological Oxygen Demand, which leads to decrease **D.O.** (dissolved oxygen) which leads to death of fishes.

179. (2) This phenomenon is known as Polyploidy, wherein the cells contain more than two paired (homologous) sets of chromosomes. Polyploidy is often seen in the case of plants. The major cause of polyploidy is the non - disjunction of sister chromatids during meiotic recombination. This condition is actually useful in development of new crop varieties.

180. (2) Fat molecules are made of atoms of carbon, hydrogen and oxygen. The hydrogen and oxygen atoms binds to the carbon, as pictorially depicted. A typical fat molecule has one glycerol and three fatty acid molecules.

H
H – C – O – H H – O – C(=O) – chain of carbons, hydrogens
H – C – O – H H – O – C(=O) – chain of carbons, hydrogens
H – C – O – H H – O – C(=O) – chain of carbons, hydrogens
H
→ $3H_2O$
glycerol 3 fatty acids

AIPMT 2015 (Retest) Solved Paper

PHYSICS

1. The position vector of a particle $\vec{R}$ as a function of time is given by:

$$\vec{R} = 4\sin(2\pi t)\hat{i} + 4\cos(2\pi t)\hat{j}$$

where R is in meter, t in seconds and $\hat{i}$ and $\hat{j}$ denote unit vectors along x-and y-directions, respectively.
Which one of the following statements is wrong for the motion of particle?

(a) Magnitude of acceleration vector is $\frac{v^2}{R}$, where v is the velocity of particle
(b) Magnitude of the velocity of particle is 8 meter/second
(c) path of the particle is a circle of radius 4 meter.
(d) Acceleration vector is along - $\vec{R}$

2. The energy of the em waves is of the oder of 15 keV. To which part of the spectrum does it belong?
(a) Infra-red rays (b) Ultraviolet rays
(c) γ-rays (d) X-rays

3. A beam of light consisting of red, green and blue colours is incident on a right angled prism. The refractive index of the material of the prism for the above red, green and blue wavelengths are 1.39, 1.44 and 1.47, respectively.

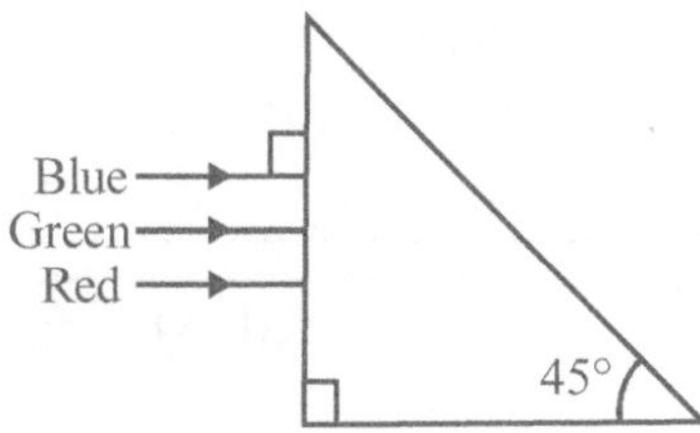

The prism will:
(a) separate all the three colours from one another
(b) not separate the three colours at all
(c) separate the red colour part from the green and blue colours
(d) separate the blue colour part from the red and green colours

4. Two particles A and B, move with constant velocities $\vec{v}_1$ and $\vec{v}_2$. At the initial moment their position vectors are $\vec{r}_1$ and $\vec{r}_2$ respectively. The condition for particles A and B for their collision is:

(a) $\vec{r}_1.\vec{v}_1 = \vec{r}_2.\vec{v}_2$
(b) $\vec{r}_1 \times \vec{v}_1 = \vec{r}_2 \times \vec{v}_2$
(c) $\vec{r}_1 - \vec{r}_2 = \vec{v}_1 - \vec{v}_2$
(d) $\frac{\vec{r}_1 - \vec{r}_2}{|\vec{r}_1 - \vec{r}_2|} = \frac{\vec{v}_2 - \vec{v}_1}{|\vec{v}_2 - \vec{v}_1|}$

5. At the first minimum adjacent to the central maximum of a single-slit diffraction pattern, the phase difference between the Huygen's wavelet from the edge of the slit and the wavelet from the midpoint of the slit is :

(a) $\frac{\pi}{2}$ radian (b) π radian
(c) $\frac{\pi}{8}$ radian (d) $\frac{\pi}{4}$ radian

6. A proton and an alpha particle both enter a region of uniform magnetic field B, moving at right angles to field B. If the radius of circular orbits for both the particles is equal and the kinetic energy acquired by proton is 1 MeV the energy acquired by the alpha particle will be:
(a) 0.5 MeV (b) 1.5 MeV
(c) 1 MeV (d) 4 MeV

7. A circuit contains an ammeter, a battery of 30V and a resistance 40.8Ω all connected in series. If the ammeter has a coil of resistance 480Ω and a shunt of 20Ω, the reading in the ammeter will be:
(a) 0.25 A (b) 2A
(c) 1 A (d) 0.5 A

8. The value of coefficient of volume expansion of glycerine is $5 \times 10^{-4}\ K^{-1}$. The fractional change in the density of glycerine for a rise of 40°C in its temperature, is:
(a) 0.020 (b) 0.025
(c) 0.010 (d) 0.015

9. An ideal gas is compressed to half its initial volume by means of several processes. Which of the process results in the maximum work done on the gas?
(a) Isobaric (b) Isochoric
(c) Isothermal (d) Adiabatic

10. A series R-C circuit is connected to an alternating voltage source. Consider two situations:
(a) When capacitor is air filled.
(b) When capacitor is mica filled.
Current through resistor is i and voltage across capacitor is V then :
(a) $V_a > V_b$ (b) $i_a > i_b$
(c) $V_a = V_b$ (d) $V_a < V_b$

11. Light of wavelength 500 nm is incident on a metal with work function 2.28 eV. The de Broglie wavelength of the emitted electron is:
(a) $< 2.8 \times 10^{-9}\ m$ (b) $\geq 2.8 \times 10^{-9}\ m$
(c) $\leq 2.8 \times 10^{-12}\ m$ (d) $< 2.8 \times 10^{-10}\ m$

12. Two metal wires of identical dimension are connected in series. If σ_1 and σ_2 are the conductivities of the metal wires respectively, the effective conductivity of the combination is :
(a) $\frac{\sigma_1 + \sigma_2}{2\sigma_1\sigma_2}$ (b) $\frac{\sigma_1 + \sigma_2}{\sigma_1\sigma_2}$
(c) $\frac{\sigma_1\sigma_2}{\sigma_1 + \sigma_2}$ (d) $\frac{2\sigma_1\sigma_2}{\sigma_1 + \sigma_2}$

13. An automobile moves on a road with a speed of 54 km h^{-1}. The radius of its wheels is 0.45 m and the moment of inertia of the wheel about its axis of rotation is 3 kg m^2. If the vehicle is brought to rest in 15s, the magnitude of average torque transmitted by its brakes to the wheel is :
(a) 8.58 kg m^2 s^{-2} (b) 10.86 kg m^2 s^{-2}
(c) 2.86 kg m^2 s^{-2} (d) 6.66 kg m^2 s^{-2}

14. A source of sound S emitting waves of frequency 100 Hz and an observor O are located at some distance from each other. The source is moving with a speed of 19.4 ms^{-1} at an angle of 60° with the source observer line as shown in the figure. The observor is at rest. The apparent frequency observed by the observer is (velocity of sound in air 330 ms^{-1})

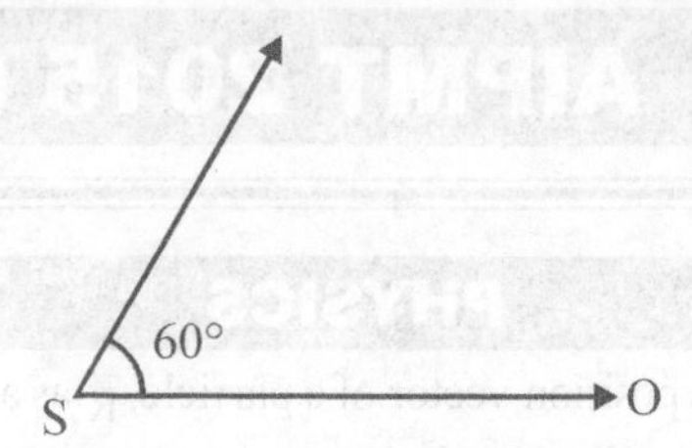

(a) 103 Hz (b) 106 Hz
(c) 97 Hz (d) 100 Hz

15. On a frictionless surface a block of mass M moving at speed v collides elastically with another block of same mass M which is initially at rest. After collision the first block moves at an angle θ to its initial direction and has a speed $\frac{v}{3}$. The second block's speed after the collision is :
(a) $\frac{3}{4}v$ (b) $\frac{3}{\sqrt{2}}v$
(c) $\frac{\sqrt{3}}{2}v$ (d) $\frac{2\sqrt{2}}{3}v$

16. Point masses m_1 and m_2 are placed at the opposite ends of a rigid rod of length L, and negligible mass. The rod is to be set rotating about an axis perpendicular to it. The position of point P on this rod through which the axis should pass so that the work required to set the rod rotating with angular velocity ω_0 is minimum, is given by :

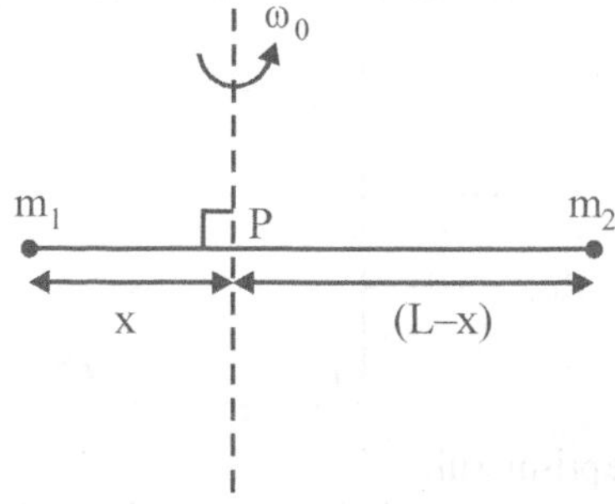

(a) $x = \frac{m_1}{m_2}L$ (b) $x = \frac{m_2}{m_1}L$
(c) $x = \frac{m_2 L}{m_1 + m_2}$ (d) $x = \frac{m_1 L}{m_1 + m_2}$

17. A ball is thrown vertically downwards from a height of 20 m with an initial velocity v_0. It collides with the ground loses 50 percent of its energy in collision and rebounds to the same height. The initial velocity v_0 is :
(Take $g = 10\ ms^{-2}$)
(a) $20\ ms^{-1}$ (b) $28\ ms^{-1}$
(c) $10\ ms^{-1}$ (d) $14\ ms^{-1}$

18. A nucleus of uranium decays at rest into nuclei of thorium and helium. Then :
(a) the helium nucleus has less momentum than the thorium nucleus.
(b) the helium nucleus has more momentum than the thorium nucleus.
(c) the helium nucleus has less kinetic energy than the thorium nucleus.
(d) the helium nucleus has more kinetic energy than the thorium nucleus.

19. An electron moves on a straight line path XY as shown. The abcd is a coil adjacent to the path of electron.
What will be the direction of current if any, induced in the coil?

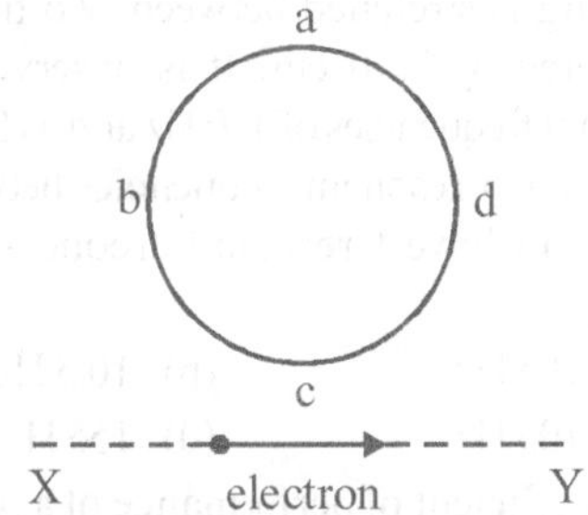

(a) adcb
(b) The current will reverse its direction as the electron goes past the coil
(c) No current induced
(d) abcd

20. A particle is executing a simple harmonic motion. Its maximum acceleration is α and maximum velocity is β.
Then its time period of vibration will be :
(a) $\frac{\alpha}{\beta}$ (b) $\frac{\beta^2}{\alpha}$
(c) $\frac{2\pi\beta}{\alpha}$ (d) $\frac{\beta^2}{\alpha^2}$

21. Two slits in Young's experiment have widths in the ratio 1 : 25. The ratio of intensity at the maxima and minima in the interference pattern, $\frac{I_{max}}{I_{min}}$ is:
(a) $\frac{121}{49}$ (b) $\frac{49}{121}$
(c) $\frac{4}{9}$ (d) $\frac{9}{4}$

22. If potential (in volts) in a region is expressed as $V(x, y, z) = 6\ xy - y + 2yz$, the electric field (in N/C) at point (1, 1, 0) is :
(a) $-(6\hat{i}+5\hat{j}+2\hat{k})$ (b) $-(2\hat{i}+3\hat{j}+\hat{k})$
(c) $-(6\hat{i}+9\hat{j}+\hat{k})$ (d) $-(3\hat{i}+5\hat{j}+3\hat{k})$

23. A parallel plate air capacitor has capacity 'C' distance of separation between plates is 'd' and potential difference 'V' is applied between the plates. Force of attraction between the plates of the parallel plate air capacitor is :
(a) $\frac{CV^2}{2d}$ (b) $\frac{CV^2}{d}$
(c) $\frac{C^2V^2}{2d^2}$ (d) $\frac{C^2V^2}{2d^2}$

24. A plank with a box on it at one end is gradually raised about the other end. As the angle of inclination with the horizontal reaches 30° the box starts to slip and slides 4.0 m down the plank in 4.0s. The coefficients of static and kinetic friction between the box and the plank will be, respectively :

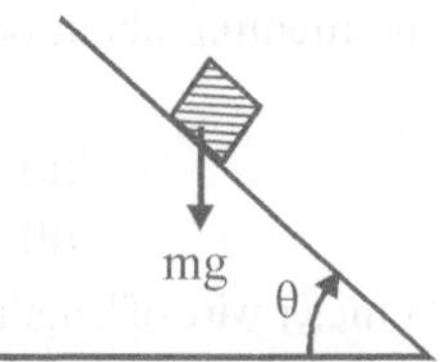

(a) 0.6 and 0.5 (b) 0.5 and 0.6
(c) 0.4 and 0.3 (d) 0.6 and 0.6

25. In the spectrum of hydrogen, the ratio of the longest wavelength in the Lyman series to the longest wavelength in the Balmer series is
(a) $\frac{9}{4}$ (b) $\frac{27}{5}$
(c) $\frac{5}{27}$ (d) $\frac{4}{9}$

26. In the given figure, a diode D is connected to an external resistance R = 100 Ω and an e.m.f. of 3.5 V. If the barrier potential developed across the diode is 0.5 V, the current in the circuit will be:

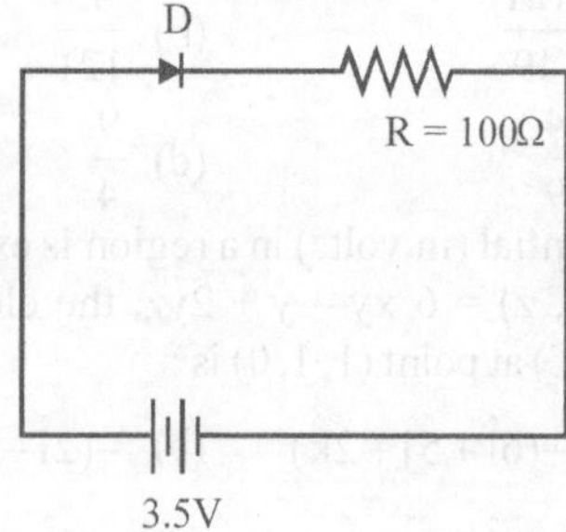

(a) 40 mA (b) 20 mA
(c) 35 mA (d) 30 mA

27. A satellite S is moving in an elliptical orbit around the earth. The mass of the satellite is very small compared to the mass of the earth. Then,

(a) the total mechanical energy of S varies periodically with time.
(b) the linear momentum of S remains constant in magnitude.
(c) the acceleration of S is always directed towards the centre of the earth.
(d) the angular momentum of S about the centre of the earth changes in direction, but its magnitude remains constant.

28. A force $\vec{F} = \alpha\hat{i} + 3\hat{j} + 6\hat{k}$ is acting at a point $\vec{r} = 2\hat{i} - 6\hat{j} - 12\hat{k}$. The value of α for which angular momentum about origin is conserved is :

(a) 2 (b) zero
(c) 1 (d) -1

29. A potentiometer wire of length L and a resistance r are connected in series with a battery of e.m.f. E_0 and a resistance r_1. An unknown e.m.f. E is balanced at a length l of the potentiometer wire. The e.m.f. E will be given by :

(a) $\frac{E_0 r}{(r + r_1)} \cdot \frac{l}{L}$ (b) $\frac{E_0 l}{L}$

(c) $\frac{LE_0 r}{(r + r_1)l}$ (d) $\frac{LE_0 r}{l r_1}$

30. 4.0 g of a gas occupies 22.4 litres at NTP. The specific heat capacity of the gas at constant volume is $5.0 JK^{-1}$. If the speed of sound in this gas at NTP is 952 ms^{-1}, then the heat capacity at constant pressure is (Take gas constant R = 8.3 $JK^{-1} mol^{-1}$)

(a) 7.5 $JK^{-1} mol^{-1}$ (b) 7.0 $JK^{-1} mol^{-1}$
(c) 8.5 $JK^{-1} mol^{-1}$ (d) 8.0 $JK^{-1} mol^{-1}$

31. Two stones of masses m and 2 m are whirled in horizontal circles, the heavier one in radius $\frac{r}{2}$ and the lighter one in radius r. The tangential speed of lighter stone is n times that of the value of heavier stone when they experience same centripetal forces. The value of n is :

(a) 3 (b) 4
(c) 1 (d) 2

32. A remote - sensing satellite of earth revolves in a circular orbit at a height of 0.25×10^6 m above the surface of earth. If earth's radius is 6.38×10^6 m and g = 9.8 ms^{-2}, then the orbital speed of the satellite is:

(a) 8.56 km s^{-1} (b) 9.13 km s^{-1}
(c) 6.67 km s^{-1} (d) 7.76 km s^{-1}

33. A string is stretched between two fixed points separated by 75.0 cm. It is observed to have resonant frequencies of 420 Hz and 315 Hz. There are no other resonant frequencies between these two. The lowest resonant frequency for this string is :

(a) 205 Hz (b) 10.5 Hz
(c) 105 Hz (d) 155 Hz

34. The coefficient of performance of a refrigerator is 5. If the inside temperature of freezer is -20°C, then the temperature of the surroundings to which it rejects heat is

(a) 41°C (b) 11°C
(c) 21°C (d) 31°C

35. Water rises to a height 'h' in a capillary tube. If the length of capaillary tube above the surface of water is made less than 'h' then :

(a) water rises upto the top of capillary tube and stays there without overflowing
(b) water rises upto a point a little below the top and stays there
(c) water does not rise at all.
(d) Water rises upto the tip of capillary tube and then starts overflowing like fountain.

36. Two vessels separately contain two ideal gases A and B at the same temperature. The pressure of A being twice that of B. Under such conditions, the density of A is found to be 1.5 times the density of B. The ratio of molecular weight of A and B is :

(a) $\frac{3}{4}$ (b) 2

(c) $\frac{1}{2}$ (d) $\frac{2}{3}$

37. The Young's modulus of steel is twice that of brass. Two wires of same length and of same area of cross section, one of steel and another of brass are suspended from the same roof. If we want the lower ends of the wires to be at the same level, then the weights added to the steel and brass wires must be in the ratio of :

(a) 2 : 1 (b) 4 : 1

(c) 1 : 1 (d) 1 : 2

38. The input signal given to a CE amplifier having a voltage gain of 150 is $V_i = 2\cos\left(15t+\frac{\pi}{3}\right)$. The corresponding output signal will be :

(a) $75\cos\left(15t+\frac{2\pi}{3}\right)$

(b) $2\cos\left(15t+\frac{5\pi}{6}\right)$

(c) $300\cos\left(15t+\frac{4\pi}{3}\right)$

(d) $300\cos\left(15t+\frac{\pi}{3}\right)$

39. In an astronomical telescope in normal adjustment a straight black line of lenght L is drawn on inside part of objective lens. The eye-piece forms a real image of this line. The length of this image is *l*. The magnification of the telescope is :

(a) $\frac{L}{I}-1$ (b) $\frac{L+I}{L-I}$

(c) $\frac{L}{I}$ (d) $\frac{L}{I}+1$

40. The heart of man pumps 5 litres of blood through the arteries per minute at a pressure of 150 mm of mercury. If the density of mercury be 13.6×10^3 kg/m^3 and g = 10m/s^2 then the power of heart in watt is :

(a) 2.35 (b) 3.0

(c) 1.50 (d) 1.70

41. If dimensions of critical velocity υ_c of a liquid flowing through a tube are expressed as $[\eta^x \rho^y r^x]$, where η, ρ and r are the coefficient of viscosity of liquid, density of liquid and radius of the tube respectively, then the values of x, y and z are given by :

(a) -1, -1, 1 (b) -1, -1, -1

(c) 1,1,1 (d) 1, -1, -1

42. A photoelectric surface is illuminated successively by monochromatic light of wavelength λ and $\frac{\lambda}{2}$. If the maximum kinetic energy of the emitted photoelectrons in the second case is 3 times that in the first case, the work function of the surface of the material is : (h = Planck's constant, c = speed of light)

(a) $\frac{hc}{\lambda}$ (b) $\frac{2hc}{\lambda}$

(c) $\frac{hc}{3\lambda}$ (d) $\frac{hc}{2\lambda}$

43. The cylindrical tube of a spray pump has radius, R, one end of which has n fine holes, each of radius r. If the speed of the liquid in the tube is V, the speed of the ejection of the liquid through the holes is :

(a) $\frac{VR^2}{nr^2}$ (b) $\frac{VR^2}{n^3r^2}$

(c) $\frac{V^2R}{nr}$ (d) $\frac{VR^2}{n^2r^2}$

44. If vectors $\vec{A} = \cos\omega t\hat{i} + \sin\omega t\hat{j}$ and $\vec{B} = \cos\frac{\omega t}{2}\hat{i} + \sin\frac{\omega t}{2}\hat{j}$ are functions of time, then the value of t at which they are orthogonal to each other is :

(a) $t = \frac{\pi}{2\omega}$ (b) $t = \frac{\pi}{\omega}$

(c) $t = 0$ (d) $t = \frac{\pi}{4\omega}$

45. A rectangular coil of length 0.12 m and width 0.1 m having 50 turns of wire is suspended vertically in a uniform magnetic field of strength 0.2 weber/m^2. The coil carries a current of 2A. If the plane of the coil is inclined at an angle of 30° with the direction of the field, the torque required to keep the coil in stable equilibrium will be :

(a) 0.20 Nm (b) 0.24 Nm

(c) 0.12 Nm (d) 0.15 Nm

CHEMISTRY

46. Reaction of phenol with chloroform in presence of dilute sodium hydroxide finally introduces which one of the following functional group ?

(a) $-CH_2Cl$ (b) $-COOH$

(c) $-CHCl_2$ (d) $-CHO$

47. If the equilibrium constant for

$N_2(g) + O_2(g) \rightleftharpoons 2NO(g)$ is K,

the equilibrium constant for

$\frac{1}{2}N_2(g) + \frac{1}{2}O_2(g) \rightleftharpoons NO(g)$ will be:

(a) $K^{\frac{1}{2}}$ (b) $\frac{1}{2}K$

(c) K (d) K^2

48. 20.0 g of a magnesium carbonate sample decomposes on heating to give carbon dioxide and 8.0 g magnesium oxide. What will be the percentage purity of magnesium carbonate in the sample ?

(a) 75 (b) 96

(c) 60 (d) 84

49. The number of water molecules is maximum in :

(a) 18 molecules of water

(b) 1.8 gram of water

(c) 18 gram of water

(d) 18 moles of water

50. The formation of the oxide ion $O^{2-}(g)$, from oxygen atom requires first an exothermic and then an endothermic step as shown below :

$O(g) + e^- \rightarrow O^-(g);\ \Delta_f H^\ominus = -141\ kJ\ mol^{-1}$

$O^-(g) + e^- \rightarrow O^{2-}(g);\ \Delta_f H^\ominus = +780\ kJ\ mol^{-1}$

Thus process of formation of O^{2-} in gas phase is unfavourable even though O^{2-} is isoelectronic with neon. It is due to the fact that

(a) Electron repulsion outweighs the stability gained by achieving noble gas configuration

(b) O^- ion has comparatively smaller size than oxygen atom

(c) Oxygen is more electronegative

(d) Addition of electron in oxygen results in larger size of the ion.

51. What is the mole fraction of the solute in a 1.00 m aqueous solution ?

(a) 0.177 (b) 1.770

(c) 0.0354 (d) 0.0177

52. The rate constant of the reaction $A \rightarrow B$ is 0.6×10^{-3} mole per second. If the concentration of A is 5 M then concentration of B after 20 minutes is :

(a) 1.08 M (b) 3.60 M

(c) 0.36 M (d) 0.72 M

53. Decreasing order of stability of O_2, O_2^-, O_2^+ and O_2^{2-} is :

(a) $O_2^+ > O_2 > O_2^- > O_2^{2-}$

(b) $O_2^{2-} > O_2^- > O_2 > O_2^+$

(c) $O_2 > O_2^+ > O_2^{2-} > O_2^-$

(d) $O_2^- > O_2^{2-} > O_2^+ > O_2$

54. Which one of the following esters gets hydrolysed most easily under alkaline conditions?

(a) O_2N–C$_6$H$_4$–$OCOCH_3$

(b) H_3CO–C$_6$H$_4$–$OCOCH_3$

(c) C$_6$H$_5$–$OCOCH_3$

(d) 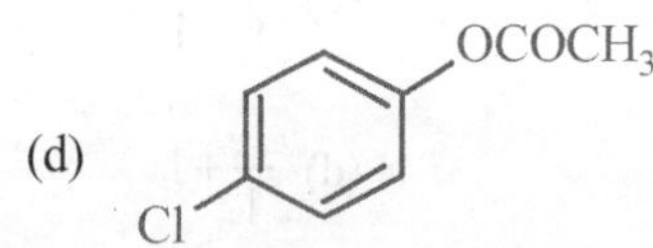

55. On heating which of the following releases CO_2 most easily ?
(a) K_2CO_3 (b) Na_2CO_3
(c) $MgCO_3$ (d) $CaCO_3$

56. Which one of the following pairs of solution is not an acidic buffer ?
(a) $HClO_4$ and $NaClO_4$
(b) CH_3COOH and CH_3COONa
(c) H_2CO_3 and Na_2CO_3
(d) H_3PO_4 and Na_3PO_4

57. The sum of coordination number and oxidation number of the metal M in the complex $[M(en)_2(C_2O_4)]Cl$ (where *en* is ethylenediamine) is:
(a) 9 (b) 6
(c) 7 (d) 8

58. Which of the statements given below is incorrect?
(a) Cl_2O_7 is an anhydride of perchloric acid
(b) O_3 molecule is bent
(c) ONF is isoelectronic with O_2N^-.
(d) OF_2 is an oxide of fluorine

59. In the reaction with HCl, an alkene reacts in accordance with the Markovnikov's rule, to give a product 1-chloro-1-methylcyclohexane. The possible alkene is :

(a) [methylenecyclohexane: CH_2 exocyclic double bond] (b) [1-methylcyclohexene: CH_3]

(c) (a) and (b) (d) [3-methylcyclohexene: CH_3]

60. 2,3-Dimethyl-2-butene can be prepared by heating which of the following compounds with a strong acid ?
(a) $(CH_3)_2CH-\underset{\underset{CH_3}{|}}{CH}-CH=CH_2$
(b) $(CH_3)_3C-CH=CH_2$
(c) $(CH_3)_2C=CH-CH_2-CH_3$
(d) $(CH_3)_2CH-CH_2-CH=CH_2$

61. The following reaction

$$C_6H_5NH_2 + Cl-CO-C_6H_5 \xrightarrow{NaOH} C_6H_5-NH-CO-C_6H_5$$

is known by the name :
(a) Friedel-Craft's reaction
(b) Perkin's reaction
(c) Acetylation reaction
(d) Schotten-Baumen reaction

62. In the extraction of copper from its sulphide ore, the metal finally obtained by the reduction of cuprous oxide with :
(a) iron (II) sulphide (b) carbon monoxide
(c) copper (I) sulphide (d) sulphur dioxide

63. If Avogadro number N_A, is changed from 6.022×10^{23} mol^{-1} to 6.022×10^{20} mol^{-1} this would change :
(a) the definition of mass in units of grams
(b) the mass of one mole of carbon
(c) the ratio of chemical species to each other in a balanced equation.
(d) the ratio of elements to each other in a compound

64. The variation of the boiling point of the hydrogen halides is in the order HF > HI > HBr > HCl.
What explains the higher boiling point of hydrogen fluoride?
(a) The electronegativity of fluorine is much higher than for other elements in the group.
(b) There is strong hydrogen bonding between HF molecules
(c) The bond energy of HF molecules is greater than in other hydrogen halides.
(d) The effect of nuclear shielding is much reduced in fluorine which polarises the HF molecule.

65. Which of the following reaction (s) can be used for the preparation of alkyl halides ?
(I) $CH_3CH_2OH + HCl \xrightarrow{anh.ZnCl_2}$
(II) $CH_3CH_2OH + HCl \longrightarrow$

(III) $(CH_3)_3COH + HCl \longrightarrow$

(IV) $(CH_3)_2CHOH + HCl \xrightarrow{\text{anh.}ZnCl_2}$

(a) (I), (III) and (IV) only
(b) (I) and (II) only
(c) (IV) only
(d) (III) and (IV) only

66. The name of complex ion, $[Fe(CN)_6]^{3-}$ is :
(a) Hexacyanoiron (III) ion
(b) Hexacyanitoferrate (III) ion
(c) Tricyanoferrate (III) ion
(d) Hexacyanidoferrate (III) ion

67. Assuming complete ionization, same moles of which of the following compounds will require the least amount of acidified $KMnO_4$ for complete oxidation
(a) $FeSO_4$ (b) $FeSO_3$
(c) FeC_2O_4 (d) $Fe(NO_2)_2$

68. In which of the following pairs, both the species are not isostructural ?
(a) $SiCl_4, PCl_4^+$
(b) diamond, silicon carbide
(c) NH_3, PH_3
(d) XeF_4, XeO_4

69. Caprolactum is used for the manufacture of :
(a) Nylon - 6 (b) Teflon
(c) Terylene (d) Nylon - 6,6

70. The hybridization involved in complex $[Ni(CN)_4]^{2-}$. is (At. No. Ni = 28)
(a) dsp^2 (b) sp^3
(c) d^2sp^2 (d) d^2sp^3

71. What is the mass of precipitate formed when 50 mL of 16.9% solution of $AgNO_3$ is mixed with 50 mL of 5.8% NaCl solution ?
(Ag = 107.8, N = 14, O = 16, Na = 23, Cl = 35.5)
(a) 28 g (b) 3.5 g
(c) 7 g (d) 14 g

72. Gadolinium belongs to 4f series. It's atomic number is 64. Which of the following is the correct electronic configuration of gadolinium ?
(a) $[Xe]4f^86d^2$ (b) $[Xe]4f^95s^1$
(c) $[Xe]4f^75d^16s^2$ (d) $[Xe]4f^65d^26s^2$

73. Which of the following is not the product of dehydration of OH ?

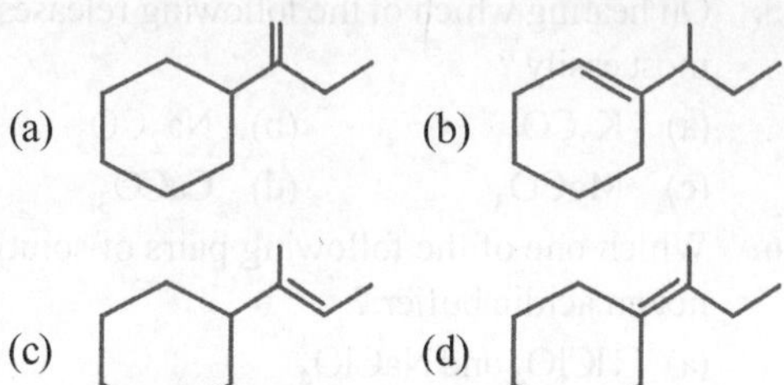

74. A gas such as carbon monoxide would be most likely to obey the ideal gas law at :
(a) high temperatures and low pressures.
(b) low temperatures and high pressures.
(c) high temperatures and low pressures.
(d) low temperatures and low pressures.

75. The stability of +1 oxidation state among Al, Ga, In and Tl increases in the sequence :
(a) Ga < In < Al < Tl (b) Al < Ga < In < Tl
(c) Tl < In < Ga < Al (d) In < Tl < Ga < Al

76. What is the pH of the resulting solution when equal volumes of 0.1 M NaOH and 0.01 M HCl are mixed ?
(a) 12.65 (b) 2.0
(c) 7.0 (d) 1.04

77. Strong reducing behaviour of H_3PO_2 is due to
(a) presence of one –OH group and two P–H bonds
(b) high electron gain enthalpy of phosphorus
(c) high oxidation state of phosphorus
(d) presence of two –OH groups and one P–H bond.

78. The number of structural isomers possible from the molecular formula C_3H_9N is :
(a) 4 (b) 5
(c) 2 (d) 3

79. Which of the following statements is not correct for a nucleophile ?
(a) Nucleophile is a Lewis acid
(b) Ammonia is a nucleophile
(c) Nucleophiles attack low e^- density sites
(d) Nucleophiles are not electron seeking.

80. Number of possible isomers for the complex $[Co(en)_2Cl_2]Cl$ will be (en = ethylenediamine)
(a) 2 (b) 1
(c) 3 (d) 4

81. Which is the correct order of increasing energy of the listed orbitals in the atom of titanium ?
(a) 3s 4s 3p 3d (b) 4s 3s 3p 3d
(c) 3s 3p 3d 4s (d) 3s 3p 4s 3d

82. In an S_N1 reaction on chiral centres there is :
(a) 100 % racemization
(b) inversion more than retention leading to partial racemization
(c) 100 % retention
(d) 100 % inversion

83. The vacant space in bcc lattice cell is :
(a) 26 % (b) 48 %
(c) 23 % (d) 32 %

84. The heat of combustion of carbon to CO_2 is –393.5 kJ/mol. The heat released upon formation of 35.2 g of CO_2 from carbon and oxygen gas is
(a) –315 kJ (b) +315kJ
(c) –630 kJ (d) –3.15 kJ

85. Aqueous solution of which of the following compounds is the best conductor of electric current ?
(a) Acetic acid, $C_2H_4O_2$
(b) Hydrochloric acid, HCl
(c) Ammonia, NH_3
(d) Fructose, $C_6H_{12}O_6$

86. The oxidation of benzene by V_2O_5 in presence of air produces :
(a) benzoic anhydride (b) maleic anhydride
(c) benzoic acid (d) benzaldehyde

87. Reaction of a carbonyl compound with one of the following reagents involves nucleophilic addition followed by elimination of water. The reagent is :
(a) a Grignard reagent
(b) hydrazine in presence of feebly acidic solution
(c) hydrocyanic acid
(d) sodium hydrogen sulphite

88. Method by which Aniline cannot be prepared is:
(a) hydrolysis of phenylisocyanide with acidic solution
(b) degradation of benzamide with bromine in alkaline solution
(c) reduction of nitrobenzene with H_2/Pd in ethanol
(d) potassium salt of phthalimide treated with chlorobenzene followed by hydrolysis with aqueous NaOH solution.

89. Two possible stereo-structures of $CH_3CHOH.COOH$, which are optically active, are called.
(a) Diastereomers (b) Atropisomers
(c) Enantiomers (d) Mesomers

90. The correct statement regarding defects in crystalline solid is :
(a) Schottky defects have no effect on the density of crystalline solids
(b) Frenkel defects decrease the density of crystalline solids
(c) Frenkel defect is a dislocation defect.
(d) Frenkel defect is found in halides of alkaline metals.

BIOLOGY

91. In his classic experiments on sea plants, Mendel did not use
(a) Pod length (b) Seed shape
(c) Flower position (d) Seed colour

92. Which one of the following is not applicable to RNA?
(a) 5' phosphoryl and 3' hydroxyl ends
(b) Heterocyclic nitrogenous bases
(c) Chargaff's rule
(d) Complementary base pairing

93. Male gametophyte in angiosperms produces:
(a) Single sperm and vegetative cell
(b) Single sperm and two vegetative cells
(c) Three sperms
(d) Two sperms and a vegetative cell

94. Which of the following are not membrane-bound?
(a) Ribosomes (b) Lysosomes
(c) Mesosomes (d) Vacuoles

95. The chitinous exoskeleton of arthropods is formed by the polymerisation of:
(a) D-glucosamine
(b) N-acetyl glucosamine
(c) lipoglycans
(d) Keratin sulphate and chondroitin sulphate

96. Among china rose, mustard, brinjal, potato, guava, cucumber, onion and tulip, how many plants have superior ovary?
(a) Six (b) Three
(c) Four (d) Five

97. The function of the gap junction is to
(a) facilitate communication between adjoining cells by connecting the cytoplasm for rapid transfer of ions, small molecules and some large molecules.

(b) separate two cells from each other.
(c) stop substance from leading across a tissue
(d) performing cementing to keep neighbouring cells together

98. Which one of the following immunoglobulins does constitute the largest percentage in human milk?
(a) lg M (b) lg A
(c) lg G (d) lg D

99. In mammalian eye, the 'fovea' is the center of the visual field, where:
(a) the optic nerve leaves the eye
(b) only rods are present
(c) more rods than cones are found
(d) high density of cones occur, but has no rods.

100. Doctors use stethoscope to hear the sounds produced during each cardiac cycle. The second sound is heard when:
(a) Ventricular wall vibrate due to gushing in of blood from atria
(b) Semilunar valves close down after the blood flows into vessels from ventricles
(c) AV node receives signal from SA node
(d) AV valves open up

101. Coconut water from a tender coconut is
(a) Free nuclear endosperm
(b) Innermost layers of the seed coar
(c) Degenerated nucellus
(d) Immature emryo

102. The cutting of DNA at specific locations became possible with the discovery of
(a) Probes
(b) Selectable markers
(c) Ligases
(d) Restriction enzymes

103. Which of the following structures is not found in a prokaryotic cell?
(a) Ribosome (b) Mesosome
(c) Plasma membrane (d) Nuclear envelope

104. Choose the correct option for the following events of meiosis in correct sequence:
(A) Crossing over
(B) Synapsis
(C) Terminalisation of chiasmata
(D) Disappearance of nucleolus
(a) $B \rightarrow A \rightarrow C \rightarrow D$ (b) $A \rightarrow B \rightarrow C \rightarrow D$
(c) $A \rightarrow B \rightarrow D \rightarrow C$ (d) $D \rightarrow C \rightarrow B \rightarrow A$

105. A column of water within xylem vessels of tall trees does not break under its weight because of:
(a) Tensile strength of water
(b) Lignification of xylem vessels
(c) Positive root pressure
(d) Dissolved sugars in water

106. The imperfect fungi which are decomposer of litter and help in mineral cycling belong to:
(a) Basidiomycetes (b) Phycomycetes
(c) Ascomycetes (d) Deuteromycetes

107. The structures that help some bacteria to attach to rocks and / or host tissues are:
(a) Fimbriae (b) Mesosomes
(c) Holdfast (d) Rhizoids

108. The DNA molecule to which the gene of interest is integrated for cloning is called:
(a) Vector (b) Template
(c) Carrier (d) Transformer

109. Pick up the wrong statement
(a) Prostista have photosynthetic and heterotrophic modes of nutrition
(b) Some fungi are edible
(c) Nuclear membrane is present Monera
(d) Cell wall is absent in Animalia

110. Metagenesis refers to:
(a) Alteration of generation between asexual and sexual phases of an organisms
(b) Occurrence of a drastic change in form during post-embroyonic development
(c) Presence of a segmented body and parthenogenetic mode of reproduction
(d) Presence of different morphic forms

111. Which of the following events is not associated with ovulation in human female?
(a) Full development of Graffin follicle
(b) Release of secondary oocyte
(c) LH surge
(d) Decrease in estradiol

112. Which of the following joints would allow no movement?
(a) Cartilaginoius joint
(b) Synovial joint
(c) Ball and Socket joint
(d) Fibrous joint

113. Match the following list of microbes and their importance:

(1)	*Sacharomyces cerevisiae*	(i)	Production of immunosuppressive agents
(2)	*Monascus Purpureus*	(ii)	Ripening of Swiss cheese
(3)	*Trichoderma polysporum*	(iii)	Commercial prodcution of ethanol
(4)	*Propionibacterium sharmanii*	(iv)	Production of blood cholesterol lowering agents

	(1)	(2)	(3)	(4)
(a)	(iv)	(iii)	(ii)	(i)
(b)	(iv)	(ii)	(i)	(iii)
(c)	(iii)	(i)	(iv)	(ii)
(d)	(iii)	(iv)	(i)	(ii)

114. The UN conference of Parties on climate change in the year 2012 was held at:
(a) Doha (b) Lima
(c) Warsaw (d) Durban

115. If you suspect major deficiency of antibodies in a person, to which of the following would you look for confirmatory evidences?
(a) Serum albumins
(b) Haemocytes
(c) Serum globulins
(d) Fibrinogin in plasma

116. Chromatophores take part in:
(a) Growth (b) Movement
(c) Respiration (d) Photosynthesis

117. Acid rain is caused by increase in the atmospheric concentration of:
(a) SO_3 and CO (b) CO_2 and CO
(c) O_3 and dust (d) SO_2 and NO_2

118. During ecological succession:
(a) the establishment of a new biotic community is very fast in its primary phase.
(b) the numbers and types of animals remain constant.
(c) the changes lead to a community that is in near equlibrium with the environment and is called pioneer community.
(d) the gradual and predictable change in species composition occurs in a given area.

119. The oxygen evolved during photosynthesis comes from water molecules. Which one of the following pairs of elements is involved in this reaction?
(a) Manganese and Potassium
(b) Magnesium and Molybdenum
(c) Magnesium and Chlorine
(d) Manganese and Chlorine

120. Which of the following pairs is not correctly matched?

	Mode of reproduction	**Example**
(a)	Rhizome	Banana
(b)	Binary fission	*Sargassum*
(c)	Conidia	*Penicillium*
(d)	Offset	Water hyacinth

121. In the following human pedigree, the filled symbols represent the affected individuals. Identify the type of given pedigree.

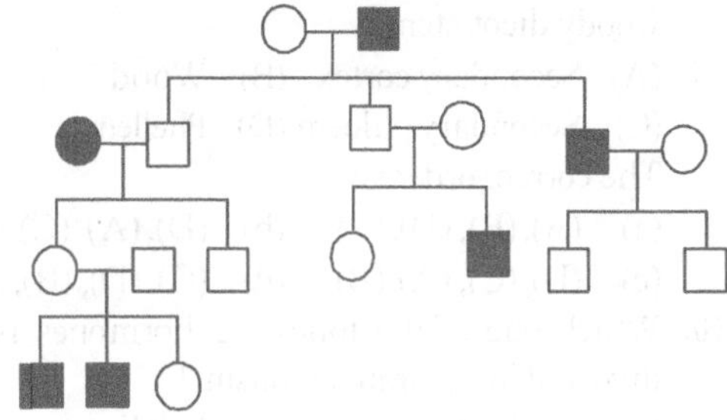

(a) X- linked recessive
(b) Autosomal recessive
(c) X-linked dominant
(d) Autosomal dominant

122. Which one of the following animals has two separate circulatory pathways?
(a) Lizard (b) Whale
(c) Shark (d) Frog

123. Flowers are unisexual in:
(a) Cucumber (b) China rose
(c) Onion (d) Pea

124. Which one of the following fruits is parthenocarpic?
(a) Apple (b) Jackfruit
(c) Banana (d) Brinjal

125. A pleiotropic gene:
(a) is a gene evolved during Pliocene.
(b) controls a trait only in combination with another gene
(c) controls multiple traits in an individual.
(d) is expressed only in primitive plants

126. Which of the following is not a function of the skeletal system?
(a) Storage of minerals
(b) Production of body heat
(c) Locomotion
(d) Production of erythrocytes

127. A jawless fish, which lays eggs in fresh water and whose ammocoetes larvae after metamorphosis return to the ocean is:
(a) *myxine* (b) *Neomyxine*
(c) *Petromyzon* (d) *Eptatretus*

128. Filiform apparatus is characteristic feature of:
(a) Nucellar embryo (b) Aleurone cell
(c) Synergids (d) Generative cell

129. Read the different components from (A) to (D) in the list given below and tell the correct order of the components with reference to their arrangement from outer side to inner side in a woody dicot stem.
(A) Secondary cortex (B) Wood
(C) Secondary phloem (D) Phellem
The correct order is:
(a) (A), (B), (D), (C) (b) (D), (A), (C), (B)
(c) (D), (C), (A), (B) (d) (C), (D), (B), (A)

130. Which one of the following hormones is not involved in sugar metabolism ?
(a) Aldosterone (b) Insulin
(c) Glucagon (d) Cortisone

131. Golden rice is a genetically modified crop plant where the incorporated gene is meant for biosynthesis of :
(a) Vitamin C (b) Omega 3
(c) Vitamin A (d) Vitamin B

132. Outbreeding is an important strategy of animal husbandry because it :
(a) is useful in producing purelines of animals.
(b) is useful in overcoming inbreeding depression.
(c) exposes harmful recessive genes that are eliminated by selection.
(d) helps in accumulation of superior genes.

133. A gene showing codominance has:
(a) alleles tightly linked on the same chromosome
(b) alleles that are recessive to each other
(c) both alleles independently expressed in the heterozygote
(d) one allele dominant on the other

134. Which one of the following hormones though synthesised elsewhere is stored and released by the master gland ?
(a) Luteinizing hormone
(b) Prolactin
(c) Melanocyte stimulating hormone
(d) Antidiuretic hormone

135. Increase in concentration of the toxicant at successive trophic levels is known as :
(a) Biodeterioration
(b) Biotransformation
(c) Biogeochemical
(d) Biomagnification

136. Industrial melanism is an example of :
(a) Natural selection (b) Mutation
(c) Neo Lamarckism (d) Neo Darwinism

137. The primary dentition in human differs from permanent dentition in not having one of the following type of teeth :
(a) Premolars (b) Molars
(c) Incisors (d) Canine

138. The wheat grain has an embryo with one, large, shield-shaped cotyledon known as:
(a) Coleorrhiza (b) Scutellum
(c) Coleoptile (d) Epiblast

139. The body cells in cockroach discharge their nitrogenous waste in the haemolymph mainly in the form of :
(a) Potassium urate (b) Urea
(c) Calcium carbonate (d) Ammonia

140. Which of the following biomolecules does have phosphodiester bond ?
(a) Monosaccharides in a polysaccharide
(b) Amino acids in a polypeptide
(c) Nucleic acids in a nucleotide
(d) Fatty acids in a diglyceride

141. The term 'linkage' was coined by :
(a) T. Boveri (b) G. Mendel
(c) W. Sutton (d) T.H. Morgan

142. Which one is wrong statement ?
(a) Mucor has biflagellate zoospores
(b) Haploid endosperm is typical feature of gymnosperms
(c) Brown algae have chlorophyll *a* and *c* and fucoxanthin
(d) Archegonia are found in Bryophyta, Pteridophyta and Gymnosperms.

(a) Implantation of embryo at site other than uterus.
(b) Implantation of defective embryo in the uterus
(c) Pregnancies terminated due to hormonal imbalance.
(d) Pregnancies with genetic abnormality.

144. Most animals that live in deep oceanic waters are:
(a) secondary consumers
(b) tertiary consumers
(c) detritivores
(d) primary consumers

145. Which of the following diseases is caused by a protozoan ?
(a) Influenza (b) Babesiosis
(c) Blastomycosis (d) Syphilis

146. In which of the following interactions both partners are adversely affected ?
(a) Predation (b) Parasitism
(c) Mutualism (d) Competition

147. Identify the correct order of organisation of genetic material from largest to smallest :
(a) Genome, chromosome, nucleotide, gene
(b) Genome, chromosome, gene, nucleotide
(c) Chromosome, genome, nucleotide, gene
(d) Chromosome, gene, genome, nucleotide

148. A colour blind man marries a woman with normal sight who has no history of colour blindness in her family.
What is the probability of their grandson being colour blind ?
(a) 1 (b) Nil
(c) 0.25 (d) 0.5

149. In photosynthesis the light-independent reactions take place at :
(a) Photosystem-I (b) Photosystem-II
(c) Stromal matrix (d) Thylakoid lumen

150. In which of the following both pairs have correct combination ?

(a)	Gaseous nutrient cycle	Carbon and sulphur
	Sedimentary nutrient cycle	Nitrogen and Phosphorus
(b)	Gaseous nutrient cycle	Nitrogen and sulphur
	Sedimentary nutrient cycle	Carbon and Phosphorus
(c)	Gaseous nutrient cycle	Sulphur and Phosphorus
	Sedimentary nutrient cycle	Carbon and Nitrogen
(d)	Gaseous nutrient cycle	Carbon and Nitrogen
	Sedimentary nutrient cycle	Sulphur and Phosphorus

151. The introduction of t-DNA into plants involves:
(a) Altering the pH of the soil, then heat shocking the plants
(b) Exposing the plants to cold for a brief period
(c) Allowing the plant roots to stand in water
(d) Infection of the plant by *Agrobacterium tumefaciens*

152. The wings of a bird and the wings of an insect are :
(a) analogous structures and represent convergent evolution
(b) phylogenetic structures and represent divergent evolution
(c) homologous structures and represent convergent evolution
(d) homologous structures and represent divergent evolution

153. Root pressure develops due to :
(a) Low osmotic potential in soil
(b) Passive absorption
(c) Increase in transpiration
(d) Active absorption

154. Human urine is usually acidic because :
(a) excreted plasma proteins are acidic
(b) potassium and sodium exchange generates acidity
(c) hydrogen ions are actively secreted into the filtrate.
(d) the sodium transporter exchanges one hydrogen ion for each sodium ion, in peritubular capillaries.

155. A protoplast is a cell :
(a) without nucleus
(b) undergoing division
(c) without cell wall
(d) without plasma membrane

156. The species confined to a particular region and not found elsewhere is termed as :
(a) Alien (b) Endemic
(c) Rare (d) Keystone

157. Select the wrong statements :
(a) W.M. Stanley showed that viruses could be crystallized
(b) The term '*contagium vivum fluidum*' was coined by M.W. Bejerinek
(c) Mosaic disease in tobacco and AIDS in human being are caused by viruses
(d) The viroids were dicovered by D.J. Ivanowski

158. Axile placentation is present in
(a) Lemon (b) Pea
(c) *Argemone* (d) *Dianthus*

159. A childless couple can be assisted to have a child through a technique called GIFT. The full form of this technique is :
(a) Gamete intra fallopian transfer
(b) Gamete internal fertillization and transfer
(c) Germ cell internal fallopian transfer
(d) Gemete inseminated fallopian transfer

160. Destruction of the anterior horn cell of the spinal cord would result in loss of :
(a) voluntary motor impulses
(b) commissural impulses
(c) integrating impulses
(d) sensory impulses

161. During biological nitrogen fixation, inactivation of nitrogenase by oxygen poisoning is prevented by :
(a) Xanthophyll (b) Carotene
(c) Cytochrome (d) Leghemoglobin

162. An association of individuals of different species living in the same habitat and having functional interactions is:
(a) Biotic community (b) Ecosystem
(c) Population (d) Ecological niche

163. Name the pulmonary disease in which alveolar surface area involved in gas exchange is drastically reduced due to damage in the alveolar walls.
(a) Emphysema (b) Pneumonia
(c) Asthma (d) Pleurisy

164. Balbiani rings are sites of :
(a) Nucleotide synthesis
(b) Polysaccharide synthesis
(c) RNA and protein synthesis
(d) Lipid synthesis

165. Match the columns and identify the correct option.

	Column-I		Column-II
(1)	Thylakoids	(i)	Disc-shaped sacs in Golgi apparatus
(2)	Cristae	(ii)	Condensed structure of DNA
(3)	Cisternae	(iii)	Flat membranous sacs in stroma
(4)	Chromatin	(iv)	Infoldings in mitochondria

	(1)	(2)	(3)	(4)
(a)	(iii)	(iv)	(i)	(ii)
(b)	(iii)	(i)	(iv)	(ii)
(c)	(iii)	(iv)	(ii)	(i)
(d)	(iv)	(iii)	(i)	(ii)

166. Cellular organelles with membranes are :
(a) chormosomes, ribosomes and endoplasmic reticulum
(b) endoplasmic reticulum, ribosomes and nuclei
(c) lysosomes, Golgi apparatus and mitochondria
(d) nuclei, ribosomes and mitochondria

167. Auxin can be bioassayed by :
(a) Hydroponics
(b) Potometer
(c) Lettuce hypocotyl elongation
(d) Avena coleoptile curvature

168. Which of the following layers in an antral follicle is acelluar ?
(a) Theca interna (b) Stroma
(c) Zona pellucida (d) Granulosa

169. Satellite DNA is important because it :
(a) shows high degree of polymorphism in population and also the same degree of polymorphism in an individual, which is heritable from parents to children.
(b) does not code for proteins and is same in all members of the population
(c) codes for enzymes needed for DNA replication
(d) codes for proteins needed in cell cycle.

170. Cell wall is absent in :
(a) *Funaria* (b) Mycoplasma
(c) *Nostoc* (d) *Aspergillus*

171. In angiosperms, microsporogenesis and megasporogenesis :
(a) form gametes without further divisions
(b) Involve meiosis
(c) occur in ovule
(d) occur in anther

172. Roots play insignificant role in absorption of water in :
(a) Pistia (b) Pea
(c) Wheat (d) Sunflower

173. Which of the following are most suitable indicators of SO_2 pollution in the environment?
(a) Conifers (b) Algae
(c) Fungi (d) Lichens

174. Grafted kidney may be rejected in a patient due to :
(a) Cell-midiated immune response
(b) Passive immune response
(c) Innate immune response
(d) Humoral immune response

175. Body having meshwork of cell, internal cavities lined with food filtering flagellated cells and indirect development are the characteristics of phylum.
(a) Porifera (b) Mollusca
(c) Protozoa (d) Coelenterate

176. In which group of organisms the cell walls form two thin overlapping shells which fit together?
(a) Euglenoids (b) Dinoflagellates
(c) Slime moulds (d) Chrysophytes

177. Choose the wrong statements:
(a) Neurospora is used in the study of biochemical genetics
(b) Morels and truffles are poisonoues mushrooms
(c) Yeast is unicellular and useful in fermentation
(d) *Penicillium* is multicellular and produces antibiotics

178. In human females, meiosis-II is not complete until?
(a) fertilization
(b) uterine implantation
(c) birth
(d) puberty

179. Eutrophication of water bodies leading to killing of fishes is mainly due to non-availability of :
(a) light (b) essential minerals
(c) oxygen (d) food

180. The enzyme that is not present in succus entericus is :
(a) nucleases (b) nucleosidase
(c) lipase (d) maltase

HINTS & SOLUTIONS

PHYSICS

1. (b) Here, $x = 4\sin(2\pi t)$ ...(*i*)

$y = 4\cos(2\pi t)$ ...(*ii*)

Squaring and adding equation (*i*) and (*ii*)

$x^2 + y^2 = 4^2 \Rightarrow R = 4$

Motion of the particle is circular motion, acceleration vector is along $-\vec{R}$ and its magnitude $= \frac{V^2}{R}$

Velocity of particle, $V = \omega R = (2\pi)(4) = 8\pi$

2. (d) Energy of x-ray is (100 ev to 100 kev) Hence energy of the order of 15 kev belongs to x-rays.

3. (c) For total internal reflection, incident angle (i) > critical angle (i_c)

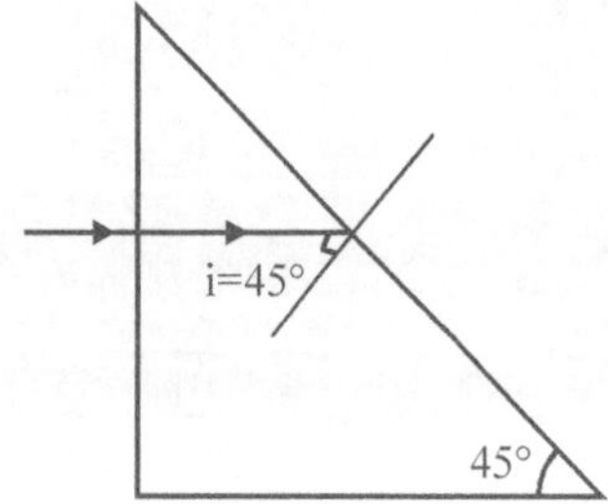

So, $\sin i > \sin i_c$

$\sin 45° > \frac{1}{\mu} \quad \Rightarrow \mu > \sqrt{2} \Rightarrow 1.414$

Since refractive index μ of green and voilet are greater than 1.414 so they will total internal reflected. But red colour will be refracted.

4. (d) For collision $\vec{V}_{B/A}$ should be along $\overline{B \rightarrow A}\left(\hat{r}_{A/B}\right)$

So, $\frac{\vec{V}_2 - \vec{V}_1}{|V_2 - V_1|} = \frac{\vec{r}_1 - \vec{r}_2}{|r_1 - r_2|}$

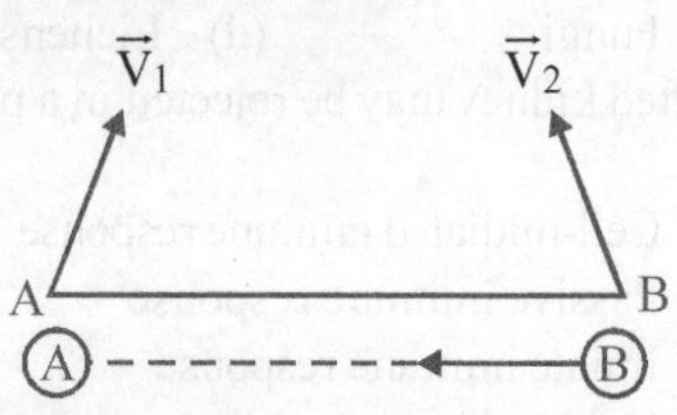

5. (b) For first minima at P

$AP - BP = \lambda$

$AP - MP = \frac{\lambda}{2}$

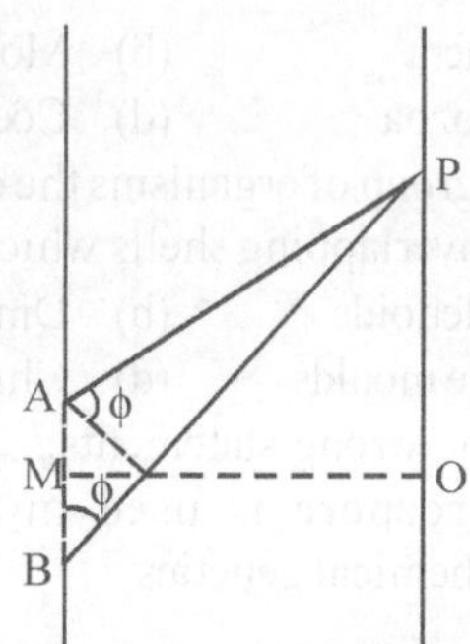

So phase difference, $\phi = \frac{2\pi}{\lambda} \times \frac{\lambda}{2} = \pi$ radian

6. (c) As we know, $F = qvB = \frac{mv^2}{R}$

$\therefore R = \frac{mV^2}{qB} = \frac{\sqrt{2m(kE)}}{qB}$

Since R is same so, $KE \propto \frac{q^2}{m}$

Therefore KE of α particle

$= \frac{q^2}{m} = \frac{(2)^2}{4} = 1$ MeV

7. (d) From circuit diagram

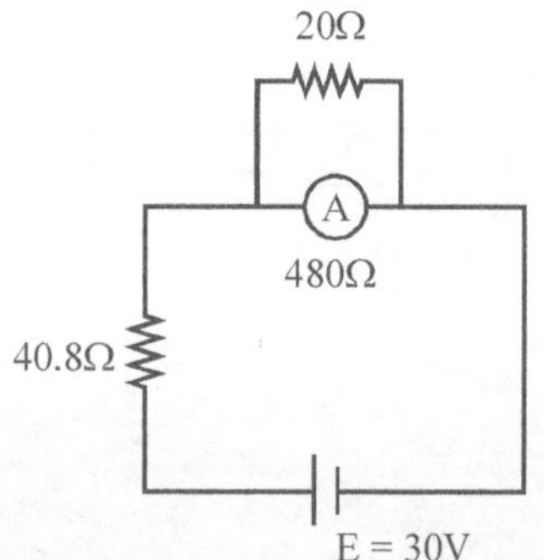

Resistance of ammeter $= \frac{480 \times 20}{480+20} = 19.2\Omega$.

Total resistance $R = 40.8 + 19.2 = 60\Omega$

Reading in the ammeter $i = \frac{V}{R}$

$= \frac{30}{40.8+19.2} = 0.5A$

8. (a) From question,

Rise in temperature $\Delta t = 40°C$

Fractional change in the density $\frac{\Delta\rho}{\rho_0} = ?$

Coefficient of volume expansion

$\gamma = 5 \times 10^{-4} K^{-1}$

$\rho = \rho_0 (1 - \gamma\Delta t)$

$\Rightarrow \frac{\Delta\rho}{\rho_0} = \gamma\Delta T = (5 \times 10^{-4})(40) = 0.02$

9. (d) Since area under the curve is maximum for adiabatic process so, work done (W = PdV) on the gas will be maximum for adiabatic process

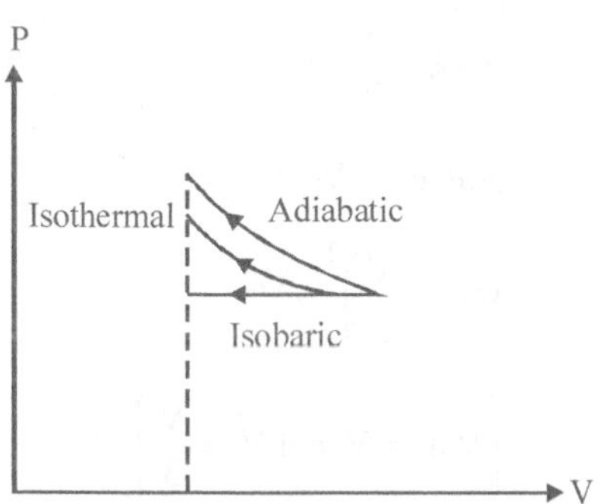

10. (a) For series R – C circuit, capacitive reactance,

$$Z_c = \sqrt{R^2 + \left(\frac{1}{C\omega}\right)^2}$$

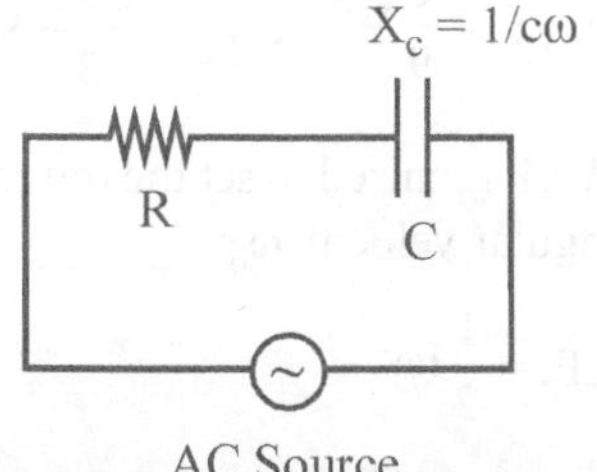

Current $i = \frac{V}{Z_c} = \frac{V}{\sqrt{R^2 + \left(\frac{1}{C\omega}\right)^2}}$

$$V_c = iX_c = \frac{V}{\sqrt{R^2 + \left(\frac{1}{C\omega}\right)^2}} \times \frac{1}{C\omega}$$

$$V_c = \frac{V}{\sqrt{(RC\omega)^2 + 1}}$$

If we fill a di-electric material like mica instead of air then capacitance $C\uparrow \Rightarrow V_c\downarrow$

So, $V_a > V_b$

11. (b) Given : work function ϕ of metal = 2.28 eV

Wavelength of light $\lambda = 500$ nm $= 500 \times 10^{-9}$m

$$KE_{max} = \frac{hc}{\lambda} - \phi$$

$$KE_{max} = \frac{6.6 \times 10^{-34} \times 3 \times 10^8}{5 \times 10^{-7}} - 2.82$$

$$KE_{max} = 2.48 - 2.28 = 0.2 \text{ ev}$$

$$\lambda_{min} = \frac{h}{p} = \frac{h}{\sqrt{2m(KE)_{max}}}$$

$$= \frac{\frac{20}{3} \times 10^{-34}}{\sqrt{2 \times 9 \times 10^{-31} \times 0.2 \times 1.6 \times 10^{-19}}}$$

$$\lambda_{min} = \frac{25}{9} \times 10^{-9}$$

$= 2.80 \times 10^{-9}$ nm $\quad \therefore \lambda \geq 2.8 \times 10^{-9}$ m

12. (d) In figure, two metal wires of identical dimension are connected in series

A σ_1 σ_2

l l

$$R_{eq} = \frac{l}{\sigma_1 A} + \frac{l}{\sigma_2 A} = \frac{l_{eq}}{\sigma_{eq} A_{eq}}$$

$$\frac{2l}{\sigma_{eq}A} = \frac{l}{A}\left(\frac{\sigma_1+\sigma_2}{\sigma_1\sigma_2}\right)$$

$$\therefore \sigma_{eq} = \frac{2\sigma_1\sigma_2}{\sigma_1+\sigma_2}$$

13. **(d)** Given : speed $V = 54\ kmh^{-1} = 15\ ms^{-1}$
Moment of inertia, $I = 3\ kgm^2$
Time $t = 15s$

$$\omega_i = \frac{V}{r} = \frac{15}{0.45} = \frac{100}{3} \quad \omega_f = 0$$

$$\omega_f = \omega_i + \alpha t$$

$$0 = \frac{100}{3} + (-\alpha)(15) \qquad \Rightarrow \alpha = \frac{100}{45}$$

Average torque transmitted by brakes to the wheel

$$\tau = (I)(\alpha) = 3 \times \frac{100}{45} = 6.66\ kgm^2s^{-2}$$

14. **(a)** Here, original frequency of sound, $f_0 = 100$ Hz
Speed of source $V_s = 19.4 \cos 60° = 9.7$

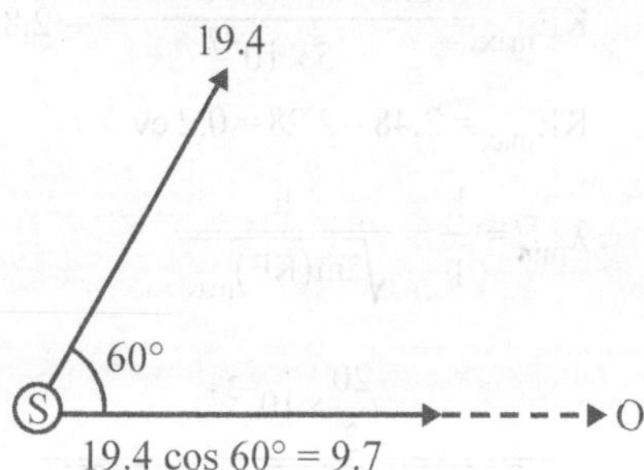

From Doppler's formula

$$f^1 = f_0\left(\frac{V-V_0}{V-V_s}\right)$$

$$f^1 = 100\left(\frac{V-0}{V-(+9.7)}\right)$$

$$f^1 = 100\frac{V}{V\left(1-\frac{9.7}{V}\right)}$$

$$f^1 = 100\left(1+\frac{9.7}{330}\right) = 103\text{Hz}$$

Apparent frequency $f^1 = 103$ Hz

15. **(d)** Here, $M_1 = M_2$ and $u_2 = 0$

$$u_1 = V, \quad V_1 = \frac{V}{3}; \qquad V_2 = ?$$

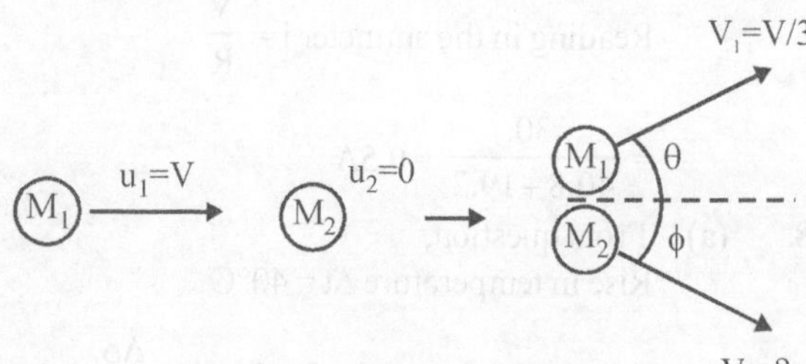

From figure, along x-axis,
$$M_1u_1 + M_2u_2 = M_1V_1\cos\theta + M_2V_2\cos\phi \qquad ...(i)$$

Along y-axis
$$0 = M_1V_1\sin\theta - M_2V_s\sin\phi \qquad ...(ii)$$

By law of conservation of kinetic energy

$$\frac{1}{2}M_1u_1^2 + \frac{1}{2}M_2u_2^2 = \frac{1}{2}M_1V_1^2 + \frac{1}{2}M_2V_2^2 \qquad ...(iii)$$

Putting $M_1 = M_2$ and $u_2 = 0$ in equation (*i*), (*ii*) and (*iii*) we get

$$\theta + \phi = \frac{\pi}{2} = 90°$$

and $u_1^2 = V_1^2 + V_2^2$

$$V^2 = \left(\frac{V}{3}\right)^2 + V_2^2$$

$$\left[\because u_1 = V \text{ and } V_1 = \frac{V}{3}\right]$$

$$\text{or, } V^2 - \left(\frac{V}{3}\right)^2 = V_2^2$$

$$V^2 - \frac{V^2}{9} = V_2^2$$

$$\text{or } V_2^2 = \frac{8}{9}V^2 \Rightarrow V_2 = \frac{2\sqrt{2}}{3}V$$

16. **(c)** Work required to set the rod rotating with angular velocity ω_0

$$K.E. = \frac{1}{2}I\omega^2$$

Work is minimum when I is minimum.
I is minimum about the centre of mass
So, $(m_1)(x) = (m_2)(L - x)$
or, $m_1x = m_2L - m_2x$

$$\therefore x = \frac{m_2L}{m_1 + m_2}$$

17. (a) When ball collides with the ground it loses its 50% of energy

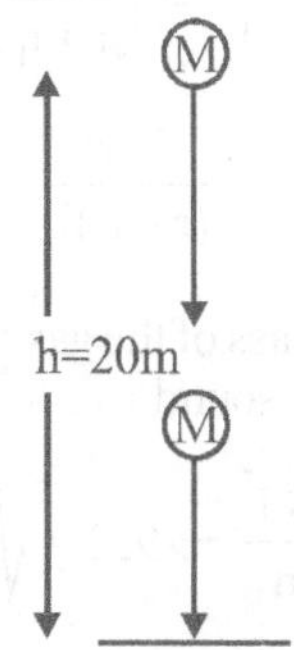

$$\therefore \frac{KE_f}{KE_i} = \frac{1}{2} \Rightarrow \frac{\frac{1}{2}mV_f^2}{\frac{1}{2}mV_i^2} = \frac{1}{2}$$

$$\text{or } \frac{V_f}{V_i} = \frac{1}{\sqrt{2}}$$

$$\text{or, } \frac{\sqrt{2gh}}{\sqrt{V_0^2 + 2gh}} = \frac{1}{\sqrt{2}}$$

or, $4gh = V_0^2 + 2gh$
$\therefore V_0 = 20ms^{-1}$

18. (d) In an explosion a body breaks up into two pieces of unequal masses both part will have numerically equal momentum and lighter part will have more velocity.
$U \rightarrow Th + He$

$$KE_{Th} = \frac{P^2}{2m_{Th}}, KE_{He} = \frac{P^2}{2m_{He}}$$

sinc m_{He} is less so KE_{He} will be more.

19. (b) Current will be induced,
when e^- comes closer the induced current will be anticlockwise
when e^- comes farther induced current will be clockwise

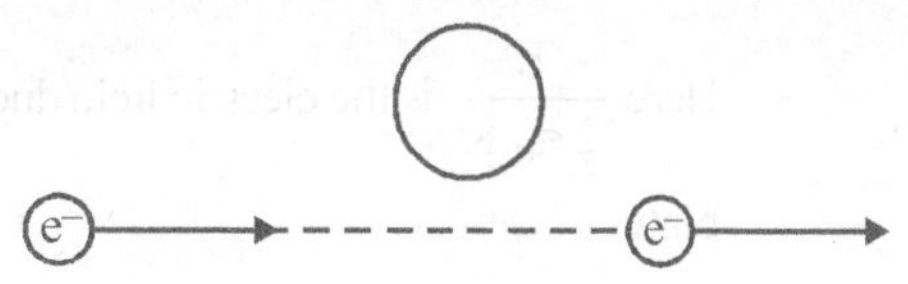

20. (c) As, we know, in SHM
Maximum acceleration of the particle, $\alpha = A\omega^2$
Maximum velocity, $\beta = A\omega$

$$\Rightarrow \omega = \frac{\alpha}{\beta}$$

$$\Rightarrow T = \frac{2\pi}{\omega} = \frac{2\pi\beta}{\alpha} \quad \left[\because \omega = \frac{2\pi}{T}\right]$$

21. (d) The ratio of slits width $= \frac{1}{25}$ (given)

$$\therefore \frac{I_1}{I_2} = \frac{25}{1}$$

$$I \propto A^2 \Rightarrow \frac{I_1}{I_2} = \frac{A_1^2}{A_2^2} = \frac{25}{1} \text{ or } \frac{A_1}{A_2} = \frac{5}{1}$$

$$\frac{A_{max}}{A_{min}} = \frac{A_1 + A_2}{A_1 - A_2} = \frac{5+1}{5-1} = \frac{6}{4} = \frac{3}{2}$$

$$\therefore \frac{I_{max}}{I_{min}} = \frac{A_{max}^2}{A_{min}^2} = \left(\frac{3}{2}\right)^2 = \frac{9}{4}$$

22. (a) Potential in a region
$V = 6xy - y + 2yz$
As we know the relation between electric potential and electric field is $\vec{E} = \frac{-dV}{dx}$

$$\vec{E} = \left(\frac{\partial V}{\partial x}\hat{i} + \frac{\partial V}{\partial y}\hat{j} + \frac{\partial V}{\partial z}\hat{k}\right)$$

$$\vec{E} = \left[(6y\hat{i} + (6x - 1 + 2z)\hat{j} + (2y)\hat{k}\right]$$

$$\vec{E}_{(1,1,0)} = -(6\hat{i} + 5\hat{j} + 2\hat{k})$$

23. (a) Force of attraction between the plates,
$F = qE$

$$= q \times \frac{\sigma}{2\in_0 K} = q\left(\frac{q}{2dK \in_0}\right) = \frac{q^2}{2dK \in_0}$$

Here $\frac{\sigma}{2\epsilon_0 K}$ is the electric field due to one plate

$q = CV \quad \therefore F = \frac{C^2V^2}{2dK\epsilon_0}$

24. (a) Coefficient of static friction,

$\mu_s = \tan 30° = \frac{1}{\sqrt{3}} = 0.577 \cong 0.6$

$S = ut + \frac{1}{2}at^2$

$4 = \frac{1}{2}a(4)^2 \Rightarrow a = \frac{1}{2} = 0.5$ [$\because$ s = 4m and t = 4s given]

$a = g\sin\theta - \mu_k(g)\cos\theta$

$\Rightarrow \mu_k = \frac{0.9}{\sqrt{3}} = 0.5$

25. (c) As we know, $\frac{1}{\lambda} = R_e\left(\frac{1}{n_1^2} - \frac{1}{n_2^2}\right)$

For Lyman series, $n_1 = 1$ and $n_2 = 2, 3, 4...$
For Balmer series $n_1 = 2$ and $n_2 = 3, 4, 5...$

$\frac{1}{\lambda_1} = R_e\left(\frac{1}{1^2} - \frac{1}{2^2}\right)$

$\frac{1}{\lambda_2} = R_e\left(\frac{1}{2^2} - \frac{1}{3^2}\right)$

$\therefore \frac{\lambda_1}{\lambda_2} = \frac{5}{27}$

26. (d) Current $I = \frac{V}{R} = \frac{(3.5 - 0.5)}{100}A$

[$\because$ Barrier potential $V_B = 0.5V$]

$= \frac{3}{100} = 30mA$

27. (c) The gravitational force on the satellite will be aiming towards the centre of the earth so acceleration of the satellite will also be aiming towards the centre of the earth.

28. (d) From Newton's second law for rotational motion,

$\vec{\tau} = \frac{d\vec{L}}{dt}$, if $\vec{L}$ = constant then $\vec{\tau} = 0$

So, $\vec{\tau} = \vec{r} \times \vec{F} = 0$

$(2\hat{i} - 6\hat{j} - 12\hat{k}) \times (\alpha\hat{i} + 3\hat{j} + 6\hat{k}) = 0$

Solving we get $\alpha = -1$

29. (a) EMF, $E = Kl$ where $K = \frac{V}{L}$ potential gradient

$K = \frac{V}{L} = \frac{iR}{L} = \left(\frac{E_0 r}{r + r_1}\right)\frac{l}{L}$

So, $E = Kl = \frac{E_0 rl}{(r + r_1)L}$

30. (d) Molar mass of the gas = 4g/mol
Speed of sound

$V = \sqrt{\frac{\gamma RT}{m}} \Rightarrow 952 = \sqrt{\frac{\gamma \times 3.3 \times 273}{4 \times 10^{-3}}}$

$\Rightarrow \gamma = 1.6 = \frac{16}{10} = \frac{8}{5}$

Also, $\gamma = \frac{C_P}{C_V} = \frac{8}{5}$

So, $C_P = \frac{8 \times 5}{5} = 8JK^{-1}mol^{-1}$ [$C_V = 5.0\ JK^{-1}$ given]

31. (d) According to question, two stones experience same centripetal force

i.e. $F_{C_1} = F_{C_2}$

or, $\frac{mv_1^2}{r} = \frac{2mv_2^2}{(r/2)}$ or, $V_1^2 = 4V_2^2$

So, $V_1 = 2V_2$ i.e., n = 2

32. (d) **Given:** Height of the satellite from the earth's surface $h = 0.25 \times 10^6 m$
Radius of the earth $R = 6.38 \times 10^6 m$
Acceleration due to gravity $g = 9.8\ m/s^2$
Orbital velocity, $V_0 = ?$

$V_0 = \sqrt{\frac{GM}{(R+h)}} = \sqrt{\frac{GM}{R^2}.\frac{R^2}{(R+h)}}$

$= \sqrt{\frac{9.8 \times 6.38 \times 6.38}{6.63 \times 10^6}}$

$= 7.76\ km/s \quad \left[\because \frac{GM}{R^2} = g\right]$

33. (c) In a stretched string all multiples of frequencies can be obtained i.e., if fundamental frequency is n then higher frequencies will be 2n, 3n, 4n ...

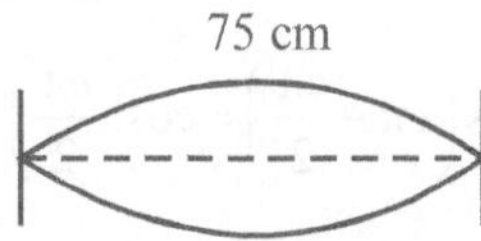

So, the difference between any two successive frequencies will be 'n'
According to question, n = 420 – 315
= 105 Hz
So the lowest frequency of the string is 105 Hz.

34. (d) Coefficient of performance,

$$Cop = \frac{T_2}{T_1 - T_2}$$

$$5 = \frac{273-20}{T_1-(273-20)} = \frac{253}{T_1-253}$$

$5T_1 - (5 \times 253) = 253$
$5T_1 = 253 + (5 \times 253) = 1518$

$$\therefore T_1 = \frac{1518}{5} = 303.6$$

or, $T_1 = 303.6 - 273 = 30.6 \cong 31°C$

35. (a) Water rises upto the top of capillary tube and stays there without overflowing.

36. (a) From PV = nRT

$$P_A = \frac{\rho_A M_A}{RT} \text{ and } P_B = \frac{\rho_B M_B}{RT}$$

From question,

$$\frac{P_A}{P_B} = \frac{\rho_A}{\rho_B}\frac{M_A}{M_B} = 2\frac{M_A}{M_B} = \frac{3}{2}$$

$$\text{So, } \frac{M_A}{M_B} = \frac{3}{4}$$

37. (a) Young's modulus $Y = \frac{W}{A} \cdot \frac{l}{\Delta l}$

$$\frac{W_1}{Y_1} = \frac{W_2}{Y_2}$$

[$\because$ A, l, Δl same for both brass and steel]

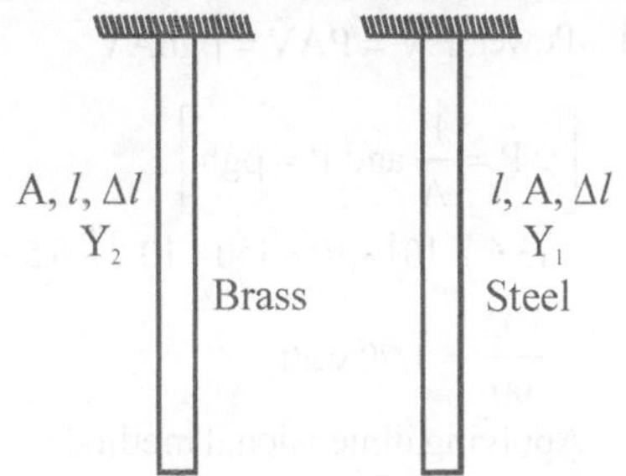

$$\frac{W_1}{W_2} = \frac{Y_1}{Y_2} = 2 \qquad [Y_{steel}/Y_{brass} = 2 \text{ given}]$$

38. (c) **Given :** Voltage gain $A_V = 150$

$$V_i = 2\cos\left(15t + \frac{\pi}{3}\right); V_0 = ?$$

For CE transistor phase difference between input and output signal is $\pi = 180°$

Using formula, $A_V = \frac{V_0}{V_i}$

$\Rightarrow V_0 = A_V \times V_i$

$$= 150 \times 2\cos\left(15t + \frac{\pi}{3}\right)$$

$$\text{or } V_0 = 300\cos\left(15t + \frac{\pi}{3} + \pi\right)$$

$$V_0 = 300\cos\left(15t + \frac{4}{3}\pi\right)$$

39. (c)

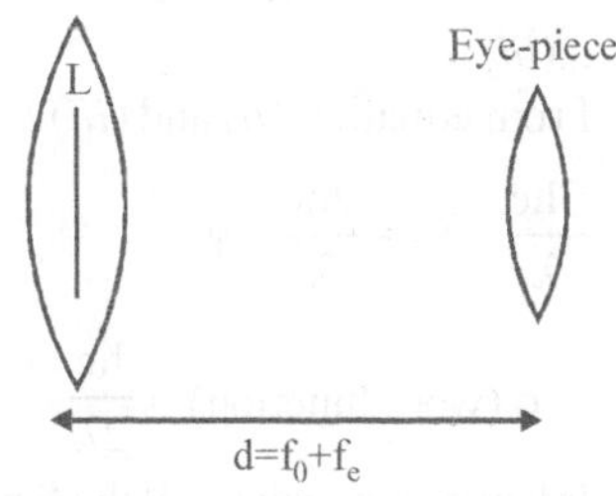

Magnification by eye piece

$$m = \frac{f}{f+u}$$

$$-\frac{I}{L} = \frac{f_e}{f_e + [-(f_0 + f_e)]} = -\frac{f_e}{f_0} \text{ or, } \frac{I}{L} = \frac{f_e}{f_0}$$

$$\text{Magnification, } M = \frac{f_0}{f_e} = \frac{L}{I}$$

40. (d) Power $\vec{F}.\vec{V} = PA\vec{V} = \rho ghAV$

$$\left[\because P = \frac{F}{A} \text{ and } P = \rho gh\right]$$

$= 13.6 \times 10^3 \times 10 \times 150 \times 10^{-3} \times 0.5 \times 10^{-3}/60$

$= \frac{102}{60} = 1.70$ watt

41. (d) Applying dimensional method :

$V_C = \eta^x \rho^y r^z$

$[M^0LT^{-1}] = [ML^{-1}T^{-1}]^x [ML^{-3}T^0]^y [M^0LT^0]^z$

Equating powers both sides

$x + y = 0; -x = -1 \therefore x = 1$

$1 + y = 0 \therefore y = -1$

$-x - 3y + z = 1$

$-1 - 3(-1) + z = 1$

$-1 + 3 + z = 1$

$\therefore z = -1$

42. (d) Photoelectric equations

$Ek_{1max} = \frac{hc}{\lambda} - \phi$ ...(*i*)

and $Ek_{2max} = \frac{hc}{\lambda/2} - \phi$

$EK_{2max} = \frac{2hc}{\lambda} - \phi$ ...(ii)

From question, $Ek_{2max} = 3Ek_{1max}$

Multiplying equation (i) by 3

$3Ek_{1max} = 3\left(\frac{hc}{\lambda} - \phi\right)$

...(*iii*)

From equation (*ii*) and (*iii*)

$\frac{3hc}{\lambda} - 3\phi = \frac{2hc}{\lambda} - \phi$

$\therefore \phi$ (work function) $= \frac{hc}{2\lambda}$

43. (a) Inflow rate of volume of the liquid = Outflow rate of volume of the liquid

$\pi R^2 V = n\pi r^2(v) \Rightarrow v = \frac{\pi R^2 V}{n\pi r^2} = \frac{VR^2}{nr^2}$

44. (b) Two vectors are

$\vec{A} = \cos\omega t\hat{i} + \sin\omega t\hat{j}$

$\vec{B} = \cos\frac{\omega t}{2}\hat{i} + \sin\frac{\omega t}{2}\hat{j}$

For two vectors $\vec{A}$ and $\vec{B}$ to be orthogonal A.B = 0

$\vec{A}.\vec{B} = 0 = \cos\omega t.\cos\frac{\omega t}{2} + \sin\omega t.\sin\frac{\omega t}{2}$

$= \cos\left(\omega t - \frac{\omega t}{2}\right) = \cos\left(\frac{\omega t}{2}\right)$

So, $\frac{\omega t}{2} = \frac{\pi}{2}$ $\quad\therefore\ t = \frac{\pi}{\omega}$

45. (a) Here, number of turns of coil, N = 50

Current through the coil I = 2A

Area $A = l \times b = 0.12 \times 0.1 m^2$

Magnetic field $\vec{B} = 0.2\ \omega/m^2$

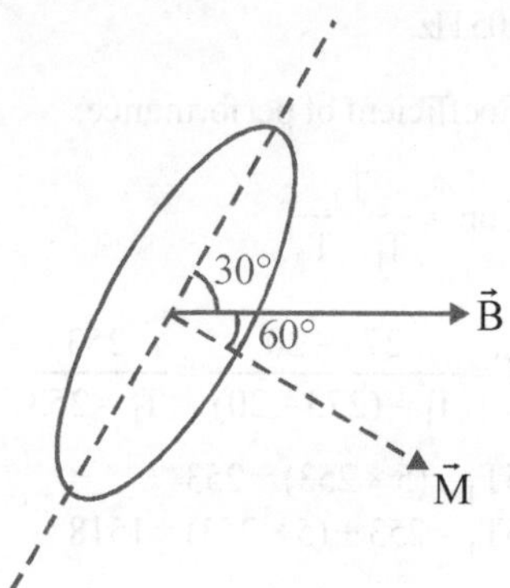

Torque required to keep the coil in stable equilibrium.

$\tau = \vec{M} \times \vec{B} = MB \sin 60°$

$= NiAB \sin 60°$

$= 50 \times 2 \times 0.12 \times 0.1 \times 0.2 \times \frac{\sqrt{3}}{2}$

$= 12\sqrt{3} \times 10^{-2} = 0.20784$ Nm

CHEMISTRY

46. (d) $CHCl_3 + NaOH \rightleftharpoons :\overset{\ominus}{C}Cl_3 + H_2O$

$:\overset{\ominus}{C}Cl_3 \longrightarrow :CCl_2 + Cl^{\ominus}$

(Reaction scheme: phenoxide + :CCl_2 → o-$\overset{\ominus}{C}Cl_2$ phenoxide; + H_2O, $-OH^{\ominus}$ → o-$CHCl_2$ phenoxide; $OH^{\ominus}$, $-2HCl$ → o-CHO phenoxide)

Therefore functional group – CHO is introduced.

47. (a) $N_2(g) + O_2(g) \rightleftharpoons 2NO(g)$

$$K = \frac{[NO]^2}{[N_2][O_2]}$$

$$\frac{1}{2}N_2 + \frac{1}{2}O_2 \rightleftharpoons NO$$

$$K' = \frac{[NO]}{[N_2]^{1/2}[O_2]^{1/2}} = K^{1/2}$$

48. (d) $MgCO_3 \longrightarrow MgO + CO_2$

84 g of $MgCO_3$ form 40 g of MgO

$\therefore$ 20g of $MgCO_3$ form $\frac{40 \times 20}{84}$ g of MgO

= 9.52 g of MgO

Since 8.0 g of MgO is formed

Purity of sample $= \frac{8}{9.52} \times 100 = 84.0\%$

49. (d) No. of moles of water

In 1.8 g of H_2O = 0.1 moles

In 18 g of H_2O = 1 moles

1 mole contain 6.022×10^{23} molecules of water

therefore maximum number of molecules is in 18 moles of water.

50. (a) Incoming electrons occupies the smaller n = 2 shell, also negative charge on oxygen (O^-) is another factor due to which incoming electron feel repulsion.

Hence electron repulsion outweigh the stability gained by achieving noble gas configuration.

51. (d) Molality $= \frac{W}{M} \times \frac{1}{V(kg)}$

$= \frac{n}{V(kg)}$ (where n_{solute} = W/M)

$1.00\,m = \frac{n}{V(kg)}$ i.e., 1 mole in 1 kg of water

Moles of 1 kg H_2O

$$= \frac{1000g}{18g/mol} = 55.55 \text{ mole}$$

Moles of solute = 1

Mole fraction

$$= \frac{n_{solute}}{n_{solute} + n_{water}} = \frac{1}{(1 + 55.55)}$$

= 0.01768 = 0.0177

52. (d) Rate constant k = 0.6×10^{-3} mole per second. (unit mole per second shows zero order reaction)

For a zero order reaction

$[A] = [A]_0 - kt$

and $[A_0] - [A] = [B] = kt$

$= 0.6 \times 10^{-3} \times 20 \times 60 = 0.72$ M

53. (a) According to molecular orbital theory as bond order decreases stability of the molecule decreases

Bond order $= \frac{1}{2}(N_b - N_a)$

Bond order for $O_2^+ = \frac{1}{2}(10-5) = 2.5$

Bond order for $O_2 = \frac{1}{2}(10-6) = 2$

Bond order for $O_2^- = \frac{1}{2}(10-7) = 1.5$

Bond order for $O_2^{2-} = \frac{1}{2}(10-8) = 1.0$

hence the correct order is

$O_2^+ > O_2 > O_2^- > O_2^{2-}$

54. (a) Among the substituent attached to the benzene ring, $-NO_2$ group is the most electron withdrawing, thus withdraws electron density from carbonyl carbon thus facilitate the attack of $OH^{\ominus}$ ion.

(–R effect of $-NO_2$ group)

55. (c) Carbonates becomes more thermally stable down the group, therefore $MgCO_3$ will leave CO_2 easily.

56. (a) Among the given acids, $HClO_4$ is a very strong acid, completely dissociates

$\therefore [HA] \rightarrow 0$ in this case, hence cannot be used for acidic buffer.

57. (a) $[M(en)_2(C_2O_4)]Cl$

C_2O_4 = bidentate ligand, carry – 2 charge

en = bidentate ligand, carry 0 charge

$\therefore$ M carry + 3 charge;

coordination number = 6

$\therefore$ Sum = +3 + 6 = 9

58. (d) OF_2; among the following O and F, F is more electronegative than oxygen. So OF_2 cannot be called oxide because in that case fluorine is in +1 oxidation state which is not possible, so OF_2 is called oxygen difluoride.

59. (c) 1-chloro-1-methylcyclohexane.

$$+ H-Cl \rightarrow \; + Cl^{\ominus} \rightarrow$$

$$+ H-Cl \rightarrow \; + Cl^{\ominus} \rightarrow$$

60. (b) $CH_3C(CH_3) = C(CH_3) - CH_3$

2,3-Dimethyl-2-butene

$(CH_3)_2CH-CH(CH_3)-CH=CH_2$ contain 7 carbon atoms

$$CH_3-C(CH_3)_2-CH=CH_2 + H^{\oplus} \longrightarrow CH_3-C(CH_3)_2-\overset{\oplus}{C}H-CH_3$$

Rearrangement

$$(CH_3)_2C=C(CH_3)_2 \longleftarrow CH_3-\overset{\oplus}{C}(CH_3)-C(H)(CH_3)_2$$

2, 3 - dimethyl – 2 - butene (Stable 3° carbocation)

61. (d)

$$C_6H_6 \xrightarrow[CH_3Cl]{AlCl_3} C_6H_5CH_3$$ (Friedel Craft's Reaction)

$$C_6H_5CHO + (RCH_2CO)_2O \xrightarrow{Base} ArCH=C(R)COOH + HOOCCH_2R$$

Perkin Reaction

$$C_6H_5OH + CH_3-\overset{O}{\overset{\|}{C}}-Cl \longrightarrow C_6H_5-O-\overset{O}{\overset{\|}{C}}-CH_3$$

Acetylation Reaction

$$C_6H_5NH_2 + Cl-\overset{O}{\overset{\|}{C}}-C_6H_5 \xrightarrow{NaOH} C_6H_5NH-\overset{O}{\overset{\|}{C}}-C_6H_5$$

Schotten-Baumen Reaction

62. (c) $2Cu_2S(s) + 3O_2(g) \rightarrow 2Cu_2O(s) + 2SO_2(g)$
The unchanged Cu_2S, mixed with Cu_2O and heated strongly in absence of air
$2Cu_2S + 2Cu_2O \rightarrow 6Cu + SO_2$

63. (b) If 6.022×10^{23} changes to 6.022×10^{20}/mol than this would change mass of one mole of carbon.

64. (b) The H-bonding is present in HF due to high electronegativity of fluorine atom. While H-bonding is not present in HI, HBr and HCl.

65. (a) $ZnCl_2$ is a lewis acid and interact with alcohol.

$$CH_3-CH_2OH + ZnCl_2 \rightarrow R-\overset{\oplus}{\underset{H}{\underset{|}{O}}}-ZnCl_2 \quad (I)$$
$(R = CH_3-CH_2-)$

$$R-\overset{\oplus}{\underset{H}{\underset{|}{O}}}-ZnCl_2 \rightarrow R^{\oplus} + [HOZnCl_2]^{\ominus}$$
(I)

1 mole of carbon = 12gm of carbon
$= 6.023 \times 10^{23}$ C – atoms.
Carbocation is formed as intermediate in the S_N^1 mechanism which these reaction undergoes.
In the absence of $ZnCl_2$ formation of primary carbocation is difficult which is the case with (ii) while (i) undergoes reaction.
(iii) Tertiary carbocation easily formed due to the stability.

$$CH_3-\underset{CH_3}{\underset{|}{\overset{OH}{\overset{|}{C}}}}-CH_3 \longrightarrow$$

$$CH_3-\underset{CH_3}{\underset{|}{\overset{\oplus}{C}}}-CH_3 + H_2O$$

(iv) In the presence of $ZnCl_2$, 2° carbocation is formed from $(CH_3)_2-\underset{H}{\underset{|}{C}}-OH$

i.e., $CH_3-\overset{\oplus}{C}H-CH_3$

66. (d) Carbocation
Hexacyanidoterrate (III) ion.

67. (a) Considering the same moles of compounds $Fe^{2+}\ SO_4^{2\ominus}$ only $Fe^{2\oplus}$ is oxidised by $KMnO_4$. $\overset{(+6)}{SO_4^{2-}}$ in which sulphur is in highest oxidation state cannot be oxidised.

$Fe^{2+}\ SO_3^{2-} \longrightarrow Fe^{2\oplus}$

$\overset{(+4)}{SO_3^{2-}} \longrightarrow \overset{(+6)}{SO_4^{2-}} + 2e^-$

Sulphur can be oxidised to (+6) i.e to SO_4^{2-}

In $FeC_2O_4 \longrightarrow Fe^{2\oplus}$

Carbon can be oxidised to (+IV) i.e to CO_2

$Fe(NO_2)_2 \longrightarrow Fe^{2\oplus}$

$$+\overset{(+3)}{2NO_2^{\ominus}} \downarrow 2NO_3^- + 2e^- \uparrow (+5)$$
Nitrogen can be oxidised to (+V) state

68. (d) XeF_4, XeO_4

XeF_4 (Square planar) $\quad XeO_4$ [Tetrahedral]

69. (a)

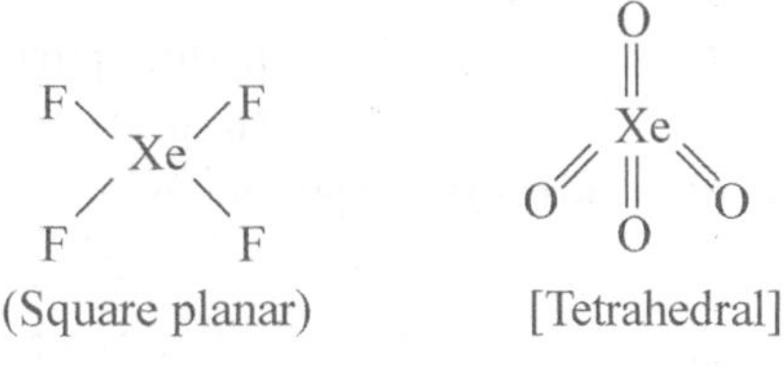

$\left[-\overset{H}{N}-(CH_2)_5-\overset{O}{\overset{||}{C}}- \right]_n$
Nylon–6

70. **(a)** $Ni^{2\oplus} = [Ar]^{18} 4s^0 3d^8$

Valence bond theory can be used to predict shape.

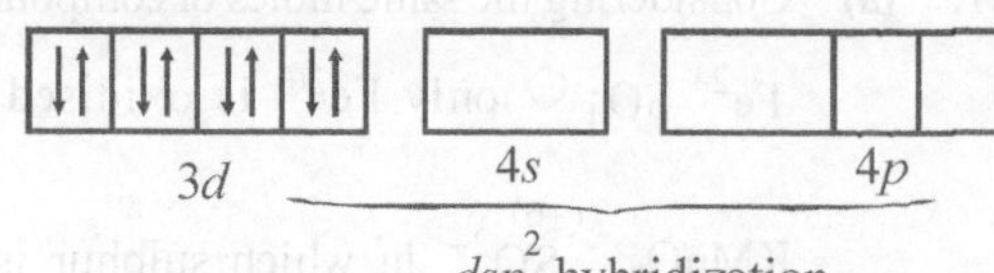

dsp^2 hybridization

(In presence of ligand, pairing of electron occurs)

$\therefore$ Square planar.

71. **(c)** 50 ml of 16.9% solution of $AgNO_3$

$$\left(\frac{16.9}{100} \times 50\right) = 8.45 \text{ g of } AgNO_3$$

$$n_{mole} = \frac{8.45g}{(107.8 + 14 + 16 \times 3)\ g/mol}$$

$$= \left(\frac{8.45\ g}{169.8 g/mol}\right) = 0.0497 \text{ moles}$$

50 ml of 5.8% solution of NaCl contain

$$NaCl = \left(\frac{5.8}{100} \times 50\right) = 2.9\ g$$

$$n_{NaCl} = \frac{2.9g}{(23 + 35.5) g/mol} = 0.0495 \text{ moles}$$

$$AgNO_3 + NaCl \rightarrow AgCl\downarrow + Na^{\ominus} + Cl^{\ominus}$$

1 mole 1 mole 1 mole

$\therefore$ 0.049 mole 0.049 mole 0.049 mole of AgCl

$$n = \frac{w}{M} \rightarrow w = (n_{AgCl}) \times \text{Molecular Mass}$$

$$= (0.049) \times (107.8 + 35.5)$$

$$= 7.02\ g$$

72. **(c)** $Gd(64) = [Xe]4f^7 5d^1 6s^2$

73. **(b)** [Reaction scheme: alcohol + $H^{\oplus}$ → protonated alcohol → carbocation → alkene products]

74. **(a)** At high temperature and low pressure.

75. **(b)** Lower oxidation state become more stable on moving down the group

Al < Ga < In < Tl

76. **(a)** gm eq of NaOH = 0.1 × V = 0.1V

gm eq of HCl = 0.01 × V = 0.01V

gm eq of NaOH > gm eq. HCl

hence resultant solution should be basic, hence from the eqn

$$M_1V_1 - M_2V_2 = MV$$

$$0.1V - 0.01V = MV$$

$$M = \frac{0.09}{2} = 0.045 = 4.5 \times 10^{-2}$$

Now $pOH = -\log[OH^-]$

$= -\log 4.5 \times 10^{-2} = 1.34$

$\because$ pH + pOH = 14

$\therefore$ pH = 14 – 1.34 = 12.65

77. **(a)** The acids which contain P-H bond have strong reducing properties. Thus H_3PO_2 acid is good reducing agent as it contains two P–H bonds and reduces, for example, $AgNO_3$ to metallic silver.

$$4\,AgNO_3 + 2H_2O + H_3PO_2 \longrightarrow 4Ag + 4HNO_3 + H_3PO_4$$

78. **(a)** C_3H_9N

$CH_3-CH_2-CH_2-NH_2$

$CH_3-NH-CH_2-CH_3$

$CH_3-CH(NH_2)-CH_3$, $CH_3-N(CH_3)-CH_3$

79. **(a)** Nucleophile is a species that provide electron while species which are deficient of electrons termed as lewis acid, hence nucleophiles are usually lewis bases.

80. **(c)** $[Co(en)_2 Cl_2]Cl$

[Structures of trans (I) and cis (II) isomers of $[Co(en)_2Cl_2]^+$]

trans (I) and cis (II)

Mirror image of (II)

(Structure: Co bonded to N, N, N, N (two bidentate N–N ligands), Cl, Cl)

81. **(d)** (n + 1) rule can be used. Titanium is a multi electron system

(n + 1)	3s	<	3p	<	4s	<	3d
	↓		↓		↓		↓
	(3 + 0)		(3 + 1)		(4 + 0)		(3 + 2)
	‖		‖		‖		‖
	3		4		4		5

If (n + 1) values are same, than value of "n" has to be considered.

82. **(b)** In S_N1 reaction, carbocation a planar species as intermediate is formed.

(Structure: planar carbocation $\overset{\oplus}{C}$ with R_1, R_2, R_3; Nu attacking from above and below)

So attack from below or above the plane can take place.

If 50% attack below and above the plane of carbocation take place than 100% racemization occurs but it may not be highly probable.

i.e. if inversion occurs more than retention leading to partial racemization.

83. **(d)** Number of atoms per unit cell

$= 8 \times (1/8) + 1 = 2$

$$r = \frac{\sqrt{3}a}{4}$$

$$\text{Packing Fraction (P.F.)} = \frac{2 \times \frac{4}{3}\pi r^3}{(4r/\sqrt{3})^3}$$

$$= \frac{\sqrt{3}\pi}{8} = 0.68$$

% P.F. = 68% ; % of vacant space

= 100 – 68 = 32%

84. **(a)** $C + O_2 \rightarrow CO_2 + 393.5$ kJ/mol

12g 44g

44g is formed from 12g of carbon

35.2g is formed from $\frac{12 \times 35.2}{44}$ g of C

$= 9.6$ g of C $= 9.6/12 = 0.8$ mole

1 mole release heat 393.5 kJ

0.8 mole release heat $= 393.5 \times 0.8$

$= 314.8 \text{ kJ} \approx 315$ kg

85. **(b)** HCl completely dissociates to give H⊕ and Cl⊖ ions, hence act as very good electrolyte. While rest are non- electrolytes.

86. **(b)** $2C_6H_{6(g)} + 9O_{2(g)} \xrightarrow[410°C]{V_2O_5}$

(Structure: maleic anhydride, CH–C(=O)–O–C(=O)–CH with CH=CH) $+ 4CO_{2(g)} + 4H_2O_{(g)}$

Maleic anhydride

87. **(b)** $R - \overset{O}{\overset{\|}{C}} - R^1 + NH_2 - NH_2 \xrightarrow{H^\oplus}$

$R - \underset{NH-NH_2}{\underset{|}{\overset{OH}{\overset{|}{C}}}} - R^1 \xrightarrow[\text{Elimination}]{H_2O} R - \overset{N-NH_2}{\overset{\|}{C}} - R'$

(Addition)

While in all other case no elimination take place.

88. **(d)** Because arylhalides does not undergoes nucleophilic substitution easily.

89. **(c)** (Structure: $CH_3 - \underset{OH}{\underset{|}{\overset{H}{\overset{|}{C}}}} - COOH$)

Four different Substituent, only one chiral centre. Hence only enantiomers arc possible.

90. **(c)** (Diagram: lattice of A and B ions with A^+ moved to interstitial site)

Frenkel defect is dislocation effect

BIOLOGY

91. (a) Mendel did not use pod length for his experiment.
92. (c) Chargaff's rule is not applicable to RNA
93. (d) Two sperms and a vegetative cell are produced by male gametophyte in angiosperms.
94. (a) Ribosomes are not membrance-bound cell organelle.
95. (b) The chitinous exoskeleton of arthropods is formed by the polymerisation of N-acetyl glucosamine.
96. (a) Superior ovary is found in china rose, mustard brinjal, potato, onion and tulip. Guava and cucumber have inferior ovary.
97. (a) The function of the gap junction is to facilitate communication between adjoining cells by connecting the cytoplasm for rapid transfer of ions, small molecules and some large molecules.
98. (b) Colostrum contains high levels of lgA, which gives passive immunity to foetus.
99. (d) Fovea centralis is the most sensitive part of retina. It has high density of cones, but rods are not found.
100. (b) The second sound heard by doctors using stethoscope occurs when semilunar valves close down after the blood flows into vesels from ventricles.
101. (a) Coconut water is the free nuclear endosperm which is used by us.
102. (d) Restriction enzymes are used to cut DNA at specific locations.
103. (d) Nuclear envelope is not found in a prokaryotic cell.
104. (a) Synapsis $\rightarrow$ Crossing over $\rightarrow$ Terminalisation of chiasmata $\rightarrow$ Disappearance of nucleolus
105. (a) Due to tensile strength of water, a column of water within xylem vessels of tall trees does not break under its weight.
106. (d) Class- deuteromycetes contains imperfect fungi which play role in decomposition of organic wastes.
107. (a) Fimbriae assist to some bacteria to get attachment with rocks or host body to get establishment and nutrition.
108. (a) A vector is a DNA molecule which is used as a vehicle to carry the gene of interest to another cell.
109. (c) The kingdom Monera possesses unicellular organisms (e.g - bacteria) having no nuclear membrane.
110. (a) Metagenesis is defined as alternation of generation found in phylum cnidaria (eg. Obelia). In this phenomenon one generation of an organism reproduces asexually, followed by a sexually reproducing generation.
111. (d) In human female, during ovulation, LH surge takes place due to hypersecretion of estrogen, which induces positive feed back to anterior pituitary for secretion of LH.
112. (d) Fibrous joint is an attachment between bones, being held together by fibrous connective tissue. For example- suture between skull bones.
113. (d) a - (iii), b - (iv), c - (i), d - (ii)
114. (a) This conference occured from 20 Nov to 8 December, 2012 to extend the life of the Kyoto Protocal at Doha in Qatar.
115. (c) Serum globulins test provides confirmatory evidence measuring the deficiency of antibodies (γ -globulin) in a person.
116. (d) Chromatophores play an important role in the process of photo synthesis. They contain pigments and are found in blue green algae.
117. (d) 60 - 70% SO_2 and 20 - 30% NO_2 contribute in the acid rain.
118. (d) The gradual and predictable change, in the composition of species takes place in a given area during ecological succession.
119. (d) During photosynthesis photolysis of water is induced by Mn^{++} and Cl^- ions.

120. (b) Binary fission usually takes place in *Amoeba, Paramoecium* and *Euglena.*

121. (b) Autosomal recessive is a type of disorder in which two copies of an abnormal gene must be found for the disease in the affected person.

122. (b) Whale is a mammal. It has four chambered heart having two atria and two ventricles. Oxygenated and deoxygenated blood flow in separate circulatory pathways.

123. (a) The flowers of cucumber are unisexual, it means they have only male flowers or only female flowers.

124. (c) Parthenocarpic fruits (eg - banana) are produced without fertilization of ovule.

125. (c) A pleiotropic gene regulates multiple traits (characte- ristics) in an individual.

126. (b) Production of body heat is caused by the process of metabolism (respiration).

127. (c) *Petrormyzon marinus*, commonly known as sea lamprey lays eggs in fresh water and its larvae after metamorphosis return to the ocean (saline water).

128. (c) Synergids have filiform apparatus.

129. (b) The correct sequence from outerside to inner side in a woody dicot stem is as follows:

Phellem $\rightarrow$ Secondary cortex $\rightarrow$ Secondary phloem $\rightarrow$ Wood , [(d) $\rightarrow$ (a) $\rightarrow$ (c) $\rightarrow$ (b)]

130. (a) Aldosterne is produced by adrenal cortex and plays an important role in the regulation of Na^+ and K^+ levels in body.

131. (c) Golden rice (*Oryza sativa*) is a genetically modified crop. It biosynthesizes β - carotene which is the precursor of vitamin-A.

132. (b) Outbreeding is useful in the problem of inbreeding depression.

133. (c) In codominance both alleles are independently expressed in the heterozygote.

134. (d) ADH (Antidiuretic hormone) and oxytocin are produced by hypothalamus and stored in posterior pituitary.

135. (d) Biomagnification means an increase in concentration of toxins through the trophic levels of a food chain.

136. (a) Industrial melanism occured in England due to industrial revolution. It caused black coating of moths for the adaptation in the present environment.

137. (a) The dental formula for milk teeth is $\frac{2102}{2102}$, so premolars are absent in the primary dentition.

138. (b) Scutellum is large, shield shaped cotyledon that helps in the absorption of nutrients from endosperm during germination.

139. (a) The body cells in cockroach discharge their nitrogenous waste in the haemolymph as a solution of sodium or potassium urate (a weak alkaline fluid).

140. (c) Nucleic acids have phophodiester bond in a nucleotide.

141. (d) Thomas Hunt Morgan won the Nobel Prize (1933) in physiology or medicine for the function of chromosomes in heredity.

142. (a) The spores are non motile in Mucor.

143. (a) In ectopic pregnancies, the implantation of embryo does not occur in uterus but at other site.

144. (c) Detritivores are the organisms which feed on dead plants and animal residues.

145. (b) Babesiosis is malaria like disease caused by protozoon-*Babesia*. In this desease haemoglobinuric fever takes place.

146. (d) Competition occurs for light, food and space among organisms in which all partners are adversely affected in an ecosystom.

147. (b) Genome > chromosomes > gene > nucleotide.

148. (c) XX × X^cY

Normal women Colourblind man

♀ \ ♂	X^c	Y
x	XX^c	XY
x	XX^c	XY

The daughters of this couple will have normal eye sight and carrier if one of the carrier daughter marries with normal eyed man.

XX^c × XY

Normal women Normal man

♀ \ ♂	X	Y
x	XX	XY
x^c	XX	X^cY

Only 25% grandson will show colourblindness.

149. (c) Stromal matrix contains a number of flattened membranous sacs called thylakoid or lamellae.

150. (d) Carbon and Nitrogen are gaseous nutrient cycle. Sulphur and phosphorus are sedimentary nutrient cycle.

151. (d) *Agrobacterium tumefaciens* introduces t-DNA into the plant.

152. (a) The bird's wings and insect's wings are analogous in structure. They show convergent evolution having similar adaptive functional structure developing in different species.

153. (d) Active absorption creates root pressure. In this process, the expenditure of energy takes place for the movement of substances against concentration gradient.

154. (c) Urine has acidic nature because hydrogen ions(H^+) are components of an acid which are secreted into the filtrate.

155. (c) Cell wall is absent in a protoplast.

156. (b) Endemic species is confined to a particular area due to special type of adaptation according to that region.

157. (d) T.O. Dinear (1971) discovered the viroids which are smaller than viruses.

158. (a) When the placenta is axial and ovules are attached to it in a multilocular ovary, it is known as axile placentation.

159. (a) GIFT (Gamete Intra Fallopian Transfer) is a technique helping to have a child.

160. (a) In poliomyelitis, anterior horn cells of spinal cord are destructed which causes loss of motor activities of limbs.

161. (d) Leghemoglobin prevents oxygen poisoning which causes inactivation of nitrogenase.

162. (a) Biotic community is an association of individuals of different species living in the same habitat and showing functional interactions.

163. (a) In the disease, emphysema, alveolar surface area is reduced due to destruction of alveolar walls.

164. (c) RNA and protein synthesis occur in Balbiani rings.

165. (a) (a) → (iii), (b) → (iv), (c) → (i), (d) → (ii)

166. (c) Lysosomes, golgi apparatus and mitochondria are the cell organelles which have membranes.

167. (d) Avena coleoptile curvature is used for the bioassay of auxin.

168. (c) Zona pellucida is formed as a new membrane by secondary oocyte around itself.

169. (a) Satellite DNA displays high degree of polymorphism in population and also the same degree of polymorphism in an individual, which is inherited from parents to children (offsprings).

170. (b) Mycoplasma lacks cell wall.

171. (b) In meiosis, the number of chromosomes are reduced by half producing haploid daughter cells. The microspore mother cell and the megaspore mother cell undergo meiosis to produce haploid microspore and megaspore respectively.

172. (a) Pistia a hydrophyte plant where absorption of water by root is not important.

173. (d) Lichens cannot grow in the place where sulpher dioxide, pollutant is available in the environment.

174. (a) Cell-midiated immune response is a system which is specialized for a particular person.

175. (a) The given characteristic features define the phylum porifera.

176. (d) In chrysophytes, the cell walls form two thin overplapping shells holding together. The body of Diatoms appears like soap box due to overlapping shells.

177. (b) Morel and truffles are used as food and they are members of Ascomycetes fungi.

178. (a) Meiosis-II does not complete untill fertilization occurs in females (in human being).

179. (c) Due to eutrophication water bodies lack oxygen causing death of aquatic life.

180. (a) Succus entericus lacks enzyme nucleases.

AIPMT 2015

PHYSICS

1. If energy (E), velocity (V) and time (T) are chosen as the fundamental quantities, the dimensional formula of surface tension will be :
   (a) $[EV^{-1}T^{-2}]$ (b) $[EV^{-2}T^{-2}]$
   (c) $[E^{-2}V^{-1}T^{-3}]$ (d) $[EV^{-2}T^{-1}]$

2. A particle of unit mass undergoes one-dimensional motion such that its velocity varies according to $v(x) = bx^{-2n}$ where b and n are constants and x is the position of the particle. The acceleration of the particle as d function of x, is given by:
   (a) $-2nb^2x^{-4n-1}$ (b) $-2b^2x^{-2n+1}$
   (c) $-2nb^2e^{-4n+1}$ (d) $-2nb^2x^{-2n-1}$

3. A ship A is moving Westwards with a speed of 10 km h^{-1} and a ship B 100 km South of A, is moving Northwards with a speed of 10 km h^{-1}. The time after which the distance between them becomes shortest, is :
   (a) 5 h (b) $5\sqrt{2}$ h
   (c) $10\sqrt{2}$ h (d) 0 h

4. A block A of mass m_1 rests on a horizontal table. A light string connected to it passes over a frictionless pulley at the edge of table and from its other end another block B of mass m_2 is suspended. The coefficient of kinetic friction between the block and the table is μ_k. When the block A is sliding on the table, the tension in the string is
   (a) $\dfrac{(m_2 - \mu k\, m_1)g}{(m_1 + m_2)}$ (b) $\dfrac{m_1m_2(1 + \mu_k)g}{(m_1 + m_2)}$
   (c) $\dfrac{m_1m_2(1 - \mu_k)g}{(m_1 + m_2)}$ (d) $\dfrac{(m_2 + \mu_k\, m_1)g}{(m_1 + m_2)}$

5. Three blocks A, B and C of masses 4 kg, 2 kg and 1 kg respectively, are in contact on a frictionless surface, as shown. If a force of 14 N is applied on the 4 kg block then the contact force between A and B is

   (a) 6 N (b) 8 N
   (c) 18 N (d) 2 N

6. A block of mass 10 kg, moving in x direction with a constant speed of 10 ms^{-1}, is subject to a retarding force F = 0.1 × J/m during its travel from x = 20 m to 30 m. Its final KE will be :
   (a) 450 J (b) 275 J
   (c) 250 J (d) 475 J

7. Two similar springs P and Q have spring constants K_P and K_Q, such that $K_P > K_Q$. They are stretched, first by the same amount (case a,) then by the same force (case b). The work done by the springs W_P and W_Q are related as, in case (a) and case (b), respectively
   (a) $W_P = W_Q$; $W_P = W_Q$
   (b) $W_P > W_Q$; $W_Q > W_P$
   (c) $W_P < W_Q$; $W_Q < W_P$
   (d) $W_P = W_Q$; $W_P > W_Q$

8. A particle of mass m is driven by a machine that delivers a constant power of k watts. If the particle starts from rest the force on the particle at time t is
   (a) $\sqrt{mk}\ t^{-1/2}$ (b) $\sqrt{2mk}\ t^{-1/2}$
   (c) $\dfrac{1}{2}\sqrt{mk}\ t^{-1/2}$ (d) $\sqrt{\dfrac{mk}{2}}t^{-1/2}$

9. A mass m moves in a circle on a smooth horizontal plane with velocity v_0 at a radius R_0. The mass is attached to string which passes through a smooth hole in the plane as shown.

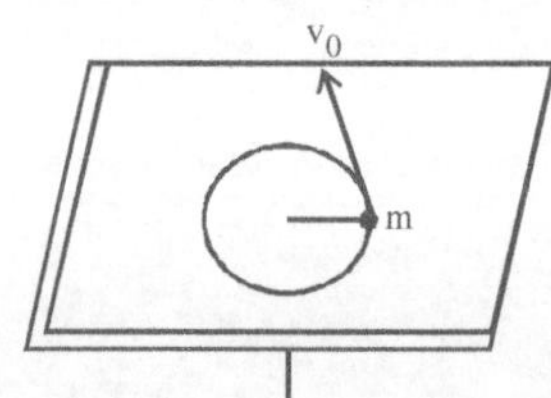

The tension in the string is increased gradually and finally m moves in a circle of radius $\frac{R_0}{2}$. The final value of the kinetic energy is

(a) $\frac{1}{4}mv_0^2$ (b) $2mv_0^2$

(c) $\frac{1}{2}mv_0^2$ (d) mv_0^2

10. A rod of weight W is supported by two parallel knife edges A and B and is in equilibrium in a horizontal position. The knives are at a distance d from each other. The centre of mass of the rod is at distance x from A. The normal reaction on A is

(a) $\frac{Wd}{x}$ (b) $\frac{W(d-x)}{x}$

(c) $\frac{W(d-x)}{d}$ (d) $\frac{Wx}{d}$

11. Three identical spherical shells, each of mass m and radius r are placed as shown in figure. Consider an axis XX' which is touching to two shells and passing through diameter of third shell. Moment of inertia of the system consisting of these three spherical shells about XX' axis is

(a) $3mr^2$

(b) $\frac{16}{5}mr^2$

(c) $4mr^2$

(d) $\frac{11}{5}mr^2$

12. Kepler's third law states that square of period of revolution (T) of a planet around the sun, is proportional to third power of average distance r between sun and planet i.e. $T^2 = Kr^3$ here K is constant. If the masses of sun and planet are M and m respectively then as per Newton's law of gravitation force of attraction between them is $F = \frac{GMm}{r^2}$, here G is gravitational constant. The relation between G and K is described as

(a) $GMK = 4\pi^2$ (b) $K = G$

(c) $K = \frac{1}{G}$ (d) $GK = 4\pi^2$

13. Two spherical bodies of mass M and 5 M and radii R and 2 R released in free space with initial separation between their centres equal to 12 R. If they attract each other due to gravitational force only, then the distance covered by the smaller body before collision is

(a) 4.5 R (b) 7.5 R

(c) 1.5 R (d) 2.5 R

14. The approximate depth of an ocean is 2700 m. The compressibility of water is 45.4×10^{-11} Pa^{-1} and density of water is 10^3 kg/m^3. What fractional compression of water will be obtained at the bottom of the ocean ?

(a) 1.0×10^{-2} (b) 1.2×10^{-2}

(c) 1.4×10^{-2} (d) 0.8×10^{-2}

15. A wind with speed 40 m/s blows parallel to the roof of a house. The area of the roof is 250 m^2. Assuming that the pressure inside the house is atmospheric pressure, the force exerted by the wind on the roof and the direction of the force will be ($\rho_{air} = 1.2$ kg/m^3)

(a) 4.8×10^5 N, upwards

(b) 2.4×10^5 N, upwards

(c) 2.4×10^5 N, downwards

(d) 4.8×10^5 N, downwards

16. On observing light from three different stars P, Q and R, it was found that intensity of violet colour is maximum in the spectrum of P, the intensity of green colour is maximum in the spectrum of R and the intensity of red colour is maximum in the spectrum of Q. If T_P, T_Q and T_R are the respective absolute temperature of P, Q and R, then it can be concluded from the above observations that

(a) $T_P > T_R > T_Q$ (b) $T_P < T_R < T_Q$

(c) $T_P < T_Q < T_R$ (d) $T_P > T_Q > T_R$

17. The two ends of a metal rod are maintained at temperatures 100°C and 110°C. The rate of heat flow in the rod is found to be 4.0 J/s. If the ends are maintained at temperatures 200°C and 210°C, the rate of heat flow will be

(a) 16.8 J/s (b) 8.0 J/s

(c) 4.0 J/s (d) 44.0 J/s

18. A Carnot engine, having an efficiency of $\eta = \frac{1}{10}$ as heat engine, is used as a refrigerator. If the work done on the system is 10 J, the amount of

energy absorbed from the reservoir at lower temperature is

(a) 99 J (b) 90 J
(c) 1 J (d) 100 J

19. Figure below shows two paths that may be taken by a gas to go from a state A to a state C.

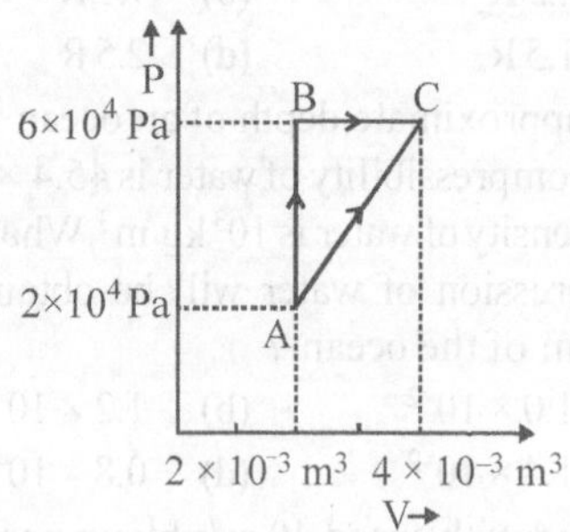

In process AB, 400 J of heat is added to the system and in process BC, 100 J of heat is added to the system. The heat absorbed by the system in the process AC will be

(a) 500 J (b) 460 J
(c) 300 J (d) 380 J

20. One mole of an ideal diatomic gas undergoes a transition from A to B along a path AB as shown in the figure.

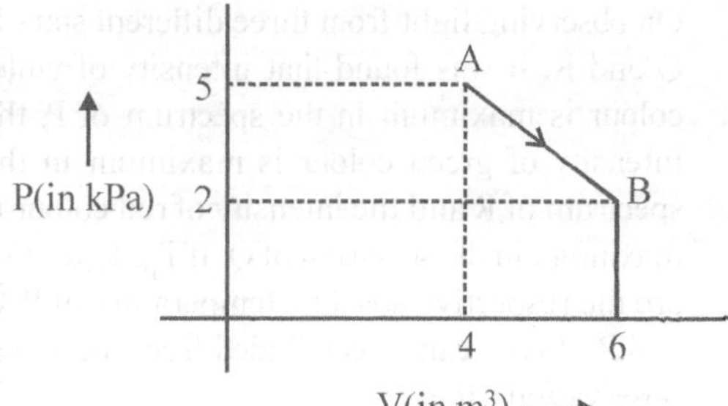

The change in internal energy of the gas during the transition is:

(a) –20 kJ (b) 20 J
(c) –12 kJ (d) 20 kJ

21. The ratio of the specific heats $\frac{C_p}{C_v} = \gamma$ in terms of degrees of freedom (n) is given by

(a) $\left(1+\frac{n}{3}\right)$ (b) $\left(1+\frac{2}{n}\right)$

(c) $\left(1+\frac{n}{2}\right)$ (d) $\left(1+\frac{1}{n}\right)$

22. When two displacements represented by $y_1 = a\sin(\omega t)$ and $y_2 = b\cos(\omega t)$ are superimposed the motion is:

(a) simple harmonic with amplitude $\frac{a}{b}$

(b) simple harmonic with amplitude $\sqrt{a^2+b^2}$

(c) simple harmonic with amplitude $\frac{(a+b)}{2}$

(d) not a simple harmonic

23. A particle is executing SHM along a straight line. Its velocities at distances x_1 and x_2 from the mean position are V_1 and V_2, respectively. Its time period is

(a) $2\pi\sqrt{\frac{x_2^2-x_1^2}{V_1^2-V_2^2}}$ (b) $2\pi\sqrt{\frac{V_1^2+V_2^2}{x_1^2+x_2^2}}$

(c) $2\pi\sqrt{\frac{V_1^2-V_2^2}{x_1^2-x_2^2}}$ (d) $2\pi\sqrt{\frac{x_1^2-x_2^2}{V_1^2-V_2^2}}$

24. The fundamental frequency of a closed organ pipe of length 20 cm is equal to the second overtone of an organ pipe open at both the ends. The length of organ pipe open at both the ends is

(a) 100 cm (b) 120 cm
(c) 140 cm (d) 80 cm

25. The electric field in a certain region is acting radially outward and is given by E = Ar. A charge contained in a sphere of radius 'a' centred at the origin of the field, will be given by

(a) $A\varepsilon_0 a^2$ (b) $4\pi\varepsilon_0 Aa^3$
(c) $\varepsilon_0 Aa^3$ (d) $4\pi\varepsilon_0 Aa^2$

26. A parallel plate air capacitor of capacitance C is connected to a cell of emf V and then disconnected from it. A dielectric slab of dielectric constant K, which can just fill the air gap of the capacitor, is now inserted in it. Which of the following is incorrect ?

(a) The energy stored in the capacitor decreases K times.

(b) The chance in energy stored is $\frac{1}{2}CV^2\left(\frac{1}{K}-1\right)$

(c) The charge on the capacitor is not conserved.

(d) The potential difference between the plates decreases K times.

27. Across a metallic conductor of non-uniform cross section a constant potential difference is applied. The quantity which remains constant along the conductor is :
(a) current (b) drift velocity
(c) electric field (d) current density

28. A potentiometer wire has length 4 m and resistance 8Ω. The resistance that must be connected in series with the wire and an accumulator of e.m.f. 2V, so as to get a potential gradient 1 mV per cm on the wire is
(a) 40 Ω (b) 44 Ω
(c) 48 Ω (d) 32 Ω

29. A, B and C are voltmeters of resistance R, 1.5 R and 3R respectively as shown in the figure. When some potential difference is applied between X and Y, the voltmeter readings are V_A, V_B and V_C respectively. Then

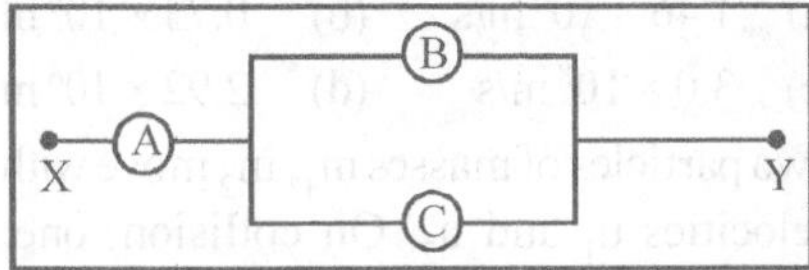

(a) $V_A \neq V_B = V_C$ (b) $V_A = V_B \neq V_C$
(c) $V_A \neq V_B \neq V_C$ (d) $V_A = V_B = V_C$

30. An electron moving in a circular orbit of radius r makes n rotations per second. The magnetic field produced at the centre has magnitude:
(a) Zero (b) $\frac{\mu_0 n^2 e}{r}$
(c) $\frac{\mu_0 ne}{2r}$ (d) $\frac{\mu_0 ne}{2\pi r}$

31. A wire carrying current I has the shape as shown in adjoining figure. Linear parts of the wire are very long and parallel to X-axis while semicircular portion of radius R is lying in Y-Z plane. Magnetic field at point O is :

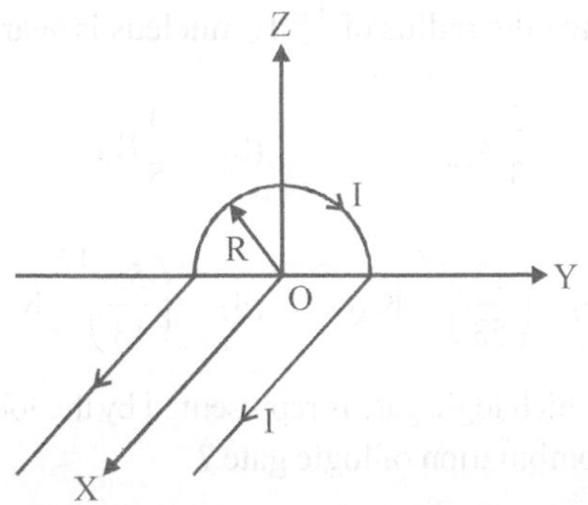

(a) $\vec{B} = -\frac{\mu_0}{4\pi}\frac{I}{R}\left(\mu\hat{i} \times 2\hat{k}\right)$
(b) $\vec{B} = -\frac{\mu_0}{4\pi}\frac{I}{R}\left(\pi\hat{i} + 2\hat{k}\right)$
(c) $\vec{B} = \frac{\mu_0}{4\pi}\frac{I}{R}\left(\pi\hat{i} - 2\hat{k}\right)$
(d) $\vec{B} = \frac{\mu_0}{4\pi}\frac{I}{R}\left(\pi\hat{i} + 2\hat{k}\right)$

32. A conducting square frame of side 'a' and a long staight wire carrying current I are located in the same plane as shown in the figure. The frame moves to the right with a constant velocity 'V'. The emf induced in the frame will be proportional to

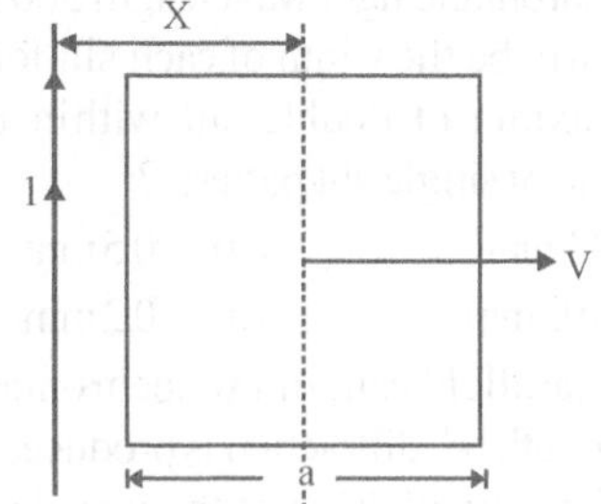

(a) $\frac{1}{(2x-a)^2}$ (b) $\frac{1}{(2x+a)^2}$
(c) $\frac{1}{(2x-a)(2x+a)}$ (d) $\frac{1}{x^2}$

33. A resistance 'R' draws power 'P' when connected to an AC source. If an inductance is now placed in series with the resistance, such that the impedance of the circuit becomes 'Z', the power drawn will be
(a) $P\sqrt{\frac{R}{Z}}$ (b) $P\left(\frac{R}{Z}\right)$
(c) P (d) $P\left(\frac{R}{Z}\right)^2$

34. A radiation of energy 'E' falls normally on a perfectly reflecting surface. The momentum transferred to the surface is (C = Velocity of light)
(a) $\frac{2E}{C}$ (b) $\frac{2E}{C^2}$
(c) $\frac{E}{C^2}$ (d) $\frac{E}{C}$

35. The refracting angle of a prism is 'A', and refractive index of the material of the prism is cot(A/2). The angle of minimum deviation is :
(a) $180° - 2A$ (b) $90° - A$
(c) $180° + 2A$ (d) $180° - 3A$

36. Two identical thin plano-convex glass lenses (refractive index 1.5) each having radius of curvature of 20 cm are placed with their convex surfaces in contact at the centre. The intervening space is filled with oil of refractive index 1.7. The focal length of the combination is
(a) –25 cm (b) –50 cm
(c) 50 cm (d) –20 cm

37. In a double slit experiment, the two slits are 1 mm apart and the screen is placed 1 m away. A monochromatic light wavelength 500 nm is used. What will be the width of each slit for obtaining ten maxima of double slit within the central maxima of single slit pattern ?
(a) 0.1 mm (b) 0.5 mm
(c) 0.02 mm (d) 0.2 mm

38. For a parallel beam of monochromatic light of wavelength 'λ', diffraction is produced by a single slit whose width 'a' is of the wavelength of the light. If 'D' is the distance of the screen from the slit, the width of the central maxima will be :
(a) $\frac{D\lambda}{a}$ (b) $\frac{Da}{\lambda}$
(c) $\frac{2Da}{\lambda}$ (d) $\frac{2D\lambda}{a}$

39. Which of the following figures represent the variation of particle momentum and the associated de-Broglie wavelength?

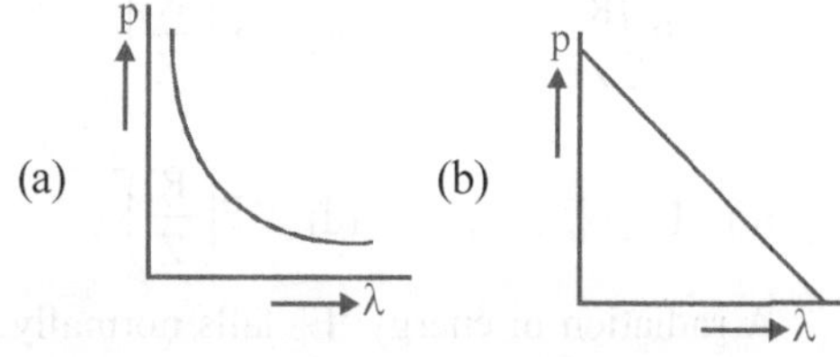

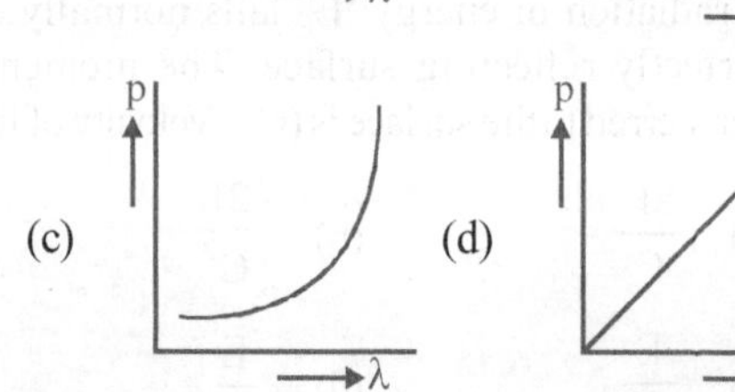

40. A certain metallic surface is illuminated with monochromatic light of wavelength λ. The stopping potential for photo-electric current for this light is $3V_0$. If the same surface is illuminated with light of wavelength 2λ, the stopping potential is V_0. The threshold wavelength for this surface for photo-electric effect is
(a) 4λ (b) $\frac{\lambda}{4}$
(c) $\frac{\lambda}{6}$ (d) 6λ

41. Consider 3rd orbit of He^+ (Helium), using non-relativistic approach, the speed of electron in this orbit will be [given $K = 9 \times 10^9$ constant, Z = 2 and h (Plank's Constant) $= 6.6 \times 10^{-34}$ J s]
(a) 1.46×10^6 m/s (b) 0.73×10^6 m/s
(c) 3.0×10^8 m/s (d) 2.92×10^6 m/s

42. Two particles of masses m_1, m_2 move with initial velocities u_1 and u_2. On collision, one of the particles get excited to higher level, after absorbing energy ε. If final velocities of particles be v_1 and v_2 then we must have
(a) $\frac{1}{2}m_1u_1^2 + \frac{1}{2}m_2u_2^2 = \frac{1}{2}m_1v_1^2 + \frac{1}{2}m_2v_2^2 - \varepsilon$
(b) $\frac{1}{2}m_1u_1^2 + \frac{1}{2}m_2u_2^2 - \varepsilon = \frac{1}{2}m_1v_1^2 + \frac{1}{2}m_2V_2^2$
(c) $\frac{1}{2}m_1^2u_1^2 + \frac{1}{2}m_2^2u_2^2 + \varepsilon = \frac{1}{2}m_1^2v_1^2 + \frac{1}{2}m_2^2v_2^2$
(d) $m_1^2u_1 + m_2^2u_2 - \varepsilon = m_1^2v_1 + m_2^2v_2$

43. If radius of the $^{27}_{12}Al$ nucleus is taken to be R_{Al}, then the radius of $^{125}_{53}Te$ nucleus is nearly:
(a) $\frac{5}{3}R_{Al}$ (b) $\frac{3}{5}R_{Al}$
(c) $\left(\frac{13}{53}\right)^{1/3} R_{Al}$ (d) $\left(\frac{53}{13}\right)^{1/3} R_{Al}$

44. Which logic gate is represented by the following combination of logic gate ?

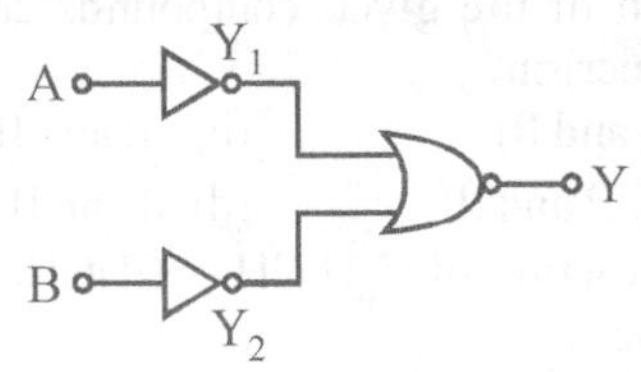

(a) NAND (b) AND
(c) NOR (d) OR

45. If in a p-n junction, a square input signal of 10 V is applied as shown, then the output across R_L will be

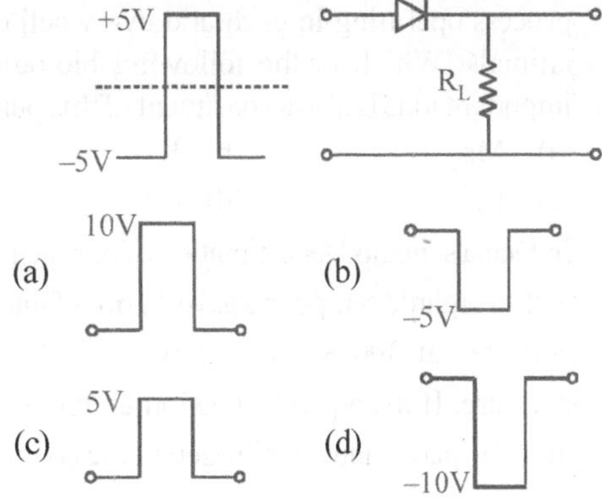

CHEMISTRY

46. A mixture of gases contains H_2 and O_2 gases in the ratio of 1 : 4 (w/w). What is the molar ratio of the two gases in the mixture ?
(a) 4 : 1 (b) 16 : 1
(c) 2 : 1 (d) 1 : 4

47. The species Ar, K^+ and Ca^{2+} contain the same number of electrons. In which order do their radii increase ?
(a) $Ca^{2+} < Ar < K^+$ (b) $Ca^{2+} < K^+ < Ar$
(c) $K^+ < Ar < Ca^{2+}$ (d) $Ar < K^+ < Ca^{2+}$

48. The correct bond order in the following species is:
(a) $O_2^{2+} < O_2^- < O_2^+$ (b) $O_2^+ < O_2^- < O_2^{2+}$
(c) $O_2^- < O_2^+ < O_2^{2+}$ (d) $O_2^{2+} < O_2^+ < O_2^-$

49. Which of the following pairs of ions are isoelectronic and isostructural ?
(a) ClO_3^-, CO_3^{2-} (b) SO_3^{2-}, NO_3^-
(c) ClO_3^-, SO_3^{2-} (d) CO_3^{2-}, SO_3^{2-}

50. Which of the following options represents the correct bond order ?
(a) $O_2^- < O_2 < O_2^+$ (b) $O_2^- > O_2 < O_2^+$
(c) $O_2^- < O_2 > O_2^+$ (d) $O_2^- > O_2 > O_2^+$

51. Maximum bond angle at nitrogen is present in which of the following ?
(a) NO_2^- (b) NO_2^+
(c) NO_3^- (d) NO_2

52. Which of the following species contains equal number of σ- and π-bonds :
(a) XeO_4 (b) $(CN)_2$
(c) $CH_2(CN)_2$ (d) HCO_3^-

53. The K_{sp} of Ag_2CrO_4, AgCl, AgBr and AgI are respectively, 1.1×10^{-12}, 1.8×10^{-10}, 5.0×10^{-13}, 8.3×10^{-17}. Which one of the following salts will precipitate last if $AgNO_3$ solution is added to the solution containing equal moles of NaCl, NaBr, NaI and Na_2CrO_4?
(a) AgCl (b) AgBr
(c) Ag_2CrO_4 (d) AgI

54. If the value of an equilibrium constant for a particular reaction is 1.6×10^{12}, then at equilibrium the system will contain :-
(a) mostly reactants
(b) mostly products
(c) similar amounts of reactants and products
(d) all reactants

55. Which of the following statements is correct for a reversible process in a state of equilibrium ?
(a) $\Delta G = 2.30\ RT \log K$
(b) $\Delta G^o = -2.30\ RT \log K$
(c) $\Delta G^o = 2.30\ RT \log K$
(d) $\Delta G = -2.30\ RT \log K$

56. Solubility of the alkaline earth's metal sulphates in water decreases in the sequence :-
(a) Ca > Sr > Ba > Mg (b) Sr > Ca > Mg > Ba
(c) Ba > Mg > Sr > Ca (d) Mg > Ca > Sr > Ba

57. The function of "Sodium pump" is a biological process operating in each and every cell of all animals. Which of the following biologically important ions is also a consituent of this pump :-
(a) Mg^{2+} (b) K^{+}
(c) Fe^{2+} (d) Ca^{2+}

58. In Duma's method for estimation of nitrogen, 0.25 g of an organic compound gave 40 mL of nitrogen collected at 300 K temperature and 725 mm pressure. If the aqueous tension at 300 K is 25 mm, the percentage of nitrogen in the compound is :
(a) 18.20 (b) 16.76
(c) 15.76 (d) 17.36

59. Which of the following is the most correct electron displacement for a nucleophilic reaction to take place?

(a) 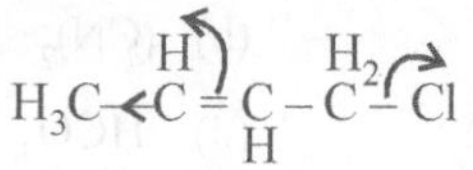 $H_3C \leftarrow \overset{H}{C} = \underset{H}{C} - \overset{H_2}{C} - Cl$

(b) $H_3C \rightarrow \overset{H}{C} = \underset{H}{C} - \overset{H_2}{C} - Cl$

(c) $H_3C \rightarrow \overset{H}{C} = \underset{H}{C} - \overset{H_2}{C} - Cl$

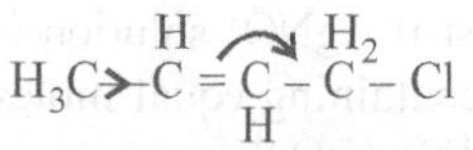

(d) 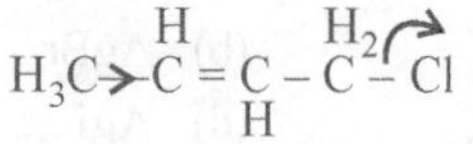 $H_3C \rightarrow \overset{H}{C} = \underset{H}{C} - \overset{H_2}{C} - Cl$

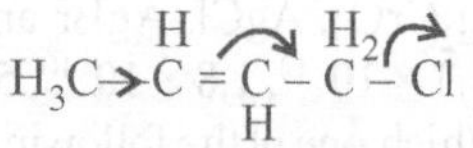

60. Consider the following compounds

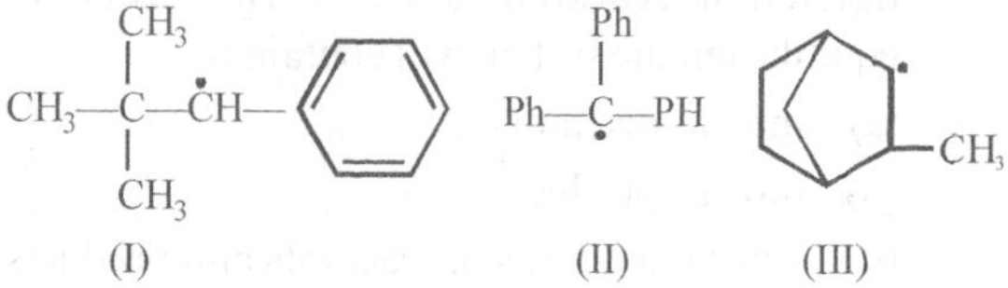

$CH_3-C(CH_3)_2-\dot{C}H-C_6H_5$ (I) $Ph_2\dot{C}-Ph$ (II) (III) CH_3

Hyperconjugation occurs in :
(a) II only (b) III only
(c) I and III (d) I only

61. Given

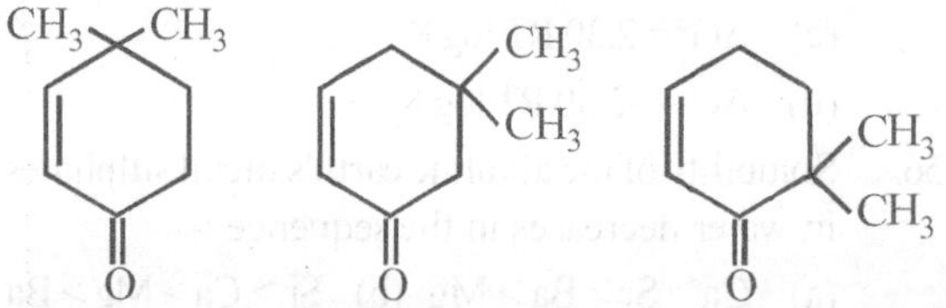

(I) (II) (III)

Which of the given compounds can exhibit tautomerism?
(a) I and III (b) II and III
(c) I, II and III (d) I and II

62. The reaction of $C_6H_5CH = CHCH_3$ with HBr produces:

(a) $C_6H_5CH_2\underset{Br}{C}HCH_3$

(b) $C_6H_5CH_2CH_2CH_2Br$

(c) 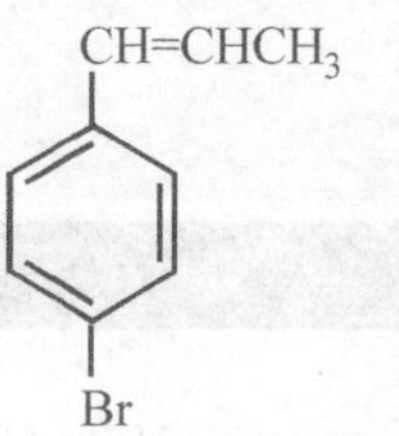 $CH{=}CHCH_3$ (para) Br

(d) $C_6H_5\underset{Br}{C}HCH_2CH_3$

63. The total number of π-bond electrons in the following structure is :-

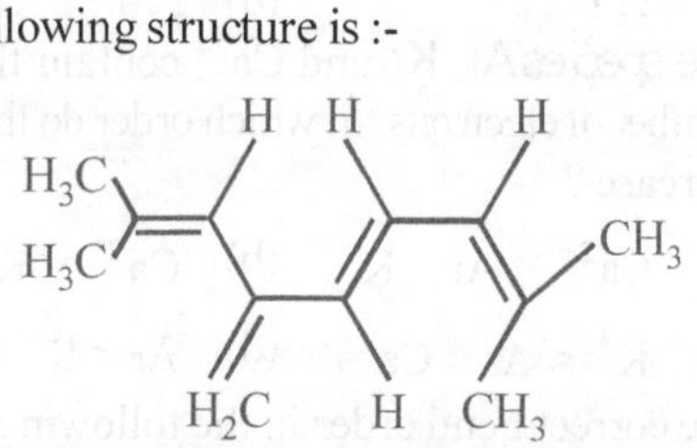

(a) 8 (b) 12
(c) 16 (d) 4

64. Given:

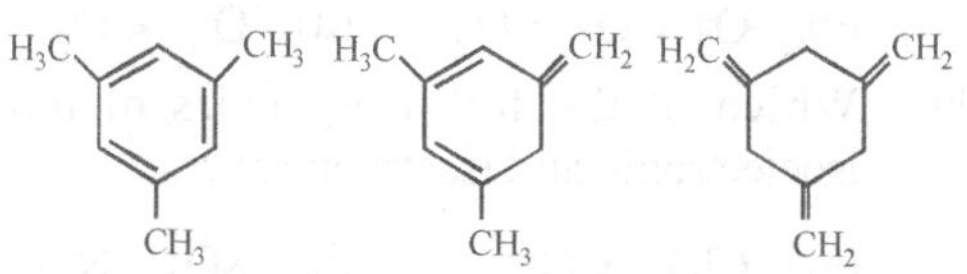

(I) (II) (III)

The enthalpy of the hydrogenation of these compounds will be in the order as :-
(a) III > II > I (b) II > III > I
(c) II > I > III (d) I > II > III

65. A single compound of the structure :

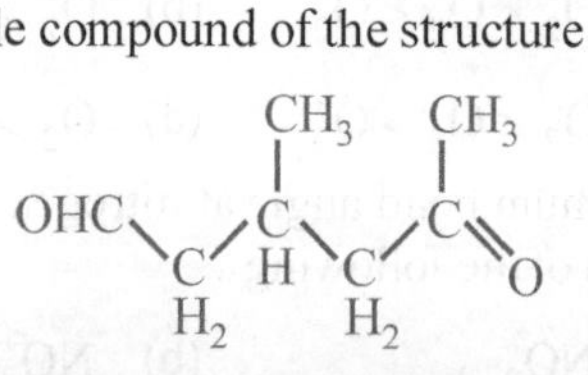

$OHC-CH_2-CH(CH_3)-CH_2-C(CH_3)=O$

is obtainable from ozonolysis of which of the following cyclic compounds ?

(a) 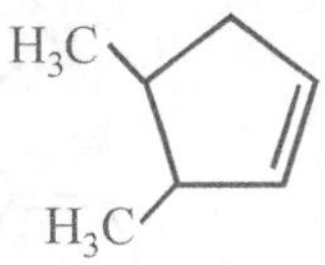 H_3C, H_3C

(b) 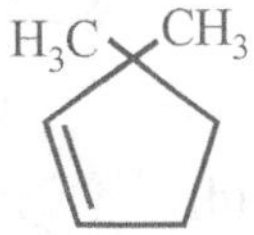 H_3C CH_3

(c) 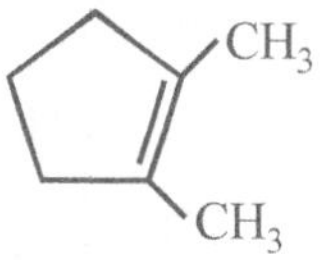 CH_3, CH_3

(d) 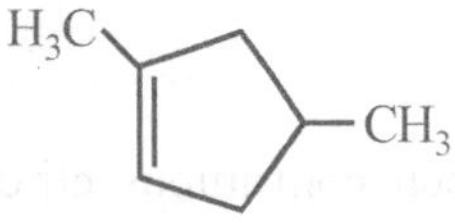 H_3C, CH_3

66. A given metal crystallizes out with a cubic structure having edge length of 361 pm. If there are four metal atoms in one unit cell, what is the radius of one atom?

(a) 127 pm (b) 80 pm
(c) 108 pm (d) 40 pm

67. Which one of the following electrolytes has the same value of van't Hoff's factor (i) as that of the $Al_2(SO_4)_3$ (if all are 100% ionised) ?

(a) $K_3[Fe(CN)_6]$ (b) $Al(NO_3)_3$
(c) $K_4[Fe(CN)_6]$ (d) K_2SO_4

68. The boiling point of 0.2 mol kg^{-1} solution of X in water is greater than equimolal solution of Y in water. Which one of the following statements is true in this case ?

(a) Molecular mass of X is greater than the molecular mass of Y.
(b) Molecular mass of X is less than the molecular mass of Y.
(c) Y is undergoing dissociation in water while X undergoes no change.
(d) X is undergoing dissociation in water.

69. Which one is not equal to zero for an ideal solution:

(a) ΔS_{mix}
(b) ΔV_{mix}
(c) $\Delta P = P_{observed} - P_{Raoult}$
(d) ΔH_{mix}

70. Which property of colloidal solution is independent of charge on the colloidal particles:-

(a) Electrophoresis (b) Electro-osmosis
(c) Tyndall effect (d) Coagulation

71. A device that converts energy of combustion of fuels like hydrogen and methane, directly into electrical energy is known as :

(a) Electrolytic cell (b) Dynamo
(c) Ni-Cd cell (d) Fuel Cell

72. The activation energy of a reaction can be determined from the slope of which of the following graphs ?

(a) $\frac{\ln K}{T}$ vs. T (b) $\ln K$ vs. $\frac{1}{T}$
(c) $\frac{T}{\ln K}$ vs. $\frac{1}{T}$ (d) $\ln K$ vs . T

73. When initial concentration of a reactant is doubled in a reaction, its half-life period is not affected. The order of the reaction is :

(a) First
(b) Second
(c) More than zero but less than first
(d) Zero

74. "Metals are usually not found as nitrates in their ores".

Out of the following two ((i) and (ii)) reasons which is/are true for the above observation ?

(i) Metal nitrates are highly unstable.
(ii) Metal nitrates are highly soluble in water.

(a) (i) and (ii) are false
(b) (i) is false but (ii) is true
(c) (i) is true but (ii) is false
(d) (i) and (ii) are true

75. Nitrogen dioxide and sulphur dioxide have some properties in common. Which property is shown by one of these compounds, but not by the other?

(a) is a reducing agent
(b) is soluble in water
(c) is used as a food-preservative
(d) forms 'acid-rain'

76. Magnetic moment 2.84 B.M. is given by :-
(At. nos, Ni = 28, Ti = 22, Cr = 24, Co = 27)

(a) Ti^{3+} (b) Cr^{2+}
(c) Co^{2+} (d) Ni^{2+}

77. The number of *d*-electrons in Fe^{2+} (Z = 26) is not equal to the number of electrons in which one of the following?
(a) *p*-electrons in Cl (Z = 17)
(b) *d*-electrons in Fe (Z = 26)
(c) *p*-electrons in Ne (Z = 10)
(d) *s*-electrons in Mg (Z = 12)

78. Which of the following processes does not involve oxidation of iron ?
(a) Decolourization of blue $CuSO_4$ solutution by iron
(b) Formation of $Fe(CO)_5$ from Fe
(c) Liberation of H_2 from steam by iron at high temperature
(d) Rusting of iron sheets

79. Because of lanthanoid contraction, which of the following pairs of elements have nearly same atomic radii ? (Numbers in the parenthesis are atomic numbers).
(a) Zr (40) and Nb (41) (b) Zr (40) and Hf (72)
(c) Zr (40) and Ta (73) (d) Ti (22) and Zr (40)

80. The angular momentum of electron in 'd' orbital is equal to :
(a) $\sqrt{2}\,\hbar$ (b) $2\sqrt{3}\,\hbar$
(c) $0\hbar$ (d) $\sqrt{6}\,\hbar$

81. Cobalt (III) chloride forms several octahedral complexes with ammonia. Which of the following will not give test of chloride ions with silver nitrate at 25ºC ?
(a) $CoCl_3{\cdot}4NH_3$ (b) $CoCl_3{\cdot}5NH_3$
(c) $CoCl_3{\cdot}6NH_3$ (d) $CoCl_3{\cdot}3NH_3$

82. Which of these statements about $[Co(CN)_6]^{3-}$ is true ?
(a) $[Co(CN)_6]^{3-}$ has four unpaired electrons and will be in a low-spin configuration.
(b) $[Co(CN)_6]^{3-}$ has four unpaired electrons and will be in a high spin configuration.
(c) $[Co(CN)_6]^{3-}$ has no unpaired electrons and will be in a high-spin configurtion.
(d) $[Co(CN)_6]^{3-}$ has no unpaired electrons and will be in a low-spin configuration.

83. In which of the following compounds, the C - Cl bond ionisation shall give most stable carbonium ion?
(a) $(H_3C)_2C(CH_3)-Cl$ (b) $C_6H_5-CH(H)-Cl$
(c) $O_2NH_2C-CH(H)-Cl$ (d) $(H_3C)_2CH-Cl$

84. The reaction

$$CH_3-\underset{CH_3}{\overset{CH_3}{C}}-ONa + CH_3CH_2Cl \xrightarrow[-NaCl]{} CH_3-\underset{CH_3}{\overset{CH_3}{C}}-O-CH_2-CH_3$$

is called :-
(a) Williamson continuous etherification process
(b) Etard reaction
(c) Gatterman - Koch reaction
(d) Williamson Synthesis

85. An organic compound 'X' having molecular formula $C_5H_{10}O$ yields phenyl hydrazone and gives negative response to the iodoform test and Tollen's test. It produces *n*-pentane on reduction. 'X' could be :-
(a) 2-pentanone
(b) 3-pentanone
(c) *n*-amyl alcohol
(d) pentanal

86. Treatment of cyclopentanone (cyclopentane ring =O) with methyl lithium gives which of the following species?
(a) Cyclopentanonyl cation
(b) Cyclopentanonyl radical
(c) Cyclopentanonyl biradical
(d) Cyclopentanonyl anion

87. The enolic form of ethyl acetoacetate as below has:

$$H_3C-C(OH)=CH-C(=O)-OC_2H_5 \rightleftharpoons H_3C-C(=O)-CH_2-C(=O)-OC_2H_5$$

(a) 16 sigma bonds and 1 pi - bond
(b) 9 sigma bonds and 2 pi - bonds
(c) 9 sigma bonds and 1 pi - bond
(d) 18 sigma bonds and 2 pi - bonds

88. The electrolytic reduction of nitrobenzene in strongly acidic medium produces :-
(a) Azoxybenzene
(b) Azobenzene
(c) Aniline
(d) p-Aminophenol

89. Biodegradable polymer which can be produced from glycine and aminocaproic acid is :
(a) PHBV (b) Buna - N
(c) Nylon 6, 6 (d) Nylon 2- nylon 6

90. Bithional is generally added to the soaps as an additive to function as a/an :
(a) Dryer
(b) Buffering agent
(c) Antiseptic
(d) Softner

BIOLOGY

91. Which one one of the following matches is correct ?

1	*Alternaria*	Sexual reproduction absent	Deuteromycetes
2	*Mucor*	Reproduction by Conjugation	Ascomycetes
3	*Agaricus*	Parasitic fungus	Basidiomycetes
4	*Phytophthora*	Aseptate mycelium	Basidiomycetes

92. True nucleus is absent in :
(a) *Mucor* (b) *Vaucheria*
(c) *Volvox* (d) *Anabaena*

93. In which of the following gametophyte is not independent free living ?
(a) *Marchantia* (b) *Pteris*
(c) *Pinus* (d) *Funaria*

94. Read the following five statements (A to E) and select the option with all correct statements :
(A) Mosses and Lichens are the first organisms to colonise a bare rock.
(B) *Selaginella* is a homosporous pteridophyte
(C) Coralloid roots in Cycas have VAM
(D) Main plant body in bryophytes is gametophytic, whereas in pteridophytes it is sporophytic
(E) In gymnosperms, male and female gametophytes are present within sporangia located on sporophyte
(a) (B), (C) and (D) (b) (A), (D) and (E)
(c) (B), (C) and (E) (d) (A), (C) and (D)

95. Male gemetes are flagellated in :
(a) *Anabaena* (b) *Ectocarpus*
(c) *Spirogyra* (d) *Polysiphonia*

96. Which one of the following statements is wrong?
(a) Agar - agar is obtained from *Gelidium* and *Gracilaria*
(b) *Chlorella* and *Spirulina* are used as space food
(c) Mannitol is stored food in Rhodophyceae
(d) Algin and carragen are products of algae

97. Which of the following animals is not viviparous?
(a) Elephant (b) Platypus
(c) Whale (d) Flying fox (Bat)

98. The terga, sterna and pleura of cockroach body are joined by :
(a) Muscular tissue
(b) Arthrodial membrane
(c) Cartilage
(d) Cementing glue

99. Which of the following represents the correct combination without any exception?

	Characteristics	Class
(a)	Mouth ventral, gills without operculum; skin with placoid scales; persistent notochord	Chondrichthyes
(b)	Sucking and circular mouth; jaws absent, integument without scales; paired appendages	Cyclostomata
(c)	Body covered with feathers; skinmoist and glandular; fore-limbsform wings; lungs with air sacs	Aves
(d)	Mammary gland; hair on body;pinnae; two pairs of Limbs	Mammalia

100. Leaves become modified into spines in :-
(a) Pea (b) Onion
(c) Silk Cotton (d) *Opuntia*

101. $\oplus \, ⚥ \, K_{(5)} \, \widehat{C_{(5)} A_5} \, G_{(2)}$ is the floral formula of
(a) *Sesbania* (b) *Petunia*
(c) *Brassica* (d) *Allium*

102. Keel is the characteristic feature of flower of :
(a) *Indigofera* (b) *Aloe*
(c) Tomato (d) Tulip

103. Perigynous flowers are found in :-
(a) Cucumber (b) China rose
(c) Rose (d) Guava

104. In ginger vegetative propagation occurs through:
(a) Offsets (b) Bulbils
(c) Runners (d) Rhizome

105. Transmission tissue is characteristic feature of :
(a) Solid style (b) Dry stigma
(c) Wet stigma (d) Hollow style

106. A major characteristic of the monocot root is the presence of :
(a) Scattered vascular bundles
(b) Vasculature without cambium
(c) Cambium sandwiched between phloem and xylem along the radius
(d) Open vascular bundles

107. In a ring girdled plant:
(a) The root dies first
(b) The shoot and root die together
(c) Neither root nor shoot will die
(d) The shoot dies first

108. Vascular bundles in monocotyledons are considered closed because:
(a) Cambium is absent
(b) There are no vessels with perforations
(c) Xylem is surrounded all around by phloem
(d) A bundle sheath surrounds each bundle

109. Which of the following characteristics is mainly responsible for diversification of insects on land?
(a) Bilateral symmetry (b) Exoskeleton
(c) Eyes (d) Segmentation

110. The guts of cow and buffalo possess:
(a) *Chlorella spp.* (b) Methanogens
(c) Cyanobacteria (d) *Fucus spp.*

111. DNA is not present in :
(a) Ribosomes (b) Nucleus
(c) Mitochondria (d) Chloroplast

112. Nuclear envelope is a derivative of :
(a) Membrane of Golgi complex
(b) Microtubules
(c) Rough endoplasmic reticulum
(d) Smooth endoplasmic reticulum

113. Which one of the following is not an inclusion body found in prokaryotes ?
(a) Cyanophycean granule
(b) Glycogen granule
(c) Polysome
(d) Phosphate granule

114. Select the correct matching in the following pairs:
(a) Smooth ER – Synthesis of lipids
(b) Rough ER – Synthesis of glycogen
(c) Rough ER – Oxidation of fatty acids
(d) Smooth ER – Oxidation of phospholipids

115. Which one of the following statements in incorrect?
(a) In competitive inhibition, the inhibitor molecule is not chemically changed by the enzyme.
(b) The competitive inhibitor does not affect the rate of breakdown of the enzyme-substrate complex.
(c) The presence of the competitive inhibitor decreases the Km of the enzyme for the substrate.
(d) A competitive inhibitor reacts reversibly with the enzyme to form an enzyme-inhibitor complex.

116. Select the correct option :

	I		II
A	Synapsis aligns homologous chromosomes	(i)	Anaphase-II
B	Synthesis of RNA and protein	(ii)	Zygotene
C	Action of enzyme recombinase	(iii)	G2-phase
D	Centromeres do not separate but chromatids move towards opposite poles	(iv)	Anaphase-I
		(v)	Pachytene

	A	B	C	D
(a)	(ii)	(iii)	(v)	(iv)
(b)	(i)	(ii)	(v)	(iv)
(c)	(ii)	(iii)	(iv)	(v)
(d)	(ii)	(i)	(iii)	(iv)

117. A somatic cell that has just completed the S phase of its cell cycle, as compared to gamete of the same species, has :

(a) same number of chromosomes but twice the amount of DNA
(b) twice the number of chromosomes and four times the amount of DNA
(c) four times the number of chromosomes and twice the amount of DNA
(d) twice the number of chromosomes and twice the amount of DNA

118. Transpiration and root pressure cause water to rise in plants by :
(a) Pulling and pushing it, respectively
(b) Pushing it upward
(c) Pushing and pulling it, respectively
(d) Pushing it upward

119. Minerals known to be required in large amounts for plant growth include :
(a) calcium, magnesium, manganese, copper
(b) potassium, phosphorus, selenium, boron
(c) magnesium, sulphur, iron, zinc
(d) phosphorus, potassium, sulphur, calcium

120. Which one gives the most valid and recent explanation for stomatal movements ?
(a) Potassium influx and efflux
(b) Starch hydrolysis
(c) Guard cell photosynthesis
(d) Transpiration

121. The structures that are formed by stacking of organized flattened membranous sacs in the chloroplasts are :
(a) Grana (b) Stroma lamellae
(c) Stroma (d) Cristae

122. Typical growth curve in plants is :
(a) Linear
(b) Stair-steps shaped
(c) Parabolic
(d) Sigmoid

123. What causes a green plant exposed to the light on only one side, to bend toward the source of light as it grows ?
(a) Green plants seek light because they are phototropic
(b) Light stimulates plant cells on the lighted side to grow faster
(c) Auxin accumulates on the shaded side, stimulating greater cell elongation there.
(d) Green plants need light to perform photosynthesis

124. Which of the following statements is not correct?
(a) Goblet cells are present in the mucosa of intestine and secrete mucus
(b) Oxyntic cells are present in the mucosa of stomach and secrete HCl.
(c) Acini are present in the pancreas and secrete carboxypeptidase
(d) Brunner's glands are present in the submucosa of stomach and secrete pepsinogen

125. Gastric juice of infants contains :
(a) nuclease, pepsinogen, lipase
(b) pepsinogen, lipase, rennin
(c) amylase, rennin, pepsinogen
(d) maltase, pepsinogen, rennin

126. Cytochromes are found in :
(a) Outer wall of mitochondria
(b) Cristae of mitochondria
(c) Lysosomes
(d) Matrix of mitochondria

127. When you hold your breath, which of the following gas changes in blood would first lead to the urge to breathe?
(a) rising CO_2 concentration
(b) falling CO_2 concentration
(c) rising CO_2 and falling O_2 concentration
(d) falling O_2 concentration

128. Which one of the following is correct ?
(a) Serum = Blood + Fibrinogen
(b) Lymph = Plasma + RBC + WBC
(c) Blood = Plasma + RBC + WBC
(d) Plasma = Blood – Lymphocytes

129. Erythropoiesis starts in :
(a) Liver (b) Spleen
(c) Red bone marrow (d) Kidney

130. Blood pressure in the mammalian aorta is maximum during :
(a) Diastole of the right ventricle
(b) Systole of the left ventricle
(c) Diastole of the right atrium
(d) Systole of the left atrium

131. Which of the following does not favour the formation of large quantities of dilute urine?
(a) Caffeine
(b) Renin
(c) Atrial-natriuretic factor
(d) Alcohol

132. Removal of proximal convoluted tubule from the nephron will result in:
   (a) More concentrated urine
   (b) No change in quality and quantity of urine
   (c) No urine formation
   (d) More diluted urine
133. Sliding filament theory can be best explained as:
   (a) Actin and Myosin filaments shorten and slide pass each other
   (b) Actin and Myosin filaments do not shorten but rather slide pass each other
   (c) When myofilaments slide pass each other, Myosin filaments shorten while Actin filaments do not shorten
   (d) When myofilaments slide pass each other Actin filaments shorten while Myosin filament do not shorten
134. Glenoid cavity articulates :
   (a) Scapula with acromion
   (b) Clavicle with scapula
   (c) Humerus with scapula
   (d) Clavicle with acromion
135. A gymnast is able to balance his body upside down even in the total darkness because of :
   (a) Vestibular apparatus
   (b) Tectorial membrane
   (c) Organ of corti
   (d) Cochlea
136. Which of the following regions of the brain is incorrectly paired with its function?
   (a) Cerebellum – language comprehension
   (b) Corpus callosum – communication between the left and right cerebral cortices
   (c) Cerebrum – calculation and contemplation
   (d) Medulla oblongata – homeostatic control
137. A chemical signal that has both endocrine and neural roles is ?
   (a) Calcitonin (b) Epinephrine
   (c) Cortisol (d) Melatonin
138. Which one of the following may require pollinators, but is genetically similar to autogamy?
   (a) Xenogamy (b) Apogamy
   (c) Cleistogamy (d) Geitonogamy
139. Which one of the following statements is not true?
   (a) Pollen grains of some plants cause severe allergies and bronchial afflictions in some people
   (b) The flowers pollinated by flies and bats secrete foul odour to attract them
   (c) Honey is made by bees by digesting pollen collected from flowers
   (d) Pollen grains are rich in nutrients, and they are used in the form of tablets and syrups
140. The hilum is a scar on the :
   (a) Fruit, where it was attached to pedicel
   (b) Fruit, where style was present
   (c) Seed, where micropyle was present
   (d) Seed, where funicle was attached
141. Which of the following are the important floral rewards to the animal pollinators?
   (a) Nectar and pollen grains
   (b) Floral fragrance and calcium crystals
   (c) Protein pellicle and stigmatic exudates
   (d) Colour and large size flower
142. Which of these is not an important component of initiation of parturition in humans ?
   (a) Synthesis of prostaglandins
   (b) Release of oxytocin
   (c) Release of prolactin
   (d) Increase in estrogen and progesterone ratio
143. Capacitation refers to changes in the :
   (a) Ovum before fertilization
   (b) Ovum after fertilization
   (c) Sperm after fertilization
   (d) Sperm before fertilization
144. Which of the following cells during gametogenesis is normally dipoid?
   (a) Spermatid
   (b) Spermatogonia
   (c) Secondary polar body
   (d) Primary polar body
145. Hysteresctomy is surgical removal of :
   (a) Prostate gland (b) Vas-deference
   (c) Mammary glands (d) Uterus
146. How many pairs of contrasting characters in pea plants were studied by Mendel in his experiments ?
   (a) Six (b) Eight
   (c) Seven (d) Five
147. Which is the most common mechanism of genetic variation in the population of sexually reproducing organism?
   (a) Chromosomal aberrations
   (b) Genetic drift
   (c) Recombination
   (d) Transduction

148. Alleles are :
(a) true breeding homozygotes
(b) different molecular forms of a gene
(c) heterozygotes
(d) different phenotype

149. A man with blood group 'A' marries a woman with blood group 'B'. What are all the possible blood groups of their offsprings ?
(a) A,B and AB only (b) A,B,AB and O
(c) O only (d) A and B only

150. In sea urchin DNA, which is double stranded, 17% of the bases were shown to be cytosine. The percentages of the other three bases expected to be present in this DNA are :-
(a) G 17%, A 16.5%, T 32.5%
(b) G 17%, A 33%, T 33%
(c) G 8.5%, A 50%, T 24.5%
(d) G 34%, A 24.5%, T 24.5%

151. The movement of a gene from one linkage group to another is called :
(a) Duplication (b) Translocation
(c) Crossing over (d) Inversion

152. Gene regulation governing lactose operon of *E.coli* that involves the lac I gene product is :
(a) Negative and inducible because repressor protein prevents transcription
(b) Negative and repressible because repressor protein prevents transcription
(c) Feedback inhibition because excess of b-galactosidase can switch off trascription
(d) Positive and inducible because it can be induced by lactose

153. The chromosomes in which centromere is situated close to one end are:
(a) Acrocentric (b) Telocentric
(c) Sub-metacentric (d) Metacentric

154. Multiple alleles are present :
(a) At different loci on the same chromosome
(b) At the same locus of the chromosome
(c) On non-sister chromatids
(d) On different chromosomes

155. An abnormal human baby with 'XXX' sex chromosomes was born due to :
(a) formation of abnormal ova in the mother
(b) fusion of two ova and one sperm
(c) fusion of two sperms and one ovum
(d) formation of abnormal sperms in the father

156. Which of the following had the smallest brain capacity ?
(a) *Homo sapiens*
(b) *Homo neanderthalensis*
(c) *Homo habilis*
(d) *Homo erectus*

157. Which of the following is not a sexually transmitted disease ?
(a) Acquired Immuno Deficiency Syndrome (AIDS)
(b) Trichomoniasis
(c) Encephalitis
(d) Syphilis

158. HIV that causes AIDS, first starts destroying:
(a) Leucocytes
(b) Helper T- Lymphocytes
(c) Thrombocytes
(d) B- Lymphocytes

159. The active form of *Entamoeba histolytica* feeds upon:
(a) mucosa and submucosa of colon only
(b) food in intestine
(c) blood only
(d) erythrocytes; mucosa and submucosa of colon

160. Which of the following viruses is not transferred through semen of an infected male?
(a) Human immunodeficiency virus
(b) Chikungunya virus
(c) Ebola virus
(d) Hepatitis B virus

161. Match each disease with its correct type of vaccine:

(a)	tuberculosis	(i)	harmless virus
(b)	whooping cough	(ii)	inactivated toxin
(c)	diphtheria	(iii)	killed bacteria
(d)	polio	(iv)	harmless bacteria

	(a)	(b)	(c)	(d)
(a)	(iii)	(ii)	(iv)	(i)
(b)	(iv)	(iii)	(ii)	(i)
(c)	(i)	(ii)	(iv)	(iii)
(d)	(ii)	(i)	(iii)	(iv)

162. Which of the following endoparasites of humans does show viviparity ?
(a) *Enterobius vermicularis*
(b) *Trichinella spiralis*
(c) *Ascaris lumbricoides*
(d) *Ancylostoma duodenale*

163. Which of the following enhances or induces fusion of protoplasts ?
(a) Polyethylene glycol and sodium nitrate
(b) IAA and kinetin
(c) IAA and gibberellins
(d) Sodium chloride and potassium chloride

164. A technique of micropropagation is :-
(a) Somatic embryogenesis
(b) Protoplast fusion
(c) Embryo rescue
(d) Somatic hybridization

165. The crops engineered for glyphosate are resistant/ tolerant to :
(a) Bacteria (b) Insects
(c) Herbicides (d) Fungi

166. In Bt cotton, the Bt toxin present in plant tissue as pro-toxin is converted into active toxin due to:
(a) Acidic pH of the insect gut
(b) Action of gut micro-organisms
(c) Presence of conversion factors in insect gut
(d) Alkaline pH of the insect gut

167. Which body of the Government of India regulates GM research and safety of introducing GM organisms for public services ?
(a) Indian Council of Agricultural Research
(b) Genetic Engineering Approval Committee
(c) Research Committee on Genetic Manipulation
(d) Bio-safety committee

168. Most animals are tree dwellers in a:
(a) Thorn woodland
(b) Temperate deciduous forest
(c) Tropical rain forest
(d) Coniferous forest

169. A population will not exist in Hardy - Weinberg equilibrium if :
(a) There are no mutations
(b) There is no migration
(c) The population is large
(d) Individuals mate selectively

170. Rachel Carson's famous book "Silent Spring" is related to :
(a) Noise pollution
(b) Population explosion
(c) Ecosystem management
(d) Pesticide pollution

171. Vertical distribution of different species occupying different levels in a biotic community is known as:
(a) Stratification (b) Zonation
(c) Pyramid (d) Divergence

172. The mass of living material at a trophic level at a particular time is called :
(a) Standing state
(b) Net primary productivity
(c) Standing crop
(d) Gross primary productivity

173. In an ecosystem the rate of production of organic matter during photosynthesis is termed as:
(a) Gross primary productivity
(b) Secondary productivity
(c) Net productivity
(d) Net primary productivity

174. Secondary Succession takes place on/in :
(a) Degraded forest
(b) Newly created pond
(c) Newly cooled lava
(d) Bare rock

175. The following graph depicts changes in two populations (A and B) of herbivores in a grassy field A possible reason for these changes is that:

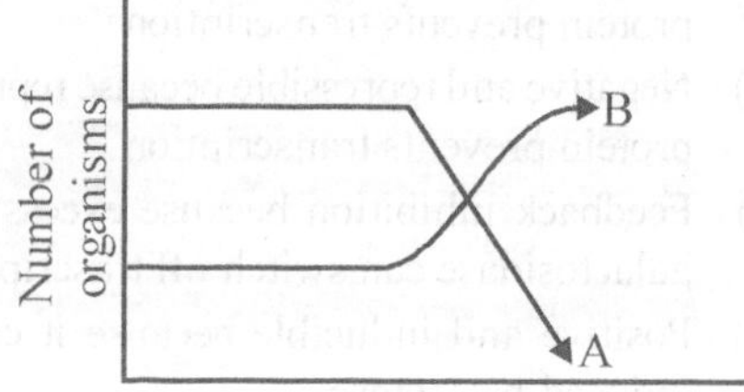

(a) Population B competed more successfully for food than population A
(b) Population A produced more offspring than population B
(c) Population A consumed the members of population B
(d) Both plant populations in this habitat decreased

176. Cryopreservation of gametes of threatened species in viable and fertile condition can be referred to as:-
(a) Advanced *exsitu* conservation of biodiversity
(b) *In situ* conservation by sacred groves
(c) *In situ* cryo-conservation of biodiversity
(d) *In situ* conservation of biodiversity

177. In which of the following both pairs have correct combination:
   (a) *In situ* conservation : Cryopreservation
   Ex situ conservation : Wildlife Sanctuary
   (b) *In situ* conservation : Seed Bank
   Ex situ conservation : National Park
   (c) *In situ* conservation : Tissue culture
   Ex situ conservation : Sacred groves
   (d) *In situ* conservation : National Park
   Ex situ conservation : Botanical Garden

178. High value of BOD (Biochemical Oxygen Demand) indicates that :
   (a) Water is highly polluted
   (b) Water is less polluted
   (c) Consumption of organic matter in the water is higher by the microbes
   (d) Water is pure

179. The UN Conference of Parties on climate change in the year 2011 was held in :
   (a) South Africa (b) Peru
   (c) Qatar (d) Poland

180. Which of the following is not one of the prime health risks associated with greater UV radiation through the atmosphere due to depletion of stratospheric ozone ?
   (a) Reduced Immune System
   (b) Damage to eyes
   (c) Increased liver cancer
   (d) Increased skin cancer

HINTS & SOLUTIONS

PHYSICS

1. (b) Let surface tension

$s = E^a V^b T^c$

$$\frac{MLT^{-2}}{L} = (ML^2T^{-2})^a \left(\frac{L}{T}\right)^b (T)^C$$

Equating the dimension of LHS and RHS

$ML^0T^{-2} = M^aL^{2a+b}T^{-2a-b+c}$

$\Rightarrow \quad a = 1, 2a + b = 0, \quad -2a - b + c = -2$

$\Rightarrow \quad a = 1, b = -2, c = -2$

Hence, the dimensions of surface tension are $[E V^{-2} T^{-2}]$

2. (a) According to question,

$V(x) = bx^{-2n}$

So, $\frac{dv}{dx} = -2\,nb\,x^{-2n-1}$

Acceleration of the particle as function of x,

$$a = v\frac{dv}{dx} = bx^{-2n}\left\{b(-2n)x^{-2n-1}\right\}$$

$$= -2nb^2x^{-4n-1}$$

3. (a) $\vec{V}_A = 10\left(-\hat{i}\right)$

$\vec{V}_B = 10\left(\hat{j}\right)$

$\vec{V}_{BA} = 10\,\hat{j} + 10\,\hat{i} = 10\sqrt{2}$ km/h

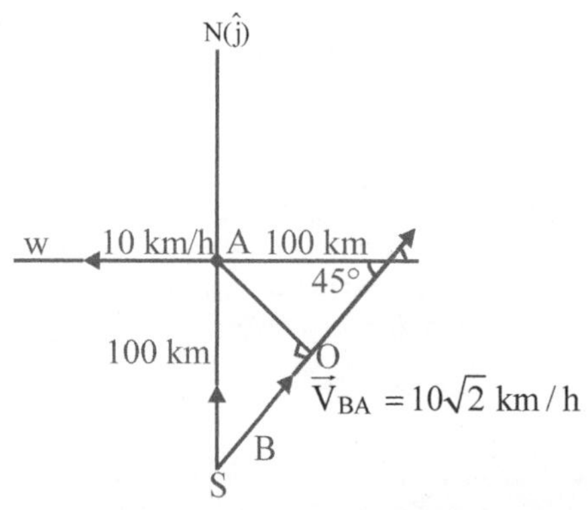

Distance OB = 100 cos 45° = $50\sqrt{2}$ km

Time taken to each the shortest distance between

A and B $= \frac{OB}{V_{BA}} = \frac{50\sqrt{2}}{10\sqrt{2}} = 5h$

4. (b) For the motion of both the blocks

$m_1a = T - \mu_k m_1 g$

$m_2g - T = m_2a$

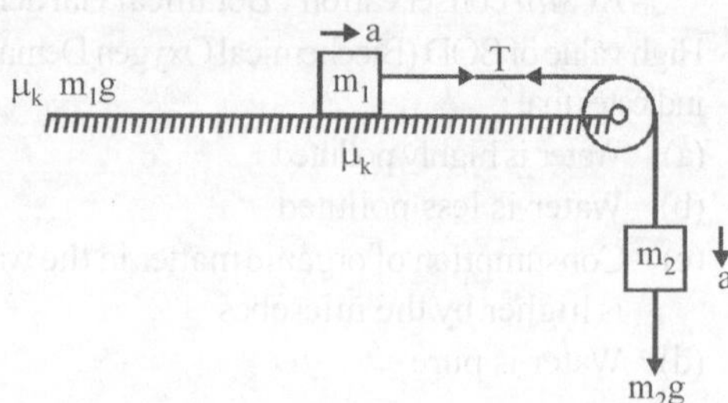

$$a = \frac{m_2g - \mu_k\, m_1g}{m_1 + m_2}$$

$$m_2g - T = (m_2)\left(\frac{m_2g - \mu_k\, m_1g}{m_1 + m_2}\right)$$

solving we get tension in the string

$$T = \frac{m_1mg\,(1+\mu_k)g}{m_1 + m_2}$$

5. (a) Acceleration of system $a = \frac{F_{net}}{M_{total}}$

$$= \frac{14}{4+2+1} = \frac{14}{7} = 2\ m/s^2$$

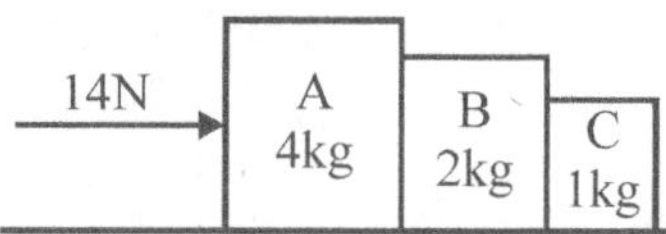

The contact force between A and B

$= (m_B + m_C) \times a = (2+1) \times 2 = 6N$

6. (d) From, $F = ma$

$$a = \frac{F}{m} = \frac{0.1x}{10} = 0.01x = V\frac{dV}{dx}$$

So, $\int_{v_1}^{v_2} vdV = \int_{20}^{30} \frac{x}{100}dx$

$$-\frac{V^2}{2}\Bigg|_{V_2}^{V_2} = \frac{x^2}{200}\Bigg|_{20}^{30} = \frac{30\times 30}{200} - \frac{20\times 20}{200}$$

$= 4.5 - 2 = 2.5$

$\frac{1}{2}m\left(V_2^2 - V_1^2\right) = 10 \times 2.5\text{ J} = -25\text{J}$

Final K.E. $= \frac{1}{2}mv_2^2 = \frac{1}{2}mv_1^2 - 25$

$= \frac{1}{2} \times 10 \times 10 \times 10 - 25$

$= 500 - 25 = 475$ J

7. (b) As we know work done in stretching spring

$w = \frac{1}{2}kx^2$

where k = spring constant
x = extension

Case (a) If extension (x) is same,

$W = \frac{1}{2}K x^2$

So, $W_P > W_Q$ $(\because K_P > K_Q)$

Case (b) If spring force (F) is same $W = \frac{F^2}{2K}$

So, $W_Q > W_P$

8. (d) As we know power $P = \frac{dw}{dt}$

$\Rightarrow \quad w = Pt = \frac{1}{2}mV^2$

So, $v = \sqrt{\frac{2Pt}{m}}$

Hence, acceleration $a = \frac{dV}{dt} = \sqrt{\frac{2P}{m}} \cdot \frac{1}{2\sqrt{t}}$

Therefore, force on the particle at time 't'

$= ma = \sqrt{\frac{2Km^2}{m}} \cdot \frac{1}{2\sqrt{t}}$

$= \sqrt{\frac{Km}{2t}} = \sqrt{\frac{mK}{2}}\, t^{-1/2}$

9. (b) Applying angular momentum conservation

$mV_0R_0 = (m)(V^1)\left(\frac{R_0}{2}\right)$

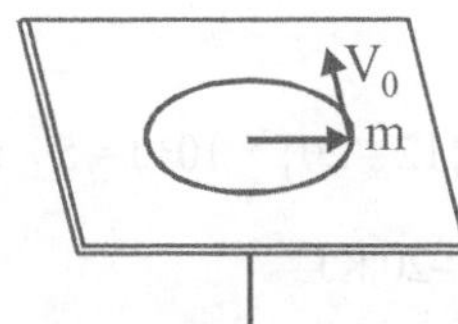

$\therefore \quad v^1 = 2V_0$

Therefore, new KE $= \frac{1}{2}m(2V_0)^2 = 2mv_0^2$

10. (c) By torque balancing about B

$N_A(d) = W(d - x)$

$N_A = \frac{W(d-x)}{d}$

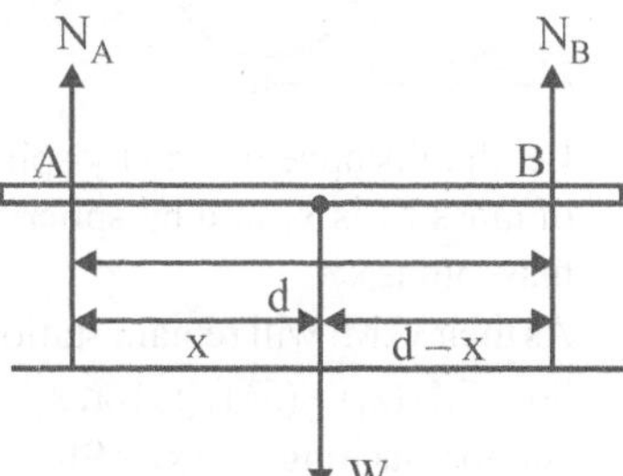

11. (c) Moment of inertia of shell 1 along diameter

$I_{diameter} = \frac{2}{3}MR^2$

Moment of inertia of shell 2 = m. i of shell 3

$= I_{tengential} = \frac{2}{3}MR^2 + MR^2 = \frac{5}{3}MR^2$

So, I of the system along x x[1]

$= I_{diameter} + (I_{tengential}) \times 2$

or, $I_{total} = \frac{2}{3}MR^2 + \left(\frac{5}{3}MR^2\right) \times 2$

$= \frac{12}{3}MR^2 = 4MR^2$

12. (a) As we know, orbital speed, $V_{orb} = \sqrt{\frac{GM}{r}}$

Time period $T = \frac{2\pi r}{v_{orb}} = \frac{2\pi r}{\sqrt{GM}}\sqrt{r}$

Squarring both sides,

$T^2 = \left(\frac{2\pi r\sqrt{r}}{\sqrt{GM}}\right)^2 = \frac{4\pi^2}{GM} \cdot r^3$

$$\Rightarrow \quad \frac{T^2}{r^3} = \frac{4\pi^2}{GM} = K$$

$\Rightarrow GMK = 4\pi^2$.

13. (b)

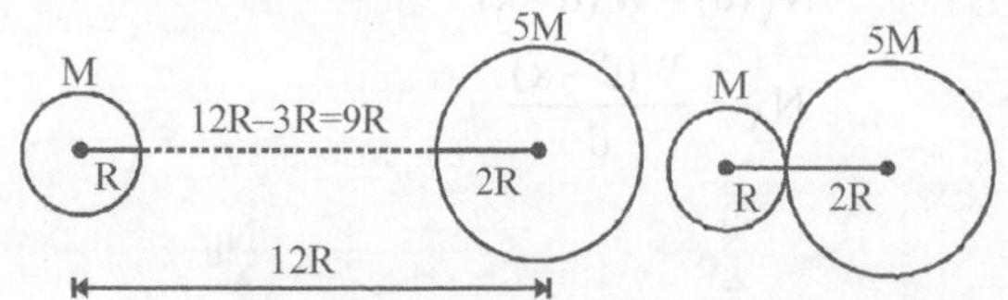

Let the distance moved by spherical body of mass M is x_1 and by spherical body of mass 5m is x_2

As their C.M. will remain stationary

So, $(M)(x_1) = (5M)(x_2)$ or, $x_1 = 5x_2$

and for tauching $x_1 + x_2 = 9R$

So, $x_1 = 7.5R$

14. (b) Compressibility of water,

$K = 45.4 \times 10^{-11}\ Pa^{-1}$

density of water $P = 10^3\ kg/m^3$

depth of ocean, $h = 2700$ m

We have to find $\frac{\Delta V}{V} = ?$

As we know, compressibility,

$$K = \frac{1}{B} = \frac{(\Delta V / V)}{P}\ (P = Pgh)$$

So, $(\Delta V/V) = KPgh$

$= 45.4 \times 10^{-11} \times 10^3 \times 10 \times 2700$

$= 1.2258 \times 10^{-2}$

15. (b) According to Bernoulli's theorem,

$$P + \tfrac{1}{2}\rho v^2 = P_0 + 0$$

So, $\Delta P = \frac{1}{2}\rho v^2$

$F = \Delta PA = \frac{1}{2}\rho v^2 A$

$= \frac{1}{2} \times 1.2 \times 40 \times 40 \times 250$

$= 2.4 \times 10^5$ N (upwards)

16. (a) From Wein's displacement law,

$\lambda_m \times T = \text{constant}$

P – max. intensity is at violet

$\Rightarrow \lambda_m$ is minimum $\Rightarrow$ temp maximum

R – max. intensity is at green $\Rightarrow \lambda_m$ is moderate $\Rightarrow$ temp moderate

Q – max. intensity is at red $\Rightarrow \lambda_m$ is maximum $\Rightarrow$ temp minimum i.e., $T_p > T_R > T_Q$

17. (c) As the temperature difference $\Delta T = 10°C$ as well as the thermal resistance is same for both the cases, so thermal current or rate of heat flow will also be same for both the cases.

18. (b) Efficiency of carnot engine

$$n = 1 - \frac{T_2}{T_1} \text{ i.e., } \frac{1}{10} = 1 - \frac{T_2}{T_1}$$

$$\Rightarrow \frac{T_2}{T_1} = 1 - \frac{1}{10} = \frac{9}{10} \Rightarrow \frac{T_1}{T_2} = \frac{10}{9}$$

$$\therefore \quad w = Q_2 \cdot \left(\frac{T_1}{T_2} - 1\right)$$

$$\text{i.e., } 10 = Q_2\left(\frac{10}{9} - 1\right) \quad 10 = Q_2\left(\frac{1}{9}\right)$$

$\Rightarrow Q_2 = 90J$

So, 90 J heat is absorbed at lower temperature.

19. (b) In cyclic process ABCA

$Q_{cycle} = W_{cycle}$

$Q_{AB} + Q_{BC} + Q_{CA} = \text{ar. of } \Delta ABC$

$$+400 + 100 + Q_{C\to A} = \frac{1}{2}(2 \times 10^{-3})(4 \times 10^4)$$

$\Rightarrow \quad Q_{C\to A} = -460$ J

$\Rightarrow \quad Q_{A\to C} = +460$ J

20. (a) Change in internal energy from $A \to B$

$$\Delta U = \frac{f}{2} nR\Delta T = \frac{f}{2} nR(T_f - T_i)$$

$$= \frac{5}{2}\{P_fV_f - P_iV_i\}$$

(As gas is diatomic $\therefore f = 5$)

$$= \frac{5}{2}\{2 \times 10^3 \times 6 - 5 \times 10^3 \times 4\}$$

$$= \frac{5}{2}\{12 - 20\} \times 10^3\ J = 5 \times (-4) \times 10^3\ J$$

$\Delta U = -20$ KJ

21. (b) Let 'n' be the degree of freedom

$$\gamma = \frac{C_p}{C_v} = \frac{\left(\frac{n}{2}+1\right)R}{\left(\frac{n}{2}\right)R} = \left(1+\frac{2}{n}\right)$$

22. (b) The two displacements equations are $y_1 = a\sin(\omega t)$

and $y_2 = b\cos(\omega t) = b\sin\left(\omega t + \frac{\pi}{2}\right)$

$y_{eq} = y_1 + y_2$

$= a\sin\omega t + b\cos\omega t = a\sin\omega t + b\sin\left(\omega t + \frac{\pi}{2}\right)$

Since the frequencies for both SHMs are same, resultant motion will be SHM.

Now $A_{eq} = \sqrt{a^2 + b^2 + 2ab\cos\frac{\pi}{2}}$

$\Rightarrow A_{eq} = \sqrt{a^2 + b^2}$

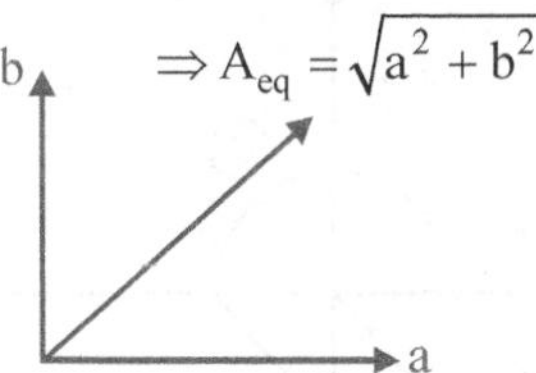

23. (a) As we know, for particle undergoing SHM,

$V = \omega\sqrt{A^2 - X^2}$

$V_1^2 = \omega^2(A^2 - x_1^2)$

$V_2^2 = \omega^2(A^2 - x_2^2)$

Substructing we get,

$$\frac{V_1^2}{\omega^2} + x_1^2 = \frac{V_2^2}{\omega^2} + x_2^2$$

$$\Rightarrow \frac{V_1^2 - V_2^2}{\omega^2} = x_2^2 - x_1^2$$

$$\Rightarrow w = \sqrt{\frac{V_1^2 - V_2^2}{x_2^2 - x_1^2}}$$

$$\Rightarrow T = 2\pi\sqrt{\frac{x_2^2 - x_1^2}{V_1^2 - V_2^2}}$$

24. (b) Fundamental frequency of closed organ pipe

$$V_c = \frac{V}{4l_c}$$

Fundamental frequency of open organ pipe

$$V_0 = \frac{V}{2l_0}$$

Second overtone frequency of open organ pipe $= \frac{3V}{2l_0}$

From question,

$$\frac{V}{4l_c} = \frac{3V}{2l_0}$$

$\Rightarrow l_0 = 6l_c = 6 \times 20 = 120$ cm

25. (b) Net flux emmited from a spherical surface of radius a according to Gauss's theorem

$\phi_{net} = \frac{q_{in}}{\varepsilon_0}$

or, $(Aa)(4\pi a^2) = \frac{q_{in}}{\varepsilon_0}$

So, $q_{in} = 4\pi\varepsilon_0 A a^3$

26. (c) Capacitance of the capacitor, $C = \frac{Q}{V}$

After inserting the dielectric, new capacitance

$C^1 = K.C$

New potential difference

$V^1 = \frac{V}{K}$

$u_i = \frac{1}{2}cv^2 = \frac{Q^2}{2C}$ $(\because Q = cv)$

$u_f = \frac{Q^2}{2f} = \frac{Q^2}{2kc} = \frac{C^2V^2}{2KC} = \left(\frac{u_i}{k}\right)$

$\Delta u = u_f - u_i = \frac{1}{2}cv^2\left\{\frac{1}{k} - 1\right\}$

As the capacitor is isolated, so change will remain conserved p.d. between two plates of the capacitor

$L = \frac{Q}{KC} = \frac{V}{K}$

27. (a) Here, metallic conductor can be considered as the combination of various conductors connected in series. And in series combination current remains same.

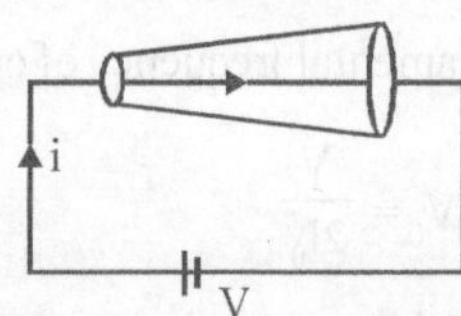

28. (d) Total potential difference across potentiometer wire = $10^{-3} \times 400$ volt = 0.4 volt

potential gradient = $\frac{\text{lmv}}{\text{cm}}$

$= 10^{-3}$ v/cm $= 10^{-1}\frac{\text{v}}{\text{m}}$

Let resistance of RΩ connected in series.

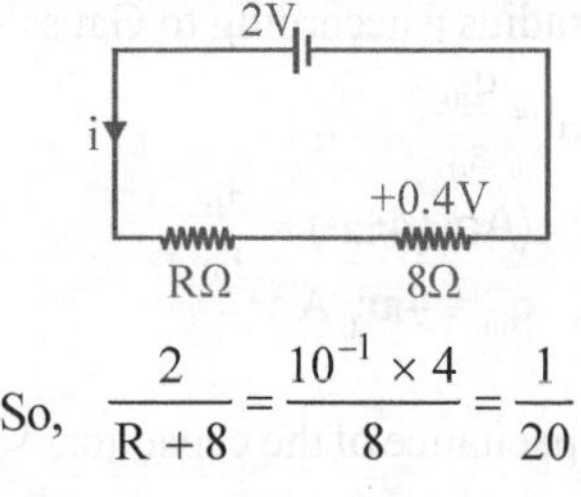

So, $\frac{2}{R+8} = \frac{10^{-1} \times 4}{8} = \frac{1}{20}$

$\Rightarrow R + 8 = 40$ or, $R = 32\,\Omega$

29. (d) Effective resistance of B and

$= \frac{R_B \cdot R_C}{R_B + R_C} = \frac{1.5R \times 3R}{1.5R + 3R} = \frac{4.5R^2}{4.5R} = R$

i.e., equal to resistance of voltmeter A.

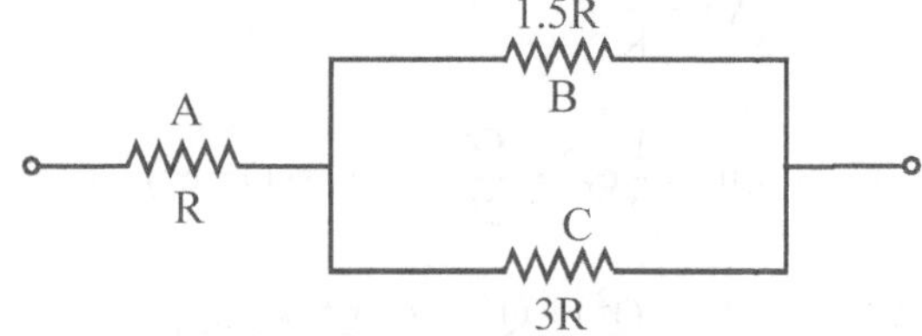

In parallel potential difference is same so, $V_B = V_C$ and in series current is same

So, $V_A = V_B = V_C$

30. (c) Radius of circular orbit = r

No. of rotations per second = n

i.e., $T = \frac{1}{n}$

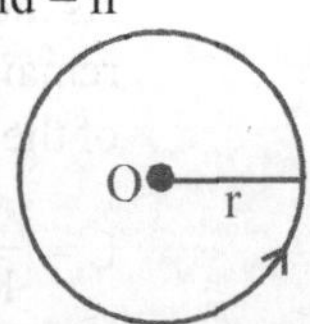

Magnetic field at its centre, B_c =?

As we know, current

$i = \frac{e}{T} = \frac{e}{(1/n)}$ = en = equivalent current

Magnetic field at the centre of circular orbit,

$B_c = \frac{\mu_0 i}{2r} = \frac{\mu_0 ne}{2r}$

31. (b) Megnetic field due to segment '1'

$\overrightarrow{B_1} = \frac{\mu_0 I}{4\pi R}[\sin 90° + \sin 0°](-\hat{k})$

$= \frac{-\mu_0 I}{4\pi R}(\hat{k}) = \vec{B}_3$

Magnetic field due to segment 2

$B_2 = \frac{\mu_0 I}{4R}(-\hat{i}) = \frac{-\mu_0 I}{4\pi R}(\pi\hat{i})$

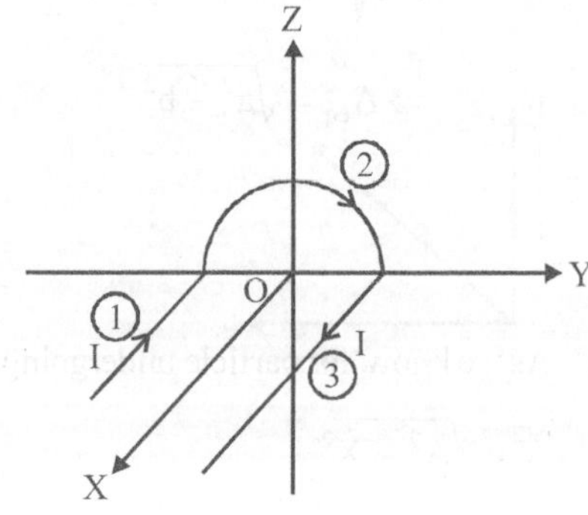

$\therefore$ $\vec{B}$ at centre

$\vec{B}_c = \vec{B}_1 + \vec{B}_2 + \vec{B}_3$

$= \frac{-\mu_0 I}{4\pi R}(\pi\hat{i} + 2\hat{k})$

32. (c) Emf induced in side 1 of frame $e_1 = B_1 V\ell$

$B_1 = \frac{\mu_0 I}{2\pi(x - a/2)}$

Emf induced in side 2 of frame $e_2 = B_2 V\ell$

$B_2 = \frac{\mu_0 I}{2\pi(x + a/2)}$

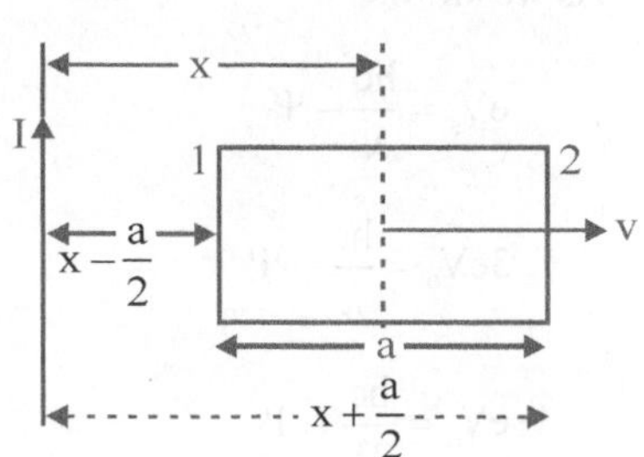

Emf induced in square frame

$e = B_1 V\ell - B_2 V\ell$

$$= \frac{\mu_0 I}{2\pi(x - a/2)}\ell v - \frac{\mu_0 I}{2\pi(x + a/2)}\ell v$$

$$\text{or, } e \propto \frac{1}{(2x - a)(2x + a)}$$

33. (d)

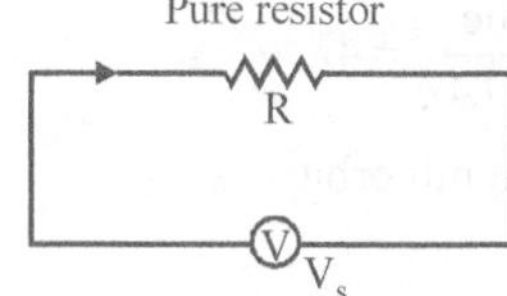

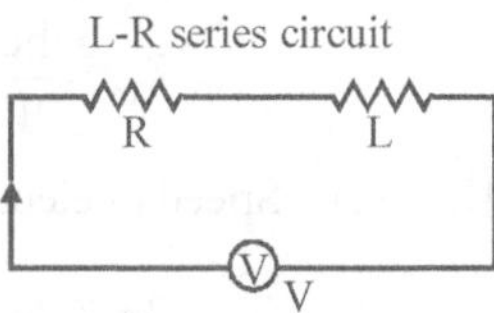

Phasor diagram

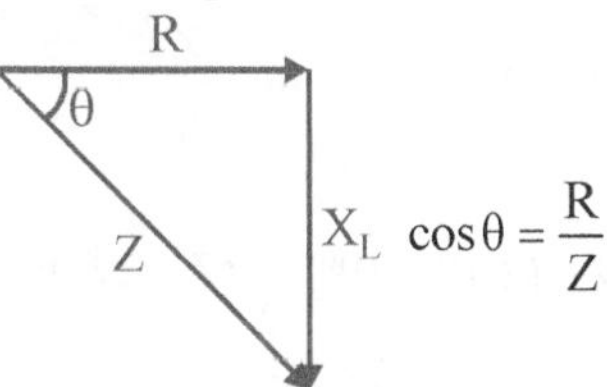

Z = impedance

For pure resistor circuit, power

$$P = \frac{V^2}{R} \Rightarrow V^2 = PR$$

For L-R series circuit, power

$$P^1 = \frac{V^2}{Z}\cos\theta = \frac{V^2}{Z}.\frac{R}{Z} = \frac{PR}{Z^2}.R = P\left(\frac{R}{Z}\right)^2$$

34. (a) Momentum of light falling on reflecting surface $p = \frac{E}{C}$

As surface is perfectly reflecting so momentum reflect $p^1 = -\frac{E}{C}$

$\frac{E}{C}$ $\frac{E}{C} = P$

So, momentum transferred

$$= P - P^1 = \frac{E}{C} - \left(-\frac{E}{C}\right) = \frac{2E}{C}$$

35. (a) As we know, the refractive index of the material of the prism

$$\mu = \frac{\sin\left(\frac{\delta_m + A}{2}\right)}{\sin(A/2)}$$

$$\cot A/2 = \frac{\sin\left(\frac{A + \delta_m}{2}\right)}{\sin A/2} = \frac{\cos(A/2)}{\sin(A/2)}$$

$[\because \mu = \cot(A/2)]$

$$\Rightarrow \text{Sin}\left(\frac{\delta_m + A}{2}\right) = \sin(90° + A/2)$$

$$\Rightarrow \delta_{min} = 180° - 2A$$

36. (b) Using lens maker's formula,

$$\frac{1}{f} = (\mu - 1)\left(\frac{1}{R_1} - \frac{1}{R_2}\right)$$

n = 1.5 n = 1.7 n = 1.5

$$\frac{1}{f_1} = \left(\frac{1.5}{1} - 1\right)\left(\frac{1}{\infty} - \frac{1}{-20}\right)$$

$$\Rightarrow f_1 = 40\text{cm}$$

$$\frac{1}{f_2} = \left(\frac{1.7}{1} - 1\right)\left(\frac{1}{-20} - \frac{1}{+20}\right) \Rightarrow f_2 = -\frac{100}{7}\text{ cm}$$

and $\frac{1}{f_3} = \left(\frac{1.5}{1} - 1\right)\left(\frac{1}{\infty} - \frac{1}{-20}\right) \Rightarrow f_3 = 40\text{ cm}$

$$\frac{1}{f_{eq}}=\frac{1}{f_1}+\frac{1}{f_2}+\frac{1}{f_3}\Rightarrow\frac{1}{f_{eq}}=\frac{1}{40}+\frac{1}{-100/7}+\frac{1}{40}$$

$\therefore \quad f_{eq}=-50$ cm

Therefore, the focal length of the combination is – 50 cm.

37. (d) Here, distance between two slits,

$d=1mm=10^{-3}m$

distance of screen from slits, D = 1 m

wavelength of monochromatic light used,

$\lambda=500nm=500\times10^{-9}m$

width of each slit a = ?

Width of central maxima in single slit pattern

$$=\frac{2\lambda D}{a}$$

Fringe width in double slit experiment

$$\beta=\frac{\lambda D}{d}$$

So, required condition $\frac{10\lambda D}{d}=\frac{2\lambda D}{a}$

$$\Rightarrow \quad a=\frac{d}{5D}=\frac{1}{5}\times10^{-3}\,m=0.2\,mm$$

38. (d) Linear width of central maxima

$$=D(2q)=2Dq\;\frac{2D\lambda}{a}$$

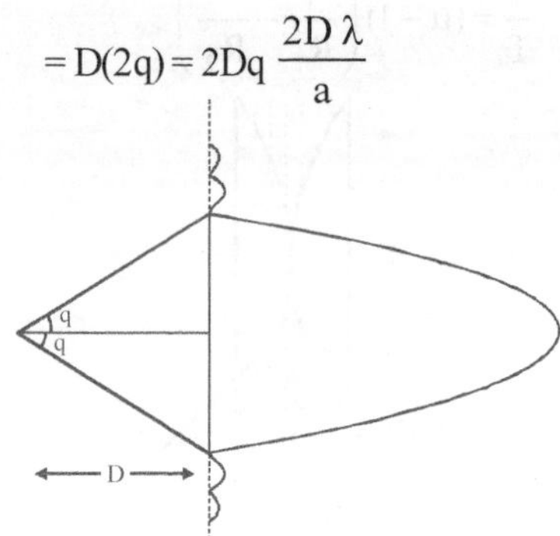

39. (a) According to De-broglie $p=\frac{h}{\lambda}$ or $P\propto\frac{1}{\lambda}$

where

P = particle momentum; λ = de-Broglie wavelength

h = Plank's constant

$P\propto\frac{1}{\lambda}$ represents rectangular hyperbola.

40. (a) As we know,

$$eV_s=\frac{hc}{\lambda}-\Psi$$

$$3eV_o=\frac{hc}{\lambda}-\Psi \quad ...(1)$$

$$eV_o=\frac{hc}{2\lambda}-\Psi \quad ...(2)$$

$$3eV_o=\frac{3hc}{2\lambda}-3\Psi \quad ...(3)$$

Multiplying eqn. (2) by (3) and subtracting it from eqn (1)

$$\Psi=\frac{hc}{4\lambda}$$

So, threshold wavelength,

$$\lambda_{th}=\frac{hc}{\Psi}=\frac{hc}{hc/4\lambda}=4\lambda$$

41. (a) Speed of electron in nth orbit

$$V_n=\frac{2\pi KZe^2}{nh}$$

$$V=(2.19\times10^6\,m/s)\,\frac{Z}{n}$$

$$V=(2.19\times10^6)\,\frac{2}{3}\;(Z=2\;\&\;n=3)$$

$V=1.46\times10^6$ m/s

42. (b) By law of conservation of energy,

$K.E_f=K.E_i-$ excitation energy (ε)

or $\frac{1}{2}mv_1^2+\frac{1}{2}mv_2^2=\frac{1}{2}m_1u_1^2+\frac{1}{2}m_2u_2^2=\varepsilon$

43. (a) As we know, $R=R_0\,(A)^{1/3}$

where A = mass number

$R_{Al}=R_0\,(27)^{1/3}=3R_0$

$R_{Te}=R_0\,(125)^{1/3}=5R_0=\frac{5}{3}R_{Al}$

44. (b) First two gates are NOT gates and the last gate is NOR gate.

Thus, $y_1=\overline{A},\ y_2=\overline{B}$ and $y=\overline{y_1+y_2}$

The truth table corresponding to this is as follows:

A	B	$y_1 = \overline{A}$	$y_2 = \overline{B}$	$y_1 + y_2$	$y = \overline{y_1 + y_2}$	A . B
0	0	1	1	1	0	0
0	1	1	0	1	0	0
1	0	0	1	1	0	0
1	1	0	0	0	1	1

Thus the combination of gate represents AND gate.

45. (c) Here P-N junction diode rectifies half of the ac wave i.e., acts as half wave rectifier. During + ve half cycle

Diode → forward biased output across will be

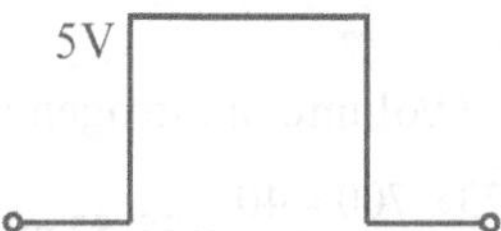

During –ve half cycle Diode → reverse biased output will not obtained.

CHEMISTRY

46. (a) Ratio of weight of gases = $w_{H_2} : w_{O_2} = 1 : 4$

Ratio of moles of gases = $n_{H_2} : n_{O_2} = \frac{1}{2} : \frac{4}{32}$

$\therefore$ Molar Ratio = $\frac{1}{2} \times \frac{32}{4} = 4 : 1$

47. (b) In isoelectronic species the radius decrease with increase in nuclear charge hence increasing order of radius is $Ca^{+2} < K^+ < Ar$

48. (c) O_2^+ ion - Total number of electrons (16 – 1) = 15. Electronic configuration

$$\sigma 1s^2 < \sigma^* 1s^2 < \sigma 2s^2 < \sigma^* 2s^2 < \sigma 2p_x^2$$
$$< \pi 2p_y^2 = \pi 2p_z^2 < \pi^* 2p_y^1$$

Bond order = $\frac{N_b - N_a}{2} = \frac{10-5}{2} = \frac{5}{2} = 2\frac{1}{2}$

O_2^- (Super oxide ion): Total number of electrons $(16+1) = 17$. Electronic configuration

$$\sigma 1s^2 < \sigma^* 1s^2 < \sigma 2s^2 < \sigma^* 2s^2 < \sigma 2p_x^2$$
$$< \pi 2p_y^2 = \pi 2p_z^2 < \pi^* 2p_y^2 = \pi^* 2p_z^1$$

Bond order

$= \frac{(N_b - N_a)}{2} = \frac{10-7}{2} = \frac{3}{2} = 1\frac{1}{2}$

O_2^{+2} ion: Total number of electrons = (16 – 2) = 14 Electronic configuration
$\sigma 1s^2 < \sigma^* 1s^2 < \sigma 2s^2 < \sigma^* 2s^2 < \sigma 2p_x^2 < \pi 2p_y^2 = \pi 2p_z^2$

Bond order =

$\frac{(N_b - N_a)}{2} = \frac{10-4}{2} = \frac{6}{2} = 3$

So bond order: $O_2^- < O_2^+ < O_2^{2+}$

49. (c) ClO_3^- and SO_3^{-2} both have same number of electrons (42) and central atom in each being sp^3 hybridised. Both are having one lone pair on central atom hence they are pyramidal.

50. (a) Oxygen molecule (O_2) - Total number of electrons = 16 and electronic configuration is

$$\sigma 1s^2 < \sigma^* 1s^2 < \sigma 2s^2 < \sigma^* 2s^2 < \sigma 2p_x^2$$
$$< \pi 2p_y^2 = \pi 2p_z^2 < \pi^* 2p_y^1 = \pi^* 2p_z^1$$

Bond order = $\frac{N_b - N_a}{2} = \frac{10-6}{2} = \frac{4}{2} = 2$

O_2^+ ion - Total number of electrons (16 – 1) = 15. Electronic configuration

$$\sigma 1s^2 < \sigma^* 1s^2 < \sigma 2s^2 < \sigma^* 2s^2 < \sigma 2p_x^2$$
$$< \pi 2p_y^2 = \pi 2p_z^2 < \pi^* 2p_y^1$$

Bond order = $\frac{N_b - N_a}{2} = \frac{10-5}{2} = \frac{5}{2} = 2\frac{1}{2}$

O_2^- (Super oxide ion) Total number of electrons $(16+1) = 17$. Electronic configuration

$$\sigma 1s^2 < \sigma^* 1s^2 < \sigma 2s^2 < \sigma^* 2s^2 < \sigma 2p_x^2$$
$$< \pi 2p_y^2 = \pi 2p_z^2 < \pi^* 2p_y^2 = \pi^* 2p_z^1$$

Bond order $= \frac{(N_b - N_a)}{2} = \frac{10-7}{2} = \frac{3}{2} = 1\frac{1}{2}$

51. (b) NO_2^+ has *sp* hybridisation so it is linear with bond angle = 180°.

52. (a) [Structure of XeO_4: Xe with four O atoms]

Number of σ bonds = 4

Number of π bonds = 4

53. (c) Ag_2CrO_4

$K_{sp} = [Ag^+]^2 \left[Cr_2O_4^{-2} \right] = 1.1 \times 10^{-12}$

$[Ag^+] = \sqrt{\frac{1.1 \times 10^{-12}}{\left[Cr_2O_4^{-2}\right]}}$

AgCl

$K_{sp} = [Ag^+][Cl^-] = 1.8 \times 10^{-10}$

$[Ag^+] = \frac{1.8 \times 10^{-10}}{[Cl^-]}$

AgBr

$K_{sp} = [Ag^+][Br^-] = 5.0 \times 10^{-13}$

$[Ag^+] = \frac{5.3 \times 10^{-13}}{[Br^-]}$

AgI

$K_{sp} = [Ag^+][I^-] = 8.3 \times 10^{-17}$

$[Ag^+] = \frac{8.3 \times 10^{-17}}{[I^-]}$

If we take

$\left[Cr_2O_4^{-2}\right] = [Cl^-] = [Br^-] = [I^-] = 1$ than maximum $[Ag^+]$ will be required in case of Ag_2CrO_4.

54. (b) Equilibrium constant for reaction:

$K = 1.6 \times 10^{12} = \frac{[Product]}{[Reactant]}$

The value of K is very high so the system will contain mostly products at equilibrium.

55. (b) $\Delta G° = -2.30RT\log K$
because at equilibrium $\Delta G = 0$

56. (d) Solubility of alkaline earth metal sulphates decreases down the group due decrease in hydration energy.

$MgSO_4 > CaSO_4 > SrSO_4 > BaSO_4$ →

Hydration energy ↓ Solubility ↓

57. (b) K^+ ion is a constituent of sodium pump.

58. (b) Wt. of organic substance = 0.25 g

$V_1 = 40\text{ mL}, T_1 = 300\text{ K}$

$P_1 = 725 - 25 = 700$ mm of Hg

$P_2 = 760$ mm of Hg (at STP)

$T_2 = 273$ K

$$\frac{P_1V_1}{T_1} = \frac{P_2V_2}{T_2}$$

V_2 (Volume of nitrogen at STP)

$= \frac{273 \times 700 \times 40}{300 \times 760} = 33.52$ mL

Percentage of nitrogen

$= \frac{28 \times \text{volume of } N_2 \text{ at STP} \times 100}{22400 \times \text{wt. of organic substance}}$

$= \frac{28 \times 33.52 \times 100}{22400 \times 0.25} = 16.76\%$

59. (b) π bond is transferred after leaving Cl^-

$CH_3-CH=CH.CH_2-Cl \rightarrow CH_3-CH=CH-\overset{\oplus}{C}H_2 \rightarrow$

$CH_3-\overset{\oplus}{C}H-CH=CH_2$

60. (b) Only structure (III) has H in conjugation with free radical.
So, hyperconjugation is possible in III only.

61. (c) All of these compounds show tautomerism

[Keto–enol tautomer structures: 4,4-dimethylcyclohex-2-enone ⇌ enol (OH); 6,6-dimethylcyclohex-2-enone ⇌ enol (OH)]

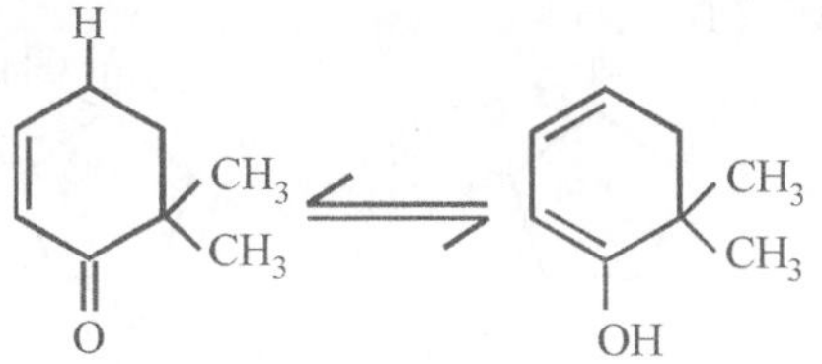

H / O (cyclohexenone with CH_3, CH_3) $\rightleftharpoons$ OH (enol with CH_3, CH_3)

62. (d) $C_6H_5—CH=CH—CH_3 + HBr$

$\downarrow$

$C_6H_5—\overset{\oplus}{C}H—CH_2—CH_3$
(Benzyl carbocation)

$\downarrow Br^{\ominus}$

$C_6H_5—\underset{Br}{CH}—CH_2—CH_3$

63. (a) No. of double bonds = 4
No. of π bond electrons
= 2 × no. of double bond
$= 2 \times 4 = 8$

64. (a) Enthalpy of hydrogenation

$\propto \frac{1}{\text{stability of alkene}}$

$\therefore$ III > II > I

65. (d) 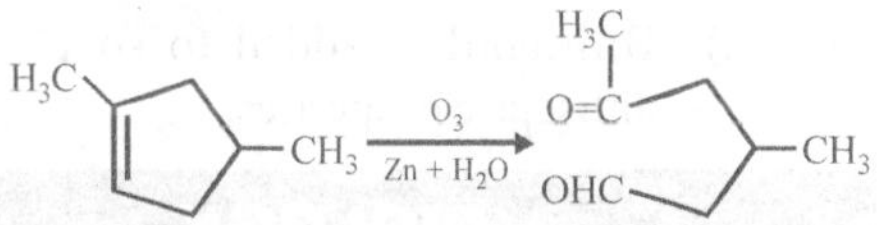 H_3C–(cyclopentene)–CH_3 $\xrightarrow[Zn + H_2O]{O_3}$ H_3C–C(=O)–…–CH_3, OHC

66. (a) For FCC : $r = \frac{\sqrt{2}a}{4} = \frac{1.41 \times 361}{4} = 127\text{ pm}$

67. (c) $K_4[Fe(CN)_6] \rightleftharpoons 4K^+ + [Fe(CN)_6]^-$

and $Al_2(SO_4)_3 \rightarrow 2Al^{3+} + 3SO_4^{2-}$

$\therefore$ van't Hoff factor is 5 for both $Al_2(SO_4)_3$ and $K_4[Fe(CN)_6]$

68. (d) $\Delta T_b = iK_b m$
Given, $(\Delta T_b)_x > (\Delta T_b)_y$
$\therefore$ $i_x K_b m > i_y K_b m$
(K_b is same for same solvent)
$i_x > i_y$
So, x is undergoing dissociation in water.

69. (a) For an ideal solution $\Delta S_{mix} > 0$

70. (c) Tyndall effect is an optical property, and it is independent of charge on colloidal particles.

71. (d) A device that converts energy of combustion of fuels, directly into electrical energy is known as fuel cell.

72. (b) Arrhenius equation

$K = A.e^{-E_a/RT} \Rightarrow \ln K = \ln A - \frac{E_a}{RT}$

$\text{slope} = \frac{-E_a}{R}$

so, activation energy of reaction can be determined from the slope of ln K vs $\frac{1}{T}$

73. (a) $t_{1/2} = \frac{0.693}{k}$

For first order $t_{1/2}$ is independent of initial concentration of reactant.

74. (b) Metal nitrates are highly soluble in water and are very stable for e.g. $NaNO_3$ and KNO_3.

75. (c) SO_2 is widely used in food and drinks industries for its property as a preservative and antioxidant while NO_2 is not used as food preservative.

76. (d) Magnetic moment = 2.84 B.M. This indicates that 2 unpaired electrons are present.

For $Ni^{+2} = 4s^0\ 3d^8$ | ⇅ | ⇅ | ⇅ | ↑ | ↑ |

Number of unpaired electrons (n) = 2
Hence Ni^{2+} gives magnetic moment 2.84 B.M.

77. (a) $Fe^{+2} = 3d^6$ (number of '*d*' electrons = 6)
In $Cl = 1s^2\ 2s^2\ 2p^6\ 3s^2\ 3p^5$
total *p* electrons = 11, which are not equal to number of '*d*' electrons in Fe^{+2}
p-electrons in $Ne = 1s^2\ 2s^2\ 2p^6 = 6$
s-electrons in $Mg = 1s^2\ 2s^2\ 2p^6\ 3s^2 = 6$

78. (b) Formation of $Fe(CO)_5$ from Fe involves no change in oxidation state of iron.

79. (b) Due to lanthanoid contraction atomic radii of Zr and Hf is almost similar.

80. (d) Orbital angular momentum = $\sqrt{\ell(\ell+1)}.\hbar$

For *d*-orbital $\ell = 2$

angular momentum = $\sqrt{2(2+1)}\hbar = \sqrt{6}\hbar$

81. (d) $CoCl_3 . 3NH_3$ will not give test for chloride ions with silver nitrate due to absence of ionisable chloride atoms.

$CoCl_3 . 3NH_3 \Rightarrow [Co(NH_3)_3 Cl_3]$

$[Co(NH_3)_3 Cl_3] \xrightarrow{AgNO_3}$ no ppt

82. (d) In $[Co(CN)_6]^{-3}$ O.N. of Co is +3

∴ $Co^{+3} = 3d^6 4s^0$

CN^- is a strong field ligand

∴ Pairing of electrons occurs so in this complex no unpaired electron is present and it is low spin complex.

83. (a) Tertiary butyl chloride will give the most stable tertiary carbonium ion among the other given compounds

$(H_3C)_3C{-}Cl \xrightarrow{-Cl^{\ominus}} (H_3C)_3C^+$

t-butyl carbonium ion

84. (d) Williamson synthesis is one of the best methods for the preparation of symmetrical and unsymmetrical ethers. In this method, an alkyl halide is allowed to react with sodium alkoxide.

85. (b) $H_3C{-}CH_2{-}\underset{\underset{O}{\|}}{C}{-}CH_2{-}CH_3$

does not give iodoform test due to absence of $(CH_3{-}\overset{\overset{O}{\|}}{C}{-})$ group. It also does, not give Tollen's test.

On reduction it gives n-pentane

$H_3C{-}CH_2{-}\underset{\underset{O}{\|}}{C}{-}CH_2{-}CH_3 \xrightarrow[\text{Zn-Hg/HCl or } NH_2\text{-}NH_2/OH^-]{\text{reduction}}$

$CH_3{-}CH_2{-}CH_2{-}CH_2{-}CH_3$

n-pentane

86. (d) Cyclopentanone $+ \overset{\ominus}{C}H_3 \overset{\oplus}{Li} \rightarrow$ 1-methylcyclopentan-1-olate ($\overset{\ominus}{O}$, CH_3)

Cylopentanoyl anion

87. (d) Enolic form of ethyl acetoacetate has 18 sigma and 2 pi-bonds as shown below:

$H_3C{-}C(OH){=}CH{-}C({=}O){-}O{-}CH_2{-}CH_3$ (each single bond σ; each double bond one σ and one π)

88. (d)

nitrobenzene (NO_2) $\xrightarrow[\text{(Strongly acidic medium)}]{\text{Electrolytic reduction}}$ Phenyl hydroxylamine (HN—OH) $\xrightarrow{\text{rearrangement}}$ p-aminophenol (NH_2, OH)

89. (d)

$H_2N{-}CH_2{-}COOH + H_2N{-}(CH_2)_5{-}COOH$

Glycine Aminocapric acid

$\rightarrow (HN{-}CH_2{-}CO{-}NH{-}(CH_2)_5{-}CO){-}$

Nylon-2-nylon-6

90. (c) Bithionol is added to soaps to impart antiseptic properties.

BIOLOGY

91. (a) *Alternaria* belongs to class - Deuteromycetes, showing absence of sexual reproduction. Asexual reproduction takes place by conidia produced on conidiophores.

92. (d) *Anabaena* is a cyanobacteria which lack a true nucleus because of absence of nuclear membrane.

93. (c) Pinus belongs to gymnosperms in which male and female gametophytes do not have an independent free living existance. They remain within the sporangia which are of two types — microsporangia and megasporangia.

94. (b) *Selaginella* is a heterosporus pteridophyte containing micro & megaspores. In *Cycas,* corolloid root has cyanobacteria - Anabaena.

95. (b) Male gametes are flagellated in *Ectocarpus* (phaeophyceae). They possess heterokont and lateral flagella.

96. (c) Mannitol or laminarin is the stored food in phaeophyceae (brown algae).

97. (b) *Platypus* is oviparous (egg laying animal). It belongs to class-mammalia.

98. (b) Between the various sclerite, a flexible membrane exists which is known as arthrodial membrane.

99. (a) (i) Aves possess dry skin, without glands except oil gland near the base of tail.
(ii) Pinnae are not found in aquatic animals and egg laying mammals.
(iii) In cyclostomes, unpaired appendages (joints) are found.

100. (d) In *Opuntia*, leaves become modified into spines to avoid transpiration so as to save water for desert adaptation. The stems of this plant appear like-leaf, termed as-phylloclades which perform the process of photosynthesis.

101. (b) Floral formula of *Petunia* (solanaccac) is

$\oplus ⚥ K_{(5)} \overset{\frown}{C_{(5)} A_5} G_{(2)}$.

102. (a) *Indigofera* is a member of family fabaceae. It has keel type of floral structure in which two anterior fused petals are present.

103. (c) Ovary is half superior and half inferior in perigynous flower.

104. (d) Vegetative propagation takes place through rhizome in ginger.

105. (a) A solid style has transmission tissue which has large intercellular spaces. It allows growth of pollen tube in pistil.

106. (b) Monocot root shows radial and closed vascular bundle (xylem & phloem).

107. (a) In a ring girdled plants roots die first due to damage of phloem causing inhibition of translocation of food.

108. (a) Vascular bundles in monocotyledons are considered closed due to absence of cambium.

109. (b) The exoskeleton of insects consists of chitinous cuticle. It gets hardened due to deposition of calcium. It prevents dessication and gives protection.

110. (b) Methanogens (microorganisms producing methane) are found in the guts of ruminant animals e.g. cows and buffalloes.

111. (a) Ribosomes are composed of ribonucleic acid and proteins and are not surrounded by any membrane. These are site for protein synthesis.

112. (c) In late prophase, nuclear envelope disappears and reappears in late telophase from rough endoplasmic reticulum (RER).

113. (c) Polysomes are found in eukaryotes and are defined as a cluster of ribosomes attached to a mRNA molecule. Polysomes are number of ribosomal complexes situated on mRNA.

114. (a) Lipid synthesis takes place in smooth endoplasmic reticulum.

115. (c) K_m increases but V_{max} remains the same in competitive inhibition. This is because, on slightly increasing the substrate concentration, the effect of inhibitor is removed.

116. (a) Pachytene - Crossing over mediated by recombinase enzyme.

Zygotene - Pairing of homologous chromosomes

G_2 phase - Synthesis of RNA and protein

Anaphase I - Contromeres do not separate but chromatids move towards opposite poles.

117. (b) When s-phase completes, a somatic cell contains 2n number of chromosomes and 4C content of DNA.

118. (a) Transpiration creates pulling (Negative pressure) force. Root pressure creates positive pressure developed in xylem. It is measured by manometer.

119. (d) Phosphorus, potassium, sulphur and calcium play an important role in plants growth. These minerals are required in large amount. Phosphorus is required for all phosphorylation reactions and constitution of cell membrane, nucleic acids and some proteins. Potassium is related with protein synthesis, closing and opening

of stomata and activation of enzyme. Calcium regulates metaolic activities, function of cell membrane and stabilizes the structure of chromosomes. Sulphur is the main constituent of the amino acids, cystine and methionine, coenzymes and vitamins.

120. (a) The opening and closing of stomata are caused by influx and efflux of potassium ions (K^+). The increase of K^+ results in opening of stoma and decrease of K^+ causes closing of stoma. The turgidity of guard cells induces to open the pores of stomata found on the surface of leaves.

121. (a) In chloroplast which are green coloured plastids, thylakoids are arranged in stacks like the pile of coins called grana.

122. (d) Sigmoid growth curve maintains equilibrium with carrying capacity, it forms asymptote stage in plants.

123. (c) Auxin stimulates cell elongation. It accumulates on shaded side which results in more elongation of cells towards shaded side of the plant. This causes bending of the plant towards source of light.

124. (d) Duodenum contains Brunner's glands which secrete mucus and digestive juices.

125. (b) Gastric juice of infants contains pepsinogen, lipase and rennin. Gastric juice does not contain amylase, maltase & nuclease.

126. (b) Cytochromes are fond in mitochondria. These are located on the inner membrane of mitochondria and are related with phosphorylation.

127. (a) Chemoreceptors in the medulla get stimulated by increase in CO_2 concentration in the blood of arteries. Decrease in O_2 concentration has no significant effect on chemoreceptors.

128. (c) Blood consists of plasma, RBC, WBC and platelets.

129. (c) Erythropoiesis is the process of formation of R.B.C. In the first month of pregnancy, yolk sac is the heamopoietic tissue. After 5 weeks, it is followed by the liver. Red bone marrow from 6 months onwards becomes the principal site of erythropoiesis.

130. (b) The wall of left ventricle is the thickest of all the Chambers. It supplies oxygenated blood throughout the body. Aorta arises from left ventricle and the musculature is best developed.

131. (b) The secretion of Renin is induced by decrease in blood pressure and blood volume. It maintains the proper BP and BV. It has no role in the formation of large quantities of urine.

132. (d) Gnerally all of the essential nutrients and 70% to 80% of electrolytes and water are reabsorbed by this segment. Removal of proximal convolnted tubule from the nephron will result in dilution of urine.

133. (b) Sliding filament theory was given by Huxley and Huxley (1954). It states that Actin and Myosin filaments do not become short but rather slide pass each other. Because of sliding of actin filaments over myosin the length of I-band will change.

134. (c) Glenoid cavity articulates humerus with scapula.

135. (a) Vestibular apparatus has specific receptors called crista and macula to maintain the balance and posture of body

136. (a) Cerebellum maintains the balance and body posture. It is not concerned with logical part.

137. (b) Epinephrine has two role as a hormone and as a neurotransmitter.

138. (d) Geitonogamy, genetically, shows similarity with autogamy because the pollen grains are born on the same plant.

139. (c) Honey is produced by worker bees using nectar (fructose) of flowering plants.

140. (d) The hilum is a scar found on the seed coat through which the developing seeds are attached to the fruit.

141. (a) Nectar and pollen grains are the usual floral rewards which the animal pollinators get.

142. (c) The role of prolactin is not concerned with initiation of parturition in humans. Prolactin is responsible for milk synthesis in mammary glands. When the levels of estrogen and progesterone become low at parturition, then anterior pituitary activates to release prolactin.

143. (d) Capacitation and acrosomal reaction occur before fertilization. The process of capacitation happens when sperms enter vagina. During this period, cholesterol vesicles are removed from the acrosome.

144. (b) During gametogenesis, spermatogonia are diploid because of undergoing the process of meiosis.

145. (d) A hysterectomy is a surgery in which woman's uterus is removed. It is adopted for the treatment of uterine fibroids and cancer of uterine zone.

146. (c) Seven pairs of contrasting characters were selected in pea plant and studied by Mendel in his experiment.

147. (c) The most common cause of variations is recombination in the organism which are reproduced by sexual way.

148. (b) Alleles are defined as alternative form of same gene.

149. (b) Possible genotype of man with blood group A × Possible genotype of woman with blood group B

$I^A I^A, I^A I^O \times I^B I^B, I^B I^O$

If the genotype is

$I^A I^O \times I^B I^O$

The possibility of resultant blood group may be A, B, AB and O.

150. (b) Chargaff's rule states that A = T and G ≡ C. The molar amount of adenine = molar amount of thymine. The molar amount of guanine – molar amount of cytosine.

Hence, G is 17%, so, C = 17%
A = 33%, so, T = 33%

151. (b) In translocation, the movement of a gene takes place from one linkage group to another between non-homologous chromosomes.

152. (a) Lac operon under control of repressor shows a negative regulation. Operon has inducible nature.

153. (a) In acrocentric chromosomes, one arm is very short but another is long.

154. (b) All alleles of a gene are situated on the same loci of chromosome in organisms.

155. (a) A human baby having abnormality with 'XXX' sex chromosomes is born due to evolution of abnormal ova in mother's ovary. This is caused due to non-disjunction of X chromosome in the mother.

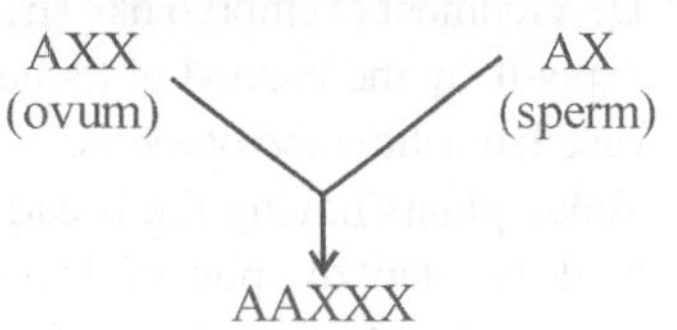

156. (c) Brain capacity in decreasing order:-
Homo sapiens (1300 - 1600 c.c)
Homo neander thalensis (approx. 1400 c.c.)
Homo eractus (600 - 700 c.c.)
Homo habilis (550 - 687 c.c.)
Thus the lowest brain capacity is of *Homo habilis* as shown above. *Homo habilis* lived in early pleistocene period and used tools of chipped stones.

157. (c) Encephalitis is a disease of inflammation of the brain. It is not transmitted sexually. Most commonly it is caused by a virus.

158. (b) After infection, HIV starts to destroy the T-cells (T-helper lymphocytes). T. cells are very important for the immune system. In the early stage of infection, the decline in numbers of T.cells is observed.

159. (d) *Entamoeba histolytica* is found in the colon. It feeds on mucosa and submucosa as well as phagocytose RBCs.

160. (b) The virus of chikunguniya is *Arbovirus* transmitted by *Aedes* mosquito. In this disease the patient feels fever lasting 2-7 days.

161. (b) Tuberculosis vaccine (BCG) has inactivated bacteria. In whooping cough vaccine, there are killed pathogens of *Bordetella pertussis* which cause whooping cough. In DPT diphtheria toxoid is present. Sabin polio vaccine contains inactivated virus.

162. (b) *Trichinella spiralis* is an endoparasite found in human intestine. It is the smallest nematode producing larvae in large number. These larvae bore the intestine of human being and enter the blood and lymphatic systems.

163. (a) Polyethylene glycol and sodium nitrate play an important role in the fusion of protoplasts from the same or different species. It is done for the formation of somatic hybrid cells. This process is adopted when normal sexual reproduction is not possible for the production of hybrids

164. (a) Development of embryo like structure from explant by the method of tissue culture, is called somatic embryogenesis.

165. (c) Today plants having the broad leaves are made resistant to a powerful biodegradable herbicide glyphosate. It is an active ingredient of Round Up ready plant. It disturbs the working of EPSP synthetase enzyme. If it is taken up by crop plants they will die. So, the bioengineers have transferred gene for synthesis of EPSP synthetase enzyme to crop plant.

166. (d) Bt toxin are solubilised in alkaline pH of the insect gut causing death.

167. (b) Genetic Engineering Approval Committee is the body of Government of India. It regulates GM research and safety of introducing GM organisms for public service in the country.

168. (c) In tropical rain forest zone, most of the animals prefer to live on trees. The reason is that the floor of forest is full of humidity, decomposing leaves and other organic matters and is the habitat of insects etc.

169. (d) The theory of Hardy- Weinberg equilibrium states that in the absence of disturbance on the level of genetic structure, a population's existance will not continue. Thus mating should be in a random way.

170. (d) Rachel Carson's book 'Silent Spring' reveals the harmful effect of DDT, a fatal pesticide. The writer opposed to chemical industry in U.S.A. This book influenced the national policy over pesticide use in many countries all over the world.

171. (a) Stratification is the occurrence of vertical zonation in the ecosystem & indicates the presence of favorable environmental conditions, for e.g., trees occupy top vertical strata or layer of a forest, shrubs the second. Herbs & grasses occupy the bottom layers. It is absent or poor where environmental conditions are unfavorable, e.g. desert ecosystems have very few trees & shrubs.

172. (c) Standing crop is the amount of living biomass in an ecosystem. It indicates the productivity & luxuriance of growth. It is expressed in the form of number or biomass of organisms per unit area.

173. (a) Gross Primary Productivity (GPP) is the rate of production of biomass or accumulation of energy by green plants per unit area per unit time. GPP depends on the chlorophyll content.

174. (a) Secondary succession takes place in disturbed area having poor vegetation.

175. (a) The given graph illustrates that population B got success in the grassy field in comparison to population A.

176. (a) In such type of conservation, the threatened animals and plants are taken out of their natural habitat and protected in special areas like zoological parks and wild life sanctuaries.

177. (d) *In-situ* (on site) conservation is within the natural habitat eg- sea, river and forest. *Ex-situ* (off site) conservation is outside the natural habitat e.g. zoological park.

178. (a) BOD (Biochemical Oxygen Demand) is a method which measures the level of water pollution. The polluted water has bacteria and many other microorganisms which consume the dissolved oxygen of water. The higher the BOD, the lower would be the dissolved oxygen in polluted water. The high value of BOD indicates the highly polluted water.

179. (a) The UN Conference of Parties on climate change in 2011, was held in Durban, South Africa. It aimed at decrease of pollutant emission.

180. (c) Liver cancer is caused due to aflatoxin released by *Aspergillus fungus* and Vinyl chloride. UV-rays are not related with liver cancer.

AIPMT 2014

PHYSICS

1. If force (F), velocity (V) and time (T) are taken as fundamental units, then the dimensions of mass are:
(a) $[F V T^{-1}]$ (b) $[F V T^{-2}]$
(c) $[F V^{-1} T^{-1}]$ (d) $[F V^{-1} T]$

2. A projectile is fired from the surface of the earth with a velocity of $5\ ms^{-1}$ and angle θ with the horizontal. Another projectile fired from another planet with a velocity of $3\ ms^{-1}$ at the same angle follows a trajectory which is identical with the trajectory of the projectile fired from the earth. The value of the acceleration due to gravity on the planet is (in ms^{-2}) given $g = 9.8\ m/s^2$
(a) 3.5 (b) 5.9
(c) 16.3 (d) 110.8

3. A particle is moving such that its position coordinate (x, y) are
(2m, 3m) at time t = 0
(6m, 7m) at time t = 2 s and
(13m, 14m) at time t = 5s.
Average velocity vector $(\vec{V}_{av})$ from t = 0 to t = 5s is :
(a) $\frac{1}{5}(13\hat{i} + 14\hat{j})$ (b) $\frac{7}{3}(\hat{i} + \hat{j})$
(c) $2(\hat{i} + \hat{j})$ (d) $\frac{11}{5}(\hat{i} + \hat{j})$

4. A system consists of three masses m_1, m_2 and m_3 connected by a string passing over a pulley P. The mass m_1 hangs freely and m_2 and m_3 are on a rough horizontal table (the coefficient of friction = μ). The pulley is frictionless and of negligible mass. The downward acceleration of mass m_1 is :
(Assume $m_1 = m_2 = m_3 = m$)
(a) $\frac{g(1 - g\mu)}{g}$
(b) $\frac{2g\mu}{3}$
(c) $\frac{g(1 - 2\mu)}{3}$
(d) $\frac{g(1 - 2\mu)}{2}$

5. The force 'F' acting on a particle of mass 'm' is indicated by the force-time graph shown below. The change in momentum of the particle over the time interval from zero to 8 s is :

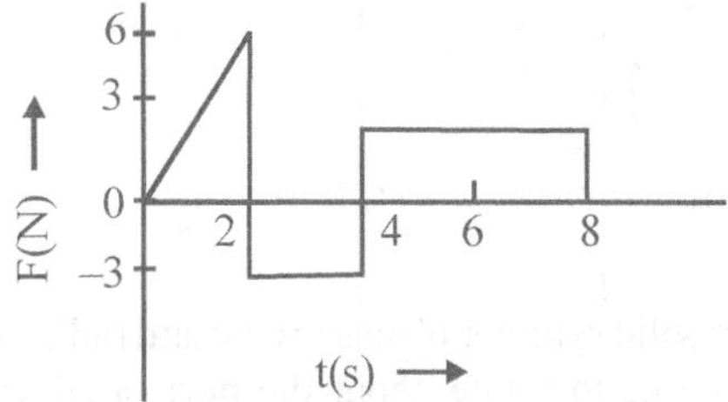

(a) 24 Ns (b) 20 Ns
(c) 12 Ns (d) 6 Ns

6. A balloon with mass 'm' is descending down with an acceleration 'a' (where a < g). How much mass should be removed from it so that it starts moving up with an acceleration 'a'?
(a) $\frac{2ma}{g + a}$ (b) $\frac{2ma}{g - a}$
(c) $\frac{ma}{g + a}$ (d) $\frac{ma}{g - a}$

7. A body of mass (4m) is lying in x-y plane at rest. It suddenly explodes into three pieces. Two pieces, each of mass (m) move perpendicular to each other with equal speeds (v). The total kinetic energy generated due to explosion is :
(a) mv^2 (b) $\frac{3}{2}mv^2$
(c) $2mv^2$ (d) $4mv^2$

8. The oscillation of a body on a smooth horizontal surface is represented by the equation,
$X = A \cos(\omega t)$
where X = displacement at time t
ω = frequency of oscillation
Which one of the following graphs shows correctly the variation of 'a' with 't'?

(a)
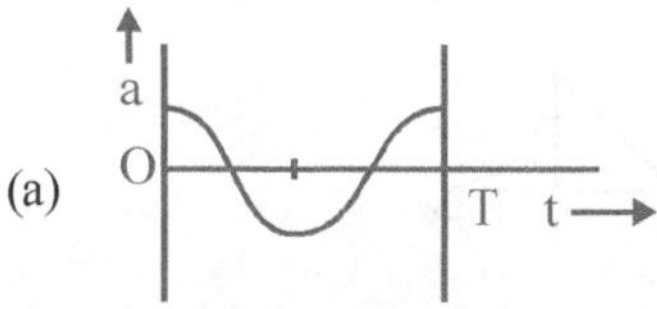

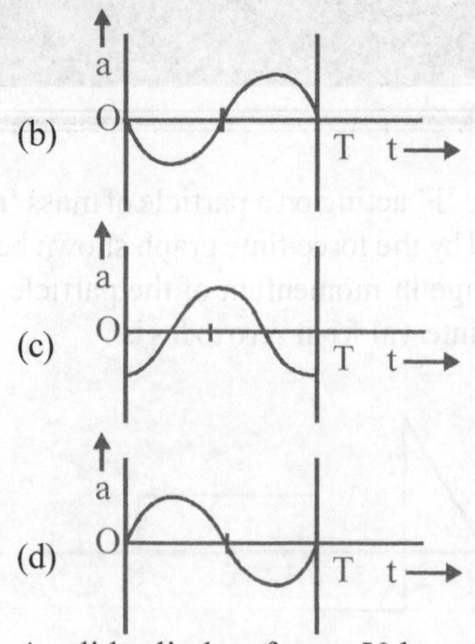

9. A solid cylinder of mass 50 kg and radius 0.5 m is free to rotate about the horizontal axis. A massless string is wound round the cylinder with one end attached to it and other hanging freely. Tension in the string required to produce an angular acceleration of 2 revolutions s^{-2} is :
(a) 25 N (b) 50 N
(c) 78.5 N (d) 157 N

10. The ratio of the accelerations for a solid sphere (mass 'm' and radius 'R') rolling down an incline of angle 'θ' without slipping and slipping down the incline without rolling is :
(a) 5 : 7 (b) 2 : 3
(c) 2 : 5 (d) 7 : 5

11. A black hole is an object whose gravitational field is so strong that even light cannot escape from it. To what approximate radius would earth (mass = 5.98×10^{24} kg) have to be compressed to be a black hole?
(a) 10^{-9} m (b) 10^{-6} m
(c) 10^{-2} m (d) 100 m

12. Dependence of intensity of gravitational field (E) of earth with distance (r) from centre of earth is correctly represented by:

(a) 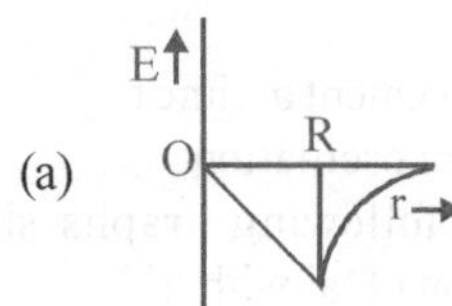

(b) 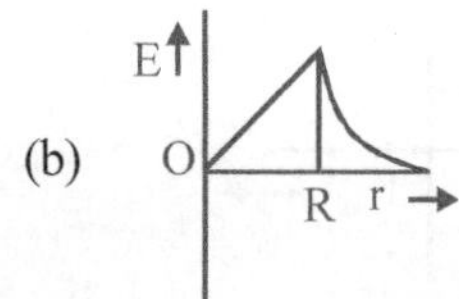

(c) 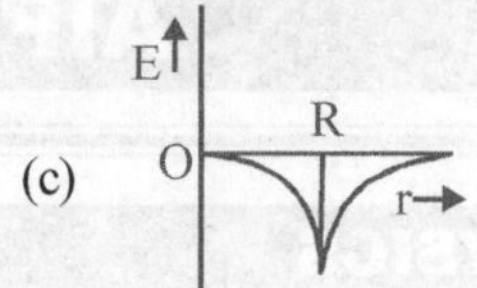

(d) 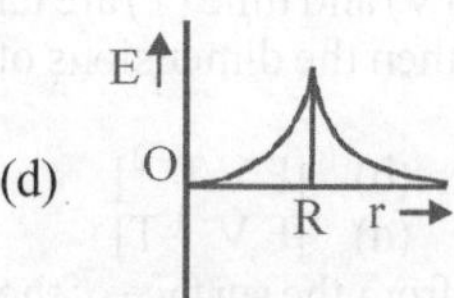

13. Copper of fixed volume 'V; is drawn into wire of length 'l'. When this wire is subjected to a constant force 'F', the extension produced in the wire is 'Δl'. Which of the following graphs is a straight line?
(a) Δl versus $\frac{1}{l}$ (b) Δl versus l^2
(c) Δl versus $\frac{1}{l^2}$ (d) Δl versus l

14. A certain number of spherical drops of a liquid of radius 'r' coalesce to form a single drop of radius 'R' and volume 'V'. If 'T' is the surface tension of the liquid, then :
(a) energy $= 4VT\left(\frac{1}{r}-\frac{1}{R}\right)$ is released
(b) energy $= 3VT\left(\frac{1}{r}+\frac{1}{R}\right)$ is absorbed
(c) energy $= 3VT\left(\frac{1}{r}-\frac{1}{R}\right)$ is released
(d) energy is neither released nor absorbed

15. Steam at 100°C is passed into 20 g of water at 10°C. When water acquires a temperature of 80°C, the mass of water present will be :
[Take specific heat of water = 1 cal g^{-1} °C^{-1} and latent heat of steam = 540 cal g^{-1}]
(a) 24 g (b) 31.5 g
(c) 42.5 g (d) 22.5 g

16. Certain quantity of water cools from 70°C to 60°C in the first 5 minutes and to 54°C in the next 5 minutes. The temperature of the surroundings is:
(a) 45°C (b) 20°C
(c) 42°C (d) 10°C

17. A monoatomic gas at a pressure P, having a volume V expands isothermally to a volume 2V and then adiabatically to a volume 16V. The final pressure of the gas is : (take $\gamma = \frac{5}{3}$)

(a) 64P (b) 32P

(c) $\frac{P}{64}$ (d) 16P

18. A thermodynamic system undergoes cyclic process ABCDA as shown in fig. The work done by the system in the cycle is :

(a) P_0V_0

(b) $2P_0V_0$

(c) $\frac{P_0V_0}{2}$

(d) Zero

19. The mean free path of molecules of a gas, (radius 'r') is inversely proportional to :

(a) r^3 (b) r^2

(c) r (d) $\sqrt{r}$

20. If n_1, n_2 and n_3 are the fundamental frequencies of three segments into which a string is divided, then the original fundamental frequency n of the string is given by :

(a) $\frac{1}{n} = \frac{1}{n_1} + \frac{1}{n_2} + \frac{1}{n_3}$

(b) $\frac{1}{\sqrt{n}} = \frac{1}{\sqrt{n_1}} + \frac{1}{\sqrt{n_2}} + \frac{1}{\sqrt{n_3}}$

(c) $\sqrt{n} = \sqrt{n_1} + \sqrt{n_2} + \sqrt{n_3}$

(d) $n = n_1 + n_2 + n_3$

21. The number of possible natural oscillation of air column in a pipe closed at one end of length 85 cm whose frequencies lie below 1250 Hz are : (velocity of sound = 340 ms^{-1})

(a) 4 (b) 5

(c) 7 (d) 6

22. A speeding motorcyclist sees trafic jam ahead of him. He slows down to 36 km/hour. He finds that traffic has eased and a car moving ahead of him at 18 km/hour is honking at a frequency of 1392 Hz. If the speeds of sound is 343 m/s, the frequency of the honk as heard by him will be :

(a) 1332 Hz (b) 1372 Hz

(c) 1412 Hz (d) 1464 Hz

23. Two thin dielectric slabs of dielectric constants K_1 and K_2 ($K_1 < K_2$) are inserted between plates of a parallel plate capacitor, as shown in the figure. The variation of electric field 'E' between the plates with distance 'd' as measured from plate P is correctly shown by :

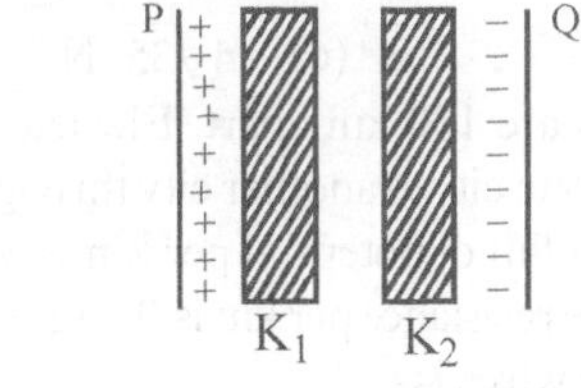

(a)

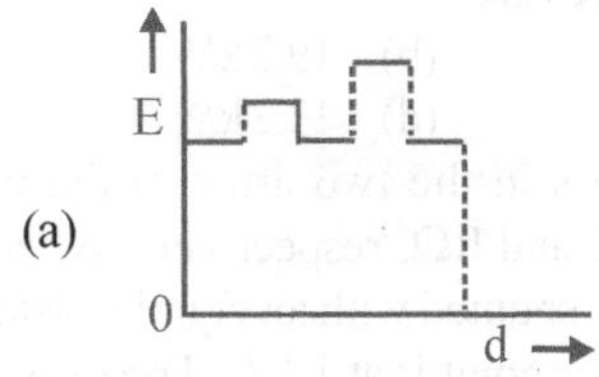

(b)

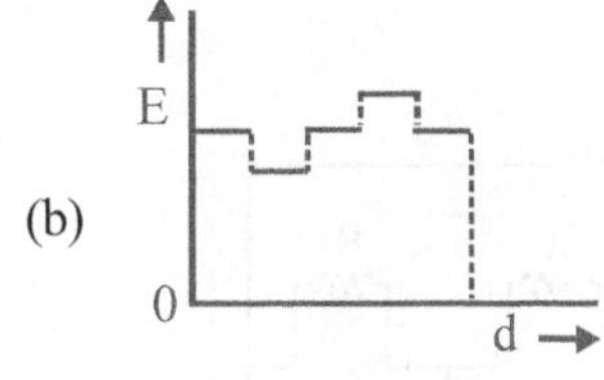

(c)

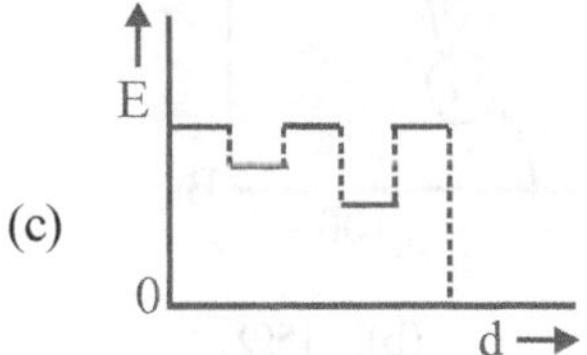

(d)

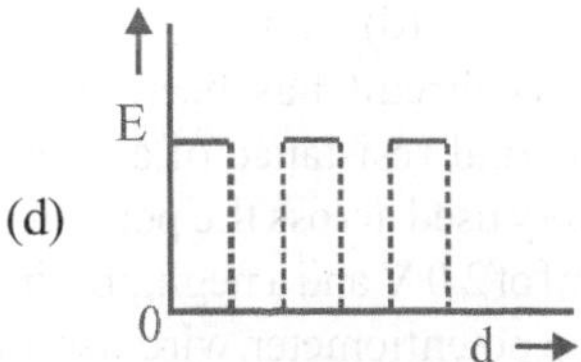

24. A conducting sphere of radius R is given a charge Q. The electric potential and the electric field at the centre of the sphere respectively are:

(a) Zero and $\frac{Q}{4\pi\varepsilon_0 R^2}$

(b) $\frac{Q}{4\pi\varepsilon_0 R}$ and Zero

(c) $\frac{Q}{4\pi\varepsilon_0 R}$ and $\frac{Q}{4\pi\varepsilon_0 R^2}$

(d) Both are zero

25. In a region, the potential is represented by V(x, y, z) = 6x – 8xy – 8y + 6yz, where V is in volts and x, y, z are in metres. The electric force experienced by a charge of 2 coulomb situated at point (1, 1, 1) is :

(a) $6\sqrt{5}$ N (b) 30 N

(c) 24 N (d) $4\sqrt{35}$ N

26. Two cities are 150 km apart. Electric power is sent from one city to another city through copper wires. The fall of potential per km is 8 volt and the average resistance per km is 0.5 Ω. The power loss in the wires is :

(a) 19.2 W (b) 19.2 kW
(c) 19.2 J (d) 12.2 kW

27. The resistances in the two arms of the meter bridge are 5Ω and RΩ, respectively. When the resistance R is shunted with an equal resistance, the new balance point is at 1.6 l_1. The resistance 'R' is :

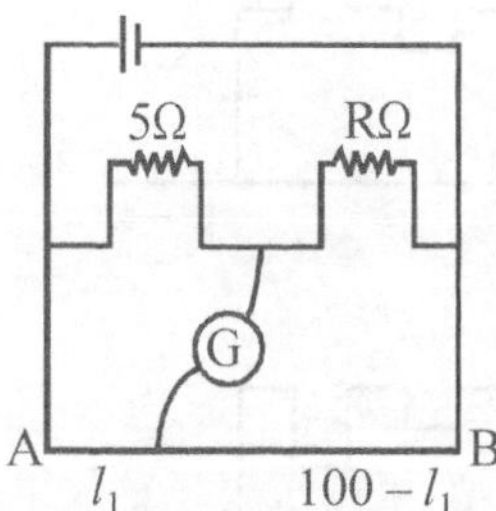

(a) 10Ω (b) 15Ω
(c) 20Ω (d) 25Ω

28. A potentiometer circuit has been set up for finding the internal resistance of a given cell. The main battery used across the potentiometer wire, has an emf of 2.0 V and a negligible internal resistance. The potentiometer wire itself is 4m long, When the resistace R, connected across the given cell, has values of
(i) infinity (ii) 9.5Ω
The balancing lengths', on the potentiometer wire are found to be 3 m and 2.85 m, respectively. The value of internal resistance of the cell is

(a) 0.25Ω (b) 0.95Ω
(c) 0.5Ω (d) 0.75Ω

29. Following figures show the arrangement of bar magnets in different configurations. Each magnet has magnetic dipole moment $\vec{m}$. Which configuration has highest net magnetic dipole moment ?

A.
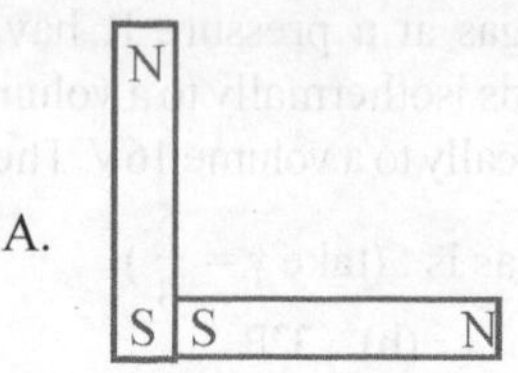

B.
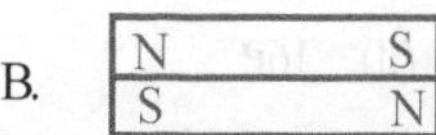

C.
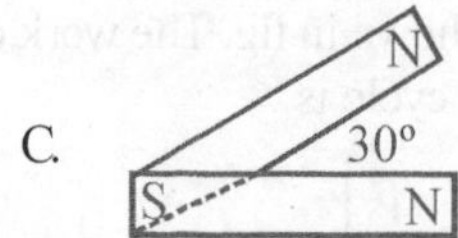

D.
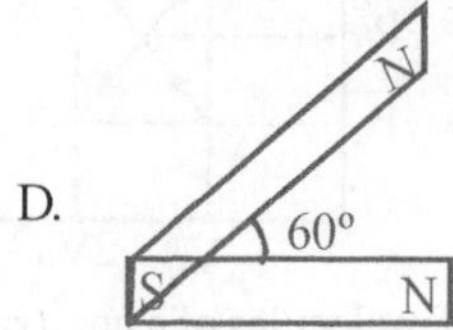

(a) A (b) B
(c) C (d) D

30. In an ammeter 0.2% of main current passes through the galvanometer. If resistance of galvanometer is G, the resistance of ammeter will be :

(a) $\frac{1}{499}G$ (b) $\frac{499}{500}G$

(c) $\frac{1}{500}G$ (d) $\frac{500}{499}G$

31. Two identical long conducting wires AOB and COD are placed at right angle to each other, with one above other such that 'O' is their common point for the two. The wires carry I_1 and I_2 currents respectively. Point 'P' is lying at distance 'd' from 'O' along a direction perpendicular to the plane containing the wires. The magnetic field at the point 'P' will be :

(a) $\frac{\mu_0}{2\pi d}\left(\frac{I_1}{I_2}\right)$ (b) $\frac{\mu_0}{2\pi d}(I_1 + I_2)$

(c) $\frac{\mu_0}{2\pi d}(I_1^2 - I_2^2)$ (d) $\frac{\mu_0}{2\pi d}(I_1^2 \times I_2^2)^{1/2}$

32. A thin semicircular conducting ring (PQR) of radius 'r' is falling with its plane vertical in a horizontal magnetic field B, as shown in figure. The potential difference developed across the ring when its speed is v, is :

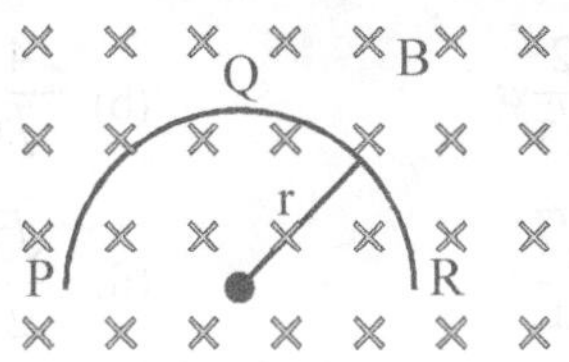

(a) Zero
(b) $Bv\pi r^2/2$ and P is at higher potnetial
(c) πrBv and R is at higher potnetial
(d) 2rBv and R is at higher potential

33. A transformer having efficiency of 90% is working on 200V and 3kW power supply. If the current in the secondary coil is 6A, the voltage across the secondary coil and the current in the primary coil respectively are :
(a) 300 V, 15A (b) 450 V, 15A
(c) 450V, 13.5A (d) 600V, 15A

34. Light with an energy flux of 25×10^4 Wm^{-2} falls on a perfectly reflecting surface at normal incidence. If the surface area is 15 cm^2, the average force exerted on the surface is :
(a) 1.25×10^{-6} N (b) 2.50×10^{-6} N
(c) 1.20×10^{-6} N (d) 3.0×10^{-6} N

35. A beam of light of $\lambda = 600$ nm from a distant source falls on a single slit 1 mm wide and the resulting diffraction pattern is observed on a screen 2 m away. The distance between first dark fringes on either side of the central bright fringe is:
(a) 1.2 cm (b) 1.2 mm
(c) 2.4 cm (d) 2.4 mm

36. In the Young's double-slit experiment, the intensity of light at a point on the screen where the path difference is λ is K, (λ being the wave length of light used). The intensity at a point where the path difference is $\lambda/4$, will be:
(a) K (b) K/4
(c) K/2 (d) Zero

37. If the focal length of objective lens is increased then magnefying power of :
(a) microscope will increase but that of telescope decrease.
(b) microscope and telescope both will increase.
(c) microscope and telescope both will decrease
(d) microscope will decrease but that of telescope increase.

38. The angle of a prism is 'A'. One of its refracting surfaces is silvered. Light rays falling at an angle of incidence 2A on the first surface returns back through the same path after suffering reflection at the silvered surface. The refractive index μ, of the prism is :
(a) 2 sin A (b) 2 cos A
(c) $\frac{1}{2}\cos A$ (d) tan A

39. When the energy of the incident radiation is incredased by 20%, the kinetic energy of the photoelectrons emitted from a metal surface increased from 0.5 eV to 0.8 eV. The work function of the metal is :
(a) 0.65 eV (b) 1.0 eV
(c) 1.3 eV (d) 1.5 eV

40. If the kinetic energy of the particle is increased to 16 times its previous value, the percentage change in the de-Broglie wavelength of the particle is :
(a) 25 (b) 75
(c) 60 (d) 50

41. Hydrogen atom in ground state is excited by a monochromatic radiation of $\lambda = 975$ Å. Number of spectral lines in the resulting spectrum emitted will be
(a) 3 (b) 2
(c) 6 (d) 10

42. The Binding energy per nucleon of ${}_3^7\text{Li}$ and ${}_2^4\text{He}$ nuclei are 5.60 MeV and 7.06 MeV, respectively.

In the nuclear reaction ${}_3^7\text{Li} + {}_1^1\text{H} \rightarrow {}_2^4\text{He} + Q$, the value of energy Q released is :
(a) 19.6 MeV (b) −2.4 MeV
(c) 8.4 MeV (d) 17.3 MeV

43. A radio isotope 'X' with a half life 1.4×10^9 years decays to 'Y' which is stable. A sample of the rock from a cave was found to contain 'X' and 'Y' in the ratio 1 : 7. The age of the rock is :
(a) 1.96×10^9 years (b) 3.92×10^9 years
(c) 4.20×10^9 years (d) 8.40×10^9 years

44. The given graph represents V - I characteristic for a semiconductor device.

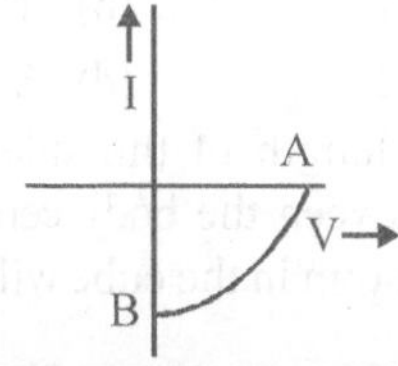

Which of the following statement is **correct** ?

(a) It is V - I characteristic for solar cell where, point A represents open circuit voltage and point B short circuit current.

(b) It is a for a solar cell and point A and B represent open circuit voltage and current, respectively.

(c) It is for a photodiode and points A and B represent open circuit voltage and current, respectively.

(d) It is for a LED and points A and B represent open circuit voltage and short circuit current, respectively.

45. The barrier potential of a p-n junction depends on:

(A) type of semi conductor material

(B) amount of doping

(C) temperature

Which one of the following is correct ?

(a) (A) and (B) only

(b) (B) only

(c) (B) and (C) only

(d) (A), (B) and (C)

CHEMISTRY

46. What is the maximum number of orbitals that can be identified with the following quantum numbers?

$n = 3, \ell = 1, m_\ell = 0$

(a) 1 (b) 2

(c) 3 (d) 4

47. Calculate the energy in joule corresponding to light of wavelength 45 nm :

(Planck's constant $h = 6.63 \times 10^{-34}$ Js; speed of light $c = 3 \times 10^8\ ms^{-1}$)

(a) 6.67×10^{15} (b) 6.67×10^{11}

(c) 4.42×10^{-15} (d) 4.42×10^{-18}

48. Equal masses of H_2, O_2 and methane have been taken in a container of volume V at temperature 27°C in identical conditions. The ratio of the volumes of gases $H_2 : O_2$: methane would be :

(a) 8 : 16 : 1 (b) 16 : 8 : 1

(c) 16 : 1 : 2 (d) 8 : 1 : 2

49. If a is the length of the side of a cube, the distance between the body centered atom and one corner atom in the cube will be :

(a) $\frac{2}{\sqrt{3}}a$ (b) $\frac{4}{\sqrt{3}}a$

(c) $\frac{\sqrt{3}}{4}a$ (d) $\frac{\sqrt{3}}{2}a$

50. Which property of colloids is not dependent on the charge on colloidal particles ?

(a) Coagulation (b) Electrophoresis

(c) Electro - osmosis (d) Tyndall effect

51. Which of the following salts will give highest pH in water ?

(a) KCl (b) NaCl

(c) Na_2CO_3 (d) $CuSO_4$

52. Of the following 0.10m aqueous solutions, which one will exhibit the largest freezing point depression?

(a) KCl (b) $C_6H_{12}O_6$

(c) $Al_2(SO_4)_3$ (d) K_2SO_4

53. When 22.4 litres of H_2(g) is mixed with 11.2 litres of Cl_2(g), each at S.T.P., the moles of HCl(g) formed is equal to:-

(a) 1 mol of HCl(g) (b) 2 mol of HCl(g)

(c) 0.5 mol of HCl(g) (d) 1.5 mol of HCl(g)

54. When 0.1 mol MnO_4^{2-} is oxidised the quantity of electricity required to completely oxidise MnO_4^{2-} to MnO_4^- is:

(a) 96500 C (b) 2 × 96500 C

(c) 9650 C (d) 96.50 C

55. Using the Gibbs energy change, $\Delta G° = + 63.3$kJ, for the following raction,

$Ag_2CO_3 \rightleftharpoons 2Ag^+(aq) + CO_3^{2-}(aq)$

the K_{sp} of Ag_2CO_3(s) in water at 25°C is:-

$(R = 8.314\ JK^{-1}\ mol^{-1})$

(a) 3.2×10^{-26} (b) 8.0×10^{-12}

(c) 2.9×10^{-3} (d) 7.9×10^{-2}

56. The weight of silver (at wt. = 108) displaced by a quantity of electricity which displaces 5600 mL of O_2 at STP will be :-

(a) 5.4 g (b) 10.8 g

(c) 54.9 g (d) 108.0 g

57. Which of the following statements is correct for the spontaneous adsorption of a gas ?

(a) ΔS is negative and, therefore, ΔH should be highly positive

(b) ΔS is negative and therefore, ΔH should be highly negative

(c) ΔS is positive and, therefore, ΔH should be negative
(d) ΔS is positive and, therefore, ΔH should also be highly positive

58. For the reversible reaction,
$N_2(g) + 3H_2(g) \rightleftharpoons 2NH_3(g) + \text{heat}$
The equilibrium shifts in forward direction
(a) By increasing the concentration of $NH_3(g)$
(b) By decreasing the pressure
(c) By decreasing concentration of $N_2(g)$ and $H_2(g)$
(d) By increasing pressure and decreasing temperature.

59. For the reaction :
$X_2O_4(\ell) \rightarrow 2XO_2(g)$
$\Delta U = 2.1$ k cal, $\Delta S = 20$ cal K^{-1} at 300 K
Hence ΔG is:-
(a) 2.7 k cal (b) −2.7 k cal
(c) 9.3 k cal (d) −9.3 k cal

60. For a given exothermic reaction, K_p and K_p' are the equilibrium constants at temperatures T_1 and T_2, respectively. Assuming that heat of reaction is constant in temperature range between T_1 and T_2, it is readily observed that:
(a) $K_p > K_p'$ (b) $K_p < K_p'$
(c) $K_p = K_p'$ (d) $K_p = \frac{1}{K_p'}$

61. Which of the following orders of ionic radii is correctly represented ?
(a) $H^- > H^+ > H$ (b) $Na^+ > F^- > O^{2-}$
(c) $F^- > O^{2-} > Na^+$ (d) $Al^{3+} > Mg^{2+} > N^{3-}$

62. 1.0 g of magnesium is burnt with 0.56 g O_2 in a closed vessel. Which reactant is left in excess and how much ?
(At. wt. Mg = 24 ; O = 16)
(a) Mg, 0.16 g (b) O_2, 0.16 g
(c) Mg, 0.44 g (d) O_2, 0.28 g

63. The pair of compounds that can exist together is:
(a) $FeCl_3, SnCl_2$ (b) $HgCl_2, SnCl_2$
(c) $FeCl_2, SnCl_2$ (d) $FeCl_3$, KI

64. Be^{2+} is isoelectronic with which of the following ions?
(a) H^+ (b) Li^+
(c) Na^+ (d) Mg^{2+}

65. Which of the following molecules has the maximum dipole moment ?
(a) CO_2 (b) CH_4
(c) NH_3 (d) NF_3

66. Which one of the following species has plane triangular shape ?
(a) N_3^- (b) NO_3^-
(c) NO_2^- (d) CO_2

67. Acidity of diprotic acids in aqueous solutions increases in the order:-
(a) $H_2S < H_2Se < H_2Te$
(b) $H_2Se < H_2S < H_2Te$
(c) $H_2Te < H_2S < H_2Se$
(d) $H_2Se < H_2Te < H_2S$

68. (i) $H_2O_2 + O_3 \rightarrow H_2O + 2O_2$
(ii) $H_2O_2 + Ag_2O \rightarrow 2Ag + H_2O + O_2$
Role of hydrogen peroxide in the above reactions is respectively -
(a) Oxidizing in (i) and reducing in (ii)
(b) Reducing in (i) and oxidizing in (ii)
(c) Reducing in (i) and (ii)
(d) Oxidizing in (i) and (ii)

69. Artificial sweetner which is stable under cold conditions only is :-
(a) Saccharine (b) Sucralose
(c) Aspartame (d) Alitame

70. In acidic medium, H_2O_2 changes $Cr_2O_7^{-2}$ to CrO_5 which has two (–O–O) bonds. Oxidation state of Cr in CrO_5 is:-
(a) +5 (b) +3
(c) +6 (d) −10

71. The reaction of aqueous $KMnO_4$ with H_2O_2 in acidic conditions gives:
(a) Mn^{4+} and O_2 (b) Mn^{2+} and O_2
(c) Mn^{2+} and O_3 (d) Mn^{4+} and MnO_2

72. Among the following complexes the one which shows zero crystal field stabilization energy (CFSE):
(a) $[Mn(H_2O)_6]^{3+}$ (b) $[Fe(H_2O)_6]^{3+}$
(c) $[Co(H_2O)_6]^{2+}$ (d) $[Co(H_2O)_6]^{3+}$

73. Magnetic moment 2.83 BM is given by which of the following ions ?
(At. nos. Ti = 22, Cr = 24, Mn = 25, Ni = 28):-
(a) Ti^{3+} (b) Ni^{2+}
(c) Cr^{3+} (d) Mn^{2+}

74. Which of the following complexes is used as an anti-cancer agent:
(a) *mer*-$[Co(NH_3)_3Cl_3]$ (b) *cis*-$[PtCl_2(NH_3)_2]$
(c) *cis*-$K_2[PtCl_2Br_2]$ (d) Na_2CoCl_4

75. Reason of lanthanoid contraction is:-
(a) Negligible screening effect of '*f*' orbitals
(b) Increasing nuclear charge
(c) Decreasing nuclear charge
(d) Decreasing screening effect

76. In the following reaction, the product (A)

$C_6H_5\overset{+}{N}\equiv NCl^- + C_6H_5NH_2 \xrightarrow{H^+} (A)$ (Yellow dye) is :

(a) $C_6H_5-N=N-NH-C_6H_5$
(b) $C_6H_5-N=N-C_6H_4-NH_2$ (NH₂ ortho to N=N)
(c) $C_6H_5-N=N-C_6H_4-NH_2$ (NH₂ meta to N=N)
(d) $C_6H_5-N=N-C_6H_4-NH_2$ (NH₂ para to N=N)

77. Which of the following will be most stable diazonium salt $R\overset{+}{N_2}X^-$?
(a) $CH_3\overset{+}{N_2}X^-$ (b) $C_6H_5\overset{+}{N_2}X^-$
(c) $CH_3CH_2\overset{+}{N_2}X^-$ (d) $C_6H_5CH_2\overset{+}{N_2}X^-$

78. D (+) glucose reacts with hydroxylamine and yields an oxime. The structure of the oxime would be :

(a)
CH=NOH
H–C–OH
HO–C–H
HO–C–H
H–C–OH
CH₂OH

(b)
CH=NOH
HO–C–H
HO–C–H
H–C–OH
H–C–OH
CH₂OH

(c)
CH=NOH
HO–C–H
H–C–OH
HO–C–H
H–C–OH
CH₂OH

(d)
CH=NOH
H–C–OH
HO–C–H
H–C–OH
H–C–OH
CH₂OH

79. Which of the following hormones is produced under the condition of stress which stimulates glycogenolysis in the liver of human beings?
(a) Thyroxin (b) Insulin
(c) Adrenaline (d) Estradiol

80. Which one of the following is an example of a thermosetting polymer?
(a) $\left(CH_2-\underset{Cl}{\underset{|}{C}}=CH-CH_2\right)_n$
(b) $\left(CH_2-\underset{Cl}{\underset{|}{CH}}\right)_n$
(c) $\left(\underset{H}{\overset{}{N}}-(CH_2)_6-\underset{H}{N}-\overset{O}{\overset{||}{C}}-(CH_2)_4-\overset{O}{\overset{||}{C}}\right)_n$
(d) Phenol–formaldehyde polymer: $\left(C_6H_3(OH)-CH_2-C_6H_3(OH)-CH_2\right)_n$

81. Which of the following organic compounds polymerizes to form the polyester Dacron?
(a) Propylene and para HO—(C_6H_4)— OH
(b) Benzoic acid and ethanol
(c) Terephthalic acid and ethylene glycol
(d) Benzoic acid and para HO–(C_6H_4)–OH

82. Which of the following is not a common component of Photochemical Smog?
(a) Ozone
(b) Acrolein
(c) Peroxyacetyl nitrate
(d) Chlorofluorocarbons

83. In the Kjeldahl's method for estimation of nitrogen present in a soil sample, ammonia evolved from 0.75 g of sample neutralized 10 mL of 1 M H_2SO_4. The percentage of nitrogen in the soil is :
(a) 37.33 (b) 45.33
(c) 35.33 (d) 43.33

84. What products are formed when the following compounds is treated with Br_2 in the presence of $FeBr_3$?

CH_3, CH_3

(a) CH_3, Br, CH_3 and CH_3, CH_3, Br
(b) Br, CH_3, CH_3 and CH_3, Br, CH_3
(c) Br, CH_3, CH_3 and CH_3, CH_3, Br
(d) CH_3, CH_3, Br and CH_3, Br, CH_3

85. Which of the following compounds will undergo racemisation when solution of KOH hydrolyses?
(i) $C_6H_5CH_2Cl$ (ii) $CH_3CH_2CH_2Cl$
(iii) $H_3C-CH(CH_3)-CH_2Cl$ (iv) $C(CH_3)(H)(Cl)(C_2H_5)$
(a) (i) and (ii) (b) (ii) and (iv)
(c) (iii) and (iv) (d) (i) and (iv)

86. Among the following sets of reactants which one produces anisole?
(a) CH_3CHO; RMgX
(b) C_6H_5OH; NaOH ; CH_3I
(c) C_6H_5OH; neutral $FeCl_3$
(d) C_6H_5–CH_3; CH_3COCl; $AlCl_3$

87. Which of the following will not be soluble in sodium hydrogen carbonate?
(a) 2, 4, 6-trinitrophenol
(b) Benzoic acid
(c) *o*-Nitrophenol
(d) Benzenesulphonic acid

88. Which one is most reactive towards Nucleophilic addition reaction?
(a) C_6H_5–CHO (b) C_6H_5–$COCH_3$

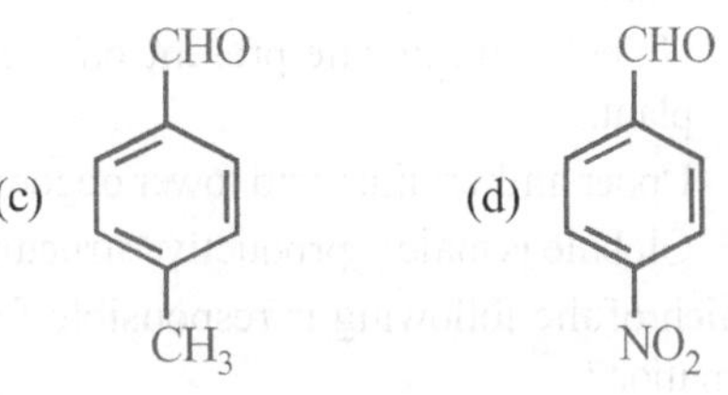

89. Identify Z in the sequence of reactions:
$$CH_3CH_2CH{=}CH_2 \xrightarrow{HBr/H_2O_2} Y \xrightarrow{C_2H_5ONa} Z$$
(a) CH_3—$(CH_2)_3$—O—CH_2CH_3
(b) $(CH_3)_2CH_2$—O—CH_2CH_3
(c) $CH_3(CH_2)_4$—O—CH_3
(d) CH_3CH_2—$CH(CH_3)$—O—CH_2CH_3

90. Which of the following organic compounds has same hybridization as its combustion product (CO_2)?
(a) Ethane (b) Ethyne
(c) Ethene (d) Ethanol

BIOLOGY

BOTANY

91. Which one of the following shows isogamy with non-flagellated gametes?
    (a) *Sargassum* (b) *Ectocarpus*
    (c) *Ulothrix* (d) *Spirogyra*
92. Five kingdom system of classification suggested by R.H. Whittaker is **not** based on:
    (a) Presence or absence of a well defined nucleus.
    (b) Mode of reproduction.
    (c) Mode of nutrition.
    (d) Complexity of body organisation.
93. Which one of the following fungi contains hallucinogens?
    (a) *Morchella esculenta*
    (b) *Amanita muscaria*
    (c) *Neurospora sp.*
    (d) *Ustilago sp.*
94. Archaebacteria differ from eubacteria in:
    (a) Cell membrane
    (b) Mode of nutrition
    (c) Cell shape
    (d) Mode of reproduction
95. Which one of the following is **wrong** about *Chara*?
    (a) Upper oogonium and lower round antheridium.
    (b) Globule and nucule present on the same plant.
    (c) Upper antheridium and lower oogonium.
    (d) Globule is male reproductive structure.
96. Which of the following is responsible for peat formation?
    (a) *Marchanita* (b) *Riccia*
    (c) *Funaria* (d) *Sphagnum*
97. Placenta and pericarp are both edible portions in:
    (a) Apple (b) Banana
    (c) Tomato (d) Potato
98. When the margins of sepals or petals overlap one another without any particular direction, the condition is termed as:
    (a) Vexillary (b) Imbricate
    (c) Twisted (d) Valvate
99. You are given a fairly old piece of dicot stem and a dicot root. Which of the following anatomical structures will you use to distinguish between the two?
    (a) Secondary xylem
    (b) Secondary phloem
    (c) Protoxylem
    (d) Cortical cells
100. Which one of the following statements is correct?
    (a) The seed in grasses is not endospermic.
    (b) Mango is a parthenocarpic fruit.
    (c) A proteinaceous aleurone layer is present in maize grain.
    (d) A sterile pistil is called a staminode.
101. Tracheids differ from other tracheary elements in:
    (a) having casparian strips
    (b) being imperforate
    (c) lacking nucleus
    (d) being lignified
102. An example of edible underground stem is:
    (a) Carrot (b) Groundnut
    (c) Sweet potato (d) Potato
103. Which structures perform the function of mitochondria in bacteria?
    (a) Nucleoid (b) Ribosomes
    (c) Cell wall (d) Mesosomes
104. The solid linear cytoskeletal elements having a diameter of 6 nm and made up of a single type of monomer are known as:
    (a) Microtubules
    (b) Microfilaments
    (c) Intermediate filaments
    (d) Lamins
105. The osmotic expansion of a cell kept in water is chiefly regulated by:
    (a) Mitochondria (b) Vacuoles
    (c) Plastids (d) Ribosomes
106. During which phase(s) of cell cycle, amount of DNA in a cell remains at 4C level if the initial amount is denoted as 2C?
    (a) G_0 and G_1 (b) G_1 and S
    (c) Only G_2 (d) G_2 and M
107. Match the following and select the correct answer:

(A) Centriole	(i) Infoldings in mitochondria		
(B) Chlorophyll	(ii) Thylakoids		
(C) Cristae	(iii) Nucleic acids		
(D) Ribozymes	(iv) Basal body cilia or fiagella		

	(A)	(B)	(C)	(D)
(a)	(iv)	(ii)	(i)	(iii)
(b)	(i)	(ii)	(iv)	(iii)
(c)	(i)	(iii)	(ii)	(iv)
(d)	(iv)	(iii)	(i)	(ii)

108. Dr. F. Went noted that if coleoptile tips were removed and placed on agar for one hour, the agar would produce a bending when placed on one side of freshly-cut coleoptile stumps. Of what significance is this experiment?
   (a) It made possible the isolation and exact identification of auxin.
   (b) It is the basis for quantitative determination of small amounts of growth-promoting substances.
   (c) It supports the hypothesis that IAA is auxin.
   (d) It demonstrated polar movement of auxins.

109. Deficiency symptoms of nitrogen and potassium are visible first in:
   (a) Senescent leaves (b) Young leaves
   (c) Roots (d) Buds

110. In which one of the following processes CO_2 in not released?
   (a) Aerobic respiration in plants
   (b) Aerobic respiration in animals
   (c) Alcoholic fermentation
   (d) Lactate fermentation

111. Anoxygenic photosynthesis is characteristic of:
   (a) *Rhodospirillum* (b) *Spirogyra*
   (c) *Chlamydomonas* (d) *Ulva*

112. A few normal seedlings of tomato were kept in a dark room. After a few days they were found to have become white-coloured like albinos. Which of the following terms will you use to describe them?
   (a) Mutated (b) Embolised
   (c) Etiolated (d) Defoliated

113. Which one of the following growth regulators is known as 'stress hormone'?
   (a) Abscissic acid (b) Ethylene
   (c) GA_3 (d) Indole acetic acid

114. Geitonogamy involves:
   (a) fertilization of a flower by the pollen from another flower of the same plant.
   (b) fertilization of a flower by the pollen from the same flower.
   (c) fertilization of a flower by the pollen from a flower of another plant in the same population.
   (d) fertilization of a flower by the pollen from a flower of another plant belonging to a distant population.

115. Male gametopyte with least number of cell is present in:
   (a) *Pteris* (b) *Funaria*
   (c) *Lilium* (d) *Pinus*

116. An aggregate fruit is one which develops from:
   (a) Multicarpellary syncarpous gynoecium
   (b) Multicarpellary apocarpus gynoecium
   (c) Complete inflorescence
   (d) Multicarpellary superior ovary

117. Pollen tablets are available in the market for:
   (a) In vitro fertilization
   (b) Breeding programmes
   (c) Supplementing food
   (d) *Ex situ* conservation

118. Function of filiform apparatus is to:
   (a) Recognize the suitable pollen at stigma
   (b) Stimulate division of generative cell
   (c) Produce nectar
   (d) Guide the entry of pollen tube

119. Non-albuminous seed is produced in:
   (a) Maize (b) Castor
   (c) Wheat (d) Pea

120. Which of the following shows coiled RNA strand and capsomeres?
   (a) Polio virus
   (b) Tobacco masaic virus
   (c) Measles virus
   (d) Retrovirus

121. Which one of the following is **wrongly** matched?
   (a) Transcription – Writing information from DNA to tRNA.
   (b) Translation – Using information in mRNA to make protein
   (c) Repressor protein – Binds to operator to stop enzyme synthesis.
   (d) Operon – Structural genes, operator and promoter.

122. Transformation was discovered by:
(a) Meselson and Stahl
(b) Hershey and Chase
(c) Griffith
(d) Watson and Crick

123. Fruit colour in squash in an example of:
(a) Recessive epistasis
(b) Dominant epistasis
(c) Complementary genes
(d) Inhibitory genes

124. Viruses have:
(a) DNA enclosed in a protein coat
(b) Prokaryotic nucleus
(c) Single chromosome
(d) Both DNA and RNA

125. The first human hormone produced by recombinant DNA technology is:
(a) Insulin (b) Estrogen
(c) Thyroxin (d) Progesterone

126. An analysis of chromosomal DNA using the Southern hybridization technique **does not** use:-
(a) Electrophoresis (b) Blotting
(c) Autoradiography (d) PCR

127. *In vitro* clonal propagation in plants is characterized by:
(a) PCR and RAPD
(b) Northern blotting
(c) Electrophoresis and HPLC
(d) Microscopy

128. An alga which can be employed as food for human being is:
(a) *Ulothrix* (b) *Chlorella*
(c) *Spirogyra* (d) *Polysiphonia*

129. Which vector can clone only a small fragment of DNA?
(a) Bacterial artificial chromosome
(b) Yeast artificial chromosome
(c) Plasmid
(d) Cosmid

130. An example of *ex situ* conservation is:
(a) National Park
(b) Seed Bank
(c) Wildlife Sanctuary
(d) Sacred Grove

131. A location with luxuriant growth of lichens on the trees indicates that the:
(a) Trees are very healthy
(b) Trees are heavily infested
(c) Location is highly polluted
(d) Location is not polluted

132. Match the following and select the correct option:
(A) Earthworm (i) Pioneer species
(B) Succession (ii) Detritivore
(C) Ecosystem service (iii) Natality
(D) Population growth (iv) Pollination

	(A)	(B)	(C)	(D)
(a)	(i)	(ii)	(iii)	(iv)
(b)	(iv)	(i)	(iii)	(ii)
(c)	(iii)	(ii)	(iv)	(i)
(d)	(ii)	(i)	(iv)	(iii)

133. A species facing extremely high risk of extinction in the immediate future is called:
(a) Vulnerable
(b) Endemic
(c) Critically Endangered
(d) Extinct

134. The zone of atmosphere in which the ozone layer is present is called:
(a) Ionosphere (b) Mesosphere
(c) Stratosphere (d) Troposphere

135. The organization which publishes the Red List of species is:
(a) ICFRE (b) IUCN
(c) UNEP (d) WWF

ZOOLOGY

136. Select the Taxon mentioned that represents both marine and fresh water species:
(a) Echinoderms (b) Ctenophora
(c) Cephalochordata (d) Cnidaria

137. Which one of the following living organisms completely lacks a cell wall?
(a) Cyanobacteria
(b) Sea – fan(*Gorgonia*)
(c) *Saccharomyces*
(d) Blue–green algae

138. *Planaria* possess high capacity of:
(a) Metamorphosis
(b) Regeneration
(c) Alternation of generation
(d) Bioluminescence

139. A marine cartilaginous fish that can produce electric current is:
(a) *Pristis* (b) *Torpedo*
(c) *Trygon* (d) *Scoliodon*

140. Choose the correctly matched pair:
(a) Tendon–Specialized connective tissue
(b) Adipose tissue – Dense connective tissue
(c) Areolar tissue – Loose connective tissue
(d) Cartilage–Loose connective tissue

141. Choose the correctly matched pair:-
(a) Inner lining of salivary ducts – Ciliated epithelim
(b) Moist surface of buccal cavity – Glandular epithelium
(c) Tubular parts of nephrons – Cuboidal epithelium
(d) Inner surface of bronchioles – Squamous epithelium

142. In 'S' phase of the cell cycle:
(a) Amount of DNA doubles in each cell.
(b) Amount of DNA remains same in each cell.
(c) Chromosome number is increased.
(d) Amount of DNA is reduced to half in each cell.

143. The motile bacteria are able to move by:
(a) Fimbriae (b) Flagella
(c) Cilia (d) Pili

144. Select the option which is **not correct** with respect to enzyme action:
(a) Substrate binds with enzyme at its active site.
(b) Addition of lot of succinate does not reverse the inhibition of succinic dehydrogenase by malonate.
(c) A non-competitive inhibitor binds the enzyme at a site distinct from that which binds the substrate.
(d) Malonate is a competitive inhibitor of succinic dehydrogenase.

145. Which one of the following is a non - reducing carbohydrate?
(a) Maltose
(b) Sucrose
(c) Lactose
(d) Ribose 5 - phosphate

146. The enzyme recombinase is required at which stage of meiosis:
(a) Pachytene (b) Zygotene
(c) Diplotene (d) Diakinesis

147. The initial step in the digestion of milk in humans is carried out by
(a) Lipase (b) Trypsin
(c) Rennin (d) Pepsin

148. Fructose is absorbed into the blood through mucosa cells of intestine by the process called:
(a) active transport
(b) facilitated transport
(c) simple diffusion
(d) co-transport mechanism

149. Approximately seventy percent of carbon-dioxide absorbed by the blood will be transported to the lungs:
(a) as bicarbonate ions
(b) in the form of dissolved gas molecules
(c) by binding to R.B.C.
(d) as carbamino - haemoglobin

150. Person with blood group AB is considered as universal recipient because he has:
(a) both A and B antigens on RBC but no antibodies in the plasma.
(b) both A and B antibodies in the plasma.
(c) no antigen on RBC and no antibody in the plasma.
(d) both A and B antigens in the plasma but no antibodies.

151. How do parasympathetic neural signals affect the working of the heart?
(a) Reduce both heart rate and cardiac output.
(b) Heart rate is increased without affecting the cardiac output.
(c) Both heart rate and cardiac output increase.
(d) Heart rate decreases but cardiac output increases.

152. Which of the following causes an increase in sodium reabsorption in the distal convoluted tubule?
(a) Increase in aldosterone levels
(b) Increase in antidiuretic hormone levels
(c) Decrease in aldosterone levels
(d) Decrease in antidiuretic hormone levels

153. Select the correct matching of the type of the joint with the example in human skeletal system:

Type of joint	Example
(a) Cartilaginous joint	between frontal and pariental
(b) Pivot joint	between third and fourth cervical vertebrae
(c) Hinge joint	between humerus and pectoral girdle
(d) Gliding joint	between carpals

154. Stimulation of a muscle fiber by a motor neuron occurs at:
(a) the neuromuscular junction
(b) the transverse tubules
(c) the myofibril
(d) the sacroplasmic reticulum

155. Injury localized to the hypothalamus would most likely disrupt:
(a) short - term memory.
(b) co-ordination during locomotion.
(c) executive functions, such as decision making.
(d) regulation of body temperature.

156. Which one of the following statements is **not correct**?
(a) Retinal is the light absorbing portion of visual photo pigments.
(b) In retina the rods have the photopigment rhodopsin while cones have three different photopigments.
(c) Retinal is a derivative of Vitamin C.
(d) Rhodopsin is the purplish red protein present in rods only.

157. Identify the hormone with its **correct** matching of source and function:
(a) Oxytocin - posterior pituitary, growth and maintenance of mammary glands.
(b) Melatonin - pineal gland, regulates the normal rhythm of sleepwake cycle.
(c) Progesterone - corpus-luteum, stimulatiuon of growth and activities of female secondary sex organs.
(d) Atrial natriuretic factor - ventricular wall increases the blood pressure.

158. Fight-or-flight reactions cause activation of:
(a) the parathyroid glands, leading to increased metabolic rate.
(b) the kidney, leading to suppression of reninangiotensin-aldosterone pathway.
(c) the adrenal medulla, leading to increased secretion of epinephrine and norepinephrene.
(d) the pancreas leading to a reduction in the blood sugar levels.

159. The shared terminal duct of the reproductive and urinary system in the human male is:
(a) Urethra (b) Ureter
(c) Vas deferens (d) Vasa efferentia

160. The main function of mammalian corpus luteum is to produce:
(a) estrogen only
(b) progesterone
(c) human chorionic gonadotropin
(d) relaxin only

161. Select the correct option describing gonadotropin activity in a normal pregnant female:
(a) High level of FSH and LH stimulates the thickening of endometrium.
(b) High level of FSH and LH facilitate implantation of the embryo.
(c) High level of hCG stimulates the synthesis of estrogen and progesterone.
(d) High level of hCG stimulates the thickening of endometrium.

162. Tubectomy is a method of sterilization in which:
(a) small part of the fallopian tube is removed or tied up.
(b) ovaries are removed surgically.
(c) small part of vas deferens is removed or tied up.
(d) uterus is removed surgically.

163. Which of the following is a hormone releasing Intra Uterine Device (IUD)?
(a) Multiload 375 (b) LNG-20
(c) Cervical cap (d) Vault

164. Assisted reproductive technology, IVF involves transfer of:
(a) Ovum into the fallopian tube.
(b) Zygote into the fallopian tube.
(c) Zygote into the uterus.
(d) Embryo with 16 blastomeres into the fallopian tube.

165. A man whose father was colour blind marries a woman who had a colour blind mother and normal father. What percentage of male children of this couple will be colour blind?
    (a) 25% (b) 0%
    (c) 50% (d) 75%
166. In a population of 1000 individuals 360 belong to genotype AA, 480 to Aa and the remaining 160 to aa. Based on this data, the frequency of allele A in the population is:
    (a) 0.4 (b) 0.5
    (c) 0.6 (d) 0.7
167. A human female with Turner's syndrome:
    (a) has 45 chromosomes with XO.
    (b) has one additional X chromosome.
    (c) exhibits male characters.
    (d) is able to produce children with normal husband.
168. Select the correct option:

	Direction of RNA synthesis	Direction of reading of the template DNA strand
(a)	5′—3′	3′—5′
(b)	3′—5′	5′—3′
(c)	5′—3′	5′—3′
(d)	3′—5′	3′—5′

169. Commonly used vectors for human genome sequencing are:
    (a) T-DNA
    (b) BAC and YAC
    (c) Expression Vectors
    (d) T/A Cloning Vectors
170. Forelimbs of cat, lizard used in walking; forelimbs of whale used in swimming and forelimbs of bats used in flying are an example of:
    (a) Analogous organs
    (b) Adaptive radiation
    (c) Homologous organs
    (d) Convergent evolution
171. Which one of the following are analogous structures?
    (a) Wings of Bat and Wings of Pigeon.
    (b) Gills of Prawn and Lungs of Man.
    (c) Thorns of *Bougainvillea* and Tendrils of *Cucurbita*
    (d) Flippers of Dolphin and Legs of Horse
172. Which is the particular type of drug that is obtained from the plant whose one flowering branch is shown below?

    (a) Hallucinogen (b) Depressant
    (c) Stimulant (d) Pain - killer
173. At which stage of HIV infection does one usually show symptoms of AIDS:-
    (a) Within 15 days of sexual contact with an infected person.
    (b) When the infected retro virus enters host cells.
    (c) When HIV damages large number of helper T-Lymphocytes.
    (d) When the viral DNA is produced by reverse transcriptase.
174. To obtain virus - free healthy plants from a diseased one by tissue culture technique, which part/parts of the diseased plant will be taken:
    (a) Apical meristem only
    (b) Palisade parenchyma
    (c) Both apical and axillary meristems
    (d) Epidermis only
175. What gases are produced in anaerobic sludge digesters?
    (a) Methane and CO_2 only
    (b) Methane, Hydrogen Sulphide and CO_2
    (c) Methane, Hydrogen Sulphide and O_2
    (d) Hydrogen Sulphide and CO_2
176. Just as a person moving from Delhi to Shimla to escape the heat for the duration of hot summer, thousands of migratory birds from Siberia and other extremely cold northern regions move to:
    (a) Western Chat
    (b) Meghalaya
    (c) Corbett National Park
    (d) Keoladeo National Park

177. Given below is a simplified model of phosphorus cycling in a terrestrial ecosystem with four blanks (A-D). Identify the blanks:-

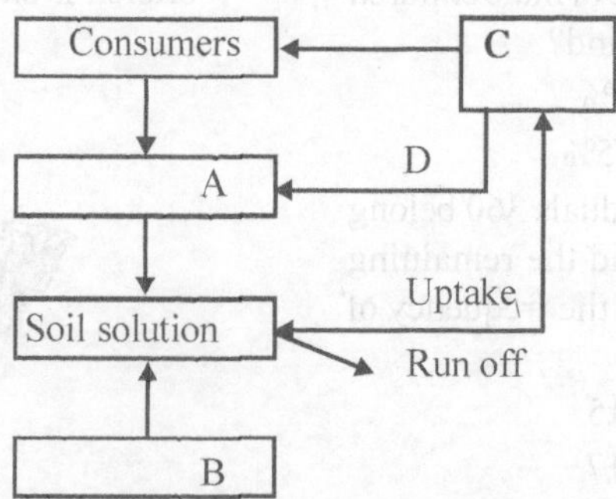

	A	B	C	D
(a)	Rock minerals	Detritus	Litter fall	Producers
(b)	Litter fall	Producers	Rock minerals	Detritus
(c)	Detritus	Rock minerals	Producer	Litter fall
(d)	Producers	Litter fall	Rock minerals	Detritus

178. Given below is the representation of the extent of global diversity of *invertebrates*. What groups the four portions (A-D) represent respectively?

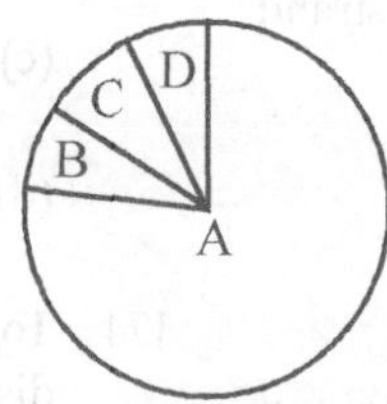

	A	B	C	D
(a)	Insects	Crustaceans	Other animal groups	Molluscs
(b)	Crustaceans	Insects	Molluscs	Other animal groups
(c)	Molluscs	Other animal groups	Crustaceans	Insects
(d)	Insects	Molluscs	Crustaceans	Other animal groups

179. A scrubber in the exhaust of a chemical industrial plant removes:-
(a) gases like sulphur dioxide
(b) particulate matter of the size 5 micrometer or above
(c) gases like ozone and methane
(d) particularte matter of the size 2.5 micrometer or less

180. If 20 J of energy is trapped at producer level, then how much energy will be available to peacock as food in the following chain?
plant $\rightarrow$ mice $\rightarrow$ snake $\rightarrow$ peacock:
(a) 0.02 J
(b) 0.002 J
(c) 0.2 J
(d) 0.0002 J

HINTS & SOLUTIONS

PHYSICS

1. (d) Force = mass × acceleration

$\Rightarrow \quad [\text{Mass}] = \left[\dfrac{\text{force}}{\text{acceleration}}\right]$

$= \left[\dfrac{\text{force}}{\text{velocity / time}}\right] = [\text{F V}^{-1}\text{ T}]$

2. (a) Horizontal range $= \dfrac{u^2 \sin 2\theta}{g}$ so $g \propto u^2$

or $\dfrac{g_{planet}}{g_{earth}} = \dfrac{(u_{planet})^2}{(u_{earth})^2}$

Therefore $g_{planet} = \left(\dfrac{3}{5}\right)^2 (9.8\ \text{m/s}^2)$

$= 3.5\ \text{m/s}^2$

3. (d) $\vec{v}_{av} = \dfrac{\Delta\vec{r}\ (\text{displacement})}{\Delta t\ (\text{time taken})}$

$= \dfrac{(13-2)\hat{i} + (14-3)\hat{j}}{5-0} = \dfrac{11}{5}(\hat{i}+\hat{j})$

4. (c) Acceleration

$= \dfrac{\text{Net force in the direction of motion}}{\text{Total mass of system}}$

$= \dfrac{m_1 g - \mu(m_2 + m_3)g}{m_1 + m_2 + m_3} = \dfrac{g}{3}(1 - 2\mu)$

$(\because\ m_1 = m_2 = m_3 = m \text{ given})$

5. (c) Change in momentum,

$\Delta p = \int F dt$

= Area of F-t graph

= ar of Δ – ar of ▭ + ar of ▭

$= \dfrac{1}{2} \times 2 \times 6 - 3 \times 2 + 4 \times 3$

= 12 N-s

6. (a) Let upthrust of air be F_a then

For downward motion of balloon

$F_a = mg - ma$

$mg - F_a = ma$

For upward motion

$F_a - (m - \Delta m)g = (m - \Delta m)a$

Therefore $\Delta m = \dfrac{2ma}{g + a}$

7. (b) By conservation of linear momentum

$2mv_1 = \sqrt{2}mv \Rightarrow v_1 = \dfrac{v}{\sqrt{2}}$

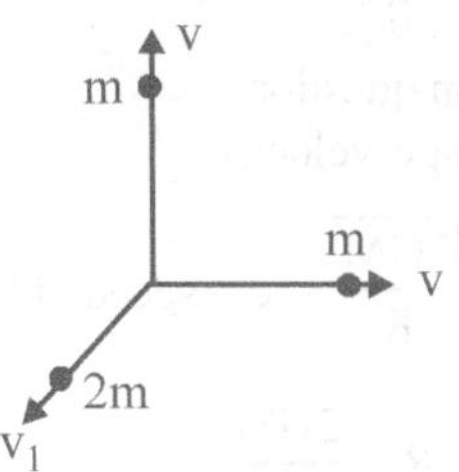

As two masses of each of mass m move perpendicular to each other.

Total KE generated

$= \dfrac{1}{2}mv^2 + \dfrac{1}{2}mv^2 + \dfrac{1}{2}(2m)v_1^2$

$= mv^2 + \dfrac{mv^2}{2} = \dfrac{3}{2}mv^2$

8. (c) Displacement, x = A cos (ωt) (given)

Velocity, $v = \dfrac{dx}{dt} = -A\omega \sin(\omega t)$

Acceleration, $a = \dfrac{dv}{dt} = -Ag\omega^2 \cos(\omega t)$

Hence graph (c) correctly dipicts the variation of a with t.

9. (d) Here α = 2 revolutions/s² = 4π rad/s² (given)

$I_{cylinder} = \dfrac{1}{2}MR^2 = \dfrac{1}{2}(50)(0.5)^2$

$= \dfrac{25}{4}\ \text{Kg-m}^2$

As τ = Iα so TR = Iα

$\Rightarrow T = \dfrac{I\alpha}{R} = \dfrac{\left(\frac{25}{4}\right)(4\pi)}{(0.5)} N = 50\pi N = 157\ N$

10. (a) For solid sphere rolling without slipping on inclined plane, acceleration

$a_1 = \dfrac{g \sin\theta}{1 + \dfrac{K^2}{R^2}}$

For solid sphere slipping on inclined plane without rolling, acceleration

$a_2 = g \sin\theta$

Therefore required ratio $= \frac{a_1}{a_2}$

$$= \frac{1}{1+\frac{K^2}{R^2}} = \frac{1}{1+\frac{2}{5}} = \frac{5}{7}$$

11. (c) From question,
Escape velocity

$$= \sqrt{\frac{2GM}{R}} = c = \text{speed of light}$$

$$\Rightarrow \quad R = \frac{2GM}{c^2}$$

$$= \frac{2 \times 6.6 \times 10^{-11} \times 5.98 \times 10^{24}}{(3 \times 10^8)^2} m$$

$$= 10^{-2} m$$

12. (a) First when $(r < R)$ $E \propto r$ and then when $r > R$

$E \propto \frac{1}{r^2}$. Hence graph (c) correctly dipicts.

13. (b) As $Y = \frac{\frac{F}{A}}{\frac{\Delta l}{l}} \Rightarrow \Delta l = \frac{Fl}{AY}$

But $V = Al$ so $A = \frac{V}{l}$

Therefore $\Delta l = \frac{Fl^2}{VY} \propto l^2$

Hence graph of Δl versus l^2 will give a straight line.

14. (c) As surface area decreases so energy is released.
Energy released $= 4\pi R^2 T[n^{1/3} - 1]$
where $R = n^{1/3} r$

$$= 4\pi R^3 T\left[\frac{1}{r} - \frac{1}{R}\right] = 3VT\left[\frac{1}{r} - \frac{1}{R}\right]$$

15. (d) According to the principle of calorimetry.
Heat lost = Heat gained
$mL_v + ms_w \Delta\theta = m_w s_w \Delta\theta$

$\Rightarrow \quad m \times 540 + m \times 1 \times (100 - 80)$
$= 20 \times 1 \times (80 - 10)$

$\Rightarrow m = 2.5$ g
Therefore total mass of water at 80°C
$= (20 + 2.5)$ g $= 22.5$ g

16. (a) Let the temperature of surroundings be θ_0
By Newton's law of cooling

$$\frac{\theta_1 - \theta_2}{t} = k\left[\frac{\theta_1 + \theta_2}{2} - \theta_0\right]$$

$$\Rightarrow \quad \frac{70 - 60}{5} = k\left[\frac{70 + 60}{2} - \theta_0\right]$$

$\Rightarrow \quad 2 = k[65 - \theta_0]$ ...(i)

Similarly,

$$\frac{60 - 54}{5} = k\left[\frac{60 + 54}{2} - \theta_0\right]$$

$\Rightarrow \quad \frac{6}{5} = k[57 - \theta_0]$ ...(ii)

By dividing (i) by (ii) we have

$$\frac{10}{6} = \frac{65 - \theta_0}{57 - \theta_0} \quad \Rightarrow \quad \theta_0 = 45^\circ$$

17. (c) For isothermal process $P_1V_1 = P_2V_2$

$$\Rightarrow \quad PV = P_2(2V) \Rightarrow P_2 = \frac{P}{2}$$

For adiabatic process

$$P_2V_2^\gamma = P_3V_3^\gamma$$

$$\Rightarrow \left(\frac{P}{2}\right)(2v)^\gamma = P_3 16v)^\gamma$$

$$\Rightarrow \quad P_3 = \frac{3}{2}\left(\frac{1}{8}\right)^{5/3} = \frac{P}{64}$$

18. (d) Work done by the system in the cycle
= Area under P-V curve and V-axis

$$= \frac{1}{2}(2P_0 - P_0)(2V_0 - V_0) +$$

$$\left[-\left(\frac{1}{2}\right)(3P_0 - 2P_0)(2V_0 - V_0)\right]$$

$$= \frac{P_0V_0}{2} - \frac{P_0V_0}{2} = 0$$

19. (b) Mean free path $\lambda_m = \frac{1}{\sqrt{2}\pi d^2 n}$

where d = diameter of molecule and d = 2r

$$\therefore \quad \lambda_m \propto \frac{1}{r^2}$$

20. (a) Total length of string $\ell = \ell_1 + \ell_2 + \ell_3$
(As string is divided into three segments)

But frequency $\propto \frac{1}{\text{length}}$ $\left(\because f = \frac{1}{2\ell}\sqrt{\frac{T}{m}}\right)$

so $\frac{1}{n} = \frac{1}{n_1} + \frac{1}{n_2} + \frac{1}{n_3}$

21. (d) In case of closed organ pipe frequency,

$$f_n = (2n+1)\frac{v}{4l}$$

for $n = 0, f_0 = 100$ Hz
$n = 1, f_1 = 300$ Hz
$n = 2, f_2 = 500$ Hz
$n = 3, f_3 = 700$ Hz
$n = 4, f_4 = 900$ Hz
$n = 5, f_5 = 1100$ Hz
$n = 6, f_6 = 1300$ Hz

Hence possible natural oscillation whose frequencies < 1250 Hz = 6(n = 0, 1, 2, 3, 4, 5)

22. (c) According to Doppler's effect
Apparent frequency

$$n' = n\left(\frac{v + v_0}{v + v_s}\right) = 1392\left(\frac{343 + 10}{343 + 5}\right).$$

$= 1412$ Hz

23. (c) Electric field, $E \propto \frac{1}{K}$

As $K_1 < K_2$ so $E_1 > E_2$
Hence graph (c) correctly dipicts the variation of electric field E with distance d.

24. (b) Due to conducting sphere
At centre, electric field E = 0

And electric potential $V = \frac{Q}{4\pi \epsilon_0 R}$

25. (d) $\vec{E} = -\frac{\partial V}{\partial x}\hat{i} - \frac{\partial V}{\partial y}\hat{j} - \frac{\partial V}{\partial z}\hat{k}$

$= -[(6-8y)\hat{i} + (-8x - 8 + 6z)\hat{j} + (6y)\hat{k}]$

At (1, 1, 1), $\vec{E} = 2\hat{i} + 10\hat{j} - 6\hat{k}$

$\Rightarrow$ $(\vec{E}) = \sqrt{2^2 + 10^2 + 6^2} = \sqrt{140} = 2\sqrt{35}$

$\therefore$ $F = q\vec{E} = 2 \times 2\sqrt{35} = 4\sqrt{35}$

26. (b) Total resistance R = (0.5 Ω/km) × (150 km) = 75 Ω
Total voltage drop = (8 V/km) × (150 km) = 1200 V

Power loss = $\frac{(\Delta V)^2}{R} = \frac{(1200)^2}{75}$ W

= 19200 W = 19.2 kW

27. (b) This is a balanced wheatstone bridge condition,

$$\frac{5}{R} = \frac{\ell_1}{100 - \ell_1} \text{ and } \frac{5}{R/2} = \frac{1.6\ell_1}{100 - 1.6\ell_1}$$

$\Rightarrow$ R = 15 Ω

28. (c) Internal resistance of the cell,

$$r = \left(\frac{E - V}{V}\right)R = \left(\frac{\ell_1 - \ell_2}{\ell_2}\right)R$$

$$= \left(\frac{3 - 2.85}{2.85}\right) \times (9.5)\ \Omega = 0.5\ \Omega$$

29. (c) Net magnetic dipole moment = $2\,M\cos\frac{\theta}{2}$

As value of $\cos\frac{\theta}{2}$ is maximum in case (c) hence net magnetic dipole moment is maximum for option (c).

30. (c) As 0.2% of main current passes through the galvanometer hence $\frac{998}{1000}I$ current through the shunt.

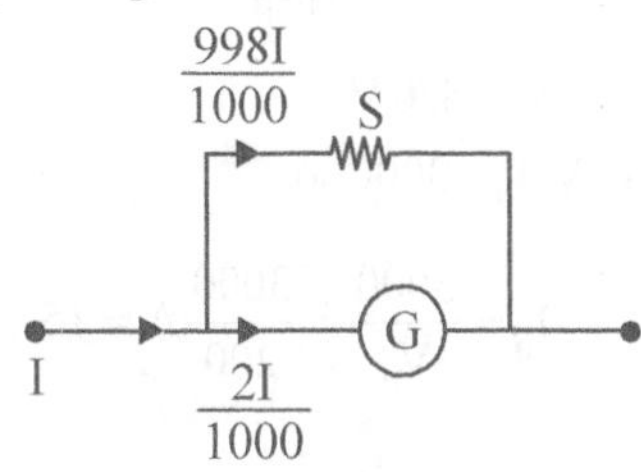

$$\left(\frac{2I}{1000}\right)G = \left(\frac{998I}{1000}\right)S$$

$\Rightarrow$ $S = \frac{G}{499}$

Total resistance of Ammeter

$$R = \frac{SG}{S+G} = \frac{\left(\frac{G}{499}\right)G}{\left(\frac{G}{499}\right) + G} = \frac{G}{500}$$

31. (d) Net magnetic field, $B = \sqrt{B_1^2 + B_2^2}$

$$= \sqrt{\left(\frac{\mu_0 I_1}{2\pi d}\right)^2 + \left(\frac{\mu_0 I_2}{2\pi d}\right)^2}$$

$$\left(\because B_1 = \frac{\mu_0 I_1}{2\pi d} \text{ and } B_2 = \frac{\mu_0 I_2}{2\pi d}\right)$$

$$= \frac{\mu_0}{2\pi d}\sqrt{I_1^2 + I_2^2}$$

32. (d) Rate of decreasing of area of semicircular ring $= \frac{dA}{dt} = (2r)V$

From Faraday's law of electromagnetic induction

$$e = -\frac{d\theta}{dt} = -B\frac{dA}{dt} = -B(2rV)$$

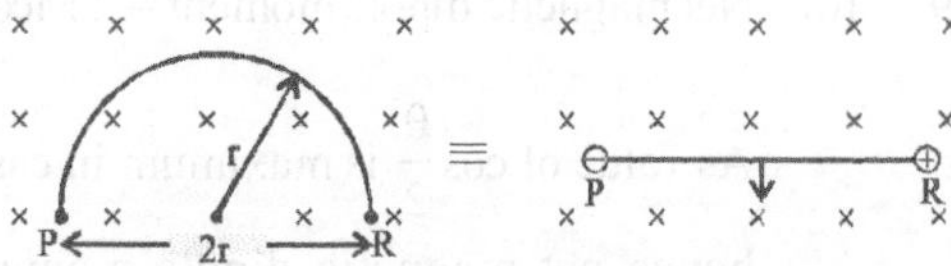

As induced current in ring produces magnetic field in upward direction hence R is at higher potential.

33. (b) Efficiency $\eta = \frac{V_s I_s}{V_p I_p} \Rightarrow 0.9 = \frac{V_s(6)}{3 \times 10^3}$

$\Rightarrow V_s = 450$ V

As $V_p I_p = 3000$ so

$$I_p = \frac{3000}{V_p} = \frac{3000}{200} A = 15A$$

34. (b) Average force $F_{av} = \frac{\Delta p}{\Delta t} = \frac{2IA}{c}$

($\because$ Power = F.V)

$$= \frac{2 \times 25 \times 10^4 \times 15 \times 10^{-4}}{3 \times 10^8}$$

$= 2.50 \times 10^{-6}$ N

35. (d) Given: D = 2m; d = 1 mm = 1×10^{-3} m

$\lambda = 600$ nm $= 600 \times 10^{-6}$ m

Width of central bright fringe (= 2β)

$$= \frac{2\lambda D}{d} = \frac{2 \times 600 \times 10^{-6} \times 2}{1 \times 10^{-3}} m$$

$= 2.4 \times 10^{-3}$ m = 2.4 mm

36. (c) For path difference λ, phase difference = 2π rad.

For path difference $\frac{\lambda}{4}$, phase difference

$$= \frac{\pi}{2} \text{ rad.}$$

As $K = 4I_0$ so intensity at given point where path difference is $\frac{\lambda}{4}$

$$K' = 4I_0 \cos^2\left(\frac{\pi}{4}\right) \left(\cos\frac{\pi}{4} = \cos 45^\circ\right)$$

$$= 2I_0 = \frac{K}{2}$$

37. (d) Magnifying power of microscope

$$= \frac{LD}{f_0 f_e} \propto \frac{1}{f_0}$$

Hence with increase f_0 magnifyig power of microscope decreases.

Magnifying power of telescope $= \frac{f_0}{f_e} \propto f_0$

Hence with increase f_0 magnifying power of telescope increases.

38. (b)

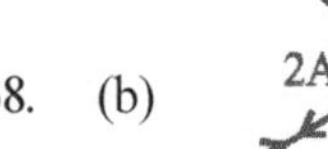
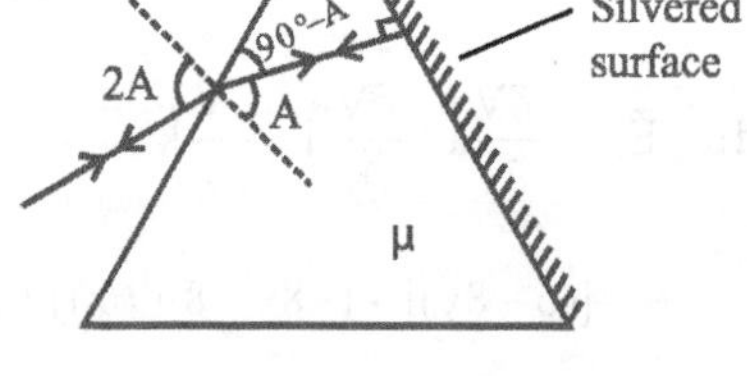

According to Snell's law $\mu = \frac{\sin i}{\sin r}$

$\Rightarrow (1) \sin 2A = (\mu) \sin A \Rightarrow \mu = 2 \cos A$

39. (b) According to Einstein's photoelectric equation, $hv = \phi_0 + K_{max}$

We have

$h\nu = \phi_0 + 0.5$ ...(i)

and $1.2h\nu = \phi_0 + 0.8$ ...(ii)

Therefore, from above two equations $\phi_0 = 1.0$ eV.

40. (b) As we know $\lambda = \frac{h}{P}$

$= \frac{h}{\sqrt{2mK}}$

$(\because\ P = \sqrt{2mKE})$

or $\frac{\lambda_1}{\lambda_2} = \sqrt{\frac{K_2}{K_1}} = \sqrt{\frac{16K}{K}} = \frac{4}{1}$

Therefore the percentage chanbe in de-Broglie wavelength $= \frac{1-4}{4} \times 100 = -75\%$

41. (c) For the $\lambda = 975$ Å

$\frac{1}{\lambda} = R\left(\frac{1}{n_1^2} - \frac{1}{n_2^2}\right)$

where R is the Rydberg constant

Solving we get $n_2 = n = 4$ ($\because n_1 = 1$ ground state)

Therefore number of spectral lines

$= \frac{n(n-1)}{2} = \frac{4(4-1)}{2} = 6$

42. (d) BE of $_2He^4 = 4 \times 7.06 = 28.24$ MeV

BE of $^7_3Li = 7 \times 5.60 = 39.20$ MeV

$^7_3Li + ^1_1H \rightarrow {_2He^4} + {_2He^4} + Q$

39.20 $\quad$ 28.24×2(= 56.48 MeV)

Therefore, Q = 56.48 – 39.20 = 17.28 MeV.

43. (c) As $\frac{N_x}{N_y} = \frac{1}{7}$ (Given)

$\Rightarrow \frac{N_x}{N_x + N_y} = \frac{1}{8} = \left(\frac{1}{2}\right)^3$

Therefore, age of the rock

$t = 3T_{1/2} = 3 \times 1.4 \times 10^9$ yrs $= 4.2 \times 10^9$ yrs.

44. (a) The given graph represents V-I characteristics of solar cell.

45. (d) The barrier potential of a p-n junction depends on amount of doping, type of semiconductor material and temperature.

CHEMISTRY

46. (a) Given: $n = 3, l = 1, m = 0$

Hence orbital is 3*p*

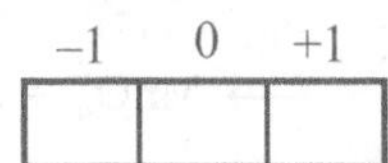

hence the number of orbital identified by m = 0 can be one only.

47. (d) $E = \frac{hc}{\lambda} = \frac{6.63 \times 10^{-34} \times 3 \times 10^8}{45 \times 10^{-9}}$

$= 4.42 \times 10^{-18}$ J

48. (c) According to Avogadro's law "At same temperature and pressure Volume ∝ no. of moles"

$n_{H_2} = \frac{w}{2};\ n_{O_2} = \frac{w}{32};\ n_{CH_4} = \frac{w}{16}$

$\because V_{H_2} : V_{O_2} : V_{CH_4} = n_{H_2} : n_{O_2} : n_{CH_4}$

$= \frac{w}{2} : \frac{w}{32} : \frac{w}{16} = 16 : 1 : 2$

49. (d) For bcc lattice body diagonal $= a\sqrt{3}$

The distance between the body centered atom and one corner atom in cube will be

$= \frac{\sqrt{3}a}{2}$

50. (d) Tyndall effect is optical property, whereas other properties are electrical properties. Hence dependent on the charge on colloids.

51. (c) Na_2CO_3 is a salt of strong base (NaOH) and weak acid (H_2CO_3). On hydrolysis this salt will produce strongly basic solution. i.e. pH will be highest (pH > 7) for this sotluion. Others are combination of

KCl = Strong acid + Strong base → neutral solution (pH ≈ 7)

NaCl = Strong acid + Strong base → neutral solution (pH ≈ 7)

$CuSO_4$ = Strong acid + wake base → Acidic solution (pH < 7)

52. (c) Colligative properties ∝ no. of particles. Since $Al_2(SO_4)_3$ contains maximum number of particles, hence will have the largest value of freezing point depression.

53. (a)

	H_2	+	Cl_2	⟶	2HCl
t = 0	22.4 lit		11.2 lit		0
t = 0 or	1 mole		0.5 mole		0
at time t	(1 – 0.5) = 0.5				0.5 × 2 = 1 mole

54. (c) $\overset{(+6)}{Mn}O_4^{2-} \longrightarrow \overset{(+7)}{Mn}O_4^- + e^-$

0.1 mole

Quantity of electricity required
$= 0.1F = 0.1 \times 96500 = 9650\,C$

55. (b) $\Delta G = -2.303 \log K$

here $K = [Ag^+]^2[CO_3^{--}] = K_{sp}$

$\therefore \quad 63.3 \times 10^{-3} = -2.303 \times 8.314 \times 298 \log K_{sp}$

$\therefore \quad \log K_{sp} = -\dfrac{63.3 \times 10^{-3}}{5705.8} = -11.09$

$\therefore \quad K_{sp} = \text{Antilog}(-11.09) = 8 \times 10^{-12}$

56. (d) $w_{O_2} = n_{O_2} \times 32$

$w_{O_2} = \dfrac{5600}{22400} \times 32 = 8g$

= 1 equivalent of O_2

= 1 equivalent of Ag = 108

57. (b) For adsorption $\Delta S < 0$ and for a spontaneous change $\Delta G = -$ ve

hence ΔH should be highly negative which is clear from the equation

$\Delta G = \Delta H - T\Delta S$

$= -\Delta H - T(-\Delta S) = -\Delta H + T\Delta S$

So if ΔH is highly negative ΔG will also be (– ve)

58. (d) Given reaction is exothermic reaction. Hence according to Le-Chatelier's principle low temperature favours the forward reaction and on increasing pressure equilibrium will shift, towards lesser number of moles i.e. forward direction.

59. (b) Given $\Delta U = 2.1$ k cal., $\Delta S = 20$ cal. K^{-1}

T = 300 K

$\because \quad \Delta H = \Delta U + \Delta n_g RT$

Putting the values given in the equation

$\Delta H = 2.1 + 2 \times \dfrac{2}{1000} \times 300$

$= 2.1 + 1.2 = 3.3$ k cal.

Now, $\Delta G = \Delta H - T\Delta S$

$= 3.3 - 300 \times \dfrac{20}{1000} = -2.7$ k cal

60. (a) In exothermic reactions on increasing temperature value of K_p decreases

$\therefore \; K_p > K_p'$ (Assuming $T_1 < T_2$)

61. (N) All answers are incorrect.

62. (a) Initially $Mg \;+\; \frac{1}{2}O_2 \longrightarrow MgO$

Mg: 1g; O_2: 0.56

or $\dfrac{1}{24}$ mole, $\dfrac{0.56}{32}$ mole

0.0416 mole, 0.0175 mole

$(0.0416 - 2 \times 0.0175)(2 \times 0.0175)$ mole
$= 0.0066$ mole

$\therefore$ Mass of Mg $= 0.0066 \times 24$
$= 0.158 \approx 0.16g$

63. (c) Both are reducing agents.

64. (b) $Be^{2+} = (4 - 2) = 2$

is isoelectronic with Li^+ $(3 - 1 = 2)$

Since both have same number of electrons in their outermost shell.

65. (c) Dipole moment of $NH_3 > NF_3$

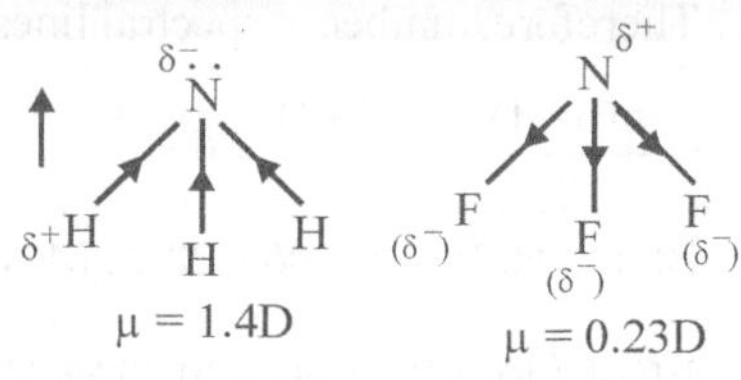

(F is more electronegative than N)

66. (b) Hybridization of $NO_3^- = \frac{1}{2}(5 + 0 + 1 - 0)$

$= \dfrac{6}{2} = 3 = sp^2$ hence geometry is trigonal planner.

NO_2^- (nitrite ion) also has sp^2 hybridization and gives a trigonal planner geometry but because there are only two outer atoms, the molecular geometry is bent with ∠120° bond angles.

67. (a) The weaking of M—H bond with increase in size of M (where M = S, Se, Te) explains the acid character of hydrides. Since on moving down the group atomic size increases hence bond length increases and hence removal tendency of H also increases.

68. (c) (i)

Reduction

$$\overset{-1}{H_2O_2} + O_3 \longrightarrow \overset{-2}{H_2O_2} + 2O_2$$

Oxidation

(ii)

Oxidation

$$\overset{-1}{H_2O_2} + Ag_2O \longrightarrow 2Ag + H_2\overset{-2}{O} + 2O_2$$

Reduction

Hence in both the reactions H_2O_2 is acting as an oxidising agent.

69. (c) Aspartame is stable under cold conditions.

70. (c) The structure of CrO_5 is

Hence CrO_5 has two proxy linkage.

Now suppose the oxi. no. of Cr is x then

$x + (-1 \times 4) + (-2) = 0$

$\therefore \; x = +6$

71. (b) $2\,KMnO_4 + 3H_2SO_4 + 5H_2O_2 \longrightarrow K_2SO_4 + 2MnSO_4 + 8H_2O + 5O_2$

i.e. Mn^{++} ion and O_2.

72. (b) Due to d^5 configuration CFSE is zero.

73. (b) Magnetic moment

$\mu = \sqrt{n(n+2)}$

$2.83 = \sqrt{n(n+2)}$

On solving n = 2

Ni^{2+} have two unpaired electron.

74. (b)

75. (a) The shape of *f*-orbitals is very much diffused and they have poor shielding effect. The effective nuclear charge increases which causes the contractiion in the size of electron charge cloud. This contraction in size is quite regular and known as lanthanoid contraction.

76. (d) $C_6H_5-N_2Cl \xrightarrow[H^+]{C_6H_5-NH_2} C_6H_5-N=N-C_6H_4-NH_2$

77. (b) Arene diazonium salts are most stable among the given options because of the dispersal of +ve charge on the benzene ring due to resonance.

78. (d) Glucose reacts with hydroxyl amine to form an oxime.

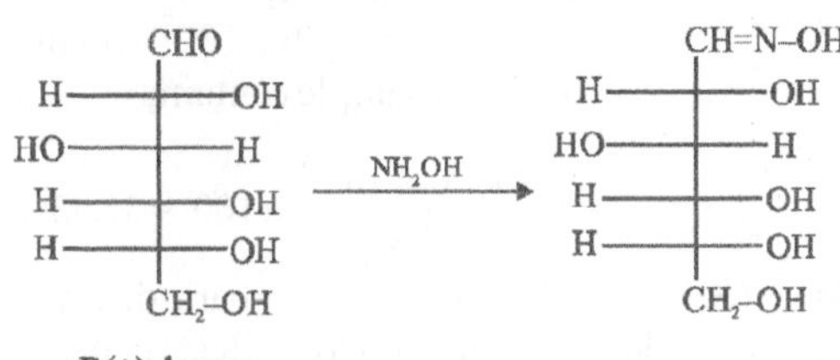

D(+)glucose

79. (c) Adrenaline is a hormone produced by adrenal glands during high stress or exciting situatiions. This powerful hormone is part of the human body's acute stress response system, also called the fight or flight response.

80. (d) Thermosetting polymers undergo chemical changes when heated and set to hard mass when cooled e.g. Bakelite.

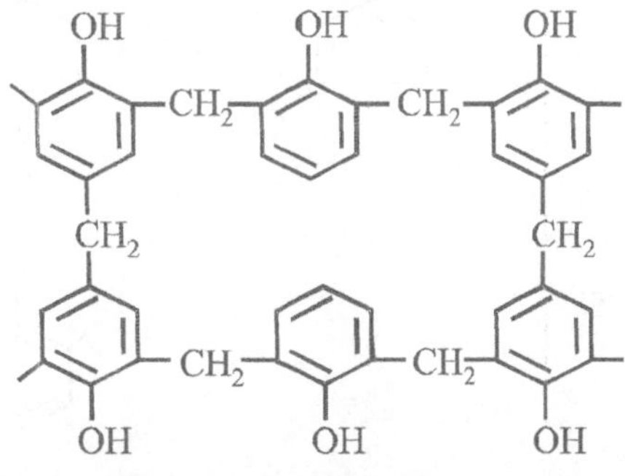

Bakelite

81. (c)

$$n\,HO.CH_2CH_2OH + n\,HOOC-C_6H_4-COOH \xrightarrow{\Delta}$$

Ethylene glycol Terephthalic acid

$$-\!\!(OCH_2CH_2-O-\overset{O}{\overset{\|}{C}}-C_6H_4-\overset{O}{\overset{\|}{C}})\!\!-_n$$

Terylene

It is resistant to mineral and organic acids. It is used for blending with wool to provide better crease, in safety helmets and aircraft battery boxes.

82. (d) The oxidised hydrocarbons and ozone in presence of humidity cause photochemical smong.

Hydrocarbons + O_2, NO_2, NO, O, O_3 → Peroxides, formaldehyde, peroxyacetylnitrate (PAN), acrodein etc. Hence chlorofluoro carbons are not common component of photochemical smog.

83. (a) 10 ml, $1M H_2SO_4$ = 20 ml, $1M NH_3$

∵ wt of N in one mole NH_3 = 14

∴ 20×10^{-3} mol $NH_3 \longrightarrow 20 \times 10^{-3} \times 14$ nitrogen

∴ 0.75 g of sample contains

$$= \frac{14 \times 20 \times 10^{-3}}{0.75} \times 100 = 37.33\%$$

84. (c) Methyl group is ortho para directing but due to steric hinderce effect, generated by two CH_3 groups substitution will not take place on position (I). Hence only two products are possible.

(m-xylene with positions III, I, II marked; CH_3, CH_3)

i.e. m-xylene $\xrightarrow[FeBr_2]{Br_2}$ (4-bromo-m-xylene: CH_3, CH_3, Br) + (2-bromo product: Br, CH_3, CH_3)

85. (N) Out of the given four compounds only (iv) compound is chiral and hence only this compound will undergo racemisation. Therefore none of the given option is correct.

86. (b) Phenols react with alkyl halides in alkaline medium to form ethers. Therefore,

$$C_6H_5OH \xrightarrow[(ii)\ CH_3I]{(i)\ NaOH} C_6H_5OCH_3$$

87. (c) *o*-nitrophenol will not be soluble in $NaHCO_3$. Due to intramolecular hydrogen bonding hydrogen on OH is strongly bound. So it can not be have as an acid and can not react with sodium bicarbonate.

88. (d) Any substituent in the carbonyl compound that increases the positive charge on the carbonyl carbon will increase reactivity towards nucleophilic addition.

$-NO_2$ shows $-M$ effect hence

(p-nitrobenzaldehyde: CHO, NO_2) is most reactive towards nucleophilic addition reaction.

89. (a)

$$CH_3-CH_2-CH=CH_2 \xrightarrow[\text{(Peroxide effect)}]{HBr/H_2O_2}$$

$$\underset{(Y)}{CH_3-CH_2-CH_2-CH_2Br}$$

↓

$$\underset{(Z)}{CH_3-(CH_2)_3-O-CH_2-CH_3}$$

90. (b) The combustion reaction of ethylene is

$$C_2H_2 + \frac{5}{2}O_2 \rightarrow 2CO_2 + H_2O$$

Both $HC\equiv CH$ and CO_2 have *sp* hybridization.

BIOLOGY

Botany

91. (d) In *Spirogyra,* sexual reproduction occurs through conjugation. Gametes are non-flagellated morphologically similar. But physiologically different (isogamy with physiological anisogamy).

92. (a) **Five kingdom system of classification** was proposed by **R.H. Whittaker** (1969). The five kingdom classification is based on the following criteria :
   - Complexity of cell structure – Prokaryotes or Eukaryotes
   - Complexity of organisms body – Unicellular or Multicellular
   - Mode of obtaining nutrition – Autotrophs or Heterotrophs
   - Phylogenetic relationships

93. (b) Several mushrooms such as *Amanita muscaria*, *Psilocybe mexicana* and *Panaeolus* spp. secrete hallucinogenic substances like psilocybin and psilocin. These substances may destroy brain cells and perception power of human beings.

94. (a) Archaebacteria differ from other bacteria in having a different cell wall structure. They lack peptidoglyan in cell wall and possess a monolayer of branched fatty acids attached to glycerol by ether bonds in their cell membranes.

95. (c) *Chara* is a green alga found attached to bottoms of shallow water of ponds, pools and lakes. Male sex organ is called antheridium. Female sex organ is called oogonium. Oogonium is borne at the top of four celled filament.

96. (d) *Sphagnum,* a moss, provides peat that have long been used as fuel. It has the capacity to retain water for long periods and as such used to cover the plant roots during transportation

97. (c) Tomato — edible part is pericarp and placenta.

98. (b) If the margins of sepals or petals overlap one another but not in any particular direction as in *Cassia* and gulmohur, the aestivation is called imbricate.

99. (c) Anatomically fairly old dicotyledonous root is distinguished from the dicotyledonous stem by position of protoxylem. In dicot root the protoxylem is located near the periphery of the vascular cylinder while in dicot stem the protoxylem is located near the centre of vascular bundle *i.e.,* the xylem is endarch.

100. (c) Generally, monocotyledonous seeds (*e.g.,* grasses) are endospermic but some as in orchids are non-endospermic. Mango is a drupe fruit develop from multicarpellary, syncarpous, superior ovary having one or many seeded. A sterile stamen is called a staminode. In monocots, outer covering of endosperm separates the embryo by a proteinous layer called aleurone layer.

101. (b) The walls of vessels (tracheary elements) are lignified and less thicker than tracheids. The lumen is wider. Vessels differ from tracheids in being cell fusions arising through the dissolution of end walls.

102. (d) Potato is an example of edible underground stem.
   - It is differnetiated into nodes and internodes. The 'eyes' on potato tuber indicate axillary buds at the nodes.
   - It bears scaly leaves. The leaves do not store food.
   - Axillary buds arise at the nodes.
   - Does not bear roots.

103. (d) In some bacteria (*e.g., Bacillus subtilis*) the plasma membrane form certain invaginations or in foldings called mesosomes in the cytoplasm. The mesosomes have various functions, viz., respiratory, secretory etc.

104. (b) Microtubule, microfilament and intermediate filaments along with ER form cytoskeleton. Microfilaments are non-living, solid and consists of actin protein. They are 4-6 nm in diameter.

105. (b) The vacuole is bound by a single membrane called tonoplast. It also functions as semipermeable membrane. It segregates vacuolar contents from cytoplasm, allows osmotic entry or exit of water, concentration and storage of nutrients as well as wastes.

106. (c) In M-phase both 4C and 2C of DNA are present in different stages.

107. (a) Centrosome is an organelle usually containing two cylindrical structures called centrioles. The centrioles form the basal body of cilia or flagella. In chloroplast a number of organised flattened membranous sacs called the thylakoids are present in the stroma. Chlorophyll pigments are present in the thylakoids. Each mitochondrion is a double membrane bound structure. The inner membrane forms a number of infoldings called the cristae towards the matrix. The cristae increase the surface area.

108. (b) **Charles Darwin and his son Francis Darwin** observed that the coleoptiles of Oat (*Avena sativa*) and canary grass (*Phalaris canariensis*) responded to unilateral illumination by growing towards the light source **(phototropic curvature or phototropism)**. After a series of experiments, it was concluded that the tip of the coleoptile was the site of production of a substance, that caused the bending of coleoptile.

109. (a) The parts of the plants that show the deficiency symptoms also depend on the mobility of the element in the plant. For element that are actively mobilised within the plants and exported to young developing tissues, the deficiency symptoms tend to appear first in the older tissues. *For example,* nitrogen, potassium and magnesium are visible first in the senescent leaves.

110. (d) **Lactic acid fermentation :** It occurs in lactic acid bacteria (*Lactobacillus*) and muscles (Human). Pyruvic acid produced in glycolysis is reduced by $NADH_2$ to form lactic acid without producing carbon dioxide.

$$\underset{\text{Pyruvic acid}}{CH_3COCOOH} + NADH_2 \xrightarrow[\text{FMN}, Zn^{2+}]{\textit{Lactate dehydrogenase}} \underset{\text{Lactic acid}}{CH_3CHOHCOOH} + NAD$$

111. (a) In *Rhodospirillum*, electron donor is organic compound during photosynthesis.

112. (c) Etiolation is depigmentation of leaf when plant is placed in dark for more than 36 hrs.

113. (a) Abscisic Acid (ABA) is called stress hormone which works in adverse environmental condition when there is low water content in atmosphere or in drought conditions. ABA causes the stomatal closure of leaves due to which the water loss by the plant is minimized.

114. (a) **Geitonogamy** is the transfer of pollen grains in different flowers of same plant.

115. (c) Male gametophyte is highly reduced in angiosperm and is known as pollen grain. It is 2 or 3-celled.

116. (b) Aggregate fruits (Etaerio) develop from the multicarpellary apocarpous ovary. They are of following types; Etaerio of follicles, etaerio of achenes, etaerio of berries, etaerio of drupes.

117. (c) Pollen grains are rich in nutrients therefore used as food supplements. Athletes and race horses use these as tablets to enhance performance.

118. (d) Filiform apparatus helps in the entry of pollen tube into a synergid in ovule. Filiform apparatus is in form of finger like projection comprising a core of micro fibrils enclosed in a sheath. It resembles transfer cells meant for short distance movement of metabolites. It is responsible for the absorption of food from the nucleus.

119. (d) **Exalbuminous (non-endospermic) seeds** usually store reserve food material in cotyledons. In these seeds, the endosperm is used up and not present in mature seeds, *e.g.*, bean, gram and pea.

120. (b) **TMV (Tobacco Mosaic Virus)** is a rod-shaped virus. The rod has a core which contains helically coiled single stranded **RNA.** There is a protective covering of protein called **capsid** around the infective part. Capsid consists of small subunits called **capsomeres** and has antigenic property.

121. (a) Process of copying genetic information from DNA to RNA is called **transcription**. At a time only one DNA strand is being transcribed into RNA. The strand of DNA with polarity 3' $\rightarrow$ 5' act as **template strand** and the DNA strand with polarity 5' $\rightarrow$ 3' act as **coding strand**.

122. (c) Frederick Griffith (in 1928), a British Medical officer described the phenomenon of **bacterial transformation**. He carried out experiment with ***Streptococcus pneumoniae*** (bacterium causing pneumonia) which is used to infect mice. By using S Strain (heat killed) and R strain (live) it was concluded that R strain has been transformed by some material of S strain which makes R strain virulent and enable to synthesize smooth polysachharide.

123. (b) **Epistasis** is the phenomenon of suppression of phenotypic expression of gene by a nonallelic gene which shows its own effect. A dominant epistatic allele suppresses the expression ofa nonallelic gene whether the latter is dominant or recessive. For example, fruit colour of Summer Squash (Cucurbita pepo) is governed by a gene which pruduces yellow colour in dominant state (Y-) and green colour in recessive state (yy).

124. (a) All viruses are nucleoprotein (Nucleic acid + Protein) in the structure. The nucleic acid (DNA and RNA) is genetic material. In a particular virus either DNA or RNA is genetic material never both are present in a virus.

Single stranded RNA or ss RNA - Tobacco mosaic virus (TMV)

Virus envelope is known as capsid. The capsid is composed of protein subunits called capsomere.

125. (a) Mammalian hormones were among the first products prepared in bacteria by r-DNA technology. Human insulin and human growth hormone are earliest examples.

126. (d) PCR is a technique for enzymatically replicating DNA without using a living organism such as *E. coli* or Yeast. It is commonly used in medical and biological research labs for a variety of tasks like detection of hereditary diseases, identification of genetic fingerprints etc.

127. (a) Now a days PCR and RAPD technique are used for the characterisation of *in vitro* clonal propagation in plants.

128. (b) *Chlorella* and *Spirullina* are unicellular algae, rich in proteins and are used as food supplements by space travellers.

129. (c) Plasmids are small extranuclear circular DNAs which carry extrachromosomal genes in bacteria and some fungi. They replicate independently. The best known vectors which are also available commerciallly are pBR322 and pUC-18.

130. (b) In *Ex-situ* conservation the threatened animals and plants are taken out of their natural habitat and are protected in special parks or areas like, **Zoological parks, Wild life safari parks** and **Botanical gardens** etc. The *ex situ* conservation also includes **cryopreservation**, fertilization of eggs *in vitro* and propagation of plants through **'Tissue culture methods'**, preservation of seeds through **Seed banks.**

131. (d) Lichens are very good pollution indicators they do not grow in pollutted areas. They are sensitive to sulphur dioxode pollution.

132. (d) Detrivores, (*e.g.* earthworm) break down detritus into smaller particles. The species that invade a bare area in succession is called pioneer species. The products of ecosystem processes are termed as ecosystem services, *e.g.*, healthy forest ecosystems purify air and water, mitigate droughts and floods etc. Natality refers to number of births during a given period in the population.

133. (c) **Critically Endangered :** The taxon facing very high risk of extinction in the wild can become extinct any moment in the immediate future, e.g. *Sus salvinus* (Pigmy Hog), *Berberis nilghiriensis*.

134. (c) The ozone (O_3) found in upper part of the atmosphere, *i.e.*, Stratosphere, is **Good ozone**, since, it acts as a shield for absorbing UV-radiations from sun.

135. (b) IUCN or WCN maintains a red data book which is a catalogue of threatened plants and animals facing risk of extinction. The IUCN red list (2004) documents the extinction of 784 species (including 338 vertebrates, 359 invertebrates and 87 plants) in the last 500 years.

Zoology

136. (d) Members of Ctenophora, Cephalochordata and Echinodermata are exclusively marine.

137. (b) *Gorgonia* (sea-fan) is an animal. All animals lack cell wall.

138. (b) *Planaria* is a flatworm which possess a high power of regeneration.

139. (b) *Torpedo* is a sluggish fish. It is carnivorous. The prey is first killed by electric shock. The shock can also be harmful for human beings.

140. (c) Tendon is dense regular connective tissue. Adipose tissue is a type of loose connective tissue located mainly beneath the cells. Cartilage is a type of specialised connetive tissue.

141. (c) (i) Inner lining of salivary ducts - Compound epithelum
(ii) Moist surface of buccal cavity - Compound epithelium
(iii) Tubular parts of nephorns - Cuboidal epithelium.
(iv) Inner surface of bronchioles – Ciliated epithelium.

142. (a) During S or synthesis phase, replication or duplication of chromosomal DNA and synthesis of histone proteins takes place. During this time the amount of DNA per cell doubles.

143. (b) Motile bacteria have thin filamentous extensions from their cell wall called flagella.

144. (b) Inhibitions of succinic dehydrogenase by malonate is an example of competitive inhibition. Thus it is reversible reaction. On increasing the substrate (succinate) concentration the effect of inhibitor is removed and Vmax remain same.

145. (b) Sucrose is classified under non reducing sugars because it does not have any free aldehyde or keto group.

146. (a) The enzyme recombinase is required at Pachytene stage of meiosis. It catalyzes the exchange of short pieces of DNA between two long DNA strands, particularly the exchange of homologous regions between the paired maternal and paternal chromosomes.

147. (c) Rennin (also called chymosin) is an enzyme that occurs in gastric juice and is a constituent of rennet. It coagulates milk by converting caseinogen to casein. The initial step in the digestion of milk in humans is carried out by rennin.

148. (b) Facilitated transport is a form of passive transport in which materials are moved across the plasma membrane by a transport protein down their concentration gradient; hence, it does not require energy.

149. (a) CO_2 from the respiratory tissues to the lungs is transported by the blood in 3 ways:
(i) In dissolved state or as a physical solution: Very small amount physically dissolved in plasma (7% i.e. $\cong$ 0.3 ml of CO_2 by each 100 ml of blood).
(ii) Bicarbonate ions: $\cong$ 70% (i.e. $\cong$ 2.5 ml per 100 ml of blood) CO_2 diffuses in plasma & then into RBCs where it (in the presence of carbonic anhydrase) combines with H_2O to form carbonic acid which is almost spontaneously dissociated into hydrogen ion and bicarbonate ions.
(iii) Carbaminohaemoglobin : $\cong$ 23% (i.e. $\cong$ 1 ml of CO_2 per 100 ml of blood) combines with haemoglobin forming an unstable compound.

150. (a) Karl Landsteiner reported first time ABO blood groups in human beings. A, B and O blood groups were discovered by Landsteiner (1900) while AB blood group was found out by de Castello and Steini (1902). ABO blood groups are determined by the gene I (isoagglutinin). There are three alleles. I^A, I^B and I^O of this gene. Proteins produced by the I^A and I^B alleles are called A antigen and B antigen. Individuals with AB blood group have both antigen A and antigen B on their RBCs, and no antibodies for either of the antigens in their plasma.

151. (a) **Control of Heart Beat by Nervous System:** Medulla oblongata has two regulatory centre:
(i) Accelerator centre - It functions through sympathetic nervous system (SNS) and increases heart beat by the secretion of epinephrine or adrenaline.

(ii) *Depressor centre* - It functions through parasympathetic nervous system (PSNS) by the secretion of acetylcholine. It decreases heart beat, speed of conduction of action potential and thereby the cardiac output.

152. (a) Aldosterone hormone, released by the adrenal glands, helps the body regulate blood pressure. Aldosterone causes the tubules of the kidneys to increase the reabsorption of sodium and water into the blood. This increases the volume of fluid in the body, which also increases blood pressure.

153. (d) A gliding joint is a common type of synovial joint formed between bones that meet at flat or nearly flat articular surfaces. Gliding joints allow the bones to glide past one another in any direction along the plane of the joint - up and down, left and right, and diagonally. Many gliding joints are formed in the appendicular skeleton between the carpal bones of the wrist; between the carpals and the metacarpals of the palm; between the tarsal bones of the ankle; and between the tarsals and the metatarsals of the foot.

154. (a) The neuromuscular junction connects the nervous system to the muscular system *via* synapses between efferent nerve fibres and muscle fibres.

155. (d) The hypothalamus is a highly complex structure in the brain that regulates many important brain chemicals. The hypothalamus is responsible for hormone production. The hormones produced by this area govern body temperature, thirst, hunger, sleep, circadian rhythm, moods, sex drive, and the release of other hormones in the body. This area of the brain controls the pituitary gland and other glands in the body.

156. (c) Retinal is a derivative of vitamin A. Retinal is a polyene chromophore, and bound to proteins called opsins, is the chemical basis of animal vision. Bound to proteins called type 1 rhodopsins, retinal allows certain microorganisms to convert light into metabolic energy.

157. (b) Oxytocin, a neurohypophysial hormone, is produced by the hypothalamus and stored and secreted by the posterior pituitary gland. It stimulates contractions of the smooth muscle tissue in the wall of the uterus during childbirth. Progesterone is a female hormone produced by the ovaries during release of a mature egg from an ovary (ovulation). Progesterone helps prepare the lining of the uterus (endometrium) to receive the egg if it becomes fertilized by a sperm. If the egg is not fertilized, progesterone levels drop and menstrual bleeding begins. Atrial natriuretic factor is a hormonal substance produced by the right atrium of the heart that stimulates the excretion of sodium and water by the kidneys and helps regulate blood pressure.

158. (c) **Adrenal Medulla** is called *emergency gland* and secretes **adrenaline** (Epinephrine) and **nor-adrenaline** (nor-epinephrine) hormones.

Adrenaline (epinephrine) is commonly called as 'emergency hormone' **or 3F – hormone (For fear, fight & flight).** It stimulates sweating, heart beat and breathing rate. It causes the dilation of coronary artery (supplying blood to the heart muscles), bronchioles (for increasing inspiratory volume) and pupil (for better vision).

159. (a) Urethra is a tube that connects the urinary bladder to the genitals for the removal of fluids from the body. The urethra travels through the penis, and carries semen as well as urine.

160. (a) The corpus luteum is a temporary endocrine structure involved in ovulation and early pregnancy. The main secretory product of corpus luteum is progesterone, which is required for the establishment and maintenance of pregnancy.

161. (c) Synthesis of estrogen and progesterone due to high level of hCG is a normal gonadotropic activity in a normal pregnant female.

162. (a) Tubectomy also referred to, as tubal sterilization is a surgical procedure done on women as a permanent method of contraception. In this method fallopian tubes are blocked so that the ova or eggs are prevented from travelling to the uterus from the ovary. This prevents any future pregnancies to occur after the surgical procedure.

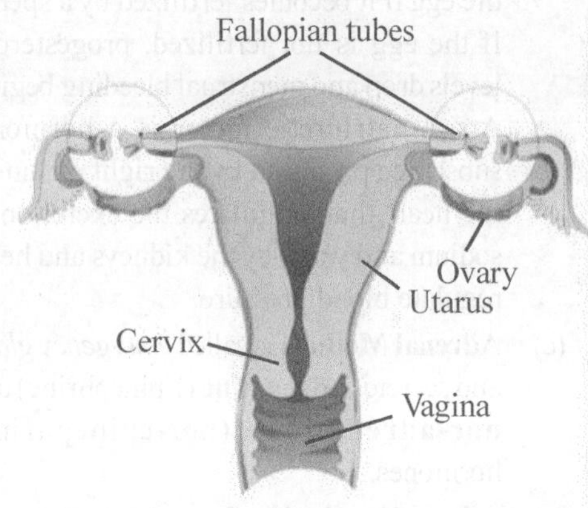

163. (b) An intrauterine device (IUD) is a small contraceptive device, often 'T'-shaped, often containing either copper or levonorgestrel, which is inserted into the uterus. They are one form of long-acting reversible contraception which is the most effective types of reversible birth control. The LNG-20 is so-called because it releases 20 micrograms of levonorgestrel per 24 hours, compared with 30 micrograms of levonorgestrel in some mini-Pills.

164. (b) IVF (*In vitro* Fertilization) is the process of fertilization by manually combining an egg and sperm in a laboratory dish. The process involves monitoring and stimulating a woman's ovulatory process, removing ovum or ova (egg or eggs) from the woman's ovaries and letting sperm fertilise them in a fluid medium in a laboratory. The fertilised egg (zygote) is cultured for 2-6 days in a growth medium and is then transferred to the mother's uterus with the intention of establishing a successful pregnancy.

165. (c) Colour blindness is a X-chromosome linked character.

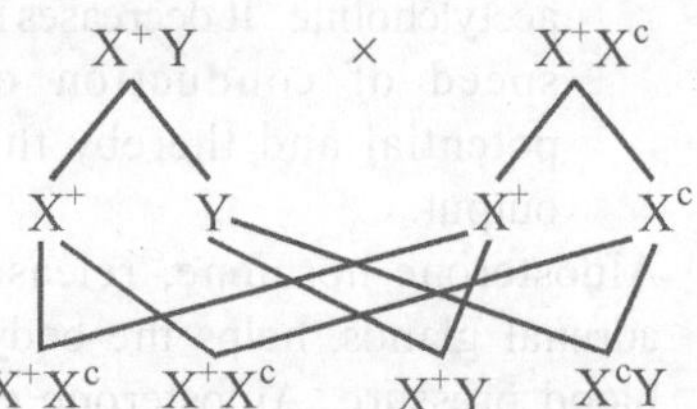

∴ Colourblind male = 50%.

166. (c) According to Hardy Weinberg principle.

$p^2 + 2pq + q^2 = 1; (p + q)^2 = 1$

(AA) $p^2 = 360$ out of 1000 individual

or $p^2 = 36$ out of 100.

$q^2 = 160$ out of 1000 or $q^2 = 16$ out of 100

so $q = \sqrt{.16} = .4$. As $p + q = 1$

so p is 0.6.

167. (a) Turner's syndrome is a chromosomal condition that affects development in females. A human female with Turner's syndrome has 45 chromosomes with XO. The most common feature of Turner syndrome is short stature, which becomes evident by about age 5.

168. (a) Synthesis of RNA exhibits several features that are synonymous with DNA replication. RNA synthesis requires accurate and efficient initiation, elongation proceeds in the 5'->3' direction (i.e. the polymerase moves along the template strand of DNA in the 3'->5' direction), and RNA synthesis requires distinct and accurate termination. Transcription exhibits several features that are distinct from replication.

169. (b) Human genome sequencing is a process that determines the complete DNA sequence of an organism's genome at a single time. This requires sequencing all of an organism's chromosomal DNA as well as DNA contained in the mitochondria and, for plants, in the chloroplast. Commonly used vectors for human genome sequencing are BAC (Bacterial artificial chromosomes) and YAC (Yeast artificial chromosomes).

170. (c) Organs which have a common fundamental anatomical plan and similar embryonic origin whatever varied functions they may perform are regarded as homologous organs. For examples the flippers of a whale, a bats wing, fore-limb of a horse, a bird's wing and forelimbs of human are structurally as well as functionally different.

171. (b) Analogous organs are those that perform similar functions but have entirely different embryonic origins.

172. (a) The plant illustrated in diagram is *Datura* which has hallucinogenic properties. Hallucinogen is a substance that produces psychological effects normally associated only with dreams, schizophrenia, or religious visions. It produces changes in perception (ranging from distortions in what is sensed to perceptions of objects where there are none), thought, and feeling.

173. (c) HIV infection does usually show symptoms of AIDS When HIV damages large number of helper T-Lymphocytes (CD4 cells). AIDS is the stage of HIV infection that occurs when one immune system is badly damaged and one become vulnerable to infections and infection-related cancers called opportunistic infections. When the number of ones CD4 cells falls below 200 cells per cubic millimetre of blood (200 cells/mm3), one is considered to have progressed to AIDS.

174. (c) To obtain virus - free healthy plants from a diseased one by tissue culture technique, both apical and axillary meristems of the diseased plant will be taken. Plant tissue culture is used to maintain or grow plant cells, tissues or organs under sterile conditions on a nutrient culture medium of known composition. Plant tissue culture is widely used to produce clones of a plant in a method known as micro propagation.

175. (b) Anaerobic digestion is a series of processes in which microorganisms break down biodegradable material in the absence of oxygen, used for industrial or domestic purposes to manage waste and/or to release energy. The process of anaerobic digestion produces a biogas, consisting of methane (it will burn), carbon dioxide (it does not burn) and traces of other contaminant gases.

176. (d) The Keoladeo National Park is a famous avifauna sanctuary in India that plays host to thousands of birds especially during the summer season. It is formerly known as the Bharatpur Bird Sanctuary in Bharatpur, Rajasthan. October to February is the best time, for the migratory birds are there as well as the residents. In August local birds start their nest building and rear their young for the next few months. October/November is when the migrants arrive. Most stay till March including the Siberian Crane.

177. (c) Phosphorus is an important element for living beings. Consumers obtain phosphorus directly or indirectly from plants. Phosphorus is also present in phosphatic rocks. It is released during the decomposition of plant and animal remains. The released phosphorus may reach the deeper layers of soil and gets deposited as phosphate rocks. All plants and animals eventually die and in due time, their organic remains or debris decay through the action of micro-organism and the phosphates are release into the water for recycling.

178. (d) Arthropoda is the largest phylum of Animalia which includes insects. Over two-thirds of all named species on earth are arthropods. They have organ system level of organisation. Mollusca is the second largest animal phylum. They are terrestrial or aquatic (marine or fresh).

179. (a) Scrubbers are air pollution control devices that use liquid to remove particulate matter or gases from an industrial exhaust or flue gas stream. This atomized liquid (typically water) entrains particles and pollutant gases in order to effectively wash them out of the gas flow. It removes or scrubs SO_2 emissions from the exhaust of coal-fired power plants.

180. (a) According to 10 percent law, during the transfer of energy from organic food from one trophic level to the next, only about ten percent of energy from organic matter is stored as flesh. The remaining is lost during transfer, broken down in respiration, or lost to incomplete digestion by higher trophic levels. If 20 J of energy is trapped at producer level, then 0.02 J of energy will be available to peacock as food in the given food chain.

NEET 2013

PHYSICS

1. A uniform force of $(3\hat{i}+\hat{j})$ newton acts on a particle of mass 2 kg. The particle is displaced from position $(2\hat{i}+\hat{k})$ meter to position $(4\hat{i}+3\hat{j}-\hat{k})$ meter. The work done by the force on the particle is
(a) 6 J (b) 13 J (c) 15 J (d) 9 J

2. A, B and C are three points in a uniform electric field. The electric potential is

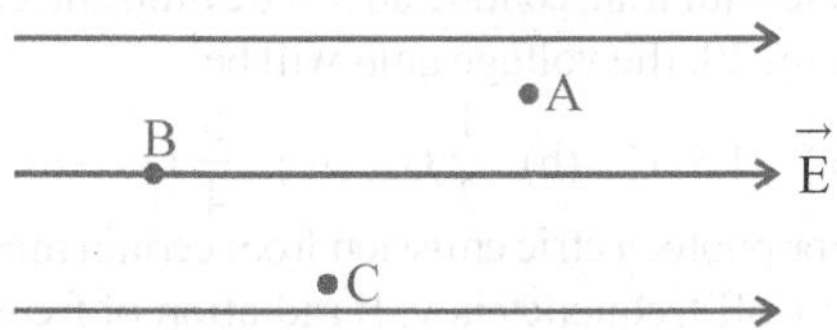

(a) maximum at B
(b) maximum at C
(c) same at all the three points A, B and C
(d) maximum at A

3. A coil of self-inductance L is connected in series with a bulb B and an AC source. Brightness of the bulb decreases when
(a) number of turns in the coil is reduced
(b) a capacitance of reactance $X_C = X_L$ is included in the same circuit
(c) an iron rod is inserted in the coil
(d) frequency of the AC source is decreased

4. The upper half of an inclined plane of inclination θ is perfectly smooth while lower half is rough. A block starting from rest at the top of the plane will again come to rest at the bottom, if the coefficient of friction between the block and lower half of the plane is given by
(a) $\mu = \frac{2}{\tan\theta}$ (b) $\mu = 2\tan\theta$
(c) $\mu = \tan\theta$ (d) $\mu = \frac{1}{\tan\theta}$

5. The wetability of a surface by a liquid depends primarily on
(a) surface tension
(b) density
(c) angle of contact between the surface and the liquid
(d) viscosity

6. The condition under which a microwave oven heats up a food item containing water molecules most efficiently is
(a) the frequency of the microwaves has no relation with natural frequency of water molecules.
(b) microwaves are heat waves, so always produce heating.
(c) infra-red waves produce heating in a microwave oven.
(d) the frequency of the microwaves must match the resonant frequency of the water molecules.

7. A gas is taken through the cycle A → B → C → A, as shown in figure. What is the net work done by the gas ?

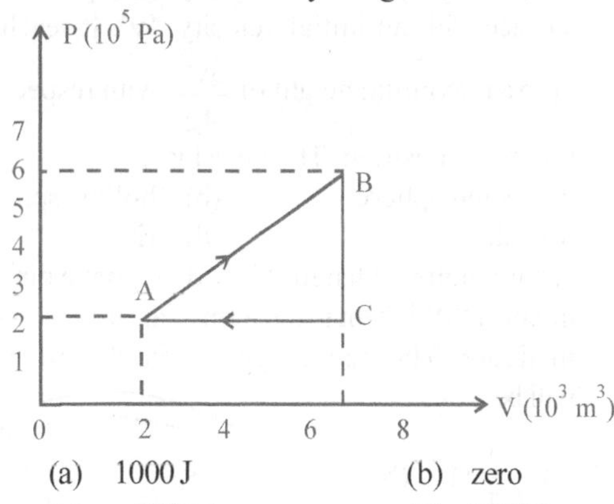

(a) 1000 J (b) zero
(c) –2000 J (d) 2000 J

8. A wire loop is rotated in a magnetic field. The frequency of change of direction of the induced e.m.f. is
(a) twice per revolution
(b) four times per revolution
(c) six times per revolution
(d) once per revolution

9. The velocity of a projectile at the initial point A is $(2\hat{i}+3\hat{j})$ m/s. It's velocity (in m/s) at point B is
(a) $-2\hat{i}+3\hat{j}$
(b) $2\hat{i}-3\hat{j}$
(c) $2\hat{i}+3\hat{j}$
(d) $-2\hat{i}-3\hat{j}$

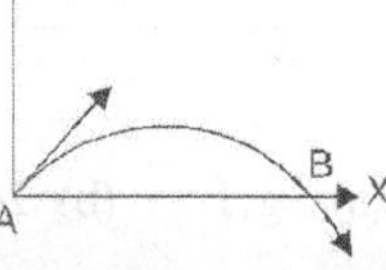

10. The following four wires are made of the same material. Which of these will have the largest extension when the same tension is applied ?
(a) Length = 100 cm, diameter = 1 mm
(b) Length = 200 cm, diameter = 2 mm
(c) Length = 300 cm, diameter = 3 mm
(d) Length = 50 cm, diameter = 0.5 mm

11. A wire of resistance 4 Ω is stretched to twice its original length. The resistance of stretched wire would be
(a) 4 Ω (b) 8 Ω (c) 16 Ω (d) 2 Ω

12. A piece of iron is heated in a flame. It first becomes dull red then becomes reddish yellow and finally turns to white hot. The correct explanation for the above observation is possible by using
(a) Wien's displacement law
(b) Kirchoff's law
(c) Newton's law of cooling
(d) Stefan's law

13. A small object of uniform density rolls up a curved surface with an initial velocity 'v'. It reaches upto a maximum height of $\frac{3v^2}{4g}$ with respect to the initial position. The object is a
(a) solid sphere (b) hollow sphere
(c) disc (d) ring

14. A bar magnet of length 'ℓ' and magnetic dipole moment 'M' is bent in the form of an arc as shown in figure. The new magnetic dipole moment will be
(a) $\frac{3}{\pi}$ M (b) $\frac{2}{\pi}$ M
(c) $\frac{M}{2}$ (d) M

15. A rod PQ of mass M and length L is hinged at end P. The rod is kept horizontal by a massless string tied to point Q as shown in figure. When string is cut, the initial angular acceleration of the rod is

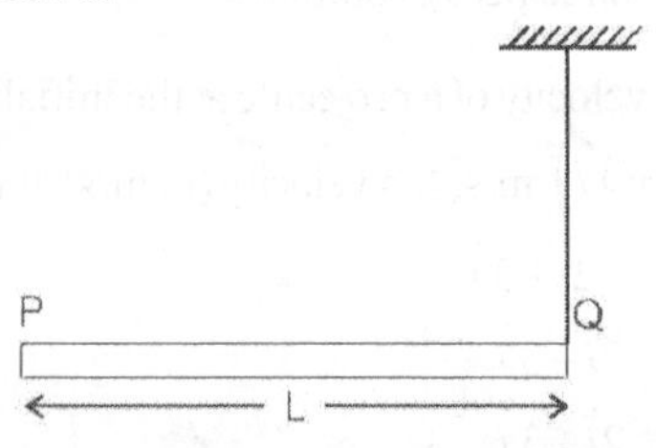

(a) g/L (b) 2g/L (c) $\frac{2g}{3L}$ (d) $\frac{3g}{2L}$

16. In a n-type semiconductor, which of the following statement is true?
(a) Electrons are minority carriers and pentavalent atoms are dopants.
(b) Holes are minority carriers and pentavalent atoms are dopants.
(c) Holes are majority carriers and trivalent atoms are dopants.
(d) Electrons are majority carriers and trivalent atoms are dopants.

17. In a common emitter (CE) amplifier having a voltage gain G, the transistor used has transconductance 0.03 mho and current gain 25. If the above transistor is replaced with another one with transconductance 0.02 mho and current gain 20, the voltage gain will be
(a) 1.5G (b) $\frac{1}{3}$G (c) $\frac{5}{4}$G (d) $\frac{2}{3}$G

18. For photoelectric emission from certain metal the cut-off frequency is ν. If radiation of frequency 2ν impinges on the metal plate, the maximum possible velocity of the emitted electron will be (m is the electron mass)
(a) $\sqrt{h\nu / m}$ (b) $\sqrt{2h\nu / m}$
(c) $2\sqrt{h\nu / m}$ (d) $\sqrt{h\nu / (2m)}$

19. In Young's double slit experiment, the slits are 2 mm apart and are illuminated by photons of two wavelengths $\lambda_1 = 12000$Å and $\lambda_2 = 10000$Å. At what minimum distance from the common central bright fringe on the screen 2 m from the slit will a bright fringe from one interference pattern coincide with a bright fringe from the other ?
(a) 6mm (b) 4mm (c) 3mm (d) 8mm

20. Three blocks with masses m, 2 m and 3 m are connected by strings as shown in the figure. After an upward force F is applied on block m, the masses move upward at constant speed v. What is the net force on the block of mass 2m? (g is the acceleration due to gravity)

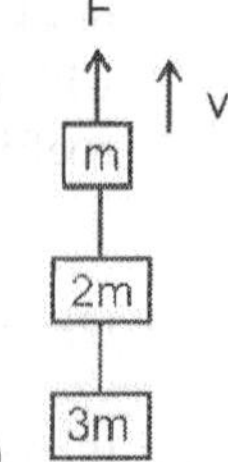

(a) 2 mg (b) 3 mg (c) 6 mg (d) zero

21. A certain mass of Hydrogen is changed to Helium by the process of fusion. The mass defect in fusion reaction is 0.02866 a.m.u. The energy liberated per a.m.u. is
(Given : 1 a.m.u = 931 MeV)
(a) 26.7 MeV (b) 6.675 MeV
(c) 13.35 MeV (d) 2.67 MeV

22. If we study the vibration of a pipe open at both ends, then which of the following statements is not true ?
(a) Odd harmonics of the fundamental frequency will be generated
(b) All harmonics of the fundamental frequency will be generated
(c) Pressure change will be maximum at both ends
(d) Antinode will be at open end

23. An explosion breaks a rock into three parts in a horizontal plane. Two of them go off at right angles to each other. The first part of mass 1 kg moves with a speed of 12 ms^{-1} and the second part of mass 2 kg moves with speed 8 ms^{-1}. If the third part flies off with speed 4 ms^{-1} then its mass is
(a) 5 kg (b) 7 kg (c) 17 kg (d) 3 kg

24. In an experiment four quantities a, b, c and d are measured with percentage error 1%, 2%, 3% and 4% respectively. Quantity P is calculated as follows $P = \frac{a^3 b^2}{cd}$ % error in P is
(a) 10% (b) 7% (c) 4% (d) 14%

25. A source of unknown frequency gives 4 beats/s, when sounded with a source of known frequency 250 Hz. The second harmonic of the source of unknown frequency gives five beats per second, when sounded with a source of frequency 513 Hz. The unknown frequency is
(a) 246 Hz (b) 240 Hz
(c) 260 Hz (d) 254 Hz

26. The internal resistance of a 2.1 V cell which gives a current of 0.2 A through a resistance of 10 Ω is
(a) 0.5 Ω (b) 0.8 Ω (c) 1.0 Ω (d) 0.2 Ω

27. A current loop in a magnetic field
(a) can be in equilibrium in one orientation
(b) can be in equilibrium in two orientations, both the equilibrium states are unstable
(c) can be in equilibrium in two orientations, one stable while the other is unstable
(d) experiences a torque whether the field is uniform or non-uniform in all orientations

28. The wavelength λ_e of an electron and λ_p of a photon are of same energy E are related by
(a) $\lambda_p \propto \lambda_e$ (b) $\lambda_p \propto \sqrt{\lambda_e}$
(c) $\lambda_p \propto \frac{1}{\sqrt{\lambda_e}}$ (d) $\lambda_p \propto \lambda_e^2$

29. The half life of a radioactive isotope 'X' is 20 years. It decays to another element 'Y' which is stable. The two elements 'X' and 'Y' were found to be in the ratio of 1 : 7 in a sample of a the given rock. The age of the rock is estimated to be
(a) 60 years (b) 80 years
(c) 100 years (d) 40 years

30. The resistance of the four arms P, Q, R and S in a Wheatstone's bridge are 10 ohm, 30 ohm, 30 ohm and 90 ohm, respectively. The e.m.f. and internal resistance of the cell are 7 volt and 5 ohm respectively. If the galvanometer resistance is 50 ohm, the current drawn from the cell will be
(a) 0.2 A (b) 0.1 A (c) 2.0 A (d) 1.0 A

31. In the given (V–T) diagram, what is the relation between pressure P_1 and P_2 ?

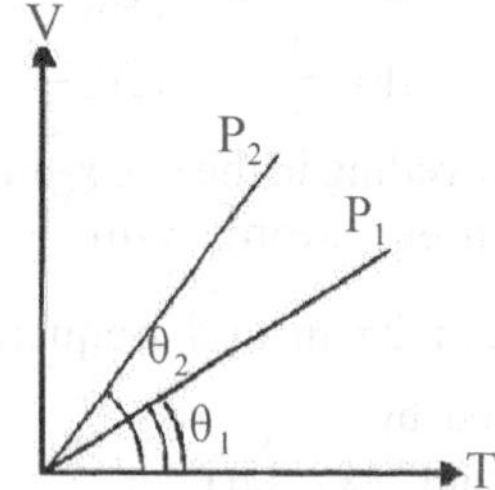

(a) $P_2 > P_1$ (b) $P_2 < P_1$
(c) Cannot be predicted (d) $P_2 = P_1$

32. The molar specific heats of an ideal gas at constant pressure and volume are denoted by C_p and C_v, respectively. If $\gamma = \frac{C_p}{C_v}$ and R is the universal gas constant, then C_v is equal to
(a) $\frac{R}{(\gamma - 1)}$ (b) $\frac{(\gamma - 1)}{R}$
(c) γR (d) $\frac{1+\gamma}{1-\gamma}$

33. The amount of heat energy required to raise the temperature of 1g of Helium at NTP, from T_1K to T_2K is
(a) $\frac{3}{2} N_a k_B (T_2 - T_1)$
(b) $\frac{3}{4} N_a k_B (T_2 - T_1)$
(c) $\frac{3}{4} N_a k_B \frac{T_2}{T_1}$
(d) $\frac{3}{8} N_a k_B (T_2 - T_1)$

34. A plano convex lens fits exactly into a plano concave lens. Their plane surfaces are parallel to each other. If lenses are made of different materials of refractive indices μ_1 and μ_2 and R is the radius of curvature of the curved surface of the lenses, then the focal length of the combination is

(a) $\frac{R}{2(\mu_1-\mu_2)}$ (b) $\frac{R}{(\mu_1-\mu_2)}$

(c) $\frac{2R}{(\mu_2-\mu_1)}$ (d) $\frac{R}{2(\mu_1+\mu_2)}$

35. During an adiabatic process, the pressure of a gas is found to be proportional to the cube of its temperature. The ratio of $\frac{C_p}{C_v}$ for the gas is

(a) 2 (b) $\frac{5}{3}$ (c) $\frac{3}{2}$ (d) $\frac{4}{3}$

36. A wave travelling in the +ve x-direction having displacement along y-direction as 1m, wavelength 2π m and frequency $\frac{1}{\pi}$ Hz is represented by

(a) y = sin (2πx − 2πt)
(b) y = sin (10πx − 20πt)
(c) y = sin (2πx + 2πt)
(d) y = sin (x − 2t)

37. The output(X) of the logic circuit shown in figure will be

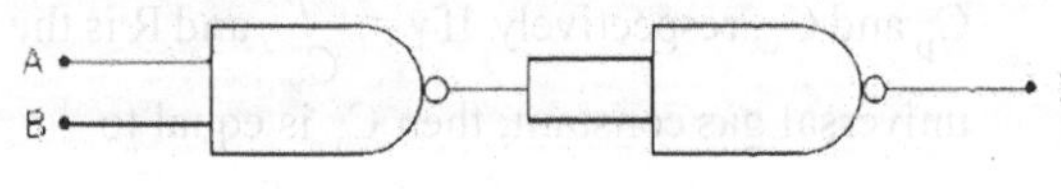

(a) $X=\overline{A.B}$ (b) X = A.B

(c) $X=\overline{A+B}$ (d) $X=\overline{\overline{A},\overline{B}}$

38. A body of mass 'm' is taken from the earth's surface to the height equal to twice the radius (R) of the earth. The change in potential energy of body will be

(a) $\frac{2}{3}$ mgR (b) 3 mgR

(c) $\frac{1}{3}$ mgR (d) mg2R

39. The ratio of longest wavelengths corresponding to Lyman and Blamer series in hydrogen spectrum is

(a) $\frac{3}{23}$ (b) $\frac{7}{29}$ (c) $\frac{9}{31}$ (d) $\frac{5}{27}$

40. Infinite number of bodies, each of mass 2 kg are situated on x-axis at distances 1m, 2m, 4m, 8m, ... respectively, from the origin. The resulting gravitational potential due to this system at the origin will be

(a) $-\frac{8}{3}$ G (b) $-\frac{4}{3}$ G (c) − 4G (d) −G

41. When a proton is released from rest in a room, it starts with an initial acceleration a_0 towards west. When it is projected towards north with a speed v_0 it moves with an initial acceleration $3a_0$ towards west. The electric and magnetic fields in the room are respectively

(a) $\frac{ma_0}{e}$ west, $\frac{2ma_0}{ev_0}$ down

(b) $\frac{ma_0}{e}$ east, $\frac{3ma_0}{ev_0}$ up

(c) $\frac{ma_0}{e}$ east, $\frac{3ma_0}{ev_0}$ down

(d) $\frac{ma_0}{e}$ west, $\frac{2ma_0}{ev_0}$ up

42. For a normal eye, the cornea of eye provides a converging power of 40D and the least converging power of the eye lens behind the cornea is 20D. Using this information, the distance between the retina and the eye lens of the eye can be estimated to be

(a) 2.5 cm (b) 1.67 cm (c) 1.5 cm (d) 5 cm

43. A parallel beam of fast moving electrons is incident normally on a narrow slit. A fluorescent screen is placed at a large distance from the slit. If the speed of the electrons is increased, which of the following statements is correct ?

(a) The angular width of the central maximum of the diffraction pattern will increase.
(b) The angular width of the central maximum will decrease.
(c) The angular width of the central maximum will be unaffected.
(d) Diffraction pattern is not observed on the screen in case of electrons.

44. Two pith balls carrying equal charges are suspended from a common point by strings of equal length. The equilibrium separation between them is r. Now the strings are rigidly clamped at half the height. The equilibrium separation between the balls now become

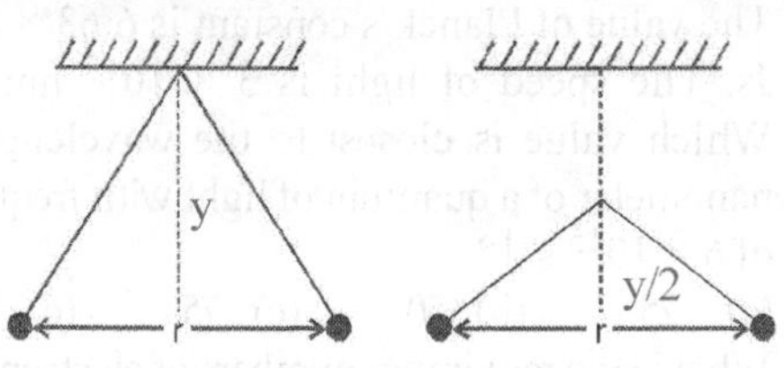

(a) $\left(\frac{r}{\sqrt[3]{2}}\right)$ (b) $\left(\frac{2r}{\sqrt{3}}\right)$

(c) $\left(\frac{2r}{3}\right)$ (d) $\left(\frac{r}{\sqrt{2}}\right)^2$

45. A stone falls freely under gravity. It covers distances h_1, h_2 and h_3 in the first 5 seconds, the next 5 seconds and the next 5 seconds respectively. The relation between h_1, h_2 and h_3 is

(a) $h_1 = \frac{h_2}{3} = \frac{h_3}{5}$

(b) $h_2 = 3h_1$ and $h_3 = 3h_2$

(c) $h_1 = h_2 = h_3$

(d) $h_1 = 2h_2 = 3h_3$

CHEMISTRY

46. Which is the monomer of Neoprene in the following ?

(a) $CH_2 = C(CH_3) — CH = CH_2$

(b) $CH_2 = C(Cl) — CH = CH_2$

(c) $CH_2 = CH — C \equiv CH$

(d) $CH_2 = CH — CH = CH_2$

47. A magnetic moment of 1.73 BM will be shown by one among the following :

(a) $[Ni(CN)_4]^{2-}$ (b) $TiCl_4$

(c) $[CoCl_6]^{4-}$ (d) $[Cu(NH_3)_4]^{2+}$

48. A metal has a *fcc* lattice. The edge length of the unit cell is 404 pm. The density of the metal is 2.72 g cm^{-3}. The molar mass of the metal is : (N_A Avogadro's constant = 6.02×10^{23} mol^{-1})

(a) 30 g mol^{-1} (b) 27 g mol^{-1}

(c) 20 g mol^{-1} (d) 40 g mol^{-1}

49. Structure of the compound whose IUPAC name is 3-ethyl-2-hydroxy-4-methylhex-3-en-5-ynoic acid is :

(a) OH COOH

(b) OH COOH

(c) COOH OH

(d) OH COOH

50. Which of the following structure is similar to graphite?

(a) B (b) B_4C (c) B_2H_6 (d) BN

51. Some *meta*-directing substituents in aromatic substitution are given. Which one is most deactivating?

(a) $-SO_3H$ (b) –COOH

(c) $-NO_2$ (d) –C°N

52. How many grams of concentrated nitric acid solution should be used to prepare 250 mL of 2.0M HNO_3 ? The concentrated acid is 70% HNO_3

(a) 90.0 g (b) 70.0 g (c) 54.0 g (d) 45.0 g

53. The order of stability of the following tautomeric compounds is :

$$\underset{\text{I}}{CH_2 = \overset{OH}{\overset{|}{C}} - CH_2 - \overset{O}{\overset{\|}{C}} - CH_3} \rightleftharpoons \underset{\text{II}}{CH_3 - \overset{O}{\overset{\|}{C}} - CH_2 - \overset{O}{\overset{\|}{C}} - CH_3} \rightleftharpoons \underset{\text{III}}{CH_3 - \overset{OH}{\overset{|}{C}} = CH - \overset{O}{\overset{\|}{C}} - CH_3}$$

(a) III > II > I (b) II > I > III

(c) II > III > I (d) I > II > III

54. Antiseptics and disinfectants either kill or prevent growth of microorganisms. Identify which of the following statements **is not true:**
(a) Chlorine and iodine are used as strong disinfectants.
(b) Dilute solutions of Boric acid and Hydrogen Peroxide are strong antiseptics.
(c) Disinfectants harm the living tissues.
(d) A 0.2% solution of phenol is an antiseptic while 1% solution acts as a disinfectant.

55. Nylon is an example of:
(a) Polysaccharide (b) Polyamide
(c) Polythene (d) Polyester

56. Among the following ethers, which one will produce methyl alcohol on treatment with hot concentrated HI?
(a) $CH_3-CH_2-\underset{\displaystyle CH_3}{\underset{|}{CH}}-O-CH_3$
(b) $CH_3-\overset{\displaystyle CH_3}{\overset{|}{\underset{\displaystyle CH_3}{\underset{|}{C}}}}-O-CH_3$
(c) $CH_3-\underset{\displaystyle CH_3}{\underset{|}{CH}}-CH_2-O-CH_3$
(d) $CH_3-CH_2-CH_2-CH_2-O-CH_3$

57. Which of these is not a monomer for a high molecular mass silicone polymer?
(a) Me_2SiCl_2 (b) Me_3SiCl
(c) $PhSiCl_3$ (d) $MeSiCl_3$

58. Identify the correct order of solubility in aqueous medium:
(a) $ZnS > Na_2S > CuS$
(b) $Na_2S > CuS > ZnS$
(c) $Na_2S > ZnS > CuS$
(d) $CuS > ZnS > Na_2S$

59. What is the activation energy for a reaction if its rate doubles when the temperature is raised from 20°C to 35°C? ($R = 8.314\ J\,mol^{-1}\,K^{-1}$)
(a) $269\ kJ\,mol^{-1}$ (b) $34.7\ kJ\,mol^{-1}$
(c) $15.1\ kJ\,mol^{-1}$ (d) $342\ kJ\,mol^{-1}$

60. A hydrogen gas electrode is made by dipping platinum wire in a solution of HCl of pH = 10 and by passing hydrogen gas around the platinum wire at one atm pressure. The oxidation potential of electrode would be ?
(a) 0.59 V (b) 0.118 V
(c) 1.18 V (d) 0.059 V

61. The value of Planck's constant is 6.63×10^{-34} Js. The speed of light is 3×10^{17} nm s^{-1}. Which value is closest to the wavelength in nanometer of a quantum of light with frequency of $6 \times 10^{15}\ s^{-1}$?
(a) 25 (b) 50 (c) 75 (d) 10

62. What is the maximum numbers of electrons that can be associated with the following set of quantum numbers?
$n = 3, l = 1$ and $m = -1$
(a) 6 (b) 4 (c) 2 (d) 10

63. Which of the following lanthanoid ions is diamagnetic ?
(At nos. Ce = 58, Sm = 62, Eu = 63, Yb = 70)
(a) Sm^{2+} (b) Eu^{2+} (c) Yb^{2+} (d) Ce^{2+}

64. 6.02×10^{20} molecules of urea are present in 100 mL of its solution. The concentration of solution is :
(a) 0.01 M (b) 0.001 M
(c) 0.1 M (d) 0.02 M

65. Based on equation $E = -2.178 \times 10^{-18}\ J\left(\frac{Z^2}{n^2}\right)$, certain conclusions are written. Which of them is not correct ?
(a) Larger the value of n, the larger is the orbit radius.
(b) Equation can be used to calculate the change in energy when the electron changes orbit.
(c) For n = 1, the electron has a more negative energy than it does for n = 6 which mean that the electron is more loosely bound in the smallest allowed orbit.
(d) The negative sign in equation simply means that the energy or electron bound to the nucleus is lower than it would be if the electrons were at the infinite distance from the nucleus.

66. An excess of $AgNO_3$ is added to 100 mL of a 0.01 M solution of dichlorotetraaquachromium (iii) chloride. The number of moles of AgCl precipitated would be :
(a) 0.002 (b) 0.003 (c) 0.01 (d) 0.001

67. $KMnO_4$ can be prepared from K_2MnO_4 as per the reaction:

$$3MnO_4^{2-} + 2H_2O \rightleftharpoons 2MnO_4^{-} + MnO_2 + 4OH^-$$

The reaction can go to completion by removing OH^- ions by adding.

(a) KOH (b) CO_2 (c) SO_2 (d) HCl

68. Which of the following compounds will not undergo Friedal-Craft's reaction easily :

(a) Xylene (b) Nitrobenzene
(c) Toluene (d) Cumene

69. Which of these is least likely to act as Lewis base?

(a) F^- (b) BF_3 (c) PF_3 (d) CO

70. The basic structural unit of silicates is :

(a) SiO_4^{4-} (b) SiO_3^{2-} (c) SiO_4^{2-} (d) SiO

71. Maximum deviation from ideal gas is expected from

(a) $N_2(g)$ (b) $CH_4(g)$ (c) $NH_3(g)$ (d) $H_2(g)$

72. Which is the strongest acid in the following :

(a) $HClO_3$ (b) $HClO_4$
(c) H_2SO_3 (d) H_2SO_4

73. Reaction by which Benzaldehyde cannot be prepared:

(a) C_6H_5–COCl + H_2 in presence of Pd - $BaSO_4$

(b) C_6H_6 + CO + HCl in presence of anhydrous $AlCl_3$

(c) C_6H_5–COOH + Zn/Hg and conc. HCl

(d) C_6H_5–CH_3 + CrO_2Cl_2 in CS_2 followed by H_3O^+

74. The radical, C_6H_5–$\dot{C}H_2$ is aromatic because it has :

(a) 7 p-orbitals and 6 unpaired electrons
(b) 7 p-orbitals and 7 unpaired electrons
(c) 6 p-orbitals and 7 unpaired electrons
(d) 6 p-orbitals and 6 unpaired electrons

75. Roasting of sulphides gives the gas X as a by product. This is colorless gas with choking smell of burnt sulphur and caused great damage to respiratory organs as a result of acid rain. Its aqueous solution is acidic, acts as a reducing agent and its acid has never been isolated. The gas X is :

(a) SO_2 (b) CO_2 (c) SO_3 (d) H_2S

76. At 25°C molar conductance of 0.1 molar aqueous solution of ammonium hydroxide is 9.54 $ohm^{-1}\ cm^2 mol^{-1}$ and at infinite dilution its molar conductance is 238 $ohm^{-1}\ cm^2\ mol^{-1}$. The degree or ionisation of ammonium hydroxide at the same concentration and temperature is :

(a) 20.800% (b) 4.008%
(c) 40.800% (d) 2.080%

77. Which of the following statements about the interstitial compounds is incorrect ?

(a) They are chemically reactive.
(b) They are much harder then the pure metal.
(c) They have higher melting points than the pure metal.
(d) They retain metallic conductivity.

78. In the reaction

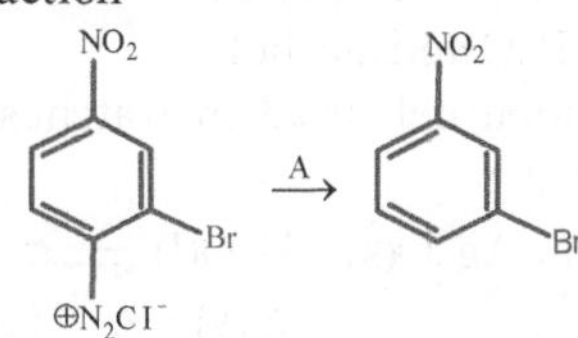

A is :

(a) Cu_2Cl_2 (b) H_3PO_2 and H_2O
(c) H^+/H_2O (d) $HgSO_4/H_2SO_4$

79. Which of the following is electron - deficient ?

(a) $(SiH_3)_2$ (b) $(BH_3)_2$
(c) PH_3 (d) $(CH_3)_2$

80. Which one of the following molecules contains no π bond?

(a) H_2O (b) SO_2 (c) NO_2 (d) CO_2

81. Which of the following does not give oxygen on heating?

(a) $Zn(ClO_3)_2$ (b) $K_2Cr_2O_7$
(c) $(NH_4)_2Cr_2O_7$ (d) $KClO_3$

82. Which of the following is a polar molecule ?

(a) SF_4 (b) SiF_4 (c) XeF_4 (d) BF_3

83. The structure of isobutyl group in an organic compound is :

(a) $CH_3-\underset{|}{C}H-CH_2-CH_3$

(b) $CH_3-CH_2-CH_2-CH_2-$

(c) $CH_3-\overset{CH_3}{\underset{CH_3}{C}}-$

(d) $(CH_3)_2CH-CH_2-$

84. Which of the following is paramagnetic ?
(a) O_2^- (b) CN^- (c) NO^+ (d) CO

85. The number of carbon atoms per unit cell of diamond unit cell is :
(a) 8 (b) 6 (c) 1 (d) 4

86. XeF_2 is isostructural with :
(a) ICl_2^- (b) $SbCl_3$ (c) $BaCl_2$ (d) TeF_2

87. A reaction having equal energies of activation for forward and reverse reaction has :
(a) $\Delta G = 0$ (b) $\Delta H = 0$
(c) $\Delta H = \Delta G = \Delta S = 0$ (d) $\Delta S = 0$

88. Dipole-induced dipole interactions are present in which of the following pairs :
(a) Cl_2 and CCl_4
(b) HCl and He atoms
(c) SiF_4 and He atoms
(d) H_2O and alcohol

89. A button cell used in watches functions as following

$$Zn(s) + Ag_2O(s) + H_2O(l) \rightleftharpoons 2Ag(s) + Zn^{2+}(aq) + 2OH^-(aq)$$

If half cell potentials are :
$Zn^{2+}(aq) + 2e^- \rightarrow Zn(s); E^o = -0.76$ V
$Ag_2O(s) + H_2O(l) + 2e^- \rightarrow 2Ag(s) + 2OH^-(aq)$; $E^o = 0.34$ V
The cell potential will be :
(a) 0.42 V (b) 0.84 V (c) 1.34 V (d) 1.10 V

90. Nitrobenzene on reaction with conc. HNO_3/H_2SO_4 at 80 - 100°C forms which one of the following products ?
(a) 1, 3 - Dinitrobenzene
(b) 1, 4 - Dinitrobenzene
(c) 1, 2, 4 - Trinitrobenzene
(d) 1, 2 - Dinitrobenzene

BIOLOGY

91. The diagram shows an important concept in the genetic implication of DNA. Fill in the blanks A to C.

$$DNA \xrightarrow{A} mRNA \xrightarrow{B} protein \xrightarrow[C]{\text{Proposed by}}$$

(a) A-translation B - transciption C-Erwin Chargaff
(b) A-transcription B - translation C-Francis Crick
(c) A-translation B - extension C-Rosalind Franklin
(d) A-transcription B - replication C-James Watson

92. Perisperm differs from endosperm in;
(a) having no reserve food
(b) being a diploid tissue
(c) its formatting by fusion of secondary nucleus with several sperms
(d) being a haploid tissue

93. Besides paddy fields cyanobacteria are also found inside vegetative part of :
(a) *Cycas* (b) *Equisetum*
(c) *Psilotum* (d) *Pinus*

94. Which of the following statements is correct in relation to the endocrine system?
(a) Organs in the body like gastrointestinal tract, heart, kidney and liver do not produce any hormones.
(b) Non-nutrient chemicals produced by the body in trace amount that act as intercellular messenger are known as hormones.
(c) Releasing and inhibitory hormones are produced by the pituitary gland .
(d) Adenohypophysis is under direct neural regulation of the hypothalamus.

95. Megasporangium is equivalent to :
(a) Fruit (b) Nucellus
(c) Ovule (d) Embryo sac

96. If two persons with 'AB' blood group marry and have sufficiently large number of children these children could be classified as 'A' blood group: 'AB' blood group: 'B' blood group in 1 : 2 : 1 ratio. Modern technique of protein electrophoresis reveals presence of both 'A' and 'B' type proteins in 'AB' blood group individuals. This is an example of :
(a) incomplete dominance
(b) Partial dominance
(c) Complete cominance
(d) Codominance

97. A pregnant female deliver a baby who suffers from stunted growth, mental retardation/low intelligence quotient and abnormal skin. This is the result of :
(a) Low secretion of growth hormone
(b) Cancer of the thyroid gland
(c) Over secretion of pars distalis
(d) Deficiency of iodine in diet

98. Which one of the following organelle in the figure correctly matches with its function ?

(a) Golgi apparatus, protein synthesis
(b) Golgi apparatus, formation of glycolipids
(c) Rough endoplasmic reticulum, protein synthesis
(d) Rough endoplasmic reticulum, formation of glycoproteins

99. A phosphoglycerate is always made up of :
(a) only an unsaturated fatty acid esterified to a glycerol molecule to which a phosphate group is also attached
(b) a saturated or unsaturated fatty acid esterified to a glycerol molecule to which a phosphate group is also attached
(c) a saturated or unsaturated fatty acid esterified to a phosphate group which is also attached to a glycerol molecule.
(d) only a saturated fatty acid esterified to a glycerol molecule to which a phosphate group is also attached

100. During sewage treatment, biogases are produced which include :
(a) methane, oxygen, hydrogen sulphide
(b) hydrogen sulphide, methane, sulphur dioxide
(c) hydrogen sulphide, nitrogen, methane
(d) methane, hydrogen sulphide, carbon dioxide

101. The eye of octopus and eye of cat show different patterns of structure, yet they perform similar function. This is an example of :
(a) Homologous organs that have evolved due to divergent evolution.
(b) Analogous organs that have evolved due to convergent evolution.
(c) Analogous organs that have evolved due to divergent evolution.
(d) Homologous organs that have evolved due to convergent evolution.

102. Which of the following criteria does not pertain to facilitated transport?
(a) High selectivity
(b) Transport saturation
(c) Uphill transport
(d) Requirement of special membrane proteins

103. The process by which organisms with different evolutionary history evolve similar phenotypic adaptations in response to a common environmental challenge, is called :
(a) Convergent evolution
(b) Non-random evolution
(c) Adaptive radiation
(d) Natural selection

104. Infection of *Ascaris* usually occurs by :
(a) Eating imperfectly cooked pork.
(b) Tse-tse fly.
(c) Mosquito bite.
(d) Drinking water containing eggs of *Ascaris*.

105. The Air Prevention and Control of Pollution Act came into force in :
(a) 1981 (b) 1985 (c) 1990 (d) 1975

106. Which group of animals belong to the same phylum?
(a) Earthworm , Pinworm, Tapeworm
(b) Prawn, Scorpion, Locusta
(c) Sponge, Sea anemone, Starfish
(d) Malarial parasite, *Amoeba*, Mosquito

107. Which of the following cannot be detected in a developing foetus by amniocentesis?
(a) Sex of the foetus
(b) Down syndrome
(c) Jaundice
(d) Klinefelter syndrome

108. The Golgi complex plays a major role
(a) in digesting proteins and carbohydrates
(b) as energy transferring organelles
(c) in post translational modification of proteins and glycosidation of lipids
(d) in trapping the light and transforming it into chemical energy

109. Select the correct match of the digested products in humans given in column -I with their absorption site and mechanism in column-II

	Column I	Column II
(a)	Fructose, Na^+	Small intestine, passive absorption
(b)	Glycerol, fatty acids	Duodenum, move as chylomicrons
(c)	Cholesterol, maltose	Large intestine, active absorption
(d)	Glycine, glucose	Small intestine, active absorption

110. Menstrual flow occurs due to lack of :
(a) FSH (b) Oxytocin
(c) Vasopressin (d) Progesterone

111. The characteristics and an example of a synovial joint in humans is :

	Characteristics	Examples
(a)	Fluid filled between two joints, provides cushion	Skull bones
(b)	Fluid filled synovial cavity between two bones	Joint between atlas and axis
(c)	Lymph filled between two bones, limited movement	gliding joint between carpals
(d)	Fluid cartilage between two bones, limited movements	Knee joint

112. Isogamous condition with non-flagellated gametes is found in :
(a) *Spirogyra* (b) *Volvox*
(c) *Fucus* (d) *Chlamydomonas*

113. A stage in cell division is shown in the figure. Select the answer which gives correct identification of the stage with its characteristics.

(a)	Late anaphase	Chromosomes move away from equatorial plate, Golgi complex not present
(b)	Cytokinesis	Cell plate formed, mitochondria distributed between two daughter cells
(c)	Telophase	Endoplasmic reticulum and nucleolus not reformed yet
(d)	Telophase	Nuclear envelop reforms, golgi complex reforms

114. Seed coat is not thin, membranous in :
(a) Coconut (b) Groundnut
(c) Gram (d) Maize

115. The diagram given here is the standard ECG of a normal person, the P-wave represents the :

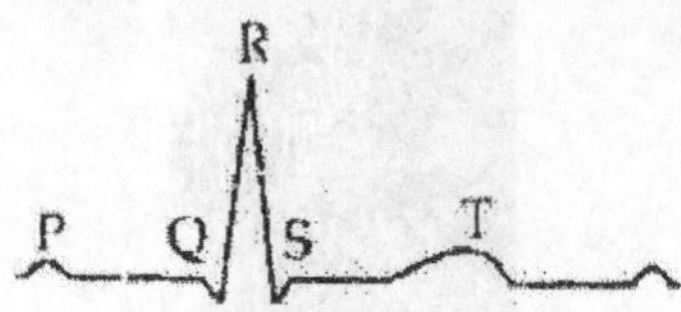

(a) Initiation of the ventricular contraction
(b) Beginning of the systole
(c) End of systole
(d) Contraction of both the atria

116. Which Mendelian idea is depicted by a cross in which the F_1 generation resembles both the parents?
(a) Law of dominance
(b) Inheritance of one gene
(c) Co-dominance
(d) Incomplete dominance

117. The tendency of population to remain in genetic equilibrium may be disturbed by :
(a) lack of migration
(b) lack of mutations
(c) lack of random mating
(d) random mating

118. If both parents are carriers for thalassemia, which is an autosomal recessive disorder, what are the chances of pregnancy resulting in an affected child?
(a) 50% (b) 25%
(c) 100% (d) no chance

119. In plant breeding programmes, the entire collection (of plants/seeds) having all the diverse alleles for all genes in a given crop is called :
(a) cross-hybridisation among the selected parents.
(b) evaluation and selection of parents.
(c) germplasm collection
(d) selection of superior recombinants.

120. The cell-mediated immunity inside the human body is carried out by :
(a) B-lymphocytes (b) Thrombocytes
(c) Erythrocytes (d) T-lymphocytes

121. Match the name of the animal (column I), with one characteristics (column II), and the phylum/class (column III) to which it belongs :

	Column I	Column II	Column III
(a)	*Ichthyophis*	terrestrial	Reptilia
(b)	*Limulus*	body covered by chitinous exoskeleton	Pisces

(c)	*Adamsia*	radially symmetrical	Porifera
(d)	*Petromyzon*	ectoparasite	Cyclostomata

122. Pigment-containing membranous extensions in some cyanobacteria are :
(a) Basal bodies (b) Pneumatophores
(c) Chromatophores (d) Heterocysts

123. Kyoto Protocol was endorsed at :
(a) CoP - 5 (b) CoP - 6
(c) CoP - 4 (d) CoP - 3

124. Select the answer which correctly matches the endocrine gland with the hormone it secrets and its function/deficiency symptom:

	Endocrine gland	Hormone	Function/ deficiency symptom
(a)	Posterior pituitary	Growth Hormone (GH)	Oversecretion Stimulates abnormal growth
(b)	Thyroid gland	Thyroxine	Lack of iodine in diet results in goitre
(c)	Corpus luteum	Testosterone	Stimulates spermato-genesis
(d)	Anterior pituitary	Oxytocin	Stimulates uterus contraction during child birth

125. The first stable product of fixation of atmospheric nitrogen in leguminous plants is :
(a) Ammonia (b) NO_3^-
(c) Glutamate (d) NO_2^-

126. Natural reservoir of phosphorus is :
(a) Animal bones (b) Rock
(c) Fossils (d) Sea water

127. What external changes are visible after the last moult of a cockroach nymph?
(a) Anal cerci develop
(b) Both fore wings and hind wings develop
(c) Labium develops
(d) Mandibles become harder

128. What is the correct sequence of sperm formation?
(a) Spermatogonia, spermatocyte, spermatozoa, spermatid
(b) Spermatogonia; spermatozoa, spermatocyte, spermatid
(c) Spermatogonia, spermatocyte, spermatid, spermatozoa
(d) Spermatid, spermatocyte, spermatogonia, spermatozoa

129. Select the wrong statement :
(a) Anisogametes differ either in structure, function or behaviour.
(b) In Oomycetes female gamete is smaller and motile, while male gamete is larger and non-motile.
(c) *Chalmydomonas* exhibits both isogamy and anisogamy and *Fucus* shows oogamy.
(d) Isogametes are similar in structure, function and behaviour.

130. Monoecious plant of *Chara* shows occurrence of :
(a) stamen and carpel of the same plant
(b) upper antheridium and lower oogonium on the same plant
(c) upper oogonium and lower antheridium on the same plant
(d) antheridiophore and archegoniophore on the same plant

131. The essential chemical components of many coenzymes are :
(a) Nucleic acids (b) Carbohydrates
(c) Vitamins (d) Proteins

132. Which of the following statements is not true of two genes that show 50% recombination frequency?
(a) The genes are tightly linked
(b) The genes show independent assortment
(c) If the genes are present on the same chromosome, they undergo more than one crossovers in every meiosis
(d) The genes may be on different chromosomes

133. Read the following statements (A-E) and answer the question which follows them.
1. In liverworts, mosses and ferns gametophytes are free-living
2. Gymnosperms and some ferns are heterosporous.
3. Sexual reproduction in *Fucus*, *Volvox* and *Albugo* is oogamous
4. The sporophyte in liveworts is more elaborate than that in mosses
5. Both, *Pinus* and *Marchantia* are dioecious

How many of the above statements are correct?
(a) Two (b) Three (c) Four (d) One

134. The incorrect statement with regard to Haemophilia is :
(a) It is a recessive disease
(b) It is a dominant disease
(c) A single protein involved in the clotting of blood is affected
(d) It is a sex-linked disease

135. Advantage of cleistogamy is :
(a) More vigorous offspring
(b) No dependence of pollinators
(c) Vivipary
(d) Higher genetic variability

136. Transition state structure of the substrate formed during an enzymatic reaction is :
(a) permanent but unstable
(b) transient and unstable
(c) permanent and stable
(d) transient but stable

137. In china rose the flowers are :
(a) Actinomorphic, epigynous with valvate aestivation
(b) Zygomorphic, hypogynous with imbricate aestivation
(c) Zygomorphic, epigynous with twisted aestivation
(d) Actinomorphic, hypogynous with twisted aestivation

138. Age of a tree can be estimated by :
(a) biomass
(b) number of annual rings
(c) diameter of its heartwood
(d) its height and girth

139. Which of the following are likely to be present in deep sea water ?
(a) Eubacteria (b) Blue-green algae
(c) Saprophytic fungi (d) rchaebacteria

140. Variation in gene frequencies within populations can occur by chance rather than by natural section. The is referred to as :
(a) Genetic drift (b) Random mating
(c) Genetic load (d) Genetic flow

141. A sedentary sea anemone gets attached to the shell lining of hermit crab. The association is :
(a) Symbiosis (b) Commensalism
(c) Amensalism (d) Ectoparasitism

142. Which of the following is not correctly matched for the organism and its cell wall degrading enzyme?
(a) Plant cells-Cellulase
(b) Algae-Methylase
(c) Fungi-Chitinase
(d) Bacteria-Lysozyme

143. Product of sexual reproduction generally generates:
(a) Prolonged dormancy
(b) New genetic combination leading to variation
(c) Large biomass
(d) Longer viability of seeds

144. Which of the following represent maximum number of species among global biodiversity?
(a) Lichens (b) Fungi
(c) Mosses and Ferns (d) Algae

145. One of the legal methods of birth control is :
(a) by abstaining from coitus from day 10 to 17 of the menstrual cycle
(b) by having coitus at the time of day break
(c) by a premature ejaculation during coitus
(d) abortion by taking an appropriate medicine

146. Which one of the following processes during decomposition is correctly described?
(a) Humification-Leads to the accumulation of a dark coloured substance humus which undergoes microbial action at every fast rate
(b) Catabolism-Last step decomposition under fully anaerobic condition
(c) Leaching-Water soluble inorganic nutrients rise to the top layers of soil
(d) Fragmentation-Carried out by organisms such as earthworm

147. DNA fragments generated by the restriction endonucleases in a chemical reaction can be separated by :
(a) Polymerase chain reaction
(b) Electrophoresis
(c) Restriction mapping
(d) Centrifugation

148. The figure shows a diagrammatic view of human respiratory system with labels A, B, C and D. Select the option which gives correct identification and main function and/or characteristics.

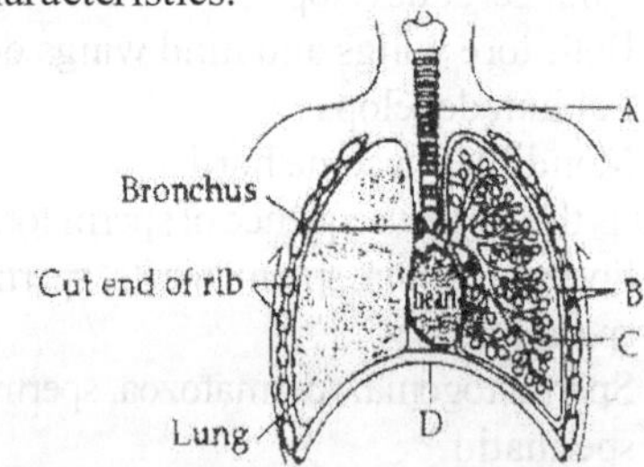

(a) B-pleural membrane-surrounds ribs on both sides to provides cushion against rubbing.
(b) C-Alveoli-thin walled vascular bag like structures for exchange of gases.
(c) D-Lower end of lungs-diaphragm pulls it down during inspiration
(d) A-trachea-long tube supported by complete cartilaginous rings for conducting inspired air.

149. Which one of the following is not used for ex-situ plant conservation ?
(a) Seed banks
(b) Shifting cultivation
(c) Botanical Gardens
(d) Field gene banks

150. Lenticels are involved in :
(a) Gaseous exchange (b) Food transport
(c) Photosynthesis (d) Transpiration

151. Among bitter gourd, mustard, brinjal, pumpkin, china rose, lupin, cucumber, sunhemp, gram, guava, bean, chilli, plum, petunia, tomato, rose, withania, potato, onion, aloe and tulip how many plants have hypogynous flower ?
(a) Ten (b) Fifteen
(c) Eighteen (d) Six

152. The complex formed by a pair of synapsed homologous chromosomes is called
(a) Kinetochore (b) Bivalent
(c) Axoneme (d) Equatorial plate

153. Which one of the following statements is correct?
(a) Sporogenous tissue is haploid
(b) Endothecium produces the microspores
(c) Tapetum nourishes the developing pollen
(d) Hard outer layer of pollen is called intine

154. A major site for synthesis of lipids is :
(a) SER (b) Symplast
(c) Nucleoplasm (d) RER

155. Select the correct statement with respect to locomotion in humans :
(a) Accumulation of uric acid crystals in joints causes their inflammation
(b) The vertebral column has 10 thoracic vertebrae.
(c) The joint between adjacent vertebrae is a fibrous joint
(d) The decreased level of progesterone causes osteoporosis in old people

156. A biologist studied the population of rats in a barn. He found that the average natality was 250, average mortality 240, immigration 20 and emigration 30. The net increase in population is :
(a) 15 (b) 05 (c) zero (d) 10

157. Parts A, B, C and D of the human eye are shown in the diagram. Select the option which gives correct identification along with its functions/ characteristics :

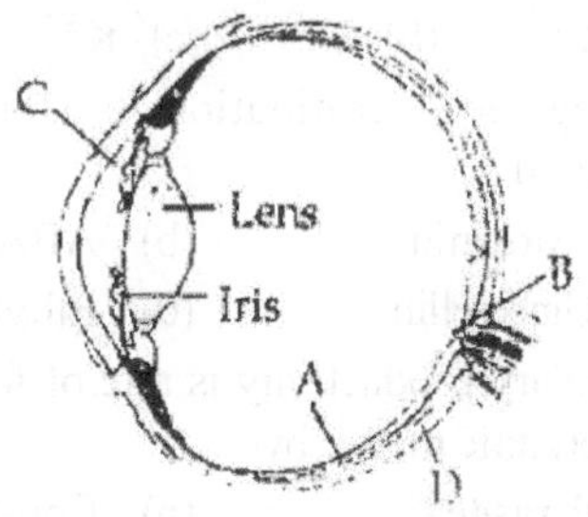

(a) B-Blind spot-has only a few rods and cones
(b) C-Aqueous chamber-reflects the light which does not pass through the lens.
(c) D-Choroid-is anterior part forms ciliary body
(d) A-Retina-contains photoreceptors-rods and cones.

158. Which of the following are correctly matched with respect to their taxonomic classification?
(a) Centipede, millipede, spider, scorpion-Insecta
(b) House fly, butterfly, tse tse fly, silverfish-Insecta
(c) Spiny anteater, sea urchin, sea cucumber-Echinodermata
(d) Flying fish, cuttlefish, silverfish-Pisces

159. Figure shown schematic plan of blood circulation in humans with labels A to D. Identify the label and give its function's.

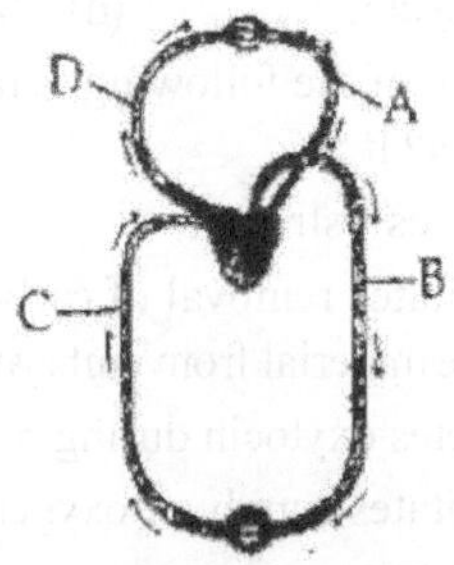

(a) B-Pulmonary artery-takes blood from heart to lungs, $PO_2 = 90$ mm Hg
(b) C-Vena Cava - takes blood from body parts to right auricle, $PCO_2 = 45$ mm Hg
(c) D - Dorsal aorta - takes blood from Heart to body Part $PO_2 = 95$ mm Hg
(d) A-Pulmonary vein - takes impure blood from body parts, $PO_2 = 60$ mm Hg

160. The most abundant intracellular cation is :
(a) Ca^{++} (b) H^+ (c) K^+ (d) Na^+

161. During seed germination its stored food is mobilized
(a) Cytokinin (b) ABA
(c) Gibberellin (d) Ethylene

162. Secondary productivity is rate of formation of new organic matter by
(a) Parasite (b) Consumer
(c) Decomposer (d) Producer

163. The colonies of recombinant bacteria appear white in contrast to blue colonies of non-recombinant bateria because of :
(a) Insertional inactivate of alphagalactosidase in non-recombinant bacteria
(b) Insertional inactivation of alpha-galactosidase in recombinant bacteria
(c) Inactivation of glycosidase enzyme in recombinant bacteria
(d) Non-recombinant bacteria containing beta-galactosidase

164. Which of the following Bt crops is being grown in India by the farmers?
(a) Cotton (b) Brinjal
(c) Soyabean (d) Maize

165. Interfascicular cambium develops from the cells of :
(a) Xylem parenchyma (b) Endomermis
(c) Pericycle (d) Medullary rays

166. Which one of the following is not the function of placenta? It :
(a) secretes estrogen
(b) facilitates removal of carbon dioxide and waste material from embryo.
(c) secretes oxytocin during parturition
(d) facilitates supply of oxygen and nutrients to embyo

167. Which of the metabolites is common to respiration mediated breakdown of fats, carbohydrates and proteins?
(a) Fructose 1, 6 - bisphosphate
(b) Pyruvic acid
(c) Acetyl CoA
(d) Glucose - 6 - phosphate

168. According to Darwin, The organic evolution is due to :
(a) Interspecific competition
(b) Competition within closely related species
(c) Reduced feeding efficiency in one species
(d) Intraspecific competition

169. Which enzymes will be produced in a cell in which there is a nonsense mutation in the lac Y gene?
(a) Lactose permease
(b) Transacetylase
(c) Lactose permease and transcetylase
(d) β- galactosidase

170. A good producer of citric acid is :
(a) *Pseudomonas* (b) *Clostridium*
(c) *Saccharomyces* (d) *Aspergillus*

171. Macro molecule chitin is :
(a) Phosphorus containing polysaccharide
(b) Sulphur containing polysaccharide
(c) Simple polysaccharide
(d) Nitrogen containing polysaccharide

172. The H-zone in the skeletal muscle fibre is due to :
(a) The central gap between myosin filaments in the A-band.
(b) The central gap between actin filaments extending through myosin filaments in the A-band
(c) Extension of myosin filaments in the central portion of the A-band.
(d) The absence of myofibrils in the central portion of A-band.

173. Meiosis takes place in :
(a) Conidia (b) Gemmule
(c) Megaspore (d) Meiocyte

174. A diagram showing axon terminal and synapse is given. Identify correctly at least two of A-D.

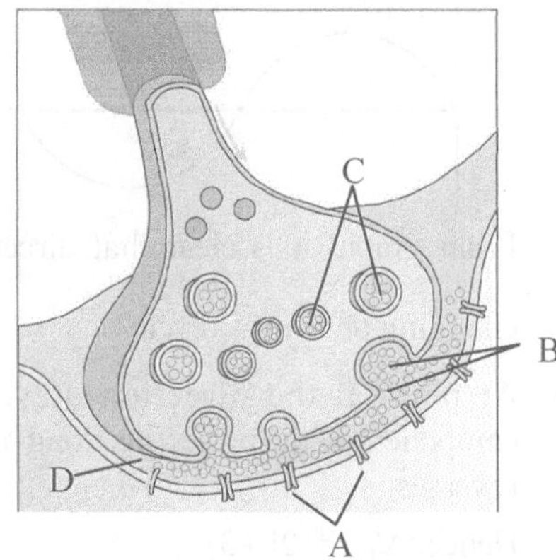

(a) B-Synaptic connection, D-K^+
(b) A-Neurotransmitter, B-Synaptic cleft
(c) C- Neurotransmitter, D-Ca^{++}
(d) A-Receptor, C-Synaptic vesicles

175. Which one of the following is not a correct statement ?
(a) Botanical gardens have collection of living plants for reference.
(b) A museum has collection of photographs of plants and animals
(c) Key is taxonomic aid for identification of specimens.
(d) Herbarium houses dried, pressed and preserved plant specimens.

176. Global warming can be controlled by :
(a) Reducing reforestation, increasing the use of fossil fuel.
(b) Increasing deforestation, slowing down the growth of human population.
(c) Increasing deforestation, reducing efficiency of energy usage.
(d) Reducing deforestation, cutting down use of fossil fuel.

177. The three boxes in this diagram represents the three major biosynthetic pathways in aerobic respiration. Arrows represents net reactants or products.

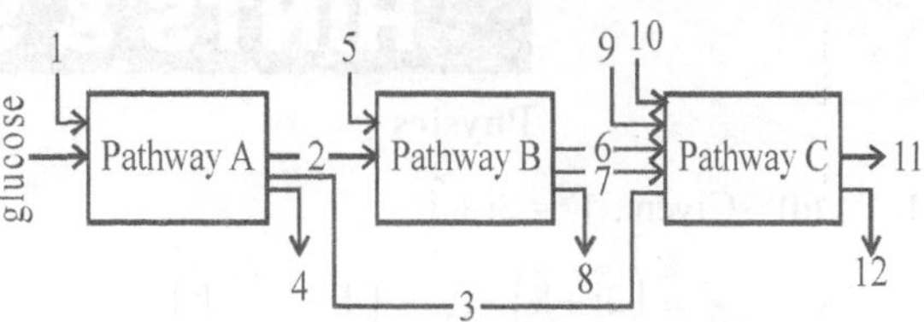

Arrows numbered 4, 8 and 12 can all be :
(a) ATP (b) H_2O
(c) FAD^+ or $FADH_2$ (d) NADH

178. Artificial insemination mean:
(a) Transfer of sperms of husband to a test tube containing ova
(b) Artificial introduction of sperms of a healthy donor into the vagina
(c) Introduction of sperms of a healthy donor directly into the ovary
(d) Transfer of sperms of a healthy donor to a test tube containing ova

179. One of the representatives of phylum Arthropoda is :
(a) Silverfish (b) Pufferfish
(c) Flying fish (d) Cuttlefish

180. Figure shown human urinary system with structures labelled A to D. Select option which correctly identifies them and gives their characteristics and /or functions.

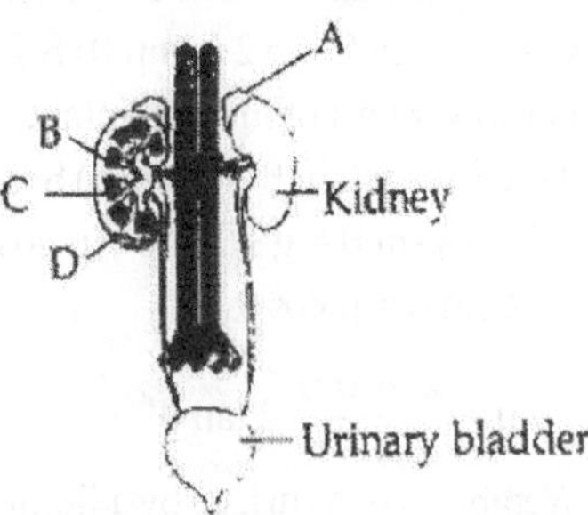

(a) B-pelvis-broad funnel shaped space inner to hilum, directly connected to loops of Henle.
(b) C-Medulla-inner zone of kidney and contains complex nephrons.
(c) D - Cortex - outer part of kidney and do not contain any part of nephrons
(d) A-Adrenal gland - located at the anterior part of kidney. Secrete Catecholamines which stimulate glycogen breakdown.

HINTS & SOLUTIONS

Physics

1. (d) Given : $\vec{F} = 3\hat{i} + \hat{j}$

$\vec{r_1} = (2\hat{i} + \hat{k})$, $\vec{r_2} = (4\hat{i} + 3\hat{j} - \hat{k})$

$\vec{r} = \vec{r_2} - \vec{r_1} = (4\hat{i} + 3\hat{j} - \hat{k}) - (2\hat{i} + \hat{k})$

or $\vec{r} = 2\hat{i} + 3\hat{j} - 2\hat{k}$

So work done by the given force $w = \vec{f}.\vec{r}$

$= (3\hat{i} + \hat{j}).(2\hat{i} + 3\hat{j} - 2\hat{k}) = 6 + 3 = 9J$

2. (a) Potential at B, V_B is maximum

$V_B > V_C > V_A$

As in the direction of electric field potential decreases.

3. (c) By inserting iron rod in the coil,

$L \uparrow z \uparrow I \downarrow$ so brightness $\downarrow$

4. (b)

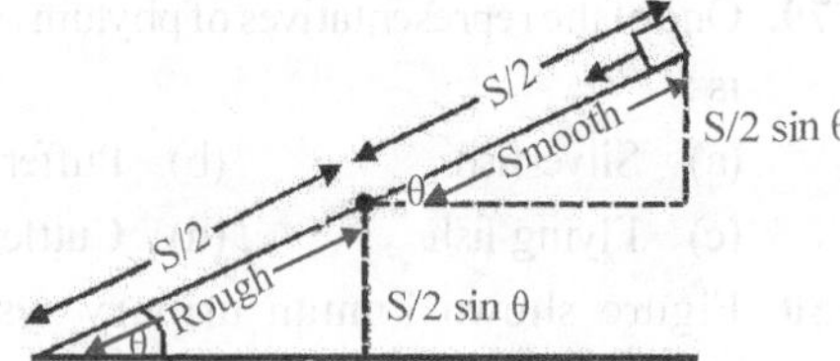

For upper half of inclined plane

$v^2 = u^2 + 2a\, S/2 = 2\,(g \sin\theta)\, S/2 = gS \sin\theta$

For lower half of inclined plane

$0 = u^2 + 2\,g\,(\sin\theta - \mu\cos\theta)\, S/2$

$\Rightarrow -gS\sin\theta = gS\,(\sin\theta - \mu\cos\theta)$

$\Rightarrow 2\sin\theta = \mu\cos\theta$

$\Rightarrow \mu = \dfrac{2\sin\theta}{\cos\theta} = 2\tan\theta$

5. (c) Wetability of a surface by a liquid primarily depends on angle of contact between the surface and liquid.

If angle of contact is acute liquids wet the solid and vice-versa.

6. (d) Required condition : Frequency of microwaves = Resonant frequency of water molecules.

7. (a) W_{net} = Area of triangle ABC $= \dfrac{1}{2}$ AC × BC

$= \dfrac{1}{2} \times 5 \times 10^{-3} \times 4 \times 10^5 = 1000\,J$

8. (a) This is the case of periodic EMI

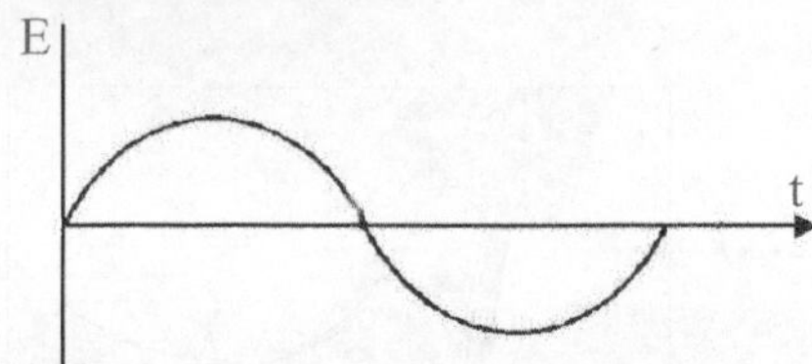

From graph, it is clear that direction is changing once in $\dfrac{1}{2}$ cycle.

9. (b) At point B the direction of velocity component of the projectile along Y - axis reverses.

Hence, $\vec{V_B} = 2\hat{i} - 3\hat{j}$

10. (d) $F = \dfrac{YA}{L} \times \ell$

So, extension, $\ell \propto \dfrac{L}{A} \propto \dfrac{L}{D^2}$

[∵ F and Y are constant]

$\ell_1 \propto \dfrac{100}{1^2} \propto 100;$ $\quad \ell_2 \propto \dfrac{200}{2^2} \propto 50;$

$\ell_3 \propto \dfrac{300}{3^2} \propto \dfrac{100}{3};$ $\quad \ell_4 \propto \dfrac{50}{\frac{1}{4}} \propto 200$

The ratio of $\dfrac{L}{D^2}$ is maximum for case (d).

Hence, option (d) is correct.

11. (c) Resistance $R = \dfrac{\rho\ell}{A}$; $\quad \because \ell' = 2\ell$

$\therefore A' = \dfrac{A}{2}$ $\therefore R' \rho\dfrac{2\ell}{\frac{A}{2}} = 4R = 4 \times 4\,\Omega = 16\,\Omega$

Therefore the resistance of new wire becomes 16 Ω

12. (a) Wein's displacement law

According to this law $\lambda_{max} \propto \dfrac{1}{T}$

or, $\lambda_{max} \times T$ = constant

So, as the temperature increases λ decreases.

13. (c)

$\dfrac{3V^2}{4g}$

V

From law of conservation of mechanical energy

$$\frac{1}{2}I\omega^2 + 0 + \frac{1}{2}mv^2 = mg \times \frac{3v^2}{4g}$$

$$\Rightarrow \frac{1}{2}I\omega^2 = \frac{3}{4}mv^2 - \frac{1}{2}mv^2$$

$$= \frac{mv^2}{2}\left(\frac{3}{2}-1\right)$$

or, $\frac{1}{2}I\frac{V^2}{R^2} = \frac{mv^2}{4}$ or, $I = \frac{1}{2}mR^2$

Hence, object is a disc.

14. (a) Magnetic dipole moment

$M = m \times \ell$ $M' = m \times r$

From figure

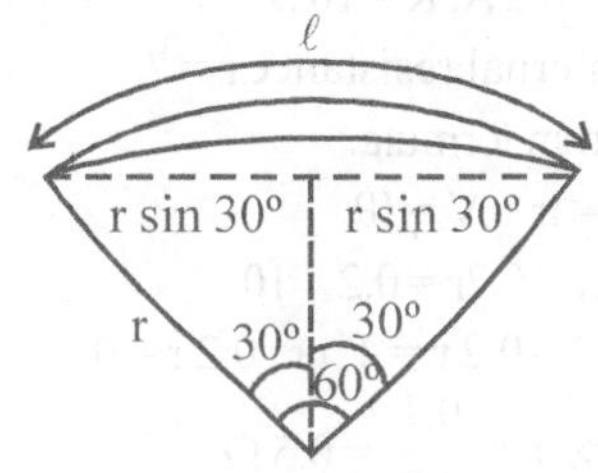

$\ell = \frac{\pi r}{3}$ or $r = \frac{3\ell}{\pi}$

so, $M' = m \times r = \frac{m \times 3\ell}{\pi} = \frac{3}{\pi}M$

15. (d)

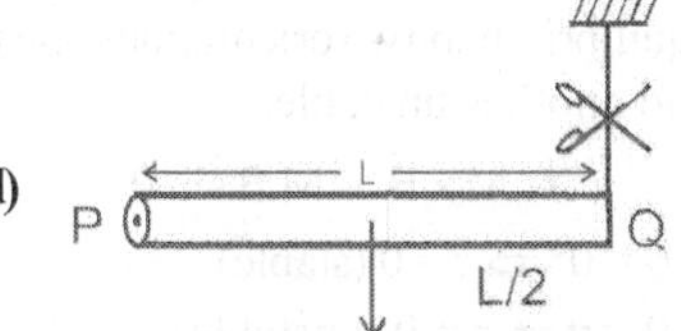

Weight of the rod will produce the torque

$$\tau = mg\frac{L}{2} = I\alpha = \frac{mL^2}{3}\alpha \left[\because I_{rod} = \frac{ML^2}{3}\right]$$

Hence, angular acceleration $\alpha = \frac{3g}{2L}$

16. (b) In a n-type semiconductor holes are minority carriers and pentavalent atoms are dopants.

17. (d) Voltage gain $\Delta_V = \beta \frac{R_{out}}{R_{in}}$

$$\Rightarrow G = 25 \frac{R_{out}}{R_{in}} \quad ...(i)$$

Transconductance $g_m = \frac{\beta}{R_{in}}$

$$\Rightarrow R_{in} = \frac{\beta}{g_m} = \frac{25}{0.03}$$

Putting this value of R_{in} in eqn. (i)

$$G = 25\frac{R_{out}}{25} \times 0.03 \quad ...(ii)$$

$$\therefore G' = 20\frac{R_{out}}{20} \times 0.02 \quad ...(iii)$$

From eqs. (ii) and (iii)

Voltage gain of new transistor $G' = \frac{2}{3}G$

18. (b) From photo-electric equation,

$$h\nu' = h\nu + K_{max} \quad ...(i)$$

$$h.2\nu = h\nu + \frac{1}{2}mV^2_{max}$$

$[\therefore \nu' = 2\nu]$

$$\Rightarrow h\nu = \frac{1}{2}mV^2_{max} \Rightarrow V_{max} = \sqrt{\frac{2h\nu}{m}}$$

19. (a) $\because y = \frac{n\lambda D}{d}$

$\therefore n_1\lambda_1 = n_2\lambda_2$

$\Rightarrow n_1 \times 12000 \times 10^{-10} = n_2 \times 10000 \times 10^{-10}$

or, $n(12000 \times 10^{-10}) = (n+1)(10000 \times 10^{-10})$

$\Rightarrow n = 5$

$\left(\because \lambda_1 = 12000 \times 10^{-10}\text{m}; \lambda_2 = 10000 \times 10^{-10}\text{m}\right)$

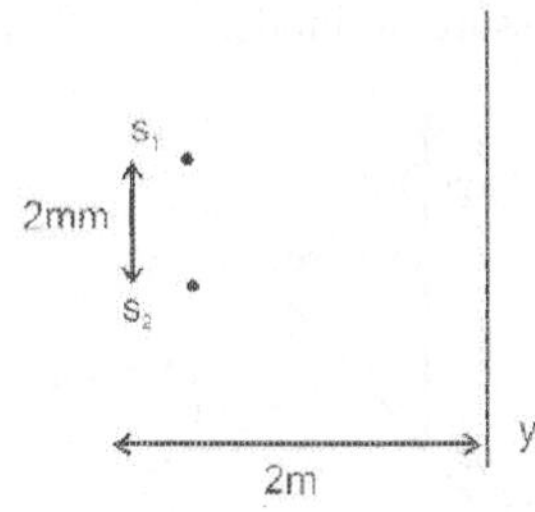

$$y_{common} = \frac{n\lambda_1 D}{d} = \frac{5\left(12000 \times 10^{-10}\right) \times 2}{2 \times 10^{-3}}$$

$(\because d = 2 \text{ mm and } D = 2\text{m})$

$= 5 \times 12 \times 10^{-4}\text{ m} = 60 \times 10^{-4}\text{ m}$

$= 6 \times 10^{-3}\text{m} = 6\text{ mm}$

20. (d)

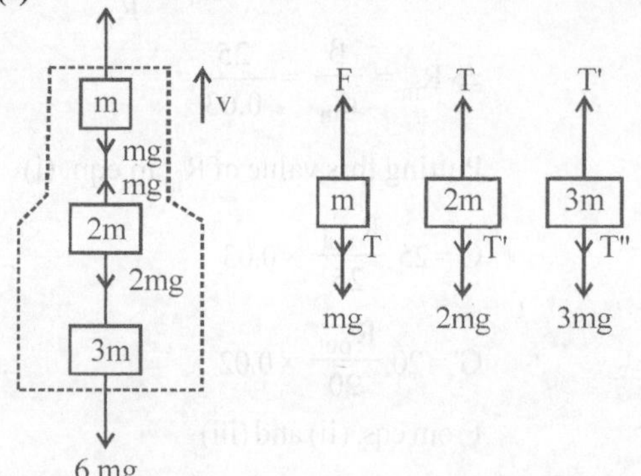

From figure

$F = 6\,mg$,

As speed is constant, acceleration $a = 0$

$\therefore 6\,mg = 6ma = 0, F = 6\,mg$

$\therefore T = 5\,mg, T' = 3\,mg$

$T'' = 0$

F_{net} on block of mass 2 m

$= T - T' - 2\,mg = 0$

ALTERNATE:

$\because$ v = constant

so, $a = 0$, Hence, $F_{net} = ma = 0$

21. (b) Mass defect $\Delta m = 0.02866$ a.m.u.

Energy $= 0.02866 \times 931 = 26.7$ MeV

As ${}_1H^2 + {}_1H^2 \longrightarrow {}_2He^4$

Energy liberated per a.m.u = 13.35/2 MeV = 6.675 MeV

22. (c) Pressure change will be minimum at both ends. In fact, pressure variation is maximum at $\ell/2$ because the displacement node is pressure antinode.

23. (a)

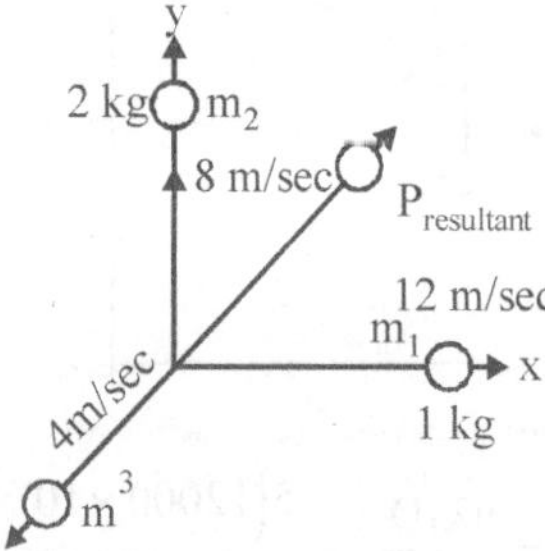

$$P_{resultant} = \sqrt{12^2 + 16^2} = \sqrt{144 + 256} = 20$$

$m_3 v_3 = 20$ (momentum of third part)

or, $m_3 = \frac{20}{4} = 5$ kg

24. (d) $P = \frac{a^3 b^2}{cd}, \frac{\Delta P}{P} \times 100\% = 3\frac{\Delta a}{a} \times 100\% + 2\frac{\Delta b}{b} \times 100\% + \frac{\Delta c}{c} \times 100\% + \frac{\Delta d}{d} \times 100\%.$

$= 3 \times 1\% + 2 \times 2\% + 3\% + 4\% = 14\%$

25. (d) When sounded with a source of known frequency fundamental frequency

$= 250 \pm 4$ Hz $= 254$ Hz or 246 Hz

2nd harmonic if unknown frequency (suppose) 254 Hz $= 2 \times 254 = 508$ Hz

As it gives 5 beats

$\therefore 508 + 5 = 513$ Hz

Hence, unknown frequency is 254 Hz

26. (a) Given : emf $\varepsilon = 2.1$ V

$I = 0.2$ A, $R = 10\Omega$

Internal resistance r = ?

From formula.

$\varepsilon - Ir = V = IR$

$2.1 - 0.2r = 0.2 \times 10$

$2.1 - 0.2\,r = 2$ or $0.2\,r = 0.1$

$\Rightarrow r = \frac{0.1}{0.2} = 0.5\,\Omega$

ALTERNATE : $i = \frac{\varepsilon}{r + R} \Rightarrow 0.2 = \frac{2.1}{r + 10}$

$\Rightarrow 2.1 = 0.2\,r + 2 \Rightarrow r = \frac{1}{2} = 0.5\,\Omega$

27. (c) A current loop in a magnetic field is in equilibrium in two orientations one is stable and another unstable.

$\because \quad \tau = \vec{M} \times \vec{B} = MB\sin\theta$

If $\theta = 0° \Rightarrow \tau = 0$ (stable)

If $\theta = \pi \Rightarrow \tau = 0$ (unstable)

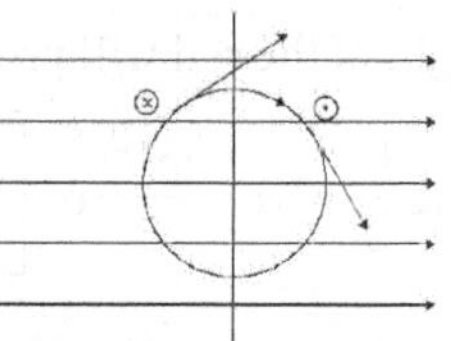

Do not experience a torque in some orientations

Hence option (c) is correct.

28. (d) As $P = \frac{E}{C}$

$\lambda_p = \frac{\hbar C}{E}$ ...(i) $\qquad \lambda_e^2 = \frac{\hbar h}{\sqrt{2mE}}$ ...(ii)

From equations (i) and (ii)

$\lambda_p \propto \lambda_e^2$

29. **(a)** The value of x is $\frac{1}{8}$

$= \frac{x_0}{8} = \frac{x_0}{2^3} \Rightarrow t = 3T = 3 \times 20 = 60$ years

Hence the estimated age of the rock is 60 years

ALTERNATE:

	X →	Y_0
at t = 0	N_0	0
at t = t	N	$N_0 - N$

$$\frac{N}{N_0 - N} = \frac{1}{7} = \frac{N}{N_0} = \frac{1}{8}$$

t = 3T

= 3 × 20 = 60 years

30. **(a)** Given : V = 7 V

r = 5Ω

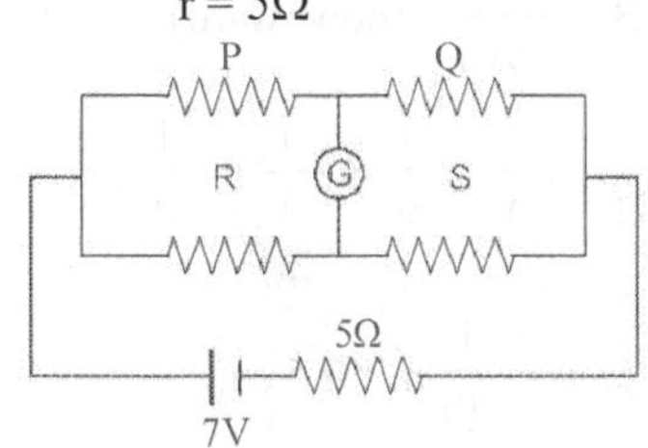

$$R_{eq} = \frac{40 \times 120}{40 + 120}\,\Omega$$

$$I = \frac{V}{R} = \frac{7}{5 + \frac{40 \times 120}{40 + 120}}$$

$$= \frac{7}{5 + 30} = \frac{1}{5} = 0.2\,A.$$

31. **(b)** $P_1 > P_2$

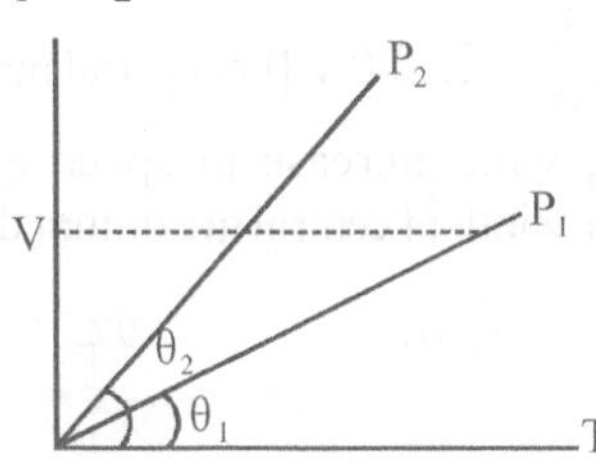

As V = constant ⇒ P ∝ T

Hence from V–T graph $P_1 > P_2$

32. **(a)** $C_p - C_v = R \Rightarrow C_p = C_v + R$

$$\because \gamma = \frac{C_p}{C_v} = \frac{C_v + R}{C_v} = \frac{C_v}{C_v} + \frac{R}{C_v}$$

$$\Rightarrow \gamma = 1 + \frac{R}{C_v} \Rightarrow \frac{R}{C_v} = \gamma - 1 \Rightarrow C_v = \frac{R}{\gamma - 1}$$

33. **(d)** From first law of thermodynamics

$\Delta Q = \Delta U + \Delta W$

$$= \frac{3}{2} \cdot \frac{1}{4} R(T_2 - T_1) + 0$$

$$= \frac{3}{8} N_a K_B (T_2 - T_1) \; [\because K = \frac{R}{N}]$$

34. **(b)**

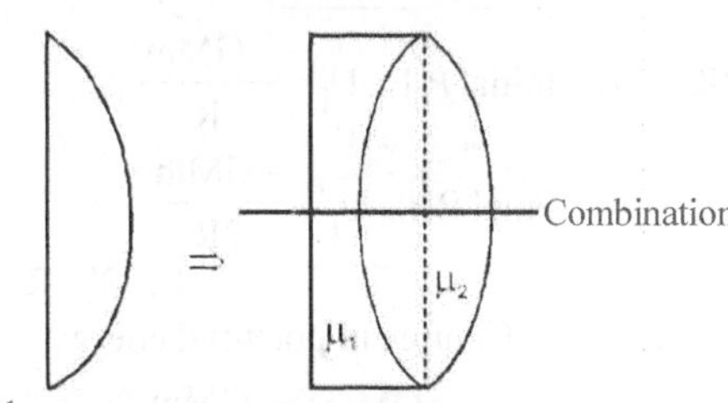

$$\frac{1}{f} = \frac{1}{f_1} + \frac{1}{f_2}$$

$$= (\mu_1 - 1)\left(\frac{1}{\infty} - \frac{1}{-R}\right) + (\mu_2 - 1)\left(\frac{1}{\infty} - \frac{1}{R}\right)$$

$$= \frac{(\mu_1 - 1)}{R} - \frac{(\mu_2 - 1)}{R} \Rightarrow \frac{1}{f} = \frac{\mu_1 - \mu_2}{R}$$

$$\Rightarrow f = \frac{R}{\mu_1 - \mu_2}$$

Hence, focal length of the combination is $\frac{R}{\mu_1 - \mu_2}$.

35. **(c)** According to question $P \propto T^3$

But as we know for an adiabatic process the pressure $P \propto T^{\frac{\gamma}{\gamma - 1}}$.

So, $\frac{\gamma}{\gamma - 1} = 3 \Rightarrow \gamma = \frac{3}{2}$ or, $\frac{C_p}{C_v} = \frac{3}{2}$

36. **(d)** As $Y = A \sin(\omega t - kx + \phi)$

$$\omega = 2\pi f = \frac{2\pi}{\pi} = 2 \quad [\because f = \frac{1}{\pi}]$$

$$k = \frac{2\pi}{\lambda} = \frac{2\pi}{2\pi} = 1 \quad [\because \lambda = 2\pi]$$

$\therefore Y = 1 \sin(2t - x + \phi) \quad [\because A = 1\,m]$

37. **(b)**

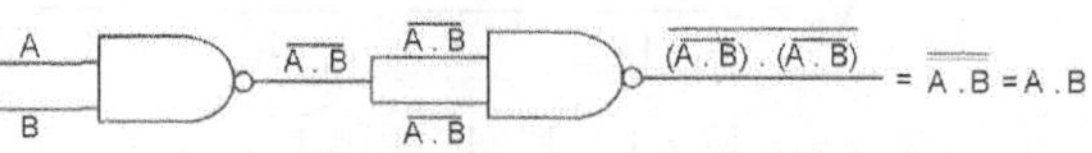

i.e., output X = A.B

ALTERNATE:

A	B	X
0	0	0
1	0	0
0	1	0
1	1	1

38. (a) Initial P. E., $U_i = \dfrac{-GMm}{R}$,

Final P.E., $U_f = \dfrac{-GMm}{3R}$

$[\because R' = R + 2R = 3R]$

$\therefore$ Change in potential energy,

$$\Delta U = \frac{-GMm}{3R} + \frac{GMm}{R}$$

$$= \frac{GMm}{R}\left(1 - \frac{1}{3}\right) = \frac{2}{3}\frac{GMm}{R} = \frac{2}{3}mgR$$

$$\left(\because \frac{GMm}{R} = mgR\right)$$

ALTERNATE : $\Delta U = \dfrac{mgh}{1 + \dfrac{h}{R}}$

By placing the value of h = 2R we get

$\Delta U = \dfrac{2}{3}mgR.$

39. (d) For Lyman series $(2 \to 1)$

$$\frac{1}{\lambda_L} = R\left[1 - \frac{1}{2^2}\right] = \frac{3R}{4}$$

For Balmer series $(3 \to 2)$

$$\frac{1}{\lambda_B} = R\left[\frac{1}{4} - \frac{1}{9}\right] = \frac{5R}{36}$$

$$\Rightarrow \frac{\lambda_L}{\lambda_B} = \frac{\dfrac{4}{3R}}{\dfrac{36}{5R}} = \frac{4}{36}\left(\frac{5}{3}\right) = \frac{5}{27}$$

40. (c)

2 kg 2 kg 2 kg

0 1 2 3 4 5 6 7 8 m

Gravitational potential $V = \dfrac{-Gm}{r}$

$$V_0 = -\frac{G \times 2}{1} - \frac{G \times 2}{2} - \frac{G \times 2}{4} - \frac{G \times 2}{8}$$

$$-2G\left[1 + \frac{1}{2} + \frac{1}{4} + \frac{1}{8} + ...\infty\right]$$

$$= -2G \times \frac{1}{1 - \frac{1}{2}} = -2G \times \frac{1}{\frac{1}{2}} = -4G.$$

41. (a)

N, W, E, S, $\vec{E}$

When moves with an acceleration a_0 towards west, electric field

$$E = \frac{F}{q} = \frac{ma_0}{e} \text{ (West)}$$

When moves with an acceleration $3a_0$ towards east, magnetic field

$$B = \frac{2ma_0}{ev_0} \text{ (downward)}$$

42. (a) $\dfrac{1}{f} = \dfrac{1}{v} - \dfrac{1}{u} = \dfrac{1}{\frac{1}{40}} = \dfrac{1}{u}$

$$\Rightarrow u = \frac{1}{40} m = 2.5 \text{ cm}$$

43. (b)

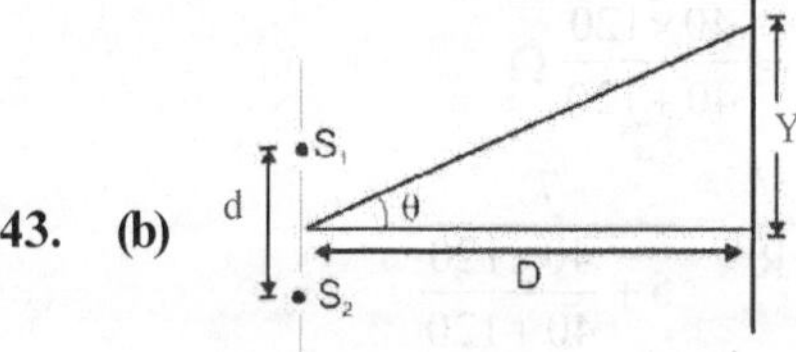

Angular width, $\theta = \dfrac{Y}{D} = \dfrac{n\lambda D}{dD}$ $\left[\because Y = \dfrac{D\lambda}{d}\right]$

so, $\theta = \dfrac{\lambda}{d}$, $v \uparrow \lambda \downarrow \theta \downarrow$ [For central maxima n = 1]

Hence, with increase in speed of electrons angular width of central maximum decreases.

44. (a) 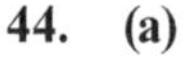

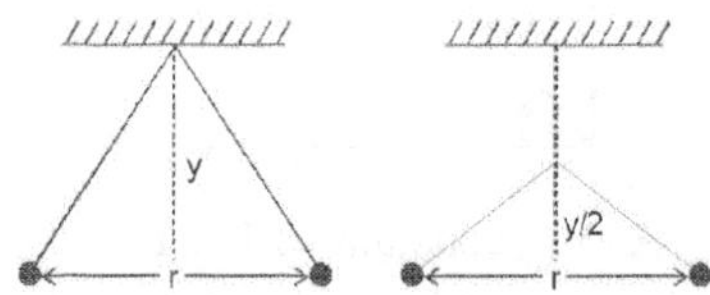

From figure, $\tan\theta = \dfrac{F_e}{mg} \Rightarrow \dfrac{r/2}{y} = \dfrac{\dfrac{kq^2}{r^2}}{mg}$

$\left[\because F = \dfrac{kq^2}{r^2} \text{ from coulomb's law}\right]$

$$\Rightarrow r^3 \propto y \Rightarrow r'^3 \propto \frac{y}{2} \Rightarrow \frac{r'}{r} = \frac{1}{2^{1/3}}$$

$$\Rightarrow r' = \frac{r}{\sqrt[3]{2}}$$

45. (a) $\because \quad h = \frac{1}{2}gt^2$

$$\therefore \quad h_1 = \frac{1}{2}g(5)^2 = 125$$

$$h_1 + h_2 = \frac{1}{2}g(10)^2 = 500$$

$$\Rightarrow h_2 = 375$$

$$h_1 + h_2 + h_3 = \frac{1}{2}g(15)^2 = 1125$$

$$\Rightarrow h_3 = 625$$

$$h_2 = 3h_1,\ h_3 = 5h_1$$

$$\text{or } h_1 = \frac{h_2}{3} = \frac{h_3}{5}$$

Chemistry

46. (b) $CH_2 = \overset{Cl}{\overset{|}{C}} - CH = CH_2$ (chloroprene) is the monomer of neoprene.

$$\underset{\text{Chloroprene}}{CH_2 = \overset{Cl}{\overset{|}{C}} - CH = CH_2} \xrightarrow[K_2S_2O_8]{\text{Polymerisation}} \underset{\text{Neoprene}}{\left[CH_2 - \overset{Cl}{\overset{|}{C}} = CH - CH_2\right]_n}$$

47. (d) $[Cu(NH_3)_4]^{2+}$ hybridisation dsp^2

$Cu^{+2} - 3d^9$ has one unpaired e^-

↑↓	↑↓	↑↓	↑↓	↑

So magnetic moment

$$\mu = \sqrt{n(n+2)} = \sqrt{1(1+2)} = \sqrt{3} = 1.73$$

48. (b) Density is given by

$d = \frac{Z \times M}{N_A a^3}$; where Z = number of formula units present in unit cell, which is 4 for *fcc*

a = edge length of unit cell. M = Molecular mass

$$2.72 = \frac{4 \times M}{6.02 \times 10^{23} \times (404 \times 10^{-10})^3}$$

$(\because 1\text{pm} = 10^{-10}\text{ cm})$

$$M = \frac{2.72 \times 6.02 \times (404)^3}{4 \times 10^7} = 26.99$$

$= 27 \text{ gm mole}^{-1}$

49. (a) [Structure: H_3C–CH_2–C3(=C4(–CH_3)–C5≡C6H)–C2H(OH)–C1OOH]

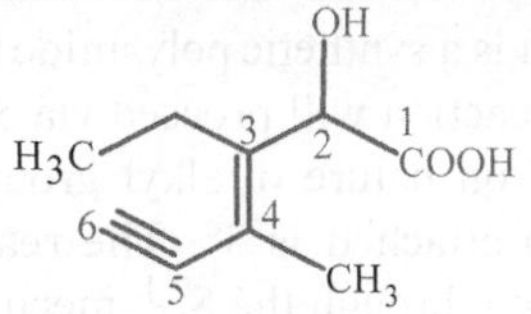

IUPAC name of the structure is 3-ethyl-2-hydroxy-4-methylhex-3-en-5-ynoic acid

50. (d) Boron nitride (BN) is known as inorganic graphite. The most stable form is hexagonal one. It has layered structure similar to graphite.

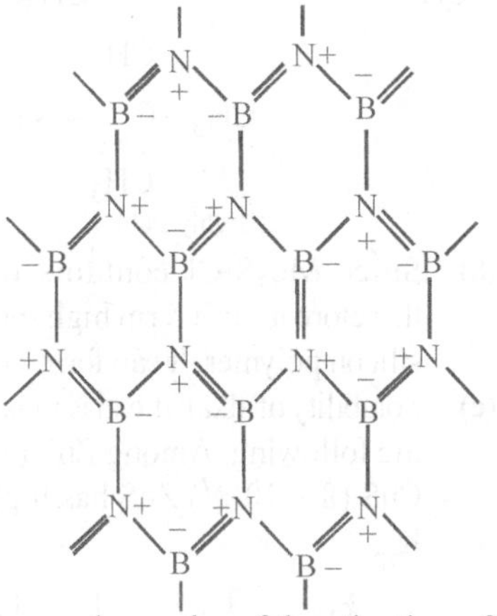

51. (c) Decreasing order of deactivating effect of the given *m*-directing group is

$> NO_2 > -CN > -SO_3H > -COOH$

$-NO_2$ group is most deactivating group due to strong – E, – I and – M effects.

52. (d) $\text{Molarity (M)} = \frac{\text{wt} \times 1000}{\text{mol. wt.} \times \text{vol (ml)}}$

$$2 = \frac{\text{wt.}}{63} \times \frac{1000}{250} \quad \text{wt.} = \frac{63}{2}\text{ gm}$$

$$\text{wt. of 70\% acid} = \frac{100}{70} \times 31.5 = 45 \text{ gm}$$

53. (a) Enolic form predominates in compounds containing two carbonyl groups separated by a $-CH_2$ group. This is due to following two factors.

(i) Presence of conjugation which increases stability.

(ii) Formation of intramolecular hydrogen bond between enolic hydroxyl group and second carbonyl group which leads to stablisation of the molecule.

Hence the correct answer is III > II > I.

54. (b) Dilute solutions of boric acid and hydrogen peroxide are weak antiseptics.

55. (b) Nylon is a synthetic polyamide polymer.

56. (b) The reaction will proceed via S_N^1 or S_N^2 based on nature of alkyl group. If alkyl group attached is 3°. The reaction will proceed through the S_N^1 mechanism and if alkyl group is primary reaction will proceed through S_N^2 mechanism.

$$CH_3-\underset{CH_3}{\overset{CH_3}{\underset{|}{\overset{|}{C}}}}-O-CH_3 \xrightarrow[S_N1]{HI} CH_3-\underset{CH_3}{\overset{CH_3}{\underset{|}{\overset{|}{C}}}}-\overset{+}{O}\langle^{H}_{CH_3} \longrightarrow$$

$$CH_3-\underset{CH_3}{\overset{CH_3}{\underset{|}{\overset{+}{C}}}} \xrightarrow{I^-} CH_3-\underset{CH_3}{\overset{CH_3}{\underset{|}{\overset{|}{C}}}}-I$$

(Highly stable)

57. (b) Since Me_3SiCl contains only one Cl, therefore it can't form high molecular mass silicon polymer. It can form only dimer.

58. (c) Solubility of alkali metal is maximum among the following. Among ZnS (1.7×10^{-5}) & CuS (8×10^{-37}) ZnS has higher value of K_{sp}.

59. (b) $\log \frac{k_2}{k_1} = \frac{E_a}{2.303R}\left(\frac{1}{T_1} - \frac{1}{T_2}\right)$

$\log 2 = \frac{E_a}{2.303 \times 8.314}\left[\frac{1}{293} - \frac{1}{308}\right]$

$0.3 = \frac{E_a}{2.303 \times 8.314} \times \frac{15}{293 \times 308}$

$E_a = \frac{0.3 \times 2.303 \times 8.314 \times 293 \times 308}{15}$.

$= 34673 \text{ J mole}^{-1} = 34.7 \text{ J mole}^{-1}$

60. (a) $H_2 \longrightarrow 2H^+ + 2e^-$

1 atm $\quad 10^{-10}$

$E_{H_2/H^+} = 0 - \frac{0.059}{2}\log\frac{(10^{-10})^2}{1}$

$E_{H_2/H^+} = +0.59 \text{ V}$

61. (c) $C = \nu\lambda$

$\lambda = \frac{C}{\nu} = \frac{3 \times 10^{17}}{6 \times 10^{15}} = 50 \text{ nm}$

62. (c) $n = 3 \rightarrow 3^{rd}$ shell

$l = 1 \rightarrow$ p sub shell.

m = – 1 is possible for two electrons present in an orbital.

63. (c) $Sm^{2+}(Z = 62)$

$[Xe]4f^6\, 6s^2$ – 6 unpaired e^-

$Eu^{2+}(Z = 63)$

$[Xe]4f^7\, 6s^2$ – 7 unpaired e^-

$Yb^{2+}(Z = 70)$

$[Xe]4f^{14}\, 6s^2$ – 0 unpaired e^-

$Ce^{2+}(Z = 58)$

$[Xe]4f^1\, 5d^1\, 6s^2$ – 2 unpaired e^-

Only Yb^{2+} is diamagnetic.

64. (a) $M = \frac{6.02 \times 10^{20} \times 1000}{100 \times 6.02 \times 10^{23}} = \frac{6.02 \times 10^{21}}{6.02 \times 10^{23}}$

$= 0.01 \text{ M}$

65. (c) Energy of an electron at infinite distance from the nucleus is zero. As an electron approaches the nucleus, the electron attraction increases and hence the energy of electron decreases and thus becomes negative. Thus as the value of *n* decreases, *i.e.* lower the orbit is, more negative is the energy of the electron in it.

66. (d) $[Cr(H_2O)_4Cl_2]Cl + AgNO_3 \longrightarrow AgCl + [Cr(H_2O)_4Cl_2]NO_3$

$\text{Molarity} = \frac{\text{wt}}{\text{mol. mass}} \times \frac{1000}{\text{vol.}}$

$\frac{\text{wt}}{\text{mol. mass}} = \text{molarity} \times \frac{\text{vol.}}{1000} = \frac{0.01 \times 100}{1000}$

$= 0.001$

67. (b) HCl and SO_2 are reducing agents and can reduce MnO_4^-. CO_2 which is neither oxidising and nor reducing will provide only acidic medium. It can shift reaction in forward direction and reaction can go to completion.

68. (b) $-NO_2$ is a powerful electron withdrawing group. Its presence on ring makes the ring less active

69. (b) BF_3 is Lewis acid(e^- pair acceptor)

70. (a) SiO_4^{4-} is basic structural unit of silicates.

71. (c) Higher the critical temperature more easily will be the gas liquify. Now since most easily liquifiable gas show larger deviation, NH_3 will show maximum deviation from ideal behaviour.

72. (b) $HClO_4$ is the strongest acid amongst all because the oxidation state or Cl is maximum (+7).

73. (c) Zn/Hg and HCl reduce carboxyl group to methylene group (Clemmensen reduction).

74. (d) Presence of $6p$ orbitals, each containing one unpaired electron, in a six membered cyclic structure is in accordance with Huckel rule of aromaticity.

75. (a) Based on the features given above the gas must be SO_2.

76. (b) $\alpha = \frac{\Lambda_M}{\Lambda_M^{\infty}} = \frac{9.54}{238} = 0.04008 = 4.008\,\%$.

77. (a) In interstitial compounds small atoms like H, B & C enter into the void sites between the packed atoms of crystalline metal. They retain metallic conductivity and are chemically inert.

78. (b) [Reaction scheme: 2-nitro-5-bromobenzenediazonium chloride (NO_2, Br, $\oplus N_2Cl^-$) + $H_3PO_2 + H_2O$ → 3-bromonitrobenzene (NO_2, Br) $+ H_3PO_3 + HCl$]

79. (b) $(BH_3)_2$ or (B_2H_6)

It contains two 3 centre - 2 electron bonds and present above and below the plane of molecules compounds which do not have sufficient number of electrons to form normal covalent bonds are called electron deficient molecules.

80. (a) $H \overset{\sigma}{-} \ddot{O}: \overset{\sigma}{-} H$

$O \leftarrow \overset{\sigma}{\underset{\pi}{S{=}O}}$ $\quad O \leftarrow \overset{\sigma}{\underset{\pi}{N{=}O}}$ $\quad O \overset{\sigma}{\underset{\pi}{=}} C \overset{\sigma}{\underset{\pi}{=}} O$

81. (c) $(NH_4)_2Cr_2O_7 \xrightarrow{\Delta} N_2 + Cr_2O_3 + 4H_2O$

$Zn(ClO_3)_2 \xrightarrow{\Delta} ZnCl_2 + 3O_2$

$2\,KClO_3 \xrightarrow{\Delta} 2KCl + 3O_2$

$4K_2Cr_2O_7 \xrightarrow{\Delta} 4K_2CrO_4 + 2Cr_2O_3 + 3O_2$

82. (a) SF_4 has 4 bond pairs and 1 lone pair of electrons, sp^3d hybridisation leads to irregular shape :SF_4 [structure: S with four F atoms and a lone pair] and resultant $\mu \neq 0$.

83. (d) $\begin{matrix} CH_3 \\ CH_3 \end{matrix} \!\!>\! CH - CH_2 -$ (iso-butyl group)

84. (a) Molecular orbital configuration of O_2^- is

$O_2^-\,(17) = \sigma 1s^2, \sigma^* 1s^2, \sigma 2s^2, \sigma^* 2s^2, \sigma 2p_z^2, \pi 2p_x^2 = \pi 2p_y^2, \pi^* 2p_x^2 = \pi^* 2p_y^1$

85. (a) Diamond is like ZnS. In diamond cubic unit cell, there are eight corner atoms, six face centered atoms and four more atoms inside the structure.

Number of atoms present in a diamond cubic cell

$= 8 \times \frac{1}{8} + 6 \times \frac{1}{2} + 4 = 8$

(corners) (face centered) (inside body)

86. (a) $F{-}\ddot{\underset{\cdot\cdot}{Xe}}{:}{-}F$ $\quad sp^3d$ and Linear

$^-Cl{-}\underset{\cdot\cdot}{I}{:}{-}Cl$ $\quad sp^3d$ and Linear

87. (b) $\Delta H = E_{a_f} - E_{a_b} = 0$

88. (b) This type of attractive force operates between the polar molecules having permanent dipole and the molecules lacking permanent dipole.

HCl is polar ($\mu \neq 0$) and He is non polar ($\mu = 0$), thus gives dipole-induced dipole interaction.

89. (d) $E^\circ_{Cell} = E^\circ_{OP} + E^\circ_{RP}$
$= 0.76 + 0.314 = 1.10\,V$

90. (a) Nitrobenzene $\xrightarrow[80-100°C]{HNO_3/H_2SO_4}$ 1,3 - Dinitrobenzene

Biology

91. (b) DNA $\xrightarrow{A}$ mRNA $\xrightarrow{B}$ protein $\xrightarrow[C]{\text{Proposed by}}$

In this question A is transcription, B - translation C - Francis Crick (central dogma) It is unidirectional flow of information DNA to mRNA (transcription) and then decoding the information present in mRNA in the formation of polypeptide chain or protein (translation).

92. (b) Perisperm is remnants of nucellus which is diploid (2n) but endosperm is triploid (3n). Perisperm occurs in the seeds of Black pepper, coffee, castor, cardamum,

Nymphaea. Endosperm is the food laden tissue which is meant for nourishing the embryo in seed plants. In angiosperms the endosperm is formed as a result of vegetative fertilization, triple fusion or fusion of a male gamete with diploid secondary nucleus of the central cell.

93. (a) In *Cycas* specialised root called coralloid roots are associated with N_2 – fixing cyanobacteria either *Nostoc* or *Anabaena.* Coralloid roots lie near the soil surface. They are irregular and often dichotomously branched. Root hair and root cap are absent in these roots.

94. (b) Hormones are non-nutrient chemicals which act as intercellular messengers and are produced in trace amounts.
Gastrointestinal tracts secretes four major peptide hormones – gastrin, secretin, cholecystokinin (CCK) and gastric inhibitory peptide (GIP) while juxtaglomerular cells of kidney secretes erythropoietin, a peptide hormone. Releasing and inhibitory hormones are produced by hypothalamus. Neurohypophysis or posterior pituitary is under direct neural regulation of the hypothalamus.

95. (c) Ovule is also called integumented megasporangium. It develops into seed after fertilisation in spermatophytes. It occurs singly or in a cluster inside ovary with parenchymatous cushions called placenta.

96. (d) ABO blood group system in human beings is an example of codominant, dominant recessive and multiple alletes. Blood groups are controlled by the gene I located on 9th chromosome that has 3 multiple alleles, out of which any two are found in a person. In codominance both gene express it self completely.

97. (d) Iodine is essential for the normal rate of hormone synthesis in the thyroid. Deficiency of iodine in our diet results in **hypothyroidism** and enlargement of the thyroid gland, commonly called **goitre.** Hypothyroidism during pregnancy causes defective development and maturation of the growing baby leading to stunted growth (cretinism), mental retardation, low intelligence quotient, abnormal skin, deaf-mutism, etc.

98. (c) Rough endoplasmic reticulum is a network or reticulum of tiny tubular structures scattered in the cytoplasm and bear ribosomes on their outer surface. These are involved in protein synthesis and secretion. They are extensive and continuous with the outer membrane of the nucleus.

99. (b) A saturated or unsaturated fatty acid esterified to a glycerol molecule to which a phosphate group is also attached forms phosphoglycerides.
Phosphoglycerides are major structural components of cell membranes. They are, therefore, also called membrane lipids. Lecithin is one example.

100. (d) The major component of biogas is methane (about 50-68%) which is highly inflammable. The other gases are carbon dioxide (25 - 35%), hydrogen (1 - 5%), nitrogen (2 - 7%), Oxygen (0 - 0.1%) and rarely hydrogen sulfide. Biogas is a "mixture of gases" produced from degradable organic matter by the activity of various anaerobic bacteria that offers a low cost alternative for energy requirements.

101. (b) The eye of octopus and eye of cat show different patterns of structure, yet they perform similar functions. This is an example of analogous organs.
Analogous organs have evolved due to convergent evolution . Analogous organs have developed in the evolutionary process through adaptation of quite different organisms to similar mode of life.

102. (c) Substance that have a hydrophilic moiety, find it difficult to pass through the membrane, their movement has to be facilitated by the membrane protein without expenditure of ATP energy. Facilitated diffusion cannot cause net transport of molecules from a low to a high concentration or uphill transport. Transport rate reaches a maximum when all of the protein transporters are being used (saturation). Facilitated diffusion is very specific.

103. (a) The wings is a classic example of convergent evolution. Flying insects, birds and bats same all evolved the capacity of flight independently. All four serve the same function and are similar in structure but anatomically different.

104. (d) *Ascaris*, an intestinal parasite causes **ascariasis.** Symptoms of these disease include internal bleeding, muscular pain, fever, anemia and blockage of the intestinal passage. The eggs of the parasite are excreted along with the faeces of infected persons which contaminate soil, water, plants, etc. A healthy person acquires this infection through contaminated water vegetables, fruits, etc.

105. (a) In India, the Air (Prevention and control of pollution) Act came into force in 1981 but was amended in 1987 to include noise as an air pollutant.

106. (b) Prawn , Scorpion and Locusta belong to phylum Arthropoda. This is the largest phylum of Animalia, over two-thirds of all named species on earth are Arthropoda. Phylum is named so due to presence of jointed appendages (arthros = jointed, poda = appendages) in body of members of this phylum.

107. (c) Jaundice cannot be detected in a developing foetus by amniocentesis. It is a foetal sex determination test based on chromosomal pattern in the amniotic fluid surrounding the developing embryo. It is now legally ban in India.

108. (c) Golgi apparatus plays a major role in post translational modification of proteins forming glycoprotein and glycosidation of lipid forming glycolipids. A number of proteins and lipids synthesised on endoplasmic reticulum (rough and smooth respectively) are modified in the cisternae of the Golgi apparatus before they are released from it trans face.

109. (d) Small intestine is major area of absorbtion of nutrients. Approximately 80% of absorbtion take place here. Glucose, fructose, fatty acids, amino acids (Glycine etc.) are absorbed through mucosa into blood and lymph by active absorbtion.

110. (d)

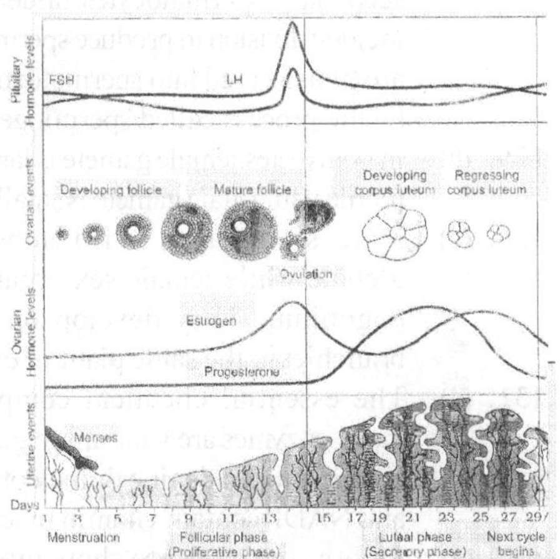

During menstrual or bleeding phase which comprises of 3 - 5 days out of 28 days menstrual cycle, the production of LH from the anterior lobe of the pituitary gland is considerably reduced . The withdrawal of this hormone causes degeneration of corpus luteum and therefore, progesterone production is reduced. Production of estrogens is also reduced in this phase.

111. (b) **Synovial joints** are characterised by the presence of a fluid filled synovial cavity between the articulating surfaces of the two bones. Such an arrangement allows considerable movement. These joints help in locomotion and many other movements. Ball and socket joint (between humerus and pectoral girdle), Hinge joint (knee joint), Pivot joint (between atlas and axis), Gliding joint (between the carpals) and Saddle joint (between carpal and metacarpal of thumb) are some examples.

112. (a) In *Spirogyra,* sexual reproduction occurs through conjugation. Gametes are non-flagellated morphologically similar. But physiologically different (isogamy with physiological anisogamy). *Volvox* and *fucus* are examples of oogamous and *Chlamydomonas* contains isogamous flagellated gametes .

113. (d) Telophase is the stage of reconstitution of nuclei. The chromosomes that have reached their respective poles decondense and lose their individuality and collect in a mass in the two poles. Nuclear envelope assemble around chromatin mass. Nucleolus, Golgi complex and ER reform.

114. (a) The seed coat develops from integuments originally surrounding the ovule. It is thick and hard in coconut which protect the embryo from mechanical injury and from drying out.

115. (d)

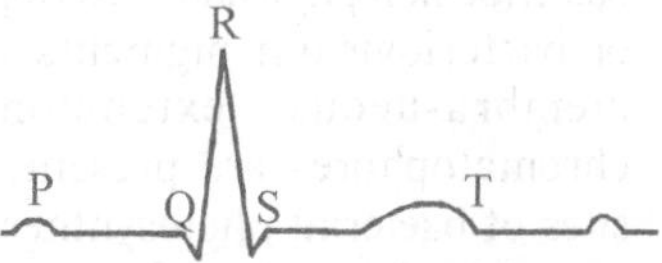

The P-wave represents the electrical **excitation (or depolarisation) of the atria,** which leads to the contraction of both the atria. The QRS complex represents the **depolarisation of the ventricles,** which initiates the ventricular contraction. The contraction starts shortly after Q and marks the beginning of the systole.

116. (c) In Co-dominance F_1 generation resemble both the parents. Ex: Blood group inheritance.

117. (c) The tendency of population to remain in genetic equilibrium may be disturbed by five factors. These are gene migration or gene flow, genetic drift, mutation, genetic recombination and natural selection.

118. (b) Genotype of carrier parents is – Aa (male parent) × Aa (female parent)

	A	a ♂
A	AA	Aa
♀ a	Aa	*aa

AA → normal child (25%)
Aa → carriers child (50%)
aa → affected child (25%)

119. (c) Germplasm collection is the first step of plant breeding programmes. As genetic variability is the root of any breeding programme. In many crops pre-existing genetic variability is available from wild relatives of the crop. Collection and preservation of all the different wild varieties, species and relatives of the cultivated species (followed by their characteristics) is a pre requisite for effective exploitation of natural genes available in the population.

120. (d) T – lymphocytes mediates cell mediated immunity (CMI). It is one of the two types of acquired immunity responsible for graft or transplant rejection.

121. (d)

Ichthyophis	–	Amphibian
Limulus	–	Arthropoda
Adamsia	–	Cnidaria
Petramyzon	–	Jawless vertebrate & ectoparasite, cyclostomate

122. (c) In some photoautotrophic bacteria bacteriochlorophyll and bacteriophaeophytin or bacterioviridin pigments containing membra-neous extensions called chromatophores are present. These are sites of bacterial photosynthesis .

123. (d) In year 1997, at the conference of parties III (COP - 3) , Kyoto, Japan the Kyoto conference on climate change took place. In that conference developed countries agreed to specific targets for cutting their emissions of green house gases. A general framework has defined for this with specifics to be detailed in next few years. This become known as the kyoto protocol.

124. (b)
- Growth hormone secreted by Anterior pituitary
- Corpus leutum secreted Progesterone
- Oxytocin is secreted by Posterior pituitary

125. (a) Ammonia is first stable compound produce in fixation of atmospheric nitrogen in leguminous plant. A molecule of nitrogen is change into two molecules of ammonia. It requires 16 ATP, a source of electron donor ferredoxin, hydrogen donor NADPH or $FMNH_2$ and dinitrogenase enzyme.

126. (b) The natural reservoir of phosphate is rock which contains phosphorus in the form of phosphates when rocks are weathered, minute amounts of these phosphate dissolve in soil solution and are absorbed by the roots of the plants . Herbivores and other animals obtain this element from plants.

127. (b) The development of cockroach is paurometabolous meaning that their development is through nymphal stage. The nymphs look very much like adults. The nymph grows by moulting about 13 times to reach adult form. The last nymphal stage has wingpads while adult cockroaches have wings.

128. (c) In testis, the immature male germ cells or spermatogonia (2n) multiply by mitotic divide and increase in number. Some spermatogonia (2n) known as primary spermatocytes divide meiotic division to form secondary spermatocytes (n). The secondary spermatocytes undergo second meiotic division to produce spermatid which are transformed into spermatozoa (sperms) by the process called spermiogenesis.

129. (b) In oomycetes female gamete is large and non motile while male gamete is small & motile.

130. (c) Male sex organ is called antheridium or globule while female sex organ is called oogonium. They develop on the same branchlet in the same plant in *chara.*

131. (c) The essential chemical components of many enzymes are vitamins, e.g., coenzyme nicotinamide adenine dinucleotide (NAD) and NADP contain vitamin niacin.

132. (a) Tightly linked genes show more linkage then crossing over.

133. (b) In liverworts and ferns gametophytes are free living while in fern sporophytes are free living. Gymnosperms and genera like *Selaginella* and *Salvinia* are heterosporous. The sporophyte in mosses are more elaborate than that of liverworts, *Pinus* is monoecious and heterosporous. *Marchantia* is dioecious.

134. (b) Haemophilia is sex linked recessive disease in which a simple protein that is a part of protein cascade involved in clotting of blood is affected. Due to this, in an affected individual a simple cut will result in non stop bleeding .

135. (b) Cleistogamy favours no dependence on pollinator because flowers never open. In such flowers, the anthers and stigma lie close to each other. When anthers dehisce in flower buds pollen grains come in contact with the stigma to effect pollination.

136. (b) Transition state structure formed during an enzymatic reaction is transient and unstable.

137. (d) In china rose (*Hibiscus rose-sinensis*) flowers are actinomorphic , hypogynous and twisted aestivation in corolla.

138. (b) Age of a tree can be estimated by number of annual rings. Annual ring constitute alternate concentric rings of spring wood and autumn wood.

139. (d) Archaebactera live in some of the most harsh habitats such as extreme salty areas (halophiles), hot springs (thermoacidophiles) and marshy areas (methanogens) and in deep sea water.

140. (a) Genetic drift is variation in gene frequencies within populations can occur by chance rather than by natural selection, sometimes the change in allele frequency is so different in the new sample of population that they become a different species. The original drifted population becomes founders and the effect is called founder effect.

141. (b) Commensalism is a interaction where one species is benefitted and other is neither benefitted nor harmed.

142. (b) Algae is a plant and so its cell wall is made up of cellulose. Cellulase enzyme is needed for degradation of its cell wall.

143. (b) Sexual reproduction leads to formation of new combination and appearance of variations. Genetic recombination, interaction etc. during sexual reproduction provides vigour and vitality to the offsprings. They better adapt themselves to changing environmental conditions and also plays an important role in evolution.

144. (b) The no. of species of fungi is 72000 that is maximum in respect to other options. The number of fungus species in the world is more than the combined total of species of fishes, amphibians, reptiles and mammals.

145. (d) Medical termination of pregnancy (MTP) or induced abortion become legal in India from 1971 with some strict conditions to avoid its misuse. Such restrictions are important to check indiscriminate and illegal female foeticides which are reported to be high in India. All the other options are of natural methods of contraception.

146. (d) Humification, catabolism, leaching and fragmentation are the steps of decomposition which operate simultaneously on the detritus. Fragmentation is breaking down detritus into smaller particles by detritivores like earthworm. By the process of leaching, water soluble inorganic nutrients go down into soil horizon and get precipitated as unavailable salts. Humification occurs at a very slow rate.

147. (b) DNA fragments generated by restriction endonucleases in a chemical reaction can be separated by gel electrophoresis. Since DNA fragments are negatively charged molecules they can be separated by forcing them to move towards the anode under an electric field through a medium/matrix. The DNA fragments separate according to their size through sieving effect provided by matrix.

148. (b) Alveoli are very thin, irregular walled bag like structures for gaseous exchange. Tracheae bronchi and bronchioles are supported by incomplete cartilaginous rings. Double layered pleural membrane surrounds the lungs with pleural fluid between them. It reduces friction on the lung surface.

149. (b) Shifting or Jhum cultivation contributes to deforestation. Ex-situ conservation is a biodiversity conservation method in which threatened animals and plants are taken out of their natural habitat and placed in protected and carring environment zoological parks, botanical gardens, wildlife safari parks, seed banks and gene banks.

150. (a) Lenticels are lens shaped openings occurs in stems of most of the woody trees. It permits the exchange of gases between the outer atmosphere and the internal tissue of the stem.

151. (b) Mustard (cruciferae) , brinjal (solanaceae) china rose (malvaceae), lupin (leguminosae), gram (leguminosae), bean (leguminosae), chilli, *Petunia*, potato, tomato, *Withania* (solanaceae), onion, *Aloe,* tulip (liliaceae) are having hypogynous flower while bittergourd, cucumber (cucurbitaceae), guava (myrtaceae) are having epigynous flower and rose are having perigynous flower.

152. (b) During zygotene, second stage of meiosis I, homologous chromosomes start pairing together forming a complex structure called synaptonemal complex. The complex formed by a pair of synapsed homologous chromosome is called a bivalent or a tetrad. Crossing over occurs between non sister chromatids of bivalent in the next stage.

153. (c) Sporogenous tissue is always diploid, endothecium is second layer of anther wall and perform the function of protection and help in dehiscence of anther to release the pollen. Hard outer layer of pollen is called exine but tapetum always nourishes the developing pollen.

Cells of the tapetum possess dense cytoplasm and generally have more than one nucleus (polypoid).

154. (a) The smooth endoplasmic reticulum is the major site for synthesis of lipid. In animal cells lipid like steroidal hormones are synthesised in SER.

155. (a) Vertebral column has 12 thoracic vertebrate. The joints between adjacent vertebrae is cartilaginous joint which permits limited movements. Progesterone is secreted by corpus luteum which supports in pregnancy in females.

156. (c) Net increase in population : (Natality + Immigration) – (Mortality + Emigration)
$(250+20)-(240+30)=270-270=0$

157. (d) Retina is the innermost layer of eyeball and it contains three layers of cells - ganglion cells, bipolar cells and photoreceptor cells. No photoreceptor cells are present in blind spot . The space between the cornea and the lens is called the aqueous chamber and contains a thin watery fluid called aqueous humor.

Choroid is anterior part of external sclera layer of eyeball while ciliary body is anterior part of middle choroid layer.

158. (b) House fly, butterfly, tse tse fly, silverfish all belongs to insecta.

159. (b) A is Pulmonary vein which takes pure blood from lungs to heart. B is dorsal aorta which takes pure blood from heart to various body parts. C is Vena cava which takes impure blood from various body pars to right auricle of heart. D is pulmonary artery which takes impure blood from heart to lungs.

160. (c) The most abundant intracellular cation is K^+.

161. (c) Gibberellins stimulate the synthesis of α-amylase and proteases enzyme in germinating grains of cereals. They are involved in the conversion of starch into sugar. The proteases convert an inactive β-amylase to the active form. The active β-amylase and α-amylase together digest starch to glucose which is mobilized to meet the metabolic demands of embryo.

162. (b) The rate of resynthesis of organic matter by the consumers is known as secondary productivity . It depends upon the loss while transferring energy containing organic matter from the previous trophic level plus the consumption due to respiration and predation. Respiration loss is about 20% for autotrophs, 30% for herbivores and upto 60% in case of carnivores. Therefore net productivity decreases with each trophic level.

163. (d) Alternative selectable markers have been developed which differentiate recombinant from non-recombinants on the basis of their ability to produce colour in the presence of chromogenic substrate. In this , a recombinant DNA is inserted within the coding sequence of an enzyme β-galactosidase. This results into inactivation of the enzyme, which is referred to as insertional inactivation. The presence of chromogenic substrate give blue coloured colonies of the plasmid in the bacteria does not have an insert. Presence of insert results into insertional inactivation of the galactosidase and the colonies do not produce any colour, these are identified as recombinant colonies.

164. (a) Bt cotton is being grown in India by the farmers. Bt cotton is pest resistant plant which could decreased the pesticides used. Bt toxin is produced by a bacterium *Bacillus thusringiensis* (Bt for short). Bt toxin gene has been cloned from the bacteria and been expressed in plants to provide resistance to insects without the need for insecticides.

165. (d) At the time of secondary growth interfascicular cambium is formed by parenchymatous medullary rays.
Interfascicular cambium along with intrafascicular cambium (formed from cambium cells present between xylem and phloem) constitute continuous cambium ring. If cut off new cells in both directions cause secondary growth in most dicotyledonous plants.

166. (c) **Function of placenta** – The placenta facilitate the supply of oxygen and nutrients to the embryo and also removal of carbon dioxide and excretory/waste materials produced by the embryo. The placenta is connected to the embryo through an umbilical cord which helps in the transport of substances to and from the embryo. Placenta also acts as an endocrine tissue and produces several hormones like **human chorionic gonadotropin** (hCG), **human placental lactogen** (hPL) , **estrogens, progestogens** etc.

167. (c) Acetyl CoA is common to respiration mediated breakdown of fats, carbohydrates and proteins. Fats are broken down to fatty acid and glycerol and again fatty acid degraded to acetyl Co - A. Protein first degraded by proteases to indiviual amino acids which deaminated to pyruric acid and further decarboxylised to acetyl Co- A.

168. (d) According to Darwin the organic evolution is due to intraspecific competition. Individuals of same species have variation in characteristics. Those characteristics which enable some to survive better in natural conditions (climate, food, physical factors etc.) would outbreed others that are less-endowed to survive under such natural conditions. There, therefore, will survive more and hence are selected by nature . He called it natural selection and implied it as a mechanism of evolution.

169. (d) A nonsense mutation is the one which stops polypeptide synthesis due to formation of a terminating or non sense codon. e.g. ATT(UAA), ATC (UAG), ACT(UGA). The lactose or lac operon of *Esherichia coli* contains structural genes (Z, Y, A) . If Y codes for termination of polypeptide chain then only the product of 'Z' gence teranscribe to form β galactosidase.

170. (d) A good source of citric acid is *Aspergillus niger* (a fungus). Apart from citric acid, oxalic acid, gallic acid, gluconic acid are extracted from fungus.

171. (d) Chitin or fungal cellulose that is nitrogen containing polysaccharide and hetero-polymer of NAG.

172. (b) Central part of thick filament, not overlapped by thin filaments is called the 'H' zone. 'H' zone is also called Hensen's Line.

173. (d) Meiosis takes place in meiocyte while Conidia and Gemmule are asexual structures and megaspore is haploid.

174. (d) A-Receptor, C-Synaptic vesicles
B is synaptic cleft. A synapse is formed by the membranes of a synaptic neuron and post synaptic neuron, which may or may not separated by a gap called synaptic cleft. It is filled by fluid called neurotransmitter which are involved in transmission of impulse at these synapses.

175. (b) **Museum** – Biological museums are generally set up in educational institutes such as schools and colleges. Museums have collections of preserved plant and animal specimens for study and reference. Specimens are preserved in the containers or jars in preservative solutions. Plant and animal specimens may also be preserved as dry specimens. Insects are preserved in insect boxes after collections, killing and pinning. Larger animals like birds and mammals are usually stuffed and preserved. Museums often have collections of skeletons of animals too.

176. (d) In first three options CO_2 concentration increases but it is decreases in option (d). Global warming is increase of earth's average temperature. Global warming is due to green house effect which is naturally occurring phenomenon that is responsible for heating of Earth's surface and atmosphere. The main green house gas is carbon-di-acid (contribute 60%).

177. (a) In a given diagram of aerobic respiration – pathway A is glycolysis, pathway B is kreb's cycle and pathway C is ETS , thus 4, 8 & 12 are ATP. ATP act as energy currency. The energy trapped in form of ATP and it broken down whenever and wherever it needs to be utilised.

178. (b) In **Artificial Insemination** (AI) technique, the semen collected either from the husband or a healthy donor is artificially introduced either into the vagina or into the uterus (IUI – **intrauterine insemination**) of the female.

179. (a) Representive of Phylum Arthropoda is silverfish. Arthropoda is the largest phylum of Animalia, which covers two-thirds of all named species.

180. (d) A – Adrenal gland – located at the anterior part of kidney, secrete Catecholamines which stimulate glycogen breakdown.

NEET (Karnataka) 2013 - PHYSICS

PHYSICS

1. The pair of quantities having same dimensions is
   (a) Young's modulus and energy
   (b) impulse and surface tension
   (c) angular momentum and work
   (d) work and torque

2. Vectors $\vec{A}, \vec{B}$ and $\vec{C}$ are such that $\vec{A}\cdot\vec{B}=0$ and $\vec{A}\cdot\vec{C}=0$. Then the vector parallel to $\vec{A}$ is
   (a) $\vec{B}$ and $\vec{C}$ (b) $\vec{A}\times\vec{B}$
   (c) $\vec{B}+\vec{C}$ (d) $\vec{B}\times\vec{C}$

3. The displacement 'x' (in meter) of a particle of mass 'm' (in kg) moving in one dimension under the action of a force, is related to time 't' (in sec) by $t=\sqrt{x}+3$. The displacement of the particle when its velocity is zero, will be
   (a) 2m (b) 4m (c) zero (d) 6m

4. A person holding a rifle (mass of person and rifle together is 100 kg) stands on a smooth surface and fires 10 shots horizontally, in 5 s. Each bullet has a mass of 10 g with a muzzle velocity of 800 ms^{-1}. The final velocity acquired by the person and the average force exerted on the person are
   (a) $-1.6\ ms^{-1}$; 8 N (b) $-0.08\ ms^{-1}$; 16 N
   (c) $-0.8\ ms^{-1}$; 8 N (d) $-1.6\ ms^{-1}$; 16 N

5. A car is moving in a circular horizontal track of radius 10 m with a constant speed of 10 m/s. A bob is suspended from the roof of the car by a light wire of length 1.0 m. The angle made by the wire with the vertical is
   (a) $0°$ (b) $\frac{\pi}{3}$ (c) $\frac{\pi}{6}$ (d) $\frac{\pi}{4}$

6. A particle with total energy E is moving in a potential energy region $U(x)$. Motion of the particle is restricted to the region when
   (a) $U(x)>E$ (b) $U(x)<E$
   (c) $U(x)=O$ (d) $U(x)\le E$

7. One coolie takes 1 minute to raise a suitcase through a height of 2 m but the second coolie takes 30 s to raise the same suitcase to the same height. The powers of two coolies are in the ratio of
   (a) 1 : 2 (b) 1 : 3 (c) 2 : 1 (d) 3 : 1

8. Two discs are rotating about their axes, normal to the discs and passing through the centres of the discs. Disc D_1 has 2 kg mass and 0.2 m radius and initial angular velocity of 50 rad s^{-1}. Disc D_2 has 4kg mass, 0.1 m radius and initial angular velocity of 200 rad s^{-1}. The two discs are brought in contact face to face, with their axes of rotation coincident. The final angular velocity (in rad s^{-1}) of the system is
   (a) 40 (b) 60 (c) 100 (d) 120

9. The ratio of radii of gyration of a circular ring and a circular disc, of the same mass and radius, about an axis passing through their centres and perpendicular to their planes are
   (a) $\sqrt{2}:1$ (b) $1:\sqrt{2}$ (c) $3:2$ (d) $2:1$

10. The radius of a planet is twice the radius of earth. Both have almost equal average mass-densities. If V_P and V_E are escape velocities of the planet and the earth, respectively, then
    (a) $V_E=1.5V_P$ (b) $V_P=1.5V_E$
    (c) $V_P=2V_E$ (d) $V_E=3V_P$

11. A particle of mass 'm' is kept at rest at a height $3R$ from the surface of earth, where 'R' is radius of earth and 'M' is mass of earth. The minimum speed with which it should be projected, so that it does not return back, is (g is acceleration due to gravity on the surface of earth)
    (a) $\left(\frac{GM}{R}\right)^{\frac{1}{2}}$ (b) $\left(\frac{GM}{2R}\right)^{\frac{1}{2}}$
    (c) $\left(\frac{gR}{4}\right)^{\frac{1}{2}}$ (d) $\left(\frac{2g}{4}\right)^{\frac{1}{2}}$

12. If the ratio of diameters, lengths and Young's modulus of steel and copper wires shown in the figure are p, q and s respectively, then the corresponding ratio of increase in their lengths would be
    (a) $\frac{7q}{(5sp)}$
    (b) $\frac{5q}{(7sp^2)}$
    (c) $\frac{7q}{(5sp^2)}$
    (d) $\frac{2q}{(5sp)}$

Steel
2m
Copper
5m

13. A fluid is in streamline flow across a horizontal pipe of variable area of cross section. For this which of the following statements is correct?

(a) The velocity is minimum at the narrowest part of the pipe and the pressure is minimum at the widest part of the pipe

(b) The velocity is maximum at the narrowest part of the pipe and pressure is maximum at the widest part of the pipe

(c) Velocity and pressure both are maximum at the narrowest part of the pipe

(d) Velocity and pressure both are maximum at the widest part of the pipe

14. The density of water at 20°C is 998 kg/m^3 and at 40°C 992 kg/m^3. The coefficient of volume expansion of water is

(a) 10^{-4}/°C (b) 3×10^{-4}/°C

(c) 2×10^{-4}/°C (d) 6×10^{-4}/°C

15. Two metal rods 1 and 2 of same lengths have same temperature difference between their ends. Their thermal conductivities are K_1 and K_2 and cross sectional areas A_1 and A_2, respectively. If the rate of heat conduction in rod 1 is four times that in rod 2, then

(a) $K_1A_1 = K_2A_2$ (b) $K_1A_1 = 4K_2A_2$

(c) $K_1A_1 = 2K_2A_2$ (d) $4K_1A_1 = K_2A_2$

16. A system is taken from state a to state c by two paths adc and abc as shown in the figure. The internal energy at a is $U_a = 10$ J. Along the path adc the amount of heat absorbed $\delta Q_1 = 50$ J and the work done $\delta W_1 = 20$ J whereas along the path abc the heat absorbed $\delta Q_2 = 36$ J. The amount of work done along the path abc is

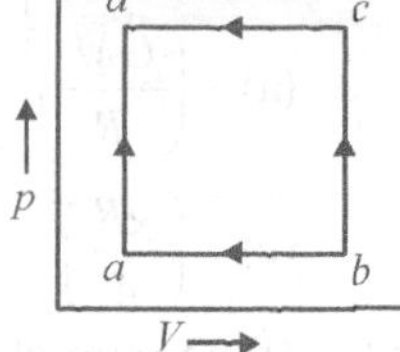

(a) 6 J (b) 10 J (c) 12 J (d) 36 J

17. Which of the following relations does not give the equation of an adiabatic process, where terms have their usual meaning?

(a) $P^{\gamma}T^{1-\gamma}$ = constant

(b) $P^{1-\gamma}\,T^{\gamma}$ = constant

(c) PV^{γ} = constant

(d) $TV^{\gamma-1}$ = constant

18. Two Carnot engines A and B are operated in series. The engine A receives heat from the source at temperature T_1 and rejects the heat to the sink at temperature T. The second engine B receives the heat at temperature T and rejects to its sink at temperature T_2. For what value of T the efficiencies of the two engines are equal?

(a) $\frac{T_1 + T_2}{2}$ (b) $\frac{T_1 - T_2}{2}$

(c) T_1T_2 (d) $\sqrt{T_1T_2}$

19. In a vessel, the gas is at a pressure P. If the mass of all the molecules is halved and their speed is doubled, then the resultant pressure will be

(a) $4P$ (b) $2P$ (c) P (d) $P/2$

20. A particle of mass m oscillates along x-axis according to equation $x = a \sin \omega t$. The nature of the graph between momentum and displacement of the particle is

(a) straight line passing through origin

(b) circle

(c) hyperbola (d) ellipse

21. The length of the wire between two ends of a sonometer is 100 cm. What should be the positions of two bridges below the wire so that the three segments of the wire have their fundamental frequencies in the ratio of 1 : 3 : 5?

(a) $\frac{1500}{23}$ cm, $\frac{2000}{23}$ cm

(b) $\frac{1500}{23}$ cm, $\frac{500}{23}$ cm

(c) $\frac{1500}{23}$ cm, $\frac{300}{23}$ cm

(d) $\frac{300}{23}$ cm, $\frac{1500}{23}$ cm

22. Two sources P and Q produce notes of frequency 660 Hz each. A listener moves from P to Q with a speed of 1 ms^{-1}. If the speed of sound is 330 m/s, then the number of beats heard by the listener per second will be

(a) zero (b) 4 (c) 8 (d) 2

23. A charge 'q' is placed at the centre of the line joining two equal charges 'Q'. The system of the three charges will be in equilibrium if 'q' is equal to

(a) $Q/2$ (b) $-Q/4$ (c) $Q/4$ (d) $-Q/2$

24. An electric dipole of dipole moment p is aligned parallel to a uniform electric field E. The energy required to rotate the dipole by 90° is

(a) pE^2 (b) p^2E

(c) pE (d) infinity

25. A 12 cm wire is given a shape of a right angled triangle ABC having sides 3 cm, 4 cm and 5 cm as shown in the figure. The resistance between two ends (AB, BC, CA) of the respective sides are measured one by one by a multi-meter. The resistances will be in the ratio of

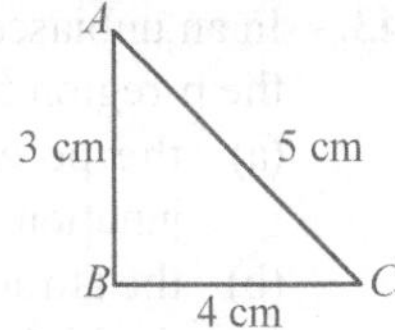

(a) 3 : 4 : 5 (b) 9 : 16 : 25
(c) 27 : 32 : 35 (d) 21 : 24 : 25

26. Two rods are joined end to end, as shown. Both have a cross-sectional area of 0.01 cm^2. Each is 1 meter long. One rod is of copper with a resistivity of 1.7×10^{-6} ohm-centimeter, the other is of iron with a resistivity of 10^{-5} ohm-centimeter.
How much voltage is required to produce a current of 1 ampere in the rods?

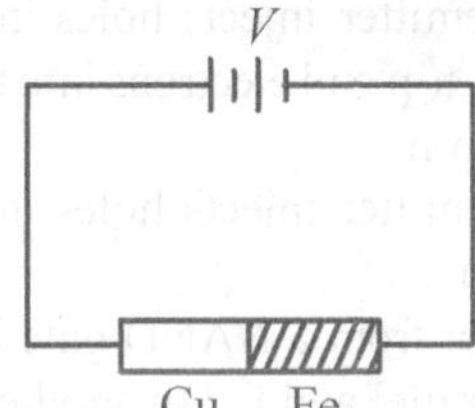

(a) 0.117V (b) 0.00145 V
(c) 0.0145 V (d) 1.7×10^{-6} V

27. Ten identical cells connected in series are needed to heat a wire of length one meter and radius 'r' by 10°C in time 't'. How many cells will be required to heat the wire of length two meter of the same radius by the same temperature in time 't'?
(a) 10 (b) 20 (c) 30 (d) 40

28. A long straight wire carries a certain current and produces a magnetic field of $2 \times 10^{-4} \frac{\text{weber}}{\text{m}^2}$ at a perpendicular distance of 5 cm from the wire. An electron situated at 5 cm from the wire moves with a velocity 10^7 m/s towards the wire along perpendicular to it. The force experienced by the electron will be
(charge on electron $= 1.6 \times 10^{-19}$ C)
(a) Zero (b) 3.2 N
(c) 3.2×10^{-16} N (d) 1.6×10^{-16} N

29. A circular coil $ABCD$ carrying a current i is placed in a uniform magnetic field. If the magnetic force on the segment AB is $\vec{F}$, the force on the remaining segment $BCDA$ is

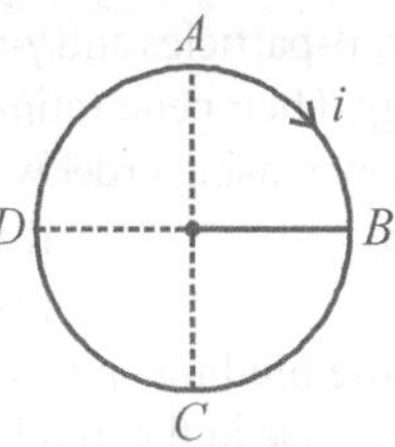

(a) $\vec{F}$ (b) $-\vec{F}$ (c) $3\vec{F}$ (d) $-3\vec{F}$

30. A bar magnet of magnetic moment M is placed at right angles to a magnetic induction B. If a force F is experienced by each pole of the magnet, the length of the magnet will be
(a) F/MB (b) MB/F
(c) BF/M (d) MF/B

31. A current of 2.5 A flows through a coil of inductance 5 H. The magnetic flux linked with the coil is
(a) 2 Wb (b) 0.5 Wb
(c) 12.5 Wb (d) Zero

32. The primary of a transformer when connected to a dc battery of 10 volt draws a current of 1 mA. The number of turns of the primary and secondary windings are 50 and 100 respectively. The voltage in the secondary and the current drawn by the circuit in the secondary are respectively
(a) 20 V and 0.5 mA (b) 20 V and 2.0 mA
(c) 10 V and 0.5 mA
(d) Zero and therefore no current

33. An electromagnetic wave of frequency $\nu = 3.0$ MHz passes from vacuum into a dielectric medium with relative permittivity $\varepsilon = 4.0$. Then
(a) wavelength is doubled and frequency is unchanged
(b) wavelength is doubled and frequency becomes half
(c) wavelength is halved and frequency remains unchanged
(d) wavelength and frequency both remain unchanged

34. An electron in hydrogen atom makes a transition $n_1 \to n_2$ where n_1 and n_2 are principal quantum numbers of the two states. Assuming Bohr's model to be valid the time period of the electron in the initial state is eight times that in the final state. The possible values of n_1 and n_2 are
(a) $n_1 = 4$ and $n_2 = 2$
(b) $n_1 = 6$ and $n_2 = 2$
(c) $n_1 = 8$ and $n_2 = 1$
(d) $n_1 = 8$ and $n_2 = 2$

35. α-particles, β-particles and γ-rays are all having same energy. Their penetrating power in a given medium in increasing order will be
(a) β, γ, α (b) γ, α, β
(c) α, β, γ (d) β, α, γ

36. How does the binding energy per nucleon vary with the increase in the number of nucleons?
(a) Increases continuously with mass number
(b) Decreases continuously with mass number
(c) First decreases and then increases with increase in mass number
(d) First increases and then decreases with increase in mass number

37. A source of light is placed at a distance of 50 cm from a photocell and the stopping potential is found to be V_0. If the distance between the light source and photocell is made 25 cm, the new stopping potential will be
(a) $2V_0$ (b) $V_0/2$ (c) V_0 (d) $4V_0$

38. The de-Broglie wavelength of neutron in thermal equilibrium at temperature T is
(a) $\frac{30.8}{\sqrt{T}}$Å (b) $\frac{3.08}{\sqrt{T}}$Å
(c) $\frac{0.308}{\sqrt{T}}$Å (d) $\frac{0.0308}{\sqrt{T}}$Å

39. Two plane mirrors are inclined at 70°. A ray incident on one mirror at angle θ after reflection falls on second mirror and is reflected from there parallel to first mirror. The value of θ is
(a) 50° (b) 45° (c) 30° (d) 55°

40. The reddish appearance of the sun at sunrise and sunset is due to
(a) the colour of the sky
(b) the scattering of light
(c) the polarisation of light
(d) the colour of the sun

41. In Young's double slit experiment the distance between the slits and the screen is doubled. The separation between the slits is reduced to half. As a result the fringe width
(a) is doubled (b) is halved
(c) becomes four times
(d) remains unchanged

42. A parallel beam of light of wavelength λ is incident normally on a narrow slit. A diffraction pattern is formed on a screen placed perpendicular to the direction of the incident beam. At the second minimum of the diffraction pattern, the phase difference between the rays coming from the two edges of slit is
(a) πλ (b) 2π (c) 3π (d) 4π

43. In an unbiased p-n junction, holes diffuse from the p-region to n-region because of
(a) the potential difference across the p-n junction
(b) the attraction of free electrons of n-region
(c) the higher hole concentration in p-region than that in n-region
(d) the higher concentration of electrons in the n-region than that in the p-region

44. One way in which the operation of a n-p-n transistor differs from that of a p-n-p
(a) the emitter junction is reversed biased in n-p-n
(b) the emitter junction injects minority carriers into the base region of the p-n-p
(c) the emitter injects holes into the base of the p-n-p and electrons into the base region of n-p-n
(d) the emitter injects holes into the base of n-p-n

45. The output from a NAND gate is divided into two in parallel and fed to another NAND gate. The resulting gate is a

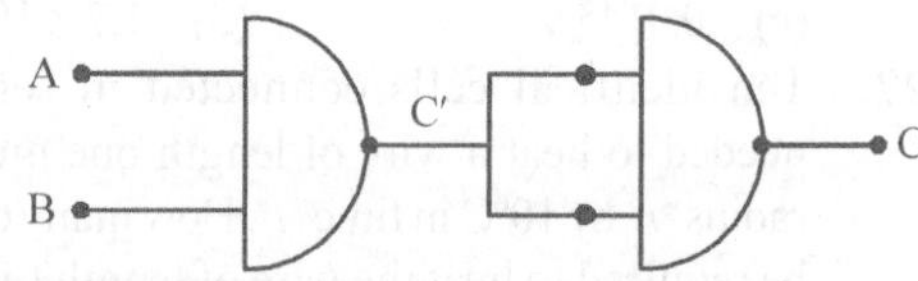

(a) NOT gate (b) AND gate
(c) NOR gate (d) OR gate

CHEMISTRY

46. The values of K_{sp} of $CaCO_3$ and CaC_2O_4 are 4.7×10^{-9} and 1.3×10^{-9} respectively at 25°C. If the mixture of these two is washed with water, what is the concentration of Ca^{2+} ions in water?
(a) 7.746×10^{-5} M (b) 5.831×10^{-5} M
(c) 6.856×10^{-5} M (d) 3.606×10^{-5} M

47. What is the density of N_2 gas at 227°C and 5.00 atm pressure? (R = 0.0821 atm K^{-1} mol^{-1})
(a) 0.29 g/ml (b) 1.40 g/ml
(c) 2.81 g/ml (d) 3.41 g/ml

48. At 100°C the K_w of water is 55 times its value at 25°C. What will be the pH of neutral solution? (log 55 = 1.74)
(a) 6.13 (b) 7.00 (c) 7.87 (d) 5.13

49. When 5 litres of a gas mixture of methane and propane is perfectly combusted at 0°C and 1 atmosphere, 16 litre of oxygen at the same temperature and pressure is consumed. The amount of heat released from this combustion in kJ ($\Delta H_{comb}(CH_4) = 890$ kJ mol^{-1}, $\Delta H_{comb}(C_3H_8) = 2220$ kJ mol^{-1}) is
(a) 32 (b) 38 (c) 317 (d) 477

50. According to law of photochemical equivalence the energy absorbed (in ergs/mole) is given as ($h = 6.62 \times 10^{-27}$ ergs, $c = 3 \times 10^{10}$ cm s^{-1}, $N_A = 6.02 \times 10^{23}$ mol^{-1})
(a) $\frac{1.956\times10^{16}}{\lambda}$ (b) $\frac{1.956\times10^{8}}{\lambda}$
(c) $\frac{2.859\times10^{5}}{\lambda}$ (d) $\frac{2.859\times10^{16}}{\lambda}$

51. The dissociation constant of a weak acid is 1×10^{-4}. In order to prepare a buffer solution with a pH = 5 the [Salt]/[Acid] ratio should be
(a) 1 : 10 (b) 4 : 5 (c) 10 : 1 (d) 5 : 4

52. A reaction is 50% completed in 2 hours and 75% completed in 4 hours. The order of reaction is
(a) 0 (b) 1 (c) 2 (d) 3

53. Accumulation of lactic acid ($HC_3H_5O_3$), a monobasic acid in tissues leads to pain and a feeling of fatigue. In a 0.10 M aqueous solution, lactic acid is 3.7% dissociated. The value of dissociation constant, K_a, for this acid will be:
(a) 2.8×10^{-4} (b) 1.4×10^{-5}
(c) 1.4×10^{-4} (d) 3.7×10^{-4}

54. Three thermochemical equations are given below:
(i) $C_{(graphite)} + O_2(g) \rightarrow CO_2(g)$; $\Delta_r H° = x$ kJ mol^{-1}
(ii) $C_{(graphite)} + \frac{1}{2}O_2(g) \rightarrow CO(g)$; $\Delta_r H° = y$ kJ mol^{-1}
(iii) $CO(g) + \frac{1}{2}O_2(g) \rightarrow CO_2(g)$; $\Delta_r H° = z$ kJ mol^{-1}
Based on the above equations, find out which of the relationship given below is correct?
(a) $x = y - z$ (b) $z = x + y$
(c) $x = y + z$ (d) $y = 2z - x$

55. In which of the following pair both the species have sp^3 hybridization?
(a) H_2S, BF_3 (b) SiF_4, BeH_2
(c) NF_3, H_2O (d) NF_3, BF_3

56. In an experiment it showed that 10 mL of 0.05 M solution of chloride required 10 mL of 0.1 M solution of $AgNO_3$, which of the following will be the formula of the chloride (X stands for the symbol of the element other than chlorine):
(a) X_2Cl (b) X_2Cl_2 (c) XCl_2 (d) XCl_4

57. How many grams of cobalt metal will be deposited when a solution of cobalt (II) chloride is electrolyzed with a current of 10 amperes for 109 minutes (1 Faraday = 96,500 C; Atomic mass of Co = 59 u)
(a) 0.66 (b) 4.0 (c) 20.0 (d) 40.0

58. Which condition is not satisfied by an ideal solution?
(a) $\Delta_{mix} H = 0$
(b) $\Delta_{mix} V = 0$
(c) $\Delta_{mix} S = 0$
(d) Obeyance to Raoult's Law

59. For a reaction between A and B the order with respect to A is 2 and the order with respect to B is 3. The concentrations of both A and B are doubled, the rate will increase by a factor of:
(a) 10 (b) 12 (c) 16 (d) 32

60. The pair of species that has the same bond order in the following is:
(a) O_2, B_2 (b) CO, NO^+
(c) NO^-, CN^- (d) O_2, N_2

61. The correct IUPAC name for $[CrF_2(en)_2]Cl$ is:
(a) Chloro difluoridobis (ethylene diamine) chromium (III)
(b) Chloro difluoridoethylenediaminechromium (III) chloride
(c) Difluoridobis (ethylene diamine) chromium (III) chloride
(d) Difluorobis-(ethylene diamine) chromium (III) chloride

62. The outer electronic configuration of Gd (At. No. 64) is:
(a) $4f^4 5d^5 6s^1$ (b) $4f^5 5d^4 6s^1$
(c) $4f^7 5d^1 6s^2$ (d) $4f^3 5d^5 6s^2$

63. Identify the incorrect statement, regarding the molecule XeO_4:
(a) XeO_4 molecule is tetrahedral
(b) XeO_4 molecule is square planar
(c) There are four $p\pi - d\pi$ bonds
(d) There are four $sp^3 - p$, σ bonds

64. Crystal field splitting energy for high spin d^4 octahedral complex is:
(a) $-1.6\,\Delta_0$ (b) $-1.2\,\Delta_0$
(c) $-0.6\,\Delta_0$ (d) $-0.8\,\Delta_0$

65. In which of the following ionisation processes the bond energy increases and the magnetic behaviour changes from paramagnetic to diamagnetic?

(a) $N_2 \rightarrow N_2^+$ (b) $O_2 \rightarrow O_2^+$
(c) $C_2 \rightarrow C_2^+$ (d) $NO \rightarrow NO^+$

66. Which one of the following arrangements represents the correct order of least negative to most negative electron gain enthalpy for C, Ca, Al, F and O?

(a) $Ca < Al < C < O < F$
(b) $Al < Ca < O < C < F$
(c) $Al < O < C < Ca < F$
(d) $C < F < O < Al < Ca$

67. Sc (Z = 21) is a transition element but Zn (Z = 30) is not because

(a) both Sc and Zn do not exhibit variable oxidation states
(b) both Sc^{3+} and Zn^{2+} ions are colourless and form white compounds
(c) in case of Sc, 3*d* orbitals are partially filled but in Zn these are completely filled
(d) last electron is assumed to be added to 4s level in case of Zn

68. In Castner-Kellner cell for production of sodium hydroxide:

(a) Brine is electrolyzed with Pt electrodes
(b) Brine is electrolyzed using graphite electrodes
(c) Molten sodium chloride is electrolysed
(d) Sodium amalgam is formed at mercury cathode

69. Which statement is wrong?

(a) Feldspars are not aluminosilicates
(b) Beryl is an example of cyclic silicate
(c) Mg_2SiO_4 is orthosilicate
(d) Basic structural unit in silicates is the SiO_4 tetrahedron

70. Which is diamagnetic?

(a) $[Fe(CN)_6]^{3-}$ (b) $[Co(F_6)]^{3-}$
(c) $[Ni(CN)_4]^{2-}$ (d) $[NiCl_4]^{2-}$

71. Consider the half-cell reduction reaction :
$Mn^{2+} + 2e^- \rightarrow Mn, E° = -1.18$ V
$Mn^{2+} \rightarrow Mn^{3+} + e^-, E° = -1.51$ V
The E° for the reaction $3Mn^{2+} \rightarrow Mn^0 + 2Mn^{3+}$, and possibility of the forward reaction are, respectively

(a) −2.69 V and no (b) −4.18 V and yes
(c) +0.33 V and yes (d) +2.69 V and no

72. The outer orbitals of C in ethene molecule can be considered to be hybridized to give three equivalent sp^2 orbitals. The total number of sigma (σ) and pi (π) bonds in ethene molecule is

(a) 1 sigma (σ) and 2 pi (π) bonds
(b) 3 sigma (σ) and 2 pi (π) bonds
(c) 4 sigma (σ) and 1 pi (π) bonds
(d) 5 sigma (σ) and 1 pi (π) bonds

73. In a particular isomer of $[Co(NH_3)_4Cl_2]^0$, the Cl-Co-Cl angle is 90°, the isomer is known as

(a) Linkage isomer (b) Optical isomer
(c) *cis*-isomer (d) Position isomer

74. The anion of acetylacetone (acac) forms $Co(acac)_3$ chelate with Co^{3+}. The rings of the chelate are

(a) three membered (b) five membered
(c) four membered (d) six membered

75. The metal oxide which cannot be reduced to metal by carbon is

(a) Fe_2O_3 (b) Al_2O_3 (c) PbO (d) ZnO

76. Which among the following is a paramagnetic complex?

(a) $Mo(CO)_6$ (b) $[Co(NH_3)_6]^{3+}$
(c) $[Pt(en)Cl_2]$ (d) $[CoBr_4]^{2-}$
(At. No. of Mo = 42, Pt = 78)

77. Dettol is the mixture of

(a) Terpineol and Bithionol
(b) Chloroxylenol and Bithionol
(c) Chloroxylenol and Terpineol
(d) Phenol and Iodine

78. Arrange the following in increasing order of stability.

(A) $(CH_3)_2\overset{\oplus}{C}-CH_2CH_3$ (B) $(CH_3)_3\overset{\oplus}{C}$
(C) $(CH_3)_2\overset{\oplus}{CH}$ (D) $CH_3\overset{\oplus}{CH_2}$
(E) $\overset{\oplus}{CH_3}$

(a) E < D < C < B < A (b) E < D < C < A < B
(c) D < E < C < A < B (d) A < E < D < C < B

79. Given

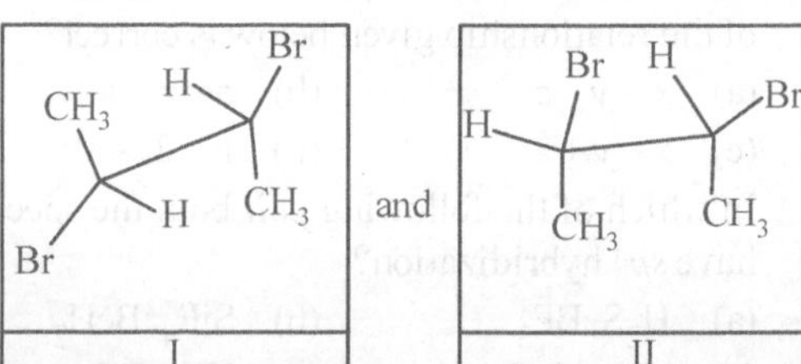

I and II are

(a) A pair of optical isomers
(b) Identical
(c) A pair of conformers
(d) A pair of geometrical isomers

80. In the following reaction :

$$HC \equiv CH \xrightarrow[Hg^{2+}]{H_2SO_4} 'P'$$

Product 'P' will not give

(a) Iodoform test
(b) Tollen's reagent test
(c) Brady's reagent test
(d) Victor Meyer test

81. On hydrolysis of a "compound", two compounds are obtained. One of which on treatment with sodium nitrite and hydrochloric acid gives a product which does not respond to iodoform test. The second one reduces Tollen's reagent and Fehling's solution. The "compound" is

(a) $CH_3CH_2CH_2CON(CH_3)_2$
(b) $CH_3CH_2CH_2NC$
(c) $CH_3CH_2CH_2CN$
(d) $CH_3CH_2CH_2ON=O$

82. Homolytic fission of the following alkanes forms free radicals CH_3-CH_3, $CH_3-CH_2-CH_3$, $(CH_3)_2CH-CH_3$, $CH_3-CH_2-CH(CH_3)_2$. Increasing order of stability of the radicals is

(a) $(CH_3)_3\dot{C} < (CH_3)_2\dot{C}-CH_2CH_3 < CH_3-\dot{C}H-CH_3 < CH_3-\dot{C}H_2$
(b) $(CH_3)_2\dot{C}-CH_2CH_3 < CH_3-\dot{C}H-CH_3 < CH_3-\dot{C}H_2 < (CH_3)_3\dot{C}$
(c) $CH_3-\dot{C}H_2 < CH_3-\dot{C}H-CH_3 < (CH_3)_2\dot{C}-CH_2-CH_3 < (CH_3)_3\dot{C}$
(d) $CH_3-\dot{C}H_2 < CH_3-\dot{C}H-CH_3 < (CH_3)_3\dot{C} < (CH_3)_2\dot{C}-CH_2CH_3$

83. Some reactions of amines are given. Which one is not correct ?

(a) $(CH_3)_2NH + NaNO_2 + HCl \rightarrow (CH_3)_2N-N=O$
(b) $(CH_3)_2N-C_6H_5 + NaNO_2 + HCl \rightarrow (CH_3)_2N-C_6H_4-N=NCl$
(c) $CH_3CH_2NH_2 + HNO_2 \rightarrow CH_3CH_2OH + N_2$
(d) $CH_3NH_2 + C_6H_5SO_2Cl \rightarrow CH_3NHSO_2C_6H_5$.

84. Which one of the following statements is not true?

(a) Dissolved oxygen (DO) in cold water can reach a concentration upto 10 ppm.
(b) Clean water would have a BOD value of 5 ppm.
(c) Fluoride deficiency in drinking water is harmful. Soluble fluoride is often used to bring its concentration upto 1 ppm.
(d) When the pH of rain water is higher than 6.5, it is called acid rain.

85. What is the hybridisation state of benzyl carbonium ion $C_6H_5-\overset{+}{C}H_2$?

(a) sp^3 (b) sp^2 (c) spd^2 (d) sp^2d

86. Nitrogen detection in an organic compound is carried out by Lassaigne's test. The blue colour formed corresponds to which of the following formulae?

(a) $Fe_3[Fe(CN)_6]_3$ (b) $Fe_3[Fe(CN)_6]_2$
(c) $Fe_4[Fe(CN)_6]_3$ (d) $Fe_4[Fe(CN)_6]_2$

87. Phenol is distilled with Zn dust followed by Friedel Crafts alkylation with propyl chloride in the presence of $AlCl_3$ to give a compound (B). (B) is oxidised in the presence of air to form the compound (C). The structural formula of (C) is

(a) C_6H_5-COOH (benzoic acid)

(b) $C_6H_4(COOH)(OH)$ (COOH with ortho OH)

(c) $(H_3C)_2C(-O-O-H)-C_6H_5$

(d) $H_3C-C(CH_3)(-O-O-H)-C_6H_4OH$ (OH ortho)

88. In DNA the linkages between different nitrogenous bases are:
(a) peptide linkage
(b) phosphate linkage
(c) H-bonding
(d) glycosidic linkage

89. Which of the following chemical system is non aromatic?
(a) 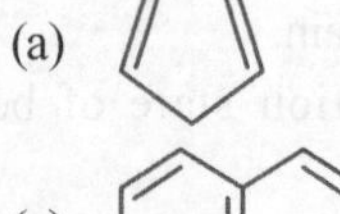 (b) 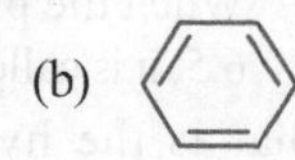
(c) 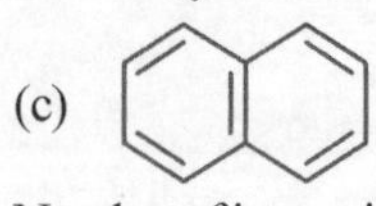 (d) 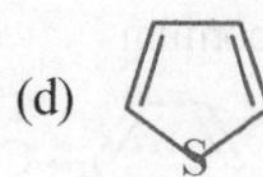

90. Number of isomeric alcohols of molecular formula $C_6H_{14}O$ which give positive iodoform test is
(a) two (b) three
(c) four (d) five

BIOLOGY

91. The plant body is thalloid in
(a) *Funaria* (b) *Sphagnum*
(c) *Salvinia* (d) *Marchantia*

92. Which one of the following is true for fungi?
(a) They are phagotrophs
(b) They lack a rigid cell wall
(c) They are heterotrophs
(d) They lack nuclear membrane

93. Specialized cells for fixing atmospheric nitrogen in *Nostoc* are
(a) Akinetes (b) Heterocysts
(c) Hormogonia (d) Nodules

94. The common characteristics between tomato and potato will be maximum at the level of their
(a) Genus (b) Family
(c) Order (d) Division

95. What is common in all the three, *Funaria, Dryopteris* and *Ginkgo?*
(a) Independent sporophyte
(b) Presence of archegonia
(c) Well developed vascular tissues
(d) Independent gametophyte

96. Which one of the following is wrongly matched?
(a) *Nostoc*-Water blooms
(b) *Spirogyra*-Motile gametes
(c) *Sargassum*-Chlorophyll *c*
(d) Basidiomycetes-Puffballs

97. Among flowers of *Calotropis*, tulip, *Sesbania, Asparagus,* Colchicine, Sweet pea, *Petunia, Indigofera,* Mustard, Soyabean, Tobacco and groundnut how many plants have corolla with valvate aestivation?
(a) Five (b) Six
(c) Seven (d) Eight

98. How many plants among China rose, *Ocimum*, sunflower, mustard, *Alstonia, guava, Calotropis* and *Nerium* (Oleander) have opposite phyllotaxy?
(a) Two (b) Three
(c) Four (d) Five

99. In a cymose inflorescence the main axis
(a) Terminates in a flower
(b) Has unlimited growth
(c) Bears a solitary flower
(d) Has unlimited growth but lateral branches end in flowers

100. Which of the following statements is **not** true for stomatal apparatus?
(a) Inner walls of guard cells are thick
(b) Guard cells invariably possess chloroplasts and mitochondria
(c) Guard cells are always surrounded by subsidiary cells
(d) Stomata are involved in gaseous exchange

101. Meristematic tissue responsible for increase in girth of tree trunk is
(a) Apical meristem
(b) Intercalary meristem
(c) Lateral meristem
(d) Phellogen

102. Inflorescence is racemose in
(a) Soyabean (b) Brinjal
(c) Tulip (d) *Aloe*

103. During the metaphase stage of mitosis spindle fibres attach to chromosomes at
(a) Centromere
(b) Kinetochore
(c) Both centromere and kinetochore
(d) Centromere, kinetochore and areas adjoining centromere

104. Uridine, present only in RNA is a
(a) Pyrimidine (b) Nucleoside
(c) Nucleotide (d) Purine

105. The term 'glycocalyx' is used for
(a) A layer surrounding the cell wall of bacteria
(b) A layer present between cell wall and membrane of bacteria
(c) Cell wall of bacteria
(d) Bacterial cell glyco-engineered to possess N-glycosylated proteins

106. Which of the following type of plastids does not contain stored food material?
(a) Amyloplasts (b) Chromoplasts
(c) Elaioplasts (d) Aleuroplasts

107. During meiosis I, the chromosomes start pairing at
(a) Leptotene (b) Zygotene
(c) Pachytene (d) Diplotene

108. Which two distinct microbial processes are responsible for the release of fixed nitrogen as dinitrogen gas (N_2) to the atmosphere?
(a) Anaerobic ammonium oxidation, and denitrification
(b) Aerobic nitrate oxidation, and nitrite reduction
(c) Decomposition of organic nitrogen, and conversion of dinitrogen to ammonium compounds
(d) Enteric fermentation in cattle, and nitrogen fixation by *Rhizobium* in root nodules of legumes

109. Which of the following best illustrates FEEDBACK in development?
(a) As tissue (X) develops, it secretes something that slows down the growth of tissue (Y)
(b) Tissue (X) secretes RNA which changes the development of tissue (Y)
(c) As tissue (X) develops, it secretes enzymes that inhibit the development of tissue (Y)
(d) As tissue (X) develops, it secretes something that induces tissue (Y) to develop

110. The viability of seeds is tested by
(a) Safranine
(b) 2, 6 dichlorophenol indophenols
(c) 2, 3, 5 triphenyl tetrazolium chloride
(d) DMSO

111. The pineapple which under natural conditions is difficult to blossom has been made to produce fruits throughout the year by application of
(a) IAA, IBA (b) NAA, 2, 4-D
(c) Phenyl acetic acid (d) Cytokinin

112. Which of the following elements is a constituent of biotin?
(a) Sulphur (b) Magnesium
(c) Calcium (d) Phosphorus

113. Bundle sheath cells
(a) Are rich in RuBisCO
(b) Are rich in PEP carboxylase
(c) Lack RuBisCO
(d) Lack both RuBisCO and PEP carboxylase

114. Which one of the following statements is **correct**?
(a) Geitonogamy involves the pollen and stigma of flowers of different plants
(b) Cleistogamous flowers are always autogamous
(c) Xenogamy occurs only by wind pollination
(d) Chasmogamous flowers do not open at all

115. Megaspores are produced from the megaspore mother cells after
(a) Meiotic division
(b) Mitotic division
(c) Formation of a thick wall
(d) Differentiation

116. Syngamy can occur outside the body of the organism in
(a) Fungi (b) Mosses
(c) Algae (d) Ferns

117. Animal vectors are required for pollination in
(a) Maize (b) *Vallisneria*
(c) Mulberry (d) Cucumber

118. Which of the following statements is **correct?**
(a) Sporopollenin can withstand high temperatures but not strong acids
(b) Sporopollenin can be degraded by enzymes
(c) Sporopollenin is made up of inorganic materials
(d) Sporopollenin can withstand high temperatures as well as strong acids and alkalis

119. Albuminous seeds store their reserve food mainly in
(a) Perisperm (b) Endosperm
(c) Cotyledons (d) Hypocotyl

120. Random unidirectional change in allele frequencies that occurs by chance in all populations and especially in small populations is known as
(a) Mutation (b) Migration
(c) Natural selection (d) Genetic drift

121. Which of the following is **not** a property of the genetic code?
(a) Universal (b) Non-overlapping
(c) Ambiguous (d) Degeneracy

122. Genetic variation in a population arises due to
(a) Mutations only
(b) Recombination only
(c) Mutations as well as recombination
(d) Reproductive isolation and selection

123. Satellite RNAs are present in some
(a) Plant viruses (b) Viroids
(c) Prions (d) Bacteriophages

124. One of the most frequently used techniques in DNA fingerprinting is
(a) AFLP (b) VNTR
(c) SSCP (d) SCAR

125. Which one of the following vectors is used to replace the defective gene in gene therapy?
(a) Ti plasmid (b) Adenovirus
(c) Cosmid (d) Ri plasmid

126. Which of the following statements is not true about somatic embryogenesis?
(a) A somatic embryo develops from a somatic cell
(b) The pattern of development of a somatic embryo is comparable to that of a zygotic embryo
(c) Somatic embryos can develop from microspores
(d) Somatic embryo is induced usually by an auxin such as 2, 4-D

127. Genes of interest can be selected from a genomic library by using
(a) Restriction enzymes (b) Cloning vectors
(c) DNA probes (d) Gene targets

128. During the process of isolation of DNA, chilled ethanol is added to
(a) Remove proteins such as histones
(b) Precipitate DNA
(c) Break open the cell to release DNA
(d) Facilitate action of restriction enzymes

129. RNA interference involves
(a) Synthesis of mRNA from DNA
(b) Synthesis of cDNA from RNA using reverse transcriptase
(c) Silencing of specific mRNA due to complementary RNA
(d) Interference of RNA in synthesis of DNA

130. Benthic organisms are affected most by
(a) Water-holding capacity of soil
(b) Light reaching the forest floor
(c) Surface turbulence of water
(d) Sediment characteristics of aquatic ecosystems

131. Which of the following has maximum genetic diversity in India?
(a) Rice (b) Mango
(c) Wheat (d) Groundnut

132. Which organization publishes the Red Data Book?
(a) GEF (b) IUCN
(c) UNEP (d) WWF

133. The largest tiger reserve in India is
(a) Nagarhole
(b) Valmiki
(c) Nagarjunsagar-Srisailam
(d) Periyar

134. The second commitment period for Kyoto Protocol was decided at
(a) Cancun (b) Durban
(c) Bali (d) Doha

135. Which one of the following is a primary consumer in maize field ecosystem?
(a) Lion (b) Grasshopper
(c) Wolf (d) Phytoplankton

136. The characteristics of class Reptilia are :
(a) Body covered with dry and cornified skin, scales over the body are epidermal, they do not have external ears
(b) Body covered with moist skin which is devoid of scales, the ear is represented by a tympanum, alimentary canal, urinary and reproductive tracts open into a common cloaca
(c) Fresh water animals with bony endoskeleton, air-bladder to regulate buoyancy
(d) Marine animals with cartilaginous endoskeleton, body covered with placoid scales

137. Which one of the following animals is correctly matched with its one characteristics and the taxon?

	Animal	Characteristic	Taxon
(a)	Duckbilled platypus	Oviparous	Mammalian
(b)	Millipede	Ventral nerve cord	Arachnida
(c)	Sea Anemone	Triploblastic	Cnidaria
(d)	Silverfish	Pectoral and Pelvic fins	Chordata

138. Which one of the following groups of animals reproduces only by sexual means?
(a) Ctenophora (b) Cnidaria
(c) Porifera (d) Protozoa

139. Sharks and dogfishes differ from skates and rays by
(a) Their pectoral fins distinctly marked off from cylindrical bodies
(b) Gill slits are ventrally placed
(c) Head and trunk are widened considerably
(d) Distinct demarcation between body and tail

140. Identify the tissue shown in the diagram and match with its characteristcs and its location

(a) Skeletal muscle, shows striations and closely attached with the bones of the limbs
(b) Smooth muscles, show branching, found in the walls of the heart
(c) Cardiac muscles, unbranched muscles, found in the walls of the heart
(d) Striated muscles, tapering at both-ends, attached with the bones of the ribs

141. Select the correct option with respect to cockroaches
(a) The fore wings are tegmina which are used in flight
(b) Malpighian tubules convert nitrogenous wastes into urea
(c) Males bear short anal styles not present in females
(d) Nervous system comprises of a dorsal nerve cord and ten pairs of ganglion

142. Select the alternative giving correct identifica-tion and function of the organelle 'A' in the diagram

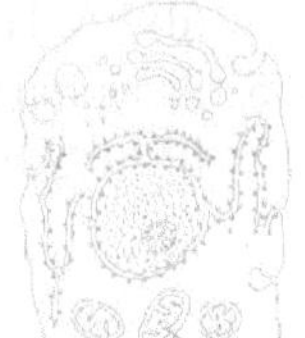

(a) Endoplasmic reticulum-synthesis of lipids
(b) Mitochondria-produce cellular energy in the form of ATP
(c) Golgi body-provides packaging material
(d) Lysosomes - secrete hydrolytic enzymes

143. Why is a capsule advantageous to a bacterium?
(a) It allows the bacterium to attach to the surface
(b) It protects the bacterium from desiccation
(c) It provides means of locomotion
(d) It allows bacterium to "hide" from host's immune system

144. Which of the following statements about enzymes is **wrong**?
(a) Enzymes require optimum pH and temperature for maximum activity
(b) Enzymes are denatured at high temperatures
(c) Enzymes are mostly proteins but some are lipids also
(d) Enzymes are highly specific

145. The figure shows a hypothetical tetrapeptide portion of a protein with parts labelled A-D. Which one of the following option is correct?

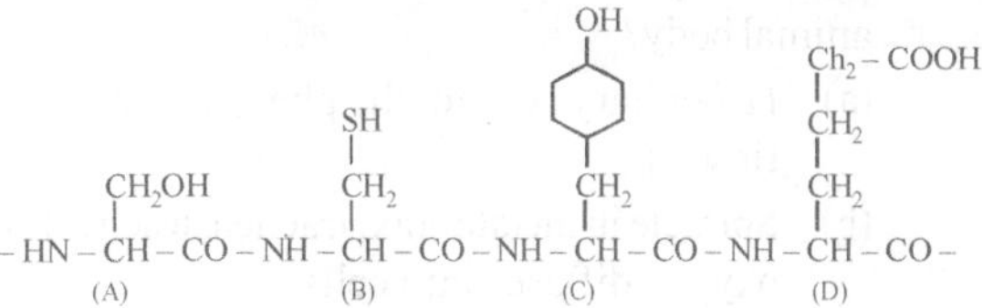

(a) A is the sulphur containing amino acid - methionine
(b) D is the acidic amino acid - glutamic acid`
(c) C is an aromatic amino acid - tryptophan
(d) A is the C - terminal amino acid and D is N terminal amino acid

146. A stage of mitosis is shown in the diagram. Which stage is it and what are its characteristics?

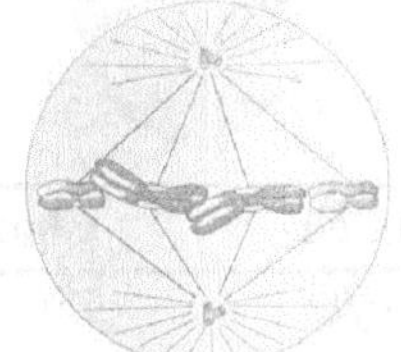

(a) Late prophase – chromosomes move to spindle equator
(b) Metaphase – spindle fibres attached to kinetochores, centromeres split and chromatids separate
(c) Metaphase – chromosomes moved to spindle equator chromosomes made up of two sister chromatids
(d) Anaphase – centromeres split and chromatids separate and start moving away

147. A healthy person eats the following diet - 5 gm raw sugar, 4 gm albumin, 10 gm pure buffalo ghee adultrated with 2 gm vegetable ghee (hydrogenated vegetable oil) and 5 gm lignin. How many calories he is likely to get?
(a) 144 (b) 126 (c) 164 (d) 112

148. Which enzymes are likely to act on the baked potatoes eaten by a man, starting from the mouth and as it moves down the alimentary canal?
(a) Salivary maltase → carboxypeptidase → trypsinogen
(b) Pancreatic amylase → salivary amylase → lipases
(c) Disaccharidase like maltase → lipases → nucleases
(d) Salivary amylase → pancreatic amylase → disaccharidases

149. Which one of the following is one of the paths followed by air/O_2 during respiration in an adult male *Periplaneta americana* as it enters the animal body?

(a) Hypopharynx, mouth, pharynx, trachea, tissues
(b) Spiracle in metathorax, trachea, tracheoles, oxygen diffuses into cells
(c) Mouth, bronchial tube, trachea, oxygen enters cells
(d) Spiracles in prothorax, tracheoles, trachea, oxygen diffuses into cells

150. The figure shows a human blood cell. Identify it and give its characteristics.

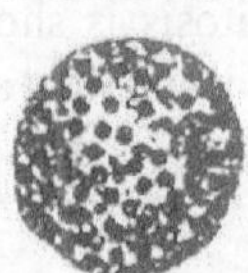

	Blood Cell	Characteristics
(a)	Monocyte	Life span 3 days, produce antibodies
(b)	Basophil	Secrete serotonin, inflammatory response
(c)	B-lymphocyte	Form about 20% of blood cells involved in immune response
(d)	Neutrophil	Most abundant blood cell, phagocytic

151. Figure shows blood circulation in humans with labels A to D. Select the option which gives correct identification of label and functions of the part:

(a) A – Artery - thick walled and blood flows evenly
(b) B – Capillary - thin without muscle layers and wall two cell thick
(c) C – Vein - thin walled and blood flows in jerks/spurts
(d) D – Pulmonary vein - takes oxygenated blood to heart $PO_2 = 95$ mmHg

152. Select the option which shows correct matching of animal with excretory organs and excretory product

	Animal	Excretory organs	Excretory product
(a)	Housefly	Renal tubules	Uric acid
(b)	*Labeo* (Rohu)	Nephridial tubes	Ammonia
(c)	Salamander	Kidney	Urea
(d)	Peacock	Kidney	Urea

153. Select the correct statement with respect to disorders of muscles in humans

(a) Rapid contractions of skeletal muscles causes muscle dystrophy
(b) Failure of neuromuscular transmission in myasthenia gravis can prevent normal swallowing
(c) Accumulation of urea and creatine in the joints cause their inflammation
(d) An overdose of vitamin D causes osteoporosis

154. During muscle contraction in humans the

(a) Actin filaments shorten
(b) Sarcomere *does not* shorten
(c) A band remain same
(d) A, H and I bands shorten

155. A sagittal section of human-brain is shown here. Identify at least two labels from A-D.

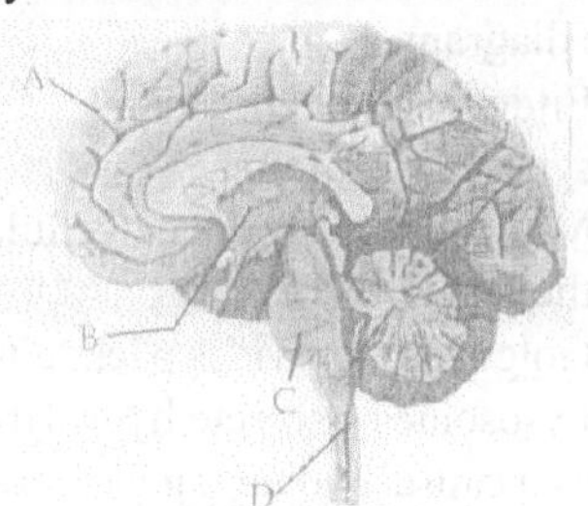

(a) A – Cerebral hemispheres; B – Cerebellum
(b) C – Mid brain; D – Cerebellum
(c) A – Cerebrum; C – Pons
(d) B – Corpus callosum; D – Medulla

156. The figure shows an axon terminal and synapse. Select the option giving correct identifications of tables A-D

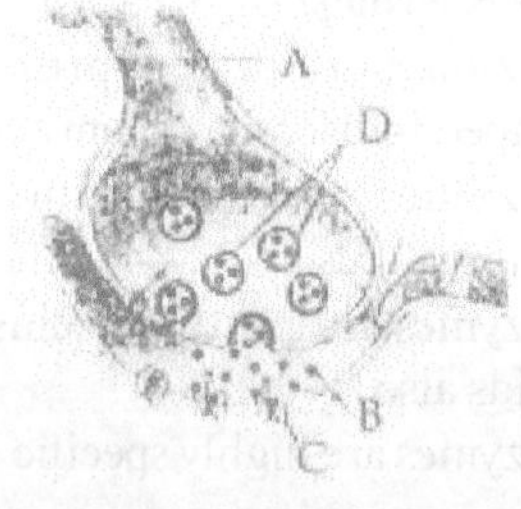

(a) A – Axon terminal; B – Serotonin complex
(b) A – Action potential; C – Neurotransmitter
(c) B – Neurotransmitter; D – Receptor capsules
(d) C – Receptor; D – Synaptic vesicles

157. Which of the following represents the action of insulin?
(a) Increases blood glucose levels by hydrolysis of glycogen
(b) Increases blood glucose levels by stimulating glucagon production
(c) Decreases blood glucose levels of forming glycogen
(d) Increases blood glucose level by promoting cellular uptake of glucose

158. Norepinephrine :
(1) Is released by sympathetic fibres
(2) Is released by parasympathetic fibres
(3) Increases the heart rate
(4) Decreases blood pressure
Which of the above said statements are correct?
(a) (1) and (4) (b) (1) and (3)
(c) (2) and (3) (d) (2) and (4)

159. The figure shows a section of human ovary. Select the option which gives the correct identification of A and B with function/ characteristic

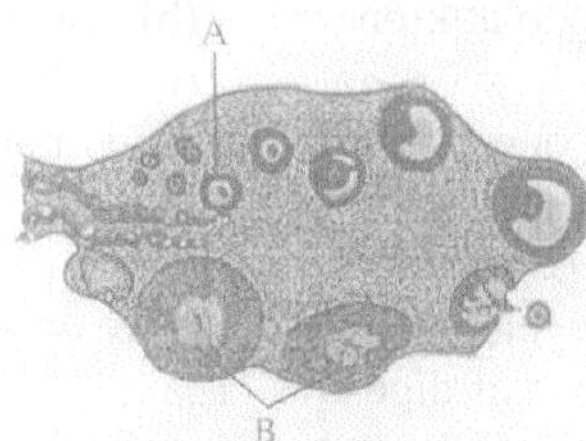

(a) A – Primary oocyte – it is the prophase – I of the meiotic division
(b) B – Corpus luteum – secretes progesterone
(c) A – Tertiary follicle – forms Graafian follicle
(d) B – Corpus luteum – secretes estrogen

160. Select the option which correctly matches the endocrine gland with its hormone and its function

	Endocrine gland	Hormone	Function
(a)	Ovary	FSH	stimulates follicular development and the secretion of estrogens.
(b)	Placenta	estrogen	initiates secretion of the milk.
(c)	Corpus luteum	estrogen	essential for maintenance of endometerium
(d)	Leydig cells	androgen	initiates the production of sperms.

161. The foetal ejection reflex in humans triggers release of
(a) Oxytocin from maternal pituitary
(b) Oxytocin from foetal pituitary
(c) Human Chorionic Gonadotropin (hCG) from placenta
(d) Human Placental Lactogen (hPL) from placenta

162. One of the following is **not** a method of contraception – which one?
(a) Tubectomy
(b) Condoms
(c) Pills of a combination of oxytocin and vasopressin
(d) Lippes loop

163. Which one of the following statements is correct regarding Sexually Transmitted Diseases (STD)?
(a) The chances of a 5 year boy contacting a STD are very little
(b) A person may contact syphilis by sharing milk with one already suffering from the disease
(c) Haemophilia is one of the STD
(d) Genital herpes and sickle-cell anaemia are both STD

164. The stage transferred into the uterus after induced fertilization of ova in the laboratory is
(a) Zygote
(b) Embryo at 4 blastomere stage
(c) Embryo at 2 blastomere stage
(d) Morula

165. Which one is the **incorrect** statement with regards to the importance of pedigree analysis?
(a) It helps to trace the inheritance of a specific trait
(b) It confirms that DNA is the carrier of genetic information
(c) It helps to understand whether the trait in question is dominant or recessive
(d) It confirms that the trait is linked to one of the autosome

166. In our society women are blamed for producing female children. Choose the correct answer for the sex-determination in humans
(a) Due to some defect in the women
(b) Due to some defect like aspermia in man
(c) Due to the genetic make up of the particular sperm which fertilizes the egg
(d) Due to the genetic make up of the egg

167. Down's syndrome in humans is due to
(a) Two 'Y' chromosomes
(b) Three 'X' chromosomes
(c) Three copies of chromosome 21
(d) Monosomy

168.

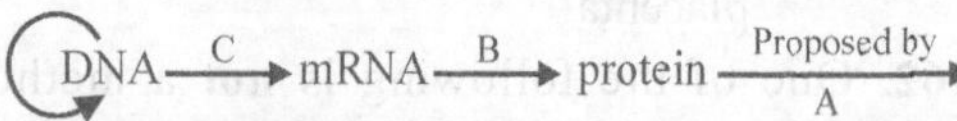

The figure gives an important concept in the genetic implication of DNA. Fill the blanks A, B and C.
(a) A – Francis Crick; B – translation; C – transcription
(b) A – Maurice Wilkins; B – transcription; C – translation
(c) A – James Watson; B – replication; C – extension
(d) A – Erwin Chargaff; B – translation; C – replication

169. In an inducible operon, the genes are
(a) Always expressed
(b) Usually not expressed unless a signal turns them "on"
(c) Usually expressed unless a signal turns them "off"
(d) Never expressed

170. The finch species of Galapagos Islands are grouped according to their food sources. Which of the following is not a finch food?
(a) Seeds (b) Carrion
(c) Insects (d) Tree buds

171. Dinosaurs dominated the world in which of the following geological era?
(a) Devonion (b) Coenozoic
(c) Jurassic (d) Mesozoic

172. Identify the site where *Wuchereria bancrofti* is normally found on human body
(a) Lymphatic vessels of the lower limbs
(b) Muscles of the legs
(c) Blood vessels of the thigh region
(d) Skin between the fingers

173. Which one of the following is a hallucinogenic drug?
(a) Opium (b) Caffeine
(c) Morphine
(d) Lysergic acid diethylamide

174. Tissue culture technique can produce infinite number of new plants from a small parental tissue. The economic importance of the technique is in raising
(a) Variants through picking up somaclonal variations
(b) Genetically uniform population identical to the original parent
(c) Homozygous diploid plants
(d) Development of new species

175. Microbe used for biocontrol of pest butterfly caterpillars is
(a) *Trichoderma sp.*
(b) *Saccharomyces cerevisiae*
(c) *Bacillus thuringiensis*
(d) *Streptococcus sp.*

176. The age pyramid with broad base indicates
(a) High percentage of young individuals
(b) High percentage of old individuals
(c) Low percentage of young individuals
(d) A stable population

177. When man eats fish which feeds on zooplankton which have eaten small plants, the producer in the chain is
(a) Zooplankton (b) Small plants
(c) Fish (d) Man

178. Which one of the following is **not** a parasitic adaptation?
(a) Loss of unnecessary sense organs
(b) Development of adhesive organs
(c) Loss of digestive organs
(d) Loss of reproductive capacity

179. Climate of the world is threatened by
(a) Increasing concentration of atmospheric oxygen
(b) Decreasing amount of atmospheric oxygen
(c) Increasing amount of atmospheric carbondioxide
(d) Decreasing amount of atmospheric carbondioxide

180. Which one of the following is **not correct** as regards the harmful effects of particulate matter of the size 2.5 micro meters or less?
(a) It can be inhaled into the lungs
(b) It can cause respiratory problems
(c) It can directly enter into our circulatory system
(d) It can cause inflammation and damage to the lungs

HINTS & SOLUTIONS

Physics

1. (d) Work = Force × displacement
Torque = Force × force arm
= mass × acceleration × length
= $[M] \times [LT^{-2}] \times [L] = [M L^2 T^{-2}]$

2. (d) Vector triple product
$\vec{A} \times (\vec{B} \times \vec{C}) = \vec{B}(\vec{A} \cdot \vec{C}) - \vec{C}(\vec{A} \cdot \vec{B}) = 0$
$\Rightarrow \vec{A} \parallel (\vec{B} \times \vec{C})$
$[\because \vec{A} \cdot \vec{B} = 0 \text{ and } \vec{A} \cdot \vec{C} = 0]$

3. (c) $\because t = \sqrt{x} + 3$
$\Rightarrow \sqrt{x} = t - 3 \Rightarrow x = (t-3)^2$
$v = \frac{dx}{dt} = 2(t-3) = 0 \Rightarrow t = 3$
$\therefore x = (3-3)^2 \quad \Rightarrow x = 0.$

4. (c) According to law of conservation of momentum
$MV + mnv = 0$
$$\Rightarrow V = \frac{-mNv}{M} = \frac{-0.01\text{ kg} \times 10 \times 800\text{ m/s}}{100}$$
$\Rightarrow -0.8$ m/s
According to work energy theorem,
Average work done = Change in average kinetic energy
i.e., $F_{av} \times S_{av} = \frac{1}{2} m V_{rms}^2$
$$\Rightarrow \frac{F_{av} V_{max} t}{2} = \frac{1}{2} m \frac{V_{rms}^2}{2} \Rightarrow F_{av} = 8\text{ N}$$

5. (d) Given; speed = 10 m/s; radius r = 10 m
Angle made by the wire with the vertical
$$\tan\theta = \frac{v^2}{rg} = \frac{10^2}{10 \times 10} = 1 \Rightarrow \theta = 45° = \frac{\pi}{4}$$

6. (d) As the particle is moving in a potential energy region.
$\therefore$ Kinetic energy ≥ 0
And, total energy E = K.E. + P.E.
$\Rightarrow U(x) \leq E$

7. (a) $\because$ Power $P = \frac{w}{t}$
$$\Rightarrow \frac{P_1}{P_2} = \frac{t_2}{t_1} = \frac{30s}{1\text{ minute}} = \frac{30s}{60s} = \frac{1}{2}$$
(t_1 = 1 minute; t_2 = 30 second given)

8. (c) Given: $m_1 = 2$ kg $\quad m_2 = 4$ kg
$r_1 = 0.2$ m $\quad r_2 = 0.1$ m
$w_1 = 50$ rad s^{-1} $\quad w_2 = 200$ rad s^{-1}
As, $I_1 W_1 = I_2 W_2$ = Constant
$$\therefore W_f = \frac{I_1 W_1 + I_2 W_2}{I_1 + I_2} = \frac{\frac{1}{2} m_1 r_1^2 w_1 + \frac{1}{2} m_2 r_2^2 w_2}{\frac{1}{2} m_1 r_1^2 + \frac{1}{2} m_2 r_2^2}$$
By putting the value of m_1, m_2, r_1, r_2 and solving we get
= 100 rad s^{-1}

9. (a) $\because I = MK^2 \therefore K = \sqrt{\frac{I}{M}}$
$I_{ring} = MR^2$ and $I_{disc} = \frac{1}{2} MR^2$
$$\frac{K_1}{K_2} = \sqrt{\frac{I_1}{I_2}} = \sqrt{\frac{MR^2}{\left(\frac{MR^2}{2}\right)}} = \sqrt{2} : 1$$

10. (c) Escape velocity, $V_e = R\sqrt{\frac{8}{3}\pi GP}$
$$\Rightarrow V_e \propto R \Rightarrow \frac{V_P}{V_E} = \frac{R_P}{R_E} = 2$$
$\Rightarrow V_P = 2V_E.$

11. (b) As we know, the minimum speed with which a body is projected so that it does not return back is called escape speed.
$$V_e = \sqrt{\frac{2GM}{r}} = \sqrt{\frac{2GM}{R+h}} = \sqrt{\frac{2GM}{4R}}$$
$$= \left(\frac{GM}{2R}\right)^{\frac{1}{2}} \quad (\because h = 3R)$$

12. (c) From formula,
Increase in length
$$\Delta L = \frac{FL}{AY} = \frac{4FL}{\pi D^2 Y}$$
$$\frac{\Delta L_S}{\Delta L_C} = \frac{F_S}{F_C}\left(\frac{D_C}{D_S}\right)^2 \frac{Y_C}{Y_S} \frac{L_S}{L_C}$$
$$= \frac{7}{5} \times \left(\frac{1}{p}\right)^2 \left(\frac{1}{s}\right) q = \frac{7q}{(5sp^2)}$$

13. **(b)** According to Bernoulli's theorem,

$$P+\frac{1}{2}\rho v^2 = \text{constant and } Av = \text{constant}$$

If A is minimum, v is maximum, P is minimum.

14. **(b)** From question,

$\Delta\rho = (998-992)\text{ kg/m}^3 = 6\text{ kg/m}^3$

$$\rho = \frac{998+992}{2}\text{ kg/m}^3 = 995\text{ kg/m}^3$$

$$\rho = \frac{m}{V}$$

$$\Rightarrow \frac{\Delta\rho}{\rho} = -\frac{\Delta V}{V} \Rightarrow \left|\frac{\Delta\rho}{\rho}\right| = \left|\frac{\Delta V}{V}\right|$$

$\therefore$ Coefficient of volume expansion of water,

$$\frac{1}{V}\frac{\Delta V}{\Delta t} = \frac{1}{\rho}\frac{\Delta\rho}{\Delta t}$$

$$= \frac{6}{995\times 20} \approx 3\times10^{-4}\,/°\text{C}$$

15. **(b)** $Q_1 = 4Q_2$ (Given)

$$\Rightarrow \frac{K_1A_1\Delta t}{L} = 4\frac{K_2A_2\Delta t}{L}$$

$$\Rightarrow K_1A_1 = 4K_2A_2.$$

16. **(a)** From first law of thermodynamics

$$Q_{adc} = \Delta U_{adc} + W_{adc}$$
$$50\text{ J} = \Delta U_{adc} + 20\text{ J}$$
$$\Delta U_{adc} = 30\text{ J}$$

Again, $Q_{abc} = \Delta U_{abc} + W_{abc}$

$$W_{abc} = Q_{abc} - \Delta U_{abc}$$
$$= Q_{abc} - \Delta U_{adc}$$
$$= 36\text{ J} - 30\text{ J}$$
$$= 6\text{ J}$$

17. **(a)** Adiabatic equations of state are

$PV^\gamma = \text{constant}$

$TV^{\gamma-1} = \text{constant}$

$P^{1-\gamma}T^\gamma = \text{constant}.$

18. **(d)** Efficiency of engine A, $\eta_1 = 1-\frac{T}{T_1}$,

Efficiency of engine B, $\eta_2 = 1-\frac{T_2}{T}$

Here, $\eta_1 = \eta_2$

$$\therefore \frac{T}{T_1} = \frac{T_2}{T} \Rightarrow T = \sqrt{T_1T_2}$$

19. **(b)** $\because P = \frac{1}{3}\frac{mn}{V}V_{\text{rms}}^2$

When mass is halved and speed is doubled then

Resultant pressure,

$$P' = \frac{1}{3}\times\frac{m}{2}\times\frac{n}{V}(2v_{\text{rms}})^2 = 2P.$$

20. **(d)** As $\frac{v^2}{a^2\omega^2}+\frac{y^2}{a^2} = 1$ This is the equation of ellipse. Hence the graph is an ellipse. P versus x graph is similar to V versus x graph.

21. **(a)** From formula, $f = \frac{1}{x}\sqrt{\frac{T}{m}}$

$$\Rightarrow \frac{1}{f} \propto l$$

$$\therefore l_1 : l_2 : l_3 = \frac{1}{f_1} : \frac{1}{f_2} : \frac{1}{f_3}$$

$$= f_2f_3 : f_1f_3 : f_1f_2$$

[Given: $f_1 : f_2 : f_3 = 1:3:5$]

$= 15:5:3$

Therefore the positions of two bridges below the wire are

$$\frac{15\times100}{15+5+3}\text{ cm and } \frac{15\times100+5\times100}{15+5+3}\text{ cm}$$

i.e., $\frac{1500}{23}$ cm, $\frac{2000}{23}$ cm

22. **(b)** $\frac{\Delta f}{f} = \frac{v}{C}$

$$\Rightarrow \frac{(\text{Beats})/2}{f} = \frac{v}{C} \Rightarrow \text{Beats} = \frac{2fv}{C} = 4.$$

23. **(b)** The system of three charges will be in equilibrium.

A ——— q ——— B

$Q \leftarrow r \rightarrow O \leftarrow r \rightarrow Q$

For this, force between charge at A and B + force between charge at point O and either at A or B is zero.

i.e., $\frac{KQ^2}{r^2} + \frac{KQq}{(r/2)^2} = 0$

By solving we get,

$$q = -\frac{Q}{4}.$$

24. **(c)** When electric dipole is aligned parallel $\theta = 0°$ and the dipole is rotated by $90°$ i.e., $\theta = 90°$.

Energy required to rotate the dipole

$W = U_f - U_i = (-pE\cos 90°) - (-pE\cos 0°)$

$= pE.$

25. **(c)** Resistance is directly proportional to length

$$\frac{1}{R_{AB}} = \frac{1}{3} + \frac{1}{4+5} = \frac{(4+5)+3}{(3)(4+5)}$$

$$R_{AB} = \frac{3\times(4+5)}{3+(4+5)} = \frac{27}{12}$$

Similarly,

$$R_{BC} = \frac{4\times(3+5)}{4+(3+5)} = \frac{32}{12}$$

$$R_{AC} = \frac{5\times(3+4)}{5+(3+4)} = \frac{35}{12}$$

$\therefore R_{AB} : R_{BC} : R_{AC} = 27 : 32 : 35$

26. **(a)** Copper rod and iron rod are joined in series.

$$\therefore R = R_{Cu} + R_{Fe} = (\rho_1 + \rho_2)\frac{\ell}{A}$$

$$\left(\because R = \rho\frac{\ell}{A}\right)$$

From ohm's law $V = RI$

$= (1.7 \times 10^{-6} \times 10^{-2} + 10^{-5} \times 10^{-2}) \div 0.01 \times 10^{-4}$ volt

$= 0.117$ volt $(\because I = 1A)$

27. **(b)** Resistance is directly proportionl to length of the wire. As length is doubled so mass is doubled and resistance is doubled.

We have

$$\frac{(10E)^2}{R}t = mS\Delta T, \text{ Now } \frac{(nE)^2 t}{2R} = (2m)S\Delta T$$

$$\Rightarrow \frac{n^2E^2t}{2R} = 2\frac{10^2E^2t}{R}$$

$\Rightarrow n = 20$

28. **(c)** Given:

Magnetic field B = 2×10^{-4} weber/m^2

Velocity of electron, v = 10^7 m/s

Lorentz force $F = qvB \sin\theta$

$= 1.6 \times 10^{-19} \times 10^7 \times 2 \times 10^{-4} \, (\because \theta = 90°)$

$= 3.2 \times 10^{-16}$ N

29. Here, $\vec{F}_{AB} + \vec{F}_{BCDA} = \vec{0}$

$$\Rightarrow \vec{F}_{BCDA} = -\vec{F}_{AB} = -\vec{F}$$

$$(\because \vec{F}_{AB} = \vec{F})$$

30. **(b)** $FL = MB$ (= Torque) $\Rightarrow L = \frac{MB}{F}$

31. **(c)** Given: current I = 2.5 A

Inductance, L = 5H

Magnatic flux, ϕ = ?

We know, $\phi = LI \Rightarrow 5 \times 2.5$ Wb = 12.5 Wb

32. **(d)** A transformer is essentially an AC device. DC source so no mutual induction between coils

$\Rightarrow E_2 = 0$ and $I_2 = 0$

33. **(c)** Given: frequency f = 2MHz, relative permittivity $\epsilon_r = 4$

From formula,

$$\text{velocity } v = \frac{c}{\sqrt{\epsilon_r}} = \frac{c}{2} \Rightarrow \lambda' = \frac{\lambda}{2}$$

[Since frequency remains unchanged]

34. **(a)** $\because T \propto n^3$

$Tn_1 = 8\,Tn_2$ (given)

Hence, $n_1 = 2n_2$

35. **(c)** Increasing order of penetrating power : $\alpha < \beta < \gamma$.

For same energy, lighter particle has higher penentrating power.

36. **(d)**

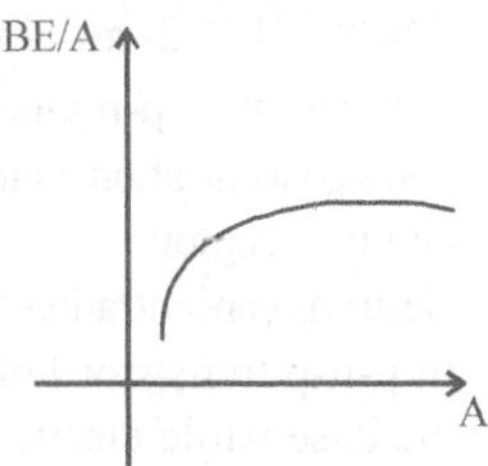

From the graph of BE/A versus mass number A it is clear that, BE/A first increases and then decreases with increase in mass number.

37. **(c)** Since, stopping potential is independent of distance hence new stopping potential will remain unchanged i.e., new stopping potential = V_0.

38. **(a)** From formula

$$\lambda = \frac{h}{\sqrt{2mKT}}$$

$$= \frac{6.63\times10^{-34}}{\sqrt{2\times1.67\times10^{-27}\times1.38\times10^{-23}T}}m$$

[By placing value of h, m and k)

$$= \frac{30.8}{\sqrt{T}}\text{Å}$$

39. (a)

From fig. $40° + \theta = 90° \therefore \theta = 90° - 40° = 50°$

40. (b) It is due to scattering of light. Scattering $\propto \frac{1}{\lambda^4}$. Hence the light reaches us is rich in red.

41. (c) Fringe width $\beta = \frac{\lambda D}{d}$;

From question $D' = 2D$ and $d' = \frac{d}{2}$

$\therefore \beta' = \frac{\lambda D^1}{d^1} = 4\beta$

42. (d) Conditions for diffraction minima are

Path diff. $\Delta x = n\lambda$ and Phase diff. $\delta\phi = 2n\pi$

Path diff. $= n\lambda = 2\lambda$

Phase diff. $= 2n\pi = 4\pi \quad (\because n = 2)$

43. (c) In p-region of p-n junction holes concentration > electrons concentration and in n-region electrons concentration > holes concentration.

44. (c) In p-n-p transistor holes are injected into the base while electrons are injected into the base of n-p-n transistor. Emitter-base junction is forward biased.

45. (b) $C' = \overline{A \cdot B} \Rightarrow C = \overline{\overline{A \cdot B}} = A \cdot B$

Hence the resultant gate is AND gate.

Chemistry

46. (a) $CaCO_3 \longrightarrow \underset{x}{Ca^{2+}} + \underset{x}{CO_3^{2-}}$

$CaC_2O_4 \longrightarrow \underset{y}{Ca^{2+}} + \underset{y}{C_2O_4^{2-}}$

$\therefore [Ca^{2+}] = x + y$

Now, $K_{sp}(CaCO_3) = [Ca^{2+}][CO_3^{2-}]$

or $4.7 \times 10^{-9} = (x+y)x$

similarly, $K_{sp}(CaC_2O_4) = [Ca^{2+}][C_2O_4^{2-}]$

or $1.3 \times 10^{-9} = (x+y)y$

On solving, we get

$[Ca^{2+}] = 7.746 \times 10^{-5}$ M

47. (d) Density $= \frac{PM}{RT} = \frac{5 \times 28}{0.0821 \times 500} = 3.41$ g/ml

48. (a) K_w at 25°C $= 1 \times 10^{-14}$

At 25°C

$K_w = [H^+][OH^-] = 10^{-14}$

At 100°C (given)

$K_w = [H^+][OH^-] = 55 \times 10^{-14}$

$\because$ for a neutral solution

$[H^+] = [OH^-]$

$\therefore \quad [H^+]^2 = 55 \times 10^{-14}$

or $\quad [H^+] = (55 \times 10^{-14})^{1/2}$

$\because \quad pH = -\log[H^+]$

On taking log on both side

$-\log[H^+] = -\log(55 \times 10^{-14})^{1/2}$

$pH = -\frac{1}{2}\log 55 + 7\log 10$

$pH = -0.87 + 7 = 6.13$

49. (c) $\underset{x}{CH_4} + \underset{2x}{2O_2} \longrightarrow CO_2 + 2H_2O$

$\underset{(5-x)}{C_3H_8} + \underset{5(5-x)}{5O_2} \longrightarrow 3CO_2 + 4H_2O$

$2x + 5(5-x) = 16 \quad \Rightarrow x = 3L$

$\therefore$ Heat released

$= \frac{3}{22.4} \times 890 + \frac{2}{22.4} \times 2220 = 317$

50. (N) $E = \frac{hc}{\lambda} \times N_A$

$= \frac{6.62 \times 10^{-27} \times 3 \times 10^{10} \times 6.02 \times 10^{23}}{\lambda}$

$= \frac{1.19 \times 10^8}{\lambda}$

Hence none of the given option is correct.

51. (c) $pH = pK_a + \log\frac{[Salt]}{[Acid]}$

$5 = 4 + \log\frac{[Salt]}{[Acid]} \qquad \because pK_a = -\log K_a$

Given $K_a = 1 \times 10^{-4}$

$\therefore \quad pK_a = -\log(1 \times 10^{-4}) = 4$

Now from Handerson equation

$pH = pK_a + \log\frac{[Salt]}{[Acid]}$

Putting the values

$5 = 4 + \log\frac{[Salt]}{[Acid]}$

$\log\frac{[\text{Salt}]}{[\text{Acid}]} = 5-4=1$

Taking antilog

$[\text{Salt}]/[\text{Acid}] = 10 = \frac{10}{1}$

52. **(b)** For a first order reaction,

$t_{75\%} = 2 \times t_{50\%}$

53. **(c)** $\alpha = \sqrt{\frac{K_a}{c}}; 0.037 = \sqrt{\frac{K_a}{0.10}};$

$K_a = (0.037)^2 \times 0.10 = 1.37 \times 10^{-4}$

54. **(c)** Applying Hess's law, equation (i) can be obtained by adding equations (ii) and (iii).

$\therefore x = y + z$

55. **(c)** Applying VSEPR theory, both NF_3 and H_2O are sp^3 hybridized.

56. **(c)** Millimoles of solution of chloride

$= 0.05 \times 10 = 0.5$

Millimoles of $AgNO_3$ solution

$= 10 \times 0.1 = 1$

So, the millimoles of $AgNO_3$ are double than the chloride solution.

$\therefore XCl_2 + 2AgNO_3 \longrightarrow 2AgCl + X(NO_3)_2$

57. **(c)** Applying,

$w = Zit = \frac{Eit}{96500}$

Equivalent weight of cobalt (II) = 59/2

I = 10 A

Time (t) = 109 min = 109 × 60 sec

Substituting these values we get,

$w = \frac{59 \times 10 \times 109 \times 60}{2 \times 96500} = 20.0$

58. **(c)** An ideal solution is that solution in which each component obeys Raoult's law under all conditions of temperatures and concentrations. For an ideal solution.

$\Delta H_{mix} = 0$ and $\Delta V_{mix} = 0$.

59. **(d)** Rate $= k[A]^2[B]^3$

when concentrations of both A and B are doubled then

rate $= k[2A]^2[2B]^3$

$= 32\, k[A]^2[B]^3$

$\therefore$ rate will increase by a factor of 32.

60. **(b)** No. of electrons in CO = 6 + 8 = 14

No. of electrons in $NO^+ = 7 + 8 - 1 = 14$

$\therefore$ CO and NO^+ are isoelectronic species.

Isoelectronic species have identical bond order.

61. **(c)** IUPAC name of $[CrF_2(en)_2]Cl$ is Difluoridobis(ethylenediamine)chromium (III) chloride.

62. **(c)** Gd [64] = [Xe] $4f^7 5d^1 6s^2$

63. **(b)** XeO_4 (structure: Xe with four O atoms, one double-bonded at top and three below)

64. **(c)** CFSE = $(-0.4x + 0.6y)\Delta_0$ where

$x \rightarrow e^-$s in t_{2g} orbital

$y \rightarrow e^-$s in e_g orbital

$\therefore -0.4 \times 3 + 0.6 \times 1 = -0.6\,\Delta_0$

65. **(d)** (a) $N_2 \longrightarrow N_2^+$

B.O. 3 → 2.5

Bond energy decreases

Magnetic behaviour changes from diamagnetic to paramagnetic

(b) $O_2 \longrightarrow O_2^+$

B.O. 2 → 2.5

Bond energy increases

Magnetic behaviour does not change.

(c) $C_2 \longrightarrow C_2^+$

B.O. 2 → 2.5

Bond energy decreases

Magnetic behaviour changes from diamagnetic to paramagnetic

(d) $NO \longrightarrow NO^+$

B.O. 2 → 2.5

bond energy increases

Magnetic behaviour changes from paramagnetic to diamagnetic

66. **(a)** As the nuclear charge increases, the force of attraction between the nucleus and the incoming electron increses and hence the elecron gain enthalpy becomes more negative, hence the correct order is

Ca < Al < C < O < F

67. **(c)** A transition element must have incomplete *d*-subshell. Zinc have completely filled *d* subshell having $3d^{10}$ configuration. Hence do not show properties of transition elements to any appreciable extent except for their ability to form complexes.

68. **(d)** At the cathode, since the discharge potential of Na^+ ions is lower than that of H^+ ions on the mercury, cathode, so Na^+ ions are discharged while H^+ ions remain in the solution.

$Na^+ + e^- \longrightarrow Na$
$2Na + xHg \longrightarrow Na_2Hgx$
Sodium Amalgam

69. (a) Feldspars are aluminosilicates.

70. (c) $Ni^{+2} \rightarrow 3d^8$ = ⇅ ⇅ ⇅ ↑ ↑
CN^- is a strong ligand and causes pairing of $3d$ electrons of Ni^{2+}.
∴ It is diamagnetic.

71. (a) $\Delta E^\circ = E^\circ_{red} + E^\circ_{oxd}$
$= -1.81 - 1.51 = -2.69$
Since ΔE° is negative
∴ $\Delta G = -nFE^\circ$, ΔG will have positive value so, forward reaction is not possible.

72. (d) $H \overset{\sigma}{-} C \overset{\sigma}{\underset{\pi}{=}} C \overset{\sigma}{-} H$ (each C bonded to H by σ bond)

73. (c) [Octahedral structure of Co with Cl, Cl, NH_3, NH_3, NH_3, NH_3; 90° between the two Cl]

74. (d) Acetylacetone forms six membered stable ring complexes.

75. (b) Al_2O_3 cannot be reduced by carbon.

76. (d) $Co^{2+} \Rightarrow [Ar]4s^03d^7$, Br^- is weak ligand.

77. (c) Dettol is a mixture of chloroxylenol and terpineol.

78. (b) Stability depends on number of hyperconjugative structures.

79. (c) Conformers are form of stereoisomers in which isomers can be interconverted by rotations about single bonds. I and II are staggered and eclipsed conformers respectively.

80. (d) $CH \equiv CH \xrightarrow{Hg^{+2}/H_2SO_4} CH_3 - \overset{O}{\overset{||}{C}} - H$

CH_3CHO does not give Victor Meyer test.

81. (b) Hydrolysis gives $CH_3CH_2CH_2NH_2$ + HCOOH. On treatment with $NaNO_2$ and HCl former gives $CH_3CH_2CH_2OH$ which does not give iodoform test. HCOOH gives Tollen's reagent test.

82. (c) Stability depends on number of hyperconjugative structure.

83. (b) Secondary amine react with nitrous acid to give N-Nitrosoamines.

$(CH_3)_2N - C_6H_5 + NaNO_2 + HCl \longrightarrow C_6H_5 - N(CH_3) - N = O + NaCl + CH_3OH$

84. (d) Acid rain is the rain water containing sulphuric acid and nitric acid which are formed from the oxides of sulphur and nitrogen present in the air as pollutants and rain water has a pH of 4-5.

85. (b) In the carbonium ion the carbon atom carrying the positive charge is sp^2 hybridized.

86. (c) The blue colour is of $Fe_4[Fe(CN)_6]_3$

87. (c)

Phenol $\xrightarrow[\Delta]{Zn}$ Benzene $\xrightarrow[CH_3CH_2CH_2Cl]{AlCl_3}$ $CH_3-CH(C_6H_5)-CH_3$ $\xrightarrow{O_2}$ $(CH_3)_2C(C_6H_5)-O-O-H$

88. (c) The base pairs of the two strands of DNA are linked together through H-bonds.

89. (a) Huckel rule is not obeyed. It has only four electrons. Further it does not have continous conjugation.

90. (c) $CH_3\,\overset{OH}{\overset{|}{CH}}-CH_2CH_2CH_2CH_3$

$CH_3-\overset{OH}{\overset{|}{CH}}-\underset{CH_3}{\underset{|}{CH}}\,CH_2CH_3$,

$CH_3-\overset{OH}{\overset{|}{CH}}-CH_2-\underset{CH_3}{\underset{|}{CH}}-CH_3$,

$CH_3-\overset{OH}{\overset{|}{CH}}-\underset{CH_3}{\overset{CH_3}{\overset{|}{\underset{|}{C}}}}-CH_3$

Biology

91. **(d)** The plant body of a liverwort is haploid (n), gametophytic, small, dorsoventrally flattened, thallose, dichotomously branched fixed by unicellular and unbranched rhizoids, *e.g., Marchantia.*

92. **(c)** Fungi lack chlorophyll, hence, they do not prepare their food by photosynthesis. They can grow where organic material is available. So, they are heterotrophs that acquire their nutrient by absorption and store in the form of glycogen.

93. **(b)** Heterocysts are large sized, thick-walled specialised cells which occur in terminal, intercalary or lateral position in **filamentous cyanobacteria,** *e.g., Nostoc.* They have enzyme **nitrogenase** and specialised to perform biological nitrogen fixation.

94. **(b)** Families are characterised on the basis of both vegetative and reproductive features of plant species. Tomato *(Lycopersicon esculentum)* and potato *(Solanum tuberosum)* belong to the same family Solanaceae.

95. **(b)** The female sex organ archegonium is formed in bryophytes (*Funaria),* pteridophytes (*Dryopteris*) and gymnosperms (*Ginkgo*).

96. **(b)** Cyanobacteria, e.g., *Nostoc*, grow in such abundance as to form water blooms. *Sargassum* belongs to brown algae which possess chl *a*, *c*, carotenoids, xanthophyll and a characteristic brown pigment, fucoxanthin. Commonly known forms of basidiomycetes are mushrooms, bracket fungi or puffballs. In *Spirogyra* gametes are non-flagellated (non-motile) but similar in size. They show amoeboid movements.

97. **(c)** *Calotropis,* tulip, *Asparagus,* colchicine, *Petunia,* mustard, tobacco having valvate aestivation while sweet pea, groundnut, soyabean, *Indigofera, Sesbania* having vexillary aestivation.

98. **(b)** Phyllotaxy is the pattern of arrangement of leaves on the stem or branch.

- *Ocimum,* Guava, *Calotropis* → Opposite phyllotaxy
- China rose, Sunflower, Mustard → Alternate phyllotaxy
- *Alstonia, Nerium* (Oleander) → Whorled phyllotaxy

99. **(a)** In cymose inflorescence, the growth of the main axis is definite because the growing point of peduncle is used up in the formation of a flower. Further growth of flowering axis is continued by one or more lateral branches which also end in flower.

100. **(c)** Sometimes, a few epidermal cells in the vicinity of the guard cells become specialised in their shape and size and are known as subsidiary cells or accessory cells.

101. **(c)** Lateral meristems are meristems which occur parallel to the circumference of the organs in which they develop. They undergo periclinal divisions producing secondary tissues on the outer and inner sides and increase girth of the plant organs. Examples are vascular cambium and cork-cambium.

102. **(a)** • Inflorescence in family fabaceae (e.g., soyabean) is racemose.

• Inflorescence in family liliaceae (e.g., tulip, *Aloe*) is solitary/cymose; often umbellate.

• Inflorescence in family solanaceae (e.g., brinjal) is solitary, axillary or cymose as in *Solanum.*

103. **(b)** Kinetochores are small disc-shaped structures at the surface of the centromeres which serve as the sites of attachment of spindle fibres to the chromosomes that are moved into position at the centre of the cell. Hence, the metaphase is characterised by all the chromosomes coming to lie at the equator.

104. **(b)** The combination of pentose sugar with nitrogenous bases (purines or pyrimidines) is called nucleoside. Examples are adenosine, guanosine, cytidine, thymidine and uridine.

105. **(a)** Glycocalyx or mucilage is the outermost coating of bacterial cells/cell wall which is rich in polysaccharides. A thick and tougher mucilage is called capsule which gives gummy or sticky trait to cells. It protects the cells from dessication, toxins and preventing attachment to foreign invaders.

106. **(b)** Chromoplasts are non-photosynthetic coloured plastides which synthesise and stored carotenoid pigmentes. They, therefore, appear orange red and yellow where as amyloplast (store starch), aleuroplast (store proteins) and elaioplast (store oil droplets and fats) are leucoplasts, colourless plastids.

107. **(b)** During zygotene, a substage of Prophase I of meiosis I, chromosomes start pairing together called synapsis. Such paired chromosomes are called as homologous chromosomes. A complex structure *i.e.*, synaptonemal complex is formed by a pair of synapsed homologous chromosomes called a bivalent or a tetrad.

108. (a) In the global nitrogen cycle, bacterial denitrification is recognised as important process that converts fixed nitrogen to atmospheric nitrogen gas, N_2. In marine nitrogen cycle, anaerobic oxidation of ammonium coupled to nitrate reduction, contributes substantially to N_2–production in marine sediments.

109. (d) During embryonic development, the primary organiser signals the development of another organ or tissue by secreting chemical factors.

110. (c) Dehydrogenase enzymes present in living tissue reduce the tetrazolium chloride to formazan, a reddish, water insoluble compound. This reaction occurs in or near living cells which are releasing hydrogen in respiration processes. Viable tissues produce a normal red-colour, weak living tissue produce an abnormal colour. Dead tissues do not stain, remaining usually white.

111. (b) Plants which are difficult to flower can be made to do so by spraying them with 2, 4, –D (2, 4 – dichlorophenoxy acetic acid) and NAA (napthalene acetic acid) which are synthetic auxins, *e.g.,* litchi, pineapple.

112. (a) Biotin, also known as vitamin H, is a water-soluble B-vitamin. It is a coenzyme for carboxylase enzymes, involved in the synthesis of fatty acids, isoleucine, valine and in gluconeogenesis. Sulphur is a constituent of biotin.

113. (a) In C_4 plants, bundle sheath cells are rich in RuBisCO, but lack PEPcase. The CO_2 released in the bundle sheath cells enters the C_3 or the Calvin pathway, common to all plants. Thus, the basic pathway that results in the formation of sugars, the calvin pathway is common to the C_3 and C_4 plants.

114. (b) Cleistogamous flowers do not expose their reproductive parts. Anthers and stigma lie close to each other. Pure autogamy occurs since there is no chance of cross-pollination. Cleistogamy is the most efficient floral adaptation for promoting self-pollination. E.g., *Viola mirabilis* and *Oxalis autosella.*

115. (a) Single Megaspore Mother Cell (MMC) with dense cytoplasm and a prominent nucleus gets differentiated from nucellus near the micropylar region. This Megaspore Mother Cell (MMC) undergoes meiosis to form '4' haploid cells called **megaspores** and the process of formation is known as **megasporogenesis.**

116. (c) In most aquatic organisms, such as a majority of algae and fishes as well as amphibians, syngamy occurs in the external medium (water), *i.e.,* outside the body of the organism. This type of gametic fusion is called external fertilisation.

117. (d) There are different types of vectors involved in pollination.

For example, Maize, mulberry → pollination by wind. *Vallisneria* → pollination through water (hydrophily). Cucumber → Bees are brought for the commercial plantings of cucumber.

118. (d) Pollen grains are generally spherical and a prominent two-layered wall. The hard outer layer called the exine is made up of sporopollenin which is one of the most resistant organic material known. It can withstand high temperatures and strong acids and alkali.

119. (b) Endosperm is the nutritive tissue which provides nourishment to the embryo in seed plant. Albuminous seeds retain a part of endosperm as it is not completely used up during embryo development (*e.g.*, wheat, maize, barley, castor, sunflower).

120. (d) Genetic drift is random change in allele number and frequency in a gene pool due to chance (e.g., small size of population). It is caused by sampling error or error in gene pool sample that is to form the next generation. The sampling gene pool is generally small in size. Variability is also limited.

121. (c) Genetic code is the relationship of amino acid sequence in a polypeptide and nucleotide/ base sequence in mRNA/antisense strand of DNA.

It is **universal,** *i.e.*, a codon specifies the same amino acid in all organisms, **non-overlapping**, *i.e.*, adjacent codons are independent with no base being member of two codons, **degeneracy**, *i.e.*, some amino acids are coded by more than one codon, hence the code is degenerate, **unambiguous,** *i.e.*, one codon codes for only one amino acid.

122. (c) Crossing over leads to recombination of genetic material on the two chromosomes. Mutation results in alternation of DNA sequences and consequently results in change in the genotype and the phenotype of an organism. In addition to recombination, mutation is another phenomenon that leads to variation in DNA.

123. (a) Plant viruses often contain parasites of their own, referred to as satellites. Satellite RNAs are dependent on their associated (helper) virus for both replication and encapsidation. Example—Tobacco Necrosis Virus (TNV). Viroids are infectious agents smaller than viruses. Bacteriophages are viruses that infect the bacteria. A prion is an infectious agent that is composed primarily of protein.

124. (b) The technique of DNA fingerprinting was developed by Dr. Alec Jeffrey in 1984. It is a technique generally using repeated sequences (repetitive DNA) in the human genome that produces a pattern of band that is unique for every individuals. These short nucleotide repeats vary in number from person to person and are called **variable number of tandem repeat (VNTR).** VNTR belongs to class of satellite DNA referred to as minisatellite.

125. (b) Adenovirus is non-enveloped dsDNA virus which cause respiratory diseases. Adenovirus is used as a vehicle to administer targeted therapy in the form of recombinant DNA or protein. Specific modifications on fibre proteins are used to target adenovirus to certain cell types, a major effort is made to limit hepatotoxicity and prevent multiple organ failure.

126. (c) Somatic embryogenesis is a process where a plant or embryo is derived from a single somatic cell or group of somatic cells. Somatic embryos are formed from plant cells that are not normally involved in the development of embryos, *i.e.*, ordinary plant tissue. No endosperm or seed coat is formed around a somatic embryo.

127. (c) A hybridization probe is a fragment of DNA of variable length which is used in DNA samples to detect the presence of nucleotide sequence (the DNA target) that are complementary to the sequence in the probe. The probe hybridize to single–stranded DNA whose base sequence allow probe target base-pairing due to complementary between the probe and target.

128. (b) Ethanol is non-polar molecule while DNA is a polar molecule, it has a net negative charge due to the phosphate groups attached to it. When DNA is placed into a solution of 100% ethanol, it becomes insoluble and precipitates out of solution.

129. (c) RNAi is a method of cellular defense in all eukaryotes. It is a system within living cells that helps to control the activity of specific genes. This method involves silencing of mRNA due to complementary double stranded RNA that prevents translation of target gene or mRNA [silencing]. Source of ds RNA is **retrovirus** (having RNA genome) or **transposons** (mobile genetic material).

130. (d) The sediment characteristics often determine the type of benthic animals that can thrive there.

131. (a) During the period 1960 to 2000 rice production went up from 35 million tonnes to 89.5 million tonnes. This was due to the development of semi-dwarf varieties of rice. There are 2,00,000 varieties of rice in India.

132. (b) IUCN or WCN maintains a red data book which is a catalogue of threatened plants and animals facing risk of extinction. The IUCN red list (2004) documents the extinction of 784 species (including 338 vertebrates, 359 invertebrates and 87 plants) in the last 500 years.

133. (c) Nagarjunsagar-Srisailam Tiger Reserve is the largest Tiger reserve in India and the only Tiger Reserve in Andhra Pradesh state.

134. (d) The Kyoto Protocol is an international agreement linked to the United Nations Framework Convention on climate change which commits its parties by setting internationally binding emissions reduction targets. In Doha, Qatar on 8 December 2012, the "Doha Amendment to the Kyoto Protocol" was adopted. The second commitment period was from 1st Jan. 2013 to 31st Dec. 2020.

135. (b) In an ecosystem, producers (e.g., plants/phytoplankton) belong to first trophic level, herbivores or primary consumer (e.g., grasshopper) to the second and carnivores or secondary consumer (e.g., lion, wolf) to the third trophic level.

136. (a) In option (a) all the characteristics belong to class Reptilia. In options (b), (c) and (d) the characteristics belong to the classes Amphibia, Osteichthyes and Chondrichthyes respectively.

137. (a) Duckbilled platypus is oviparous and comes under phylum mammalia. Millipede belongs to the phylum arthropoda. Sea-anemone is diploblastic belongs to phylum cnidaria. Silver-fish is an insect belongs to phylum arthropoda, having long antennae, no wings and move in a wiggling motion that resembles the movement of a fish.

138. (a) In **Ctenophora,** sexes are not separate. Reproduction takes place only by sexual means. In **Cnidaria,** which exist in both forms, polyps produce medusae asexually and medusae form the polyps sexually. **Porifera (sponge)** reproduce asexually by fragmentation and sexually by formation of gametes. **Protozoans** come under protista which reproduce asexually and sexually by cell fusion and zygote formation.

139. (a) Sharks and dogfishes have cylindrical body while skates and rays have flattened body with winglike pectoral fins which are not distinct from body.

140. (a) Skeletal muscle tissue is closely attached to sketetal bones. In a typical muscle such as the biceps, striated (striped) skeletal muscle fibres are bundled together in a parallel fashion.

141. (c) In cockroaches, males bear a pair of short, thread like anal styles which are absent in females.

142. (b) In fig., (A) shows the cell organelle mitochondria. The mitochondria are bounded by two membranes, *i.e.,* outer membrane and inner membrane. Mitochondria are referred as "powerhouse" of the cell as they produce 95% of ATP. This energy is produced during the break down of food molecules which involve glycolysis, oxidative decarboxylation and oxidative phosphorylation (krebs cycle and respiratory chain).

143. (d) Capsule is a layer that lies outside the cell wall of bacteria. The capsule can protect cells from engulfment by eukaryotic cells, such as macrophages. They also exclude bacterial viruses and most hydrophobic toxic materials such as detergents.

144. (c) Almost all enzymes are proteins. There are some nucleic acids that behave like enzymes. There are called ribozyme (also called RNA enzyme or catalytic RNA).

145. (b) Glutamic acid and aspartic acid are acidic amino acids with two carboxylic groups and one amino group.

146. (c) Metaphase chromosome is made up of two sister chromatides which are held together by the centromere. Chromosomes are moved to spindle equator and gets aligned along metaphase plate through spindle fibres to both poles.

147. (a) Physiological value of carbohydrates is 4.0 kcal/g, proteins 4.0 kcal/g and of fats is 9.0 kcal/g. Hence,

5 g raw sugar will yield $5 \times 4.0 = 20.0$ kcal
4 g albumin (protein) will yield $4 \times 4.0 = 16.0$ kcal
$10 + 2$ g of fat will yield $12 \times 9.0 = 108.0$ kcal
Total yield = 144 kcal.

148. (d) Chemical process of digestion started in the oral cavity by the hydrolytic action of the carbohydrate (potato contains starch) splitting enzyme, the salivary amylase. Carbohydrates in the chyme are hydrolysed by pancreatic amylase into disaccharides.

149. (b) The number of spiracles in cockroach is 10-pairs (2-pairs in thoracic and 8-pairs in abdominal region). The thoracic pairs of spiracles are present on pleuron between prothorax - mesothorax and metathorax. Each spiracle opens into a chamber called atrium. After atrium the tracheal tube ramifies into fine branches of tracheae, and then tracheoles.

150. (b) Basophils secrete histamine, serotonin, heparin, etc., and are involved in inflammatory reactions.

151. (d) Artery is blood vessel which receives blood from heart, has elastic thick wall shows jerky movements due to pumping activity of heart. Vein is a blood vessel which carries blood towards the heart, has wider lumen with internal valves where blood flows smoothly and slowly. Blood capillary is a very fine blood vessel which has a single layered wall. Pulmonary veins are the only veins which carry oxygenated blood. ($_{P}O_2 = 95$ mmHg, $_{P}CO_2 = 40$ mmHg)

152. (c) Salamander (Amphibia; Caudata) excrete urea by help of kidneys.

153. (b) Myasthenia gravis is a chronic autoimmune muscular disease. It causes breakdown of neuromuscular junction due to which the brain loses control over muscles. The symptoms may include drooping eyelids, difficulty in swallowing muscle fatigue, difficult breathing and inability to control facial expressions.

154. **(c)** Increase in Ca^{++} level into the sarcoplasm leads to the binding of calcium with a subunit of troponin on actin filaments and there by remove the masking of active sites for myosin. Utilising the energy from ATP hydrolysis, the myosin head now binds to the exposed active sites on actin to form a cross bridge. This pulls the attached actin filaments towards the centre of 'A' band. The 'Z' line attached to these actins are also pulled inwards thereby causing a shortening of the sarcomere, *i.e.*, contraction. It is clear from the above steps, that during shortening of the muscle, *i.e.*, contraction, the 'I' bands get reduced, whereas the 'A' bands retain the length.

155. **(c)** Cerebrum is the first and most developed part of fore brain. It makes 2/3 part of total brain. Pons is a small spherical projection, which is situated below the midbrain and upper side of the medulla oblongata. It acts as a relay centre among different parts of brain. B and D are thalamus and spinal cord respectively.

156. **(d)** In the given figure, A, B, C and D are axon, neurotransmitters, receptors and synaptic vesicles respectively.

157. **(c)** Insulin is a peptide hormone, which plays a major role in the regulation of glucose homeostasis. Insulin acts mainly on hepatocytes and adipocytes (cells of adipose tissue), and enhances cellular glucose uptake and utilization. Insulin also stimulates conversion of glucose to glycogen (glycogenesis) in the target cells.

158. **(b)** Norepinephrine is released by sympathetic fibres *i.e*, rapidly secreted in response to stress of any kind and during emergency situations. It increases the heart beat, the strength of heart contraction and the rate of respiration.

159. **(b)** The corpus luteum is a temporary endocrine structure in female mammals that is involved in the production of relatively high levels of progesterone. A marked in the figure shows primary follicle, a layer of granulosa cells, surrounds each primary oocyte. A large number of these follicles degenerate during the phase from birth to puberty.

160. **(d)** Leydig cells or interstitial cells, which are present in the intertubular spaces produce a group of hormones called androgens mainly testosterone. Androgens play a major stimulatory role in the process of spermatogenesis (formation of spermatozoa).

161. **(a)** Parturition is induced by a complex neuroendocrine mechanism. The signals for parturition originate from the fully developed foetus and the placenta which induce mild uterine contractions called foetal ejection reflex. This triggers release of oxytocin from the maternal pituitary.

162. **(c)** Oxytocin is birth hormone and acts on the smooth muscles of our body and stimulates their contraction. Vasopressin acts mainly at the kidney and stimulates resorption of water and electrolytes and reduces loss of water through urine. Hence, it is also called as anti-diuretic hormone (ADH).

163. **(a)** STDs are more common in 15-24 years age group. Gonorrhoea, syphilis, genital herpes, chlamydiasis, genital warts, trichomoniasis, hepatitis-B and AIDS are some of the common STDs. Haemophilia is sex-linked recessive disease, which show its transmission from unaffected carrier female to some of male progeny. Sickle cell anaemia is an autoimmune linked recessive trait in which both the partners are carrier for the gene.

164. **(d)** In Intra-Uterine Transfer (IUT) embryo with more than 8-blastomeres stage (morula) is used for transfer into the uterus.

165. **(b)** Pedigree is a chart showing the record of inheritance of certain genetic traits for two or more ancestral generations of an individual, abnormality or disease.

166. **(c)** In case of humans, the sex determining mechanism is XY type. Out of 23 pairs of chromosomes, 22 pairs are exactly same in both males and females called autosomes. A pair of X-chromosomes present in the female, whereas the presence of an X and Y chromosome are determinant of male characteristic. In case the ovum fertilises with a sperm carrying X-chromosome the zygote develops into a female (XX) and the fertilisation of ovum with Y-chromosome carrying sperm results into a male offspring.

167. **(c)** Down's syndrome is the chromosomal disorders due to the presence of an additional copy of the chromosome number 21 (trisomy of 21). The affected individual is short statured with small round head, furrowed tongue and partially open mouth and mental development is retarted.

168. (a) In this question A is Franis Crick, B is translation and C is transcription. It is unidirectional flow of information DNA to mRNA (transcription) and then decoding the information present in mRNA in the formation of polypeptide chain or protein (translation).

169. (b) Inducible operons are usually switched off. This is a type of operon which is switched on when a chemical called inducer is present. The inducer is almost always a substrate.

170. (b) Darwin observed an amazing diversity of creatures on galapagos islands. He realised that there were many varieties of finches in the same island like seed-eating, with altered beaks insectivorous and vegetarian finches. Carrion are dead bodies. No finches feed on carrion.

171. (d) Dinosaurs dominated the world in mesozoic era. In Triassic period, giant reptiles (dinosaurs) emerged. In Jurassic period, dinosaurs diversified and abundant (golden age of dinosaurs) and in cretaceous period, mass extinction of dinosaurs occurred.

172. (a) *Wuchereria (W. bancrofti* and *W. malayi),* filarial worms causing chronic inflammation of the organs in which they live for many years, usually the lymphatic vessels of the lower limbs and the disease caused by them known as **elephantiasis** or **filariasis.** The genital organs are mainly affected, resulting in gross deformities. The pathogens are transmitted to a healthy person through the bite by the female mosquito vectors.

173. (d) Hallucinogens are drugs that cause illusions and delusions (hallucinations) and change the feelings or perception. Examples are Bhang, Charas, Hashish, Marijuana (all cannabinoids) and LSD (Lysergic acid diethylamide). Opium and morphine are opiate narcotics (opioids) having analgesic (pain relieving effect) while caffeine is a stimulant that increase the activity of nervous system, and cause awakening.

174. (b) The method of producing thousands of plants through tissue culture is called micro-propagation. Each of these plants will be genetically identical to the original plant from which they were grown, *i.e.*, they are somaclones. Many important food plants like tomato, banana, apple have been produed on commercial scale using this method.

175. (c) Microbial biocontrol agent that can be introduced in order to control butterfly caterpillars is the bacteria *Bacillus thuringiensis (Bt).* They sprayed onto vulnerable plants, where they eaten by the insect larvae. In the gut of the larvae, the toxin is released and the larvae get killed. The bacterial disease will kill the caterpillars, but leave other insects unharmed.

176. (a) Age pyramid is a graphic representation of different age groups of population with pre-reproductive groups at the base, reproductive ones in the middle and post-reproductive groups at the top. In **Triangular age pyramid,** the number of pre-reproduction individual is very large while reproductive are moderate and post-reproductive are fewer.

177. (b) Plants are producers which can prepare their food by the process of photosynthesis. Zooplanktons, fish and man are primary, secondary and tertiary carnivores respectively.

178. (d) In Parasitism, one species (parasite) is benefitted and the other (host) is harmed. Parasites have very high reproduction capacity. The life cycles of parasites are often complex, involving one or two intermediate hosts or vectors to facilitate parasitisation of its primary host.

179. (c) The increase in the level of green house gases (e.g., CO_2, CH_4, CFC, N_2O etc.) has led to the considerable heating of earth surface leading to **global warming.** The relative contributions of various green house gases to global warming is

$CO_2(60\%) > CH_4(20\%) > CFC(14\%) > N_2O(6\%)$

180. (c) According to Centrol Pollution Control Board (CPCB) particulate size 2.5 micrometers or less in diameter are responsible for causing various breathing and respiratory problems. They can also cause irritation, inflammations, damage to the lungs and premature deaths.

CBSE PMT SCREENING 2012

PHYSICS

1. Electron in hydrogen atom first jumps from third excited state to second excited state and then from second excited state to the first excited state. The ratio of the wavelength $\lambda_1 : \lambda_2$ emitted in the two cases is
(a) 7 : 5 (b) 27 : 20 (c) 27 : 5 (d) 20 : 7
2. When a string is divided into three segments of length l_1, l_2, and l_3 the fundamental frequencies of these three segments are v_1, v_2 and v_3 respectively. The original fundamental frequency (v) of the string is
(a) $\sqrt{v} = \sqrt{v_1} + \sqrt{v_2} + \sqrt{v_3}$
(b) $v = v_1 + v_2 + v_3$
(c) $\frac{1}{v} = \frac{1}{v_1} + \frac{1}{v_2} + \frac{1}{v_3}$
(d) $\frac{1}{\sqrt{v}} = \frac{1}{\sqrt{v_1}} + \frac{1}{\sqrt{v_2}} + \frac{1}{\sqrt{v_3}}$
3. A 200 W sodium street lamp emits yellow light of wavelength 0.6 μm. Assuming it to be 25% efficient in converting electrical energy to light, the number of photons of yellow light it emits per second is
(a) 1.5×10^{20} (b) 6×10^{18}
(c) 62×10^{20} (d) 3×10^{19}
4. Two ideal diodes are connected to a battery as shown in the circuit. The current supplied by the battery is :

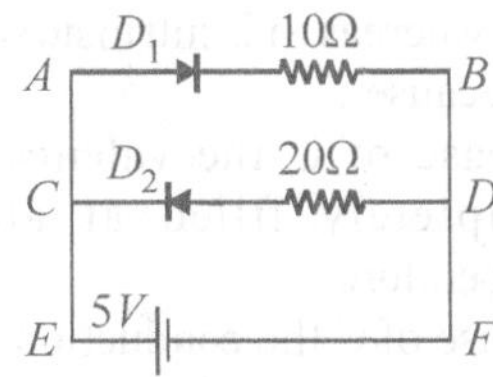

(a) 0.75 A (b) zero
(c) 0.25 A (d) 0.5 A
5. When a mass is rotating in a plane about a fixed point, its angular momentum is directed along :
(a) a line perpendicular to the plane of rotation
(b) the line making an angle of 45° to the plane of rotation
(c) the radius
(d) the tangent to the orbit
6. An electric dipole of moment 'p' is placed in an electric field of intensity 'E'. The dipole acquires a position such that the axis of the dipole makes an angle θ with the direction of the field. Assuming that the potential energy of the dipole to be zero when θ = 90°, the torque and the potential energy of the dipole will respectively be :
(a) $pE\sin\theta, -pE\cos\theta$
(b) $pE\sin\theta, -2pE\cos\theta$
(c) $pE\sin\theta, 2pE\cos\theta$
(d) $pE\cos\theta, -pE\cos\theta$
7. In a *CE* transistor amplifier, the audio signal voltage across the collector resistance of 2kΩ is 2V. If the base resistance is 1kΩ and the current amplification of the transistor is 100, the input signal voltage is :
(a) 0.1 V (b) 1.0 V (c) 1 mV (d) 10 mV
8. A coil of resistance 400Ω is placed in a magnetic field. If the magnetic flux φ (wb) linked with the coil varies with time *t* (sec) as $\phi = 50t^2 + 4$. The current in the coil at $t = 2$ sec is :
(a) 0.5 A (b) 0.1 A (c) 2 A (d) 1 A
9. A thermodynamic system is taken through the cycle *ABCD* as shown in figure. Heat rejected by the gas during the cycle is :

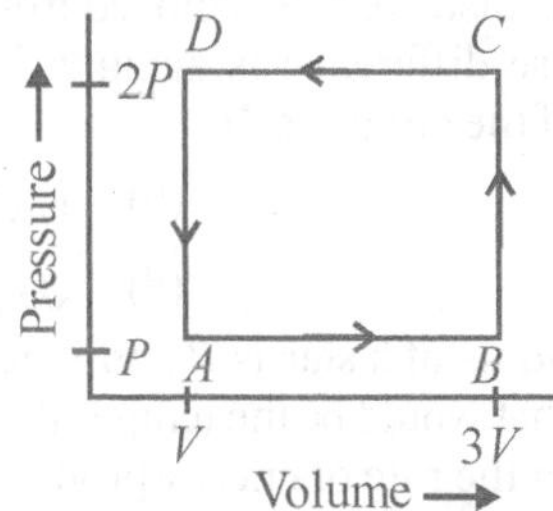

(a) 2PV (b) 4PV (c) $\frac{1}{2}$PV (d) PV
10. If the nuclear radius of ^{27}Al is 3.6 fermi, the approximate nuclear radius of ^{64}Cu in fermi is :
(a) 2.4 (b) 1.2 (c) 4.8 (d) 3.6
11. Two similar coils of radius *R* are lying concentrically with their planes at right angles to each other. The currents flowing in them are *I* and 2 *I*, respectively. The resultant magnetic field induction at the centre will be:
(a) $\frac{\sqrt{5}\mu_0 I}{2R}$ (b) $\frac{3\mu_0 I}{2R}$
(c) $\frac{\mu_0 I}{2R}$ (d) $\frac{\mu_0 I}{R}$

12. The potential energy of particle in a force field is $U = \frac{A}{r^2} - \frac{B}{r}$, where A and B are positive constants and r is the distance of particle from the centre of the field. For stable equilibrium, the distance of the particle is :
(a) $B/2A$ (b) $2A/B$
(c) A/B (d) B/A

13. When a biconvex lens of glass having refractive index 1.47 is dipped in a liquid, it acts as a plane sheet of glass. This implies that the liquid must have refractive index.
(a) equal to that of glass
(b) less than one
(c) greater than that of glass
(d) less than that of glass

14. The horizontal range and the maximum height of a projectile are equal. The angle of projection of the projectile is :
(a) $\theta = \tan^{-1}\left(\frac{1}{4}\right)$ (b) $\theta = \tan^{-1}(4)$
(c) $\theta = \tan^{-1}(2)$ (d) $\theta = 45°$

15. In an electrical circuit R, L, C and an a.c. voltage source are all connected in series. When L is removed from the circuit, the phase difference between the voltage and the current in the circuit is $\pi/3$. If instead, C is removed from the circuit, the phase difference is again $\pi/3$. The power factor of the circuit is :
(a) 1/2 (b) $1/\sqrt{2}$
(c) 1 (d) $\sqrt{3}/2$

16. If the radius of a star is R and it acts as a black body, what would be the temperature of the star, in which the rate of energy production is Q?
(a) $Q/4\pi R^2\sigma$ (b) $(Q/4\pi R^2\sigma)^{-1/2}$
(c) $(4\pi R^2 Q/\sigma)^{1/4}$ (d) $(Q/4\pi R^2\sigma)^{1/4}$
(σ stands for Stefan's constant)

17. The current (I) in the inductance is varying with time according to the plot shown in figure.

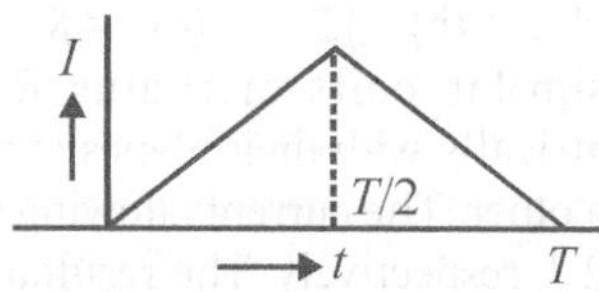

Which one of the following is the correct variation of voltage with time in the coil?

(a)
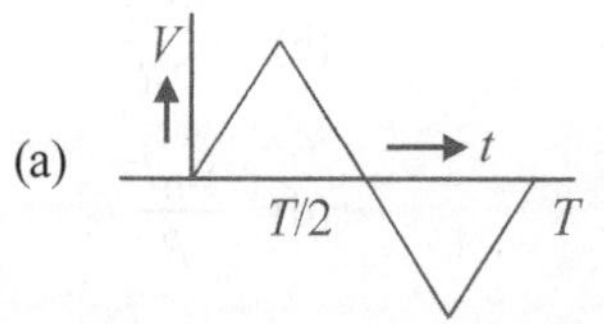

(b)
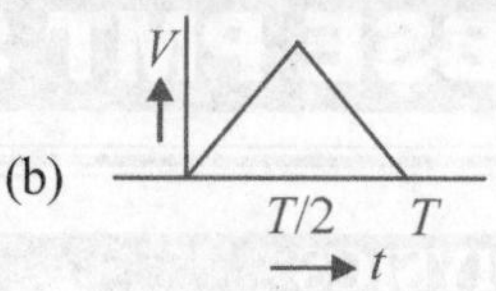

(c)
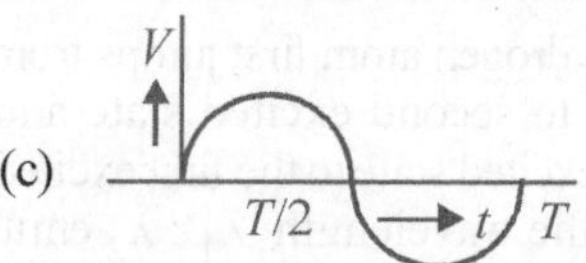

(d)
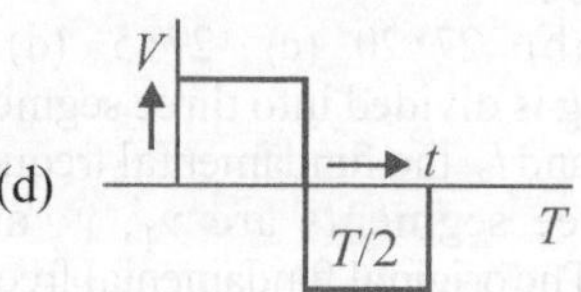

18. A milli voltmeter of 25 milli volt range is to be converted into an ammeter of 25 ampere range. The value (in ohm) of necessary shunt will be :
(a) 0.001 (b) 0.01 (c) 1 (d) 0.05

19. Two persons of masses 55 kg and 65 kg respectively, are at the opposite ends of a boat. The length of the boat is 3.0 m and weighs 100 kg. The 55 kg man walks up to the 65 kg man and sits with him. If the boat is in still water the centre of mass of the system shifts by :
(a) 3.0 m (b) 2.3 m (c) zero (d) 0.75 m

20. A mixture consists of two radioactive materials A_1 and A_2 with half lives of 20 s and 10 s respectively. Initially the mixture has 40 g of A_1 and 160 g of A_2. The amount of the two in the mixture will become equal after :
(a) 60 s (b) 80 s (c) 20 s (d) 40 s

21. C and Si both have same lattice structure, having 4 bonding electrons in each. However, C is insulator whereas Si is intrinsic semiconductor. This is because :
(a) In case of C the valence band is not completely filled at absolute zero temperature.
(b) In case of C the conduction band is partly filled even at absolute zero temperature.
(c) The four bonding electrons in the case of C lie in the second orbit, whereas in the case of Si they lie in the third.
(d) The four bonding electrons in the case of C lie in the third orbit, whereas for Si they lie in the fourth orbit.

22. The height at which the weight of a body becomes 1/16th, its weight on the surface of earth (radius R), is :
(a) $5R$ (b) $15R$ (c) $3R$ (d) $4R$

23. An electron of a stationary hydrogen atom passes from the fifth energy level to the ground level. The velocity that the atom acquired as a result of photon emission will be :

(a) $\frac{24hR}{25m}$ (b) $\frac{25hR}{24m}$

(c) $\frac{25m}{24hR}$ (d) $\frac{24m}{25hR}$

(m is the mass of the electron, R, Rydberg constant and h, Planck's constant)

24. A compass needle which is allowed to move in a horizontal plane is taken to a geomagnetic pole. It :

(a) will become rigid showing no movement
(b) will stay in any position
(c) will stay in north-south direction only
(d) will stay in east-west direction only

25. In the circuit shown, the cells A and B have negligible resistances. For $V_A = 12\text{V}$, $R_1 = 500\Omega$ and $R = 100\Omega$ the galvanometer (G) shows no deflection. The value of V_B is :

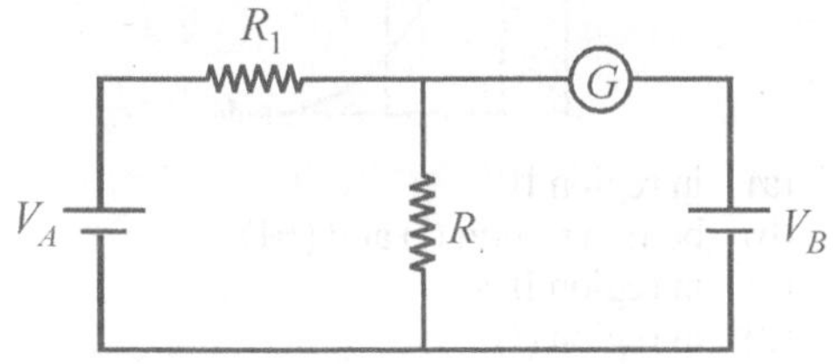

(a) 4 V (b) 2 V (c) 12 V (d) 6 V

26. Four point charges $-Q$, $-q$, $2q$ and $2Q$ are placed, one at each corner of the square. The relation between Q and q for which the potential at the centre of the square is zero is :

(a) $Q = -q$ (b) $Q = -\frac{1}{q}$

(c) $Q = q$ (d) $Q = \frac{1}{q}$

27. A car of mass 1000 kg negotiates a banked curve of radius 90 m on a frictionless road. If the banking angle is 45°, the speed of the car is :

(a) 20 ms^{-1} (b) 30 ms^{-1}
(c) 5 ms^{-1} (d) 10 ms^{-1}

28. A solid cylinder of mass 3 kg is rolling on a horizontal surface with velocity 4 ms^{-1}. It collides with a horizontal spring of force constant 200 Nm^{-1}. The maximum compression produced in the spring will be :

(a) 0.5m (b) 0.6m
(c) 0.7m (d) 0.2m

29. One mole of an ideal gas goes from an initial state A to final state B via two processes : It first undergoes isothermal expansion from volume V to $3V$ and then its volume is reduced from $3V$ to V at constant pressure. The correct P-V diagram representing the two processes is :

(a)

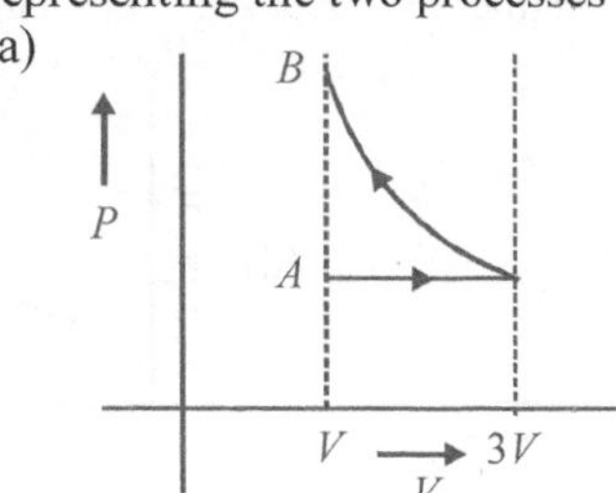

(b)

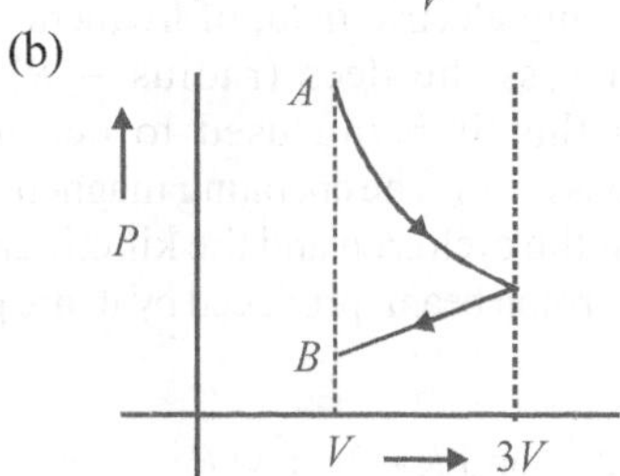

(c)

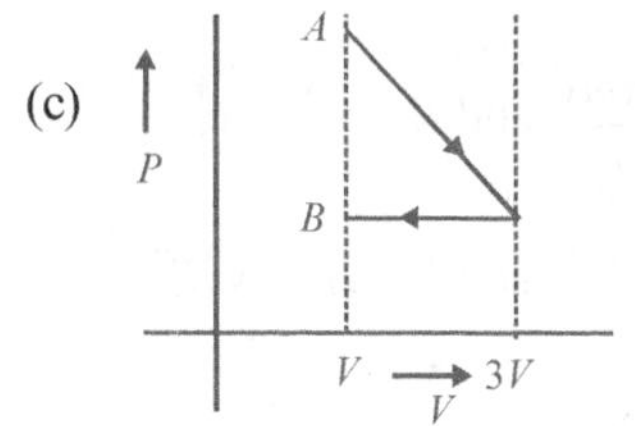

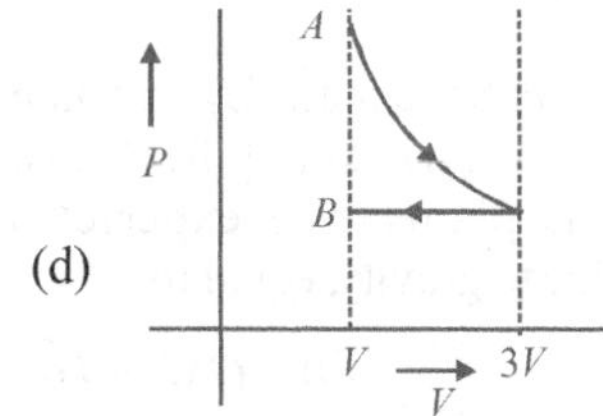

(d)

30. Two spheres A and B of masses m_1 and m_2 respectively collide. A is at rest initially and B is moving with velocity v along x-axis. After collision B has a velocity $\frac{v}{2}$ in a direction perpendicular to the original direction. The mass A moves after collision in the direction

(a) same as that of B
(b) opposite to that of B
(c) $\theta = \tan^{-1}(1/2)$ to the x-axis
(d) $\theta = \tan^{-1}(-1/2)$ to the x-axis

31. Liquid oxygen at 50 K is heated to 300 K at constant pressure of 1 atm. The rate of heating is constant. Which one of the following graphs represents the variation of temperature with time?

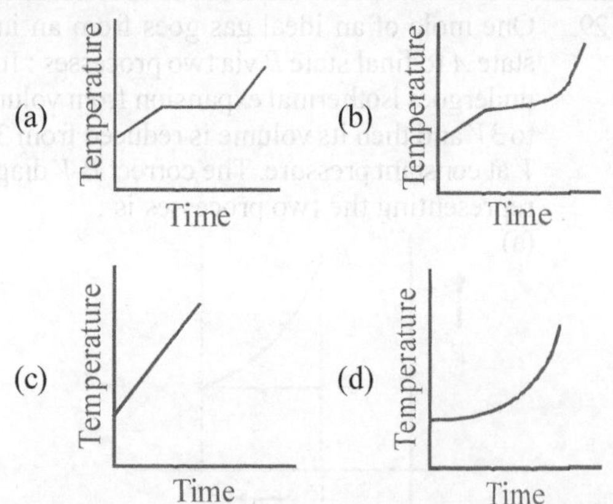

32. An alternating electric field, of frequency ν, is applied across the dees (radius = R) of a cyclotron that is being used to accelerate protons (mass = m). The operating magnetic field (B) used in the cyclotron and the kinetic energy (K) of the proton beam, produced by it, are given by :

(a) $B = \frac{m\nu}{e}$ and $K = 2m\pi^2\nu^2R^2$

(b) $B = \frac{2\pi m\nu}{e}$ and $K = m^2\pi\nu R^2$

(c) $B = \frac{2\pi m\nu}{e}$ and $K = 2m\pi^2\nu^2R^2$

(d) $B = \frac{m\nu}{e}$ and $K = m^2\pi\nu R^2$

33. A spherical planet has a mass M_P and diameter D_P. A particle of mass m falling freely near the surface of this planet will experience an acceleration due to gravity, equal to :

(a) $4GM_P/D_P^2$ (b) GM_Pm/D_P^2
(c) GM_P/D_P^2 (d) $4GM_Pm/D_P^2$

34. A ray of light is incident at an angle of incidence, i, on one face of prism of angle A (assumed to be small) and emerges normally from the opposite face. If the refractive index of the prism is μ, the angle of incidence i, is nearly equal to :

(a) μA (b) $\frac{\mu A}{2}$ (c) $\frac{A}{\mu}$ (d) $\frac{A}{2\mu}$

35. The damping force on an oscillator is directly proportional to the velocity. The unit of the constant of proportionality is
(a) $kgms^{-1}$ (b) $kgms^{-2}$
(c) kgs^{-1} (d) kgs

36. A concave mirror of focal length 'f_1' is placed at a distance of 'd' from a convex lens of focal length 'f_2'. A beam of light coming from infinity and falling on this convex lens-concave mirror combination returns to infinity.
The distance 'd' must equal to :
(a) $f_1 + f_2$ (b) $-f_1 + f_2$
(c) $f_1 + 2f_2$ (d) $-2f_1 + f_2$

37. A geostationary satellite is orbiting the earth at a height of $5R$ above that surface of the earth, R being the radius of the earth. The time period of another satellite in hours at a height of $2R$ from the surface of the earth is :

(a) 5 (b) 10 (c) $6\sqrt{2}$ (d) $\frac{6}{\sqrt{2}}$

38. Transfer characteristics [output voltage (V_0) vs input voltage (V_1)] for a base biased transistor in CE configuration is as shown in the figure. For using transistor as a switch, it is used :

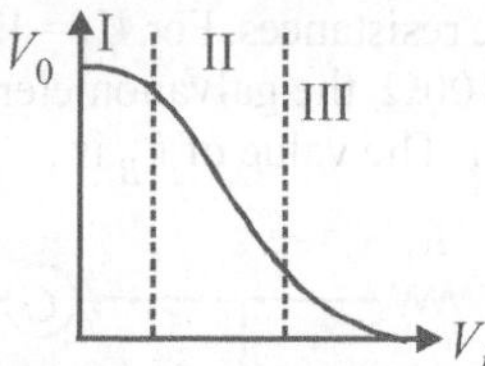

(a) in region III
(b) both in region (I) and (III)
(c) in region II
(d) in region (I)

39. If voltage across a bulb rated 220 Volt-100 Watt drops by 2.5% of its rated value, the percentage of the rated value by which the power would decrease is :
(a) 20% (b) 2.5% (c) 5% (d) 10%

40. A particle has initial velocity $(2\vec{i} + 3\vec{j})$ and acceleration $(0.3\vec{i} + 0.2\vec{j})$. The magnitude of velocity after 10 seconds will be :

(a) $9\sqrt{2}$ units (b) $5\sqrt{2}$ units
(c) 5 units (d) 9 units

41. Monochromatic radiation emitted when electron in hydrogen atom jumps from first excited to the ground state irradiates a photosensitive material. The stopping potential is measured to be 3.57 V. The threshold frequency of the materials is :
(a) 4×10^{15} Hz (b) 5×10^{15} Hz
(c) 1.6×10^{15} Hz (d) 2.5×10^{15} Hz

42. ABC is an equilateral triangle with O as its centre. $\vec{F}_1$, $\vec{F}_2$ and $\vec{F}_3$ represent three forces acting along the sides AB, BC and AC respectively. If the total torque about O is zero the magnitude of

$\vec{F}_3$ is :

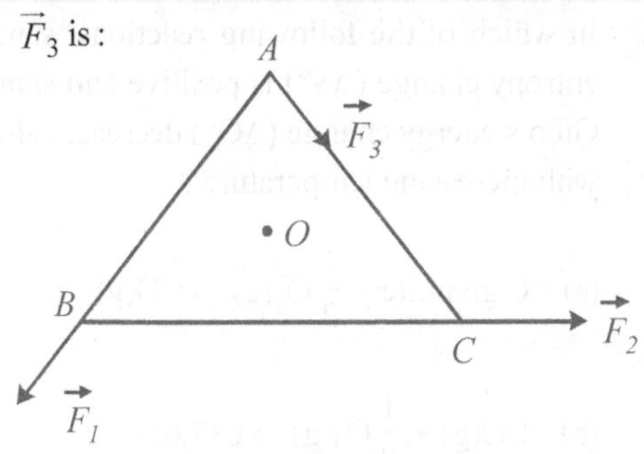

(a) $\vec{F}_1 + \vec{F}_2$ (b) $\vec{F}_1 - \vec{F}_2$

(c) $\frac{\vec{F}_1 + \vec{F}_2}{2}$ (d) $2(\vec{F}_1 + \vec{F}_2)$

43. The figure shows a logic circuit with two inputs *A* and *B* and the output *C*. The voltage wave forms across *A*, *B* and *C* are as given. The logic gate is :

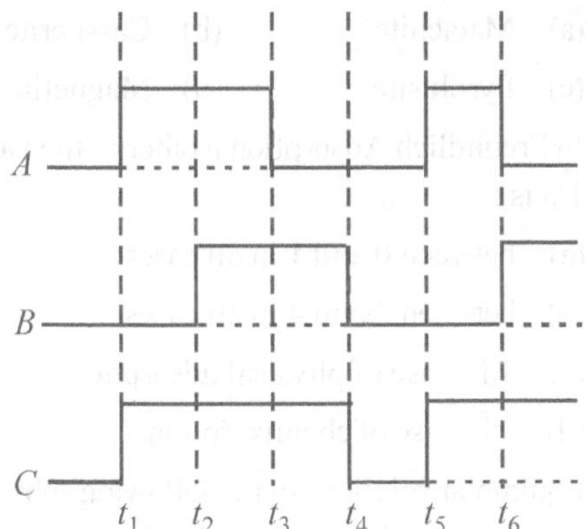

(a) OR gate (b) NOR gate
(c) AND gate (d) NAND gate

44. What is the flux through a cube of side '*a*' if a point charge of *q* is at one of its corner :

(a) $\frac{2q}{\varepsilon_0}$ (b) $\frac{q}{8\varepsilon_0}$

(c) $\frac{q}{\varepsilon_0}$ (d) $\frac{q}{2\varepsilon_0}6a^2$

45. An α-particle moves in a circular path of radius 0.83 cm in the presence of a magnetic field of 0.25 Wb/m^2. The de-Broglie wavelength associated with the particle will be :
(a) 1 Å (b) 0.1 Å
(c) 10 Å (d) 0.01 Å

46. The electric field associated with an e.m. wave in vacuum is given by $\vec{E} = \hat{i}\, 40 \cos (kz - 6 \times 10^8 t)$, where *E*, *z* and *t* are in volt/m, metre and seconds respectively. The value of wave vector *k* is :
(a) 2 m^{-1} (b) 0.5 m^{-1}
(c) 6 m^{-1} (d) 3 m^{-1}

47. The motion of a particle along a straight line is described by equation :
$x = 8 + 12t - t^3$
where *x* is in metre and *t* in second. The retardation of the particle when its velocity becomes zero, is :
(a) 24 ms^{-2} (b) zero
(c) 6 ms^{-2} (d) 12 ms^{-2}

48. The magnifying power of a telescope is 9. When it is adjusted for parallel rays the distance between the objective and eyepiece is 20 cm. The focal length of lenses are :
(a) 10 cm, 10 cm (b) 15 cm, 5 cm
(c) 18 cm, 2 cm (d) 11 cm, 9 cm

49. Two sources of sound placed close to each other are emitting progressive waves given by $y_1 = 4 \sin 600\,\pi t$ and $y_2 = 5 \sin 608\,\pi t$. An observer located near these two sources of sound will hear :
(a) 4 beats per second with intensity ratio 25 : 16 between waxing and waning.
(b) 8 beats per second with intensity ratio 25 : 16 between waxing and waning
(c) 8 beats per second with intensity ratio 81 : 1 between waxing and waning
(d) 4 beats per second with intensity ratio 81 : 1 between waxing and waning

50. A ring is made of a wire having a resistance $R_0 = 12\ \Omega$. Find the points *A* and *B* as shown in the figure, at which a current carrying conductor should be connected so that the resistance *R* of the sub-circuit between these points is equal to $\frac{8}{3}\Omega$.

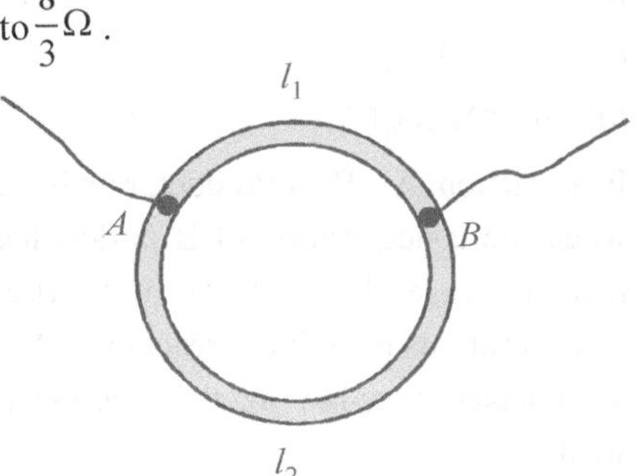

(a) $\frac{l_1}{l_2} = \frac{5}{8}$ (b) $\frac{l_1}{l_2} = \frac{1}{3}$

(c) $\frac{l_1}{l_2} = \frac{3}{8}$ (d) $\frac{l_1}{l_2} = \frac{1}{2}$

CHEMISTRY

51. Aluminium is extracted from alumina (Al_2O_3) by electrolysis of a molten mixture of :
(a) $Al_2O_3 + HF + NaAlF_4$
(b) $Al_2O_3 + CaF_2 + NaAlF_4$
(c) $Al_2O_3 + Na_3AlF_6 + CaF_2$
(d) $Al_2O_3 + KF + Na_3AlF_6$

52. pH of a saturated solution of $Ba(OH)_2$ is 12. The value of solubility product (K_{SP}) of $Ba(OH)_2$ is :
(a) 3.3×10^{-7} (b) 5.0×10^{-7}
(c) 4.0×10^{-6} (d) 5.0×10^{-6}

53. When Cl_2 gas reacts with hot and concentrated sodium hydroxide solution, the oxidation number of chlorine changes from :
(a) zero to +1 and zero to –5
(b) zero to –1 and zero to +5
(c) zero to –1 and zero to +3
(d) zero to +1 and zero to –3

54. Maximum number of electrons in a subshell with : $l = 3$ and $n = 4$ is :
(a) 14 (b) 16
(c) 10 (d) 12

55. Which one of the following is an outer orbital complex and exhibits paramagnetic behaviour ?
(a) $[Ni(NH_3)_6]^{2+}$
(b) $[Zn(NH_3)_6)]^{2+}$
(c) $[Cr(NH_3)_6]^{3+}$
(d) $[CO(NH_3)_6]^{3+}$

56. In a reaction, A + B → Product, rate is doubled when the concentration of B is doubled, and rate increases by a factor of 8 when the concentrations of both the reactants (A and B) are doubled rate, law for the reaction can be written as :
(a) Rate = $k[A][B]^2$
(b) Rate = $k[A]^2[B]^2$
(c) Rate = $k[A][B]$
(d) Rate = $k[A]^2B]$

57. In which of the following reactions, standard entropy change (ΔS°) is positive and standard Gibb's energy change (ΔG°) decreases sharply with increasing temperature ?
(a) $C\text{ graphite} + \frac{1}{2}O_2(g) \rightarrow CO(g)$
(b) $CO(g) + \frac{1}{2}O_2(g) \rightarrow CO_2(g)$
(c) $Mg(s) + \frac{1}{2}O_2(g) \rightarrow MgO(s)$
(d) $\frac{1}{2}C\text{ graphite} + \frac{1}{2}O_2(g) \rightarrow \frac{1}{2}CO_2(g)$

58. Which one of the following is a mineral of iron ?
(a) Malachite (b) Cassiterite
(c) Pyrolusite (d) Magnetite

59. In Freundlich Adsorption isotherm, the value of 1/n is :
(a) between 0 and 1 in all cases
(b) between 2 and 4 in all cases
(c) 1 in case of physical adsorption
(d) 1 in case of chemisorption

60. Equimolar solutions of the following substances were prepared separately. Which one of these will record the highest pH value ?
(a) $BaCl_2$ (b) $AlCl_3$
(c) LiCl (d) $BeCl_2$

61. 50 mL of each gas A and of gas B takes 150 and 200 seconds respectively for effusing through a pin hole under the similar condition. If molecular mass of gas B is 36, the molecular mass of gas A will be :
(a) 96 (b) 128
(c) 32 (d) 64

62. The correct set of four quantum numbers for the valence electron of rubidium atom (Z = 37) is :
(a) 5, 1, + 1/2 (b) 6, 0, 0 + 1/2
(c) 5, 0, 0 + 1/2 (d) 5, 1, 0 + 1/2

63. In which of the following compounds, nitrogen exhibits highest oxidation state ?
(a) N_2H_4 (b) NH_3
(c) N_3H (d) NH_2OH

64. Predict the product in the given reaction.

m-chlorobenzaldehyde (CHO, Cl) $\xrightarrow{50\%\ KOH}$

(a) m-$ClC_6H_4CH_2OH$ + m-$ClC_6H_4CH_2COO^-$

(b) m-$HOC_6H_4CH_2OH$ + m-$C_6H_4(OH)_2$

(c) m-$ClC_6H_4CH_2OH$ + m-$ClC_6H_4COO^-$

(d) m-$HOC_6H_4CH_2OH$ + m-$HOC_6H_4COO^-$

65. Acetone is treated with excess of ethanol in the presence of hydrochloric acid. The product obtained is :

(a) $CH_3CH_2CH_2-\overset{O}{\overset{\|}{C}}-CH_3$

(b) $CH_3CH_2CH_2-\overset{O}{\overset{\|}{C}}-CH_2CH_2CH_3$

(c) $(CH_3)_2C\begin{matrix}OH\\OC_2H_5\end{matrix}$

(d) $(CH_3)_2C\begin{matrix}OC_2H_5\\OC_2H_5\end{matrix}$

66. A metal crystallizes with a face-centered cubic lattice. The edge length of the unit cell is 408 pm. The diameter of the metal atom is :
(a) 288 pm (b) 408 pm
(c) 144 pm (d) 204 pm

67. Which one of the following, statements is incorrect about enzyme catalysis?
(a) Enzymes are mostly proteinous in nature.
(b) Enzyme action is specific.
(c) Enzymes are denaturated by ultraviolet rays and at high temperature.
(d) Enzymes are least reactive at optimum temperature.

68. In a zero-order reaction for every 10° rise of temperature, the rate is doubled. If the temperature is increased from 10°C to 100°C, the rate of the reaction will become :
(a) 256 times (b) 512 times
(c) 64 times (d) 128 times

69. Deficiency of vitamin B_1 causes the disease
(a) Convulsions (b) Beri-Beri
(c) Cheilosis (d) Sterility

70. Among the following compounds the one that is most reactive towards electrophilic nitration is:
(a) Benzoic acid (b) Nitrobenzene
(c) Toluene (d) Benzene

71. Buffer solutions have constant acidity and alkalinity because
(a) these give unionised acid or base on reaction with added acid or alkali.
(b) acids and alkalies in these solutions are shielded from attack by other ions.
(c) they have large excess of H^+ or OH^- ions
(d) they have fixed value of pH

72. The correct order of decreasing acid strength of trichloroacetic acid (*A*), trifluoroacetic acid (*A*), acetic acid (*C*) and formic acid (*D*) is :
(a) B>A>D>C
(b) B>D>C>A
(c) A>B>C>D
(d) A>C>B>D

73. Which one of the following sets of monosaccharides forms sucrose?
(a) α–D-Galactopyranose and α–D-Glucopyranose
(b) α–D-Glucopyranose and β–D-Fructofuranose
(c) β–D-Glucopyranose and α–D- Fructofuranose
(d) α–D-Glucopyranose and β–D-Fructopyranose

74. The enthalpy of fusion of water is 1.435 kCal/mol. The molar entropy change for the melting of ice at 0°C is :
(a) 10.52 cal / (mol K)
(b) 21.04 cal / (mol K)
(c) 5.260 cal / (mol K)
(d) 0.526 cal / (mol K)

75. Which one of the following pairs is isostructural (i.e., having the same shape and hybridization) ?
(a) $\left[BCl_3 \text{ and } BrCl_3^-\right]$
(b) $\left[NH_3 \text{ and } NO_3^-\right]$
(c) $\left[NF_3 \text{ and } BF_3\right]$
(d) $\left[BF_4^- \text{ and } NH_4^+\right]$

76. Bond order of 1.5 is shown by :
(a) O_2^+ (b) O_2^-
(c) O_2^{2-} (d) O_2

77. Which one of the following is not a condensation polymer ?
(a) Melamine (b) Glyptal
(c) Dacron (d) Neoprene

78. In the following sequence of reactions

$CH_3 - Br \xrightarrow{KCN} A \xrightarrow{H_3O^+}$

$B \xrightarrow[\text{ether}]{LiAlH_4} C$, the end product (C) is :
(a) Acetone (b) Methane
(c) Acetaldehyde (d) Ethyl alcohol

79. Which nomenclature is not according to IUPAC system?
(a) $Br-CH_2-CH=CH_2$,
1-Bromo-prop-2-ene
(b) $CH_3-CH_2-C(CH_3)(Br)-CH_2-CH(CH_3)CH_3$
4 -Bromo, 2, 4 -di -methylhexane
(c) $CH_3-CH(CH_3)-CH(C_6H_5)-CH_2-CH_3$,
2-Methyl-3-phenylpentane
(d) $CH_3-C(=O)-CH_2-CH_2-CH_2COOH$
5-oxohexanoic acid

80. The number of octahedral void(s) per atom present in a cubic close-packed structure is :
(a) 1 (b) 3
(c) 2 (d) 4

81. In the extraction of copper from its sulphide ore, the metal is finally obtained by the reduction of cuprous oxide with :
(a) Copper (I) sulphide (Cu_2S)
(b) Sulphur dioxide (SO_2)
(c) Iron sulphide (FeS)
(d) Carbon monoxide (CO)

82. Identify the alloy containing a non-metal as a constituent in it.
(a) Invar (b) Steel
(c) Bell metal (d) Bronze

83. A mixture of potassium chlorate, oxalic acid and sulphuric acid is heated. During the reaction which element undergoes maximum change in the oxidation number ?
(a) S (b) H
(c) Cl (d) C

84. Which one of the alkali metals, forms only, the normal oxide, M_2O on heating in air ?
(a) Rb (b) K
(c) Li (d) Na

85. The ease of adsorption of the hydrated alkali metal ions on an ion-exchange resins follows the order :
(a) $Li^+ < K^+ < Na^+ < Rb^+$
(b) $Rb^+ < K^+ < Na^+ < Li^+$
(c) $K^+ < Na^+ < Rb^+ < Li^+$
(d) $Na^+ < Li^+ < K^+ < Rb^+$

86. Which one of the following statements regarding photochemical smog is not correct?
(a) Carbon monoxide does not play any role in photochemical smog formation.
(b) Photochemical smog is an oxidising agent in character.
(c) Photochemical smog is formed through photochemical reaction involving solar energy.
(d) Photochemical smog does not cause irritation in eyes and throat.

87. CH_3CHO and $C_6H_5CH_2CHO$ can be distinguished chemically by :
(a) Benedict test
(b) Iodoform test
(c) Tollen's reagent test
(d) Fehling solution test

88. Which of the statements is not true?
(a) On passing H_2S through acidified $K_2Cr_2O_7$ solution, a milky colour is observed.
(b) $Na_2Cr_2O_7$ is preferred over $K_2Cr_2O_7$ in volumetric analysis.
(c) $K_2Cr_2O_7$ solution in acidic medium is orange.
(d) $K_2Cr_2O_7$ solution becomes yellow on increasing the pH beyond 7.

89. Standard enthalpy of vapourisation $\Delta_{vap} H°$ for water at 100°C is 40.66 kJ mol^{-1}. The internal energy of vaporisation of water at 100°C (in kJ mol^{-1}) is :
(a) +37.56 (b) −43.76
(c) +43.76 (d) +40.66
(Assume water vapour to behave like an ideal gas).

90. Identify the wrong statement in the following :
(a) Amongst isoelectronic species, smaller the positive charge on the cation, smaller is the ionic radius.
(b) Amongst isoelectronic species, greater the negative charge on the anion, larger is the ionic radius.
(c) Atomic radius of the elements increases as one moves down the first group of the periodic table.
(d) Atomic radius of the elements decreases as one moves across from left to right in the 2nd period of the periodic table.

91. Which of the following statements is not valid for oxoacids of phosphorus?
(a) Orthophosphoric acid is used in the manufacture of triple superphosphate.
(b) Hypophosphorous acid is a diprotic acid.
(c) All oxoacids contain tetrahedral four coordinated phosphorus.
(d) All oxoacids contain atleast one P = O and one P — OH group.

92. The protecting power of lyophilic colloidal sol is expressed in terms of :
(a) Coagulation value
(b) Gold number
(c) Critical miscelle concentration
(d) Oxidation number

93. Sulphur trioxide can be obtained by which of the following reaction :
(a) $CaSO_4 + C \xrightarrow{\Delta}$
(b) $Fe_2(SO_4)_3 \xrightarrow{\Delta}$
(c) $S + H_2SO_4 \xrightarrow{\Delta}$
(d) $H_2SO_4 + PCl_5 \xrightarrow{\Delta}$

94. P_A and P_B are the vapour pressure of pure liquid components, A and B, respectively of an ideal binary solution. If X_A represents the mole fraction of component A, the total pressure of the solution will be.
(a) $P_A + X_A(P_B - P_A)$
(b) $P_A + X_A(P_A - P_B)$
(c) $P_B + X_A(P_B - P_A)$
(d) $P_B + X_A(P_A - P_B)$

95. Which of the following acids does not exhibit optical isomerism ?
(a) Maleic acid (b) α-amino acids
(c) Lactic acid (d) Tartaric acid

96. Which of the following species contains three bond pairs and one lone pair around the central atom ?
(a) H_2O (b) BF_3
(c) NH_2^- (d) PCl_3

97. Limiting molar conductivity of NH_4OH $\left(\text{i.e., } \overset{0}{\Lambda}_{m(NH_4OH)}\right)$ is equal to :
(a) $\overset{0}{\Lambda}_{m(NH_4Cl)} + \overset{0}{\Lambda}_{m(NaCl)} - \overset{0}{\Lambda}_{m(NaOH)}$
(b) $\overset{0}{\Lambda}_{m(NaOH)} + \overset{0}{\Lambda}_{m(NaCl)} - \overset{0}{\Lambda}_{m(NH_4Cl)}$
(c) $\overset{0}{\Lambda}_{m(NH_4OH)} + \overset{0}{\Lambda}_{m(NH_4Cl)} - \overset{0}{\Lambda}_{m(HCl)}$
(d) $\overset{0}{\Lambda}_{m(NH_4Cl)} + \overset{0}{\Lambda}_{m(NaOH)} - \overset{0}{\Lambda}_{m(NaCl)}$

98. The pair of species with the same bond order is :
(a) O_2^{2-}, B_2 (b) O_2^+, NO^+
(c) NO, CO (d) N_2, O_2

99. Which of the following statements is false?
(a) Artificial silk is derived from cellulose.
(b) Nylon-66 is an example of elastomer.
(c) The repeat unit in natural rubber is isoprene.
(d) Both starch and cellulose are polymers of glucose.

100. In the following reaction :

$$H_3C-\underset{CH_3}{\overset{CH_3}{|}}{C}-CH=CH_2 \xrightarrow{H_2O/H^+} \underset{\text{Minor Product}}{A} + \underset{\text{Minor Product}}{B}$$

The major product is :

(a) $H_3C-C(CH_3)(OH)-CH(CH_3)-CH_3$

(b) $CH_2(OH)-C(CH_3)_2-CH_2-CH_3$

(c) $H_3C-C(CH_3)_2-CH(OH)-CH_3$

(d) $H_3C-C(CH_3)_2-CH_2-CH_2OH$

BIOLOGY

101. Motile zygote of *Plasmodium* occurs in :
(a) Gut of female *Anopheles*
(b) Salivary glands of *Anopheles*
(c) Human RBCs
(d) Human liver

102. The human hind brain comprises three parts, one of which is :
(a) Spinal cord (b) Corpus callosum
(c) Cerebellum (d) Hypothalamus

103. Which part of the human ear plays no role in hearing as such but is otherwise very much required?
(a) Eustachian tube
(b) Organ of corti
(c) Vestibular apparatus
(d) Ear ossicles

104. The most abundant prokaryotes helpful to humans in making curd from milk and in production of antibiotics are the ones categorised as :
(a) Cyanobacteria
(b) Archaebacteria
(c) Chemosynthetic autotrophs
(d) Heterotrophic bacteria

105. Which one of the following statements is false in respect of viability of mammalian sperm ?
(a) Sperm is viable for only up to 24 hours.
(b) Survival of sperm depends on the pH of the medium and is more active in alkaline medium.
(c) Viability of sperm is determined by its motility.
(d) Sperms must be concentrated in a thick suspension.

106. Evolution of different species in a given area starting from a point and spreading to other geographical areas is known as :
(a) Adaptive radiation
(b) Natural selection
(c) Migration
(d) Divergent evolution

107. What is the figure given below showing in particular ?

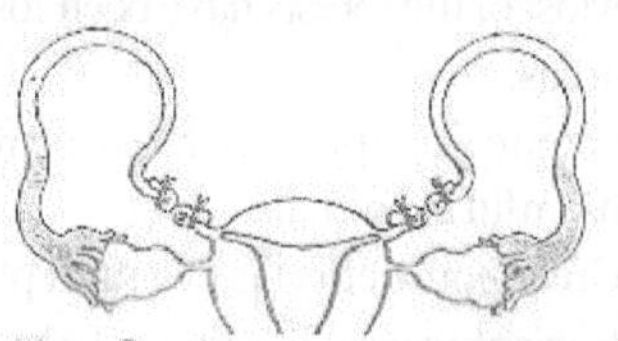

(a) Ovarian cancer (b) Uterine cancer
(c) Tubectomy (d) Vasectomy

108. In an area where DDT had been used extensively, the population of birds declined significantly because:
(a) birds stopped laying eggs
(b) earthworms in the area got eradicated
(c) cobras were feeding exclusively on birds
(d) many of the birds laid, did not hatch

109. The figure below is the diagrammatic representation of the *E.Coli* vector pBR 322. Which one of the given options correctly identifies its certain component (s) ?

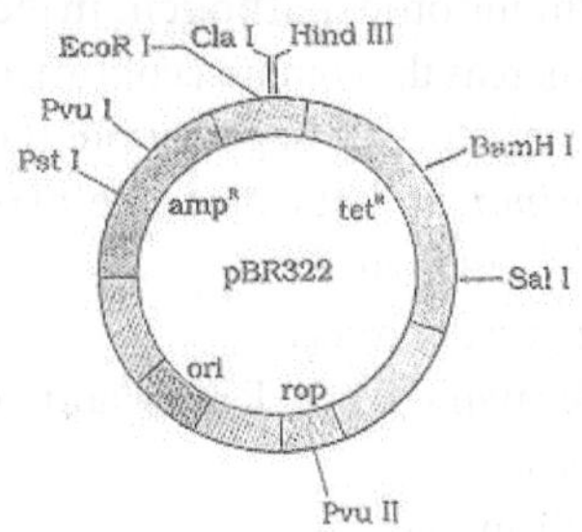

(a) ori - original restriction enzyme
(b) rop-reduced osmotic pressure
(c) Hind III, EcoRI - selectable markers
(d) amp^R, tet^R - antibiotic resistance genes

110. The common bottle cork is a product of :
(a) Dermatogen (b) Phellogen
(c) Xylem (d) Vascular Cambium

111. Widal Test is carried out to test :
(a) Malaria (b) Diabetes mellitus
(c) HIV/AIDS (d) Typhoid fever

112. Which part would be most suitable for raising virus-free plants for micropropagation ?
(a) Bark (b) Vascular tissue
(c) Meristem (d) Node

113. Which one of the following is a wrong statement?

(a) Most of the forests have been lost in tropical areas.

(b) Ozone in upper part of atmosphere is harmful to animals.

(c) Greenhouse effect is a natural phenomenon.

(d) Eutrophication is a natural phenomenon in freshwater bodies.

114. Companion cells are closely associated with :

(a) Sieve elements (b) Vessel elements

(c) Trichomes (d) Guard cells

115. Common cold differs from pneumonia in, that :

(a) Pneumonia is a communicable disease whereas the common cold is a nutritional deficiency disease.

(b) Pneumonia can be prevented by a live attenuated bacterial vaccine whereas the common cold has no effective vaccine.

(c) Pneumonia is caused by a virus while the common cold is caused by the bacterium *Haemophilus influenzae*.

(d) Pneumonia pathogen infects alveoli whereas the common cold affects nose and respiratory passage but not the lungs.

116. *Pheretima* and its close relatives derive nourishment from :

(a) sugarcane roots

(b) decaying fallen leaves and soil organic matter.

(c) soil insects

(d) small pieces of fresh fallen leaves of maize, etc.

117. Removal of RNA polymerase III from nucleoplasm will affect the synthesis of :

(a) t RNA (b) hn RNA

(c) m RNA (d) r RNA

118. A process that makes important difference between C_3 and C_4 plants is :

(a) Transpiration (b) Glycolysis

(c) Photosynthesis (d) Photorespiration

119. PCR and Restriction Fragment Length Polymorphism are the methods for :

(a) Study of enzymes

(b) Genetic transformation

(c) DNA sequencing

(d) Genetic Fingerprinting

120. Best defined function of Manganese in green plants is :

(a) Photolysis of water

(b) Calvin cycle

(c) Nitrogen fixation

(d) Water absorption

121. Measuring Biochemical Oxygen Demand (BOD) is a method used for :

(a) estimating the amount of organic matter in sewage water.

(b) working out the efficiency of oil driven automobile engines.

(c) measuring the activity of *Saccharomyces cerevisae* in producing curd on a commercial scale.

(d) working out the efficiency of RBCs about their capacity to carry oxygen.

122. Which one of the following is not a part of a transcription unit in DNA ?

(a) The inducer (b) A terminator

(c) A promoter (d) The structural gene

123. A certain road accident patient with unknown blood group needs immediate blood transfusion. His one doctor friend at once offers his blood. What was the blood group of the donor?

(a) Blood group B (b) Blood group AB

(c) Blood group O (d) Blood group A

124. Consumption of which one of the following foods can prevent the kind of blindness associated with vitamin 'A' deficiency ?

(a) '*Flaver Savr*' tomato

(b) Canolla

(c) Golden rice

(d) Bt-Brinjal

125. The maximum amount of electrolytes and water (70 - 80 percent) from the glomerular filtrate is reabsorbed in which part of the nephron ?

(a) Ascending limb of loop of Henle

(b) Distal convoluted tubule

(c) Proximal convoluted tubule

(d) Descending limb of loop of Henle

126. Both, autogamy and geitonogamy are prevented in -

(a) Papaya (b) Cucumber

(c) Castor (d) Maize

127. Placentation in tomato and lemon is

(a) Parietal (b) Free central

(c) Marginal (d) Axile

128. A person entering an empty room suddenly finds a snake right in front on opening the door. Which one of the following is likely to happen in his neuro-hormonal control system ?
(a) Sympathetic nervous system is activated releasing epinephrin and norepinephrin from adrenal medulla.
(b) Neurotransmitters diffuse rapidly across the cleft and transmit a nerve impulse.
(c) Hypothalamus activates the parasympathetic division of brain.
(d) Sympathetic nervous system is activated releasing epinephrin and norepinephrin from adrenal cortex.

129. Which one of the following is not a gaseous biogeochemical cycle in ecosystem ?
(a) Sulphur cycle (b) Phosphorus cycle
(c) Nitrogen cycle (d) Carbon cycle

130. A single strand of nucleic acid tagged with a radioactive molecule is called :
(a) Vector (b) Selectablemarker
(c) Plasmid (d) Probe

131. Which one of the following options gives one correct example each of convergent evolution and divergent evolution?

	Convergent evolution	Divergent evolution
(a)	Eyes of octopus and mammals	Bones of forelimbs of vertebrates
(b)	Thorns of Bougainvillia and tendrils of *Cucurbita*	Wings of butterflies and birds
(c)	Bones of forelimbs of vertebrates	Wings of butterfly and birds
(d)	Thorns of Bougainvillia and tendrils of *Cucurbita*	Eyes of Octopus and mammals

132. An organic substance that can withstand environmental extremes and cannot be degraded by any enzyme is :
(a) Cuticle (b) Sporopollenin
(c) Lignin (d) Cellulose

133. *Cycas* and *Adiantum* resemble each other in having:
(a) Seeds (b) Motile Sperms
(c) Cambium (d) Vessels

134. What was the most significant trend in the evolution of modern man (*Homo sapiens*) from his ancestors ?
(a) Shortening of jaws
(b) Binocular vision
(c) Increasing cranial capacity
(d) Upright posture

135. Cymose inflorescence is present in :
(a) *Solanum* (b) *Sesbania*
(c) *Trifolium* (d) *Brassica*

136. Ribosomal RNA is actively synthesized in
(a) Lysosomes (b) Nucleolus
(c) Nucleoplasm (d) Ribosomes

137. During gamete formation, the enzyme recombinase participates during
(a) Metaphase - I (b) Anaphase - II
(c) Prophase - I (d) Prophase - II

138. Identify the possible link "A" in the following food chain :
Plant → insect - frog → "A" → Eagle
(a) Rabbit (b) Wolf
(c) Cobra (d) Parrot

139. Phyllode is present in :
(a) *Asparagus*
(b) *Euphorbia*
(c) *Australian Acacia*
(d) *Opuntia*

140. Given below is an imaginary pyramid of numbers. What could be one of the possibilities about certain organisms at some of the different levels ?

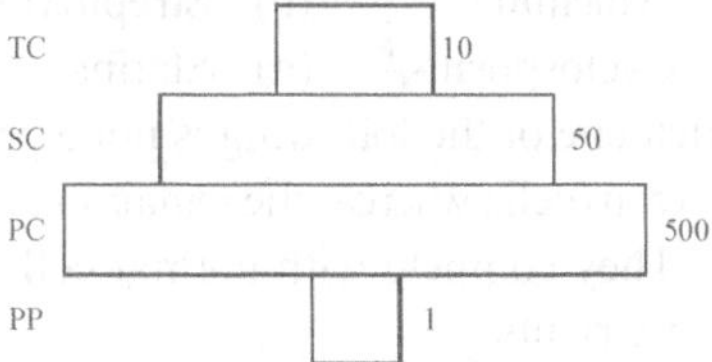

(a) Level PC is "insects" and level SC is "small insectivorous birds".
(b) Level PP is "phytoplanktons" in sea and "Whale" on top level TC
(c) Level one PP is "pipal trees" and the level SC is "sheep".
(d) Level PC is "rats" and level SC is "cats".

141. *Monascus purpureus* is a yeast used commercially in the production of:
(a) ethanol
(b) streptokinase for removing clots from the blood vessels.
(c) Citric acid
(d) blood cholesterol lowering statins

142. The correct sequence of cell organelles during photorespiration is :
(a) Chloroplast-Golgibodies-mitochondria

(b) Chloroplast-Rough Endoplasmic reticulum, Dictyosomes
(c) Chloroplast-peroxisome-mitochondria
(d) Chloroplast-vacuole-peroxisome

143. Which one of the following is correctly matched ?
(a) Passive transport of nutrients - ATP
(b) Apoplast - Plasmodesmata
(c) Potassium - Readily immobilisation
(d) Bakane of rice seedlings - F. Skoog

144. Anormal- visioned man whose father was colour. blind, marries a woman whose father was also colour blind. They have their first child as a daughter. what are the chance that this child would be colour blind ?
(a) 100% (b) zero percent
(c) 25% (d) 50%

145. Signals for parturition originate from :
(a) Both placenta as well as fully developed foetus
(b) Oxytocin released from maternal pituitary
(c) Placenta only
(d) Fully developed foetus only

146. A patient brought to a hospital with myocardial infarction is normally immediately given :
(a) Penicillin (b) Streptokinase
(c) Cyclosporin-A (d) Statins

147. Which one of the following is not a property of cancerous cells whereas the remaining three are ?
(a) They compete with normal cells for vital nutrients.
(b) They do not remain confined in the area of formation.
(c) They divide in an uncontrolled manner
(d) They show contact inhibition.

148. The gynoecium consists of many free pistils in flowers of
(a) *Aloe* (b) *Tomato*
(c) *Papaver* (4) *Michelia*

149. Which one of the following is not a function of an ecosystem
(a) Energy flow (b) Decomposition
(c) Productivity (d) Stratification

150. In a normal pregnant woman, the amount of total gonadotropin activity was assessed. The result expected was
(a) High level of circulating FSH and LH in the uterus to stimulate implantation of the embyro
(b) High level of circulatting HCG to stimulate endometrial thickening
(c) High level of FSH and LH in uterus to stimulate endometrical thickening
(d) High level of circulating HCG to stimulate estrogen and progesterone synthesis

151. Which one of the following areas in India, is a hotspot of biodiversity?
(a) Eastern Ghats (b) Gangetic Plain
(c) Sunderbans (d) Western Ghats

152. Which one of the follwing is a correct statement?
(a) Pteridophyte gametophyte has a protonemal and leafy stage
(b) In gymnosperms female gametophyte is free-living
(c) Antheridiophores and archegoniophores are present in pteridophytes.
(d) Origin of seed habit can be traced in pteridophytes

153. Which one of the following does not differ in *E.coli* and *Chlamydomonas*?
(a) Ribosomes
(b) Chromosomal Organization
(c) Cellwall
(d) Cellmembrane

154. The cyanobacteria are also referred to as
(a) protists (b) golden algae
(c) Slime moulds (d) blue green algae

155. The test-tube Baby Programme employs which one of the following techniques
(a) Intra cytoplasmic sperm injection (ICSI)
(b) Intra uterine insemination (IUI)
(c) Gamete intra fallopian transfer (GIET)
(d) Zygote intra fallopian transfer (ZIFT)

156. Which one of the following is a case of wrong matching?
(a) Somatic hybridization - Fusion of two diverse cells
(b) Vector DNA -Site for t-RNA synthesis.
(c) Micropropagation - *in vitro* production of plants in large numbers.
(d) Callus - Unorganised mass of cell produced in tissue culture

157. The highest number of species in the world is represented by
(a) Fungi (b) Mosses
(c) Algae (d) Lichens

158. In which one of the following options the two examples are correctly matched with their particular type of immunity.

	Examples	Types of immunity
(a)	Polymorphonuclear leukocytes and monocytes	Cellular barriers
(b)	Anti-tetanus and anti-snake bite injection	Active immunity
(c)	Saliva in mouth and Tear in eyes	Physical barriers
(d)	Mucus coating of epithelium lining the urinogenital tract-and the HCl in stomach	Physiological barriers

159. Which one of the following is wrong statement?
(a) *Anabaena* and *Nostoc* are capable of fixing nitrogen in free living state also.
(b) Root nodule forming nitrogen fixers live as aerobes under free-living conditions.
(c) Phosphorus is a constituent of cell membranes, certain nucleic acids and cell proteins.
(d) *Nitrosomonas* and *Nitrobacter* are chemoautotrophs.

160. Anxiety and eating spicy food together in an otherwise normal human, may lead to
(a) Indigestion (b) Jaundice
(c) Diarrhoea (d) Vomiting

161. The Leydig cells found in the human body are the secretory source of
(a) Progesterone (b) intestinal mucus
(c) glucagon (d) androgens

162. Compared to those of humans, the erythrocytes in frog are
(a) Without nucleus but with haemoglobin
(b) nucleated and with haemoglobin
(c) very much smaller and fewer
(d) nucleated and without haemoglobin.

163. In which one of the following the genus name, its two charcters and its phylum are not correctly matched, whereas the remaining three are correct

	Genus name	Two characters	Phylum
(a)	*Pila*	(a) Body Segmented Mouth with radula	Mollusca
(b)	*Asterias*	(b) Spiny Skinned Water vascular system	Echinoermata
(c)	*Sycon*	(c) Pore bearing Canal system	Porifera
(d)	*Periplaneta*	(d) Jointed appendages Chitinous exoskeleton	Arthropoda

164. What is true about ribosomes?
(a) The prokaryotic ribosomes are 80S, where "S" stands for sedimentation coefficient
(b) These are composed of ribonucleic acid and proteins
(c) These are found only in eukaryotic cells
(d) These are self-splicing introns of some RNAs.

165. Cirrhosis of liver is caused by the chronic intake of
(a) Opium
(b) Alcohol
(c) Tobacco (Chewing)
(d) Cocaine

166. Which one is a **true** statement regarding DNA polymerase used in PCR?
(a) It is used to ligate introduced DNA in recipient cell
(b) It serves as a selectable marker
(c) It is isolated from a virus
(d) It remains active at high temperature

167. Which statement is **wrong** for viruses?
(a) All are parasites
(b) All of them have helical symmetry
(c) They have ability to synthesize nucleic acids and proteins
(d) Antibiotics have no effect on them

168. Which one of the following is **correctly** matched?
(a) Onion - Bulb
(b) Ginger - Sucker
(c) *Chlamydomonas* - Conidia
(d) Yeast - Zoospores

169. How many plants in the list given below have composite fruits that develop from an inflorescence Walnut, poppy, radish, fig, pineapple, apple, tomato, mulberry?
(a) Four (b) Five
(c) Two (d) Three

170. Given below is the diagrammatic representation of one of the categories of small molecular weight organic compounds in the living tissues. Identify the category shown and the one blank component "X" in it.

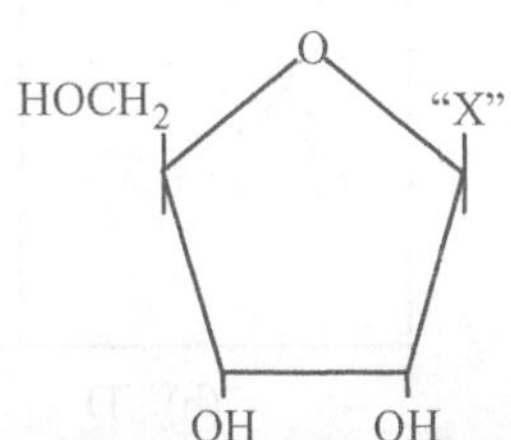

	Category	Component
(a)	Cholesterol	Guanin
(b)	Amino acid	NH_2
(c)	Nucleotide	Adenine
(d)	Nucleoside	Uracil

171. Which one of the following microbes forms symbiotic association with plants and helps them in their nutrition?
(a) *Azotobacter* (b) *Aspergillus*
(c) *Glomus* (d) *Trichoderma*

172. The extinct human who lived 1,00,000 to 40,000 years ago, in Europe, Asia and parts of Africa, With short stature, heavy eyebrows, retreating fore heads, large jaws with heavy teeth, stocky bodies, a lumbering gait and stooped posture was
(a) *Homo habilis*
(b) Neanderthal human
(c) Cro-magnan humans
(d) *Ramapithecus*

173. If one strand of DNA has the nitrogenous base sequence at ATCTG, what would be the complementary RNA strand sequence
(a) TTAGU (b) UAGAC
(c) AACTG (d) ATCGU

174. Which one of the following pairs of hormones are the examples of those that can easily pass through the cell membrane of the target cell and bind to a receptor inside it (Mostly in the nucleus)
(a) Insulin, glucagon
(b) Thyroxin, insulin
(c) Somatostain, oxytocin
(d) Cortisol, testosterone

175. Nuclear mebrane is absent in
(a) *Penicillium* (b) *Agaricus*
(c) *Volvox* (d) *Nostoc*

176. Which one is the most abundant protein in the animal world?
(a) Trypsin (b) Haemoglobin
(c) Collagen (d) Insulin

177. Which one of the following is common to multicellular fungi, filamentous algae and protonema of mosses?
(a) Diplontic life cycle
(b) Members of kingdom plantae
(c) Mode of Nutrition
(d) Multiplication by fragmentation

178. Which one single organism or the pair of organisms is **correctly** assigned to its taxonomic group?
(a) *Paramoecium* and *Plasmodium* belong to the same kingdom as that of *Penicillium*
(b) Lichen is a composite organism formed from the symbiotic association of an algae and a protozoan
(c) Yeast used in making bread and beer is a fungus
(d) *Nostoc* and *Anabaena* are examples of protista

179. Even in absence of pollinating agents seed setting is assured in
(a) *Commellina* (b) *Zostera*
(c) *Salvia* (d) Fig

180. Yeast is used in the production of
(a) Citric acid and lactic acid
(b) Lipase and pectinase
(c) Bread and beer
(d) Cheese and butter

181. Which one out of A-D given below correctly respresents the structural formula of the basic amino acid

A	B	C	D
NH_2 | H—C—COOH | CH_2 | CH_2 | C // \\ O OH	NH_2 | H—C—COOH | CH_2 | OH	CH_2OH | CH_2 | CH_2 | NH_2	NH_2 | H—C—COOH | CH_2 | CH_2 | CH_2 | CH_2 | NH_2

(a) C (b) D (c) A (d) B

182. The upright pyramid of number is absent in
(a) Pond (b) Forest
(c) Lake (d) Grassland

183. What is correct to say about the hormone action in humans
(a) Glucagon is secreted by β-cells of islets of Langerhans and stimulates glycogenolysis
(b) Secretion of thymosins is stimulated with ageing
(c) In females FSH first binds with specific receptors on ovarian cell membrane
(d) FSH stimulates the secretion of estrogen and progesterone

184. Closed vascular bundles lack
(a) Ground tissue (b) conjunctive tissue
(c) Cambium (d) Pith

185. Which of the following are the correct statement for respiration in human?
(a) Cigarette smoking may lead of inflammation of bronchi
(b) Neural signals from pneumotaxic centre in pons region of brain can increase the duration of inspiration
(c) Workers in grinding and stone - breaking industries may suffer from lung fibrosis
(d) About 90% of carbon dioxide (CO_2) is carried by haemoglobin as carbamino haemoglobin.

186. Removal of introns and joining of exons in a defined order during transcription is called :
(a) Looping (b) Inducing
(c) Slicing (d) Splicing

187. F_2 generation in a Mendelian cross showed that both genotypic and phenotypic ratios are same as 1 : 2 : 1. It represents a case of :
(a) Co-dominance
(b) Dihybrid cross
(c) Monohybrid cross with complete dominance
(d) Monohybrid cross with incomplete dominance

188. Given below is the representation of a certain event at a particular stage of a type of cell division. Which is this stage?

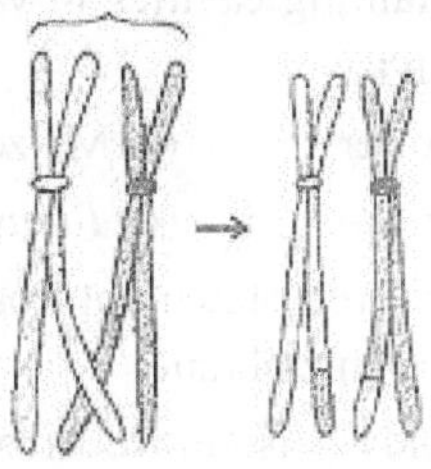

(a) Prophase I during meiosis
(b) Prophase II during meiosis
(c) Prophase of Mitosis
(d) Both prophase and metaphase of mitosis

189. People who have migrated from the planes to an area adjoining Rohatang Pass about six months back :
(a) have more RBCs and their haemoglobin has a lower binding affinity to O_2.
(b) are not physically fit to play games like football.
(c) suffer from altitude sickness with symptoms like nausea, fatigue, etc.
(d) have the usual RBC count but their haemoglobin has very high binding affinity to O_2.

190. For transformation, micro-particles coated with DNA to be bombarded with gene gun are madc up of :
(a) Silver or Platinum
(b) Platinum or Zinc
(c) Silicon or Platinum
(d) Gold or Tungsten

191. A nitrogen-fixing microbe associated with Azolla in rice fields is :
(a) *Spirulina* (b) *Anabaena*
(c) *Frankia* (d) *Tolypothrix*

192. Select the correct statement regarding the specific disorder of muscular or skeletal system :-
(a) Muscular dystrophy - age related shortening or muscles.
(b) Osteoporosis - decrease in bone mass and higher chance of fractures with advancing age.
(c) Myasthenia gravis - Auto immune disorder which inhibits sliding of myosin filaments
(d) Gout - inflammation of joints due to extra deposition of calcium.

193. Water containing cavities in vascular bundles are found in :

(a) Sunflower (b) Maize

(c) *Cycas* (d) *Pinus*

194. Select the correct statement from the following regarding cell membrane.

(a) Na^+ and K^+ ions move across cell membrane by passive transport

(b) Proteins make up 60 to 70% of the cell membrane.

(c) Lipids are arranged in a bilayer with polar heads towards the inner part.

(d) Fluid mosaic model of cell membrane was proposed by Singer and Nicolson

195. Gymnosperms are also called soft wood spermatophytes because they lack :

(a) Cambium

(b) Phloem fibres

(c) Thick-walled tracheids

(d) Xylem fibres

196. The coconut water and the edible part of coconut are equivalent to :

(a) Endosperm (b) Endocarp

(c) Mesocarp (d) Embryo

197. Vexillary aestivation is characteristic of the family

(a) Fabaceae (b) Asteraceae

(c) Solanaceae (d) Brassicaceae

198. Which one of the following is an example of carrying out biological control of pests/diseases using microbes ?

(a) *Trichoderma* sp. against certain plant pathogens

(b) Nucleopolyhedrovirus against white rust in *Brassica*

(c) Bt - cotton to increase cotton yield

(d) Lady bird beetle against aphids in mustard

199. Select the correct statement from the ones given below with respect to *Periplaneta americana*.

(a) Nervous system located dorsally, consists of segmentally arranged ganglia joined by a pair of longitudinal connectives.

(b) Males bear a pair of short thread like anal styles.

(c) There are 16 very long Malpighian tubules present at the junctions of midgut and hindgut.

(d) Grinding of food is carried out only by the mouth parts.

200. Maximum nutritional diversity is found in the group.

(a) Fungi (2) Animalia

(c) Monera (4) Plantae

CBSE PMT MAINS 2012

PHYSICS

1. The dimensions of $(\mu_0\varepsilon_0)^{-1/2}$ are :
   (a) $[L^{1/2}T^{-1/2}]$ (b) $[L^{-1}T]$
   (c) $[LT^{-1}]$ (d) $[L^{1/2}T^{1/2}]$

2. A stone is dropped from a height h. It hits the ground with a certain momentum P. If the same stone is dropped from a height 100% more than the previous height, the momentum when it hits the ground will change by :
   (a) 68% (b) 41%
   (c) 200% (d) 100%

3. A car of mass m is moving on a level circular track of radius R. If μ_s represents the static friction between the road and tyres of the car, the maximum speed of the car in circular motion is given by :
   (a) $\sqrt{\mu_s mRg}$ (b) $\sqrt{Rg/\mu_s}$
   (c) $\sqrt{mRg/\mu_s}$ (d) $\sqrt{\mu_s Rg}$

4. A car of mass m starts from rest and accelerates so that the instantaneous power delivered to the car has a constant magnitude P_0. The instantaneous velocity of this car is proportional to :
   (a) $t^2 P_0$ (b) $t^{1/2}$
   (c) $t^{-1/2}$ (d) $\frac{t}{\sqrt{m}}$

5. A circular platform is mounted on a frictionless vertical axle. Its radius $R = 2$ m and its moment of inertia about the axle is 200 kgm^2. It is initially at rest. A 50 kg man stands on the edge of the platform and begins to walk along the edge at the speed of 1 ms^{-1} relative to the ground. Time taken by the man to complete one revolution is :
   (a) π s (b) $\frac{3\pi}{2}$s
   (c) 2π s (d) $\frac{\pi}{2}$s

6. The moment of inertia of a uniform circular disc is maximum about an axis perpendicular to the disc and passing through :

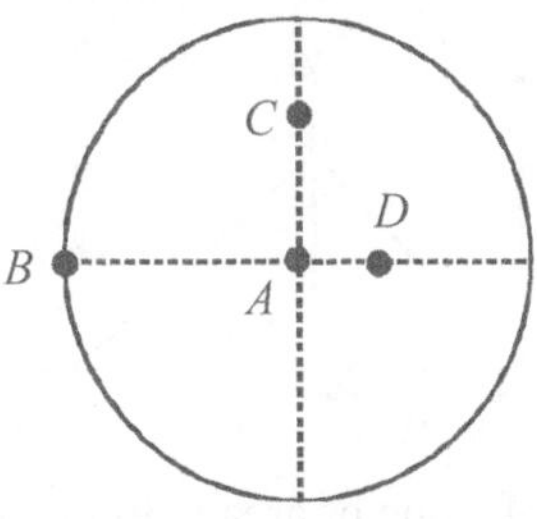

   (a) B (b) C
   (c) D (d) A

7. Three masses are placed on the x-axis : 300 g at origin, 500 g at $x = 40$ cm and 400 g at $x = 70$ cm. The distance of the centre of mass from the origin is :
   (a) 40 cm (b) 45 cm
   (c) 50 cm (d) 30 cm

8. If v_e is escape velocity and v_0 is orbital velocity of a satellite for orbit close to the earth's surface, then these are related as :
   (a) $v_0 = \sqrt{2}v_e$ (b) $v_0 = v_e$
   (c) $v_e = \sqrt{2v_0}$ (d) $v_e = \sqrt{2}v_0$

9. Which one of the following plots represents the variation of gravitational field on a particle with distance r due to a thin spherical shell of radius R ? (r is measured from the centre of the spherical shell)

(a)

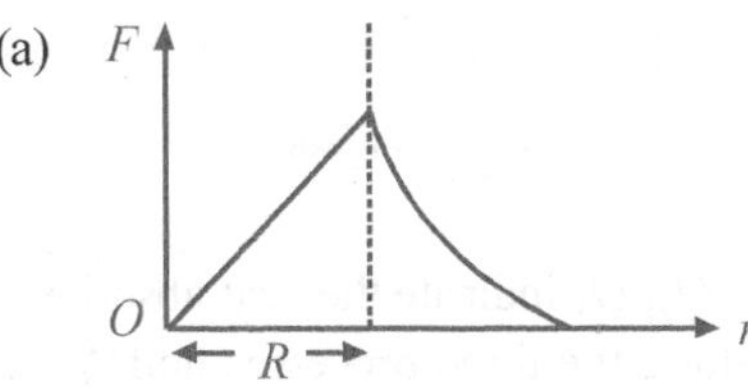

(b)

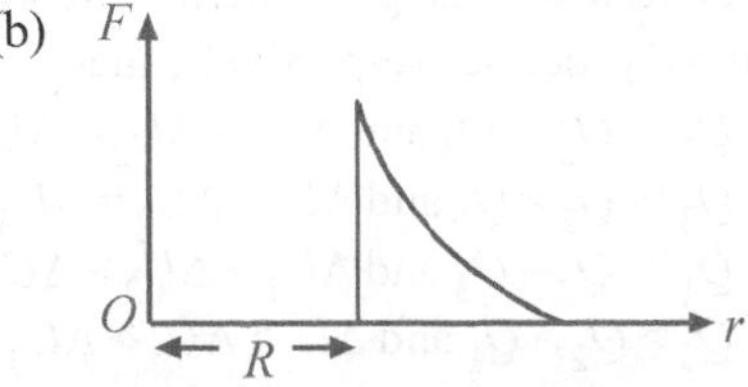

(c)

(d)

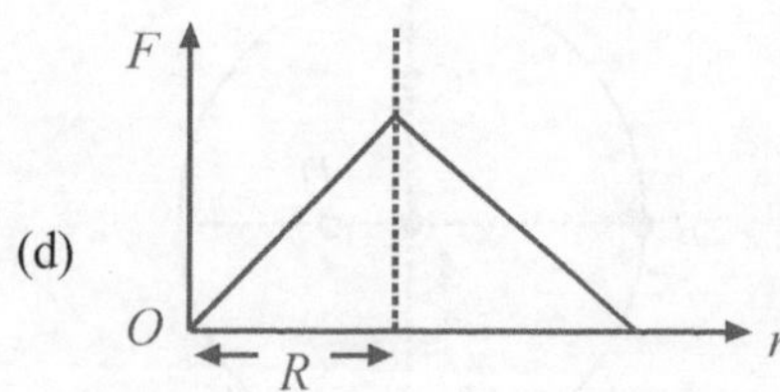

10. A slab of stone of area 0.36 m^2 and thickness 0.1 m is exposed on the lower surface to steam at 100°C. A block of ice at 0°C rests on the upper surface of the slab. In one hour 4.8 kg of ice is melted. The thermal conductivity of slab is : (Given latent heat of fusion of ice = 3.36 × 10^5 Jkg^{-1}.) :
(a) 1.24 J/m-s-°C
(b) 1.29 J/m-s-°C
(c) 2.05 J/m-s-°C
(d) 1.02 J/m-s-°C

11. An ideal gas goes from state A to state B via three different processes as indicated in the P-V diagram :

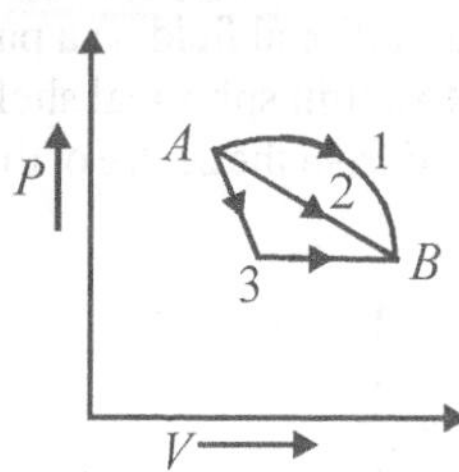

If Q_1, Q_2, Q_3 indicate the heat absorbed by the gas along the three processes and ΔU_1, ΔU_2, ΔU_3 indicate the change in internal energy along the three processes respectively, then
(a) $Q_1 > Q_2 > Q_3$ and $\Delta U_1 = \Delta U_2 = \Delta U_3$
(b) $Q_3 > Q_2 > Q_1$ and $\Delta U_1 = \Delta U_2 = \Delta U_3$
(c) $Q_1 = Q_2 = Q_3$ and $\Delta U_1 > \Delta U_2 > \Delta U_3$
(d) $Q_3 > Q_2 > Q_1$ and $\Delta U_1 > \Delta U_2 > \Delta U_3$

12. The equation of a simple harmonic wave is given by

$$y = 3\sin\frac{\pi}{2}(50t - x)$$

Where x and y are in metres and t is in seconds. The ratio of maximum particle velocity to the wave velocity is
(a) 2π (b) $\frac{3}{2}\pi$
(c) 3π (d) $\frac{2}{3}\pi$

13. A train moving at a speed of 220 ms^{-1} towards a stationary object, emits a sound of frequency 1000 Hz. Some of the sound reaching the object gets reflected back to the train as echo. The frequency of the echo as detected by the driver of the train is : (speed of sound in air is 330 ms^{-1})
(a) 3500 Hz (b) 4000 Hz
(c) 5000 Hz (d) 3000 Hz

14. A parallel plate capacitor has a uniform electric field E in the space between the plates. If the distance between the plates is d and area of each plate is A, the energy stored in the capacitor is :
(a) $\frac{1}{2}\varepsilon_0 E^2$ (b) $E^2 Ad/\varepsilon_0$
(c) $\frac{1}{2}\varepsilon_0 E^2 Ad$ (d) $\varepsilon_0 EAd$

15. Two metallic spheres of radii 1 cm and 3 cm are given charges of -1×10^{-2} C and 5×10^{-2} C, respectively. If these are connected by a conducting wire, the final charge on the bigger sphere is :
(a) 2×10^{-2} C (b) 3×10^{-2} C
(c) 4×10^{-2} C (d) 1×10^{-2} C

16. The power dissipated in the circuit shown in the figure is 30 Watts. The value of R is :

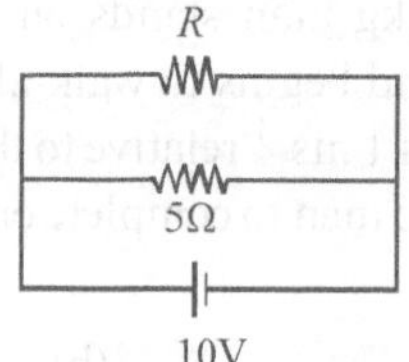

(a) 20 Ω (b) 15 Ω
(c) 10 Ω (d) 30 Ω

17. Cell having an emf ε and internal resistance r is connected across a variable external resistance R. As the resistance R is increased, the plot of potential difference V across R is given by :

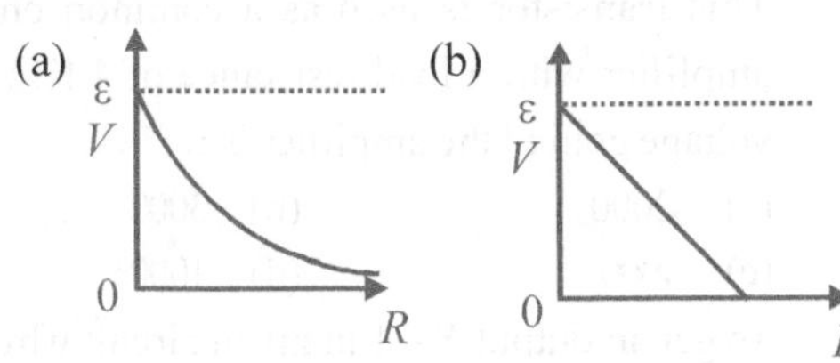

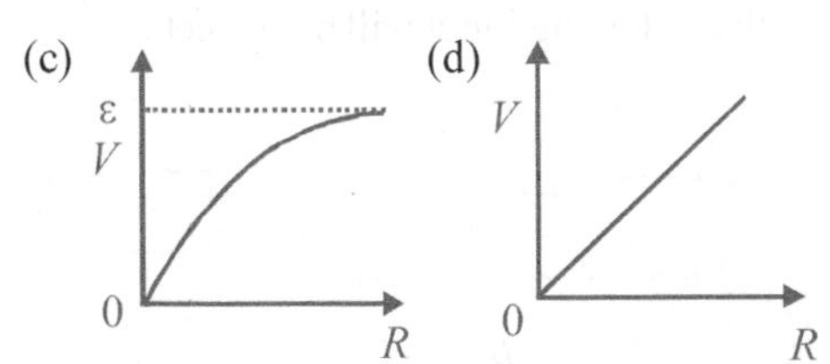

18. A proton carrying 1 MeV kinetic energy is moving in a circular path of radius R in uniform magnetic field. What should be the energy of an α-particle to describe a circle of same radius in the same field?
(a) 2 MeV (b) 1 MeV
(c) 0.5 MeV (d) 4 MeV

19. A magnetic needle suspended parallel to a magnetic field requires $\sqrt{3}$ J of work to turn it through 60°. The torque needed to maintain the needle in this position will be :
(a) $2\sqrt{3}$J (b) 3J
(c) $\sqrt{3}$J (d) $\frac{3}{2}$J

20. The instantaneous values of alternating current and voltages in a circuit are given as

$$i = \frac{1}{\sqrt{2}}\sin(100\pi t) \text{ ampere}$$

$$e = \frac{1}{\sqrt{2}}\sin(100\pi t + \pi/3) \text{ Volt}$$

The average power in Watts consumed in the circuit is :
(a) $\frac{1}{4}$ (b) $\frac{\sqrt{3}}{4}$
(c) $\frac{1}{2}$ (d) $\frac{1}{8}$

21. In a coil of resistance 10 Ω, the induced current developed by changing magnetic flux through it, is shown in figure as a function of time. The magnitude of change in flux through the coil in Weber is :

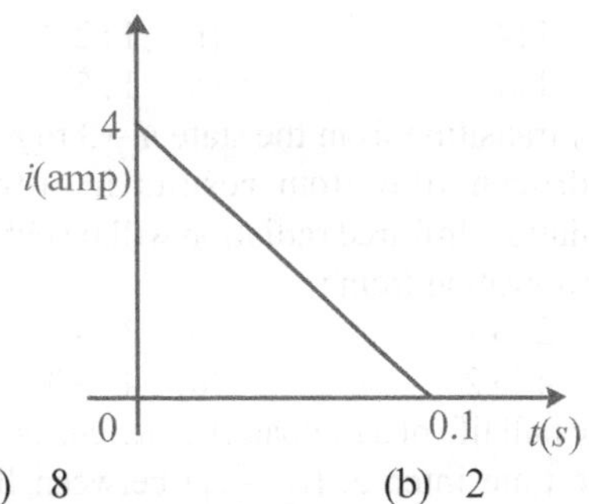

(a) 8 (b) 2
(c) 6 (d) 4

22. The ratio of amplitude of magnetic field to the amplitude of electric field for an electromagnetic wave propagating in vacuum is equal to :
(a) the speed of light in vacuum
(b) reciprocal of speed of light in vacuum
(c) the ratio of magnetic permeability to the electric susceptibility of vacuum
(d) unity

23. For the angle of minimum deviation of a prism to be equal to its refracting angle, the prism must be made of a material whose refractive index :
(a) lies between $\sqrt{2}$ and 1
(b) lies between 2 and $\sqrt{2}$
(c) is less than 1
(d) is greater than 2

24. A rod of length 10 cm lies along the principal axis of a concave mirror of focal length 10 cm in such a way that its end closer to the pole is 20 cm away from the mirror. The length of the image is :
(a) 10 cm (b) 15 cm
(c) 2.5 cm (d) 5 cm

25. If the momentum of electron is changed by P, then the de Broglie wavelength associated with it changes by 0.5%. The initial momentum of electron will be :
(a) 200 P (b) 400 P
(c) $\frac{P}{200}$ (d) 100 P

26. Two radiations of photons energies 1 eV and 2.5 eV, successively illuminate a photosensitive metallic surface of work function 0.5 eV. The ratio of the maximum speeds of the emitted electrons is :

(a) 1 : 4 (b) 1 : 2
(c) 1 : 1 (d) 1 : 5

27. The transition from the state $n = 3$ to $n = 1$ in a hydrogen like atom results in ultraviolet radiation. Infrared radiation will be obtained in the transition from :

(a) $2 \rightarrow 1$ (b) $3 \rightarrow 2$
(c) $4 \rightarrow 2$ (d) $4 \rightarrow 3$

28. The half life of a radioactive nucleus is 50 days. The time interval $(t_2 - t_1)$ between the time t_2 when $\frac{2}{3}$ of it has decayed and the time t_1 when $\frac{1}{3}$ of it had decayed is :

(a) 30 days (b) 50 days
(c) 60 days (d) 15 days

29. The input resistance of a silicon transistor is 100 Ω. Base current is changed by 40 μA which results in a change in collector current by 2 mA. This transistor is used as a common emitter amplifier with a load resistance of 4 KΩ. The voltage gain of the amplifier is :

(a) 2000 (b) 3000
(c) 4000 (d) 1000

30. To get an output $Y = 1$ in given circuit which of the following input will be correct :

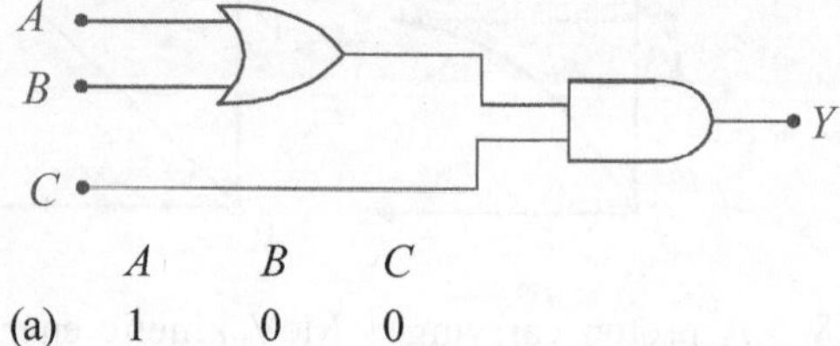

	A	B	C
(a)	1	0	0
(b)	1	0	1
(c)	1	1	0
(d)	0	1	0

CHEMISTRY

31. Given that the equilibrium constant for the reaction $2SO_2(g) + O_2(g) \rightleftharpoons 2SO_3(g)$ has a value of 278 at a particular temperature. What is the value of the equilibrium constant for the following reaction at the same temperature ?

$$SO_3(g) \rightleftharpoons SO_2(g) + \frac{1}{2}O_2(g)$$

(a) 1.8×10^{-3} (b) 3.6×10^{-3}

(c) 6.0×10^{-2} (d) 1.3×10^{-5}

32. Structure of a mixed oxide is cubic close-packed (c.c.p). The cubic unit cell of mixed oxide is composed of oxide ions. One fourth of the tetrahedral voids are occupied by divalent metal A and the octahedral voids are occupied by a monovalent metal B. The formula of the oxide is :

(a) ABO_2 (b) A_2BO_2

(c) $A_2B_3O_4$ (d) AB_2O_2

33. Given the reaction between 2 gases represented by A_2 and B_2 to give the compound AB(g).

$A_2(g) + B_2(g) \rightleftharpoons 2AB(g)$.

At equilibrium, the concentration

of $A_2 = 3.0 \times 10^{-3}$ M

of $B_2 = 4.2 \times 10^{-3}$ M

of $AB = 2.8 \times 10^{-3}$ M

If the reaction takes place in a sealed vessel at 527°C, then the value of K_C will be :

(a) 2.0 (b) 1.9

(c) 0.62 (d) 4.5

34. Activation energy (E_a) and rate constants (k_1 and k_2) of a chemical reaction at two different temperatures (T_1 and T_2) are related by :

(a) $\ln\frac{K_2}{K_1} = -\frac{E_a}{R}\left(\frac{1}{T_1} - \frac{1}{T_2}\right)$

(b) $\ln\frac{K_2}{K_1} = -\frac{E_a}{R}\left(\frac{1}{T_2} - \frac{1}{T_1}\right)$

(c) $\ln\frac{K_2}{K_1} = -\frac{E_a}{R}\left(\frac{1}{T_2} + \frac{1}{T_1}\right)$

(d) $\ln\frac{K_2}{K_1} = \frac{E_a}{R}\left(\frac{1}{T_1} - \frac{1}{T_2}\right)$

35. During change of O_2 to O_2^- ion, the electron adds on which one of the following orbitals ?

(a) π^* orbital (b) π orbital

(c) σ^* orbital (d) σ orbital

36. Standard reduction potentials of the half reactions are given below :

$F_2(g) + 2e^- \rightarrow 2F^-(aq)$; $E° = +2.85$ V

$Cl_2(g) + 2e^- \rightarrow 2Cl^-(aq)$; $E° = +1.36$ V

$Br_2(l) + 2e^- \rightarrow 2Br^-(aq)$; $E° = +1.06$ V

$I_2(s) + 2e^- \rightarrow 2I^-(aq)$; $E° = +0.53$ V

The strongest oxidising and reducing agents respectively are :

(a) F_2 and I^- (b) Br_2 and Cl^-

(c) Cl_2 and Br^- (d) Cl_2 and I_2

37. A certain gas takes three times as long to effuse out as helium. Its molecular mass will be :

(a) 27 u (b) 36 u

(c) 64 u (d) 9 u

38. The orbital angular momentum of a p-electron is given as :

(a) $\frac{h}{\sqrt{2}\pi}$ (b) $\sqrt{3}\frac{h}{2\pi}$

(c) $\sqrt{\frac{3}{2}}\frac{h}{\pi}$ (d) $\sqrt{6}.\frac{h}{2\pi}$

39. Vapour pressure of chloroform ($CHCl_3$) and dichloromethane (CH_2Cl_2) at 25°C are 200 mm Hg and 41.5 mm Hg respectively. Vapour pressure of the solution obtained by mixing 25.5 g of $CHCl_3$ and 40 g of CH_2Cl_2 at the same temperature will be : (Molecular mass of $CHCl_3$ = 119.5 u and molecular mass of CH_2Cl_2 = 85 u).

(a) 173.9 mm Hg (b) 615.0 mm Hg

(c) 347.9 mm Hg (d) 285.5 mm Hg

40. Molar conductivities ($\Lambda°_m$) at infinite dilution of NaCl, HCl and CH_3COONa are 126.4, 425.9 and 91.0 S cm^2 mol^{-1} respectively. $\Lambda°_m$ for CH_3COOH will be :

(a) 425.5 S cm^2 mol^{-1}

(b) 180.5 S cm^2 mol^{-1}

(c) 290.8 S cm^2 mol^{-1}

(d) 390.5 S cm^2 mol^{-1}

41. For real gases van der Waals equation is written as

$$\left(p+\frac{an^2}{V^2}\right)(V-nb)=nRT$$

where 'a' and 'b' are van der Waals constants.
Two sets of gases are :
(I) O_2, CO_2, H_2 and He
(II) CH_4, O_2 and H_2
The gases given in set-I in increasing order of 'b' and gases given in set-II in decreasing order of 'a', are arranged below. Select the correct order from the following :
(a) (I) He < H_2 < CO_2 < O_2 (II) CH_4 > H_2 > O_2
(b) (I) O_2 < He < H_2 < CO_2 (II) H_2 > O_2 > CH_4
(c) (I) H_2 < He < O_2 < CO_2 (II) CH_4 > O_2 > H_2
(d) (I) H_2 < O_2 < He < CO_2 (II) O_2 > CH_4 > H_2

42. Equal volumes of two monoatomic gases, A and B, at same temperature and pressure are mixed. The ratio of specific heats (C_p/C_v) of the mixture will be :
(a) 0.83 (b) 1.50
(c) 3.3 (d) 1.67

43. Red precipitate is obtained when ethanol solution of dimethylglyoxime is added to ammoniacal Ni(II). Which of the following statements is not true ?
(a) Red complex has a square planar geometry.
(b) Complex has symmetrical H-bonding
(c) Red complex has a tetrahedral geometry.
(d) Dimethylglyoxime functions as bidentate ligand.

[dimethylglyoxime = $H_3C-C=N-OH$ | $H_3C-C=N-OH$]

44. Low spin complex of d^6-cation in an octahedral field will have the following energy :
(a) $\frac{-12}{5}\Delta_0+P$ (b) $\frac{-12}{5}\Delta_0+3P$
(c) $\frac{-2}{5}\Delta_0+2P$ (d) $\frac{-2}{5}\Delta_0+P$
(Δ_0 = Crystal Field Splitting Energy in an octahedral field, P = Electron pairing energy)

45. Which one of the following does not correctly represent the correct order of the property indicated against it?
(a) Ti < V < Cr < Mn : increasing number of oxidation states
(b) $Ti^{3+} < V^{3+} < Cr^{3+} < Mn^{3+}$: increasing magnetic moment
(c) Ti < V < Cr < Mn : increasing melting points
(d) Ti < V < Mn < Cr : increasing 2nd ionization enthalpy

46. Four successive members of the first series of the transition metals are listed below. For which one of them the standard potential $\left(E^0_{M^{2+}/M}\right)$ value has a positive sign?
(a) Co (Z = 27) (b) Ni (Z = 28)
(c) Cu (Z = 29) (d) Fe (Z = 26)

47. In the replacement reaction

$$\rangle C-I + MF \longrightarrow \rangle C-F + MI$$

The reaction will be most favourable if M happens to be :
(a) Na (b) K
(c) Rb (d) Li

48. In which of the following arrangements the given sequence is not strictly according to the property indicated against it ?
(a) HF < HCl < HBr < HI : increasing acidic strength
(b) $H_2O < H_2S < H_2Se < H_2Te$: increasing pK_a values
(c) $NH_3 < PH_3 < AsH_3 < SbH_3$: increasing acidic character
(d) $CO_2 < SiO_2 < SnO_2 < PbO_2$: increasing oxidising power

49. Four diatomic species are listed below. Identify the correct order in which the bond order is increasing in them:
(a) $NO < O_2^- < C_2^{2-} < He_2^+$
(b) $O_2^- < NO < C_2^{2-} < He_2^+$
(c) $C_2^{2-} < He_2^+ < O_2^- < NO$
(d) $He_2^+ < O_2^- < NO < C_2^{2-}$

50. The catalytic activity of transition metals and their compounds is ascribed mainly to :
(a) their magnetic behaviour
(b) their unfilled d-orbitals
(c) their ability to adopt variable oxidation state
(d) their chemical reactivity

51. Which of the following exhibit only + 3 oxidation state ?
(a) U (b) Th
(c) Ac (d) Pa

52. The Gibbs' energy for the decomposition of Al_2O_3 at 500°C is as follows :
$$\frac{2}{3}Al_2O_3 \longrightarrow \frac{4}{3}Al + O_2;\ \Delta_rG = +960\,kJ\,mol^{-1}$$
The potential difference needed for the electrolytic reduction of aluminium oxide (Al_2O_3) at 500°C is at least :
(a) 4.5 V (b) 3.0 V
(c) 2.5 V (d) 5.0 V

53. Chloroamphenicol is an :
(a) antifertility drug
(b) antihistaminic
(c) antiseptic and disinfectant
(d) antibiotic-broad spectrum

54. Consider the following reaction :
$$C_6H_5COCl \xrightarrow[Pd\text{-}BaSO_4]{H_2} \text{'A'}$$
The product 'A' is :
(a) C_6H_5CHO
(b) C_6H_5OH
(c) $C_6H_5COCH_3$
(d) C_6H_5Cl

55. Which one of the following sets forms the biodegradable polymer?
(a) $CH_2=CH-CN$ and $CH_2=CH-CH=CH_2$
(b) H_2N-CH_2-COOH and $H_2N-(CH_2)_5-COOH$
(c) $HO-CH_2-CH_2-OH$ and $HOOC-C_6H_4-COOH$
(d) $C_6H_5-CH=CH_2$ and $CH_2=CH-CH=CH_2$

56. An organic compound (C_3H_9N) (A), when treated with nitrous acid, gave an alcohol and N_2 gas was evolved. (A) on warming with $CHCl_3$ and caustic potash gave (C) which on reduction gave isopropylmethylamine. Predict the structure of (A).
(a) $(CH_3)_2CH-NH_2$
(b) $CH_3CH_2-NH-CH_3$
(c) $CH_3-N(CH_3)-CH_3$
(d) $CH_3CH_2CH_2-NH_2$

57. Which of the following reagents will be able to distinguish between 1-butyne and 2-butyne?
(a) $NaNH_2$ (b) HCl
(c) O_2 (d) Br_2

58. Consider the reaction :
$$RCHO + NH_2NH_2 \rightarrow RCH=N-NH_2$$
What sort of reaction is it ?
(a) Electrophilic addition – elimination reaction
(b) Free radical addition – elimination reaction
(c) Electrophilic substitution – elimination reaction
(d) Nucleophilic addition – elimination reaction

59. Which of the following compounds will give a yellow precipitate with iodine and alkali ?
(a) Acetophenone
(c) Methyl acetate
(b) Acetamide
(d) 2-Hydroxypropane

60. Which of the following compounds can be used as antifreeze in automobile radiators ?
(a) Methyl alcohol (b) Glycol
(c) Nitrophenol (d) Ethyl alcohol

CBSE PMT MAINS 2012- BIOLOGY

BIOLOGY

61. How many organisms in the list given below are autotrophs?
Lactobacillus, Nostoc, Chara, Nitrosomonas, Nitrobacter, Streptomyces, Saccharomyces, Trypanosomes, Porphyra, Wolffia
(a) Four (b) Five
(c) Six (d) Three

62. Read the following five statements (A - E) and answer the question.
(A) In *Equisetum* the female gametophyte is retained on the parent sporophyte.
(B) In *Ginkgo* male gametophyte is not independent.
(C) The sporophyte in *Riccia* is more developed than that in *Polytrichum.*
(D) Sexual reproduction in *Volvox* is isogamous.
(E) The spores of slime molds lack cell walls.
How many of the above statements are correct?
(a) Two (b) Three
(c) Four (d) One

63. Which one of the following pairs is wrongly matched?
(a) *Ginkgo* -Archegonia
(b) *Salvinia* - Prothallus
(c) Viroids - RNA
(d) Mustard - Synergids

64. In the five-kingdom classification, *Chlamydomonas* and *Chlorella* have been included in
(a) protista (b) algae
(c) plantae (d) monera

65. For its activity, carboxypeptidase requires
(a) zinc (b) iron
(c) niacin (d) copper

66. Which one of the following structures is an organelle within an organelle?
(a) Ribosome (b) Peroxisome
(c) ER (d) Mesosome

67. Which one of the following is a wrong statement regarding mutations?
(a) Deletion and insertion of base pairs cause frame-shift mutations.
(b) Cancer cells commonly show chromosomal aberrations.
(c) UV and Gamma rays are mutagens.
(d) Change in a single base pair of DNA does not cause mutation.

68. A test cross is carried out to
(a) determine the genotype of a plant at F_2.
(b) predict whether two traits are linked.
(c) assess the number of alleles of a gene.
(d) determine whether two species or varieties will breed successfully.

69. Read the following four statements (A-D).
(A) In transcription, adenosine pairs with uracil.
(B) Regulation of *lac* operon by repressor is referred to as positive regulation.
(C) The human genome has approximately 50,000 genes.
(D) Haemophilia is a sex-linked recessive disease.
How many of the above statements are correct?
(a) Two (b) Three
(c) Four (d) One

70. Which one of the following organisms is correctly matched with its three characteristics?
(a) Pea: C_3 pathway, Endospermic seed, Vexillary aestivation
(b) Tomato: Twisted aestivation, Axile placentation, Berry
(c) Onion: Bulb, Imbricate aestivation, Axile placentation
(d) Maize: C_3 pathway, Closed vascular bundles, Scutellum

71. How many plants in the list given below have marginal placentation?
Mustard, Gram, Tulip, Asparagus, Arhar, Sun hemp, Chilli, Colchicine, Onion, Moong, Pea, Tobacco, Lupin
(a) Four (b) Five
(c) Six (d) Three

72. Read the following four statements (A-D).
(A) Both, photophosphorylation and oxidative phosphorylation involve uphill transport of protons across the membrane.
(B) In dicot stems, a new cambium originates from cells of pericycle at the time of secondary growth.
(C) Stamens in flowers of *Gloriosa* and *Petunia* are polyandrous.
(D) Symbiotic nitrogen-fixers occur in free-living state also in soil.

How many of the above statements are correct?
(a) Two (b) Three
(c) Four (d) One

73. Through their effect on plant growth regulators, what do the temperature and light control in the plants?
(a) Apical dominance (b) Flowering
(c) Closure of stomata (d) Fruit elongation

74. Which one of the following generally acts as an antagonist to gibberellins?
(a) Zeatin (b) Ethylene
(c) ABA (d) IAA

75. As compared to a dicot root, a monocot root has
(a) many xylem bundles.
(b) relatively thicker periderm.
(c) inconspicuous annual rings.
(d) more abundant secondary xylem.

76. For its action, nitrogenase requires
(a) Mn^{2+}
(b) light
(c) high input of energy
(d) super oxygen radicals

77. Vernalization stimulates flowering in
(a) zamikand (b) turmeric
(c) carrot (d) ginger

78. What is the function of germ pore?
(a) Emergence of radicle
(b) Absorption of water for seed germination
(c) Initiation of pollen tube
(d) Release of male gametes

79. Which one of the following statements is wrong?
(a) When pollen is shed at two-celled stage, double fertilization does not take place.
(b) Vegetative cell is larger than generative cell.
(c) Pollen grains in some plants remain viable for months.
(d) Intine is made up of cellulose and - pectin.

80. Plants with ovaries having only one or a few ovules, are generally pollinated by
(a) bees (b) butterflies
(c) birds (d) wind

81. Sacred groves are specially useful in
(a) preventing soil erosion.
(b) year-round flow of water in rivers.
(c) generating environmental awareness.
(d) conserving rare and threatened species.

82. The rate of formation of new organic matter by rabbit in a grassland, is called
(a) net productivity
(b) secondary productivity
(c) net primary productivity
(d) gross primary productivity

83. *Cuscuta* is an example of
(a) ectoparasitism (b) brood parasitism
(c) predation (d) endoparasitism

84. The second stage of hydrosere is occupied by plants like
(a) *Azolla* (b) *Typha*
(c) *Salix* (d) *Vallisneria*

85. Green revolution in India occurred during
(a) 1960.s (b) 1970.s
(c) 1980.s (d) 1950.s

86. In gobar gas, the maximum amount is that of
(a) butane (b) methane
(c) propane (d) carbon dioxide

87. Read the following four statements (A-D).
(A) Colostrum is recommended for the new born because it is rich in antigens.
(B) Chikungunya is caused by a Gram negative bacterium.
(C) Tissue culture has proved useful in obtaining virus-free plants.
(D) Beer is manufactured by distillation of fermented grape juice.

How many of the above statements are wrong?
(a) Two (b) Three
(c) Four (d) One

88. Tobacco plants resistant to a nematode have been developed by the introduction of DNA that produced (in the host cells)
(a) both sense and anti-sense RNA
(b) a particular hormone
(c) an antifeedant
(d) a toxic protein

89. Biolistics (gene-gun) is suitable for
(a) DNA finger printing.
(b) Disarming pathogen vectors.
(c) Transformation of plant cells.
(d) Constructing recombinant DNA by joining with vectors.

90. In genetic engineering, the antibiotics are used
(a) as selectable markers.
(b) to select healthy vectors.
(c) to keep the cultures free of infection.
(d) as sequences from where replication starts.

91. Which one of the following pairs of animals are similar to each other pertaining to the feature stated against them?
(a) *Pteropus* and *Ornithorhyncus* - Viviparity
(b) Garden lizard and Crocodile - Three chambered heart
(c) *Ascaris* and *Ancylostoma* - Metameric segmentation
(d) Sea horse and Flying fish - Cold blooded (poikilothermal)

92. Which one of the following categories of animals, is correctly described with no single exception in it?
(a) All sponges are marine and have collared cells.
(b) All mammals are viviparous and possess diaphragm for breathing.
(c) All bony fishes have four pairs of gills and an operculum on each side.
(d) All reptiles possess scales, have a three chambered heart and are cold blooded (poikilothermal).

93. Which one of the following organisms is scientifically correctly named, correctly printed according to the International Rules of Nomenclature and correctly described?
(a) *Musca domestica* - The common house lizard, a reptile.
(b) *Plasmodium falciparum* – A protozoan pathogen causing the most serious type of malaria.
(c) *Felis tigris* - The Indian tiger, well protected in Gir forests.
(d) *E.coli* - Full name *Entamoeba coli, a* commonly occurring bacterium in human intestine.

94. Which one of the following cellular parts is correctly described?
(a) Centrioles - Sites for active RNA synthesis.
(b) Lysosomes - Optimally active at a pH of about 8.5.
(c) Thylakoids - Flattened membranous sacs forming the grana of chloroplasts.
(d) Ribosomes - Those on chloroplasts are larger (80s) while those in the cytoplasm are smaller (70s).

95. Identify the meiotic stage in which the homologous chromosomes separate while the sister chromatids remain associated at their centromeres?
(a) Metaphase I (b) Metaphase II
(c) Anaphase I (d) Anaphase II

96. Which one of the following biomolecules is correctly characterized?
(a) Lecithin - A phosphorylated glyceride found in cell membrane.
(b) Palmitic acid - An unsaturated fatty acid with 18 carbon atoms.
(c) Adenylic acid - Adenosine with a glucose phosphate molecule.
(d) Alanine amino acid - Contains an amino group and an acidic group anywhere in the molecule.

97. The idea of mutations was brought forth by
(a) Gregor Mendel, who worked on *Pisum sativum.*
(b) Hugo de Vries, who worked on evening primrose.
(c) Hardy Weinberg, who worked on allele frequencies in a population.
(d) Charles Darwin, who observed a wide variety of organisms during sea voyage.

98. Which of the following forms the basis of DNA Finger printing?
(a) The relative proportions of purines and pyrimidines in DNA.
(b) Satellite DNA occurring as highly repeated short DNA segments.
(c) The relative difference in the DNA occurrence in blood, skin and saliva.
(d) The relative amount of DNA in the ridges and grooves of the fingerprints.

99. Represented below is the inheritance pattern of a certain type of traits in humans.

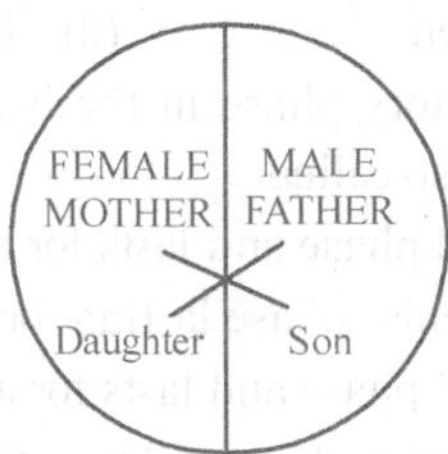

Which one of the following conditions could be an example of this pattern?

(a) Thalassemia

(b) Haemophilia

(c) Phenylketonuria

(d) Sickle cell anaemia

100. Given below is the diagrammatic sketch of a certain type of connective tissue. Identify the parts labeled A, B, C and D, and select the correct answer from the given option.

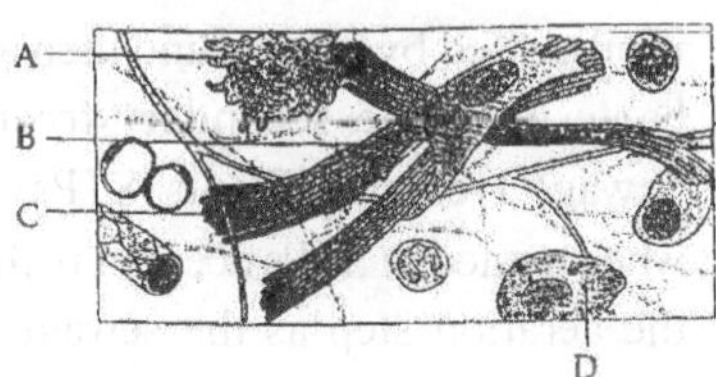

	Part-A	Part-B	Part-C	Part - D
(a)	Macrophage	Fibroblast	Collagen fibres	Mast cells
(b)	Mast cell	Macrophage	Fibroblast	Collagen fibres
(c)	Macrophage	Collagen fibres	Fibroblast	Mast cell
(d)	Mast cell	Collagen fibres	Fibroblast	Macrophage

101. Which one of the following options gives the correct categorization of six animals according to the type of nitrogenous wastes (A, B, C), they give out?

	A- Ammonotelic	B-Ureotelic	C- Uricotelic
(a)	Pigeon, Humans	Aquatic Amphibia, Lizards	Cockroach, Frog
(b)	Frog, Lizards	Aquatic Amphibia, Humans	Cockroach, Pigeon
(c)	Aquatic Amphibia	Frog, Humans	Pigeon, Lizards, Cockroach
(d)	Aquatic Amphibia	Cockroach, Humans	Frog, Pigeon, Lizards

102. Where do certain symbiotic micro-organisms normally occur in human body?

(a) Caecum

(b) Oral lining and tongue surface

(c) Vermiform appendix and rectum

(d) Duodenum

103. Which one of the following pairs of chemical substances is correctly categorized?

(a) Calcitonin and thymosin - Thyroid hormones

(b) Pepsin and prolactin - Two digestive enzymes secreted in stomach

(c) Troponin and myosin - Complex proteins in striated muscles

(d) Secretin and rhodopsin - Polypeptide hormones

104. The supportive skeletal structures in the human external ears and in the nose tip are examples of

(a) ligament (b) areolar tissue

(c) bone (d) cartilage

105. The four sketches (1, 2, 3 and 4) given below, represent four different types of animal tissues. Which one of these is correctly identified in the options given, along with its correct location and function?

(1) 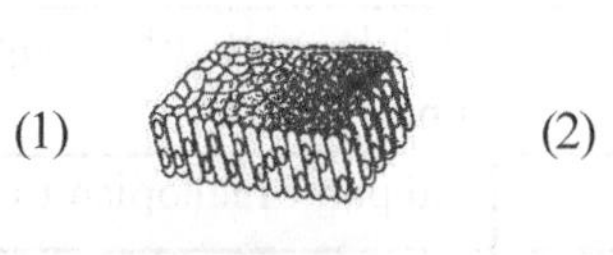 (2) 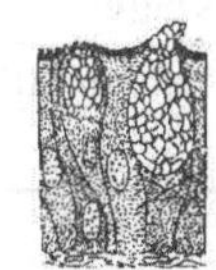

(3)  (4) 

		Tissue	Location	Function
(a)	(1)	Glandular epithelium	Intestine	Secretion
(b)	(3)	Collagen fibres	Cartilage	Attach skeletal muscles to bones
(c)	(4)	Smooth muscle tissue	Heart	Heart contraction
(d)	(1)	Columnar epithelium	Nephron	Secretion and absorption

106. A fall in glomerular filtration rate (GFR) activates
(a) adrenal cortex to release aldosterone.
(b) adrenal medulla to release adrenaline.
(c) juxta - glomerular cells to release renin.
(d) posterior pituitary to release vasopressin.

107. Which one of the following characteristics is common to both in humans and adult frogs?
(a) Four - chambered heart
(b) Internal fertilization
(c) Nucleated RBCs
(d) Ureotelic mode of excretion

108. Identify the human developmental stage shown below as well as the related right place of its occurrence in a normal pregnant woman, and select the right option for the two together.

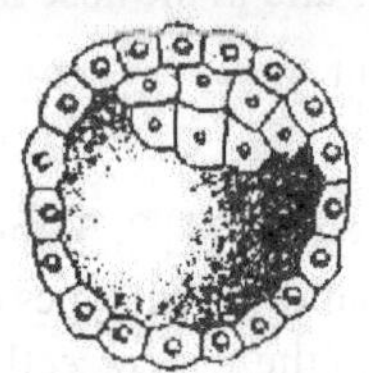

	Develeopmental stage	Site of occurrence
(a)	Late morula	Middle part of fallopian tube
(b)	Blastula	End part of fallopian tube
(c)	Blastocyst	Uterine wall
(d)	8 celled morula	Starting point of fallopian tube

109. Which one of the following human organs is often called the graveyard of RBCs?
(a) Gall bladder (b) Kidney
(c) Spleen (d) Liver

110. The secretory phase in the human menstrual cycle is also called
(a) luteal phase and lasts for about 6 days
(b) follicular phase lasting for about 6 days
(c) luteal phase and lasts for about 13 days
(d) follicular phase and lasts for about 13 days

111. Select the correct statement about biodiversity.
(a) Large scale planting of Bt cotton has no adverse effect on biodiversity.
(b) Western Ghats have a very high degree of species richness and endemism.
(c) Conservation of biodiversity is just a fad pursued by the developed countries.
(d) The desert areas of Rajasthan and Gujarat have a very high level of desert animal species as well as numerous rare animals.

112. The domestic sewage in large cities
(a) has a high BOD as it contains both aerobic and anaerobic bacteria.
(b) is processed by aerobic and then anaerobic bacteria in the secondary treatment in Sewage Treatment Plants (STPs).
(c) when treated in STPs does not really require the aeration step as the sewage contains adequate oxygen.
(d) has very high amounts of suspended solids and dissolved salts.

113. Which one of the following sets of items in the options 1 - 4 are correctly categorized with one exception in it?

	Items	Category	Exception
(a)	UAA, UAG, UGA	Stop codons	UAG
(b)	Kangaroo, Koala, Wombat	Australian marsupials	Wombat
(c)	*Plasmodium, Cuscuta, Trypanosoma*	Protozoan parasites	*Cuscuta*
(d)	Typhoid, Pneumonia, Diphtheria	Bacterial diseases	Diphtheria

114. Identify the likely organisms (1), (2), (3) and (4) in the food web shown below.

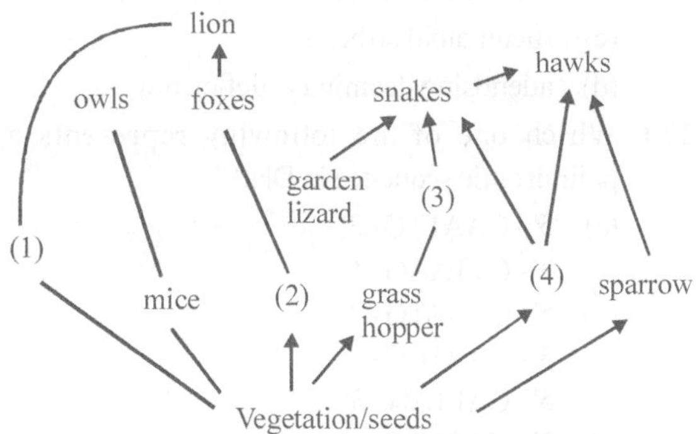

Options

	(1)	(2)	(3)	(4)
(a)	Deer	Rabbit	Frog	Rat
(b)	Dog	Squirrel	Bat	Deer
(c)	Rat	Dog	Tortoise	Crow
(d)	Squirrel	Cat	Rat	Pigeon

115. Consider the following four statements (A-D) and select the option which includes all the correct ones only.

(1) Single cell *Spirulina* can produce large quantities of food rich in protein, minerals, vitamins etc.

(2) Body weight-wise the micro-organism *Methylophilus methylotrophus* may be able to produce several times more proteins than the cows per day.

(3) Common button mushrooms are a very rich source of vitamin C.

(4) A rice variety has been developed which is *very* rich in calcium.

Options :

(a) Statements (3), (4)

(b) Statements (1), (3) and (4)

(c) Statements (2), (3) and (4)

(d) Statements (1), (2)

116. Identify the molecules (a) and (b) shown below and select the right option giving their source and use.

(a)

(b)

Options

	Molecule	Source	Use
(a)	(a) Cocaine	*Erythroxylum coca*	Accelerates the transport of dopamine
(b)	(b) Heroin	*Cannabis sativa*	Depressant and slows down body functions
(c)	(b) Cannabinoid	*Atropa belladona*	Produces hallucinations
(d)	(a) Morphine	*Papaver somniferum*	Sedative and pain killer

117. Which one of the following statements is correct with respect to immunity?

(a) Preformed antibodies need to be injected to treat the bite by a viper snake.

(b) The antibodies against small pox pathogen are produced by T – lymphocytes.

(c) Antibodies are protein molecules, each of which has four light chains.

(d) Rejection of a kidney graft is the function of B-lymphocytes.

118. The figure below shows three steps (A, B, C) of Polymerase Chain Reaction (PCR). Select the option giving correct identification together with what it represents?

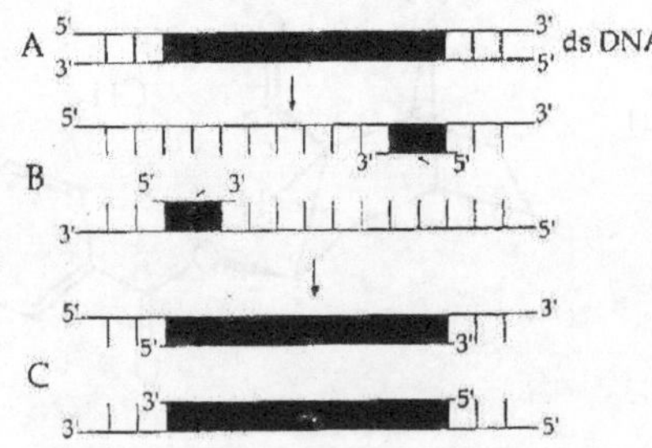

(a) B - Denaturation at a temperature of about 98°C separating the two DNA strands.

(b) A - Denaturation at a temperature of about 50°C.

(c) C - Extension in the presence of heat stable DNA polymerase.

(d) A - Annealing with two sets of primers.

119. The first clinical gene therapy was given for treating

(a) chicken pox

(b) diabetes mellitus

(c) rheumatoid arthritis

(d) adenosine deaminase deficiency

120. Which one of the following represents a palindromic sequence in DNA?

(a) 5' - GAATTC - 3'
3' - CTTAAG - 5'

(b) 5' - CCAATG - 3'
3' - GAATCC - 5'

(c) 5' - CATTAG - 3'
3' - GATAAC - 5'

(d) 5' - GATACC - 3'
3' - CCTAAG - 5'

HINTS & SOLUTIONS (SCREENING)

PHYSICS

1. (c)

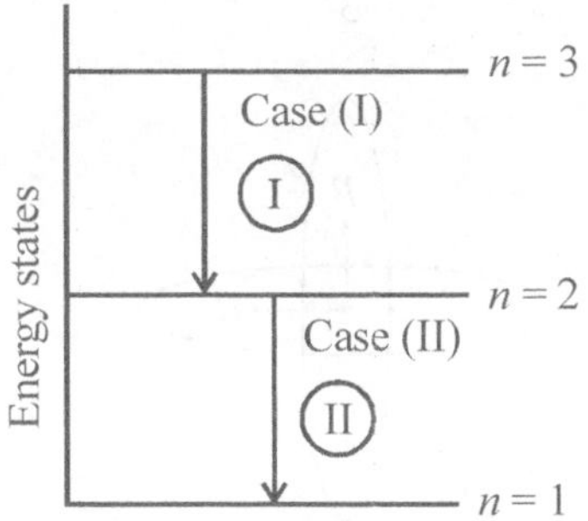

The wave number $(\bar{v})$ of the radiation $= \frac{1}{\lambda}$

$$= R_{\infty}\left[\frac{1}{n_1^2} - \frac{1}{n_2^2}\right]$$

Now for case (I) $n_1 = 3, n_2 = 2$

$$\frac{1}{\lambda_1} = R_{\infty}\left[\frac{1}{9} \frac{-1}{4}\right], R_{\infty} = \text{Rydberg constant}$$

$$\frac{1}{\lambda_1} = R_{\infty}\left[\frac{4-9}{36}\right] = \frac{-5R_{\infty}}{36}$$

$$\Rightarrow \quad \lambda_1 = \frac{-36}{5R_{\infty}}$$

$$\frac{1}{\lambda_2} = R_{\infty}\left[\frac{1}{4} - \frac{1}{1}\right] = \frac{-3R_{\infty}}{4}$$

$$\lambda_2 = \frac{-4}{3R_{\infty}}$$

$$\Rightarrow \quad \frac{\lambda_1}{\lambda_2} = \frac{-36}{5R_{\infty}} \times \frac{3R_{\infty}}{-4}$$

$$\frac{\lambda_1}{\lambda_2} = \frac{27}{5}$$

2. (c) Fundamental frequency is given by

$$v = \frac{1}{2l}\sqrt{\frac{T}{\mu}} \Rightarrow v \propto \frac{1}{l} \Rightarrow P \propto \frac{1}{v}$$

Since, P divided into l_1, l_2 and l_3 segments

Here $l = l_1 + l_2 + l_3$

$$\text{So} \quad \frac{1}{v} = \frac{1}{v_1} + \frac{1}{v_2} + \frac{1}{v_3}$$

3. (a) Given that, only 25% of 200W electrical energy converted into light of yellow colour

$$\left(\frac{hc}{\lambda}\right) \times N = 200 \times \frac{25}{100}$$

Where N is the no. of photons emitted per second, h = Plank's constant, c, speed of light.

$$N = \frac{200 \times 25}{100} \times \frac{\lambda}{hc}$$

$$= \frac{200 \times 25 \times 0.6 \times 10^{-6}}{100 \times 6.2 \times 10^{-34} \times 3 \times 10^{8}} = 1.5 \times 10^{20}$$

4. (d) Here D_1 is in forward bias and D_2 is in reverse bias so, D_1 will conduct and D_2 will not conduct. Thus, no current will flow through DC.

$$I = \frac{V}{R} = \frac{5}{10} = \frac{1}{2} \text{amp.}$$

5. (a) $\because \vec{L} = \vec{r} \times \vec{p}$

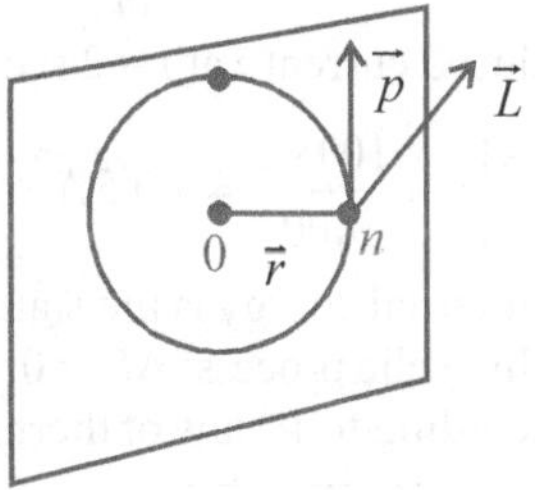

By right hand screw rule, the direction of $\vec{L}$ is $\perp$ to the plane containing $\vec{r}$ & $\vec{p}$.

The mass is rotating in the plane, about a fixed point, thus this plane will contain $\vec{r}$ & $\vec{p}$ and the direction of $\vec{L}$' will be $\perp$ to the plane.

6. (a) The torque on the dipole is given as

$\tau = PE \sin\theta$

The potential energy of the dipole in the electric field is given as

$U = -PE \cos\theta$

7. (d)

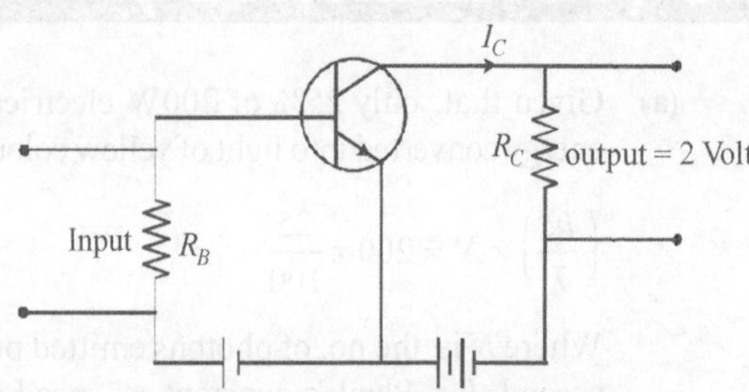

The output voltage, across the load R_C
$V_0 = I_C R_C = 2$ volt
The collector current (I_C)

$$I_C = \frac{2}{2\times10^3} = 10^{-3}\text{ A}$$

$$(\beta)\text{ current gain} = \frac{I_C}{I_B} = 100$$

$$I_B = \frac{I_C}{100} = \frac{10^{-3}}{100} = 10^{-5}\text{ A}$$

Input voltage (V_i)
$V_i = R_B I_B = 1\times10^3\times10^{-5} = 10^{-2}$ Volt
$V_i = 10$mV

8. (a) According, to Faraday's law of induction

Induced e.m.f. $\varepsilon = -\frac{d\phi}{dt} = -(100t)$

Induced current i at $t = 2$ sec

$$= \left|\frac{\varepsilon}{R}\right| = +\frac{100\times2}{400} = +0.5\text{ A}$$

9. (a) ∵ Internal energy is the state function.
∴ In cyclic process; $\Delta U = 0$
According to 1st law of thermodynamics

$$\boxed{\Delta Q = \Delta U + W}$$

So heat absorbed
$\Delta Q = W =$ Area under the curve
$= -(2V)(P) = -2PV$
So heat rejected $= 2PV$

10. (c) The radius of the nucleus is directly proportional to cube root of atomic number i.e. $R \propto A^{1/3}$
$\Rightarrow \quad R = R_0 A^{1/3}$, where R_0 is a constant of proportionality

$$\frac{R_2}{R_1} = \left(\frac{A_2}{A_1}\right)^{1/3} = \left(\frac{64}{27}\right)^{1/3} = \frac{4}{3}$$

where R_1 = the radius of ^{27}Al, and A_1 = atomic mass number of Al
R_2 = the radius of ^{64}Cu and A_2 = atomic mass number of Cu

$$R_2 = 3.6\times\frac{4}{3} = 4.8\text{ fermi}$$

11. (a)

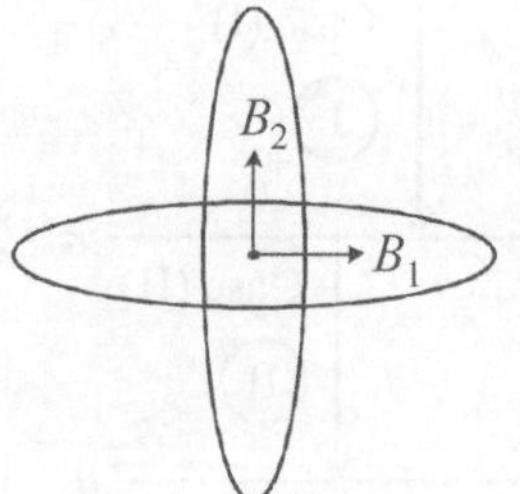

The magnetic field, due to the coil carrying current I ampere

$$B_1 = \frac{\mu_0 I}{2R}$$

The magnetic field due to the coil, carrying current $2I$ ampere

$$B_2 = \frac{\mu_0(2I)}{2R}$$

The resultant B

$$B_{net} = \sqrt{B_1^2 + B_2^2 + 2B_1B_2\cos\theta},\ \theta = 90°$$

$$B_{net} = \sqrt{B_1^2 + B_2^2} = \frac{\mu_0(2I)}{2R}\sqrt{1+4}$$

$$= \frac{\sqrt{5}\mu_0 I}{2R}$$

12. (b) For equilibrium

$$\frac{dU}{dr} = 0 \quad\Rightarrow\quad \frac{-2A}{r^3} + \frac{B}{r^2} = 0$$

$$r = \frac{2A}{B}$$

For stable equilibrium

$\frac{d^2U}{dr^2}$ should be positive for the value of r.

Here $\frac{d^2U}{dr^2} = \frac{6A}{r^4} - \frac{2B}{r^3}$ is +ve value

So, $r = \frac{2A}{B}$

13. (a) $\frac{1}{f}=\left(\frac{\mu_g}{\mu_m}-1\right)\left(\frac{1}{R_1}-\frac{1}{R_2}\right)$

If $\mu_g=\mu_m$, then $\frac{1}{f}=(1-1)\left(\frac{1}{R_1}-\frac{1}{R_2}\right)$

$\Rightarrow \quad \frac{1}{f}=0$

$$\boxed{f=\frac{1}{0}=\infty}$$

This implies that the liquid must have refractive index equal to glass.

14. (b) Horizontal range

$$R=\frac{u^2\sin 2\theta}{g} \quad ....(1)$$

Maximum height

$$H=\frac{u^2\sin^2\theta}{2g} \quad ....(2)$$

According to the problem

$R=H$

$$\frac{u^2\sin 2\theta}{g}=\frac{u^2\sin^2\theta}{2g}$$

$\Rightarrow \quad 2\sin\theta\cos\theta=\frac{\sin^2\theta}{2}$

$2\cos\theta=\frac{\sin\theta}{2}$

$\Rightarrow \quad \cot\theta=\frac{1}{4}$

$\Rightarrow \quad \tan\theta=4$

$\Rightarrow \quad \theta=\left[\tan^{-1}(4)\right]$

15. (c)

When L is removed from the circuit

$$\frac{X_C}{R}=\tan\frac{\pi}{3}$$

$$X_C=R\tan\frac{\pi}{3} \quad ....(1)$$

When C is remove from the circuit

$$\frac{X_L}{R}=\tan\frac{\pi}{3}$$

$$X_C=R\tan\frac{\pi}{3} \quad ...(2)$$

Net impedence $Z=\sqrt{R^2+(X_L-X_C)^2}=R$

Power factor $\cos\phi=\frac{R}{Z}=1$

16. (d) Stefan's law for black body radiation

$Q=\sigma e A T^4$

$$T=\left[\frac{Q}{\sigma(4\pi R^2)}\right]^{1/4}$$

Here $e=1$

$A=4\pi R^2$

17. (d) $V=-L\frac{di}{dt}$

Here $\frac{di}{dt}$ is + ve for $\frac{T}{2}$ time and

$\frac{di}{dt}$ is – ve for next $\frac{T}{2}$ time

18. (a) Galvanometer is converted into ammeter, by connected a shunt, in parallel with it.

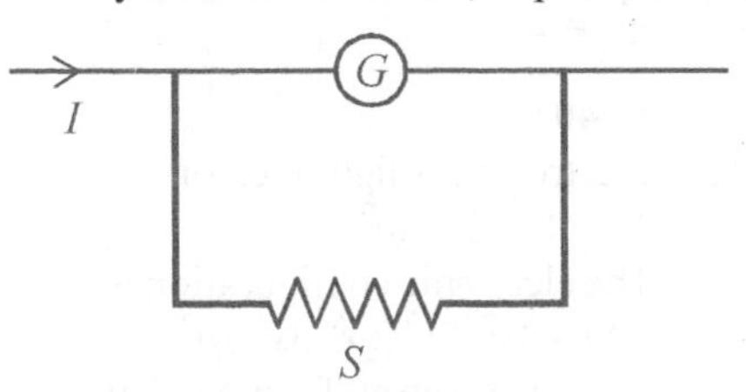

$$\frac{GS}{G+S}=\frac{V_G}{I}=\frac{25\times10^{-3}}{25}$$

$$\frac{GS}{G+S}=0.001\Omega$$

Here $S<<G$

So $S=0.001\ \Omega$

19. (c)

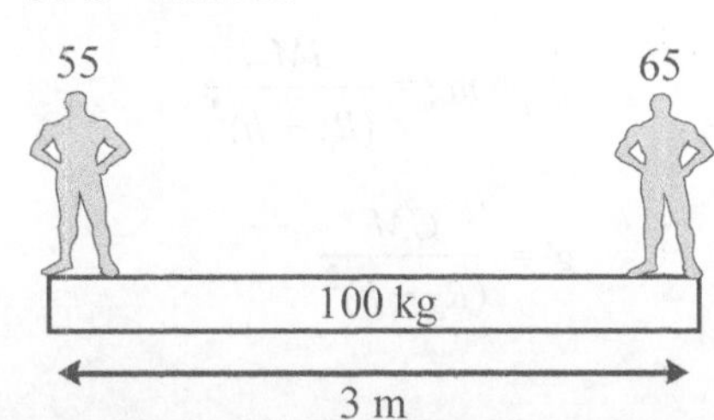

There is no external force so centre of mass of the system will not shift

20. (d) Let, the amount of the two in the mixture will become equal after t years.
The amount of A_1, which remains after t years

$$N_1 = \frac{N_{01}}{(2)^{t/20}}$$

The amount of A_2, which remains, after t years

$$N_2 = \frac{N_{02}}{(2)^{t/10}}$$

According to the problem
$N_1 = N_2$

$$\frac{40}{(2)^{t/20}} = \frac{160}{(2)^{t/10}}$$

$$2^{t/20} = 2^{\left(\frac{t}{10}-2\right)}$$

$$\frac{t}{20} = \frac{t}{10} - 2$$

$$\frac{t}{20} - \frac{t}{10} = 2$$

$$\frac{t}{20} = 2$$

$t = 40$ s

21. (c) Electronic configuration of
$^6C = 1s^2, 2s^2\, 2p^2$
The electronic configuration of
$^{14}Si = 1s^2, 2s^2\, 2p^6, 3s^2\, 3p^2$
As they are away from Nucleus, so effect of nucleus is low for Si even for Sn and Pb are almost mettalic.

22. (c) Let at h height, the weight of a body becomes $\frac{1}{16}$ of its weight on the surface.

$$W_h = 16\, W_s$$

$$W_h = mg = \frac{GM_m}{(R_e + h)^2}$$

$$g' = \frac{GM}{(R_e + h)^2}$$

Similarly, $g = \frac{GM}{R_e^2}$

$$\frac{g'}{g} = \frac{R_e^2}{(R_e + h)^2} \Rightarrow g' = g\left(1 + \frac{h}{R}\right)^{-2}$$

$\Rightarrow \quad W_h = 16\, W_s$

$\Rightarrow \quad g' = \frac{g}{16}$

$$g' = 16g\left(1 + \frac{h}{R}\right)^{-2}$$

$$\frac{1}{16} = \left(1 + \frac{h}{R}\right)^{-2} \Rightarrow 4 = 1 + \frac{h}{R}$$

$h = 3R$

23. (a) For emission, the wave number of the radiation is given as

$$\frac{1}{\lambda} = Rz^2\left(\frac{1}{n_1^2} - \frac{1}{n_2^2}\right)$$

R = Rydberg constant, Z = atomic number

$$= R\left(\frac{1}{1^2} - \frac{1}{5^2}\right) = R\left(1 - \frac{1}{25}\right) \Rightarrow \frac{1}{\lambda} = R\frac{24}{25}$$

linear momentum

$$P = \frac{h}{\lambda} = h \times R \times \frac{24}{25} \text{ (de-Broglie hypothesis)}$$

$$\Rightarrow mv = \frac{24hR}{25} \Rightarrow V = \frac{24hR}{25m}$$

24. (b) Since magnetic field is in vertical direction and needle is free to totate in horizontal plane only so magnetic force cannot rotate the needle in horizontal plane. Hence needle can stay in any position.

25. (b)

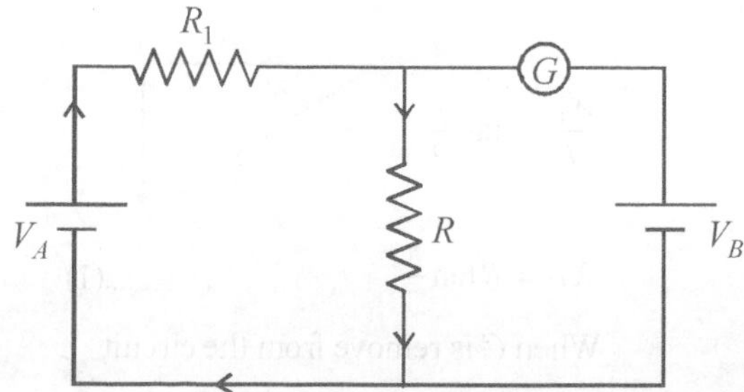

Since deflection in galvanometer is zero so current will flow as shown in the above diagram.

Current $I = \dfrac{V_A}{R_1 + R} = \dfrac{12}{500+100} = \dfrac{12}{600}$

So $V_B = IR = \dfrac{12}{600} \times 100 = 2V$

26. (a) Let the side length of square be 'a' then potential at centre O is

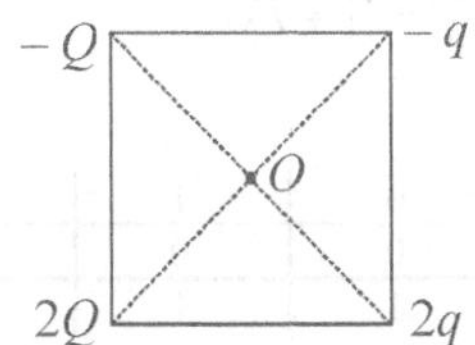

$$V = \frac{k(-Q)}{\left(\frac{a}{\sqrt{2}}\right)} + \frac{k(-q)}{\frac{a}{\sqrt{2}}} + \frac{k(2q)}{\frac{a}{\sqrt{2}}} + \frac{k(2Q)}{\frac{a}{\sqrt{2}}} = 0$$

(Given)

$$-Q - q + 2q + 2Q = 0 = Q + q = 0$$

$\therefore\ Q = -q$

27. (c) For banking, $\tan\theta = \dfrac{V^2}{Rg}$

$$\tan 45° = \frac{V^2}{90 \times 10} = 1$$

$V = 30$ m/s

28. (b) At maximum compression the solid cylinder will stop, so loss in K.E. of cylinder = gain in P.E. of spring

$$\Rightarrow \quad \frac{1}{2}mv^2 + \frac{1}{2}I\omega^2 = \frac{1}{2}kx^2$$

$$\Rightarrow \quad \frac{1}{2}mv^2 + \frac{1}{2}\frac{mR^2}{2}\left(\frac{v}{R}\right)^2 = \frac{1}{2}kx^2$$

$$\Rightarrow \quad \frac{3}{4}mv^2 = \frac{1}{2}kx^2$$

$$\Rightarrow \quad \frac{3}{4} \times 3 \times (4)^2 = \frac{1}{2} \times 200\, x^2$$

$$\Rightarrow \quad \frac{36}{100} = x^2 \ \Rightarrow\ x = 0.6\ \text{m}$$

29. (d) 1st process is isothermal expansion which is only correct shown in option (d)
2nd process is isobaric compression which is correctly shown in option (d)

30. (c) $m_2 \quad m_1$

(B) → v (A)

$u = 0$

Conservation of linear momentum along x-direction

$$m_2 v = m_1 v_x \ \Rightarrow\ \frac{m_2 v}{m_1} = v_x$$

along y-direction

$$m_2 \times \frac{v}{2} = m_1 v_y \ \Rightarrow\ v_y = \frac{m_2 v}{2m_1}$$

Note: Let A moves in the direction, which makes an angle θ with initial direction i.e.,

$$\tan\theta = \frac{v_y}{v_x} = \frac{m_2 v}{2m_1} \Big/ \frac{m_2 v}{2m_1}$$

$$\tan\theta = \frac{1}{2}$$

$\Rightarrow \quad \theta = \tan^{-1}\left(\dfrac{1}{2}\right)$ to the x-axis.

31. (a) Initially liquid oxygen will gain the temperature up to its boiling temperature then it change its state to gas. After this again its temperature will increase, so corresponding graph will be (a)

32. (c) Time period of cyclotron is

$$T = \frac{1}{\upsilon} = \frac{2\pi m}{eB};\ B = \frac{2\pi m}{e}\upsilon;\ R = \frac{m\upsilon}{eB} = \frac{p}{eB}$$

$$\Rightarrow \quad P = eBR = e \times \frac{2\pi m\upsilon}{e} R = 2\pi m\upsilon R$$

$$\text{K.E.} = \frac{p^2}{2m} = \frac{(2\pi m\upsilon R)^2}{2m} = 2\pi^2 m\upsilon^2 R^2$$

33. (a) Gravitational attraction force on particle B,

$$F_g = \frac{GM_P m}{(D_P/2)^2}$$

Acceleration of particle due to gravity

$$a = \frac{F_g}{m} = \frac{4GM_P}{D_P^2}$$

34. (a) For normally emerge $e = 0$
Therefore $r_2 = 0$ and $r_1 = A$
Snell's law for incident ray
$1 \sin i = \mu \sin r_1 = \mu \sin A$
For small angle
$i = \mu A$

35. (c) $F \propto v \Rightarrow F = kv$

$$k = \frac{F}{v} \Rightarrow [k] = \frac{[\text{kgms}^{-2}]}{[\text{ms}^{-1}]} = \text{kgs}^{-1}$$

36. (c)

$d = f_1 + 2f_2$

37. (c) According to Kepler's law of period
$T^2 \propto R^3$

$$\frac{T_1^2}{T_2^2} = \frac{R_1^3}{R_2^3} = \frac{(6R)^3}{(3R)^3} = 8$$

$$\frac{24 \times 24}{T_2^2} = 8$$

$$T_2^2 = \frac{24 \times 24}{8} = 72 = 36 \times 2$$

$$T_2 = 6\sqrt{2}$$

38. (b) I $\rightarrow$ ON
II $\rightarrow$ OFF
In IIIrd state it is used as a amplifier, it is active region.

39. (c) Resistance of bulb is constant

$$P = \frac{V^2}{R} \Rightarrow \frac{\Delta p}{p} = \frac{2\Delta V}{V} + \frac{\Delta R}{R}$$

$$\frac{\Delta p}{p} = 2 \times 2.5 + 0 = 5\%$$

40. (b) $\vec{v} = \vec{u} + \vec{a}t$

$$v = (2\hat{i} + 3\hat{j}) + (0.3\hat{i} + 0.2\hat{j}) \times 10 = 5\hat{i} + 5\hat{j}$$

$$|\vec{v}| = \sqrt{5^2 + 5^2}$$

$$|\vec{v}| = 5\sqrt{2}$$

41. (c) $n \rightarrow 2 - 1$
$E = 10.2$ eV
$kE = E - \phi$
$Q = 10.20 - 3.57$
$h\upsilon_0 = 6.63$ eV

$$\upsilon_0 = \frac{6.63 \times 1.6 \times 10^{-19}}{6.67 \times 10^{-34}} = 1.6 \times 10^{15} \text{ Hz}$$

42. (a) $\vec{F}_1 x + \vec{F}_2 x = \vec{F}_3 x$

$$\vec{F}_3 = \vec{F}_1 + \vec{F}_2$$

43. (a)

A	0	1	1	0
B	0	0	1	1
C	1	1	1	1

OR gate

44. (b) Eight identical cubes are required to arrange so that this charge is at centre of the cube

So flux $\phi = \dfrac{q}{8\varepsilon_0}$

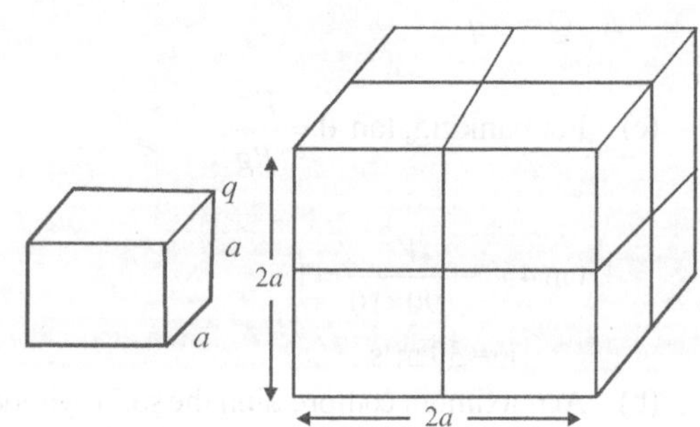

45. (d) The de-Broglie wavelength associated with the particle will be

$$\lambda = \frac{h}{p} \Rightarrow \lambda = \frac{h}{mv},$$

h = Plank's constant = 6.63×10^{-34} J.s
For circular motion, $F_c = qvB$

$$\Rightarrow \frac{mv^2}{r} = qvB$$

$$\Rightarrow \frac{mv}{qB} = r$$

$$r = \frac{mv}{qB} \Rightarrow mv = qrB$$

$\Rightarrow \quad (2e)(0.83\times10^{-2})\left(\frac{1}{4}\right)$

$$\lambda=\frac{6.6\times10^{-34}\times4}{2\times1.6\times10^{-19}\times0.83\times10^{-2}}$$

$\lambda=9.93\times10^{-34+21}=0.01$ Å

46. (a) On comparing the given equation to

$\vec{E}=a_0\hat{i}\cos(\omega t-kz)$

$\omega=6\times10^{8z}$,

$$k=\frac{2\pi}{r}=\frac{\omega}{c}$$

$$k=\frac{\omega}{c}=\frac{6\times10^{8}}{3\times10^{8}}=2\,\text{m}^{-1}$$

47. (d) $x=8+12t-t^3$

The final velocity of the particle will be zero, because it retarded.

$V=0+12-3t^2=0$

$3t^2=12$

$t=2$ sec

Now the retardation

$$a=\frac{dv}{dt}=0-6t$$

$a[t=2]=-12\text{ m/s}^2$

retardation $=12\text{ m/s}^2$

48. (c) M.P. $=9=\dfrac{f_0}{f_e}$

$\Rightarrow \quad f_0=9f_e$ ...(1)

$f_0+f_e=20$ ...(2)

On solving

$f_0=18$ cm = focal length of the objective

$f_e=2$ cm = focal length of the eyepiece

49. (d) $2\pi f_1=600\pi$

$f_1=300$ ... (1)

$2\pi f_2=608\pi$

$f_2=304$ ...(2)

$|f_1-f_2|=4$ beats

$$\frac{I_{\max}}{I_{\text{m.n}}}=\frac{(A_1+A_2)^2}{(A_1+A_2)^2}=\frac{(5+4)^2}{(5-4)^2}=\frac{81}{1},$$

where A_1 and A_2 are amplitudes of given two sound waves.

50. (d) Let x is the resistance per unit length then

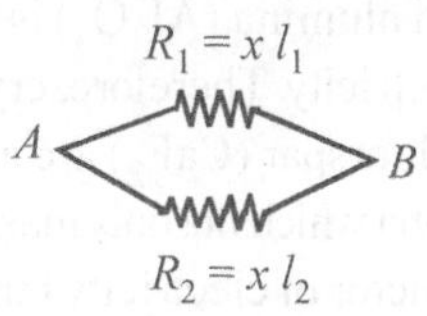

Equivalent resistance $R=\dfrac{R_1R_2}{R_1+R_2}$

$$=\frac{(x_1l_1)(x_2l_2)}{xl_1+xl_2}$$

$$\Rightarrow \quad \frac{8}{3}=x\frac{l_1l_2}{l_1+l_2}$$

$$\frac{8}{3}=x\frac{l_1}{\frac{l_1}{l_2}+1} \quad ...(i)$$

Also, $R_0=xl_1+xl_2$

$12=x(l_1+l_2)$

$$12=xl_2\left(\frac{l_1}{l_2}+1\right) \quad ...(ii)$$

$$\frac{Eq.(i)}{Eq.(ii)}\Rightarrow\frac{\frac{8}{3}}{\frac{12}{1}}=\frac{\dfrac{xl_1}{\left(\frac{l_1}{l_2}+1\right)}}{xl_2\left(\frac{l_1}{l_2}+1\right)}=\frac{l_1}{l_2\left(\frac{l_1}{l_2}+1\right)^2}$$

$$\Rightarrow\left(\frac{l_1}{l_2}+1\right)^2\times\frac{8}{36}=\frac{l_1}{l_2}$$

$$(y^2+1+2y)\times\frac{8}{36}=y\ (\text{where } y=\frac{l_1}{l_2})$$

$8y^2+8+16y=36y$

$\Rightarrow \quad 8y^2-20y+8=0$

$\Rightarrow \quad 2y^2-5y+2=0$

$\Rightarrow \quad 2y^2-4y-y+2=0$

$\Rightarrow \quad 2y(y-2)-1(y-2)=0$

$\Rightarrow \quad (2y-1)(y-2)=0$

$$\Rightarrow \quad y=\frac{l_1}{l_2}=\frac{1}{2}$$

CHEMISTRY

51. (c) Fused alumina (Al_2O_3) is a bad conductor of electricity. Therefore, cryolite (Na_3AlF_6) and fluorspar (CaF_2) are added to purified alumina which not only make alumina a good conductor of electricity but also reduce the melting point of the mixture to around 1140 K.

52. (b) Given pH = 12

or $[H^+] = 10^{-12}$

Since, $[H^+][OH^-] = 10^{-14}$

$$\therefore \quad [OH^-] = \frac{10^{-14}}{10^{-12}} = 10^{-2}$$

$$Ba(OH)_2 \rightleftharpoons \underset{s}{Ba^{2+}} + \underset{2s}{2OH^-}$$

$[OH^-] = 10^{-2}$

$2s = 10^{-2}$

$$s = \frac{10^{-2}}{2}$$

$$K_{sp} = 4s^3$$

$$= 4 \times \left(\frac{10^{-2}}{2}\right)^3$$

$$= 5 \times 10^{-7}$$

53. (b) On reaction with hot and concentrated alkali a mixture of chloride and chlorate is formed

$$3Cl_2 + 3\,NaOH_{(excess)} \xrightarrow{Hot} 5Na\overset{-1}{Cl} + Na\overset{+5}{Cl}O_3 + 3H_2O$$

54. (a) $(n = 4, l = 3) \Rightarrow 4f$ subshell

Since, maximum no. of electrons in a subshell $= 2(2l + 1)$

So, total no. of electron in $4f$ subshell

$= 2(2 \times 3 + 1) = 14$ electrons.

55. (a) $[Ni(NH_3)_6]^{2+}$

$Ni^{2+} = 3d^8$, according to CFT $= t_{2g}^6\, e_g^2$

therefore, hybridisation is sp^3d^2 & complex is paramagnetic.

56. (d) According to statement given in the question, it is clear that

$r \propto [A]^2$ and $r \propto [B]$

That means order of reaction with respect to B is 1 and w.r.t A is 2

Hence, Rate $= k[A]^2[B]^1$

57. (a) Since, in the first reaction gaseous products are forming from solid carbon hence entropy will increase i.e. $\Delta s = +ve$.

$$C\,(gr.) + \frac{1}{2}\,O_2(g) \rightarrow CO(g);\ \Delta S° = +\,ve$$

Since, $\Delta G° = \Delta H° - T\Delta S$ hence the value of ΔG decrease on increasing temperature.

58. (d) [$Fe_3O_4 \Rightarrow$ Magnetite]

[$CuCO_3\,Cu(OH)_2$ – Malachite]

Pyrolusite – MnO_2 Cassiterite – SnO_2.

59. (a) According to Freundlich Adsorption isotherm

$$\frac{x}{m} = KP^{\frac{1}{n}}$$

at low pressure $\frac{1}{n} = 1$

$$\therefore \quad \frac{x}{m} \propto P^1$$

at high pressure $\frac{1}{n} = 0$

$$\frac{x}{m} \propto P°$$

i.e. the value of n varies between 0 to 1

60. (a) ($AlCl_3$, LiCl & $BeCl_2$)) all these solutions are acidic due to cationic hydrolysis, where as $BaCl_2$, is salt of strong base and strong acid, hence its solution will almost neutral i.e., pH ≈ 7.

61. (c) $$\frac{V_A}{t_A} \Big/ \frac{V_B}{t_B} = \sqrt{\frac{M_B}{M_A}}$$

$$\Rightarrow \frac{200}{150} = \sqrt{\frac{36}{M_A}} \Rightarrow \frac{4}{3} = \sqrt{\frac{36}{M_A}}$$

$$\Rightarrow \frac{16}{9} = \frac{36}{M_A} \Rightarrow M_A = \frac{81}{4} = 20.25$$

So, most approximate answer is 32

62. (c) Electronic configuration of Rb = [Kr] $5s^{-1}$
Set of quantum numbers, n = 5
$\ell = 0, \therefore$ *s* orbital
m = 0, s = + 1/2

63. (c)

Compound	Oxidation number of nitrogen
N_2H_4	= –2
NH_3	= –3
N_3H	= –1/3
NH_2OH	= –1

64. (c) Cannizzaro reaction - when an aldehyde containing no α – H undergo reaction in presence of 50% KOH. It disproportionates to form a molecule of carboxylic acid and a molecule of alcohol.

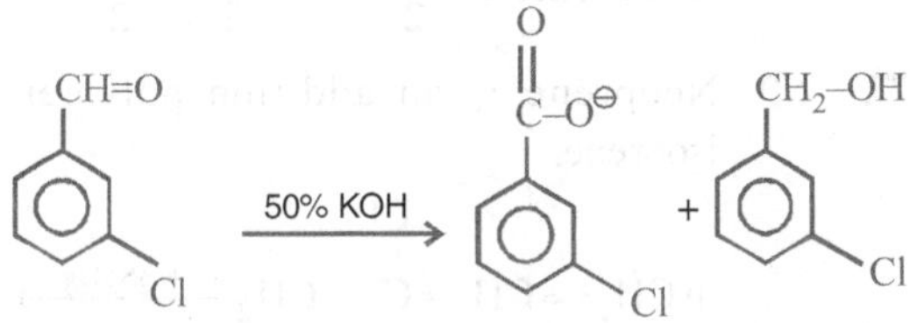

m-chlorobenzaldehyde (CH=O, Cl) $\xrightarrow{50\%\ KOH}$ m-chlorobenzoate ($C(=O)-O^{\ominus}$, Cl) + m-chlorobenzyl alcohol (CH_2-OH, Cl)

65. (d) Anhydrous alcohols add to the carbonyl group of aldehydes in the presence of anhydrous hydrogen chloride to form acetals via hemiacetals.

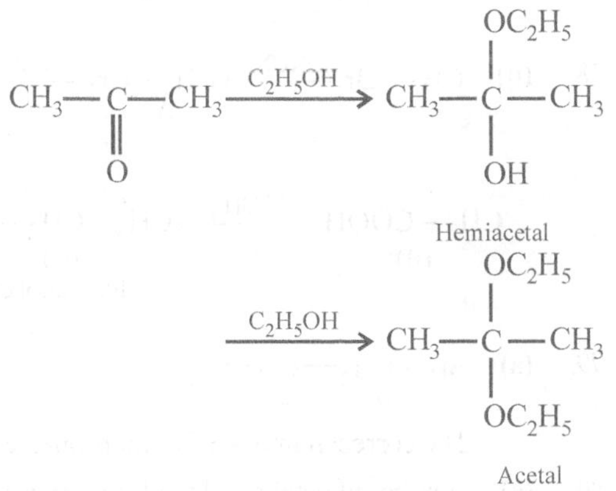

$$CH_3-\underset{\underset{O}{\|}}{C}-CH_3 \xrightarrow{C_2H_5OH} CH_3-\underset{OH}{\overset{OC_2H_5}{C}}-CH_3$$

Hemiacetal

$$\xrightarrow{C_2H_5OH} CH_3-\underset{OC_2H_5}{\overset{OC_2H_5}{C}}-CH_3$$

Acetal

66. (a) For CCP $\sqrt{2}a = 4r$

$$\frac{\sqrt{2}\times 408}{2} = 2r \qquad (2r = \text{Diameter})$$

Diameter = 288.5

67. (d) Enzymes are most reactive at optimum temperature. The optimum temperature for enzyme activity lies between 40°C to 60°C.

68. (b) $\dfrac{r_{100°C}}{r_{10°C}} = 2^{\left(\frac{100-10}{10}\right)} = 2^9 = 512$

69. (b) Beri-Beri.

70. (c) Electrophilic rate order

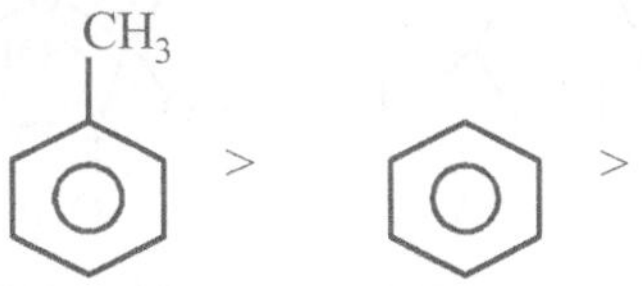

Toluene (CH_3) > Benzene 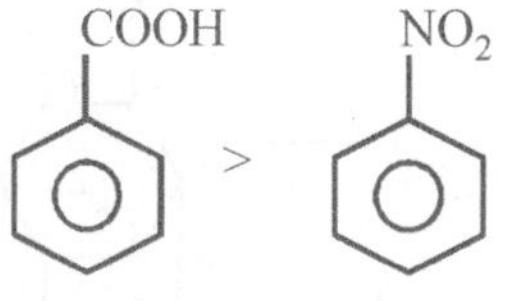 > Benzoic acid (COOH) > Nitrobenzene (NO_2)

Since nitration is an electrophilic substitution hence presence of electron releasing group like CH_3 in the nucleus facilitates nitration.

71. (a) Lets take an example of an acidic buffer CH_3COOH and CH_3COONa.
$CH_3COOH \rightleftharpoons CH_3COO^- + H^+$;
$CH_3COONa \rightleftharpoons CH_3COO^- + Na^+$
when few drops of HCl are added to this buffer, the H^+ of HCl immediatly combine with CH_3COO^- ions to form undissociated acetic acid molecules. Thus there will be no appreciable change in its pH value. Like wise if few drops of NAOH are added, the OH^- ions will combine with H^+ ions to form unionised water molecule. Thus pH of solution will remain constant.

72. (a) $CF_3COOH > CCl_3COOH > HCOOH > CH_3COOH$ (K_a order)
The halogenated fatty acids are much stronger acids than the parent fatty acid and more over the acidity among the halogenated fatty acid is increased almost

proportionately with the increase in electronegativity of the halogen present. Further formic acid having no alky group is more acidic than acetic acid.

73. (b)

α-D-Glucopyranose + β-D-Fructofuranose

$\longrightarrow$ Fructose $+ H_2O$

Sucrose is a disaccharide of α–D-Glucopyranose and β–D-Fructofuranose.

74. (c) $\Delta S = \frac{\Delta H}{T} = \frac{1.435 \times 10^3}{273}$

$= 5.260 \text{ cal/mol} - K$

75. (d) BF_4^- hybridisation sp^3, tetrahedral structure.

NH_4^+ hybridisation sp^3, tetrahedral structure.

76. (b) $(O_2) =$

$$\sigma 1s^2 \sigma^* 1s^2 \ \sigma 2s^2 \sigma^* 2s^2 \ \sigma 2p_z^2 \ \pi 2p_x^2$$

$$= \pi 2p_y^2 \ \pi^* 2p_x^1 = \pi^* 2p_y^1$$

Bond order $= \frac{N_b - N_a}{2} = \frac{10-6}{2} = \frac{4}{2} = 2$

$$\left(O_2^+ \text{ion}\right) = \sigma 1s^2 \ \sigma^* 1s^2 \ \sigma 2s^2 \ \sigma^* 2s^2$$

$$\sigma 2p_z^2 \ \pi 2p_x^2 = \pi 2p_y^2 \ \pi^* 2p_x^1$$

Bond order $\frac{N_b - N_a}{2} = \frac{10-5}{2} = \frac{5}{2} = 2\frac{1}{2}$

$$\left(O_2^-\right) = \sigma 1s^2 \ \sigma^* 1s^2 \ \sigma 2s^2 \ \sigma^* 2s^2 \ \sigma 2P_z^2$$

$$\pi 2p_x^2 = \pi 2p_y^2 \ \pi^* 2p_x^2 \ \pi^* 2p_y^1$$

Bond order

$$= \frac{(N_b - N_a)}{2} = \frac{10-7}{2} = \frac{3}{2} = 1\frac{1}{2}$$

$$\left(O_2^{2-}\right) = \sigma 1s^2 \sigma^* 1s^2 \ \sigma 2s^2 \ \sigma^* 2s^2 \sigma 2p_z^2 \ \pi 2p_x^2$$

$$= \pi 2p_y^2 \ \pi^* 2p_x^2 = \pi^* 2p_y^2$$

Bond order $\frac{N_b - N_a}{2} = \frac{10-8}{2} = \frac{2}{2} = 1$

77. (d) Neoprene is an addition polymer of isoprene.

$$n\,CH_2{=}CH{-}\overset{Cl}{\overset{|}{C}}{=}CH_2 \xrightarrow{K_2S_2O_8} \left(CH_2{-}CH{=}\overset{Cl}{\overset{|}{C}}{-}CH_2\right)_n$$

Chloroprene → Neoprene

78. (d) $CH_3-Br \xrightarrow{KCN} \underset{A}{CH_3-CN} \xrightarrow{H_3O^+}$

$$\underset{(B)}{CH_3-COOH} \xrightarrow{LiAlH_4} \underset{(C)}{CH_3-CH_2-OH}$$

Ethyl alcohol

79. (a) $Br{-}\underset{3}{CH_2}{-}\underset{2}{CH}{=}\underset{1}{CH_2}$

The correct name is 3-Bromoprop-1-ene.

80. (a) Number of octahedral voids in ccp, is equal to effective number of atoms.

In ccp, effective number of atoms are 4 so, 4 octahedral voids.

So, 1 octahedral voids per atom.

81. (a) Cuprous oxide formed during roasting of cuprous sulphide is mixed with few amount of cuprous sulphide and heated in a reverberatory furnace to get metallic copper.

$$2Cu_2O + Cu_2S \rightarrow 6Cu + SO_2(g)$$

82. (b) Invar is a nickel iron alloy, Bell metal is an alloy of about 80% copper and 20% tin, Bronze is also an alloy of copper and tin. Steel : It always have few % of carbon.

83. (c) $\overset{+5}{KClO_3} + H_2\overset{+6}{C_2}O_4 + H_2\overset{+6}{SO_4} \rightarrow K_2SO_4 + \overset{-1}{KCl} + CO_2 + H_2O$

i.e. maximum change in oxidation number is observed in Cl (+5 to −1).

84. (c) All the alkali metals when heated with oxygen form different types of oxides for example lithium forms lithium oxide (Li_2O), sodium forms sodium peroxide (Na_2O_2), while K, Rb and Cs form their respective superoxides.

$$2Li + \frac{1}{2}O_2 \rightarrow Li_2O$$

85. (b) All alkali metal salts are ionic (except Lithium) and soluble in water due to the fact that cations get hydrated by water molecules. The degree of hydration depends upon the size of the cation. Smaller the size of a cation, greater is its hydration energy.

Relative ionic radii :

$$Cs^+ > Rb^+ > K^+ > Na^+ > Li^+$$

Relative ionic radii in water or relative degree of hydration:

$$Li^+ > Na^+ > K^+ > Rb^+ > Cs^+$$

86. (d) The oxidised hydrocarbons and ozone in presence of humidity cause photochemical smog.

Hydrocarbons + O_2, NO_2, NO, O, O_3 → Peroxides, formaldehyde, peroxyacetylnitrate (PAN), acrolein etc.

It is oxidising in nature and causes irritation to eyes, lungs, nose, asthamatic attack and damage plants.

87. (b) CH_3CHO gives Iodoform test but $C_6H_5CH_2CHO$ does not give Iodoform test due to absence of methyl group

88. (b) $Na_2Cr_2O_7$ is hygroscopic.

89. (d) $H_2O(\ell) \rightleftharpoons H_2O(g) + Q$

$\Delta E = 37558\ J/mol$

$\Delta E = 37.56\ kJ\ mol^{-1}$

90. (a) As the positive charge increases on metal cation, radius decreases. This is due to the fact that nuclear charge in the case of a cation is acting on lesser number of electrons and pulls them closer.

91. (b) [H—P(=O)(—H)—O—H] Hypophosphorous acid

(H_3PO_2) is a monobasic acid. i.e., it has only one ionisable hydrogen atom or one OH is present H—P(→O)(—H)—OH

92. (b) The lyophobic sols are less stable than lyophilic sols. The lyophilic sols are thus used to protect the lyophobic sols. This property of lyophilic sols is known as protective action of lyophilic sols.

Which can be represented by gold number.

93. (b) $Fe_2(SO_4)_3 \xrightarrow{\Delta} Fe_2O_3 + SO_3$

94. (d) $P = P_AX_A + P_BX_B$

$= P_AX_A + P_B(1 - X_A)$

$\Rightarrow P_AX_A + P_B - P_BX_A$

$\Rightarrow P_B + X_A(P_A - P_B)$

95. (a) $HOOC-CH=CH-COOH$ (cis) maleic acid (H and H on same side)

$HOOC(H)C=C(H)COOH$ (Trans)

It shows geometrical isomerism but does not show optical isomerism.

96. PCl_3

$:\ddot{P}(\ddot{\underset{..}{Cl}}:)_3$

97. (d) $\overset{0}{\Lambda}_{m(NH_4Cl)} = \overset{0}{\Lambda}_{mNH_4^+} + \overset{0}{\Lambda}_{mCl^-}$

$\overset{0}{\Lambda}_{m(NaOH)} = \overset{0}{\Lambda}_{mNa^+} + \overset{0}{\Lambda}_{mOH^-}$

$\overset{0}{\Lambda}_{(NaCl)} = \overset{0}{\Lambda}_{mNa^+} + \overset{0}{\Lambda}_{mCl^-}$

$\therefore\ \overset{0}{\Lambda}_{m}(NH_4^+) + \overset{0}{\Lambda}_{m}(OH^-)$

$= \overset{0}{\Lambda}_{m}(NH_4^+) + \overset{0}{\Lambda}_{m}(Cl^-) + \overset{0}{\Lambda}_{m}(Na^+)$

$+ \overset{0}{\Lambda}_{m}(OH^-) - \left[\overset{0}{\Lambda}_{m}(Na^+) + \overset{0}{\Lambda}_{m}(Cl^-)\right]$

$\overset{0}{\Lambda}_{m(NH_4OH)} = \overset{0}{\Lambda}_{m(NH_4Cl)} + \overset{0}{\Lambda}_{m(NaOH)}$

$- \overset{0}{\Lambda}_{m(NaCl)}$

98. (a) Both O_2^{2-} and B_2 has bond order equal to 1.

$B_2(10) = [\sigma 1s^2 \sigma^* 1s^2 \sigma 2s^2 \sigma^* 2s^2$
$\pi 2p^1{}_y, \pi 2p_z{}^1]$

$\text{Bondorder} = \dfrac{N_b - N_a}{2}$

$= \dfrac{6-4}{2} = \dfrac{2}{2} = 1$

B_2 is known in the gas phase

$O_2^{--} = \sigma 1s^2 \sigma^* 1s^2 \sigma 2s^2 \sigma^* 2s^2 (\sigma 2p_z)^2$
$(\pi 2p_x)^2 (\pi 2p_y)^2 (\pi^* 2p_x)^2 (\pi^* 2p_y)^2$

$\text{Bond order} = \dfrac{1}{2}(8-6) = 1$

99. (b) Nylon-66 is an example of first synthetic fibres produced from the simple molecules. It is prepared by condensation polymerisation of adipic acid and haxamethylene diamine.

100. (a) $CH_3-C(CH_3)_2-CH=CH_2 \xrightarrow{H^+}$

$CH_3-C(CH_3)_2-\overset{\oplus}{C}H-CH_3 \xrightarrow[-H^+]{H_2O}$ $CH_3-C(CH_3)(H_3C)-CH(OH)-CH_3$ (Minor)

↓ 1,2-methyl shiftting

$CH_3-\overset{\oplus}{C}(CH_3)-CH(CH_3)-CH_3$

A (Minor)

$\xrightarrow[-H^+]{H_2O}$ $CH_3-C(OH)(CH_3)-CH(CH_3)-CH_3$

(Major)

B

BIOLOGY

101. (a) Motile zygote of *Plasmodium* occurs in gut of female *Anopheles*. Zygote formed in stomach of mosquito about 9 to 10 days after sucking the blood of an infected human. *Anopheles* receives RBCs containing different stages of erthyrocytic cycle, including gametocyte. In its gut, all stages except the gametocytes are digested. The gametocytes remain unaffected by digestive enzymes of the mosquito, hatch out from the RBCs into the lumen of mosquito's stomach and form sperm and ovum by gametogenesis. Syngamy or fusion of male and female gamete.

102. (c) The human hind brain comprises of pons, cerebellum and medulla oblongata.

103. (c) Vestibular apparatus is a part of inner ear which has no role in hearing but responsible for the maintenance of balance of the body and posture.

104. (d) The most abundant prokaryotes helpful to humans in making curd from milk and in production of antibiotics are the heterotrophic bacteria. *Lactobacillis* bacteria convert milk into curd.

105. (d)

106. (a) Process of evolution of different species in a given area starting from a point and radiating to other area of geographical areas is called adaptive radiations.
Example : Darwin's finches,
Australian marsupials.

107. (c) The figure shows the tubectomy. This is a surgical method to prevent pregnancy in women. In tubectomy small part of the fallopian tube is removed or tied through a small cut in the abdomen or through vagina. It is very effective method but reversibility is very poor.

108. (d) High concentration of DDT disturbs calcium metabolism in birds which caused thinning of eggshell and their premature breaking.

109. (d) In pBR 322
ori-represents site of origin or replication
rop-represents those proteins that take part in replication of plasmid.
Hind III, EcoRI- Recoginition sites of Restriction endonucleases
amp^R and tet^R - They are antibiotic resistant gene part

110. (b) The common bottle cork is the product of phellogen. Phellogen produces cork or phellem on the outer side. It consists of dead and compactly arranged rectangular cells that possess suberised cells walls. The cork cells contain tannins. Hence, they appear brown or dark brown in colour. The cork cells of some plants are filled with air *e.g.*, *Quercus suber* (Cork Oak or Bottle Cork).

111. (d) Widel test is carried out to test typhoid fever caused by ***Salmonella typhii*** bacteria. Typhoid vaccine is available.

112. (c) Meristem is most suitable for raising virus free plants for micropropagation.

113. (b) Ozone present in upper part of the atmosphere *ie*. stratosphere is benefical for living beings. As it functions as a shield against harmful UV-radiation.

114. (a) Companion cells are narrow, elongated and thin walled **living cells**. They lie on the **sides** of the sieve tubes and are closely associated with them through **compound plasmodesmata**. It is supposed that the nuclei of the companion cells control the activities of the sieve tube through plasmodesmata. Companion cells also help in maintaining a proper **pressure gradient** in the sieve tube elements.

115. (d) Common cold is most infectious viral disease caused by ***Rhino*** viruses which is transmitted through inhalation of droplets from infected person or through contaminated objects. Pneumonia is a bacterial disease caused by *Streptococcus pneumoniae* and *Haemophillus influenzae*. This serious disease of lungs spread by sputum of the patient.

116. (b) Earthworm derives its nutrition from decaying fallen leaves and soil organic matter. Thus the type of feeding in earthworm is detritus.

117. (a) Removal of RNA polymerase III from nucleoplasm will affect the synthesis of tRNA while RNA polymerase I forms r-RNA and RNA polymerase II form m-RNA is Eukaryotes.

118. (d) Photorespiration is absent is C_4 plants. Photorespiration does not produce energy or reducing power. Rather, it consumes energy. Further, it undoes the work of photosynthesis. There is 25% loss of fixed CO_2. Therefore, photorespiration is a highly wasteful process. This happens only in case of C_3 plants.

119. (d)

120. (a) The best defined function of manganese is in the splitting of water to liberate oxygen during photosynthesis. It is absorbed in the form of manganous ions (Mn^{2+}). It activates many enzymes involved in photosynthesis, respiration and nitrogen metabolism.

121. (a) BOD is a measure of organic matter present in water. It refers to amount of O_2 consumed by microbes to decompose all the organic matter in 1 L of water at 20°C for 5 days. The greater the BOD of waste water, more is its polluting potential.

122. (a) Transcription unit consists of promoter, structural gene & terminator.

123. (c) Blood group O acts as universal donor.

124. (c) Golden rice is vitamin A rich variety developed by rDNA technology and used in the treatment of vitamin A deficiency.

125. (c) Nearly all the essential nutrients, and 70-80 percent of electrolytes and H_2O are reabsorbed by proximal convoluted tubules.

126. (a) Papaya is dioecious *i.e.* male and female flowers occurs in separate plants so that it prevents both autogamy & geitonogamy (method of self pollination).

127. (d)

Parietal	Ex -	Mustard
Free central	Ex -	*Primula & Dianthus*
Marginal	Ex -	Pea
Axile	Ex -	Tomato, lemon, Chinarose

128. (a) Epinephrine and norepinephrine are secreted by adrenalmedulla in response to stress of any kind and during emergency situations and are called emergency hormones or hormones of flight, or fight.

129. (b) Phosphorus is mostly used as phosphate. Its **reservoir pool** is phosphate rocks while cycling pool is soil for terrestrial ecosystems and water for aquatic ecosystems.

130. (d) A single strand DNA or RNA tagged with radioactive molecule that is used in of hybridization of DNA or RNA is called probe.

131. (a)

132. (b) Sporopollenin is fatty substance present in pollen wall and provides resistance against extremes conditions like high temperature, acid, bases.

133. (b) *Cycas* (a gymnosperm) and *Adiantum* known as Maiden hair fern is a pteridophyte resemble each other in having motile sperm. Seeds, cambium are common in gymno-sperms and absent in pteridophytes. True vessels are absent in both pteridophytes and gymnospems.

134. (d)

135. (a) Cymose infloresence in present in *Solanum*. Cymose inflorescence is the name of determinate or definite inflorescence in which the tip of the main axis terminates in a flower and further growth continues by one or more lateral branches which also behave like the main axis.

136. (b) Ribosomal RNA is actively synthesized in nucleolus. Nucleolus is also known as ribosomal factory.

137. (c) Recombinase enzyme catalyzes the exchange of short pieces of DNA between two long DNA strands particularly the exchange of homologous regions between the paired maternal and paternal chromosome in prophase - I.

138. (c)

139. (c) Phyllode is present in *Australian Acacia*. It is modification of leaf in which lamina is absent and petiole become flattened and perform the function of food synthesis.

140. (a)

141. (d) *Monascus purpureus* is a yeast used in the production of statins which are used in lowering blood cholestrol.

142. (c) **143. (c)**

144. (b) If a normal visioned man marries a woman whose father was also colourblind. Then his wife would be carried of this disease if her mother was normal. This trait passed in to children but daughters produce by this couple are carrier not the colourblind. 50% of sons would be colourblind.

145. (a) The signals for child birth (parturition) originate from the fully matured foetus and placenta which induce mild uterine contractions called **foetal ejection reflex**.

146. (b) Streptokinase is immediately given to dissolve the thrombus carring myocardial infarction

147. (d) Contact inhibition is a property of normal cell not of cancer cells. Due to this property they remain in contact with other cells inhibit their growth.

148. (d)

149. (d) Four important functional aspects of the ecocystem are
(i) Productivity (ii) Decomposition,
(iii) Energy flow and (iv) Nutrient cycling.

150. (d)
151. (d) Hots are the geographical area where biodiversity is maximum. Two hotspots in India are Western Ghats and North eastern himlayan region.
152. (d)
(1) Gametophyte of bryophytes bears protonemal & leafy stage.
(2) In gymnosperm female gametophyte is not free living.
(3) They are present in *Marchantia* or which is a bryophyte.
(4) Origin of seed habit started in pteridophyte *Selaginaella*.
153. (d) Cell membrane of *E.coli*, a bacteria and *Chlamydomonas* a unicellular green alga does not differ. As in both cell membrane is composed of lipids and proteins.
154. (d) Cyanobacteria are also referred as blue green algae, they perform oxygenic photosynthesis. They are most successful autotrophic organisms on earth which one found in all types of environment - fresh water, sea water, salt marshes, moist rocks, tree trunks, moist soils, hot springs, frozen waters.
155. (d) Test tube baby programme employs zygote intrafallopion transfer (ZIFT) technique. In this technique fusion of ovum and sperm is done outside the body of woman to form zygote which is allowed to divide forming 8 blastomercs, then it is transfered the fallopion tube of the woman.
156. (b) Vector DNA are the DNA molecules that can carry a foreign DNA segment and replicate inside the host cells. Vector DNA may be plasmids, a bacteriophage, cosmids, yeast artificial chromosomes.
157. (a)
158. (a) Polymorphonuclear leukoytes named so, as they have multilobed nucleus or neutrophils and monoytes are the cellular barrier provide innate or nonspecific immunity. Cellular barriers are the internal defence or second live of defence.
159. (c) Phosphorus is a constituent of nucleic acids both DNA and RNA but absent in cell membranes and cell proteins.
160. (a) Anxiety and eating spicy food together in normal healthy man can lead to indigestion which is difficulty in digestion.
161. (d) Leydig cells found in testis is the endocrine portion of the testis which are present in between seminiferous tubules secretes androgens or male sex hormone *e.g.* testosterone.
162. (b) The erythrocytes or RBCs in frog are large, oval and biconvex nucleated cell with respiratory red pigment haemoglobin in its cytoplasm.
163. (a) Molluscans are soft bodied animals. Their body is unsegmented with a distinct head, muscular foot and visceral hump. In *Pila* the buccal cavity contains a resping organ, the radula with transverse rows of teeth.
164. (b) Ribosomes are amembraneous (*ie.* without membrane) cell organelle composed of rRNA and protein. These are found in both prokaryotic and eukaryotic cells. In prokaryotes, ribosomes are 70S type while in eukaryotes, it is 80S type.
165. (b) Long term intake of alcohol causes damage to liver which is known as cirrhosis of liver with continued alcohol intake, there is destruction of hepatocytes and fibroblasts (cell which form fibres) and slimulaties of collagean protein formation.
166. (d) The name of this DNA polymerase is *Taq* polymerase extracted from a thermophilic bacterial.
167. (b) All the viruses are obligate parasite *i.e.* they remain inert outside the host cell. They have ability to syntheize nucleic acids and proteins by using host cellular machinery (ribosomes, tRNAs, aminoacids, energy). Three shapes are found in viruses helical (elongate body *e.g.* T.M.V) cuboidal (short broad body with rhombic rounded, polyhedral shape *e.g.* Poliomyelitis virus) and binal (with birth cuboidal and helical parts *e.g.* many bacteriophages like T_2). Antibiotics have no effect on them, antiviral drugs can only kill them.
168. (a) Onion - Bulb - Undeground stem , Ginger - Rhizome, *Chlamydomonas* - Zoospore Yeast - ascospores
169. (d) Fig-developed by hypanthodium inflorescence; pineapple- developed by spike inflorescence; Mulberry-developed by catkin inflorescence.
170. (d) A combination of a nitrogen base with a pentose sugar is known as **nucleoside**. The nitrogen base combines with the sugar molecule at its carbon atom 1′ in a glycosidic bond (C – N – C) by one of its nitrogen atoms (usually 1 in pyrimidines and 9 in purines). Depending upon the type of pentose sugar, nucleosides are differentiated into **ribonucleosides** and **deoxyribonucleosides**.
171. (c) *Glomus* is endomycorrhiza that helps in absorption of nutrition specially phosphorus from soil.
172. (b)

173. (b) Sequence of DNA is ATCTG then sequence of m-RNA will UAGAC. As adenine base pairs with uracil and guanine with cytosine.

174. (d) Cortisol and testosterone are steroid hormones which are lipid soluble and easily pass through the cell membrane of a target cell and bind to specific intracellular receptor.

175. (d) *Nostoc* is prokaryote while rest are eukaryotes. Nuclear membrane is absent in prokaryotes.

176. (c) Collagen is the most abundant protein of animal world. Rubisco (ribulose biphosphte carboxylase - oxygenase) is not only the most abundant protein in plants but also the whole biosphere.

177. (d)

178. (c) *Saccharomyces cervisiae* is a yeast used in making bread (Baker's yeast) and commercial production of ethanol.

Paramoecium & *Plasmodium* are of animal kingdom while *pencillium* is a fungi. Lichen is composite organism formed from the symbiotic association of an algae and a fungus. *Nostoc* & *Anabaena* are examples of kingdom monera.

179. (a)

180. (c) *Saccharomyces cervisiae* is a yeast used in making bread (Baker's yeast) and commercial production of ethanol.

181. (b) This is structural formula of basic amino acid lysine.

182. (b)

183. (c) Glucagon is secreted by α cells of islets of langerhans and stimulate glycogenolysis *i.e.* breakdown of glycogen into glucose Thymosin hormone secreted from thymus gland stimulates the development of certain kinds of white blood cells involved in producing immunity. It also hostens attainment of sexual maturity.

184. (c) In closed vascular bundle cambium is absent between xylem and phloem.

185. (c) (a) and (c). Neural signals from pneumotaxic center in pons region of brain can limit the duration of inspiration. About 25% of carbodioxide (CO_2) is carried by haemoglobin as carboxyhaemglobin.

186. (d) Remonal of introns by spliceosome in hnRNA and joining of exons by DNA liagse in a defined order during transcription is called splicing.

187. (d) Incomplete (partial or mosaic) dominance is the phenomenon where none of the two contrasting alleles or factors is dominant. The expression of the character in a hybrid or F_1 individual is intermediate or a fine mixture of the expression of the two factors (as found in homozygous state). Incomplete or mosaic inheritance is not an example of pre-mendelian concept of blending inheritance because the parental types reappear in the F_2 generation. It is however, considered by some workers to be an example of quantitative inheritance where only a single gene pair is involved. F_2 phenotypic ratio is 1 : 2 : 1, similar to genotypic ratio.

(i) Crl incomplete dominance is reported in flowers of Four O'Clock or *Mirabilis jalapa* and *Antirrhinum majus* (Snapdragon or Dog flower).

188. (a) Diagram represents crossing over that takes place in pachytene stage of prophase - I during meiosis.

189. (a) People going to hilly areas will have polycythemia (more number of RBC in blood) after six months and their Hb has a lower binding affinity to O_2.

190. (d) For gene transfer into the host cell without using vector microparticles made of tungsten and Gold coated with foregin DNA are bombarded into target cells at a very high velocity.

191. (b) *Anabaena azollae* resides in the leaf cavities of the fern *Anabaena*. It fixes nitrogen. A part of the fixed nitrogen is excreted in the cavities and becomes available to the fern. The decaying fern plants release the same for utilization of the rice plants.

192. (b) Major causative factors of osteoporosis are imbalances of hormones like calcitonin of thyroid, parathormone of parathyroids, and sex hormones and deficiencies of calcium and vitamin D.

193. (b) Stem of maize has water containing cavities in vascular bundles.

194. (d) Fluid mosaic membrane model was proposed by Singer and Nicholson in 1972. It is most accepted model of structure of biomembrane.

195. (d)

196. (a) Liquid endosperm is edible part of coconut.

197. (a) **198. (a)** **199. (b)** **200.** (c)

HINTS & SOLUTIONS (MAINS)

PHYSICS

1. (c) $(\mu_0\varepsilon_0)^{-1/2} = \dfrac{1}{\sqrt{\mu_0\varepsilon_0}} = C$: Speed of light

where, ε_0 = permittivity of free space
μ_0 = permeability of free space
So dimensions $[LT^{-1}]$

2. (b) Momentum $P = mv = m\sqrt{2gh}$
($\because v^2 = u^2 + 2gh$; Here $u = 0$)
When stone hits the ground momentum

$$P = m\sqrt{2gh}$$

when same stone dropped from $2h$ (100% of initial) then momentum

$$P' = m\sqrt{2g(2h)} = \sqrt{2}P$$

Which is changed by 41% of initial.

3. (d) For smooth driving maximum speed of car v Then

$$\frac{mv^2}{R} = \mu_s mg$$

$$\therefore \quad v = \sqrt{\mu_s Rg}$$

4. (b) Constant power of car $P_0 = F.V = ma.v$

$$P_0 = m\frac{dv}{dt}.v$$

$P_0 dt = mvdv$ Integrating

$$P_0.t = \frac{mv^2}{2}$$

$$v = \sqrt{\frac{2P_0 t}{m}}$$

$\because P_0$, m and 2 are constant

$$\therefore \quad v \propto \sqrt{t}$$

5. (c) Using conservation principle
$L_i = 0$ (Initial momentum)
$L_f = mvR - I\omega$ (Final momentum)
According to the conservation of momentum
$L_i = L_f$

$$\Rightarrow mvR - I.\omega = 0$$

$$mvR = I.\omega$$

$$\omega = \left(\frac{1}{2}\right)$$

$$(v + \omega R)t = 2\pi R$$

$$\left(1 + \frac{1}{2}\times 2\right) = 2\pi \times 2$$

$$t = 2\pi \text{ sec.}$$

6. (a) According to parallel axis theorem of the moment of Inertia

$$I = I_{cm} + md^2$$

d is maximum for point B so I_{max} about B.

7. (a) $X_{cm} = \dfrac{m_1x_1 + m_2x_2 + m_3x_3}{m_1 + m_2 + m_3}$

$$X_{cm} = \frac{300\times(0) + 500(40) + 400\times 70}{300 + 500 + 400}$$

$$X_{cm} = \frac{500\times 40 + 400\times 70}{1200}$$

$$X_{cm} = \frac{50 + 70}{3} = \frac{120}{3} = 40 \text{ cm}$$

8. (d) $v_e = \sqrt{\dfrac{2GM}{R}} \Rightarrow v_0 = \sqrt{\dfrac{GM}{R}}$

$$v_e = \sqrt{2}v_0$$

9. (b) The Gravitational field due to a thin spherical shell of radius R at distance r.

$$F = \frac{GM}{r^2} \text{ (If } r > R)$$

For $r = R$ i.e. on the surface of the shell

$$F = \frac{GM}{R^2}$$

For $r < R$ i.e. inside the shell
$F = 0$

10. (a)

Rate of heat given by steam = Rate of heat taken by ice
where K = Thermal conductivity of the slab
m = Mass of the ice
L = Latent heat of melting/fusion
A = Area of the slab

$$\frac{dQ}{dt} = \frac{KA(100-0)}{1} = m\frac{dL}{dt},$$

$$\frac{K \times 100 \times 0.36}{0.1} = \frac{4.8 \times 3.36 \times 10^5}{60 \times 60}$$

$K = 1.24$ J/ms °C

11. (a) Initial and final condition is same for all process

$\Delta U_1 = \Delta U_2 = \Delta U_3$

From first law of thermodynamics

$\Delta Q = \Delta U + \Delta W$

Work done

$\Delta W_1 > \Delta W_2 > \Delta W_3$ (Area of P.V. graph)

So $\Delta Q_1 > \Delta Q_2 > \Delta Q_3$

12. (b) $y = 3\sin\frac{\pi}{2}(50t - x)$

or $y = 3\sin\left(25\pi t - \frac{\pi}{2}x\right)$ on comparing with the standard wave equation

$y = a\sin(\omega t - kx)$

Wave velocity $v = \frac{\omega}{k} = \frac{25\pi}{\pi/2} = 50$ m/sec.

The velocity of particle

$$v_p = \frac{\partial y}{\partial t} = 75\pi\cos\left(25\pi t - \frac{\pi}{2}x\right)$$

$v_{p\,max} = 75\pi$

then $\frac{v_{p\max}}{v} = \frac{75\pi}{50} = \frac{3\pi}{2}$

13. (c) Frequency of the echo detected by the driver of the train is (According to Doppler effect in sound)

$$f' = \left(\frac{v+u}{v-u}\right)f$$

where f = Original frequency of source of sound

f' = Apparent frequency of source because of the relative motion between source and observer.

$$f' = \left(\frac{330+220}{330-220}\right)1000 = 5000 \text{ Hz}$$

14. (c) The energy stored by a capacitor

$$U = \frac{1}{2}CV^2 \quad \text{...(i)}$$

V is the p.d. between two plates of the capacitor.

The capacitance of the parallel plate capacitor

$$C = \frac{A\varepsilon_0}{d}$$

Substituting the value of C in equation (i)

$$U = \frac{1}{2}\frac{A\varepsilon_0}{d}(Ed)^2 = \frac{1}{2}\frac{A\varepsilon_0 E^2 d}{d} \quad (\because V = E.d.)$$

15. (b) At equilibrium, potential of both spheres becomes same if charge of sphere one x and other sphere $Q - x$ then (where $Q = 4 \times 10^{-2}$ C)

$v_1 = v_2$

$$\frac{kx}{1\text{cm}} = \frac{k(Q-x)}{3\text{cm}}$$

$3x = Q - x \Rightarrow 4x = Q$

$$x = \frac{Q}{4} = \frac{4 \times 10^{-2}}{4}C = 1 \times 10^{-2}$$

$Q' = Q - x = 3 \times 10^{-2}C$

16. (c) The power dissipated in the circuit.

$$P = \frac{V^2}{R_{eq}}$$

...(i)

$v = 10$ volt

$$\frac{1}{R_{eq}} = \frac{1}{R} + \frac{1}{5} = \frac{5+R}{5R}$$

$$R_{eq} = \left(\frac{5R}{5+R}\right)$$

$P = 30$ W

Substituting the values in equation (i)

$$30 = \frac{(10)^2}{\left(\frac{5R}{5+R}\right)}$$

$$\frac{15R}{5+R} = 10$$

$15R = 50 + 10R$

$5R = 50$

$R = 10\,\Omega$

17. (c) The current through the resistance R

$$I = \left(\frac{\varepsilon}{R+r}\right)$$

The potential difference across R

$$V = IR = \left(\frac{\varepsilon}{R+r}\right)R$$

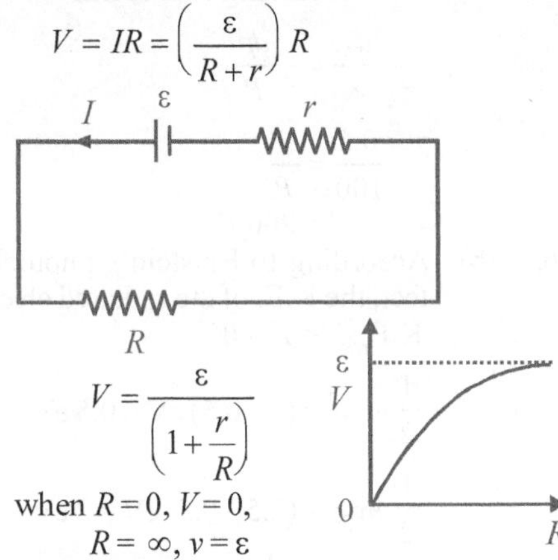

$$V = \frac{\varepsilon}{\left(1+\frac{r}{R}\right)}$$

when $R = 0, V = 0,$
$R = \infty, v = \varepsilon$

Thus V increases as R increases upto certain limit, but it does not increase further.

18. (b) According to the principle of circular motion in a magnetic field

$$F_c = F_m \Rightarrow \frac{mv^2}{R} = qVB$$

$$\Rightarrow R = \frac{mv}{qB} = \frac{P}{qB} = \frac{\sqrt{2m.k}}{qB}$$

$$R_\alpha = \frac{\sqrt{2(4m)K'}}{2qB}$$

$$\frac{R}{R_\alpha} = \sqrt{\frac{K}{K'}}$$

But $R = R_\alpha$ (given)
Thus $K = K' = 1$ MeV

19. (b) According to work energy theorem
$W = U_{final} - U_{initial} = MB(\cos 0 - \cos 60°)$

$$W = \frac{MB}{2} = \sqrt{3}\text{J} \quad ...(i)$$

$$\tau = \vec{M}\times\vec{B} = MB\sin 60° = \left(\frac{MB\sqrt{3}}{2}\right) \quad ...(ii)$$

From equation (i) and (ii)

$$\tau = \frac{2\sqrt{3}\times\sqrt{3}}{2} = 3\text{J}$$

20. (d) The average power in the circuit where $\cos\phi$ = power factor
$< P > = V_{rms} \times I_{rms}\cos\phi$

$$\phi = \pi/3 = \text{phase difference} = \frac{180}{3} = 60$$

$$V_{rms} = \frac{\frac{1}{\sqrt{2}}}{\sqrt{2}} = \frac{1}{2}\text{volt}$$

$$I_{rms} = \frac{\frac{1}{\sqrt{2}}}{\sqrt{2}} = \left(\frac{1}{2}\right)A$$

$$\cos\phi = \cos\frac{\pi}{3} = \frac{1}{2}$$

$$< P > = \frac{1}{2}\times\frac{1}{2}\times\frac{1}{2} = \frac{1}{8}W$$

21. (b) The charge through the coil = area of current-time$(i-t)$ graph

$$q = \frac{1}{2}\times 0.1\times 4 = 0.2\text{ C}$$

$$q = \frac{\Delta\phi}{R} \quad \because \text{Change in flux } (\Delta\phi) = q\times R$$

$$q = 0.2 = \frac{\Delta\phi}{10}$$

$\Delta\phi = 2$ Weber

22. (b) The average energy stored in the electric field $U_E = \frac{1}{2}\varepsilon_0 E^2$
The average energy stored in the magnetic field $= U_B = \frac{1}{2}\frac{B^2}{\mu_0}$,
According to conservation of energy
$U_E = U_B$

$$\varepsilon_0\mu_0 = \frac{B^2}{E^2}$$

$$\frac{B}{E} = \sqrt{\varepsilon_0\mu_0} = \frac{1}{C}$$

23. (b)

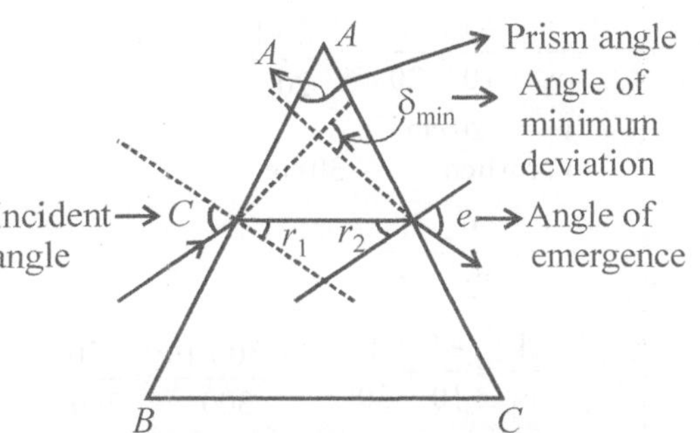

The angle of minimum deviation is given as
$\delta_{min} = i + e - A$
for minimum deviation
$\delta_{min} = A$
then $2A = i + e$

in case of δ_{min} $i = e$

$$2A = 2i \quad r_1 = r_2 = \frac{A}{2}$$

$$i = A = 90°$$

From Snell's law

$1 \sin i = n \sin r_1$

$$\sin A = n \sin \frac{A}{2}$$

$$2 \sin\frac{A}{2}\cos\frac{A}{2} = n\sin\frac{A}{2}$$

$$2 \cos\frac{A}{2} = n$$

when $A = 90° = i_{min}$

then $n_{min} = \sqrt{2}$

when $i = A = 0$ then, $n_{max} = 2$

24. (d)

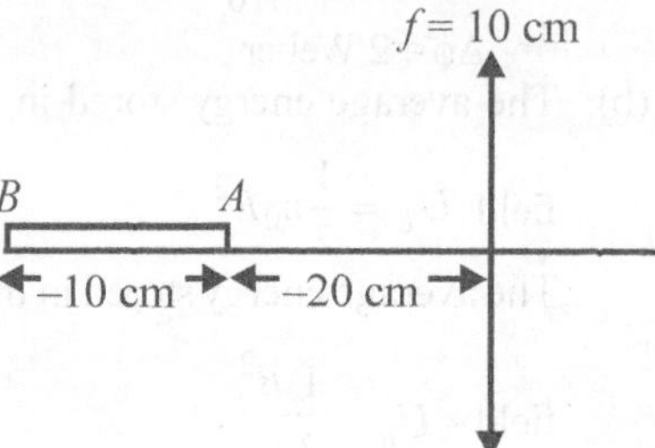

The focal length of the mirror

$$-\frac{1}{f} = \frac{1}{v} + \frac{1}{u}$$

For A end of the rod the image distance

When $u_1 = -20$ cm

$$\Rightarrow \frac{-1}{10} = \frac{1}{v_1} - \frac{1}{20}$$

$$\frac{1}{v_1} = \frac{-1}{10} + \frac{1}{20} = \frac{-2+1}{20}$$

$v_1 = -20$ cm

For when $u_2 = -30$ cm

$$\frac{1}{f} = \frac{1}{v_2} - \frac{1}{30}$$

$$\frac{1}{v_2} = \frac{-1}{10} + \frac{1}{30} = \frac{-30+10}{300} = \frac{-20}{300}$$

$v_2 = -15$ cm

$L = v_2 - v_1 = -15 - (-20)$

$L = 5$ cm

25. (a) The de-Broglie's wavelength associated with the moving electron $\lambda = \frac{h}{P}$

Now, according to problem

$$\frac{d\lambda}{\lambda} = -\frac{dp}{P}$$

$$\frac{0.5}{100} = \frac{P}{P'}$$

$$P' = 200\,P$$

26. (b) According to Einstein's photoelectric effect, the K.E. of the radiated electrons

$K.E_{max} = E - W$

$$\frac{1}{2}mv_1^2 = (1 - 0.5)\,\text{eV} = 0.5\,\text{eV}$$

$$\frac{1}{2}mv_2^2 = (2.5 - 0.5)\,\text{eV} = 2\,\text{eV}$$

$$\frac{v_1}{v_2} = \sqrt{\frac{0.5}{2}} = \frac{1}{\sqrt{4}} = \frac{1}{2}$$

27. (d) $\because$ The frequency of the transition $v \propto \frac{1}{n^2}$, where $n = 1, 2, 3$..........

28. (b) $N_1 = N_0\,e^{-\lambda t}$ $\quad N_1 = \frac{1}{3}N_0$

$$\frac{N_0}{3} = N_0 e^{-\lambda t_2}$$

$$\Rightarrow \frac{1}{3} = e^{-\lambda t^2} \quad ...(i)$$

$$N_2 = \frac{2}{3}N_0$$

$$\frac{2}{3}N_0 = N_0 e^{-\lambda t_1}$$

$$\Rightarrow \frac{2}{3} = e^{-\lambda t_1} \quad ...(ii)$$

Dividing equation (i) by equation (ii)

$$\frac{1}{2} = e^{-\lambda(t_2 - t_1)}$$

$$\lambda(t_2 - t_1) = \ln 2$$

$$t_2 - t_1 = \frac{\ln 2}{\lambda} = T_{1/2} = 50 \text{ days}$$

29. (a) Voltage gain $(A_V) = \frac{V_{out}}{V_{in}} = \frac{I_{out}}{I_{in}} \times \frac{R_{out}}{R_{in}}$

$$A_V = \frac{2\times10^{-3}}{40\times10^{-6}} \times \frac{4\times10^3}{100} = 2\times100 = 2000$$

30. (b) When $A = 1, B = 0, C = 1$ then $Y = 1$

$y^1 = 1 + 0 = 1$

$y = y^1 C = 1.1 = 1$

CHEMISTRY

31. (c) $2SO_2 + O_2 \rightleftharpoons 2SO_3 \quad K = 278 \text{ (given)}$

$$SO_3 \rightleftharpoons SO_2 + \frac{1}{2}O_2 \quad K' = \left(\sqrt{\frac{1}{K}}\right)$$

$$= \sqrt{\frac{1}{278}} = \sqrt{35.97 \times 10^{-4}} = 6 \times 10^{-2}$$

32. (d) $A^{2+} = \frac{1}{4} \times 8 = 2$

$B^+ = 4 \times 1 = 4$

$O^{2-} = 8 \times \frac{1}{8} + 6 \times \frac{1}{2} = 4$

Hence the formula of the oxide will be $A_2B_4O_4$ or AB_2O_2.

33. (c) $A_2 + B_2 \rightleftharpoons 2AB \quad K_c = \frac{[AB]^2}{[A_2][B_2]}$

$$K_c = \frac{(2.8 \times 10^{-3})^2}{3 \times 10^{-3} \times 4.2 \times 10^{-3}}$$

$$= \frac{(2.8)^2}{3 \times 4.2} = 0.62$$

34. (b, d)

According to Arrhenius equation

$$\ln\frac{K_2}{K_1} = \frac{E_a}{R}\left(\frac{1}{T_1} - \frac{1}{T_2}\right)$$

$$= -\frac{E_a}{R}\left(\frac{1}{T_2} - \frac{1}{T_1}\right)$$

$$\ln\frac{K_1}{K_2} = -\frac{E_a}{R}\left(\frac{1}{T_1} - \frac{1}{T_2}\right)$$

35. (a) $O_2 = KK(\sigma 2s)^2 (\sigma^* 2s)^2 (\sigma 2p_z)^2 (\pi 2p_x)^2 (\pi 2p_y)^2 (\pi^* 2p_x)^1 (\pi^* 2p_y)^1$

$O_2^- = KK(\sigma 2s)^2 (\sigma^* 2s)^2 (\sigma 2p_z)^2 (\pi 2p_x)^2 (\pi 2p_y)^2 (\pi^* 2p_x)^2 (\pi^* 2p_y)^1$

36. (a) Higher the value of reduction potential higher will be the oxidising power whereas the lower the value of reduction potential higher will be the reducing power.

37. (b) $r \propto \sqrt{\frac{1}{M}}$

$$\frac{r_1}{r_2} = \sqrt{\frac{M_2}{M_1}}$$

$$\frac{3r_1}{r_1} = \sqrt{\frac{M_2}{4}}$$

$$9 = \frac{M_2}{4}$$

$M_2 = 36$ g/mole

38. (a) Orbital angular momentum

$$= \frac{h}{2\pi}\sqrt{\ell(\ell+1)}$$

For p orbital $\ell = 1$

$$\text{So,} = \frac{h}{2\pi}\sqrt{2} = \frac{h}{\sqrt{2}\pi}$$

39. () $n_{CHCl_3} = \frac{25.5}{119.5} = 0.213$

$$n_{CH_2Cl_2} = \frac{40}{85} = 0.47$$

$$P_T = P_A^\circ X_A + P_B^\circ X_B$$

$$= 200 \times \frac{0.213}{0.683} + 41.5 \times \frac{0.47}{0.683}$$

$$= 62.37 + 28.55$$

$$= 90.92$$

40. (d) $\Lambda^0_{CH_3COOH} = \Lambda^0_{CH_3COONa} + \Lambda^0_{HCl} - \Lambda^0_{NaCl}$

$$= 91 + 425.9 - 126.4 = 390.5$$

41. () None

Molar mass ↑, 'a' increases

size of molecule ↑, 'b' increase

b(L/mol)	a(bar. L^2/mol^2)
$H_2 \rightarrow 0.02661$	$CH_4 \rightarrow 2.25$
$He \rightarrow 0.0237$	$O_2 \rightarrow 1.36$
$O_2 \rightarrow 0.03183$	$H_2 \rightarrow 0.244$
$CO_2 \rightarrow 0.04267$	

42. (d) $\frac{C_P}{C_V} = \frac{\frac{5}{2}R}{\frac{3}{2}R} = \frac{5}{3} = 1.67$

43. (c) Nickel ions are frequently detected by the formation of red precipitate of the complex of nickel dimethylglyoxime, when heated with dimethylglyoxime.

$$\begin{array}{l} CH_3-C=NOH \\ \quad\;\;\; | \\ CH_3-C=NOH \end{array} + Ni^{++} \longrightarrow$$

Dimethylglyoxime

$$\begin{array}{ccccc} & OH & & O & \\ & | & & \uparrow & \\ CH_3-C=N & \searrow & & \swarrow N=C-CH_3 & \\ | \quad\; | & & Ni & & | \\ CH_3-C=N & \nearrow & & \searrow N=C-CH_3 & \\ & \downarrow & & | & \\ & O & & OH & \end{array}$$

Nickel dimethylglyoxime

44. (b) $d^6 - t_{2g}^{\;2,2,2}\, e_g^{\;0,0}$ (in low spin)

C.F.S.E $= -0.4 \times 6\Delta_0 + 3P$

$= -\frac{12}{5}\Delta_0 + 3P$

45. (c) The melting points of the transition element first rise to a maximum and then fall as the atomic number increases manganese have abnormally low melting points.

46. (c) $E^0_{Cu^{+2}/Cu} = 0.34$ volt,

other has – ve $E^0_{R.P.}$

$E^0_{Co^{++}/Co} = -0.28$

$E^0_{Ni^{++}/Ni} = -0.25$

$E^0_{Fe^{++}/Fe} = -0.44$

47. (c) Tertiary halide can show ionic reaction with MF so, MF should be most ionic for reaction to proceed forward. Hence 'M' should be 'Rb'.

48. (b) If acidic nature is high, K_a is high and pk_a is low

	H_2O	H_2S
K_a	1.8×10^{-6}	1.3×10^{-7}
	H_2Se	H_2Te
K_a	1.3×10^{-4}	2.3×10^{-3}

since $pk_a = -\log k_a$

Hence the order of pk_a will be

$H_2O > H_2S > H_2Se > H_2Te$

49. (d) Bond order $= \frac{N_b - N_a}{2}$

$He_2^+ = \sigma(1s)^2\, \sigma^*(1s)$ B.O. $= 0.5$

$O_2^- = KK\sigma(2s)^2\; \sigma^*(2s)^2\, \sigma(2p_z)^2\, \pi(2p_x)^2$ $\pi(2p_y)^2\;\; \pi^*(2p_x)^2\;\; \pi^*(2p_y)^1$ B.O. = 1.5

NO $= KK\, \sigma(2s)^2\;\; \sigma^*(2s)^2\, \pi(2p_x)^2\, \pi(2p_y)^2$ $\sigma(2p_z)^2\, \pi^*(2p_x)^1$ B.O. $= 2.5$

$C_2^{2-} = KK\, \sigma(2s)^2\;\; \sigma^*(2s)^2\, \pi(2p_x)^2\, \pi(2p_y)^2$ $\sigma(2p_z)^2$ B.O. $= 3.0$

50. (c) The transition metals and their compounds are used as catalysts. Because of the variable oxidation states, due to this, they easily absorb and re-emit wide range of energy to provide the necessary activation energy.

51. (c) Ac (89) = [Rn] $[6d^1]$ $[7s^2]$

52. (c) $\Delta G = -nFE°$

for the reaction $n = \frac{2}{3} \times 2 \times 3 = 4$

$960 \times 10^3 = -4 \times 96500 \times E°$

$E° = -2.5$ volt

So, it needed 2.5 volt for reduction

53. (d) Chloroamphenicol is a broad spectrum antibiotic.

54. (a) $C_6H_5COCl \xrightarrow[\text{Pd–BaSO}_4]{H_2} C_6H_5CH{=}O$

It is Rosenmund reaction.

55. (b) Biodegradable polymer is Nylon-2-Nylon-6 which is copolymer of glycine (H_2N-CH_2-COOH) and amino caproic acid $(H_2N-(CH_2)_5-COOH)$.

$nH_2N-CH_2-COOH +$ (glycine) $nH_2N-(CH_2)_5-COOH$ (amino caproic acid)

$\downarrow$

$\left(HN-CH_2-\overset{O}{\overset{\|}{C}}-HN-(CH_2)_5-\overset{O}{\overset{\|}{C}} \right)_n$

nylon – 2 – nylon – 6

56. (a) $CH_3-\underset{\underset{CH_3\,(A)}{|}}{CH}-NH_2 \xrightarrow{HNO_2} CH_3-\underset{\underset{CH_3}{|}}{CH}-OH + N_2\uparrow$

isopropyl amine; isopropyl alcohol

$\xrightarrow{CHCl_3/KOH} CH_3-\underset{\underset{CH_3}{|}}{CH}-N\equiv C$

isopropyl isocyanide

$\xrightarrow{\text{reduction}} CH_3-\underset{\underset{CH_3}{|}}{CH}-NH-CH_3$

Isopropyl methyl amine

57. (a) 1-Butyne and 2-butyne are distinguish by $NaNH_2$ because 1-Butyne react with $NaNH_2$ due to presence of terminal hydrogen.

$CH_3-CH_2-C\equiv CH + NaNH_2 \longrightarrow$
1– Butyne

$CH_3CH_3-C\equiv CNa + NH_3$

58. (d) $R-CH=O + H_2N-NH_2 \rightarrow R-CH=N-NH_2$

Such reactions take place in slightly acidic medium and involve nucleophilic addition of the ammonia derivative.

59. (a, d)

It is iodoform reaction. Acetophenone $\left[CH_3-\underset{\underset{O}{\|}}{C}-C_6H_5 \right]$ and 2-Hydroxypropane $\left[CH_3-\underset{\underset{OH}{|}}{CH}-CH_3 \right]$ both give a yellow precipitate of CHI_3 (iodoform) with iodine and alkali.

60. (b) Glycol is used as an antifreeze in automobiles.

BIOLOGY

61. (c) Autotrophs are those organisms that are able to make energy-containing organic molecules from inorganic raw material by using basic sunlight. *Nostoc, Chara, Porphyra* and *Wolffia* are photoautotrophs while *Nitrosomonas* and *Nitrobacter* are chemoautotrophs.

62. (a) Statement (A) and (B) are correct

Riccia is liverwort in which simplest sporophyte consists of capsule only while *Polytrichum* is moss in which sporophyte consists of foot seta & capsule. *Volvox* is a fresh water green colonial alga. Reproduction is both sexual and asexual. Sexual reproduction is oogamous. Slime molds are consumer decomposer protists. They possess characters of plants (cellulosic cell wall), animals (phagotrophic nutrition) and fungi (spores).

63. (a) Archegonia are not formed in *Ginkgo.*

64. (b) *Chlamydomonas* & *Chlorella* have been included in algae. Algae are chlorophyllus, thalloid avascular plants with no cellular differentiation. Algae belong to thallophyta of plant kingdom.

65. (a) Carboxypeptidase is an exopeptidase enzyme secreted by the pancreas that acts only on the peptide linkage of a terminal amino acid containing a free carboxyl group. Zinc is a cofactor for carboxypeptidase. In digestion carboxypeptidase, requires zinc for its protein degrading action.

66. (a) Ribosome are small naked (non membrane bound) particles made of r-RNA and proteins. Ribosomes are also seen in the organelles like mitochondria and chloroplasts. They are the cell's protein factories and are found on RER and scattered in the cytoplasm as polyribosomes. Ribosomes are the sites at which information carried in the genetic code is converted into protein molecules.

67. (d) Change in single base pair of DNA is also a type of mutations called point mutations. It is a type of mutation that causes the replacement of a single base nucleotide with another nucleotide of the genetic material, DNA or RNA. For example, a point mutation is the cause of sickle cell disease.

68. (a) A test cross, first introduced by Gregor Mendel, is used to determine if an individual exhibiting a dominant trait is homozygous or heterozygous for that trait. It takes place between F_1 - Generation and recessive parent.

69. (a) Statement A & D is correct

Regulation of lac operon by repressor is referred as negative regulation. In negative regulation, a repressor molecule binds to the operator of an operon and terminates transcription. In positive regulation, an activator interacts with the RNA polymerase in the promoter region to initiate transcription.

Human genome contains some 20,000-25,000 genes billion bases.

70. (c) Onion - Bulb - Underground stems - Imbricate aestivation -Axile placentation - Member of Liliaceae.

71. (c) Marginal placentation is a placentation with ovules borne on the wall along the ventral suture of a simple ovary. Gram, Arhar, Sun hemp, Moong, Pea & Lupin belongs to fabaceae family that bears marginal placentation.

72. (c) All the four statements are correct.

73. (b) Flowering is induced by light and temperature.

74. (c) Gibberellins & ABA are antagonistic with each other. ABA counteracts many effects of gibberellins like induction of hydrolases and alpha- amylases in barley seedlings.

75. (a) The vascular bundles are arranged in a loose circle inside the endodermis of a monocot root. In a monocot root, more than six vascular bundles are present. It shows polyarch condition

76. (c) Nitrogenases are enzymes used by some organisms to fix atmospheric nitrogen gas. It is the only known family of enzymes that accomplish this process. Nitrogenase requires high input of energy & anaerobic condition.

77. (c) Carrot is a biennial plant which flowers in the second season. Vernalization is the subjection of seeds or seedlings to low temperature in order to hasten plant development and flowering. Vernalization is commonly used for crop plants such as winter rye and is possible because the seeds and buds of many plants require cold in order to break dormancy. Winter cereals are therefore sown in

the autumn for flowering the following year. However, if germinating seeds are artificially vernalized they can be sown in the spring for flowering the same year. Biennial plants, such as carrot (Daucus carota), will remain in their non - flowering rosette form until subjected to cold treatment.

78. **(c)** The germ pores are apertures in the exine layer of the pollen grain where the sporopollenin is absent. The germ pore helps in the formation of the germ tube or pollen tube and which makes its exit on germination.

79. **(a)** In over 60 per cent of angiosperms, pollen grains are shed at cell 2-celled stage. In the remaining species the generative cell divides mitotically to give rise to the two male gametes before pollen grains are shed (3-celled stage.)

80. **(d)** Wind pollinated flowers have generally single ovule in each ovary.

81. **(d)** An area with particular types of trees dedicated to local deities or ancestral spirits that are protected by local communities through social traditions and taboos incorporating spiritual and ecological values are called as sacred groves.

Sacred groves act as an ideal centre for biodiversity conservation. Several plants and animals that are threatened in the forest are still well conserved in some of the sacred groves. It has been observed that several medicinal plants that are not to be found in the forest are abundant in the sacred groves. Further, rare, endangered, threatened and endemic species are often concentrated in sacred groves.

82. **(b)** At the trophic level of consumers the rate at which food energy is assimilated is called secondary productivity. Rabbit is a consumer.

83. **(a)** *Cuscuta*, or Dodder plant, is a parasitic plant that wraps around other plants for nourishment. *Cuscuta* is found on outer side of the host and is total stem parasite. *Cuscuta* a parasite of *Acacia.*

84. **(d)** A hydrosere is a plant succession which occurs in a freshwater lake. Second stage of hydrosere is submerged stage that is represented by *Vallisneria, Hydrilla, Elodea, Utricularia etc.*

85. **(a)** Green revolution is the introduction of high-yielding varieties of seeds and the increased use of fertilizers and irrigation, which provided the increase in production needed to make India self-sufficient in food grains, thus improving agriculture in India. Green revolution in India occurred in 1960.s

86. **(b)** In gobar gas the maximum amount of methane which is produced by methanogenic bacteria. Gobar gas is a gas mixture which is generated when organic compounds are fermented in the absence of air (anaerobic fermentation). Biogas contains 50 – 70% methane, 30 – 40% carbon dioxide, and traces of hydrogen, hydrogen sulphide and nitrogen. Methane is a combustible gas, which means it can be burned. It can be used as a fuel for cooking and lighting.

87. **(a)** Statement (A) and (B) are wrong.

Colostrum is the first lacteal secretion (thin yellowish fluid) produced by the mammary gland of a mother prior to the production of milk. It is recommended for the new born because it is rich in antibodies and minerals.

Chikungunya fever is a viral disease transmitted to humans by the bite of infected *Aedes Aegypti* mosquitoes.

88. **(a)** RNA interference technique, sense & antisense RNA fused to form dsRNA that silent the expression of m- RNA of nematode. RNA interference is a novel strategy adopted to prevent infestation of nematode *Meloidegyne incognitia* in roots of tobacco plants. .

89. **(d)** Biolistic it is direct gene transferred method for constructing recombinant DNA. The gene gun was invented by John C. Sanford with Edward Wolf. A gene gun can be used to genetically infect cells or whole organisms with foreign DNA by aiming the barrel of the gun and firing. The microshot projectiles in the biolistic gene gun are made of microscopic (or nano) sized gold or platinum powders. These expensive powders are soaked in DNA or RNA (in raw or plasmid form) that are engineered for insertion into the genome of the cells or organisms under the gun.

90. **(a)** Antibiotics are powerful medicines that fight bacterial infections. They either kill bacteria or keep them from reproducing. In genetic engineering, the antibiotics are used as selectable markers.

91. (d) Sea horse and flying fish are cold blooded animals. *Ornithorhyncus* is oviparous. Crocodile has four chambered heart. *Ascaris* and *Ancylostoma* are segmented roundworms.

92. (a) All sponges are marine and have collared cells without any exception. Sponges are animals of the phylum Porifera. They are multicellular organisms which have bodies full of pores and channels allowing water to circulate through them, consisting of jelly-like mesohyl sandwiched between two thin layers of cells.

93. (c) *Plasmodium falciparum* – A protozoan pathogen causes the most serious type of malaria that is falciparum malaria.

Musca domestica – House fly, an insect belongs to arthropoda phylum.

Felis tigris - The Bengal tiger, well protected in Sundarbans (Tiger reserve)

E.coli - Full name *Escherichia coli, a* commonly occurring bacterium in human intestine.

94. (c) Centrioles serve as basal bodies for cilia and flagella. They are concerned with spindle formation during cell division. Lysosomes are membrane-bound organelles containing many hydrolytic enzymes, which are optimally active at an acid pH (near pH 5). Thylakoids are membrane-bound organelles found within chloroplasts. The thylakoid membrane, forms many flattened, fluid-filled tubules that enclose a single convoluted compartment. These tubules tend to stack on top of each other to forma structure called a granum.

95. (c) Anaphase I begins when the two chromosomes of each bivalent (tetrad) separate and start moving toward opposite poles of the cell as a result of the action of the spindle. In anaphase I the sister chromatids remain attached at their centromeres and move together toward the poles.

96. (a) Palmitic acid is one of the most common saturated fatty acids found in animals and plants. It has 16 carbons including the carboxyl carbon. Adenylic acid is a nucleotide consisting of adenine, ribose or deoxyribose, and a phosphate group. It is a constituent of DNA or RNA. It is also called adenosine monophosphate. Amino acids are organic acids (with carboxylic group COOH) having amino group ($-NH_2$) generally attached to Carbon or carbon next to carboxylic group. The carbon also bears a variable alkyl group (R) or hydrogen or hydrocarbon. In alanine is represented by methyl group.

97. (b) Mutations are alterations of genetic material. They occur frequently during DNA duplication in cell division. The idea of mutations was brought forth by Hugo de Vries. The plant on which de Vries had experimented was *Oenothera lamarckiana* (Evening primrose). The mutation observed by de Vries in *Oenothera lamarckiana* was chromosomal number variant.

98. (b) DNA fingerprinting is a test to identify and evaluate the genetic information-called DNA (deoxyribonucleic acid)-in a person's cells. DNA fingerprinting is a form of identification based on sequencing specific non-coding portions of DNA that are known to have a high degree of variability from person to person. These sections are known as Tandem repeats. The test is used to determine whether a family relationship exists between two people, to identify organisms causing a disease, and to solve crimes.

99. (b) The inheritance pattern of a particular trait hown in the above picture results in haemophilia. Haemophilia is a group of inherited blood disorders in which the blood does not clot properly. It is caused by a fault in one of the genes that determine how the body makes blood clotting factor VIII or IX. These genes are located on the X chromosome.

100. (a) Connective tissue is the most diverse of the four tissue types with a wide variety of functions. Connective tissue is made up of cells and fibers in a "jelly-like" ground substance.

111. (c) Those animals that excrete ammonia are called as ammonotelic, eg. aquatic amphibia. Those animals that excrete urea are called as ureotelic, eg. frog, humans.

Those animals that excrete uric acid are called as uricotelic, eg. pigeon, lizards, and cockroach.

102. (a) The caecum is a pouch-like portion of the large intestine which hosts some symbiotic micro-organism. The caecum absorbs water and salts from undigested foods before they continue on to the large intestine.

103. (c) Troponin is a protein which is found on actin filament and myosin protein is found in myosin filament. Both actin and myosin are complex proteins in striated muscles.

Thymosin is a hormone secreted by the thymus that stimulates development of T cells. Prolactin is a hormone released by the pituitary gland that stimulates breast development and milk production in women. Rhodopsin, also known as visual purple, is not a hormone. It is a biological pigment in photoreceptor cells of the retina that is responsible for the first events in the perception of light.

104. (d) Cartilage is a type of connective tissue which is present in human external ears and in the nose tip.

105. (a) Glandular epithelium is invaginated epithelia into connective tissue which differentiate into secretory units. Examples include sebaceous glands of the skin and glands in the intestinal lining (exocrine glands), and many endocrine glands releasing hormones, such as the thyroid follicle.

106. (c) The amount of the filtrate formed by the kidneys per minute is called glomerular filtration rate (GFR). GFR in a healthy individual is approximately 125 ml/minute, i.e., 180 litres per day. A fall in GFR can activate the JG cells to release renin which can stimulate the glomerular blood flow and thereby the GFR back to normal. Renin converts angiotensinogen in blood to angiotensin I and further angiotensin II. Angiotensin II being powerful vasoconstrictor increases the glomerular blood pressure and thereby GFR

107. (d) Adult frog and human exhibit ureotelism because there excretory waste product is urea.

108. (c) Blastocyst is a thin-walled hollow structure in early embryonic development that contains a cluster of cells called the inner cell mass from which the embryo arises. The outer layer of cells gives rise to the placenta and other supporting tissues needed for fetal development within the uterus while the inner cell mass cells give rise to the tissues of the body. The blastocyst reaches the womb (uterus) around day 5, and implants into the uterine wall on about day 6.

109. (c) Spleen is an organ of the lymphatic system located in the left side of the abdominal cavity under the diaphragm, the muscular partition between the abdomen and the chest. It is called graveyard of RBC because fragment of red blood cells, old and dead cells are constantly being removed from the blood streams by it.

110. (c) The menstrual cycle is the scientific term for the physiological changes that can occur in fertile women for the purposes of sexual reproduction and fertilization. The secretory phase in the human menstrual cycle is also called luteal phase and lasts for about 13 days. During secretory phase, the endometrium prepares for the implantation of an embryo and the corpus luteum is active and secretes progesterone hormone.

111. (b) Western ghat is biodiversity rich zone along with endemism. The forests of the Western Ghats are some of the best representatives of non-equatorial tropical evergreen forests in the world. The Western Ghats have evolved into one of the richest centers of endemism owing to their isolation from other moist areas.

112. (b) Sewage treatment is the process of removing contaminants from wastewater and household sewage, both runoff (effluents) and domestic. It includes physical, chemical, and biological processes to remove physical, chemical and biological contaminants. Sewage treatment generally involves three stages, called primary, secondary and tertiary treatment. It is processed by aerobic and then anaerobic bacteria in the secondary treatment in Sewage Treatment Plants (STPs).

113. (c) UAG is also a stop codon. Wombats are also Australian marsupials. They are short-legged, muscular quadrupeds, approximately 1 metre (40 in) in length with a short, stubby tail. Diphtheria is an acute infectious disease caused by the bacteria *Corynebacterium diphtheriae.*

Cuscuta, or Dodder plant, is not a protozoan. It is a parasitic vine that wraps around other plants for nourishment.

114. (a) Food web is a network of food chains or feeding relationships by which energy and nutrients are passed on from one species of living organisms to another.

115. (d) *Spirulina* is SCP rich in protein, vitamins & minerals. 250 gram biomass of *Methylophilus methylotrophus* produces 25tonn protein/day while cow of 250 Kg. produces only 200 gm. protein/day.

Common button mushrooms are a very rich source of vitamin D. A rice variety has been developed which is *very* rich in iron content.

116. (d) Molecule (a) represents structure of morphine. Morphine is the most abundant alkaloid found in opium, the dried sap (latex) derived from shallowly slicing the unripe seedpods of the opium, or common and/or edible, poppy Papaver *somniferum*. Morphine is a potent opiate analgesic drug that is used to relieve severe pain.

117. (a). Preformed antibodies need to be injected to treat the bite by a viper snake. It is also a type of immunization which is called as passive immunization.

Antibodies, produced by B-cells, are typically made of basic structural units—each with two large heavy chains and two small light chains. B cells differentiate into plasma cells that secrete antibodies. Antibodies are proteins that bind to specific antigens and mark them for destruction by, for example, marking them more recognizable to phagocytic cells. Rejection of a kidney graft is not a function of B lymphocyte.

118. (c) PCR is a technique for enzymatically replicating DNA without using a living organism such as *E. coli* or yeast. It is commonly used in medical and biological research labs for a variety of tasks like detection of hereditary diseases, identification of genetic fingerprints etc.

The correct steps shown in the above figure are:

A – Denaturation at a temperature of about 94° to 98°C. During the denaturation, the double strand melts open to single stranded DNA, and all enzymatic reactions stop.

B – Annealing (binding of DNA primer to the separated strands. Occurs at 50° to 65°Celsius, which is lower than the optimal temperature of the DNA polymerases)

C – Extension or elongation of the strands using the DNA primer with heat-stable DNA polymerases, most frequently Taq (*Thermus aquaticus*) at 72ºC.

119. (d) Gene therapy is an experimental technique that uses genes to treat or prevent disease. The first clinical gene therapy was given for treating adenosine deaminase deficiency. A four-year old girl became the first gene therapy patient on September 14, 1990 at the NIH Clinical Center. Adenosine deaminase deficiency, also called ADA deficiency or ADA-SCID is an autosomal recessive metabolic disorder that causes immunodeficiency. ADA deficiency is due to a lack of the enzyme adenosine deaminase.

120. (a) A palindromic sequence is a nucleic acid sequence (DNA or RNA) that is the same whether read 5' (five-prime) to 3' (three prime) on one strand or 5' to 3' on the complementary strand with which it forms a double helix.

5.- GAATTC - 3.

3.- CTTAAG - 5.

It is a palindromic sequence of DNA cut by restriction enzyme ECORI.

NEET

National Eligibility Entrance Test

Test Series **Full Test - 1**

Time : - 3 Hours **Max. Marks:- 720**

INSTRUCTIONS

1. Answers are to be given on Response Sheet provided at the end of this paper. You can tear this Response Sheet along the cut marks provided. Fill in the particulars carefully in the Response Sheet provided.
2. The test is of 3 hours duration.
3. The Test Booklet consists of **180** questions. The maximum marks are **720**.
4. There are four sections in the question paper.

   The distribution of marks subjectwise in each section is as under for each correct response.

   Section 1 – Physics (180 Marks) – 45 Questions

   Section 2 – Chemistry (180 Marks) – 45 Questions

   Section 3 – Botany (180 Marks) – 45 Questions

   Section 4 – Zoology (180 Marks) – 45 Questions
5. You will be awarded 4 Marks if you have darkened only the bubble corresponding to the correct answer and zero mark if no bubble is darkened. In all other cases, minus one (– 1) mark will be awarded.
6. Use Blue/Black Ball Point Pen only for writing particulars/marking responses on the Response Sheet.

   Use of pencil is strictly prohibited.
7. You are advised to create an actual examination hall condition at your home before appearing in the test.
8. Rough work is to be done on the space provided for this purpose in the Test Booklet only.
9. Do not fold or make any stray marks on the Response Sheet.
10. Once you have evaluated the marks in the test, you should assess it thoroughly with the help of the 'Test Assessment and Analysis sheet' given at the back of the response sheet.

SECTION 1 - PHYSICS

1. A force is given by $F = at + bt^2$, where t is time, the dimensions of a and b are respectively
(a) $[M L T^{-4}]$ and $[M L T^{-1}]$
(b) $[M L T^{-1}]$ and $[M L T^{0}]$
(c) $[M L T^{-3}]$ and $[M L T^{-4}]$
(d) $[M L T^{-3}]$ and $[M L T^{0}]$

2. Which of the following time-displacement graph is not possible in nature?

(a) 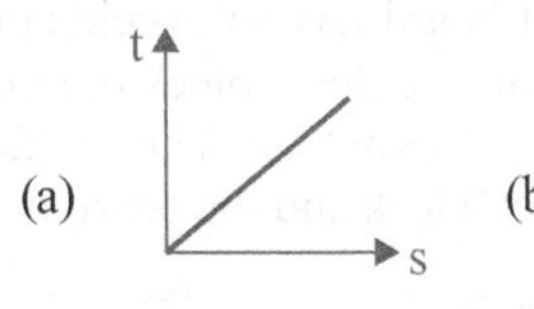

(b) 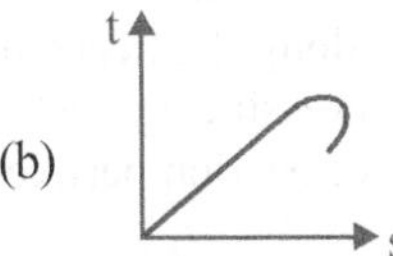

(c) 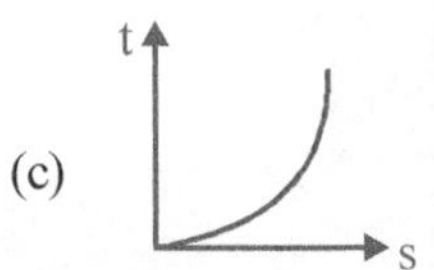

(d) 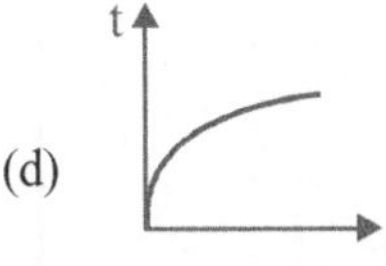

3. A particle undergoes simple harmonic motion having time period T. The time taken in 3/8th oscillation is
(a) $\frac{3}{8}T$ (b) $\frac{5}{8}T$
(c) $\frac{5}{12}T$ (d) $\frac{7}{12}T$

4. The angular velocity of a body changes from ω_1 to ω_2 without applying torque but by changing moment of inertia. The ratio of initial radius of gyration to the final radius of gyration is
(a) $\omega_2 : \omega_1$ (b) $\omega_2^2 : \omega_1^2$
(c) $\sqrt{\omega_2} : \sqrt{\omega_1}$ (d) $1/\omega_2 : 1/\omega_1$

5. For the velocity time graph shown in the figure below the distance covered by the body in the last two seconds of its motion is what fraction of the total distance travelled by it in all the seven seconds?

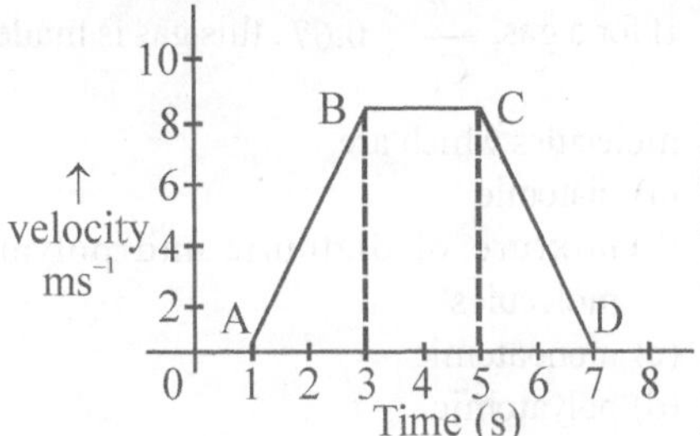

(a) $\frac{1}{2}$ (b) $\frac{1}{4}$
(c) $\frac{2}{3}$ (d) $\frac{1}{3}$

6. The work function of aluminium is 4.2 eV. If two photons each of energy 3.5 eV strike an electron of aluminium, then emission of electron will
(a) depend upon the density of the surface
(b) possible
(c) not possible
(d) None of these

7. If a capacitor of capacitance 'C' is connected in series with an inductor of inductance L, then the angular frequency will be
(a) $\sqrt{\frac{1}{LC}}$ (b) $\sqrt{\frac{L}{C}}$
(c) LC (d) $\sqrt{LC}$

8. Two wires A and B of the same material, having radii in the ratio 1 : 2 and carry currents in the ratio 4 : 1. The ratio of drift speed of electrons in A and B is
(a) 16 : 1 (b) 1 : 16
(c) 1 : 4 (d) 4 : 1

9. A stone is thrown with a velocity u making an angle θ with the horizontal. The horizontal distance covered by its fall to ground is maximum when the angle θ is equal to
(a) 0° (b) 30°
(c) 45° (d) 90°

10. If for a gas, $\frac{R}{C_V} = 0.67$, this gas is made up of molecules which are
(a) diatomic
(b) mixture of diatomic and polyatomic molecules
(c) monoatomic
(d) polyatomic

11. ABC is an equilateral triangle. Charges +q are placed at each corner as shown in fig. The electric intensity at centre O will be

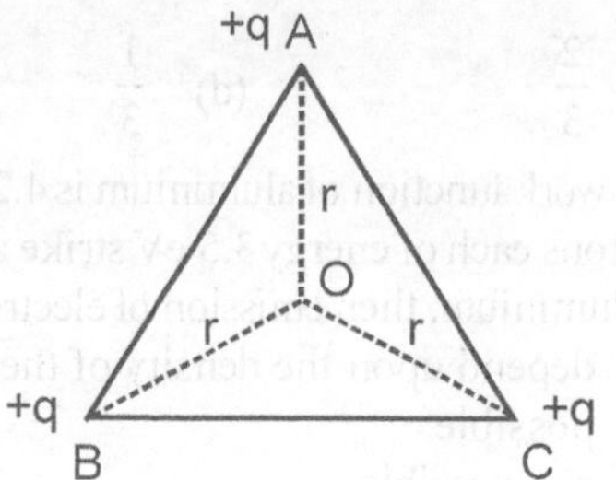

(a) $\frac{1}{4\pi\epsilon_o}\frac{q}{r}$ (b) $\frac{1}{4\pi\epsilon_o}\frac{q}{r^2}$

(c) $\frac{1}{4\pi\epsilon_o}\frac{3q}{r^2}$ (d) zero

12. Photoelectric current from a given photocell will depend upon
(a) number of photons striking per second
(b) frequency of incident photon
(c) material of the target
(d) None of these

13. A circular disc A of radius r is made from an iron plate of thickness t and another circular disc B of radius 4r is made from an iron plate of thickness t/4. The relation between the moments of inertia I_A and I_B is
(a) $I_A > I_B$
(b) $I_A = I_B$
(c) $I_A < I_B$
(d) depends on the actual value of t and r

14. Einstein's work on photoelectric effect provided support for the equation
(a) $E = h\nu$ (b) $E = mc^2$
(c) $E = \frac{-Rhc}{n^2}$ (d) $K.E. = \frac{1}{2}mv^2$

15. A metallic bar is heated from 0°C to 100°C. The coeficient of linear expansion is $10^{-5}\ K^{-1}$. What will be the percentage increase in length?
(a) 0.01% (b) 0.1%
(c) 1% (d) 10%

16. A rough vertical board has an acceleration *a* along the horizontal so that a block of mass M pressing against it does not fall. The coefficient of friction between block and the board is

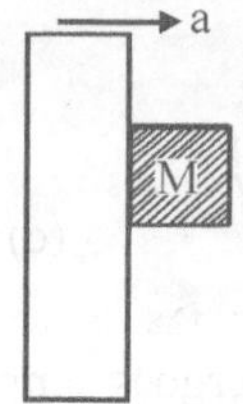

(a) $> \frac{a}{g}$ (b) $< \frac{g}{a}$

(c) $= \frac{a}{g}$ (d) $> \frac{g}{a}$

17. What should be the minimum value of refractive index of the material of the prism for the reflections to take place as shown in the figure?

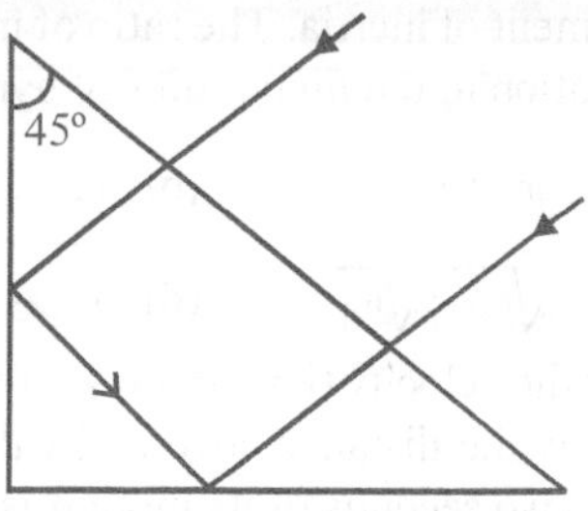

(a) 1.7 (b) 1.4
(c) 1.2 (d) 2.7

18. Electric lines of force about a negative point charge are
(a) circular, anti-clockwise
(b) circular, clockwise
(c) radial, inwards
(d) radial, outwards

19. Ohm's law is not obeyed by
(a) electrolytes (b) discharge tube
(c) vacuum tubes (d) all of the above

20. The temperature of an iron block is 140°F. Its temperature on the Celsius scale is
(a) 60° (b) 160°
(c) 140° (d) 132°

21. A gun fires two bullets at 60° and 30° with horizontal. The bullets strike at some horizontal distance. The ratio of maximum height for the two bullets is in the ratio of
(a) 2 : 1 (b) 3 : 1
(c) 4 : 1 (d) 1 : 1

22. A generator has an e.m.f. of 440 Volt and internal resistance of 400 Ohm. Its terminals are connected to a load of 4000 Ohm the voltage across the load is
(a) 220 volt (b) 440 volt
(c) 200 volt (d) 400 volt

23. Electric field inside a copper wire of length 10 metres, resistance 2 ohm connected to a 10 volt battery is
(a) $1\ Vm^{-1}$ (b) $0.5\ Vm^{-1}$
(c) $10\ Vm^{-1}$ (d) $5\ Vm^{-1}$

24. The radius vector, drawn from the sun to a planet, sweeps out equal areas in equal intervals of time. This is the statement of
(a) Kepler's first law (b) Kepler's second aw
(c) Newton's first law (d) Kepler's third law

25. The path difference between the two waves :

$$y_1 = a_1 \sin\left(\omega t - \frac{2\pi x}{\lambda}\right)$$

and $y_2 = a_2 \sin\left(\omega t - \frac{2\pi x}{\lambda} + \phi\right)$ will be

(a) $\frac{2\pi}{\lambda}\phi$ (b) $\frac{2\pi}{\lambda}\left(\phi - \frac{\pi}{2}\right)$
(c) $\frac{\lambda}{2\pi}\phi$ (d) $\frac{2\pi}{\lambda}\left(\phi + \frac{\pi}{2}\right)$

26. The diagram shows the energy levels for an electron in a certain atom. Which transition shown represents the emission of a photon with the most energy?

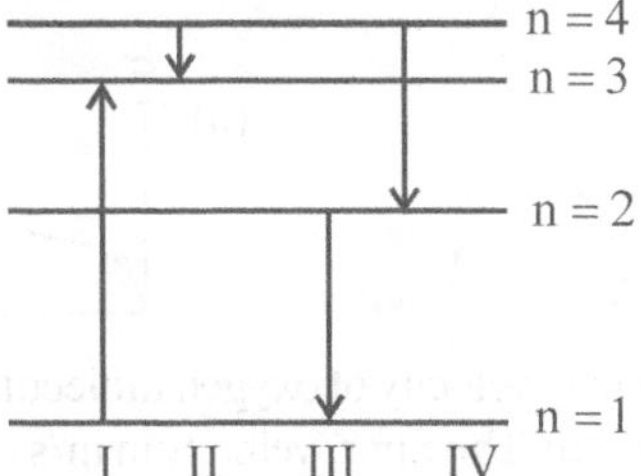

(a) IV (b) III
(c) II (d) I

27. The direction of propagation of electromagnetic waves is given by the direction of
(a) Vector $\vec{E}$ (b) Vector $\vec{B}$
(c) Vector ($\vec{E} \times \vec{B}$) (d) None of these

28. A body of mass 10 kg and velocity 10 m/s collides with a stationary body of mass 5 kg. After collision both bodies stick to each other, velocity of the bodies after collision will be
(a) $\frac{3}{10}$ m/s (b) $\frac{18}{3}$ m/s
(b) $\frac{9}{20}$ m/s (d) $\frac{20}{3}$ m/s

29. Two particles of mass m_1 and m_2 ($m_1 > m_2$) attract each other with a force inversely proportional to the square of the distance between them. If the particles are initially held at rest and then released, the centre of mass will
(a) move towards m_1 (b) move towards m_2
(c) remain at rest (d) Nothing can be said

30. Which of the following graphs represent the variation of thermo emf (E) of a thermocouple with temperature θ of hot junction (the cold junction being kept at 0°C)

(a) (b)

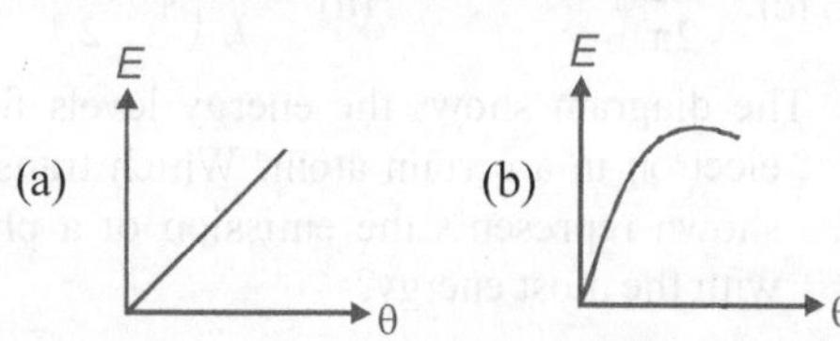

(c) (d)

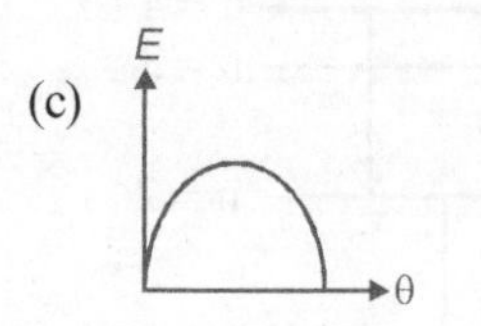

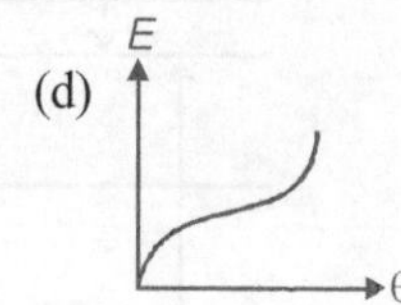

31. The r.m.s. velocity of oxygen molecule at 16°C is 474 m/sec. The r.m.s. velocity in m/s of hydrogen molecule at 127°C is

(a) 1603 (b) 1896
(c) 2230.59 (d) 2730

32. To demonstrate the phenomenon of interference, we require two sources which emit radiation

(a) of the same frequency
(b) of different wavelengths
(c) of the same frequency and having a definite phase relationship
(d) of nearly the same frequency

33. The oscillating electric and magnetic field vectors of electromagnetic wave are oriented along

(a) the same direction and in phase
(b) the same direction but have a phase difference of 90°
(c) mutually perpendicular directions and are in phase
(d) mutually perpendicular directions but has a phase difference of 90°

34. A uniform spherical shell gradually shrinks maintaining its shape. The gravitational potential at the centre

(a) increases (b) decreases
(c) remains constant (d) oscillates

35. If v_e and v_0 represent the escape velocity and orbital velocity of a satellite corresponding to a circular orbit of radius R, then

(a) $v_e = v_o$
(b) $v_e = \sqrt{2}\, v_o$
(c) $v_e = (1/\sqrt{2})\, v_o$
(d) v_e and v_o are not related

36. At 0°K which of the following properties of a gas will be zero?

(a) Kinetic energy
(b) Potential energy
(c) Vibrational energy
(d) Density

37. A galvanometer coil has a resistance of 15Ω and gives full scale deflection for a current of 4 mA. To convert it to an ammeter of range 0 to 6 A

(a) 10 mΩ resistance is to be conncected in parallel to the galvanometer
(b) 10 mΩ esistance is to be connected in series with the galvanometer
(c) 0.1 Ω resistance is to be connected in parallel to the galvanometer
(d) 0.1 Ω resistance is to be connected in series with the galvanometer

38. A uniform rod of mass m, length ℓ, area of cross-section A has Young's modulus Y. If it is hanged vertically, elongation under its own weight will be

(a) $\dfrac{mg\ell}{2AY}$ (b) $\dfrac{2mg\ell}{AY}$
(c) $\dfrac{mg\ell}{AY}$ (d) $\dfrac{mgY}{A\ell}$

39. The magnifying power of a telescope is 9. When it is adjusted for parallel rays, the distance between the objective and the eye piece is found to be 20 cm. The focal length of lenses are

(a) 18 cm, 2 cm (b) 11 cm, 9 cm
(c) 10 cm, 10 cm (d) 15 cm, 5 cm

40. If two soap bubbles of different radii are connected by a tube. Then
(a) air flows from the smaller bubble to the bigger
(b) air flows from bigger bubble to the smaller bubble till the sizes are interchanged
(c) air flows from the bigger bubble to the smaller bubble till the sizes become equal
(d) there is no flow of air.

41. An object undergoing SHM takes 0.5 s to travel from one point of zero velocity to the next such point. The distance between those points is 50 cm. The period, frequency and amplitude of the motion is
(a) 1s, 1Hz, 25 cm (b) 2s, 1Hz, 50 cm
(c) 1s, 2Hz, 25 cm (d) 2s, 2Hz, 50 cm

42. A plane wave of wavelength 6250 Å is incident normally on a slit of width 2×10^{-2} cm. The width of the principal maximum on a screen distant 50 cm will be
(a) 312.5×10^{-3} cm (b) 312.5×10^{-3} m
(c) 312.5×10^{-3} m (d) 312m

43. In a p-type semi-conductor germanium is doped with
(a) aluminium (b) boron
(c) gallium (d) all of these

44. The table given below represents the truth table for which of the following combinations of logic gates?

A	B	Y
0	0	1
0	1	0
1	0	1
1	1	0

(a) 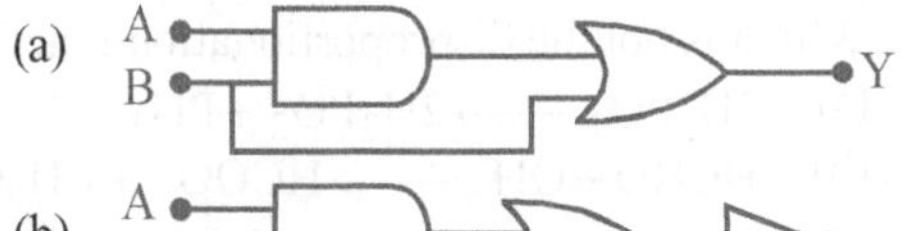

(b) 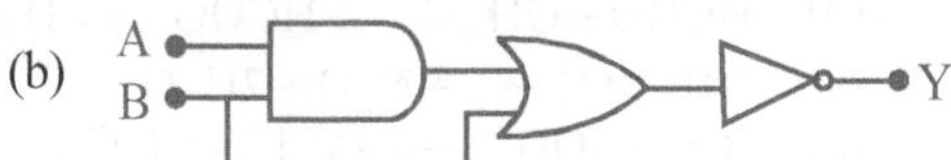

(c) 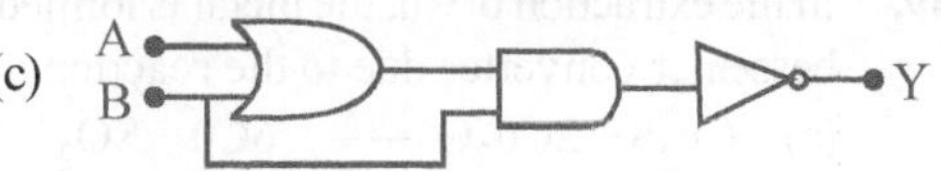

(d) None of these

45. In a transistor
(a) both emitter and collector have same length
(b) length of emitter is greater than that of collector
(c) length of collector is greater than that of emitter
(d) any one of emitter and collector can have greater length

SECTION 2 - CHEMISTRY

46. Fluorine is more electronegative than either boron or phosphorus. What conclusion can be drawn from the fact that BF_3 has no dipole moment but PF_3 does ?
(a) BF_3 is not spherically symmetrical but PF_3 is spherically symmetrical.
(b) BF_3 molecule must be linear
(c) The atomic radius of P is larger than that of B
(d) The BF_3 molecule must be planar triangular

47. At 298 K, equal volumes of SO_2, CH_4 and O_2 are mixed in empty container. The total pressure exerted is 2.1 atm. The partial pressure of CH_4 in mixture is :
(a) 0.6 atm. (b) 1.2 atm
(c) 2.4 atm (d) 3.6 atm.

48. 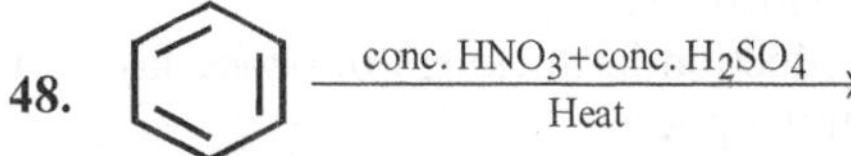

$$\xrightarrow[\text{Heat}]{\text{conc. } HNO_3 + \text{conc. } H_2SO_4} X \xrightarrow{Cl_2/FeCl_3} Y$$

The product Y is
(a) *p*-chloronitrobenzene
(b) *m*-chloronitrobenzene
(c) *o*-chloronitrobenzene
(d) *o*-*p*-dichloronitrobenzene

49. In the extraction of Cu, the metal is formed in the bessemer converter due to the reaction :
(a) $Cu_2S + 2Cu_2O \longrightarrow 6Cu + SO_2$
(b) $Cu_2S \longrightarrow 2Cu + S$
(c) $Fe + Cu_2O \longrightarrow 2Cu + FeO$
(d) $2Cu_2O \longrightarrow 4Cu + O_2$

50. An example of electrophilic substitution reaction is
(a) Chlorination of methane
(b) Conversion of methyl chloride to methyl alcohol
(c) Nitration of benzene
(d) Formation of ethylene from ethyl alcohol.

51. Which of the following molecules is most suitable to disperse benzene in water ?
(a) $\overset{O}{\parallel}$ $\bar{O}\overset{+}{Na}$
(b) $Na\overset{+}{}\,\bar{O}$ O $\bar{O}\,\overset{+}{Na}$
(c) CH_3
(d) Cl

52. Ozone hole refers to
(a) Increase in concentration of ozone
(b) Hole in ozone layer
(c) Reduction in thickness of ozone layer in troposphere
(d) Reduction in thickness of ozone layer in stratsophere

53. Which of the following does not represent the correct order of the properties indicated
(a) $Ni^{2+} > Cr^{2+} > Fe^{2+} > Mn^{2+}$ (size)
(b) Sc > Ti > Cr > Mn (size)
(c) $Mn^{2+} > Ni^{2+} < Co^{2+} < Fe^{2+}$ (unpaired electron)
(d) $Fe^{2+} > Co^{2+} > Ni^{2+} > Cu^{2+}$ (unpaired electron)

54. The major product formed in the following reaction is :

$$CH_3-\underset{H}{\underset{|}{\overset{CH_3}{\overset{|}{C}}}}-CH_2Br \xrightarrow[CH_3OH]{CH_3O^-}$$

(a) $CH_3-\underset{H}{\underset{|}{\overset{CH_3}{\overset{|}{C}}}}-CH_2OCH_3$
(b) $CH_3-\underset{OCH_3}{\underset{|}{CH}}-CH_2CH_3$
(c) $CH_3-\overset{CH_3}{\overset{|}{C}}=CH_2$
(d) $CH_3-\underset{OCH_3}{\underset{|}{\overset{CH_3}{\overset{|}{C}}}}-CH_3$

55. The electrons, identified by quantum numbers n and l (i) n = 4, l = 1 (ii) n = 4, l = 0 (iii) n = 3, l = 2 (iv) n = 3, l = 1 can be placed in order of increasing energy, from the lowest to highest, as
(a) (iv) < (ii) < (iii) < (i)
(b) (ii) < (iv) < (i) < (iii)
(c) (i) < (iii) < (ii) < (iv)
(d) (iii) < (i) < (iv) < (ii)

56. In an adiabatic process which of the following is true?
(a) $q = +w$ (b) $q = 0$
(c) $\Delta E = q$ (d) $P\Delta V = 0$

57. In which of the following cases, the stability of two oxidation states is correctly represented
(a) $Ti^{3+} > Ti^{4+}$ (b) $Mn^{2+} > Mn^{3+}$
(c) $Fe^{2+} > Fe^{3+}$ (d) $Cu^{+} > Cu^{2+}$

58. Which is not the disproportionation reaction ?
(a) $3H_3PO_2 \longrightarrow 2H_3PO_2 + PH_3$
(b) $HCHO + OH^- \longrightarrow HCOO^- + CH_3OH$
(c) $NH_4NO_3 \longrightarrow N_2O + 2H_2O$
(d) $3Cl_2 + 6OH^- \longrightarrow 5Cl^- + ClO_3^- + 3H_2O$

59. The major organic product in the reaction, $CH_3—O—CH(CH_3)_2 + HI \rightarrow$ Product is

(a) $ICH_2OCH(CH_3)_2$

(b) $CH_3OC(CH_3)_2$ with I on the central carbon

(c) $CH_3I + (CH_3)_2CHOH$

(d) $CH_3OH + (CH_3)_2CHI$

60. For the reaction $H_{2(g)} + Br_{2(g)} \rightleftharpoons 2HBr_{(g)}$, the rate law is rate $= k[H_2][Br_2]^{1/2}$. Which of the following statement is true about this reaction

(a) The reaction is of second order.

(b) Molecularity of the reaction is 3/2

(c) The unit of k is s^{-1}

(d) Molecularity of the reaction is 2

61. The correct order of solubility in water for He, Ne, Ar, Kr, Xe is

(a) He > Ne > Ar > Kr > Xe

(b) Xe > Kr > Ar > Ne > He

(c) Ne > Ar > Kr > He > Xe

(d) Ar > Ne > He > Kr > Xe

62. If uncertainty in position and velocity are equal then uncertainty in momentum will be

(a) $\frac{1}{2}\sqrt{\frac{mh}{\pi}}$ (b) $\frac{1}{2}\sqrt{\frac{h}{\pi m}}$

(c) $\frac{h}{4\pi m}$ (d) $\frac{mh}{4\pi}$

63. The correct IUPAC name for

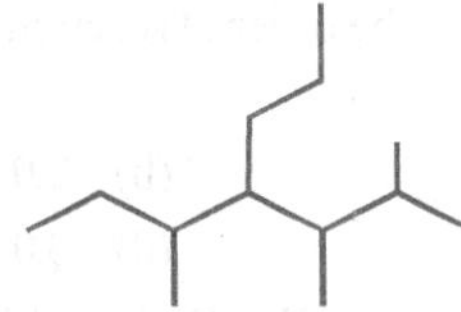

(a) 5-methyl -4-(1'-2'-demethylpropyl) heptane

(b) 3-methyl-4-(1',2'-dimethylpropyl) heptane

(c) 2,3,5-trimethyl-4-propylheptane

(d) 4-propyl-2,3,5-trimethylpeptane

64. Which of the following shows iso-structural species?

(a) NH_4^+ and NH_2^-

(b) CH_3^- and CH_3^+

(c) SO_4^{2-}, PO_4^{3-} and $[BF_4^-]$

(d) NH_4^+ and NH_3

65. When dihydroxyacetone reacts with HIO_4, the product is/are :

(a) HCHO (b) HCOOH

(c) HCHO and HCOOH (d) HCHO and CO_2

66. Hard water when passed through ion exchange resin containing R'COOH groups, becomes free from :

(a) Cl^- (b) SO_4^{2-}

(c) H_3O^+ (d) Ca^{2+}

67. A certain compound (X) when treated with copper sulphate solution yields a brown precipitate. On adding hypo solution, the precipitate turns white. The compound is

(a) K_2CO_3 (b) KI

(c) KBr (d) K_3PO_4

68. Consider the following reactions:

(i) $H^+_{(aq)} + OH^-_{(aq)} = H_2O_{(l)}$, $\Delta H = -X_1$ kJ mol^{-1}

(ii) $H_{2(g)} + \frac{1}{2}O_{2(g)} = H_2O(l)$, $\Delta H = -X_2$ kJ mol^{-1}

(iii) $CO_{2(g)} + H_{2(g)} = CO_{(g)} + H_2O$, $\Delta H = -X_3$ kJ mol^{-1}

(iv) $C_2H_{2(g)} + \frac{5}{2}O_{2(g)} = 2CO_2(g) + H_2O(l)$' $\Delta H = +4X_4$ kJ mol^{-1}

Enthalpy of formation of $H_2O_{(l)}$ is

(a) $+X_3$ kJ mol^{-1} (b) $-X_4$ kJ mol^{-1}

(c) $+X_1$ kJ mol^{-1} (d) $-X_2$ kJ mol^{-1}

69. Which of the following possesses a sp-carbon in its structure ?

(a) $CH_2=CCl-CH=CH_2$

(b) $CCl_2=CCl_2$

(c) $CH_2=C=CH_2$

(d) $CH_2=CH-CH=CH_2$.

70. A balloon has maximum capacity of 20 L. At one atmospheric pressure 10 L of air is filled in the balloon. It will burst when pressure is (assuming isothermal condition)

(a) >0.5 atm (b) <0.5 atm
(c) $=0.5$ atm (d) ≥ 0.5 atm

71. Which one of the following complexes will have four different isomers ?

(a) $[Co(en)_2Cl_2]Cl$
(b) $[Co(en)(NH_3)Cl_2]Cl$
(c) $[Co(PPh_3)_2Cl_2]Cl$
(d) $[Co(en)_3]Cl_3$

72. Solubility product of a salt AB is 1×10^{-8} in a solution in which the concentration of A^+ ions is 10^{-3} M. The salt will precipitate when the concentration of B^- ions is kept

(a) between 10^{-8} M to 10^{-7} M
(b) between 10^{-7} M to 10^{-8} M
(c) $> 10^{-5}$ M
(d) $< 10^{-8}$ M

73. Sucrose in water is dextro-rotatory, $[\alpha]_D = +66.4°$. When boiled with dilute HCl, the solution becomes leavo-rotatory, $[\alpha]_D = -20°$. In this process the sucrose molecule breaks into

(a) L-glucose + D-fructose
(b) L-glucose + L-fructose
(c) D-glucose + D-fructose
(d) D-glucose + L-fructose

74. Select correct statement(s).

(a) Cyanamide ion (CN_2^{2-}) is isoelectronic with CO_2 and has the same linear structure
(b) Mg_2C_3 reacts with water to form propyne
(c) CaC_2 has NaCl type lattice
(d) All of the above

75. For $-\overset{\overset{\Large O}{||}}{C}-\ddot{N}H-$(peptide bond)

Which statement is incorrect about peptide bond?

(a) C — N bond length in proteins is longer than usual bond length of the C — N bond
(b) Spectroscopic analysis shows planar structure of the $-\underset{\underset{\Large O}{||}}{C}-NH-$ group
(c) C — N bond length in proteins is smaller than usual bond length of the C—N bond
(d) None of the above

76. Vapour pressure of benzene at 30°C is 121.8 mm. When 15 g of a non volatile solute is dissolved in 250 g of benzene its vapour pressure decreased to 120.2 mm. The molecular weight of the solute (Mo. wt. of solvent = 78)

(a) 356.2 (b) 456.8
(c) 530.1 (d) 656.7

77. In nitrogen family, the H-M-H bond angle in the hydrides gradually becomes closer to 90° on going from N to Sb. This shows that gradually

(a) The basic strength of the hydrides increases
(b) Almost pure *p*-orbitals are used for M-H bonding
(c) The bond energies of M-H bonds increase
(d) The bond pairs of electrons become nearer to the central atom

78. For reaction $aA \rightarrow xP$, when [A] = 2.2 mM, the rate was found to be 2.4 mMs^{-}. On reducing concentration of A to half, the rate changes to 0.6 mMs^{-1}. The order of reaction with respect to A is :

(a) 1.5 (b) 2.0
(c) 2.5 (d) 3.0

79. The tendency of BF_3, BCl_3 and BBr_3 to behave as Lewis acid decreases in the sequence:

(a) $BCl_3 > BF_3 > BBr_3$
(b) $BBr_3 > BCl_3 > BF_3$
(c) $BBr_3 > BF_3 > BCl_3$
(d) $BF_3 > BCl_3 > BBr_3$

80. If 'a' stands for the edge length of the cubic systems : simple cubic, body centred cubic and face centred cubic, then the ratio of radii of the spheres in these systems will be respectively,

(a) $\frac{1}{2}a : \frac{\sqrt{3}}{4}a : \frac{1}{2\sqrt{2}}a$ (b) $\frac{1}{2}a : \sqrt{3}a : \frac{1}{\sqrt{2}}a$

(c) $\frac{1}{2}a : \frac{\sqrt{3}}{2}a : \frac{\sqrt{3}}{2}a$ (d) $1a : \sqrt{3}a : \sqrt{2}a$

81. Which of the following is correct order of acidity?
(a) $HCOOH > CH_3COOH > ClCH_2COOH > C_2H_5COOH$
(b) $ClCH_2COOH > HCOOH > CH_3COOH > C_2H_5COOH$
(c) $CH_3COOH > HCOOH > ClCH_2COOH > C_2H_5COOH$
(d) $C_2H_5COOH > CH_3COOH > HCOOH > ClCH_2COOH$

82. Penicillin is :
(a) analgesic (b) antipyretic
(c) antimalarial (d) antibiotic

83. For the reaction $2NO + Br_2 \longrightarrow 2NOBr$, the following mechanism has been given

$$NO + Br_2 \xrightarrow{\text{fast}} NOBr_2$$

$$NOBr_2 + NO \xrightarrow{\text{slow}} 2NOBr$$

Hence, rate law is
(a) $k'[NO]^2[Br_2]$ (b) $k'[NO][Br_2]$
(c) $k'[NOBr_2][NO]$ (d) $k'[NO][Br_2]^2$

84. Which of the following reactions can produce aniline as main product?
(a) $C_6H_5NO_2 + Zn/KOH$
(b) $C_6H_5NO_2 + Zn/NH_4Cl$
(c) $C_6H_5NO_2 + LiAlH_4$
(d) $C_6H_5NO_2 + Zn/HCl$

85. Which of the following reactions is used to make a fuel cell?
(a) $Cd(s) + 2Ni(OH)_3(s) \longrightarrow CdO(s) + 2Ni(OH)_2(s) + H_2O(l)$
(b) $Pb(s) + PbO_2(s) + 2H_2SO_4(aq) \longrightarrow 2PbSO_4(s) + 2H_2O(l)$
(c) $2H_2(g) + O_2(g) \longrightarrow 2H_2O(l)$
(d) $2Fe(s) + O_2(g) + 4H^+(aq) \longrightarrow 2Fe^{2+}(aq) + 2H_2O(l)$

86. Which does not exist ?
(a) $[SiCl_6]^{2-}$ (b) $[GeF_6]^{2-}$
(c) $[CCl_6]^{2-}$ (d) $[SnCl_6]^{2-}$

87. In which case, van't Hoff factor *i* remains unchanged ?
(a) $PtCl_4$ reacts with aq. KCl
(b) aq. $ZnCl_2$ reacts with aq. NH_3
(c) aq. $FeCl_3$ reacts with aq. $K_4[Fe(CN)]_6$
(d) $KMnO_4$ reduced to MnO_2 in alkaline medium

88. The unit of equivalent conductivity is
(a) $S\ cm^{-2}$
(b) ohm cm^2 (g - equivalent)
(c) ohm cm
(d) $ohm^{-1}cm^2$ (g equivalent)$^{-1}$

89. Perlon is
(a) Terylene (b) Rubber
(c) Nylon – 6 (d) Polyester

90. Tertiary nitro compounds do not tautomerise because
(a) there is no double bond
(b) there is no α-hydrogen
(c) oxygen is more electronegative than hydrogen
(d) all of the above

SECTION 3 - BOTANY

91. The hormone which controls cell division and cell differentiation is
(a) ABA (b) auxin
(c) gibberellin (d) cytokinin

92. Respiratory Quotient (R.Q.) is defined as:
(a) vol. of O_2/vol. of CO_2
(b) vol. of CO_2/vol. of O_2
(c) vol. of O_2/vol. of N_2
(d) vol. of N_2/vol. of CO_2

93. Which one of the following is a long day plant?
(a) Bajra (b) Soyabean
(c) Tobacco (d) Wheat

94. Choose the right option.

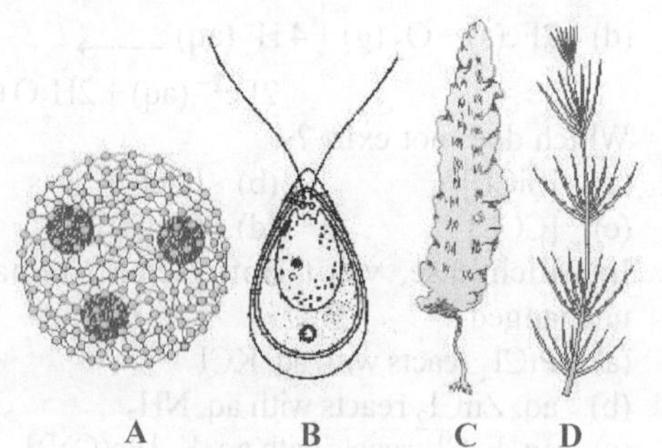

	A	B	C	D
(a)	*Volvox*	*Chlamydo-monas*	*Laminaria*	*Chara*
(b)	*Chara*	*Laminaria*	*Volvox*	*Chlamydo-monas*
(c)	*Laminaria*	*Volvox*	*Chlamyd-omonas*	*Chara*
(d)	*Chlamyd-omonas*	*Chara*	*Laminaria*	*Volvox*

95. Connecting link between glycolysis and Krebs cycle is/before entering Krebs cycle pyruvate is changed to
(a) oxaloacetate
(b) phosphoenol pyruvate
(c) pyruvate
(d) acetyl CoA

96. Differentiation of shoot is controlled by
(a) high gibberellin : cytokinin ratio
(b) high auxin : cytokinin ratio
(c) high cytokinin : auxin ratio
(d) high gibberellin : auxin ratio

97. An enzyme that can stimulate germination of barley seeds is
(a) lipase (b) protease
(c) invertase (d) α-amylase

98. The enzyme used for joining two DNA fragments is called:
(a) ligase
(b) restriction endonuclease
(c) DNA polymerase
(d) gyrase

99. Thorn is a modification of
(a) stem (b) leaf
(c) stipule (d) root

100. Dry indehiscent single-seeded fruit formed from bicarpellary syncarpous inferior ovary is
(a) Caryopsis (b) Cypsela
(c) Berry (d) Cremocarp

101. DDT residues are rapidly passed through food chain causing biomagnification because DDT is
(a) moderately toxic
(b) non-toxic to aquatic animals
(c) water soluble
(d) lipo soluble

102. Deep black soil is productive due to high proportion of
(a) sand and zinc (b) gravel and calcium
(c) clay and humus (d) silt and earthworm

103. Which the following shows the correct explanation of the given figures?

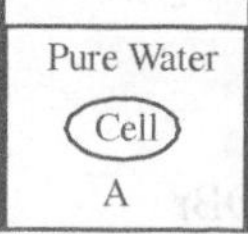

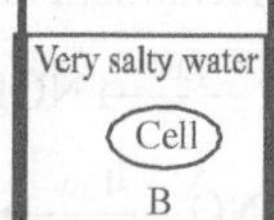

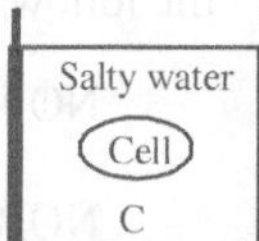

(a) Cell "A" will lose H_2O, Cell "B" will gain H_2O, Cell "C" neither gain nor loses H_2O.
(b) Cell "A" neither gain nor loses H_2O, Cell "B" will gain H_2O, Cell "C" will lose H_2O.
(c) Cell "A" will gain water, Cell "B" neither gain nor loses H_2O, Cell "C" lose H_2O.
(d) Cell "A" will gain H_2O, Cell "B" will lose H_2O, Cell "C" neither gain nor loses H_2O.

104. Female gametophyte of angiosperms is represented by
(a) Ovule
(b) Megaspore mother cell
(c) Embryo sac
(d) Nucellus

105. Chasmogamy refers to the condition where
(a) Flowers remains closed
(b) Flowers absent
(c) Flowers open
(d) Flowers gamopetalous

106. Today, concentration of green house gases is high because of
(a) Use of refrigerator
(b) Increased combustion of oil and coal
(c) Deforestation
(d) All the above

107. Cross between AaBB and aaBB will form
(a) 1 AaBB : 1aaBB (b) All AaBB
(c) 3 AaBB : 1 aaBB (d) 1 AaBB : 3 aaBB

108. Which of the following is the most stable ecosystem?
(a) Forest (b) Desert
(c) Mountain (d) Ocean

109. Parthenocarpic tomato fruits can be produced by
(a) treating the plants with low concentrations of gibberellic acid and auxins
(b) raising the plants from vernalized seeds
(c) treating the plants with phenylmercuric acetate
(d) removing androecium of flowers before pollen grains are released

110. A gene pair hides the effect of another gene. The phenomenon is
(a) epistasis (b) dominance
(c) mutation (d) None of the above

111. A mutation at one base of the first codon, of a gene, produces a non-functional protein. Such a mutation is called
(a) nonsense mutation
(b) missense mutation
(c) frameshift mutation
(d) reverse mutation

112. The rate at which light energy is converted to the chemical energy of organic molecules in the ecosystem's is
(a) net primary productivity
(b) gross primary productivity
(c) net secondary productivity
(d) gross secondary productivity

113. Consider the following statements concerning food chains :
(i) Removal of 80% tigers from an area resulted in greatly increased growth of vegetation
(ii) Removal of most of the carnivores resulted in an increased population of deers
(iii) The length of food chains is generally limited to 3-4 trophic levels due to energy loss
(iv) The length of food chains may vary from 2 to 8 trophic levels

Which two of the above statements are correct?
(a) (ii) and (iii) (b) (iii) and (iv)
(c) (i) and (iv) (d) (i) and (ii)

114. The picture below shows three different types of plants (marked as P-I, P-II and P-III) which flower on the basis of their critical photoperiod. Now identify these plants (P-I, II and III).

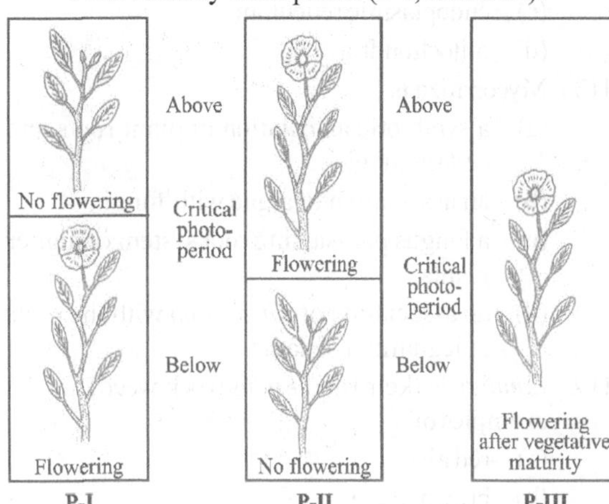

(a) P-I = Long day plant; P-II = Short day plant; P-III = Day neutral plant
(b) P-I = Short day plant; P-II = Long day plant; P-III = Day neutral plant
(c) P-I = Short day plant; P-II = Short day plant; P-III = Day neutral plant
(d) P-I =Long day plant; P-II = Long day plant; P-III = Day neutral plant

115. Binomial nomenclature was introduced by
(a) Carolus Linnaeus
(b) Charles Darwin
(c) Bentham and Hooker
(d) Aristotle

116. Pteridophytes differ from mosses/bryophytes in possessing
(a) independent gametophyte
(b) well developed vascular system
(c) archegonia structure
(d) flagellate spermatozoids

117. The cell organelle involved in glyco-sylation of protein is
(a) ribosome
(b) peroxisome
(c) endoplasmic reticulum
(d) mitochondria

118. Mycorrhiza is
(a) a symbiotic association of plant roots and certain fungi.
(b) an association of algae with fungi.
(c) a fungus parasitising root system of higher plants.
(d) an association of *Rhizobium* with the roots of leguminous plants.

119. *Laminaria* (kelp) and *Fucus* (rock weed) are the examples of
(a) red algae
(b) brown algae
(c) green algae
(d) golden brown algae

120. The outer layer of vacuole is called
(a) cell wall (b) tonoplast
(c) plasmalayer (d) leucoplast

121. A narrow layer of thin walled cells found between phloem/bark and wood of a dicot is
(a) cork cambium (b) vascular cambium
(c) endodermis (d) pericycle

122. Which of the following statements is true?
(a) Vessels are multicellular with narrow lumen.
(b) Tracheids are multicellular with narrow lumen.
(c) Vessels are unicellular with wide lumen.
(d) Tracheids are unicellular with wide lumen.

123. The linking of antibiotic resistance gene with the plasmid vector became possible with
(a) DNA ligase (b) Endonucleases
(c) DNA polymerase (d) Exonucleases

124. Which one of the following is not included under *in-situ* conservation ?
(a) Botanical garden (b) Biosphere reserve
(c) National park (d) Sanctuary

125. *Monascus purpureus* is a yeast used commercially in the production of :
(a) ethanol
(b) streptokinase for removing clots from the blood vessels.
(c) Citric acid
(d) blood cholesterol lowering statins

126. Gel electrophoresis is used for
(a) cutting of DNA into fragments
(b) separation of DNA fragments according to their size
(c) construction of recombinant DNA by joining with cloning vectors
(d) isolation of DNA molecule

127. DNA or RNA segment tagged with a radioactive molecule is called
(a) Vector (b) Probe
(c) Clone (d) Plasmid

128. Which of the following is considered a hot-spot of biodiversity in India ?
(a) Indo-Gangetic Plain (b) Eastern Ghats
(c) Aravalli Hills (d) Western Ghats

129. Which one of the following is the correct matched pair of an endangered animal and National Park ?
(a) Rhinoceros - Kaziranga National Park
(b) Wild Ass - Dudhwa National Park
(c) Great Indian Bustard - Keoladeo National Park
(d) Lion - Corbett National Park

130. The movement of mineral ions into plant root cells as a result of diffusion is called
(a) osmosis
(b) active absorption
(c) passive absorption
(d) endocytosis

131. The rate of transpiration in plants is dependent upon
(a) temperature and soil
(b) light and temperature
(c) wind, temperature and light
(d) light, temperature, atmospheric humidity and wind

132. Stomata open and close due to
(a) circadian rhythm
(b) genetic clock
(c) pressure of gases inside the leaves
(d) turgor pressure of guard cells

133. Which of the following statements about nitrification is not true?
(a) *Nitrobacter* oxidizes nitrite to nitrate.
(b) *Nitrosomonas* and *Nitrosococcus* convert ammonium ions to nitrite.
(c) Nitrification reactions are energy-producing (exergonic) reactions.
(d) Heterotrophic plants are more directly dependent on the nitrifying bacteria for usable nitrogen than autotrophic plants.

134. Grana are present inside the :
(a) mitochondria
(b) chloroplast
(c) endoplasmic reticulum
(d) ribosome

135. Chemiosmotic hypothesis given by Peter Mitchell proposes the mechanism of
(a) synthesis of ATP
(b) synthesis of $FADH_2$
(c) synthesis of NADH
(d) synthesis of NADPH

SECTION 4 - ZOOLOGY

136. Which ion is essential for muscle contraction?
(a) Na^+ (b) K^+
(c) Ca^{2+} (d) Cl^-

137. Which one of the following organs in the human body is most affected due to shortage of oxygen?
(a) Intestine (b) Skin
(c) Kidney (d) Brain

138. Elbow joint is an example of
(a) hinge joint (b) gliding joint
(c) ball and socket joint (d) pivot joint

139. Choose the correct names of the different bacteria according to their shapes.

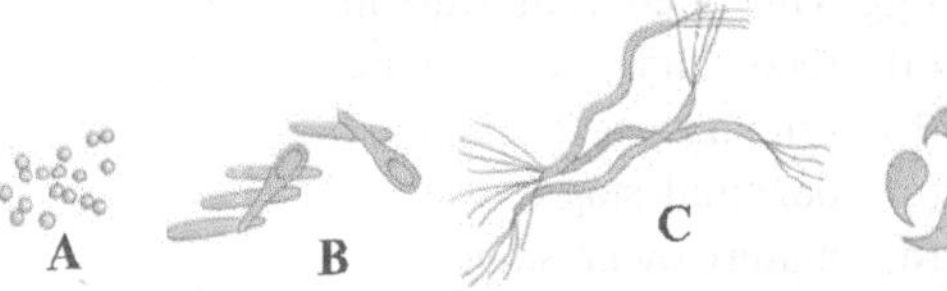

(a) A – Cocci, B – Bacilli, C – Spirilla, D – Vibrio
(b) A – Bacilli, B – Cocci, C – Spirilla, D – Vibrio
(c) A – Spirilla, B – Bacilli, C – Cocci, D – Vibrio
(d) A – Spirilla, B – Vibrio, C – Cocci, D – Bacilli

140. Match the source gland with respective hormone as well as the function correctly.

	Source gland	Hormone	Function
(a)	Anterior pituitary	Oxytocin	Contraction of uterus muscles during child birth
(b)	Posterior pituitary	Vasopressin	Stimulates reabsorption of water in the distal tubules in the nephron
(c)	Corpus luteum	Estrogen	Supports pregnancy
(d)	Thyroid	Thyroxine	Regulates blood calcium level

141. The nerve transmitter, produced at the synapse and neuromuscular junction, is
(a) GTP (b) ATP
(c) acetylcholine (d) phosphokinase

142. In the chemistry of vision in mammals, the photosensitive substance is called
(a) sclerotin (b) retinol
(c) rhodopsin (d) melanin

143. Emphysema developes mainly because of
(a) Allergy or hypersensitisation
(b) Spasm of the smooth muscles of bronchioles
(c) Cigarette smoking
(d) Inflammation of the alveoli

144. Egg is liberated from ovary in
(a) secondary oocyte stage
(b) primary oocyte stage
(c) oogonial stage
(d) mature ovum stage

145. Animals that can tolerate a narrow range of salinity are
(a) stenohaline (b) euryhaline
(c) anadromous (d) catadromous

146. The most important component of the oral contraceptive pills is
(a) Progesterone
(b) Growth hormone
(c) Thyroxine
(d) Luteinizing hormone

147. The following graph of relative concentrations of the four hormones present in the blood plasma of a woman during her menstrual cycle. Identify the hormones.

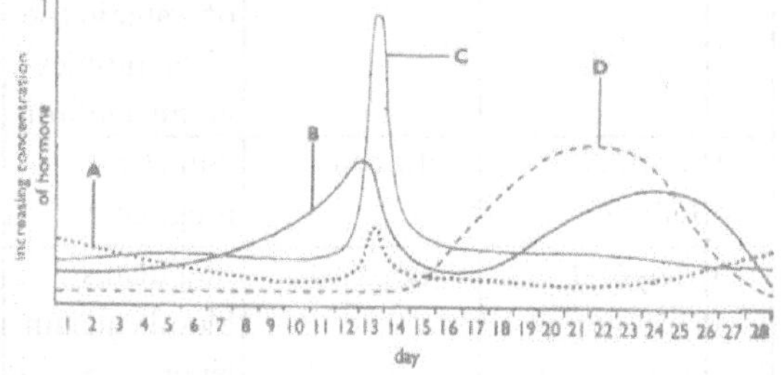

	A	B	C	D
(a)	FSH	Progesterone	LH	Oestrogen
(b)	LH	Progesterone	FSH	Oestrogen
(c)	FSH	Oestrogen	LH	Progesterone
(d)	LH	Oestrogen	FSH	Progesterone

148. Reproductive health in society can be improved by –
(i) Introduction of sex education in schools.
(ii) Increased medical assistance.
(iii) Awareness about contraception and STDs.
(iv) Equal opportunities to male and female child.
(v) Ban on aminocentesis.
(vi) Encouraging myths and misconceptions.
(a) All of these
(b) (i), (ii), (iv) & (vi)
(c) (i), (ii), (iii), (iv) & (v)
(d) (ii) and (v)

149. Foetal ejection reflex in human female is induced by
(a) release oxytocin from pituitary
(b) fully developed foetus and placenta
(c) differentiation of mammary glands
(d) pressure exerted by amniotic fluid

150. Natality is the characteristic of a population which means
(a) the total number of individuals present per unit area at a given time
(b) the increase in number of individuals in a population under given environmental conditions
(c) loss of individuals due to death in a population under given environmental conditions
(d) the movement of individuals into and out of population

151. In human female the blastocyst
(a) Forms placenta even before implantation
(b) Gets implanted into uterus 3 days after ovulation
(c) Gets nutrition from uterine endometrial secretion only after implantation
(d) Gets implanted in endometrium by the trophoblast cells

152. In humans, male XXY and female XXXX occur due to
(a) aneuploidy
(b) euploidy
(c) autosomal syndrome
(d) None of these

153. Which one of the following is an exotic Indian fish ?
(a) *Catla catla*
(b) *Heteropneustes fossilis*
(c) *Cyprinus carpio*
(d) *Labeo rohita*

154. Which of the following primate is the closest relative of humans?
(a) Rhesus monkey (b) Orangutan
(c) Gorilla (d) Gibbon

155. Adaptive radiation refers to
(a) evolution of different species from a common ancestor
(b) migration of members of a species to different geographical areas
(c) power of adaptation in an individual to a variety of environments
(d) adaptations due to geographical isolation.

156. What was the most significant trend in evolution of modern man (*Homo sapiens*) from his ancestors ?
(a) Upright posture
(b) Shortening of jaws
(c) Binocular vision
(d) Increasing brain capacity

157. Which of the following is a viral disease of poultry birds?
(a) Anthrax (b) Ranikhet
(c) Coccidiosis (d) None of these

158. Which one of the followings breeding process increases homozygosity?
(a) Inbreeding
(b) Out breeding
(c) Cross breeding
(d) Inter-specific breeding

159. Classification of Porifera is based on
(a) branching (b) spicules
(c) reproduction (d) symmetry

160. The kind of epithelium which forms the inner walls of blood vessels is
(a) cuboidal epithelium
(b) columnar epithelium
(c) ciliated columnar epithelium
(d) squamous epithelium

161. Myoglobin is present in
(a) all muscle fibres
(b) white muscle fibres only
(c) red muscle fibres only
(d) both white and red muscle fibres

162. Mast cells are associated with
(a) exocrine glands
(b) endocrine glands
(c) areolar connective tissue
(d) neural tissue

163. Which of the following animal is cold blooded and has 4 - chambered heart?
(a) Salamander (b) *Ornithorhynchus*
(c) Crocodile (d) *Calotes*

164. Man, in the life cycle of *Plasmodium*, is
(a) primary host (b) secondary host
(c) intermediate host (d) None of these

165. The most active phagocytic white blood cells are
(a) neutrophils and monocytes
(b) neutrophils and eosinophils
(c) lymphocytes and macrophages
(d) eosinophils and lymphocytes

166. Human insulin is being commercially produced from a transgenic species of
(a) *Escherichia* (b) *Mycobacterium*
(c) *Rhizobium* (d) *Saccharomyces*

167. Obstacle to large scale transplantation of organs is
(a) insufficiency of organ donors
(b) immunological rejection of foreign bodies
(c) religious or ethnic considerations
(d) lack of effective surgical techniques

168. Which one of the following is an Indian medicinal plant ?
(a) *Saccharum officinarum*
(b) *Rauwolfia serpentina*
(c) *Oryza sativa*
(d) *Solanum melongena*

169. Hypersensitivity to an allergen is associated with
(a) aberrant functioning of the immune mechanism
(b) increase in ambient temperature
(c) age of the individual
(d) food habits

170. Human immuno deficiency virus (HIV) has a protein coat and a genetic material which is
(a) single stranded DNA
(b) double stranded DNA
(c) single stranded RNA
(d) double stranded RNA

171. Dolly sheep was obtained by –
(a) Cloning the udder cell (somatic cell) fused with enucleated oocyte
(b) Cloning of gametes
(c) Tissue culture
(d) None of the above

172. Which of the following forms the basis of DNA finger printing?
(a) The relative proportions of purines and pyrimidines in DNA.
(b) Satellite DNA occurring as highly repeated short DNA segments.
(c) The relative difference in the DNA occurrence in blood, skin and saliva.
(d) The relative amount of DNA in the ridges and grooves of the fingerprints.

173. Which one of the following correctly represents the normal adult human dental formula ?
(a) $\frac{3}{3}, \frac{1}{1}, \frac{3}{2}, \frac{1}{1}$ (b) $\frac{2}{2}, \frac{1}{1}, \frac{3}{2}, \frac{3}{3}$
(c) $\frac{2}{2}, \frac{1}{1}, \frac{2}{2}, \frac{3}{3}$ (d) $\frac{3}{3}, \frac{1}{1}, \frac{3}{3}, \frac{3}{3}$

174. Toxic substances are detoxified in human body in
(a) kidney (b) lungs
(c) liver (d) stomach

175. The function of rennin is
(a) vasodilation
(b) reduce blood pressure
(c) degradation of angiotensinogen
(d) None of the above

176. Heart is covered by
(a) Peritoneum (b) Pleural membrane
(c) Pericardium (d) Visceral membrane

177. Which of the following is incorrect?
(a) Iodine is needed for thyroxine formation.
(b) Calcium regulates the excitibility of nerve fibres.
(c) Potassium plays an important role in the regulation of acid base balance in cell.
(d) Phosphorus helps to maintain the osmotic pressure of the body fluids.

178. An adult human with average health has systolic and diastolic pressures as
(a) 80mm Hg and 80 mm Hg
(b) 70 mm Hg and 120 mm Hg
(c) 120 mm Hg and 80 mm Hg
(d) 50 mm Hg and 80 mm Hg

179. Reabsorption of water in distal parts of kidney tubules/urine formation is controlled by
(a) relaxin (b) calcitonin
(c) oxytocin (d) vasopressin

180. Dark purplish gland lying on the left side of abdomen is called :
(a) liver (b) spleen
(c) gall bladder (d) appendix

RESPONSE SHEET

Test Code: ....Time taken :.......................... Date:...............

	(a)	(b)	(c)	(d)		(a)	(b)	(c)	(d)		(a)	(b)	(c)	(d)		(a)	(b)	(c)	(d)
1.					46.					91.					136.				
2.					47.					92.					137.				
3.					48.					93.					138.				
4.					49.					94.					139.				
5.					50.					95.					140.				
6.					51.					96.					141.				
7.					52.					97.					142.				
8.					53.					98.					143.				
9.					54.					99.					144.				
10.					55.					100.					145.				
11.					56.					101.					146.				
12.					57.					102.					147.				
13.					58.					103.					148.				
14.					59.					104.					149.				
15.					60.					105.					150.				
16.					61.					106.					151.				
17.					62.					107.					152.				
18.					63.					108.					153.				
19.					64.					109.					154.				
20.					65.					110.					155.				
21.					66.					111.					156.				
22.					67.					112.					157.				
23.					68.					113.					158.				
24.					69.					114.					159.				
25.					70.					115.					160.				
26.					71.					116.					161.				
27.					72.					117.					162.				
28.					73.					118.					163.				
29.					74.					119.					164.				
30.					75.					120.					165.				
31.					76.					121.					166.				
32.					77.					122.					167.				
33.					78.					123.					168.				
34.					79.					124.					169.				
35.					80.					125.					170.				
36.					81.					126.					171.				
37.					82.					127.					172.				
38.					83.					128.					173.				
39.					84.					129.					174.				
40.					85.					130.					175.				
41.					86.					131.					176.				
42.					87.					132.					177.				
43.					88.					133.					178.				
44.					89.					134.					179.				
45.					90.					135.					180.				

TEST ASSESSMENT AND ANALYSIS SHEET

How to fill this sheet : After you finish the paper and have checked it first try to attempt again all the questions which you have done wrong or have not been able to attempt. Go to detailed solutions only after you have attempted all the questions of your own.Attempting the questions on your own without seeing the solutions will help you in filling this table

ANALYSIS OF TEST

Total number of questions	
Questions Attempted	
Correct Questions	
Wrong Questions	
	Net Score

Strike rate $\left(\frac{\text{Correct questions}}{\text{Questions attempt}} \times 100\right) =$ ....

Percentage Marks $\left(\frac{\text{Correct Answers}}{\text{Total number of questions}} \times 100\right) =$ ....

Analysis of wrong questions : Divide the questions which you have done wrong in 3 broad categories.

Reasons for questions attempted wrong	**Number of Questions**
Knew but calculation mistake (solved on my own after exam)	
Knew but got confused (solved on my own after exam)	
Did not knew the question (couldn't solve even after exam)	

Questions not attempted : Divide the questions not attempted in 3 broad categories

Reasons for unattempted questions	**Number of Questions**
Knew the ques but time was less (solved on my own after exam)	
Knew the ques but got confused (solved on my own after exam)	
Did not knew the questions (couldn't solve even after exam)	

List down your weak areas you need to improve your score :

1)____

2)____

3)____

NEET

National Eligibility Entrance Test

Test Series **Full Test - 2**

Time : - 3 Hours **Max. Marks:- 720**

INSTRUCTIONS

1. Immediately fill in the particulars on this page of the Test Booklet with Blue/Black Ball Point Pen. Use of pencil is strictly prohibited.
2. Answers are to be given on Response Sheet provided at the end of this paper. You can tear this Response Sheet along the cut marks provided. Fill in the particulars carefully in the Response Sheet provided.
3. The test is of 3 hours duration.
4. The Test Booklet consists of **180** questions. The maximum marks are **720**.
5. There are four section in the question paper.

   The distribution of marks subjectwise in each part is as under for each correct response.

   Section 1 – Physics (180 Marks) – 45 Questions

   Section 2 – Chemistry (180 Marks) – 45 Questions

   Section 3 – Botany (180 Marks) – 45 Questions

   Section 4 – Zoology (180 Marks) – 45 Questions
6. You will be awarded 4 Marks if you have darkened only the bubble corresponding to the correct answer and zero mark if no bubble is darkened. In all other cases, minus one (– 1) mark will be awarded.
7. Use Blue/Black Ball Point Pen only for writing particulars/marking responses on Side-1 and Side-2 of the Response Sheet.

   Use of pencil is strictly prohibited.
8. You are advised to create an actual examination hall simulation.
9. Rough work is to be done on the space provided for this purpose in the Test Booklet only.
10. Do not fold or make any stray marks on the Response Sheet.
11. Once you have evaluated the marks in the test, you should assess it thoroughly.

SECTION 1 - PHYSICS

1. The dimensions of magnetic field in M, L, T and C (coulomb) is given as
(a) $[MLT^{-1}C^{-1}]$ (b) $[MT^2C^{-2}]$
(c) $[MT^{-1}C^{-1}]$ (d) $[MT^{-2}C^{-1}]$

2. Error in the measurement of radius of a sphere is 1%. Then error in the measurement of volume is
(a) 1% (b) 5%
(c) 3% (d) 8%

3. In the fig. S_1 and S_2 are identical springs. The oscillation frequency of the mass m is f. If one spring is removed, the frequency will become

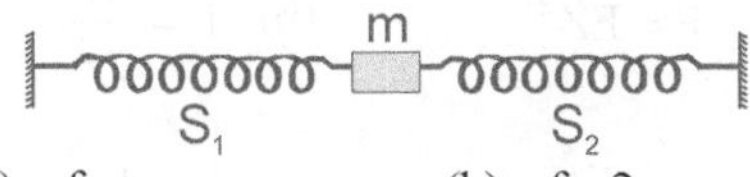

(a) f (b) $f \times 2$
(c) $f \times \sqrt{2}$ (d) $f/\sqrt{2}$

4. A body of mass 2 kg is rotating on a circular path of radius 0.8 m with an angular velocity 44 rad/sec. If radius of path becomes 1 m then value of angular velocity will be
(a) 35.28 rad/sec (b) 14.08 rad/sec
(c) 28.16 rad/sec (d) 24.08 rad/sec

5. The motion of particle is described by the equation $x = a + bt^2$, where a = 15 cm and b = 3 cm/sec^2. Its instant velocity at time 3 sec will be
(a) 36 cm/sec (b) 9 cm/sec
(c) 4.5 cm/sec (d) 18 cm/sec

6. Minimum excitation potential of Bohr's first orbit in hydrogen atom will be
(a) 3.4 eV (b) 3.6 eV
(c) 10.2 eV (d) 13.6 eV

7. A coil of insulated wire is connected to a battery. If it is taken to galvanometer, its pointer is deflected, because
(a) induced current is produced
(b) the coil acts like a magnet
(c) the number of turns in the coil of the galvanometer are changed
(d) None of these

8. A heater of 220 V heats a volume of water in 5 minutes time. A heater of 110 V heats the same volume of water in
(a) 5 minutes (b) 8 minutes
(c) 10 minutes (d) 20 minutes

9. Which of the following graph cannot possibly represent one dimensional motion of a particle?

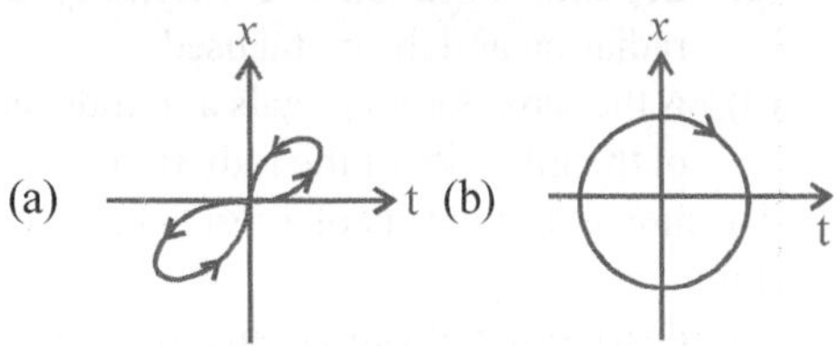

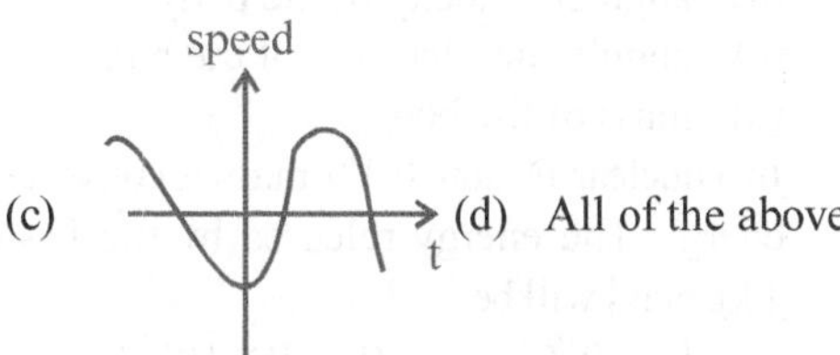

(d) All of the above

10. The perfect gas equation for 4 gram of hydrogen gas is
(a) PV=RT (b) PV=2RT
(c) $PV=\frac{1}{2}RT$ (d) PV=4RT

11. In the given figure, the capacitors C_1, C_3, C_4 and C_5 have a capacitance of 4μF each. If the capacitor C_2 has a capacitance of 10μF, then effective capacitance between A and B is

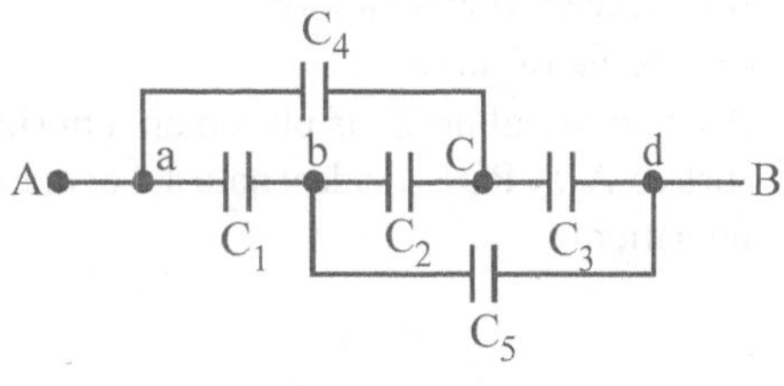

(a) 8μF (b) 6μF
(c) 4μF (d) 2μF

12. According to Einstein's photoelectric equation, the plot of the kinetic energy of the emitted photoelectrons from a metal vs the frequency of the incident radiation gives a straight line whose slope
(a) depends on the intensity of the radiation
(b) depends of the mature of the metal used
(c) depends both on the intensity of the radiation and the metal used
(d) is the same for all metals and independent of the intensity of the radiation

13. The moment of inertia of a rigid body, depends upon
(a) distribution of mass from axis of rotation
(b) angular velocity of the body
(c) angular acceleration of the body
(d) mass of the body.

14. In a nuclear fission, 0.1% mass is converted into energy. The energy released by the fission of 1 kg mass will be
(a) 9×10^{19} J (b) 9×10^{17} J
(c) 9×10^{16} J (d) 9×10^{13} J

15. If the coefficient of cubical expansion is x times of the coefficient of superficial expansion, then value of x is
(a) 3 (b) 2.5
(c) 1.5 (d) 2

16. When a bus suddenly takes a turn, the passengers are thrown outwards because of
(a) speed of motion
(b) inertia of motion
(c) acceleration of motion
(d) None of these

17. An equilateral prism is placed on a horizontal surface. A ray PQ is incident onto it. For minimum deviation

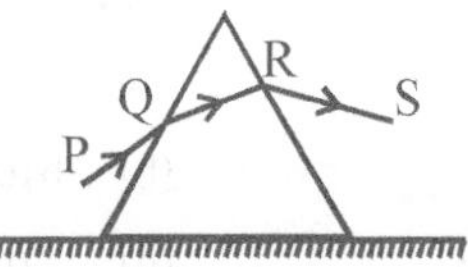

(a) PQ is horizontal
(b) QR is horizontal
(c) RS is horizontal
(d) any one will be horizontal

18. A tin nucleus (atomic number $Z = 50$) has a radius of 6.6×10^{-15} m. The potential on its surface will be (the charge on the proton = 1.6×10^{-19} C)
(a) 1.1×10^7 V (b) 2.1×10^7 V
(c) 3.1×10^7 V (d) 0.15×10^7 V

19. A relation between Faraday's constant F, chemical equivalent E and electrochemical equivalent Z is
(a) $F = EZ^2$ (b) $F = \frac{E}{Z^2}$
(c) $F = \frac{E^2}{Z}$ (d) $F = \frac{E}{Z}$

20. Two rods of the same length and areas of cross-section A_1 and A_2 have their ends at the same temperature K_1 and K_2 are the thremal conductivities of the two rods. The rate of flow of heat is same in both rods if
(a) $\frac{A_1}{A_2} = \frac{K_1}{K_2}$ (b) $\frac{A_1}{A_2} = \frac{K_2}{K_1}$
(c) $A_1A_2 = K_1K_2$ (d) $A_1K_1^2 = A_2K_2^2$

21. A stone is just released from the window of a train moving along a horizontal straight track. The stone will hit the ground following a
(a) straight line path (b) circular path
(c) parabolic path (d) hyperbolic path

22. Lenz's law is a consequence of the law of conservation of
(a) charge (b) mass
(c) energy (d) momentum

23. An equipotential surface is that surface
(a) on which each and every point has the same potential
(b) which has negative potential
(c) which has positive potential
(d) which has zero potential

24. The kinetic energy of a body becomes four times its inital value. The new momentum will be
(a) same as initial value
(b) twice the initial value
(c) thrice the initial value
(d) four times the initial value

25. An observer moves towards a stationary source of sound with a speed 1/5th ofthe speed of sound. The wavelength and frequency of the source emitted any λ and f respectively. The apparent frequency and wavelength recorded by the observer are respectively
(a) 0.8f, 0.8λ (b) 1.2f, 1.2λ
(c) 1.2f, λ (d) f, 1.2λ

26. The photoelectric effect is based upon the law of conservation of
(a) momentum
(b) energy
(c) angular momentum
(d) mass

27. A choke is preferred to a resistance for limiting current in AC circuit because
(a) choke is cheap
(b) there is no wastage of power
(c) choke is compact in size
(d) choke is a good absorber of heat

28. If a body of mass 3 kg is dropped from the top of a tower of height 25m. Then its kinetic energy after 3 sec is
(a) 557 J (b) 748 J
(c) 1050 J (d) 1296 J

29. A gymnast takes turns with her arms & legs stretched. When she pulls her arms & legs in
(a) the angular velocity decreases
(b) the moment of inertia decreases
(c) the angular velocity stays constant
(d) the angular momentum increases

30. A current of I ampere flows in a wire forming a circular arc of radius r metres subtending an angle θ at the centre as shown. The magnetic field at the centre O in tesla is

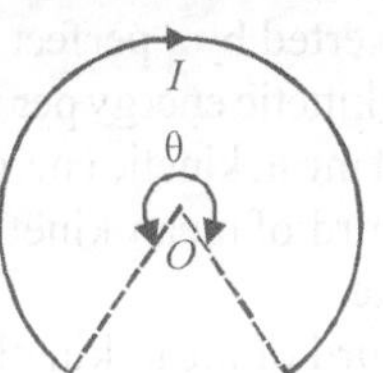

(a) $\frac{\mu_0 I\theta}{4\pi r}$ (b) $\frac{\mu_0 I\theta}{2\pi r}$
(c) $\frac{\mu_0 I\theta}{2r}$ (d) $\frac{\mu_0 I\theta}{4r}$

31. The number of degrees of freedom for each atom of a monoatomic gas is
(a) 3 (b) 5
(c) 6 (d) 1

32. Ability of the eye to see objects at all distances is called
(a) binocular vision (b) myopia
(c) hypermetropia (d) accommodation

33. The velocity of electromagnetic radiation in a medium of permittivity ε_0 and permeability μ_0 is given by
(a) $\sqrt{\frac{\epsilon_0}{\mu_0}}$ (b) $\sqrt{\mu_0 \epsilon_0}$
(c) $\frac{1}{\sqrt{\mu_0 \epsilon_0}}$ (d) $\sqrt{\frac{\mu_0}{\epsilon_0}}$

34. The numerical ratio of displacement to distance is
(a) always less than one
(b) always equal to one
(c) always more than one
(d) equal to or less than one

35. The change in the value of 'g' at a height 'h' above the surface of the earth is the same as at a depth 'd' below the surface of the earth. When both 'd' and 'h' are much smaller than the radius of earth, then which one of the following is correct?
(a) $d = \frac{3h}{2}$ (b) $d = \frac{h}{2}$
(c) $d = h$ (d) $d = 2h$

36. Pressure exerted by a perfect gas is equal to
(a) mean kinetic energy per unit volume
(b) half of mean kinetic energy per unti volume
(c) two third of mean kinetic energy per unit volume
(d) one third of mean kinetic energy per unit volume

37. Transformer is based upon the principle of
(a) self induction (b) mutual induction
(c) eddy current (d) None of these

38. A spherical ball of iron of radius 2 mm is falling through a column of glycerine. If densities of glycerine and iron are respectively 1.3×10^3 kg/m^3 and 8×10^3 kg/m^3, η for glycerine = 0.83 Nm^{-2} sec, then the terminal velocity is
(a) 0.7 m/s (b) 0.07 m/s
(c) 0.007 m/s (d) 0.0007 m/s

39. The time taken by light to pass through 4 mm thick glass slab of refractive index 1.5 will be (velocity of light in air = 3×10^8 m/s)
(a) 8×10^{-11} s (b) 2×10^{-11} s
(c) 8×10^{-8} s (d) 8×10^{-8} s

40. Surface tension of a solution is 30×10^{-2} N/m. The radius of the soap bubble is 5 cm. The surface energy of soap bubble is
(a) 1.8×10^{1} J (b) 1.8×10^{2} J
(c) 1.0×10^{-1} J (d) 1.8×10^{-2} J

41. A cylindrical resonance tube, open at both ends, has a fundamental frequency f in air. If half of the length is dipped vertically in water, the fundamental frequency of the air column will be
(a) $\frac{3f}{2}$ (b) 2f
(c) f (d) $\frac{f}{2}$

42. Which of the following waves have the maximum wavelength ?
(a) Infrared rays (b) UV rays
(c) Radio waves (d) X – rays

43. In an p-n-p transistor working as a common base amplifier current gain is 0.96 and emitter current is 7.2 mA. The base current is
(a) 0.2 mA (b) 0.29 mA
(c) 0.35 mA (d) 0.4 mA

44. The forward biasd diode is
(a) –2V (diode) +2V
(b) –4V (diode) –3V
(c) 3V (diode) 5V
(d) 0V (diode) –2V

45. In V-I characteristic of a p-n junction reverse biasing results in
(a) leakage current
(b) the current barrier across junction increases
(c) no flow of current
(d) large current

SECTION 2 - CHEMISTRY

46. IUPAC name of the following compound :

(cyclopropyl)–C(=O)–N(CH_3)(CH_3)

(a) N, N-dimethylcyclopropanecarboxamide
(b) N-methylcyclopropanamide
(c) cyclopropionamide
(d) none of the above

47 The electrical charge on a colloidal particle is observed by:
(a) Ultramicroscope
(b) Scattering
(c) Brownian movement
(d) Electrophoresis

48. A is a lighter phenol and B is an aromatic carboxylic acid. Separation of a mixture of A and B can be carried out easily by using a solution of
(a) Sodium hydroxide (b) Sodium sulphate
(c) calcium chloride (d) Sodium bicarbonate

49. $(Ag + Pb)$ alloy $\xrightarrow[\text{Zinc is added}]{\text{Melt and}}$ $(Ag + Pb + Zn)$ melt $\xrightarrow{\text{Cool}}$ $\xrightarrow[\text{Layer Y}]{\text{Layer X}}$

Select correct statement based on above scheme.

(a) Layer X contains Zn and Ag
(b) Layer Y contains Pb and Ag but amount of silver in this layer is smaller than in layer X
(c) X and Y are immiscible layers
(d) All are correct statements

50. For the reaction : $H_2 + Cl_2 \xrightarrow{\text{sunlight}} 2HCl$ the order of reaction is

(a) 0 (b) 2
(c) 1 (d) 3

51. Which of the following is not correct ?

(a) $3O_2 \underset{\text{discharge}}{\overset{\text{Silent electric}}{\rightleftharpoons}} 2O_3;\ \Delta H = -284.5\ kJ.$
(b) Ozone undergoes addition reaction with unsaturated carbon compounds.
(c) Sodium thiosulphate reacts with I_2 to form sodium tetrathionate and sodium iodide.
(d) Ozone oxidises lead sulphide to lead sulphate.

52. (I) $n = 3, l = 2, m_1 = -2$
(II) $n = 3, l = 1, m_1 = 0$
(III) $n = 3, l = 0, m_1 = -1$
(IV) $n = 3, l = 2, m_1 = 0$
(V) $n = 3, l = 3, m_1 = -2$

Of these question state designation which does not describe an allowed state for an electron in an atom ?

(a) I and IV (b) III and V
(c) II and V (d) IV and V

53. Which of the following compounds is a good conductor of electricity in solution state?

(a) covalent (b) molecular solid
(c) metallic solid (d) ionic compounds

54. Which is the most suitable reagent among the following to distinguish compound (3) from rest of the compounds ?

1. $CH_3 - C \equiv C - CH_3$
2. $CH_3 - CH_2 - CH_2 - CH_3$
3. $CH_3 - CH_2C \equiv CH$
4. $CH_3 - CH = CH_2.$

(a) Bromine in carbon tetrachloride
(b) Bromine in acetic acid
(c) Alk $KMnO_4$
(d) Ammonical silver nitrate.

55. $N_2 + 3H_2 \longrightarrow 2NH_3$. In this reaction equivalent weight of N_2 is

(a) 4.67 (b) 28
(c) 14 (d) 2.33

56. Which of the following is/are the hazardous pollutant(s) present in automobile exhaust gases?

(i) N_2 (ii) CO
(iii) CH_4 (iv) Oxides of nitrogen

(a) (ii) and (iii) (b) (i) and (ii)
(c) (ii) and (iv) (d) (i) and (iii)

57. $Zn^{2+} \to Zn_{(s)}$; $E^0 = -0.76$ V

$Cu^{2+} \to Cu_{(s)}$; $E^0 = -0.34$ V

Which of the following is spontaneous?

(a) $Zn^{2+} + Cu \to Zn + Cu^{2+}$
(b) $Cu^{2+} + Zn \to Cu + Zn^{2+}$
(c) $Zn^{2+} + Cu^{2+} \to Zn + Cu$
(d) None of these

58. Cuprous compounds such as CuCl, CuCN and CuSCN are the only salts stable in water due to

(a) high hydration energy of Cu^+ ions
(b) their inherent tendency not do disproportionate
(c) diamagnetic nature
(d) insolubility in water

59. A compound formed by elements X and Y crystallizes in a cubic structure in which the X atoms are at the corners of a cube and the Y atoms are at the face centres. The formula of the compound is

(a) XY_3 (b) X_3Y
(c) XY (d) XY_2

60. An alkene having molecular formula C_7H_{14} was subjected to ozonolysis in the presence of zinc dust. An equimolar amount of the following two compounds was obtained

$$\begin{matrix} CH_3 \\ CH_3 \end{matrix} \rangle C=O \text{ and } \begin{matrix} CH_3 \\ CH_3CH_2 \end{matrix} \rangle C=O$$

The IUPAC name of the alkene is
(a) 3, 4-dimethyl-3-pentene
(b) 3, 4-dimethyl-2-pentene
(c) 2, 3-dimethyl-3-pentene
(d) 2, 3-dimethyl-2-pentene

61. At a particular temperature, the vapour pressures of two liquids A and B are respectively 120 and 180 mm of mercury. If 2 moles of A and 3 moles of B are mixed to form an ideal solution, the vapour pressure of the solution at the same temperature will be (in mm of mercury)
(a) 156 (b) 145
(c) 150 (d) 108

62. Which of the following complex ions is expected to absorb visible light?
(a) $[Ti(en)_2(NH_3)_2]^{4+}$
(b) $[Cr(NH_3)_6]^{3+}$
(c) $[Zn(NH_3)_6]^{2+}$
(d) $[Sc(H_2O)_3(NH_3)_3]^{3+}$
(At. no. Zn = 30, Sc = 21, Ti = 22, Cr = 24)

63. According to the adsorption theory of catalysis, the speed of the reaction increases because-
(a) Adsorption lowers the activation energy of the reaction
(b) The concentration of reactant molecules at the active centres of the catalyst becomes high due to strong adsorption
(c) In the process of adsorption, the activation energy of the molecules becomes large
(d) Adsorption produces heat which increases the speed of the reaction

64. A pressure cooker reduces cooking time for food because
(a) boiling point of water involved in cooking is increased
(b) the higher pressure inside the cooker crushes the food material
(c) cooking involves chemical changes helped by a rise in temperature
(d) heat is more evenly distributed in the cooking space

65. Most reactive halide towards S_N1 reaction is
(a) *n*-Butyl chloride
(b) *sec*-Butyl chloride
(c) *tert*-Butyl chloride
(d) Allyl chloride

66. The absolute configuration of the following :

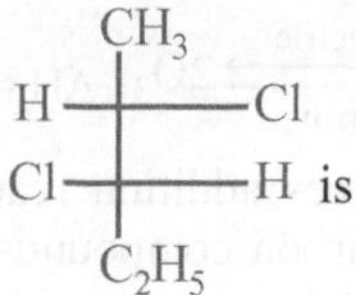

is
(a) 2S, 3R (b) 2S, 3S
(c) 2R, 3S (d) 2R, 3R

67. The rate constant *k*, for the reaction

$$N_2O_5(g) \longrightarrow 2NO_2(g) + \frac{1}{2}O_2(g)$$

is $1.3 \times 10^{-2} s^{-1}$. Which equation given below describes the change of $[N_2O_5]$ with time ? $[N_2O_5]_0$ and $[N_2O_5]_t$ corrospond to concentration of N_2O_5 initially and at time *t*.
(a) $[N_2O_5]_t = [N_2O_5]_0 + kt$
(b) $[N_2O_5]_0 = [N_2O_5]_t e^{kt}$
(c) $\log[N_2O_5]_t = \log[N_2O_5]_0 + kt$
(d) $\text{In} \dfrac{[N_2P_5]_0}{[N_2P_5]_t} = kt$

68. Among the following the most stable compound is :
(a) *cis*-1, 2-cyclohexanediol
(b) *trans*-1, 2-cyclohexanediol
(c) *cis*-1,3-cyclohexanediol
(d) *trans*-1, 3-cyclohexanediol

69. Which of the following is true?
(a) Coke is found in nature
(b) Producer gas is a mixture of CO & H_2
(c) CO is used in the extraction of Ni by Mond's process
(d) CO_2 can be prepared by dehydration of formic acid.

70. Aniline is reacted with bromine water and the resulting product is treated with an aqueous solution of sodium nitrite in presence of dilute hydrochloric acid. The compound so formed is converted into a tetrafluoroborate which is subsequently heated dry. The final product is
(a) 1,3, 5-tribromobenzene
(b) p-bromofluorobenzene
(c) p-bromoaniline
(d) 2,4, 6-tribromofluorobenzene

71. With which one of the following elements silicon should be doped so as to give p-type of semiconductor ?
(a) Germanium (b) Arsenic
(c) Selenium (d) Boron

72. 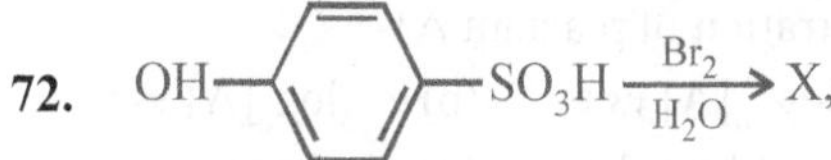

X is identified as
(a) 2, 4, 6-tribromophenol
(b) 2-bromo-4-hydroxylbenzene sulphonic acid
(c) 3, 5-dibromo-4-hydroxybenzene sulphonic acid
(d) 2-bromophenol

73. The appearance of colour in solid alkali metal halides is generally due to
(a) Schottky defect
(b) Frenkel defect
(c) Interstitial positions
(d) F-centres

74. Which of the following is a basic dye?
(a) Congo red (b) Aniline yellow
(c) Alizarin (d) Indigo

75. Aluminium displaces hydrogen from acids but copper does not. A galvanic cell prepared by combining Cu / Cu^{2+} and Al/Al^{3+} has an e.m.f of 2.0 V at 298°K. If the potential of copper electrode is + 0.34 V, that of aluminium is
(a) + 1.66 (b) – 1.66
(c) + 2.34 (d) – 2.3 V

76. An organic compound X on treatment with pyridinium chlorochromate in dichloromethane gives compound Y. Compound Y reacts with I_2 and alkali to form triiodomethane. The compound 'X' is
(a) C_2H_5OH (b) CH_3CHO
(c) CH_3COCH_3 (d) CH_3COOH

77. A sudden large jumps between the values of second and third ionization energies of an element, would be associated with which of the following electronic configuration?
(a) $1s^2, 2s^2, 2p^6, 3s^1$
(b) $1s^2, 2s^2, 2p^6, 3s^1, 3p^1$
(c) $1s^2, 2s^2, 2p^6, 3s^2$
(d) $1s^2, 2s^2, 2p^6, 3s^2, 3p^2$

78. Which of the products is formed when acetone is reacted with barium hydroxide solution?
(a) $CH_3-\overset{O}{\overset{||}{C}}-CH_2-\underset{OH}{\underset{|}{\overset{CH_3}{\overset{|}{C}}}}-CH_3$
(b) $CH_3-\overset{O}{\overset{||}{C}}-\underset{CH_3}{\underset{|}{CH}}-\underset{OH}{\underset{|}{CH}}-CH_3$
(c) $CH_3-\overset{O}{\overset{||}{C}}-\underset{OH}{\underset{|}{CH}}-\underset{CH_3}{\underset{|}{CH}}-CH_3$
(d) $CH_3-\underset{CH_3}{\underset{|}{\overset{OH}{\overset{|}{C}}}}-\underset{CH_3}{\underset{|}{\overset{OH}{\overset{|}{C}}}}-CH_3$

79. For the manufacture of NH_3 by the following reaction

$$N_2 + 3H_2 \rightleftharpoons 2NH_3 + 23kcal$$

The favourable conditions are

(a) low temp., high pressure and catalyst
(b) low temp., low pressure and catalyst
(c) high temp., high pressure and catalyst
(d) high temp., low pressure and catalyst

80. Ethanol can be prepared more easily by which reaction ?

(i) $CH_3CH_2Br + H_2O \longrightarrow CH_3CH_2OH$
(ii) $CH_3CH_2Br + Ag_2O$ (in boiling water) $\longrightarrow CH_3CH_2OH$

(a) by (i) reaction
(b) by (ii) reaction
(c) Both reactions proceed at same rate
(d) by none

81. The oxidation states of iodine in HIO_4, H_3IO_5 and H_5IO_6 are respectively

(a) +1, +3, +7 (b) +7, +7, +3
(c) +7, +7, +7 (d) +7, +5, +3

82. Which is correct statement?

(a) Starch is a polymer of α-glucose
(b) In cyclic structure of fructose, there are four carbons and one oxygen atom
(c) Amylose is a component of cellulose
(d) Proteins are composed of only one type of amino acids

83. Regarding F^- and Cl^- which of the following statements is/are correct?

(i) Cl^- can give up an electron more easily than F^-
(ii) Cl^- is a better reducing agent than F^-
(iii) Cl^- is smaller in size than F^-
(iv) F^- can be oxidized more readily than Cl^-

(a) (i) and (ii) (b) (i), (ii) and (iv)
(c) (iii) and (iv) (d) only (i)

84. Among the following, the species having square planar geometry for central atom are

(i) XeF_4 (ii) SF_4
(iii) $[NiCl_4]^{2-}$ (iv) $[PtCl_4]^{2-}$

(a) (i) and (iv) (b) (i) and (ii).
(c) (ii) and (iii) (d) (iii) and (iv)

85. Water glass is :

(a) Na_2SiO_3 (b) Mg_2Si
(c) $SiCl_4$ (d) $Ca(H_2PO_4)_2$

86. Which one of the following statement is *not true*?

(a) In vulcanization the formation of sulphur bridges between different chains make rubber harder and stronger.
(b) Natural rubber has the *trans* -configuration at every double bond
(c) Buna-S is a copolymer of butadiene and styrene
(d) Natural rubber is a 1, 4 - polymer of isoprene

87. For a first order reaction, to obtain a positive slope, we need to plot {where [A] is the concentration of reactant A}

(a) $-\log_{10}[A]$ *vs* t (b) $-\log_e[A]$ *vs* t
(c) $\log_{10}[A]$ *vs* log t (d) [A] *vs* t

88. Among the following molecules

(i) XeO_3 (ii) $XeOF_4$ (iii) XeF_6

Those having same number of lone pairs on Xe are

(a) (i) and (ii) only (b) (i) and (iii) only
(c) (ii) and (iii) only (d) (i), (ii) and (iii)

89. The de-Broglie wavelength associated with a particle of mass 10^{-6} kg moving with a velocity of 10 ms^{-1} is :

(a) 6.63×10^{-7} m (b) 6.63×10^{-16} m
(c) 6.63×10^{-21} m (d) 6.63×10^{-29} m

90. Which of the following is not a fat soluble vitamin?

(a) Vitamin B complex (b) Vitamin D
(c) Vitamin E (d) Vitamin A

SECTION 3 - BOTANY

91. The first organisms to appear on earth were
(a) photoautotrophs (b) chemoautotrophs
(c) chemoheterotrophs (d) heterotrophs

92. Floridean starch is found in
(a) chlorophyceae (b) rhodophyceae
(c) phaeophyceae (d) cyanophyceae

93. A water fern which is used as a green manure in rice fields is
(a) *Salvinia* (b) *Mucor*
(c) *Aspergillus* (d) *Azolla*

94. Which of the following does not have stomata ?
(a) Hydrophytes
(b) Mesophytes
(c) Xerophytes
(d) Submerged hydrophytes

95. Which is correct about transport or conduction of substances?
(a) Organic food moves up through phloem
(b) Organic food moves up through xylem
(c) Inorganic food moves upwardly and downwardly through xylem
(d) Organic food moves upwardly and downwardly through phloem

96. The quiescent centre in root meristem serves as a
(a) site for storage of food which is utilized during maturation.
(b) reservoir of growth hormones.
(c) reserve for replenishment of damaged cells of the meristem.
(d) region for absorption of water.

97. Which one of the following option correctly represents the type of life cycle patterns?

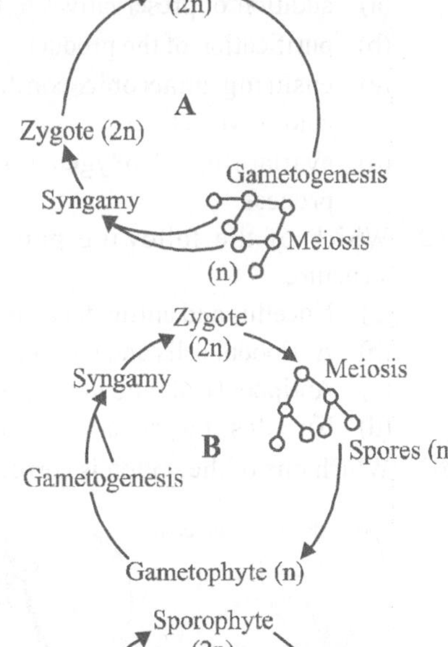

(a) A - Diplontic, B - Haplodiplontic, C - Haplontic
(b) A - Haplodiplontic, B - Haplontic, C - Diplontic
(c) A - Haplontic, B - Diplontic, C - Haplodiplontic
(d) A - Diplontic, B - Haplontic, C - Haplodiplontic

98. The "Eyes" of the potato tuber are
(a) root buds (b) flower buds
(c) shoot buds (d) axillary buds

99. Acid rains are produced by
(a) excess emissions of NO_2 and SO_2 from burning fossil fuels
(b) excess production of NH_3 by industry and coal gas
(c) excess release of carbon monoxide by incomplete combustion
(d) excess formation of CO_2 by combustion and animal respiration

100. Which of these is a pollution related to occupational health hazard disease ?
(a) Pneumoconiosis (b) Asthma
(c) Flurosis (d) Silicosis

101. Stirred tank bioreactors have been designed for
(a) addition of preservatives to the product
(b) purification of the product
(c) ensuring anaerobic conditions in the culture vessel
(d) availability of oxygen throughout the process

102. Which of the following pair has haploid structures?
(a) Nucellus and antipodal cells
(b) Antipodal cells and egg cell
(c) Antipodal cells and megaspore mother cell
(d) Nucellus and primary endosperm nucleus

103. Which one of the option is correct?

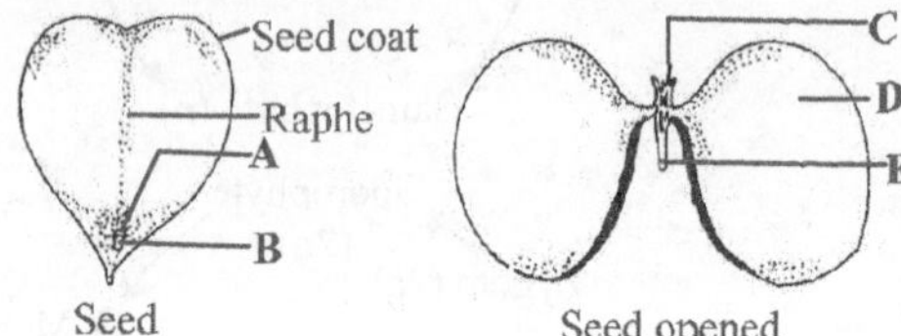

(a) A - Hilum, B - Micropyle, C - Radicle, D - Cotyledon, E - Plumule
(b) A - Hilum, B - Micropyle, C - Plumule, D - Cotyledon, E - Radicle
(c) A - Micropyle, B - Hilum, C - Plumule, D - Cotyledon, E - Radicle
(d) A - Hilum, B - Micropyle, C - Plumule, D - Radicle, E - Cotyledon

104. Double fertilization is fusion of
(a) two eggs
(b) two eggs and polar nuclei with pollen nuclei
(c) one male gamete with egg and other with synergid
(d) one male gamete with egg and other with secondary nucleus

105. Plasmolysis occurs when the cell is placed in
(a) isotonic solution
(b) hypotonic solution
(c) hypertonic solution
(d) None of the above

106. 'Whip-tail' disease in cauliflower is noted due to deficiency of
(a) manganese (b) magnesium
(c) molybdenum (d) nitrogen

107. The organelles which take part in photo-respiration are
(a) chloroplast, mitochondria, nucleus
(b) chloroplast, mitochondria, lysosome
(c) mitochondria, chloroplast, peroxisome
(d) mitochondria, lysosomes, peroxisome

108. Photosynthetically active radiation is represented by the range of wavelength of
(a) 340-450 nm (b) 400-700 nm
(c) 500-600 nm (d) 400-950 nm

109. Which one of the following is essential for photolysis of water ?
(a) Manganese (b) Zinc
(c) Copper (d) Boron

110. Which element is required in the germination of pollen grain?
(a) Chlorine (b) Potassium
(c) Boron (d) Calcium

111. Study the figure and choose the correct statement regarding this.

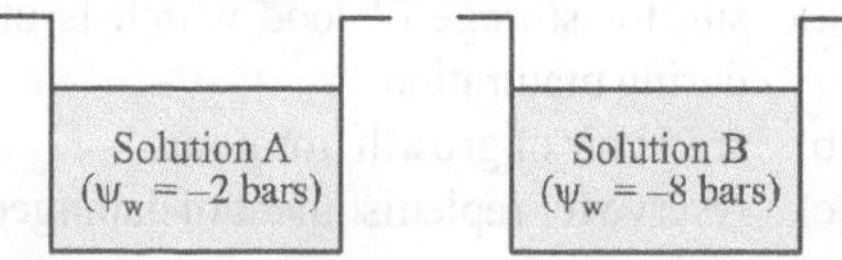

(a) Kinetic energy (K.E.) of H_2O in A solution > K.E. of water in B solution.
(b) K.E. of water in B solution > K.E. of water in A solution.
(c) K.E. of water in A solution = K.E. of H_2O in B solution.
(d) Water potential has nothing to do with K.E. of water in a solution.

112. The dihybrid ratio in F_2 - generation is
(a) 1 : 1 : 1 : 1 (b) 2 : 1 : 2 : 1
(c) 3:1 (d) 9 : 3 : 3 : 1

113. The zone of atmosphere that lies near the ground is
(a) troposphere (b) stratosphere
(c) homosphere (d) ionosphere

114. Reverse transcriptase is
(a) RNA dependent RNA polymerase
(b) DNA dependent RNA polymerase
(c) DNA dependent DNA polymerase
(d) RNA dependent DNA polymerase

115. Jaya and Ratna developed for green revolution in India are the varieties of
(a) maize (b) rice
(c) wheat (d) bajra

116. A lake near a village suffered heavy mortality of fishes within a few days. Consider the following reasons for this.
(i) Lots of urea and phosphate fertilizer were used in the crops in the vicinity
(ii) The area was sprayed with DDT by an aircraft
(iii) The lake water turned green and stinky
(iv) Phytoplankton populations in the lake declined initially thereby greatly reducing photosynthesis.
(a) (i) and (iii) (b) (i), (ii) and (iv)
(c) (iii) and (iv) (d) (ii), (iii) and (iv)

117. Which of the following ecosystems has highest rate of gross primary production?
(a) Grasslands
(b) Mangroves
(c) Coral reefs
(d) Equatorial rain forest

118. Glycogen is a polymer of
(a) galactose (b) glucose
(c) fructose (d) sucrose

119. What is the end product of glycolysis of a glucose molecule?
(a) Fructose 1, 6- diphosphate
(b) Pyruvate and ATP
(c) Phosphoglyceraldehyde
(d) Lactic acid and ATP

120. Which one of the following acids is a derivative of carotenoids ?
(a) Indole-3 -acetic acid
(b) Gibberellic acid
(c) Abscisic acid
(d) Indole butyric acid

121. Which one of the following generally acts as an antagonist to gibberellins?
(a) Zeatin (b) Ethylene
(c) ABA (d) IAA

122. Other than DNA polymerase, which is/ are the enzyme(s) involved in DNA synthesis?
(a) Topoisomerase (b) Helicase
(c) RNA primase (d) All of these

123. Which of the following is not true for oxidative phosphorylation?
(a) It uses oxygen as the initial electron donor.
(b) It involves the redox reactions of electron transport chain.
(c) It involves an ATP synthase located in the inner mitochondrial membrane.
(d) It depends on chemiosmosis.

124. Consumption of which one of the following foods can prevent the kind of blindness associated with vitamin 'A' deficiency ?
(a) '*Flavr Savr*' tomato
(b) Canolla
(c) Golden rice
(d) Bt-Brinjal

125. Biodiversity of a geographical region represents
(a) endangered species found in the region.
(b) the diversity in the organisms living in the region.
(c) genetic diversity present in the dominant species of the region.
(d) species endemic to the region.

126. Use of transgenic plants as biological factories for the production of special chemicals is called–
(a) Molecular farming
(b) Molecular genetics
(c) Molecular mapping
(d) Dry farming

127. Which animal has become extinct from India?
(a) Snow Leopard (b) Hippopotamus
(c) Wolf (d) Cheetah

128. Agarose extracted from sea weeds finds use in :
(a) Spectrophotometry
(b) Tissue culture
(c) PCR
(d) Gel electrophoresis

129. Streptomycin is obtained from
(a) *Streptomyces griseus*
(b) *S. aureofaciens*
(c) *S. venezuelae*
(d) *S. ramosus*

130. The nucleolus is the site of formation of
(a) spindle fibres (b) chromosomes
(c) ribosomes (d) peroxysomes

131. Niche overlap indicates
(a) two different parasites on the same host
(b) sharing of one or more resources between the two species
(c) mutualism between two species
(d) active cooperation between two species

132. Coconut milk (coconut water) is widely used in tissue culture because it contains
(a) auxins (b) ethylene
(c) cytokinin (d) gibberellins

133. The okazaki fragments in DNA chain growth
(a) polymerize in the 3' - to - 5' direction and forms replication fork
(b) prove semi-conservative nature of DNA replication
(c) polymerize in the 5' - to - 3' direction and explain 3' - to - 5' DNA replication
(d) result in transcription

134. Leghaemoglobin helps in
(a) nitrogen fixation
(b) protecting nitrogenase from O_2
(c) destroys bacteria
(d) transport of food in plants

135. Important objective of biotechnology in agriculture section is –
(a) To produce pest resistant varieties of plants
(b) To increase the nitrogen content
(c) To decrease the seed number
(d) To increase the plant weight

ZOOLOGY

136. A chordate character is
(a) gills
(b) spiracles
(c) postanal tail
(d) chitinous exoskeleton

137. Which bacteria is utilized in gober gas plant?
(a) Methanogens
(b) Nitrifying bacteria
(c) Ammonifying bacteria
(d) Denitrifying bacteria

138. Excretory organs of Cockroach are
(a) flame cells
(b) nephridia
(c) hreen glands
(d) malpighian tubules

139. What is common among silverfish, scorpion, crab and honey bee?
(a) Compound eyes
(b) Poison gland
(c) Jointed legs
(d) Metamorphosis

140. Tendons and ligaments are the example of
(a) areolar connective tissue
(b) adipose tissue
(c) dense regular connective tissue
(d) loose connective tissue

141. The type of epithelial cells which line inner surface of fallopian tubes, bronchioles and small bronchi are known as
(a) squamous epithelium
(b) columnar epithelium
(c) ciliated epithelium
(d) cubical epithelium

142. The figure given below shows a small part of human lung where exchange of gases takes place. In which one of the options given below, the one part A, B, C or D is correctly identified along with its function?

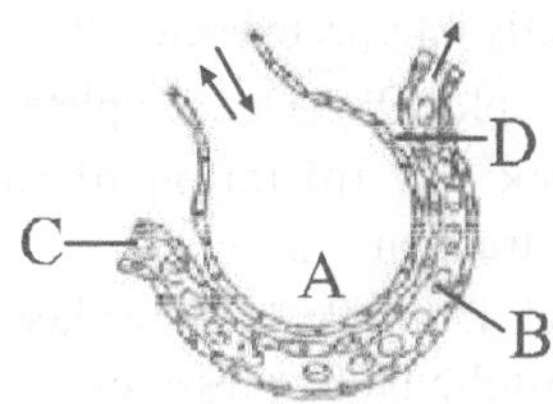

Options
(a) B : Red blood cell - transport of CO_2 mainly
(b) C : Arterial capillary - passes oxygen to tissues
(c) A : alveolar cavity - main site of exchange of respiratory gases
(d) D : Capillary wall - exchange of O_2 and CO_2 takes place here

143. Which one of the following groups of structures/organs have similar function?
(a) Typhlosole in earthworm, intestinal villi in rat and contractile vacuole in *Amoeba.*
(b) Nephridia in earthworm, malpighian tubules in cockroach and urinary tubules in rat.
(c) Antennae of cockroach, tympanum of frog and clitellum of earthworm.
(d) Incisors of rat, gizzard (proventriculus) of cockroach and tube feet of starfish.

144. The correct sequence in the process of development of human embryo is
(a) fertilization—zygote—cleavage—morula—blastula—gastrula
(b) fertilization—cleavage—morula—zygote—blastula—gastrula
(c) fertilization—zygote—blastula—morula—cleavage—gastrula
(d) cleavage—zygote—fertilization—morula—blastula—gastrula

145. Vasa efferentia are the ductules leading from
(a) testicular lobules to rete testis
(b) rete testis to vas deferens
(c) vas deferens to epididymis
(d) epididymis to urethra

146. Which one of the following is the correct matching of the events occurring during menstrual cycle?
(a) Proliferative phase: Rapid regeneration of myometrium and maturation of Graffian follicle.
(b) Development of corpus luteum : Secretory phase and increased secretion of progesterone.
(c) Menstruation: Breakdown of myometrium and ovum not fertilised.
(d) Ovulation: LH and FSH attain peak level and sharp fall in the secretion of progesterone.

147. The number of births per thousand people in the population is expressed as :
(a) Reproduction rate (b) Conception rate
(c) Crude birth rate (d) Growth rate

148. Umbilical cord has
(a) Two arteries carrying blood to placenta and one vein returning blood to foetus
(b) One artery carrying blood to placenta and two veins returning blood to foetus
(c) Two arteries bringing blood to foetus and one vein carrying blood to placenta
(d) One artery bringing blood to foetus and two veins carrying blood to placenta

149. Refer the following diagram and identify the parts of a kidney indicated

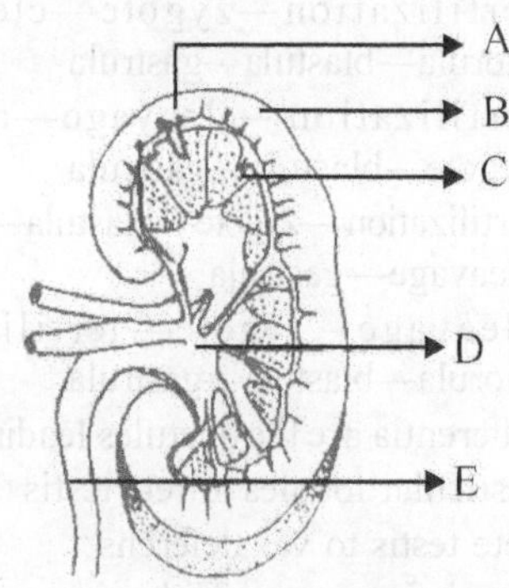

(a) A = Cortex, B = Nephron, C = Pelvis, D = Medulla, E = Ureter
(b) A = Cortex, B = Medulla, C = Nephron, D = Pelvis, E = Ureter
(c) A = Nephron, B = Cortex, C = Medulla, D = Ureter, E = Pelvis
(d) A = Nephron, B = Cortex, C = Medulla, D = Pelvis, E = Ureter

150. The logistic population growth is expressed by the equation :

(a) $dt/dN = Nr\left(\frac{K-N}{K}\right)$

(b) $dN/dt = rN\left(\frac{K-N}{K}\right)$

(c) $dN/dt = rN$

(d) $dN/dt = rN\left(\frac{N-K}{N}\right)$

151. Which one of the following pairs of food components in humans reaches the stomach totally undigested ?
(a) Starch and cellulose
(b) Protein and starch
(c) Starch and fat
(d) Fat and cellulose

152. Which one of the following is a **correct** matching pair?
(a) Lubb - Sharp closure of AV valves at the beginning of ventricular systole.
(b) Dup - Sudden opening of semilunar valves at the beginning of ventricular diastole.
(c) Pulsation of the radial artery valves in the blood vessels.
(d) Purkinje fibers - Initiation of the heart beat.

153. Proximal and distal convoluted tubules are parts of
(a) Seminiferous tubules (b) Nephron
(c) Oviduct (d) Vas deferens

154. Which one of the components of ECG in human is correctly interpreted below ?
(a) Complex QRS-One complete Pulse
(b) Peak T - Initiation of total cardiac contraction
(c) Peak P and Peak R together-Systolic and diastolic blood pressures
(d) Peak P- Initiation of left atrial contraction only

155. With reference to the blood in a normal person, which one of the following statements is correct?
(a) Compared to arteries, veins are less numerous and hold less of the body's blood at any given time.
(b) Blood cells constitute about 70 percent of the total volume of the blood.
(c) White blood cells (WBC) are made by lymph nodes only.
(d) The blood has more platelets than WBC.

156. Deposition of uric acid crystals within the synovial joint causes:
(a) osteoarthritis
(b) rheumatoid arthritis
(c) gout
(d) paralysis

157. Homologous organs are
(a) Wings of Insects and Bat
(b) Gills of Fish and lungs of Rabbit
(c) Pectoral fins of Fish and fore limbs of Horse
(d) Wings of Grosshopper and Crow

158. In Down's syndrome of a male child, the sex complement is
(a) XO (b) 45 XY
(c) 45 XX (d) XXY

159. Which of the following combinations is generally recommended for composite fish farming in India?
(a) *Catla, Labeo, Cirrhinus*
(b) *Catla, Cyprinus, Clarias*
(c) *Clarias, Channa, Cyprinus*
(d) *Cirrhinus, Cyprinus, Channa*

160. Lack of independent assortment of two genes A and B in fruit fly *Drosophila* is due to
(a) repulsion (b) recombination
(c) linkage (d) crossing over

161. Jurassic period of the mesozoic era was characterised by
(a) Radiation of reptiles and origin of mammal-like reptiles
(b) Dinosaurs become extinct and angiosperms appeared
(c) Flowering plants and first dinosaurs appeared
(d) Gymnosperms were dominant plants and first birds appeared

162. The "Cri-du-Chat" syndrome is caused by change in chromosome structure involving
(a) deletion (b) duplication
(c) inversion (d) translocation

163. In history of biology, human genome project led to the development of :
(a) biotechnology (b) biomonitoring
(c) bioinformatics (d) biosystematics

164. Look at the diagram given below.

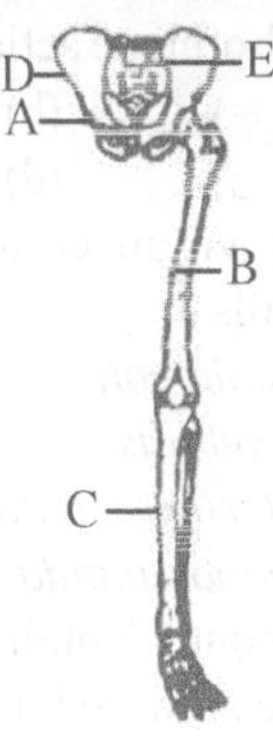

Parts labelled as 'A', 'B', 'C', 'D' and 'E' respectively indicate
(a) ilium, femur, tibia, pubis and sacrum
(b) pubis, tibia, femur, ilium and sacrum
(c) ilium, femur, tibia, pubis, and sacrum
(d) pubis, femur, tibia, ilium and sacrum

165. Oxygen dissociation curve of haemoglobin is
(a) Sigmoid (b) Hyperbolic
(c) Linear (d) Hypobolic

166. Which of the following is made up of a single bone in mammal ?
(a) Dentary (b) Hyoid
(c) Upper jaw (d) All of these

167. Intercostal muscles are found attached with
(a) diaphragm (b) ribs
(c) pleura (d) lungs

168. Eustachian tube connects
(a) External ear with middle ear
(b) External ear with internal ear
(c) Middle ear with pharynx
(d) Internal pharynx

169. Which one of the following pairs correctly matches a hormone with a disease resulting from its deficiency?
(a) Luteinizing - Failure of ovulation
(b) Insulin - Diabetes insipidus
(c) Thyroxine - Tetany
(d) Parathyroid - Diabetes mellitus

170. Which one of the following is not a second messenger in hormone action ?
(a) Calcium (b) Sodium
(c) cAMP (d) cGMP

171. A genetically engineered bacteria used for clearing oil spills is :
(a) *Escherischia coli*
(b) *Bacillus subtilis*
(c) *Agrobacterium tumifaciens*
(d) *Pseudomonas putida*

172. What is true about T-lymphocytes in mammals?
(a) These are produced in thyroid
(b) There are three main types — cytotoxic T-cells, helper T-cells and suppressor T-cells
(c) These originate in lymphoid tissues
(d) They scavenge damaged cells and cellular debris

173. The use of *Cannabis* products (bhang, ganja, charas, marijuana or hashish) causes
(a) depression of brain activity and feeling of calmness
(b) alters thoughts, perceptions and feelings
(c) suppresses brain function and relieves pain
(d) stimulates the nervous system and increases alterness and activity.

174. During anaerobic digestion of organic waste, such as in producing biogas, which one of the following is left undegraded ?
(a) Lipids
(b) Lignin
(c) Hemi-cellulose
(d) Cellulose

175. Carcinoma refers to
(a) benign tumours of the connective tissue
(b) malignant tumours of the connective tissue
(c) malignant tumours of the skin or mucous membrane
(d) malignant tumours of the colon

176. Short-lived immunity acquired from mother to foetus across placenta or through mother's milk to the infant is categorised as
(a) innate non-specific immunity
(b) active immunity
(c) passive immunity
(d) cellular immunity

177. African sleeping sickness is due to
(a) *Plasmodium vivax* transmitted by Tsetse fly
(b) *Trypanosoma lewsii* transmitted by Bed Bug
(c) *Trypanosoma gambiense* transmitted by *Glossina palpalis*
(d) *Entamoeba gingivalis* spread by Housefly

178. Which one of the following are rich in fructose, calcium and some enzymes?
(a) Male accessory glands
(b) Liver
(c) Pancreas
(d) Salivary glands

179. The number of floating ribs, in the human body, is
(a) 6 pairs (b) 5 pairs
(c) 3 pairs (d) 2 pairs

180. Which one of the following statements is correct?
(a) Cro-Magnon man's fossil has been found in Ethiopia
(b) *Homo erectus* is the ancestor of man
(c) Neanderthal man is the direct ancestor of *Homo sapiens*
(d) *Australopithecus* is the real ancestor of modern man

RESPONSE SHEET

Test Code: ....Time taken :......................Date:...........

	(a)	(b)	(c)	(d)		(a)	(b)	(c)	(d)		(a)	(b)	(c)	(d)		(a)	(b)	(c)	(d)
1.					46.					91.					136.				
2.					47.					92.					137.				
3.					48.					93.					138.				
4.					49.					94.					139.				
5.					50.					95.					140.				
6.					51.					96.					141.				
7.					52.					97.					142.				
8.					53.					98.					143.				
9.					54.					99.					144.				
10.					55.					100.					145.				
11.					56.					101.					146.				
12.					57.					102.					147.				
13.					58.					103.					148.				
14.					59.					104.					149.				
15.					60.					105.					150.				
16.					61.					106.					151.				
17.					62.					107.					152.				
18.					63.					108.					153.				
19.					64.					109.					154.				
20.					65.					110.					155.				
21.					66.					111.					156.				
22.					67.					112.					157.				
23.					68.					113.					158.				
24.					69.					114.					159.				
25.					70.					115.					160.				
26.					71.					116.					161.				
27.					72.					117.					162.				
28.					73.					118.					163.				
29.					74.					119.					164.				
30.					75.					120.					165.				
31.					76.					121.					166.				
32.					77.					122.					167.				
33.					78.					123.					168.				
34.					79.					124.					169.				
35.					80.					125.					170.				
36.					81.					126.					171.				
37.					82.					127.					172.				
38.					83.					128.					173.				
39.					84.					129.					174.				
40.					85.					130.					175.				
41.					86.					131.					176.				
42.					87.					132.					177.				
43.					88.					133.					178.				
44.					89.					134.					179.				
45.					90.					135.					180.				

TEST ASSESSMENT AND ANALYSIS SHEET

How to fill this sheet : After you finish the paper and have checked it first try to attempt again all the questions which you have done wrong or have not been able to attempt. Go to detailed solutions only after you have attempted all the questions of your own.Attempting the questions on your own without seeing the solutions will help you in filling this table

ANALYSIS OF TEST	
Total number of questions	
Questions Attempted	
Correct Questions	
Wrong Questions	
	Net Score

Strike rate $\left(\frac{\text{Correct questions}}{\text{Questions attempt}} \times 100\right) =$ ....

Percentage Marks $\left(\frac{\text{Correct Answers}}{\text{Total number of questions}} \times 100\right) =$ ....

Analysis of wrong questions : Divide the questions which you have done wrong in 3 broad categories.

Reasons for questions attempted wrong	**Number of Questions**
Knew but calculation mistake (solved on my own after exam)	
Knew but got confused (solved on my own after exam)	
Did not knew the question (couldn't solve even after exam)	

Questions not attempted : Divide the questions not attempted in 3 broad categories

Reasons for unattempted questions	**Number of Questions**
Knew the ques but time was less (solved on my own after exam)	
Knew the ques but got confused (solved on my own after exam)	
Did not knew the questions (couldn't solve even after exam)	

List down your weak areas you need to improve your score :

1)______________________________

2)______________________________

3)______________________________

NEET

National Eligibility Entrance Test

Test Series **Full Test - 3**

Time : - 3 Hours **Max. Marks:- 720**

INSTRUCTIONS

1. Immediately fill in the particulars on this page of the Test Booklet with Blue/Black Ball Point Pen. Use of pencil is strictly prohibited.
2. Answers are to be given on Response Sheet provided at the end of this paper. You can tear this Response Sheet along the cut marks provided. Fill in the particulars carefully in the Response Sheet provided.
3. The test is of 3 hours duration.
4. The Test Booklet consists of **180** questions. The maximum marks are **720**.
5. There are four section in the question paper.

   The distribution of marks subjectwise in each part is as under for each correct response.

   Section 1 – Physics (180 Marks) – 45 Questions

   Section 2 – Chemistry (180 Marks) – 45 Questions

   Section 3 – Botany (180 Marks) – 45 Questions

   Section 4 – Zoology (180 Marks) – 45 Questions
6. You will be awarded 4 Marks if you have darkened only the bubble corresponding to the correct answer and zero mark if no bubble is darkened. In all other cases, minus one (– 1) mark will be awarded.
7. Use Blue/Black Ball Point Pen only for writing particulars/marking responses on Side-1 and Side-2 of the Response Sheet.

   Use of pencil is strictly prohibited.
8. You are advised to create an actual examination hall simulation.
9. Rough work is to be done on the space provided for this purpose in the Test Booklet only.
10. Do not fold or make any stray marks on the Response Sheet.
11. Once you have evaluated the marks in the test, you should assess it thoroughly.

SECTION 1 - PHYSICS

1. Which one of the following represents the correct dimensions of the coefficient of viscosity?
(a) $[ML^{-1}T^{-1}]$ (b) $[MLT^{-1}]$
(c) $[ML^{-1}T^{-2}]$ (d) $[ML^{-2}T^{-2}]$

2. The component of vector $\bar{a} = 2i + 3j$ along the vector $i + j$ is
(a) $\frac{5}{\sqrt{2}}$ (b) $10\sqrt{2}$
(c) $5\sqrt{2}$ (d) 5

3. In pressure-volume diagram, the isochoric, isothermal, isobaric and iso-entropic parts respectively, are

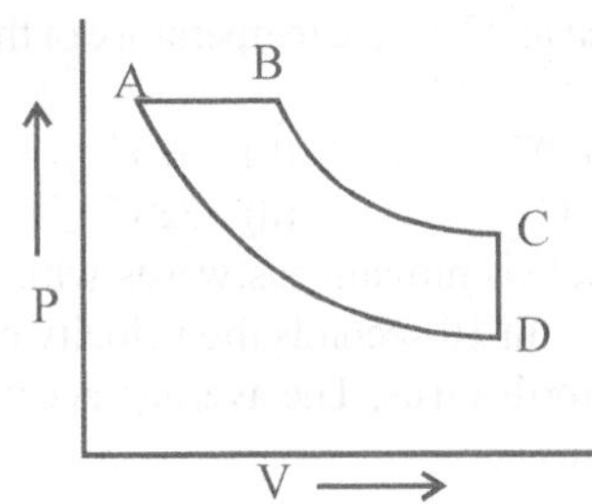

(a) BA, AD, DC, CB (b) DC, CB, BA, AD
(c) AB, BC, CD, DA (d) CD, DA, AB, BC.

4. A satellite of mass m is put into a circular orbit of height h from the surface of the earth (mass = M_e, radius = R_e). The orbital speed of satellite is
(a) $\sqrt{\frac{2GM_e h}{R_e(R_e+h)}}$ (b) $\sqrt{\frac{2GM_e}{R_e+h}}$
(c) $\sqrt{\frac{GM_e}{(R_e+h)^3}}$ (d) $\sqrt{\frac{GM_e}{2(R_e+h)}}$

5. A point initially at rest moves along x-axis. Its acceleration varies with time as $a = (6t + 5)\text{m/s}^2$. If it starts from origin, the distance covered in 2s is
(a) 20m (b) 18m
(c) 16m (d) 25m

6. What is the ratio of the shortest wavelength of the Balmer series to the shortest wavelength of the Lyman series ?
(a) 4 : 1 (b) 4 : 3
(c) 4 : 9 (d) 5 : 9

7. Use of eddy currents is done in the following except
(a) moving coil galvanometer
(b) electric brakes
(c) induction motor
(d) dynamo

8. Kirchhoff's I and II laws are based on conservation of
(a) energy and charge respectively
(b) charge and energy respectively
(c) mass and charge respectively
(d) None of these

9. A passenger travels along the straight road for half the distance with velocity v_1 and the remaining half distance with velocity v_2. Then average velocity is given by
(a) v_1v_2 (b) v_2^2/v_1^2
(c) $(v_1 + v_2)/2$ (d) $2v_1v_2/(v_1 + v_2)$

10. An ideal gas heat engine operates in a Carnot cycle between 227°C and 127°C. It absorbs 6 kcal at the higher temperature. The amount of heat (in kcal) converted into work is equal to
(a) 1.2 (b) 4.8
(c) 3.5 (d) 1.6

11. The current I and voltage V graphs for a given metallic wire at two different temperatures T_1 and T_2 are shown in the fig. It is concluded that

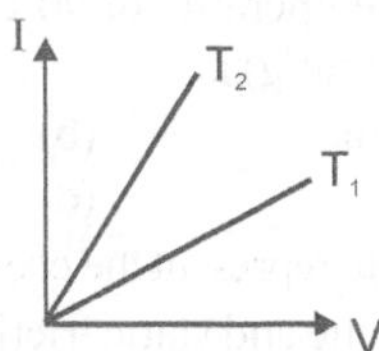

(a) $T_1 > T_2$ (b) $T_1 < T_2$
(c) $T_1 = T_2$ (d) $T_1 = 2T_1$

12. The work function of a metallic substance is 5 eV. The threshold frequency is approximately
(a) 1.6×10^7 Hz (b) 8.68×10^{15} Hz
(c) 9.68×10^{17} Hz (d) 1.2×10^{15} Hz

13. With O as the origin of the coordinate axis, the X and Y-coordinates of the centre of mass of the system of particles shown in the figure may be given as (Here m and 2m represent the masses of the particles).

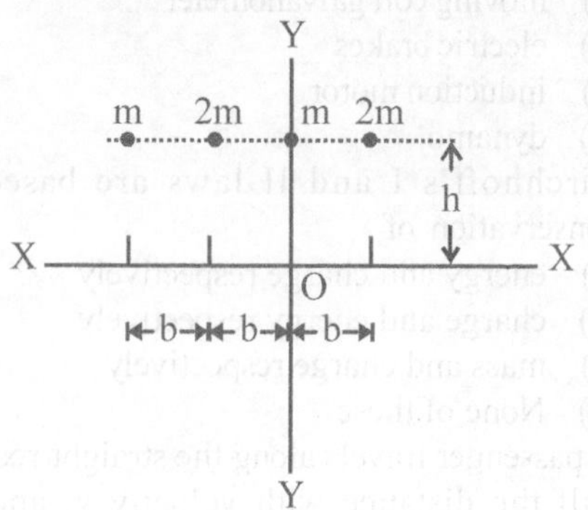

(a) $\left(-\frac{b}{2},0\right)$ (b) $\left(-\frac{b}{2},b\right)$
(c) $\left(-\frac{b}{3},b\right)$ (d) $\left(-\frac{2}{5},b,b\right)$

14. A nuclear reaction is given by
$_{Z}X^{A} \rightarrow {}_{Z+1}Y^{A} + {}_{-1}e^{0} + \bar{\nu}$, represents
(a) fission (b) β-decay
(c) α-decay (d) fusion

15.. Steam is passed into 22 gm of water at 20°C. The mass of water that will be present when the water acquires a temperatue of 90°C (Latent heat of steam is 540 cal/g) is
(a) 24.83 gm (b) 24 gm
(c) 36.6 gm (d) 30 gm

16. If μ_r, μ_k and μ_s represent the coefficients of rolling friction, kinetic and static friction, then
(a) $\mu_r = \mu_s = \mu_k$ (b) $\mu_r > \mu_k > \mu_s$
(c) $\mu_r < \mu_s < \mu_k$ (d) $\mu_r < \mu_k < \mu_s$

17. A prism of refractive index $\sqrt{2}$ has a refracting angle of 60°. At what angle a ray must be incident on it so that it suffers a minimum deviation?
(a) 45° (b) 60°
(c) 90° (d) 180°

18. At the centre of a cubical box + Q charge is placed. The value of total flux that is coming out a wall is
(a) Q/ε_o (b) $Q/3\varepsilon_o$
(c) $Q/4\varepsilon_o$ (d) $Q/6\varepsilon_o$

19. Electromotive force is the force which is able to maintain a constant
(a) potential difference
(b) power
(c) resistance
(d) current

20. 16 g of oxygen at 37°C is mixed with 14 g of nitrogen at 27°C, the temperature of the mixture will be
(a) 30.5°C (b) 37°C
(c) 27°C (d) 32°C

21. A particle is moving eastwards with a velocity of 5 m/s. In 10 seconds the velocity changes to 5 m/s northwards. The average acceleration in this time is
(a) $\frac{1}{2}$ ms^{-2} North
(b) $\frac{1}{\sqrt{2}}$ ms^{-2} North – East
(c) $-\frac{1}{\sqrt{2}}$ ms^{-2} North-West
(d) zero

22. In an AC generator, a coil with N turns, all of the same area A and total resistance R, rotates with frequency ω in a magnetic field B. The maximum value of emf generated in the coil is
(a) NABRω (b) NAB
(c) NABR (d) NABω

23. In bringing an electron towards another electron, the electrostatic potential energy of system
(a) increases
(b) decreases
(c) becomes zero
(d) remains unchanged

24. If the radius of the earth shrinks by 2% and mass remains unchanged then acceleration due to gravity will
(a) increase by 2% (b) decrease by 2%
(c) increase by 4% (d) decrease by 4%

25. The fundamental frequency of a closed end organ pipe is n. Its length is doubled and radius is halved. Its frequency will become nearly
(a) n/2 (b) n/3
(c) n (d) 2 n

26. Rectifier converts
(a) mechanical energy to electrical energy
(b) a.c. to d.c.
(c) light energy to electrical energy
(d) None of these

27. The core of any transformer is laminated so as to
(a) reduce the energy loss due to eddy currents
(b) make it light weight
(c) make it robust and strong
(d) increase the secondary voltage

28. A particle moves under the effect of a force F = cx from x = 0 to x = x_1, the work done in the process is
(a) cx_1^2 (b) $\frac{1}{2}cx_1^2$
(c) $2cx_1^2$ (d) zero

29. A boy is spinning on a rotating stool with his arms down to his sides. Suddenly he stretches his arms. Which of the following is true regarding his angular speed (ω) and the angular momentum (L)?
(a) ω decreases but L remains same
(b) Both ω and L decrease
(c) ω remains constant but L decreases
(d) ω increases but L remains constant

30. The resistance of a resistor with the following colour code

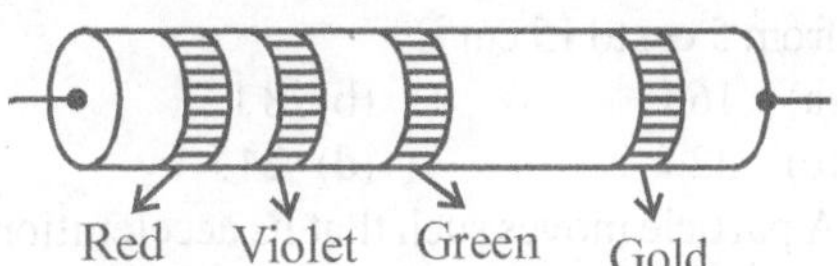

is equal to
(a) $26 \times 10^4\ \Omega \pm 5\%$ (b) $25 \times 10^4\ \Omega \pm 10\%$
(c) $35 \times 10^5\ \Omega \pm 5\%$ (d) $27 \times 10^5\ \Omega \pm 5\%$

31. According to the kinetic theory of gases, the pressure exerted by a gas on the walls is measured as
(a) rate of change of momentum imparted to the walls per second per unit area.
(b) momentum imparted to the walls per unit area
(c) change of momentum imparted to the walls per unit area.
(d) change in momentum per unit volume

32. A person who can see things most clearly at a distance of 10 cm, requires spectacles to enable to see clearly things at a distance of 30 cm. What should be the focal length of the spectacles?
(a) 15 cm (Concave) (b) 15 cm (Convex)
(c) 10 cm (d) 0

33. Electromagnetic waves are transverse in nature is evident by
(a) polarization (b) interference
(c) reflection (d) diffraction

34. A diametrical tunnel is dug across the earth. A ball is dropped into the tunnel from one side. The velocity of the ball when it reaches the centre of the earth is (Given : gravitational potential at the centre of earth = $-\frac{3}{2}\frac{GM}{R}$)
(a) $\sqrt{R}$ (b) $\sqrt{gR}$
(c) $\sqrt{2.5gR}$ (d) $\sqrt{7.1gR}$

35. A spring of force constant 800 N/m has an extension of 5 cm. The work done in extending it from 5 cm to 15 cm is
(a) 16 J (b) 8 J
(c) 32 J (d) 24 J

36. A particle moves such that its acceleration 'a' is given by a = –bx where x is the displacement from equilibrium position and b is constant. The period of oscillation is
(a) $2\pi/b$ (b) $2\pi/\sqrt{b}$
(c) $\sqrt{2\pi/b}$ (d) $2\sqrt{\pi/b}$

37. Which of the following law refers the direction of induced emf ?
(a) Biot-Savart's law
(b) Ampere's circutal law
(c) Kirchoff's law
(d) Lenz's law

38. A rain drop of radius 0.3 mm has a terminal velocity in air = 1 m/s. the viscosity of air is 8×10^{-5} poise. The viscous force on it is
(a) 45.2×10^{-4} dyne (b) 101.73×10^{-5} dyne
(c) 16.95×10^{-4} dyne (d) 16.95×10^{-5} dyne

39. Which of the following pairs can produce erect, diminished and virtual image?
(a) Concave lens and convex mirror
(b) Convex lens and convex mirror
(c) Convex lens and concave mirror
(d) Concave lens and concave mirror

40. Bernoulli's theorem is based on
(a) conservation of mass, energy and momentum
(b) conservation of momentum
(c) conservation of mass
(d) conservation of energy

41. A glass rod 20 cm long is clamped at the middle. It is set into the longitudinal vibration. If the emitted sound frequency is 4000 Hz, the velocity of sound in glass will be
(a) 2800 m/s (b) 3200 m/s
(c) 1600 m/s (d) 2000 m/s

42. In a solenoid, if number of turns is doubled, then self inductance will become
(a) half (b) double
(c) 1/4 times (d) quadruple

43. A transistor has an $\alpha = 0.95$, then β is equal to
(a) $\frac{1}{19}$ (b) 19
(c) 1.5 (d) 0.95

44. To which logic gate does the truth table given below correspond?

A	B	Y
0	0	0
0	1	1
1	0	1
1	1	0

(a) AND gate (b) NAND gate
(c) NOR gate (d) XOR gate

45. Optical fibres are used for long distance communication because
(a) it amplifies signals to be transmitted
(b) it transfer signals faster than electrical cables
(c) it pre-emphasise weak signals
(d) it provide little attenuation as compared to electrical cable for light propagation

SECTION 2 - CHEMISTRY

46. If 0.1 M solution of glucose and 0.1 M solution of urea are placed on two sides of the semipermeable membrane to equal heights, then it will be correct to say that
(a) There will be no net movement across the membrane
(b) Glucose will flow towards urea solution
(c) urea will flow towards glucose solution
(d) water will flow from urea solution to glucose

47. In the Freidel Craft's acylation reaction, the effective electrophile is
(a) $RCOCl^{\oplus}$ (b) $AlCl_3$
(c) RCOCl (d) $RCO^{\oplus}$

48. White lead is
(a) Pb_3O_4
(b) PbO
(c) $2PbCO_3.Pb(OH)_2$
(d) $Pb(CH_3COO)_2.Pb(OH)_2$

49. Specific volume of cylindrical virus particle is 6.02×10^{-2} cc/gm. whose radius and length 7 Å & 10 Å respectively. If $N_A = 6.02 \times 10^{23}$, find molecular weight of virus
(a) 3.08×10^3 kg/mol (b) 3.08×10^4 kg/mol
(c) 1.54×10^4 kg/mol (d) 15.4 kg/mol

50. Excess of ethanol and con. H_2SO_4 on heating upto 140°C produces
(a) diethyl ether
(b) diethyl sulphate
(c) ethyl hydrogen sulphate
(d) ethylene

51. Consider the following statements
I. The radius of an anion is larger than that of the parent atom.
II. The ionization energy generally increases with increasing atomic number in a period.
III. The electronegativity of an element is the tendency of an isolated atom to attract an electron.
Which of the above statements is/are correct?
(a) I alone (b) II alone
(c) I and II (d) II and III

52. Cl_2 changes to Cl^- and ClO^- in cold NaOH. Equivalent weight of Cl_2 will be
(a) M (b) $\frac{M}{2}$
(c) $\frac{M}{3}$ (d) $\frac{3M}{3}$

53. Which of the following pentafluoride is not obtained?
(a) SbF_5 (b) BiF_5
(c) PF_5 (d) AsF_5

54. In the conversion of Br_2 to BrO_3^-, the oxidation number of Br changes from
(a) zero to + 5 (b) + 1 to + 5
(c) zero to – 3 (d) + 2 to + 5

55. Arrange the following phenols in order of their increasing acidity.

OH, NO_2 (I)   OH, NO_2 (II)   OH, NO_2 (III)

(a) I > III > II (b) I > II > III
(c) III > I > II (d) III > II > I

56. The average molecular speed is greatest in which of the following gas samples ?
(a) 1.0 mol of O_2 at 560 K
(b) 0.50 mol of Ne at 500 K
(c) 0.20 mol of CO_2 at 440 K
(d) 2.0 mol of He at 140 K

57. Aniline is an activated system for electrophilic substitution. The compound formed on heating aniline with acetic anhydride is

(a) NH_2, $COCH_3$ (b) NH_2, $COCH_3$

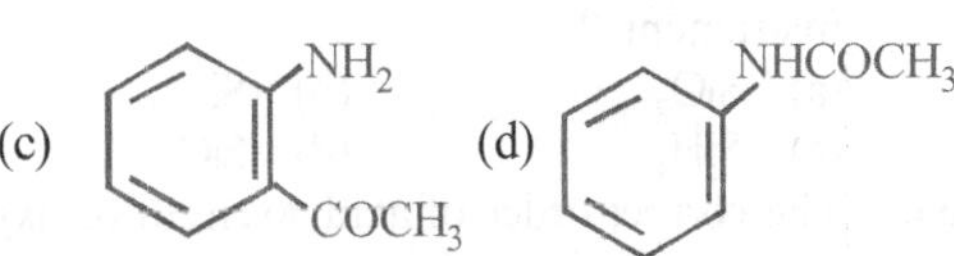

58. The formula mass of Mohr's salt is 392. The iron present in it is oxidised by $KMnO_4$ in acid medium. The equivalent mass of Mohr's salt is
(a) 392 (b) 31.6
(c) 278 (d) 156

59. Sewage mostly constitutes
(a) Non-biodegradable pollutants
(b) Biodegradable pollutants
(c) Effluents
(d) Air pollutants

60. The Langmuir adsorption isotherm is deduced using the assumption
(a) the adsorption sites are equivalent in their ability to adsorb the particles
(b) the heat of adsorption varies with coverage
(c) the adsorbed molecules interact with each other
(d) the adsorption takes place in multilayers.

61. Three dimensional molecules with cross links are formed in the case of a
(a) thermoplastic
(b) thermosetting plastic
(c) both (a) and (b)
(d) none of the above

62. The volume-temperature graphs of a given mass of an ideal gas at constant pressure are shown below.

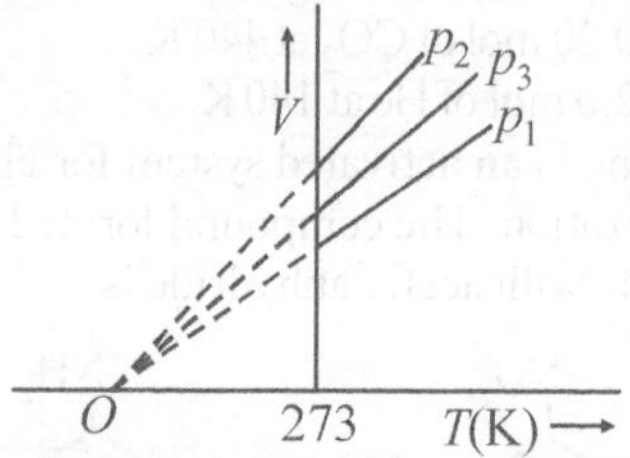

What is the correct order of pressures ?
(a) $p_1 > p_3 > p_2$ (b) $p_1 > p_2 > p_3$
(c) $p_2 > p_3 > p_1$ (d) $p_2 > p_1 > p_3$

63. Which of the following is used for making optical instruments?
(a) SiO_2 (b) Si
(c) SiH_4 (d) SiC

64. The correct order of acid strength of oxyacids is:
(a) $HClO > HClO_2 > HClO_3 > HClO_4$
(b) $HClO_2 > HClO_3 > HClO > HClO_4$
(c) $HClO_4 > HClO_3 > HClO_2 > HClO$
(d) $HClO_3 > HClO_4 > HClO > HClO_2$

65. 25.3 g of sodium carbonate, Na_2CO_3 is dissolved in enough water to make 250 mL of solution. If sodium carbonate dissociates completely, molar concentration of sodium ions, Na^+ and carbonate ions, CO_3^{2-} are respectively (Molar mass of Na_2CO_3 = 106 g mol^{-1})
(a) 0.955 M and 1.910 M
(b) 1.910 M and 0.955 M
(c) 1.90 M and 1.910 M
(d) 0.477 M and 0.477 M

66. In which alkyl halide, S_N2 mechanism is favoured maximum?
(a) CH_3Cl (b) CH_3CH_2Cl
(c) $(CH_3)_2CHCl$ (d) $(CH_3)_3C-Cl$

67. Equilibrium constant K changes with temperature. At 300 K, equilibrium constant is 25 and at 400 K it is 10. Hence, backward reaction will have energy of activation
(a) equa to that of forward reaction
(b) less than that of forward reaction
(c) greater than that of forward reaction
(d) given values are not sufficient to explain given statement

68. Which of the two ions from the list given below that have the geometry that is explained by the same hybridization of orbitals, NO_2^-, NO_3^-, NH_2^-, NH_4^+, SCN^-?
(a) NO_2^- and NO_3^- (b) NH_4^+ and NO_3^-
(c) SCN^- and NH_2^- (d) NO_2^- and NH_2^-

69. HO–C_6H_4 + C_6H_5–$\overset{+}{N_2}Cl^-$ $\xrightarrow{\text{base}}$

(a) C_6H_5–N=N–C_6H_4–OH

(b) 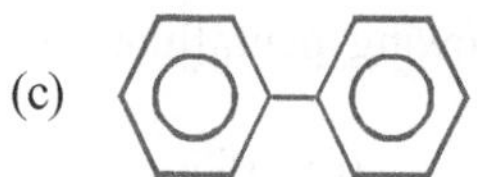

(c) 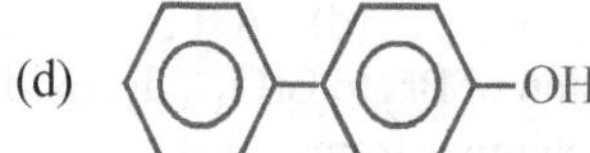

(d) C_6H_5–C_6H_4–OH

70. White lead is
(a) Pb_3O_4
(b) PbO
(c) $2PbCO_3.Pb(OH)_2$
(d) $Pb(CH_3COO)_2.Pb(OH)_2$

71. The commonest method for the extraction of metals from oxide ores involves :
(a) Reduction with carbon and carbon monoxide
(b) Reduction with aluminium
(c) Reduction with hydrogen
(d) Electrolytic method

72. With which one of the following elements silicon should be doped so as to give p-type of semiconductor ?
(a) Germanium (b) Arsenic
(c) Selenium (d) Boron

73. IUPAC name of $[Pt(NH_3)_3 (Br) (NO_2) Cl] Cl$ is
(a) Triamminechlorobromonitroplatinum (IV) chloride
(b) Triamminebromonitrochloroplatinum (IV) chloride
(c) Triamminebromochloronitroplatinum (IV) chloride
(d) Triamminenitrochlorobromoplatinum (IV) chloride

74. Van't Hoff factor is:
(a) More than one in case of association
(b) Less than one in case of dissociation
(c) $\frac{\text{normal molecular mass}}{\text{observed molecular mass}}$
(d) $\frac{\text{observed molcular mass}}{\text{normal molecular mass}}$

75. Which of the following compound will undergo self aldol condensation in the presence of cold dilute alkali ?
(a) $CH_2 = CH - CHO$
(b) $CH \equiv C - CHO$
(c) C_6H_5CHO
(d) CH_3CH_2CHO.

76. When one mole of an ideal gas is compressed to half its initial volume and simultaneously heated to twice in initial temperature, the change in entropy (ΔS) is
(a) $C_V \ln 2$ (b) $C_P \ln 2$
(c) $R \ln 2$ (d) $(C_V - R) \ln 2$

77. If x is amount of adsorbate and m is amount of adsorbent, which of the following relations is not related to adsorption process ?
(a) $x / m = K(p)$ at very low pressure.
(b) $x / m = Kp^{1/n}$ at inter mediate pressure
(c) $\frac{x}{m} = K$ at very high pressure
(d) $\frac{x}{m} = p$

78. For the reaction of one mole of zinc dust with one mole of sulphuric acid in a bomb calorimeter, ΔU and w corresponds to
(a) $\Delta U < 0, w = 0$ (b) $\Delta U < 0, w < 0$
(c) $\Delta U > 0, w = 0$ (d) $\Delta U > 0, w > 0$

79. Reactivity of hydrogen atoms attached to different carbon atoms in alkanes has the order
(a) Tertiary > Primary > Secondary
(b) Primary > Secondary > Tertiary
(c) Both (a) and (b)
(d) Tertiary > Secondary > Primary.

80. In most cases, for a rise of 10K temperature the rate constant is doubled to tripled. This is due to the reason that
(a) collision frequency increases by a factor of 2 to 3.
(b) fraction of molecules possessing threshold energy increases by a factor of 2 to 3
(c) Activation energy is lowered by a factor of 2 to 3.
(d) none of these

81. The restricted rotation about carbon carbon double bond in 2-butene is due to
(a) Overlap of one s- and sp^2-hybridized orbitals
(b) Overlap of two sp^2-hybridized orbitals
(c) Overlap of one p- and one sp^2-hybridized orbitals
(d) Sideways overlap of two p- orbitals.

82. Which of the following colligative property can provide molar mass of proteins (or polymers or colloids) with greatest precision ?
(a) Osmotic pressure
(b) Elevation of boiling point
(c) Depression of freezing point
(d) Relative lowering of vapour pressure

83. Cyclic hydrocarbon 'A' has all the carbon and hydrogen atoms in a single plane. All the carbon carbon bonds have the same length, less than 1.54 Å, but more than 1.34 Å. The C – C – C bond angle will be
(a) 109°28' (b) 100°
(c) 180° (d) 120°

84. The boiling point of three saturated hydrocarbons A, B and C are – 102°C, –43.4°C and – 0.6°C respectively. The hydrocarbon having the maximum number of carbon atoms in its molecule is
(a) A (b) B
(c) C (d) none of these

85. Which of the following electrolyte will have maximum flocculation value for $Fe(OH)_3$ sol?
(a) NaCl (b) Na_2S
(c) $(NH_4)_3PO_4$ (d) K_2SO_4

86. In DNA the complementary bases are
(a) adenine and thymine; guanine and cytosine
(b) uracil and adenine; cytosine and guanine
(c) adenine and guanine; thymine and cytosine
(d) adenine and thymine; guanine and uracil

87. What is the correct order of spin only magnetic moment (in BM) of Mn^{2+}, Cr^{2+} and V^{2+}?
(a) $Mn^{2+} > V^{2+} > Cr^{2+}$
(b) $V^{2+} > Cr^{2+} > Mn^{2+}$
(c) $Mn^{2+} > Cr^{2+} > V^{2+}$
(d) $Cr^{2+} > V^{2+} > Mn^{2+}$

88. In which of the following cases, pH is greater than 7?
(a) 50 ml of 0.1 M HCl + 50 ml of 0.1 M NaCl
(b) 50 ml of 0.1 M H_2SO_4 + 50 ml of 0.2 M NaOH
(c) 50 ml of 0.1 M CH_3COOH + 50 ml of 0.1 M KOH
(d) 50 ml of 0.1 M HNO_3 + 50 ml of 0.1 M NH_3

89. In the presence of platinum catalyst, hydrocarbon A adds hydrogen to form n-hexane. When hydrogen bromide is added to A instead of hydrogen, only a single bromo compound is formed. Which of the following is A?
(a) $CH_3-CH_2-CH=CH-CH_2-CH_3$
(b) $CH_3-CH_2-CH_2-CH=CH-CH_3$
(c) $CH_3-CH=CH-CH_2-CH_2-CH_3$
(d) $CH_2=CH-CH_2-CH_2-CH_2-CH_3$

90. $\underset{\text{Benzene}}{C_6H_6} + \underset{\text{Methyl chloride}}{CH_3Cl} \xrightarrow{AlCl_3} \underset{\text{Toluene}}{C_6H_5CH_3} + HCl$

The above reaction is:
(a) Wurtz Fittig reaction
(b) Grignard reaction
(c) Friedel-Craft's reaction
(d) Ullmann reaction

SECTION 3 - BOTANY

91. Most plants are green in colour because
(a) the atmosphere filters out all the colours of the visible light spectrum except green.
(b) green light is the most effective wavelength region of the visible spectrum in sunlight for photosynthesis.
(c) chlorophyll is least effective in absorbing green light.
(d) green light allows maximum photosynthesis.

92. Which of the following organism possesses characteristics of both a plant and an animal?
(a) Bacteria (b) *Euglena*
(c) *Mycoplasma* (d) *Paramecium*

93. Function of companion cells is
(a) providing energy to sieve elements for active transport.
(b) providing water to phloem.
(c) loading of sucrose into sieve elements by passive transport.
(d) loading of sucrose into sieve elements.

94. In cell cycle, DNA replication takes place in
(a) G_1 phase
(b) G_2 phase
(c) mitotic metaphase
(d) S phase

95. During cell division, the spindle fibres attach to the chromosome at a region called
(a) chromocentre (b) kinetochore
(c) centriole (d) chromomere

96. Organisms which are indicator of SO_2 pollution of air are
(a) Mosses (b) Lichens
(c) Mushrooms (d) Puffballs

97. Choose the correct option for A, B, C and D of a mature embryo sac .

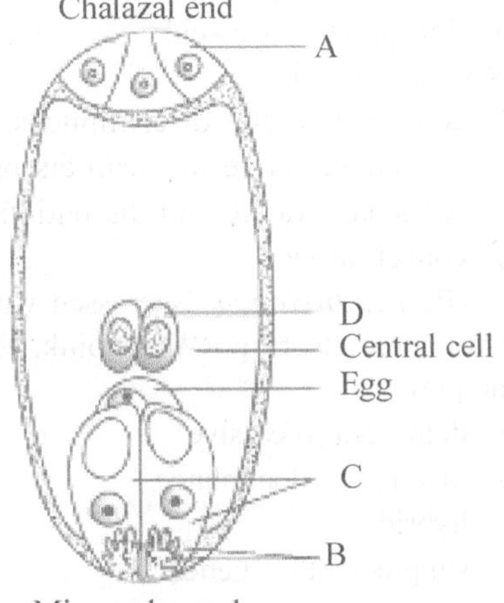

(a) A – Filiform apparatus; B – Antipodal cells; C – Synergids; D – Polar nuclei
(b) A – Antipodal cells; B – Filiform apparatus; C – Synergids; D – Polar nuclei
(c) A – Antipodal cells; B – Filiform apparatus; C – Polar nuclei; D – Synergids
(d) A – Antipodal cells; B – Polar nuclei; C – Filiform apparatus; D – Synergids

98. In which one pair both the plants can be vegetative propagated by leaf pieces?
(a) *Bryophyllum* and *Kalanchoe*
(b) *Chrysanthemum* and *Agave*
(c) *Agave* and *Kalanchoe*
(d) *Asparagus* and *Bryophyllum*

99. Syngamy means
(a) fusion of gametes
(b) fusion of cytoplasms
(c) fusion of two similar spores
(d) fusion of two dissimilar spores

100. Identify the correctly matched pair.
(a) Basal Convention - Biodiversity Conservation
(b) Kyoto Protocol - Climatic change
(c) Montreal Protocol - Global warming
(d) Ramsar Convention - Ground water pollution

101. Secondary sewage treatment is mainly a
(a) physical process
(b) mechanical process
(c) chemical process
(d) biological process

102. A mutually beneficial association necessary for survival of both partners is
(a) mutualism/symbiosis
(b) commensalism
(c) amensalism
(d) both (a) and (b)

103. Cotyledons and testa respectively are edible parts in
(a) walnut and tamarind
(b) french bean and coconut
(c) cashew nut and litchi
(d) groundnut and pomegranate

104. When pollen grains are not transferred from anthers to stigma in a flower, due to the barrier, it is called
(a) herkogamy (b) heterogamy
(c) cleistogamy (d) dichogamy

105. Water potential is equal to
(a) Ψ_s + O.P. (b) Ψ_s + T.P.
(c) $\Psi_p + \Psi_w$ (d) $\Psi_s + \Psi_p$

106. Which of the following is a free living aerobic non-photosynthetic nitrogen-fixer?
(a) *Rhizobium* (b) *Azotobacter*
(c) *Azospirillum* (d) *Nostoc*

107. What kinds of the structures of proteins are shown in the following figure ?

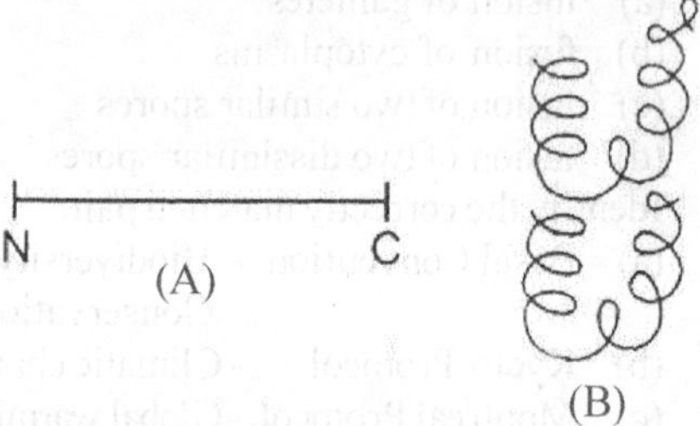

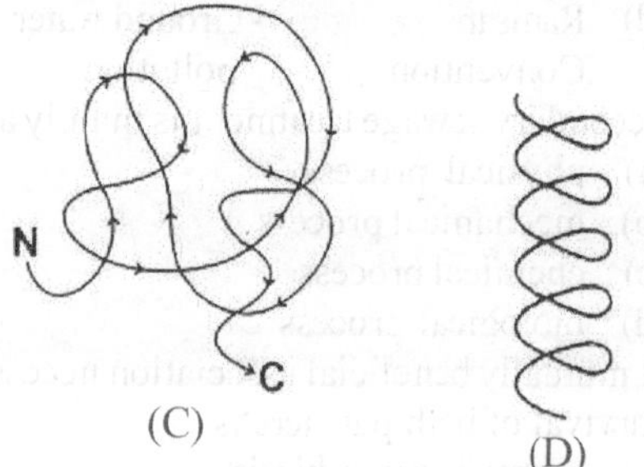

(a) A = 1° structure, B = 2° structure, C = 3° structure, D = 4° structure
(b) A = 4° structure, B = 2° structure, C = 3° structure, D = 2° structure
(c) A = 1° structure, B = 4° structure, C = 3° structure, D = 2° structure
(d) A = 4° structure, B = 3° structure, C = 2° structure, D = 1° structure

108. In soil, the water available for root absorption is
(a) gravitational water
(b) capillary water
(c) hygroscopic water
(d) combined water

109. Which one of the following mineral elements plays an important role in biological nitrogen fixation ?
(a) Molybdenum
(b) Copper
(c) Manganese
(d) Zinc

110. C_4 acid, formed in the mesophyll of C_4 plants leaf during photosynthesis is
(a) OAA or malic acid or aspartic acid
(b) pyruvic acid
(c) succinic acid
(d) fumaric acid

111. *Azotobacter* and *Beijerinckia* are the examples of
(a) symbiotic nitrogen-fixers
(b) non-symbiotic nitrogen-fixers
(c) ammonifying bacteria
(d) disease causing bacteria

112. The first carbon dioxide acceptor in C_4-plants is
(a) Phosphoenol-pyruvate
(b) Ribulose 1, 5-diphosphate
(c) Oxalo-acetic acid
(d) Phosphoglyceric acid

113. DNA replication is
(a) conservative and discontinuous
(b) semiconservative and semi discontinuous
(c) semiconservative and discontinuous
(d) conservative

114. Red (RR) *Antirrhinum* is crossed with white (WW) one. Offspring RW are pink. This is an example of
(a) dominant -recessive
(b) incomplete dominance
(c) hybrid
(d) supplementary genes

115. Food chain in which micro-organisms breakdown the food formed by primary producers is
(a) parasitic food chain
(b) detritus food chain
(c) consumer food chain
(d) predator food chain

116. Which of the following pairs is a sedimentary type of biogeochemical cycle?
(a) Oxygen and nitrogen
(b) Phosphorus and sulphur
(c) Phosphorus and nitrogen
(d) Phosphorus and carbon dioxide

117. Independent assortment of genes does not take place when
(a) genes are located on homologous chromosomes
(b) genes are linked and located on same chromosomes
(c) genes are located on non-homologous chromosomes
(d) All the above

118. Breeding of crops with high levels of minerals, vitamins and proteins is called
(a) Somatic hybridisation
(b) Biofortification
(c) Biomagnification
(d) Micropropagation

119. If we completely remove the decomposers from an ecosystem, its functioning will be adversely affected, because
(a) energy flow will be blocked
(b) herbivores will not receive solar energy
(c) mineral movement will be blocked
(d) rate of decomposition will be very high

120. A nucleotide is formed of
(a) Purine, pyrimidine and phosphate
(b) Purine, sugar and phosphate
(c) Nitrogen base, sugar and phosphate
(d) Pyrimidine, sugar and phosphate

121. A chemical believed to be involved in flowering is
(a) gibberellin (b) kinetin
(c) florigen (d) IBA

122. In RNA, thymine is replaced by
(a) Adenine (b) Guanine
(c) Cytosine (d) Uracil

123. Which of the following is a reducing sugar?
(a) Galactose
(b) Gluconic acid
(c) β-methyl galactoside
(d) Sucrose

124. Which of the following processes make direct use of oxygen ?
(a) Glycolysis
(b) Fermentation
(c) Electron transport
(d) Krebs citric acid cycle

125. Glycolysis is a
(a) redox process (b) aerobic process
(c) oxidative process (d) reductive process

126. Which of the following is a biological uncoupler of oxidative phosphorylation ?
(a) Thermogenin
(b) 2, 4 – Dichlorophenoxy acetic acid
(c) 2, 4 – Dinitrophenol
(d) Ethylene diaminotetra acetic acid

127. Wildlife is destroyed most when
(a) there is lack of proper care
(b) mass scale hunting for foreign trade
(c) its natural habitat is destroyed
(d) natural calamity occurs

128. Golden rice is a transgenic crop of the future with the following improved trait :
(a) insect resistance
(b) high lysine (essential amino acid) content
(c) high protein content
(d) high vitamin-A content

129. Largest amount of fresh water is found in
(a) lakes and streams
(b) underground
(c) polar ice caps and glaciers
(d) rivers

130. A common biocontrol agent for the control of plant diseases is
(a) Baculovirus
(b) *Bacillus thuringiensis*
(c) *Glomus*
(d) *Trichoderma*

131. A nitrogen-fixing microbe associated with *Azolla* in rice fields is :
(a) *Spirulina* (b) *Anabaena*
(c) *Frankia* (d) *Tolypothrix*

132. Which group of vertebrates comprises the highest number of endangered species ?
(a) Birds (b) Mammals
(c) Fishes (d) Reptiles

133. PCR and Restriction Fragment Length Polymorphism are the methods for :
(a) Study of enzymes
(b) Genetic transformation
(c) DNA sequencing
(d) Genetic Fingerprinting

134. The C_4 plants are photosynthetically more efficient than C_3 plants because:
(a) the CO_2 compensation point is more
(b) CO_2 generated during photorespiration is trapped and recycled through PEP carboxylase
(c) the CO_2 efflux is not prevented
(d) they have more chloroplasts

135. The enormous diversity of protein molecules is due mainly to the diversity of
(a) amino groups on the amino acids
(b) R groups on the amino acids
(c) amino acid sequences within the protein molecule
(d) peptide bonds

SECTION 4 - ZOOLOGY

136. 'Comma' shaped bacteria are known as
(a) coccus (b) spiral
(c) spirillum (d) vibrio

137. Which one of the following pairs of animals comprises 'jawless fishes'?
(a) Mackerals and Rohu
(b) Lampreys and hag fishes
(c) Guppies and hag fishes
(d) Lampreys and eels

138. Intercalated discs are the communication junctions between the cells of
(a) cardiac muscles
(b) striped muscles
(c) adipose tissue
(d) nerve and striated muscles

139. Which one of the following groups of animals is bilaterally symmetrical and triploblastic?
(a) Aschelminthes (round worms)
(b) Ctenophores
(c) Sponges
(d) Coelenterates (Cnidarians)

140. Pseudostratified epithelium is found in
(a) oesophagus (b) respiratory tract
(c) urinary tract (d) kidney

141. Male and female cockroaches can be distinguished externally through
(a) anal styles in male
(b) anal cerci in female
(c) anal style and antennae in females
(d) Both (b) and (c)

142. Nerve cells do not divide because they do not have
(a) nucleus (b) centrosome
(c) golgi body (d) mitochondria

143. Binary fission is found in –
(a) *Amoeba* (b) *Paramecium*
(c) *Planaria* (d) All of these

144. Human population growth in India
(a) tends to follow a sigmoid curve as in case of many other animal species
(b) tends to reach a zero population growth as in case of some animal species
(c) can be reduced by permitting natural calamities and enforcing birth control measures
(d) can be regulated by following the national programme of family planning

145. The diaphragm, cervical cap and vaults are
(a) Disposable contraceptive devices
(b) Reusable contraceptives
(c) IUDs
(d) Implants

146. The human embryo, with 8 to 16 blastomere is called
(a) Morula (b) Blastula
(c) Gastrula (d) Foetus

147.

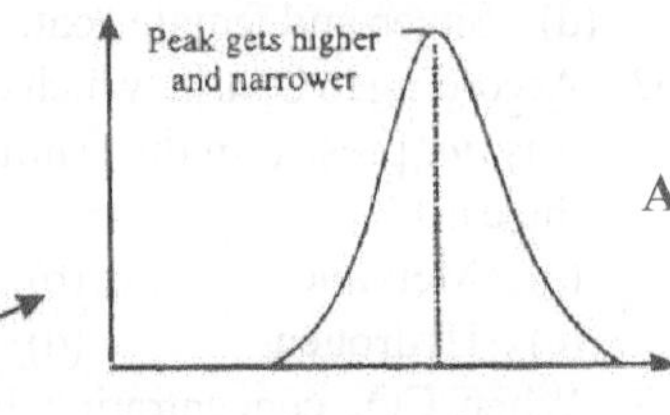

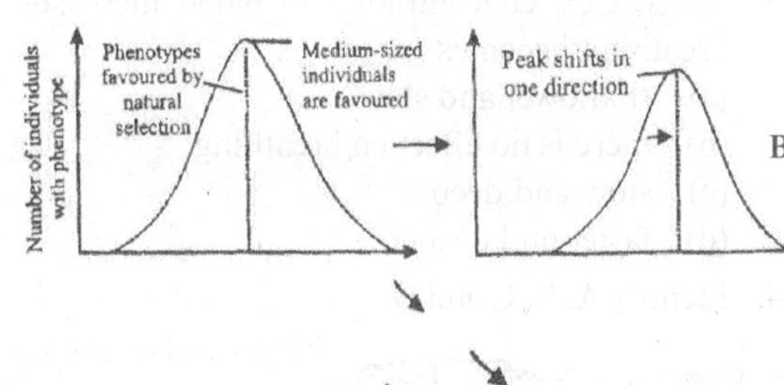

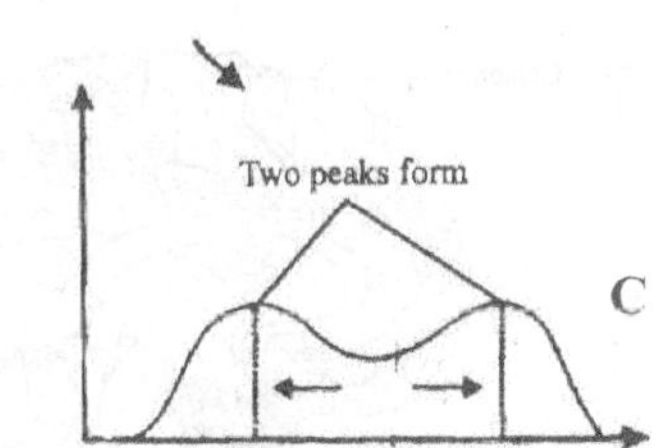

Following diagrammatic representation refers the natural selection on different traits. In which all the three graphs A, B and C are identified correctly
(a) A – Directional, B – Stabilising, C – Disruptive
(b) A – Stabilising, B – Directional, C – Disruptive
(c) A – Stabilising, B – Disruptive, C – Directional
(d) A – Directional, B – Disruptive, C – Stabilising

148. Which one of the following statements about human sperm is correct?
(a) Acrosome has a conical pointed structure used for piercing and penetrating the egg, resulting in fertilisation
(b) The sperm lysis in the acrosome dissolve the egg envelope facilitating fertilisation
(c) Acrosome serves as a sensory structure leading the sperm towards the ovum
(d) Acrosome serves no particular function

149. Bulbourethral gland is present in
(a) Males and is another name for Uterus masculina
(b) Females and is another name for Bertholin's gland
(c) Males and is another name for Cowper's gland
(d) None of these

150. Which of the following are water soluble vitamins?
(a) B, C (b) A, D, K
(b) A, D, E, K (d) K, B, A, E

151. Blood pressure is measured by
(a) sphygmomanometer
(b) stethoscope
(c) electrocardiogram
(d) phonocardiogram

152. Oxyntic cells are located in
(a) Islets of Langerhans
(b) Gastric epithelium and secrete pepsin
(c) Kidneys and secrete renin
(d) Gastric epithelium and secrete HCl

153. If Henle's loop were absent from mammalian nephron which of the following is to be expected?
(a) The urine will be more concentrated
(b) The urine will be more dilute
(c) There will be no urine formation
(d) There will be hardly any change in the quality and quantity of urine formed

154. Hepatic portal system carries
(a) Oxygenated blood from liver to intestine
(b) Deoxygenated blood from liver to intestine
(c) Oxygenated blood from intestine to liver
(d) Deoxygenated blood from intestine to liver

155. "Kwashiorkar' and 'Beri-Beri' are
(a) communicable diseases
(b) infectious diseases
(c) deficiency diseases
(d) None of the above

156. ABO blood group system is due to
(a) multifactor inheritance
(b) incomplete dominance
(c) multiple allelism
(d) epistasis

157. Evolution is
(a) progressive development of a race
(b) history and development of a race alongwith variations
(c) history of a race
(d) development of a race

158. Bull semen for the purpose of artificial insemination is stored in
(a) Ice (b) Liquid oxygen
(c) Liquid nitrogen (d) Liquid CO_2

159. The kind of evolution in which two species of different geneology come to resemble one another closely, is termed as
(a) progressive evolution
(b) convergent evolution
(c) parallel evolution
(d) retrogressive evolution

160. *Drosophila* flies with XXY genotype are females, but human beings with such genotype are abnormal males. It shows that
(a) Y-chromosome is essential for sex determination in *Drosophila.*
(b) Y-chromosome is female determinant in *Drosophila.*
(c) Y-chromosome is male determination in human beings.
(d) Y-chromosome has no role in sex determination either in *Drosophila* or in human beings.

161. Hinny is a hybrid of male
(a) Horse and female donkey
(b) Donkey and female horse
(c) Goat and female lamb
(d) Sheep and female goat.

162. According to Oparin, which one of the following was ***not*** present in the primitive atmosphere of the earth?
(a) Methane (b) Oxygen
(c) Hydrogen (d) Water vapour

163. When CO_2 concentration in blood increases, breathing becomes
(a) shallower and slow
(b) there is no effect on breathing
(c) slow and deep
(d) faster and deeper

164. Identify A, B, C and D

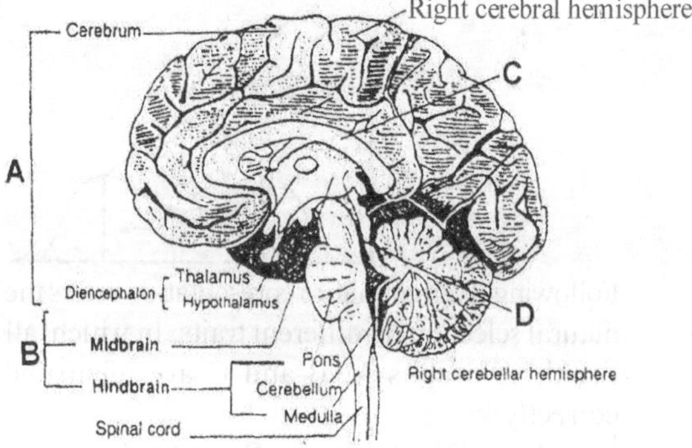

(a) A – Forebrain, B – Brainstem, C – Corpus callosum, D – Cerebral aqueduct
(b) A – Forebrain, B – Brainstem, C – Cerebral aqueduct, D – Corpus callosum
(c) A – Brainstem, B – Forebrain, C – Corpus callosum, D – Cerebral aqueduct
(d) A – Brainstem, B – Forebrain, C – Cerebral aqueduct, D – Corpus luteum

165. Afferent nerve fibres carry impulses from
   (a) effector organs to CNS
   (b) receptors to CNS
   (c) CNS to receptors
   (d) CNS to muscles

166. Which one of the following is the correct statement for respiration in humans?
   (a) Workers in grinding and stone-breaking industries may suffer from lung fibrosis.
   (b) About 90% of carbon dioxide (CO_2) is carried by haemoglobin as carbaminohaemoglobin.
   (c) Cigarette smoking may lead to inflammation of bronchi.
   (d) Neural signals from pneumotoxic centre in pons region of brain can increase the duration of inspiration.

167. The blood calcium level is lowered by the deficiency of
   (a) parathormone
   (b) thyroxine
   (c) both calcitonin and parathormone
   (d) calcitonin

168. Which of the following statements regarding glucagon is false?
   (a) It is secreted by α-cells of Langerhans.
   (b) It acts antagonistically to insulin.
   (c) It decreases blood sugar level.
   (d) The gland responsible for its secretion is heterocrine gland.

169. Which hormone possesses anti-insulin effect?
   (a) Cortisol (b) Calcitonin
   (c) Oxytocin (d) Aldosterone

170. Which is the example of conditioned reflex ?
   (a) Eyes closed when anything enter into it.
   (b) Hand took up when piercing with needle.
   (c) Salivation in a hungry dog in response to ringing of a bell.
   (d) Digestion food goes forward in alimentary canal.

171. Which one of the following techniques made it possible to genetically engineerd living organism ?
   (a) Recombinant DNA techniques
   (b) X-ray diffraction
   (c) Heavier isotope labelling
   (d) Hybridization

172. The diagram opposite shows a homologous (bivalent) pair of chromosomes during meiosis.

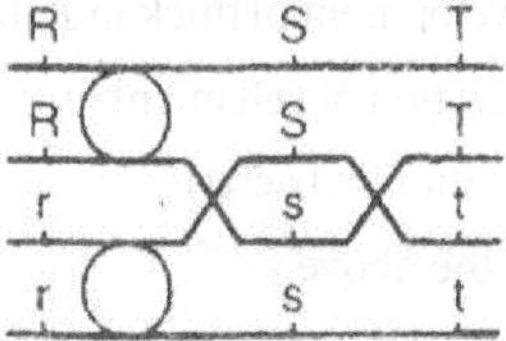

Which of the following correctly represents the final products of the second meiotic division ?

(a)
R S T
R s T
r S t
r s t

(b)
R S T
R s t
r S T
r s t

(c)
R S T
R s T
S t
r s t

(d)
R S T
R s t
r S T
r s t

173. A sexually transmitted disease caused by bacteria is

(a) Leprosy
(b) AIDS
(c) Syphilis
(d) Pertussis

174. What is mode of bacterial resistance against antibiotics ?

(a) Development of thick mucilaginous layer
(b) Alteration of cell membrane
(c) Mutation in bacteria
(d) All the above

175. Which one of the following is a wrong matching of a microbe and its industrial product, while the remaining three are correct ?

(a) Yeast - statins
(b) *Acetobacter aceti* - acetic acid
(c) *Clostridium butylicum* - lactic acid
(d) *Aspergillus niger* - citric acid

176. To which type of barriers under innate immunity, do the saliva in the mouth and the tears from the eyes, belong?

(a) Cytokine barriers
(b) Cellular barriers
(c) Physiological barriers
(d) Physical barriers

177. Select the correct statement with respect to diseases and immunisation?

(a) If due to some reason B-and T-lymphocytes are damaged, the body will not produce antibodies against a pathogen
(b) Injection of dead / inactivated pathogens causes passive immunity
(c) Certain protozoans have been used to mass produce hepatitis B vaccine.
(d) Injection of snake antivenom against snake bite is an example of active immunisation

178. Which one of the following is a matching pair of a vitamin and the deficiency disease related with it ?

(a) Riboflavin — beri beri
(b) Thiamine — xerophthalmia
(c) Niacin — pellagra
(d) Calciferol — scurvy

179. Which one of the following is the correct matching of three items and their grouping category?

	Items	Group
(a)	Ilium, ischium, pubis	- coxal bones of pelvic girdle
(b)	Actin, myosin, rhodopsin.	- muscle proteins
(c)	Cytosine, uracil, thiamine	- pyrimidines
(d)	Malleus, incus, cochlea	- ear ossicles

180. What is the first step in the Southern blot technique?

(a) Denaturation of DNA on the gel for hybridization with specific probe.
(b) Production of a group of genetically identical cells.
(c) Digestion of DNA by restriction enzyme.
(d) Denaturation of DNA from a nucleated cell such as the one from the scene of crime.

RESPONSE SHEET

Test Code: ....Time taken :......................Date:..............

	(a)	(b)	(c)	(d)		(a)	(b)	(c)	(d)		(a)	(b)	(c)	(d)		(a)	(b)	(c)	(d)
1.					46.					91.					136.				
2.					47.					92.					137.				
3.					48.					93.					138.				
4.					49.					94.					139.				
5.					50.					95.					140.				
6.					51.					96.					141.				
7.					52.					97.					142.				
8.					53.					98.					143.				
9.					54.					99.					144.				
10.					55.					100.					145.				
11.					56.					101.					146.				
12.					57.					102.					147.				
13.					58.					103.					148.				
14.					59.					104.					149.				
15.					60.					105.					150.				
16.					61.					106.					151.				
17.					62.					107.					152.				
18.					63.					108.					153.				
19.					64.					109.					154.				
20.					65.					110.					155.				
21.					66.					111.					156.				
22.					67.					112.					157.				
23.					68.					113.					158.				
24.					69.					114.					159.				
25.					70.					115.					160.				
26.					71.					116.					161.				
27.					72.					117.					162.				
28.					73.					118.					163.				
29.					74.					119.					164.				
30.					75.					120.					165.				
31.					76.					121.					166.				
32.					77.					122.					167.				
33.					78.					123.					168.				
34.					79.					124.					169.				
35.					80.					125.					170.				
36.					81.					126.					171.				
37.					82.					127.					172.				
38.					83.					128.					173.				
39.					84.					129.					174.				
40.					85.					130.					175.				
41.					86.					131.					176.				
42.					87.					132.					177.				
43.					88.					133.					178.				
44.					89.					134.					179.				
45.					90.					135.					180.				

TEST ASSESSMENT AND ANALYSIS SHEET

How to fill this sheet : After you finish the paper and have checked it first try to attempt again all the questions which you have done wrong or have not been able to attempt. Go to detailed solutions only after you have attempted all the questions of your own.Attempting the questions on your own without seeing the solutions will help you in filling this table

ANALYSIS OF TEST	
Total number of questions	
Questions Attempted	
Correct Questions	
Wrong Questions	
	Net Score

Strike rate $\left(\frac{\text{Correct questions}}{\text{Questions attempt}} \times 100\right) =$ ....

Percentage Marks $\left(\frac{\text{Correct Answers}}{\text{Total number of questions}} \times 100\right) =$ ....

Analysis of wrong questions : Divide the questions which you have done wrong in 3 broad categories.

Reasons for questions attempted wrong	**Number of Questions**
Knew but calculation mistake (solved on my own after exam)	
Knew but got confused (solved on my own after exam)	
Did not knew the question (couldn't solve even after exam)	

Questions not attempted : Divide the questions not attempted in 3 broad categories

Reasons for unattempted questions	**Number of Questions**
Knew the ques but time was less (solved on my own after exam)	
Knew the ques but got confused (solved on my own after exam)	
Did not knew the questions (couldn't solve even after exam)	

List down your weak areas you need to improve your score :

1)______________________________

2)______________________________

3)______________________________

NEET

National Eligibility Entrance Test

Test Series **Full Test - 4**

Time : - 3 Hours **Max. Marks:- 720**

INSTRUCTIONS

1. Immediately fill in the particulars on this page of the Test Booklet with Blue/Black Ball Point Pen. Use of pencil is strictly prohibited.
2. Answers are to be given on Response Sheet provided at the end of this paper. You can tear this Response Sheet along the cut marks provided. Fill in the particulars carefully in the Response Sheet provided.
3. The test is of 3 hours duration.
4. The Test Booklet consists of **180** questions. The maximum marks are **720**.
5. There are four section in the question paper.

   The distribution of marks subjectwise in each part is as under for each correct response.

   Section 1 – Physics (180 Marks) – 45 Questions

   Section 2 – Chemistry (180 Marks) – 45 Questions

   Section 3 – Botany (180 Marks) – 45 Questions

   Section 4 – Zoology (180 Marks) – 45 Questions
6. You will be awarded 4 Marks if you have darkened only the bubble corresponding to the correct answer and zero mark if no bubble is darkened. In all other cases, minus one (– 1) mark will be awarded.
7. Use Blue/Black Ball Point Pen only for writing particulars/marking responses on Side-1 and Side-2 of the Response Sheet.

   Use of pencil is strictly prohibited.
8. You are advised to create an actual examination hall simulation.
9. Rough work is to be done on the space provided for this purpose in the Test Booklet only.
10. Do not fold or make any stray marks on the Response Sheet.
11. Once you have evaluated the test, you should assess it thoroughly.

SECTION 1 - PHYSICS

1. The resistance R of a wire is given by the relation $R = \frac{\rho \ell}{\pi r^2}$. Percentage error in the measurement of r, ℓ and r is 1%, 2% and 3% respectively. Then the percentage error in the measurement of R is

(a) 6 (b) 9
(c) 8 (d) 10

2. Figure shows the v-t graph for two particles P and Q. Which of the following statements regarding their relative motion is true? Their relative velocity

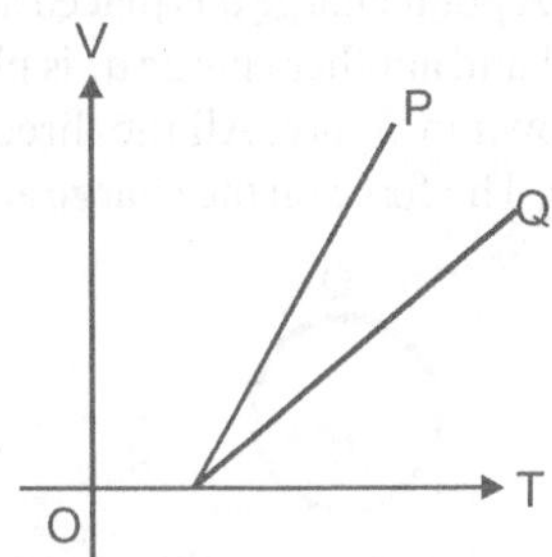

(a) is zero
(b) is non-zero but constant
(c) continuously decreases
(d) continuously increases

3. Time required to boil 2 litres of water initially at 20°C by a heater coil which works at 80% efficiency spending 500 joule/s is

(a) 82 minutes (b) 50 minutes
(c) 28 minutes (d) 37 minutes

4. A mass is tied to a string and rotated in a vertical circle, the minimum velocity of the body at the top is

(a) $\sqrt{gr}$ (b) g/r
(c) $\left(\frac{g}{r}\right)^{3/2}$ (d) gr

5. A man projects a coin upwards from the gate of a uniformly moving train. The path of coin for the man will be

(a) parabolic
(b) inclined straight line
(c) vertical straight line
(d) horizontal straight line

6. What is the disintegration constant of radon, if the number of its atoms diminishes by 18% in 24 h?

(a) $2.1 \times 10^{-3} s^{-1}$ (b) $2.1 \times 10^{-4} s^{-1}$
(c) $2.1 \times 10^{-5} s^{-1}$ (d) $2.1 \times 10^{-6} s^{-1}$

7. The frequencies of X-rays, γ-rays and ultraviolet rays are respectively *a*, *b*, and *c*. Then

(a) $a < b, b < c$ (b) $a < b, b > c$
(c) $a > b, b > c$ (d) $a > b, b < c$

8. A galvanometer can be changed into an ammeter by using

(a) low resistance shunt in series
(b) low resistance shunt in parallel
(c) high resistance shunt in series
(d) high resistance shunt in parallel

9. Which of the following statements is **FALSE** for a particle moving in a circle with a constant angular speed ?

(a) The acceleration vector points to the centre of the circle
(b) The acceleration vector is tangent to the circle
(c) The velocity vector is tangent to the circle
(d) The velocity and acceleration vectors are perpendicular to each other.

10. A monoatomic gas at 27°C is compressed adiabatically to $\frac{8}{27}$ of its original volume. The rise in temperature will be

(a) 300°C (b) 350°C
(c) 375°C (d) 400°C

11. Each of the resistance in the network shown is equal to R. The resistance between the terminals A and B is

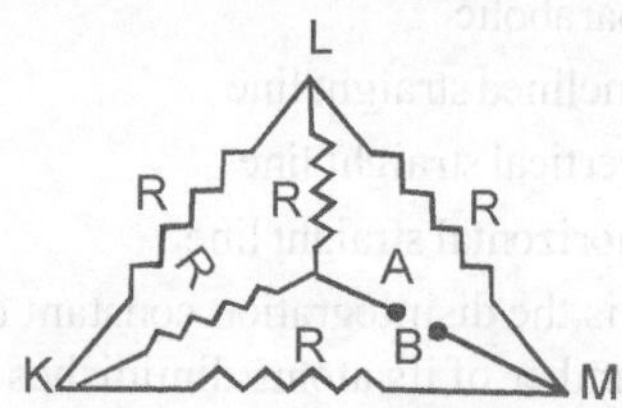

(a) R (b) 5R
(c) 3R (d) 6R

12. The energy of hydrogen atom in the nth orbit is E_n, then the energy in the nth orbit of single ionised helium atom is

(a) $\frac{E_n}{2}$ (b) $2E_n$
(c) $4E_n$ (d) $\frac{E_n}{4}$

13. Two identical particles move towards each other with velocity 2v and v respectively. The velocity of centre of mass is
(a) v (b) v/3
(c) v/2 (d) zero.

14. The mass number of He is 4 and that for sulphur is 32. The radius of sulphur nuclei is larger than that of helium by
(a) $\sqrt{8}$ (b) 4
(c) 2 (d) 8

15. According to Newton's law of cooling, the rate of cooling of a body is proportional to $(\Delta\theta)^n$, where $\Delta\theta$ is the difference of the temperature of the body and the surroundings, then n is equal to
(a) two (b) three
(c) four (d) one

16. A body having initial velocity of 10 m/s moving on a rough surface comes to rest after moving 50 m. What is coefficient of friction between the body and surface? ($g = 10\ m/s^2$)
(a) 0.5 (b) 0.2
(c) 0.3 (d) 0.1

17. The separation between successive fringes in a double slit arrangement is x. If the whole arrangement is dipped under water what will be the new fringe separation? [The wavelenght of light being used is 5000 Å]
(a) 1.5x (b) x
(c) 0.75x (d) 2x

18. A thin, metallic spherical shell contains a charge Q on it. A point charge q is placed at the cente of the shell and another charge q_1 is placed outside it as shown in figure. All the three charges are positive. The force on the charge at the centre is

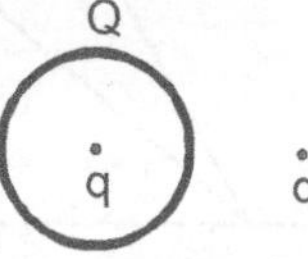

(a) towards left (b) towards right
(c) upward (d) zero

19. A charge *q* is moving with a velocity *v* parallel to a magnetic field *B*. Force on the charge due to magnetic field is
(a) qvB (b) qB/v
(c) zero (d) Bv/q

20. A bucket full of hot water is kept in a room and it cools from 75°C to 70°C in T_1 minutes, from 70°C to 65°C in T_2 minutes and from 65°C to 60°C in T_3 minutes. Then
(a) $T_1 = T_2 = T_3$ (b) $T_1 < T_2 < T_3$
(c) $T_1 > T_2 > T_3$ (d) $T_1 < T_2 > T_3$

21. In uniform circular motion, the velocity vector and acceleration vector are
(a) perpendicular to each other
(b) in same direction
(c) in opposite direction
(d) not related to each other

22. The current in a coil of L = 40 mH is to be increased uniformly from 1A to 11A in 4 milli sec. The induced e.m.f. will be
(a) 100 V (b) 0.4 V
(c) 440 V (d) 40 V

23. Two capacitors when connected in series have a capacitance of 3 μF, and when connected in parallel have a capacitance of 16 μF. Their individual capacities are
(a) 1 μF, 2 μF (b) 6 μF, 2 μF
(c) 12 μF, 4 μF (d) 3 μF, 16 μF

24. The weight of a body will be the least at
(a) poles
(b) equator
(c) at height equal to R
(d) centre of the earth

25. When a tuning fork produces sound waves in air, which one of the following is same in the material of tuning fork as well as in air?
(a) Wavelength (b) Frequency
(c) Velocity (d) Amplitude

26. The fermi energy for a substance is
(a) independent of T
(b) directly proportional to $\sqrt{T}$
(c) directly proportional to T
(d) directly proportional to T^2

27. ABC is a triangular plate of uniform thickness. The sides are in the ratio shown in the figure. I_{AB}, I_{BC} and I_{CA} are the moments of inertia of the plate about AB, BC and CA as axes respectively. Which one of the following relations is correct?

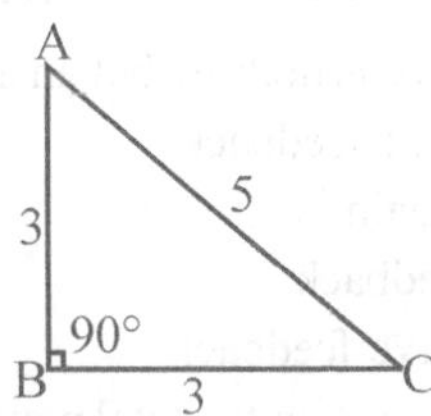

(a) $I_{AB} > I_{BC}$ (b) $I_{BC} > I_{AB}$
(c) $I_{AB} + I_{BC} = I_{CA}$ (d) I_{CA} is maximum

28. Consider the following statement:
When jumping from some height, you should bend your knees as you come to rest, instead of keeping your legs stiff. Which of the following relations can be useful in explaining the statement? Where symbols have their usual meanings.
(a) $\Delta\vec{p}_1 = -\Delta\vec{p}_2$ (b) $\Delta E = \Delta(PE + KE) = 0$
(c) $\vec{F}\Delta t = m\Delta\vec{v}$ (d) $\Delta x \propto \Delta F$

29. The amplitude of magnetic field of an electromagnetic wave is 2×10^{-7}T. It's electric field amplitude if the wave is travelling in free space is
(a) 6 Vm^{-1} (b) 60 Vm^{-1}
(c) 10/6 Vm^{-1} (d) None of these

30. The magnetic flux density B at a distance r from a long straight wire carrying a steady current varies with distance r as

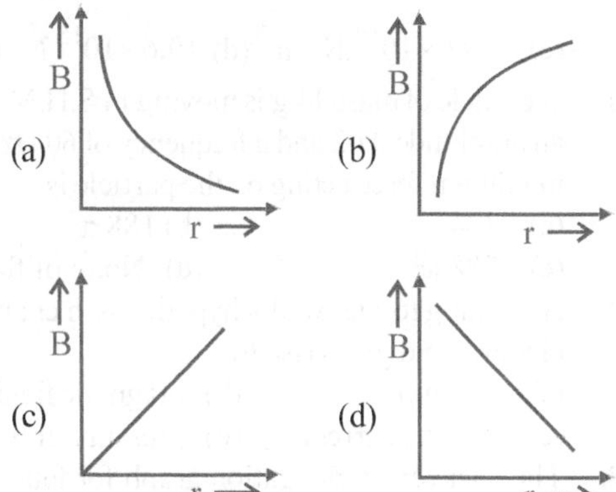

31. A metal piece is heated upto T° abs. The temperature of the surrounding is t° abs. The heat in the surrounding due to radiation is proportional to
(a) $(T - T)^4$ (b) $T^4 - t^4$
(c) $(T - t)^{1/4}$ (d) $T^2 - t^2$

32. Given, ${}_a\mu_g = \frac{3}{2}$, ${}_a\mu_w = \frac{4}{3}$, if a convex lens of focal length 10 cm is placed in water, then its focal length in water is
(a) equal to 40 cm (b) equal to 20 cm
(c) equal to 10 cm (d) None of these

33. If an alternating current is flowing in a spring, then the spring will be changing
(a) in a straight line (b) periodically
(c) elliptically (d) first (c) then (a)

34. The unit vector along $2i - 3j + k$ is
(a) $\frac{2i-3j+k}{\sqrt{14}}$ (b) $\frac{2i-3j+k}{5}$
(c) $\frac{2i-3j+k}{\sqrt{15}}$ (d) None of these

35. An iron rod of length 2m and cross-sectional area of 50 mm^2 is stretched by 0.5 mm, when a mass of 250 kg is hung from its lower end. Young's modulus of iron rod is
(a) $19.6\times10^{20}\ N/m^2$ (b) $19.6\times10^{18}\ N/m^2$
(c) $19.6\times10^{10}\ N/m^2$ (d) $19.6\times10^{15}\ N/m^2$

36. A particle of mass 1 kg is moving in S.H.M. with an amplitude 0.02 and a frequency of 60 Hz. The maximum force acting on the particle is
(a) $144\ \pi^2$ (b) $188\ \pi^2$
(c) $288\ \pi^2$ (d) None of these

37. According to Maxwell's hypothesis, a changing electirc field gives rise to
(a) an e.m.f (b) magnetic field
(c) electric current (d) pressure gradient.

38. The load versus elongation graph for four wires has been shown in the figure. The thinest wire is

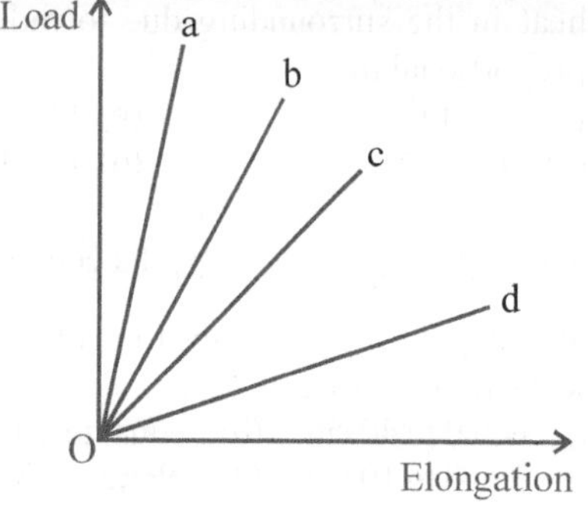

(a) a (b) b
(c) c (d) d

39. A ray of light is incident on the surface of separation of a medium with the velocity of light at an angle 45° and is refracted in the medium at an angle 30°. Velocity of light in the medium will be (velocity of light in air = 3×10^8 m/s)
(a) 3.8×10^8 m/s (b) 3.38×10^8 m/s
(c) 2.12×10^8 m/s (d) 1.56×10^8 m/s

40. The rain drops are in spherical shape due to
(a) residual pressure (b) thrust on drop
(c) surface tension (d) viscosity

41. A dip circle is so set that its needle moves freely in the magnetic meridian. In this position, the angle of dip is 40º. Now the dip circle is rotated so that the plane in which the needle moves makes an angle of 30º with the magnetic meridian. In this position, the needle will dip by an angle
(a) 40º (b) 30º
(c) more than 40º (d) less than 40º

42. If the critical angle for total internal reflection from a medium to vacuum is 30°. Then velocity of light in the medium is
(a) $1.5\times10^8\ m/s$ (b) $2\times10^8\ m/s$
(c) $3\times10^8\ m/s$ (d) $0.75\times10^8\ m/s$

43. An oscillator is nothing but an amplifier with
(a) positive feedback
(b) large gain
(c) no feedback
(d) negative feedback

44. For an AM-system the total power of modulated signal is 600 W and that of carrier is 400 W, the modulation index is
(a) 0.25 (b) 0.36
(c) 0.54 (d) 1

45. The drift current in a p-n junction is
(a) from the n-side to the p-side
(b) from the p-side to the n-side
(c) from the n-side to the p-side if the junction is forward-baised and in the opposite direction if it is reverse biased
(d) from the p-side to the n-side if the junction is forward-baised and in the opposite direction if it is reverse-baised

SECTION 2 - CHEMISTRY

46. An acidic solution of 'X' does not give precipitate on passing H_2S through it. 'X' gives white precipitate when NH_4OH is added to it. The white precipitate dissolves in excess of NaOH solution. Pure 'X' fumes in air and dense white fumes are obtained when a glass rod dipped in NH_4OH is put in the fumes. Compound 'X' can be
(a) $ZnCl_2$ (b) $FeCl_3$
(c) $AlCl_3$ (d) $SnCl_2$

47. CN^- is a strong field ligand. This is due to the fact that
(a) it carries negative charge
(b) it is a pseudohalide
(c) it can accept electrons from metal species
(d) it forms high spin complexes with metal species

48. The weight of NaCl decomposed by 4.9g of H_2SO_4, if 6 g of sodium hydrogen sulphate and 1.825 g of HCl, were produced in the reaction is:
(a) 6.921 g (b) 4.65 g
(c) 2.925 g (d) 1.4 g

49. Which one of the following statement is **not** true ?
(a) pH of drinking water should be between 5.5–9.5.
(b) Concentration of DO below 6 ppm is good for the growth of fish.
(c) Clean water would have a BOD value of less than 5 ppm.
(d) Oxides of sulphur, nitrogen and carbon are the most widespread air pollutant.

50. Which of the following statements is not correct for nitrogen ?
(a) Its electronegativity is very high
(b) d-orbitals are available for bonding
(c) It is a typical non-metal
(d) Its molecular size is small

51. Heat of dissociation of CH_3COOH is 0.005 kcal g^{-1}, hence enthalpy change when 1 mol of $Ca(OH)_2$ is completely neutralised by CH_3COOH is
(a) –27.4 kcal (b) –13.6 kcal
(c) –26.8 kcal (d) –27.1 kcal

52. 0.4 moles of HCl and 0.2 moles of $CaCl_2$ were dissolved in water to have 500 mL of solution, the molarity of Cl^- ion is:
(a) 0.8 M (b) 1.6 M
(c) 1.2 M (d) 10.0 M

53. In sodium fusion test of organic compounds, the nitrogen of the organic compound is converted into
(a) Sodamide (b) Sodium cyanide
(c) Sodium nitrite (d) Sodium nitrate

54. Specific volume of cylindrical virus particle is 6.02×10^{-2} cc/gm. whose radius and length 7 Å & 10 Å respectively. If $N_A = 6.02 \times 10^{23}$, find molecular weight of virus
(a) 3.08×10^3 kg/mol (b) 3.08×10^4 kg/mol
(c) 1.54×10^4 kg/mol (d) 15.4 kg/mol

55. Inductive effect involves
(a) displacement of σ-electrons
(b) delocalisation of π-electrons
(c) delocalisation of σ-electrons
(d) displacement of π-electrons

56. The energy of a photon is 3×10^{-12} erg. What is its wavelength in nm ?
($h = 6.62 \times 10^{-27}$ erg-sec; $c = 3 \times 10^{10}$ cm/s)
(a) 662 (b) 1324
(c) 66.2 (d) 6.62

57. The decreasing order of stability of alkyl carbonium ion is in the order of :

(a) $R-\overset{R}{\underset{R}{C^+}} > R-\overset{R}{\underset{H}{C^+}} > R-\overset{H}{\underset{H}{C^+}} > H-\overset{H}{\underset{H}{C^+}}$

(b) $H-\overset{H}{\underset{H}{C^+}} > R-\overset{H}{\underset{H}{C^+}} > R-\overset{R}{\underset{H}{C^+}} > R-\overset{R}{\underset{R}{C^+}}$

(c) $R-\overset{R}{\underset{H}{C^+}} > R-\overset{R}{\underset{R}{C^+}} > R-\overset{H}{\underset{H}{C^+}} > H-\overset{H}{\underset{H}{C}}$

(d) $R-\overset{H}{\underset{H}{C^+}} > R-\overset{R}{\underset{H}{C^+}} > R-\overset{R}{\underset{R}{C^+}} > H-\overset{H}{\underset{H}{C}}$

58. Aluminium vessels should not be washed with materials containing washing soda since
(a) washing soda is expensive
(b) washing soda is easily decomposed
(c) washing soda reacts with Al to form soluble aluminate
(d) washing soda reacts with Al to form insoluble aluminium oxide

59. The following data are for the decomposition of ammonium nitrite in aqueous solution :

Vol. of N_2 in cc	Time (min)
6.25	10
9.00	15
11.40	20
13.65	25
35.65	Infinity

The order of rection is :
(a) Zero (b) One
(c) Two (d) Three

60. Which of the following reagents convert propene to 1-propanol?
(a) H_2O, H_2SO_4
(b) aqueous KOH
(c) $MgSO_4, NaBH_4/H_2O$
(d) B_2H_6, H_2O_2, OH^-

61. A closed container contains equal number of oxygen and hydrogen molecules at a total pressure of 740 mm. If oxygen is removed form the system then pressure will
(a) Become double of 740 mm
(b) Become half of 740 mm
(c) Become 1/9 of 740 mm
(d) Remains unchanged

62. Compound X of molecular formula C_4H_6 takes up one equivalent of hydrogen in presence of Pt to form another compound Y which on ozonolysis gives only ethanoic acid. The compound X can be
(a) $CH_2 = CH - CH = CH_2$
(b) $CH_2 = C = CHCH_3$
(c) $CH_3C \equiv CCH_3$
(d) All the three

63. 1 M solution of CH_3COOH should be diluted to .... times so that pH is doubled.
(a) four times (b) 5.55×10^4 times
(c) 5.55×10^6 times (d) 10^{-2} times

64. The following reactions take place in the blast furnace in the preparation of impure iron. Identify the reaction pertaining to the formation of the slag.
(a) $Fe_2O_3(s) + 3\,CO(g) \rightarrow 2\,Fe\,(l) + 3\,CO_2\,(g)$
(b) $CaCO_3\,(s) \rightarrow CaO\,(s) + CO_2\,(g)$
(c) $CaO\,(s) + SiO_2(s) \rightarrow CaSiO_3\,(s)$
(d) $2C(s) + O_2\,(g) \rightarrow 2\,CO(g)$

65. Which one of the following is NOT a buffer solution?
(a) $0.8\,M\,H_2S + 0.8\,M\,KHS$
(b) $2M\,C_6H_5NH_2 + 2M\,C_6H_5\overset{+}{N}H_3Br^-$
(c) $3M\,H_2CO_3 + 3M\,KHCO_3$
(d) $0.05\,M\,KClO_4 + 0.05\,M\,HClO_4$

66. Which of the following statements is false ?
(a) Radon is obtained from the decay of radium
(b) Helium is inert gas
(c) Xenon is the most reactive among the rare gases
(d) The most abundant rare gas found in the atmosphere is helium

67. Which one of the following is expected to exhibit optical isomerism?
(en = ethylenediamine)
(a) cis-$[Pt(NH3)_2Cl_2]$
(b) trans-$[Pt(NH_3)_2Cl_2]$
(c) cis-$[Co(en)_2Cl_2]$
(d) trans-$[Co(en)_2Cl_2]$

68. For the reaction,
$$2N_2O_5 \longrightarrow 4NO_2 + O_2$$
the rate of reaction is :
(a) $\frac{1}{2}\frac{d}{dt}[N_2O_5]$ (b) $2\frac{d}{dt}[N_2O_5]$
(c) $\frac{1}{4}\frac{d}{dt}[NO_2]$ (d) $4\frac{d}{dt}[NO_2]$

69. Four successive members of the first row transition elements are listed below with their atomic numbers. Which one of them is expected to have the highest third ionization enthalpy?
(a) Vanadium (Z = 23)
(b) Chromium (Z = 24)
(c) Manganese (Z = 25)
(d) Iron (Z = 26)

70. $[Cr(H_2O)_6]Cl_3$ (at no. of Cr = 24) has a magnetic moment of 3.83 B. M. The correct distribution of 3d electrons in the Chromium of the complex is
(a) $3d_{xy}{}^1, (3d_{x^2-y^2})^1, 3d_{yz}{}^1$
(b) $3d_{xy}{}^1, 3d_{yz}{}^1, 3d_{xz}{}^1$
(c) $3d_{xy}{}^1, 3d_{yz}{}^1, 3d_{dz^2}$
(d) $(3d_{x^2-y^2})^1, 3d_{z^2}, 3d_{xz}{}^1$

71. In the balanced chemical reaction
$$IO_3^- + aI^- + bH^+ \longrightarrow cH_2O + dI_2$$
a, b, c and d respectively corresponds to
(a) 5, 6, 3, 3 (b) 5, 3, 6, 3
(c) 3, 5, 3, 6 (d) 5, 6, 5, 5

72. Which of the following statements is true?
(a) Silicon exhibits 4 coordination number in its compound
(b) Bond energy of F_2 is less than Cl_2
(c) Mn(III) oxidation state is more stable than Mn(II) in aqueous state
(d) Elements of 15th group shows only +3 and +5 oxidation states

73. Which of the following compounds has the highest boiling point?
(a) $CH_3CH_2CH_2Cl$
(b) $CH_3CH_2CH_2CH_2Cl$
(c) $CH_3CH(CH_3)CH_2Cl$
(d) $(CH_3)_3CCl$

74. Which one of the following statements is not correct ?
(a) Nickel forms $Ni(CO)_4$
(b) All the transition metals form monometallic carbonyls
(c) Carbonyls are formed by transition metals
(d) Transition metals form complexes

75. Hydrogen has an ionisation energy of 1311 kJ mol^{-1} and for chlorine it is 1256 kJ mol^{-1}. Hydrogen forms H^+ (aq) ions but chlorine does not form Cl^+ (aq) ions because
(a) H^+ has lower hydration enthalpy
(b) Cl^+ has lower hydration enthalpy
(c) Cl has high electron affinity
(d) Cl has high electronegativity

76. The number of enantiomers of the compound $CH_3CHBrCHBrCOOH$ is :
(a) 2 (b) 3
(c) 4 (d) 6

77. Equivalent weighs of $KMnO_4$ acidic medium, neutral medium and concentrated alkaline medium respectively are $\frac{M}{5}, \frac{M}{1}, \frac{M}{3}$. Reduced products can be
(a) $MnO_2, MnO_4^{2-}, Mn^{2+}$
(b) $MnO_2, Mn^{2+}, MnO_4^{2-}$
(c) $Mn^{2+}, MnO_4^{2-}, MnO_2$
(d) $Mn^{2+}, MnO_2, MnO_4^{2-}$

78. Which of these have no unit?
(a) Electronegativity
(b) Electron affinity
(c) Ionisation energy
(d) Excitation potential

79. Which of the following statements is not correct for sigma and pi-bonds formed between two carbon atoms?
(a) Sigma-bond determines the direction between carbon atoms but a pi-bond has no primary effect in this regard
(b) Sigma-bond is stronger than a pi-bond
(c) Bond energies of sigma- and pi-bonds are of the order of 264 kJ/mol and 347 kJ/mol, respectively
(d) Free rotation of atoms about a sigma-bond is allowed but not in case of a pi-bond

80. The reactivity of metals with water is in the order of
(a) Na > Mg > Zn > Fe > Cu
(b) Cu > Fe > Zn > Mg > Na
(c) Mg > Zn > Na > Fe > Cu
(d) Zn > Na > Mg > Fe > Cu

81. The emf of Daniell cell at 298 K is E_1
$Zn \mid ZnSO_4\,(0.01\,M) \mid\mid CuSO_4\,(1.0\,M) \mid Cu$
When the concentration of $ZnSO_4$ is 1.0 M and that of $CuSO_4$ is 0.01 M, the emf changed to E_2
What is the relation between E_1 and E_2?
(a) $E_1 = E_2$ (b) $E_2 = 0 \neq E_2$
(c) $E_1 > E_2$ (d) $E_1 < E_2$

82. 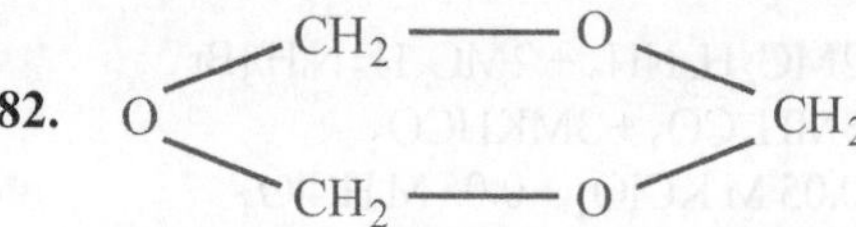

The above shown polymer is obtained when a carbonyl compound is allowed to stand. It is a white solid. The polymer is
(a) Trioxane (b) Formose
(c) Paraformaldehyde (d) Metaldehyde.

83. The correct order of atomic/ionic sizes is
(a) N < Li < B
(b) $F^- < O^{2-} < N^{3-}$
(c) $Ca^{2+} < S^{2-} < Cl^-$
(d) $Na^+ < Mg^{2+} < Cl^-$

84. The rapid change of pH near the stoichiometric point of an acid-base titration is the basis of indicator detection. pH of the solution is related to ratio of the concentrations of the conjugate acid (HIn) and base (In^-) forms of the indicator by the expression
(a) $\log\frac{[In^-]}{[HIn]} = pK_{In} - pH$
(b) $\log\frac{[HIn]}{[In^-]} = pK_{In} - pH$
(c) $\log\frac{[HIn]}{[In^-]} = pH - pK_{In}$
(d) $\log\frac{[In^-]}{[HIn]} = pH - pK_{In}$

85. In the diazotization of arylamines with sodium nitrite and hydrochloric acid, an excess of hydrochloric acid is used primarily to
(a) Supress the concentration of free aniline available for coupling
(b) Supress hydrolysis of phenol
(c) Ensure a stoichiometric amount of nitrous acid
(d) Neutralise the base liberated

86. In lake test of Al^{3+} ion, there is formation of coloured floating lake. It is due to
(a) adsorption of litmus by H_2O
(b) adsorption of litmus by $Al(OH)_3$
(c) adsorption of litmus by $Al(OH)_4^-$
(d) none of these

87. An ideal gas expands in volume from 1×10^{-3} to 1×10^{-2} m^3 at 300 K against a constant pressure of 1×10^5 Nm^{-2}. The work done is
(a) 270 kJ (b) -900 kJ
(c) -900 kJ (d) 900 kJ

88. Ethanol and dimethyl ether form a pair of functional isomers. The boiling point of ethanol is higher than that of dimethyl ether, due to the presence of
(a) H-bonding in ethanol
(b) H-bonding in dimethyl ether
(c) CH_3 group in ethanol
(d) CH_3 group in dimethyl ether

89. Which of the following reactions will not result in the formation of anisole?
(a) Phenol + dimethyl sulphate in presence of a base
(b) Sodium phenoxide is treated with methyl iodide
(c) Reaction of diazomethane with phenol
(d) Reaction of methylmagnesium iodide with phenol

90. What will be the heat of formation of methane, if the heat of combustion of carbon is '–x' kJ, heat of formation of water is '–y' kJ and heat of combustion of methane is 'z' kJ ?
(a) $(-x-y+z)$ kJ (b) $(-z-x+2y)$ kJ
(c) $(-x-2y-z)$ kJ (d) $(-x-2y+z)$ kJ

SECTION 3 - BOTANY

91. A taxon is
(a) a group of related families
(b) a group of related species
(c) a type of living organisms
(d) a taxonomic group of any ranking

92. Peat moss is another name of
(a) *Sphagnum* (b) *Marchantia*
(c) *Riccia* (d) *Dryopteris*

93. In fern, spores are formed in
(a) sporangium (b) oogonium
(c) archegonium (d) stomium

94. Which one of the following is a true fruit?
(a) Apple (b) Pear
(c) Cashew nut (d) Coconut

95. The cork cambium, cork and secondary cortex are collectively called
(a) phelloderm (b) phellogen
(c) periderm (d) phellem

96. Which of the following algae are suitable for human consumption?
(a) *Laminaria* and *Fucus*
(b) *Gracilaria* and *Chondrus*
(c) *Porphyra* and *Spirogyra*
(d) *Rhodymania* and *Porphyra*

97. Choose the correct option.
(i) Lysosomes are double membranous vesicles budded off from Golgi apparatus and contain digestive enzymes.
(ii) Endoplasmic reticulum consists of a network of membranous tubule and helps in transport, synthesis and secretion.
(iii) Leucoplasts are bound by two membranes, lack pigment but contain their own DNA and protein synthesising machinery.
(iv) Sphaerosomes are single membrane bound organelle which are associated with synthesis and storage of lipids.
(a) (i) only (b) (i) and (ii)
(c) (ii), (iii) and (iv) (d) All of these

98. Chosse the correct combinations.

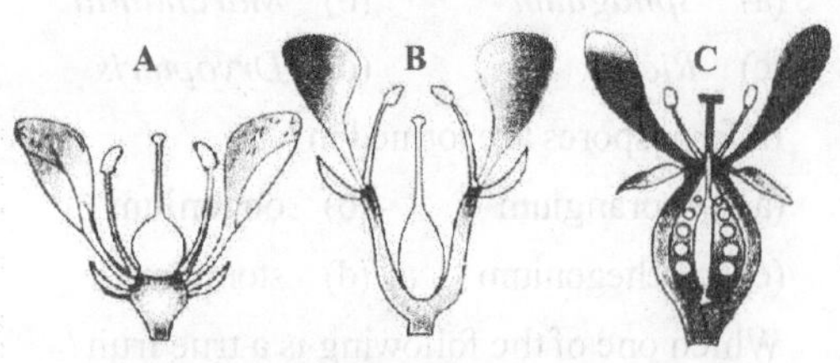

I. Hypogynous flower
II. Perigynous flower
III. Epigynous flower
(a) A - I, B - II, C - III (b) A - I, B - III, C - II
(c) A - III, B - II, C - I (d) A - III, B - I, C - II

99. Which one of the following is not a method of vegetative propagation?
(a) Budding (b) Layering
(c) Sowing (d) Tissue culture

100. Entry of pollen tube through micropyle is
(a) Chalazogamy (b) Mesogamy
(c) Porogamy (d) Pseudogamy

101. Competition for light, nutrients and space is most severe between
(a) closely related organism growing in different habitats
(b) closely related organisms growing in the same habitat
(c) distantly related organisms growing in the same habitat
(d) distantly related organisms growing in different habitats

102. In oogamy, fertilization involves
(a) a small non-motile female gamete and a large motile male gamete
(b) a large non-motile female gamete and a small motile male gamete
(c) a large non-motile female gamete and a small nonmotile male gamete
(d) a large motile female gamete and a small nonmotile male gamete

103. Photochemical smog formed in congested metropolitan cities mainly consists of
(a) Ozone, peroxyacetyl nitrate and NO_X
(b) Smoke, peroxyacetyl nitrate and SO_2
(c) Hydrocarbon, SO_2 and CO_2
(d) Hydrocarbon, ozone and SO_X

104. The electrostatic precipitator is used for removing particulate matter from
(a) Exhaust of the thermal power plant
(b) Exhaust from the automobiles
(c) Industrial effluents
(d) Kitchen waste

105. Keystone species in an ecosystem are those
(a) present in maximum number
(b) that are most frequent
(c) attaining a large biomass
(d) contributing to ecosystem properties

106. Initiation codon of protein synthesis (in eukaryotes) is
(a) GUA (b) GCA
(c) CCA (d) AUG

107. What is the best pH of soil for cultivation of plants ?
(a) 3.4 – 5.4 (b) 6.5 – 7.5
(c) 4.5 – 8.5 (d) 5.6 – 6.5

108. Telomerase is an enzyme which is a
(a) simple protein (b) RNA
(c) ribonucleoprotein (d) repetitive DNA

109. Mass of living matter at a trophic level in an area at any time is called
(a) standing crop (b) deteritus
(c) humus (d) standing state

110. The *Triticale* is an intergeneric hybrid between :
(a) wheat and maize (b) maize and rye
(c) wheat and rye (d) bajra and wheat

111. Which one is a neem product used as insect repellent?
(a) Azadirachtin (b) Rotenone
(c) Parathione (d) Endrin

112. Choose the correct option.

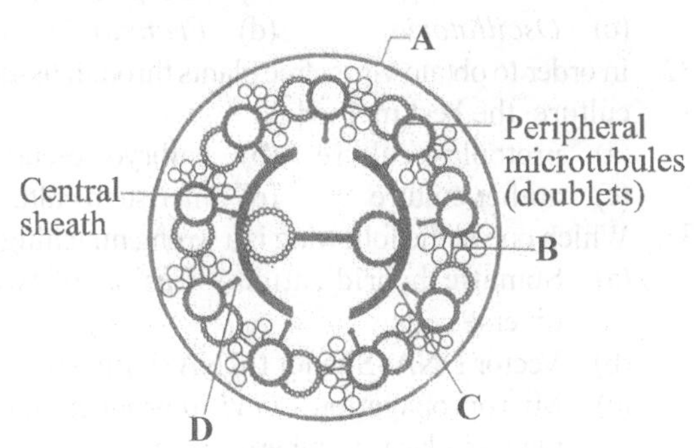

(a) A – Plasma membrane, B – Interdoublet bridge, C – Central microtubule, D – Radial spoke
(b) A – Plasma membrane, B – Arm, C – Central microtubule, D – Radial spoke
(c) A – Plasma membrane, B – Interdoublet bridge, C – Hub, D – Radial spoke
(d) A – Plasma membrane, B – Interdoublet bridge, C – Hub, D – Arm

113. An ecosystem which can be easily damaged but can recover after some time if damaging effect stops will be having
(a) low stability and high resilience
(b) high stability and low resilience
(c) low stability and low resilience
(d) high stability and high resilience

114. The mode of catching insects in *Drosera* plants is by means of
(a) sensitive glandular hairs which secrete a sweet, viscous, shining substance.
(b) specially sensitive trigger hairs.
(c) leaves which are modified into pitcher.
(d) leaf segments modified into bladder.

115. Quantasomes are present in
(a) chloroplast (b) mitochondria
(c) golgi body (d) lysosome

116. The water potential and osmotic potential of pure water are
(a) zero and zero (b) 100 and 100
(c) zero and 100 (d) 100 and zero

117. Photorespiration is favoured by
(a) high O_2 and low CO_2
(b) low light and high O_2
(c) low temperature and high O_2
(d) low O_2 and high CO_2

118. A free living nitrogen-fixing cyanobacterium which can also form symbiotic association with the water fern *Azolla* is
(a) *Tolypothrix* (b) *Chlorella*
(c) *Nostoc* (d) *Anabaena*

119. Hydroponics is
(a) nutrient less culture
(b) water less culture
(c) soilless culture
(d) None of these

120. Krebs cycle occurs in
(a) mitochondria (b) cytoplasm
(c) chloroplasts (d) ribosomes

121. Most abundant organic compound on earth is
(a) Protein (b) Cellulose
(c) Lipids (d) Steroids

122. Terminal cytochrome of respiratory chain which donates electrons to oxygen is
(a) Cyt. b (b) Cyt. c
(c) Cyt. a_1 (d) Cyt. a_3

123. To avoid excessive water loss during severe drought stress, the closure of stomata is signalled by the production of
(a) IAA (b) NAA
(c) ABA (d) IBA

124. In short day plants, flowering is induced by
(a) photoperiod less than 12 hours.
(b) photoperiod below a critical length and uninterrupted long night.
(c) long night.
(d) short photoperiod and interrupted long night.

125. The major reason that glycolysis is not as energy productive as respiration is that
(a) NAD^+ is regenerated by alcohol or lactate production, without the high-energy electrons passing through the electron transport chain.
(b) it is the pathway common to fermentation and respiration.
(c) it does not take place in a specialized membrane-bound organelle.
(d) pyruvate is more reduced than CO_2; it still contains much of the energy from glucose.

126. The catalytic efficiency of two different enzymes can be compared by the
(a) formation of the product
(b) pH optimum value
(c) K_m value
(d) molecular size of the enzyme

127. Biodiversity Act of India was passed by the parliament in the year
(a) 1992 (b) 1996
(c) 2000 (d) 2002

128. 'Axenic culture' is
(a) culture of tissue
(b) growing of shrubs
(c) growing of tall trees
(d) culture of tissue free from contamination

129. Which one thing is not true about antibiotics?
(a) The term "antibiotic" was coined by Selman Waksman in 1942
(b) First antibiotic was discovered by Alexander Flemming
(c) Each antibiotic is effective only against one particular kind of germ
(d) Some persons can be allergic to a particular antibiotic

130. Main objective of production/use of herbicide resistant GM crops is to
(a) eliminate weeds from the field without the use of manual labour
(b) eliminate weeds from the field without the use of herbicides
(c) encourage eco-friendly herbicides
(d) reduce herbicide accumulation in food articles for health safety

131. The common nitrogen fixer in paddy fields is
(a) *Rhizobium* (b) *Azospirillum*
(c) *Oscillatoria* (d) *Frankia*

132. In order to obtain virus- free plants through tissue culture, the best method is
(a) protoplast culture (b) embryo rescue
(c) anther culture (d) meristem culture

133. Which one of the following is a wrong matching?
(a) Somatic hybridization - Fusion of two diverse cells
(b) Vector DNA -Site for t-RNA synthesis
(c) Micropropagation - *in vitro* production of plants in large numbers
(d) Callus - Unorganised mass of cell produced in tissue culture

134. 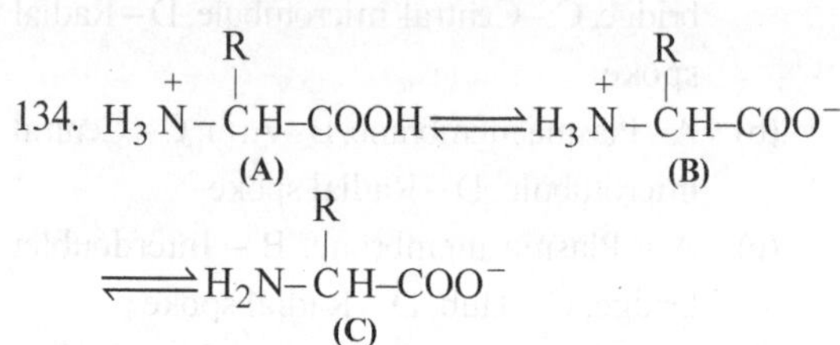

Which of the above is Zwitterionic form?
(a) A (b) C
(c) B (d) All of these

135. Restriction endonucleases are enzymes which
(a) make cuts at specific positions within the DNA molecule
(b) recognize a specific nucleotide sequence for binding of DNA ligase
(c) restrict the action of the enzyme DNA polymerase
(d) remove nucleotides from the ends of the DNA molecule

SECTION 4 - ZOOLOGY

136. Sex factor in bacteria is
(a) Chromosomal replicon
(b) F-replicon
(c) RNA
(d) Sex-pilus

137. Animals/organisms floating on the surface of water are
(a) plankton (b) pelagic
(c) benthos (d) neritic

138. The cell junctions called tight, adhering and gap junctions are found in
(a) connective tissue (b) epithelial tissue
(c) neural tissue (d) muscular tissue

139. Spleen is referred to as
(a) temporary endocrine gland
(b) graveyard of RBC
(c) largest gland
(d) store house of WBC

140. Given below are four matchings of an animal and its kind of respiratory organ :
(i) Silver fish – trachea
(ii) Scorpion – book lung
(iii) Sea squirt – pharyngeal gills
(iv) Dolphin – skin
The correct matchings are
(a) (iii) and (iv) (b) (i) and (iv)
(c) (i), (ii) and (iii) (d) (ii) and (iv)

141. In the mouthparts of the cockroach, the organ of mastication is
(a) labium (b) maxillae
(c) mandibles (d) labrum

142. Which one of the following characters is not typical of the class Mammalia?
(a) Thecodont dentition
(b) Alveolar lungs
(c) Ten pairs of cranial nerves
(d) Seven cervical vertebrae

143. Natural parthenogenesis occurs in:
(a) Protozoans (b) Earthworm
(c) All insects (d) Honeybee

144. Consider the statements given below regarding contraception and answer as directed thereafter:
(i) Medical Termination of Pregnancy (MTP) during first trimester is generally safe
(ii) Generally chances of conception are nil until mother breast-feeds the infant upto two years
(iii) Intrauterine devices like copper-T are effective contraceptives
(iv) Contraception pills may be taken upto one week after coitus to prevent conception
Which two of the above statements are correct?
(a) ii and iii (b) iii and iv
(c) i and iii (d) i and ii

145. Identify the blood vessels A to D.

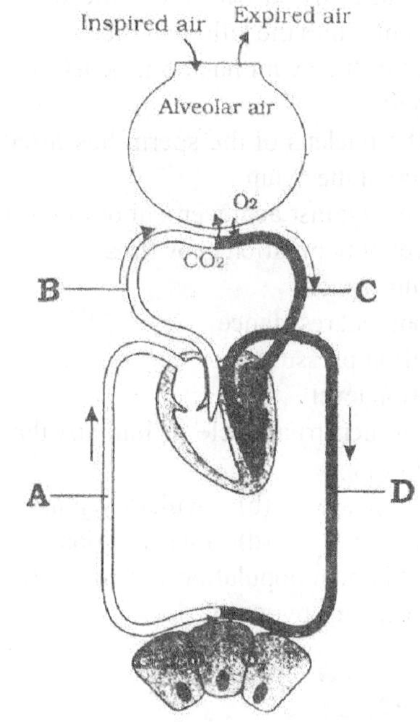

(a) A- Systemic vein, B-Pulmonary artery, C-Pulmonary vein, D-Systemic artery
(b) A-Systemic artery, B-Pulmonary artery, C- Pulmonary vein, D- Systemic vein
(c) A-Pulmonary artery, B-Systemic vein, C-Pulmonary vein, D- Systemic artery
(d) A-Systemic vein, B-Pulmonary vein, C- Pulmonary artery, D- Systemic artery

146. Which pathway of the male reproductive system is correct for the sperms transportation?
(a) Vas efferentia→Vas deferens→Epididymis
(b) Vas deferens→Epididymis→Seminal vesicle
(c) Epididymis→Vas deferens→Urethra
(d) Rete testis→Epididymis→Vas efferentia

147. The second maturation division of the mammalian ovum occurs:
(a) in the Graafian follicle following the first maturation division
(b) Shortly after ovulation before the ovum makes entry into the fallopian tube
(c) Until after the ovum has been penetrated by a sperm
(d) Until the nucleus of the sperm has fused with that of the ovum

148. A force acting against achievement of highest possible level of population growth is
(a) Carrying capacity
(b) Environment resistance
(c) Population pressure
(d) Saturation level

149. The phase of menstrual cycle in humans that lasts for 7-8 days, is
(a) follicular phase (b) ovulatory phase
(c) luteal phase (d) menstruation

150. What type of human population is represented by the following age pyramid?

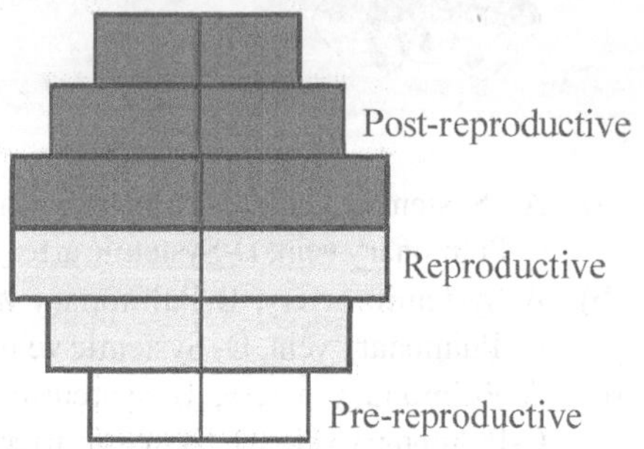

(a) Vanishing population
(b) Stable population
(c) Declining population
(d) Expanding population

151. The 'Mule' is the result of
(a) Inbreeding depression
(b) Out breeding
(c) Cross breeding
(d) Inter-specific hybridization

152. Haemophilia is more common in males because it is a
(a) Recessive character carried by Y–chromosome
(b) Dominant character carried by Y-chromosome
(c) Dominant trait carried by X-chromosome
(d) Recessive trait carried by X-chromosome

153. Theory of inheritance of acquired characters was given by
(a) Wallace (b) Lamarck
(c) Darwin (d) De Vries

154. The animal husbandry deals with the care, breeding and management of
(a) Domesticated animals
(b) Fishes
(c) Honey bees and silk worms
(d) All of these

155. 'Inland fishery' refers to
(a) Culturing fish in fresh water
(b) Trapping and capturing fishes from sea coast
(c) Deep sea fishing
(d) Extraction of oil from fishes

156. The most popular breed of fowl in India is
(a) White leg horn (b) Aseel
(c) Plymouth (d) Langshan

157. Which of following teeth are lophodont?
(a) Incisor and canine
(b) Premolar and molar
(c) Canine and premolar
(d) Premolar and incisor

158. Pacemaker of heart is
(a) AV node (b) Bundle of His
(c) SA node (d) Purkinje fibres

159. Uricotelism is found in
(a) Frogs and toads
(b) Mammals and birds
(c) Birds, reptiles and insects
(d) Fishes and fresh water protozoans

160. A large proportion of oxygen is left unused in the human blood even after its uptake by the body tissues. This O_2
(a) acts as a reserve during muscular exercise
(b) raise the pCO_2 of blood to 75 mm of Hg.
(c) is enough to keep oxyhaemoglobin saturation at 96%
(d) helps in releasing more O_2 to the epithelial tissues.

161. The basic functional unit of the human kidney is
(a) nephron (b) nephridia
(c) pyramid (d) Henle's loop

162. Urea from the blood can be removed by
(a) Uremia (b) Diuresis
(c) Dialysis (d) Micturition

163. Which one of the following correctly explains the function of a specific part of a human nephron ?
(a) Podocytes : create minute spaces (slite pores) for the filtration of blood into the Bowman's capsule.
(b) Henle's loop : most reabsorption of the major substances from the glomerular filtrate.
(c) Distal convoluted tubule : reabsorption of K^+ ions into the surrounding blood capillaries.
(d) Afferent arteriole : carries the blood away from the glomerular towards renal vein.

164. The nerve centres which control the body temperature and the urge for eating are contained in:
(a) hypothalamus (b) pons
(c) cerebellum (d) thalamus

165. Rods and cones of eyes are modified
(a) multipolar neuron
(b) unipolar neuron
(c) bipolar neuron
(d) None of these

166. Which of the following is both exocrine and endocrine gland ?
(a) Liver (b) Pancreas
(c) Thyroid (d) Adrenal

167. The sensation of fatigue in the muscles after prolonged strenuous physical work, is caused by
(a) a decrease in the supply of oxygen
(b) minor wear and tear of muscle fibres
(c) the depletion of glucose
(d) the accumulation of lactic acid

168. Consider the diagram given below

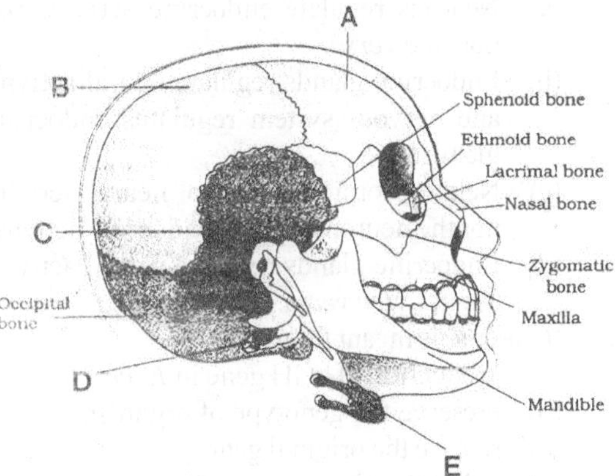

Identify the labelled parts as A, B, C, D and E respectively.
(a) Frontal bone, Parietal bone, Temporal bone, Occipital condyle and Hyoid bone
(b) Frontal bone, Temporal bone, Parietal bone, Occipital condyle and Hyoid bone
(c) Frontal bone, Parietal bone, Temporal bone, Hyoid bone and Occipital condyle
(d) Parietal bone, Frontal bone, Temporal bone, Occipital condyle and Hyoid bone

169. Which one of the following statements is incorrect ?
(a) The presence of non-respiratory air sacs, increases the efficiency of respiration in birds.
(b) In insects, circulating body fluids serve to distribute oxygen to tissues.
(c) The principle of countercurrent flow facilitates efficient respiration in gills of fishes.
(d) The residual air in lungs slightly decreases the efficiency of respriation in mammals.

170. Which one of the following does not act as a neurotransmitter ?
(a) Epinephrine (b) Norepinephrine
(c) Cortisone (d) Acetylcholine

171. Which one of the following statements is correct?
(a) Neurons regulate endocrine activity, but not *vice versa.*
(b) Endocrine glands regulate neural activity and nervous system regulates endocrine glands.
(c) Neither hormones control neural activity nor the neurons control endocrine activity.
(d) Endocrine glands regulate neural activity but not *vice versa.*

172. 'Cloning' is meant for/to
(a) production of hGH gene in *E. coli*
(b) preserve the genotype of organism
(c) replace the original gene
(d) All of the above

173. A cell coded protein formed in response to infection with most animal viruses is
(a) Antigen (b) Antibody
(c) Interferon (d) Histone

174. Which one of the following is not used in organic farming?
(a) *Glomus* (b) Earthworm
(c) *Oscillatoria* (d) Snail

175. ELISA is used to detect viruses where the key reagent is
(a) RNase
(b) alkaline phosphatase
(c) catalase
(d) DNA probe

176. Vitamin B_{12} is formed during fermentation of
(a) *Ashloya gossipii*
(b) *Rhizopus stolonifer*
(c) *Propionibacteria*
(d) *Saccharomyces cerevisiae*

177. Which one is a correctly match sexually transmitted disease with its pathogen?
(a) Syphilis – *Leishmania donovani*
(b) AIDS – *Bacillus anthracis*
(c) Urethritis – *Entamoeba gingivalis*
(d) Gonorrhoea – *Neisseria gonorrhoeae*

178. Which one of the following depresses brain activity and produced feelings of calmness, relaxation and drowsiness?
(a) Morphine (b) Valium
(c) Amphetamines (d) Hashish

179. Which one of the following is correctly matched pair of the given secretion and its primary role in human physiology?
(a) Sebum — Sexual attraction
(b) Sweat — Thermoregulation
(c) Saliva — Tasting food
(d) Tears — Excretion of salts

180. Consider the following four statements (i-iv) and select the option which includes all the correct ones only.
(i) Single cell *Spirulina* can produce large quantities of food rich in protein, minerals, vitamins etc.
(ii) Body weight-wise the micro-organism *Methylophilus methylotrophus* may be able to produce several times more proteins than the cows per day.
(iii) Common button mushrooms are a very rich source of vitamin C.
(iv) A rice variety has been developed which is very rich in calcium.
(a) Statements (ii) and (iv)
(b) Statements (i), (iii) and (iv)
(c) Statements (ii), (iii) and (iv)
(d) Statements (i) and (ii)

RESPONSE SHEET

Test Code: ....Time taken :...................... Date:...............

	(a)	(b)	(c)	(d)		(a)	(b)	(c)	(d)		(a)	(b)	(c)	(d)		(a)	(b)	(c)	(d)
1.					46.					91.					136.				
2.					47.					92.					137.				
3.					48.					93.					138.				
4.					49.					94.					139.				
5.					50.					95.					140.				
6.					51.					96.					141.				
7.					52.					97.					142.				
8.					53.					98.					143.				
9.					54.					99.					144.				
10.					55.					100.					145.				
11.					56.					101.					146.				
12.					57.					102.					147.				
13.					58.					103.					148.				
14.					59.					104.					149.				
15.					60.					105.					150.				
16.					61.					106.					151.				
17.					62.					107.					152.				
18.					63.					108.					153.				
19.					64.					109.					154.				
20.					65.					110.					155.				
21.					66.					111.					156.				
22.					67.					112.					157.				
23.					68.					113.					158.				
24.					69.					114.					159.				
25.					70.					115.					160.				
26.					71.					116.					161.				
27.					72.					117.					162.				
28.					73.					118.					163.				
29.					74.					119.					164.				
30.					75.					120.					165.				
31.					76.					121.					166.				
32.					77.					122.					167.				
33.					78.					123.					168.				
34.					79.					124.					169.				
35.					80.					125.					170.				
36.					81.					126.					171.				
37.					82.					127.					172.				
38.					83.					128.					173.				
39.					84.					129.					174.				
40.					85.					130.					175.				
41.					86.					131.					176.				
42.					87.					132.					177.				
43.					88.					133.					178.				
44.					89.					134.					179.				
45.					90.					135.					180.				

TEST ASSESSMENT AND ANALYSIS SHEET

How to fill this sheet : After you finish the paper and have checked it first try to attempt again all the questions which you have done wrong or have not been able to attempt. Go to detailed solutions only after you have attempted all the questions of your own.Attempting the questions on your own without seeing the solutions will help you in filling this table

ANALYSIS OF TEST	
Total number of questions	
Questions Attempted	
Correct Questions	
Wrong Questions	
	Net Score

Strike rate $\left(\frac{\text{Correct questions}}{\text{Questions attempt}} \times 100\right) =$ ....

Percentage Marks $\left(\frac{\text{Correct Answers}}{\text{Total number of questions}} \times 100\right) =$ ....

Analysis of wrong questions : Divide the questions which you have done wrong in 3 broad categories.

Reasons for questions attempted wrong	Number of Questions
Knew but calculation mistake (solved on my own after exam)	
Knew but got confused (solved on my own after exam)	
Did not knew the question (couldn't solve even after exam)	

Questions not attempted : Divide the questions not attempted in 3 broad categories

Reasons for unattempted questions	Number of Questions
Knew the ques but time was less (solved on my own after exam)	
Knew the ques but got confused (solved on my own after exam)	
Did not knew the questions (couldn't solve even after exam)	

List down your weak areas you need to improve your score :

1)____

2)____

3)____

NEET

National Eligibility Entrance Test

Test Series **Full Test - 5**

Time : - 3 Hours **Max. Marks:- 720**

INSTRUCTIONS

1. Immediately fill in the particulars on this page of the Test Booklet with Blue/Black Ball Point Pen. Use of pencil is strictly prohibited.
2. Answers are to be given on Response Sheet provided at the end of this paper. You can tear this Response Sheet along the cut marks provided. Fill in the particulars carefully in the Response Sheet provided.
3. The test is of 3 hours duration.
4. The Test Booklet consists of **180** questions. The maximum marks are **720**.
5. There are four section in the question paper.

   The distribution of marks subjectwise in each part is as under for each correct response.

   Section 1 – Physics (180 Marks) – 45 Questions

   Section 2 – Chemistry (180 Marks) – 45 Questions

   Section 3 – Botany (180 Marks) – 45 Questions

   Section 4 – Zoology (180 Marks) – 45 Questions
6. You will be awarded 4 Marks if you have darkened only the bubble corresponding to the correct answer and zero mark if no bubble is darkened. In all other cases, minus one (– 1) mark will be awarded.
7. Use Blue/Black Ball Point Pen only for writing particulars/marking responses on Side-1 and Side-2 of the Response Sheet.

   Use of pencil is strictly prohibited.
8. You are advised to create an actual examination hall simulation.
9. Rough work is to be done on the space provided for this purpose in the Test Booklet only.
10. Do not fold or make any stray marks on the Response Sheet.
11. Once you have evaluated the test, you should assess it thoroughly.

SECTION 1 - PHYSICS

1. The time of oscillation T of a small drop of liquid depends on radius r, density ρ and surface tension S. The relation between them is given by

(a) $T \propto \sqrt{\frac{S}{\rho r^3}}$ (b) $T \propto \sqrt{\frac{\rho r^3}{S}}$

(c) $T \propto \sqrt{\frac{S^2 r^3}{\rho}}$ (d) $T \propto \sqrt{\frac{\rho r^3}{S}}$

2. The number of significant figures in 3400 is
(a) 7 (b) 6
(c) 12 (d) 2

3. A sample of ideal monoatomic gas is taken round the cycle *ABCA* as shown in the figure. The work done during the cycle is

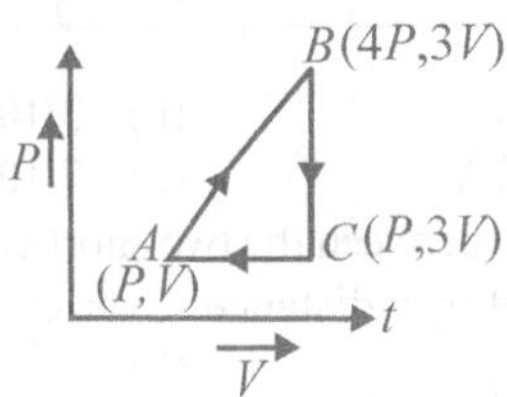

(a) 3 *PV* (b) zero
(c) 9 *PV* (d) 6 *PV*

4. The escape velocity for a body projected vertically upwards from the surface of earth is 11 km/s. If the body is projected at an angle of 45° with the vertical, the escape velocity will be
(a) 22 km/s (b) 11 km/s
(c) $\frac{11}{\sqrt{2}}$ km/s (d) $11\sqrt{2}$ km/s

5. A projectile is projected at an angle of 45° with speed u. The radius of curvature of its trajectory at the maximum height is

(a) $\frac{u^2}{g}$ (b) $\frac{u^2}{4g}$

(c) $\frac{2u^2}{g}$ (d) $\frac{u^2}{2g}$

6. As intensity of incident light increases
(a) photoelectric current increase
(b) K.E. of emitted photoelectron increases
(c) photo electric current decreases
(d) K.E. of emitted photoelectrons decreases

7. Which is true of the following in a purely resistive ac circuit?
(a) The current leads the driving voltage
(b) The driving voltage leads the current.
(c) The current and driving voltage are in the same phase.
(d) Any of the above may be true depending on the value of resistance

8. Two resistors A and B have resistances R_A and R_B respectively with $R_A < R_B$. The resistivities of their materials are ρ_A and ρ_B. Then
(a) $\rho_A > \rho_B$
(b) $\rho_A = \rho_B$
(c) $\rho_A < \rho_B$
(d) The information is not sufficient to find the relation between ρ_A and ρ_B

9. A block of mass 60 kg just slides over a horizontal distance of 0.9 m. If the coefficient of friction between their surface is 0.15 then work done against friction will be
(a) 79.4 J (b) 97.54 J
(c) 105.25 J (d) None of these

10. The efficiency of carnot engine when source temperature is T_1 and sink temperature is T_2 will be

(a) $\frac{T_1 - T_2}{T_1}$ (b) $\frac{T_2 - T_1}{T_2}$

(c) $\frac{T_1 - T_2}{T_2}$ (d) $\frac{T_1}{T_2}$

11. A capacitor of the capacitance 10 μF is charged by connecting through a resistance of 20Ω to a battery of 20V, as shown. How much energy is supplied by the battery?

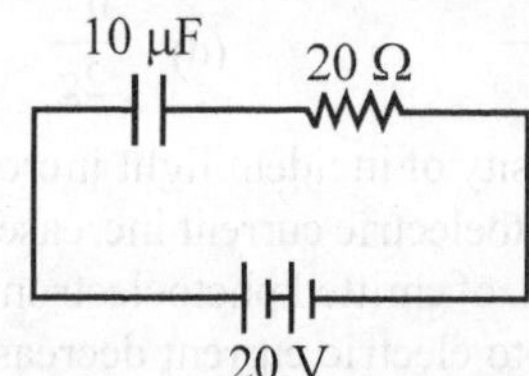

(a) <2 mJ (b) 2mJ
(c) >2 mJ (d) None of these

12. In which of the following Bohr's orbit n in a hydrogen atom emits the photons of lowest frequency ?
(a) n = 4 to n = 3 (b) n = 2 to n = 1
(c) n = 4 to n = 2 (d) n = 3 to n = 1

13. A wheel having moment of inertia 2 kg-m^2 about its vertical axis, rotates at the rate of 60 rpm about this axis, The torque which can stop the wheel's rotation in one minute would be

(a) $\frac{\pi}{18}$ N − m (b) $\frac{2\pi}{15}$ N − m

(c) $\frac{\pi}{12}$ N − m (d) $\frac{\pi}{15}$ N − m

14. Photoelectric effect shows
(a) wave like behaviour of light
(b) paritcle like behaviour of light
(c) both wave like and particle like behaviour
(d) Neither wave like nor particle like behaviour of light

15. The K.E. of one mole of an ideal gas is E = (3/2) RT. Then C_p will be
(a) 0.5 R (b) 0.1 R
(c) 1.5 R (d) 2.5 R

16. A triangular block of mass M with angles 30°, 60° and 90° rest with its 30° – 90° side on a horizontal table. A cubical block of mass m rests on 60° – 30° side. The acceleration which M must have relative to the table to keep m stationary relative to the triangular block is (assuming frictionless contact)
(a) g (b) $g/\sqrt{2}$
(c) $g/\sqrt{3}$ (d) $g/\sqrt{5}$

17. The condition for obtaining secondary maxima in the diffraction pattern due to single slit is
(a) $a \sin\theta = n\lambda$ (b) $a \sin\theta = (2n-1)\frac{\lambda}{2}$
(c) $a \sin\theta = (2n-1)\lambda$ (d) $a \sin\theta = \frac{n\lambda}{2}$

18. Seven resistances, each of value 20 Ω, are connected to a 2 V battery as shown in the figure. The ammeter reading will be

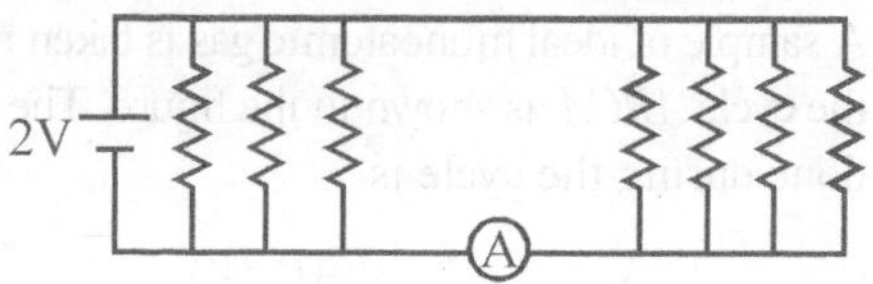

(a) 1/10 A (b) 3/10 A
(c) 4/10 A (d) 7/10 A.

19. The force between the two short electric dipoles, separated by a distance *r*, varies as
(a) r^2 (b) r^{-3}
(c) r^4 (d) r^{-4}

20. If *R* is universal gas constant, the amount of heat needed to raise the temperature of 2 moles of an ideal monoatomic gas from 273 K to 373 K, when no work is done, is
(a) 100 *R* (b) 150 *R*
(c) 300 *R* (d) 500 *R*

21. A block B of mass 4 kg is placed on a rough horizontal plane. A time dependent horizontal force f = kt acts on the block, k = 2 N/s^2. The frictional force between the block and the plane at t = 2 sec is (μ = 0.2)
(a) 4N (b) 8N
(c) 12N (d) zero

22. In a transformer, number of turns in the primary coil are 140 and that in the secondary coil are 280. If current in primary coil is 4 A, then that in the secondary coil is

(a) 4 A (b) 2 A
(c) 6 A (d) 10 A

23. A closed spherical surface of radius R encloses an electric dipole. The net electric flux through the surface is

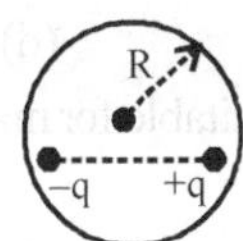

(a) zero (b) $2q.4\pi R^2$
(c) $\frac{q}{\varepsilon_0}$ (d) $\frac{2q}{\varepsilon_0}$

24. How much energy is required for a body of mass 1000 kg to escape from the earth? ($g = 9.8\ m/s^2$, R = 6400 km)

(a) about 3.2×10^{10} J (b) about 6.4×10^{10} J
(c) about 6.4×10^6 J (d) None of these

25. A phase difference between two points separated by 0.8 m in a wave of frequency 120 Hz is $\pi/2$. The wave velocity is

(a) 384 m/s (b) 768 m/s
(c) 250 m/s (d) 154 m/s

26. For a single side band transmission a balanced modulator is used to

(a) increase power of carrier wave
(b) increase amplitude of carrier wave
(c) suppress audio signal
(d) suppress carrier component

27. The amplitude of an alternate voltage is 120 V. Its rms value will be

(a) 107.3 V (b) 84.8 V
(c) 70.7 V (d) 68.8 V

28. A bag of mass M hangs by a long thread and a bullet (mass m) comes horizontally with velocity V and gets caught in the bag. Then for the combined (bag + bullet) system

(a) momentum $= \frac{mvM}{M+m}$

(b) kinetic energy $= \frac{mV^2}{2}$

(c) momentum $= \frac{mV(M+m)}{M}$

(d) kinetic energy $= \frac{m^2 v^2}{2(M+m)}$

29. A simple pendulum is executing simple harmonic motion with a time period T. If the length of the pendulum is increased by 21% the increase in the time period of the pendulum of increased length is

(a) 50% (b) 30%
(c) 21% (d) 10%

30. A wheel with ten metallic spokes each 0.50m long is rotated with a speed of 120 rev/min in a plane normal to the earth's magnetic field at the place. If the magnitude of the field is 0.40 G, the induced emf between the axle and the rim of the wheel is equal to

(a) 1.256×10^{-3} V (b) 6.28×10^{-4} V
(c) 1.256×10^{-4} V (d) 6.28×10^{-5} V

31. The average kinetic energy of a molecule of a perfect gas is

(a) (2/3) KT (b) 1.5 KT
(c) 2.5 KT (d) None of these

32. A ray of monochromatic light suffers minimum deviation of 38°, while passing through a prism of refracting angle 60°. Refractive index of the prism material is

(a) 2.4 (b) 0.8
(c) 1.3 (d) 1.5

33. In a circuit, containing a resistance R and a capacitance C, if the frequency f of impressed ac increases, the impedance Z of circuit will

(a) decrease
(b) increase
(c) remain constant
(d) first decrease and then increase

34. Two bodies of masses 10 kg and 100 kg are separated by a distance of 2m ($G = 6.67 \times 10^{-11}$ $Nm^2 kg^{-2}$). The gravitational potential at the mid point on the line joining the two is

(a) 7.3×10^{-7} J/kg (b) 7.3×10^{-9} J/kg
(c) -7.3×10^{-9} J/kg (d) 7.3×10^{-6} J/kg

35. For a given material, the Young's modulus is 2.4 times that of rigidity modulus, the Poission's ratio is

(a) 0.2 (b) 0.4
(c) 1.2 (d) 2.4

36. A particle executes S.H.M. having time period T, then the time period with which the potential energy changes is

(a) T (b) 2 T
(c) T/2 (d) ∞

37. A bar magnet of magnetic moment $\vec{M}$ is placed in the magnetic field $\vec{B}$. The torque acting on the magnet is

(a) $\vec{M} \times \vec{B}$ (b) $\vec{M} - \vec{B}$
(c) $\frac{1}{2} \vec{M} \times \vec{B}$ (d) $\vec{M} + \vec{B}$

38. Figure shows a capillary rise H. If the air is blown through the horizontal tube in the direction as shown, then rise in capillary tube will be

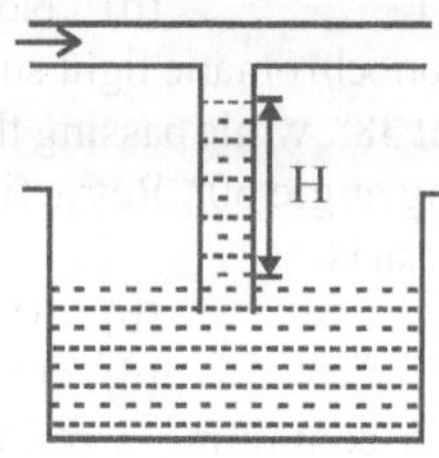

(a) = H (b) > H
(c) < H (d) zero

39. If red light and violet light rays are of focal lengths f_R and f_V respectively, then which one of the following is true ?

(a) $\lambda_R \leq \lambda_V$ (b) $\mu_R > \mu_V$
(c) $\mu_R = \mu_V$ (d) $\mu_R < \mu_V$

40. A soap bubble has radius r and volume V. If the excess pressure inside the bubble is P. Then PV is proportional to

(a) r (b) r^2
(c) r^3 (d) r^4

41. The materials suitable for making electromagnets should have

(a) high retentivity and low coercivity
(b) low retentivity and low coercivity
(c) high retentivity and high coercivity
(d) low retentivity and high coercivity

42. The photo electric work function for a metal surface is 4.125 eV. The cut-off wavelength for this surface is

(a) 4125 A° (b) 3000 A°
(c) 6000 A° (d) 2062 A°

43. In a n-p-n transistor

(a) holes move from emitter to base
(b) negative charge moves from emitter to base
(c) holes move from base to collector
(d) negative charge moves from collector to base

44. Ground waves are polarised

(a) Parallel to the earth's surface
(b) normal to the earth's surface
(c) at an angle 45° from earth's surface
(d) in any direction.

45. The logic behind 'NOR' gate is that it gives

(a) high output when both inputs are low
(b) high output when both inputs are high
(c) low output when both inputs are low
(d) None of these

SECTION 2 - CHEMISTRY

46. Huckel's rule states that a monocyclic conjugated compound will be aromatic if it contains
(a) $(4n + 2\pi)$ electrons
(b) $(4\pi + 2n)$ electrons
(c) 4π electrons
(d) $(4n + 2)\pi$ electrons

47. Which of the following is optically inactive?

(a) 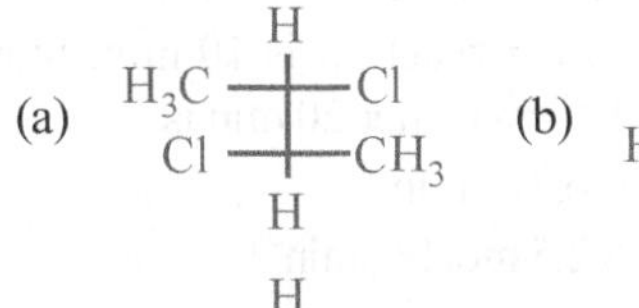

(b) H at top; Cl — C — CH_3; H_3C — C — Cl; H at bottom

(c) H at top; H_3C — C — Cl; H_3C — C — Cl; H at bottom

(d) none of these.

48. EAN of cobalt is 36 in $[Co(NH_3)_2O_2(en)Cl]$. Thus, O_2 is
(a) dioxide (b) superoxide ion
(c) peroxide ion (d) oxide

49. Chromyl chloride oxidises toluene to benzaldehyde, this reaction is known as
(a) Rosenmund reaction
(b) Wurtz reaction
(c) Etard reaction
(d) Fittig reaction

50. If 0.1 M of a weak acid is taken, and its percentage of degree of ionization is 1.34%, then its ionization constant will be :
(a) 0.8×10^{-5} (b) 1.79×10^{-5}
(c) 0.182×10^{-5} (d) none of the above

51. An organic compound X (molecular formula $C_6H_7O_2N$) has six carbon atoms in a ring system, two double bonds and a nitro group as substituent, X is
(a) Homocyclic but not aromatic
(b) Aromatic but not homocyclic
(c) Homocyclic and aromatic
(d) Heterocyclic and aromatic

52. In which of the following regions hydrogen and helium are found
(a) Stratosphere (b) Mesosphere
(c) Exosphere (d) Troposphere

53. Milk of magnesia is
(a) $Mg(OH)_2$ (b) $Ca(OH)_2$
(c) $CaCO_3$ (d) $MgCO_3$

54. Which of the following has highest knocking property ?
(a) Aromatic hydrocarbons
(b) Olefins
(c) Branched chain paraffins
(d) Straight chain paraffins

55. Among the following compounds (I - III), the ease of their reaction with electrophiles is,

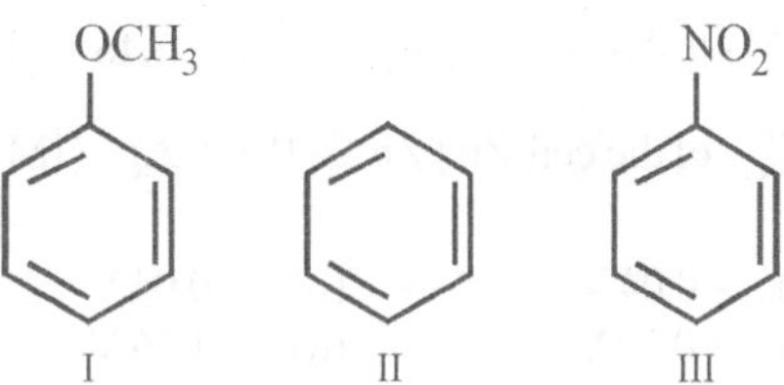

(a) II > III > I (b) III > II > I
(c) II > I > III (d) I > II > III

56. Haemoglobin contains 0.334% of iron by weight. The molecular weight of haemoglobin is approximately 67200. The number of iron atoms (at. wt. of Fe is 56) present in one molecule of haemoglobin are
(a) 1 (b) 6
(c) 4 (d) 2

57. Nessler's reagent is
(a) $KHgI_4$ (b) $K_2HgI_4 + NH_4OH$
(c) $K_2HgI_4 + KOH$ (d) $KHgI_4 + NH_4OH$

58. The solubility of hydroxides, fluorides or oxalates of the metals of Group II A
(a) increases down the group
(b) decreases down the group
(c) varies randomly
(d) is constant

59. AB; crystallizes in a body centred cubic lattice with edge length '*a*' equal to 387 pm. The distance between two oppositely charged ions in the lattice is :
(a) 335 pm (b) 250 pm
(c) 200 pm (d) 300 pm

60. The reagent (s) which can be used to distinguish acetophenone from benzophenone is (are)
(a) 2,4- Dinitrophenylhydrazine
(b) Aqueous solution of $NaHSO_3$
(c) Benedict reagent
(d) I_2 and NaOH.

61. For the cell
$Zn \underset{\ominus}{|} Zn^{2+}(1M) || Cu^{2+}(1M) \underset{\oplus}{|} Cu$, E^0_{cell} is 1.10 V,
$E^0_{cell^{2+}/Cu} = 0.34\,V$ and for the cell $Cu | Cu^{2+}$
$(1M) || Ag^+ (1M) | Ag$, $E^0_{cell} = 0.46\,V$ hence,
E^0_{cell} of the cell $Zn | Zn^{2+}(1M) || Ag^+(1M) | Ag$ is
(a) −0.04 V (b) +0.04 V
(c) +0.30 V (d) +1.56 V

62. Among the following complexes, optical activity is possible in
(a) $[Co(NH_3)_6]^{3+}$
(b) $[Co(H_2O)_2(NH_3)_2Cl_2]^+$
(c) $[Cr(H_2O)_2Cl_2]^+$
(d) $[Co(CN)_5NC]$

63. The r.m.s velocity of hydrogen is $\sqrt{7}$ times the r.m.s velocity of nitrogen. If T is the temperature of the gas , then
(a) $T_{(H_2)} = T_{(N_2)}$ (b) $T_{(H_2)} > T_{(N_2)}$
(c) $T_{(H_2)} < T_{(N_2)}$ (d) $T_{(H_2)} = \sqrt{7}T_{(N_2)}$

64. The correct order of ionic radii of Y^{3+}, La^{3+}, Eu^{3+} and Lu^{3+} is
(a) $Y^{3+} < La^{3+} < Eu^{3+} < Lu^{3+}$
(b) $Lu^{3+} < Eu^{3+} < La^{3+} < Y^{3+}$
(c) $La^{3+} < Eu^{3+} < Lu^{3+} < Y^{3+}$
(d) $Y^{3+} < Lu^{3+} < Eu^{3+} < La^{3+}$

65. The pK_a of an amino acid is 9.15. At what pH amino acid is 5% dissociated ?
(a) 9.15 (b) 4.85
(c) 9.44 (d) 7.87

66. In the froth floatation process for the purification of ores, the ore particles float because
(a) They are light
(b) Their surface is hydro dphobic i.e. not easily wetted by water
(c) They bear electrostatic charge
(d) They are insoluble

67. T_{50} of first -order reaction is 10 min. Starting with 10 mol L^{-1}, rate after 20 min is
(a) 0.0693 mol L^{-1} min^{-1}
(b) 0.0693 × 2.5 mol L^{-1} min^{-1}
(c) 0.0693 × 5 mol L^{-2} min^{-1}
(d) 0.0693 × 10 mol L^{-1} min^{-1}

68. Which of the following organometallic compound is σ and π bonded?
(a) $[Fe(\eta^5 - C_5H_5)_2]$
(b) $Fe(CH_3)_3$
(c) $K[PtCl_3(\eta^2 - C_2H_4)]$
(d) $[Co(CO)_5NH_3]^{2+}$

69. Which of the following statement is false ?
(a) For 1 mole of an ideal gas, $C_p - C_v = R$
(b) $\left(\frac{\partial E}{\partial T}\right)_T = 0$ for an ideal gas
(c) $\Delta q = \Delta w + p\Delta v$
(d) For reversible isothermal expansion of 1 mole of an ideal gas from volume V_1 to V_2, work done is equal to RT ln (V_2/V_1)

70. The correct order of the decreasing ionic radii among the following isoelectronic species are :
(a) $Ca^{2+} > K^+ > S^{2-} > Cl^{-1}$
(b) $Cl^- > S^{2-} > Ca^{2+} > K^+$
(c) $S^{2-} > Cl^- > K^+ > Ca^{2+}$
(d) $K^+ > Ca^{2+} > Cl^- > S^{2-}$

71. When two halogens are attached to same carbon atom, it is known as :
(a) *vic*-dihalide (b) *gem*-dihalide
(c) α, ω -dihalide (d) α, β-dihalide

72. SnO_2 is taken in basic medium and current is passed. Colloidal sol migrates towards
(a) anode (+ plate) (b) cathode (– plate)
(c) both (a) and (b) (d) none of these

73. 0.45 g of acid of molecular weight 90 was neutralized by 20 ml. of a 0.5N caustic potash. The basicity of an acid is
(a) 1 (b) 2
(c) 3 (d) 4

74. Which of the following is a broad spectrum drug?
(a) Plasmoquine (b) Chloroquine
(c) Chloramphenicol (d) D.D.T.

75. A compound on treatment with NaOH followed by addition of $AgNO_3$ produces white precipitate at room temperature. The precipitate is soluble in NH_4OH. The compound is identified as
(a) vinyl chloride (b) benzyl chloride
(c) chlorobenzene (d) ethyl bromide

76. At what angles for the first order diffraction, spacing between two planes respectively are λ and $\frac{\lambda}{2}$?
(a) $0°, 90°$ (b) $90°, 0°$
(c) $30°, 90°$ (d) $90°, 30°$

77. Equivalent conductance of an electrolyte containing NaF at infinite dilution is 90.1 $Ohm^{-1}cm^2$. If NaF is replaced by KF what is the value of equivalent conductance?
(a) $90.1\ Ohm^{-1}cm^2$ (b) $111.2\ Ohm^{-1}cm^2$
(c) 0 (d) $222.4\ Ohm^{-1}cm^2$

78. The enthalpy of combustion of C_6H_6 is –3250 kJ, when 0.39 gm of C_6H_6 is burnt in excess of oxygen in an open vessel, the amount of heat evolved is
(a) 8.32 kJ (b) 12.36 kJ
(c) 16.25 kJ (d) 20.74 kJ

79. In which of the following is there a consistent decrease in atomic radius as the atomic number increases?
(a) halogens
(b) representative elements
(c) transition elements
(d) lanthanides

80. The basicity of aniline is less than that of cyclohexylamine. This is due to
(a) + R effect of —NH_2 group
(b) – I effect of —NH_2 group
(c) – R effect of —NH_2 group
(d) hyperconjugation effect

81. $E° = \frac{RT}{nF} \ln K_{eq}$
The above equation is called :
(a) Gibb's equation
(b) Gibb's–Helmholtz equation
(c) Nernst equation
(d) Van der Waal's equation

82. Mac Arthur process is used for the extraction of:
(a) Au (b) Pt
(c) Cu (d) Zn

83. H_3BO_2 is
(a) monobasic and weak Lewis acid
(b) monobasic and weak Bronsted acid
(c) monobasic and strong Lewis acid
(d) tribasic and weak Bronsted acid

84. The following equilibrium constants are given:
$N_2 + 3H_2 \rightleftharpoons 2NH_3; K_1$
$N_2 + O_2 \rightleftharpoons 2NO; K_2$
$H_2 + \frac{1}{2}O_2 \rightleftharpoons H_2O; K_2$
The equilibrium constant for the oxidation of NH_3 by oxygen to give NO is
(a) $\frac{K_2 K_3^2}{K_1}$ (b) $\frac{K_2^2 K_3}{K_1}$
(b) $\frac{K_1 K_2}{K_3}$ (d) $\frac{K_2 K_3^3}{K_1}$

85. The ortho/para directing group among the following is :

(a) COOH (b) CN
(c) $COCH_3$ (d) $NHCONH_2$

86. Benzoic acid gives benzene on being heated with X and phenol gives benzene on being heated with Y. Therefore X and Y are respectively

(a) Soda-lime and copper
(b) Zn dust and NaOH
(c) Zn dust and soda-lime
(d) Soda-lime and zinc dust.

87. The enthalpies of the following reactions are shown below.

$$\frac{1}{2}H_2(g) + \frac{1}{2}O_2(g) \rightarrow OH(g); \quad \Delta H = 42.09 \text{ kJ mol}^{-1}$$

$H_2(g) \rightarrow 2H(g)$; $\Delta H = 435.89 \text{ kJ mol}^{-1}$

$O_2(g) \rightarrow 2O(g)$; $\Delta H = 495.05 \text{ kJ mol}^{-1}$

Calculate the O — H bond energy for the hydroxyl radical.

(a) $223.18 \text{ kJ mol}^{-1}$ (b) $423.38 \text{ kJ mol}^{-1}$
(c) $513.28 \text{ kJ mol}^{-1}$ (d) $113.38 \text{ kJ mol}^{-1}$

88. Elements X, Y and Z have atomic numbers 19, 37 and 55 respectively. Which of the following statements is true about them ?

(a) Their ionization potential would increase with increasing atomic number
(b) 'Y' would have an ionisation potential between those of 'X' and 'Z'
(c) 'Z' would have the highest ionization potential
(d) 'Y' would have the highest ionization potential.

89. MnO_4^{2-} (1 mole) in neutral aqueous medium disproportionates to

(a) 2/3 mole of MnO_4^- and 1/3 mole of MnO_2
(b) 1/3 mole of MnO_4^- and 2/3 mole of MnO_2
(c) 1/3 mole of Mn_2O_7 and 1/3 mole of MnO_2
(d) 2/3 mole of Mn_2O_7 and 1/3 mole of MnO_2

90. Formation of a solution from two components can be considered as

(i) pure solvent $\rightarrow$ separated solvent molecules, ΔH_1
(ii) pure solute $\rightarrow$ separated solute molecules, ΔH_2
(iii) separated solvent and solute molecules $\rightarrow$ solution, ΔH_3

Solution so formed will be ideal if

(a) $\Delta H_{soln} = \Delta H_1 + \Delta H_2 - \Delta H_3$
(b) $\Delta H_{soln} = \Delta H_1 - \Delta H_2 - \Delta H_3$
(c) $\Delta H_{soln} = \Delta H_3 - \Delta H_1 - \Delta H_2$
(d) $\Delta H_{soln} = \Delta H_1 + \Delta H_2 + \Delta H_3$

SECTION 3 - BOTANY

91. Botanical name of banana is

(a) *Musa paradisica*
(b) *Phaseolus vulgaris*
(c) *Ricinus communis*
(d) *Ananas sativus*

92. Which meristem helps in increasing girth?

(a) Lateral meristem
(b) Intercalary meristem
(c) Primary meristem
(d) Apical meristem

93. The correct sequence in cell cycle is

(a) S - G_1 - G_2 - M (b) S - M - G_1 - G_2
(c) G_1 - S - G_2 - M (d) M_1 - G_1 - G_2 - S

94. In Chlorophyceae, sexual reproduction occurs by

(a) isogamy and anisogamy
(b) isogamy, anisogamy and oogamy
(c) oogamy only
(d) anisogamy and oogamy

95. Algae have cell wall made up of

(a) cellulose, galactans and mannans
(b) hemicellulose, pectins and proteins
(c) pectins, cellulose and proteins
(d) cellulose, hemicellulose and pectins.

96. Which of the following cell organelle remains enveloped by a single unit membrane?
(a) Mitochondria (b) Lysosomes
(c) Nucleus (d) Chloroplast

97. Which one of the following animals is correctly matched with its particular named taxonomic category ?
(a) Tiger - tigris, the species
(b) Cuttle fish - mollusca, a class
(c) Humans - primata, the family
(d) Housefly - musca, an order

98. The endosperm of gymnosperm is
(a) triploid (b) haploid
(c) diploid (d) polyploid

99. When funiculum, chalaza, and micropyle lie in one straight line then ovule is called –
(a) Amphitropous (b) Orthotropous
(c) Campylotropous (d) Anatropous

100. Which one of the following statements is correct for secondary succession ?
(a) It begins on a bare rock
(b) It occurs on a deforested site
(c) It follows primary succession
(d) It is similar to primary succession except that it has a relatively fast pace

101. Identified A, B, C and D.

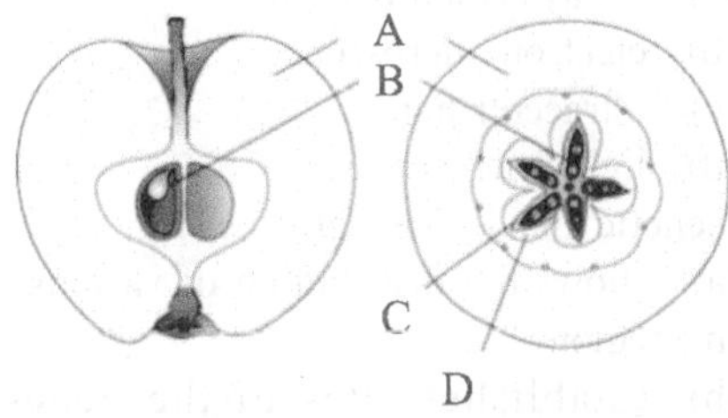

(a) A – Mesocarp; B – Endocarp; C – Seed; D – Thalamus
(b) A – Seed; B – Thalamus; C – Mesocarp; D – Endocarp
(c) A – Thalamus; B – Seed; C – Endocarp; D – Mesocarp
(d) A – Mesocarp; B – Endocarp; C – Seed; D – Thalamus

102. Transfer of pollen to the stigma of another flower of the same plant is
(a) Autogamy (b) Allogamy
(c) Xenogamy (d) Geitonogamy

103. Biochemical Oxygen Demand *(BOD)* in a river water
(a) has no relationship with concentration of oxygen in the water.
(b) gives a measure of *Salmonella* in the water.
(c) increases when sewage gets mixed with river water.
(d) remains unchanged when algal bloom occurs.

104. A drupe develops in
(a) mango (b) wheat
(c) pea (d) tomato

105. The parenchyma tissue which forms the bulk of ovule where the sporogenous tissue is produced is –
(a) Megaspore mother cell
(b) Nucellus
(c) Ovule
(d) Embryo sac

106. Which element is located at the centre of the porphyrin ring in chlorophyll ?
(a) Manganese (b) Calcium
(c) Magnesium (d) Potassium

107. Movement of ions or molecules in a direction opposite to that of prevailing electrochemical gradient is known as
(a) diffusion
(b) active transport
(c) pinocytosis
(d) brownian movement

108. Which one is the correct summary equation of photosynthesis?
(a) $C_6H_{12}O_6 + 6O_2 \longrightarrow 6CO_2 + 6H_2O + \text{energy}$
(b) $C_6H_{12}O_6 + 6O_2 + 6H_2O \longrightarrow 6CO_2 + 12H_2O + \text{energy}$
(c) $6CO_2 + 6H_2O \longrightarrow 6H_2O + C_6H_{12}O_6$
(d) $6CO_2 + 12H_2O \xrightarrow[\text{Chlorophyll}]{\text{Light}} 6O_2 + C_6H_{12}O_6 + 6H_2O$

109. Which one is a long day plant?
(a) Tobacco (b) *Glycine max*
(c) *Mirabilis jalapa* (d) Spinach

110. Which of the following is used as antitranspirant?
(a) Calcium carbonate
(b) Phenyl mercuric acetate
(c) Cobalt chloride
(d) Naphthol acetic acid

111. RuBisCO stands for
(a) Ribulose Biphosphate Carboxylase Oxygenase
(b) Ribulose Phosphate Carboxylase Oxygenase
(c) Ribulose Phosphate Carboxylic Oxygenase
(d) None of the above

112. Turgor pressure become equal to the wall pressure when
(a) water leaves the cell
(b) water enters the cell
(c) no exchange of water takes place
(d) solute goes from cell into water

113. The figure given below shows the conversion of a substrate into product by an enzyme. In which one of the four options (a-d) the components or reaction labelled as A, B, C and D are identified correctly?

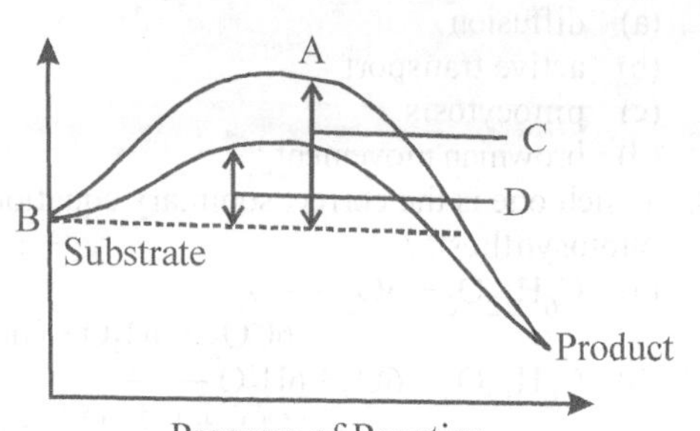

(a) A –Potential energy, B –Transition state, C –Activation energy with enzyme, D –Activation energy without enzyme
(b) A –Transition state, B –Potential energy, C –Activation energy without enzyme, D –Activation energy with enzyme
(c) A –Potential energy, B –Transition state, C –Activation energy with enzyme, D –Activation energy with enzyme
(d) A –Activation energy with enzyme, B –Transition state, C –Activation energy without enzyme, D –Potential energy

114. Plant obtains nitrogen from soil in the form of
(a) nitrite (b) nitrate
(c) ammonia (d) hydroxylamine

115. Which of the following statement is incorrect ?
(a) C_3 plants respond to higher temperature, show higher photosynthetic rate while C_4 plants have lower optimum temperature.
(b) Tropical plants have higher temperature optimum than the plants adapted to temperate climate.
(c) Light reaction is less temperature sensitive than dark reaction.
(d) The effect of water as a factor is more through its effect on plant, rather than directly on photosynthesis.

116. Nucleotide arrangement in DNA can be seen by
(a) X-ray crystallography
(b) electron microscope
(c) ultracentrifuge
(d) light microscope

117. Genetic map is one that
(a) shows the distribution of various species in a region
(b) establishes sites of the genes on a chromosome
(c) establishes the various stages in gene evolution
(d) show the stages during the cell division

118. DNA template sequence of CTGATAGC is transcribed over mRNA as
(a) GUCTUTCG (b) GACUAUCG
(c) GAUTATUG (d) UACTATCU

119. Humus is essential for plant growth because:
   (a) it is rich in nutrients and increases the water holding capacity of soil
   (b) it increases aeration of soil
   (c) it increases porocity of soil
   (d) All of the above
120. Which one of the following pesticides is banned now a-days?
   (a) DDT (b) Eldrin
   (c) Aldrin (d) Toxaphene
121. Ozone hole means
   (a) hole in the ozone layer in stratosphere.
   (b) decrease in the thickness of ozone in the stratosphere.
   (c) decrease in concentration of ozone in trophosphere.
   (d) increase in the concentration of ozone in trophosphere.
122. The new varieties of plants are produced by
   (a) selection and hybridization
   (b) mutation and selection
   (c) introduction and mutation
   (d) selection and introduction
123. Which form of RNA has a structure resembling clover leaf?
   (a) rRNA (b) hn RNA
   (c) m RNA (d) t RNA
124. Site of respiration in bacteria is
   (a) episome (b) ribosome
   (c) mesosome (d) microsome
125. The following diagram is the polynucleotide chain. Identify A, B, C, D and E.

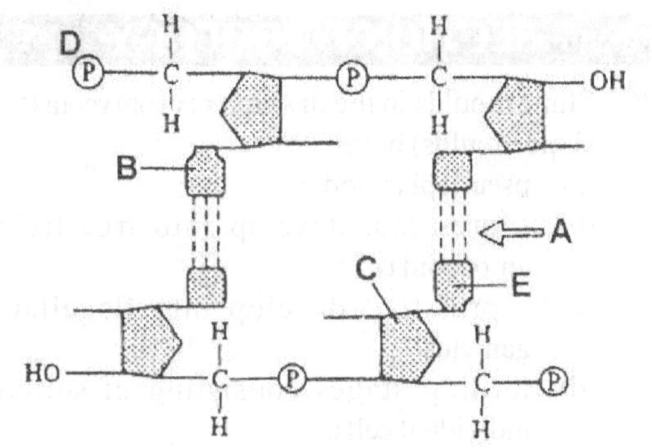

   (a) A – Hydrogen bonds, B – Pyrimidine, C – Hexose (deoxyribose) sugar, D – 5' end, E – Purine base.
   (b) A – Hydrogen bonds, B – Purine base, C – Hexose (deoxyribose) sugar, D – 5' end, E – Pyrimidine.
   (c) A – Hydrogen bonds, B – Pyrimidine, C – Pentose (deoxyribose) sugar, D – 5' end, E – Purine base
   (d) A – Hydrogen bonds, B – Purine base, C – Pentose (deoxyribose) sugar, D – 5' end, E – Pyrimidine
126. Carrier ions like Na^+ facilitate the absorption of substances like:
   (a) amino acids and glucose
   (b) glucose and fatty acids
   (c) fatty acids and glycerol
   (d) fructose and some amino acids
127. The method that renders the seed coat permeable to water so that embryo expansion is not physically retarded is
   (a) vernalization (b) stratification
   (c) denudation (d) scarification
128. In Kreb's cycle, the FAD precipitates as electron acceptor during the conversion of
   (a) succinyl CoA to succinic acid
   (b) α - ketoglutarate to succinyl CoA
   (c) fumaric acid to malic acid
   (d) succinic acid to fumaric acid
129. In the DNA molecule :
   (a) there are two strands which run antiparallel- one in $5' \rightarrow 3'$ direction and other in $3' \rightarrow 5'$ direction.
   (b) the total amount of purine nucleotides and pyrimidine nucleotides is not always equal.
   (c) there are two strands which run parallel in the $5' \rightarrow 3'$ direction.
   (d) the proportion of Adenine in relation to thymine varies with the organism.

130. Parthenocarpic fruit can be produced by the application of which of the following auxin?
(a) IBA (Indole butyric acid)
(b) IAA (Indole acetic acid)
(c) 2, 4-D (2-4 dichlorophenoxy acetic acid)
(d) All of the above.

131. MAB stands for
(a) Man and Biology Programme
(b) Man and Biosphere Programme
(c) Mammals and Biosphere Programme
(d) Mammals and Biology Programme

132. Which one of the following help in absorption of phosphorus from soil by plants?
(a) *Glomus*
(b) *Rhizobium*
(c) *Frankia*
(d) *Anabaena*

133. The area where wild populations, traditional life styles and genetic resources are protected is :
(a) Core Zone
(b) Buffer Zone
(c) Biosphere reserve
(d) Manipulation Zone

134. The figure below is the diagrammatic representation of the *E.Coli* vector pBR 322. Which one of the given options correctly identifies its certain component (s) ?

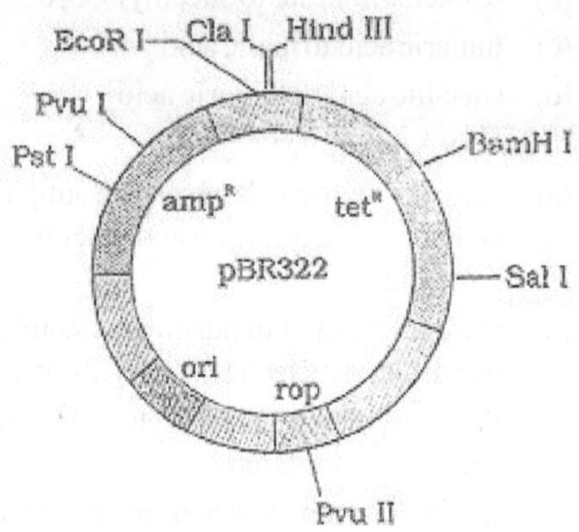

(a) ori - original restriction enzyme
(b) rop-reduced osmotic pressure
(c) Hind III, EcoRI - selectable markers
(d) amp^R, tet^R - antibiotic resistance genes

135. Which one of the following option is correct for A, B, C and D?

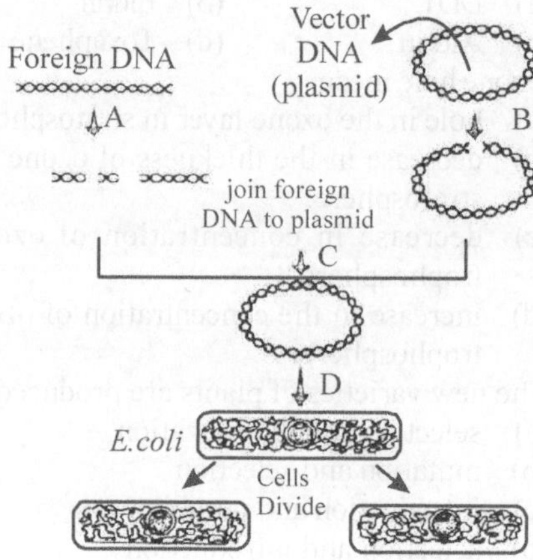

(a) A – Exonuclease, B – Endonuclease, C – DNA ligase, D – Transformation
(b) A – Exonuclease, B – Exonuclease, C – DNA ligase, D – Transformation
(c) A – Exonuclease, B – Endonuclease, C – Hydrolase, D – Transduction
(d) A – Restriction endonuclease, B – Restriction endonuclease, C – DNA ligase, D – transformation

SECTION 4 - ZOOLOGY

136. Slime moulds in the division myxomycota (true slime moulds) have
(a) pseudoplasmodia.
(b) spores that develop into free living amoeboid cells.
(c) spores that develop into flagellated gametes.
(d) feeding stages consisting of solitary individual cells.

137. Which of the following substances would cause coagulation of blood at the site of its introduction?
(a) Prothrombin (b) Fibrinogen
(c) Thromboplastin (d) Heparin

138. Leech is
(a) carnivorous (b) sanguivorous
(c) ectoparasite (d) Both (b) and (c)

139. RBCs in human are
(a) biconcave and nucleated
(b) biconvex and nucleated
(c) biconcave and enucleated
(d) biconvex and enucleated

140. Lymph differs from blood in possessing
(a) only WBCs.
(b) more RBCs and WBCs.
(c) more RBCs and few WBCs.
(d) more WBCs and few RBCs.

141. Which one of the following is NOT a characteristic of phylum Annelida?
(a) Closed circulatory system
(b) Segmentation
(c) Pseudocoelom
(d) Ventral nerve cord

142. Fish which can be used in biological control of mosquitoes/Larvicidal fish is
(a) *Eel* (b) *Carp*
(c) Cat Fish (d) *Gambusia*

143. A person which shows the secondary sexual characters of both male and female is called –
(a) Intersex (b) Hermaphrodite
(c) Bisexual (d) Gynandromorph

144. Amniocentesis is a technique
(a) by which the essential amino acids in the body can be estimated
(b) by which any chromosomal anomalies in the foetus can be detected
(c) in which the sex of the foetus can be reversed
(d) that can be used for correcting genetic disorders of the foetus

145. The diagram shows an antibody molecule. Identify A to F.

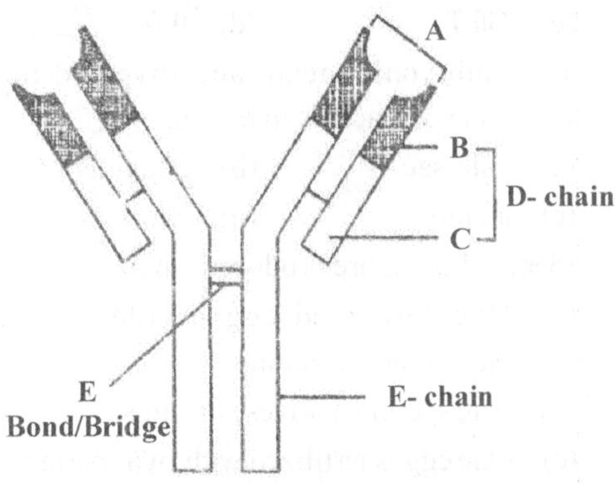

(a) **A** - Antigen binding site; **B** - Variable region (of L-Chain); **C** - Constant region (of L-Chain); **D** - Light polypeptide chain (L-Chain); **E** - Heavy polypeptide chain (H-Chain); **F** - Disulfide (bond)
(b) **A** - Antigen binding site; **B** - Constant region (of L-Chain); **C** - Variable region (of L-Chain); **D** - Light polypeptide chain (L-Chain); **E** - Heavy polypeptide chain (H-Chain); **F** - Disulfide (bond)
(c) **A**- Antigen binding site; **B** - Variable region (of L-Chain); **C** - Constant region (of L-Chain); **D** - Heavy polypeptide chain (L-Chain); **E** - Light polypeptide chain (H-Chain); **F** - Hydrogen (bond)
(d) **A** - Antigen binding site; **B** - Variable region (of L-Chain); **C** - Constant region (of L-Chain); **D** - Light polypeptide chain (L-Chain); **E** - Heavy polypeptide chain (H-Chain); **F** - Hydrogen (bond)

146. The copper ions of IUDs
(a) Suppress the motility and fertilization capacity of sperms
(b) Make the uterus unsuitable for implantation
(c) Increase phagocytosis of sperms
(d) Make cervix hostile to sperms

147. The transfer of zygote or early embryo (up to 8 blastomere) into fallopian tube is
(a) IVF and ET (b) ZIFT
(c) GIFT (d) IUT

148. The embryonic membrane involved in the formation of placenta in human is
(a) yolk sac (b) allantois
(c) amnion (d) chorion

149. Identical twins are produced when
(a) One fertilized egg divides and two blastomeres separate
(b) One sperm fertilizes two eggs
(c) One egg is fertilized with two sperms
(d) Two eggs are fertilized

150. Which of the following induces parturition ?
(a) Vasopressin
(b) Oxytocin
(c) Growth hormone
(d) Thyroid stimulating hormone

151. Jaundice is a disorder of
(a) Excretory system
(b) Skin and eyes
(c) Digestive system
(d) Circulatory system

152. Child death may occur in the marriage between
(a) Rh^+ man and Rh^+ woman
(b) Rh^+ man and Rh^- woman
(c) Rh^- man and Rh^- woman
(d) Rh^- man and Rh^+ woman

153. Continued consumption of a diet rich in butter, red meat and eggs for a long period may lead to
(a) vitamin A toxicity
(b) kidney stones
(c) hypercholesterolemia
(d) urine laden with ketone bodies

154. A person who is one along hunger strike and is surviving only on water, will have
(a) less amino acids in his urine
(b) more glucose in his blood
(c) less urea in his urine
(d) more sodium in his urine

155. Haemoglobin is having maximum affinity with:
(a) NH_3 (b) O_2
(c) CO_2 (d) CO

156. Thickening of arteries due to cholesterol deposition is
(a) arteriosclerosis (b) rheumatic heart
(c) blood pressure (d) cardiac arrest

157. Which of the following components of blood does not enter into the nephron?
(a) Urea (b) Water
(c) Glucose (d) Plasma protein

158. The most striking example of point mutation is found in a disease called
(a) thalassemia (b) night blindness
(c) down's syndrome (d) sickle cell anaemia

159. In live stock breeding experiments the following stage is transferred to surrogate mothers
(a) Unfertilized eggs
(b) Fertilized eggs
(c) 8 to 32 celled embryo
(d) Frozen semen

160. The presence of gill slits, in the embryos of all vertebrates, supports the theory of
(a) biogenesis
(b) recapitulation
(c) metamorphosis
(d) organic evolution

161. Which of the following has been recently used for increasing productivity of super milk cows?
(a) Artificial insemination by a pedigreed bull only
(b) Superovulation of a high production cow only
(c) Embryo transplantation only
(d) A combination of superovulation, artificial insemination and embryo transplantation into a 'carrier cow' (surrogate mother)

162. The change of the light-coloured variety of peppered moth *(Biston betularia)* to its darker variety *(Biston carbonaria)* is due to
(a) mutation (b) regeneration
(c) genetic isolation (d) temporal isolation

163. A person with the sex chromosomes XXY suffers from
(a) Down's syndrome
(b) Klinefelter's syndrome
(c) Turner's syndrome
(d) Gynandromorphism

164. There is no life on moon due to the absence of
(a) O_2 (b) water
(c) light (d) temperature

165. The density of a population in a given habitat during a given period, fluctuates due to changes in four basic process. On this basis fill up A and B boxes in the given diagram with correct options

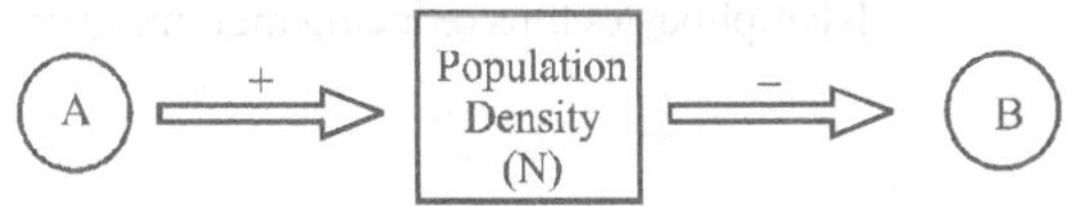

(a) A = Natality + Immigration, B = Mortality + Emigration
(b) A = Natality + Mortality, B = Immigration + Emigration
(c) A = Birth rate + Death rate, B = Migration + Emigration
(d) A = Natality + Emigration, B = Mortality + Immigration

166. Two of the body parts which do not appear in MRI may be
(a) molar teeth and eye lens
(b) scapula and canines
(c) ligaments and ribs
(d) tendons and premolars

167. Chemically hormones are
(a) biogenic amines only
(b) proteins, steroids and biogenic amines
(c) proteins only
(d) steroids only

168. The largest quantity of air that can be expired after a maximal inspiratory effort is called
(a) residual volume (b) tidal volume
(c) vital capacity (d) total lung volume

169. The black pigment in the eye which reduces the internal reflection is located in
(a) retina (b) iris
(c) cornea (d) sclerotic

170. Myofibrils are made up of
(a) Myosin and actin
(b) Myosin and troponin
(c) Actin and tropomyosin
(d) All the above components

171. During the propagation of a nerve impulse, the action potential results from the movement of:
(a) K^+ ions from extracellular fluid to intracellular fluid.
(b) Na^+ ions from intracellular fluid to extracellular fluid.
(c) K^+ ions from intracellular fluid to extracellular fluid.
(d) Na^+ ions from extracellular fluid to intracellular fluid.

172. Testosterone is produced by
(a) sertoli cells (b) leydig's cells
(c) oxyntic cells (d) pituitary gland

173. Which one of the following is used as vector for cloning genes into higher organisms?
(a) Baculovirus
(b) *Salmonella typhimurium*
(c) *Rhizopus nigricans*
(d) Retrovirus

174. Which one of the following pairs is incorrect ?
(a) Plasmid - small piece of extrachromosomal DNA in bacteria
(b) Interferon - an enzyme that interferes with DNA replication
(c) Cosmid - A vector for carrying large DNA fragments into host cells
(d) Myeloma - antibody producing tumor cells.

175. In which one of the following pairs of diseases both are caused by viruses ?
(a) Tetanus and typhoid
(b) Whooping cough and sleeping sickness
(c) Syphilis and AIDS
(d) Measles and Rabies

176. Electroporation procedure involves
(a) fast passage of food through sieve pores in phloem elements with the help of electric stimulation.
(b) opening of stomatal pores during night by artificial light.
(c) making transient pores in the cell membrane to introduce gene constructs.
(d) purification of saline water with the help of a membrane system.

177. Widal test is used for the diagnosis of
(a) Malaria (b) Pneumonia
(c) Tuberculosis (d) Typhoid

178. Increased asthmatics attacks in certain seasons are related to
(a) eating fruits preserved in tin containers
(b) inhalation of seasonal pollen
(c) low temperature
(d) hot and humid environment.

179. For transformation, micro-particles coated with DNA to be bombarded with gene gun are made up of :
(a) Silver or Platinum
(b) Platinum or Zinc
(c) Silicon or Platinum
(d) Gold or Tungsten

180. Which one of the following statements is correct with respect to AIDS?
(a) The HIV can be transmitted through eating food together with an infected person
(b) Drug addicts are least susceptible to HIV infection .
(c) AIDS patients are being fully cured with proper care and nutrition
(d) The causative HIV retrovirus enters helper T-lymphocytes thus reducing their numbers

RESPONSE SHEET

Test Code: ....Time taken : .... Date:..............

	(a)	(b)	(c)	(d)		(a)	(b)	(c)	(d)		(a)	(b)	(c)	(d)		(a)	(b)	(c)	(d)
1.					46.					91.					136.				
2.					47.					92.					137.				
3.					48.					93.					138.				
4.					49.					94.					139.				
5.					50.					95.					140.				
6.					51.					96.					141.				
7.					52.					97.					142.				
8.					53.					98.					143.				
9.					54.					99.					144.				
10.					55.					100.					145.				
11.					56.					101.					146.				
12.					57.					102.					147.				
13.					58.					103.					148.				
14.					59.					104.					149.				
15.					60.					105.					150.				
16.					61.					106.					151.				
17.					62.					107.					152.				
18.					63.					108.					153.				
19.					64.					109.					154.				
20.					65.					110.					155.				
21.					66.					111.					156.				
22.					67.					112.					157.				
23.					68.					113.					158.				
24.					69.					114.					159.				
25.					70.					115.					160.				
26.					71.					116.					161.				
27.					72.					117.					162.				
28.					73.					118.					163.				
29.					74.					119.					164.				
30.					75.					120.					165.				
31.					76.					121.					166.				
32.					77.					122.					167.				
33.					78.					123.					168.				
34.					79.					124.					169.				
35.					80.					125.					170.				
36.					81.					126.					171.				
37.					82.					127.					172.				
38.					83.					128.					173.				
39.					84.					129.					174.				
40.					85.					130.					175.				
41.					86.					131.					176.				
42.					87.					132.					177.				
43.					88.					133.					178.				
44.					89.					134.					179.				
45.					90.					135.					180.				

TEST ASSESSMENT AND ANALYSIS SHEET

How to fill this sheet : After you finish the paper and have checked it first try to attempt again all the questions which you have done wrong or have not been able to attempt. Go to detailed solutions only after you have attempted all the questions of your own.Attempting the questions on your own without seeing the solutions will help you in filling this table

ANALYSIS OF TEST

Total number of questions	
Questions Attempted	
Correct Questions	
Wrong Questions	
	Net Score

Strike rate $\left(\dfrac{\text{Correct questions}}{\text{Questions attempt}} \times 100\right) =$ ....

Percentage Marks $\left(\dfrac{\text{Correct Answers}}{\text{Total number of questions}} \times 100\right) =$ ....

Analysis of wrong questions : Divide the questions which you have done wrong in 3 broad categories.

Reasons for questions attempted wrong	**Number of Questions**
Knew but calculation mistake (solved on my own after exam)	
Knew but got confused (solved on my own after exam)	
Did not knew the question (couldn't solve even after exam)	

Questions not attempted : Divide the questions not attempted in 3 broad categories

Reasons for unattempted questions	**Number of Questions**
Knew the ques but time was less (solved on my own after exam)	
Knew the ques but got confused (solved on my own after exam)	
Did not knew the questions (couldn't solve even after exam)	

List down your weak areas you need to improve your score :

1)________________________________

2)________________________________

3)________________________________

NEET

National Eligibility Entrance Test

Test Series **Full Test - 6**

Time : - 3 Hours **Max. Marks:- 720**

INSTRUCTIONS

1. Immediately fill in the particulars on this page of the Test Booklet with Blue/Black Ball Point Pen. Use of pencil is strictly prohibited.
2. Answers are to be given on Response Sheet provided at the end of this paper. You can tear this Response Sheet along the cut marks provided. Fill in the particulars carefully in the Response Sheet provided.
3. The test is of 3 hours duration.
4. The Test Booklet consists of **180** questions. The maximum marks are **720**.
5. There are four section in the question paper.

   The distribution of marks subjectwise in each part is as under for each correct response.

   Section 1 – Physics (180 Marks) – 45 Questions

   Section 2 – Chemistry (180 Marks) – 45 Questions

   Section 3 – Botany (180 Marks) – 45 Questions

   Section 4 – Zoology (180 Marks) – 45 Questions
6. You will be awarded 4 Marks if you have darkened only the bubble corresponding to the correct answer and zero mark if no bubble is darkened. In all other cases, minus one (– 1) mark will be awarded.
7. Use Blue/Black Ball Point Pen only for writing particulars/marking responses on Side-1 and Side-2 of the Response Sheet.

   Use of pencil is strictly prohibited.
8. You are advised to create an actual examination hall simulation.
9. Rough work is to be done on the space provided for this purpose in the Test Booklet only.
10. Do not fold or make any stray marks on the Response Sheet.
11. Once you have evaluated the test, you should assess it thoroughly.

SECTION 1 - PHYSICS

1. Which of the following quantities has a unit but dimensionless?
(a) Strain
(b) Reynolds number
(c) Angular displacement
(d) Poisson's ratio

2. The velocity versus time curve of moving point is as given in the figure.

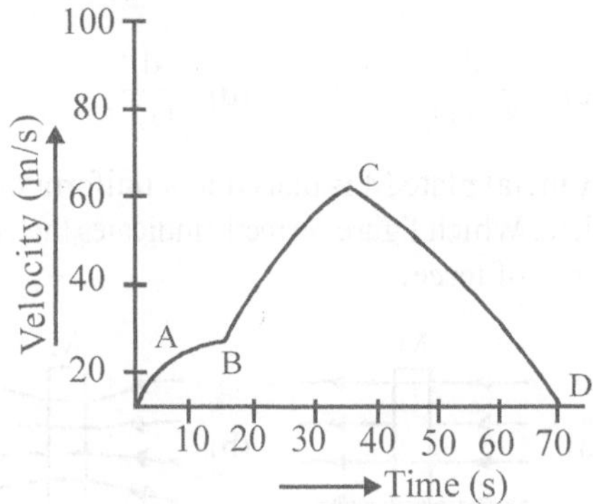

The maximum acceleration will be
(a) $4 m/s^2$ (b) $3 m/s^2$
(c) $2 m/s^2$ (d) $1 m/s^2$

3. When the amount of work done is 300 J and change in internal energy is 100 J, then the heat supplied is
(a) 400 J (b) 350 J
(c) 200 J (d) 150 J

4. If 'S' is stress and 'Y' is young's modulus of material of a wire, the energy stored in the wire per unit volume is
(a) $\frac{S^2}{2Y}$ (b) $2S^2Y$
(c) $\frac{S}{2Y}$ (d) $\frac{2Y}{S^2}$

5. A gun is aimed at a horizontal target. It takes $\frac{1}{2}$ s for the bullet to reach the target. The bullet hits the target x metre below the aim. Then, x is equal to
(a) $\frac{9.8}{4}$ m (b) $\frac{9.8}{8}$ m
(c) 9.8 m (d) 19.6 m.

6. Control rods used in nuclear reactors are made of
(a) Stainless steel (b) Graphite
(c) Cadmium (d) Plutonium

7. If the number of turns per unit length of a coil of a solenoid is doubled, the self inductance of the solenoid will
(a) remain unchanged
(b) be halved
(c) be doubled
(d) become four times

8. The internal resistance of a cell is the resistance of
(a) material used in cell
(b) electrolyte used in cell
(c) electrodes of the cell
(d) vessel of the cell

9. A mass of 1 kg is suspended by a thread. It is
(i) lifted with an acceleration $4.9 m/s^2$
(ii) lowered with an acceleration $4.9 m/s^2$.
The ratio of the tension is
(a) 3 : 1 (b) 1 : 3
(c) 1 : 2 (d) 2 : 1

10. A Carnot engine is working between 127°C and 27°C. The increase in efficiency will be maximum when the temperature of
(a) the source is increased by 50°C
(b) the sink is decreased by 50°C
(c) source is increased by 25°C and that of sink is decreased by 25°C
(d) both source and sink are decreased by 25°C each

11. Four capacitors each of $25\mu F$ are connected as shown in diagram. The DC voltmeter reads 200 volt. The charge on each plate of capacitor will be

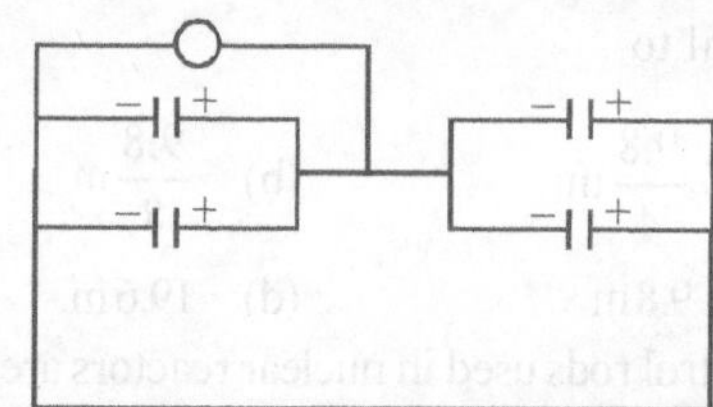

(a) 5×10^{-2} C (b) 2×10^{-2} C
(c) 5×10^{-3} C (d) 2×10^{-2} C

12. If the momentum of a particle is doubled, then its de-Broglie wavelength will become
(a) two times (b) four times
(c) unchanged (d) half times

13. Angular momentum of a system of particles changes when
(a) force acts on a body
(b) torque acts on a body
(c) direction of velocity changes
(d) None of these

14. The minimum energy required to eject an electron, from the metal surface is called
(a) atomic energy (b) mechanical energy
(c) electrical energy (d) workfunction

15. A sample of gas expands form volume V_1 to V_2. The amount of work done by the gas is greatest, when the expansion is
(a) isothermal (b) adiabatic
(c) isobaric (d) all of these

16. A monkey is accelerating down a string whose breaking strength is two third of his weight. The minimum acceleration of the monkey should be
(a) 0 (b) $\frac{2}{3}g$
(c) g (d) $\frac{g}{3}$

17. In Young's expt., the distance between two slits is $\frac{d}{3}$ and the distance between the screen and the slit is 3 D. The number of fringes in $\frac{1}{3}$m on the screen, formed by monochromatic light of wavelength 3λ, will be
(a) $\frac{d}{9D\lambda}$ (b) $\frac{d}{27D\lambda}$
(c) $\frac{d}{81D\lambda}$ (d) $\frac{d}{D\lambda}$

18. A metal plate M is placed in a uniform electric field. Which figure correctly indicates the electric lines of force?

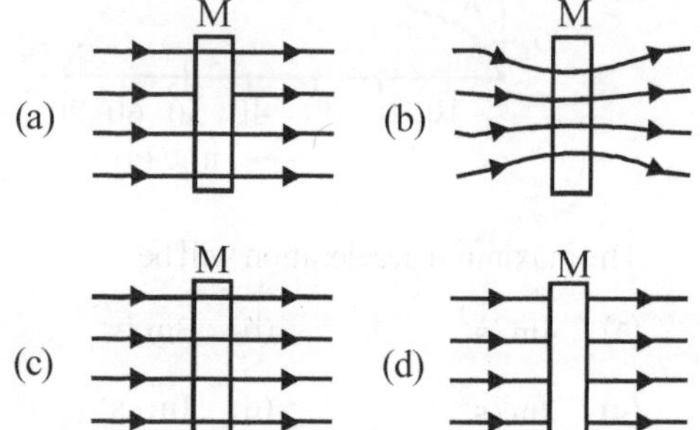

19. What will happen when a 40 watt, 220 volt lamp and 100 watt, 220 volt lamp are connected in series across 40 volt supply?
(a) 100 watt lamp will fuse
(b) 40 watt lamp will fuse
(c) Both lamps will fuse
(d) Neither lamp will fuse

20. Work is done by 0.1 mole of a gas at 27°C. To double its volume at constant pressure, required work done is (R = 2 cal/mol. K)
(a) 546 cal (b) 60 cal
(c) 600 cal (d) 54 cal

21. The coefficient of friction between two surfaces is 1 then angle of friction is
(a) 60° (b) 45°
(c) 30° (d) None of these

22. If a current increases from zero to one ampere in 0.1 second in a coil of 5 mH, then the magnitude of the induced e.m.f. will be
(a) 0.005 volt (b) 0.5 volt
(c) 0.05 volt (d) 5 volt

23. Potentiometer measures potential more accurately because
(a) it measures potential in open circuit
(b) it uses sensitive galvanometer for null deflection
(c) it uses high resistance potentiometer wire
(d) it measures potential in closed circuit

24. Weightlessness in satellite is due to
(a) zero gravitational acceleration
(b) zero acceleration
(c) zero mass
(d) None of these

25. Two waves whose intensities are 9 and 16 are made to interfere. The ratio of maximum and minimum intensities in the interference pattern is
(a) 49 : 1 (b) 25 : 7
(c) 10 : 9 (d) 4 : 3

26. In an AM wave, the information is contained within
(a) r.f. carrier wave
(b) only lower and upper side frequencies
(c) both r.f. carrier and side frequencies
(d) None of these

27. When the current in an ac circuit is wattless, the phase difference between the applied voltage and circuit current will be
(a) 45° (b) 60°
(c) 90° (d) 180°

28. A crane is used to lift 1000 kg of coal from a mine 100 m deep. The time taken by the crane is 1 hour. The efficiency of the crane is 80%. If g = 10 ms^{-2}, then the power of the crane is
(a) 10^4 W (b) 10^5 W
(c) $\frac{10^4}{36\times 8}$ W (d) $\frac{10^5}{36\times 8}$ W

29. The escape velocity of a planet, having mass 6 times and radius 2 times of the earth, is
(a) $\sqrt{2}v_{(es)}$ (b) $2v_{(es)}$
(c) $3v_{(es)}$ (d) $\sqrt{3}v_{(es)}$

30. A coil in the shape of an equilateral triangle of side l is suspended between the pole pieces of a permanent magnet such that $\vec{B}$ is in plane of the coil. If due to a current i in the triangle a torque τ acts on it, the side l of the triangle is
(a) $\frac{2}{\sqrt{3}}\left(\frac{\tau}{Bi}\right)^{\frac{1}{2}}$ (b) $\frac{2}{3}\left(\frac{\tau}{Bi}\right)$
(c) $2\left(\frac{\tau}{\sqrt{3}Bi}\right)^{\frac{1}{2}}$ (d) $\frac{1}{\sqrt{3}}\frac{\tau}{Bi}$

31. A fixed mass of gas at constant pressure occupies a volume V. The gas undergoes a rise in temperature so that the root mean square velocity of its molecules is doubled. The new volume will be
(a) V/2 (b) $V/\sqrt{2}$
(c) 2V (d) 4V

32. The sky appears blue, because
(a) blue light is scattered the most
(b) blue light is absorbed
(c) red light is absorbed
(d) it is its natural colour

33. The tuning circuit of a radio receiver has a resistance of 50 Ω, an inductance of 10 mH and a variable capacitance C. A 1 MHz radio wave produces a potential difference of 0.1 mV. The value of capacitance to produce the resonance will be

(a) 2.5 pF (b) 5.0 pF
(c) 25 pF (d) 50 pF

34. A force of 10 N acts on a body of mass 20 kg for 10 seconds. Change in its momentum is

(a) 5 kg m/s (b) 100 kg m/s
(c) 200 kg m/s (d) 1000 kg m/s

35. A steel rod of radius R = 10 mm and length L= 100 cm is stretched along its length by a force $F = 6.28 \times 10^4$ N. If the Young's modulus of steel is $Y = 2 \times 10^{11}$ N/m^2, the percentage elongation in the length of the rod is

(a) 0.100 (b) 0.314
(c) 2.015 (d) 1.549

36. Energy of simple harmonic motion depends upon

(a) $\frac{1}{\omega^2}$. (b) ω
(c) a^2 (d) $\frac{1}{a^2}$

37. The relation between B, H and I in S.I. units is

(a) $B = \mu_0(H+I)$ (b) $B = H + 4\pi I$
(c) $H = \mu_0(B+I)$ (d) None of these

38. A tank is filled with water upto a height H. Water is allowed to come out of a hole P in one of the walls at a depth h below the surface of water (see fig.) Express the horizontal distance X in terms of H and h.

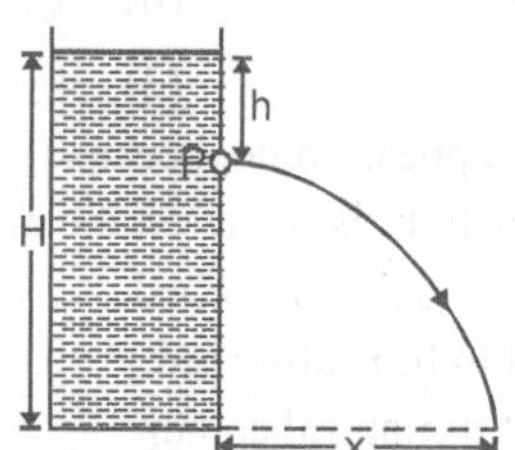

(a) $X = \sqrt{h(H-h)}$ (b) $X = \sqrt{\frac{h}{2}(H-h)}$
(c) $X = 2\sqrt{h(H-h)}$ (d) $X = 4\sqrt{h(H-h)}$

39. A point object is 24 cm above the surface of water ($\mu = 4/3$) in lake. A fish inside the water will observe the image to be at a point

(a) 6 cm above the surface of water
(b) 6 cm below the surface of water
(c) 18 cm above the surface of water
(d) 32 cm above the surface of water

40. According to Hook's law of elasticity, if stress is increased, then the ratio of stress to strain

(a) remains constant (b) becomes zero
(c) decreases (d) increases

41. Current gain in common emitter configuration is more than 1 because

(a) $I_c < I_b$ (b) $I_c < I_e$
(c) $I_c > I_e$ (d) $I_c > I_b$

42. Logic gates are the building blocks of a

(a) analog system
(b) digital system
(c) abacus system
(d) None of these

43. The forbidden energy gap for germanium crystal at 0 K is

(a) 0.071 eV (b) 0.71 eV
(c) 2.57 eV (d) 6.57 eV

44. Let n_p and n_e be the number of holes and conduction electrons in an intrinsic semiconductor. Then

(a) $n_p > n_e$. (b) $n_p = n_e$.
(c) $n_p < n_e$. (d) $n_p \neq n_e$.

45. Forward baising is that in which applied voltage

(a) increases potential barrier
(b) cancels the potential barrier
(c) is equal to 1.5 volts
(d) None of these

SECTION 2 - CHEMISTRY

46. Among the following pairs of ions, the lower oxidation state in aqueous solution is more stable than the other, in :

(a) Ti^+, Ti^{3+} (b) Cu^+, Cu^{2+}

(c) Cr^{2+}, Cr^{3+} (d) V^{2+}, VO^{2+}

47. If NaCl is doped with 10^{-4} mol % of $SrCl_2$, the concentration of cation vacancies will be ($N_A = 6.02 \times 10^{23}\ mol^{-1}$)

(a) $6.02 \times 10^{16}\ mol^{-1}$ (b) $6.02 \times 10^{17}\ mol^{-1}$

(c) $6.02 \times 10^{14}\ mol^{-1}$ (d) $6.02 \times 10^{15}\ mol^{-1}$

48. Which of the following isomerism is exhibited by $CH_3 - O - C_3H_7$ and $C_2H_5OC_2H_5$

(a) optical isomerism

(b) chain isomerism

(c) metamerism

(d) position isomerism

49. Which of the following statements, about the advantage of roasting of sulphide ore before reduction is not true?

(a) The ΔG_f^o of the sulphide is greater than those for CS_2 and H_2S.

(b) The ΔG_f^o is negative for roasting of sulphide ore to oxide.

(c) Roasting of the sulphide to the oxide is thermodynamically feasible.

(d) Carbon and hydrogen are suitable reducing agents for metal sulphides.

50. Graph between $\log\left(\frac{x}{m}\right)$ and log p is a straight line at angle 45° with intercept OA as shown. Hence, $\left(\frac{x}{m}\right)$ at a pressure of 0.2 atm is

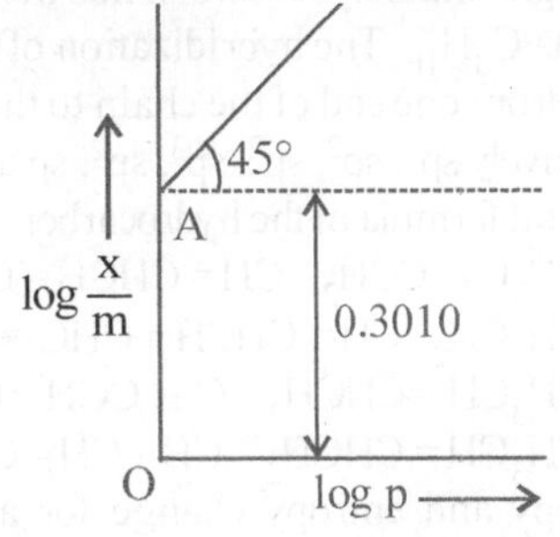

(a) 0.2 (b) 0.4

(c) 0.6 (d) 0.8

51. The correct order of increasing $[H_3O^+]$ in the following aqueous solutions is

(a) $0.01\ M\ H_2S < 0.01 M\ H_2SO_4 < 0.01\ M\ NaCl < 0.01 M\ NaNO_2$

(b) $0.01\ M\ NaCl < 0.01\ M\ NaNO_2 < 0.01\ M\ H_2S < 0.01 M\ H_2SO_4$

(c) $0.01\ M\ NaNO_2 < 0.01\ M\ NaCl < 0.01\ M\ H_2S < 0.01\ M\ H_2SO_4$

(d) $0.01\ M\ H_2S < 0.01\ M\ NaNO_2 < 0.01\ M\ NaCl < 0.01\ M\ H_2SO_4$

52. Pollution in large cities can be checked only by

(a) Shifting of factories out of the residential area

(b) Less use of insecticides

(c) Proper disposal of organic wastes, sewage and industrial effluents

(d) All the above

53. All form ideal solution except

(a) C_6H_6 and $C_6H_5\ CH_3$

(b) C_2H_6 and C_2H_5I

(c) C_6H_5Cl and $C_6H_5\ Br$

(d) $C_2H_5\ I$ and $C_2H_5\ OH$.

54. Which of the following metal ions will form complexes with the same magnetic moment and geometry irrespective of the nature of ligands?

(a) Ni^{2+} (b) Fe^{2+}

(c) Cu^{2+} (d) Co^{2+}

55. A straight chain hydrocarbon has the molecular formula C_8H_{10}. The hybridization of the carbon atoms from one end of the chain to the other are respectively sp^3, sp^2, sp^2, sp^3, sp^2, sp and sp. The structural formula of the hydrocarbon would be :
(a) $CH_3C \equiv CCH_2-CH=CHCH=CH_2$
(b) $CH_3CH_2-CH=CHCH=CHC \equiv CH$
(c) $CH_3CH=CHCH_2-C \equiv CCH=CH_2$
(d) $CH_3CH=CHCH_2-CH-CH-C \equiv CH$.

56. Enthalpy and entropy change for a chemical reaction are -2.5×10^3 cals and 7.4 cals deg^{-1} respectively. At 25°C the reaction is
(a) Reversible (b) Irreversible
(c) Spontaneous (d) Non spontaneous

57. Which of the following has correct increasing basic strength ?
(a) $MgO < BeO < CaO < BaO$
(b) $BeO < MgO < CaO < BaO$
(c) $BaO < CaO < MgO < BeO$
(d) $CaO < BaO < BeO < MgO$

58. One litre of 1 M $CuSO_4$ solution is electrolysed. After passing 2 F of electricity, molarity of $CuSO_4$ solution will be
(a) $\frac{M}{2}$ (b) $\frac{M}{4}$
(c) M (d) 0

59. Which of the following is commercially known as **oxone** ?
(a) $Na_2O_2 + HCl$ (b) $Na_2O + HCl$
(c) $Na_2O_2 + Na_2O$ (d) none of these

60. $M(OH)_x$ has $K_{sp} = 4 \times 10^{-4}$ M. Hence, x is
(a) 1 (b) 2
(c) 3 (d) 4

61. Which is not correct regarding the adsorption of a gas on surface of solid?
(a) On increasing temperature, adsorption increases contineuously
(b) Enthalpy and entropy changes are –ve
(c) Adsorption is more for some specific substance
(d) This Phenomenon is reversible

62. At its melting point ice is lighter than water because
(a) H_2O molecules are more closely packed in solid state
(b) Ice crystals have hollow hexagonal arrangement of H_2O molecules.
(c) On melting of ice the H_2O molecule shrinks in size
(d) Ice froms mostly heavy water on first melting.

63. The most suitable method for separtion of a 1 : 1 mixture of ortho and para nitrophenols is
(a) Sublimation (b) Chromatography
(c) Crystallization (d) Steam distillation

64. A and B in the following reactions are

$$R-\underset{\underset{O}{\|}}{C}-R' \xrightarrow[KCN]{HCN/} A \xrightarrow{B} R-\underset{R'}{C}\begin{matrix} OH \\ CH_2NH_2 \end{matrix}$$

(a) $A = RR'C\begin{matrix} CN \\ OH \end{matrix}$, $B = LiAlH_4$
(b) $A = RR'C\begin{matrix} OH \\ COOH \end{matrix}$, $B = NH_3$
(c) $A = RR'C\begin{matrix} CN \\ OH \end{matrix}$, $B = H_3O^{\oplus}$
(d) $A = RR'CH_2CN$, $B = NaOH$

65. On adding 0.1 M solution each of $[Ag^+]$, $[Ba^{2+}]$, $[Ca^{2+}]$ in a Na_2SO_4 solution, species first precipitated is
$[K_{sp}\, BaSO_4 = 10^{-11}, K_{sp}\, CaSO_4 = 10^{-6}, K_{sp}\, Ag_2SO_4 = 10^{-5}]$
(a) Ag_2SO_4 (b) $BaSO_4$
(c) $CaSO_4$ (d) All of these

66. Which one of the following statements concerning lanthanide elements is false?
(a) Lanthanides are separated from one another by ion exchange method
(b) The ionic radii of trivalent lanthanides steadily increase with increase in atomic number
(c) All lanthanides are highly dense metals
(d) Most characteristic oxidation state of lanthanides is +3

67. The fraction of total volume occupied by the atoms present in a simple cube is

(a) $\frac{\pi}{3\sqrt{2}}$ (b) $\frac{\pi}{4\sqrt{2}}$

(c) $\frac{\pi}{4}$ (d) $\frac{\pi}{6}$

68. The hypothetical complex chlorodiaquotriammine cobalt (III) chloride can be represented as

(a) $[CoCl(NH_3)_3(H_2O)_2]Cl_2$
(b) $[Co(NH_3)_3(H_2O)Cl_3]$
(c) $[Co(NH_2)_3(H_2O)_2Cl]$
(d) $[Co(NH_3)_3(H_2O)_3]Cl_3$

69. Which of the following may be considered to be an organometallic compound?

(a) Nickel tetracarbonyl
(b) Chlorophyll
(c) $K_3[Fe(C_2O_4)_3]$
(d) $[Co(en)_3]Cl_3$

70. If K_1 and K_2 are respective equilibrium constants for the two reactions

$XeF_6(g) + H_2O(g) \rightleftharpoons XeOF_4(g) + 2HF(g)$

$XeO_4(g) + XeF_6(g) \rightleftharpoons XeOF_4(g) + XeO_3F_2(g)$

the equilibrium constant for the reaction

$XeO_4(g) + 2HF(g) \rightleftharpoons XeO_3F_2(g) + H_2O(g)$

will be

(a) $\frac{K_1}{K_2^2}$ (b) $K_1.K_2$

(c) $\frac{K_1}{K_2}$ (d) $\frac{K_2}{K_1}$

71. Alkyl halide on heating with dry Ag_2O produces

(a) ether (b) ester
(c) ketone (d) hydrocarbon

72. In the following reaction 'A' is

$$C_2H_5MgBr + H_2C-CH_2 \xrightarrow{H_2O} A$$
(with O bridging the two CH_2 groups, i.e. an epoxide)

(a) $C_2H_5CH_2CHO$ (b) $C_2H_5CH_2CH_2OH$
(c) $C_2H_5CH_2OH$ (d) C_2H_5CHO

73. The standard electrode potentials of four elements A, B, C and D are –3.05, –1.66, –0.40 and +0.80. The highest chemical reactivity will be exhibited by :

(a) A (b) B
(c) C (d) D

74. Causticisation is a process in which

(a) soap is prepared by treating vegetable oils with caustic soda
(b) paper is formed
(c) petroleum is treated with caustic soda before fractional distillation
(d) caustic soda is manufactured by treating sodium carbonate with lime.

75. Which of the following statements is incorrect regarding benzyl chloride?

(a) It gives white precipitate with alcoholic $AgNO_3$
(b) It is an aromatic compound with substitution in the side chain
(c) It undergoes nucleophilic substitution reaction
(d) It is less reactive than vinyl chloride

76. One mole of an ideal gas is allowed to expand reversibly and adiabatically from a temperature of 27°C. If the work done during the process is 3 kJ, then final temperature of the gas is (C_v = 20 J/K)

(a) 100 K (b) 150 K
(c) 195 K (d) 255 K

77. Chemically borax is

(a) Sodium metaborate
(b) Sodium orthoborate
(c) Sodium tetraborate
(d) Sodium tetraborate decahydrate

78. Acetylenic hydrogens are acidic because

(a) Sigma electron density of C – H bond in acetylene is nearer to carbon, which has 50% s-character
(b) Acetylene has only open hydrogen in each carbon
(c) Acetylene contains least number of hydrogens among the possible hydrocarbons having two carbons
(d) Acetylene belongs to the class of alkynes with molecular formula, C_nH_{2n-2}.

79. $H_2S(g) \longrightarrow HS^-(g) + H^+(g),\ \Delta H^\circ = x_1$

$\Delta H_f^\circ[H_2S(g)] = x_2,\ \Delta H_f^\circ[H(g)] = x_3$

hence, $\Delta H_f^\circ[HS]^-$ is

(a) $x_1 + x_2 - x_3$ (b) $x_3 - x_1 - x_2$
(c) $x_1 - x_2 - x_3$ (d) $x_3 - x_1 + x_2$

80. In qualitative analysis, the metals of Group I can be separated from other ions by precipitating them as chloride salts. A solution initially contains Ag^+ and Pb^{2+} at a concentration of 0.10 M. Aqueous HCl is added to this solution until the Cl^- concentration is 0.10 M. What will the concentrations of Ag^+ and Pb^{2+} be at equilibrium?
(K_{sp} for $AgCl = 1.8 \times 10^{-10}$,
K_{sp} for $PbCl_2 = 1.7 \times 10^{-5}$)

(a) $[Ag^+] = 1.8 \times 10^{-7}$ M; $[Pb^{2+}] = 1.7 \times 10^{-6}$ M
(b) $[Ag^+] = 1.8 \times 10^{-11}$ M; $[Pb^{2+}] = 8.5 \times 10^{-5}$ M
(c) $[Ag^+] = 1.8 \times 10^{-9}$ M; $[Pb^{2+}] = 1.7 \times 10^{-3}$ M
(d) $[Ag^+] = 1.8 \times 10^{-11}$ M; $[Pb^{2+}] = 8.5 \times 10^{-4}$ M

81. Carbon and CO gas are used to reduce which of the following pairs of metal oxides for extraction of metals ?

(a) FeO, SnO (b) SnO, ZnO
(c) BaO, Na_2O_2 (d) FeO, ZnO

82. Which one of the alkaline earth metals shows some anomalous behaviour and has same electronegativity as aluminium?

(a) Sr (b) Ca
(c) Ba (d) Be

83. Identify the correct order of electron affinity for O^-, O, F^- and Na^+.

(a) $Na^+ < F^- < O^- < O$
(b) $O < O^- < F^- < Na^+$
(c) $O^- < O << F^- < Na^+$
(d) $F^- < Na^+ < O^- < O$

84. For the reaction, $2Cl(g) \longrightarrow Cl_2(g)$, the signs of ΔH and ΔS respectively, are:

(a) +, – (b) +, +
(c) –, – (d) –, +

85. Which of the following can not be isoelectronic?

(a) two different cations
(b) two different anions
(c) cation and anion
(d) two different atoms

86. When CH_3Cl and $AlCl_3$ are used in Friedel-Crafts reaction, the electrophile is

(a) Cl^+ (b) $AlCl_4^-$
(c) CH_3^+ (d) $AlCl_2^+$

87. An endothermic reaction with high activation energy for the forward reaction is given by the diagram :

(a)

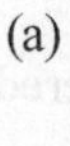

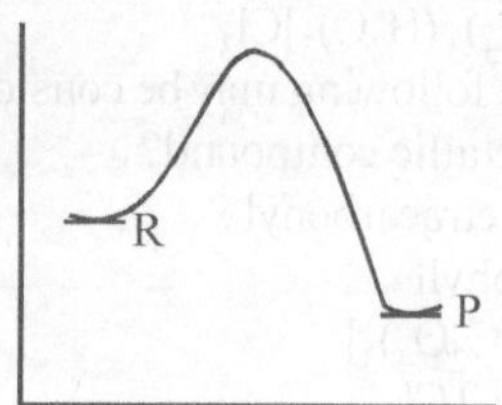

Reaction co-ordinate

(b)

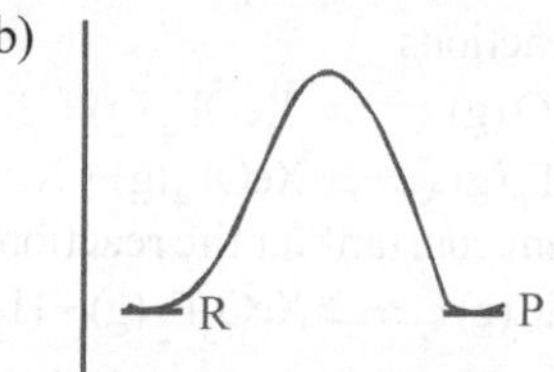

Reaction co-ordinate

(c)

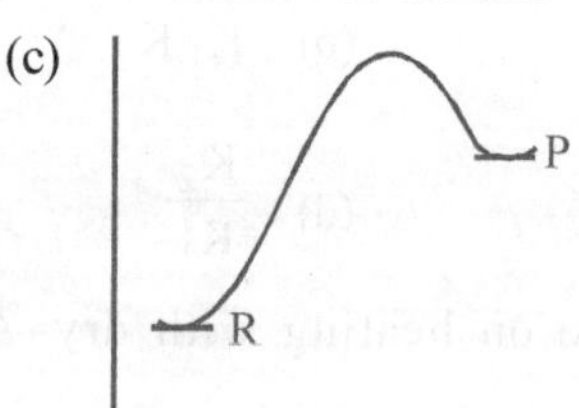

Reaction co-ordinate

(d)

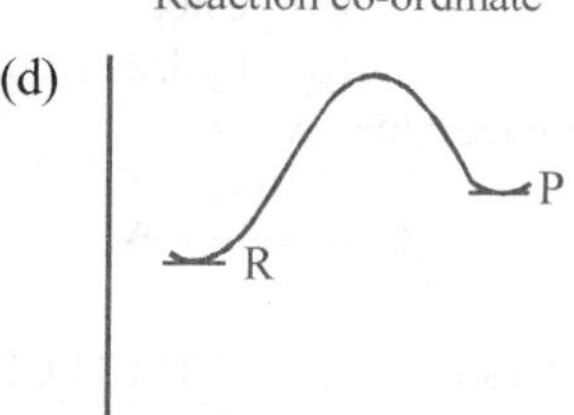

Reaction co-ordinate

88. The alkaline earth metals Ba, Sr, Ca and Mg may be arranged in the order of their decreasing first ionisation potential as

(a) Mg, Ca, Sr, Ba (b) Ca, Sr, Ba, Mg
(c) Sr, Ba, Mg, Ca (d) Ba, Mg, Ca, Sr

89. The monomer of the polymer;

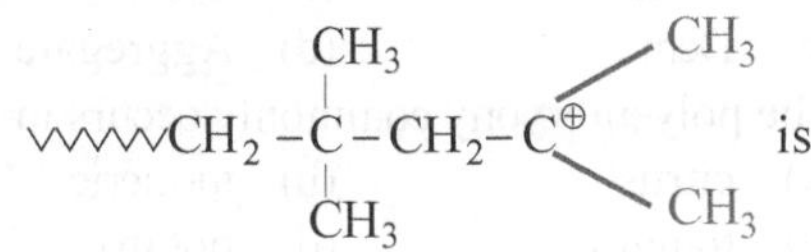

(a) $H_2C=C(CH_3)_2$
(b) $CH_3CH=CHCH_3$
(c) $CH_3CH=CH_2$
(d) $(CH_3)_2C=C(CH_3)_2$

90. Two electrolytic cells, one containing acidified ferrous chloride and another acidified ferric chloride, are connected in series. The ratio of iron deposited at cathodes in the two cells will be :

(a) 3 : 1 (b) 2 : 1
(c) 1 : 1 (d) 3 : 2

SECTION 3 - BOTANY

91. The living organisms can be unexceptionally distinguished from the non-living things on the basis of their ability for

(a) interaction with the environment and progressive evolution
(b) reproduction
(c) growth and movement
(d) responsiveness to touch.

92. The largest flower found is known as

(a) *Rafflesia* (b) *Tecoma*
(c) *Musa* (d) Cauliflower

93. Pulses are belong to the family

(a) fabaceae (b) asteraceae
(c) poaceae (d) solanaceae

94. People recovering from long illness are often advised to include the alga *Spirulina* in their diet because it

(a) makes the food easy to digest.
(b) is rich in proteins.
(c) has antibiotic properties.
(d) restores the intestinal microflora.

95. In a cereal grain the single cotyledon of embryo is represented by

(a) scutellum (b) prophyll
(c) coleoptile (d) coleorrhiza

96. The chief water conducting elements of xylem in gymnosperms are :

(a) vessels (b) fibres
(c) transfusion tissue (d) tracheids

97. Consider the following statements regarding the major pigments and stored food in the different groups of algae and choose the correct option.

(i) In Chlorophyceae, the stored food material is starch and the major pigments are chlorophyll-*a* and *d*.
(ii) In Phaeophyceae, laminarin is the stored food and major pigments are chlorophyll-*a* and *b*.
(iii) In Rhodophyceae, floridean starch is the stored food and the major pigments are chlorophyll-*a*, *d* and phycoerythrin.

(a) (i) is correct, but (ii) and (iii) are wrong.
(b) (i) and (ii) are correct, but (iii) is wrong.
(c) (i) and (iii) are correct, but (ii) is wrong.
(d) (iii) is correct, but (i) and (ii) are wrong.

98. Chromosome duplication without nuclear division refers to

(a) Meiosis (b) Mitosis
(c) Androgenesis (d) Endomitosis

99. During which stages (or prophase I substages) of meiosis do you expect to find the bivalents and DNA replication respectively?
(a) Pachytene and interphase (between two meiotic divisions)
(b) Pachytene and interphase (just prior to prophase I)
(c) Pachytene and S phase (of interphase just prior to prophase I)
(d) Zygotene and S phase (of interphase prior to prophase I)

100. Perisperm is
(a) remnant of endosperm
(b) persistant nucellus
(c) remnant of embryo
(d) part of endosperm

101. Identified A, B, C and D of a seed.

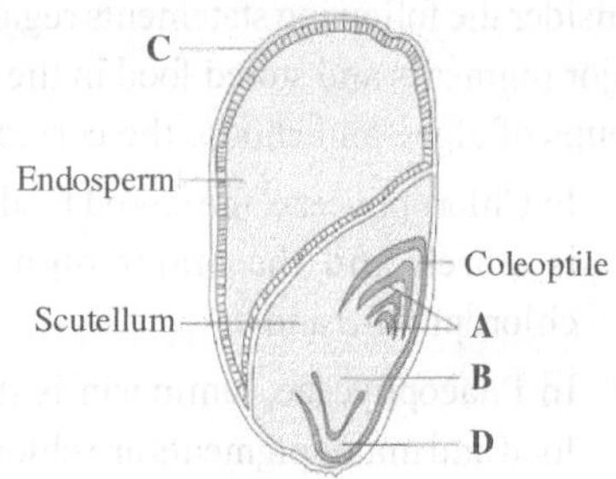

(a) A – Plumule; B – Radicle; C – Pericarp; D – Coleorhiza
(b) A – Radicle; B – Plumule; C – Pericarp; D – Coleorhiza
(c) A – Radicle; B – Plumule; C – Coleorhiza; D – Pericarp
(d) A – Radicle; B – Coleorhiza; C – Plumule; D – Pericarp

102. One of the most resistant biological material is
(a) lignin (b) hemicellulose
(c) lignocellulose (d) sporopollenin

103. Which of the following is a pollution related occupational health hazard disease ?
(a) Pneumoconiosis (b) Asthma
(c) Flurosis (d) Silicosis

104. A population of genetically identical individuals, obtained from asexual reproduction is
(a) Callus (b) Clone
(c) Deme (d) Aggregate

105. The polyembryony commonly occurs in
(a) citrus (b) turmeric
(c) tomato (d) potato

106. In which one of the following habitats does the diurnal temperature of soil surface vary most?
(a) Shrub land (b) Forest
(c) Desert (d) Grassland

107. In increasing order of organizational complexity, which one of the following is the correct sequence?
(a) Population, species, community, ecosystem
(b) Population, variety, species, ecosystem
(c) Population, ecosystem, species, community
(d) Species, variety, ecosystem, community

108. In dicot embryo the radicle is formed by
(a) epibasal tier of embryo
(b) hypobasal tier of embryo
(c) hypophysis of suspensor
(d) terminal cell of suspensor

109. Osmosis means movement of
(a) solute from lower concentration to higher concentration.
(b) solute from higher concentration to lower concentration.
(c) solvent from lower concentration of solution to higher concentration of solution.
(d) solvent from higher concentration of solution to lower concentration of solution.

110. Solution outside a cell has higher concentration than cytoplasm, then the solution is
(a) isotonic (b) acidic
(c) hypotonic (d) hypertonic

111. The figure below shows the structure of a mitochondrion with its four parts labelled (A), (B), (C) and (D).

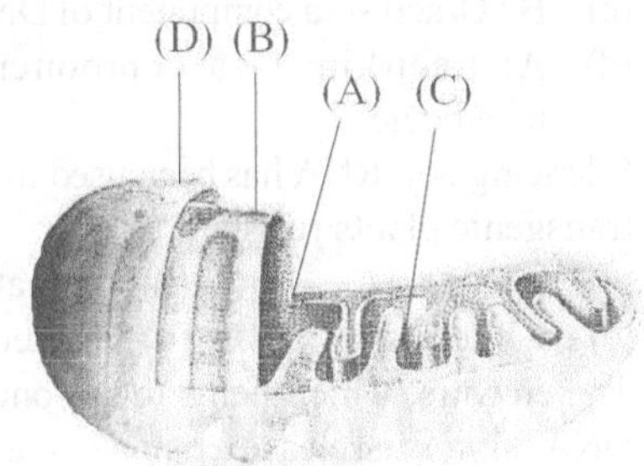

Select the part correctly matched with its function.
(a) Part (D): Outer membrane – gives rise to inner membrane by splitting
(b) Part (B): Inner membrane – forms infoldings called cristae
(c) Part (C): Cristae – possess single circular DNA molecule and ribosomes
(d) Part (A): Matrix – major site for respiratory chain enzymes

112. Which of the following is correct set of micronutrient for plants?
(a) Mg, Si, Fe, Cu, Ca
(b) Cu, Fe, Zn, B, Mn
(c) Mg, Fe, Zn, B, Mn
(d) Mo, Zn, Cl, Mg, Ca

113. During $Na^+ - K^+$ pump
(a) $3Na^+$ and $2K^+$ are transported
(b) $1Na^+$ and $2K^+$ are transported
(c) $3Na^+$ and $3K^+$ are transported
(d) Depends on requirement of cell

114. The technique of obtaining large number of plantlets by tissue culture method is called
(a) Plantlet culture
(b) Organ culture
(c) Micropropagation
(d) Macropropagation

115. Extranuclear inheritance occurs in
(a) peroxisome and ribosome
(b) chloroplast and mitochondria
(c) mitochondria and ribosome
(d) chloroplast and lysosome

116. Which one of the following pairs, is not correctly matched?
(a) Gibberellic acid - Leaf fall
(b) Cytokinin - Cell wall elongation
(c) IAA - Cell wall elongation
(d) Abscissic acid - Stomatal closure.

117. Test cross involves
(a) crossing between two F_1 hybrids
(b) crossing the F_1 hybrid with a double recessive genotype
(c) crossing between two genotypes with dominant trait
(d) crossing between two genotypes with recessive trait

118. What base is responsible for hot spots for spontaneous point mutations?
(a) Adenine (b) 5-bromouracil
(c) 5-methylcytosine (d) Guanine

119. When one gene controls two or more different characters simultaneously, the phenomenon is called
(a) apomixis (b) pleiotropy
(c) polyploidy (d) polyteny

120. Pneumatophores are characteristic of plants growing in
(a) saline soils
(b) sandy soils
(c) marshy places and salt lakes
(d) dryland regions

121. What is antisense technology ?
(a) A cell displaying a foreign antigen used for synthesis of antigens
(b) Production of somaclonal variants in tissue cultures
(c) When a piece of RNA that is complementary in sequence is used to stop expression of a specific gene
(d) RNA polymerase producing DNA

122. An enzyme brings about
   (a) decrease in reaction time
   (b) increase in reaction time
   (c) increase in activation energy
   (d) reduction in activation energy

123. The enzymes hexokinase which catalyses glucose to glucose-6-phosphate in glycolysis is inhibited by glucose-6-phosphate. This is an example of
   (a) competitive inhibition
   (b) non-competitive inhibition
   (c) feedback allosteric inhibition
   (d) positive feedback

124. Apical dominance in plants is caused by
   (a) high concentration of auxins in the terminal bud.
   (b) high concentration of gibberellins in the apical bud.
   (c) high concentration of auxins in the lateral bud.
   (d) absence of auxins and gibberellins in apical bud.

125. Primary precursor of IAA is
   (a) phenylalanine (b) tyrosine
   (c) tryptophan (d) leucine

126. In a plant growing under dark condition, the leaves turn light coloured, internodes become much elongated and it is termed as
   (a) vernalization (b) etiolation
   (c) phyllotaxy (d) chlorosis

127. Which one of the following structural formulae of two organic compounds is correctly identified along with its related function ?

A. $R_2-\overset{O}{\overset{\|}{C}}-O-CH$; $CH_2-O-\overset{O}{\overset{\|}{C}}-R$ (above); $CH_2-O-P(OH)-O-CH_2-CH_2-N(CH_3)_3$ (below)

B. (purine ring structure bearing NH_2, with N, N, N and NH in the rings)

   (a) B : Adenine — a nucleotide that makes up nucleic acids
   (b) A: Triglyceride — major source of energy
   (c) B : Uracil — a component of DNA
   (d) A: Lecithin — a component of cell membrane

128. Silencing of mRNA has been used in producing transgenic plants resistant to:
   (a) bollworms (b) nematodes
   (c) white rusts (d) bacterial blights

129. The enzyme, which helps to cut one strand of DNA duplex to release tension of coiling of two strands is
   (a) DNA ligase
   (b) DNA polymerase I
   (c) topo-isomerase
   (d) swielases (helicase or unwindases)

130. Continuous addition of sugars in 'fed batch' fermentation is done to:
   (a) produce methane
   (b) obtain antibiotics
   (c) purify enzymes
   (d) degrade sewage

131. Cell elongation in internodal regions of the green plants takes place due to
   (a) indole acetic acid (b) cytokinins
   (c) gibberellins (d) ethylene

132. The 'Earth Summit' held at Rio de Janerio in 1992 resulted into
   (a) Compilation of Red list
   (b) Establishment of Biosphere Reserves
   (c) Convention on Biodiversity
   (d) Development of Hot Spots of Biodiversity

133. Tobacco plants resistant to a nematode have been developed by the introduction of DNA that produced (in the host cells)
   (a) both sense and anti-sense RNA
   (b) a particular hormone
   (c) an antifeedant
   (d) a toxic protein

134. Chloramphenicol and erythromycin (broad spectrum antibiotics) are produced by
(a) *Streptomyces* (b) *Nitrobacter*
(c) *Rhizobium* (d) *Penicillium*

135. For biogas production besides dung an extensive use of which weed is recommended in our country–
(a) *Mangifera indica*
(b) *Hydrilla*
(c) *Eicchornia crassipes*
(d) *Solanum*

SECTION 4 - ZOOLOGY

136. Which one of the following statements about Mycoplasma is wrong ?
(a) They are pleomorphic.
(b) They are sensitive to penicillin.
(c) They cause diseases in plants.
(d) They are also called (Pleuro pneumonia like organisms) PPLO.

137. Housefly possesses
(a) two pairs of wings
(b) one pair of wings
(c) three pairs of wings
(d) four pair of wings

138. The longest cells in human body are
(a) nerve cells (b) bone cells
(c) leg muscle cells (d) heart-muscle cells

139. Major inorganic component of vertebrate bone is
(a) calcium carbonate
(b) calcium phosphate
(c) sodium hydroxide
(d) potassium hydroxide

140. Histamine is secreted by
(a) mast cells (b) fibroblast
(c) histiocytes (d) melanocytes

141. Which of the following is not a granulocyte?
(a) Lymphocyte (b) Neutrophil
(c) Basophil (d) Eosinophil

142. The largest gland of human body is
(a) thyroid (b) pituitary
(c) pancreas (d) liver

143. Which one of the following statements is correct with respect to salt water balance inside the body of living organisms?
(a) When water is not available, camels do not produce urine but store urea in tissues.
(b) Salmon fish excretes lot of stored salt through gill membrane in fresh water.
(c) *Paramecium* discharges concentrated salt solution by contractile vacuoles.
(d) The body fluids of fresh water animals are generally hypotonic to surrounding water.

144. Identify A to E.

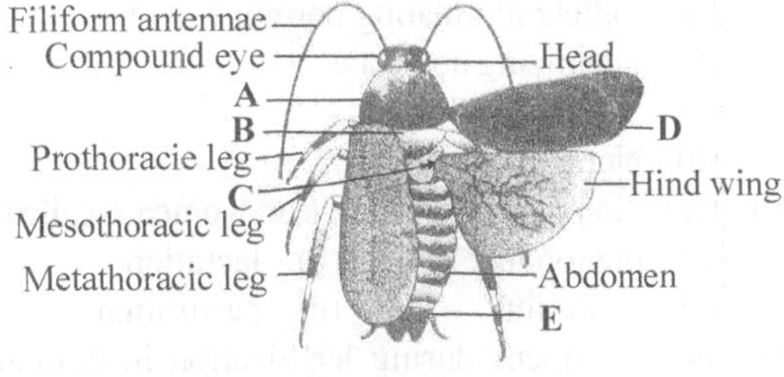

Figure - External features of cockroach

(a) A- Pronotum, B- Mesothorax, C- Metathorax, D- Tegmina, E- Pleura
(b) A-Pronotum, B-Mesothorax, C-Metathorax, D-Tegmina, E- Sterna
(c) A-Pronotum, B-Mesothorax, C-Metathorax, D- Tegmina, E-Anal cerci
(d) A-Pronotum, B-Mesothorax, C- Metathorax, D-Tegmina, D- Anal style.

145. Tying up or removing a small part of fallopian duct is called
(a) Vasectomy (b) Ductus arteriosus
(c) Archidectomy (d) Tubcctomy

146. The technique called gamete intrafallopian transfer (GIFT) is recommended for those females:
(a) who cannot produce an ovum
(b) who cannot retain the foetus inside uterus.
(c) whose cervical canal is too narrow to allow passage for the sperms
(d) who cannot provide suitable environment for fertilisation

147. Fertilization occurs in, human, rabbit and other placental mammals in –
(a) Ovary (b) Uterus
(c) Fallopian tubes (d) Vagina

148. Sertoli cells are found in
(a) ovaries and secrete progesterone
(b) adrenal cortex and secrete adrenaline
(c) seminiferous tubules and provide nutrition to germ cells
(d) pancreas and secrete cholecystokinin

149. Birth control pills check ovulation in female by inhibiting the secretion of –
(a) follicle stimulating hormone
(b) luteinizing hormone
(c) Both (a) and (b)
(d) None of the above

150. Cessation of menstrual cycle in women is called
(a) menopause (b) lactation
(c) ovulation (d) parturition

151. What happens during fertilisation in humans after many sperms reach close to the ovum?
(a) Cells of corona radiata trap all the sperms except one
(b) Only two sperms nearest the ovum penetrate zona pellucida
(c) Secretions of acrosome helps one sperm enter cytoplasm of ovum through zona pellucida
(d) All sperms except the one nearest to the ovum lose their tails

152. Coronary artery disease (CAD) is often referred to as
(a) Heart failure (b) Cardiac arrest
(c) Atherosclerosis (d) Thrombosis

153. Antidiuretic hormone
(a) Secretion is determined by plasma osmolarity
(b) Increases permeability of renal collecting duct cells to water
(c) Is secreted by nerve cells with their cell bodies in hypothalamus
(d) All the above

154. Which one of the following is a matching pair of a substrate and its particular digestive enzyme?
(a) Starch — maltase
(b) Lactose — rennin
(c) Maltose — steapsin
(d) Casein — chymotrypsin

155. Which one of the following pairs of the cells with their secretion is **correctly** matched?
(a) Oxyntic cells - A secretion with pH between 2.0 and 3.0.
(b) Alpha cells of Islets of Langerhans - Secretion that decreases blood sugar level.
(c) Kupffer cells - A digestive enzyme that hydrolysis nucleic acids.
(d) Sebaceous glands - A digestive enzyme that hydrolysis nucleic acids

156. Bulk of carbon dioxide (CO_2) released from body tissues into the blood is present as
(a) bicarbonate in blood plasma and RBCs
(b) free CO_2 in blood plasma
(c) 70% carbamino- haemoglobin and 30% as bicarbonate
(d) carbamino-haemoglobin in RBCs

157. 'Bundle of His' is a part of which one of the following organs in humans?
(a) Brain (b) Heart
(c) Kidney (d) Pancreas

158. In mammals, the urinary bladder opens into
(a) Uterus (b) Urethra
(c) Vestibule (d) Ureter

159. Barr body in mammals represents
(a) all the heterochromatin in female cells
(b) Y-chromosomes in somatic cells of male
(c) all heterochromatin in male and female cells
(d) one of the two X-chromosomes in somatic cells of females

160. Genetic drift is change of
(a) gene frequency in same generation
(b) appearance of recessive genes
(c) gene frequency from one generation to next
(d) None of the above

161. Two different species can not live for long duration in the same niche or habitat. This law is
(a) Allen's law
(b) Gause's hypothesis
(c) Dollo's rule
(d) Weisman's theory

162. Which one of the following statements is correct in relation to honey bees?
(a) *Apis indica* is the largest wild honey bee in India
(b) Honey is predominantly sucrose and arabinose
(c) Beewax is a waste product of honey bees
(d) Communication in honey bees was discovered by Karl Von Frisch

163. Select the **incorrect** statement from the following:
(a) Galactosemia is an inborn error of metabolism
(b) Small population size results in random genetic drift in a population
(c) Baldness is a sex -limited trait
(d) Linkage is an exception to the principle of independent assortment in heredity

164. Biometric genetics deals with :
(a) the biochemical explanations of various genetical phenomena
(b) the effect of environment on genetic set up organisms
(c) the genetical radiations on the living organisms
(d) the inheritance of quantitative traits

165. Murrah, Mehsana, Jaffarbadi are breeds of
(a) Buffalo (b) Cow
(c) Cattle (d) Horse

166. Identify the molecules (A) and (B) shown below and select the right option giving their source and use.

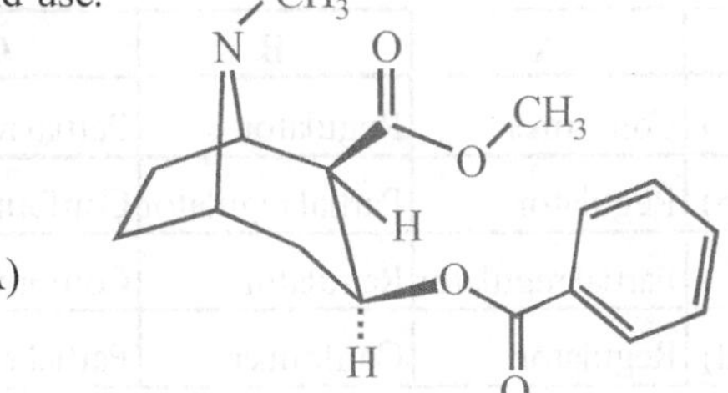

(B)
OH
O
H

Options

	Molecule	Source	Use
(a)	(A) Cocaine	*Erythroxylum coca*	Accelerates the transport of dopamine
(b)	(B) Heroin	*Cannabis sativa*	Depressant and slows down body functions
(c)	(C) Cannabinoid	*Atropa belladona*	Produces hallucinations
(d)	(A) Morphine	*Papaver somniferum*	Sedative and pain killer

167. The technique used for estimation of minute amounts of hormones and drugs is called
(a) electrophoresis
(b) electroencephalogram
(c) fractionation
(d) radioimmunoassay

168. Which of the following is the contractile protein of a muscle?
(a) Myosin (b) Tropomyosin
(c) Actin (d) Tubulin

169. During oxygen transport the oxyhaemoglobin at the tissue level liberates oxygen to the cells because in tissue
(a) O_2 concentration is high and CO_2 is low
(b) O_2 concentration is low and CO_2 is high
(c) O_2 tension is high and CO_2 tension is low
(d) O_2 tension is low and CO_2 tension is high

170. Select the **correct** statement regarding the specific disorder of muscular or skeletal system :
(a) *Myasthenia gravis* - Autoimmune disorder which inhibits sliding of myosin filaments.
(b) *Gout* - inflammation of joints due to extra deposition of calcium.
(c) *Muscular dystrophy* - age related shortening of muscles.
(d) *Osteoporosis* - decrease in bone mass and higher chances of fractures with advancing age.

171. Which one of the following pairs is incorrectly matched?
(a) Glucagon - Beta cells (source)
(b) Somatostatin - Delta cells (source)
(c) Corpus luteum - Relaxin (secretion)
(d) Insulin - Diabetes mellitus (disease)

172. Melanocyte stimulating hormone (MSH) is produced by
(a) parathyroid
(b) pars intermedia of pituitary
(c) anterior pituitary
(d) posterior pituitary

173. Nicotine acts as a stimulant, because it mimics the effect of
(a) thyroxine (b) acetylcholine
(c) testosterone (d) dopamine

174. A major component of gobar gas is –
(a) Butane (b) Ammonia
(c) Methane (d) Ethane

175. Rennin used in cheese industry is –
(a) Antibiotic (b) Enzyme
(c) Alkaloid (d) Inhibitor

176. The polymerase chain reaction (PCR) technology was discovered by
(a) Karry Mullis (b) Saiki *et al*
(c) Craig Venter (d) Maxam and Gilbert

177. Antibodies in our body are complex
(a) steroids (b) prostaglandins
(c) glycoproteins (d) lipoproteins

178. A person likely to develop tetanus is immunised by administering
(a) preformed antibodies
(b) wide spectrum antibiotics
(c) weakened germs
(d) dead germs

179. cDNA probes are copied from the messenger RNA molecules with the help of–
(a) Restriction enzymes
(b) Reverse transcriptase
(c) DNA polymerase
(d) Adenosine deaminase

180. The figure given below is a diagrammatic representation of response of organisms to abiotic factors. What do A, B and C represent respectively?

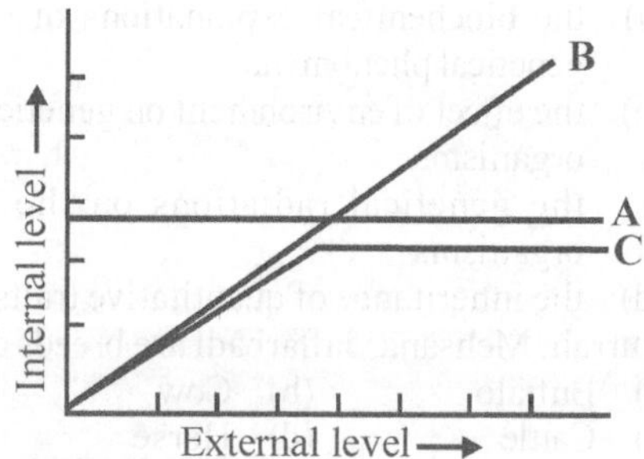

	A	B	C
(a)	Conformer	Regulator	Partial regulator
(b)	Regulator	Partial regulator	Conformer
(c)	Partial regulator	Regulator	Conformer
(d)	Regulator	Conformer	Partial regulator

RESPONSE SHEET

Test Code: ....Time taken :...................... Date:..............

	(a)	(b)	(c)	(d)		(a)	(b)	(c)	(d)		(a)	(b)	(c)	(d)		(a)	(b)	(c)	(d)
1.					46.					91.					136.				
2.					47.					92.					137.				
3.					48.					93.					138.				
4.					49.					94.					139.				
5.					50.					95.					140.				
6.					51.					96.					141.				
7.					52.					97.					142.				
8.					53.					98.					143.				
9.					54.					99.					144.				
10.					55.					100.					145.				
11.					56.					101.					146.				
12.					57.					102.					147.				
13.					58.					103.					148.				
14.					59.					104.					149.				
15.					60.					105.					150.				
16.					61.					106.					151.				
17.					62.					107.					152.				
18.					63.					108.					153.				
19.					64.					109.					154.				
20.					65.					110.					155.				
21.					66.					111.					156.				
22.					67.					112.					157.				
23.					68.					113.					158.				
24.					69.					114.					159.				
25.					70.					115.					160.				
26.					71.					116.					161.				
27.					72.					117.					162.				
28.					73.					118.					163.				
29.					74.					119.					164.				
30.					75.					120.					165.				
31.					76.					121.					166.				
32.					77.					122.					167.				
33.					78.					123.					168.				
34.					79.					124.					169.				
35.					80.					125.					170.				
36.					81.					126.					171.				
37.					82.					127.					172.				
38.					83.					128.					173.				
39.					84.					129.					174.				
40.					85.					130.					175.				
41.					86.					131.					176.				
42.					87.					132.					177.				
43.					88.					133.					178.				
44.					89.					134.					179.				
45.					90.					135.					180.				

TEST ASSESSMENT AND ANALYSIS SHEET

How to fill this sheet : After you finish the paper and have checked it first try to attempt again all the questions which you have done wrong or have not been able to attempt. Go to detailed solutions only after you have attempted all the questions of your own.Attempting the questions on your own without seeing the solutions will help you in filling this table

ANALYSIS OF TEST	
Total number of questions	
Questions Attempted	
Correct Questions	
Wrong Questions	
	Net Score

Strike rate $\left(\frac{\text{Correct questions}}{\text{Questions attempt}} \times 100\right)$ = ...

Percentage Marks $\left(\frac{\text{Correct Answers}}{\text{Total number of questions}} \times 100\right)$ = ...

Analysis of wrong questions : Divide the questions which you have done wrong in 3 broad categories.

Reasons for questions attempted wrong	Number of Questions
Knew but calculation mistake (solved on my own after exam)	
Knew but got confused (solved on my own after exam)	
Did not knew the question (couldn't solve even after exam)	

Questions not attempted : Divide the questions not attempted in 3 broad categories

Reasons for unattempted questions	Number of Questions
Knew the ques but time was less (solved on my own after exam)	
Knew the ques but got confused (solved on my own after exam)	
Did not knew the questions (couldn't solve even after exam)	

List down your weak areas you need to improve your score :

1)________________________________

2)________________________________

3)________________________________

NEET

National Eligibility Entrance Test

Test Series **Full Test - 7**

Time : - 3 Hours **Max. Marks:- 720**

INSTRUCTIONS

1. Immediately fill in the particulars on this page of the Test Booklet with Blue/Black Ball Point Pen. Use of pencil is strictly prohibited.
2. Answers are to be given on Response Sheet provided at the end of this paper. You can tear this Response Sheet along the cut marks provided. Fill in the particulars carefully in the Response Sheet provided.
3. The test is of 3 hours duration.
4. The Test Booklet consists of **180** questions. The maximum marks are **720**.
5. There are four section in the question paper.

   The distribution of marks subjectwise in each part is as under for each correct response.

   Section 1 – Physics (180 Marks) – 45 Questions

   Section 2 – Chemistry (180 Marks) – 45 Questions

   Section 3 – Botany (180 Marks) – 45 Questions

   Section 4 – Zoology (180 Marks) – 45 Questions
6. You will be awarded 4 Marks if you have darkened only the bubble corresponding to the correct answer and zero mark if no bubble is darkened. In all other cases, minus one (– 1) mark will be awarded.
7. Use Blue/Black Ball Point Pen only for writing particulars/marking responses on Side-1 and Side-2 of the Response Sheet.

   Use of pencil is strictly prohibited.
8. You are advised to create an actual examination hall simulation.
9. Rough work is to be done on the space provided for this purpose in the Test Booklet only.
10. Do not fold or make any stray marks on the Response Sheet.
11. Once you have evaluated the marks in the test, you should assess it thoroughly.

SECTION 1 - PHYSICS

1. Identify the pair whose dimensions are equal
(a) torque and work (b) stress and energy
(c) force and stress (d) force and work

2. A body is dropped from a tower, another body is dropped from the same tower after one second. What is the separation between the two bodies two second after the release of the first body? ($g = 9.8\ m/s^2$)
(a) 4.9m (b) 24.5m
(c) 9.8m (d) 19.6m

3. The linear velocity of a rotating body is given by
$$\vec{v} = \vec{\omega} \times \vec{r}$$
If $\vec{\omega} = \hat{i} - 2\hat{j} + 2\hat{k}$ and $\vec{r} = 4\hat{j} - 3\hat{k}$, then the magnitude of $\vec{v}$ is
(a) $\sqrt{29}$ units (b) $\sqrt{31}$ units
(c) $\sqrt{37}$ units (d) $\sqrt{41}$ units

4. A body falls on the ground from a height of 10 m and rebounds to a height of 2.5 m. Then the ratio of velocities of the body just before and just after the collision will be
(a) 2 : 1 (b) 1 : 2
(c) 4 : 1 (d) 1 : 3

5. A fan is making 600 revolutions per minute. If after some time it makes 1200 revolutions per minute, then increases in its angular velocity will be
(a) 60π rad/sec (b) 40π rad/sec
(c) 20π rad/sec (d) 10π rad/sec

6. Two bullets are fired horizontally with different velocities from the same height. Which will reach the ground first?
(a) Slower one
(b) Faster one
(c) Both will reach simultaneously
(d) It cannot be predicted

7. Two blocks $m_1 = 5$ gm and $m_2 = 10$ gm are hung vertically over a light frictionless pulley as shown here. What is the acceleration of the masses when they are left free?

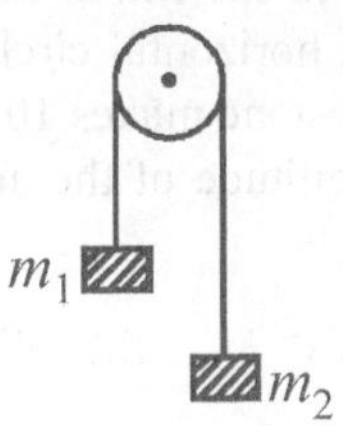

(a) $g/3\ ms^{-2}$ (b) $g/2\ ms^{-2}$
(c) $g\ ms^{-2}$ (d) $g/5\ ms^{-2}$
(where g is acceleration due to gravity)

8. A man weighing 80 kg, stands on a weighing scale in a lift which is moving upwards with a uniform acceleration of $5 m/s^2$. What would be the reading on the scale ? ($g = 10\ m/s^2$)
(a) 1200 N (b) Zero
(c) 400 N (d) 800 N

9. Two bodies moving towards each other collide and move away in opposite direction. There is some rise in temperature of bodies because a part of the kinetic energy is converted into
(a) heat energy (b) nuclear energy
(c) electriccal energy (d) mechanical energy

10. A body is initially at rest. It undergoes one-dimensional motion with constant acceleration. The power delivered to it at time t is proportional to :
(a) $t^{1/2}$ (b) t
(c) $t^{3/2}$ (d) t^2

11. A cord is wound round the circumference of the wheel of radius r. The axis of the wheel is horizontal and the moment of inertia about its centre is I. A weight mg is attached to the cord at the end. The weight falls from rest. After falling through a distance h, the angular velocity of the wheel will be
(a) $\sqrt{2gh}$ (b) $\sqrt{\frac{2mgh}{I+2mr^2}}$
(c) $\sqrt{\frac{2mgh}{I+mr^2}}$ (d) $\sqrt{\frac{2gh}{I+mr}}$

12. A stone tied to one end of string of 1m long is whirled in a horizontal circle with a constant speed. If the stone makes 10 revolutions in 20 sec, the magnitude of the acceleration of the stone is

(a) 988 cm/s^2 (b) 490 cm/s^2

(c) 986 cm/s^2 (d) 996 cm/s^2

13. If the distance between the earth and the sun were half its present value, the number of days in a year would have been

(a) 64.5 (b) 129

(c) 182.5 (d) 730

14. A planet in a distant solar system is 10 times more massive than the earth and its radius is 10 times smaller. Given that the escape velocity from the earth is 11 km s^{-1}, the escape velocity from the surface of the planet would be

(a) 1.1 km s^{-1} (b) 11 km s^{-1}

(c) 110 km s^{-1} (d) 0.11 km s^{-1}

15. The diagram shown below represents the applied forces per unit area with the corresponding change X (per unit length) produced in a thin wire of uniform cross section in the curve shown. The region in which the wire behaves like a liquid is

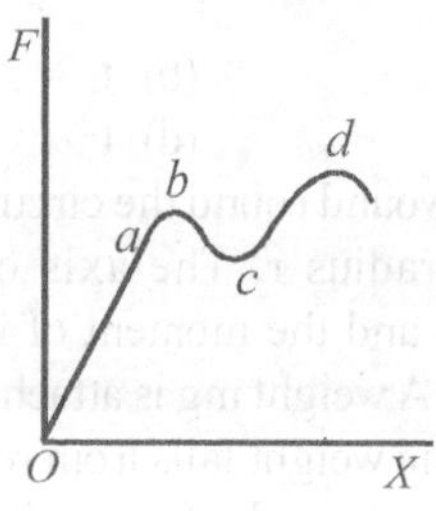

(a) ab (b) bc

(c) cd (d) Oa

16. What is the work dorre in blowing a soap bubble of radius 0.2 m ? (The surface tension of soap solution is 0.06 N/m)

(a) $192\pi \times 10^{-4}$ J (b) $96\pi \times 10^{-4}$ J

(c) $194.2\pi \times 10^{-4}$ J (d) None of these

17. Steam at 100°C is passed into 1.1 kg of water contained in a calorimeter of water equivalent 0.02 kg at 15°C till the temperature of the calorimeter and its contents rises to 80°C. The mass of the steam condensed in kg is

(a) 0.130 (b) 0.065

(c) 0.260 (d) 0.135

18. Thermal capacity of 40 g of aluminium ($s = 0.2$ cal/g K) is

(a) 168 joule /°C (b) 672 joule/°C

(c) 840 joule/°C (d) 33.6 joule/°C

19. The molar specific heat at constant pressure of an ideal gas is (7/2) R. The ratio of specific heat at constant pressure to that at constant volume is

(a) 8/7 (b) 5/7

(c) 9/7 (d) 7/5

20. In thermodynamic processes, which of the following statements is not true?

(a) In an isochoric process pressure remains constant

(b) In an isothermal process the temperature remains constant

(c) In an adiabatic process PV^{γ} = constant

(d) In an adiabatic process the system is insulated from the surroundings

21. The correct statement of the law of equipartition of energy is

(a) the total energy of a gas is equally divided among all the molecules

(b) the gas possess equal energies in all the three directions x, y and z-axis

(c) the total energy of a gas is equally divided between kinetic and potential energies

(d) the total kinetic energy of a gas molecules is equally divided among translational and rotational kinetic energies

22. A simple pendulum of length l has a maximum angular displacement θ. The maximum kinetic energy of the bob is

(a) $mgl(1 - \cos\theta)$ (b) $0.5\,mgl$

(c) mgl (d) $5\,mgl$

23. A tuning fork of frequency 392 Hz, resonates with 50 cm length of a string under tension (T). If length of the string is decreased by 2%, keeping the tension constant, the number of beats heard when the string and the tuning fork made to vibrate simultaneously is
(a) 4 (b) 6
(c) 8 (d) 12

24. Two parallel large thin metal sheets have equal surface charge densities $(\sigma = 26.4\times10^{-12} c/m^2)$ of opposite signs. The electric field between these sheets is
(a) 1.5 N/C (b) 1.5×10^{-10} N/C
(c) 3 N/C (d) 3×10^{-10} N/C

25. An electric dipole placed in a non-uniform electric field experiences
(a) both, a torque and a net force
(b) only a force but no torque
(c) only a torque but no net force
(d) no torque and no net force

26. Equipotential surfaces associated with an electric field which is increasing in magnitude along the x-direction are
(a) planes parallel to yz-plane
(b) planes parallel to xy-plane
(c) planes parallel to xz -plane
(d) coaxial cylinders of increasing radii around the x-axis

27. Five capacitor, each of the capacitance C, are connected, as shown. What is the equivalent capacitance between the points Aand B?

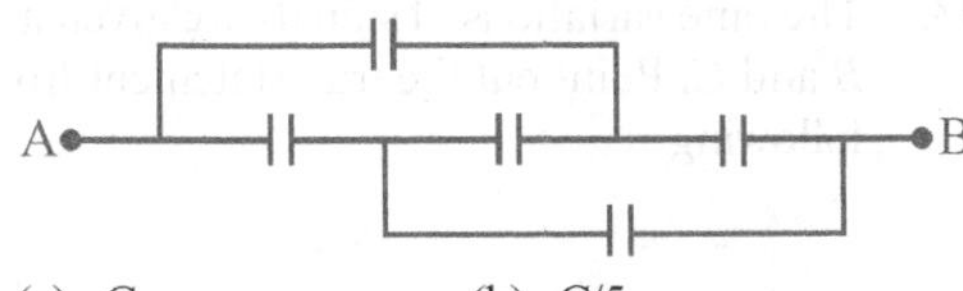

(a) C (b) C/5
(c) 2C (d) 5C

28. A piece of copper and another of germanium are cooled from room temperature to 80° K. The resistance of
(a) each of them increases
(b) each of them decreases
(c) copper increases and germanium decreases
(d) copper decreases and germanium increases

29. Six identical resistors are connected as shown in the figure. The equivalent resistance will be

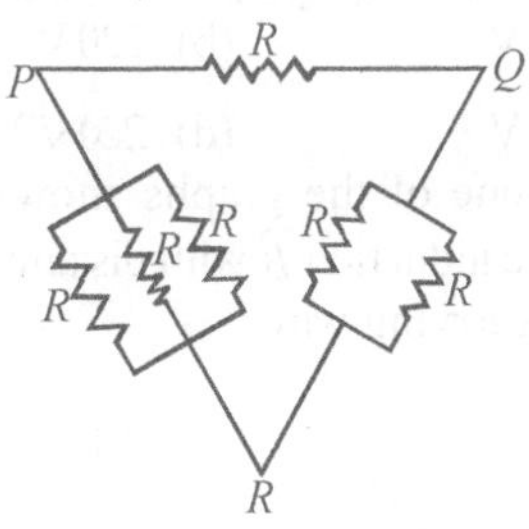

(a) maximum between P and R
(b) maximum between Q and R
(c) maximum between P and Q
(d) all are equal

30. If an ammeter is to be used in place of a voltmeter, then we must connect with the ammeter a
(a) low resistance in parallel
(b) high resistance in parallel
(c) high resistance in series
(d) low resistance in series.

31. The current flows from A to B as shown in figure, then the direction of the induced current in the loop will be

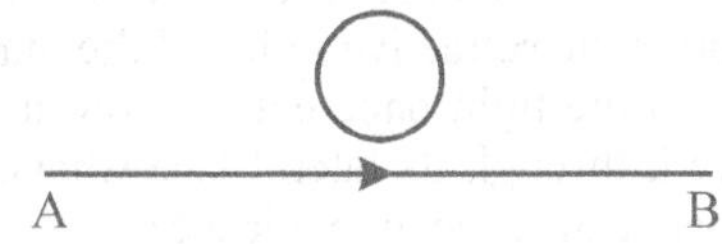

(a) straight line (b) anti-clockwise
(c) clockwise (d) None of these

32. The earth's magnetic field at a given point is 0.5×10^{-5} Wb m^{-2}. This field is to be annulled by magnetic induction at the centre of a circular conducting loop of radius 5.0 cm . The current required to be flown in the loop is nearly
(a) 0.2 A (b) 0.4 A
(c) 4A (d) 40 A

33. The turn ratio in a step up transformer is 1 : 2 if a Lechlanche cell of 1.5 V is connected across the input, then the voltage across the output will be

(a) 0.1 V (b) 1.5 V
(c) 0.75 V (d) zero

34. If an AC main supply is given to be 220 V. The average emf during a positive half cycle will be

(a) 198 V (b) 220 V
(c) 240 V (d) $220\sqrt{2}$ V

35. Which one of the graphs shows variation of magnetic induction $\vec{B}$ with distance r from a long current carrying wire?

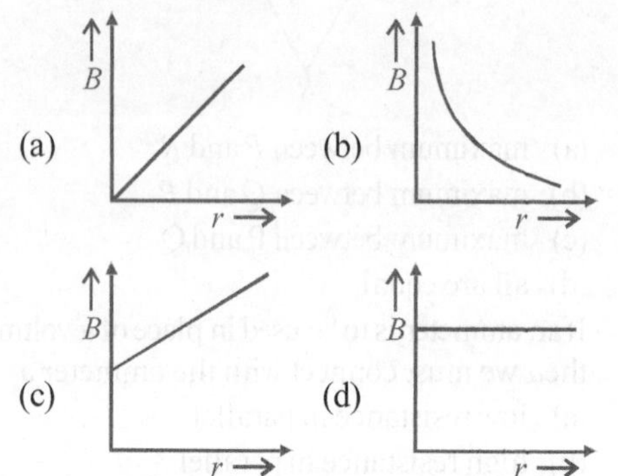

36. Which of the following waves have the maximum wavelength ?

(a) Infrared rays (b) UV rays
(c) Radiowaves (d) X-rays

37. Light enters at an angle of incidence in a transparent rod of refractive index n. For what value of the refractive index of the material of the rod the light once entered into it will not leave it through its lateral face whatsoever be the value of angle of incidence?

(a) $n > \sqrt{2}$ (b) $n = 1$
(c) $n = 1.1$ (d) $n = 1.3$

38. The condition for an achromatic doublet, is (f_1 and f_2 are the focal lengths of individual lenses)

(a) $f_1 - f_2 = \frac{\omega_1\omega_2}{2} f_2$ (b) $\omega_1 f_1 + \omega_2 f_2 = 0$

(c) $\frac{\omega_1}{f_1} + \frac{\omega_2}{f_2} = 0$ (d) $\frac{f_1}{f_2} + \frac{\omega_1}{\omega_1} = 0$

39. If yellow light emitted by sodium lamp in Young's double slit experiment is replaced by a monochromatic blue light of the same intensity

(a) fringe width will decrease
(b) fringe width will increase
(c) fringe width will remain unchanged
(d) fringes will become less intense

40. The binding energy of deuteron ($_1^2$H) is 1.15 MeV per nucleon and an alpha particle ($_2^4$He) has a binding energy of 7.1 MeV per nucleon. Then in the reaction

$$_1^2\text{H} + _1^2\text{H} \longrightarrow _2^2\text{He} + Q$$

the energy released Q is

(a) 5.95 MeV (b) 26.1 MeV
(c) 23.8 MeV (d) 289.4 MeV

41. When light of wavelength 300 nm (nanometre) falls on a photoelectric emitter, photoelectrons are liberated. For another emitter, however, light of 600 nm wavelength is sufficient for creating photoemission. What is the ratio of the work functions of the two emitters?

(a) 1 : 2 (b) 2 : 1
(c) 4 : 1 (d) 1 : 4

42. The transfer ratio β of a transistor is 50. The input resistance of the transistor when used in the common emitter configuration is 1kΩ. The peak value of the collector A.C. current for an A.C. input voltage of 0.01 V, is

(a) 500 μA (b) 0.25 μA
(c) 0.01 μA (d) 100 μA

43. The time variations of signals are given as in A, B and C. Point out the true statement from the following

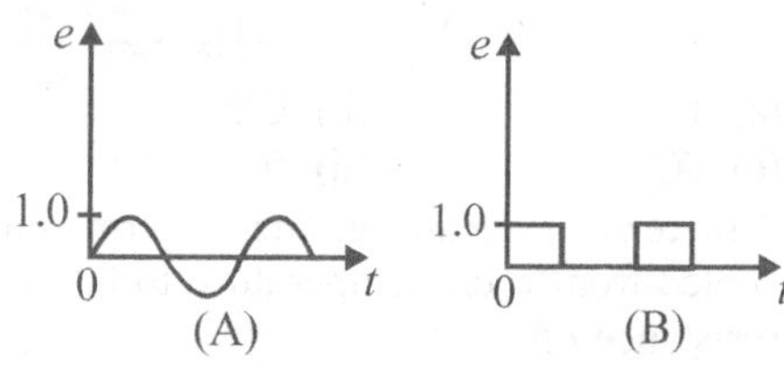

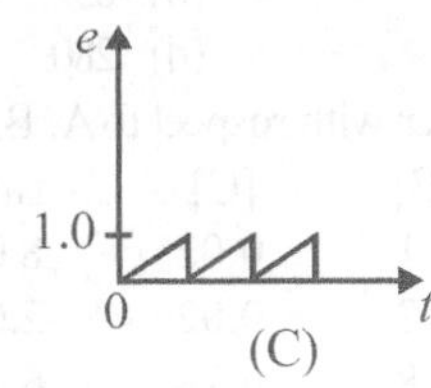

(C)

(a) A, B and C are analogue signals
(b) A and B are analogue, but C is digital signal
(c) A and C are digital, but B is analogue signal
(d) A and C are analogue, but B is digital signal

44. For a common emitter circuit if $\frac{I_C}{I_E} = 0.98$ then current gain for common emitter circuit will be
(a) 49 (b) 98
(c) 4.9 (d) 25.5

45. The maximum and minimum amplitude of an AM wave are 90 mV and 30 mV respectively. The depth of modulation is
(a) 0.6 (b) 0.5
(c) 0.4 (d) 0.3

SECTION 2 - CHEMISTRY

46. The angular momentum of the electron in first excited energy state of hydrogen atom is
(a) $\frac{h}{\pi}$ (b) $\frac{h}{2\pi}$
(c) $\sqrt{2(2+1)}\frac{h}{2\pi}$ (d) None of these

47. When NaCl is dopped with 1.0×10^{-3} mole of $SrCl_2$, the number of cation vacancy is
(a) 6.023×10^{18} (b) 6.023×10^{20}
(c) $2 \times 6.023 \times 10^{20}$ (d) 3.011×10^{20}

48. A 0.5 M NaOH solution offers a resistance of 31.6 ohm in a conductivity cell at room temperature. What shall be the approximate molar conductance of this NaOH solution if cell constant of the cell is 0.367 cm^{-1}.
(a) 23.4 S cm^2 $mole^{-1}$ (b) 23.2 S cm^2 $mole^{-1}$
(c) 46.45 S cm^2 $mole^{-1}$ (d) 54.64 S cm^2 $mole^{-1}$

49. Ammonium dichromate on heating gives
(a) chromic acid & ammonia
(b) chromium sesquioxide & nitrogen
(c) chromium sesquioxide & ammonia
(d) chromic acid and N_2

50. Predict the relative acidic strength among the following
(a) H_2O, H_2S, H_2Se, H_2Te
(b) $H_2O < H_2S < H_2Se < H_2Te$
(c) $H_2Te < H_2Se < H_2S < H_2O$
(d) $H_2O < H_2Se < H_2S < H_2Te$

51. The catalyst used in the preparation of an alkyl chloride by the action of dry HCl on an alcohol is
(a) anhydrous $AlCl_3$ (b) $FeCl_3$
(c) anhydrous $ZnCl_2$ (d) Cu

52. Among the following the aromatic compound is

(a) 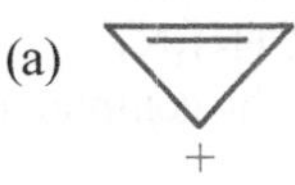 (b) 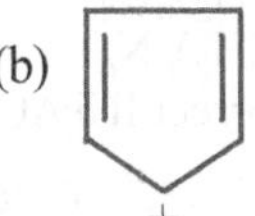

(c)  (d)

53. The four quantum numbers that could identify the third $3p$ electron in sulphur are
(a) $n = 3, l = 0, m = +1, s = +\frac{1}{2}$
(b) $n = 2, l = 2, m = -1, s = +\frac{1}{2}$
(c) $n = 3, l = 2, m = +1, s = -\frac{1}{2}$
(d) $n = 3, l = 1, m = -1, s = +\frac{1}{2}$

54. $Sn^{4+} + 2e^- \longrightarrow Sn^{2+}$ $E° = 0.13$ V
$Br_2 + 2e^- \longrightarrow 2Br^-$ $E° = 1.08$ V
Calculate K_{eq} for the cell at 20°C formed by two electrodes
(a) 10^{41} (b) 10^{32}
(c) 10^{-32} (d) 10^{-42}

55. Calculate the pH of a solution obtained by mixing 2 ml of HCl of pH 2 and 3 ml of solution of KOH of pH = 12
(a) 10.30 (b) 3.70
(c) 11.30 (d) None of these

56. Which of the following represents a correct sequence of reducing power of the following elements?
(a) Li > Cs > Rb (b) Rb > Cs > Li
(c) Cs > Li > Rb (d) Li > Rb > Cs

57. Paramagnetism of Cr (Z = 24), Mn^{2+} (Z = 25) and Fe^{3+} (Z = 26) are x, y and z respectively. They are in the order
(a) $x = y = z$ (b) $x > y > z$
(c) $x = y > z$ (d) $x > y = z$

58. Formaldehyde reacts with ammonia to give urotropine is
(a) $(CH_2)_6N_4$ (b) $(CH_2)_4N_3$
(c) $(CH_2)_6N_6$ (d) $(CH_2)_3N_3$

59. The correct IUPAC name of the compound

CHO, NO_2, OCH_3 (benzene ring with CHO at C-1, NO_2 at C-2, OCH_3 at C-4) is

(a) 4-methoxy-2-nitrobenzaldehyde
(b) 4-formyl-3 nitroanisole
(c) 4- methoxy-6-nitrobenzaldehyde
(d) 2-formyl-5-methoxynitrobenzene

60. The favourable condition for a process to be spontaneous is :
(a) $T\Delta S > \Delta H, \quad \Delta H = +\text{ive}, \ \Delta S = -\text{ive}$
(b) $T\Delta S > \Delta H, \quad \Delta H = +\text{ive}, \ \Delta S = +\text{ive}$
(c) $T\Delta S = \Delta H, \quad \Delta H = -\text{ive}, \ \Delta S = -\text{ive}$
(d) $T\Delta S = \Delta H, \quad \Delta H = +\text{ive}, \ \Delta S = +\text{ive}$

61. Vapour pressure (in torr) of an ideal solution of two liquids A and B is given by : $P = 52X_A + 114$ where X_A is the mole fraction of A in the mixture. The vapour pressure (in torr) of equimolar mixture of the two liquids will be :
(a) 166 (b) 83
(c) 140 (d) 280

62. What is order with respect to A, B, C, respectively

[A]	[B]	[C]	rate (M/sec.)
0.2	0.1	0.02	8.08×10^{-3}
0.1	0.2	0.02	2.01×10^{-3}
0.1	1.8	0.18	6.03×10^{-3}
0.2	0.1	0.08	6.464×10^{-2}

(a) –1, 1, 3/2 (b) –1, 1, 1/2
(c) 1, 3/2, –1 (d) 1, –1, 3/2

63. Fluorine does not show highest oxidation state opposite to other halogens, because
(a) it is most electronegative
(b) it has no *d*-orbital
(c) its atomic radius is very small
(d) F^- ion is stable and isoelectronic with neon

64. Glucose contains in addition to aldehyde group.
(a) one secondary –OH and four primary –OH groups
(b) one primary –OH and four secondary –OH groups
(c) two primary –OH and three secondary –OH groups
(d) three primary –OH and two secondary –OH groups

65. Which of the following product is obtained by treating 1-butyne with $HgSO_4$ and H_2SO_4?
(a) $CH_3CH_2COCH_3$
(b) $CH_3CH_2CH_2CHO$
(c) $CH_3CH_2CH_2COOH$
(d) $CH_3CH_2CH=CH_2$

66. Which of the following compound is expected to be optically active?
(a) $(CH_3)_2CHCHO$
(b) $CH_3CH_2CH_2CHO$
(c) $CH_3CH_2CHBrCHO$
(d) $CH_3CH_2CBr_2CHO$

67. The product obtained on reaction of C_2H_5Cl with hydrogen over palladium carbon is :
(a) C_3H_8 (b) C_4H_{10}
(c) C_2H_6 (d) C_2H_4

68. Which of the following can be predicted from electronegativity values of elements ?
(a) Dipole moment of a molecule
(b) Valency of elements
(c) Polarity of bonds
(d) Position in electrochemical series

69. The reaction
$3ClO^-_{(aq)} \longrightarrow ClO^-_{3(aq)} + 2Cl^-_{(aq)}$ is an example of
(a) oxidation reaction
(b) reduction reaction
(c) disproportionation reaction
(d) decomposition reaction

70. The concentration of a reactant X decreases from 0.1 M to 0.005 M in 40 min. If the reaction follows first order kinetics, the rate of the reaction when the concentration of X is 0.01 M will be
(a) 1.73×10^{-4} M min^{-1}
(b) 3.47×10^{-4} M min^{-1}
(c) 3.47×10^{-5} M min^{-1}
(d) 7.5×10^{-4} M min^{-1}

71. Mark the false statement?
(a) A salt bridge is used to eliminate liquid junction potential
(b) The Gibbs free energy change, ΔG is related with electromotive force E as $\Delta G = -nFE$
(c) Nernst equation for single electrode potential is $E = E^o - \frac{RT}{nF} \log a_{M^{n+}}$
(d) The efficiency of a hydrogen-oxygen fuel cell is 23%

72. The paramagnetism of transition element compounds is due to
(a) paired eletrons spining in opposite directions
(b) unpaired eletrons in *d* and *f*-orbitals
(c) shared valance electrons
(d) unpaired electrons in *s* or *p*-orbitals.

73. Aniline, chloroform and alcoholic KOH react to produce a bad smelling substance which is
(a) phenyl isocyanide
(b) phenyl cyanide
(c) chlorobenzene
(d) benzyl alcohol.

74. The species with a radius less than that of Ne is
(a) Mg^{2+} (b) F^-
(c) O^{2-} (d) K^+

75. Vapour density of the equilibrium mixture of the reaction $SO_2Cl_{2(g)} \rightleftharpoons SO_{2(g)} + Cl_{2(g)}$ is 50.0. Percent dissociation of SO_2Cl_2 is :
(a) 33.33 (b) 35.0
(c) 30.0 (d) 66.67

76. The chemical reaction,
$2AgCl_{(s)} + H_{2(g)} \longrightarrow 2HCl_{(aq)} + 2Ag(s)$
taking place in a galvanic cell is represented by the notation
(a) $Pt_{(s)} | H_{2(g)}, 1\text{ bar} | 1M\,KCl_{(aq)} | AgCl_{(s)} | Ag_{(s)}$
(b) $Pt_{(s)} | H_{2(g)}, 1\text{ bar} | 1M\,HCl_{(aq)} | 1M\,Ag^+_{(aq)} | Ag_{(s)}$
(c) $Pt_{(s)} | H_{2(g)}, 1\text{ bar} | 1M\,HCl_{(aq)} | AgCl_{(s)} | Ag_{(s)}$
(d) $Pt_{(s)} | H_{2(g)}, 1\text{ bar} | 1M\,HCl_{(aq)} | Ag_{(s)} | AgCl_{(s)}$

77. H_3PO_3 is
(a) neutral (b) basic
(c) a tribasic acid (d) a dibasic acid

78. Of the following which is diamagnetic in nature?
(a) $[CoF_6]^{3+}$ (b) $[NiCl_4]^{2-}$
(c) $[CuCl_4]^{2-}$ (d) $[Ni(CN)_4]^{2-}$

79. Which of the following products is formed when benzaldehyde is treated with CH_3MgBr and the addition product so obtained is subjected to acid hydrolysis ?
(a) A secondary alcohol
(b) A primary alcohol
(c) Phenol
(d) tert-Butyl alcohol

80. Mole fraction of methanol in its aqueous solution is 0.5. The concentration of solution in terms of percent by mass of methanol is
(a) 36 (b) 50
(c) 64 (d) 72

81. The unit cell of an ionic compound is a cube in which cations (*A*) occupy each of the corners and anions (*B*) are at the centres of each face . The simplest formula of the ionic compound is
(a) AB_2 (b) A_3B
(c) AB_3 (d) A_4B_3

82. Select the process that represents smelting.

(a) $2Al_2O_3 + 6H_2O \xrightarrow{\Delta} 4Al(OH)_3$

(b) $ZnCO_3 \xrightarrow{\Delta} ZnO + CO_2$

(c) $Fe_2O_3 + 3C \xrightarrow{\Delta} 2Fe + 3CO$

(d) $2Pb + O_2 \xrightarrow{\Delta} 2PbO + 2SO_2$

83. Electrolytic reduction of alumina to aluminium by Hall- Heroult process is carried out

(a) in the presence of NaCl

(b) in the presence of fluorite

(c) in the presence of cryolite which forms a melt with lower melting temperature

(d) in the presence of cryolite which forms a melt with higher melting temperature

84. Complexes $[Co(NH_3)_5SO_4]$ Br and $[Co(NH_3)_5Br]SO_4$ can be distinguished by

(a) conductance measurement

(b) using $BaCl_2$

(c) using $AgNO_3$

(d) both b and c

85. Nylon is a :

(a) polysaccharide (b) polyester

(c) polyamide (d) all of the above

86. 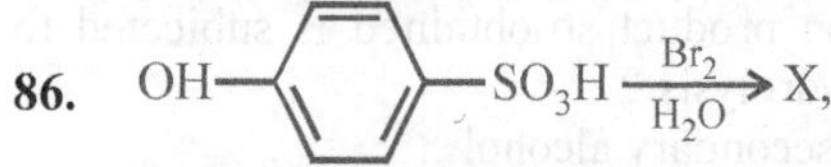

X is identified as

(a) 2, 4, 6-tribromophenol

(b) 2-bromo-4-hydroxylbenzene sulphonic acid

(c) 3,5-dibromo-4-hydroxybenzene sulphonic acid

(d) 2-bromophenol

87. The non-polar molecule is :

(a) NF_3 (b) SO_3

(c) $CHCl_3$ (d) ClO_2

88. The hybridization of P in PO_4^{3-} is the same as of

(a) S in SO_3 (b) N in NO^-_3

(c) S in SO_4^{--} (d) I in ICl_4^-

89. Solution of potash alum is acidic in nature. This is due to hydrolysis of

(a) SO_4^{2-} (b) K^+

(c) $Al_2(SO_4)_3$ (d) Al^{3+}

90. When conc. HNO_3 acts on our skin, the skin becomes yellow, because

(a) HNO_3 acts as an oxidising agent

(b) HNO_3 acts as a dehydrating agent

(c) Nitro-cellulose is formed

(d) The proteins are converted into xanthoproteins

SECTION 3 - BOTANY

91. Basic unit of taxonomy is

(a) Genus (b) Species

(c) Family (d) Order

92. Which one of the following is a living fossil?

(a) *Pinus* (b) *Opuntia*

(c) *Ginkgo* (d) *Thuja*

93. Alginates (alginin), used as highly efficient gauze in internal operations are obtained from cell walls of

(a) Cyanophyceae (b) Phaeophyceae

(c) Rhodophyceae (d) All of these

94. Bryophytes resemble algae in the following aspects

(a) Filamentous body, presence of vascular tissues and autotrophic nutrition

(b) Differentiation of plant body into root, stem and leaves and autotrophic nutrition

(c) Thallus like plant body, presence of root and autotrophic nutrition

(d) Thallus like plant body, lack of vascular tissues and autotrophic nutrition

95. Clove is

(a) flower bud (b) axillary bud

(c) thalamus (d) ovule

96. When gynoecium is present in the top most position of thalamus, the flower is known as

(a) inferior (b) epigynous

(c) perigynous (d) hypogynous

97. Which is not a stem modification ?
(a) Rhizome of ginger
(b) Corm of *Colocasia*
(c) Pitcher of *Nepenthes*
(d) Tuber of potato

98. In land plants, the guard cells differ from other epidermal cells in having
(a) cytoskeleton
(b) mitochondria
(c) endoplasmic reticulum
(d) chloroplasts

99. Which option is correctly matched with the diagrams?

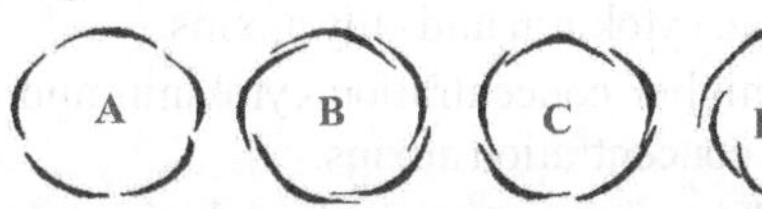

(a) A-Valvate B-Twisted, C-Imbricate, D-Vexillary
(b) A-Vexillary, B-Valvate, C-Twisted, D-Imbricate
(c) A-Imbricate, B-Vexillary, C-Valvate, D-Twisted
(d) A-Twisted, B-Imbricate, C-Vexillary, D-Valvate

100. Most abundant lipid in the cell membrane is
(a) cholesterol (b) phospholipids
(b) glycolipids (d) cerebrosides

101. Histones are
(a) glycoproteins (b) mucoproteins
(c) basic proteins (d) acid proteins

102. The given graph shows the effect of substrate concentration on the rate of reaction of the enzyme green -gram -phosphatase. What does the graph indicate ?

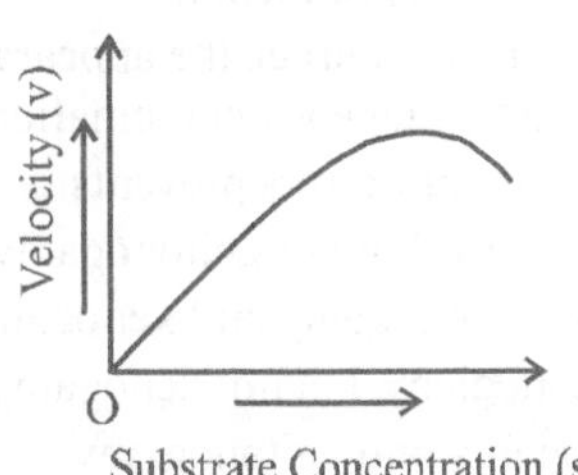

(a) The rate of enzyme reaction is directly proportional to the substrate concentration
(b) Presence of an enzyme inhibitor in the reaction mixture
(c) Formation of an enzyme-substrate complex
(d) At higher substrate-concentration the pH increases.

103. The Km value of the enzyme is the value of the substrate concentration at which the reaction reaches to
(a) Zero (b) 2 Vmax
(c) ½ Vmax (d) ¼ Vmax

104. A plant cell placed in pure water will
(a) expand until the osmotic potential or solute potential reaches that of water.
(b) becomes more turgid until the pressure potential of cell reaches its osmotic potential.
(c) become more turgid until the osmotic potential reaches that of pure water.
(d) becomes less turgid until the osmotic potential reaches that of pure water.

105. During cell division, sometimes there will be failure of separation of sister chromatids. This event is called
(a) interference (b) complementation
(c) non-disjunction (d) coincidence

106. In which stage of the cell cycle histone proteins synthesized in a eukaryotic cell?
(a) During G_2 stage of prophase
(b) During S-phase
(c) During entire prophase
(d) During telophase

107. Which of the following is not a correct pairing of a macronutrient and the major functions it performs in the life of a plant ?
(a) Potassium-enzyme activation, water balance, ion balance.
(b) Calcium-activity of membranes and cytoskeleton, second messenger.
(c) Sulphur-in proteins and coenzymes.
(d) Iron-in active sites of many redox enzymes and electron carriers.

108. Which of the following statement is false?
(a) H_2S, not H_2O, is involved in photosynthesis of purple sulphur bacteria.
(b) Light and dark reactions are stopped in the absence of light.
(c) Calvin cycle occurs in the grana of chloroplast.
(d) ATP is produced during light reaction *via* chemiosmosis.

109. The casparian strip prevents water and minerals from entering the stele through the
(a) plasmodesmata (b) apoplast
(c) symplast (d) xylem vessel

110. 'Whip-tail' disease in cauliflower is due to deficiency of
(a) manganese (b) magnesium
(c) molybdenum (d) nitrogen

111. R.Q. is highest when respiratory substance is
(a) fat (b) malic acid
(c) glucose (d) protein

112. Electron transport chain is inhibited by
(a) rotenone and amytal
(b) antimycin-A
(c) cyanide (CN^-), azide (N^{3-}) and carbon monoxide (CO)
(d) All of the above

113. Given below is the pathway of light reaction. Identify the given blanks indicated by A, B, C, D and E.

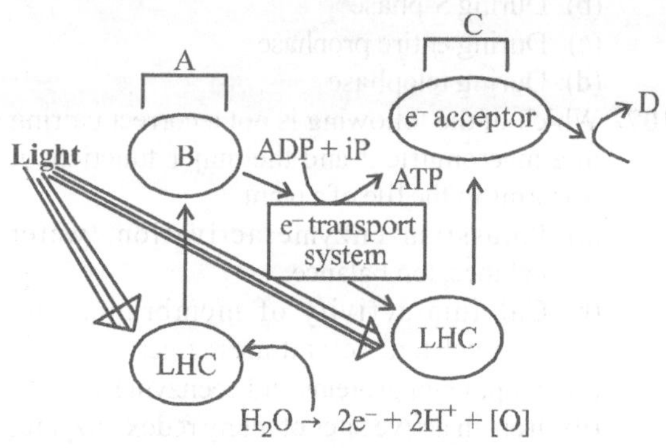

	A	B	C	D
(a)	P 700	H^+ acceptor	P680	$NADP^+$
(b)	Photosystem I	e^- acceptor	Photosystem II	$NADPH_2$ + ATP
(c)	Photosystem II	H^+ acceptor	P700	NADPH
(d)	Photosystem II	e^- acceptor	Photosystem I	NADPH + H^+

114. In tissue culture roots can be induced by
(a) no cytokinin and only auxins.
(b) higher concentration cytokinin and lower concentration auxins.
(c) lower concentration of cytokinin and higher concentration of auxins.
(d) only cytokinin and no auxins.

115. Phytohormones are
(a) chemicals regulating flowering.
(b) chemicals regulating secondary growth.
(c) hormones regulating growth from seed to adulthood.
(d) regulators synthesized by plants and influencing physiological processes.

116. Which one of the following is a natural growth inhibitor?
(a) NAA (b) ABA
(c) IAA (d) GA

117. The disadvantage of parthenogenesis is –
(a) Establishment of polyploid generation
(b) Elimination of variety in population
(c) Means of reproduction
(d) Does not encourage the appearance of new and advantageous combinations of genes

118. Unisexuality of flowers prevents
(a) autogamy, but not geitonogamy
(b) both geitonogamy and xenogamy
(c) geitonogamy, but not xenogamy
(d) autogamy and geitonogamy

119. One of the important consequences of geographical isolation is
(a) preventing speciation
(b) speciation through reproductive isolation
(c) random creation of new species
(d) no change in the isolated fauna.

120. Genetic drift operates only in
(a) smaller populations
(b) larger populations
(c) mendelian populations
(d) island populations

121. Three crops that contribute maximum to global food grain production are
(a) Wheat, rice and maize
(b) Wheat, rice and barley
(c) Wheat, maize and sorghum
(d) Rice, maize and sorghum

122. Farmers have reported over 50% higher yields of rice by using the biofertilizer
(a) *Azolla pinnata*
(b) *Cyanobacteria*
(c) *Legume-Rhizobium* symbiosis
(d) Mycorrhiza

123. Himgiri developed by hybridisation and selection for disease resistance against rust pathogens is a variety of
(a) chilli (b) maize
(c) sugarcane (d) wheat

124. Which one of the following is not used for construction of ecological pyramids ?
(a) Number of individuals
(b) Rate of energy flow
(c) Fresh weight
(d) Dry weight

125. In agarose gel electrophoresis
(a) DNA migrates towards the negative electrode
(b) supercoiled plamids migrate slower than their nicked counterparts
(c) larger molecules migrate faster than smaller molecules
(d) ethidium bromide can be used to visualize the DNA

126. The genetically-modified (GM) brinjal in India has been developed for:
(a) insect-resistance
(b) enhancing shelf life
(c) enhancing mineral content
(d) drought-resistance

127. Which one of the following is not observed in biodiversity hotspots ?
(a) Endemism
(b) Accelerated species loss
(c) Lesser inter-specific competition
(d) Species richness

128. Which one of the following expanded forms of the following acronyms is correct?
(a) IPCC = International Panel for Climate Change
(b) UNEP = United Nations Environmental Policy
(c) EPA = Environmental Pollution Agency
(d) IUCN = International Union for Conservation of Nature and Natural Resources

129. Photochemical smog pollution does not contain
(a) Nitrogen dioxide
(b) Carbon dioxide
(c) PAN (peroxy acetyl nitrate)
(d) Ozone

130. When huge amount of sewage is dumped into a river, its B.O.D. will
(a) increase (b) decrease
(c) sharply decrease (d) remain unchanged

131. Which one of the following shows maximum genetic diversity in India ?
(a) Groundnut (b) Rice
(c) Maize (d) Mango

132. The greatest biomass of autotrophs in the oceans is that of
(a) sea grasses and slime moulds
(b) free floating microalgae, cyanobacteria and nanoplankton
(c) benthic brown algae, coastal red algae and daphnids
(d) benthic diatoms and marine viruses

133. Which of the following is based upon the principle of antigen-antibody interaction?
(a) PCR (b) ELISA
(c) R DNA technology (d) RNA

134. The figure below shows three steps (A, B, C) of Polymerase Chain Reaction (PCR). Select the option giving correct identification together with what it represents?

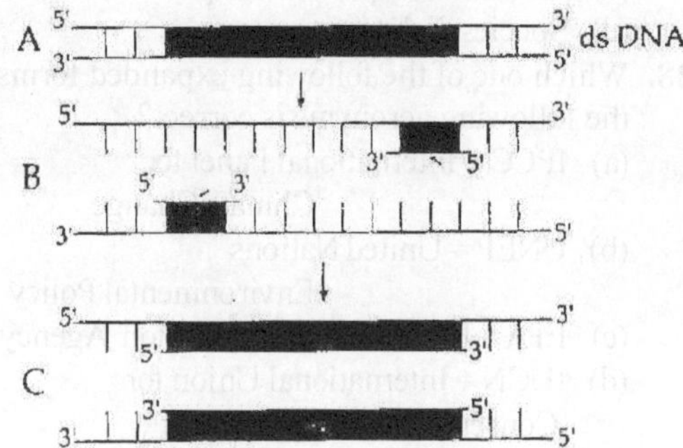

(a) B - Denaturation at a temperature of about 98°C separating the two DNA strands.
(b) A - Denaturation at a temperature of about 50°C.
(c) C - Extension in the presence of heat stable DNA polymerase.
(d) A - Annealing with two sets of primers.

135. Which of the following statement(s) is/are true?
(i) Uneven thickening of cell wall is characteristic of sclerenchyma.
(ii) Periblem forms cortex of the stem and the root.
(iii) Tracheids are the chief water transporting elements in gymnosperms.
(iv) Companion cell is devoid of nucleus at maturity.
(v) The Commercial cork is obtained from *Quercus suber.*
(a) (i) and (iv) only (b) (ii) and (v) only
(c) (iii) and (iv) only (d) (ii), (iii) and (v) only

SECTION 4 - ZOOLOGY

136. In prokaryotes, the genetic material is
(a) linear DNA with histones
(b) circular DNA with histones
(c) linear DNA without histones
(d) circular DNA without histones

137. In *Amoeba* and *Paramecium* osmoregulation occurs through
(a) pseudopodia (b) nucleus
(c) contractile vacuole (d) general surface

138. Animals with metameric segmentation, bilateral symmetry and closed circulatory system belong to phylum
(a) Annelida (b) Echinodermata
(c) Arthropoda (d) Mollusca

139. Which one of the following characters is **not** typical of the class Mammalia?
(a) Thecodont dentition
(b) Alveolar lungs
(c) Ten pairs of cranial nerves
(d) Seven cervical vertebrae

140. Which of the following type of cell junction is not found in animal tissues ?
(a) Desmosome (b) Tight junction
(c) Gap junction (d) Plasmodesmata

141. Which one of the following cellular components of the blood is responsible for the production of antibodies?
(a) Thrombocyte (b) Lymphocyte
(c) Monocyte (d) Erythrocyte

142. Sprain is caused due to the excessive pulling of
(a) Muscles (b) Tendons
(c) Ligaments (d) All of the above

143. Which one of the following pairs of the kind of cells and their secretion is correctly matched?
(a) Oxyntic cells—a secretion with pH between 2.0 and 3.0
(b) Alpha cells of islets of Langerhans—secretion that decreases blood sugar level
(c) Kupffer cell—a digestive enzyme that hydrolyses nucleic acids
(d) Sebaceous glands—a secretion that evaporates for cooling

144. In cockroach, the corpora allata is
(a) A sense organ
(b) A tactile organ
(c) An endocrine organ
(d) A digestive gland

145. Which group contains biocatalysts?
(a) Myosin, oxytocin, adrenalin
(b) Peptidase, amylase, rennin
(c) Glucose, amino acids, fatty acids
(d) Rhodopsin, pepsin, steapsin

146. Both the crown and root of a tooth is covered by a layer of bony hard substance. It is called
(a) enamel (b) dentine
(c) bony socket (d) cementum

147. The respiratory centre, which regulates respiration, is located in
(a) Cerebral peduncle
(b) Vagus nerve
(c) Pons
(d) Medulla oblongata

148. Even when there is no air in it, human trachea does not collapse due to presence of
(a) Bony rings (b) Turgid pressure
(c) Chitinous rings (d) Cartilaginous rings

149. The structure which does not contribute to the breathing movements in mammals is
(a) Larynx (b) Ribs
(c) Diaphragm (d) Intercostal muscles

150. Find out the incorrect answer from the following?
(a) Veins are typically larger in diameter than arteries
(b) Because of their small size, capillaries contain blood that is moving more quickly than in other parts of the circulatory system
(c) The walls of arteries are elastic, enabling them to stretch and shrink during changes in blood pressure
(d) Veins contain more blood than any other part of the circulatory system

151. Given below are four statements (i-iv) regarding human blood circulatory system
(i) Arteries are thick-walled and have narrow lumen as compared to veins
(ii) Angina is acute chest pain when the blood circulation to the brain is reduced
(iii) Persons with blood group AB can donate blood to any person with any blood group under ABO system
(iv) Calcium ions play a very important role in blood clotting
Which two of the above statements are correct?
(a) (i) and (iv) (b) (i) and (iv)
(c) (ii) and (iii) (d) (iii) and (iv)

152. In Prawn, excretion is carried out by
(a) Nephrons (b) Malpighian tubules
(c) Flame cells (d) Green glands

153. Columns of Bertini in the kidneys of mammals are formed as extensions of
(a) Cortex into medulla
(b) Cortex into pelvis
(c) Medulla into pelvis
(d) Pelvis into ureter

154. Consider the following four statements (i - iv) about certain desert animals such as kangaroo rat
(i) They have dark colour and high rate of reproduction and excrete solid urine
(ii) They do not drink water, breathe at a slow rate to conserve water and have their body covered with thick hairs
(iii) They feed on dry seeds and do not require drinking water
(iv) They excrete very concentrated urine and do not use water to regulate body temperature.
Which two of the above statements for such animals are true?
(a) (i) and (ii) (b) (iii) and (iv)
(c) (ii) and (iii) (d) (iii) and (i)

155. Which of the following is an autoimmune disorder ?
(a) Myasthenia gravis (b) Osteoporosis
(c) Muscular dystrophy (d) Gout

156. Thin filaments of skeletal muscle fibres contain
(a) Actin protein only
(b) Actin and myosin proteins
(c) Actin, Troponin and Tropomyosin proteins
(d) Actin, Troponin, Myosin and Tropomyosin proteins

157. The joint in our neck which allows us to rotate our head left to right is
(a) pivot joint (b) hinge joint
(c) saddle joint (d) ellipsoid joint

158. Among which one of the following groups of chemicals, all are neurotransmitters?
(a) Glycine, dopamine, melatonin
(b) Somatostatin, serotonin, acetylcholine
(c) Noradrenaline, somatostatin, threonine
(d) Acetylcholine, noradrenaline, dopamine

159. The enzyme required for the conduction of nerve impulse across synapse is:
(a) peroxidase
(b) choline acetylase
(c) ascorbic acid oxidase
(d) succinic dehydrogenase

160. The purplish red pigment rhodopsin contained in the rods type of photoreceptor cells of the human eye, is a derivative of
(a) Vitamin A (b) Vitamin B_1
(c) Vitamin C (d) Vitamin D

161. Which hormone interacts with membrane bound receptor and does not normally enter the target cell ?
(a) Follicle stimulating hormone
(b) Estrogen
(c) Thyroxin
(d) Cortisol

162. The 24 hour (diurnal) rhythm of our body such as the sleep-wake cycle is regulated by the hormone
(a) Adrenaline (b) Melatonin
(c) Calcitonin (d) Prolactin

163. Steroid hormones transmit their information by
(a) stimulating the receptors present on cell membrance
(b) entering into the cell and modifying cellular contents
(c) entering into the cell and modifying nuclear organization
(d) the help of an intracellular second messenger

164. During regeneration, modification of an organ to other organ is known as
(a) Morphogenesis (b) Epimorphosis
(c) Morphallaxis (d) Accretionary growth

165. In oogamy, fertilization involves
(a) a small non-motile female gamete and a large motile male gamete
(b) a large non-motile female gamete and a small motile male gamete
(c) a large non-motile female gamete and a small non motile male gamete
(d) a large motile female gamete and a small nonmotile male gamete

166. The part of fallopian tube closest to the ovary is
(a) isthmus (b) infundibulum
(c) cervix (d) ampulla

167. Correct sequence of cell stages in spermatogenesis is:
(a) spermatocytes, spermatids, spermatogonia, spermatozoa
(b) spermatogonia, spermatocytes, spermatids, spermatozoa
(c) spermatocytes, spermatogonia, spermatids, spermatozoa
(d) spermatogonia, spermatids, spermatocytes, spermatozoa

168. Progestasert and LNG-20 are
(a) Implants
(b) Copper releasing IUDs
(c) Non-medicated IUDs
(d) Hormone releasing IUDs

169. Which one of the following conditions correctly describes the manner of determining the sex in the given example?
(a) Homozygous sex chromosomes (ZZ) determine female sex in birds.
(b) XO type of sex chromosomes determine male sex in grasshopper.
(c) XO condition in human as found in Turner syndrome, determines female sex.
(d) Homozygous sex chromosomes (XX) produce male in *Drosophila.*

170. Select the correct statement from the ones given below with respect to dihybrid cross.
(a) Tightly linked genes on the same chromosome show higher recombinations
(b) Genes far apart on the same chromosome show very few recombinations
(c) Genes loosely linked on the same chromosome show similar recombinations as the tightly linked ones
(d) Tightly linked genes on the same chromosome show very few recombinations

171. One gene-one enzyme relationship was established for the first time in
(a) *Salmonella typhimurium*
(b) *Escherichia coli*
(c) *Diplocococcus pneumoniae*
(d) *Neurospora crassa.*

172. The transforming principle of *Pneumococcus* as found out by Avery, MacLeod and McCarty was
(a) mRNA (b) DNA
(c) protein (d) polysaccharide

173. In emphysema –
(a) Gas exchange area of lungs is reduced
(b) Gas exchange area of lungs is increased
(c) Trachea gets narrowed
(d) Larynx is permanently closed

174. Damage to thymus in a child may lead to –
(a) a reduction in haemoglobin content of blood
(b) a reduction in stem cell production
(c) loss of antibody mediated immunity
(d) loss of cell mediated immunity

175. Two microbes found to be very useful in genetic engineering are
(a) *Vibrio cholerae* and a tailed bacteriophage
(b) *Diplococcus* sp. and *Pseudomonas* sp.
(c) Crown gall bacterium and *Caenorhabditis elegans*
(d) *Escherichia coli* and *Agrobacterium tumefaciens*

176. An "Urn" shaped population age pyramid represents
(a) Growing population
(b) Static population
(c) Declining population
(d) Threatenped population

177. Study the four statements (i–iv) given below and select the two correct ones out of them:
(i) A lion eating a deer and a sparrow feeding on grain are ecologically similar in being consumers
(ii) Predator star fish *Pisaster* helps in maintaining species diversity of some invertebrates
(iii) Predators ultimately lead to the extinction of prey species
(iv) Production of chemicals such as nicotine, strychnine by the plants are metabolic disorders

The two correct statements are:
(a) (ii) and (iii) (b) (iii) and (iv)
(c) (i) and (iv) (d) (i) and (ii)

178. Reason of diversity in living beings is due to
(a) mutation
(b) long term evolutionary change
(c) gradual change
(d) short term evolutionary change

179. Part of alimentary canal meant for reabsorption of the maximum H_2O is
(a) Small intestine (b) large intestine
(c) rectum (d) colon

180. Which one of the following pairs of structures is correctly matched with their correct description?

	Structures	Description
(a)	Tibia and fibula	Both form parts of knee joint
(b)	Cartilage and cornea	No blood supply but do require oxygen for respiratory need
(c)	Shoulder joint and elbow joint	Ball and socket type of joint
(d)	Premolars and molars	20 in all and 3– rooted

RESPONSE SHEET

Test Code: ....Time taken :.......................Date:...........

	(a)	(b)	(c)	(d)		(a)	(b)	(c)	(d)		(a)	(b)	(c)	(d)		(a)	(b)	(c)	(d)
1.					46.					91.					136.				
2.					47.					92.					137.				
3.					48.					93.					138.				
4.					49.					94.					139.				
5.					50.					95.					140.				
6.					51.					96.					141.				
7.					52.					97.					142.				
8.					53.					98.					143.				
9.					54.					99.					144.				
10.					55.					100.					145.				
11.					56.					101.					146.				
12.					57.					102.					147.				
13.					58.					103.					148.				
14.					59.					104.					149.				
15.					60.					105.					150.				
16.					61.					106.					151.				
17.					62.					107.					152.				
18.					63.					108.					153.				
19.					64.					109.					154.				
20.					65.					110.					155.				
21.					66.					111.					156.				
22.					67.					112.					157.				
23.					68.					113.					158.				
24.					69.					114.					159.				
25.					70.					115.					160.				
26.					71.					116.					161.				
27.					72.					117.					162.				
28.					73.					118.					163.				
29.					74.					119.					164.				
30.					75.					120.					165.				
31.					76.					121.					166.				
32.					77.					122.					167.				
33.					78.					123.					168.				
34.					79.					124.					169.				
35.					80.					125.					170.				
36.					81.					126.					171.				
37.					82.					127.					172.				
38.					83.					128.					173.				
39.					84.					129.					174.				
40.					85.					130.					175.				
41.					86.					131.					176.				
42.					87.					132.					177.				
43.					88.					133.					178.				
44.					89.					134.					179.				
45.					90.					135.					180.				

TEST ASSESSMENT AND ANALYSIS SHEET

How to fill this sheet : After you finish the paper and have checked it first try to attempt again all the questions which you have done wrong or have not been able to attempt. Go to detailed solutions only after you have attempted all the questions of your own.Attempting the questions on your own without seeing the solutions will help you in filling this table

ANALYSIS OF TEST	
Total number of questions	
Questions Attempted	
Correct Questions	
Wrong Questions	
	Net Score

Strike rate $\left(\frac{\text{Correct questions}}{\text{Questions attempt}} \times 100\right) =$ ....

Percentage Marks $\left(\frac{\text{Correct Answers}}{\text{Total number of questions}} \times 100\right) =$ ....

Analysis of wrong questions : Divide the questions which you have done wrong in 3 broad categories.

Reasons for questions attempted wrong	Number of Questions
Knew but calculation mistake (solved on my own after exam)	
Knew but got confused (solved on my own after exam)	
Did not knew the question (couldn't solve even after exam)	

Questions not attempted : Divide the questions not attempted in 3 broad categories

Reasons for unattempted questions	Number of Questions
Knew the ques but time was less (solved on my own after exam)	
Knew the ques but got confused (solved on my own after exam)	
Did not knew the questions (couldn't solve even after exam)	

List down your weak areas you need to improve your score :

1)______________________________

2)______________________________

3)______________________________

NEET

National Eligibility Entrance Test

Test Series **Full Test - 8**

Time : - 3 Hours **Max. Marks:- 720**

INSTRUCTIONS

1. Immediately fill in the particulars on this page of the Test Booklet with Blue/Black Ball Point Pen. Use of pencil is strictly prohibited.
2. Answers are to be given on Response Sheet provided at the end of this paper. You can tear this Response Sheet along the cut marks provided. Fill in the particulars carefully in the Response Sheet provided.
3. The test is of 3 hours duration.
4. The Test Booklet consists of **180** questions. The maximum marks are **720**.
5. There are four section in the question paper.

   The distribution of marks subjectwise in each part is as under for each correct response.

   Section 1 – Physics (180 Marks) – 45 Questions

   Section 2 – Chemistry (180 Marks) – 45 Questions

   Section 3 – Botany (180 Marks) – 45 Questions

   Section 4 – Zoology (180 Marks) – 45 Questions
6. You will be awarded 4 Marks if you have darkened only the bubble corresponding to the correct answer and zero mark if no bubble is darkened. In all other cases, minus one (– 1) mark will be awarded.
7. Use Blue/Black Ball Point Pen only for writing particulars/marking responses on Side-1 and Side-2 of the Response Sheet.

   Use of pencil is strictly prohibited.
8. You are advised to create an actual examination hall simulation.
9. Rough work is to be done on the space provided for this purpose in the Test Booklet only.
10. Do not fold or make any stray marks on the Response Sheet.
11. Once you have evaluated the marks in the test, you should assess it thoroughly.

SECTION 1 - PHYSICS

1. Out of the following pair , which one does NOT have identical dimensions
(a) impulse and momentum
(b) angular momentum and planck's constant
(c) work and torque
(d) moment of inertia and moment of a force

2. An eye specialist prescribes spectacles having combination of convex lens of focal length 40 cm in contact with a concave lens of focal length 25 cm. The power of this lens combination in diopter is
(a) +1.5 (b) −1.5
(c) +6.67 (d) −6.67

3. For the velocity time graph shown in the figure below the distance covered by the body in the last two seconds of its motion is what fraction of the total distance travelled by it in all the seven seconds?

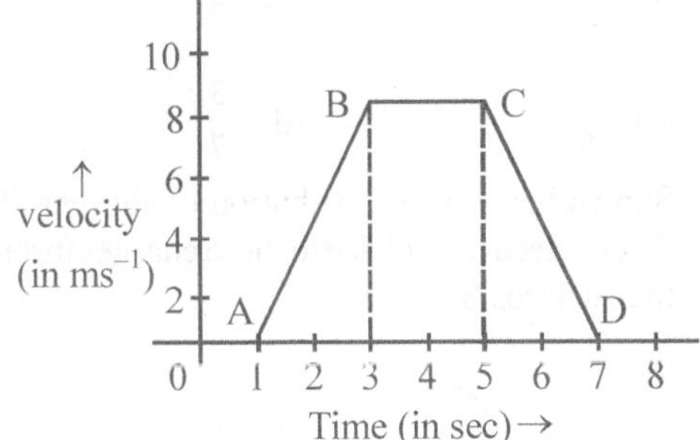

(a) $\frac{1}{2}$ (b) $\frac{1}{4}$
(c) $\frac{2}{3}$ (d) $\frac{1}{3}$

4. The angles which the vector $\vec{A} = 3\hat{i} + 6\hat{j} + 2\hat{k}$ makes with the co-ordinate axes are
(a) $\cos^{-1}\frac{3}{7}, \cos^{-1}\frac{4}{7}, \cos^{-1}\frac{1}{7}$
(b) $\cos^{-1}\frac{3}{7}, \cos^{-1}\frac{6}{7}, \cos^{-1}\frac{2}{7}$
(c) $\cos^{-1}\frac{4}{7}, \cos^{-1}\frac{5}{7}, \cos^{-1}\frac{3}{7}$
(d) None of these

5. The effective acceleration of a body when thrown upwards with acceleration a is
(a) $(g + a)$ (b) $(a - g)$
(c) $\sqrt{(g^2 + a^2)}$ (d) $(g - a^2/g)$

6. Which one of the following statements is false ?
(a) A vector has only magnitude, whereas a scalar has both magnitude and direction
(b) Distance is a scalar quantity but displacement is a vector quantity
(c) Momentum, force and torque are vector quantities
(d) Mass, speed and energy are scalar quantities

7. A motor car is travelling at 30 m/s on a circular road of radius 500 m. It is increasing in speed at the rate of 2 m/s^2. Then the acceleration of the car will be
(a) 4 m/s^2 (b) 3 m/s^2
(c) 2.7 m/s^2 (d) 8 m/s^2

8. If Q, E and W denote respectively the heat added, change in internal energy and the work done in a closed cyclic process, then
(a) $W = 0$ (b) $Q = W = 0$
(c) $E = 0$ (d) $Q = 0$

9. The moment of inertia of disc about a tangent axis in its plane is
(a) $\frac{MR^2}{4}$ (b) $\frac{3MR^2}{2}$
(c) $\frac{5}{4}MR^2$ (d) $\frac{7MR^2}{4}$

10. A boat is sent across a river with a velocity of 8 km h^{-1}. If the resultant velocity of boat is 10 km h^{-1} , then the velocity of the river is
(a) 12.8 km h^{-1} (b) 6 km h^{-1}
(c) 8 km h^{-1} (d) 10 km h^{-1}

11. How many electrons make up a charge of 20 μC?
(a) 1.25×10^{14} (b) 2.23×10^{14}
(c) 3.25×10^{14} (d) 5.25×10^{14}

12. Two masses *A* and *B* each of mass *M* are fixed together by a massless spring. A force acts on the mass *B* as shown in fig. At the instant the mass *A* has acceleration *'a'*. What is the acceleration of mass *B* ?

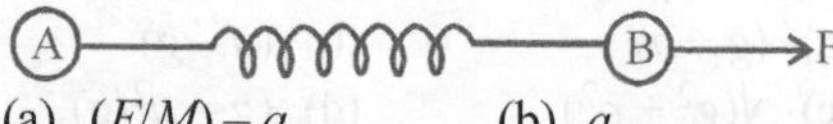

(a) $(F/M) - a$ (b) a
(c) $-a$ (d) (F/M)

13. When sound waves travel from air to water, which one of the following remains constant?
(a) Time period (b) Frequency
(c) Velocity (d) Wavelength

14. An ideal gas *A* and a real gas *B* have their volumes increased from *V* to 2*V* under isothermal conditions. The increase in internal energy
(a) will be same in both *A* and *B*
(b) will be zero in both the gases
(c) of *B* will be more than that of *A*
(d) of *A* will be more than that of *B*

15. What is the effective capacitance between points *X* and *Y*?

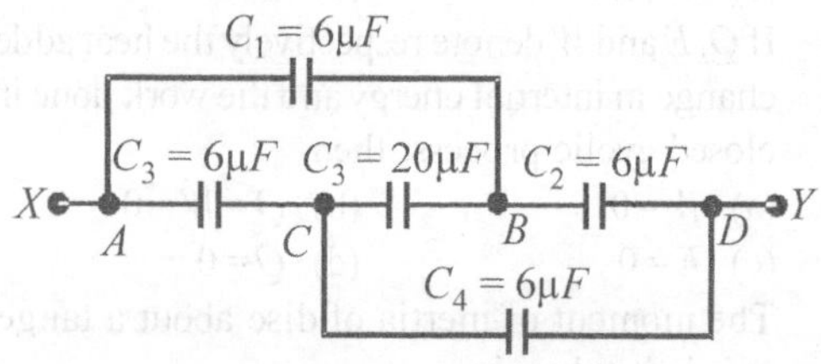

(a) 24 μ*F* (b) 18 μ*F*
(c) 12 μ*F* (d) 6 μ*F*

16. A particle of mass m_1 is moving with a velocity v_1 and another particle of mass m_2 is moving with a velocity v_2. Both of them have the same momentum but different kinetic energies are E_1 and E_2 respectively. If $m_1 > m_2$ then
(a) $E_1 = E_2$ (b) $E_1 < E_2$
(c) $\frac{E_1}{E_2} = \frac{m_1}{m_2}$ (d) $E_1 > E_2$

17. A force F acting on an object varies with distance *x* as shown here. The force is in N and *x* in *m*. The work done by the force in moving the object from *x* = 0 to *x* = 6 m is

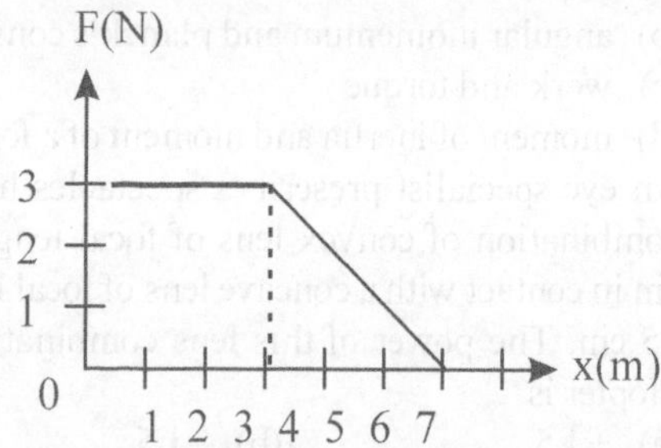

(a) 18.0 J (b) 13.5 J
(c) 9.0 J (d) 4.5 J

18. If a spherical ball rolls on a table without slipping the fraction of its total energy associated with rotational energy is
(a) $\frac{3}{5}$ (b) $\frac{2}{7}$
(c) $\frac{2}{5}$ (d) $\frac{3}{7}$

19. Shown below is a distribution of charges. The flux of electric field due to these charges through the surfaces S is

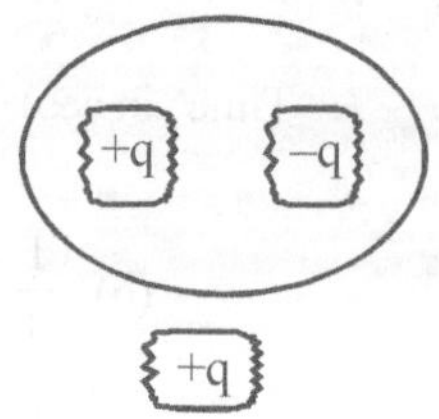

(a) $3q/\varepsilon_0$ (b) $2q/\varepsilon_0$
(c) q/ε_0 (d) zero

20. 10 cm is a wavelength corresponding to the spectrum of
(a) infrared rays (b) ultraviolet rays
(c) microwaves (d) γ-rays

21. The transformer voltage induced in the secondary coil of a transformer is mainly due to
(a) a varying electric field
(b) a varying magnetic field
(c) the vibrations of the primary coil
(d) the iron core of the transformer

22. The escape velocity for a body projected vertically upwards from the surface of earth is 11 km/s. If the body is projected at an angle of 45°with the vertical, the escape velocity will be
(a) $11\sqrt{2}$ km/s (b) 22 km/s
(c) 11 km/s (d) $\frac{11}{\sqrt{2}}$ km/s

23. An open U-tube contains mercury. When 11.2 cm of water is poured into one of the arms of the tube, how high does the mercury rise in the other arm from its initial level?
(a) 0.56 cm (b) 1.35 cm
(c) 0.41 cm (d) 2.32 cm

24. A steel wire of length 20 cm and uniform cross-section 1 mm^2 is tied rigidly at both the ends. The temperature of the wire is altered from 40ºC to 20ºC. Coefficient of linear expansion for steel $\alpha = 1.1 \times 10^{-5}$/ºC and Y for steel is 2.0×10^{11} N/m^2. The change in tension of the wire is
(a) 2.2×10^6 newton (b) 16 newton
(c) 8 newton (d) 44 newton

25. The fastest mode of transfer of heat is
(a) conduction (b) convection
(c) radiation (d) None of these

26. The B – H curve (i) and (ii) shown in fig associated with

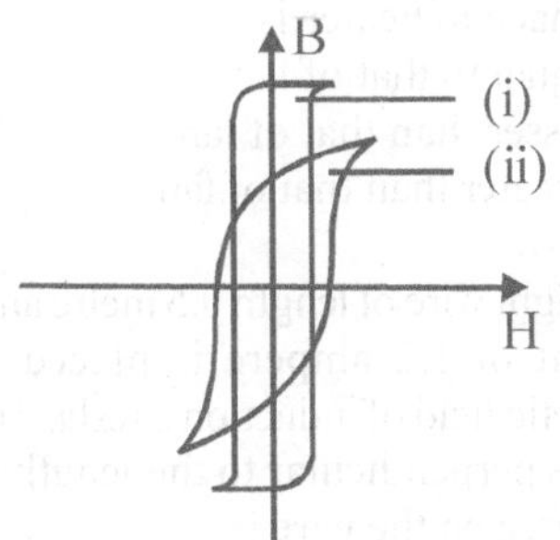

(a) (i) diamagnetic and (ii) paramagnetic substance
(b) (i) paramagnetic and (ii) ferromagnetic substance
(c) (i) soft iron and (ii) steel
(d) (i) steel and (ii) soft iron

27. The pressure P of an ideal gas and its mean K.E. per unit volume are related as
(a) $P = \frac{E}{2}$ (b) $P = E$
(c) $P = \frac{3E}{2}$ (d) $P = \frac{2E}{3}$

28. A body cools from 50.0°C to 48°C in 5s. How long will it take to cool from 40.0°C to 39°C? Assume the temperature of surroundings to be 30.0°C and Newton's law of cooling to be valid.
(a) 2.5 s (b) 10 s
(c) 20 s (d) 5 s

29. The spatial distribution of the electric field due to two charges (A, B) is shown in figure. Which one of the following statements is correct ?

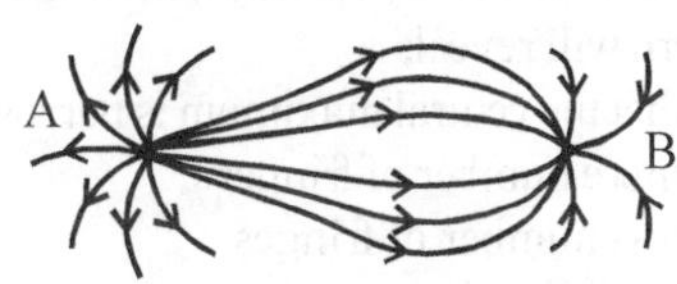

(a) A is +ve and B –ve; $|A| > |B|$
(b) A is –ve and B +ve; $|A| = |B|$:
(c) Both are +ve but A > B
(d) Both are –ve but A > B

30. The equation of plane wave is given by
$$y = 2\sin\pi\left(200t - \frac{x}{15}\right)$$
where displacement y is given in cm and time t in second, then the velocity of the wave is
(a) 3000 cm/sec (b) 200 cm/sec
(c) 150 cm/sec (d) 2 cm /sec

31. A primary cell has an e.m.f. of 1.5 volt. When short-circuited it gives a current of 3 ampere. The internal resistance of the cell is
(a) 4.5 ohm (b) 2 ohm
(c) 0.5 ohm (d) (1/4.5) ohm

32. The magnetic lines of force inside a bar magnet
(a) are from *N*-pole to *S*-pole of magnet
(b) do not exist
(c) depend upon the area of cross-section of bar magnet
(d) are from *S*-pole to *N*-pole of magnet

33. Five resistances have been connected as shown in the figure. The effective resistance between A and B is

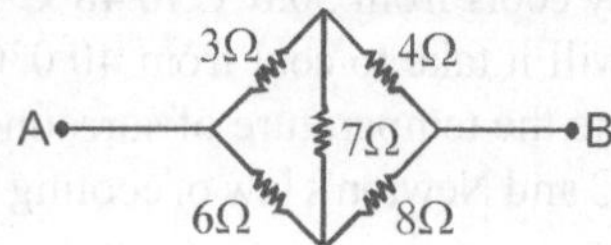

(a) 14/3Ω (b) 20/3Ω
(c) 14Ω (d) 21Ω

34. Yellow light is used in a single slit diffraction experiment with slit width of 0.6 mm. If yellow light is replaced by X– rays, then the observed pattern will reveal,
(a) that the central maximum is narrower
(b) more number of fringes
(c) less number of fringes
(d) no diffraction pattern

35. Which one is possible?
(a) ${}_7N^{19} + {}_0n^1 \rightarrow {}_7N^{16} + {}_1H^1$
(b) ${}_{16}S^{32} + {}_1H^1 \rightarrow {}_{17}Cl^{35} + {}_2He^4$
(c) ${}_8O^{16} + {}_0n^1 \rightarrow {}_7N^{14} + 3\,{}_1H^1 + 2\,{}_{-1}\beta^0$
(d) ${}_1H^1 + {}_1H^1 \rightarrow {}_2He^4$

36. A conducting square loop of side *L* and resistance *R* moves in its plane with a uniform velocity *v* perpendicular to one of its side. A magnetic induction *B* constant in time and space, pointing perpendicular and into the plane of the loop exists everywhere. The current induced in the loop is
(a) $\frac{Blv}{R}$ clockwise
(b) $\frac{Blv}{R}$ anti-clockwise
(c) $\frac{2Blv}{R}$ anti – clockwise
(d) zero

37. When a ray of light enters a glass slab from air,
(a) its wavelength decreases
(b) its wavelength increases
(c) its frequency decreases
(d) neither its wavelength nor its frequency changes.

38. Main function of the RF amplifiers in a superheterodyne receiver is to
(a) amplify signal
(b) reject unwanted signal
(c) discriminate against image frequency signal and IF-signal
(d) All the above

39. An electric fan and a heater are marked as 100 W, 220 V and 1000 W, 220 V respectively. The resistnace of heater is
(a) equal to that of fan
(b) lesser than that of fan
(c) greater than that of fan
(d) zero

40. A straight wire of length 0.5 metre and carrying a current of 1.2 ampere is placed in uniform magnetic field of induction 2 tesla. The magnetic field is perpendicular to the length of the wire. The force on the wire is

(a) 2.4 N (b) 1.2 N
(c) 3.0 N (d) 2.0 N

41. Monochromatic light of frequency 6.0×10^{14} Hz is produced by a laser. The power emitted is 2×10^{-3} W. The number of photons emitted, on the average, by the source per second is
(a) 5×10^{16} (b) 5×10^{17}
(c) 5×10^{14} (d) 5×10^{15}

42. Of the various series of the hydrogen spectrum, the one which lies wholly in the ultraviolet region is
(a) Lyman series (b) Balmer series
(c) Paschen series (d) Bracket series

43. An artificial satellite moving in a circular orbit around the earth has a total (kinetic + potential) energy E_0. Its potential energy is
(a) $-E_0$ (b) $1.5 E_0$
(c) $2E_0$ (d) E_0

44. Of the diodes shown in the following diagrams, which one is reverse biased ?

(a)

(b)
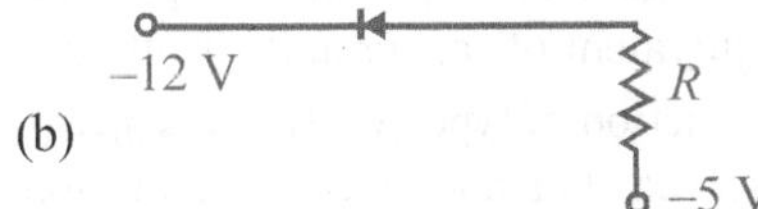

(c)
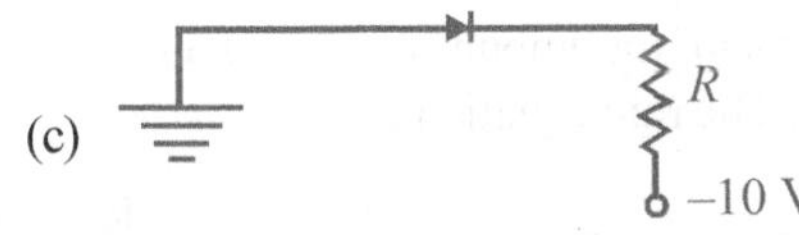

(d)
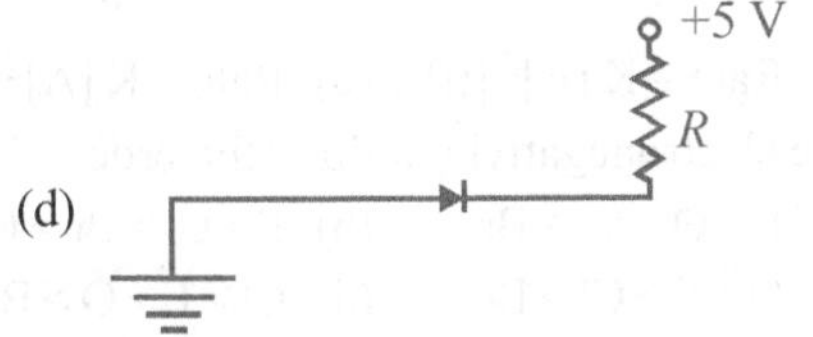

45. Which of the following gates will have an output of 1?

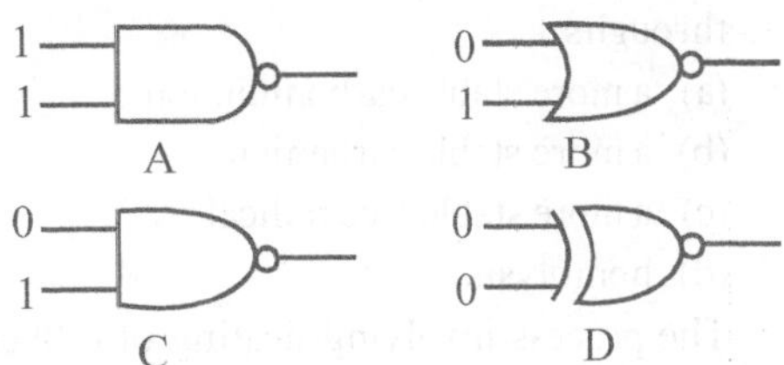

(a) D (b) A
(c) B (d) C

SECTION 2 - CHEMISTRY

46. Which of the following contains maximum number of atoms?
(a) 6.023×10^{21} molecules of CO_2
(b) 0.44 g of CO_2
(c) 22.4 L of CO_2 at STP
(d) None of these

47. Which of the following statements is true ?
(a) ΔG may be lesser or greater or equal to ΔH
(b) ΔG is always proportional to ΔH
(c) ΔG is always greater than ΔH
(d) ΔG is always less than ΔH

48. 1.08 g of pure silver was converted into silver nitrate and its solution was taken in a beaker. It was electrolysed using platinum cathode and silver anode. 0.01 Faraday of electricity was passed using 0.15 volt above the decomposition potential of silver. The silver content of the beaker after the above shall be
(a) 0 g (b) 0.108 g
(c) 0.108 g (d) 1.08 g

49. Number of electrons transfered in each case when $KMnO_4$ acts as an oxidising agent to give MnO_2, Mn^{2+}, $Mn(OH)_3$ and MnO_4^{2-} are respectively
(a) 3,5,4 and 1 (b) 4,3,1 and 5
(c) 1,3,4 and 5 (d) 5,4,3 and 1

50. Addition of HI to double bond of propene yields isopropyl iodide and not n-propyl iodide as the major product, because addition proceeds throughs
(a) a more stable carbonium ion
(b) a more stable carbanion
(c) a more stable free radical
(d) homolysis

51. The process involving heating of rubber with sulphur is called
(a) Galvanisation (b) Vulcanization
(c) Bessemerisaion (d) Sulphonation

52. Which of the following is a redox reaction ?
(a) Reaction of H_2SO_4 with NaOH
(b) In atomosphere, O_3 from O_2 by lightning
(c) Nitrogen oxides from nitrogen and oxygen decomposition by lightning
(d) Evaporation of H_2O

53. Standard entropy of X_2, Y_2 and XY_3 are 60, 40 and 50 J K^{-1} mol^{-1}, respectively. For the reaction,

$\frac{1}{2}X_2 + \frac{3}{2}Y_2 \rightarrow XY_3, \Delta H = -30kJ$, to be at equilibrium, the temperature will be
(a) 1250 K (b) 500 K
(c) 1000 K (d) 750 K

54. The cell constant of a given cell is 0.47 cm^{-1}. The resistance of a solution placed in this cell is measured to be 31.6 ohm. The conductivity of the solution (in S cm^{-1} where S has usual meaning) is
(a) 0.15 (b) 1.5
(c) 0.015 (d) 150

55. Consider the following changes

$A \rightarrow A^+ + e^- : E_1$ and $A^+ \rightarrow A^{2+} + e^- : E_2$

The energy required to pull out the two electrons are E_1 and E_2 respectively. The correct relationship between two energies would be
(a) $E_1 < E_2$ (b) $E_1 = E_2$
(c) $E_1 > E_2$ (d) $E_1 \geq E_2$

56. Pure chloroform is prepared by
(a) distilling chloral hydrate with aqueous sodium hydroxide
(b) heating ethanol with bleaching powder
(c) heating acetone with bleaching powder
(d) reducing carbon tetrachloride

57. The end product (C) in the following sequence of reactions is

$$HC \equiv CH \xrightarrow[20\%\ H_2SO_4]{1\%\ HgSO_4} A \xrightarrow[H_2O]{CH_3MgX} B \xrightarrow{[O]} (C)$$

(a) Acetic acid (b) Isopropyl alcohol
(c) Acetone (d) Ethanol

58. At 0^0 C and one atm pressure, a gas occupies 100 cc. If the pressure is increased to one and a half-time and temperature is increased by one third of absolute temperature, then final volume of the gas will be
(a) 80 cc (b) 88.9 cc
(c) 66.7 cc (d) 100 cc

59. Which one of the following statement is not true
(a) The conjugate base of $H_2PO_4^-$ is HPO_4^{2-}
(b) pH + pOH = 14 for all aqueous solutions
(c) the pH of 1×10^{-8} M HCl is 8
(d) 96,500 coulombs of electricity when passed through a $CuSO_4$ solution deposits 1 gram equivalent of copper at the Cathode

60. For a reaction of type $A + B \longrightarrow$ products, it is observed that doubling concentration of A causes the reaction rate to be four times as great, but doubling amount of B does not affect the rate. The rate equation is
(a) Rate = K [A] [B] (b) Rate = $\frac{K}{4}[A]^2$
(c) Rate = K $[A]^2$ [B] (d) Rate = K $[A]^2$ $[B]^2$

61. The electronegativity follows the order
(a) F > O > Cl > Br (b) F > Cl > Br > O
(c) O > F > Cl > Br (d) Cl > F > O > Br

62. Which of the following is not formed when glycerol reacts with HI?
(a) $CH_2=CH-CH_2I$
(b) $CH_2OH-CHI-CH_2OH$
(c) $CH_3-CH=CH_2$
(d) $CH_3-CHI-CH_3$

63. The amine that does not react with acetyl chloride is
(a) CH_3NH_2 (b) $(CH_3)_2NH$
(c) $(CH_3)_3N$ (d) None of these

64. In what manner increase of pressure will affect the following equation?
$C(s) + H_2O(g) \rightleftharpoons CO(g) + H_2(g)$
(a) Shift in the forward direction
(b) Shift in the reverse direction
(c) Increase in the yield of hydrogen
(d) No effect

65. For preparing a buffer solution of pH 6 by mixing sodium acetate and acetic acid, the ratio of the concentration of salt and acid should be ($K_a = 10^{-5}$)
(a) 1 : 10 (b) 10 : 1
(c) 100 : 1 (d) 1 : 100

66. The activation energy for a hypothetical reaction, A $\rightarrow$ Product, is 12.49 kcal/mole. If temperature is raised from 295 to 305, the rate of reaction increased by
(a) 60% (b) 100%
(c) 50% (d) 20%

67. BCl_3 is a planar molecule whereas NCl_3 is pyramidal because
(a) BCl_3 has no lone pair of electrons but NCl_3 has a lone pair of electrons
(b) B—Cl bond is more polar than N—Cl bond
(c) Nitrogen atom is smaller than boron atom
(d) N—Cl bond is more covalent than B—Cl bond

68. Indicate the wrongly named compound
(a) $CH_3-\underset{\displaystyle CH_3}{\underset{|}{CH}}-CH_2-CH_2-CHO$
(4-methyl -1- pentanal)
(b) $CH_3-\underset{\displaystyle CH_3}{\underset{|}{CH}}-C\equiv C-COOH$
(4- methyl -2- pentyn -1- oic acid)
(c) $CH_3CH_2CH_2-\underset{\displaystyle CH_3}{\underset{|}{CH}}-COOH$
(2- methyl -1- pentanoic acid)

(d) $CH_3-CH_2-CH=CH-\overset{\displaystyle O}{\overset{||}{C}}-CH_3$
(3- hexen -5- one)

69. An organic compound, C_3H_6O does no give a precipitate with 2, 4 dinitrophenyl-hydrazine reagent and does not react with metallic sodium. It could be
(a) CH_3-CH_2-CHO (b) $CH_2=CH-CH_2OH$
(c) $CH_3-CO-CH_3$ (d) $CH_2=CH-O-CH_3$

70. Which one of the following pairs of molecules will have permanent dipole moments for both members?
(a) SiF_4 and NO_2 (b) NO_2 and CO_2
(c) NO_2 and O_3 (d) SiF_4 and CO_2

71. An element has bcc strucutre having unit cells 12.08×10^{23}. The number of atoms in these cells is
(a) 12.08×10^{23} (b) 24.16×10^{23}
(c) 48.38×10^{23} (d) 12.08×10^{22}

72. An unknown metal M displaces nickel from nickel (II) sulphate solution but does not displace manganese from manganese sulphate solution. Which order represents the correct order of reducing power?
(a) Mn > Ni > M (b) Ni > Mn > M
(c) Mn > M > Ni (d) M > Ni > Mn

73. In which of the following reactions, H_2O_2 is acting as a reducing agent
(a) $H_2O_2 + SO_2 \rightarrow H_2SO_4$
(b) $2KI + H_2O_2 \rightarrow 2KOH + I_2$
(c) $PbS + 4H_2O_2 \rightarrow PbSO_4 + 4H_2O$
(d) $Ag_2O + H_2O_2 \rightarrow 2Ag + H_2O + O_2$

74. Beilstein's test is given by which of the following
(a) Halogens (b) Thio urea
(c) Pyridine (d) All

75. Glucose molecule reacts with X number of molecules of phenylhydrazine to yield osazone. The value of X is
(a) Three (b) Two
(c) One (d) Four

76. At STP, the order of root mean square speed of molecules H_2, N_2, O_2 and HBr is
(a) $H_2 > N_2 > O_2 > HBr$
(b) $HBr > O_2 > N_2 > H_2$
(c) $HBr > H_2 > O_2 > N_2$
(d) $N_2 > O_2 > H_2 > HBr$

77. Which of the following is not true ?
(a) Some disinfectants can be used as antispetics
(b) sulphadiazine is a synthetic antibacterial
(c) aspirin is analgesic as well as antipyretic
(d) polystyrene is used to make non-stick cookware

78. Vapour pressure of CCl_4 at 25°C is 143 mm Hg. 0.5 g of a non-volatile solute (mol. wt. 65) is dissolved in 100 ml of CCl_4. Find the vapour pressure of the solution.
(Density of CCl_4 = 1.58 g/cm^3)
(a) 141.93 mm (b) 94.39 mm
(c) 199.34 mm (d) 143.99 mm

79. Match list I with list II and select the correct answer using the codes given below the lists:

List I	***List II***
I. Cyanide process	A. Ultrapure Ge
II. Floatation process	B. Pine oil
III. Electrolytic reduction	C. Extraction of Al
IV. Zone refining	D. Extraction of Au

Codes:
(a) I-C, II-A, III-D, IV-B (b) I-D,II-B,III-C,IV-A
(c) I-C,II-B,III-D, IV-A (d) I-D,II-A,III-C,IV-B

80. In which of the following, resonance will be possible?
(a) $CH_2-CH_2-CH_2-CHO$
(b) $CH_2 = CH - CH = O$
(c) CH_3COCH_3
(d) $CH_2 = CH - CH_2 - CH = CH_2$

81. In Bohr series of lines of hydrogen spectrum the third line from the red end corresponds to which one of the following inter-orbit jumps of the electron for Bohr orbits in an atom of hydrogen?
(a) $3 \rightarrow 2$ (b) $5 \rightarrow 2$
(c) $4 \rightarrow 1$ (d) $2 \rightarrow 5$

82. Point out the false statement
(a) Brownian movement and Tyndall effect are shown by colloidal systems.
(b) Gold number is a measure of the protective power of a lyophillic colloid
(c) The colloidal solution of a liquid in liquid is called gel
(d) Hardy - Schulze rule is related with coagulation.

83. The compound A on heating gives a colourless gas. and a residue which dissolves in water to give B. When excess of CO_2 is passed into B and gently heated, A is formed. The compound A is
(a) $CaSO_4.2H_2O$ (b) Na_2CO_3
(c) $CaCO_3$ (d) K_2CO_3

84. The value of the 'spin only' magnetic moment for one of the following configurations is 2.84 BM. The correct one is
(a) d^5 (in strong ligand field)
(b) d^3 (in weak as well as in strong fields)
(c) d^4 (in weak ligand fields)
(d) d^4 (in strong ligand fields)

85. N_2 and O_2 are converted into monoanions, N_2^- and O_2^- respectively. Which of the following statements is wrong ?

(a) In N_2^-, N – N bond weakens

(b) In O_2^-, O - O bond order increases

(c) In O_2^-, O - O bond order decreases

(d) N_2^- becomes paramagnetic

86. The structure and hybridization of $Si(CH_3)_4$ is

(a) bent, sp (b) trigonal, sp^2

(c) octahedral, sp^3d (d) tetrahedral, sp^3

87. In a reaction $A + B \rightleftharpoons C + D$, the initial concentrations, of A and B were 0.9 mol. dm^{-3} each. At equilibrium the concentration of D was found to be 0.6 mol dm^{-3}. What is the value of equilibrium constant for the reaction

(a) 8 (b) 4

(c) 9 (d) 3

88. Lithium crystallizes as body-centred cubic cyrstals. If the length of the side of unit cell is 350 pm, the atomic radius of lithium is

(a) 303.1 pm (b) 606.2 pm

(c) 151.5 pm (d) 123.7 pm

89. An example of Perkin's reaction is

(a) $C_6H_5CHO + CH_3NO_2 \xrightarrow{KOH} C_6H_5CHCHNO_2$

(b) $C_6H_5CHO + (CH_3CO)_2O \xrightarrow{CH_3COONa} C_6H_5CH = CHCOOH$

(c) $C_6H_5CHO + CH_3CHO \xrightarrow{NaOH} C_6H_5CH= CHCHO$

(d) $C_6H_5CHO + CH_2(COOH)_2 \xrightarrow{Alc.NH_3} C_6H_5CH = CHCO_2H$

90. The ether that undergoes electrophilic substitution reactions is

(a) $CH_3OC_2H_5$ (b) $C_6H_5OCH_3$

(c) CH_3OCH_3 (d) $C_2H_5OC_2H_5$

SECTION 3 - BOTANY

91. Classification of organisms based on evolutionary as well as genetic relationships is called

(a) Biosystematics

(b) Phenetics

(c) Numerical taxonomy

(d) Cladistics

92. In gymnosperms, the ovule is naked because

(a) ovary wall is absent

(b) integuments are absent

(c) perianth is absent

(d) nucellus is absent

93. Epidermis is absent in

(a) root tip and shoot tip

(b) shoot bud and floral bud

(c) ovule and seed

(d) petiole and pedicel

94. Which pair of structures are usually found in both plant and animal cells?

(a) Cell membrane and nucleolus

(b) Cell membrane and cell wall

(c) Nucleolus and chloroplast

(d) Nucleus and cell wall

95. How many meiotic division would be required to produce 101 female gametophytes in an angiosperm?

(a) 101 (b) 26

(c) 127 (d) None of these

96. Which of the following element is necessary for translocation of sugars in plants ?

(a) Boron (b) Molybdenum

(c) Manganese (d) Iron

97. In C_4 plants, agranal chloroplasts are found in

(a) mesophyll cells

(b) epidermal cell chloroplasts of green stem

(c) bundle sheath cells

(d) chloroplasts of guard cells

98. Out of 36 ATP molecules produced per glucose molecule during respiration
(a) 2 are produced outside glycolysis and 34 during respiratory chain.
(b) 2 are produced outside mitochondria and 34 inside mitochondria.
(c) 2 during glycolysis and 34 during Krebs cycle.
(d) all are formed inside mitochondria.

99. Which of the following hormone prevents ripening of fruit?
(a) Gibberellin (b) Ethylene
(c) Cytokinin (d) ABA

100. Which one of the following represents an ovule, where the embryo sac becomes horse-shoe shaped and the funiculus and micropyle are close to each other?
(a) Amphitropous (b) Circinotropous
(c) Atropous (d) Anotropous

101. When the members of one population do not breed at the same time of the year as the members of another population it is called
(a) geographic isolation
(b) sexual isolation
(c) seasonal isolation
(d) habitat isolation

102. Diversity of habitat over the total landscape is called :
(a) β diversity
(b) γ diveristy (gamma)
(c) landscape diversity
(d) ecosystem diversity

103. DDT causes egg shell thinning in birds because it inhibits
(a) Calmodulin
(b) Calcium ATPase
(c) Magnesium ATPase
(d) Carbonic anhydrase

104. The first clinical gene therapy was given for treating
(a) chicken pox
(b) diabetes mellitus
(c) rheumatoid arthritis
(d) adenosine deaminase deficiency

105. Taxonomic hierarchy refers to
(a) Step-wise arrangement of all categories for classification of plants and animals
(b) A group of senior taxonomists who decide the nomenclature of plants and animals
(c) A list of botanists or zoologists who have worked on taxonomy of a species or group
(d) Classification of a species based on fossil record

106. A group of fungi with septate mycelium in which sexual reproduction is either unknown or lacking are classified under
(a) phycomycetes (b) deuteromycetes
(c) ascomycetes (d) basidiomycetes

107. Oxysomes or $F_0 - F_1$ particles occur on
(a) thylakoids
(b) mitochondrial surface
(c) inner mitochondrial membrane
(d) chloroplast surface

108. Which of the following layer is present nearest of plasma membrane in plant cell?
(a) Secondary wall (b) Middle lamella
(c) Primary wall (d) Tonoplast

109. Cellulose is a polymer of
(a) α-glucose (b) α-D-glucose
(c) β-D-glucose (d) β-L-glucose

110. The number of chromatids in a chromosome at anaphase is
(a) 2 in mitosis and 1 in meiosis
(b) 1 in mitosis and 2 in meiosis
(c) 2 each in mitosis and meiosis
(d) 2 in mitosis and 4 in meiosis

111. Plants die from prolonged water-logging because
(a) soil nutrients become very dilute.
(b) root respiration stops.
(c) cell sap in the plants becomes too dilute.
(d) nutrients leach down due to excess water.

112. Gibberellin was first extracted from
(a) *Gibberella fujikuroi* (b) *Gelidium*
(c) *Gracelaria* (d) *Aspergillus*

113. Atropine, an alkaloid, is obtained from
(a) *Datura anaroxia* (b) *Atropa belladona*
(c) *Hyocyamus niger* (d) *Withania somnifera*

114. If an angiospermic male plant is diploid and female plant tetraploid, the ploidy level of endosperm will be

(a) haploid (b) triploid
(c) tetraploid (d) pentaploid

115. Species diversity increase as one proceeeds from

(a) high altitude to low altitude and high latitude to low latitude
(b) low altitude to high altitude and high latitude to low latitude
(c) low altitude to high altitude and low latitude to high latitude
(d) high altitude to low altitude and low latitude to high latitude

116. Sacred groves are specially useful in

(a) preventing soil erosion.
(b) year-round flow of water in rivers.
(c) generating environmental awareness.
(d) conserving rare and threatened species.

117. Milky water of green coconut is

(a) liquid nucellus
(b) liquid of female gametophyte
(c) liquid endosperm
(d) liquid embryo

118. Which one of the following is the major difference between mosses and ferns ?

(a) Ferns lack alternation of generation while mosses show the same.
(b) Mosses are facultative aerobes while ferns are obligate aerobes.
(c) Vascular bundles of ferns show xylem vessels while those of mosses lack it.
(d) Sporophytes of ferns live much longer as compared to the sporophytes of mosses.

119. Red snow causing alga is

(a) *Chlamydomonas nivalis*
(b) *Chlamydomonas reinhardtii*
(c) *Chlamydomonas debaryanum*
(d) *Chalmydomonas media*

120. Which of the following statement (s) is/are incorrect?

(i) Calyx and corolla are reproductive organs of a flower.
(ii) Zygomorphic flower can be divided into two equal radial halves in any radial plane.
(iii) Flowers without bracts are termed as bracteate.
(iv) Parthenocarpic fruit is formed after fertilization of the ovary.
(v) In legumes, seed is non-endospermic.
(vi) Radical buds develop on roots.

(a) (i), (ii), (iii) and (iv) (b) (i), (ii) and (v)
(c) (iii), (iv) and (vi) (d) (i), (iv) and (v)

121. How many pair of nucleotides are present in one helix of B- DNA?

(a) 10 (b) 12
(c) 5 (d) 6

122. Conversion of ammonia into nitrates through *Nitrosomonas* is called

(a) nitrogen fixation (b) nitrification
(c) denitrification (d) ammonification

123. Photorespiration is favoured by

(a) high O_2 and low CO_2
(b) low light and high O_2
(c) low temperature and high O_2
(d) low O_2 and high CO_2

124. In the leaves of C_4 plants, malic acid formation during CO_2 fixation occurs in the cells of

(a) bundle sheath (b) phloem
(c) epidermis (d) mesophyll

125. A plant completing its life cycle before the onset of dry condition is said to be

(a) short day plant (b) long day plant
(c) drought escaping (d) All of these

126. The graph shows the relation between light intensity and the giving off and taking up of carbon dioxide by the leaves of a plant. Why is most carbon dioxide given off when the light intensity is zero units ?

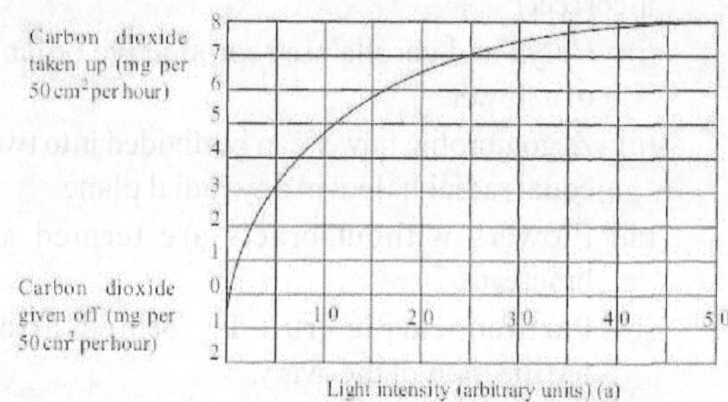

(a) Because it is just the start of the experiment.
(b) Only respiration is taking place at this intensity of light.
(c) Only photosynthesis is taking place at this intensity of light.
(d) The rate of photosynthesis is equivalent to the rate of respiration.

127. Which one of the following statement is false?
(i) Epidermal cell has small amount of cytoplasm and a large vacuole.
(ii) Waxy cuticle layer is absent in roots.
(iii) Root hairs are unicellular, while stem hairs / trichomes are multicellular.
(iv) Trichomes may be branched or unbranched, soft or stiff and prevent transpiration.
(v) Guard cells are dumbell shaped in dicots and bean-shaped in monocots.
(a) (i) only (b) (iv) only
(c) (iii) only (d) (v) only

128. In photosystem-I the first electron acceptor is
(a) cytochrome
(b) plastocyanin
(c) an iron-sulphur protein
(d) ferredoxin

129. Which one of the following graphs most closely represents the relationship between the rate of transpiration of a mesophytic leaf and the atmospheric humidity ?

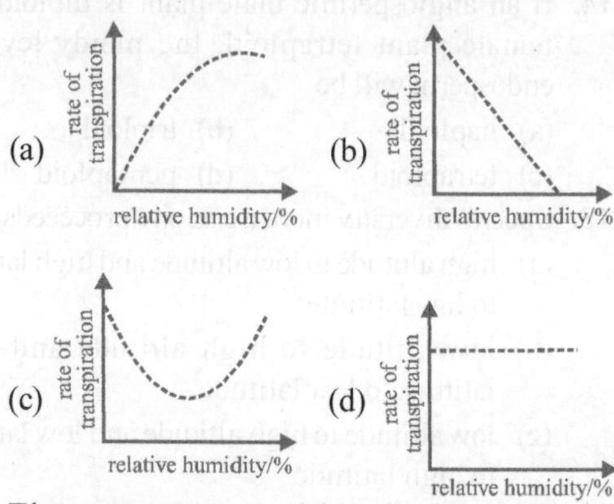

130. The two gases making highest relative contribution to the greenhouse gases are
(a) CO_2 and CH_4
(b) CH_4 and N_2O
(c) CFCs and N_2O
(d) CO_2 and N_2O

131. The upright pyramid of number is absent in
(a) Pond (b) Forest
(c) Lake (d) Grassland

132. Biolistics (gene-gun) is suitable for
(a) DNA finger printing.
(b) Disarming pathogen vectors.
(c) Transformation of plant cells.
(d) Constructing recombinant DNA by joining with vectors.

133. Read the following four statements (i - iv).
(i) In transcription, adenosine pairs with uracil.
(ii) Regulation of *lac* operon by repressor is referred to as positive regulation.
(iii) The human genome has approximately 50,000 genes.
(iv) Haemophilia is a sex-linked recessive disease.
How many of the above statements are correct?
(a) Two (b) Three
(c) Four (d) One

134. The interaction is determental to both the species, in
(a) Predation (b) Commensalism
(c) Ammensalism (d) Competition

135. Which option is true about heart wood/ duramen?
(i) It does not help in water and mineral conduction.
(ii) It is dark coloured but soft.
(iii) It has tracheary elements filled with tannins, resins, gums, oil, etc.
(iv) It is a peripheral part.
(v) Sensitive to microbes and insects, hence least durable.
(a) (i) and (iii) (b) (ii) and (iii)
(c) (iv) and (v) (d) (iii) and (iv)

SECTION 4 - ZOOLOGY

136. A few organisms are known to grow and multiply at temperatures of 100–105°C. They belong to
(a) marine archaebacteria
(b) thermophilic sulphur bacteria
(c) blue-green algae (cyanobacteria)
(d) thermophilic, subaerial fungi

137. *Archaeopteryx* is connecting link between
(a) Reptiles and birds
(b) Reptiles and mammals
(c) Fishes and reptiles
(d) Chordates and nonchordates

138. The fibres of the following muscles are fusiform and do not show striations
(a) Skeletal muscles (b) Cardiac muscles
(c) Both (a) and (b) (d) Smooth muscles

139. The most abundant mineral in human body is
(a) Magnesium (b) Sodium
(c) Calcium (d) Potasium

140. Cholecystokinin is secretion of
(a) Duodenum that causes contraction of gall bladder
(b) Globlet cells of ileum, stimulates secretion of succus entericus
(c) Liver and controls secondary sex characters
(d) Stomach that stimulates pancreas to release juice

141. The volume of air breathed in and out during normal breathing is called
(a) Vital capacity
(b) Inspiratory reserve volume
(c) Explratory reserve volume
(d) Tidal volume

142. Solenocytes and metanephridia are excretory organs of
(a) Annelida and Arthropoda
(b) Platyhelminthes and Annelida
(c) Coelenterata and Mollusca
(d) Aschelminthes and Annelida

143. Atrial natriuretic factor (ANF) is released in response to the increase in blood volume and blood pressure. Which of the followings is not the function of ANF?
(a) Stimulates aldosterone secretion
(b) Inhibits the release of renin from JGA
(c) Stimulates salt loss in urine
(d) Inhibits sodium reabsorption from collecting duct

144. Which of the following diseases is caused by the under secreation of cortisol?
(a) Anaemia
(b) Addison's disease
(c) Hyperglycemia
(d) Mental illness or retardation

145. Glycogen is converted to glucose by
(a) Insulin
(b) Glucagon
(c) Galactose
(d) Both glucagons and insulin

146. Which of the following sets represents vestigial organs?
(a) Vermiform appendix, body hair and patella
(b) Wisdom teeth, body hair and atlas vertebra
(c) Ear muscles, cochlea and coccyx
(d) Vermiform appendix, ear muscles and coccyx.

147. Hepatitis B is transmitted through
(a) Sexual contact
(b) Blood transfusion
(c) Contaminated food and water
(d) All of the above

148. Which one of the following pairs of hormones are the examples of those that can easily pass through the cell membrane of the target cell and bind to a receptor inside it (Mostly in the nucleus)?
(a) Insulin, glucagon
(b) Thyroxin, insulin
(c) Somatostain, oxytocin
(d) Cortisol, testosterone

149. Where do certain symbiotic micro-organisms normally occur in human body?
(a) Caecum
(b) Oral lining and tongue surface
(c) Vermiform appendix and rectum
(d) Duodenum

150. Radial symmetry occurs in
(a) Porifera and Coelenterata
(b) Coelenterata and Echinodermata
(c) Coelenterata and Platyhelminthes
(d) Arthropoda and Mollusca

151. Which of the following statements is/are not true?
(i) In Urochordata, notochord is present in larval tail.
(ii) In Cephalochordata, notochord extends from head to tail region.
(iii) *Branchiostoma* belongs to hemichordata.
(iv) Only one class of living members, class Cyclostomata represents the super class agnatha
(a) (ii) and (iv) (b) (i), (iii) and (iv)
(c) (iii) only (d) (i) and (iv)

152. Mycoplasma is pleuromorphic due to
(a) absence of cell wall
(b) presence of three layered cell membrane
(c) the presence of sterol
(d) None of these

153. Life span of human RBCs is
(a) 120 days (b) 20 days
(c) 9 days (d) 90 days

154. Newly born child is more resistant to cold because of the presence of
(a) Thick dermis (b) Blubber
(c) Brown fat (d) White fat

155. Which part of the alimentary canal does not secrete any enzyme?
(a) Mouth (b) Oesophagus
(c) Stomach (d) Duodenum

156. The protein coated, water soluble fat globules are called
(a) Chylomicrons (b) Micelles
(c) Chyle (d) Monoglycerides

157. What would happen if human blood becomes acidic (low pH)
(a) Oxygen carying capacity of haemoglobin increases
(b) Oxygen carrying capacity of haemoglobin decreases
(c) RBCs count increases
(d) RBCs count decreases

158. Which one of the following is a matching pair?
(a) Lubb-Sharp closure of AV valves at the beginning of ventricular systole
(b) Dup-Sudden opening of semilunar valves at the beginning of ventricular diastole
(c) Pulsation of the radial artery-Valves in the blood vessels
(d) Initiation of the heart beat -Purkinje fibres

159. Reabsorption of chloride ions from glomerular filtrate in kidney tubule occurs by
(a) Active transport
(b) Diffusion
(c) Osmosis
(d) Brownian movement

160. Ankle joint is
(a) Pivot Joint (b) Ball and socket joint
(c) Hinge joint (d) Gliding joint

161. The major function of the intervertebral disc is to
(a) Absorb shock
(b) String the vertebrae together
(c) Prevent injuries
(d) Prevent hyperextension

162. A decrease in the level of oestrogen and progesterone causes
(a) Growth and dilation of myometrium
(b) Growth of endometrium
(c) Constriction of uterine blood vessels leading to sloughing of endometrium or uterine epithelium
(d) Release of ovum from the ovary.

163. Structure connecting the foetus to placenta is
(a) umbilical cord (b) amnion
(c) yolk sac (d) chorion

164. In India which of the following species of honey bee is reared in artificial hives ?
(a) *Apis indica* (b) *Apis dorsata*
(c) *Apis florae* (d) None of these

165. A person entering an empty room suddenly finds a snake right in front on opening the door. Which one of the following is likely to happen in his neuro-hormonal control system ?
(a) Sympathetic nervous system is activated releasing epinephrine and norepinephrine from adrenal medulla.
(b) Neurotransmitters diffuse rapidly across the cleft and transmit a nerve impulse.
(c) Hypothalamus activates the parasympathetic division of brain.
(d) Sympathetic nervous system is activated releasing epinephrine and norepinephrine from adrenal cortex.

166. Which one of the following pairs of chemical substances is correctly categorized?
(a) Calcitonin and thymosin - Thyroid hormones
(b) Pepsin and prolactin - Two digestive enzymes secreted in stomach
(c) Troponin and myosin - Complex proteins in striated muscles
(d) Secretin and rhodopsin - Polypeptide hormones

167. The blind spot is the region where
(a) Image is formed
(b) Cones are numerous
(c) The optic nerve leaves out
(d) Image is formed during the dark

168. The chemical substance found in the surface layer of cytoplasm of spermatozoa is:
(a) fertilizin (b) agglutinin
(c) antifertilizin (d) hyaluronidase

169. Read the following 4-statements and mark the option that has both correct statements
(i) MTP was legalized in 1971
(ii) Inability to conceive or produce children even after 2 years of unprotected sexual cohabitation is called infertility
(iii) Surgical method of contraception prevents gamete formation
(iv) MTPs are relatively safe up to 12 weeks of pregnancy
(a) (i) and (ii) (b) (ii) and (iii)
(c) (iii) and (iv) (d) (i) and (iii)

170. Erythroblastosis foetalis occurs when
(a) Mother is Rh negative and father is Rh positive
(b) Father is Rh negative and mother is Rh positive
(c) Both are Rh positive
(d) Both are Rh negative

171. Chromosome complement with 2n – 1 is called
(a) Monosomy (b) Trisomy
(c) Nullisomy (d) Tetrasomy

172. Which of the following is a minerelocorticoid?
(a) Calciferol (b) Progesterone
(c) Adrenalin (d) Aldosterone

173. In myopia or short sightedness
(a) Image is formed slightly in front of retina because eye ball is longer
(b) Eye ball is normal but image is formed over blind spot
(c) Eye ball is normal but images is formed slightly behind the retina due to faulty lens
(d) Curvature of cornea becomes irregular

174. Sequence of meninges from inner to outside is
(a) Duramater – Archnoid – Piamater
(b) Duramater – Piamater – Arachnoid
(c) Arachnoid – Duramater - Piamater
(d) Piamater- Arachnoid - Duramater

175. The functional unit of contractile system of a striated muscle is
(a) Sarcomere (b) Z-band
(c) Sarcosome (d) Myofibril

176. Joint between bones of human skull is
(a) Hinge joint (b) Synovial joint
(c) Cartilaginous joint (d) Fibrous joint

177. Which one of the following is the *correct description* of a certain part of the normal human skelection ?
(a) Parietal bone and the temporal bone of the skull are joined by fibrous joint
(b) First vertebra is axis which articulates with the occipital condyles
(c) The 9th and 10th pairs of ribs are called the floating ribs
(d) Glenoid cavity is a depression to which the thigh bone articulates.

178. Metanephric kidneys are found in
(a) Reptiles only (b) Birds only
(c) mammals only (d) All of these

179. Which of the following set of animals has an incomplete double circulation system?
(a) Frog and crocodile (b) Shark and whale
(c) Lizard and pigeon (d) Toad and lizard

180. Study the pedigree chart given below:

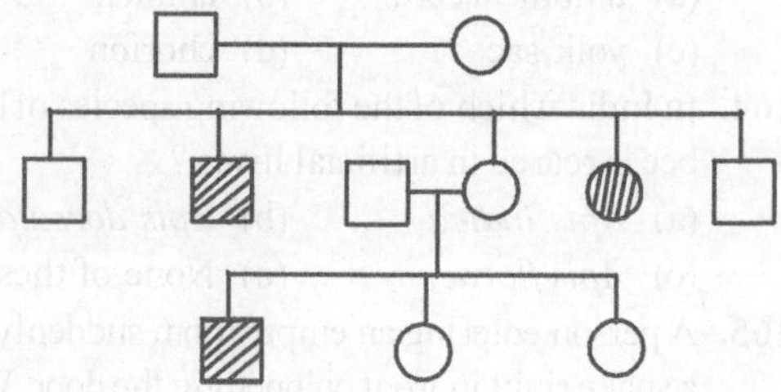

What does it show?
(a) Inheritance of a recessive sex-linked disease like haemophilia
(b) Inheritance of a sex-linked inborn error of metabolism like phenylketonuria
(c) Inheritance of a condition like phenylketonuria as an autosomal recessive trait
(d) The pedigree chart is wrong as this is not possible

RESPONSE SHEET

Test Code: .... Time taken :.................... Date:..............

	(a)	(b)	(c)	(d)		(a)	(b)	(c)	(d)		(a)	(b)	(c)	(d)		(a)	(b)	(c)	(d)
1.					46.					91.					136.				
2.					47.					92.					137.				
3.					48.					93.					138.				
4.					49.					94.					139.				
5.					50.					95.					140.				
6.					51.					96.					141.				
7.					52.					97.					142.				
8.					53.					98.					143.				
9.					54.					99.					144.				
10.					55.					100.					145.				
11.					56.					101.					146.				
12.					57.					102.					147.				
13.					58.					103.					148.				
14.					59.					104.					149.				
15.					60.					105.					150.				
16.					61.					106.					151.				
17.					62.					107.					152.				
18.					63.					108.					153.				
19.					64.					109.					154.				
20.					65.					110.					155.				
21.					66.					111.					156.				
22.					67.					112.					157.				
23.					68.					113.					158.				
24.					69.					114.					159.				
25.					70.					115.					160.				
26.					71.					116.					161.				
27.					72.					117.					162.				
28.					73.					118.					163.				
29.					74.					119.					164.				
30.					75.					120.					165.				
31.					76.					121.					166.				
32.					77.					122.					167.				
33.					78.					123.					168.				
34.					79.					124.					169.				
35.					80.					125.					170.				
36.					81.					126.					171.				
37.					82.					127.					172.				
38.					83.					128.					173.				
39.					84.					129.					174.				
40.					85.					130.					175.				
41.					86.					131.					176.				
42.					87.					132.					177.				
43.					88.					133.					178.				
44.					89.					134.					179.				
45.					90.					135.					180.				

TEST ASSESSMENT AND ANALYSIS SHEET

How to fill this sheet : After you finish the paper and have checked it first try to attempt again all the questions which you have done wrong or have not been able to attempt. Go to detailed solutions only after you have attempted all the questions of your own.Attempting the questions on your own without seeing the solutions will help you in filling this table

ANALYSIS OF TEST	
Total number of questions	
Questions Attempted	
Correct Questions	
Wrong Questions	
	Net Score

Strike rate $\left(\frac{\text{Correct questions}}{\text{Questions attempt}} \times 100\right) =$ ....

Percentage Marks $\left(\frac{\text{Correct Answers}}{\text{Total number of questions}} \times 100\right) =$ ....

Analysis of wrong questions : Divide the questions which you have done wrong in 3 broad categories.

Reasons for questions attempted wrong	Number of Questions
Knew but calculation mistake (solved on my own after exam)	
Knew but got confused (solved on my own after exam)	
Did not knew the question (couldn't solve even after exam)	

Questions not attempted : Divide the questions not attempted in 3 broad categories

Reasons for unattempted questions	Number of Questions
Knew the ques but time was less (solved on my own after exam)	
Knew the ques but got confused (solved on my own after exam)	
Did not knew the questions (couldn't solve even after exam)	

List down your weak areas you need to improve your score :

1)____________________

2)____________________

3)____________________

NEET

National Eligibility Entrance Test

Test Series **Full Test - 9**

Time : - 3 Hours **Max. Marks:- 720**

INSTRUCTIONS

1. Immediately fill in the particulars on this page of the Test Booklet with Blue/Black Ball Point Pen. Use of pencil is strictly prohibited.
2. Answers are to be given on Response Sheet provided at the end of this paper. You can tear this Response Sheet along the cut marks provided. Fill in the particulars carefully in the Response Sheet provided.
3. The test is of 3 hours duration.
4. The Test Booklet consists of **180** questions. The maximum marks are **720**.
5. There are four section in the question paper.

   The distribution of marks subjectwise in each part is as under for each correct response.

   Section 1 – Physics (180 Marks) – 45 Questions

   Section 2 – Chemistry (180 Marks) – 45 Questions

   Section 3 – Botany (180 Marks) – 45 Questions

   Section 4 – Zoology (180 Marks) – 45 Questions
6. You will be awarded 4 Marks if you have darkened only the bubble corresponding to the correct answer and zero mark if no bubble is darkened. In all other cases, minus one (– 1) mark will be awarded.
7. Use Blue/Black Ball Point Pen only for writing particulars/marking responses on Side-1 and Side-2 of the Response Sheet.

   Use of pencil is strictly prohibited.
8. You are advised to create an actual examination hall simulation.
9. Rough work is to be done on the space provided for this purpose in the Test Booklet only.
10. Do not fold or make any stray marks on the Response Sheet.
11. Once you have evaluated the marks in the test, you should assess it thoroughly.

SECTION 1 - PHYSICS

1. The respective number of significant figures for the numbers 23.023, 0.0003 and 2.1×10^{-3} are
(a) 5, 1, 2 (b) 5, 1, 5
(c) 5, 5, 2 (d) 4, 4, 2

2. If the kinetic energy of a particle is increased by 300%, the momentum of the particle will increase by
(a) 20% (b) 200%
(c) 100% (d) 50%

3. The pulleys and strings shown in the figure are smooth and of negligible mass. For the system to remain in equilibrium. The angle θ should be

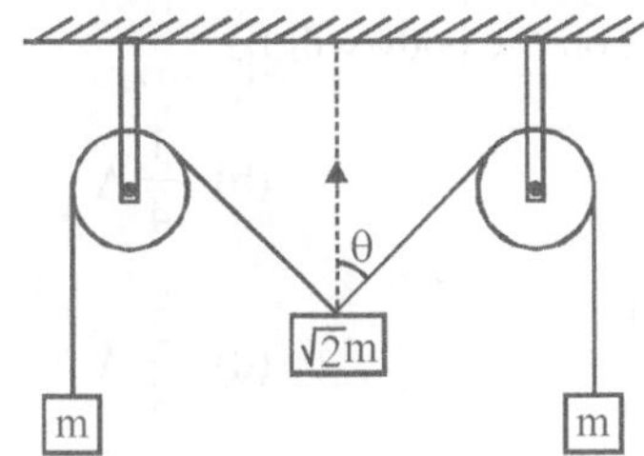

(a) 0° (b) 30°
(c) 45° (d) 60°

4. A particle starting from rest falls from a certain height. Assuming that the acceleration due to gravity remain the same throughout the motion, its displacements in three successive half second intervals are S_1, S_2 and S_3 then
(a) $S_1 : S_2 : S_3 = 1 : 5 : 9$
(b) $S_1 : S_2 : S_3 = 1 : 3 : 5$
(c) $S_1 : S_2 : S_3 = 9 : 2 : 3$
(d) $S_1 : S_2 : S_3 = 1 : 1 : 1$

5. The Bulk modulus for an incompressible liquid is
(a) zero (b) unity
(c) infinity (d) between 0 and 1

6. If $\hat{n}$ is a unit vector in the direction of the vector $\vec{A}$, then
(a) $\hat{n} = \frac{\vec{A}}{|\vec{A}|}$ (b) $\hat{n} = \vec{A}|\vec{A}|$
(c) $\hat{n} = \frac{|\vec{A}|}{\vec{A}}$ (d) $\hat{n} = \hat{n} \times \vec{A}$

7. A transformer is employed to
(a) convert A.C. into D.C.
(b) convert D.C. into A.C.
(c) obtain a suitable A.C. voltage
(d) obtain a suitable D.C. voltage

8. As a result of change in the magnetic flux linked to the closed loop shown in the fig, an e.m.f. V volt is induced in the loop.

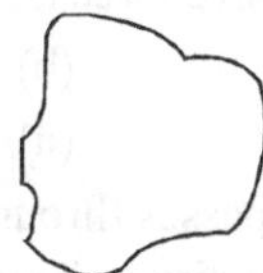

The work done (in joule) in taking a charge Q coulomb once along the loop is
(a) QV (b) $2QV$
(c) $QV/2$ (d) zero

9. What is the acceleration of a projectile at its heighest point
(a) maximum (b) minimum
(c) zero (d) g

10. A mass of 2kg is whirled in a horizontal circle by means of a string at initial speed of 5 revolution per minute. Keeping the radius constant, the tension in the string is doubled. The new speed is nearly
(a) 2.25 rpm (b) 7 rpm
(c) 10 rpm (d) 14 rpm

11. In a two dimensional motion, instantaneous speed v_0 is a positive constant. Then which of the following is necessarily true?

(a) The average velocity is not zero at any time
(b) Average acceleration must always vanish
(c) Displacements in equal time intervals are equal
(d) Equal plane lengths are traversed in equal intervals

12. A sphere of radius R has uniform volume charge density. The electric potential at a point ($r < R$) is
(a) due to the charge inside a sphere of radius r only
(b) due to the entire charge of the sphere
(c) due to the charge in the spherical shell of inner and outer radii r and R, only
(d) independent of r

13. A ball of mass 150 g, moving with an acceleration 20 m/s^2, is hit by a force, which acts on it for 0.1 sec. The impulsive force is
(a) 0.5 N (b) 0.1 N
(c) 0.3 N (d) 1.2 N

14. A ray of light passes through four transparent media with refractive indices μ_1, μ_2, μ_3 and μ_4 as shown in the figure. The surfaces of all media are parallel. If the emergent ray CD is parallel to the incident ray AB, we must have

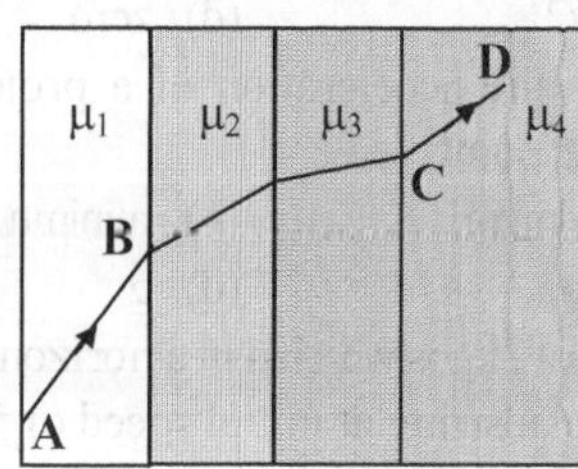

(a) $\mu_1 = \mu_2$ (b) $\mu_2 = \mu_3$
(c) $\mu_3 = \mu_4$ (d) $\mu_4 = \mu_1$

15. The meniscus of a liquid contained in one of the limbs of a narrow U-tube is held in an electromagnet with the meniscus in line with the field. The liquid is seen to rise. This indicates that the liquid is
(a) ferromagnetic (b) paramagnetic
(c) diamagnetic (d) non-magnetic

16. A body of mass 3 kg is under a constant force which causes a displacement s in meter in it, given by the relation $s = \frac{1}{3}t^2$, where t is in second. Work done by the force in 2 second is
(a) $\frac{3}{8}$ J (b) $\frac{8}{3}$ J
(c) $\frac{19}{5}$ J (d) $\frac{5}{19}$ J

17. Under a constant torque the angular momentum of a body changes from A to 4A in 4 second. The torque on the body will be
(a) 1A (b) $\frac{1}{4}$ A
(c) $\frac{4}{3}$ A (d) $\frac{3}{4}$ A

18. The kinetic theory of gases
(a) explains the behaviour of an ideal gas
(b) describes the motion of a single atom or molecule
(c) relates the temperature of the gas with K.E. of atoms of the gas
(d) all of the above

19. An electric bulb marked 40 W and 200V, is used in a circuit of supply voltage 100V. Now its power is
(a) 10 W (b) 20 W
(c) 40 W (d) 100 W

20. Two wires of same material and length but cross-sections in the ratio 1 : 2 are used to suspend the same loads. The extensions in them will be in the ratio of
(a) 1 : 2 (b) 2 : 1
(c) 4 : 1 (d) 1 : 4

21. The moment of inertia of a disc of mass m and radius R about an axis, which is tangential to the circumference of the disc and parallel to its diameter is

(a) $\frac{3}{2}mR^2$ (b) $\frac{2}{3}mR^2$

(c) $\frac{5}{4}mR^2$ (d) $\frac{4}{5}mR^2$

22. The height at which the acceleration due to gravity becomes $\frac{g}{9}$ (where g = the acceleration due to gravity on the surface of the earth) in terms of R, the radius of the earth, is

(a) $\frac{R}{\sqrt{2}}$ (b) $R/2$

(c) $\sqrt{2}R$ (d) $2R$

23. A thermodynamic system is taken from state A to B along ACB and is brought back to A along BDA as shown in the PV diagram. The net work done during the complete cycle is given by the area

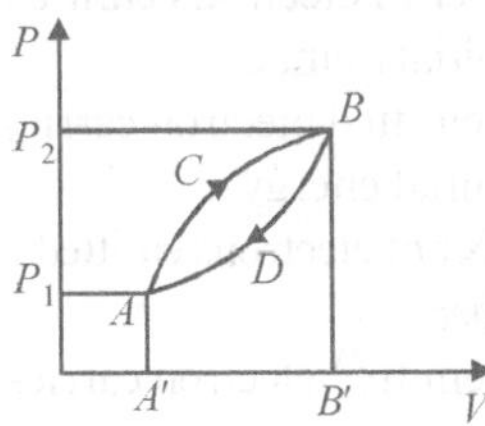

(a) $P_1ACBP_2P_1$ (b) $ACBB'A'A$

(c) $ACBDA$ (d) $ADBB'A'A$

24. A man is watching two trains, one leaving and the other coming with equal speed of 4 m/s. If they sound their whistles each of frequency 240 Hz, the number of beats heard by man (velocity of sound in air = 320 m/s) will be equal to

(a) 12 (b) 0

(c) 3 (d) 6

25. Two identical galvanometers are taken, one is to be converted into an ammeter and other into a milliammeter. Shunt of milliammeter compared to ammeter is

(a) less (b) more

(c) zero (d) None of these

26. When a wave travels in a medium the particles displacement is given by the equation y = 0.03 sin π(2t – 0.01x), where x and y are in seconds. The wavelength of the wave is

(a) 200m (b) 100m

(c) 20m (d) 10m

27. The value of current I in the circuit shown in figure is

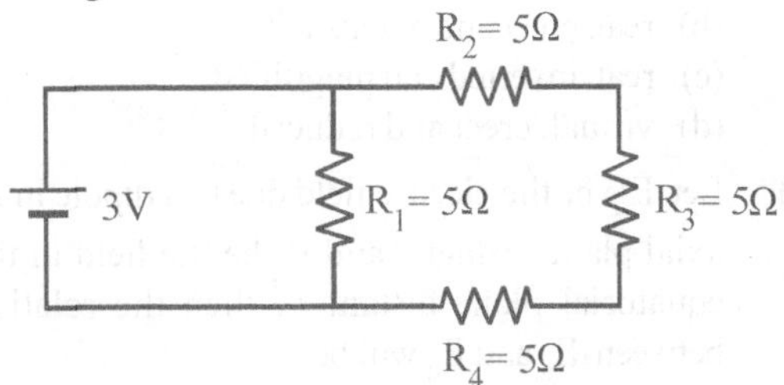

(a) 1.8 A (b) 0.8 A

(c) 0.2 A (d) 1.6 A

28. The oscillating electric and magnetic field vectors of electromagnetic wave are oriented along

(a) the same direction and in phase

(b) the same direction but have a phase difference of 90°

(c) mutually perpendicular directions and are in phase

(d) mutually perpendicular directions but has a phase difference of 90°

29. A particle of mass 10 g is kept on the surface of a uniform sphere of mass 100 kg and radius 10 cm. Find the work to be done against the gravitational force between them to take the particle far away from the sphere (Take $G = 6.67 \times 10^{-11} Nm^2/kg^2$)

(a) 3.33×10^{-10} J (b) 13.34×10^{-10} J

(c) 6.67×10^{-10} J (d) 6.67×10^{-9} J

30. If the cold junction of a thermo-couple is kept at 0°C and the hot junction is kept at T°C then the relation between neutral temperature (T_n) and temperature of inversion (T_i) is

(a) $T_n = 2T_i$ (b) $T_n = T_i - T$
(c) $T_n = T_i + T$ (d) $T_n = T_i/2$

31. An engine has an efficiency of 1/6. When the temperature of sink is reduced by 62°C, its efficiency is doubled. Temperature of the source is

(a) 37°C (b) 62°C
(c) 99°C (d) 124°C

32. In a compound microscope, the intermediate image is

(a) virtual, erect and magnified
(b) real, erect and magnified
(c) real, inverted and magnified
(d) virtual, erect and reduced

33. Let E_a be the electric field due to a dipole in its axial plane distant ℓ and E_q be the field in the equatorial plane distant ℓ', then the relation between E_a and E_q will be

(a) $E_a = 4E_q$ (b) $E_q = 2E_a$
(c) $E_a = 2E_q$ (d) $E_q = 3E_a$

34. The capacitance between points *P* and *Q* in circuit shown in fig is

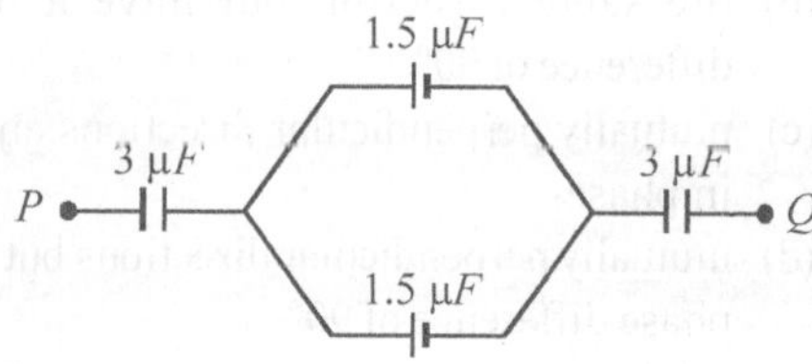

(a) 1μF (b) 1.5 μF
(c) 9 μF (d) 6.75 μF

35. Two coherent monochromatic light beams of intensities *I* and 4 *I* are superimposed. The maximum and minimum possible intensities in the resulting beam are

(a) 5*I* and *I* (b) 5*I* and 3*I*
(c) 9*I* and *I* (d) 9*I* and 3*I*

36. When an electron jumps from the fourth orbit to the second orbit, one gets the

(a) second line of Lyman series
(b) second line of Paschen series
(c) second line of Balmer series
(d) first line of Pfund series

37. Solar radiation emitted by sun resembles that emitted by a black body at a temperature of 6000 K. Maximum intensity is emitted at a wavelength of about 4800 Å. If the sun were to cool down from 6000 K to 3000K, then the peak intensity would occur at a wavelength

(a) 4800 Å (b) 9600 Å
(c) 7200 Å (d) 6400 Å

38. A uniform magnetic field acts at right angles to the direction of motion of electron. As a result, the electron moves in a circular path of radius 2 cm. If the speed of electron is doubled, then the radius of the circular path will be

(a) 2.0 cm (b) 0.5 cm
(c) 4.0 cm (d) 1.0 cm

39. A photoelectric cell is illuminated by a point source of light 1 m away. When the source is shifted to 2 m, then

(a) number of electrons emitted is a quarter of the initial number
(b) each emitted electron carries one quarter of the initial energy
(c) number of electrons emitted is half the initial number
(d) each emitted electron carries half the initial energy

40. A hollow cylinder has *a* charge *q* coulomb within *i*. If ϕ is the electric flux in units of voltmeter associated with the curved surface *B*, the flux linked with the plane surface *A* in units of voltmeter will be

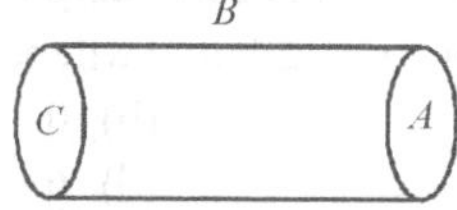

(a) $\frac{q}{2\varepsilon_0}$ (b) $\frac{\phi}{3}$

(c) $\frac{q}{\varepsilon_0}-\phi$ (d) $\frac{1}{2}\left(\frac{q}{\varepsilon_0}-\phi\right)$

41. Two flat circular coils have a common center, but their planes are at right angles to each other. The inner coil has 150 turns and radius of π cm. The outer coil has 400 turns and a radius of 2π cm. The magnitude of the resultant magnetic induction at the common centers of the coils when a current of 200 mA is sent through each of them is

(a) 10^{-3} Wb/m^2

(b) 2×10^{-3} Wb/m^2

(c) 5×10^{-3} Wb/m^2

(d) 7×10^{-3} Wb/m^2

42. The cathode of a photoelectric cell is changed such that the work function changes from W_1 to W_2 $(W_2 > W_1)$. If the current before and after changes are I_1 and I_2, all other conditions remaining unchanged, then (assuming $h\nu > W_2$)

(a) $I_1 = I_2$ (b) $I_1 < I_2$

(c) $I_1 > I_2$ (d) $I_1 < I_2 < 2I_1$

43. Optical fibres transmit light along its axis, by the process of

(a) total internal reflection

(b) refraction

(c) interference

(d) diffraction

44. When a *p-n* junction diode is reverse biased the flow of current across the junction is mainly due to

(a) diffusion of charges

(b) drift of charges

(c) depends on the nature of material

(d) both drift and diffusion of charges

45. If in a p-n junction diode, a square input signal of 10 V is applied as shown

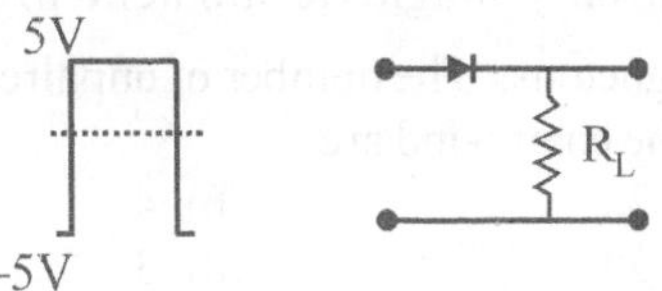

Then the output signal across R_L will be

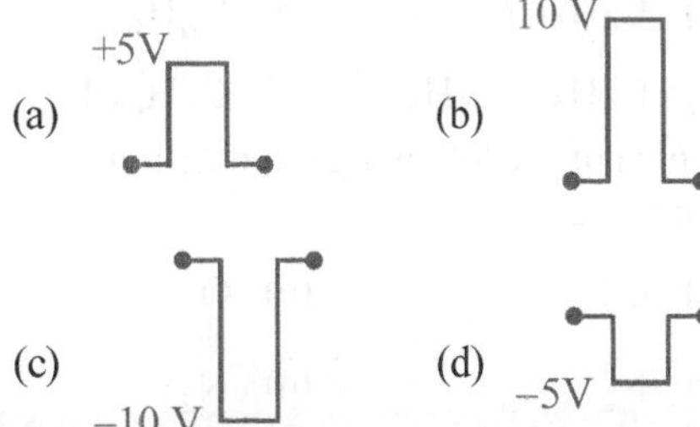

SECTION 2 - CHEMISTRY

46. The values of Planck's constant is 6.63×10^{-34} Js. The velocity of light is 3.0×10^{8} m s^{-1}. Which value is closest to the wavelength in nanometres of a quantum of light with frequency of 8×10^{15} s^{-1}?

(a) 5×10^{-18} (b) 4×10^{1}

(c) 3×10^{7} (d) 2×10^{-25}

47. Activation energy of the reaction is

(a) the energy released during the reaction

(b) the energy evolved when activated complex is formed

(c) minimum amount of energy needed to overcome the potential barrier

(d) the energy needed to form one mole of the product

48. Which of the following order is wrong?

(a) $NH_3 < PH_3 < AsH_3$ — Acidic

(b) Li < Be < B < C — IE_1

(c) $Al_2O_3 < MgO < Na_2O < K_2O$ — Basic

(d) $Li^+ < Na^+ < K^+ < Cs^+$ — Ionic radius

49. A compound of a metal ion $M^{x+}(Z = 24)$ has a spin only magnetic moment of $\sqrt{15}$ Bohr Magnetons. The number of unpaired electrons in the compound are

(a) 2 (b) 4
(c) 5 (d) 3

50. The reaction $C_6H_5N_2{}^+Cl^-$ with CuCl gives

(a) C_6H_5Cl (b) C_6H_6
(c) $C_6H_5-C_6H_5$ (d) $C_6H_4Cl_2$

51. Which of the following ion does not have bond order of 2.5 ?

(a) O_2^- (b) O_2^+
(c) N_2^+ (d) N_2^-

52. An ideal gas expands in volume from 1×10^{-3} to $1\times10^{-2}m^3$ at 300 K against a constant pressure of 1×10^5 Nm^{-2}. The work done is

(a) 270 kJ (b) −900 kJ
(c) −900 J (d) 900 kJ

53. On the basis of the information available from the reaction $\frac{4}{3}Al + O_2 \longrightarrow \frac{2}{3}Al_2O_3$, $\Delta G = -827\ kJmol^{-1}$ of O_2, the minimum e.m.f. required to carry out electrolysis of Al_2O_3 is ($F = 96500\ C\ mol^{-1}$)

(a) 4.28 V (b) 6.42 V
(c) 8.56 V (d) 2.14 V

54. Iron sulphide is heated in air to form A, an oxide of sulphur. A is dissolved in water to give an acid. The basicity of this acid is

(a) 2 (b) 3
(c) 1 (d) Zero

55. Which of the following cannot be made by using Williamson's synthesis?

(a) Methoxybenzene
(b) Benzyl p-nitrophenyl ether
(c) Methyl tertiary butyl ether
(d) Di-tert-butyl ether

56. Solubility of a substance which dissolves with a decrease in volume and absorption of heat will be favoured by

(a) High P and High T (b) Low P and Low T
(c) High P and Low T (d) Low P and High T

57. Which of the following is an expression of Raoults law if P_A is the partial pressure of the solvent in a solution. $P^0{}_A$ is the partial pressure of pure solvent and if X_A and X_B are the mole fraction of the solute and the solvent respectively?

(a) $P_A = P^o_A X_A$ (b) $P_A = P^o_A X_A\left(\frac{1}{X_B}\right)$
(c) $P_A = P^o_A X_B$ (d) $P^o_A = P\left(\frac{X_A}{X_B}\right)$

58. Standard potentials (E°) for some half-reactions are given below :

(I) $Sn^{4+} + 2e \longrightarrow Sn^{2+}$; $E° = +0.15V$

(II) $2Hg^{2+} + 2e \longrightarrow Hg_2^{2+}$; $E° = 0.92V$

(III) $PbO_2 + 4H^+ + 2e \longrightarrow Pb^{2+} + 2H_2O$; $E° = +1.45V$

based on the above, which one of the following statements is correct?

(a) Sn^{4+} is a stronger oxidising agent than Pb^{4+}
(b) Sn^{2+} is a stronger reducing agent than $Hg_2{}^{2+}$
(c) Pb^{2+} is a stronger oxidisng agent than Pb^{4+}
(d) Pb^{2+} is a stronger reducing agent than Sn^{2+}

59. A metal which is not affected by conc. H_2SO_4, HNO_3 or alkalies forms a compound X This compound X can be used to give a complex which finds its application for toning in photography. The metal is

(a) Au (b) Ag
(c) Hg (d) Cu

60. Phenol on reaction with Br_2 in non-polar aprotic solvent furnishes
(a) 2, 4, 6-Tribromophenol
(b) *p*-Bromophenol
(c) *o- and p*-Bromophenol
(d) *m*-Bromophenol

61. The rate of forward reaction is two times that of the reverse reaction at a given temperature and identical concentration. $K_{equilibrium}$ is
(a) 0.5 (b) 1.5
(c) 2.5 (d) 2.0

62. A metallic crystal crystallizes into a lattice containing a sequence of layers AB AB AB......Any packing of spheres leaves out voids in the lattice. What percentage of volume of this lattice is empty space?
(a) 74% (b) 26%
(c) 50% (d) none of these.

63. Sodium thiosulphate, $Na_2S_2O_3 . 5H_2O$ is used in photography to
(a) Reduce the silver bromide grains to metallic silver
(b) Convert the metallic silver to silver salt
(c) Remove undecomposed AgBr as soluble silver thiosulphate complex.
(d) Remove reduced silver

64. Which among the following factors is the most important in making fluorine the strongest oxidizing halogen ?
(a) Hydration enthalpy
(b) Ionization enthalpy
(c) Electron affinity
(d) Bond dissociation energy

65. α-D- Glucose and β- D-glucose differ from each other due to difference in one carbon with respect to its
(a) Size of hemiacetal ring
(b) Number of OH groups
(c) Configuration
(d) Conformation

66. The K_{sp} of CuS, Ag_2S and HgS are 10^{-31}, 10^{-44} and 10^{-54} respectively. The solubility of these sulphides are in the order
(a) $Ag_2S > CuS > HgS$ (b) $AgS > HgS > CuS$
(c) $HgS > Ag_2S > CuS$ (d) $CuS > Ag_2S > HgS$

67. In which of the following Tyndall effect is not observed ?
(a) Suspensions (b) Emulsions
(c) Sugar solution (d) Gold sol

68. The most electropositive metals are isolated from their ores by
(a) High temperature reduction with carbon
(b) Self reduction
(c) Thermal decomposition
(d) Electrolysis of fused ionic salts

69. The hypothetical complex chloro diaquatriamminecobalt (III) chloride can be represented as
(a) $[CoCl(NH_3)_3(H_2O)_2]Cl_2$
(b) $[Co(NH_3)_3(H_2O)Cl_3]$
(c) $[Co(NH_3)_3(H_2O)_2Cl]$
(d) $[Co(NH_3)_3(H_2O)_3]Cl_3$

70. Which of the following is a polyamide molecule?
(a) Terylene (b) Rayon
(c) Nylon – 6 (d) Polystyrene

71. The electronegativities of F, Cl, Br and I are 4.0, 3.0, 2.8, 2.5 respectively. The hydrogen halide with a high percentage of ionic character is
(a) HF (b) HCl
(c) HBr (d) HI

72. The pH of a solution is increased from 3 to 6; its H^+ ion concentration will be
(a) reduced to half
(b) doubled
(c) reduced by 1000 times
(d) increased by 1000 times

73. The first ionization energies of alkaline earth metals are higher than those of the alkali metals. This is because

(a) There is an increase in the nuclear charge of the alkaline earth metals
(b) There is a decrease in the nuclear charge of the alkaline earth metals
(c) There is no change in the nuclear charge
(d) None of the above

74. Identify the statement that is not correct as far as structure of diborane is concerned
(a) There are two bridging hydrogen atoms and four terminal hydrogen atoms in diborane
(b) Each boron atom forms four bonds in diborane
(c) The hydrogen atoms are not in the same plane in diborane
(d) All, B – H bonds in diborane are similar

75. Which of the following is correct order of acidity?
(a) $HCOOH > CH_3COOH > ClCH_2COOH > C_2H_5COOH$
(b) $ClCH_2COOH > HCOOH > CH_3COOH > C_2H_5COOH$
(c) $CH_3COOH > HCOOH > ClCH_2COOH > C_2H_5COOH$
(d) $C_2H_5COOH > CH_3COOH > HCOOH > ClCH_2COOH$

76. 12 gm of Mg (atomic mass 24) will react completely with hydrochloric acid to give
(a) One mol of H_2
(b) 1/2 mol of H_2
(c) 2/3 mol of O_2
(d) both 1/2 mol of H_2 and 1/2 mol of O_2

77. The enthalpy of formation for C_2H_4 (g) CO_2 (g) and H_2O (l) at 25° C and 1 atm pressure are 52, –394 and – 286 kJ mol^{-1} respectively. The enthalpy of combustion of C_2H_4 (g) will be
(a) +1412 kJ mol^{-1} (b) –1412 kJ mol^{-1}
(c) 141.2 kJ mol^{-1} (d) –141.2 kJ mol^{-1}

78. What will be the emf for the given cell
$Pt \mid H_2(P_1) \mid H^+(aq) \mid H_2(P_2) \mid Pt$
(a) $\frac{RT}{F}\ln\frac{P_1}{P_2}$ (b) $\frac{RT}{2F}\ln\frac{P_1}{P_2}$
(c) $\frac{RT}{F}\ln\frac{P_2}{P_1}$ (d) None of these

79. Ozone is an important constituent of stratosphere because it
(a) Destroys bacteria which are harmful to human life
(b) Prevents the formation of smog over large cities
(c) Removes poisonous gases of the atmosphere by reacting with them
(d) Absorbs ultraviolet radiation which is harmful to human life

80. Which of the following products is formed when benzaldehyde is treated with CH_3MgBr and the addition product so obtained is subjected to acid hydrolysis ?
(a) A secondary alcohol
(b) A primary alcohol
(c) Phenol
(d) tert-Butyl alcohol

81. In which of the following pairs, there is greatest difference in the oxidation number of the underlined elements ?
(a) $\underline{N}O_2$ and $\underline{N}_2O_4$ (b) $\underline{P}_2O_5$ and $\underline{P}_4O_{10}$
(c) $\underline{N}_2O$ and $\underline{N}O$ (d) $\underline{S}O_2$ and $\underline{S}O_3$

82. The boiling point of a solution of 0.11 g of a substance in 15 g of ether was found to be 0.1°C higher than that of pure ether. The molecular weight of the substance will be (K_b = 2.16°K kg mol^{-1})
(a) 148 (b) 158
(c) 168 (d) 178

83. Which one of the following is an example of homogeneous catalysis ?
(a) Haber's process of synthesis of ammonia
(b) Catalytic conversion of SO_2 to SO_3 in contact process
(c) Catalytic hydrogenation of oils
(d) Acid hydrolysis of methyl acetate

84. The decomposition of a substance follows first order kinetics. Its concentration is reduced to 1/8th of its initial value in 24 minutes. The rate constant of the decomposition process is
(a) $1/24\ min^{-1}$
(b) $\frac{0.692}{24}\ min^{-1}$
(c) $\frac{2.303}{24}\log\left(\frac{1}{8}\right) min^{-1}$
(d) $\frac{2.303}{24}\log\left(\frac{8}{1}\right) min^{-1}$

85. The electronic configuration of elements A, B and C are [He] $2s^1$, [Ne] $3s^1$ and [Ar] $4s^1$ respectively. Which one of the following order is correct for IE_1 (in $kJ\ mol^{-1}$) of A, B and C?
(a) A > B > C (b) C > B > A
(c) B > C > A (d) C > A > B

86. An element (X) forms compounds of the formula XCl_3, X_2O_5 and Ca_3X_2 but does not form XCl_5. Which of the following is the element X ?
(a) B (b) Al
(c) N (d) P

87. Which metal is present in brass, bronze and German silver?
(a) Zn (b) Mg
(c) Cu (d) Al

88. Both geometrical and optical isomerisms are shown by
(a) $[Co(en)_2Cl_2]^+$ (b) $[Co(NH_3)_5Cl]^{2+}$
(c) $[Co(NH_3)_4Cl_2]^+$ (d) $[Cr(ox)_3]^{3-}$

89. Which one of the following is most reactive towards nucleophilic substitution reaction?
(a) $CH_2=CH-Cl$
(b) C_6H_5Cl
(c) $CH_3CH=CH-Cl$
(d) $ClCH_2-CH=CH_2$

90. Which of the following is strongest Lewis base?
(a) CH_3^- (b) NH_2^-
(c) OH^- (d) F^-

SECTION 3 - BOTANY

91. The cell as a basic unit of structure of living things was discovered by
(a) Aristotle
(b) Robert Hooke
(c) Schleiden and Schwann
(d) Gregore Mendel

92. What is the similarity between gymnosperms and angiosperms ?
(a) Phloem of both have companian cells.
(b) Endosperm is formed before fertilization in both.
(c) Origin of ovule and seed is similar in both.
(d) Both have leaves, stem and roots.

93. Fibrous root system is better adopted than tap root system for
(a) transport of organic matter
(b) absorption of water and minerals
(c) storage of food
(d) anchorage of plant to soil

94. Element necessary for middle lamella is
(a) Ca (b) Zn
(c) K (d) Cu

95. Recombination of genes occur at
(a) prophase in mitosis
(b) prophase I in meiosis
(c) prophase II in meiosis
(d) metaphase II in meiosis

96. Proton pumps in the plasma membranes of plant cells may
(a) generate a membrane potential that helps drive cations into the cell through their specific carriers.
(b) be coupled to the movement of K^+ into guard cells.
(c) drive the accumulation of sucrose in sieve-tube members.
(d) be involved in all of the above

97. The Z-scheme refers to
(a) the type of photosynthesis occurs in plants found in areas with minimal precipitation.
(b) the pattern of grana within the chloroplasts of photosynthetic plants.
(c) the carbon-fixation process which is also known as the Calvin cycle.
(d) an energy diagram for the transfer of electrons in the light reactions of photosynthesis in plants.

98. Which one of the following statement regarding auxin is/are correct?
(a) IAA and IBA are natural but NAA, 2, 4-D and 2, 4, 5-T are synthetic auxins.
(b) IAA and NAA are natural but IBA, 2, 4, 5-T and 2, 4-D are synthetic auxin.
(c) NAA and 2, 4, 5-T are natural but IAA, IBA and 2, 4-D are synthetic auxins.
(d) IAA, NAA, IBA, 2, 4-D and 2, 4, 5-T are synthetic auxins.

99. Male gametes in angiosperms are formed by the division of
(a) generative cell
(b) vegetative cell
(c) microspore mother cell
(d) megaspore mother cell

100. In a cross between AABB × aabb, the ratio of F_2 genotypes between AABB, AaBB, Aabb and aabb would be
(a) 9 : 3 : 3 : 1 (b) 2 : 1 : 1 : 2
(c) 1 : 2 : 2 : 1 (d) 7 : 5 : 3 : 1

101. In negative operon
(a) co-repressor binds with repressor
(b) co-repressor does not bind with repressor
(c) co-repressor binds with inducer
(d) cAMP have negative effect on *lac* operon

102. Which of the following does not have stomata?
(a) Hydrophytes
(b) Mesophytes
(c) Xerophytes
(d) Submerged hydrophytes

103. Which of the following is pair of bio-fertilizers?
(a) *Azolla* and BGA
(b) *Nostoc* and legumes
(c) *Rhizobium* and grasses
(d) *Salmonella* and *E. coli*

104. Transgenic plants are the ones:
(a) generated by introducing foreign DNA into a cell and regenerating a plant from that cell.
(b) produced after protoplast fusion in artificial medium.
(c) grown in artificial medium after hybridization in the field.
(d) produced by a somatic embryo in artificial medium.

105. Montreal protocol aims at
(a) Biodiversity conservation
(b) Control of water pollution
(c) Control of CO_2 emission
(d) Reduction of ozone depleting substances

106. The basis of karyotaxonomy is
(a) Number of nucleoli
(b) Sedimentation rate of ribosomes
(c) Chromosome banding
(d) Chromosome number

107. In Chlorophyceae, sexual reproduction occurs by
(a) isogamy and anisogamy
(b) isogamy, anisogamy and oogamy
(c) oogamy only
(d) anisogamy and oogamy

108. Velamen is found in
(a) roots of screwpine
(b) aerial and terrestrial roots of orchids
(c) leaves of *Ficus elastica*
(d) only aerial roots of orchids

109. Collenchyma differs from parenchyma in having
(a) living protoplasm
(b) cellulose walls
(c) vacuoles
(d) pectin deposits at corners

110. Which of the following is not an aromatic amino acid?
(a) Valine (b) Tyrosine
(c) Phenylalanine (d) Tryptophan

111. Aquaporins are
(a) Cytoplasmic connections between cortex cells.
(b) Openings in the lower epidermis of leaves through which water vapour escapes
(c) Openings into root hairs through which water enters
(d) Water specific channels in membranes that may regulate the rate of osmosis.

112. Which one is a C_4-plant?
(a) Papaya (b) Pea
(c) Potato (d) Maize/Corn

113. Which of the following is an important intermediate found in all the types of respiration?
(a) Acetyl CoA (b) Pyruvic acid
(c) Oxaloacetate (d) Tricarboxylic acid

114. Haploid plant cultures are got from
(a) leaves (b) root tip
(c) pollen grain (d) buds

115. In the DNA molecule
(a) the total amount of purine nucleotides and pyrimidine nucleotides is not always equal
(b) there are two strands which run parallel in the $5' \rightarrow 3'$ direction
(c) the proportion of adenine in relation to thymine varies with the organism
(d) there are two strands which run anti-parallel one in $5' \rightarrow 3'$ direction and other in $3' \rightarrow 5'$

116. Which one among the following chemicals is used for causing defoliation of forest trees?
(a) Phosphon-D
(b) Malic hydrazide
(c) 2, 4-dichlorophenoxy acetic acid
(d) Amo-1618

117. If the forest cover is reduced to half, what is most likely to happen on a long term basis ?
(a) Tribals living in these areas will starve to death
(b) Cattle in these and adjoining areas will die due to lack of fodder
(c) Large areas will become deserts
(d) Crop breeding programmes will suffer due to a reduced availability of variety of germplasm

118. Major aerosol pollutant in jet plane emission is
(a) Sulphur dioxide (b) Carbon monoxide
(c) Methane (d) Fluorocarbon

119. A renewable exhaustible natural resource is:
(a) coal (b) petroleum
(c) minerals (d) forest

120. Two animals which are the members of the same order must also be the members of the same :
(a) Class (b) Family
(c) Genus (d) Species

121. Sexual reproduction in *Spirogyra* is an advanced feature because it shows
(a) physiologically differentiated sex organs
(b) different sizes of motile sex organs
(c) same size of motile sex organs
(d) morphologically different sex organs

122. Hypanthodium is
(a) thalamus (b) fruit
(c) inflorescence (d) ovary

123. Which of the following plant shows multiple epidermis?
(a) *Croton* (b) *Allium*
(c) *Nerium* (d) *Cucurbita*

124. Four daughter cells formed after meiosis are
(a) genetically similar
(b) genetically different
(c) anucleate
(d) multinucleate

125. In an experiment, mature leaves on the plant were enclosed for a fixed amount of time in a transparent bag that had radioactive CO_2. In which part of the plant will maximum radioactivity be found after some time?
(a) Actively growing leaves.
(b) Guard cells of all the leaves.
(c) In mature leaves.
(d) Senescing leaves and roots.

126. The closely related morphologically similar sympatric populations, but reproductively isolated, are designated as
(a) clines (b) demes
(c) clones (d) sibling species

127. Which one of the following insecticides is more stable in the environment?
(a) DDT (b) Diazinon
(c) Camphechlor (d) Malathion

128. If by radiation all nitrogenase enzymes are inactivated, then there will be no
(a) fixation of nitrogen in legumes
(b) fixation of atmospheric nitrogen
(c) conversion from nitrate to nitrite in legumes
(d) conversion from ammonium to nitrate in soil

129. Which one of the following is an example of *Ex-situ* conservation?
(a) Wildlife sanctuary (b) Seed bank
(c) Sacred groves (d) National park

130. The class of enzymes that catalyze the removal of a group from a substrate without addition of water, leaving double bonds, is
(a) Transferases (b) Dehydrogenases
(c) Hydrolases (d) Lyases

131. The main organelle involved in modification and routing of newly synthesized proteins to their destinations is
(a) chloroplast
(b) mitochondria
(c) lysosome
(d) endoplasmic reticulum

132. Which of the following statements about nitrogen fixation is **false** ?
(a) Nitrogenase is only catalytic under anaerobic conditions.
(b) The energy for nitrogen fixation can be provided by either photosynthesis or respiration.
(c) In nitrogen fixation, nitrogen is reduced by the addition of three successive pairs of hydrogen atoms.
(d) Most nitrogen fixing microbes are aerobic.

133. When a photosynthetic plant is transferred to an atmosphere of enriched O_2, its rate of
(a) photosynthesis would increase.
(b) photosynthesis would decrease.
(c) respiration would decrease.
(d) osmosis would increase.

134. Incomplete oxidation of glucose into pyruvic acid with several intermediate steps is known as
(a) TCA-pathway (b) Glycolysis
(c) HMS-pathway (d) Krebs cycle

135. Common enzyme in glycolysis and pentose phosphate pathways is
(a) hexokinase (b) aconitase
(c) fumarase (d) dehydrogenase

SECTION 4 - ZOOLOGY

136. Bacteria lack alternation of generation because there is
(a) neither syngamy nor reduction division.
(b) distinct chromosomes are absent.
(c) no conjugation.
(d) no exchange of genetic material.

137. Which one of the following categories of animals, is correctly described with no single exception in it?
(a) All sponges are marine and have collared cells.
(b) All mammals are viviparous and possess diaphragm for breathing.
(c) All bony fishes have four pairs of gills and an operculum on each side.
(d) All reptiles possess scales, have a three chambered heart and are cold blooded (poikilothermal).

138. Cilia, Flagella and microvilli are associated with the
(a) Connective tissue (b) Epithelial tissue
(c) Nervous tissue (d) Muscular tissue

139. Which of the following is a component of vitamin (thiamine, biotin), Acetyl CoA, cysteine, methionine and ferrerdoxin?
(a) Fe (b) S
(c) Co (d) K

140. About 70% of CO_2 is transported as
(a) Carbonic acid
(b) Carboxyhaemoglobin
(c) Bicarbonates
(d) Carbamino compounds

141. In human heart, which of the following valve remains in contact of the oxygenated blood only?
(a) Tricuspid valve (b) Semi lunar valve
(c) Eustachian valve (d) Mitral valve

142. The Urine under normal conditions does not contain glucose because
(a) The normal blood sugar is fructose
(b) Glucose of blood is not filtered in the glomerulus
(c) Glucose in glomerular filtrate is reabsorbed in the uriniferous tubules
(d) Glucose in glomerular filtrate is converted into glycogen.

143. Blood glucose level in man is regulated by:
(a) insulin
(b) adrenaline
(c) glucagon and insulin
(d) All of the above

144. Which one of the following statements is incorrect about menstruation?
(a) During normal menstruation about 40 ml. blood is lost
(b) The menstrual fluid can easily clot
(c) At menopause in the female, there is especially abrupt increase in gonadotropic hormones
(d) The beginning of the cycle of menstruation is called menarche

145. Which one is a hereditary disease ?
(a) Cataract (b) Leprosy
(c) Blindness (d) Phenylketonuria

146. In which condition the gene ratio remains constant for any species?
(a) Sexual selection (b) Random mating
(c) Mutation (d) Gene flow

147. Geometric representation of age structure is a characteristic of
(a) population (b) landscape
(c) ecosystem (d) biotic community

148. What is true for monoclonal antibodies?
(a) These antibodies obtained from one parent and for one antigen
(b) These antibodies obtained from one parent and for two antigens
(c) These antibodies obtained from one parent and for many antigens
(d) These antibodies obtained from many parents and for many antigens

149. Decomposers are organisms that
(a) secrete chemical substances, causing death of tissues.
(b) operate in living body and disintegrate organic substances of cells step by step.
(c) attack and kill plants as well as animals.
(d) operate in relay terms, disintegrate step by step the organic constituents of dead body.

150. Capsid is
(a) genetic material of virus
(b) protein cover of virus
(c) extra genetic material of bacterium
(d) house keeping genome of bacterium

151. Transfer of genetic material from one bacterium to another in the transduction process is through
(a) conjugation
(b) bacteriophages
(c) another bacterium
(d) physical contact between donor and recipient strain

152. In which of the following animals, respiration occurs without any respiratory organ?
(a) Frog (b) Fish
(c) Cockroach (d) Earthworm

153. Spending the dry hot period in an inactive state by an animal is
(a) camouflage (b) hibernation
(c) aestivation (d) mullerian mimicry

154. In animal cells, cytokinesis involves
(a) the separation of sister chromatids
(b) the contraction of the contractile ring of microfilament
(c) depolymerisation of kinetochore microtubules
(d) a protein kinase that phosphorylates other enzymes

155. Myocardial Infarction is caused by
(a) hardening of arteries
(b) lumpy thickness develop in the inner walls of arteries
(c) clot may occur in the lumen of a coronary artery
(d) sudden interruption in blood flow towards a portion of heart

156. Continued consumption of a diet rich in butter, red meat and eggs for a long period may lead to
(a) vitamin toxicity
(b) kidney stones
(c) hyperchloesterolemia
(d) urine laden with ketone bodies

157. Synovial fluid is found in
(a) cranial cavity
(b) spinal cavity
(c) immovable joints
(d) freely movable joints

158. In an egg, the type of cleavage is determined by
(a) shape and size of the sperm
(b) size and location of the nucleus
(c) amount and distribution of yolk
(d) number of egg membranes

159. If for some reason, the vasa efferentia in the human reproductive system gets blocked, the gametes will not be transported from
(a) testes to epididymis
(b) epididymis to vas deferens
(c) ovary to uterus
(d) vagina to uterus

160. Medical Termination of Pregnancy (MTP) is considered safe up to how many weeks of pregnancy?
(a) Eight weeks (b) Twelve weeks
(c) Eighteen weeks (d) Six weeks

161. During blood typing agglutination indicates that the
(a) RBCs carry certain antigens
(b) Plasma contains certain antigens
(c) RBCs carry certain antibodies
(d) Plasma contains certain antibodies

162. Ultrasound of how much frequency is beamed into human body for sonography ?
(a) 15 – 30 MHZ
(b) 1 – 15 MHZ
(b) 45 – 70 MHZ
(d) 30 – 45 MHZ.

163. **Cry** I endotoxins obtained from *Bacillus thuringiensis* are effective against
(a) mosquitoes (b) flies
(c) nematodes (d) boll worms

164. Retention of larval characters even after sexual maturity is called
(a) Parthenogenesis (b) Ontogenesis
(c) Paedogenesis (d) Neoteny

165. The helical structure of protein is stabilized by
(a) Ester (b) Peptide bonds
(c) Disulphide bonds (d) Hydrogen bonds

166. The affinity of CO with Hb is more than oxygen by
(a) 2 times (b) 20 times
(c) 200 times (d) 2000 times

167. The structure which prevents the entry of food into the windpipe is
(a) Gullet (b) Glottis
(c) Tonsil (d) Epiglottis

168. Layer of cells that secrete enamel of tooth is
(a) Osteoblast (b) Ameloblast
(c) Odontoblast (d) Dentoblast

169. Cow's milk is slightly yellowish in colour due to the presence of
(a) Carotene
(b) Riboflavin
(c) Xanthophyll
(d) Xanthophyll and Carotene

170. Accessory excretory organs of man are
(a) only skin
(b) only skin and liver
(c) only skin and lungs
(d) skin, lungs, liver and intestine

171. Which is not correct with respect to human kidney?
(a) The peripheral region is called cortex and central medulla
(b) Malpighian capsules are present in the cortex region
(c) Blood enters glomerulus through efferent arterioles
(d) The concave part of kidney is called hilus

172. Frequency of sound is discriminated by
(a) The intensity of movement of basilar fibres of cochlea
(b) The type of fluid - perilymph or endolymph
(c) The site at the cochlear coil
(d) All of these

173. Which one of the following is the correct difference between rod cells and cone cells of our retina?

		Rod Cells	Cone Cells
(a)	Distribution	More concentrated in centre of retina	Evenly distributed all over retina
(b)	Visual acuity	High	Low
(c)	Visual pigment contained	Iodopsin	Rhodopsin
(d)	Over all function	Vision in poor light	Colour vision and detailed vision in bright light

174. Which of the following glands grows to the maximum size at puberty and then diminishes gradually?
(a) Thymus (b) Pituitary
(c) Thyroid (d) Adrenal

175. Hypoglycemic hormone is
(a) Insulin (b) Glucagon
(c) Thyroxine (d) Calcitonin

176. Saltatory conduction of nerve impulse takes place through :
(a) Myelinated fibre
(b) Non-myelinated fibre
(c) Gray fibres
(d) None of these

177. In which part of the brain corpora quadrigemina is located?
(a) Diencephalon (b) Mesencephalon
(c) Prosencephalon (d) Rhombencephalon

178. Humerus differs from the femur in having:
(a) Sigmoid notch (b) Trochanter
(c) Deltoid ridge (d) None of these

179. Diploid chromosome number in humans is
(a) 46 (b) 44
(c) 48 (d) 42

180. Which of the following shows the correct graph of arithmetic growth?

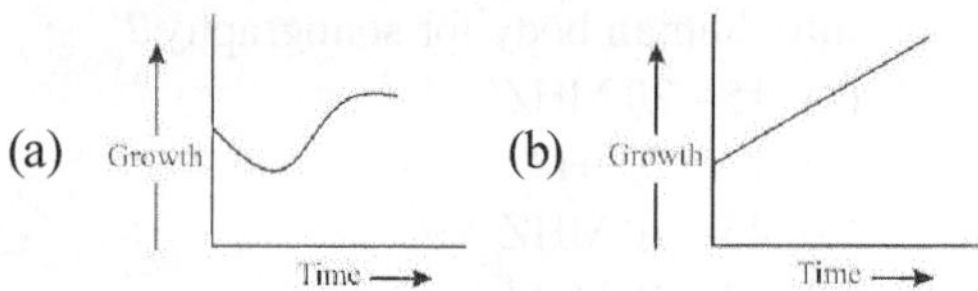

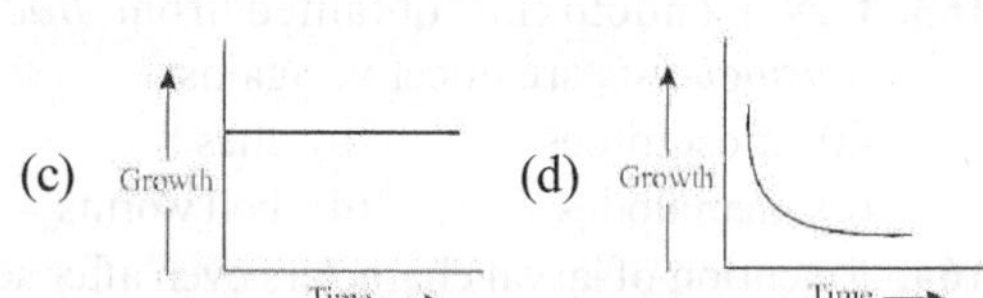

RESPONSE SHEET

Test Code: ....Time taken :....................... Date:.............

	(a)	(b)	(c)	(d)		(a)	(b)	(c)	(d)		(a)	(b)	(c)	(d)		(a)	(b)	(c)	(d)
1.					46.					91.					136.				
2.					47.					92.					137.				
3.					48.					93.					138.				
4.					49.					94.					139.				
5.					50.					95.					140.				
6.					51.					96.					141.				
7.					52.					97.					142.				
8.					53.					98.					143.				
9.					54.					99.					144.				
10.					55.					100.					145.				
11.					56.					101.					146.				
12.					57.					102.					147.				
13.					58.					103.					148.				
14.					59.					104.					149.				
15.					60.					105.					150.				
16.					61.					106.					151.				
17.					62.					107.					152.				
18.					63.					108.					153.				
19.					64.					109.					154.				
20.					65.					110.					155.				
21.					66.					111.					156.				
22.					67.					112.					157.				
23.					68.					113.					158.				
24.					69.					114.					159.				
25.					70.					115.					160.				
26.					71.					116.					161.				
27.					72.					117.					162.				
28.					73.					118.					163.				
29.					74.					119.					164.				
30.					75.					120.					165.				
31.					76.					121.					166.				
32.					77.					122.					167.				
33.					78.					123.					168.				
34.					79.					124.					169.				
35.					80.					125.					170.				
36.					81.					126.					171.				
37.					82.					127.					172.				
38.					83.					128.					173.				
39.					84.					129.					174.				
40.					85.					130.					175.				
41.					86.					131.					176.				
42.					87.					132.					177.				
43.					88.					133.					178.				
44.					89.					134.					179.				
45.					90.					135.					180.				

TEST ASSESSMENT AND ANALYSIS SHEET

How to fill this sheet : After you finish the paper and have checked it first try to attempt again all the questions which you have done wrong or have not been able to attempt. Go to detailed solutions only after you have attempted all the questions of your own.Attempting the questions on your own without seeing the solutions will help you in filling this table

ANALYSIS OF TEST	
Total number of questions	
Questions Attempted	
Correct Questions	
Wrong Questions	
	Net Score

Strike rate $\left(\frac{\text{Correct questions}}{\text{Questions attempt}} \times 100\right) =$ ....

Percentage Marks $\left(\frac{\text{Correct Answers}}{\text{Total number of questions}} \times 100\right) =$ ....

Analysis of wrong questions : Divide the questions which you have done wrong in 3 broad categories.

Reasons for questions attempted wrong	Number of Questions
Knew but calculation mistake (solved on my own after exam)	
Knew but got confused (solved on my own after exam)	
Did not knew the question (couldn't solve even after exam)	

Questions not attempted : Divide the questions not attempted in 3 broad categories

Reasons for unattempted questions	Number of Questions
Knew the ques but time was less (solved on my own after exam)	
Knew the ques but got confused (solved on my own after exam)	
Did not knew the questions (couldn't solve even after exam)	

List down your weak areas you need to improve your score :

1)____________________

2)____________________

3)____________________

NEET

National Eligibility Entrance Test

Test Series **Full Test - 10**

Time : - 3 Hours **Max. Marks:- 720**

INSTRUCTIONS

1. Immediately fill in the particulars on this page of the Test Booklet with Blue/Black Ball Point Pen. Use of pencil is strictly prohibited.
2. Answers are to be given on Response Sheet provided at the end of this paper. You can tear this Response Sheet along the cut marks provided. Fill in the particulars carefully in the Response Sheet provided.
3. The test is of 3 hours duration.
4. The Test Booklet consists of **180** questions. The maximum marks are **720**.
5. There are four section in the question paper.

   The distribution of marks subjectwise in each part is as under for each correct response.

   Section 1 – Physics (180 Marks) – 45 Questions

   Section 2 – Chemistry (180 Marks) – 45 Questions

   Section 3 – Botany (180 Marks) – 45 Questions

   Section 4 – Zoology (180 Marks) – 45 Questions
6. You will be awarded 4 Marks if you have darkened only the bubble corresponding to the correct answer and zero mark if no bubble is darkened. In all other cases, minus one (– 1) mark will be awarded.
7. Use Blue/Black Ball Point Pen only for writing particulars/marking responses on Side-1 and Side-2 of the Response Sheet.

   Use of pencil is strictly prohibited.
8. You are advised to create an actual examination hall simulation.
9. Rough work is to be done on the space provided for this purpose in the Test Booklet only.
10. Do not fold or make any stray marks on the Response Sheet.
11. Once you have evaluated the marks in the test, you should assess it thoroughly.

SECTION 1 - PHYSICS

1. The frequency of vibration f of a mass m suspended from a spring of spring constant k is given by a relation $f = c\ m^x\ k^y$, where c is a dimensionless constant. The values of x and y are

(a) $x=\frac{1}{2}, y=\frac{1}{2}$ (b) $x=-\frac{1}{2}, y=-\frac{1}{2}$

(c) $x=\frac{1}{2}, y=-\frac{1}{2}$ (d) $x=-\frac{1}{2}, y=\frac{1}{2}$

2. The sprinkling of water reduces slightly the temperature of a closed room because
   (a) temperature of water is less than that of the room
   (b) specific heat of water is high
   (c) water has large latent heat of vaporisation
   (d) water is a bad conductor of heat

3. Figure shows the graphical variation of displacement with time for the case of a particle moving along a straight line. The accelerations of the particle during the intervals OA, AB, BC and CD are respectively

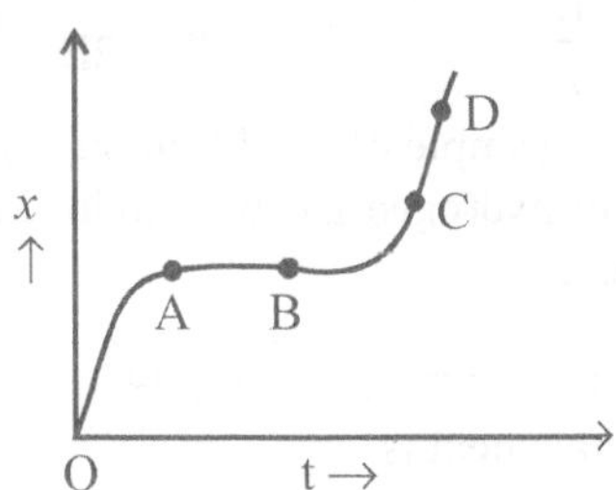

	OA	AB	BC	CD
(a)	–	0	+	0
(b)	+	0	+	+
(c)	–	0	–	0
(d)	+	0	–	+

4. Consider a vector F = 4 i – 3 j. Another vector that is perpendicular to F is
   (a) $4i+3j$ (b) $6i$
   (c) $7k$ (d) $3i-4j$

5. Weightlessness experienced while orbiting the earth in spaceships is the result of
   (a) inertia (b) acceleration
   (c) zero gravity (d) centre of gravity

6. 10 g of ice at 0°C is mixed with 100 g of water at 50°C. What is the resultant temperature of mixture?
   (a) 31.2°C (b) 32.8°C
   (c) 36.7°C (d) 38.2°C

7. The electric field at a distance r from the centre in the space between two concentric metallic spherical shells of radii r_1 and r_2 carrying charge Q_1 and Q_2 is $(r_1 < r < r_2)$

(a) $\frac{Q_1+Q_2}{4\pi\epsilon_0 (r_1+r_2)^2}$ (b) $\frac{Q_1+Q_2}{4\pi\epsilon_0 r^2}$

(c) $\frac{Q_1}{4\pi\epsilon_0 r^2}$ (d) $\frac{Q_2}{4\pi\epsilon_0 r^2}$

8. An aeroplane flying at a constant speed releases a bomb. As the bomb moves away from the aeroplane, it will
   (a) always be vertically below the aeroplane only if the plane is flying horizontally
   (b) always be vertically below the plane only if the aeroplane was flying at an angle of 45° to the horizontal
   (c) always be vertically below the plane
   (d) gradually fall behind the aeroplane if the aeroplane was flying horizontally

9. Two bodies of masses m and 4m are moving with equal kinetic energies. The ratio of their linear momenta will be
   (a) 1 : 4 (b) 4 : 1
   (c) 1 : 2 (d) 2 : 1

10. Figure shows the elliptical path of a planet around the sun. The two shaded parts have equal area. If t_1 and t_2 be the time taken by the planet to go from a to b and from c to d respectively, then

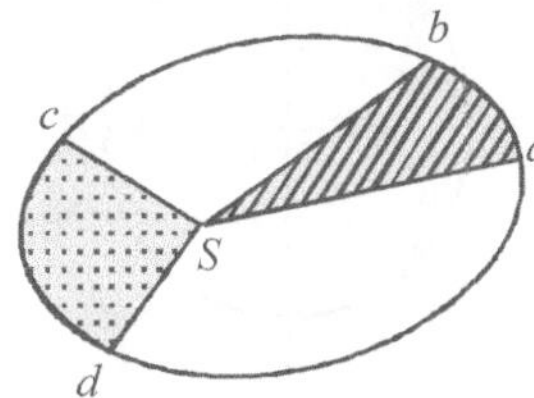

(a) $t_1 < t_2$
(b) $t_1 = t_2$
(c) $t_1 > t_2$
(d) Insufficient information to deduce the relation between t_1 and t_2

11. Which of the statements is false for a particle moving in circle with constant angular speed ?
(a) The velocity vector is tangent to the circle
(b) The acceleration vector is tangent to the circle
(c) The acceleration vector points to the centre of circle
(d) The velocity and acceleration vector are perpendicular to each other

12. A closed argon pipe and an open argon pipe are tuned to the same fundamental frequency. What is the ratio of their lengths?
(a) 1 : 2 (b) 2 : 1
(c) 2 : 3 (d) 4 : 3

13. The internal energy of an ideal gas increases during an isothermal process when the gas is
(a) expanded by adding more molecules to it
(b) expanded by adding more heat to it
(c) expanded against zero pressure
(d) compressed by doing work on it

14. A cell of internal resistance r is connected to an external resistance *R*. The current will be maximum in *R*, if
(a) $R = r$ (b) $R < r$
(c) $R > r$ (d) $R = r/2$

15. Figure below shows a body of mass *m* moving with a uniform speed *v* along a circle of radius *r*.

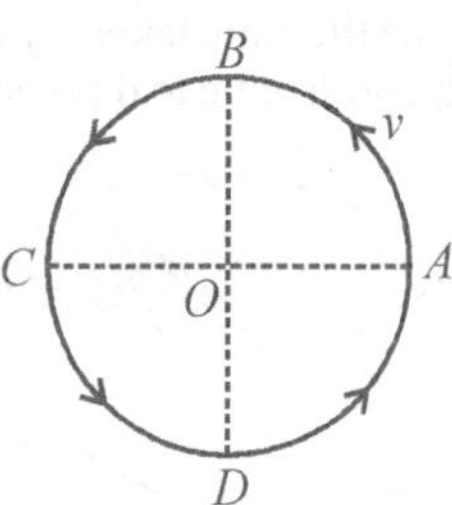

The change in velocity in going from *A* to *B* is
(a) $v\sqrt{2}$ (b) $v/\sqrt{2}$
(c) v (d) zero

16. The moment of inertia of a rigid body, depends upon
(a) distribution of mass from axis of rotation
(b) angular velocity of the body
(c) angular acceleration of the body
(d) mass of the body

17. The moment of inertia of a regular circular disc of mass 0.4 kg and radius 100 cm about the axis perpendicular to the plane of the disc and passing through its centre is
(a) 0.002 kg m^2 (b) 0.02 kg m^2
(c) 2 kg m^2 (d) 0.2 kg m^2

18. The point charges Q and –2Q are placed at some distance apart. If the electric field at the location of Q is E. Then, the electric field at the location of –2Q will be
(a) $-\frac{3E}{2}$ (b) $-E$
(c) $-\frac{E}{2}$ (d) $-2E$

19. When a sample of solid lithium is placed in a flask of hydrogen gas then following reaction happened

$${}_1^1H + {}_3Li^7 \longrightarrow {}_2He^4 + {}_2He^4$$

This statement is
(a) true
(b) false
(c) may be true at a particular pressure
(d) None of these

20. A mass of 0.5 kg moving with a speed of 1.5 m/s on a horizontal smooth surface, collides with a nearly weightless spring of force constant k = 50 N/m. The maximum compression of the spring would be

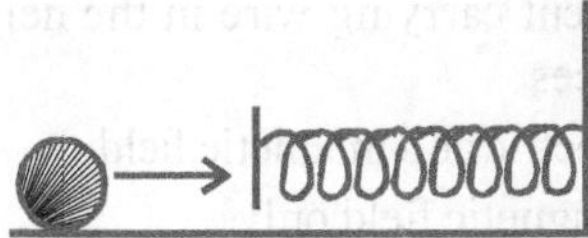

(a) 0.5m (b) 0.15m
(c) 0.12m (d) 1.5m

21. The radius R of the soap bubble is doubled under isothermal condition. If T be the surface tension of soap bubble, the work done in doing so is given by
(a) $32\pi R^2T$ (b) $24\pi R^2T$
(c) $8\pi R^2T$ (d) $4\pi R^2T$

22. The maximum electric field that can be held in air without producing ionisation of air is 10^7 V/m. The maximum potential therefore, to which a conducting sphere of radius 0.10 m can be charged in air is
(a) 10^9 V (b) 10^8 V
(c) 10^7 V (d) 10^6 V

23. The angular velocities of three bodies in simple harmonic motion are ω_1, ω_2, ω_3 with their respective amplitudes as A_1, A_2, A_3. If all the three bodies have same mass and velocity, then
(a) $A_1\omega_1 = A_2\omega_2 = A_3\omega_3$
(b) $A_1^2\omega_1 = A_2^2\omega_2 = A_3^2\omega_3$
(c) $A_1\omega_1^2 = A_2\omega_2^2 = A_3\omega_3^2$
(d) $A_1^2\omega_1^2 = A_2^2\omega_2^2 = A_3^2\omega_3^2$

24. The masses of the three wires of copper are in the ratio of 1 : 3 : 5 and their lengths are in the ratio of 5 : 3 : 1. The ratio of their electrical resistance is
(a) 1 : 3 : 5 (b) 5 : 3 : 1
(c) 1 : 25 : 125 (d) 125 : 15 : 1

25. Two condensers, one of capacity *C* and other of capacity *C*/2 are connected to a *V*-volt battery, as shown.

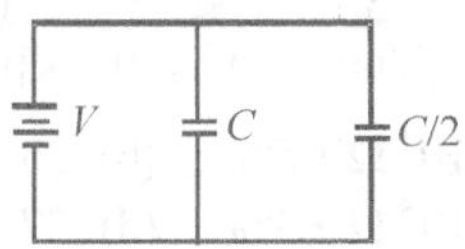

The work done in charging fully both the condensers is
(a) $\frac{1}{4}CV^2$ (b) $\frac{3}{4}CV^2$
(c) $\frac{1}{2}CV^2$ (d) $2CV^2$

26. What is relation between Y, K, η for some isotropic solid material ?
(a) $\eta = \frac{3KY}{3K+Y}$ (b) $\eta = \frac{3KY}{9K-Y}$
(c) $\eta = \frac{3KY}{K-Y}$ (d) $Y = \frac{3K\eta}{3K-\eta}$

27. Two equal masses m_1 and m_2 moving along the same straight line with velocities + 3 m/s and – 5m/s respectively, collide elastically. Their velocities after the collision will be respectively
(a) –3 m/s & +5 m/s (b) + 4 m/s for both
(c) –4 m/s & +4 m/s (d) –5m/s & +3 m/s

28. The coefficient of static friction, μ_s, between block *A* of mass 2 kg and the table as shown in the figure is 0.2. What would be the maximum mass value of block *B* so that the two blocks do not move? The string and the pulley are assumed to be smooth and massless. ($g = 10$ m/s^2)

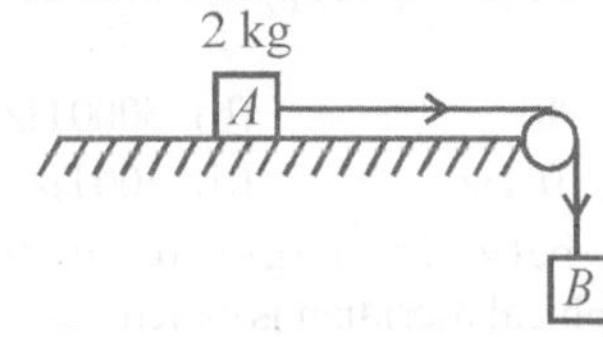

(a) 0.4 kg (b) 2.0 kg
(c) 4.0 kg (d) 0.2 kg

29. Consider a thermodynamic system. If ΔU represents the increase in its internal energy and W the work done by the system, which of the following statements is true ?
(a) $\Delta U = -W$ in an adiabatic process
(b) $\Delta U = W$ in an isothermal process
(c) $\Delta U = -W$ in an isothermal process
(d) $\Delta U = W$ in an adiabatic process

30. We have a galvanometer of resistance 25 Ω . It is shunted by a 2.5 Ω wire. The part of total current that flows through the galvanometer is

(a) $\frac{I}{I_o} = \frac{2}{11}$ (b) $\frac{I}{I_o} = \frac{1}{11}$

(c) $\frac{I}{I_o} = \frac{4}{11}$ (d) $\frac{I}{I_o} = \frac{3}{11}$

31. Two lenses of focal length $f_1 = 10$ cm and $f_2 = -20$ cm are kept as shown. The resultant power of combination will be

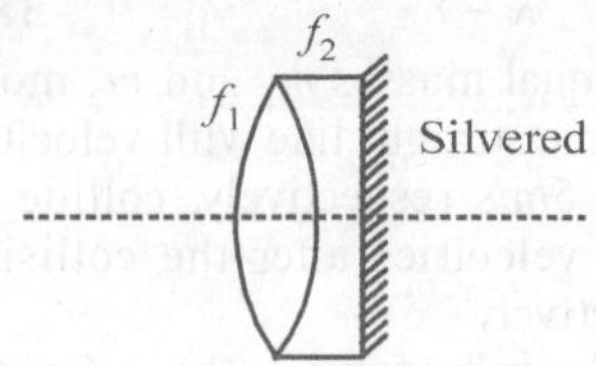

(a) –10D (b) 5D
(c) 0 (d) 10D

32. In A.C. circuit in which inductance and capacitance are joined in series, current is found to be maximum when the value of inductance is 0.5 H and the value of capacitance is 8 μF. The angular frequency of applied alternating voltage will be

(a) 400 Hz (b) 5000 Hz
(c) 2×10^5 Hz (d) 500 Hz

33. Angle between magnetic meridian and geographical meridian is called

(a) angle of dip
(b) angle of declination
(c) angle of depression
(d) both (a) and (b)

34. In Young's double-slit experiment, the separation between the slits is halved and the distance between the slits and the screen is doubled. The fringe width is

(a) unchanged (b) halved
(c) doubled (d) quadrupled

35. A current carrying wire in the neighbourhood produces

(a) electric and magnetic field
(b) magnetic field only
(c) no field
(d) electric field only

36. Consider the YDSE arrangement shown in figure. If $d = 10\lambda$ then position of 8th maxima is

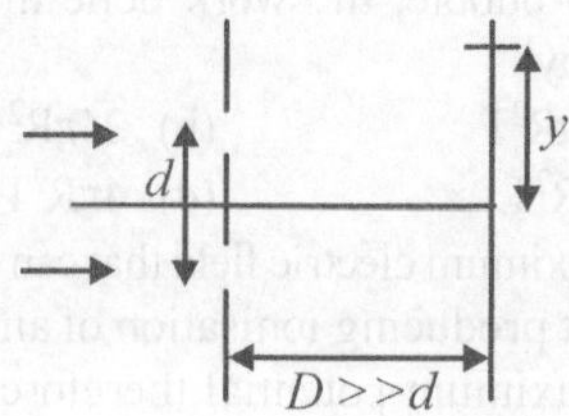

(a) $y = \frac{D}{10}$ (b) $y = \frac{D}{3}$

(c) $y = \frac{4}{5}D$ (d) $y = \frac{4D}{3}$

37. Light of wavelength 200 Å fall on aluminium surface. Work function of aluminium is 4.2 eV. What is the kinetic energy of the fastest emitted photoelectrons?

(a) 2 eV (b) 1 eV
(c) 4 eV (d) 0.2 eV

38. Which of the following waves have the maximum wavelength?

(a) Microwaves (b) UV rays
(c) Radio waves (d) X – rays

39. The resistance of a resistor with the following colour code is equal to

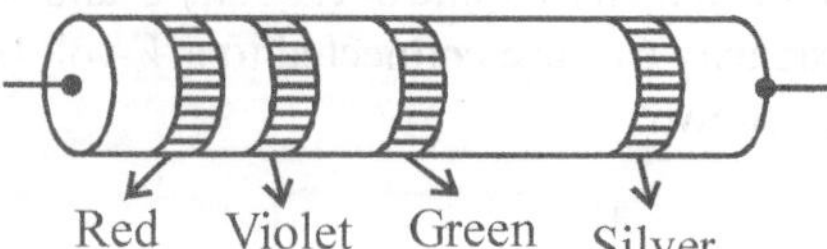

(a) $26 \times 10^4\ \Omega \pm 5\%$ (b) $25 \times 10^4\ \Omega \pm 10\%$
(c) $35 \times 10^5\ \Omega \pm 5\%$ (d) $27 \times 10^5\ \Omega \pm 10\%$

40. Behind a thin converging lens having both the surfaces of the same radius 10cm, a plane mirror has been placed. The image of an object at a distance 40cm from the lens is formed at the same position. What is the refractive index of the lens?

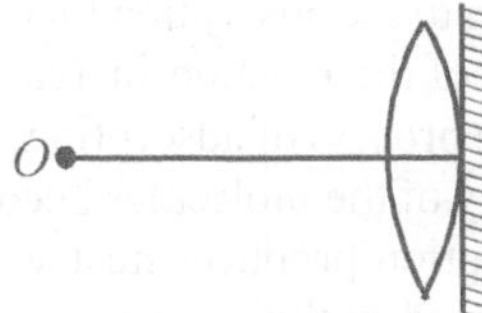

(a) 1.5 (b) 5/3
(c) 9/8 (d) None of these

41. If the wavelength of the first line of the Balmer series in the hydrogen spectrum is λ, then the wavelength of the first line of the Lyman series is
(a) $(27/5)\lambda$ (b) $(5/27)\lambda$
(c) $(32/27)\lambda$ (d) $(27/32)\lambda$

42. Communication on ground is through electromagnetic waves of wavelength
(a) larger than 600 m
(b) between 200 and 600 m
(c) between 1 and 5 m
(d) between 10^{-3} and 0.1 m

43. Impedance of circuit when a resistance R and an inductor of inductance L are connected in series in an A. C. circuit of frequency ν, is
(a) $\sqrt{R^2+2\pi^2\nu^2L^2}$ (b) $\sqrt{R^2+4\pi^2\nu^2L^2}$
(c) $\sqrt{R+4\pi^2\nu^2L^2}$ (d) $\sqrt{R+2\pi^2\nu^2L^2}$

44. Which one of the following is NOT a correct statement about semi-conductors?
(a) The electrons and holes have different mobilities in a semi-conductor
(b) In an *n*-type semi-conductor, the Fermi level lies closer to the conduction band edge
(c) Silicon is a direct band gap semi-conductor
(d) Silicon has diamond structure

45. Which of the following gates corresponds to the truth table given below?

A	*B*	*Y*
1	1	0
1	0	1
0	1	1
0	0	1

(a) NAND (b) OR
(c) AND (d) XOR

SECTION 2 - CHEMISTRY

46. The number of antibonding electron pairs in O_2^{2-} molecular ion on the basis of molecular orbital theory are (at. no. O = 8)
(a) 2 (b) 3
(c) 4 (d) 5

47. The differential rate law for the reaction $H_2 + I_2 \rightarrow 2HI$ is

(a) $-\frac{d[H_2]}{dt} = -\frac{d[I_2]}{dt} = -\frac{d[HI]}{dt}$

(b) $\frac{d[H_2]}{dt} = \frac{d[I_2]}{dt} = \frac{1}{2}\frac{d[HI]}{dt}$

(c) $\frac{1}{2}\frac{d[H_2]}{dt} = \frac{1}{2}\frac{d[I_2]}{dt} = -\frac{d[HI]}{dt}$

(d) $-2\frac{d[H_2]}{dt} = -2\frac{d[I_2]}{dt} = \frac{d[HI]}{dt}$

48. In a salt bridge KCl is used because
(a) It is an electrolyte
(b) It is good conductor of electricity
(c) The transport number of K^+ and Cl^- ions are nearly same or both have same ionic mobility
(d) It is ionic compound

49. The transition metals have a less tendency to form ions due to
(a) High ionisation energy
(b) Low heat of hydration of ions
(c) High heat of sublimation
(d) All of the above

50. Carboxylic acids are more acidic than phenol and alcohol because of
(a) intermolecular hydrogen bonding
(b) formation of dimers
(c) highly acidic hydrogen
(d) resonance stabilization of their conjugate base

51. The degree of dissociation of dinitrogen tetroxide $N_2O_4(g) \longrightarrow 2NO_2(g)$ at temperature T and total pressure P is α. Which one of the following is the correct expression for the equilibrium constant (K_p) at this temperature?
(a) $\frac{2\alpha}{(1-\alpha^2)}$ (b) $\frac{\alpha^2 P}{1-\alpha}$
(c) $\frac{4\alpha^2}{(1-\alpha^2)}$ (d) $\frac{4\alpha^2 P}{(1-\alpha^2)}$

52. The important antioxidant used in food is
(a) BHT (b) BHC
(c) BTX (d) All the three

53. Electromagnetic separation is used in the concentration of
(a) Copper pyrites (b) Bauxite
(c) Cassiterite (d) Cinnabar

54. Geometrical isomerism is not shown by
(a) $CH_3CH_2\overset{CH_3}{\overset{|}{C}}=\underset{CH_3}{\underset{|}{C}}CH_2CH_3$
(b) $C_2H_5-\underset{H}{\underset{|}{C}}=\underset{H}{\underset{|}{C}}-CH_2I$
(c) $CH_2=C(Cl)CH_3$
(d) $CH_3-CH=CH-CH=CH_2$

55. Schotten-Baumann reaction is a reaction of phenols with
(a) Benzoyl chloride and sodium hydroxide
(b) Acetyl chloride and sodium hydroxide
(c) Salicylic acid and conc. H_2SO_4
(d) Acetyl chloride and conc. H_2SO_4

56. 1 M NaCl and 1 M HCl are present in an aqueous solution. The solution is
(a) not a buffer solution with pH < 7
(b) not a buffer solution with pH > 7
(c) a buffer solution with pH < 7
(d) a buffer solution with pH > 7

57. According to the adsorption theory of catalysis, the speed of the reaction increases because
(a) in the process of adsorption, the activation energy of the molecules becomes large
(b) adsorption produces heat which increases the speed of the reaction
(c) adsorption lowers the activation energy of the reaction
(d) the concentration of product molecules at the active centres of the catalyst becomes high due to adsorption.

58. Consider the following reaction at 1000°C
A. $Zn(s)+\frac{1}{2}O_2(g) \rightarrow ZnO(s);\ \Delta G° = -360\,kJ\,mol^{-1}$
B. $C(gr)+\frac{1}{2}O_2(g) \rightarrow CO(g);\ \Delta G° = -460\,kJ\,mol^{-1}$
Choose the correct statement at 1000°C
(a) zinc can be oxidised by carbon monoxide.
(b) zinc oxide can be reduced by graphite
(c) both statements (a) and (b) are true
(d) carbon monoxide can be reduced by zinc.

59. Consider the following complex $[Co(NH_3)_5CO_3]ClO_4$.
The coordination number, oxidation number, number of *d*-electrons and number of unpaired *d*-electrons on the meal are respectively
(a) 6, 3, 6, 0 (b) 7, 2, 7, 1
(c) 7, 1, 6, 4 (d) 6, 2, 7, 3

60. Match List I (Reaction) with List II (Reagent) and select the correct answer using the codes given below the lists:

List I	List II
I. Etard reaction	A. Alcoholic KOH
II. Hydroxylation	B. Anhydrous $AlCl_3$
III. Dehydro-halogenation	C. Chromyl chloride
IV. Friedel-Crafts reaction	D. Dilute alkaline $KMnO_4$

Codes :

(a) I-A, II-B, III-C, IV-B
(b) I-D, II-C, III-A, IV-B
(c) I-C, II-D, III-A, IV-B
(d) I-B, II-A, III-D, IV-C

61. The solubility product of AgI at 25°C is 1.0×10^{-16} mol^2 L^{-2}. The solubility of AgI in 10^{-4}N solution of KI at 25°C is approximately (in mol L^{-1})

(a) 1.0×10^{-12} (b) 1.0×10^{-10}
(c) 1.0×10^{-8} (d) 1.0×10^{-16}

62. The migration of colloidal solute particles in a colloidal solution, when an electric current is applied to the solution is known as

(a) Brownian movement
(b) Electroosmosis
(c) Electrophoresis
(d) Electrodialysis

63. Which one of the following is the correct statement?

(a) $B_2H_6.2NH_3$ is known as 'inorganic benzene'
(b) Boric acid is a protonic acid
(c) Beryllium exhibits coordination number of six
(d) Chlorides of both beryllium and aluminium have bridged chloride structures in solid phase

64. The ionization constant of phenol is higher than that of ethanol because

(a) Phenoxide ion is a stronger base than ethoxide ion
(b) Phenoxide ion is stabilized through delocalization
(c) Phenoxide ion is less stable than ethoxide ion
(d) Phenoxide ion is bulkier than ethoxide ion

65. Teflon, styron and neoprene are all

(a) Copolymers
(b) Condensation polymers
(c) Homopolymers
(d) Monomers

66. For reaction $PCl_3(g) + Cl_2(g) \rightleftharpoons PCl_5(g)$, the value of K_c at 250°C is 26 mol^{-1} litre^1. The value of K_p at this temperature will be

(a) 0.61 atm^{-1} (b) 0.57 atm^{-1}
(c) 0.83 atm^{-1} (d) 0.46 atm^{-1}

67. Which of the following statement is correct if the intermolecular forces in liquids A, B and C are in the order A < B < C?

(a) B evaporates more readily than A
(b) B evaporates less readily than C
(c) A and B evaporate at the same rate
(d) A evaporates more readily than C

68. The reducing power of divalent species decreases in the order

(a) Ge > Sn > Pb (b) Sn > Ge > Pb
(c) Pb > Sn > Ge (d) None.

69. When potassium ferrocyanide crystals are heated with conc. H_2SO_4 the gas evolved is

(a) SO_2 (b) NH_3
(c) CO_2 (d) CO

70. Which of the following statements is not correct regarding aniline?

(a) It is less basic than ethylamine
(b) It can be steam-distilled
(c) It reacts with sodium to give hydrogen
(d) It is soluble in water

71. Which of the following equilibria will shift to right side on increasing the temperature?

(a) $CO(g) + H_2O(g) \rightleftharpoons CO_2(g) + H_2(g)$
(b) $2SO_2(g) + O_2(g) \rightleftharpoons 2SO_3(g)$
(c) $H_2O(g) \rightleftharpoons H_2(g) + \frac{1}{2}O_2(g)$
(d) $4HCl(g) + O_2(g) \rightleftharpoons 2H_2O(g) + 2Cl_2(g)$

72. An aqueous solution of sucrose, $C_{12}H_{22}O_{11}$, containing 34.2 g/l has an osmotic pressure of 2.38 atmospheres at 17°C. For an aqueous solution of glucose, $C_6H_{12}O_6$ to be isotonic with this solution, it would have

(a) 34.2 g/lit (b) 17.1 g/lit
(c) 18.0 g/lit (d) 36.0 g/lit of glucose

73. Prevention of corrosion of iron by Zn coating is called
(a) Galvanization
(b) Cathodic protection
(c) Electrolysis
(d) Photoelectrolysis

74. Aromatic character of benzene is proved by
(a) Resonance theory
(b) Aromatic sextet theory
(c) Orbital theory
(d) All the above

75. RNA is different from DNA because RNA contains
(a) ribose sugar and thymine
(b) ribose sugar and uracil
(c) deoxyribose sugar and thymine
(d) deoxyribose sugar and uracil

76. The de Broglie wavelength of a tennis ball of mass 60 g moving with a velocity of 10 m/s is approximately
(a) 10^{-33} metres (b) 10^{-31} metres
(c) 10^{-16} metres (d) 10^{-25} metres
Planck's constant, $h = 6.63 \times 10^{-34}$ Js

77. AB, A_2 and B_2 are diatomic molecules. If the bond enthalpies of A_2, AB and B_2 are in the ratio 1:1 :0.5 and enthalpy of formation of AB from A_2 and B_2 is $-100\ \text{kJ mol}^{-1}$. What is the bond energy of A_2 :
(a) $200\ \text{kJ mol}^{-1}$ (b) $100\ \text{kJ mol}^{-1}$
(c) $300\ \text{kJ mol}^{-1}$ (d) $400\ \text{kJ mol}^{-1}$

78. Beryllium shows diagonal relationship with aluminium. Which of the following similarity is incorrect :
(a) Be forms beryllates and Al forms aluminates
(b) $Be(OH)_2$ like $Al(OH)_3$ is basic.
(c) Be like Al is rendered passive by HNO_3.
(d) Be_2C like Al_4C_3 yields methane on hydrolysis.

79. $^{-}\!:CH_2 - \underset{\underset{\ddot{O}:}{\|}}{C} - CH_3$ and $CH_2 = \underset{\underset{:\ddot{O}:^{-}}{|}}{C} - CH_3$ are
(a) Resonating structures
(b) Tautomers
(c) Geometrical isomers
(d) Optical isomers

80. The fatty acid which shows reducing property is
(a) acetic acid (b) ethanoic acid
(c) oxalic acid (d) formic acid

81. If the nitrogen atom had electronic configuration $1s^7$ it would have energy lower than that of the normal ground state configuration $1s^2\ 2s^2\ 2p^3$ because the electrons would be closer to the nucleus. Yet $1s^7$ is not observed. It violates
(a) Heisenberg's uncertainty principle
(b) Hund's rule
(c) Pauli exclusion principle
(d) Bohr postulate of stationary orbits

82. For orthorhombic system axial ratios are $a \neq b \neq c$ and the axial angles are
(a) $\alpha = \beta = \gamma \neq 90°$
(b) $\alpha = \beta = \gamma = 90°$
(c) $\alpha = \beta = \gamma = 90°,\ \beta \neq 90°$
(d) $\alpha \neq \beta \neq \gamma = 90°$

83. When cyclohexane is poured on water, it floats, because
(a) cyclohexane is in 'boat' form
(b) cyclohexane is in 'chair' form
(c) cyclohexane is in 'crown' form
(d) cyclohexane is less dense than water

84. $(n-1)d^{10}ns^2$ is the general electronic configuration of
(a) Fe, Co, Ni (b) Cu, Ag, Au
(c) Zn, Cd, Hg (d) Se, Y, La

85. Which is colourless in water?
(a) Ti^{3+} (b) V^{3+}
(c) Cu^{3+} (d) Sc^{3+}

86. Structurally biodegrdable detergent should contain
(a) normal alkyl chain
(b) branched alkyl chain
(c) phenyl side chain
(d) cyclohexyl side chain

87. Which of the following crystals does not exhibit Frenkel defect?
(a) AgBr (b) AgCl
(c) KBr (d) ZnS

88. A reaction which is of first order w.r.t. reactant A, has a rate constant 6 min^{-1}. If we start with [A] = 0.5 mol L^{-1}, when would [A] reach the value of 0.05 mol L^{-1}
(a) 0.384 min (b) 0.15 min
(c) 3 min (d) 3.84 min

89. The correct order of radii is
(a) N < Be < B (b) $F^- < O^{2-} < N^{3-}$
(c) N < Li < K (d) $Fe^{3+} < Fe^{2+} < Fe^{4+}$

90. In Zeigler- Natta polymerisation of ethylene, the active species is
(a) $AlCl_3$ (b) Et_3Al
(c) CH_2CH_2 (d) Ti^{III}

SECTION 3 - BOTANY

91. Which plays an important role in the dispersal of spores in *Funaria*?
(a) Operculum
(b) Capsule
(c) Peristome and annulus
(d) Sporogonium

92. Read the following five statements (i – v) and answer the question.
(i) In *Equisetum* the female gametophyte is retained on the parent sporophyte.
(ii) In *Ginkgo* male gametophyte is not independent.
(iii) The sporophyte in *Riccia* is more developed than that in *Polytrichum.*
(iv) Sexual reproduction in *Volvox* is isogamous.
(v) The spores of slime molds lack cell walls.

How many of the above statements are correct?
(a) Two (b) Three
(c) Four (d) One

93. The fluidity of membranes in a plant in cold weather may be maintained by
(a) increasing the number of phospholipids with unsaturated hydrocarbon tails
(b) increasing the proportion of integral proteins
(c) increasing concentration of cholesterol in membrane
(d) increasing the number of phospholipids with saturated hydrocarbon tail

94. Addition of a solute to pure water causes
(a) negative water potential.
(b) more negative water potential.
(c) positive water potential.
(d) more positive water potential.

95. Which one of the following process help the water- soluble inorganic nutrients go down into the soil horizon and get precipitated as unavailable salts?
(a) Fragmentation (b) Leaching
(c) Catabolism (d) Humification

96. Which one of the following species of plant is considered as the world's most problematic aquatic weed?
(a) *Lantana*
(b) *Eichhornia*
(c) *Parthenium* (carrot grass)
(d) Brown algae

97. The first dicarboxylic acid in Krebs' cycle is
(a) isocitric acid (b) pyruvic acid
(c) oxalo acetic acid (d) α-ketoglutaric acid

98. $N_2 + 8e^- + 8H^+ + 16\,ATP \rightarrow 2NH_4 + H_2 + 16ADP + 16Pi$
The above equation refers to
(a) ammonification (b) nitrification
(c) nitrogen fixation (d) denitrification

99. If turgor pressure becomes equal to osmotic pressure
(a) water leaves the cell.
(b) water enters the cells.
(c) no exchange of water takes place.
(d) solute pass out of the cell.

100. The two chromatids of a metaphase chromosome represent
(a) replicated chromosomes to be separated at anaphase
(b) homologous chromosomes of a diploid set
(c) non-homologous chromosomes joined at the centromere
(d) maternal and paternal chromosomes joined at the centromere

101. A vascular bundle in which the protoxylem is pointing to the periphery is called
(a) endarch (b) exarch
(c) radial (d) closed

102. Which one of the following micro-organisms is used for production of citric acid in industries?
(a) *Penicillium citrinum*
(b) *Aspergillus niger*
(c) *Rhizopus nigricans*
(d) *Lactobacillus bulgaris*

103. Emasculation is not required when flowers are
(a) bisexual (b) intersexual
(c) unisexual (d) either (a) or (b)

104. Which one of the following is common to multicellular fungi, filamentous algae and protonema of mosses?
(a) Diplontic life cycle
(b) Members of kingdom plantae
(c) Mode of Nutrition
(d) Multiplication by fragmentation

105. Which part of the coconut produces coir?
(a) Seed coat (b) Mesocarp
(c) Epicarp (d) Pericarp

106. Root hair arises from
(a) pericycle (b) endodermis
(c) cortex (d) epiblema

107. The productivity of a crop declines when leaves begin to wilt mainly because
(a) the chlorophyll of wilting leaves decomposes.
(b) flaccid mesophyll cells are incapable of photosynthesis.
(c) stomata close, preventing CO_2 from entering the leaf.
(d) photolysis, the water-splitting step of photosynthesis, cannot occur when there is a water deficiency.

108. Which one of the following is not a function of an ecosystem?
(a) Energy flow (b) Decomposition
(c) Productivity (d) Stratification

109. The process in which mature differentiated cells reverse to meristematic activity to form callus is called
(a) dedifferentiation (b) differentiation
(c) redifferentiation (d) None of the above

110. Which one of the following statement is true?
(a) The greater the Biological Oxygen Demand (BOD) of waste water, more is its polluting potential.
(b) The greater the BOD of waste water, less is its polluting potential.
(c) The lesser the BOD of waste water, more is its polluting potential.
(d) The lesser the BOD of waste water, less is its polluting potential.

111. Ecosystem is
(a) always open
(b) always closed
(c) both open and closed depending upon community
(d) both open and closed depending upon biomass

112. Transition zone between two vegetations is
(a) ecotone (b) ecotype
(c) ecocline (d) ecosystem

113. The conditions necessary for vernalization are
(a) high temperature and water
(b) low temperature and oxygen
(c) water and carbon dioxide
(d) oxygen and water.

114. Final electron acceptor in oxidative phosphorylation is
(a) hydrogen (b) dehydrogenase
(c) cytochrome (d) oxygen

115. In a CAM plant, the concentration of organic acid
(a) increases during the day.
(b) decreases during the day.
(c) increases during night.
(d) decreases or increases during day.

116. The phenomenon of plasmolysis is evident when cells are kept in
(a) hypotonic solution
(b) hypertonic solution
(c) isotonic solution
(d) None of the above

117. Quantasomes are found in
(a) mitochondria
(b) chloroplast
(c) lysosome
(d) endoplasmic reticulum

118. Phellogen and phellem respectively denote
(a) cork and cork cambium
(b) cork cambium and cork
(c) secondary cortex and cork
(d) cork and secondary cortex

119. Which one of the following is a correct statement ?
(a) Pteridophyte gametophyte has a protonemal and leafy stage
(b) In gymnosperms female gametophyte is free-living
(c) Antheridiophores and archegoniophores are present in pteridophytes
(d) Origin of seed habit can be traced in pteridophytes

120. Function of suspensor of embryo is
(a) absorption of nourishment.
(b) push the embryo into nutritive endosperm region.
(c) formation of secondary embryos.
(d) All of the above

121. Pineapple (ananas) fruit develops from
(a) a multipistillate syncarpous flower
(b) a cluster of compactly borne flowers on a common axis
(c) a multilocular monocarpellary flower
(d) a unilocular polycarpellary flower

122. Scutellum is a/an
(a) protective covering of radicle
(b) protective covering of plumule
(c) endosperm of gymnosperms
(d) shield-shaped cotyledon

123. The common bottle cork is a product of :
(a) Dermatogen (b) Phellogen
(c) Xylem (d) Vascular cambium

124. Telomere and eukaryotic chromosome possesses short segments of
(a) guanine rich repeats.
(b) thymine rich repeats.
(c) cytosine rich repeats.
(d) adenine rich repeats.

125. What mechanism explains the movement of sucrose from source to sink ?
(a) Evaporation of water and active transport of sucrose from sink.
(b) Osmotic movement of water into the sucrose loaded sieve tube cells creating a higher hydrostatic pressure into the source than in the sink.
(c) Tension created by differences in hydrostatic pressure in the source and sink.
(d) Active transport of sucrose through the sieve tube membranes driven by proton pump.

126. Which one of the following areas in India, is a hotspot of biodiversity?
(a) Eastern Ghats (b) Gangetic Plain
(c) Sunderbans (d) Western Ghats

127. A taxon facing an extremely high risk of extinction in wild in the immediate future is called :
(a) critical endangered
(b) endangered
(c) vulnerable
(d) extinct in wild

128. Compensation point is
(a) where there is neither photosynthesis nor respiration.
(b) when rate of photosynthesis is equal to the rate of respiration.

(c) when entire food synthesized into photosynthesis remain utilized.
(d) when there is enough water just to meet the requirements of plant.

129. Which of the following hormones does not naturally occur in plants?
(a) 2,4–D (2, 4–dichloropheoxy acetic acid)
(b) IAA
(c) ABA
(d) GA

130. Which of the following statements regarding photorespiration are **true**?
(a) Photorespiration is a metabolically expensive pathway.
(b) Photorespiration is avoided when CO_2 is abundant.
(c) Photorespiration results in a loss of usable carbon dioxide.
(d) All of the above

131. A trace element essential for plant growth and radio-isotope, which is used in cancer therapy is
(a) cobalt (b) calcium
(c) sodium (d) iron

132. Which one of the following elements in plants is not remobilised?
(a) Phosphorus (b) Calcium
(c) Potassium (d) Sulphur

133. Water will be absorbed by root hairs when
(a) concentration of salt in the soil is high.
(b) concentration of solutes in the cell sap is high.
(c) plant is rapidly respiring.
(d) they are separated from soil by a permeable membrane.

134. Which of the following types of phytohormones resemble the nucleic acids in some structural aspects ?
(a) Auxin (b) Cytokinin
(c) Gibberellin (d) ABA

135. Which of the following statement(s) is/are incorrect ?
(i) Proton channel of oxysome / complex V / ATP synthase is located in F_1.
(ii) Metabolic water is produced in terminal oxidation / produced in respiration.
(iii) CoQ accepts electron from NADH dehydrogenase (complex I) and also can accept electron from $FADH_2$/ succinate Q-reductase / complex II.
(iv) Cytochrome *c* is a small protein attached to outer surface of the inner mitochondrial membrane and acts as mobile carrier for transfer of electrons between complex I (Cyt *bc*, complex) and III.
(v) Complex IV refers to cytochrome *c* oxidase (cyt *a*, a_3 and 2 Cu per centre).
(vi) If a cell is treated with a drug that inhibits ATP synthase, the pH of mitochondrial matrix will increase.
(a) (i), (ii) and (iii) (b) (iii), (v) and (vi)
(c) (i) and (iv) (d) Only (iii)

SECTION 4 - ZOOLOGY

136. Which one of following feature is possessed by Crustaceans and not by insects?
(a) Paired limbs
(b) Two pairs of antenna
(c) Chitinous exoskeleton
(d) Bilateral symmetry

137. Which period is called the "Golden age of fishes"?
(a) Jurassic period (b) Devonian period
(c) Permian period (d) Ordovician period

138. During inspiration the diaphragm
(a) relaxes to become dome-shaped
(b) contracts and flattens
(c) expands
(d) shows no change

139. Cirrhosis of liver is caused by the chronic intake of
(a) Opium
(b) Alcohol
(c) Tobacco (Chewing)
(d) Cocaine

140. Which of the following sets contains polysaccharides?
(a) Glucose, fructose, lactose
(b) Starch, glycogen, cellulose
(c) Sucrose, maltose, cellulose
(d) Galactose, starch, sucrose

141. Streptokinase which is used as a 'clot buster' obtained from
(a) *Streptococcus* (b) *Staphylococcus*
(c) *Lactobacillus* (d) *Saccharomyces*

142. Baculoviruses are excellent candidates for
(a) species-specific narrow spectrum pesticidal applications.
(b) species-specific broad spectrum pesticidal applications.
(c) species-specific narrow spectrum insecticidal applications.
(d) species-specific broad spectrum insecticidal applications.

143. Select the correct statement.
(a) Genetic engineering works only on animals and has not yet been successfully used on plants.
(b) There are no risks associated with DNA technology.
(c) The first step in PCR is heat which is used to separate both the strands of target DNA.
(d) DNA from one organism will not bond to DNA from another animal.

144. Sigmoid growth curve is represented by
(a) $dN/dt = rN$
(b) $dN/dt = rN(1 - N/K)$
(c) $Nt = N_o + B + I - D - E$
(d) $dN/dt = 1 - N/K$

145. The lactic acid generated during muscle contraction is converted to glycogen in :
(a) Muscle (b) Kidney
(c) Pancreas (d) Liver

146. Thromboplastin is secreted by
(a) platelets (b) lymphocytes
(c) helper T-cells (d) mast cells

147. A coenzyme is
(a) Same enzyme that occurs in different tissues such as heart and muscle
(b) One that shares the function of other enzyme
(c) Organic or inorganic in nature and helps activate metabolic enzymes
(d) Organic non-protein in nature and helps to activate metabolic enzymes

148. The sphincter of Oddi is present between
(a) Oesophagus and cardiac stomach
(b) Pyloric stomach and duodenum
(c) Hepatic duct and cystic duct
(d) Hepatopancreatic duct and duodenum

149. Which phases of cell division is arrested in oogonia of vertebrate ?
(a) Anaphase II
(b) Interphase
(c) Diplotene
(d) Both prophase I and II

150. Which one of the following combination is mismatched?
(a) Glycocalyx - may be capsule or slime layer
(b) Pili - Reproduction
(c) Cell wall - Protective, determines shape, prevents from bursting
(d) Flagella, Pili and Fimbriae - Surface structures of bacterial cell

151. The most basic amino acid is
(a) Arginine (b) Histidine
(c) Glycine (d) Glutamine

152. Poison glands of snake are modified
(a) Linguals (b) Sublinguals
(c) Maxillaries (d) Parotids

153. Which of the following statements are wrong?
(i) Leucocytes disintegrate in the spleen and liver.
(ii) RBC, WBC and blood platelets are produced by bone marrow.

(iii) Neutrophils bring about destruction and detoxification of toxins of protein origin.
(iv) The important function of lymphocytes is to produce antibodies.

(a) (i) and (ii) only (b) (i) and (iv) only
(c) (i) and (iii) only (d) (ii) and (iii) only

154. Which one single organism or the pair of organisms is **correctly** assigned to its taxonomic group?
(a) *Paramecium* and *Plasmodium* belong to the same kingdom as that of *Penicillium*
(b) Lichen is a composite organism formed from the symbiotic association of an algae and a protozoan
(c) Yeast used in making bread and beer is a fungus
(d) *Nostoc* and *Anabaena* are examples of protista

155. Bee dances are meant for
(a) Courtship (b) Communication
(c) Recreation (d) Instinct

156. The adults are radially symmetrical but larvae exhibit bilateral symmetry in
(a) Mollusca (b) Hemichordata
(c) Echinodermata (d) Cephalochordata

157. Which one of the following techniques is safest for the detection of cancers?
(a) Magnetic resonance imaging (MRI)
(b) Radiography (X-ray)
(c) Computed tomography (CT)
(d) Histopathological studies

158. Common cold differs from pneumonia in, that
(a) Pneumonia is caused by a virus while the common cold is caused by the bacterium *Haemophilus influenzae*
(b) Pneumonia pathogen infects alveoli whereas the common cold affects nose and respiratory passage but not the lungs
(c) Pneumonia is a communicable disease whereas the common cold is a nutritional deficiency disease
(d) Pneumonia can be prevented by a live attenuated bacterial vaccine whereas the common cold has no effective vaccine

159. Which of the following technique is used for the separation of DNA fragments ?
(a) Gel electrophoresis
(b) Chromatography
(c) Transformation
(d) Transduction

160. Plasmids are suitable vectors for gene cloning because
(a) these are small circular DNA molecules which can integrate with host chromosomal DNA.
(b) these are small circular DNA molecules with their own replication origin site.
(c) these can shuttle between prokaryotic and eukaryotic cells.
(d) these often carry antibiotic resistance genes.

161. Which one of the following statement is correct?
(a) Warm and moist environment favour decomposition whereas low temperature and anaerobiosis inhibit decomposition
(b) Warm and moist environment inhibit decomposition whereas low temperature and anaerobiosis favour decomposition
(c) Warm and anaerobiosis favour decomposition whereas low temperature favours decomposition
(d) Warm and low temperature inhibit decomposition whereas anaerobiosis favours decomposition

162. Which of the following statement(s) is/are true?
(i) Biowar is the use of biological weapons against humans and /or their crops and animals.
(ii) Bioethics is the unauthorized use of bioresources and traditional knowledge related to bioresources for commercial benefits.
(iii) Biopatent is exploitation of bioresources of other nations without proper authorisation.

(a) (ii) only (b) (i) only
(c) (i) and (ii) only (d) (i) and (iii) only

163. The species that invade a bare area are called
(a) keystone species (b) extinct species
(c) pioneer species (d) rare species

164. If receptor molecule is removed from target organ for hormone action, the target organ will
(a) continue to respond but require higher concentration of hormone
(b) continue to respond but in opposite away
(c) continue to respond without any difference
(d) not respond to hormone

165. Osteoporosis is caused by
(a) Ca^{2+} deficiency
(b) Hypersecretion of calcitonin
(c) K^+ deficiency
(d) Hypersecretion of Parathormone

166. Which hormone is related to mineral metabolism but is not a peptide / protein in nature ?
(a) Parathyroid hormone (PTH)
(b) Atrial natriuretic (ANF)
(c) Aldosterone
(d) All of these

167. Mark the incorrect statement
(a) The ear ossicle attached to tympanic membrane is malleus
(b) Opsin (of Rhodopsin) develops from vitamin A
(c) The pressure on ear drum is equalized by eustachian tube
(d) Otolith organ consists of saccule and utricle

168. The contractile protein of skeletal muscle involving ATPase activity is
(a) Myosin (b) Actin
(c) Troponin (d) Tropomyosin

169. Malpighian tubules are the excretory organs in
(a) Cockroach (b) Platyhelminthes
(c) *Ascaris* (d) *Pila*

170. The functional unit of mammalian kidney is
(a) ureter (b) urinary bladder
(c) urethra (d) nephron

171. The condition of excess urea in blood is known as
(a) Polyuria (b) Haematuria
(c) Uraemia (d) Diuresis

172. Chordae tendinae are found in
(a) joints of legs (b) atria of heart
(c) ventricles of brain (d) ventricles of heart

173. The pattern of contraction and relaxation of the heart is referred to as
(a) blood pressure (b) arterial flow
(c) blood flow (d) cardiac cycle

174. Rate of breathing in mammals largely depends upon
(a) Oxygen concentration in blood
(b) CO_2 concentration in blood
(c) Volume of O_2 in trachea
(d) Extent of depression of diaphragm

175. Which of the following statements are true ?
(i) The blood transports CO_2 comparatively easily because of its higher solubility.
(ii) Approximately 8-9% of CO_2 is transported being dissolved in the plasma of blood.
(iii) The carbon dioxide produced by the tissues, diffuses passively into the blood stream and passes into red blood corpuscles and react with water to form H_2CO_3.
(iv) The oxyhaemoglobin (HbO_2) of the erythrocytes is basic.
(v) The chloride ions diffuse from plasma into the erythrocytes to maintain ionic balance.
(a) (i), (iii) and (v) are true, (ii) and (iv) are false
(b) (i), (iii) and (v) are false, (ii) and (iv) are true
(c) (i), (ii) and (iv) are true, (iii) and (v) are false
(d) (i), (ii) and (iv) are false, (iii) and (v) are true

176. Young one of the cockroach is called
(a) Naid (b) Grub
(c) Nymph (d) Maggot

177. What is the figure given below showing in particular ?

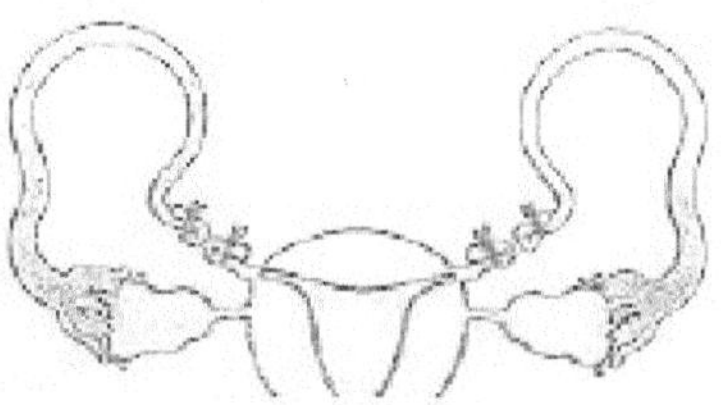

(a) Ovarian cancer (b) Uterine cancer
(c) Tubectomy (d) Vasectomy

178. Identify the human developmental stage shown below as well as the related right place of its occurrence in a normal pregnant woman, and select the right option for the two together.

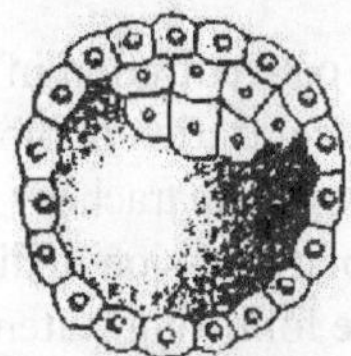

Develeopmental stage	Site of occurrence
(a) Late morula	Middle part of fallopian tube
(b) Blastula	End part of fallopian tube
(c) Blastocyst	Uterine wall
(d) 8 celled morula	Starting point of fallopian tube

179. Which one of the following statements is correct with respect to immunity?

(a) Preformed antibodies need to be injected to treat the bite by a viper snake.

(b) The antibodies against small pox pathogen are produced by T – lymphocytes.

(c) Antibodies are protein molecules, each of which has four light chains.

(d) Rejection of a kidney graft is the function of B-lymphocytes.

180. Given below is a pedigree chart showing the inheritance of a certain sex-linked trait in humans

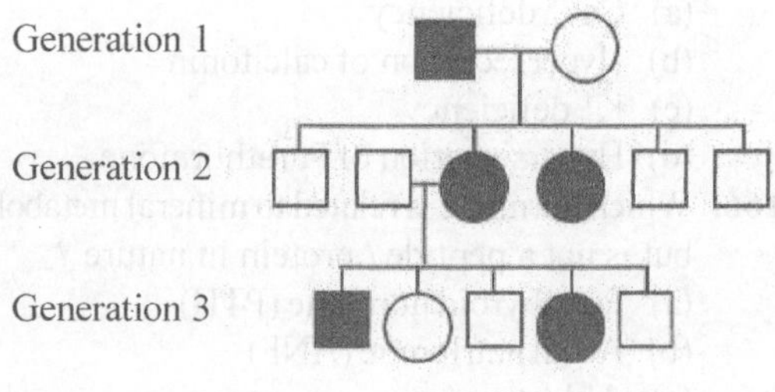

The trait traced in the above pedigree chart is

(a) dominant X - linked

(b) recessive X-linked

(c) dominant Y - linked

(d) recessive Y-linked

RESPONSE SHEET

Test Code: .... Time taken .... Date:..............

	(a)	(b)	(c)	(d)		(a)	(b)	(c)	(d)		(a)	(b)	(c)	(d)		(a)	(b)	(c)	(d)
1.					46.					91.					136.				
2.					47.					92.					137.				
3.					48.					93.					138.				
4.					49.					94.					139.				
5.					50.					95.					140.				
6.					51.					96.					141.				
7.					52.					97.					142.				
8.					53.					98.					143.				
9.					54.					99.					144.				
10.					55.					100.					145.				
11.					56.					101.					146.				
12.					57.					102.					147.				
13.					58.					103.					148.				
14.					59.					104.					149.				
15.					60.					105.					150.				
16.					61.					106.					151.				
17.					62.					107.					152.				
18.					63.					108.					153.				
19.					64.					109.					154.				
20.					65.					110.					155.				
21.					66.					111.					156.				
22.					67.					112.					157.				
23.					68.					113.					158.				
24.					69.					114.					159.				
25.					70.					115.					160.				
26.					71.					116.					161.				
27.					72.					117.					162.				
28.					73.					118.					163.				
29.					74.					119.					164.				
30.					75.					120.					165.				
31.					76.					121.					166.				
32.					77.					122.					167.				
33.					78.					123.					168.				
34.					79.					124.					169.				
35.					80.					125.					170.				
36.					81.					126.					171.				
37.					82.					127.					172.				
38.					83.					128.					173.				
39.					84.					129.					174.				
40.					85.					130.					175.				
41.					86.					131.					176.				
42.					87.					132.					177.				
43.					88.					133.					178.				
44.					89.					134.					179.				
45.					90.					135.					180.				

TEST ASSESSMENT AND ANALYSIS SHEET

How to fill this sheet : After you finish the paper and have checked it first try to attempt again all the questions which you have done wrong or have not been able to attempt. Go to detailed solutions only after you have attempted all the questions of your own.Attempting the questions on your own without seeing the solutions will help you in filling this table

ANALYSIS OF TEST	
Total number of questions	
Questions Attempted	
Correct Questions	
Wrong Questions	
	Net Score

Strike rate $\left(\frac{\text{Correct questions}}{\text{Questions attempt}} \times 100\right) =$ ....

Percentage Marks $\left(\frac{\text{Correct Answers}}{\text{Total number of questions}} \times 100\right) =$ ....

Analysis of wrong questions : Divide the questions which you have done wrong in 3 broad categories.

Reasons for questions attempted wrong	**Number of Questions**
Knew but calculation mistake (solved on my own after exam)	
Knew but got confused (solved on my own after exam)	
Did not knew the question (couldn't solve even after exam)	

Questions not attempted : Divide the questions not attempted in 3 broad categories

Reasons for unattempted questions	**Number of Questions**
Knew the ques but time was less (solved on my own after exam)	
Knew the ques but got confused (solved on my own after exam)	
Did not knew the questions (couldn't solve even after exam)	

List down your weak areas you need to improve your score :

1)____

2)____

3)____

NEET (NATIONAL ELIGIBILITY ENTRANCE TEST) Full Test - 1

ANSWER KEY																	
1	(c)	**21**	(b)	**41**	(a)	**61**	(b)	**81**	(b)	**101**	(d)	**121**	(b)	**141**	(c)	**161**	(d)
2	(b)	**22**	(d)	**42**	(a)	**62**	(a)	**82**	(d)	**102**	(c)	**122**	(d)	**142**	(c)	**162**	(c)
3	(c)	**23**	(a)	**43**	(d)	**63**	(c)	**83**	(a)	**103**	(d)	**123**	(a)	**143**	(c)	**163**	(c)
4	(c)	**24**	(b)	**44**	(a)	**64**	(c)	**84**	(d)	**104**	(c)	**124**	(a)	**144**	(a)	**164**	(b)
5	(b)	**25**	(d)	**45**	(c)	**65**	(d)	**85**	(c)	**105**	(c)	**125**	(d)	**145**	(a)	**165**	(a)
6	(c)	**26**	(b)	**46**	(d)	**66**	(d)	**86**	(c)	**106**	(d)	**126**	(b)	**146**	(a)	**166**	(a)
7	(a)	**27**	(c)	**47**	(b)	**67**	(b)	**87**	(b)	**107**	(a)	**127**	(b)	**147**	(c)	**167**	(b)
8	(a)	**28**	(d)	**48**	(b)	**68**	(d)	**88**	(d)	**108**	(d)	**128**	(d)	**148**	(c)	**168**	(b)
9	(c)	**29**	(c)	**49**	(d)	**69**	(c)	**89**	(c)	**109**	(a)	**129**	(a)	**149**	(b)	**169**	(a)
10	(c)	**30**	(c)	**50**	(c)	**70**	(b)	**90**	(b)	**110**	(a)	**130**	(c)	**150**	(b)	**170**	(c)
11	(d)	**31**	(c)	**51**	(c)	**71**	(b)	**91**	(d)	**111**	(b)	**131**	(d)	**151**	(d)	**171**	(a)
12	(a)	**32**	(c)	**52**	(d)	**72**	(c)	**92**	(b)	**112**	(b)	**132**	(d)	**152**	(b)	**172**	(b)
13	(c)	**33**	(c)	**53**	(a)	**73**	(c)	**93**	(d)	**113**	(a)	**133**	(d)	**153**	(c)	**173**	(c)
14	(a)	**34**	(a)	**54**	(d)	**74**	(d)	**94**	(a)	**114**	(b)	**134**	(b)	**154**	(c)	**174**	(c)
15	(b)	**35**	(b)	**55**	(a)	**75**	(a)	**95**	(d)	**115**	(a)	**135**	(a)	**155**	(a)	**175**	(d)
16	(d)	**36**	(a)	**56**	(b)	**76**	(a)	**96**	(c)	**116**	(b)	**136**	(c)	**156**	(d)	**176**	(c)
17	(b)	**37**	(a)	**57**	(b)	**77**	(b)	**97**	(d)	**117**	(c)	**137**	(d)	**157**	(b)	**177**	(d)
18	(c)	**38**	(c)	**58**	(c)	**78**	(b)	**98**	(a)	**118**	(a)	**138**	(a)	**158**	(a)	**178**	(c)
19	(c)	**39**	(a)	**59**	(c)	**79**	(b)	**99**	(a)	**119**	(b)	**139**	(a)	**159**	(b)	**179**	(d)
20	(a)	**40**	(a)	**60**	(d)	**80**	(a)	**100**	(b)	**120**	(b)	**140**	(b)	**160**	(d)	**180**	(b)

HINTS & SOLUTIONS

SECTION 1 -PHYSICS

1. (c) $[at] = [F]$ amd $[bt^2] = [F]$

$\Rightarrow [a] = [MLT^{-3}]$ and $[b] = [MLT^{-4}]$

2. (b) At a particular time, two values of displacement are not possible.

3. (c) Time to complete 1/4th oscillation is $\frac{T}{4}$ s.

Time to complete $\frac{1}{8}$th vibration from extreme position is obtained from

$$y = \frac{a}{2} = a\cos\omega t = a\cos\frac{2\pi}{T}t \text{ or } t = \frac{T}{6}\text{s}$$

So time to complete 3/8th oscillation

$$= \frac{T}{4} + \frac{T}{6} = \frac{5T}{12}$$

4. (c) $I_1\omega_1 : I_2\omega_2$ or $MK_1^2\omega_1 : MK_2^2\omega_2$

$$\frac{K_1}{K_2} = \sqrt{\left(\frac{\omega_2}{\omega_1}\right)}$$

5. (b) Distance in last two second

$$= \frac{1}{2} \times 10 \times 2 = 10\,\text{m}.$$

$$\text{Total distance} = \frac{1}{2} \times 10 \times (6+2) = 40\,\text{m}.$$

6. (c) Work function of aluminium is 4.2 eV. The energy of two photons can not be added at the moment photons collide with electron all its energy will be dissipated or wasted as this energy is not sufficient to knock it out.
Hence emmission of electron is not possible.

7. (a) In case of L-C circuit, angular frequency of oscillation

$$\omega = \sqrt{\frac{1}{LC}}$$

8. (a) Current flowing through the conductor, I = n e v A. Hence

$$\frac{4}{1} = \frac{nev_{d_1}\pi(1)^2}{nev_{d_2}\pi(2)^2} \text{ or } \frac{v_{d_1}}{v_{d_2}} = \frac{4\times 4}{1} = \frac{16}{1}.$$

9. (c) Since Range on horizontal plane is

$$R = \frac{u^2 \sin 2\theta}{g}$$

so it is max when $\sin 2\theta = 1 \Rightarrow \theta = \frac{\pi}{4}$

10. (c) Since $\frac{R}{C_V} = 6.7 \Rightarrow C_V = \frac{3}{2}R$, hence gas is monoatomic.

11. (d) Unit positive charge at O will be repelled equally by three charges at the three corners of triangle.

By symmetry, resultant $\vec{E}$ at O would be zero.

12. (a) The photoelectric current from a given photocell depends upon the number of photons striking per second. As the photocurrent is directly proportional to number of photoelectrons which, in turn, are proportional to the number of photons of light striking per second. Only the kinetic energy of the ejected photoelectrons depends on the frequency of radiation.

13. (c) Since moment of inertia $I = mr^2$, where m is the mass of body & r is distance of it from the fixed axis.

mass = density × volume $\{\because \rho = \text{density}\}$

$M_A = \rho\pi r^2 t$; $I_A = \frac{\rho\pi r^2 t}{2}$

$M_B = \rho\pi r^2\, 16 \times t/4 = 4\rho\pi r^2 t$

$$I_B = \frac{M_B(4r)^2}{2} = \frac{4\rho\pi r^2 t.16t^2}{2}$$

$$= \frac{64\rho\pi r^2 4t}{2}$$

$\therefore I_B > I_A$

14. (a) Einstein's photoelectric effect & compton effect established particle nature of light. These effects can be explained only, when we assume that the light has particle nature (To explain, Interference & Diffraction the light must have wave nature. It means that light has both particle and wave nature, so it is called dual nature of light)

15. (b) $\frac{\Delta \ell}{\ell} = \alpha \Delta T = 10^{-5} \times 100 = 10^3$

$$\frac{\Delta \ell}{\ell} \times 100\% = 10^{-3} \times 100$$

$$= 10^{-1} = 0.1\%$$

16. (d) Force on M = Mg
Reaction force = Ma
force of friction = $\mu R = \mu.Ma$

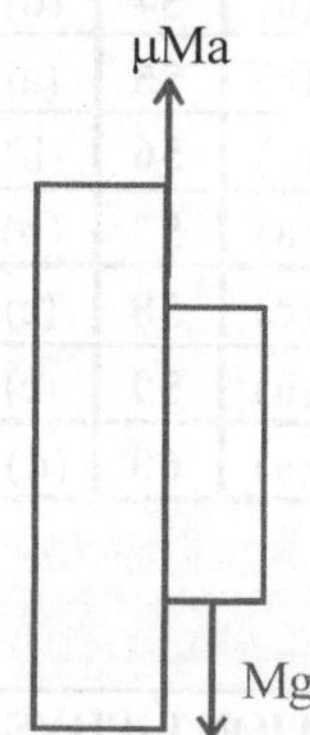

Force of friction will balance the weight. So

$\mu Ma \geq Mg$; $\mu \geq \frac{g}{a.}$

17. (b) From the figure, it is clear that angle of incidence on the reflecting surface is 45°. So, critical angle for glass air interface should be 45°.
Now, from the formula

$$\sin C = \frac{1}{\mu} \Rightarrow \sin 45° = \frac{1}{\mu}$$

$$\mu = \sqrt{2} = 1.4$$

18. (c) For a single negative point charge, electric lines of force are radial and inwards.

19. (c) Ohm's law is not obeyed by vacuum tubes. Semiconductors, transistors or diodes do not follow Ohm's law. They are also called non-ohmic devices.

20. (a) $\frac{140-32}{9}=\frac{C}{5}$ or $\frac{700-160}{9}=C$

or C = 60°C

21. (b) The bullets are fired at the same initial speed

$$\frac{H}{H'}=\frac{u^2\sin^2 60°}{2g}\times\frac{2g}{u^2\sin^2 30°}=\frac{\sin^2 60°}{\sin^2 30°}$$

$$=\frac{(\sqrt{3}/2)^2}{(1/2)^3}=\frac{3}{1}$$

22. (d) Total resistance of the circuit = 4000 + 400 = 4400 Ω

Current flowing $i=\frac{V}{R}=\frac{440}{4400}=0.1\ amp.$

Voltage across load = R i = 4000 × 0.1 = 400 volt.

23. (a) $E=\frac{V}{\ell}=\frac{10}{10}=1\ Vm^{-1}$

25. (d) Phase difference = ϕ

Path diff $=\frac{\lambda}{2\pi}\times$ phase diff. $=\frac{\lambda}{2\pi}\phi$

26. (b) $E=Rhc\left[\frac{1}{n_1^2}-\frac{1}{n_2^2}\right]$

E will be maximum for the transition for which $\left[\frac{1}{n_1^2}-\frac{1}{n_2^2}\right]$ is maximum. Here n_2 is the higher energy level.

Clearly, $\left[\frac{1}{n_1^2}-\frac{1}{n_2^2}\right]$ is maximum for the third transition, i.e. $2\rightarrow 1$. I transition represents the absorption of energy.

27. (c) Vector $(\vec{E}\times\vec{B})$

28. (d) Velocity after the collision

$$=\frac{10\times 10+5\times 0}{15}=\frac{100}{15}=\frac{20}{3}\text{m/sec}.$$

29. (c) The centre of mass remains at rest because force of attraction is mutual. No external force is acting.

31. (c) $v_{oxg.}=\sqrt{\frac{3R\times 289}{32}}$

$$\left(v_{rms}=\sqrt{\frac{3RT}{M}}\right)$$

$v_H=\sqrt{\frac{3R\times 400}{2}}$ so $v_H = 2230.59$ m/sec

32. (c) For the observation of interference phenomenon, the two source must be coherent & must have same frequency.

33. (c) The direction of oscillations of *E* and *B* fields are perpendicular to each other as well as to the direction of propagation. So, electromagnetic waves are transverse in nature.

The electric and magnetic fields oscillate in same phase.

34. (a) Gravitational potential at the centre of spherical shell $=-\frac{GM}{r}$

As the shell is shrinking, r decreases.

35. (b) $v_e=\sqrt{2}v_o$ where v_e and v_o are the escape velocity and orbital velocity respectively.

36. (a) The kinetic energy is directly proportional to temperature.

37. (a) G = 15Ω, i_g = 4 mA, i = 6 A
Required shunt,

$$S=\left(\frac{i_g}{i-i_g}\right)G=\left(\frac{4\times 10^{-3}}{6-4\times 10^{-3}}\right)\times 15$$

$$=\frac{4\times 10^{-3}}{5.996}\times 15=0.01\Omega$$

= 10mΩ (in parallel)

38. (c) $Y=\frac{F\ell}{A\Delta\ell}\Rightarrow\Delta\ell=\frac{F\ell}{YA}=\frac{mg\ell}{YA}$

39. (a) $\frac{f_0}{f_e}=9$, $\therefore f_0=9f_e$

Also $f_0+f_e=20$ ($\because$ final image is at infinity)

$9f_e+f_e=20$, $f_e=2$ cm, $\therefore f_0=18$ cm

40. (a) Let pressure outside be P_0

$\therefore P_1 \text{ (in smaller bubble)} = P_0 + \frac{2T}{r}$

$P_2 \text{ (in bigger bubble)} = P_0 + \frac{2T}{R} \ (R > r)$

$\therefore P_1 > P_2$

hence air moves from smaller bubble to bigger bubble.

41. (a) Given : T/2 = 0.5 s

$\therefore \quad T = 1s$

Frequency, $f = \frac{1}{T} = \frac{1}{1} = 1 \text{Hz}$

If A is the amplitude, then

$2A = 50 \text{ cm} \quad \Rightarrow \quad A = 25 \text{ cm}.$

42. (a) Width of central maximum

$= \frac{2\lambda D}{a} = \frac{2 \times 6250 \times 10^{-10} \times 0.5}{2 \times 10^{-4}}$

$= 3125 \times 10^{-6} \text{ m} = 312.5 \times 10^{-3} \text{ cm}.$

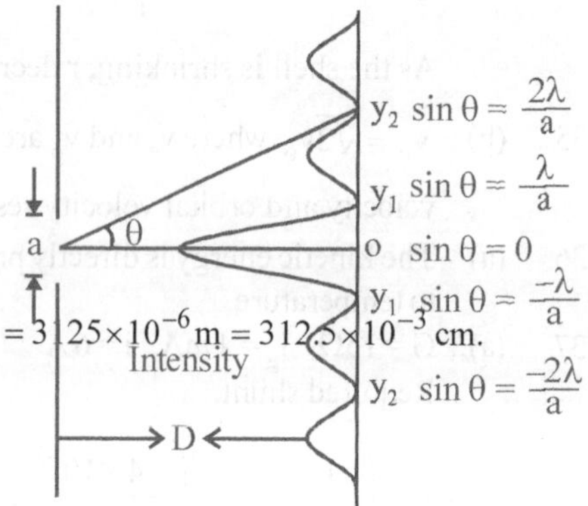

Screen position of various minima for Fraunhoffer diffraction pattern of a single slit of width a.

43. (d) In a p-type semic-onductor germanium is doped with trivalent element. All the elements, like aluminium, boron & gallium are trivalent. So, the correct alternative is (d).

45. (c) The size (or length) of collector is large in comparison to emitter (base is very small in comarison to both collector & emitter) to dissipate the heat.

SECTION 2 - CHEMISTRY

46. (d) BF_3 is planar triangular while PF_3 is pyramidal.

47. (b) Let the wt. of each gas mixed = x g

$\therefore$ mole of $SO_2 = \frac{x}{64}$

mole of $CH_4 = \frac{x}{16}$

mole of $O_2 = \frac{x}{32}$

Total number of moles of the three gases

$= \frac{x}{64} + \frac{x}{16} + \frac{x}{32} = \frac{7x}{64}$

Partial pressure exerted by a gas in the mixture of non-reacting gases (p) is given by

$p = \frac{\text{moles of that gas}}{\text{total moles}} \times \text{Total pressure}$

$\therefore p_{CH_4} = \frac{x}{16} \times \frac{64}{7x} \times 2.1 = 1.2 \text{ atm}$

48. (b) Benzene $\longrightarrow$ Nitrobenzene (NO_2) $\xrightarrow{Cl_2/FeCl_3}$ m-chloronitrobenzene (NO_2, Cl)

49. (d) Decomposition of carbonates and hydrated oxides.

50. (c) Chlorination of methane proceeds via free radical mechanism. Conversion of methyl chloride to methyl alcohol proceeds via nucleophilic substitution. Formation of ethylene from ethyl alcohol proceeds via dehydration reaction. Nitration of benzene is electrophilic substitution reaction.

51. (c) Benzene is non-polar and hence dissolves non-polar compounds (like dissolves like). Among the given compounds, only (c) is non-polar hence it dissolves in benzene.

52. (d) Ozone hole is reduction in ozone layer in stratosphere.

53. (a) In a period on moving from left to right ionic radii decreases.

(a) So order of cationic radii is $Cr^{2+} > Mn^{2+} > Fe^{2+} > Ni^{2+}$ and

(b) $Sc > Ti > Cr > Mn$

(correct order of atomic radii)

(c) For unpaired electrons

Mn^{2+} (Five) > Ni^{2+} (Two)

< Co^{2+} (Three) < Fe^{2+} (Four)

(d) For unpaired electrons

Fe^{2+} (Four) > Co^{2+} (Three)

> Ni^{2+} (Two) > Cu^{2+} (One)

54. (d) The reaction is an example of S_N1 reaction

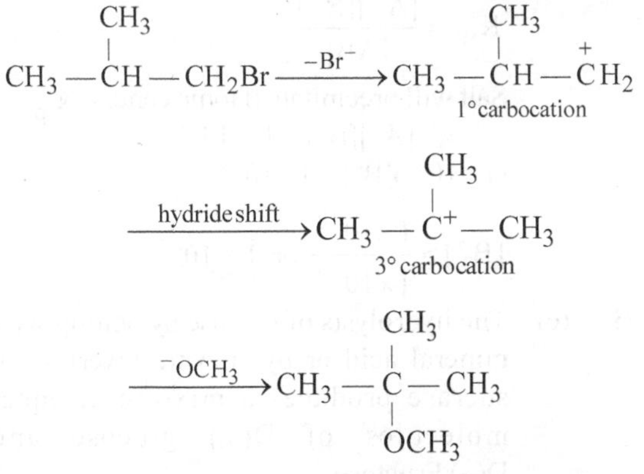

$$CH_3-\underset{|}{\overset{CH_3}{CH}}-CH_2Br \xrightarrow{-Br^-} CH_3-\overset{CH_3}{CH}-\overset{+}{C}H_2 \quad \text{(1° carbocation)}$$

$$\xrightarrow{\text{hydride shift}} CH_3-\overset{CH_3}{C^+}-CH_3 \quad \text{(3° carbocation)}$$

$$\xrightarrow{^-OCH_3} CH_3-\underset{OCH_3}{\overset{CH_3}{C}}-CH_3$$

55. (a) According to the (n + l) rule the higher the value of (n + l), the higher is the energy. When (n + l) value is the same see value of n.

	i	ii	iii	iv
(n + l)	(4 + 1)	(4 + 0)	(3 + 2)	(3 + 1)
	5	4	5	4

∴ iv < ii < iii < i

56. (b) For adiabatic process, q = 0

57. (b) Mn^{2+} ($3d^5$) is more stable than Mn^{3+} ($3d^4$).

58. (c) In NH_4NO_3, there are two different N-atoms (NH_4^+, NO_3^-) with different oxidation numbers, thus reaction is not disproportionation.

59. (c) In case of unsymmetrical ethers, the site of cleavage depends on the nature of alkyl group e.g.,

$$CH_3O-CH(CH_3)_2 + HI \xrightarrow{373K} \underset{\text{Methyl iodide}}{CH_3I} + \underset{\text{Isopropyl alcohol}}{(CH_3)_2CHOH}$$

The alkyl halide is always formed from the smaller alkyl group.

60. (d) $H_{2(g)} + Br_{2(g)} \rightleftharpoons 2HBr_{(g)}$

Rate law, $R = k[H_2][Br_2]^{1/2}$

Order of reaction = 1 + ½ = 3/2

Molecularity of reaction = 2

The unit of $k = \dfrac{R}{[H_2][Br_2]^{1/2}}$

$= \dfrac{\text{mole. lit}^{-1}\text{s}^{-1}}{[\text{mole. lit}^{-1}][\text{mole. lit}^{-1}]^{1/2}}$

$= \text{mole}^{-1/2}.\ \text{lit}^{1/2}.\ \text{s}^{-1}$

61. (b) As the molecular weight of noble gas atoms increases down the group its polarity increases due to which van-der-waal's force between them increases. Due to increased polarity of heavier inert gas, its solubility in water also increases. So, most soluble gas will be Xe and least soluble will be He. So correct order is Xe > Kr > Ar > Ne > He

62. (a) $\Delta x = \Delta v$

$\Delta x.\Delta v = \dfrac{h}{4\pi m}$

$\Delta v^2 = \dfrac{h}{4\pi m}$

∴ $\Delta v = \sqrt{\dfrac{h}{4\pi m}}$

∴ $\Delta p = m.\Delta v = m\sqrt{\dfrac{h}{4\pi m}}$

$= \dfrac{1}{2}\sqrt{\dfrac{hm}{\pi}}$

63. (c) In case two or more chains are of equal length, then the chain with greater number of side chains is selected as the principal chain.

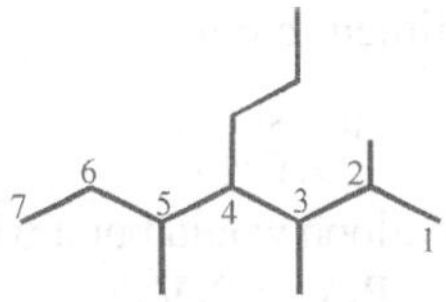

2, 3, 5-Trimethyl-4-propylheptane

64. (c) Hybridisation can be calculated by calculating the no of valence electron and dividing it by 8.

In SO_4^{2-} = Total no. of e^-

$= 6 + (6 \times 4) + 2 = 32$

So, no. of hybrid orbitals $= \dfrac{32}{8} = 4$

∴ sp^3 hybridization.

Similarly, for PO_4^{3-} ; no. of hybrid orbitals

$$=\frac{5+24+3}{8}=\frac{32}{8}=4$$

Hybridisation = sp^3

Similarly, for BF_4^-, it is sp^3.

65. (d) $\begin{matrix} CH_2OH \\ | \\ CO \\ | \\ CH_2OH \end{matrix} \xrightarrow{HIO_4} 2CH_2O + CO_2$

66. (d) An ion exchange resin containing R–COOH group exchange cations like Ca^{2+}, Mg^{2+}, Na^+, Fe^{2+} with H^+ when hard water is passed through it. This resin is called cation exchange resin.

67. (b) KI reacts with $CuSO_4$ solution to produce cuprous iodide (white precipitate) and I_2 (which gives brown colour) Iodine reacts with hypo ($Na_2S_2O_35H_2O$) solution. Decolourisaiton of solution shows the appearance of white precipitate.

$$2CuSO_4 + 4KI \rightarrow 2K_2SO_4 + \underset{\substack{\text{Cuprous iodide} \\ \text{(White ppt.)}}}{2CuI} + \underset{\substack{\text{(Brown colour} \\ \text{in solution)}}}{I_2}$$

$$2Na_2S_2O_3 + I_2 \longrightarrow \underset{\substack{\text{Sod. tetrathionate} \\ \text{(coloourless)}}}{Na_2S_4O_6} + 2NaI$$

68. (d) This reaction shows the formation of H_2O, and the X_2 represents the enthalpy of formation of H_2O because as the definition suggests that the enthalpy of formation is the heat evolved or absorbed when one mole of substance is formed from its constituent atoms.

69. (c) $\overset{sp^2}{CH_2} = \overset{sp}{C} = \overset{sp^2}{CH_2}$

70. (b) The balloon would burst when V > 20 L

$P_1V_1 = P_2V_2$

$1 \times 10 = P_2 \times 20$

$P_2 = 0.5$ atm (no bursting)

Thus, a pressure below 0.5 atm, it would burst.

71. (b) Complex $Co(en)(NH_3)_2Cl_2]Cl$ will have four different isomers.

(i) **Geometrical isomers**

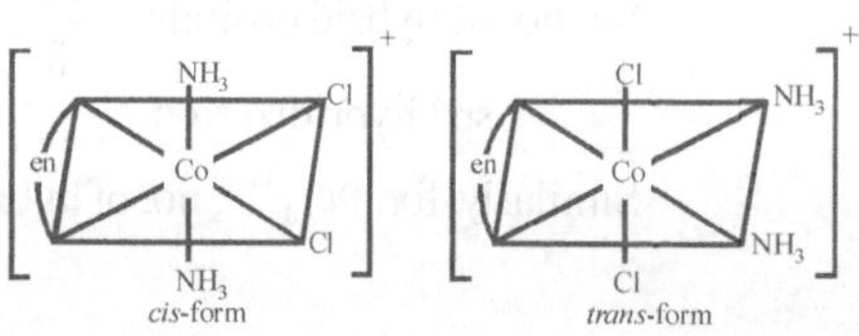

(ii) **Optical isomers**

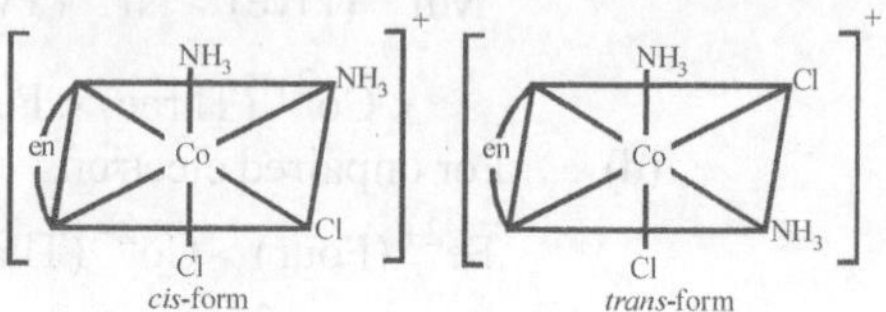

72. (c) $AB \rightleftharpoons A^+B^-$

$$K_{sp} = \frac{[A^+][B^-]}{[AB]}$$

Salt will precipitate if ionic conc. > K_{sp}

$[A^+][B^-] > 1 \times 10^{-8}$

$(1 \times 10^{-3})[B^-] > 1 \times 10^{-8}$

$$[B^-] > \frac{1 \times 10^{-8}}{1 \times 10^{-3}} \text{ or } 1 \times 10^{-5}$$

73. (c) The hydrolysis of sucrose by boiling with mineral acid or by enzyme invertase or sucrase produces a mixture of equal molecules of D(+) glucose and D(–) Fructose.

$$\underset{\substack{\text{sucrose} \\ [\alpha_D]=+66.5^\circ}}{C_{12}H_{22}O_{11}} + H_2O \xrightarrow{HCl} \underbrace{\underset{\substack{\text{D-glu cose} \\ [\alpha_D]=+52.5^\circ}}{C_6H_{12}O_6} + \underset{\substack{\text{D-Fructose} \\ [\alpha_D]=-92^\circ}}{C_6H_{12}O_6}}_{\text{Invert sugar}, [\alpha_D]=-20^\circ}$$

74. (d) In CO_2 we have 22 (6 + 8 + 8 = 22) electrons. In (CN_2^{2-}), we have 22 (6 + 7 + 7 + 2 = 22) electrons. Both CO_2 and (CN_2^{2-}) have linear structures. Thus, statement (a) is correct.

$$Mg_2C_3 + 4H_2O \longrightarrow 2Mg(OH)_2 + \underset{\text{Propyne}}{CH_3C \equiv CH}$$

i.e., statement (b) is also correct .

The structure of CaC_2 is of NaCl type i.e., statement (c) is also correct.

75. (a) Due to resonance C — N bond in protein acquires double bond character and is smaller than usual C — N bond.

$$-\overset{\overset{\ddot{O}}{\|}}{C}-\ddot{N}H- \longleftrightarrow -\overset{\overset{O^-}{|}}{C}=\underset{\oplus}{N}H-$$

76. (a) Given vapour pressure of pure solute $(P^0) = 121.8$ mm; Weight of solute (w) = 15 g

Weight of solvent (W) = 250 g; Vapour pressure of pure solvent (P) = 120.2 mm and Molecular weight of solvent (M) = 78

From Raoult's law

$$= \frac{P^o - P}{P^o} = \frac{w}{m} \times \frac{M}{W} = \frac{121.8 - 120.2}{121.8} = \frac{15}{m} \times \frac{78}{250}$$

$$\text{or } m = \frac{15 \times 78}{250} \times \frac{121.8}{1.6} = 356.2$$

77. (b) With the decrease in the electronegativity of central atom the bond angle decreases

78. (b) When the concentration of reactant is reduced to half its initial value, the rate is reduced by $\frac{2.4}{0.6} = 4$ times

It means, rate $\propto$ [reactant]^2

So, order of reaction = 2

79. (b) *p-p* overlap between B and F is maximum due to identical size and energy of *p*-orbitals, so electron deficiency in boron of BF_3 is neutralized partially to the maximum extent by back donation.

Hence BF_3 is least acidic.

80. (a) Following generalization can be easily derived for various types of lattice arrangements in cubic cells between the edge length (a) of the cell and r the radius of the sphere.

For simple cubic : $a = 2r$ or $r = \frac{a}{2}$

For body centred cubic :

$$a = \frac{4}{\sqrt{3}} r \text{ or } r = \frac{\sqrt{3}}{4} a$$

For face centred cubic :

$$a = 2\sqrt{2} r \text{ or } r = \frac{1}{2\sqrt{2}} a$$

Thus the ratio of radii of spheres for these will be

simple : bcc : fcc

$$= \frac{a}{2} : \frac{\sqrt{3}}{4} a : \frac{1}{2\sqrt{2}} a$$

i.e. option (a) is correct answer.

81. (b) Recall that presence of electron-withdrawing group increases, while presence of electron-releasing group decreases the acidity of carboxylic acids.

$$\underset{\text{(electron-withdrawing gp.)}}{ClCH_2COOH} > H-\overset{\overset{O}{||}}{C}-OH > CH_3-\overset{\overset{O}{||}}{C}-OH > C_2H_5-\overset{\overset{O}{||}}{C}-OH$$

(Electron-releasing character increasing from left to right)

82. (d) Penicillin is an antibiotic which was first obtained from a fungus, penicillium notatum by the scientist, Flemming.

83. (a) Slow step is rate determining step

$\therefore$ Rate law = k $[NOBr_2][NO]$, $[NOBr_2]$ is the intermediate

For the fast reaction, $K = \frac{[NOBr_2]}{[NO][Br_2]}$

$\therefore \quad [NOBr_2] = K[NO][Br_2]$

$\therefore \quad \text{Rate law} = Kk[NO]^2[Br_2]$

$\quad = K'[NO]^2[Br_2]$

84. (d) Various products are formed when nitroarenes are reduced. These are given below for $C_6H_5NO_2$.

Medium	Main product
In acidic medium (metal/HCl)	Aniline ($C_6H_5NH_2$)
In neutral medium (Zn/NH_4Cl)	Phenyl hydroxylamine, C_6H_5–NHOH
In alkaline medium	C_6H_5–NH–NH–C_6H_5 Hydrazobenzene

Thus, Aniline will be main product in case of (d).

85. (c) Reaction used in fuel cell is

$$2H_2(g) + O_2(g) \longrightarrow 2H_2O(\ell)$$

At anode : $[H_2 \longrightarrow 2H^+ + 2e] \times 2$

At cathode : $O_2 + 2H_2O + 4e \longrightarrow 4OH^-$

86. (c) Carbon cannot expand its coordination number beyond four due to the absence of d-orbitals, hence it cannot form $[CCl_6]^{2-}$ ion

87. (b) *i* remains unchanged when number of ions before and after complex ion remains constant.

	Solute	y	Complex	y
(a)	$PtCl_4$	5	$K_2[PtCl_6]$	3
(b)	$ZnCl_2$	3	$Zn[(NH_3)_4]Cl_2$	3

88. (d) The equivalent conductivity of a solution,

$$\Lambda_{eq} = \frac{1000}{C} \times \kappa$$

Where,

κ = specific conductance = Unit $ohm^{-1} cm^{-1}$

C = normality of the solution unit gm eq/cm^3

Hence, the unit of Λ_{eq} is $Ohm^{-1}.cm^2$ (gm equivalent)$^{-1}$.

89. (c) Nylon-6 is also called as perlon. It is a polymer of caprolactam.

Cyclo hexanone $\xrightarrow{NH_2OH}$ Oxime $\xrightarrow{\text{Beckmann rearrangement}}$

Caprolactam $\xrightarrow[\Delta]{H_2O}$ $H_2N(CH_2)_5COOH$ (ε-Amino Caproic acid)

$\xrightarrow[\text{Polymerise}]{\Delta}$ $\left(NH(CH_2)_5-\overset{O}{\overset{\|}{C}}-NH-(CH_2)_5-\overset{O}{\overset{\|}{C}}\right)_n$

Nylon-6

SECTION 3 - BOTANY

91. (d) The hormone, which controls cell division and cell differentiation, is cytokinin. Both the auxin and gibberellins cause elongation of cell and plants.

92. (b) During aerobic respiration oxygen is consumed and CO_2 is released. R.Q. is defined as ratio of volume of CO_2 released and O_2 consumed by the respiratory substrate.

$$R.Q. = \frac{\text{Volume of } CO_2}{\text{Volume of } O_2}$$

93. (d) Long-day plants usually flower in the spring or early summer; they flower only if the light periods are longer than a critical length, which is usually 9-16 hours. For example, wheat plants flower only when light periods exceed fourteen hours.

95. (d) The end product of glycolysis is pyruvate. It enters mitochondria and is oxidatively decarboxylated to acetyl CoA before entering into Krebs cycle.

96. (c) Differentiation of root is controlled by high auxin concentration. While in tissue culture auxin concentration is made high to promote rooting.

97. (d) α-amylase enzyme stimulates the germination of barley seed.

98. (a) DNA ligase is the enzyme which helps in joining two fragments of DNA. The enzyme is used in DNA replication as it joints the Okazaki segments (also in proof reading). It also finds its use in genetic engineering as it can join two or more desired nucleotide sequences of DNA.

100. (b) Dry indehiscent single-seeded fruit formed from bicarpellary syncarpous inferior ovary is cypsela. Cypsela is also called inferior, false or pseudocarpic achene, the thin fruit wall (developed from pericarp and thalamus)is attached to the seed at one point but the fruits develops from an inferior, unilocular and uniovuled ovary, *e.g.*, sunflower, marigold. Some cypsela develop pappus for dispersal *e.g.* Soncus, *Taraxacum.*

101. (d) DDT residues are rapidly passed through food chain causing biomagnification because DDT is lipo soluble. Biomagnification, also known as bioamplification or biological magnifica-tion, is the increase in concentration of a substance, such as the pesticide DDT, that occurs in a food chain as a consequence of:
Persistence (slow to be broken down by environmental processes)
Food chain energetics
Low (or non-existent) rate of internal degradation/excretion of the substance (often due to water-insolubility).

102. (c) Deep black soil has high content of clay and humus, which makes it more fertile and so productive. Humus is amorphous and colloidal mixture of complex organic substances. It provides fertility to the soil.

104. (c) Embryo sac is 7-celled structure. There is a large central cell with two polar nuclei, egg apparatus with egg cell and 2 synergids present at micropylar end and its chalazal end, 3 antipodal cells are present.

107. (a) AaBB × aaBB on crossing gives 50% individuals having genotype AaBB and 50% individuals having genotype aaBB.

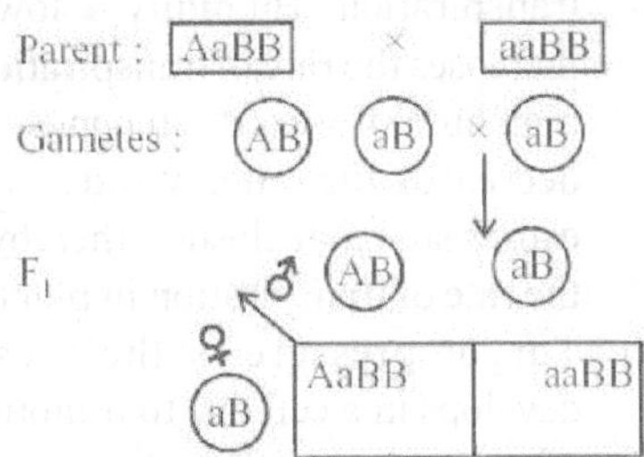

108. (d) The oceans cover about 2/3 of the whole surface of our earth. Thus it is the most stable ecosystem, because of buffering action by water.

109. (a) Parthenocarpy is the development of fruits without prior fertilization which results in the formation of seedless fruits. In some plant species, parthenocarpic fruits are produced naturally or they may be induced by treatment of the unpollinated flowers with auxin.
Removal of androecium before pollen release is called emasculation which is helpful in preventing unwanted pollination. Vernalized seeds are the chilled treated seeds for breaking dormancy. Phenyl Mercuric Acetate is an antitranspirant.
Gibberellins and Auxins are known to induce parthenocarpy in plants. If a tomato plant is treated with a low concentration of auxin and gibberellic acid it'll produce fruits without fertilization i.e. parthenocarpic fruits.

110. (a) Epistasis is the phenomenon of masking or suppressing the phenotypic impression of a gene pair by a non allelic gene pair which impresses its own effect.

111. (b) A mutation bringing about early stoppage of polypeptide formation is called nonsense mutation. Frame shift mutation is a type of gene mutation where the reading of codons is changed due to insertion or deletion of nucleotides.

112. (b) The rate at which organic molecules are formed in a green plant is called gross productivity.

113. (a) Food chain is the transfer of energy from green plants (Primary producers), through a sequence of organisms occupies in a food chain is known as its trophic level. Therefore, statements (ii) and (iii) are correct.

115. (a) Carolus Linnaeus gave the binomial system of nomenclature.

116. (b) Well developed vascular system present in the members of pteridophytes but absent in mosses as the plant body is sporophyte which is distinguished into true root, stem and leaves.

117. (c) Glycosylation is the process or result of addition of saccharides to proteins and lipids. The process is one of four principal co-translational and post-translational modification steps in the synthesis of membrane and secreted proteins and the majority of proteins synthesized in the rough ER undergo glycosylation.

118. (a) Association of algae and fungi is referred to as lichen. Symbiotic association of *Rhizobium* with roots of leguminous plants is referred to as symbiosis. Mycorrhiza is symbiotic association between fungi and roots of higher plants. The fungal partner of mycorrhiza obtains food from roots of the higher plant and in return supplies mineral elements to it.

119. (b) *Laminaria* (kelp) and *Fucus* (rock weed) are marine algae. They are the members of class- phaeophyceae (brown) algae.

120. (b) Vacuoles are present mainly in the plant cells. Each vacuole is surrounded by cytoplasmic membrane called as tonoplast which is similar to plasma membrane.

121. (b) Vascular cambium is produced by two types of meristems, fascicular and interfascicular cambium.

122. (d) Vessels are elongated, multicellular water conducting channels with wide lumen formed by end to end fusion of a large number of vessel elements. Tracheids are elongated dead cells with tapering ends having lignified walls and large or wide lumen. Their main function is conduction of water and minerals from root to leaf.

123. (a) The linking of antibiotic resistance gene with the plasmid vector became possible with DNA ligase. DNA ligase is an enzyme that is able to join together two portions of DNA and therefore plays an important role in DNA repair. DNA ligase is also used in recombinant DNA technology as it ensures that the foreign DNA is bound to the plasmid into which it is incorporated.

124. (a) *In situ* means keeping endangered species of animals or plants into natural environment and not in the environment that looks like natural but man made, like zoological & botanical gardens. *In situ* includes, national parks, sanctuaries and biosphere reserve.

125. (d) *Monascus purpureus* is a yeast used in the production of statins which are used in lowering blood cholestrol.

126. (b) Gel electrophoresis is a technique to separation of DNA fragments according to their size. DNA is negatively charged so in gel tank when electric passed, DNA move towards positive electrode.

127. (b) DNA or RNA segment tagged with a radioactive molecule is called Probe. They are used to detect the presence of complementary sequences in nucleic acid samples. Probes are used for identification and isolation of DNA and RNA.

128. (d) Hot spots are those areas which were rich in biodiversity but now under threat due to direct or indirect interference of human activites. These regions are on the edge to get some of their species extinct due to humans. Western Ghats in India are under threat due to continuous developmental activities and Doon valley is under threat due to continuous mining activities.

129. (a) Kaziranga National Park in Assam is famous as a protection environmental park for rhinoceros.

130. (c) Osmosis is the movement of solvent particles from a region of low solute concentration to a region of high solute concentration through a selectively permeable membrane.

131. (d) The factors which affect the rate of transpiration are Light – stomatal opening, Temperature – increases rate of transpiration, Humidity – low humidity increases the rate of transpiration whereas high humidity exerts an opposite effect *i.e.* decreases the rate, Wind – High wind causes stomatal closure, thereby, reducing the rate of transpiration in plants.

132. (d) Turgor pressure is the pressure that develops in a cell due to osmotic diffusion of water inside it and is responsible for pushing the membrane against cell wall. Stomata open under conditions of increased turgor pressure of guard cells and stomata get closed under conditions of decreased turgor pressure of guard cells. When turgid, they swell and bend outward. As a result, the stomatal aperture opens. When they are flaccid, the tension from the wall is released and stomatal aperture closes.

133. (d) Heterotrophic plants are less dependent on nitrogen obtained from nitrification since they receive some nitrite and nitrate through their parasitic or carnivorous nutritional modes.

134. (b) Grana are present inside the chloroplasts. Each granum may contain 10 – 100 thylakoids. Thylakoids or baggy trousers are membrane lined flattened sacs. Thylakoids (or granna) contain chloroplast proteins, photosynthetic pigements and other factors required for photosynthesis.

SECTION 4 - ZOOLOGY

136. (c) Movement of Ca^{2+} out in sarcoplasmic reticulum controls the making and breaking of actin and myosin complex actomyosin due to which muscle contraction and relaxation takes place. Albert Szent Gyorgyi worked out biochemical events of muscle contraction.

137. (d) Brain is the most vital organ. It stops functioning in the absence of O_2.

138. (a) Elbow joint is an example of hinge joint. The elbow is a hinge joint; it can open and close like a door. Hinge joint is a form of diarthrosis (freely movable joint) that allows angular movement in one plane

only, increasing or decreasing the angle between the bones e.g. elbow joint, knee joint etc.

140. (b) Posterior pituitary releases vasopressin which stimulates reabsorption of water in the distal tubules in nephron.

141. (c) The transmission of nerve impulse through synapse requires a chemical neurotransmitter. The most common neurotransmitter is acetylcholine.

142. (c) Melanin is the pigment which gives colour to the skin. Retinol is the other name for vitamin A. Sclerotin is the component of the carapace in crustaceans.

143. (c) The smooth muscles of bronchioles show spasmic contraction during expiration in asthma.

144. (a) Egg is liberated from the ovary (ovulation) at secondary oocyte stage when the meiosis II is arrested in secondary oocyte. It will be induced by the sperm.

145. (a) **Stenohaline :** Steno meaning narrow and haline meaning salt. A fish cannot handle a wide fluctuation in salt content in water. Many fresh water fish tend to be stenohaline and die in environments of high salinity such as the ocean. Fish living in coastal estuaries and tide pools are often euryhaline (tolerant to changes in salinity) as are many species which have life cycle requiring tolerance to both fresh water and sea water environments such as *Salmon*.

148. (c) Reproductive health in society can be improved by creating awareness among people about various reproduction related aspects and providing facilities and support for building up a reproductively healthy society.

149. (b) Foetal ejection reflex in human female is induced by fully developed foetus and placenta. When a woman is in a lithotomy or semi-sitting position, the foetal ejection reflex is impaired and the increased pain caused by the sacrum's inability to move as the baby descends can be intolerable.

152. (b) Aneuploidy is numerical change in the chromosome number in the genome. Euploidy is the phenomenon of having multiple or more than twice the number of genome.

153. (c) *Catla catla* and *Labeo rohita* are the two Indian major carps whereas *Heteropneustes* is a catfish. *Cyprinus* is the exotic breed.

154. (c) Chimpanzees and gorillas are our closest relatives among the living primates.

155. (a) Adaptive radiation refers to evolution of different species from a common ancestor. The mammals are adapted for different mode of life i.e. they show adaptive radiation. They can be aerial (bat), aquatic (whale and dolphins), burrowing or fossorial (rat), cursorial (horse), scantorial (squirrel) or arboreal (monkey). The adaptive radiation, the term by Osborn, is also known as Divergent evolution.

156. (d) The most significant trend in evolution of modern man (*Homo sapiens*) from his ancestors is development of brain capacity.

159. (b) The terms Porifera was given by Grant, the phylum includes animals with pores in their body. Its classification based on skeleton or spicules.

160. (d) Squamous epithelium is formed of thin discoidal and polygonal cells that fit like tiles in a floor, so is also called pavement epithelium. It is found in the walls of blood vessels, in the alveoli of lungs for exchange of gas, and in Bowman's capsule of nephron for ultra filtration.

161. (d) Myoglobin is a single-chain globular protein of 153 amino acids, containing a heme (iron-containing porphyrin) prosthetic group in the center around which the remaining apoprotein folds. It has a molecular weight of 16,700 daltons, and is the primary oxygen-carrying pigment of muscle tissues. Unlike the blood-borne hemoglobin, to which it is structurally related, this protein does not exhibit cooperative binding of oxygen, since positive cooperativity is a property of multimeric/oligomeric proteins only.

164. (b) Malaria is caused by *Plasmodium*. Asexual cycle of *Plasmodium* takes place in man hence, it is secondary host. Sexual cycle of *Plasmodium* occur in mosquito hence, it is called the primary host.

166. (a) Human insulin is being commercially produced from a transgenic species of

Escherichia coli. E. coli is a bacterium that is commonly found in the lower intestine of warm blooded animals. The bacteria can also be grown easily and its genetics are comparatively simple and easily manipulated, making it one of the best studied prokaryotic model organisms, and an important species in biotechnology.

167. (b) The major obstacle in transplantation of organs is that the recipient body does not accept the donor's organ. The body defence mechanism reject & treat the transplanted organ as a foreign particle and reacts actively.

168. (b) *Rauwolfia serpentina* belong to family *Apocynaceae*, its roots yield a chemical useful for high blood pressure.

169. (a) Allergy means all kinds of reactions manifested in the body due to hypersensitivity of substances called allergens. Allergy reactions occur due to liberation of histamine by tissue cells.

170. (c) HIV is a spherical, enveloped virus of about 90-120 nm diameter. Its nucleocapsid is icosahedral and its genome consists of a single-stranded RNA filament segmented into two identical filaments and associated with a "reverse transcriptase enzyme".

171. (a) Dolly sheep was obtained by cloning the udder cell (somatic cell) fused with enucleated oocyte.

172. (b) DNA fingerprinting is a test to identify and evaluate the genetic information-called DNA (deoxyribonucleic acid)-in a person's cells. DNA fingerprinting is a form of identification based on sequencing specific non-coding portions of DNA that are known to have a high degree of variability from person to person. These sections are known as Tandem repeats. The test is used to determine whether a family relationship exists between two people, to identify organisms causing a disease, and to solve crimes.

173. (c) The adult dental formula of human is Incisor $\frac{2}{2}$, Canine $\frac{1}{1}$, Premolar $\frac{2}{2}$, Molar $\frac{3}{3}$.

174. (c) Liver is the primary site of detoxification and elimination of body wastes and poisons. Liver detoxifies endotoxins, *e.g.* toxic NH_3 combined with CO_2 to form less toxic urea. It also detoxifies alcohol and convert them to acetaldehyde and then harmless acetyl CoA.

175. (d) Rennin (also called rennet or chymosin) is an coagulating enzyme produced from stomach of human body. It catalyzes the coagulation of milk by converting milk with soluble protein caesin into insoluble semi fluid calcium paracaesinate. This is called curdling of milk. Rennin produced in the infants immediately after birth. As the child grows, rennin production goes down and is replaced by pepsin digestive enzymes.

Renin is an enzyme which acts as hormone secreted by juxtaglomerular cells. It converts angiotensinogen into angiotensin.

177. (d) Osmotic pressure of body fluids is mainly maintained by the plasma proteins (albumins, globulins) and electrolyte ions (Na^+, K^+ etc). Phosphorus has nothing to do with the osmotic pressure of body fluids.

178. (c) An adult healthy human has average systolic and diastolic blood pressure as given below :

$$\text{B.P.} = \frac{\text{Systolic}}{\text{Diastolic}} = \frac{120 \text{ mm Hg}}{80 \text{ mm Hg}}$$

180. (b) Spleen (also known as blood bank and graveyard of RBCs) is a dark, purple gland present in the left side of abdomen against the stomach. It is internally made up of white pulp and red pulp. Its main functions are storage of blood and break down of old and senescent RBCs.

NEET (NATIONAL ELIGIBILITY ENTRANCE TEST) Full Test - 2

ANSWER KEY

1	(c)	21	(c)	41	(c)	61	(a)	81	(c)	101	(d)	121	(c)	141	(c)	161	(d)
2	(c)	22	(c)	42	(c)	62	(b)	82	(a)	102	(b)	122	(d)	142	(c)	162	(a)
3	(d)	23	(a)	43	(b)	63	(a)	83	(d)	103	(b)	123	(a)	143	(b)	163	(c)
4	(c)	24	(b)	44	(d)	64	(a)	84	(a)	104	(d)	124	(c)	144	(a)	164	(d)
5	(d)	25	(c)	45	(a)	65	(c)	85	(a)	105	(c)	125	(b)	145	(b)	165	(a)
6	(c)	26	(b)	46	(a)	66	(b)	86	(b)	106	(c)	126	(a)	146	(b)	166	(b)
7	(a)	27	(b)	47	(d)	67	(d)	87	(b)	107	(c)	127	(d)	147	(c)	167	(b)
8	(d)	28	(b)	48	(d)	68	(d)	88	(d)	108	(b)	128	(d)	148	(a)	168	(c)
9	(d)	29	(b)	49	(d)	69	(c)	89	(d)	109	(a)	129	(a)	149	(d)	169	(a)
10	(b)	30	(a)	50	(a)	70	(d)	90	(a)	110	(c)	130	(c)	150	(b)	170	(b)
11	(c)	31	(a)	51	(a)	71	(d)	91	(c)	111	(a)	131	(b)	151	(d)	171	(d)
12	(d)	32	(d)	52	(b)	72	(c)	92	(b)	112	(d)	132	(c)	152	(a)	172	(c)
13	(a)	33	(c)	53	(d)	73	(d)	93	(d)	113	(a)	133	(c)	153	(b)	173	(b)
14	(d)	34	(d)	54	(d)	74	(b)	94	(d)	114	(d)	134	(b)	154	(c)	174	(b)
15	(c)	35	(d)	55	(a)	75	(b)	95	(d)	115	(b)	135	(a)	155	(d)	175	(c)
16	(b)	36	(c)	56	(c)	76	(a)	96	(c)	116	(a)	136	(c)	156	(c)	176	(c)
17	(b)	37	(b)	57	(b)	77	(c)	97	(d)	117	(c)	137	(a)	157	(c)	177	(c)
18	(a)	38	(b)	58	(a)	78	(a)	98	(d)	118	(b)	138	(d)	158	(b)	178	(a)
19	(d)	39	(b)	59	(a)	79	(a)	99	(a)	119	(b)	139	(c)	159	(a)	179	(d)
20	(b)	40	(d)	60	(d)	80	(b)	100	(d)	120	(c)	140	(c)	160	(c)	180	(b)

HINTS & SOLUTIONS

SECTION 1 -PHYSICS

1. (c) We know that F = q v B

$$\therefore \quad B = \frac{F}{qv} = \frac{MLT^{-2}}{C \times LT^{-1}} = [MT^{-1}C^{-1}]$$

2. (c) $V = \frac{4}{3}\pi r^3$;

$$\frac{\Delta V}{V} \times 100 = 3\left(\frac{\Delta r}{r}\right) \times 100 = 3 \times 1\% = 3\%$$

3. (d) Here effective spring factor, $k_{eff} = 2\,k$;

Frequency of oscillation, $f = \frac{1}{2\pi}\sqrt{\frac{2k}{m}}$;

when one spring is removed, then spring factor = k ;

New frequency of oscillation

$$f' = \frac{1}{2\pi}\sqrt{\frac{k}{m}} = \frac{f}{\sqrt{2}}$$

4. (c) We shall apply conservation of angular momentum; that is

$$I_1\omega_1 = I_2\omega_2$$

$$\Rightarrow mv_1r_1 = mv_2r_2 \;\Rightarrow\; \omega_1 r_1^2 = \omega_2 r_2^2$$

$$\Rightarrow 44 \times (0.8)^2 = \omega_2 \times (1)^2$$

$$\Rightarrow \omega_2 = 44 \times 0.8 \times 0.8 = 28.16 \text{ radian/sec}$$

5. (d) $x = a + bt^2 = 15 + 3t^2$

$$v = \frac{dx}{dt} = 3 \times 2t = 6t$$

$$\Rightarrow \; v|_{t=3s} = 6 \times 3 = 18 \text{cm/s}$$

6. (c) Minimum excitation energy will be difference between ground level energy (n = 1) & first excited state (n = 2)

Now, $E_1 = -13.6\,eV$

$$E_2 = \frac{13.6}{2^2} eV = -3.4\,eV$$

Difference $= -3.4 + 13.6 = 10.2$ eV

8. (d) Heat produced, $H = \frac{V^2 t}{R}$. When voltage is halved, the heat produced becomes one fourth. Hence time taken to heat the water becomes four time.

9. (d) In (a), at the same time particle has two positions which is not possible. In (b), particle has two velocities at the same time. In (c), speed is negative which is not possible.

10. (b) $n = \frac{4}{2} = 2 \therefore PV = 2RT$

11. (c) The arrangement forms Wheatstone bridge as shown below :

It is balanced because $\frac{C_4}{C_1} = \frac{C_3}{C_5} = 1$

So, we can neglect C_2.

Now C_4 & C_3 are in series, C_1 & C_5 are in series so their resultant capacitance of each combination $= \frac{4 \times 4}{4+4} = \frac{16}{8} = 2\mu F$

These combinations are in parallel, so, total capacitance $= 2 + 2 = 4\mu F$

12. (d) $h\nu = \phi + \frac{1}{2}mv^2 \Rightarrow h\nu = \phi + K.E.$

$$\nu = \phi + \frac{1}{h} K.E.$$

This is equation of a straight line, if we plot ν, K-E curve. The slope of this curve is $\frac{1}{h}$ which is constant. So, it is independent of metals used.

Option (d) is correct.

13. (a) Moment of inertia of a body is given by the expression

$$I = \sum_{i=1}^{n} m_i r_i^2$$

This is the sum of product of mass of each particle and square of its distance from the axis. Naturally if distribution of mass changes, value of r will change & hence I will change.

14. (d) 0.1% of 1 kg = 0.001 kg

From the formula, $E = mc^2$

Energy released

$$= 0.001 \times (3 \times 10^8)^2 \text{ joules}$$

$$= 9 \times 10^{16} \times 10^{-3} = 9 \times 10^{13}\,J$$

15. (c) $3k = x.2k$

[Here α is coefficient of linear expansion so coefficient of cubical expansion $= 3\alpha$ & coefficient of superfacial expansion $= 2\alpha$]

$$x = \frac{3}{2} = 1.5$$

16. (b) When a bus suddenly take a turn, the passengers are thrown outwards because of inertia of motion.

17. (b) For minimum deviation, incident angle is equal to emerging angle.

$\therefore$ QR is horizontal.

18. (a) Charge on the tin nucleus = n.Q

$= (1.6 \times 10^{-19} \times 50)\,C = 8 \times 10^{-18} C$

This charge is supposed to be concentrated at its centre. Hence, potential on the surface,

$$V = \frac{1}{4\pi\varepsilon_0} \cdot \frac{q}{r}$$

$$= \frac{9 \times 10^9 \times 8 \times 10^{-18}}{6.6 \times 10^{-15}} = 1.1 \times 10^7 V$$

19. (d) One faraday charge is required to losing about electrolysis of one gram equivalent of a body. Now using this information in the relation,

$m = ZQ$

$E = ZF \Rightarrow F = \frac{E}{Z}$

21. (c) The horizontal velocity of the stone will be the same as that of the train. In this way, the horizontal motion will be uniform. The vertical motion will be controlled by the force of gravity. Hence it is accelerated motion. The resultant motion is a parabolic trajectory.

24. (b) $K = \frac{P^2}{2m} \Rightarrow P = \sqrt{2Km}$

as K is made four times P will become two times.

25. (c) $f_{apparent} = \left(\frac{u + u/5}{u}\right) f = \frac{6}{5} f = 1.2f$

wavelength remains constant (unchanged) in this case.

26. (b) Photoelectric effect is based on law of conservation of energy

27. (b) Choke coil is a device having high inductance and negligible resistance. It is used to control current in ac circuits and the power loss in a circuit containing choke coil is least.

28. (b) After 3 sec its velocity

$$v^2 = u^2 + 2gh = 0 + 2 \times 9.8 \times 25$$

$$K.E = \frac{1}{2} mv^2 = \frac{1}{2} \times 3 \times 2 \times 9.8 \times 25$$

$= 75 \times 9.8 = 748$ J.

29. (b) Since no external torque action gymnast, so angular momentum ($L = I\omega$) is conserved. After pulling her arms & legs, the angular velocity increases but moment of inertia of gymnast, decreases in, such a way that angular momentum remains constant.

30. (a) $B = \frac{\mu_0 I}{2r} \times \frac{\theta}{2\pi} = \frac{\mu_0 I\theta}{4\pi r}$

32. (d) Ability of the eye to see object at all distance is called accommodation. This is the power of the eye to change the focal length of eye lens so that it can focus rays coming from different distances.

33. (c) The velocity of electromagnetic radiation in a medium of permittivity ϵ_0 and permeability μ_0 is equal to $\frac{1}{\sqrt{\mu_0 \epsilon_0}}$.

34. (d) As displacement $\leq$ distance, therefore, the ratio is equal to or less than one.

35. (d) Variation of g with altitude is,

$$g_h = g\left[1 - \frac{2h}{R}\right];$$

variation of g with depth is,

$$g_d = g\left[1 - \frac{d}{R}\right]$$

Equating g_h and g_d, we get $d = 2h$

36. (c) $P = \frac{2/3(\frac{1}{2} mN\, v_{rms}^2)}{V}$ where m is mass of molecule of gas, N is number of molecules in gas & v_{rms} is root mean squar velocity of the molecule.

37. (b) Transformer is based upon the principle of mutual induction. Current is induced in the secondary coil due to change in current in primary coil of the transformer.

38. (b) Terminal velocity, $v_0 = \frac{2 r^2 (\rho - \rho_0) g}{9\eta}$

$$= \frac{2 \times (2 \times 10^{-3})^2 \times (8 - 1.3) \times 10^3 \times 9.8}{9 \times 0.83}$$

$= 0.07$ ms ^{1}

39. (b) Velocity of light in glass $= \frac{3 \times 10^8}{1.5}$

$= 2 \times 10^8$ m/sec

Time taken $= \frac{\text{distance}}{\text{velocity}}$

$$= \frac{4 \times 10^{-3}}{2 \times 10^8} = 2 \times 10^{-11} \text{ sec}$$

40. (d) Surface tension of solution

$(T) = 30 \times 10^{-12}$ N/m

Radius of the soap bubble (r) = 5 cm = 0.05 m

Since, the bubble has two spherical surface, hence, $\Delta A = 2 \times$ surface area $= 2 \times 4\pi r^2$

Therefore, surface energy $= T \times \Delta A$

$= 30 \times 10^{-2} \times 2 \times [4\pi (0.05)^2]$

$= 1.8 \times 10^{-2}$ J

41. (c)

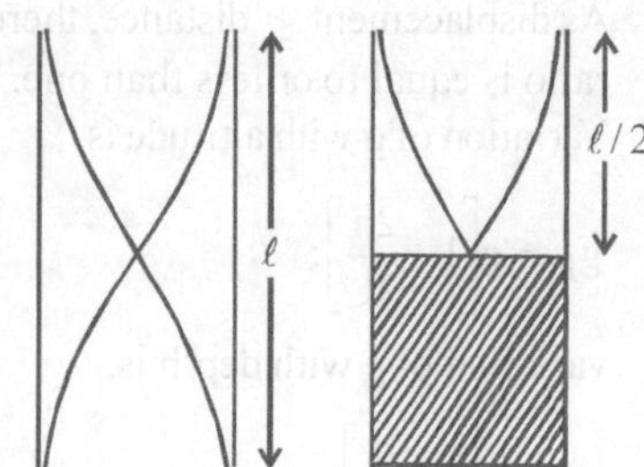

From the figure it is clear that in both the cases wavelength is same. So, frequency of fundamental tone will also be same. Hence the answer will be (c).

42. (c) Radio waves are low frequency electromagnetic waves. So, their wavelength will be comparatively large.

43. (b) For common base

Current gain $= \frac{I_c}{I_e} = 0.96$

$I_c = I_e \times 0.96 = 0.96 \times 7.2 = 6.912$

$I_b = I_e - I_c = 7.2 - 6.912$

$= 0.288 \approx 0.29$ mA

44. (d) 0V —▷|—~~ –2V It can be sketched as

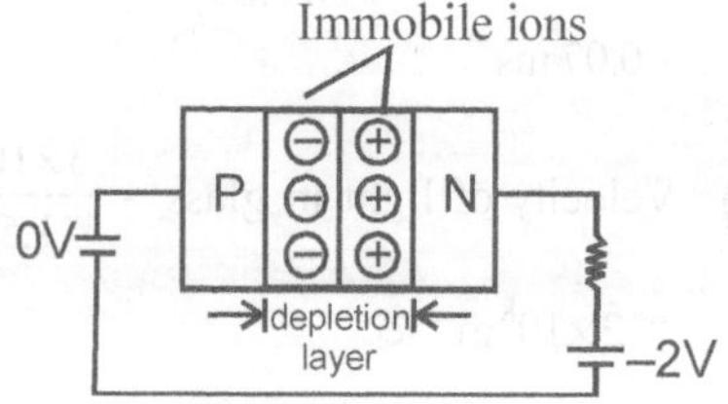

45. (a) Leakage current is the name given to the reverse current

SECTION 2 - CHEMISTRY

46. (a) It is N, N-dimethylcyclopropane – carboxamide.

47. (d) The electrical change on a colloidal particle is observed by electrophoresis. Under electric field, charged particles move in a particular direction.

48. (d) Carboxylic acids dissolve in $NaHCO_3$ but phenols do not.

$$RCOOH \xrightarrow{NaHCO_3} RCOONa + H_2O + CO_2$$

by evolving CO_2 gas.

49. (d) Zinc and lead in molten state are immiscible and form separate layers, further Zn, being ligher, forms upper layer. Silver is soluble in both Zn and Pb, and hence all statements are correct.

50. (a) The order of all photochemical reactions is zero as it does not depend upon the concentration of reactants.

51. (a) The correct reaction is

$$3O_2(g) \xrightleftharpoons[\text{Electric discharge}]{\text{Silent}} 2O_3(g);\ \Delta H = +284.5\text{ kJ.}$$

52. (b) A set of question number is valid when $l < n$ and m_l lies between $-l$ to $+l$. Thus sets I, II, IV are valid, III, V invalid.

53. (d) Ionic compounds are dissociated in solution state and form ions. Ions are good carrier of charge. So it makes solution conducting.

54. (d) Br_2 in CCl_4 (a), Br_2 in CH_3COOH (b) and alk. $KMnO_4$ (c) will react with all unsaturated compounds, i.e., 1, 3 and 4 while ammonical $AgNO_3$ (d) reacts only with terminal alkynes, i.e., 3 and hence compund 3 can be distinguished from 1, 2 and 4 by. ammonical $AgNO_3$ (d).

55. (a) $N_2 = 3H_2 = 6H$

$\therefore\ 1H = \frac{N_2}{6}$

Thus, equivalent weight of N_2

$= \frac{\text{molar mass}}{6} = \frac{28}{6} = 4.67$

56. (c) CO and oxides of Nitrogen are poisnous gases present in automobile exhaust gases.

57. (b) For spontaneous reaction E^0_{cell} must be positive.

$Zn^{2+} \longrightarrow Zn;\ E^0 = -0.76V$

$Cu^{2+} \longrightarrow Cu;\ E^0 = -0.34V$

Here Cu acts as cathode and zinc acts as anode.

$E^\circ_{cell} = E^\circ_C - E^\circ_A = (-0.34) - (-0.76)$
$= 0.42$ V.

58. (a) The smaller is the size of cation, the more is heat of hydration. Since hydration is exothermic process, thus heat evolves and hence system having lower energy are stable.

59. (a) For a cubic structure,

No. of X atoms $= 8 \times \frac{1}{8} = 1$

No. of Y atoms $= 6 \times \frac{1}{2} = 3$

$\therefore$ Formula of the compound $= XY_3$

60. (d) $(CH_3)_2C=O + O=C(CH_3)(C_2H_5)$

$\uparrow$

$(CH_3)_2C=C(CH_3)CH_2CH_3$

2,3-Dimethyl-2-pentene

61. (a) Vapour pressure of solution

$= p_A + p_B = p_A^o x_A + p_B^o x_B$

$= \frac{120 \times 2}{5} + \frac{180 \times 3}{5} = 156$ mm

62. (b) Since Cr^{3+} in the complex has unpaired electrons in the d orbital, hence will be coloured

$Ti = [Ar]3d^2 4s^2$; $Ti^{4+} = 3d^0$

$Cr = [Ar]3d^5 4s^1$; $Cr^{3+} = 3d^3$

$Zn = [Ar]3d^{10} 4s^2$; $Zn^{2+} = 3d^{10}$

$Sc = [Ar]3d^1 4s^2$; $Sc^{3+} = 3d^0$

63. (a) According to the adsorption theory of catalysis, the activity of catalysis is due to the presence of free valencies on its surface due to which surface of catalyst has chemical force of altraction. When a gas comes in contact with this surface molecules get attached attached through these valencies. Further the rate of reaction is always increases by decreases in activation energy

When any of the reactants is strongly adsorbed on the surface of catalyst, the rate becomes inversely proportional to the concentration of that reaction. The reaction is then said to be inhibited by such reactant. The lowering of activation energy always leads to the increase in speed of reaction.

64. (a) On increasing pressure, the temperature is also increased. Thus in pressure cooker due to increase in pressure the b.p. of water increases.

65. (c) More stable the carbocation, more reactive will be the parent alkyl halide towards S_{N^1} reaction.

3° > Benzyl > Allyl > 2° > 1° > methyl

66. (b) Fischer projection: C1 CH_3 (top); C2: H—C—Cl (S); C3: Cl—C—H (S); C_2H_5 (bottom)

So compound is (2S, 3S)

67. (d) As the unit of rate constant is sec^{-1}, so the reaction is first order reaction. Hence

$$k = \frac{1}{t} = \log \frac{a}{(a-x)} \text{ or } kt = \log \frac{[N_2O_5]_0}{[N_2O_5]_t}$$

68. (d) Among 1, 2- and 1, 3- configurations, 1, 3- is more stable due to less repulsion. Further among *cis* and *trans* isomers, *trans* is more stable due to less crowding.

69. (c) Option (a) is false as Coke is not found in nature. It is an artificial product. Option (d) is also false as CO_2 cannot be formed by dehydration of formic acid.

$$HCOOH \xrightarrow[H_2SO_4]{Conc.} CO + H_2O$$

Option (b) is also false as Producer gas is a mixture of CO and N_2.

70. (d) Aniline ($C_6H_5NH_2$) $+ 3Br_2 \longrightarrow$ 2, 4, 6 tribromoaniline $\xrightarrow[\text{\& dil HCl}]{NaNO_2}$

2, 4, 6 tribromoaniline

Diazonium salt (N_2Cl, 2,4,6-Br) $\xrightarrow[(-NaCl)]{NaBF_4}$ Diazonium tetra fluro borate ($\overset{+}{N_2}\overset{-}{BF_4}$, 2,4,6-Br)

Diazonium salt

Diazonium tetra fluro borate

$\xrightarrow[-N_2, -BF_3]{\Delta}$ 2,4,6 tribromofluorobenzene (F, 2,4,6-Br)

2,4,6 tribromofluorobenzene

71. (d) The semiconductors formed by the introduction of impurity atoms containing one elecron less than the parent atoms of insulators are termed as p-type semiconductors. Therefore silicon containing 14 electrons is to be doped with boron containing 13 electrons to give a p- type semi-conductor.

72. (c) $HO-C_6H_4-SO_3H \xrightarrow[H_2O]{Br_2}$ (2,6-dibromo-4-hydroxybenzenesulphonic acid: HO–$C_6H_2Br_2$–SO_3H, Br at both positions ortho to –OH)

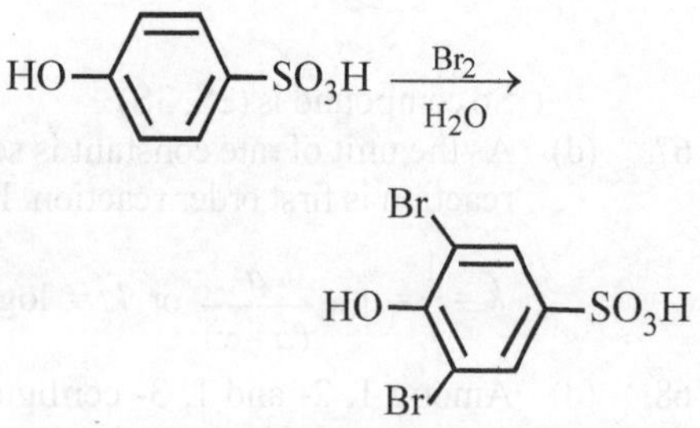

– OH group is highly activating. This is a type of electrophillic substitution reaction at ortho and para position.

73. (d) The appearance of colour in solid alkali metal halide is due to presence of F-centres found as defect in the crystal structure.

74. (b) Basic dyes have salts of amino or substituted amino group ($-NH_2$ or $-NR_2$) which in acid solution forms water soluble cations. These dyes attack the anionic sites present on the fabrics and then get attached to them. These are used to dye modified nylon and polyesters, e.g. aniline yellow, malachite green etc.

75. (b) The given statement explains that Al is placed above Cu in E.C.S. hence it can reduce Cu from its salt solution (reduction). Thus Al acts as a reducing agent now,

Given : electrode potential of copper

$(E^0_{cell}) = +0.34$ V

e.m.f of cell = 2 V

We know that,

emf of a galvanic cell (E^0_{cell})

= electrode potential of Cu (cathode) – electrode potential of Al (anode)

$2.0 = 0.34 - E^0_{Al}$

$E^0_{Al} = 0.34 - 2.00 = -1.66$ V

Hence, electrode potential of Al is – 1.66 V. Option (b) is correct.

76. (a) $X \xrightarrow{PCC} Y \xrightarrow{I_2/\text{alkali}} CHI_3$

Compund Y must give iodoform test. Further since Y is obtained by the oxidation of X which must be an alcohol (CH_3CH_2OH) and thus Y is CH_3CHO.

77. (c) On losing two electrons from $1s^2, 2s^2 2p^6, 3s^2$ electronic configuration, becomes $1s^2, 2s^2 2p^6, 3s^0$, it is more stable than $1s^2, 2s^2 2p^6, 3s^2$. Since energy gap between $2p^6$ and $3s^2$ will be more due to changing of shell from 3s to 2p. Therefore, the difference in values of IInd ionisation energy and IIIrd ionisation energy of this configuration is more.

Hence, option (c) is correct.

78. (a) Completing the given reaction.

$$CH_3-\overset{O}{\overset{\|}{C}}-CH_3 + CH_3-\overset{O}{\overset{\|}{C}}-CH_3 \xrightarrow{OH^-} CH_3-\underset{CH_3}{\overset{OH}{C}}-CH_2-\underset{CH_3}{\overset{O}{\overset{\|}{C}}}$$

Diacetonyl alcohol

[Note: Carbonyl compounds having α – H atom undergoes aldol condensation in presence of dilute base {e.g., NaOH, $Ba(OH)_2$ etc.} to form a class of compounds known as aldol].

79. (a) For exothermic reaction, the reaction goes in forward direction as temp. is lowered. If high pressure is applied, reaction will go in that direction where number of moles is less (Le-Chatelier's principle).

80. (b) Heavy metal ions, particularly Ag^+, catalyse S_N1 reaction because of presence of empty orbital.

$$CH_3CH_2Br + Ag^+ \rightleftharpoons [CH_3CH_2 \ldots Br \ldots Ag]^+$$

$$\xrightarrow[(-AgBr)]{slow} CH_3CH_2^+ \xrightarrow[\text{from } H_2O]{OH^-} CH_3CH_2OH$$

81. (c) Calculating the oxidation states of I, we get
H_5IO_6 ; $5 + x - 12 = 0$ or $x = +7$
H_3IO_5; $3 + x - 10 = 0$ or $x = +7$
HIO_4 ; $1 + x - 8 = 0$ or $x = +7$
Thus, the correct answer is +7, +7, +7

82. (a) Starch is also know as amylum which occurs in all green plants. A molecule of starch $(C_6H_{10}O_5)_n$ is built of a large number of α-glucose ring joined through oxygen-atom.

83. (d) The halide ions act as reducing agents . F^- ion does not show any reducing nature but Cl^-, Br^- & I^- ion act as reducing agents and their reducing nature is in increasing order

$$\xrightarrow[\text{Reducing nature increases}]{Cl^- \quad Br^- \quad I^-}$$

84. (a) Square Plannar Structure : XeF_4 and $[PtCl_4]^{2-}$

85. (a) Sodium silicate (Na_2SiO_3) is known as water glass. It is soluble in water. Just as plants grow in the soil, coloured crystals grow in the water glass.

86. (b)

Natural rubber
(All cis configuration)

All statements except (b) are correct

87. (b) For a first order reaction the positive slope is obtained when we plot $-\log_e [A]$ vs t.

88. (d)

XeF_6

(Xe atom in 3rd excited state)

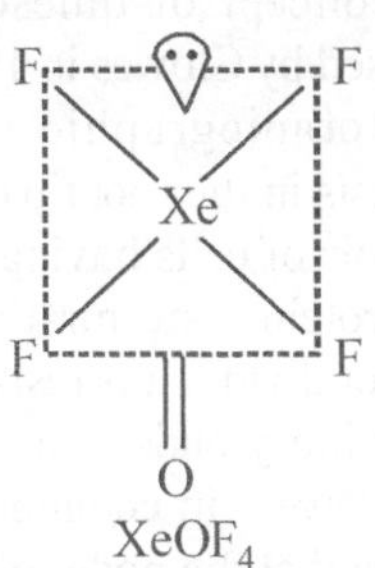

$XeOF_4$
(Xe in 3rd excited state)

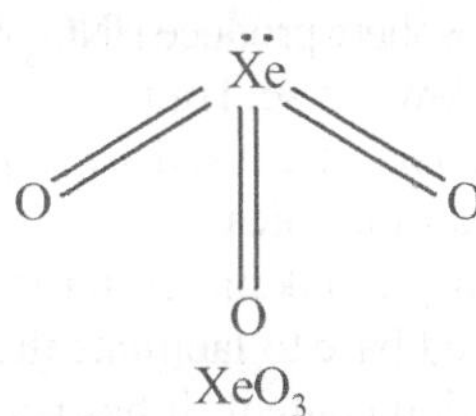

XeO_3
(Xe in 1st excited state)

Thus, all the three have same no. of lone pair of electrons i.e. one.

89. (d) $\lambda = \frac{h}{mc} = \frac{6.6 \times 10^{-34}}{10^{-6} \times 10} = 6.63 \times 10^{-29}$ m

90. (a) Vitamin B complex is water soluble vitamin whereas vitamin A, D, E and K are fat soluble vitamin.

SECTION 3-BOTANY

91. (c) During the origin of life, the first organisms evolved were chemoheterotroph. They obtained the organic material from outside which they utilized in energy production and synthesis of their own organic material.

92. (b) Reserve food in red algae (Rhodophyceae) is floridean starch. In green algae (Chlorophyceae), reserve food is starch. In brown algae Phaeophyceae, reserve food is laminarin and mannitol. In Cyanophyceae, reserve food is cyanophycean starch.

93. (d) *Azolla* has been used as a green manure crop in Vietnam and China for centuries. The ability of *Azolla* to shade out weeds in wetland rice and taro has been noted by farmers and researchers since the early 20th century.

94. (d) Submerged hydrophytes do not have stomata (a protective mechanism in aquatic plants against water logging of internal cells and tissues).

96. (c) The concept of quiescent centre was proposed by Clowes in 1961. On the basis of autoradiographic studies of DNA synthesis in the root tip of zea, he found a reservoir of cells having low DNA, RNA and protein concentration. They may or may not divide. It is resistant to damages.

98. (d) The axillary buds of the potato tuber are called "eyes" in common language. They are found at the nodes of the stem tuber.

99. (a) NO_2 and SO_2 after reacting with water in atmosphere produce HNO_3 & H_2SO_4 which fall down as acid rain.

100. (d) Silicosis is caused by exposure to dust containing silica.

101. (d) A stirred-tank bioreactor is a cylindrical or curved base to facilitate the mixing of the reaction content. It has been designed for availability of oxygen throughout the process.

102. (b) Nucellus - 2n, antipodal cells - n
Antipodal cells - n, egg cell - n
Antipodal cells - n, megaspore mother cell - 2n
Nucellus - 2n primary endosperm nucleus - 3n
Antipodal and egg cell are the product of meiotic division and rest are not.

104. (d) In angiosperms one male gamete fuses with the egg to form diploid zygote. The process is called syngamy. The other male gamete fuses with the two polar nuclei to form triploid primary endosperm nucleus. The process is called triple fusion. These two acts together known as double fertilization. The process was I[st] demonstrated by Nawaschin & Guignard in *Fritillaria* & *Lilium*.

106. (c) 'Whip-tail' disease in cauliflower is noted due to deficiency of molybdenum. Cauliflower is the most sensitive of crops to molybdenum (Mb) deficiency. Light sandy soils with pH levels close to or below pH 6 under conditions of high fertility can show severe Mb deficiency. Molybdenum is the only plant essential micronutrient that becomes less available as the soil acidifies– the opposite of copper, zinc and iron.

107. (c) The process of photorespiration requires 3 cell organelles-mitochondria, chloroplasts and peroxisomes. In chloroplasts glycolate is formed from ribulose biphosphate which passes into peroxisomes to be changed into glyoxylate. In this reaction H_2O_2 is evolved. Glyoxylate is changed to glycine which now enters mitochondria. Two molecules of glycine combine to form one molecule of serine with the evolution of CO_2.

108. (b) Photosynthesis takes place only in the visible part (400 - 700 nm wavelength) of electromagnetic radiations. Hence this component comprises the photosynthetically active radiation.

109. (a) Photolysis is catalysed by the protein-bound inorganic complex containing manganese ions (oxygen evolving complex) of photosystem II.

112. (d) When a pair of contrasting characters are studied during hybridization, it is called dihybrid cross. In F_2-generation, its ratio is 9 : 3 : 3 : 1.
It means when parents having –
Yellow, round Crossed with green wrinkled seeds seeds
(IIRR) × (iirr)
will give 9 – Yellow round seeds.
3 – Yellow wrinkled seeds.
3 – Green round seeds.
1 – Green wrinkled seeds

113. (a) A blanket of air surrounding the earth is called as atmosphere and it is divided into different strata (zone). The zone of atmosphere that lies near the ground is troposphere.

114. (d) The phenomenon of making DNA over RNA genome through enzyme reverse transcriptase is called reverse transcription or teminism.

115. (b) Jaya and Ratna are two rice varieties developed for green revolution in India.
- The scientific name of Jaya is IET-723. This paddy variety takes about 130 days to grow and the grain is long, bold and white. Its yield is 50-60 quintals per hectare.
- The scientific name of 'Ratna' is IET-1411. It takes about 130-135 days to grow. The grain is long, slender and white. Its yield is 45-50 quintal/hectare.

116. (a) Lots of urea and phosphate fertilizer were used in the crops in the vicinity and the lake water turned green and stinky. Due to this, lake near a village suffered heavy mortality of fishes within a few days.

117. (c) Coral reefs are often called "rainforest of the sea". It forms some of the most diverse ecosystem on earth. Coral reef ecosystem have the highest gross primary productivity in the sea.

118. (b) Glycogen is an animal starch, stored in the liver and is polysaccharide of– Glucose. About 30,000 glucose molecules joined by 1-4α-glycosidic bonds and its branches by 1-6α-glycosidic bonds. It gives red colour with iodine solution.
- Galactose is a monosaccharide from milk.
- Fructose is a monosaccharide from fruit.
- Sucrose is a disaccharide (Glucose + Fructose).

119. (b) As a result of glycolysis, one molecule of glucose is broken down to two molecules of pyruvate. During glycolysis 4 molecules of ATP are produced but 2 molecules of ATP are consumed in the activation of glucose molecule. Therefore, there is a net gain of 2 molecules of ATP.

120. (c) Abscisic acid (ABA), also known as abscisin II and dormin, is a plant hormone. It functions as many plant developmental processes, including bud dormancy. Abscisic acid is a derivative of carotenoids. It was called "abscisin II" originally because it was thought to play a major role in abscission of fruits. At about the same time another group was calling it "dormin" because they thought it had a major role in bud dormancy. The name abscisic acid (ABA) was coined by a compromise between the two groups.

122. (d) (i) **Helicase** also known as unwindase, these enzymes separate the two strands of DNA.

(ii) **Topoisomerases :** Tension produced by unwinding of DNA strands is reduced by these enzymes.

(iii) **RNA Primase :** This is an RNA polymerase, which helps in the synthesis of a small fragment of RNA called primer.

124. (c) Golden rice is vitamin A rich variety developed by rDNA technology and used in the treatment of vitamin A deficiency.

125. (b) Biodiversity is the number of variety of organism found within a specified geographic region.

127. (d) The 'Cheetah' has become extinct from India. The endangered subspecies *Acinonyx jubatus venaticus* lives in Asia (Iran). In 1990, there were reports in the Times of India of a cheetah sighting in eastern India. There is a chance some cheetahs remain in India, though it is doubtful. There have also been reports of Asiatic cheetahs in the Balochistan Province of Pakistan, though these continue to be unverified. The cheetah prefers to live in an open biotope, such as semi-desert, prairie, and thick brush.

128. (d) In gel electrophoresis agarose extracted from sea weed used as gel agarose, made of 0.7% gel show good resolution of large DNA and 2% gel will show good resolution of small fragments.

129. (a) 2100 antibiotics have been reported so far from actinomycetes alone. Of these maximum antibiotics have been reported from *Streptomyces* alone. Waksman isolated streptomycin from *Streptomyces griseus*.

130. (c) Nucleolus was discovered by Fontana (1781) and given name by Bowman (1840). It does not have membrane and is attached to chromatin at nucleolar organiser region (NOR). Nucleolus is site for elaboration of r-RNA and synthesis of ribsomes, hence called ribosomal factory.

131. (b) Niche indicate the habitat of a particular species and the interaction of that species with the resources present in the habitat. Niche overlap means that two or more species sharing the resources present in a particular niche.

132. (c) Coconut milk represents the non-cellular endosperm of the coconut fruit. It is widely used in tissue culture because it is extremely rich in kinetin, a type of cytokinin. Cytokinins are plant growth hormones that promote cell division, and consequently they are widely used in tissue culture practices to rapidly promote cell growth and division.

133. (c) Okazaki fragments in DNA are linked up by the enzyme DNA *ligase*. Replication always occur in 5' - 3' direction. Okazaki fragments synthesized on 3' - 5' DNA template, join to form lagging strand which grows in 3' - 5' direction.

134. (b) The root nodules of leguminous plants contain a symbiotic nitrogen fixing bacteria *Rhizobium*. Root nodules are small irregular outgrowth of the roots which are pinkish internally due to presence of a pigment called leghaemoglobin. It is related to blood pigment haemoglobin. The cells of root nodules are tetraploid and contain polyhedral bacteria called bacteroids. Leghaemoglobin is an oxygen scavenger and protects the nitrogen fixing enzyme nitrogenase of bacteroids.

SECTION 4-ZOOLOGY

136. (c) Chordates are featured by the presence of the following characters : Notochord (cephalic or head region), Pharyngeal gill slits, Postanal tail.

137. (a) *Methanobacillus* (methanogen) occurs in marshes and also in dung. It produces CH_4 gas under anaerobic condition and is utilized in gobar gas plant.

139. (c) Silver fish (Insecta), scorpion (Arachnida); crab (Crustacea) and honey bee (Insecta) all belong to phylum Arthropoda which is characterized by the presence of jointed appendages. Poison glands occur only in scorpion. Compound eyes and metamorphosis are the characteristic of insects.

143. (b) Nephridia in earthworm, malpighian tubules in cockroach and urinary tubules in the rat are excretory in function.

144. (a) Fusion of male and female gamete produces a zygote. Repeated division of the zygote is called cleavage forming a solid morula. After further division and rearrangement a fluid filled cavity surrounded by blastomeres - blastula is formed. The appearance of germ layers mark the gastrula.

145. (b) Vasa efferentia are ductules leading from rete testis to vas deferens. The rete testis is an anastomosing network of tubules located in the hilum of the testicles that carries sperm from the seminiferous tubules to the vasa efferentia.

146. (b) The corpus luteum is essential for establishing and maintaining pregnancy in females. In the ovary, the corpus luteum secretes estrogens and progesterone, which are steroid hormones responsible for the thickening of the endometrium and its development and maintenance, respectively.

150. (b) The logistic population growth is expressed by the equation $\frac{dN}{dt} = rN\left[\frac{(K-N)}{K}\right]$

where *N* is population density at time *t*, *r* is the Malthusian parameter (rate of maximum population growth) and K is the so called carrying capacity (*i.e.* maximum sustainable population). It is a type of population growth when resources are limiting.

152. (a) Lubb sound is caused partly by the closure of the bicuspid and tricuspid valves and partly by the contraction of the muscles in the ventricles. Lubb is the first heart sound.

153. (b) Nephron is the excretory unit of human excretory system. Each nephron has a Bowman's Capsule, a Proximal Convoluted Tubule (PCT), Loop of Henle (Descending & Ascending limbs) and Distal Convoluted Tubule (DCT) which then enter into collecting duct.

154. (c) Peak P-causes diastolic phase in ventricle while R-Peak causes systole in ventricle means diastolic and systolic phases represented by P & R.

155. (d) The number of blood platelets per cubic mm in human blood is 3 lacs while WBCs are 5000/cubic mm of blood. Veins are as complex as the arteries. Veins and arteries both are types of blood vessels. Arteries carry blood from heart to different organs while vein carries blood from different organs to heart. At any given time in a healthy human, the blood amount is same in both, as the circulation of blood never stops.

Blood consists of two parts:

The **plasma** (water, proteins, inorganic salts and other elements) constitutes 55-60% of blood while **cellular** part constitutes 40-45% of total blood. WBCs are produced in red bone marrow, lymph nodes and sometimes even in liver and spleen.

156. (c) In gout, there occurs a defect in uric acid metabolism resulting into its elevated level in blood (hyperuricemia). This is followed by precipitation of excessive uric acid which gets deposited in the joint spaces. These deposited crystals of uric acid causes pain in different bony joints.

157. (c) Organs that are similar in fundamental structure but different in functions are "Homologous organs", Richard Owen, introduced the term homologous. Pectoral fins of fish and fore limbs of horse similar in structure but different in functions are homologous organs.
Rest of the organs compared in the question are analogous organs.

158. (b) Down's syndrome is due to trisomy of 21st chromosome and is an autosomal abnormality.

160. (c) The lack of independent assortment in sweet pea and *Drosophila* is due to linkage.

161. (d) Jurassic period of meoszoic era was about 19-20 crore years ago & lasted for about 5.5-6 crore years. The climate was hot and damp. It is called the age of dinosaurs. Ist primitive bird *Archaeopteryx* evolved from reptiles. Ist angiosperm appeared as a dicotyledon but gymnosperms were dominant.

162. (a) Cri-du-chat/cat cry syndrome is due to the deletion of large part of the small or one of the 5th chromosome.

163. (c) Common activities in bioinformatics include mapping and analysing DNA and protein sequences, aligning different DNA *etc.* are the part of human genome project.

165. (a) It is the relationship between the percentage saturation of haemoglobin (by volume) in the blood and the oxygen tension (in partial tension) pO_2 of the blood.
It is usually a sigmoid plot. Haemoglobin molecules can bind up to four oxygen molecules in a reversible way. The shape of the curve results from the interaction of bound oxygen molecules with incoming molecules. The binding of the first molecule is difficult. However, this facilitates the binding of the second and third molecules, and it is only when the fourth molecule is to be bound that the difficulty increases, partly as a result of crowding of the haemoglobin molecule, partly as a natural tendency of oxygen to dissociate.

166. (b) Hyoid is a horse shoe shaped bone present in neck between lower law and sound box (larynx). It is not articulated to any bone, but is simply suspended, from temporal bones by means of ligaments. Hyoid provides surface for the attachment of tongue muscles.

169. (a) Luteinizing hormone (LH) stimulates ovulation. Deficiency of insulin causes diabetes mellitus. Deficiency of ADH or vasopressin causes diabetes insipidus. Deficiency of parathormone causes tetany. Deficiency of thyroxine causes cretinism in infants and myxoedema in adults.

170. (b) In heart cells AMP acts as secondary messenger which is made in the response of adrenaline and it stimulates Ca^{2+} ions to come out from the sarcoplasmic reticulum of muscle fibres which causes muscle contraction.
cGMP (Cyclic Guanosine Monophosphate) which acts as secondary messenger works on the action of acetylcholine, increase in flow of Ca^{2+} into muscle fibres & hence causes muscle relaxation. There is no role of sodium in hormonal action.

172. (c) The lymphocytes which differentiate in the thymus are known as the T - lymphocytes. T-lymphocytes are responsible for the cellular immune response.

173. (b) Tranquilizers bring about depression of brain activity and feeling of calmness. Stimulants stimulate the nervous system, and increase alertness and activity. Hallucinogens (*Cannabis* products) alters thoughts, perceptions and feelings.

175. (c) Carcinomas are malignant growths of the epithelial tissue that cover or line body organs.

179. (d) The last two pairs *i.e.* 11th and 12th pairs ribs remain free anteriorly, hence, they are called as floating ribs.

180. (b) *Cro-Magnon* fossils have been found in Europe. Neanderthal man is a transitional stage. *Australopithecus* appeared in early Pleistocene.

NEET (NATIONAL ELIGIBILITY ENTRANCE TEST) Full Test - 3

ANSWER KEY

1	(a)	21	(c)	41	(c)	61	(b)	81	(d)	101	(d)	121	(c)	141	(a)	161	(a)
2	(a)	22	(d)	42	(d)	62	(a)	82	(a)	102	(a)	122	(d)	142	(b)	162	(b)
3	(d)	23	(a)	43	(b)	63	(a)	83	(d)	103	(d)	123	(a)	143	(d)	163	(d)
4	(c)	24	(c)	44	(d)	64	(c)	84	(c)	104	(a)	124	(c)	144	(d)	164	(a)
5	(b)	25	(a)	45	(d)	65	(b)	85	(a)	105	(d)	125	(d)	145	(b)	165	(b)
6	(a)	26	(b)	46	(a)	66	(a)	86	(a)	106	(b)	126	(a)	146	(a)	166	(a)
7	(d)	27	(a)	47	(d)	67	(c)	87	(c)	107	(c)	127	(c)	147	(b)	167	(a)
8	(b)	28	(b)	48	(c)	68	(a)	88	(c)	108	(b)	128	(d)	148	(b)	168	(c)
9	(d)	29	(a)	49	(d)	69	(a)	89	(a)	109	(a)	129	(c)	149	(c)	169	(a)
10	(a)	30	(d)	50	(a)	70	(c)	90	(c)	110	(a)	130	(d)	150	(a)	170	(c)
11	(a)	31	(a)	51	(c)	71	(a)	91	(c)	111	(b)	131	(b)	151	(a)	171	(a)
12	(d)	32	(a)	52	(a)	72	(d)	92	(b)	112	(a)	132	(b)	152	(d)	172	(c)
13	(c)	33	(a)	53	(b)	73	(c)	93	(d)	113	(b)	133	(d)	153	(b)	173	(c)
14	(b)	34	(b)	54	(a)	74	(c)	94	(d)	114	(b)	134	(c)	154	(d)	174	(d)
15	(a)	35	(b)	55	(c)	75	(d)	95	(b)	115	(b)	135	(c)	155	(c)	175	(c)
16	(d)	36	(b)	56	(d)	76	(d)	96	(b)	116	(b)	136	(d)	156	(c)	176	(c)
17	(a)	37	(d)	57	(d)	77	(d)	97	(b)	117	(b)	137	(b)	157	(b)	177	(a)
18	(d)	38	(a)	58	(a)	78	(a)	98	(a)	118	(b)	138	(a)	158	(c)	178	(c)
19	(a)	39	(a)	59	(b)	79	(d)	99	(a)	119	(c)	139	(a)	159	(b)	179	(a)
20	(d)	40	(d)	60	(a)	80	(b)	100	(b)	120	(c)	140	(b)	160	(c)	180	(c)

HINTS & SOLUTIONS

SECTION 1 - PHYSICS

1. (a) From stokes law $F = 6\pi\eta rv \Rightarrow \eta = \dfrac{F}{6\pi rv}$

$$\therefore \eta = \frac{MLT^{-2}}{[L][LT^{-1}]} \Rightarrow \eta = [ML^{-1}T^{-1}]$$

2. (a) Component of $\bar{a}$ along $\bar{b} = \dfrac{\bar{a}\,.\,\bar{b}}{|\bar{b}|}$

3. (d) From C to D, V is constant. So process is isochoric. From D to A, the curve represents constant temperature. So the process is isothermal.
From A to B, pressure is constant . So, the process is isobaric.
BC represents constant entropy.

4. (c) $V = \sqrt{\dfrac{GM}{R+h}} \Rightarrow W(R+h) = \sqrt{\dfrac{GM}{(R+h)}}$

$$\Rightarrow W = \sqrt{\frac{GM}{(R+h)^3}}$$

5. (b) $\dfrac{dv}{dt} = 6t + 5 \Rightarrow dv = (6t+5)dt$

$$v = 6\frac{t^2}{2} + 5t \text{ (on integration)}$$

$$\frac{ds}{dt} = 3t^2 + 5t \Rightarrow ds = (3t^2 + 5t)dt$$

Integrating, $s = 3\dfrac{t^3}{3} + 5\dfrac{t^2}{2} = t^3 + 5\dfrac{t^2}{2}$

$\therefore$ At $t = 2$ s, distance covered,

$$s = (2)^3 + \frac{5(2)^2}{2} = 18 \text{ m}$$

6. (a) For Lyman series

$$\frac{1}{\lambda} = \left[\frac{1}{\ell^2} - \frac{1}{n^2}\right]$$

For shortest wavelength, $n = \infty$,

$$\therefore \quad \frac{1}{\lambda_1} = R \Rightarrow \lambda_1 = \frac{1}{R}$$

For Balmer series

$$\frac{1}{\lambda_1} = R\left[\frac{1}{2^2} - \frac{1}{n^2}\right]$$

For shortest wavelength, $n = \infty$,

$$\therefore \quad \frac{1}{\lambda_2} = \frac{R}{4} \Rightarrow \lambda_2 = \frac{4}{R}$$

The ratio $\frac{\lambda_2}{\lambda_1} = \frac{4}{1}$.

7. (d) Eddy current is the current produced in the body of a conductor due to electromagnetic induction. Moving coil galvanometer, electric brakes and induction motor based on electromagnetic induction. Hence they use eddy current. Dynamo doesn't use it.

8. (b) Kirchoff's Ist law states that the algebraic sum of currents (charges) meeting at the junction is zero or Σi or $\Sigma q = 0$, hence, it is based on the conservation of charge. Kirchoff's IInd law states that algebraic sum of the products of resistance and respective current in a closed mesh is zero. $\Sigma iR = \Sigma E$, hence, it is based on the conservation of energy.

9. (d) $$\frac{\frac{x}{2} + \frac{x}{2}}{\frac{x}{2v_1} + \frac{x}{2v_2}} = \frac{1}{\left(\frac{v_2 + v_1}{2v_1 v_2}\right)} = \frac{2v_1 v_2}{v_1 + v_2}$$

10. (a) Efficiency $= \frac{T_1 - T_2}{T_1}$

$T_1 = 227 + 273 = 500\text{ K}$
$T_2 = 127 + 273 = 400\text{ K}$

$$\eta = \frac{500 - 400}{500} = \frac{1}{5}$$

Hence, $V = (\eta) \times H = \frac{1}{5} \times 6 = 1.2\ kcal$

11. (a) Slope of the graph will give us rciprocal of resistance. Here resistance at temp. T_1 is greater than that of T_2. Since resistance of metallic wire is mroe at higher temp. then at lower temperature, hence $T_1 > T_2$.

12. (d) Work function (W_0) and the threshold frequency (ν_0) are related as below

$$W_0 = h\nu_0$$

$$\therefore \quad \nu_0 = \frac{W_0}{h} = \frac{(5 \times 1.6 \times 10^{-19})\,V}{(6.625 \times 10^{-34})\,Js}$$

$$= 1.2 \times 10^{15}\,\text{Hz}$$

13. (c) $$\bar{x} = \frac{m(-2b) + 2m(-b) + m \times 0 + 2m(b)}{m + 2m + m + 2m} = \frac{-b}{3}$$

14. (b) ${}_{-1}e^0$ is known as β^- particle & $\bar{\nu}$ is known as antineutrino. Since in this reaction $\bar{\nu}$ is emitted with ${}_{-1}e^0$ (β^- particle or electron), so it is known as β-decay.

15. (a) Let m be the mass of steam condensed. Then

$m \times 540 + m \times 10/2 = 22 \times 70$

$\therefore \quad m = 2.83$ gm

Now, total mass $= 22 + 2.83 = 24.83$ gm

16. (d)

17. (a) $$\mu = \frac{\sin\frac{\delta_m + A}{2}}{\sin\frac{A}{2}} \Rightarrow \sqrt{2} = \frac{\sin\frac{\delta_m + 60}{2}}{\sin 30°}$$

$$\sin\frac{\delta_m + 60}{2} = \sqrt{2} \times \frac{1}{2} = \frac{1}{\sqrt{2}} = \sin 45°$$

$$\frac{\delta_m + 60}{2} = 45°$$

$$\Rightarrow \delta_m + 60 = 90°; \quad \delta_m = 30°$$

Now $\delta_m = 2i - A$

$$\Rightarrow \quad i = \frac{\delta_m + A}{2} = \frac{30 + 60}{2} = 45°$$

18. (d) According to Gauss' Law

$$\oint E.ds = \frac{Q_{\text{enclosed by closed surface}}}{\varepsilon_0} = \text{flux}$$

so total flux $= Q/\varepsilon_0$

Since cube has six face, so flux coming out through one wall or one face is $Q/6\varepsilon_0$.

19. (a) Electromotive force is a force which is able to maintain a constant potential difference. This force comes from the chemical reaction taking place in the battery.

20. (d) Both oxygen and nitrogen will have same specific heat as they are diatomic gases.
Let θ be the temp. of mixture.
Loss of heat = gain of heat

$$\frac{16}{16}\times C_p \times (37-\theta) = \frac{14}{44}\times C_p(\theta - 27)$$

$$2\theta = 37 + 27 = 64 \Rightarrow \theta = \frac{64}{2} = 32°\,C.$$

21. (c) $|\vec{V}_2 - \vec{V}_1| = \sqrt{5^2+5^2} = 5\sqrt{2}$

$$a = \frac{5\sqrt{2}}{10} = \frac{1}{\sqrt{2}}\ m/s^2\ \text{N-W}$$

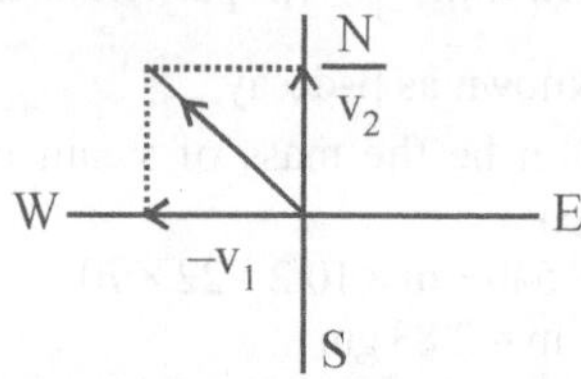

22. (d) $e = -\frac{d\phi}{dt} = -\frac{d(N\vec{B}.\vec{A})}{dt}$

$$= -N\frac{d}{dt}(BA\cos\omega t) = NBA\omega \sin \omega t$$

$$\Rightarrow e_{max} = NBA\omega$$

23. (a) The electrostatic potential energy of the two electrons, placed at the seperation r from each other, is

$$U = \left(\frac{1}{4\pi\varepsilon_0}\right)\frac{(-e)(-e)}{r} = \frac{e^2}{4\pi\varepsilon_0 r}$$

which increases when the one electron is brought towards the other and the sepeeration r decreases.

24. (c) $g = \frac{GM}{R^2} \Rightarrow \frac{dg}{g} = \frac{dM}{M} - 2\frac{dR}{R}$

$$\Rightarrow \frac{dg}{g}\times 100 = -2\left(\frac{dR}{R}\times 100\right)$$

25. (a) Frequency does not depend upon radius. As length is doubled, fundamental frequency becomes half.

26. (b) Rectifier converts a.c. current to d.c. current.

28. (b) $W = \int_0^{x_1} F\,dx = \int_0^{x_1} c\,x\,dx = \left[\frac{1}{2}cx^2\right]_0^{x_1}$

$$= \frac{1}{2}c(x_1^2 - 0) = \frac{1}{2}cx_1^2$$

29. (a) $I_1\omega_1 = I_2\omega_2$
As $I_2 > I_1$,
$\therefore \quad \omega_2 < \omega_1$

30. (d) The value of red, violet and green are : 2, 7 and 10^5. And gold represents a tolerance 5%. Thus $R = 27 \times 10^5 \pm 5\%\,\Omega$.

32. (a) The person suffers from short-sightedness because he is unable to see objects at normal distance but can see objects at nearer distance.
He will require concave lens. Things placed at 30 cm will appear to be placed at 10 cm.
So, u = 30; v = 10; f = ?
From lens formula

$$\frac{1}{v} - \frac{1}{u} = \frac{1}{f} \Rightarrow -\frac{1}{10} + \frac{1}{30} = \frac{1}{f}$$

$$\Rightarrow \frac{-3+1}{30} = \frac{1}{f} \Rightarrow f = -\frac{30}{2} = -15\text{cm}$$

Negative sign indicates that lens used is concave.

34. (b) Loss in potential energy = Gain in kinetic energy

$$-\frac{GMm}{R} - \left(-\frac{3}{2}\frac{GMm}{R}\right) = \frac{1}{2}mv^2$$

$$\Rightarrow \frac{GMm}{2R} = \frac{1}{2}mv^2 \Rightarrow v = \sqrt{\frac{GM}{R}} = \sqrt{gR}$$

35. (b) Small amount of work done in extending the spring by d_x is
dW = k x dx

$$\therefore W = k\int_{0.05}^{0.15} x\,dx$$

$$= \frac{800}{2}\left[(0.15)^2 - (0.05)^2\right]$$

$$= 400\,[(0.15+0.05)(0.15-0.05)]$$

$$= 400 \times 0.2 \times 0.1 = 8\,J$$

36. (b) $T = 2\pi\sqrt{\dfrac{\text{displacement}}{\text{acceleration}}}$

$= 2\pi\sqrt{\dfrac{x}{bx}} = 2\pi/\sqrt{b}$

37. (d) The direction of the induced emf is given by Lenz's law which states that the induced emf always opposes the change that has induced it.

38. (a) $F = 6\pi\eta r v$
$= 6 \times 3.14 \times (8 \times 10^{-5}) \times 0.03 \times 100$
$= 4.52 \times 10^{-3}$ dyne

39. (a) Both concave lens and convex mirror can produce erect, diminished and virtual image.

40. (d) Bernoulli's theorem is based on the conservation of energy of a flowing liquid.

41. (c)

→ glass rod

The length of the rod (L),

$= 0.2 = \dfrac{\lambda}{2} \Rightarrow \lambda = 0.4\,\text{m}$

∴ the velocity of the sound in gas (v),
$= \nu \times \lambda = 4000 \times 0.4 = 1600$ m/s

42. (d) Self induction of solenoid $\propto x^2$ where x is number of turns.
So, if x is doubled, self inductance will quadrupled.

43. (b) $\alpha = \dfrac{I_c}{I_e}$; $\beta = \dfrac{I_c}{I_b}$

$I_e = I_b + I_c$, $\dfrac{I_e}{I_c} = \dfrac{I_b}{I_c} + 1$

$\dfrac{1}{\alpha} = \dfrac{1}{\beta} + 1$, $\dfrac{1}{0.95} = \dfrac{1}{\beta} + 1$

$\dfrac{1}{\beta} = \dfrac{1}{0.95} - 1 = \dfrac{0.05}{0.95} = \dfrac{5}{95}$

$\beta = \dfrac{95}{5} = 19.$

44. **(d)** For this truth table, Boolean expression is $y = A.\overline{B} + \overline{A}.B$, which is for XOR gate

SECTION 2-CHEMISTRY

46. (a) As both the solutions are isotonic hence there is no net movement of the solvent through the semipermeable membrane between two solutions.

47. (d) $R\,COCl + AlCl_3 \longrightarrow R-\overset{\overset{\displaystyle O}{\|}}{C}^{\oplus} + AlCl_4^-$
Electrophile

48. (c) White lead is $2PbCO_3.Pb(OH)_2$ which is prepared by $Pb(NO_3)_2$ & Na_2CO_3. It is called as basic lead carbonate.

$3Pb(NO_3)_2 + 3Na_2CO_3 + H_2O \longrightarrow$

$Pb(OH)_2 + 2PbCO_3 + NaNO_3 + CO_2$
Basic lead carbonate

49. (d) Specific volume (volume of 1 gm) of cylindrical virus particle $= 6.02 \times 10^{-2}$ cc/gm
Radius of virus (r) = 7 Å $= 7 \times 10^{-8}$ cm
Length of virus $= 10 \times 10^{-8}$ cm
Volume of virus

$\pi r^2 l = \dfrac{22}{7} \times (7 \times 10^{-8})^2 \times 10 \times 10^{-8}$
$= 154 \times 10^{-23}$ cc

Wt. of one virus particle $= \dfrac{\text{volume}}{\text{specific volume}}$

∴ Mol. wt. of virus = Wt. of N_A particle

$= \dfrac{154 \times 10^{-23}}{6.02 \times 10^{-2}} \times 6.02 \times 10^{23}$
$= 15400$ g/mol $= 15.4$ kg/mole

50. (a) C_2H_5OH
Mechanism :
(1) Protonation :

$C_2H_5OH + H^{\oplus} \rightleftharpoons C_2H_5OH_2^{\oplus}$

(2) $C_2H_5OH_2^+ \longrightarrow C_2H_5^{\oplus} + H_2O$

$\xrightarrow{C_2H_5OH} C_2H_5-\overset{\overset{\displaystyle H}{|}}{\underset{\oplus}{O}}-C_2H_5$

↓↑

$H^{\oplus} + C_2H_5-O-C_2H_5$
Diethyl ether

51. (c) The tendency of an atom in a compound to attract a pair of bonded electrons towards itself is known as electronegativity of the atom.

52. (a) $Cl_2 + OH^- \longrightarrow Cl^- + ClO^-$

$\frac{1}{2}\underset{0}{Cl_2} \longrightarrow \underset{-1}{Cl^-} + ClO^-$ oxidation change = 1

equivalent mass of Cl_2 in reduction half-reaction.

$$= \frac{\frac{1}{2}Cl_2}{1} = \frac{M}{2}$$

$\frac{1}{2}\underset{0}{Cl_2} \longrightarrow Cl^- + \underset{+1}{ClO^-}$ reduction change = 1

equivalent mass of Cl_2 in oxidation half-reaction.

$$= \frac{\frac{1}{2}Cl_2}{1} = \frac{M}{2}$$

Thus, equivalent mass in overall reaction

$$= \frac{M}{2} + \frac{M}{2} = M$$

53. (b) BiF_5 not possible.
Bi comes in nitrogen family down the group. Hence tendency to show + 5 (co-ordination state) decreases, due to larger size of orbitals, the vaccant d orbitals can not overlap with orbitals of fluorine.

54. (a) $Br_2 \longrightarrow BrO_3^-$

O.N. of Br 0 +5

Thus, it changes from 0 to + 5.

55. (c) Observe the relative stability of their corresponding conjugate bases.

O^- (o-NO_2) I > O^- (p-NO_2) III > O^- (m-NO_2) II

I: Stabilised by both –I and – R effects of the $-NO_2$ group

III: Stabilised by both –I and – R effects, but – I is less than in I

II: Stabilised only by –I effect

However, the acidity of the corresponding phenols will be different because of H–bonding in the ortho isomer.

OH (p-NO_2) III > OH (o-NO_2) I > OH (m-NO_2) II

56. (d) Average molecular speed $= \sqrt{\frac{8RT}{\pi M}}$ is dependent on T and M. Thus, greater the value of $\sqrt{\frac{T}{M}}$, greater the molecular speed.

Thus, Gas $\sqrt{\frac{T}{M}}$

O_2 $\sqrt{\frac{560}{32}}$ 4.18

Ne $\sqrt{\frac{500}{20}}$ 5.0

CO_2 $\sqrt{\frac{440}{44}}$ 3.16

He $\sqrt{\frac{140}{4}}$ 5.91

57. (d) Aniline when treated with acetic anhydride forms acetanilide (nucleophilic substitution)

$$C_6H_5NH_2 \xrightarrow{CH_3-\overset{O}{\overset{||}{C}}-O-\overset{O}{\overset{||}{C}}-CH_3} C_6H_5NH-\overset{O}{\overset{||}{C}}-CH_3 + CH_3-\overset{O}{\overset{||}{C}}-OH$$

58. (a) Mohr's salt is $FeSO_4.(NH_4)_2SO_4.6H_2O$.
Fe^{2+} present in it is oxidized to Fe^{3+}.
So, we have, $Fe^{2+} \longrightarrow Fe^{3+} + e$

Thus, equivalent mass $= \frac{\text{Formula mass}}{1}$
= 392

59. (b) Domestic sewage constitute biodegradable pollutants.

60. (a) Langmuir adsorption isotherm is based on the assumption that every adsorption site is equivalent and the ability of a particle to bind there is independent of whether or not nearby sites are occupied.

61. (b) Thermosetting plastics consist of chain molecules which are chemically bonded, or cross-linked, with each other when heated. When thermosetting plastics cross-link, the molecules create a permanent, 3-D network that can be considered one giant molecule. Once cured, thermosetting plastic cannot

be remelted, in the same way that cured concrete cannot be reset.

62. (a) From the graph we can see the correct order of pressures $p_1 > p_3 > p_2$

63. (a) SiO_2 is used for this purpose.

64. (c) The acid strength of oxyacids of the same halogen increases with increase in oxidation state of the halogen

$$\overset{+7}{HClO_4} > \overset{+5}{HClO_3} > \overset{+3}{HClO_2} > \overset{+1}{HClO}$$

Thus Cl in $HClO_4$ is in highest oxidation state (+7) and hence strongest acid, while HClO (Cl is in +1 oxidation state) is weakest.

65. (b) Concentration of

$$Na_2CO_3 = \frac{25.3}{106} \times \frac{1000}{250} = 0.955 \text{ M}$$

$$[Na^+] = 2 \times 0.955 = 1.91 \text{ M}$$

$$\left[CO_3^{2-}\right] = 0.955 \text{ M}$$

66. (a) S_N2 mechanism is favoured maximum in methyl chloride among all the given options. Difference in rates are related to the bulk of the substituents. As the number of substituents attached to carbon bearing the halogen is increased, the reactivity towards S_N2 substitution decreases. The order of reactivity towards S_N2 mechanism is

$$CH_3Cl > CH_3CH_2Cl > (CH_3)_2CHCl > (CH_3)_3CCl$$

67. (c) K decreases with increase in temperature. Thus, forward reaction is exothermic.

E_a = activation energy of forward reaction

E_a' = ... backward ...

$= (E_a + \Delta H)$

Thus, $E_a' > E_a$

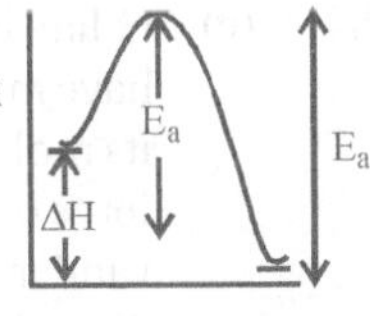

68. (a) Hybridisation $= \frac{1}{2}$ [No. of valence electrons of central atom + no. of monovalent atoms attached to it – Negative charge if any – positive charge if any]

$$NO_2^- \quad H = \frac{1}{2}[5 + 0 + 1 - 0] = 3 = sp^2$$

$$NO_3^- \quad H = \frac{1}{2}[5 + 0 + 1 - 0] = 3 = sp^2$$

$$NH_2^- \quad H = \frac{1}{2}[5 + 2 + 1 + 0] = 4 = sp^3$$

$$NH_4^+, \; H = \frac{1}{2}[5 + 4 + 0 - 1] = 4 = sp^3$$

$SCN^- = sp$

i.e., NO_2^- and NO_3^- have same hybridisation.

69. (a) $OH-C_6H_4$ + $C_6H_5-\overset{+}{N_2}\overset{-}{Cl}$ $\xrightarrow{base}$

Phenol   Phenyldiazonium chloride

$C_6H_5-N=N-C_6H_4-OH$

azo dye

70. (c) White lead is $2PbCO_3.Pb(OH)_2$ which is prepared by $Pb(NO_3)_2$ & Na_2CO_3. It is called as basic lead carbonate.

$$3Pb(NO_3)_2 + 3Na_2CO_3 + H_2O \longrightarrow$$
$$\underset{\text{Basic lead carbonate}}{Pb(OH)_2 + 2PbCO_3} + NaNO_3 + CO_2$$

71. (a) Most metals occur in their combined state. They have to be reduced to become free metals. The choice of reducing agent is decided by the factors of energetics and economics. In this regard C is the most common reducing agent used. Carbon can be used in the form of coal, coke, charcoal and carbon monoxide. e.g :-

$$SnO_2 + 2C \xrightarrow{1473-1573K} Sn + 2CO$$

$$ZnO + CO \xrightarrow{1600K} Zn + CO_2$$

$$Fe_2O_3 + CO \xrightarrow{823K} 2FeO + CO_2$$

$$FeO + CO \xrightarrow{1123K} Fe + CO_2$$

72. (d) The semiconductors formed by the introduction of impurity atoms containing one elecron less than the parent atoms of insulators are termed as p-type semiconductors. Therefore silicon containing 14 electrons is to be doped with boron containing 13 electrons to give a p- type semi-conductor.

73. (c) We know that IUPAC name of $[Pt(NH_3)_3 (Br) (NO_2)Cl]Cl$ is triamminebromochloronitroplatinum (IV) chloride.

74. (c) Van't Hoff factor

$$= \frac{\text{Normal Molecular Mass}}{\text{Observed Molecular Mass}}$$

75. (d) Aldehydes which contain a α -hydrogen on a saturated carbon, i.e., CH_3CH_2CHO undergo aldol condensation.

$$\underset{\text{propanol}}{CH_3CH_2-\overset{H}{\overset{|}{C}}=O} + H - \underset{\alpha}{\overset{CH_3}{\overset{|}{HC}}} - CHO$$

$$\xrightarrow{OH^-} CH_3CH_2-\overset{CH_3}{\underset{OH}{\overset{\beta|}{\underset{|}{C}}}} - \overset{CH_3}{\overset{\alpha|}{C}}HCHO$$

3.hydroxy,2 methyl pentanal

76. (d) When there is simultaneously change in temperature and volume (or pressure)

$$\Delta S = nC_V \log_e\left(\frac{T_2}{T_1}\right) + nR\log_e\left(\frac{V_2}{V_1}\right)$$

$$= C_V \log_e\left(\frac{2}{1}\right) + R\log_e\left(\frac{1}{2}\right)$$

$$= C_V \log_e 2 - R\log_e 2$$

$$= (C_V - R)\log_e 2$$

78. (a) $Zn + H_2SO_4 = ZnSO_4 + H_2$

In bomb calorimeter, there is no expansion in volume, so, work done will be zero. This reaction is exothermic. So, some heat will be evolved which will result in lowering of internal energy. Hence,

$$\Delta U < 0 \;\&\; w = 0$$

79. (d) The reactivity of H-atoms depends upon the stability of free radicals follows the order :

Tertiary > secondary > primary, therefore, reactivity of H-atoms follows the same order, i.e., tertiary > secondary > primary.

$$\underset{\substack{\text{Tertiary}\\\text{free radical}}}{R-\overset{R}{\underset{R}{\overset{|}{\underset{|}{C^{\bullet}}}}}} > \underset{\substack{\text{Secondary}\\\text{free radical}}}{R-\overset{R}{\underset{H}{\overset{|}{\underset{|}{C^{\bullet}}}}}} > \underset{\substack{\text{primary free}\\\text{radical}}}{R-\overset{H}{\underset{H}{\overset{|}{\underset{|}{C^{\bullet}}}}}} > \underset{\substack{\text{methyl}\\\text{free radical}}}{H-\overset{H}{\underset{H}{\overset{|}{\underset{|}{C^{\bullet}}}}}}$$

80. (b) For a 10 *K* rise in temperature, collision frequency increases merely by 1 to 2% but the number of effective collisions increases by 100 to 200%.

81. (d) Rotation around π bond is not possible. If any attempt is made to rotate one of the carbon atoms, the lobes of π-orbital will no longer remain coplanar i.e no parallel overlap will be possible and thus π-bond will break . This is known as concept of restricted rotation. In other words the presence of π-bonds makes the position of two carbon atom.

82. (a) Molecular masses of polymers are best determined by osmotic pressure method . Firstly because other colligative properties give so low values that they cannot be measured accurately and secondly, osmotic pressure measurements can be made at room temperature and do not require heating which may change the nature of the polymer.

83. (d) All the properties mentioned in the question suggest that it is a benzene molecule. Since in benzene all carbons are sp^2–hybridized, therefore, C – C – C angle is 120°

H H
1.09Å
C 1.40Å C 120°
H—C 120° C—H
C —— C
H H

84. (c) C has highest boiling point, so, it should have maximum number of carbon atoms in its molecule, because weak intermolecular force of attraction in a covalent molecule is vander waal's force of attraction which is directly proportional to the number of carbon atoms in the molecule.

85. (a) Flocculating value $\propto \frac{1}{(z)}$, where z = valency of coagulating ion.

Since, $Fe(OH)_3$ is a positively charged sol so it will be coagulated by negative ions considering the valencies of various negative ions used i.e., Cl^-, S^{2-}, PO_4^{3-} and SO_4^{2-}, we find that the valency of Cl^- is least so it will have maximum flocculation value i.e., NaCl.

86. (a) In DNA, adenine faces thymine and guanine faces cytosine.

87. (c) Number of unpaired electrons in various species.

Mn^{2+}	$3d^5$	5
Cr^{2+}	$3d^4$	4
V^{2+}	$3d^3$	3

Since, $\mu = \sqrt{n(n+2)}$
i.e., it increases with increase in number of unpaired electrons present in a species. Thus, the correct order is
$Mn^{2+} > Cr^{2+} > V^{2+}$.

88. (c) (a) $[H_3O^+] = \frac{0.1}{2} = 0.05$;
pH $= -\log 0.05 = 1.301$
(b) Complete neutralization of strong acid and strong base, pH = 7.
(c) Hydrolysis of the salt CH_3COOK, pH > 7 (salt of weak acid with strong base)
(d) Hydrolysis of the salt NH_4NO_3, pH < 7 (salt of strong acid with weak base)

89. (a) $CH_3-CH_2-CH=CH-CH_2-CH_3 \xrightarrow[\text{Platinum}]{H_2} CH_3-CH_2-CH_2-CH_2-CH_2-CH_3$

$\xrightarrow{HBr} CH_3-CH_2-\underset{Br}{\underset{|}{CH}}-CH_2-CH_2-CH_3$

90. (c) The given reaction is known as Friedel-Craft reaction.

SECTION 3-BOTANY

91. (c) The leaves appear green because of the pigment chlorophyll which does not absorb green light rather reflects it back.

92. (b) The *Euglena* is an organism, which possesses both the characteristics of plants and animals, as it can move with a flagella and also contains chlorophyll. Its nutrition is mixotrophic.

93. (d) Function of companion cell is to load sugar and amino acids into sieve elements. These cells use transmembrane proteins to take up by active transport.

94. (d) G_1 phase, also called Gap I phase is characterized by increase in cell size. In the S phase or synthetic phase DNA molecules replicate. G_2 is the second growth phase or Gap II where in there is intensive formation of RNAs and proteins. In the mitotic metaphase, the chromosomes are arranged at the equatorial plate.

95. (b) Kinetochore is the proteinaceous covering of centriole, to which spindle fibers attach.

96. (b) Lichens are composite organisms representing a symbiotic association between fungus and algae. They do not grow in the environment where pollution level is high SO_2 is strong air pollutant and lichens are very sensitive of SO_2.

99. (a) Fusion of male gamete with the egg is called syngamy.

100. (b) Kyoto Protocol (1997) : International conference held in Kyoto, Japan obtained commitments from different countries for reducing overall greenhouse gas emissions at a level 9% below 1990 level by 2008-2012.

101. (d) In secondary treatment mainly settled sewage flows to an aerobic biological treatment stage where it comes into contact with micro-organisms which remove and oxidise most of the remaining organic pollutants.

102. (a) Symbiosis or mutualism is a mutually beneficial association necessary for the survival of both the partners.

103. (d) Cotyledons and testa respectively are edible parts in groundnut and pomegranate. A cotyledon is a significant part of the embryo within the seed of a plant. Upon germination, the cotyledon may become the embryonic first leaves of a seedling. Testa is often thick or hard outer coat of a seed.

104. (a) Herkogamy – When there is some physical barrier present between the stamens and carpels avoiding any chance of self pollination.
Cleistogamy – when the flowers remain closed, self-pollination is the rule.
Dichogamy – when the two sexes mature at different times.

109. (a) Molybdenum is a component of nitrogenase and nitrate reductase involved during N_2 fixation. Copper is essential for

photosynthesis and respiration. Zinc is involved in synthesis of auxin. Manganese is activator of enzymes involved in nitrogen metabolism.

112. (a) The primary acceptor of CO_2 in C_4 plants is phosphoenol pyruvate or PEP. PEP in mesophyll cells combine with CO_2 and converted into 4 carbon compound Oxaloacetic acids by PEP carboxylase. In C_3 plant Ribulose 1, 5–diphosphate is primary acceptor of CO_2.

113. (b) In semi conservative replication of DNA in each replica one half is the old strand and the other half is a new strand synthesized over it.

114. (b) Incomplete dominance is the phenomenon of none of the alleles being dominant, with the effect that the hybrid produced by crossing two pure individuals is a mixture between the parents.

115. (b) Detritus food chain being with dead organic matter with dead plant parts, animals and their excretory product which is being acted upon by decomposers such as saprophytes to obtain energy needed for their survival.

116. (b) Biogeochemical cycles : Two types :
   (i) Gaseous : Biogenetic materials involved in circulation are gases *e.g.* N_2, O_2, CO_2 *etc.*
   (ii) Sedimentary : Biogenetic materials involved in circulation are non-gaseous *e.g.* P, Ca, S etc. forms rocks.

117. (b) Independent assortment of genes takes place only when they are located on separate non-homologus chromosomes. Where two or more than two genes are located on same chromosome, independent assortment will not be possible.

118. (b) Breeding of crops with high levels of minerals, vitamin and minerals is called biofortification. This is most practical aspect to improve the health of people.

119. (c) The decomposers act on the dead organic matter and break them down into simpler compounds and minerals.

120. (c) The nucleic acid (DNA and RNA) are repeating units of nucleotides i.e. polynucleotides. Each nucleotide comprises of nitrogenous heterocyclic bases viz. purines or pyrimidines, pentose sugar and phosphoric acid.

122. (d) RNA is a single chain of ribonucleotides. Each ribonucleotide consists of nitrogenous bases (adenine, guanine, cytosine and uracil), ribose sugar and phosphoric acid.

123. (a) Reducing sugar is any sugar that has an aldehyde group or is capable of forming one in solution through isomerisation. Reducing monosaccharides include glucose, fructose, glyceraldehyde and galactose. Many disaccharides, like lactose and maltose also have a reducing form, as one of the two units may have an open-chain form with an aldehyde group.
However, sucrose and trihalose in which the anomeric carbons of the two units are linked together, are non reducing disaccharides since neither of the rings is capable of opening.

125. (d) Glycolysis is a reductive process in which glucose is broken down in the absence of oxygen within the cytoplasm of the cell. It forms 2 pyruvate molecules, 2 net ATP molecules, 2 $NADH_2$ molecules.

127. (c) If natural habitat is destroyed, wild life is affected worstly.

128. (d) Golden rice is a transgenic crop of the future with high Vit. A content. Millions of people suffer from Vit. A deficiency which leads to vision impairment. Transgenic rice has been developed which is capable of synthesizing beta carotene, the precursor of Vitamin A. The rice variety is now being crossed into adapted varieties with field tests possible in an year or two.

129. (c) Polar ice caps & glaciers have the largest amount of pure and fresh water without any dissolved salt or impurity.

130. (d) A common biocontrol agent for control of plant diseases is *Trichoderma. Trichoderma* is a tree living fungus that exert biocontrol over several plant pathogens for the control of plant diseases. It is the natural method of pest and pathogen control.

131. (b) *Anabaena azollae* resides in the leaf cavities of the fern *Anabaena*. It fixes nitrogen. A part of the fixed nitrogen is excreted in the cavities and becomes available to the fern. The decaying fern plants release the same for utilization of the rice plants.

134. (c) The C_4 pathway allows photosynthesis to occur at very low concentrations of carbon dioxide as PEP carboxylase has an extremely high affinity for carbon dioxide. This pathway also works well at high temperatures and light intensity, enabling efficient photosynthesis in tropical plants.

135. (c) There are only 20 amino acids . These can be arranged in different order in the polypeptide chain to form a diverse array of proteins.

SECTION 4-ZOOLOGY

136. (d) Bacteria posses various forms and shapes and are of 4 different types - coccus (round), bacillus (rod shaped), vibrio (comma shaped) and spirillum (spiral like cork screw).

137. (b) Lampreys and hagfishes are unusual, jawless fish that comprise the order Cyclostomata, so named because of the circular shape of the mouth.

139. (a) Aschelminthes is bilaterally symmetrical and triploblastic. These are mostly aquatic, free living or parasitic. Their body is three layered which is ectoderm, mesoderm and endoderm.

141. (a) Sexual dimorphism is found in cockroach. A pair of anal style are present in male cockroach only. Female has broader abdomenal segments.

144. (d) Human population growth in India can be regulated by following the national programme of family planning.

148. (b) Acrosome a small pointed structure at the tip of nucleus. It breaks down just before fertilization, releasing hydrolytic enzymes that assist penetration between follicle cells that surrounds the ovum, thus facilitating fertilization.

149. (c) The gland corresponding to this in female is Bertholin.

150. (a) Vitamins are accessory food factors required in small amount for growth and metabolism. Vitamins are either water soluble $\rightarrow$ B. complex and vit.C or fat soluble $\rightarrow$ vit. A, D and K.

153. (b) It is named after its discoverer, F. G. J. Henle. In the kidney, the **loop of Henle** is the portion of the nephron that leads from the proximal convoluted tubule to the distal convoluted tubule. The loop has a hairpin bend in the renal medulla. The main function of this structure is to reabsorb water and ions from the urine. To do this, it uses a countercurrent multiplier mechanism in the medulla.

155. (c) Kwashiorkar and Beri-Beri are deficiency diseases which occur mostly in children. Kwashiorkar occurs due to deficiency of protein and Beri-Beri due to deficiency of vit-B_1 (Thiamine).

156. (c) A gene may have more than two alternative forms occupying the same locus on a chromosome such alleles are known as multiple alleles and the phenomenon is turned as multiple allelism. ABO blood group has 3 alleles. I^O, I^A or A, I^B or B.

157. (b) The term Evolution was given by Herbert Spencer that is "Descent with modifications". Evolution helps us to understand the history of life.
Evolution is a process in which something changes into a different and usually more complete or better form over time and in response to environment. This results in descendents becoming different from ancestors.

159. (b) Progressive evolution is development of organisms with more elaborate and specialized structures from those having less elaborate features e.g. amphibians from reptiles .Retrogressive or degenerative evolution is development of simpler forms from more complex ones. Such evolution has occurred in case of vestigial organs, parasitic forms, and in reduction of overspecialized structures such as wings in flightless birds. Parallel evolution is formation of similar traits in related groups of organisms independently due to similar requirement e.g. running of two toed deer and one toed Horse. Evolution of wings in insects and birds serve as example of convergent evolution.

160. (c) Sex in *Drosophila* is a function of the ratio of the number of X chromosomes to the number of autosomal sets. Therefore a *Drosophila* with a X/A = 1.0 will be a female whereas the one with a X/A ratio = 0.5 will

be male. However, in humans the presence or absence of the Y chromosome determines sex.

162. (b) According to Oparin, the atmosphere of primitive earth was reducing because its atoms were most numerous and most reactive. Free oxygen was not present in significant amount.
Large quantities of H_2, N_2, water vapour, CO_2, CH_4 and NH_3 were present.

163. (d) When the CO_2 concentration in blood increases, breathing becomes faster and deeper. As CO_2 levels increase, patients exhibit a reduction in overall level of consciousness as well as respiratory effort. Severe increases in CO_2 levels can lead to respiratory arrest.

165. (b) Afferent nerve fibres – sensory nerve fibres which carry senses from receptors to brain or spinal cord (CNS).
Efferent nerve fibres – motor nerve fibres carry orders or responses from CNS to effectors (muscles and glands).

167. (a) A peptide hormone secreted by the parathyroid gland in response to low levels of calcium in the blood. It acts to maintain normal blood levels of calcium by increasing the number of osteoclasts, which break down the bone matrix and release calcium into the blood. It also increases the reabsorption of calcium and magnesium ions in the kideny tubules, so that their concentration is maintained in the blood.

168. (c) Pancreas is a heterocrine gland comprising both endocrine and exocrine parts. Its endocrine part consists of small masses of hormone secreting cells called islets of langerhans. The α-cells of latter secrete glucagons and its β-cells secrete insulin. These two hormones have antagonistic effects on the glucose level in the blood which means that insulin decreases the blood glucose level while glucagon increases blood glucose level.

169. (a) Cortisol is secreted by the middle region of adrenal cortex. It increases the blood glucose level (which is anti-insulin effect) by converting proteins & fats into glucose.

170. (c) Conditioned reflexes are acquired reflexes and dependent on past experiences, training and learning. I.R. Pavlov demonstrated conditioned reflexes in a hungry dog. He called food and salivation in response to it as unconditioned stimulus and sound of bell and salivation in response to bell as conditioned reflexes.

171. (a) Recombinant DNA technology is the process joining together two DNA molecules from two different species that are inserted into a host organism to produce new genetic combination.

173. (c) Syphilis caused by bacteria *Treponema pallidum.*

175. (c) *Clostridium butylicum* industrially produces butyric acid.

176. (c) The saliva in the mouth and the tears from the eyes belongs to physiological barriers.

177. (a) B and T-lymphocytes produce antibodies against pathogen in the body, if due to some reason B and T-lymphocytes are damaged the body will not produce antibodies against a pathogen. Each B cell and T cell is specific for a particular antigen.

178. (c) Beri-beri is a deficiency disease caused by the lack of vitamin B1 (Thiamine) in the diet. The deficiency of riboflavin (Vit.B2) causes ariboflavinosis. Scurvy is the result of lack of vitamin C in the diet. Deficiency of calciferol (vitamin D) in the diet causes rickets in children and osteomalacia in adults.

179. (a) The pelvic girdle is formed by two innominate bones consists of three separate bones ilium, ischium and the pubis.

180. (c) The Southern blot is used to detect and identify certain DNA sequences in a sample of bodily fluid. It uses single-stranded DNA to search out their complementary strands. When a Southern blot is performed on DNA, the first step is digestion of DNA with restriction enzymes. Restriction enzymes cut DNA at known sequences, and produces DNA fragments of a certain length. Once the DNA is cut into pieces, scientists conduct electrophoresis to separate them by size.

NEET (NATIONAL ELIGIBILITY ENTRANCE TEST) Full Test - 4

ANSWER KEY																	
1	(b)	21	(a)	41	(d)	61	(b)	81	(c)	101	(b)	121	(b)	141	(c)	161	(a)
2	(d)	22	(a)	42	(a)	62	(d)	82	(a)	102	(b)	122	(d)	142	(c)	162	(c)
3	(c)	23	(c)	43	(a)	63	(b)	83	(b)	103	(a)	123	(c)	143	(d)	163	(a)
4	(a)	24	(d)	44	(d)	64	(c)	84	(d)	104	(a)	124	(b)	144	(c)	164	(a)
5	(b)	25	(b)	45	(a)	65	(d)	85	(a)	105	(d)	125	(d)	145	(a)	165	(b)
6	(d)	26	(a)	46	(a)	66	(d)	86	(b)	106	(d)	126	(c)	146	(c)	166	(b)
7	(b)	27	(b)	47	(b)	67	(c)	87	(c)	107	(b)	127	(d)	147	(c)	167	(d)
8	(b)	28	(c)	48	(c)	68	(c)	88	(a)	108	(c)	128	(d)	148	(b)	168	(a)
9	(b)	29	(b)	49	(b)	69	(c)	89	(d)	109	(a)	129	(c)	149	(b)	169	(b)
10	(c)	30	(a)	50	(b)	70	(b)	90	(d)	110	(c)	130	(d)	150	(c)	170	(c)
11	(a)	31	(b)	51	(c)	71	(a)	91	(d)	111	(a)	131	(b)	151	(d)	171	(a)
12	(c)	32	(a)	52	(b)	72	(b)	92	(a)	112	(a)	132	(d)	152	(d)	172	(b)
13	(c)	33	(b)	53	(b)	73	(b)	93	(a)	113	(a)	133	(b)	153	(b)	173	(c)
14	(c)	34	(a)	54	(d)	74	(b)	94	(d)	114	(a)	134	(c)	154	(d)	174	(d)
15	(d)	35	(c)	55	(a)	75	(b)	95	(c)	115	(a)	135	(a)	155	(a)	175	(b)
16	(d)	36	(c)	56	(a)	76	(c)	96	(d)	116	(a)	136	(b)	156	(a)	176	(c)
17	(c)	37	(b)	57	(a)	77	(c)	97	(c)	117	(a)	137	(a)	157	(b)	177	(d)
18	(d)	38	(d)	58	(d)	78	(a)	98	(a)	118	(d)	138	(b)	158	(c)	178	(b)
19	(c)	39	(c)	59	(b)	79	(c)	99	(c)	119	(c)	139	(b)	159	(c)	179	(b)
20	(b)	40	(c)	60	(d)	80	(a)	100	(c)	120	(a)	140	(c)	160	(a)	180	(d)

HINTS & SOLUTIONS

SECTION 1 - PHYSICS

1. (b) Given = $R = \frac{\rho \ell}{\pi r^2}$, then

$$\frac{\Delta R}{R} \times 100$$

$$= \frac{\Delta \rho}{\rho} \times 100 + \frac{\Delta \ell}{\ell} \times 100 + 2\frac{\Delta r}{r} \times 100$$

$$= 1\% + 2\% + 2 \times 3\% = 9\%$$

2. (d) The difference in velocities is increasing with time as both of them have more constant but different acceleration.

3. (c) Heat required to boil water $= mc\Delta\theta$

$$= 2 \times 4200 \times (100 - 20)$$
$$= 6.72 \times 10^3 \text{ J}$$

If t be the time of boil then

$\eta \times 500 \times t = 6.72 \times 10^3$

or $t = \frac{6.72 \times 10^3}{0.8 \times 500} = 28$ minutes.

4. (a) Let velocity at $A = v_A$ and velocity at $B = v_B$

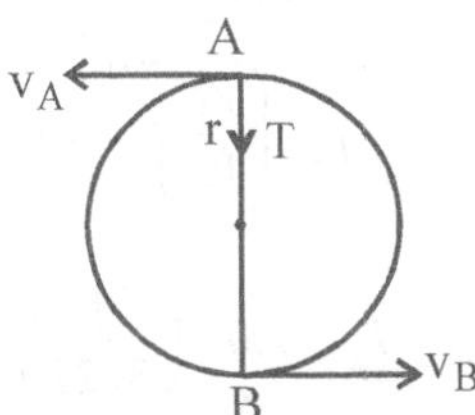

Applying conservation of energy at A & B

$$\frac{1}{2}mv_A^2 + 2gmr = \frac{1}{2}mv_B^2$$

$v_B^2 = v_A^2 + 4gr$.........(i)
Now as it is moving in circular path it has centripetal force.

At point A $\Rightarrow T + mg = \frac{mv_A^2}{r}$

for minimum velocity $T \geq 0$

or $\frac{mv_A^2}{r} \geq mg \Rightarrow v_A^2 \geq gr \Rightarrow v_A \geq \sqrt{gr}$

6. (d) For nuclear disintegration

$\lambda = \frac{2.303}{t} \log \frac{N_o}{N_o - N}$

$= \frac{2.303}{24 \times 60 \times 60} \log \frac{100}{82} = 2.1 \times 10^{-6} sec^{-1}$

7. (b) γ rays has lowest wavelength and highest frequency among them while ultraviolet ray has highest wavelength and lowest frequency.
Order of frequency : $b > a > c$

8. (b) A galvanometer can be changed into an ammeter by the use of low resistance in parallel. So that ammeter does not draw much current which may change the magnitude of main current.

9. (b) Acceleration vector is always radial (i.e . towards the center) for uniform circular motion.

10. (c) $T_1 = 27°C = 300$ K

$V_1 = V$ and $V_2 = \frac{8V}{27}$

Ratio of specific heats for monoatomic gas,

$\gamma = \frac{5}{3}$

In an adiabatic process,
$T_1 V_1^{\gamma-1} = T_2 V_2^{\gamma-1}$

or $T_2 = \left(\frac{V_1}{V_2}\right)^{\gamma-1} \times T$,

$T_2 = 300 \times \left(\frac{1}{8/27}\right)^{(5/3)-1}$

$= 300 \times \left(\frac{27}{8}\right) = 300 \times \left(\frac{9}{4}\right) = 675$ K

$= 402°C$

Hence, rise in temperature
$= T_2 - T_1 = 402 - 27 = 375°C$

11. (a) The equivalent circuit is shown in fig. Since the Wheatstone's bridge is balanced, therefore no current will flow through the arm KL. Equivalent resistance between AKM = R + R = 2 R
Equivalent resistance between ALM = R + R = 2 R
The two resistances are in parallel. Hence equivalent resistance between A and B is given by

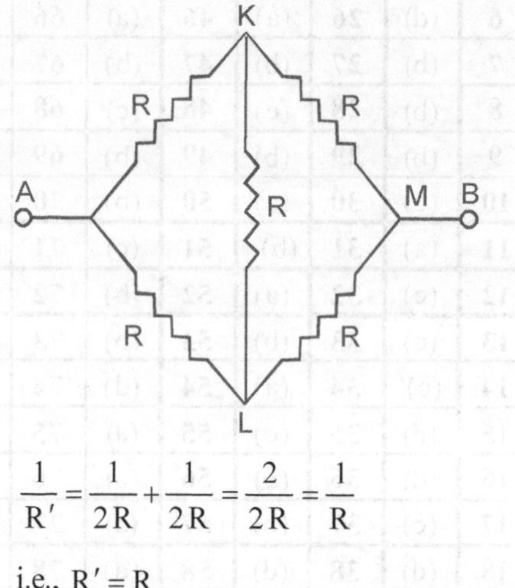

$\frac{1}{R'} = \frac{1}{2R} + \frac{1}{2R} = \frac{2}{2R} = \frac{1}{R}$

i.e., $R' = R$

12. (c) For n^{th} orbit, energy, $E_n = \frac{2\pi^2 em^4 z^2}{n^2 h^2}$

For hydrogen (z = 1), $E_n = \frac{2\pi^2 em^4}{n^2 h^2}$

For helium (z = 2),

So, $E = \frac{2\pi^2 em^4 \times 4}{n^2 h^2}$

$\frac{E}{E_n} = \frac{4}{1} \Rightarrow E = 4E_n$

13. (c) Conserving Linear Momentum
$2Mv_c = 2Mv - Mv \Rightarrow v_c = v/2$.

14. (c) $\frac{R_s}{R_\alpha} = \left(\frac{A_s}{A_\alpha}\right)^{1/3} = \left(\frac{32}{4}\right)^{1/3} = 2$

15. (d) $\frac{dH}{dt} \propto (\theta_2 - \theta_1) = (\Delta\theta)^n \Rightarrow n = 1$

16. (d) Use $a = \mu g$ and $v^2 = u^2 + 2as$

17. (c) When the arrangement is dipped in water;

$$\beta' = \beta/\mu = \frac{x}{4/3} = \frac{3}{4}x = 0.75x$$

18. (d) The charge q, which is kept at the centre of metallic spherical shell transfered to the outer surface of shell & inside the shell the electric field is zero & hence force is also zero.

19. (c) $F_n = q(\vec{v} \times \vec{B})$

$= qvB \sin\theta = 0$ (because $\theta = 0°$)

20. (b) The time of cooling increases as the difference between the temperature of body & surrounding is reduced. So $T_1 < T_2 < T_3$ (according to Newton's law of cooling).

21. (a) In uniform circular motion speed is constant. So, no tangential acceleration. It has only radial acceleration

$a_R = \frac{v^2}{R}$ [directed towards center]

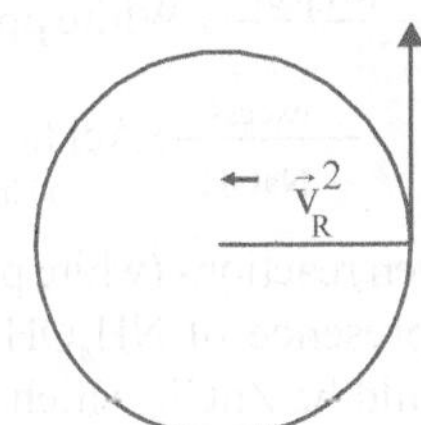

and its velocity is always in tangential direction. So these two are perpendicular to each other.

22. (a) $e = \frac{LdI}{dt} = \frac{40 \times 10^{-3}(11-1)}{4 \times 10^{-3}} = 100V$

23. (c) $C_s = \frac{C_1 C_2}{C_1 + C_2} = 3$

$C_p = C_1 + C_2 = 16 \quad \therefore \quad C_1 C_2 = 48$

$C_1 - C_2 = \sqrt{(C_1 + C_2)^2 - 4C_1 C_2}$

$= \sqrt{16^2 - 4 \times 48} = \sqrt{64} = 8$

$\therefore \quad C_1 = 12\mu F$ and $C_2 = 4\mu F$

24. (d) At the centre of the earth gravity is zero.

25. (b) Frequency does not depend upon the medium, so, it will remain same in the material of the tuning fork and in air.

26. (a) The highest energy level occupied by an electron in the energy band at zero kelvin is called Fermi level and the energy associated is known as Fermi energy. So, it is independent of T.

27. (b) The intersection of medians is the centre of mass of the triangle. Since distances of centre of mass from the sides are related as : $x_{BC} < x_{AB} < x_{AC}$ therefore $I_{BC} > I_{AB} > I_{AC}$ or $I_{BC} > I_{AB}$.

28. (c) If $\vec{F}$ force acts for short interval Δt, then

$\vec{F}\,\Delta T = m\Delta\vec{v}$.

29. (b) For electromagnetic wave,

$F_E = F_M \Rightarrow eE = BeC \Rightarrow E = B.C$

$= 2 \times 10^{-7} \times 3 \times 10^8 = 60$ V/m

30. (a) According to Biot Savart's law

$$dB = \frac{\mu_0}{4\pi} \cdot \frac{Id\ell \sin\theta}{r^2}$$

$$dB \propto \frac{1}{r^2}$$

So, graph (a) is correct.

31. **(b)** According to Stefan' law heat emitted per second per unit area $\alpha(T_1^4 - T_2^4)$ where T_1 is the temperature of body & T_2 is the temperature of surrounding (T_1 & T_2 are on Kelvin scale).

32. (a) $\frac{1}{f_\omega} = \left({}^{\omega}\mu_g - 1\right)\left(\frac{1}{r_1} - \frac{1}{r_2}\right)$

[f_ω is focal length of lens in water]

$\frac{1}{f} = ({}^{a}\mu_g - 1)\left(\frac{1}{r_1} - \frac{1}{r_2}\right)$

[f is focal length of lens in air]

Dividing,

$$\frac{f}{f_\omega} = \frac{({}^{\omega}\mu_g - 1)}{({}^{a}\mu_g - 1)};\ {}^{\omega}\mu_g = \frac{{}^{a}\mu_g}{{}^{a}\mu_w}$$

$$= \frac{3/2}{4/3} = \frac{3}{2} \times \frac{3}{4} = \frac{9}{8}$$

$$\frac{f}{f_w} = \frac{9/8 - 1}{\frac{3}{2} - 1} = \frac{1/8}{1/2} = \frac{1}{4}$$

$f_w = 4 \times f = 4 \times 10 = 40$ cm.

33. (b) If current passes through a spring, it shrinks as in two adjacent wire, current is flowing in the same direction. Even if direction of current is reversed still the spring will shrink. As AC current is a periodically changing current, the process of shrinking will also be periodic in nature.

35. (c) $Y = \frac{F/A}{\Delta\ell/\ell} = \frac{\frac{250\times10}{50\times10^{-6}}}{\frac{0.5\times10^{-3}}{2}}$

$= \frac{250\times9.8}{50\times10^{-6}} \times \frac{2}{0.5\times10^{-3}}$

$\Rightarrow 19.6\times10^{10}\,N/m^2$

36. (c) Max. force = mass × max. acceleration $= m\,4\,\pi^2\,v^2\,a = 1\times4\times\pi^2\times(60)^2\times0.02$ $= 288\,\pi^2$

37. (b) Changing electric field gives rise to displacement current which creates magnetic field around it.

39. (c)

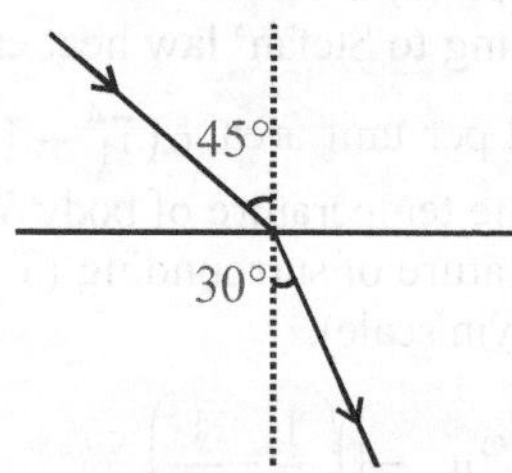

Refractive index, $\mu = \frac{\sin 45^\circ}{\sin 30^\circ}$

$\mu = \frac{1}{\sqrt{2}} \times 2 = \sqrt{2}$

Now, $\mu = \frac{\text{Velocity of light in air}}{\text{Velocity of light in medium}} = \sqrt{2}$

Velocity of light in medium

$= 3\times10^8 \times \frac{1}{\sqrt{2}} = 2.12\times10^8\,m/sec$

40. (c) Rain drops are in spherical shape due to surface tension.

41. (d) $\delta_1 = 40^\circ$, $\delta_2 = 30^\circ$, $\delta = ?$

$\cot\delta = \sqrt{\cot^2\delta_1 + \cot^2\delta_2}$

$= \sqrt{\cot^2 40^\circ + \cot^2 30^\circ}$

$\cot\delta = \sqrt{1.19^2 + 3} = 2.1$

$\therefore\ \delta = 25^\circ$ i.e. $\delta < 40^\circ$.

42. (a) $\mu = \frac{1}{\sin C} = \frac{1}{\sin 30^\circ} = \frac{1}{1/2} = 2$

Velocity of light in the medium

$= \frac{3\times10^8}{2} = 1.5\times10^8\,m/sec$

44. (d) $P_T = P_C\left(1 + \frac{m_a^2}{2}\right)$

$\therefore\ 600 = 400\left(1 + \frac{m_a^2}{2}\right) \Rightarrow \frac{3}{2} = 1 + \frac{m_a^2}{2}$

or $\frac{m_a^2}{2} = \frac{1}{2} \Rightarrow m_a = 1$

SECTION 2 - CHEMISTRY

46. (a) $X \xrightarrow{NH_4OH}$ White ppt

$\xrightarrow[NaOH]{excess}$ Acidic solution (soluble) (No. ppt with H_2S)

Given reactions (white precipitate with H_2S in presence of NH_4OH) indicate that 'X' should be $ZnCl_2$ which explains all given reactions.

$ZnCl_2 + 2H_2O \rightarrow Zn(OH)_2 + HCl\uparrow$ (White fumes)

$NH_4OH + HCl\uparrow \xrightarrow{-H_2O} NH_4Cl\uparrow$ (HCl: White fumes; NH_4Cl: Dense white fumes)

$ZnCl_2 + 2NaOH \rightarrow Zn(OH)_2 + 2NaCl$

$Zn(OH)_2 \xrightarrow[Excess]{2NaOH} Na_2ZnO_2 + 2H_2O$

47. (b) CN^- is a strong field ligand as it is a psuedohalide ion. These ions are strong coordinating ligands and hence have the tendency to form σ-bond (from the pseudo halide to the metal) and π-bond. (from the metal to pseudo halide)

48. (c) $\underset{x\,g}{NaCl} + \underset{4.9\,g}{H_2SO_4} \longrightarrow \underset{6\,g}{NaHSO_4} + \underset{1.825\,g}{HCl}$

According to law of conservation of mass "mass is neither created nor destroyed during a chemical change"

∴ Mass of the reactants = Mass of products

$$x + 4.9 = 6 + 1.825$$

or $x = 2.925$ g

49. (b) The ideal value of D.O for growth of fishes is 8 mg/ℓ. 7mg ℓ is desirable range, below this value fishes get susceptible to desease. A value of 2 mg/ℓ or below is lethal for fishes.

50. (b) In case of nitrogen, d-orbitals are not available.

51. (c) $A : CH_3COOH\text{(weak acid)} \rightleftharpoons CH_3COO^- + H^+$

$$\Delta H = 0.005 \text{ kcal g}^{-1} = 0.30 \text{ kcal mol}^{-1}$$

$B : Ca(OH)_2 + 2H^+ \rightleftharpoons Ca^{2+} + 2H_2O$

$$\Delta H = 0.005 \text{ kcal g}^{-1} = -13.17 \times 2 \text{ kcal}$$

$$Ca(OH)_2 + 2CH_3COOH \rightleftharpoons (CH_3COO)_2Ca + 2H_2O$$

$$\Delta H = 2A + B = -27.4 + 0.6 = -26.8 \text{ kcal}$$

52. (b) $\underset{0.4\text{moles}}{HCl} \rightleftharpoons H^+ + \underset{0.4\text{moles}}{Cl^-}$

$$\underset{0.2\text{moles}}{CaCl_2} \rightleftharpoons Ca^{2+} + \underset{2\times0.2=0.4\text{moles}}{2Cl^-}$$

Total Cl^- moles = 0.4 + 0.4 = 0.8 moles

$$\text{Molarity} = \frac{\text{Moles}}{\text{Vol.in L}}$$

$$\therefore \text{Molarity of } Cl^- = \frac{0.8}{0.5} = 1.6 \text{ M.}$$

53. (b) Sodium cyanide ($Na + C + N \rightarrow NaCN$). (Lassaigne's test)

54. (d) Specific volume (volume of 1 gm) of cylindrical virus particle = 6.02×10^{-2} cc/gm
Radius of virus (r) = 7 Å = 7×10^{-8} cm
Length of virus = 10×10^{-8} cm
Volume of virus

$$\pi r^2 l = \frac{22}{7} \times (7 \times 10^{-8})^2 \times 10 \times 10^{-8}$$

$$= 154 \times 10^{-23} \text{ cc}$$

$$\text{Wt. of one virus particle} = \frac{\text{volume}}{\text{specific volume}}$$

∴ Mol. wt. of virus = Wt. of N_A particle

$$= \frac{154 \times 10^{-23}}{6.02 \times 10^{-2}} \times 6.02 \times 10^{23}$$

$$= 15400 \text{ g/mol} = 15.4 \text{ kg/mole}$$

55. (a) Inductive effect involves displacement of σ-electrons.

56. (a) Using the relation $\lambda = \frac{hc}{E}$
Substituting given values, we get

$$\lambda = \frac{6.62 \times 10^{-27} \text{ erg-sec} \times 3 \times 10^{10} \text{ cm s}^{-1}}{3 \times 10^{-12} \text{ erg}}$$

$$\lambda = 6.62 \times 10^{-5} \text{ cm}$$

$$= 6.62 \times 10^{-5} \times 10^7 \text{ nm} \quad [1 \text{ cm} = 10^7 \text{ nm}]$$

$$= 662 \text{ nm}$$

57. (a) Tertiary carbonium ion has highest stability followed by secondary and then primary carbonium ion. So, option (a) is correct.

59. (b) $NH_4NO_2 \longrightarrow N_2 + 2H_2O$
Volume of N_2 formed in successive five minutes are 2.75 cc, 2.40 cc and 2.25 cc which is in decreasing order. So rate of reaction is dependent on concentration of NH_4NO_2. As decrease is not very fast so it will be first order reaction.

60. (d) We know that

$$\underset{\text{1, Propene}}{6CH_3 - CH = CH_2} \xrightarrow[\text{ether, 0°C}]{B_2H_6}$$

$$2(CH_3CH_2CH_2)_3 \xrightarrow[OH^-]{H_2O_2} \underset{\text{Propanol}}{6CH_3CH_2CH_2OH} + 2H_3BO_3$$

61. (b) $p_{tot} = 740$ mm

$$p_{tot} = p_{O_2} + p_{H_2}$$

Number of moles of O_2 and H_2 are equal

$$\therefore \quad p_{O_2} = p_{H_2}$$

$$\therefore \quad p_{O_2} + p_{H_2} = 740$$

$$p_{O_2} = 370\text{mm} = p_{H_2}$$

62. (d) Formation of only CH_3COOH by ozonolysis indicates that the compound Y should be $CH_3CH=CHCH_3$ which can be formed by all of the three given compounds

$$\underset{X}{CH_2 = CH - CH = CH_2} \xrightarrow{1H_2/Pt} \underset{Y}{CH_3 - CH = CH - CH_3}$$

$$\underset{X}{CH_3C \equiv CCH_3} \xrightarrow{1H_2/Pt} \underset{Y}{CH_3CH = CHCH_3}$$

$$\underset{X}{CH_2 = C = CHCH_3} \xrightarrow{1H_2/Pt}$$

$$\underset{Y}{CH_3CH = CHCH_3} \xrightarrow{O_3} 2CH_3COOH$$

63. (b) $pH = \frac{1}{2}[pK_a - \log 1] = \frac{pK_a}{2}$

pH' (twice of pH) = pK_a

$\therefore \quad pK_a = \frac{1}{2}[pK_a - \log C]$

$-\log C = pK_a = -\log K_a$

$C = K_a = 1.8 \times 10^{-5}$ M

dilution $= \frac{1}{C} = 5.55 \times 10^4$ times

64. (c) In blast furnace at about 1270 K, calcium carbonate is almost completely decomposed to give CaO which acts as a flux and combines with SiO_2 present as impurity (gangue) in the ore to form calcium silicate (fusible slag)

CaO(s) (basic flux) + SiO_2 (s) (acidic flux) $\longrightarrow$ $CaSiO_3$ (s) (slag)

65. (d) Buffer solution contains weak base + salt of weak base with strong acid or weak acid + salt of weak acid with strong base. In option (d) the acid used is $HClO_4$ which is strong acid and $KClO_4$ is salt of this acid with strong base. So it is not an example of buffer solution.

66. (d) The most abundant rare gas found in the atmosphere is argon and not helium.

67. (c)

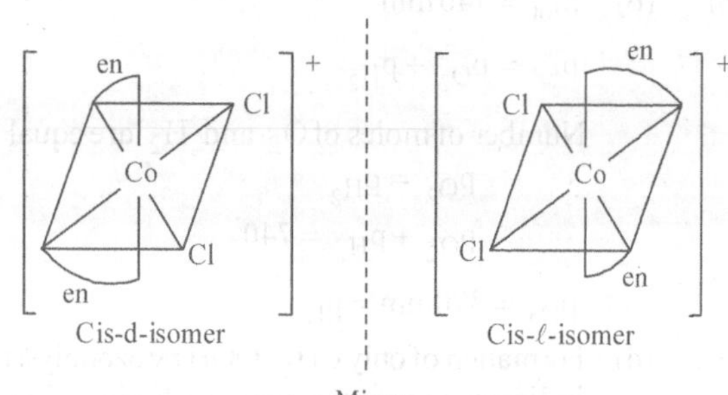

68. (c) Rate of reaction $= -\frac{1}{2}\frac{d[N_2O_5]}{dt} = \frac{1}{4}\frac{d[NO_2]}{dt}$

69. (c) For third ionization enthalpy last configuration of

		3d	4s
V	$-4s^0 3d^3$	1 1 1 _ _	_
Cr	$-4s^0 3d^4$	1 1 1 1 _	_
Mn	$-4s^0 3d^5$	1 1 1 1 1	_
Fe	$-4s^0 3d^6$	1⇂ 1 1 1 1	_

For third Ionization enthalpy Mn has stable configuration due to half filled d-orbital.

70. (b) $\mu = \sqrt{n(n+2)}$

$3.83 = \sqrt{n(n+2)}$

on solving n = 3

as per magnetic moment, it has three unpaired electron.

Cr^{3+} will have configuration as

Cr $1s^2 2s^2 2p^6 3s^2 3p^6 3d^4 4s^2$

Cr^{3+} $1s^2 2s^2 2p^6 3s^2 3p^6 3d^3$

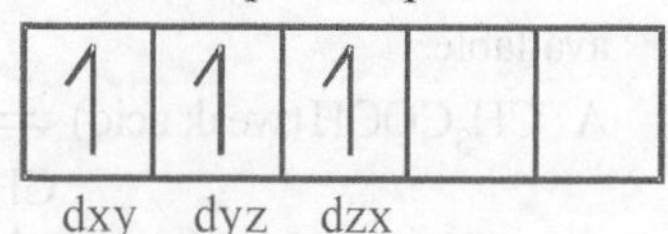

So 3dxy' 3dyz' 3dxz'

71. (a) Given reaction is

$IO_3^- + aI^- + bH^+ \longrightarrow cH_2O + dI_2$

Ist half reaction

$I^- \longrightarrow I_2$ ...(i)

$-1 \quad\quad 0$ (oxidation)

IInd half reaction

$IO_3^- \longrightarrow I_2$ ...(ii)

$+5 \quad\quad 0$ (reduction)

On balancing equation (ii) we have

$10e^- + 2IO_3^- + 12H^+ \longrightarrow I_2 + 6H_2O$ ...(iii)

Now, balance equation (i)

$2I^- \longrightarrow I_2 + 2e^-$ ....(iv)

Multiply eqn (iv) by 5 and add it to eqn (iii), we get

$2IO_3^- + 10I^- + 12H^+ \longrightarrow 6I_2 + 6H_2O$

or, $IO_3^- + 5I^- + 6H^+ \longrightarrow 3I_2 + 3H_2O$

Hence a = 5, b = 6, c = 3, d = 3

72. (b) This is because of inter-electronic replusions between lone pairs.

B.E.:	F – F	Cl – Cl
($kJ\,mol^{-1}$):	158.8	242.6

$:\ddot{F} \underset{\leftrightarrow}{\overset{\leftrightarrow}{-}} \ddot{F}:$

73. (b) Molecules having higher molecular weight and less branching have higher boiling point.

74. (b) Always transition metals combines with more than one carbonyl group.

75. (b) Hydration energy of Cl^+ is very less than H^+ hence it doesn't form Cl^+ (aq) ion.

76. (c) No. of asymmetric carbon = 2

No. of enantiomers = $2^2 = 4$.

77. (c)

	Change	Equiv. mass
$\underset{+7}{MnO_4^-} \xrightarrow{H^+} \underset{+2}{Mn^{2+}}$	5	$\frac{M}{5}$
$\underset{+7}{MnO_4^-} \xrightarrow{H_2O} \underset{6}{MnO_4^{2-}}$	1	M
$\underset{+7}{MnO_4^-} \xrightarrow{OH^-} \underset{+4}{MnO_2}$	3	$\frac{M}{3}$

78. (a) Electronegativity is the tendency of the atom to attract electrons to itself when combined in a compound as defined by Pauling. Electronegativity, is a relative term so it does not have any unit.

79. (c) As sigma bond is stronger than the π (pi) bond, so it must be having higher bond energy than π (pi) bond.

80. (a) The reactivity may be attributed to size factor, larger is the size, higher is tendency to lose electron (low I.E.). Zn is the last element of 3d series and it has larger size than Cu due to $(3d^{10}4s^2)$ configurations.
Hence, the reactivity order is
Na > Mg > Zn > Fe > Cu.

81. (c) Using the relation

$$E_{cell} = E^0_{cell} - \frac{0.0591}{n}\log\frac{[\text{anode}]}{[\text{cathode}]}$$

$$= E^0_{cell} - \frac{0.0591}{n}\log\frac{[Zn^{2+}]}{[Cu^{2+}]}$$

Substituting the given values in two cases.

$$E_1 = E^0 - \frac{0.0591}{2}\log\frac{0.01}{1.0}$$

$$= E^0 - \frac{0.0591}{2}\log 10^{-2}$$

$$= E^0 + \frac{0.0591}{2} \times 2 \text{ or } (E^0 + 0.0591)\text{V}$$

$$E_2 = E^0 - \frac{0.0591}{2}\log\frac{1}{0.01}$$

$$= E^0 - \frac{0.0591}{2}\log 10^2$$

$$= E^0 - \frac{2 \times 0.0591}{2} \text{ or } (E^0 - 0.0591)\text{V}$$

Thus, $E_1 > E_2$

82. (a) $3\,HCHO \underset{\text{aq. solution}}{\overset{\text{On keeping}}{\rightleftharpoons}}$ (cyclic: CH_2–O–CH_2–O–CH_2–O)

Trioxane
(meta formaldehyde)

83. (b) Amongst the isoelectronic species, the anion having more negative charge would have the larger size.

84. (d) For an acid-base indicator

$$HIn \rightleftharpoons H^+ + In^-$$

$$\therefore K_{In} = \frac{[H^+][In^-]}{[HIn]} \text{ or } [H^+] = K_{In} \times \frac{[HIn]}{[In^-]}$$

$$\text{or } \log H^+ = \log K_{In} + \log\frac{[HIn]}{[In^-]}$$

Taking negative on both sides

$$-\log[H^+] = -\log K_{In} - \log\frac{[HIn]}{[In^-]}$$

$$\text{or we can write } pH = pK_{In} + \log\frac{[In^-]}{[HIn]}$$

$$\text{or } \log\frac{[In^-]}{[HIn]} = pH - pK_{In}$$

85. (a) Excess of HCl is used to convert free aniline to aniline hydrochloride otherwise free aniline would undergo coupling reaction with benzenediazonium chloride.

86. (b) In lake test of Al^{3+}, there is formation of coloured floating lake. It is due to the adsorption of litmus by $Al(OH)_3$.

87. (c) $w = -P\Delta V = -10^5(1\times10^{-2} - 1\times10^{-3})$

$= -900\text{J}$

88. (a) Due to H-bonding, the boiling point of ethanol is much higher than that of the isomeric diethyl ether.

89. (d) Phenol has active (acidic) hydrogen so it reacts with CH_3MgI to give CH_4, and not anisole

$$C_6H_5OH + CH_3MgI \longrightarrow CH_4 + C_6H_5OMgI$$

90. (d) From given data, we have

$$C + O_2 \longrightarrow CO_2 - x\,kJ \ldots (i)$$

$$H_2 + \frac{1}{2}O_2 \longrightarrow H_2O - y\,kJ \quad ...(ii)$$

$$CH_4 + 2O_2 \longrightarrow CO_2 + 2H_2O + z\,kJ \;...(iii)$$

The required equation is

$$C + 2H_2 \longrightarrow CH_4 + Q$$

To get the required equation, operate

(i) + 2 × (ii) – (iii)

Thus, we get

$$C + 2H_2 \longrightarrow CH_4 + [(-x) + (-2y) - (-z)]$$

Thus, heat of formation of methane is (–x – 2y + z) kJ

SECTION 3 - BOTANY

92. (a) Peat moss is another name for *Sphagnum.* It is also known as famine food in China.

94. (d) The fruit is a mature or ripened ovary. When a fruit develops exclusively from the ovary, it is said to be true fruit. When in addition to the ovary, some other floral part also participates in the formation of fruits, then it is known as false fruit. Apple, pear, cashewnut, mulberry etc. are all false fruits.

95. (c) Phellem, phellogen and phelloderm are collectively called periderm.

100. (c) In porogamy the tip of pollen tube enters the micropyle, pushes through the nucellar tissue & finally pierces the egg-apparatus end of the embryo sac. If pollen tube enters through chalazal side it is called chalazogamy & if it enters laterally it is called mesogamy.

101. (b) Competition is most severe between the members of a population belonging to same habitat.

102. (b) In oogamy male and female gametes are morphologically as well as physiologically different. Female gametes are large and non-motile. Male gametes are small but motile.

105. (d) The keystone species in an ecosystem are those who are the main contributors to the ecosystem.

106. (d) The initiation codon is the codon which initiates the protein synthesis. They are AUG for methionine and GUG for valine.

107. (b) The best pH of soil for cultivation of plants is 6.5 – 7.5.

108. (c) Telomerase is a ribonucleoprotein which synthesize the rich strand of telomers in DNA. **Telomerase** is an enzyme that adds specific DNA sequence repeats ("TTAGGG" in all vertebrates) to the 3' ("three prime") end of DNA strands in the telomere regions, which are found at the ends of eukaryotic chromosomes. The telomeres contain condensed DNA material, giving stability to the chromosomes. The enzyme is a reverse transcriptase that carries its own RNA molecule, which is used as a template when it elongates telomeres, which are shortened after each replication cycle. Telomerase was discovered by Carol W. Greider in 1984.

109. (a) A standing crop is the quantity or total weight or energy content of the organism, which are in a particular location at a particular time.

110. (c) *Triticale* is a man made hexaploid inter genetic hybrid.

Tritium turgidum (4A) × Secale cereale (2B)

Gametes: (2A) and (B)

↓

(AAB) Sterile

↓ ← Treated with Colichicine Chromosomal doubling

Triticale (6n) (AAAABB)

111. (a) Azadirachtin is a chemical compound belonging to the limonoids. It is a secondary metabolite present in the neem tree seeds. The molecular formula is $C_{35}H_{44}O_{16}$. Azadirachtin is a highly oxidised tetranortriterpenoid which boasts a plethora of oxygen functionality, comprising an enol ether, acetal, hemiacetal, and tetra-substituted oxirane as well as a variety of carboxylic esters. It is classified among the plant secondary metabolites.

113. (a) An ecosystem having low stability can be easily damaged. An ecosystem having high resilience will take less time to recover.

116. (a) Osmotic potential is the potential of a solution to cause water movement into it across a semi- permeable membrane. Water potential is the tendency of water to leave a system.

117. (a) Photorespiratory loss of CO_2 occurs when RuBisCo starts functioning as an oxygenase instead of carboxylase under conditions of high O_2 and low CO_2. It involves three organelles chloroplast, mitochondria and peroxisomes. Half of the photosynthetically fixed carbon (in the form of RuBP) may be lost into the atmosphere through this process and no ATP formation occurs.

119. (c) Cultivation of plants by placing the roots in the nutrient solution without any soil is called hydroponics. It is also known as soilless culture/ water culture/ solution culture. It is used to determine which elements are essential for plant growth and what symptoms are produced by the absence or deficiency of essential elements.

120. (a) The enzyme involved in Krebs cycle are localized in the mitochondrial matrix.

121. (b) Cellulose is the most abundant organic compound, most abundant polysaccharide and most abundant biopolymer found on earth.

122. (d) Terminal cytochorome is cyt a_3. cyt a_3 posseses two copper centers. It helps in transfer of electrons to oxygen.

126. (c) K_m (Michealis Menten constant) is defined as that substrate concentration at which under optimum conditions the rate of an enzyme catalysed reaction reaches half the maximum rate. K_m is inversely proportional to affinity of enzyme for its substrate.

127. (d) Biodiversity Act of India was passed by the parliament in the year 2002.

128. (d) Axenic is a method of culture of isolated plant cells, tissues or organs in an artificial, nutritive medium.

129. (c) 'Each antibiotic is effective only against one particular kind of germ' is not correct.

130. (d) Main objective of production/use of herbicide resistant GM crops is to reduce herbicide accumulation in food articles for health safety. GM plants has been useful in many ways. Genetic modifications has made crops more tolerant to abiotic stresses, reduced reliance on chemical pesticides, enhanced nutritional value of food.

131. (b) The common nitrogen fixer in paddy field is *Azospirillum.* It is an anaerobic bacteria that forms loose association with roots of paddy crops.

132. (d) In order to obtain virus-free plants through tissue culture, the best method is meristem culture. Meristem tip culture is used successfully to remove viruses, bacteria, in order to produce the greatest number of plants. Meristem culture is used to produce healthy propagation stock for crops and ornamentals.

133. (b) Vector DNA are the DNA molecules that can carry a foreign DNA segment and replicate inside the host cells. Vector DNA may be plasmids, a bacteriophage, cosmids, yeast artificial chromosomes.

135. (a) Restriction endonucleases are enzymes that makes cuts at specific positions within the DNA molecule. They acts as molecular scissors. They recognise specific base sequence at palindrome sites in DNA duplex and cut its strands.

SECTION 4-ZOOLOGY

136. (b) Sex-factor or F-factor in bacteria results in high frequency conjugation. It allows bacteria to produce sex pilus necessary for conjugation.

137. (a) Organisms passively floating on the surface of water are planktons. If floating organisms are animals then it is zooplanktons and if plants– phytoplanktons.

138. (b) The cell junctions called tight, adhering and gap junctions are found in epithelial tissue. Epithelial tissue covers the whole surface of the body. It is made up of cells closely packed and ranged in one or more layers.

140. (c) Dolphin is an aquatic mammal and breathes through lungs.

142. (c) Mammals have 12 pairs of cranial nerves.

144. (c) Statements i and iii are correct.
- Medical Termination of Pregnancy (MTP) during first trimester is generally safe.
- Intrauterine device like copper-T are effective contraceptives.

146. (c) Epididymis lies between vas efferens and vas deferens.

149. (b) In menstrual cycle, menstrual phase lasts for 4 days, proliferating/ovulating phase for about 10 days and secretory phase for 14 days.

150. (c) This age pyramid represents the declining population of any organism. Population decline is the reduction over time in region's census. It can be caused for several reasons that includes heavy immigration disease, famine or sub-replacement fertility.

152. (d) Haemophilia is a sex linked recessive trait carried by X chromosome (also known as bleeder's disease).

153. (b) One of the first attempts to explain the mechanism of evolution was made by Jean Baptiste de Lamarck. His theory was Inheritance of Acquired Characters. The theory states that the characters acquired during life time are passed on to the progeny and then to subsequent generations and new species are produced.

156. (a) White leg horn is a mediterranean breed.

157. (b) Premolar and molar are lophodont teeth. Lophodont teeth with the cusps elongated to form narrow ridges. The molars in elephants and horses have cusps fused by means of intermediate masses of dentine to form ridges or lophs.

158. (c) Sino-Auricular node (S A node) present in the walls of right auricle has a myogenic initiation of heartbeat in a regular fashion and controls the pace of heartbeat called pacemaker.

160. (a) A large portion of oxygen is left unused in the human blood even after its uptake by the body tissues. This O_2 acts as a reserve during muscular exercise.

163. (a) Glome podocytes are highly specialized cells with a complex cytoarchitecture plays a major role in establishing the selective permeability of glomerular filtration barrier.

164. (a) Hypothalamus contains important nerve centres that controls the body temperature, thirst, hunger and eating, water balance and sexual function.

166. (b) Pancreas is a gland which is both exocrine and endorine. Cells of acini are exocrine which secrete enzymes and islet of Langerhans part is endocrine which secretes hormones like insulin, glucagon, *etc.*

167. (d) The sensation of fatigue in the muscles after prolonged strenuous physical work is caused by the accumulation of lactic acid.

170. (c) Epinephrine or adrenaline, norepinephrine or noradrenaline and acetylchloline are the neurotransmitters. These are released by the nerve fibres to transmit the impulse to the next neuron. Cortisone is not the neurotransmitter.

171. (a) Neurons regulate endocrine activity, but not *vice-versa*. Neurons in the hypothalamus secrete thyroid releasing hormone (TRH), which stimulates cells in the anterior pituitary to secrete thyroid-stimulating hormone (TSH).

172. (b) Cloning is the production of an organism with exactly similar genetic make up as in the mother individual. Cloning is done to preserve genotype of an individual. This is achieved by cell culture, tissue culture or genetic engineering.

173. (c) A special defence system works specially against viral infection. It has no effect on micro-organism. Cells invaded by a virus produce an antiviral protein called interferon (IFN). The latter is released from the infected cell and on reaching the nearby non-infected cells it makes them resistant to the virus infection.

174. (d) Organic farming involves use of organic wastes and other biological material along with beneficial microbes to release nutrients to crop to increase the soil fertility in an ecofriendly, and pollution free environment. *Glomus,* earthworm and *Oscillatoria* can be used in organic farming while snail cannot.

175. (b) ELISA test is a technique which can detect any amount of an antibody or antigen with the help of an enzyme. The commonly used enzymes are alkaline phosphatase and peroxidase.

177. (d) AIDS is caused by HIV virus and Gonorrhoea is caused by *Neisseria gonorrhoea.* Urethritis is inflammed of the urethra by bacteria. Syphilis is caused by *Treponema pallidum.*

178. (b) Amphetamines bring about increased alertness and sleeplessness. Hashish is a hallucinogen. Valium is a tranquilizer. Valium depresses brain activity and produces feeling of calmness, relaxation and drowsiness. Morphine is an opiate narcotic.

179. (b) Thermoregulation is the ability of an organism to keep its body temperature within certain boundaries, even when temperature surrounding is very different. In humans, sweating is primarily a means of thermoregulation.

180. (d) *Spirulina* is SCP rich in protein, vitamins & minerals. 250 gram biomass of *Methylophilus methylotrophus* produces 25tonn protein/day while cow of 250 kg. produces only 200 gm. protein/day. Common button mushrooms are a very rich source of vitamin D. A rice variety has been developed which is very rich in iron content.

NEET (NATIONAL ELIGIBILITY ENTRANCE TEST) Full Test - 5

ANSWER KEY

1	(b)	21	(a)	41	(b)	61	(d)	81	(b)	101	(c)	121	(b)	141	(c)	161	(d)
2	(d)	22	(b)	42	(b)	62	(b)	82	(a)	102	(d)	122	(a)	142	(d)	162	(a)
3	(a)	23	(a)	43	(b)	63	(c)	83	(a)	103	(c)	123	(d)	143	(d)	163	(b)
4	(b)	24	(b)	44	(b)	64	(b)	84	(d)	104	(a)	124	(c)	144	(b)	164	(b)
5	(d)	25	(a)	45	(a)	65	(d)	85	(d)	105	(b)	125	(d)	145	(a)	165	(a)
6	(a)	26	(d)	46	(d)	66	(b)	86	(d)	106	(c)	126	(a)	146	(a)	166	(b)
7	(c)	27	(b)	47	(c)	67	(b)	87	(b)	107	(b)	127	(d)	147	(a)	167	(b)
8	(d)	28	(d)	48	(c)	68	(d)	88	(b)	108	(d)	128	(d)	148	(d)	168	(c)
9	(a)	29	(d)	49	(c)	69	(c)	89	(a)	109	(d)	129	(a)	149	(a)	169	(a)
10	(a)	30	(d)	50	(c)	70	(c)	90	(d)	110	(b)	130	(d)	150	(b)	170	(d)
11	(c)	31	(b)	51	(a)	71	(b)	91	(a)	111	(a)	131	(b)	151	(c)	171	(d)
12	(a)	32	(d)	52	(c)	72	(a)	92	(a)	112	(c)	132	(a)	152	(b)	172	(b)
13	(d)	33	(a)	53	(a)	73	(b)	93	(c)	113	(b)	133	(c)	153	(c)	173	(d)
14	(b)	34	(c)	54	(d)	74	(c)	94	(b)	114	(b)	134	(d)	154	(c)	174	(b)
15	(d)	35	(a)	55	(d)	75	(b)	95	(a)	115	(a)	135	(d)	155	(d)	175	(d)
16	(c)	36	(c)	56	(c)	76	(c)	96	(b)	116	(a)	136	(c)	156	(a)	176	(c)
17	(b)	37	(a)	57	(c)	77	(a)	97	(a)	117	(b)	137	(c)	157	(d)	177	(d)
18	(c)	38	(b)	58	(a)	78	(c)	98	(b)	118	(b)	138	(d)	158	(d)	178	(b)
19	(d)	39	(d)	59	(a)	79	(b)	99	(b)	119	(d)	139	(c)	159	(c)	179	(d)
20	(c)	40	(b)	60	(d)	80	(a)	100	(b)	120	(a)	140	(a)	160	(b)	180	(d)

HINTS & SOLUTIONS

SECTION 1 -PHYSICS

1. (b) $\sqrt{\frac{\rho r^3}{S}} = \sqrt{\frac{ML^{-3}L^3}{MT^{-2}}} = T$

2. (d) The no. of significant figures in 3400 is 2.

3. (a) The work done = area of $P-V$ graph

$= \text{area of triangle } ABC$

$= \frac{1}{2} \times 3P \times 2V = 3PV$

4. (b) Since escape velocity $\left(v_e = \sqrt{2gR_e}\right)$ is independent of angle of projection, so it will not change.

5. (d) At maximum height, $v = u\cos 45° = \frac{u}{\sqrt{2}}$;

$a_c = g$

$$r = \frac{\left(\frac{u}{\sqrt{2}}\right)^2}{g} = \frac{u^2}{2g}$$

6. (a) Because when intensity of incident light increases, it means that number of photons increases in incident light. If number of incident photons increases, then number of emitted photo electrons also increases, consequently the photo electric current increases.

7. (c) In an ac circuit, the ac driving voltage v leads the circuit current i through an angle θ, given by

$$\tan\theta = \frac{X_L - X_C}{R} = \frac{0}{R} = 0 \Rightarrow \theta = 0$$

i.e., the driving voltage and current are in the same phase.

9. (a) Force of friction acting

$= \mu mg = 0.15 \times 60 \times 9.8\,N$

distance = 0.9 m

Work done = $0.15 \times 60 \times 9.8 \times 0.9$

= 79.4 joule

10. (a) Efficiency of carnot engine $= \eta = \frac{T_1 - T_2}{T_1}$

where T_1 = source temperature

T_2 = sink temperature.

11. (c) The battery supplies the equal amounts of energy to the capacitor and resistor. The total energy supplied to the two is

$U = \frac{1}{2}CV^2 + \frac{1}{2}CV^2 = CV^2$

$= 10 \times 10^{-6} \times (20)^2 = 4 \times 10^{-3}\,J = 4\,mJ$

12. (a) Energy of Bohr's orbit is given by

$$E_n = -\frac{E_1}{n^2} \Rightarrow E_2 - E_1 = E_1\left[1 - \frac{1}{4}\right] = \frac{3}{4}E_1$$

$$E_3 - E_2 = E_1\left[\frac{1}{4} - \frac{1}{9}\right] = \frac{5}{36}E_1$$

So, the gap between two successive orbitals decreases. Naturally radiations due to transition between these levels will also go on decreasing. So, transition from n = 4 to n = 3 will have lowest frequency.

13. (d) $\tau \times \Delta t = L_0$ $\{\because \text{since } L_f = 0\}$

$\Rightarrow \tau \times \Delta t = I\omega$

or $\tau \times 60 = 2 \times 2 \times 60\pi/60$

$$\left(\because\ f = 60\text{rpm} \therefore\ \omega = 2\pi f = 2\pi \times \frac{60}{60}\right)$$

$\tau = \frac{\pi}{15}\,N-m$

14. (b) Photoelectric effect is accounted by particle like bahaviour of light (i.e. by quantum theory of light)

15. (d) $C_V = dE/dT = \frac{3}{2}R$

$$C_p = \frac{3}{2}R + R = \frac{5}{2}R = 2.5R$$

16. (c)

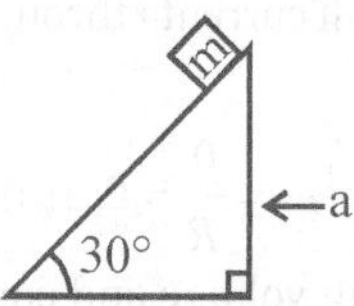

Use result, $a = g\tan\theta$, $a = g\tan 30° = g/\sqrt{3}$

20. (c) If a gas is heated at constant volume then no work is done. The heat supplied is given by

$dQ = nC_v dT$

But $C_v = \frac{f}{2}R$ where f is the degree of freedom of the gas

$\therefore\ dQ = \frac{nfRdT}{2}$

$$= \frac{2 \times 3 \times R \times (373 - 273)}{2} = 300\,R$$

21. (a) At t = 2 sec, F = 2 × 2 = 4 N, where as $f_{max} = \mu_N = 8N$. As the applied force is less than the limiting force of friction, the frictional force will be equal to applied force i.e. F = 2 × 2 = 4 N

22. (b) $I_1N_1 = I_2N_2 \Rightarrow I_2 = \frac{4 \times 140}{280} = 2A.$

23. (a) According to the Gauss's law, the net electric flux through the closed spherical surface is equal to the total chage enclosed by the surface divided by ε_0, *i.e.*,

$$\phi_E = \frac{-q + q}{\varepsilon_0} = 0$$

24. (b) $E = \frac{1}{2}mv_e^2 = \frac{1}{2} \times 1000 \times (11.2 \times 10^3)^2$

$= 6.4 \times 10^{10}\,J$ (approx.)

25. (a) Phase difference $= \frac{2\pi}{\lambda} \times$ path difference

$$\frac{\pi}{2} = \frac{2\pi}{\lambda} \times 0.8$$

$\Rightarrow \lambda = 0.8\,m \times 4 = 3.2\,m$

velocity $= \lambda \times v = 3.2 \times 120 = 384$ m/s

26. (d) Since maximum part of the power of modulated wave is contained with the carrier wave which does not transmit any desired information, hence to avoid wastage of power to suppress carrier balanced modulator is used.

27. (b) $V_{r.m.s} = \frac{V_{max}}{\sqrt{2}} = \frac{120}{\sqrt{2}} = 84.8V$

28. (d) If V is velocity of combination (bag + bullet), then from principle of conservation of linear momentum

$$(m+M)V = m\,v \text{ or } V = \frac{m\,v}{(m+M)}$$

$$K.E. = \frac{1}{2}(m+M)V^2 = \frac{m^2 v^2}{2(m+M)}$$

29. (d) $T_1 = 2\pi\sqrt{\frac{\ell}{g}}$

Length ℓ increases from ℓ to $1.21\,\ell$.

$$T_2 = 2\pi\sqrt{\frac{1.21\ell}{g}} = 1.1T_1$$

Thus % increase in time period = 10%

30. (d) $e = \frac{B\omega\ell^2}{2}$

$$= \frac{(0.4\times10^4)\times\left(2\pi\times\frac{120}{60}\right)\times(0.5)^2}{2}$$

$$= 0.628\times10^{-4}V$$

32. (d) $\mu = \frac{\sin\frac{\delta_m + A}{2}}{\sin\frac{A}{2}}$

$$= \frac{\sin\frac{98}{2}}{\sin 30} = \frac{\sin 49}{\sin 30} = 1.5$$

33. (a) The impedance of circuit is

$$Z = \sqrt{(R)^2 + (1/2\pi fc)^2}$$

Which will decrease with the increase of frequency f of impressed ac.

34. (c) $V_g = \frac{-6.67\times10^{-11}\times10}{1} - \frac{6.67\times10^{-11}\times100}{1}$

$$= -6.67\times10^{-10} - 6.67\times10^{-9}$$

$$= -6.67\times10^{-10}\times11 = -7.3\times10^{-9}\ \text{J/kg}$$

35. (a) The relation between Y, η and σ is as follows :

$$Y = 2\eta(1+\sigma)$$

$$\sigma = \frac{Y}{2\eta} - 1 = 0.2$$

36. (c) P.E. changes from zero to maximum twice in each vibration so its time period is T/2

37. (a) Torque acting on a magnet when it is rotated by θ° from its equilibrium position is $MB\sin\theta$

which is equivalent to $\vec{M}\times\vec{B}$.

38. (b) Due to increase in velocity, pressure will be low above the surface of water.

39. (d)

White light
R
V
F_V F_R

Red light less deviates than violet light.

So, $F_R > F_V$. Thus, $\mu_R < \mu_V$

Therefore, option (d) is correct.

40. (b) Excess pressure, $P = \frac{4T}{r}$

$$PV = \frac{4T}{r}\times\frac{4}{3}\pi.r^3 = \frac{16T}{3}\pi.r^2$$

$$PV \propto r^2$$

41. (b) Electro magnet should be amenable to magnetisation & demagnetization

∴ retentivity should be low & coercivity should be low

42. (b) Since work function for a metal surface is

$$W = \frac{hc}{\lambda_0}$$

where λ_0 is threshold wavelength or cut-off wavelength for a metal surface.

here $W = 4.125\,eV = 4.125\times1.6\times10^{-19}$ Joule

$$\text{so } \lambda_0 = \frac{6.6\times10^{-34}\times3\times10^8}{4.125\times1.6\times10^{-19}} = 3000\text{A}^\circ$$

43. (b)

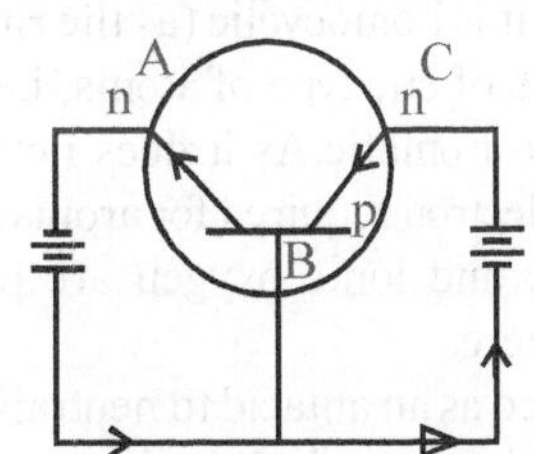

In this circuit, direction of current has been

shown. Current is flowing from base to emitter so, electrons are flowing from emitter to base.

Hence option (b) is the answer.

44. (b) Ground waves have the property of being polarized normal to the earth's surface.

45. (a) In case of 'NOR' logic circuit, when input is 0, the output is 1 that is when both the inputs are low, the output is high.

SECTION 2 - CHEMISTRY

46. (d) Huckel's rule states that for aromaticity there must be $(4n + 2)\pi$ electron present in compound where n is an integer

47. (c) It has centre of symmetry, so optically inactive.

49. (c) Toluene (CH_3) $\xrightarrow[\text{in } CCl_4]{CrO_2Cl_2}$ Benzaldehyde (CHO)

This reaction can not be given by aliphatic compounds.

50. (c) Percentage of degree of ionization = 1.34%

$\therefore$ Degree of ionization $(\alpha) = 0.0134$

$$CH_3COOH \rightleftharpoons CH_3COO^- + H^+$$

	CH_3COOH	CH_3COO^-	H^+
Initial	0.1	0	0
At equilibrium	$0.1(1-0.0134)$	0.1×0.0134	0.1×0.0134

$$K = \frac{[CH_3COO^-][H^+]}{[CH_3COOH]}$$

$$= \frac{0.1\times0.0134\times0.0134\times0.1}{0.1\times(1-0.0134)}$$

$$= 0.182\times10^{-5}$$

51. (a) [structure with NO_2] or [structure with NO_2]

Hence it is homocyclic (as the ring system is made of one type of atoms, i.e. carbon) but not aromatic. As it does not have $(4n+2)\pi$ electron required for aromaticity.

52. (c) H_2, He and ionic oxygen are present in exosphere.

53. (a) It is used as an antacid to neutralize excess of acid in stomach. It is also used in tooth pastes.

54. (d) Octane number defines the knocking tendencies of various hydrocarbons. Higher is the octane number, lesser is the knocking tendency of fuel. Experimentally it is observed that 2, 2, 4-trimethylpentane, **a branched chain** hydrocarbon has the highest anti-knocking property. It has been given an octane number 100, while *n*-heptane, a **straight chain** hydrocarbon has the highest knocking property, assigned zero octane number.

55. (d) $-OCH_3$ activates the benzene ring. $-NO_2$ deactivates the ring. Hence the reaction of the given compounds with electrophiles is in the order, I > II > III.

56. (c) Given : Percentage of the iron = 0.334%; Molecular weight of the haemoglobin = 67200 and atomic weight of the iron = 56. We know that the number of iron atoms

$$= \frac{\text{Molecular wt. of haemoglobin} \times \% \text{ of iron}}{100 \times \text{Atomic weight of iron}}$$

$$= \frac{67200\times0.334}{100\times56} = 4$$

57. (c) $K_2HgI_4 + KOH$ is called Nessler's reagent. It is used for the detection of NH_4^+ in group analysis and in quantitative determination of ammonia solution. It gives a **yellow** or **brown** precipitate with concentrations as low as 1 ppm of NH_3.

58. (a) Both lattice energy and hydrolysis energy decreases moving down the group due to gradual increase in size of M^{++} ion. So former tends to increase the solubility while latter tends to decrease it. But lattice energy has dominating role here. And therefore, solubility increases down the group.

59. (a) For BCC lattice body diagonal = $a\sqrt{3}$. The distance between the two oppositely charged ions = $\frac{a}{2}\sqrt{3}$

$$= \frac{387\times1.732}{2} = 335\text{pm}$$

60. (d) I_2 and NaOH react with acetophenone ($C_6H_5COCH_3$) to give yellow ppt. of CHI_3 but benzophenone ($C_6H_5COC_6H_5$) does not and hence can be used to distinguish between them. Oxidation of methyl ketones with sodium Hypoiodite (NaOI) or (I_2 + NaOH) gives iodoform (CHI_3) but not in case of binzophenone.

61. (d) $E^0_{cell} = E^0_{Zn/Zn^{2+}} + E^0_{Cu^{2+}/Ca}$ first cell

$1.10 = E^0_{Zn/Zn^{2+}} + 0.34$

$E^0_{Zn/Zn^{2+}} = 0.76\ V$

$E^0_{cell} = E^0_{Cu/Cu^{2+}} + E^0_{Ag^+/Ag}$ second cell

$0.46 = -0.34 + E^0_{Ag^+/Ag}$

$E^0_{Ag^+/Ag} = 0.80\ V$

For third cell

$$E^0_{cell} = E^0_{Zn/Zn^{2+}} + E^0_{Ag^+/Ag}$$
$$= 0.76 + 0.80$$
$$= 1.56\ V$$

62. (b) It is optically active when two Cl atoms are in cis position.

63. (c) $\frac{U_{H_2}}{U_{N_2}} = \sqrt{\frac{T_{H_2} \times 28}{T_{N_2} \times 2}} = \sqrt{7} \Rightarrow \frac{T_{H_2}}{T_{N_2}} = \frac{1}{2}$

$\Rightarrow T_{N_2} = 2T_{H_2}$ $\quad \therefore T_{N_2} > T_{H_2}$

64. (b) The correct order of ionic radii is

$$\underset{71}{Lu^{3+}} < \underset{63}{Eu^{3+}} < \underset{57}{La^{3+}} < \underset{39}{Y^{3+}}$$
At. No.

[In lanthanides ionic radii (M^{3+}) decreases with increase in atomic number. Y^{3+} belongs to *d*-block].

65. (d) Amino acid is 5% dissociated.

Thus, $[A^-] = [\text{conjugate base}] = 0.05$

$[HA] = [\text{Amino acid}] = 0.95$

$$pH = pK_a + \log\frac{[A^-]}{[HA]}$$

$$= 9.15 + \log\left(\frac{1}{12}\right)$$

$$= 7.87$$

66. (b) In froth floatation process, ore particles float because their surface is hydrophobic i.e., not easily wetted by oil.

67. (b) Initial concentration = 10 mol L^{-1}

$\therefore$ Conc. after 20 min (two halflives) = 2.5 mol L^{-1}

Now, $k = \frac{0.693}{t_{1/2}} = \frac{0.693}{10\,\text{min}}$

or $0.0693\ \text{min}^{-1}$

$\therefore$ rate = $k \times$ [reactant]

$= 0.0693 \times 2.5\ \text{mol}\ L^{-1}\ \text{min}^{-1}$

68. (d) $[Co(CO)_5NH_3]^{2+}$. In this complex. Co-atom attached with NH_3 through σ bonding with CO attached with dative π-bond.

69. (c) C is incorrect ; The correct is $\Delta E = \Delta q + \Delta w$

70. (c) Among the isoelectronic species, size increases with the increase in negative charge. Thus S^{2-} has the highest negative charge and hence largest in size followed by Cl^-, K^+ and Ca.

71. (b)

$CHCl_2$ | CH_3	CH_2Cl | CH_2Cl
gem-dihalide	*vic*–dihalide

72. (a) In alkaline medium SnO_2 is converted into negatively charged colloidal sol

$$SnO_2 + 2OH^- \longrightarrow SnO_3^{2-} + H_2O$$

$$SnO_2 + SnO_3^{2-} \longrightarrow \underbrace{[SnO_2]SnO_3^{2-}}_{\text{anion migrates to anode}}$$

Thus, (a)

73. (b) No. of equivalents of caustic sdoda = $20 \times 10^{-3} \times 0.5 = 10^{-2}$

$\therefore$ No. of equivalents of acid = 10^{-2}

Also, $\frac{\text{Wt.of acid} \times \text{Basicity}}{\text{MW}} = \text{No. of equiv.}$

$\Rightarrow \frac{0.45}{90} \times x = 10^{-2} \Rightarrow x = 2$

74. (c) **Chloramphenicol** is a broad spectrum drug. [Broad spectrum antibiotics are medicines effective against gram positive as well as gram negative bacteria, *e.g.*, tetracycline, chloramphenicol, etc.]

75. (b) Halides Cl^-, Br^-, I^- React with $AgNO_3$ to give

$AgCl \longrightarrow$ soluble in NH_4OH

$AgBr \longrightarrow$ sprraringly soluble in NH_4OH

$AgI \longrightarrow$ Insoluble

and the C – Cl bond is weakest in benzyl chloride [(sp^3) hybridised carbon is attached to Cl]

76. (c) $2d\sin\theta = n\lambda$

$2\lambda\sin\theta = \lambda$

$\therefore \quad \sin\theta = \frac{1}{2}$

$\theta = 30°$

$2\times\frac{\lambda}{2}\sin\theta = \lambda$

$\therefore \quad \sin\theta = 1$

$\therefore \quad \theta = 90°$

Thus, 30° and 90° as in (c).

77. (a) Because at infinite dilution the equivalent conductance of strong electrolytes furnishing same number of ions is same.

78 (c) $C_6H_6 + \frac{15}{2}O_2 \longrightarrow 6CO_2 + 3H_2O$

1 mole ≡ 3250 kJ

$\frac{0.39}{78}$ mole $\equiv 3250 \times \frac{0.39}{78} = 16.25$ kJ

80. (a) The reason for the lesser basicity of aniline as compared to that of cyclohexylamine is +R effect of $-NH_2$ group (*i.e.*, electron releasing resonance effect of $-NH_2$ group).

81. (b) Nernst equation is

$$E_{cell} = E^o_{cell} - \frac{2.303\,RT}{nF}\log\frac{[C]^c[d]^d}{[A]^a[B]^b}$$

For general cell reaction

$aA + bB \rightleftharpoons cC + dD$

When E_{cell} drops to zero the concentration of ions will be equilibrium concentrations.

$$E^o_{cell} = \frac{2.303RT}{nF}\log K_c$$

[Since at equilibrium $E_{cell} = 0$]

$$E^o_{cell} = \frac{0.0591}{n}\log K_c \text{ at 298K.}$$

82. (a) Mac Arthur process involves extraction of gold from sulphide ore with the help of KCN.

83. (a) The boron atom in boric acid, H_3BO_3 is electron deficient i.e., boric acid is a Lewis acid with one *p*-orbital vacant. There is no *d*-orbital of suitable energy in boron atom. So, it can accommodate only one additional electron pair in its outermost shell.

84. (d) Given,

$N_2 + 3H_2 \rightleftharpoons 2NH_3;\ K_1$ ....(i)

$N_2 + O_2 \rightleftharpoons 2NO;\ K_2$ ....(ii)

$H_2 + \frac{1}{2}O_2 \rightleftharpoons H_2O;\ K_3$ ....(iii)

We have to calculate

$4NH_3 + 5O_2 \longrightarrow 4NO + 6H_2O;\ K = ?$

or $2NH_3 + \frac{5}{2}O_2 \longrightarrow 2NO + 3H_2O$

For this equation, $K = \frac{[NO]^2\,[H_2O]^3}{[NH_3]^2\,[O_2]^{5/2}}$

but $K_1 = \frac{[NH_3]^2}{[N_2][H_2]^3},\ K_2 = \frac{[NO]^2}{[N_2][O_2]}$

& $K_3 = \frac{[H_2O]}{[H_2][O_2]^{½}}$ or $K_3 = \frac{[H_2O]^3}{[H_2]^3\,[O_2]^{3/2}}$

Now operate, $\frac{K_2\,.\,K_3^3}{K_1}$

$$= \frac{[NO]^2}{[N_2][O_2]} \times \frac{[H_2O]^3}{[H_2]^3\,[O_2]^{3/2}} \cdot \frac{[N_2][H_2]^3}{[NH_3]^2}$$

$$= \frac{[NO]^2\,[H_2O]^3}{[NH_3]^2\,[O_2]^{5/2}} = K$$

$$\therefore K = \frac{K_2\,.\,K_3^3}{K_1}$$

85. (d) $-NH-CONH_2$ group is ortho para directing. Nitrogen shares its lone pair with benzene ring and makes this group ortho para directing.

86. (d) $C_6H_5COOH \xrightarrow{\text{Soda-lime}(X)} C_6H_6 + Na_2CO_3 + H_2O$

$C_6H_5OH \xrightarrow{\text{Zn dust}(Y)} C_6H_6 + ZnO$

87. (b) We have to calculate the enthalpy of the reaction $OH(g) \rightarrow O(g) + H(g)$

From the given reactions, this can be obtained as follows.

$$-\left[\frac{1}{2}H_2(g) + \frac{1}{2}O_2(g) + \rightarrow OH(g)\right];\ \Delta H = -42.09\ kJ\,mol^{-1}$$

$$+\frac{1}{2}[H_2(g) \rightarrow 2H(g)];\ \Delta H = \frac{1}{2}\times 435.89\ kJ\,mol^{-1}$$

$$+\frac{1}{2}[O_2(g) \rightarrow 0\,2O(g)];\ \Delta H = \frac{1}{2}\times 495.05\ kJ\,mol^{-1}$$

Add

$$OH(g) \rightarrow H(g) + O(g)\ ;\ \Delta H = 423.38\ kJ\,mol^{-1}$$

88. (b) Elements X, Y, Z with atomic numbers 19, 37, 55 lie in group 1 (alkali metals). On moving down a group from the size of atoms increases; the outermost electrons become less strongly held. So the ionization energy decreases. Therefore, IE of Y could be between those of X and Z.

	X	Y	Z
	K (19)	Rb (37)	Cs (55)
I.E.	4.3	4.2	3.9

89. (a) $3MnO_4^{2-} + 2H_2O \rightarrow MnO_2 + 2MnO_4^- + 4OH^-$

or $MnO_4^{2-} + \frac{2}{3}H_2O \rightarrow \frac{1}{3}MnO_2 + \frac{2}{3}MnO_4^- + \frac{4}{3}OH^-$

SECTION 3 - BOTANY

92. (a) Lateral meristems occur on the sides of stem and help in increasing girth of stem and root. It divides only periclinally or radially and is responsible for increase in girth or diameter.

93. (c) The correct sequence of cell cycle is G_1, S, G_2 and M.

94. (b) In chlorophyceae, sexual reproduction takes place by all the three processes :
Isogamy – fusion of morphologically and physiologically similar gametes, Anisogamy – morphologically similar but physiologically dissimilar gametes, oogamy – fusing gametes are dissimilar in all respect.

96. (b) The membrane surrounding a lysosome allows the digestive enzymes to work at the 4.5 pH they require. They are created by the addition of hydrolytic enzymes to early endosomes from the Golgi apparatus.

97. (a) Tiger and tigris both are from same genus with particular taxonomic category.

98. (b) In gymnosperms the female gametophyte forming archegonia provides nourishment to the developing embryo. It later gets transformed into food-laden endosperm inside the seed. Endosperm provides nourishment for growth of seed at the time of seed germination. Triploid endosperm occurs in angiosperms.

100. (b) Secondary succession refers to the regrowth of a habitat in the area where disruptive event has occurred and eliminated the existing, above ground plant life of the natural habitat. So, it occurs on a deforested site.

102. (d) Autogamy = Self Pollination
Geitonogamy - (*geiton* = neighbour, *gamos* = marriage.) i.e. when pollen grains are transferred to the stigma of other flower of the same plant (i.e. neighboring flower).
Allogamy = Xenogamy – Cross pollination.

103. (c) Biochemical Oxygen Demand (BOD) in river water increases when sewage gets mixed with river water. Biochemical oxygen demand or biological oxygen demand (BOD) is a chemical procedure for determining how fast biological organisms use up oxygen in a body of water. It is used in water quality management and assessment, ecology and environmental science. BOD considered as an indication of the quality of a water source.

104. (a) Some fleshy fruits such as mango, plum etc. usually have a single hard stone that encloses a seed, called drupe.

107. (b) Active transport involves movement of ions against concentration gradient involves the expenditure of energy. Diffusion involves the movement of solute particles from region of higher concentration to a region of lower concentration. Pinocytosis is cell drinking. Brownian movement is the random to and fro movement of atoms and molecules.

109. (d) Long day plants undergo flowering after receiving light above a critical day length. *e.g.* spinach, sugarbeet. Short day plants flower only when they receive light below a critical day length, *e.g.* *Glycine max*, tobacco.

112. (c) Cell placed in a hypotonic solution becomes turgid due to endosmosis. Water exerts a pressure on the walls of the cell called Turgor pressure. Cell wall being a rigid and hard structure also exerts a pressure on Cytoplasm in response to Turgor pressure when TP = WP, DPD = 0 (No net exchange of water).

114. (b) The atmospheric nitrogen is fixed in the soil in the form of nitrate. Nitrate and ammonium can be taken up by plants or used by other soil organisms. Any nitrate or ammonium that is not used for growth is added to the pool of inorganic nitrogen in the soil.

116. (a) In 1953 Wilkins obtained very fine X-ray crystallographic pictures of DNA from which Watson and Crick developed the double helix model of DNA.

117. (b) Genetic mapping help in studying the architecture of the chromosomes and of entire genome.

118. (b) During transcription, from the DNA template complementary mRNA is formed and thymine is replaced by uracil.

119. (d) Humus is the organic material in soil is the dark brown or black colouration. In soil science, humus refers to any organic matter which has reached a point of stability, where it will break down no further and might, if conditions do not change, remain essentially as it is for centuries, if not millennia. In agriculture, humus is sometimes also used to describe mature compost, or natural compost extracted from a forest or other spontaneous source for use to amend soil. It is also used to describe a topsoil horizon that contains organic matter (humus type, humus form, humus profile).

120. (a) DDT was subsequently banned for agricultural use worldwide under the Stockholm Convention, but its limited use in disease vector control continues to these days in certain parts of the world and remains controversial. Along with the passage of the Endangered Species Act, the US ban on DDT is cited by scientists as a major factor in the comeback of the bald eagle in the contiguous US.

121. (b) Ozone layer is present in the stratosphere in the atmosphere of the earth. It protects the earth from UV rays (less than 300 nm). Hole in the ozone layer means thinning of ozone layer that allows harmful UV rays to reach earth which cause skin diseases.

122. (a) Selection is picking up only those plants for reproduction which have desired qualities. Hybridization is crossing of two or more types of plants for bringing their traits together in the progeny. Introduction is taking a plant or a new variety from an area where it grows naturally to a region where it does not occur before.

123. (d) rRNA occurs inside ribosomes. m RNA brings information from DNA to polypeptides. hnRNA are heterogenous nuclear RNA.

124. (c) Mesosomes are the invaginations of the plasma membrane that can form into vesicles. They are found to be present in both gram-positive and gram-negative bacteria. Mesosomes may play a role in cell wall formation during cell division and/ or chromosome replication and distribution and/or electron transfer systems of respiration.

126. (a) Massive ions like Na^+ facilitate the absorption of substances like amino acid and glucose through co-transport.

127. (d) Scarification is the breaking of the seed coat barriers which can be done by mechanical or chemical methods.

128. (d) In Kreb's cycle, the FAD precipitates as electron acceptor during the conversion of succinic acid to fumaric acid.

130. (d) Parthenocarpy refers the development of fruits without fertilization. Such fruits are seedless but otherwise normal in appearance, *e.g.* banana, pineapple etc. The auxins (IAA, IBA, 2, 4-D) are applied in low concentration in a lanoline paste to the stigma of flower to induce parthenocarpy.

131. (b) MAB (Man and Biosphere Programme) was launched by UNESCO in 1971. There are 14 projects under this programme, which focus on Research and Training with emphasis on ecological approach to the study of inter-relationship between man and the environment.

132. (a) *Glomus aggregatum* is a mycorrhizal fungus used as a soil inoculant in agriculture and horticulture. Its purpose is to increase the surface area of roots for nutrient absorption like phosphorus.

134. (d) In pBR 322
ori-represents site of origin or replication
rop-represents those proteins that take part in replication of plasmid.
Hind III, EcoRI- Recoginition sites of Restriction endonucleases
amp^R and tet^R - They are antibiotic resistant gene part.

SECTION 4 - ZOOLOGY

136. (c) Slime moulds in the division of myxomycota have spores that develop into flagellated gametes.

137. (c) Fibrinogen, prothrombin, thromboplastin are respectively clotting factors I, II and III. Thromboplastin released from injured platelets or injured tissue. This converts prothrombin into thrombin in the presence of calcium ions. The thrombin converts fibrinogen into fibrin and the latter forms the clot.

140. (a) Lymph differs from blood in that it contains less protein and no red blood cells. Lymph contains white blood cells.

141. (c) Pseudocoelomate is any invertebrate animal whose body cavity is a pseudocoel, a cavity between the gut and the outer body wall derived from a persistent blastocoel, rather than a true coelom. Pseudocoelomate animals include the Rotifera and Nematoda.

144. (b) Amniocentesis is the most widely used method for prenatal detection of many genetic disorders. It is also a technique used for determining the sex of the foetus.

148. (d) Chorionic villi found in the maternal blood is involved in the formation of placenta in man.

152. (b) Rh factor was discovered by Karl Landsteiner. A child of Rh^+ man will be Rh^+ whether the mother is Rh^+ or Rh^-. If the mother is Rh^+ then there will be no problem but if mother is Rh^- so when the blood of Rh^+ child (in womb) mixes with the blood of Rh^- mother then some antibodies in mother's blood are formed against Rh^+ factor which coagulate the womb blood causing death. If birth takes place then there is a possibility of child death in early years. This in known as erythroblastosis foetalis. In most cases the Ist pregnancy may succeed but after that it fails.

153. (c) Continued consumption of fat rich diet causes hypercholesterolemia. Hypercholesterolemia is the presence of high levels of cholesterol in the blood. High cholesterol raises risk for heart disease, heart attack, and stroke. Kidney stones are solid mass made up of tiny crystals. There are different types of kidney stones. The exact cause depends on the type of stone like, calcium stones, uric acid stone *etc*. Vitamin A toxicity or hypervitaminosis A is having too much of vitamin A in the body. Ketonuria is condition in which ketone bodies are present in urine. Body products excess ketone bodies as an alternate source of energy during starvation or diabetes mellitus (type 1).

154. (c) Due to a long hunger strike and survival on water, a person will have less urea in his urine because urea comes to kidney as a waste product from liver which is formed after the breakdown of protein fat, carbohydrate during hunger. It is not synthesised but the synthesised ones are catabolised.

155. (d) Haemoglobin has/having 250 times more affinity for CO as compared to oxygen. Hb readily combines with CO, forming carboxyhaemoglobin (COHb), COHb interferes with the transport of oxygen.

156. (a) High proportion of cholesterol in blood leads to deposition of cholesterol on the walls of blood vessels. This causes the arteries to lose their elasticity and get stiffened. This is called arteriosclerosis or hardening of arteries.

158. (d) Sickle cell anaemia (in which nucleotide triplet CTC is changed to CAC) affects the β-globin chain of haemoglobin. Since these changes occur at a particular locus or point of a chromosome where specific gene is located, they are called as point mutation.

160. (b) In the embryos of all vertebrates, the presence of gill slits supports the theory of recapitulation (repeating the early stages of embryogenesis in earlier evolved animals).

161. (d) In superovulation, a high milk yielding cow is induced to shed 4–6 eggs (instead of one) every 6–8 weeks (instead of 20–21 days). The superovulated donor is artificially inseminated with semen from a quality bull. The embryos developing from the eggs so fertilised are flushed out. These good quality embryos are now transfered to surrogate mother for delivery.

162. (a) The change in the colour of peppered moth is due to the mutation of single mendelian

gene for the survival in the smoke-laden industrial environment. It is called Industrial melanism.

163. (b) Down's syndrome develops due to trisomy of chromosome number 21. In Turner's syndrome, the effect appears due to fusion of a gamete without sex chromosome and a gamete with one X-chromosome (44 + X). Klinefelter individuals are phenotypically males. The defect appears due to fusion of egg having unreduced sex complement (A + X X) with a gamete carrying Y chromosome (44 + XXY). Gynandromorphism occur among *Drosophila.* In such individuals one half of the body shows male characters and the other half shows female characters.

164. (b) There is no life on moon due to absence of water. There are many other reasons for being no life on moon, life insufficient oxygens, unsuitable temperature etc.

166. (b) MRI machine does not show face image of bone and calcium, *e.g.* scapula, canine. It is also not suitable for patients with cardiac pace makers.

167. (b) Hormones are chemical messengers formed by endocrine cells. Chemically hormones are of the following types: Amines– composed of amino group e.g., Melatonin.

Amino acids	–	eg. thyroxine
Peptides	–	eg. insulin
Steroids	–	eg. aldosterone

168. (c) The maximum volume of air a person can breath in after a forced expiration or the maximum volume of air a person can breathe out after a forced inspiration is called vital capacity (VC). This includes tidal volume, inspiratory reserve volume and expiratory reserve volume.
(TV + IRV + ERV). VC varies from 3400 ml – 4800 mL, depending upon age, sex and height of the individual.

169. (a) The inner layer of the posterior two-thirds of the eyeball consists of a light sensitive layer, called retina that possesses two types of photoreceptors called the rods and the cone cells. Retina reduces the internal reflection so any damage to it leads to greater internal reflection of light often causing an increase in light sensitivity.

171. (d) Total sum of physio-electrochemical changes that takes place along the length of nerve fibre is known as nerve impulse. Change in potential due to stimulation of nerve fibre is called action potential. During propagation of nerve impulse, Na^+ enters inside so (+ve) change is formed inside the membrane. K^+ ions come out.

173. (d) Retrovirus has the ability to transform normal cells into cancerous cells. Hence, it can used as a vector for cloning desirable genes into animal cells.

175. (d) Tetanus and typhoid are caused by bacteria.
Whooping cough is caused by bacteria.
Sleeping sickness is caused by parasitic flees.
Syphilis is caused by bacteria while AIDS is final stage of HIV viral disease.
Measles and Rabies are caused by virus.

176. (c) Electroporation is the method of making cell membrane permeable for the entry of recombinant DNA into the bacteria.

177. (d) Widal test is used for the diagnosis of typhoid. It is an agglutination test for the presence of antibodies against the *Salmonella* organisms, which cause typhoid fever.

178. (b) Pollen–grains of many species are responsible for some of the severe allergies and bronchial affliction in some people often lead to chronic respiratory disorders - asthma, bronchitis.

179. (d) For gene transfer into the host cell without using vector microparticles made of tungsten and gold coated with foregin DNA are bombarded into target cells at a very high velocity.

180. (d) AIDS (Acquired Immune Deficiency Syndrome) is caused by HIV retrovirus. The virus destroys the helper T lymphocytes thus reducing their numbers.

NEET (NATIONAL ELIGIBILITY ENTRANCE TEST) Full Test - 6

ANSWER KEY

1	(c)	21	(b)	41	(d)	61	(a)	81	(b)	101	(a)	121	(c)	141	(a)	161	(b)
2	(d)	22	(c)	42	(b)	62	(b)	82	(d)	102	(d)	122	(d)	142	(d)	162	(d)
3	(a)	23	(a)	43	(b)	63	(d)	83	(c)	103	(d)	123	(c)	143	(a)	163	(c)
4	(a)	24	(d)	44	(b)	64	(a)	84	(c)	104	(b)	124	(a)	144	(c)	164	(d)
5	(b)	25	(a)	45	(b)	65	(b)	85	(d)	105	(a)	125	(c)	145	(d)	165	(a)
6	(c)	26	(b)	46	(a)	66	(b)	86	(c)	106	(c)	126	(b)	146	(a)	166	(d)
7	(d)	27	(c)	47	(b)	67	(d)	87	(c)	107	(a)	127	(d)	147	(c)	167	(d)
8	(b)	28	(d)	48	(c)	68	(a)	88	(a)	108	(c)	128	(b)	148	(c)	168	(a)
9	(a)	29	(d)	49	(d)	69	(b)	89	(a)	109	(c)	129	(c)	149	(c)	169	(d)
10	(b)	30	(c)	50	(b)	70	(d)	90	(d)	110	(d)	130	(c)	150	(a)	170	(d)
11	(c)	31	(d)	51	(c)	71	(a)	91	(b)	111	(b)	131	(c)	151	(b)	171	(a)
12	(d)	32	(a)	52	(d)	72	(b)	92	(a)	112	(b)	132	(c)	152	(c)	172	(b)
13	(b)	33	(a)	53	(d)	73	(a)	93	(a)	113	(a)	133	(a)	153	(d)	173	(b)
14	(d)	34	(b)	54	(c)	74	(d)	94	(b)	114	(a)	134	(a)	154	(a)	174	(c)
15	(a)	35	(a)	55	(d)	75	(d)	95	(a)	115	(b)	135	(c)	155	(a)	175	(b)
16	(d)	36	(c)	56	(c)	76	(b)	96	(d)	116	(a)	136	(b)	156	(a)	176	(a)
17	(c)	37	(a)	57	(b)	77	(d)	97	(d)	117	(b)	137	(b)	157	(b)	177	(c)
18	(d)	38	(c)	58	(d)	78	(a)	98	(b)	118	(c)	138	(a)	158	(b)	178	(c)
19	(b)	39	(d)	59	(a)	79	(a)	99	(d)	119	(b)	139	(b)	159	(d)	179	(b)
20	(b)	40	(a)	60	(b)	80	(c)	100	(b)	120	(c)	140	(a)	160	(c)	180	(d)

HINTS & SOLUTIONS

SECTION 1 -PHYSICS

1. (c) Angular displacement has unit (degree or radian) but it is dimensionless.

   Note : vice-versa is not possible.

2. (d) The maximum acceleration will be represented by line having maximum slope. Line BC represents maximum slope.
   Here change in velocity (Δv)
   $= 60 - 30 = 30$ m/s
   Time interval (Δt) $= 40 - 10 = 30$ s

   $$\text{acceleration (a)} = \frac{\Delta v}{\Delta t} = \frac{30}{30} = 1\text{m}/\text{s}^2.$$

3. (a) $\Delta Q = \Delta E + \Delta W = 100 + 300 = 400$ J

4. (a) Energy stored per unit volume,

   $$E = \frac{1}{2} \times \text{stress} \times \text{strain}$$

   We know that,

   $$Y = \frac{\text{stress}}{\text{strain}} \text{ or strain} = \frac{\text{stress}}{Y}$$

   $$E = \frac{1}{2} \times \text{stress} \times \frac{\text{stress}}{Y} = \frac{1}{2} \cdot \frac{S^2}{Y}$$

5. (b) $x = \frac{1}{2} g t^2$

   $$= \frac{1}{2} \times 9.8 \times \frac{1}{2} \times \frac{1}{2} = \frac{9.8}{8} \text{ m}$$

6. (c) Control rods are made of cadmium

7. (d) $L \alpha n^2$, where n is number of turn per unit length.

8. (b) The internal resistance of a cell is the resistance of electrolyte used in cell because it is the electrolyte which is responsible for conduction of electricity inside the cell.

9. (a) $T_1 = m(g+a)$; $T_2 = m(g-a) \Rightarrow \frac{T_1}{T_2} = 3:1$

10. (b) $\frac{T_1 - T_2}{T_1}$ is maximum in case (b).

11. (c) Potential over each capacitor = 200 volt
Charge on each Capacitor
$= 25\times10^{-6}\times200 = 5\times10^{-3}$ coulomb

12. (d) For de-Broglie wavelength, $\lambda = \frac{h}{mv}$
So, if mv is doubled, value of λ will be halved.

13. (b) We know that $\tau = \frac{dL}{dt}$ [rate of change of momentum]
Thus when torque acts on a body its angular momentum changes.

14. (d) The minimum energy required for the emission of electrons is called work function

15. (a) We know that,
$Q = \Delta E$ + work done by gas
Work done by gas $= Q - \Delta E$
Work done by gas is maximum when ΔE is minimum. For isothermal change $\Delta E = 0$. So for isothermal expansion work done is maximum.

16. (d) $mg - T = ma$,

given, $T = \frac{2}{3}mg \Rightarrow \frac{1}{3}mg = ma \Rightarrow a = \frac{g}{3}$

17. (c) $\beta = \frac{\lambda' D'}{d'} = \frac{3\lambda 3D}{d/3} = 27\frac{\lambda D}{d}$.

No. of fringes $= \frac{1/3}{\beta} = \frac{d}{81\lambda D}$.

19. (b) Maximum capacity of currents in 40 W
$= \frac{40}{220} = \frac{2}{11} = 1.82\text{A}$
Maximum capacity of current in 100 W
$= \frac{100}{220} = \frac{5}{11} = 0.454\text{A}$
Resistances of bulbs are
$\frac{220\times220}{40} = 1210\Omega$ & $\frac{220\times220}{100} = 484\Omega$
Now when they are joined in series with 400 volt.
Current $= \frac{400}{1210+484} = 0.236$
As current 0.236 A is more than max. capacity of 40 W bulb so, this bulb will fuse.

20. (b) Work done at constant pressure
$= P(V_2 - V_1) = nR(T_2 - T_1)$
From $\frac{V_2}{V_1} = \frac{T_2}{T_1} \Rightarrow \frac{2V}{V} = \frac{T_2}{T_1}$
$\Rightarrow T_2 = 2\times T_1 = 2\times300 = 600$
Work done $= nR(T_2 - T_1)$
$= 0.1\times2\times(600-300) = 60$ cal

21. (b) $\mu = 1 = \tan\lambda \therefore \lambda = 45°$, angle of friction = angle of repose

22. (c) $\varepsilon = (5\times10^{-3})(1/0.1) = 0.05\,\text{V}$.

24. (d)

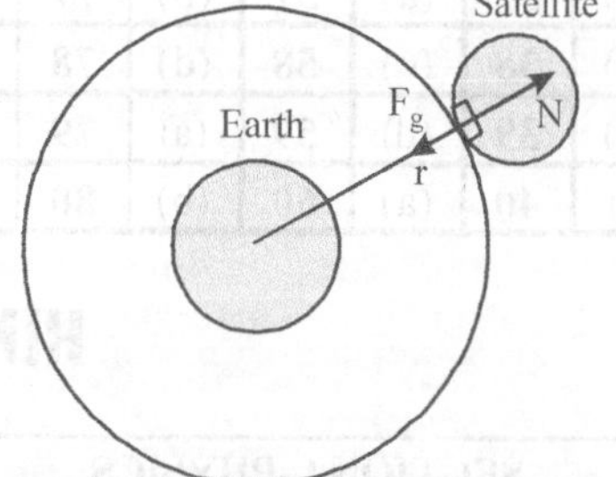

The weightlessness in a satellite is due to the fact that the normal reaction on a body on its surface is zero.

The forces acting on an object in the satellite are
(i) gravitational attraction due to earth,

$$F_g = \frac{GMm}{r^2}$$

(towards the earth)
(ii) normal reaction, N (away from the earth)
Also, the object has a centripetal acceleration a, given by

$$a = \frac{GM}{r^2}$$

By Newton's third law,

$$F_g - N = ma$$

$$\Rightarrow \frac{GMm}{r^2} - N = m\left(\frac{GM}{r^2}\right)$$

$$\Rightarrow N = 0$$

Therefore, the apparent weight of the object is zero.

25. (a) $\frac{I_1}{I_2} = \frac{9}{16}; \frac{I_1}{I_2} = \frac{A_1^2}{A_2^2}$

[A_1 & A_2 are amplitudes of waves]

$$\frac{A_1}{A_2} = \sqrt{\frac{9}{16}} = \frac{3}{4}$$

$$\frac{A_1 + A_2}{A_2 - A_1} = \frac{7}{1}$$

$$\frac{I_{max}}{I_{min}} = \frac{(A_1 + A_2)^2}{(A_2 - A_1)^2} = \left(\frac{7}{1}\right)^2 = \frac{49}{1}$$

$I_{max} : I_{min} = 49 : 1$

26. (b) The modulated voltage comprises of
(i) carrier wave of frequency ω_c
(ii) lower side frequency $(\omega_c - \omega_m)$ wave
(iii) upper side frequency $(\omega_c + \omega_m)$ wave
Thus in an AM wave information is contained in lower $(\omega_c - \omega_m)$ and upper $(\omega_c + \omega_m)$ side frequencies.

27. (c) The current in an ac circuit is said to be wattless, if the power loss in the circuit is zero, i.e.,
$P = VI\cos\theta = 0$
which gives
$\cos\theta = 0 \quad \Rightarrow \theta = 90°$

28. (d) Power supplied $= \frac{mgh}{t}$

Power used by crane $= \frac{mgh}{t} \times \frac{100}{80}$

$$= \frac{1000 \times 10 \times 100}{3600} \times \frac{100}{80} = \frac{10^5}{36 \times 8} \text{ W}$$

29. (d) Escape velocity

$$v_e = \sqrt{\frac{2GM}{R}}$$

$$v_p = \sqrt{\frac{2G(6M)}{2R}} = \sqrt{3}\sqrt{\frac{2GM}{R}}$$

$$v_p = \sqrt{3}v_e$$

30. (c) Touque (τ) on the triangle = M × B
$\tau = iAB$ (since M = iA)

$$\tau = i\frac{1}{2}\ell.\ell.\sin 60 \,.\, B = \frac{1}{2}\ell^2.\frac{\sqrt{3}}{2}B$$

$$\ell = \sqrt{\frac{4\tau}{\sqrt{3}Bi}} = 2\sqrt{\frac{\tau}{\sqrt{3}Bi}}$$

31. (d) Since v_{rms} is doubled by increasing the temp. so by $v_{rms} = \sqrt{\frac{3KT}{m}}$, the temp. increase by four times.

Now for constant pressure $\frac{V_1}{T_1} = \frac{V_2}{T_2}$

$V_1 = V, T_1 = T°K, T_2 = 4T°K, V_2 = ?$
$V_2 = 4V$

32. (a) Sky appears blue because of scattering of light when white light falls on the dust particle of air it is scattered back to the earth. Since, blue light scatters most, it enters our eyes after scattering .So sky appears blue.

33. (a) The resonant frequency

$$f_r = \frac{1}{2\pi\sqrt{LC}}$$

$$\Rightarrow 10^6 = \frac{1}{2\pi\sqrt{10 \times 10^{-3} \times C}}$$

$$\Rightarrow C = \frac{1}{4\pi^2 \times 10 \times 10^{-3} \times 10^{12}}$$

$= 2.5 \times 10^{-12}$F $= 2.5$ pF

34. (b) Change in momentum = F × t
= 10 × 10 = 100 Ns or 100 kg m/s

35. (a) $\Delta = \frac{F\ell}{AY}$

$$= \frac{6.28 \times 10^4 \times (1)}{\pi(0.01)^2 \times 2 \times 10^{11}} = \frac{1}{1000}$$

36. (c) Energy of simple harmonic motion $\propto a^2$

37. (a) In S.I. units, we have $B = \mu_0 (H + I)$

38. (c) Vertical distance covered by water before striking ground = (H – h). Time taken is,

$t = \sqrt{2(H-h)/g}$: Horizontal velocity of water coming out of hole at P, $u = \sqrt{2gh}$.

$\therefore$ Horizontal range

$= ut = \sqrt{2gh} \times \sqrt{2(H-h)/g}$

$= 2\sqrt{h(H-h)}$

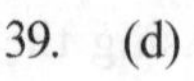

39. (d)

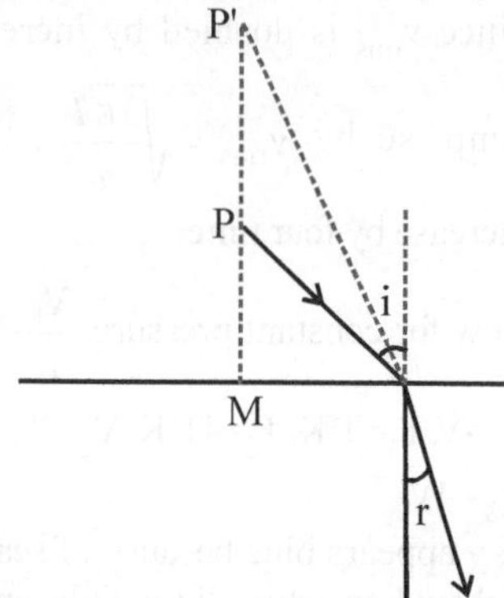

Fish inside water will see point P at P' as explained in the figure

Here $\frac{MP'}{MP} = \mu = \frac{4}{3}$

$\frac{MP'}{24} = \frac{4}{3} \Rightarrow MP' = 24 \times \frac{4}{3} = 32$ cm

above the surface of water.

41. (d) $\beta = \frac{I_c}{I_b} > 1$ or $I_c > I_b$

42. (b) Logic gates are fundamental units of digital system.

43. (b) The forbidden energy gap for germanium crystal is 0.71 eV.

44. (b) In intrinsic semi conductor, the number of holes are equal to number of electrons i.e., $n_p = n_e$

45. (b) Forward bais opposes the potential barrier and if the applied voltage is more than knee voltage it cancels the potential barrier

SECTION 2 - CHEMISTRY

46. (a) As we move down in groups 13, 14 and 15, inertness of s^2 electrons of valence shell increases (inert pair effect). Thus in metals present at the bottom of the groups 13, 14, and 15 lower oxidation state becomes more important , viz. Pb (+2 state rather +4), Bi (+3 state rather +5) and Ti (+1 rather than +3).

47 (b) Since each Sr^{++} ion provides one cation vacancy, hence
Concentration of cation vacancies
= mole % of $SrCl_2$ added = 10^{-4} mole %

$= \frac{10^{-4}}{100} \times 6.023 \times 10^{23} = 6.023 \times 10^{17}$

48. (c) $CH_3 - O - C_3H_7$ and $C_2H_5 - O - C_2H_5$ are metamers

49. (d) The sulphide ore is roasted to oxide before reduction because the ΔG_f^o of most of the sulphides are greater than those of CS_2 and H_2S, therefore neither C nor H can reduce metal sulphide to metal. Further, the standard free energies of formation of oxide are much less than those of SO_2. Hence oxidation of metal sulphides to metal oxide is thermodynamically favourable.

50. (b) By Freundlich adsorption isotherm :

$$\left(\frac{x}{m}\right) = k(p)^n$$

$$\log\left(\frac{x}{m}\right) = n\log p + \log k$$

Compare with $y = mx + c$

$\therefore$ slope = $\tan\theta = \tan 45° = 1$

$\therefore$ $n = 1$ and $\log k = 0.3010 = \log 2$

$$\log\left(\frac{x}{m}\right) = \log(0.2) + \log 2$$

$$\therefore \log\left(\frac{x}{m}\right) = \log 0.4$$

$$\therefore \left(\frac{x}{m}\right) = 0.4$$

51. (c) H_2SO_4 is a strong acid hence pH < 7 , or $[H]^+ > 10^{-7}$, the aqueous solution of NaCl is neutral hence pH ≈ 7 or $[H^+] \approx 10^{-7}$. The aqueous solution of $NaNO_2$ will be basic hence pH > 7 or $[H^+] < 10^{-7}$
H_2S is a weak acid. Therefore, the correct order will be
0.01 M $NaNO_2$ < 0.01 M NaCl < 0.01 M H_2S < 0.01 M H_2SO_4

53. (d) C_2H_5I and C_2H_5OH form non-ideal solution.

54. (c) Cu^{2+} forms complexes with the same magnetic moment and geometry irrespective of the nature of ligand. It can be explained by electronic configuration.

3d

Cu^{2+} [1L | 1L | 1L | 1L | 1]

It has 9 electrons in 3d shell. Hence any incoming ligand whether it is strong or weak will result in the formation of same geometry (if number of ligands is same) because any ligand can do nothing with this unpaired e^- in 3d shell. Moreover, the complex formed will have same magnetic moment due to this unpaired e^-.

55. (d) $\overset{sp^3}{CH_3}\,\overset{sp^2}{CH} = \overset{sp^2}{CH}\,\overset{sp^3}{CH_2} - \overset{sp^2}{CH} - \overset{sp}{C} \equiv \overset{sp}{CH}$

56. (c) $\Delta H = -2.5 \times 10^3$ cal

$\Delta S = 7.4$ cal

$\Delta G = \Delta H - T\Delta S$

$= -2.5 \times 10^3 - 298 \times 7.4 = -4705.2$

$\Delta G = -$ ive; so process is spontaneous

$\Delta G = 0$, Equilibrium

$\Delta G = +$ ive; Non-spontaneous

57. (b) $BeO < MgO < CaO < BaO$.

The basic character of the oxides increases down the group.

Among these BeO is amphoteric.

58. (d) 1 L of 1 M $CuSO_4$ = 1 mol of Cu^{2+}

$$Cu^{2+} + 2e^- \longrightarrow Cu$$

2 F of electricity will convert dil Cu^{2+} into Cu, hence $[CuSO_4] = 0$

Thus, (d)

59. (a) (Na_2O_2 + HCl) is commercially known as **oxone** and is used for bleaching of delicate fibres.

60. (b) $M(OH)_x \rightleftharpoons \underset{S}{M^{2+}} + \underset{xS}{xOH^-}$

$(S)(xS)^x = K_{sp}$

$x^x.S^{x+1} = 4 \times 10^{-12}$

$x^x (10^{-4})^{x+4} = 4 \times 10^{-12}$

x = 2

61. (a) On increasing temperature adsorption of a gas on surface of solid decreases. Solid adsorb greater amount of substances at lower temperature.

62. (b) In the structure of ice each molecule of H_2O is surrounded by three H_2O molecules in hexagonal honey comb manner. On the other hand in water, each molecule is surrounded by four neighbouring molecules randomly which results an open cage like structure. As a result there are a number of 'hole' or open spaces. In such a structure lesser number of molecules are packed per ml. When ice melts a large no. of hydrogen bonds are broken. The molecules therefore move into the holes or open spaces and come closer to each other than they were in solid state. This result sharp increase in the density. Therefore ice has lower density than water.

63. (d) The boiling point of o-nitrophenol is less than para-nitrophenol due to presence of intramolecular hydrogen bonding. Since p-nitrophenol is less volatile in than o-nitrophenol due to presence of inter molecular hydrogen bonding hence they can be separated by steam distillation.

64. (a) $R-\overset{O}{\overset{||}{C}}-R' \xrightarrow[KCN]{HCN} R-\underset{R'(A)}{\overset{OH}{C}}-CN$

$\xrightarrow[LiAlH_4(B)]{\text{Reduction by}} R-\underset{R'}{\overset{OH}{C}}-CH_2NH_2$

65. (b) The species having minimum value of K_{sp} will get precipitated first of all because ionic product will exceed the solubility product of such a species.

The K_{sp} value is minimum for $BaSO_4 (10^{-11})$, so, $BaSO_4$ will get precipitated first of all.

66. (b) The ionic radii of trivalent lanthanides decreases progressively with increase in atomic number. This decrease is known as lanthanide contraction.

67. (d) Number of atoms per unit cell = 1

Atoms touch each other along edges.

Hence $r = \frac{a}{2}$

(r = radius of atom and a = edge length)

Therefore % fraction $= \frac{\frac{4}{3}\pi r^3}{(2r)^3} = \frac{\pi}{6} = 0.52$

68. (a) Chloro diaquatriammine cobalt (III) chloride is $[CoCl(NH_3)_3(H_2O)_2]Cl_2$

69. (b) Organometallic compounds are those compounds in which there is a bond which involve metal. In chlorophyll there is bond involving metal Mg

70. (d) Reaction (II) and reverse of reaction (I) gives the desired reaction hence

$$K = K_2 \times \frac{1}{K_1} = \frac{K_2}{K_1}$$

71. (a) $2R-Cl + Ag_2O \xrightarrow[\text{ether}]{} R-O-R + 2AgCl\downarrow$

72. (b) $C_2H_5MgBr + H_2C - CH_2$ (O bridging, epoxide)

$$C_2H_5CH_2-\underset{\text{OMgBr}}{CH_2} \xrightarrow{H_2O} \underset{\text{n-butanol}}{C_2H_5CH_2CH_2OH}$$

73. (a) Standard electrode potential i.e. reduction potential of A is minimum (–3.05V) i.e. its oxidation potential is maximum which implies 'A' is maximum reactive chemically.

74. (d) $Na_2CO_3 + Ca(OH)_2 \rightarrow 2NaOH + CaCO_3$

75. (d) Benzyl chloride ($C_6H_5CH_2Cl$) is **more** reactive than vinyl chloride ($CH_2=CHCl$)

76. (b) Since the gas expands adiabatically (i.e., no change in enthalpy) so the heat is totally converted into work.

For the gas, $C_V = 20$ J/K. Thus, 20 J of heat is required for 1° change in temperature of the gas.

Heat change involved during the process i.e., work done = 3 kJ = 3000 J.

Change in temperature $= \frac{3000}{20}$ K = 150 K

Initial temperature = 300 K

Since, the gas expands so the temperature decreases and thus final temperature is

300 – 150 = 150

77. (d) Chemically borax is sodium tetraborate decahydrate and it has the formula $Na_2B_4O_7.10H_2O$.

78. (a) The acidity of acetylene or 1--alkynes can be explained on the basis of molecular orbital concept according to which formation of C—H bond in acetylene involves sp-hybridised carbon atom. Now since s electrons are closer to the nucleus than p electrons, the electrons present in a bond having more s character will be correspondingly more closer to the nucleus. Thus owing to high s character of the C—H bond in alkynes (s = 50%), the electrons constituting this bond are more strongly held by the carbon nucleus (i.e., the acetylenic carbon atom or the sp orbital acts as more electronegative species than the sp^2 and sp^3 with the result the hydrogen present on such a carbon atom (≡C—H) can be esily removed as a proton.

79. (a) $\Delta H° = \Delta H_f°(\text{products}) - \Delta H_f°(\text{reactants})$

$$\Delta H° = \Delta H_f°(HS) + \Delta H_f°(H) - \Delta H_f°(H_2S)$$

$$x_1 = x(?) + x_3 - x_2$$

$$\therefore \quad x = (x_1 + x_2 - x_3)$$

Thus, (a)

80. (c) $K_{sp} = [Ag^+][Cl^-]$

$1.8 \times 10^{-10} = [Ag^+][0.1]$

$[Ag^+] = 1.8 \times 10^{-9}$ M

$K_{sp} = [Pb^{+2}][Cl^-]^2$

$1.7 \times 10^{-5} = [Pb^{+2}][0.1]^2$

$[Pb^{+2}] = 1.7 \times 10^{-3}$ M

82. (d) Be shows diagonal relationship with Al

83. (c)

	O^-	O	F^-	Na^+
Electrons	9	8	10	10
Nuclear charge	8	8	9	11

As greater the effective nuclear charge, more is the attraction of nucleus towards electrons hence higher will be the electron affinity. Thus, Na^+ has higher E.A. than F^- and other two. Similarly, F^- has higher E.A than O^- and O. Among O^- and O, O^- has lower E.A. value than O due to its greater size. As greater the atomic radius of the atom less will be the attraction of the nucleus to electron added and hence, lower will be the value of E.A. Thus, the correct order of electron affinity is

$$O^- < O < F^- < Na^+$$

84. (c) $2Cl(g) \longrightarrow Cl_2(g)$

Entropy is decreasing (–ve) in the reaction. Further the reaction is exothermic since a bond is being formed, i.e., ΔH is also –ve.

86. (c) $CH_3Cl + AlCl_3 \longrightarrow \underset{\text{Electrophile}}{CH_3^+} + AlCl_4^-$

87. (c) Reactions which involve absorption of heat energy are called endothermic reactions. For such reactions.

$$\Sigma H_P > \Sigma H_R$$

In graphs (*c*) and (*d*), the heat of products is more than heat of reactants and thus they represent endothermic reactions. But in (*d*) only small amount of energy is absorbed (less difference between energy of reactants and products). Thus, (*c*) represents maximum activation energy.

Note : Activation energy is the excess energy that the reactant molecule must possess to cross energy barrier.

88. (a) Mg, Ca, Sr, Ba. As the size increases, the ionisation potential decreases.

89. (a) Addition of monomers follows isoprene rule

$$\begin{matrix} CH_3 \\ CH_3 \end{matrix} \!\!> C = CH_2 \quad \begin{matrix} CH_3 \\ CH_3 \end{matrix} \!\!> C = CH_2$$

H T H T

$$\downarrow$$

$$\overset{\oplus}{C}(CH_3)_2-CH_2-C(CH_3)_2-CH_2 \sim\sim$$

90. (d) At cathodes : $Fe^{+} + 2e^{-} \longrightarrow Fe$;

$Fe^{3+} + 3e^{-} \longrightarrow Fe$

$(E_{Fe})_1 = \frac{\text{At. wt.}}{2}$; $(E_{Fe})_2 = \frac{\text{At. wt.}}{3}$

Hence, $\frac{(E_{Fe})_1}{(E_{Fe})_2} = \frac{3}{2}$

SECTION 3 - BOTANY

91. (b) There are several factors and processes which differentiate living beings with non-living beings like reproduction, respiration, growth, etc. But among them reproduction is the only difference which differentiate without any exception living being with non-living beings.

92. (a) *Rafflesia* or Corpse flower is a total root parasite. It obtains the total nourishment from the root of the host plant. Its flowers only come out of the host plant and they have a diameter of 1 m and weight around 10 kg.

94. (b) Algae *Spirulina* is one of the richest sources of protein, containing 40–50 percent crude protein on dry weight basis which under favourable condition may reach upto 70 percent. So, the people recovering from long illness are advised to take it in the diet.

95. (a) Single cotyledon of embryo in cereal grain is represented by scutellum. Coleoptile represent the covering of stem. Coleorrhiza represents the covering of root.

96. (d) Tracheids are chief water conducting elements of xylem in gymnosperms. They are devoid of protoplasm and hence dead. The wall constituting the tracheids is hard, thick and lignified. These are elongated cells with tapering ends.

98. (b) Mitosis is the process in which eukaryotic cell separates the chromosomes in its cell nucleus, into two identical sets in two daughter nuclei. It is generally followed immediately by cytokinesis, which divides the nuclei, cytoplasm, organelles and cell membrane into two daughter cells containing roughly equal shares of these cellular components. Mitosis and cytokinesis together define the mitotic (M) phase of the cell cycle - the division of the mother cell into two daughter cells, genetically identical to each other and to their parent cell.

99. (d) In bivalent formation of chromosomes during meiosis, the homologous chromosomes are arranged in pairs. The phenomenon is called synapsis and it occurs during zygotene stage. DNA replication occurs during S phase or synthetic phase which is the second phase of interphase.

102. (d) Each pollen has two layered wall. The outer layer is thick, tough, cuticularised called exine which is composed of a material called "sporopollenin". It is highly resistant to biological and physical decomposition, due to which pollens are preserved for a long time in fossils.

103. (d) Silicosis is caused by exposure to dust containing silica.

104. (b) Cloning is a technique by which genetically same individuals can be produced without including any sexual reproduction e.g. Dolly sheep.

105. (a) Polyembryony refers to the formation of more than one embryo within a seed of a flowering plant. *e.g.* Citrus.

106. (c) Desert show maximum difference between day and night temperature.

107. (a) Population is an aggregation or grouping of individuals of the same species at the same time in a particular area. Species is a grouping of individuals of one or more populations resembling each other in important morphological, anatomical and biochemical characters and can potentially interbreed. Biotic community is an assemblage of interdependent, and ecosystem is a self sufficient and self regulating segment of nature comprising of a biotic community and its physical environment, both interacting and exchanging materials. The basic level of ecological organisation starts with the individual, the next more complex levels are populations, species, community and then ecosystem.

110. (d) A solution having more solutes as compared to solution having less number of solutes. Former is known as hypertonic and later hypotonic. Hence, solution out side the cytoplasm has more solutes (higher concentration) and are hypertonic. If two solutions have same solute concentration then they are isotonic.

111. (b) The mitochondrial inner membrane forms infoldings known as cristae, which allow greater surface area for protein such as cytochrome to function properly and efficiently.

112. (b) The essential elements are divided into macroelements and microelements based on the quantity in which they are required by the plants. Mn, Fe, Cu, Mo, Zn, B, and Cl are the micronutrients needed in very small quantities by the plants. C, H, O, N, P, S, K, Ca, Mg, Fe are the macronutrients required in more quantity.

113. (a) During sodium-potassium pump, the concentration of sodium ions will be about 14 times more in extra cellular fluid (outside) and concentration of potassium ions will be about 28-30 times more in axoplasm (inside). Thus, $3Na^+$ and $2K^+$ are transported during the process.

115. (b) In eukaryotic cells, two cytoplasmic organelles, mitochondria and chloroplast of green plants, contain their own genetic materials.

116. (a) Gibberllic acid stimulates cell growth of leaves and stem causing their expansion, elongation respectively and leaf fall is controlled by ABA.

117. (b) A cross of F_1 hybrid with its recessive homozygous parent is called the test cross. It is done to determine the genotype of a given plant. If the given plant has homozygous dominant traits then on test cross it gives all dominant trait plants but if it is heterozygous dominant than it gives dominant and recessive phenotypes in 1 : 1 ratio.

118. (c) The term "hot spots" was used by Benzer for the sites which are more mutable than other sites. Studies in 1978 revealed that 5-methylcytosine residues occur at the position of each hot spot.

119. (b) Pleiotropy is the condition in which a single gene influences more than one trait.
Polyploidy is a condition in which individuals have more than two complete sets of chromosomes.
Apomixis is a reproductive process in plants that superficially resembles normal sexual reproduction but there is no fusion of gametes.

120. (c) Some plants growing in salty marshes develop special roots for respiration, called pneumatophores. They are (–)vely geotropic.

121. (c) An RNA molecule whose base sequence is complementary to that of the RNA transcript of a gene, *i.e.* the sense RNA, such as a messenger RNA (m RNA). Hence, an antisense RNA can undergo base pairing with its complementary mRNA sequence. This blocks gene expression, either by preventing access for ribosome to translate the mRNA or by triggering degradation of the double stranded RNA by ribonuclease enzymes. Antisense RNA and DNA both have therapeutic potential for modifying the activity of disease causing genes.

122. (d) Enzymes accelerate chemical reaction by lowering the activation energy.

123. (c) Competitive inhibition is the reversible inhibition of enzyme activity due to presence of substrate analogues. Non competitive inhibition is the reduction of enzyme activity by a factor that has no real structural similarity with the substrate. Allosteric inhibition is reversible noncompetitive inhibition occurring in case of allosteric enzymes . Herein the inhibitors are the products or intermediates of reactions catalyzed by the enzymes. Hence it is also called end product inhibition or feedback inhibition.

127. (d) Lecithin is a fat like substance called a phospholipid, which is a part of plasma membrane.

128. (b) In this technique nematode specific genes are introduced in the host plant in such a way that it produces both sense and antisense RNA. The two RNA's being complementary to each other form a double stranded RNA (dsRNA) which is also called interfering RNA responsible for initiating RNA interference (RNA i). This (dsRNA) bind to and prevent translation of specific mRNA of nematode (gene silencing). Thus transgenic plants based on RNAi technology are resistant to nematode.

129. (c) During the process of replication the enzyme that helps to cut one strand of DNA duplex to release the tension of coiling of two strands is topo-isomerase.

130. (c) A fed batch is a biotechnological batch process which is based on feeding of a growth limiting nutrient substrate to culture. It is done for purifying enzymes.

131. (c) Gibberellins induces elongation of internodes.

133. (a) RNA interference technique, sense & antisense RNA fused to form dsRNA that silent the expression of m- RNA of nematode. RNA interference is a novel strategy adopted to prevent infestation of nematode *Meloidegyne incognitia* in roots of tobacco plants.

SECTION 4-ZOOLOGY

136. (b) While working at the Rockefeller Institute, Brown reported isolation of a PPLO from human arthritic joint tissue in 1938.
In discussing the significance of this observation, Brown reported successful treatment of arthritic patients in 1949 with a new antibiotic called aureomycin.

143. (a) When water is not available, the camels produce dry faeces and concentrated urine. One of the best sources of metabolic water in camels is oxidation of food in the hump.

146. (a) Gamete intrafallopian transfer (GIFT) is recommended for those females who cannot produce an ovum. In this process,

the eggs of the donor woman are removed and in a form of mixture with sperm transferred into fallopian tube of another woman who cannot produce ovum, but can provide suitable environment for fertilization. Thus in GIFT, site of fertilization is fallopian tube, not laboratory.

148. (c) Sertoli cells are found in the walls of seminiferous tubules of the testes. They anchor and provide nutrition to the developing germ cells especially the spermatids.

150. (a) Menopause is the period when ovulation and menstrual cycle stop in human females. The period of menopause is between 45-55 years.

154. (a) Fat is mainly digested by pancreatic lipase while protein is digested by enzymes in pancreatic juice, intestinal juice and stomach.

155. (a) Oxyntic cells or Parietal cells, are the stomach epithelium cells that secrete gastric acid intrinsic factor. These cells secrete hydrochloric acid (HCl) which makes the gastric juice acidic (pH = 2.0-3.0). Alpha cells of islets of Langerhans secretes glucagon hormone which increase the glucose level in the blood by converting glycogen to glucose in liver cells. Kupffer's cells are specialized cells in the liver that destroy bacteria, foreign proteins, and worn-out blood cells. Sebaceous glands and microscopic glands in the skin that secrete an oily/waxy matter (called sebum) to lubricate the skin and hair of mammals.

156. (a) 70% to 75% CO_2 is transported as primary buffer of the blood. Bicarbonate ion (HCO_3) in blood plasma. When CO_2 diffuses from tissues into blood then it is acted upon by the enzyme carbonic anhydrase.

157. (b) 'Bundle of His' are a typical cardiac muscle fibres, connecting the atria with ventricle.

159. (d) A barr body is one of the X-chromosomes in somatic cells of females. If there are more X-chromosomes, then there will be more barr bodies.

160. (c) Genetic drift is also known as Sewall Wright effect. It refers to the random changes in gene frequencies in a population by chance, either due to intensive inbreeding or death of a small section of population by natural calamity. Its effect is more marked in small isolated population.

161. (b) Gause's hypothesis states that two different competing species cannot coexist together for a long in the same niche, one has to get eliminated by the other competing species, if they do coexist, then they do so as a result of niche differentiation, *i.e.* differentiation of then realized niches.

162. (d) Karl Von Frisch, carried out many experiments and determined that when a foraging bee returns to the hive, it performs a waggle dance. Honey is predominantly glucose and fructose. *Dorsata* is a bigger bee than *Apis indica* (a medium sized bee). Bees wax is secreted by special wax glands to make compartments.

163. (c) Baldness is a sex influenced trait. The dominance of alleles may differ in heterozygotes of the two sexes.

164. (d) Biometric genetics is the mathematical or statistical study of genetic phenomenon. In this branch, data of various genetic traits are analysed by applying the principles of statistics. It helps in the investigation of various genetic principles and checks their correctness and probability.

166. (d) Molecule (a) represents structure of morphine. Morphine is the most abundant alkaloid found in opium, the dried sap (latex) derived from shallowly slicing the unripe seedpods of the opium, or common and/or edible, poppy *Papaver somniferum*. Morphine is a potent opiate analgesic drug that is used to relieve severe pain.

167. (d) Electroencephalogram is the recording of electric potentials originating from different parts of the brain in the form of waves. Fractionation is the technique of rupturing cells and separating their components especially cell organelles for studying their chemistry and functions. Electrophoresis is the technique of separation of charged solutes on the basis of their differential migration in an applied electric field.

168. (a) Actin and tropomyosin are part of thin filaments of skeletal muscle. Tubulin is presents in microtubules. Myosin is muscle protein.

171. (a) Glucagon is secreted by δ-cells of the islets of Langerhans in the pancreas. It increases the concentration of glucose in the blood by stimulating the breakdown of glycogen.

172. (b) Parathyroid gland secretes parathormone which regulates Ca^{2+} level in blood. Anterior lobe of pituitary secretes the following hormones growth hormone, ACTH, TSH, FSH, LH. Posterior pituitary secretes vasopressin and oxytocin.

The intermediate lobe (pars intermedium) of the pitutary gland secretes MSH which causes dispersal of pigment granules in the pigment cells which give colour to the skin.

173. (b) Nicotine and acetylcholine have the same receptors-Nicotinic cholinergic receptors and hence have the same action.

176. (a) PCR is now a common and often indispensable technique, developed in 1984 by Kary Mullis, used in medical and biological research labs for a variety of applications. These include DNA cloning for sequencing, DNA-based phylogeny, or functional analysis of genes; the diagnosis of hereditary diseases; the identification of genetic fingerprints (used in forensic sciences and paternity testing); and the detection and diagnosis of infectious diseases. In 1993 Mullis won the Nobel Prize in Chemistry for his work on PCR.

177. (c) Whenever our body gets attacked by some foreign invaders, our body's immune system produces some chemicals to kill or to react against the invader. These chemicals are actually made up of carbohydrates & proteins *i.e.* glycoproteins called antibody.

178. (c) Tetanus toxoid is a vaccine consisting of growth products of *Clostridium tetani* treated with formaldehyde serving as an active immunising agent. Hence it is weakened germs. The first tetanus toxoid (inactivated toxin) was produced in 1924 and was used successfully to prevent tetanus in the armed services during World War II. In the mid-1940s, tetanus vaccine was combined with diphtheria toxoid and inactivated pertussis vaccine to make the combination DTP vaccine for routine childhood immunization.

180. (d) In the graph, the line A represents regulator, line B represents conformer and line C represents partial regulator. Organism that are able to maintain homeostasis by physiological means that ensures constant body temperature are called regulators. Organism that are not able to maintain a constant internal temperature are called conformers.

Partial regulators are organisms that have the ability to regulate, but only over a limited range of environmental conditions, beyond which they simply conform.

NEET (NATIONAL ELIGIBILITY ENTRANCE TEST) Full Test - 7

ANSWER KEY

1	(a)	21	(b)	41	(b)	61	(c)	81	(c)	101	(c)	121	(a)	141	(b)	161	(a)
2	(b)	22	(a)	42	(a)	62	(d)	82	(c)	102	(b)	122	(a)	142	(c)	162	(b)
3	(a)	23	(c)	43	(d)	63	(b)	83	(c)	103	(c)	123	(d)	143	(a)	163	(d)
4	(a)	24	(c)	44	(a)	64	(b)	84	(d)	104	(b)	124	(c)	144	(c)	164	(b)
5	(c)	25	(a)	45	(b)	65	(a)	85	(c)	105	(c)	125	(d)	145	(b)	165	(b)
6	(c)	26	(a)	46	(a)	66	(c)	86	(c)	106	(b)	126	(a)	146	(b)	166	(b)
7	(a)	27	(a)	47	(b)	67	(c)	87	(b)	107	(d)	127	(c)	147	(d)	167	(b)
8	(a)	28	(d)	48	(b)	68	(c)	88	(c)	108	(c)	128	(d)	148	(d)	168	(d)
9	(a)	29	(c)	49	(b)	69	(c)	89	(a)	109	(b)	129	(b)	149	(a)	169	(b)
10	(b)	30	(c)	50	(a)	70	(d)	90	(d)	110	(c)	130	(a)	150	(b)	170	(d)
11	(c)	31	(c)	51	(c)	71	(c)	91	(b)	111	(b)	131	(b)	151	(a)	171	(d)
12	(c)	32	(b)	52	(a)	72	(b)	92	(c)	112	(d)	132	(b)	152	(d)	172	(b)
13	(b)	33	(d)	53	(d)	73	(a)	93	(b)	113	(d)	133	(b)	153	(a)	173	(a)
14	(c)	34	(a)	54	(b)	74	(a)	94	(d)	114	(c)	134	(c)	154	(b)	174	(d)
15	(b)	35	(b)	55	(c)	75	(b)	95	(a)	115	(d)	135	(d)	155	(a)	175	(d)
16	(a)	36	(c)	56	(a)	76	(b)	96	(d)	116	(b)	136	(d)	156	(c)	176	(c)
17	(a)	37	(a)	57	(d)	77	(d)	97	(c)	117	(d)	137	(c)	157	(a)	177	(d)
18	(d)	38	(c)	58	(a)	78	(d)	98	(d)	118	(a)	138	(a)	158	(d)	178	(b)
19	(d)	39	(a)	59	(a)	79	(a)	99	(a)	119	(b)	139	(c)	159	(b)	179	(a)
20	(a)	40	(c)	60	(b)	80	(c)	100	(b)	120	(a)	140	(d)	160	(a)	180	(b)

HINTS & SOLUTIONS

SECTION 1 - PHYSICS

1. (a) $W = \vec{F}\cdot\vec{s} = Fs\cos\theta$

$$= [MLT^{-2}][L] = [ML^2T^{-2}];$$

$$\vec{\tau} = \vec{r}\times\vec{F} \Rightarrow \tau = rF\sin\theta$$

$$= [L][MLT^{-2}] = [ML^2T^{-2}]$$

2. (b) For first body, $h_1 = \frac{1}{2}g(3)^2$

For second body, $h_2 = \frac{1}{2}g(2)^2$

3. (a) $\vec{v} = \begin{vmatrix} \hat{i} & \hat{j} & \hat{k} \\ 1 & -2 & 2 \\ 0 & 4 & -3 \end{vmatrix}$

$$\vec{v} = \hat{i}[6-8] + \hat{j}[0+3] + \hat{k}[4-0]$$

$$\vec{v} = -2\hat{i} + 3\hat{j} + 4\hat{k} \Rightarrow |\vec{v}| = \sqrt{29} \text{ units}.$$

4. (a) Let v_1 and v_2 be the velocities of the body just before the collision & and after the collision.

$$v_1^2 = \sqrt{2gh} = \sqrt{2\times10\times10}; v_1 = \sqrt{2}\times10$$

$$v_2^2 = \sqrt{2gh_1} = \sqrt{2\times10\times2.5}; v_2 = \sqrt{2}\times5$$

$$\therefore \frac{v_1}{v_2} = \frac{10\sqrt{2}}{5\sqrt{2}} = \frac{2}{1}$$

5. (c) Initial angular velocity $= 2\pi n$

$$= \frac{2\pi\times600}{60} = 20\pi/\text{sec}$$

final angular velocity

$$= \frac{2\pi\times1200}{60} = 40\pi/\text{sec}$$

change in angular velocity

$$= 40\pi - 20\pi = 20\pi \text{ rad/sec}$$

6. (c) The time taken to reach the ground depends on the height from which the projectile is fired horizontally. Here height is same for both the bullets and hence they will reach the ground simultaneously.

7. (a)

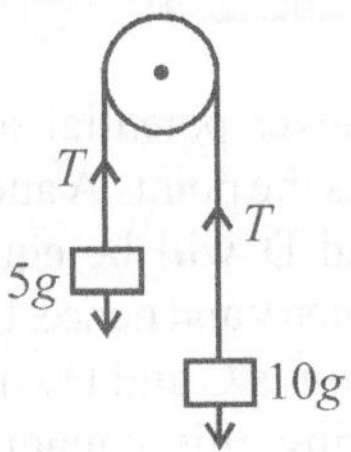

Let T be the tension in the string.

$\therefore 10g - T = 10a$ ....(i)

$T - 5g = 5a$ ....(ii)

Adding (i) and (ii),

$5g = 15a \Rightarrow a = \frac{g}{3}$ m/s^2

8. (a) Reading of the scale
= Apparent wt. of the mass = $m(g + a)$
= 80 (10 + 5) = 1200 N

9. (a) There is rise in temperature so, mechanical energy is converted into heat energy.

10. (b) Power = F.v. = m.a.v
= ma.at [v = u + at, v = at (u = 0)]
= ma^{2}t
As a and m are constant hence, Power α t

11. (c) Loss of potential energy = mgh

Gain of kinetic energy = $\frac{1}{2}I\omega^2 + \frac{1}{2}mv^2$

$= \frac{1}{2}I\omega^2 + \frac{1}{2}m\omega^2r^2 = \omega^2\left[\frac{mr^2 + I}{2}\right]$

Now, $\omega^2\left[\frac{mr^2 + I}{2}\right] = mgh$

$\omega = \left[\frac{2mgh}{I + mr^2}\right]^{1/2}$

12. (c) $n = \frac{1}{2}$

$\omega = 2\pi n = 2\pi \times \frac{1}{2} = \pi$ radian/sec

Centripetal acceleration $= \omega^2 r$

$= \pi \times \pi \times 100 = 986$ cm/sec^2

13. (b) According to Kepler's law

$\frac{T_1^2}{T_2^2} = \frac{R_1^3}{R_2^3}$

Here $T_1 = 365$ days ; $T_2 = ?$

$R_1 = R\,;\ R_2 = \frac{R}{2}$

$\Rightarrow T_2 = T_1\left(\frac{R_2}{R_1}\right)^{3/2}$

$= 365\left[\frac{R/2}{R}\right]^{3/2} = 129$ days

14. (c) $\frac{(v_e)_p}{(v_e)_e} = \frac{\sqrt{\frac{2GM_p}{R_p}}}{\sqrt{\frac{2GM_e}{R_e}}} = \sqrt{\frac{M_p}{M_e} \times \frac{R_e}{R_p}}$

$= \sqrt{\frac{10M_e}{M_e} \times \frac{R_e}{R_e/10}} = 10$

$\therefore (v_e)_p = 10 \times (v_e)_e = 10 \times 11 = 110\ km/s$

15. (b) The wire starts behaving like a liquid at point *b*. It behaves like a viscous liquid in the region *bc* of the graph.

16 (a) Work done = increase in surface area × surface tension
$= 2 \times 4\pi \times (.2)^2 \times .06$
$= 4\pi \times 4 \times 10^{-2} \times 6 \times 10^{-2}$ J
$= 192\pi \times 10^{-4}$ J [Soap bubble has two surfaces]

17. (a) $mL + m(100 - 80)$
$= 1.1 \times 1 \times (80 - 15) + 0.02 \times (80 - 15)$
$m \times 540 + 20m = 71.5 + 1.30$
$560\,m = 72.80 \quad \therefore \quad m = 0.130$

18. (d) Thermal capacity = ms = 40 × 0.2 = 8 cal/°C
= 4.2 × 8 = 33.6 joules/°C

19. (d) $C_P = \frac{7}{2}R;\ C_V = C_P - R = \frac{7}{2}R - R = \frac{5}{2}R$

$\frac{C_P}{C_V} = \frac{7/2\,R}{5/2\,R} = \frac{7}{5}$

20. (a) In an isochoric process volume remains constant whereas pressure remains constant in isobaric process.

21. (b) According to the law of equipartition of energy, the gas possesses equal energies in all the three direction x, y and z-axis.

22. (a) From the figure, it is clear that loss of height = AB = OB – OA

$= \ell - \ell\cos\theta = \ell(1-\cos\theta)$

Now, loss of P.E. = mgh = $mg\ell(1-\cos\theta)$

Gain of K.E. = $\frac{1}{2}mv^2$

$\frac{1}{2}mv^2 = mg\ell(1-\cos\theta)$

23. (c) In the first case tuning fork and string are in resonance so they have same frequency.

So, $392 = \frac{1}{2\ell}\sqrt{\frac{T}{m}} = \frac{1}{2\times 50}\sqrt{\frac{T}{m}}$.

For the second case, $n = \frac{1}{2\times 50\times 0.98}\sqrt{\frac{T}{m}}$

$\frac{n}{392} = \frac{100}{98}$; $n = 392\times\frac{100}{98} = 400$

Beats heard per sec = diff.of frequencies = 400 – 392 = 8.

24. (c) Field between two parallel sheet

$= \frac{\sigma}{\epsilon_0} = \frac{26.4\times 10^{-12}}{8.85\times 10^{-12}} = 3\ N/C$

25. (a) An electric dipole placed in a non-uniform electric field experiences a torque and a net force. In a uniform field it experiences only torque.

26. (a) Equipotential surface is always perpendicular to the direction of electric field. As the field is along x-direction, equipotential surface must be parallel to yz-plane.

27. (a) The given capacitor network an be rearranged, as shown,

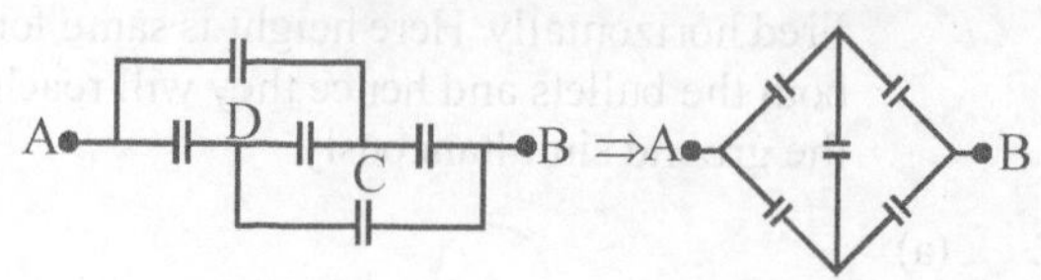

Whatever potential difference is applied across the points A and B, the potentials of C and D will be equal by the virtue of summetry and hence, the capacitor between the points C and D will never be charged. Leaving this capacitor, the remaining network consists of the two branches in parallel, each consisting of the two capacitors in series. Therefore, the equivalent capacitance between the points A and B is

$$C_{eq} = \frac{(C\times C)}{(C+C)} + \frac{(C\times C)}{(C+C)} = C$$

28. (d) Copper is a metal whereas Germanium is Semi-conductor.
Resistance of metal decreases and semiconductor increases with decrease in temperature.

29. (c) For various combinations equivalent resistance is maximum between P and Q.

30. (c) To convert a galvanometer into a voltmeter we connect a high resistance in series with the galvanometer.
The same procedure needs to be done if ammeter is to be used as a voltmeter.

31. (c)

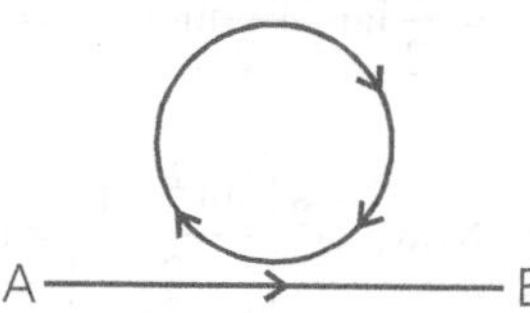

Direction of magnetic field due to indicated current will be in upward direction passing through the coil. So, a current in coil will be induced so that it decreases the flux in the coil. So, current in the clockwise direction will be induced.

32. (b) Fields created at the centre
$= 0.5\times 10^{-5}$ Wb/m^2
Due to current in circular loop, field created in the centre $= \frac{\mu_0 i}{2r}$

$$= \frac{4\pi \times 10^{-7} \times i}{2 \times 5 \times 10^{-2}} = 0.5 \times 10^{-5}$$

$$i = \frac{0.5 \times 10^{-5} \times 10 \times 10^{-2}}{4\pi \times 10^{-7}} = \frac{5}{4\pi} = \frac{35}{88}$$

$= 0.3980 = 0.4$ A

33. (d) A transformer cannot step up a d.c. input so output potential here will be zero. No potential will be induced in the secondary coil.

34. (a) $E_{rms} = 220$ V

$$\therefore E_{rms} = \frac{E_0}{\sqrt{2}} \Rightarrow E_0 = \sqrt{2}\, E_{rms}$$

Average e.m.f over half cycle

$$= \frac{2}{\pi} E_0 = 0.637 \times 1.41 \times 220 = 198.15 \text{ V}$$

35. (b) The magnetic field at a point, distant r from the straight wire conductor carrying current I is given by,

$$B = \frac{\mu_0}{4\pi} \cdot \frac{21}{r}$$

$$B \propto \frac{1}{r}$$

i.e., graph is rectangular hyperbolic.

36. (c) Radio waves are low frequency electro-magnetic waves. So, their wavelength will be comparatively large.

37. (a) Let a ray of light enter at A and the refracted beam is AB. This is incident at an angle θ. For no refraction at the lateral face, $\theta > C$ or, $\sin\theta > \sin C$ But $\theta + r = 90°$ $\Rightarrow$ $\theta = (90° - r)$

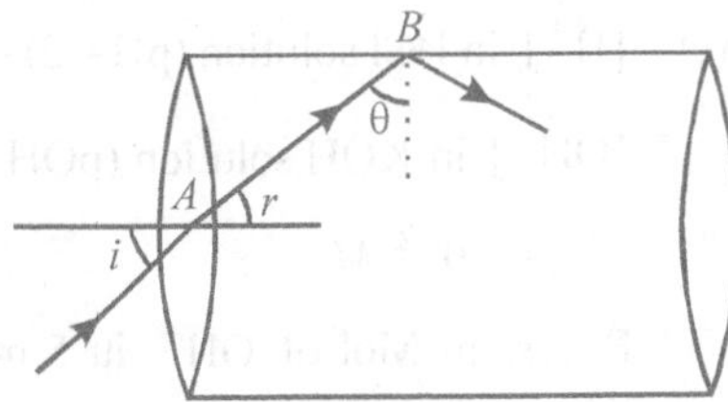

$\therefore$ $\sin(90° - r) > \sin C$
or $\cos r > \sin C$ ...(1)

From Snell's law, $n = \dfrac{\sin i}{\sin r} \Rightarrow \sin r = \dfrac{\sin i}{n}$

$$\therefore \cos r = \sqrt{1 - \sin^2 r} = \sqrt{\left(1 - \frac{\sin^2 i}{n^2}\right)}$$

$\therefore$ equation (1) gives

$$\sqrt{1 - \frac{\sin^2 i}{n^2}} > \sin C \Rightarrow 1 - \frac{\sin^2 i}{n^2} > \sin^2 C$$

Also, $\sin C = \dfrac{1}{n}$

$$\therefore 1 - \frac{\sin^2 i}{n^2} > \frac{1}{n^2} \text{ or } 1 > \frac{\sin^2 i}{n^2} + \frac{1}{n^2}$$

$$\text{or } \frac{1}{n^2}(\sin^2 i + 1) < 1 \text{ or } n^2 > (\sin^2 i + 1)$$

Maximum value of $\sin i = 1$

$$\therefore n^2 > 2 \Rightarrow n > \sqrt{2}$$

38. (c) The condition for an achromatic doublet is

$$\frac{\omega_1}{f_1} + \frac{\omega_2}{f_2} = 0$$

This gives the combination of two lenses which will be free from chromatic aberration.

39. (a) As $\beta = \dfrac{\lambda D}{d}$ and $\lambda_b < \lambda_y$,

$\therefore$ Fringe width β will decrease

40. (c) Given, ${}_1H^2 + {}_1H^2 \rightarrow {}_2H^4 + Q$

The total binding energy of the deutrons $= 4 \times 1.15 = 4.60$ MeV

The total binding energy of alpha particle $= 4 \times 7.1 = 28.4$ MeV

The energy released in the process $= 28.4 - 4.60 = 23.8$ MeV.

41. (b) $W_0 = \dfrac{hc}{\lambda_0}$ or $W_0 \propto \dfrac{1}{\lambda_0}$;

$$\Rightarrow \frac{W_1}{W_2} = \frac{\lambda_2}{\lambda_1} = \frac{600}{300} = 2$$

42. (a) $\beta = \dfrac{I_c}{I_b} \Rightarrow I_b = \dfrac{0.01}{1 \times 10^3} = 10^{-5}$

or $50 = \dfrac{I_c}{10^{-5}} \Rightarrow I_c = 50 \times 10^{-5}$

$$= 500 \times 10^{-6}$$

$$I_c = 500\,\mu A$$

43. (d) A and C are analogue but B is digital signal.

44. (a) We have, $\beta = \dfrac{\alpha}{1 - \alpha} = \dfrac{0.98}{1 - 0.98} = \dfrac{0.98}{0.02} = 49$

45. (b) $m_a = \frac{E_{max} - E_{min}}{E_{max} + E_{min}}$

$= \frac{90-30}{90+30} = \frac{60}{120} = \frac{6}{12} = 0.5$

SECTION 2 - CHEMISTRY

46. (a) Angular momentum, $mvr = n\frac{h}{2\pi}$

($n = 2$ for first excited state)

47. (b) Two Na^+ ions are replaced by one Sr^{2+} ion to maintain electrical neutrality.
Hence, number of vacancies
= Number of Sr^{2+} ions doped

$= 1.0 \times 10^{-3} \times 6.02 \times 10^{23} = 6.02 \times 10^{20}$

48. (b) Here, R = 31.6 ohm

$\therefore \quad C = \frac{1}{R} = \frac{1}{31.6} \text{ohm}^{-1} = 0.0316 \text{ ohm}^{-1}$

Specific conductance = conductance × cell constant

$= 0.0316 \text{ ohm}^{-1} \times 0.367 \text{ cm}^{-1}$

$= 0.0116 \text{ ohm}^{-1} \text{ cm}^{-1}$

Now, molar concentration = 0.5 M (given)

$= 0.5 \times 10^{-3}$ mole cm^{-3}

$\therefore$ Molar conductance $= \frac{K}{C} = \frac{0.0116}{0.5 \times 10^{-3}}$

$= 23.2 \text{ S cm}^2 \text{ mol}^{-1}$

49. (b) $(NH_4)_2Cr_2O_7 \xrightarrow{heat} Cr_2O_3 + 4H_2O + N_2$

50. (a) Assume that each has lost a proton . So we get : HO^-, HS^-, HSe^-, HTe^-
It can be easily seen that the volume available for the negative charge is increasing from HO^- to HTe^-, therefore
(i) volume available for the negative charge is increasing from left to right
(ii) charge density is decreasing from left to right
(iii) basicity is decreasing from left to right
(iv) acidity of conjugate acids is increasing from left to right

$H_2O < H_2S < H_2Se < H_2Te$

51. (c) In preparation of an alkyl chloride by the action of dry HCl, the catalyst generally used is anhydrous $ZnCl_2$.

52. (a) (a) : $4n + 2 = 2, n = 0$ (integer)

(b) : $4n + 2 = 4, n = 0.5$

(c) : $4n + 2 = 4, n = 0.5$

(d) : $4n + 2 = 4, n = 0.5$

Only (a) obeys Huckel's rule.

54. (b) $Sn^{4+} + 2e^- \longrightarrow Sn^{2+}$ E° = 0.13 V

$Br_2 + 2e^- \longrightarrow 2Br^-$ E° = 1.08 V

E° values shows Br_2 has higher reduction potential.

Hence $E_{cell} = E_R - E_L$

$= E_{Br_2/Br^-} - E_{Sn^{+4}/Sn^{+2}}$

$= 1.08 - 0.13 = 0.95$ V

Now $-\Delta G = nFE_{cell}$

n = 2, F = 96500.

$-\Delta G = 2 \times 96500 \times 0.95$ kJ/mol

Also, $\Delta G = -2.303 RT \log K_{eq}$

$\log K_{eq} = -\frac{\Delta G}{2.303 \times R \times T}$

$= \frac{-(-2 \times 96500 \times 0.95)}{2.303 \times 8.314 \times 293} = 32.6820$

K_{eq} = antilog 32.682

$= 4.78 \times 10^{32} \approx 10^{32}$

55. (c) $[H^+]$ in HCl solution (pH = 2) $= 10^{-2} M$;

$[OH^-]$ in KOH solution (pOH = 14 – 12 = 2) $= 10^{-2} M$

Excess m Mol of OH^- in 5 ml mixture

$= 3 \times 10^{-2} - 2 \times 10^{-2} = 1.0 \times 10^{-2}$;

$[OH^-]$ in mixture $= \frac{1.0 \times 10^{-2}}{5}$

$= 2 \times 10^{-3} M$;

$pOH = -\log 2 \times 10^{-3} = 3 - \log 2$;

$pH = 14 - (3 - \log 2) = 11.30$

56. (a) A reducing agent is a substance which can loose electron and hence a reducing agent should have low ionisation energy. Now since ionisation energy decreases from Li to Cs, the reducing property should increase in the opposite manner. The only exception to this is lithium. This is because the net process of converting an atom to an ion takes place in 3 steps.
(i) $M(s) \rightarrow M(g)$ ΔH = Sublimation energy
(ii) $M(g) \rightarrow M^+(g) + e^-$ ΔH = Ionisation energy
(iii) $M^+(g)+H_2O \rightarrow M^+ (aq)$ ΔH = Hydration energy
The large amount of energy liberated in hydration of Li (because of its small size) makes the overall ΔH negative. This accounts for the higher oxidation potential of lithium i.e., its high reducing power.

57. (d) Number of unpaired electrons in Cr, Mn^{2+} and Fe^{3+} are 6, 5 and 5 respectively.

58. (a) $6HCHO + 4NH_3 \longrightarrow (CH_2)_6N_4 + 6H_2O$
urotropine (hexamethylene tetramine)

59. (a) [benzene ring: 1-CHO, 2-NO_2, 4-OCH_3; positions 1–6 numbered]
4-Methoxy-2-nitrobenzaldehyde
[∵ the order of preference of substituent groups is $-CHO > -NO_2 > -OCH_3$]

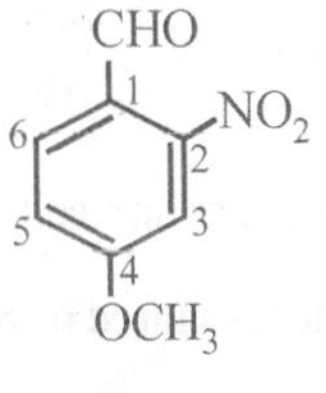

61. (c) Total V.P.,
$$P = P_A^o X_A + P_B^o X_B = P_A^o X_A + P_B^o (1 - X_A)$$
$$= (P_A^o - P_B^o) X_A + P_B^o$$
Thus, $P_B^o = 114$ torr ; $P_A^o - P_B^o = 52$
or $P_A^o = 166$ torr
Hence $P = 166 \times \frac{1}{2} + 114 \times \frac{1}{2} = 140$ torr

62. (d) If rate $= k[A]^x [B]^y [C]^z$
From first two given data
$8.08 \times 10^{-3} = k[0.2]^x [0.1]^y [0.02]^z$ .... (1)
$2.01 \times 10^{-3} = k[0.1]^x [0.2]^y [0.02]^z$ .... (2)
Divide (1) by (2) we get, $4 = 2^x (1/2)^y$
Similarly, from second and third data
$(9)^y (9)^z = 3$
$2y + 2z = 1$.
From first and fourth data $4^z = 8 = 2^3$
$2z = 3$. So $z = 3/2, y = -1, x = 1$

64. (b) Structural formula of glucose is
$$\begin{array}{c} CHO \\ | \\ (CHOH)_4 \\ | \\ CH_2OH \end{array}$$
In addition to – CHO group it contains one primary and four secondary – OH groups.

65. (a) $CH_3 - CH_2 - C \equiv CH \xrightarrow[H_3O^+]{Hg^{2+}}$
$$CH_3 - CH_2 - \underset{}{\overset{OH}{\overset{|}{C}}} = CH_2 \xrightarrow{\text{tautomerization}} CH_3CH_2COCH_3$$

66. (c) An optically active compound contains a chiral carbon atom.
(a) $H_3C - C(CH_3)(H) - CHO$ (no chiral carbon atom)
(b) $H - CH_2 - CH_2 - CH_2 - CH = O$ (each C with H above and below) (no chiral carbon atom)
(c) $H - CH_2 - CH_2 - {}^*CH(Br) - C(=O) - H$ (contains chiral carbon atom)
(d) $H - CH_2 - CH_2 - C(Br)(Br) - CH = O$ (no chiral carbon atom)

67. (c) $C_2H_5Cl + H_2 \xrightarrow{Pd} C_2H_6$
Reduction of alkyl halide by H_2 in the presence of Pd which serves as catalyst. The accumulation of hydrogen gas occurrs on the surface of active palladium so that fast reduction may be achieved.

68. (c) Greater the difference of electronegativities, more polar is the bond.

69. (c) O.N. of Cl in ClO^- changes from +1 to +5 and –1.

70. (d) For a first order reaction, we have

$$k = \frac{2.303}{t}\log\frac{N_0}{N}$$

$$\therefore\ k = \frac{2.303}{40\ \text{min}}\log\frac{0.1}{0.005}$$

$$= \frac{2.303}{40\ \text{min}} \times \log 20 = \frac{2.303}{40} \times 1.3010$$

Now rate = k × [reactant]
When [x] = 0.01 M

$$\therefore\ \text{rate} = \frac{2.303}{40} \times 1.3010 \times 0.01\ \text{M min}^{-1}$$

$$= 7.5 \times 10^{-4}\ \text{M min}^{-1}.$$

71. (c) Correct Nernst equation is

$$E = E^{o} + \frac{2.303\,RT}{nF}\log a_{M^{n+}}.$$

73. (a) This is isocyanide test

$$\underset{\text{Aniline}}{C_6H_5NH_2} + \underset{\text{Chloroform}}{CHCl_3} + 3KOH \longrightarrow \underset{\substack{\text{Phenyl isocyanide}\\ \text{(Offensive smell)}}}{C_6H_5N \overset{\rightarrow}{=} C} + 3KCl + 3H_2O$$

75. (b) $SO_2Cl_{2(g)} \rightleftharpoons SO_{2(g)} + Cl_{2(g)}$

$$\alpha\,\% = \frac{D-d}{d(y-1)} \times 100$$

$(y = 2)$

$$D = \frac{\text{molar mass of } SO_2Cl_2}{2} = \frac{135}{2} = 67.5;$$

$d = 50.0$ (given)

$$\alpha\% = \frac{67.5-50.0}{50.0(2-1)} \times 100 = 35\%$$

76. (b) $2AgCl_{(s)} + H_{2(g)} \rightarrow 2HCl_{(aq)} + 2Ag_{(s)}$
The activities of solids and liquids are taken as unity and at low concentrations, the activity of a solute is approximated to its molarity.
The cell reaction will be

$$Pt_{(s)} \mid H_{2(g)}, 1\text{bar} \mid H^{+}_{(aq)}\,1M \mid AgCl_{(aq)}\ 1M \mid Ag_{(s)}$$

77. (d) $H-\overset{\overset{\displaystyle O}{\uparrow}}{\underset{\underset{\displaystyle OH}{|}}{P}}-OH$, so dibasic acid

78. (d) $[CoF_6]^{3-}$: 3d [↑↓][↑][↑][↑][↑] 4s [:] 4p [:][:][:] 4d [:][:][ ][ ][ ]

sp^3d^2 (Paramagnetic)

$[NiCl_4]^{2-}$:
[↑↓][↑↓][↑↓][↑][↑] [:] [:][:][:]

sp^3 (paramagnetic)

$[CuCl_4]^{2-}$:
[↑↓][↑↓][↑↓][↑↓][:] [ ] [:][:][↑]

dsp^2 (paramagnetic)

$[Ni(CN)_4]^{2-}$:
[↑↓][↑↓][↑↓][↑↓][:]

(diamagnetic) [:] [:][:][ ] dsp^2 :

79. (a) Aldehydes, other than formaldehyde, when treated with RMgX give 2° alcohols.

80. (c) Mass of methanol in 1 mol solution = 0.5 × 32 = 16 g
Mass of water in solution = 0.5 × 18 = 9 g

$$\%\ \text{by mass of methanol} = \frac{16 \times 100}{16+9} = 64$$

81. (c) Number of A atoms per unit cell $= \frac{1}{8} \times 8 = 1$
(An atom at corner is shared by 8 unit cells)
Number of B atoms per unit cell $= \frac{1}{2} \times 6 = 3$
(An atom at the face centre is shared by 2 unit cells)
Hence, formula is : AB_3

82. (c) **Smelting** is the process of reduction using carbon as reducing agent.

84. (d) $[Co(NH_3)_5Br]SO_4$ gives white precipitate of $BaSO_4$ with $BaCl_{2(aq)}$ whereas $[Co(NH_3)_5SO_4]Br$ gives yellow precipitate (AgBr) with $AgNO_{3(aq)}$.

85. (c) Nylon has $-\overset{\overset{\displaystyle O}{\|}}{C}-NH-$ group linkage

86. (c) $HO-C_6H_4-SO_3H \xrightarrow[H_2O]{Br_2}$ HO–$C_6H_2(Br)_2$–SO_3H

– OH group is highly activating. This is a type of electrophillic substitution reaction at ortho and para position.

87. (b) SO_3 has trigonal planar geometry (sp^2 hybridisation of *S*) which is symmetrical.

88. (c) Number of hybrid orbitals of *P* in PO_4^{3-} =
½ [5 + 0 + 3] = 4 (sp^3)
No. of hybrid orbitals of *N* in NO^-_3
= ½ [5 + 0 + 1] = 3 (sp^2)
No. of hybrid orbitals of S in SO_4^{--}
$= \frac{1}{2}[6+0+2-0] = 4\ (sp^3)$

89. (a) $Al_2(SO_4)_3$ is a salt of weak base and a strong acid hence on hydrolysis it will produce $Al(OH)_3$ and H_2SO_4. Since H_2SO_4 is a strong acid and $Al(OH)_3$ is is a weak base hence the solution will be acidic due to SO_4^{--}.

90. (d) It is the correct answer.

SECTION 3 - BOTANY

111. (b) Because malic acid is an organic acid, which after oxidation produces more CO_2, due to which the value of R.Q. is four.

118. (a) Unisexuality of flowers prevents autogamy, but not geitonogamy. In self fertilisation , the male and female gametes are derived from the same individual. Among plants, self fertilization also called autogamy is common in many cultivated species, *eg*., wheat and oats. However, self fertilization is a form of inbreeding and does not allow for the mixing of genetic material; if it occurs over a number of generations it will result in offspring being less vigorous and productive than those resulting from cross fertilization.

119. (b) Speciation takes place via reproductive isolation which is the most important consequence of geographical isolation.

120. (a) Genetic drift can operate only on smaller populations where its fluctuation can be observed in the proportion of allele distribution in the presence of external disturbances. Genetic drift is the random change in allele number and frequency in a gene pool.

121. (a) Three crops that contribute maximum to global food grain production are Wheat, rice and maize, which belong to the family *Poaceae* (*Graminae*).

122. (a) Farmers have reported over 50% higher yields of rice by using the biofertilizer *Azolla pinnata*.

123. (d) 'Himgiri' developed by hybridisation and selection for disease resistance against rust pathogens is a variety of wheat. It is resistant to leaf/stripe rust and hill bunt.

124. (c) Fresh weight is not used for the construction of ecological pyramids because the total fresh weight does not change into energy. Hence we can say that fresh weight is not continuous in the tropic levels.

126. (a) The genetically modified brinjal in India has been developed for insect resistance. Bt brinjal is a transgenic brinjal that is developed by inserting a crystal gene from the *Bacillus thuringiensis* into the brinjal's genome. This process of insertion is accomplished using *Agrobacterium* mediated recombination.

127. (c) Lesser inter-specific competition is not observe in biodiversity hotspots. Initially 25 biodiversity hotspots were identified but subsequently nine more have of biodiversity hotspots in the world to 34. These hotspots are also regions of accelerated habitat loss.

128. (d) The correct acronym is IUCN, International Union for Conservation of Nature and Natural resources.

129. (b) Photochemical smog is made by the deposition of dust & carbon particles on the pollutant gases and water vapours. The gases found in photochemical smog are NO_2, PAN, O_3 etc.

130. (a) By dumping of huge amount of sewage, the oxygen levels are depleted, which are reflected in terms of BOD values of water. The number of microbes also increases tremendously and these also consume most

of the oxygen. Thus BOD of the river water will increase.

131. (b) In India maximum genetic diversity is seen in rice. Total of 23 variety of irrigated ecology and 8 varieties of rainfed ecology has been invented in India.

132. (b) The greatest biomass of autotrophs in the oceans is that of free floating microalgae, cyanobacteria and nanoplankton.

134. (c) PCR is a technique for enzymatically replicating DNA without using a living organism such as *E. coli* or yeast. It is commonly used in medical and biological research labs for a variety of tasks like detection of hereditary diseases, identification of genetic fingerprints etc.
The correct steps shown in the above figure are:
A – Denaturation at a temperature of about 94° to 98°C. During the denaturation, the double strand melts open to single stranded DNA, and all enzymatic reactions stop.
B – Annealing (binding of DNA primer to the separated strands. Occurs at 50° to 65°Celsius, which is lower than the optimal temperature of the DNA polymerases)
C – Extension or elongation of the strands using the DNA primer with heat-stable DNA polymerases, most frequently Taq (*Thermus aquaticus*) at 72°C.

SECTION 4 - ZOOLOGY

164. (b) Morphallaxis is a mechanism of regeneration involving reorganization of body cells. In epimorphosis new cells proliferate from the surface of the wound to form the missing structure. In accretionary growth some specialized cells retain the ability to divide and produce new cells to replace the worn-out.

165. (b) In oogamy male and female gametes are morphologically as well as physiologically different. Female gametes are large and non-motile. Male gametes are small but motile.

166. (b) The part of fallopian tube closest to the ovary is infundibulum. Infundibulum possess finger-like projections called fimbriae that help in collection of ovum after ovulation. It leads to wider part of oviduct called ampulla. The last part of oviduct is isthmus that has a narrow lumen and joins the uterus.

169. (b) In grasshopper the males lack a Y-sex chromosome and have only an X-chromosome. They produce sperm cells that contain either an X chromosome or no sex chromosome, which is designated as O.

170. (d) Linkage is the inheritance of genes of same chromosome together and capacity of these genes to retain their parental combination in subsequent generation. The strength of linkage between two genes is inversely proportional to the distance between the two. This means, two linked genes show higher frequency of recombination if the distance between them is higher and lower frequency if the distance is smaller.

171. (d) It was given by Geneticists George W. Beadle and E. L. Tatum which states that each gene in an organism controls the production of a specific enzyme. It is these enzymes that catalyze the reactions that lead to the phenotype of the organism.

172. (b) The transforming chemical discovered by Griffith in his experiments with ***Pneumococcus***, was confirmed as DNA by Avery, McLeod and McCarty.

175. (d) *Escherichia coli* is a bacterium found in human colon. On this bacterium scientists have made extensive genetic experiments to make some vital chemicals like insulin. Another bacterium is *Agrobacterium tumefaciens* which causes crown gall in plants is extensively used for genetic experiments.

177. (d) The statement (i) and (ii) are correct. The statement (iii) is incorrect because predators cannot lead to the extinction of prey species. Predator and prey evolve together. The statement (iv) is incorrect because chemical such as nicotine and strychnine produced by the plants are not metabolic disorders but are metabolic wastes.

178. (b) Diversity is due to the long term evolutionary changes. Adaptability to continuous changes in environmental conditions is important for natural selection of variants and variations generation after generation leading to emergence of diverse descendants.

NEET (NATIONAL ELIGIBILITY ENTRANCE TEST) Full Test - 8

ANSWER KEY																	
1	(d)	21	(b)	41	(d)	61	(a)	81	(b)	101	(c)	121	(a)	141	(d)	161	(a)
2	(b)	22	(c)	42	(a)	62	(b)	82	(c)	102	(b)	122	(b)	142	(b)	162	(c)
3	(b)	23	(c)	43	(c)	63	(c)	83	(c)	103	(b)	123	(a)	143	(a)	163	(a)
4	(b)	24	(d)	44	(d)	64	(b)	84	(d)	104	(d)	124	(d)	144	(b)	164	(a)
5	(b)	25	(c)	45	(d)	65	(b)	85	(c)	105	(a)	125	(c)	145	(b)	165	(a)
6	(a)	26	(c)	46	(c)	66	(b)	86	(d)	106	(b)	126	(b)	146	(d)	166	(c)
7	(c)	27	(d)	47	(a)	67	(a)	87	(b)	107	(c)	127	(d)	147	(b)	167	(c)
8	(c)	28	(b)	48	(a)	68	(d)	88	(c)	108	(a)	128	(c)	148	(d)	168	(c)
9	(c)	29	(a)	49	(a)	69	(d)	89	(b)	109	(c)	129	(b)	149	(a)	169	(a)
10	(b)	30	(a)	50	(a)	70	(c)	90	(b)	110	(b)	130	(a)	150	(b)	170	(a)
11	(a)	31	(c)	51	(b)	71	(b)	91	(d)	111	(b)	131	(b)	151	(c)	171	(a)
12	(a)	32	(d)	52	(c)	72	(c)	92	(a)	112	(a)	132	(d)	152	(a)	172	(d)
13	(b)	33	(a)	53	(d)	73	(d)	93	(a)	113	(b)	133	(a)	153	(a)	173	(a)
14	(b)	34	(d)	54	(c)	74	(d)	94	(a)	114	(d)	134	(d)	154	(c)	174	(d)
15	(d)	35	(c)	55	(a)	75	(a)	95	(a)	115	(a)	135	(a)	155	(b)	175	(a)
16	(b)	36	(d)	56	(a)	76	(a)	96	(a)	116	(d)	136	(a)	156	(a)	176	(d)
17	(b)	37	(c)	57	(c)	77	(d)	97	(c)	117	(c)	137	(a)	157	(b)	177	(a)
18	(b)	38	(d)	58	(b)	78	(a)	98	(b)	118	(d)	138	(d)	158	(a)	178	(d)
19	(d)	39	(b)	59	(c)	79	(b)	99	(d)	119	(a)	139	(c)	159	(b)	179	(d)
20	(c)	40	(b)	60	(c)	80	(b)	100	(a)	120	(a)	140	(a)	160	(c)	180	(c)

HINTS & SOLUTIONS

SECTION 1 - PHYSICS

1. (d) Moment of inertia, $I = Mr^2$

$$[I] = [ML^2]$$

Moment of force, $\vec{\tau} = \vec{r} \times \vec{F}$

$$[\vec{\tau}] = [L][MLT^{-2}] = [ML^2T^{-2}]$$

2. (b) $f_1 = +40$ cm (for convex lens) $= 0.4$ m

$f_2 = -25$ cm (for concave lens) $= -0.25$ m

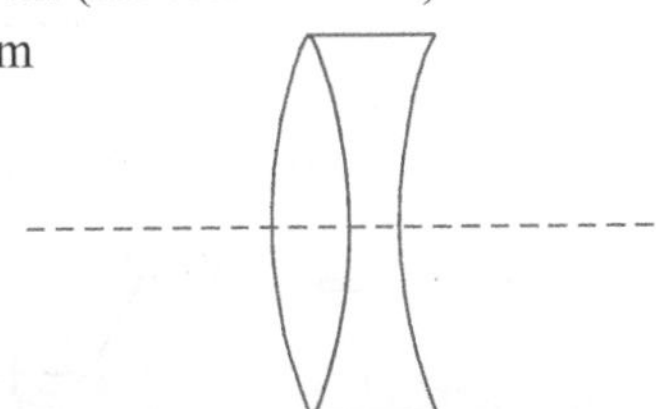

$\therefore$ Focal length (f) of the combination

$$\frac{1}{f} = \frac{1}{f_1} + \frac{1}{f_2} = \frac{1}{0.40} - \frac{1}{0.25} = \frac{0.25 - 0.4}{0.40 \times 0.25}$$

$$= -\frac{0.15}{0.1} = -1.5 \text{ dioptre.}$$

$$\Rightarrow \quad P = \frac{1}{f} = -1.5 \text{ dioptre}$$

3. (b) Distance in last two second

$$= \frac{1}{2} \times 10 \times 2 = 10 \text{ m.}$$

$$\text{Total distance} = \frac{1}{2} \times 10 \times (6 + 2) = 40 \text{ m.}$$

4. (b) $\hat{A} = \frac{3\hat{i} + +b\hat{j} + 2\hat{k}}{\sqrt{9+36+4}} = \left(\frac{3}{7}\hat{i} + \frac{6}{7}\hat{j} + \frac{2}{7}\hat{k}\right)$. If α, β and γ are angles made by $\vec{A}$ with coordinate axes, then

$\cos\alpha = \frac{3}{7}$, $\cos\beta = \frac{6}{7}$ and $\cos\gamma = \frac{2}{7}$.

5. (b) Effective acceleration of a body = $[a - g]$.
6. (a) Vector has both magnitude and direction so, statement (a) is wrong.
7. (c) Radial acceleration = centripetal acceleration

$\frac{v^2}{r} = \frac{30 \times 30}{500} = \frac{9}{5}$

Tangential acceleration $(a_t) = 2\ \text{m/sec}^2$

Total acceleration = $\sqrt{a_t^2 + a_r^2}$

$= \sqrt{2^2 + \frac{81}{25}} = \sqrt{4 + \frac{81}{25}} \sqrt{\frac{181}{25}} = 2.7\ m/\sec^2$

8. (c) In a cyclic process, the initial state coincides with the final state. Hence, the change in internal energy is zero, as it depends only on the initial and final states. But Q & W are non-zero during a cycle process.
9. (c)

Moment of inertia of a disc about a diameter

$= \frac{1}{2} \times \frac{1}{2} MR^2 = \frac{1}{4} MR^2$

$[I_x + I_y = I_z = \frac{1}{2} MR^2]$

Applying theorem of parallel axis.

$I_z = \frac{1}{4} MR^2 + MR^2 = \frac{5}{4} MR^2$

10. (b) $v_r = \sqrt{v_R^2 - v_B^2} = \sqrt{10^2 - 8^2} = 6\ \text{km h}^{-1}$
11. (a) Charge on an electron

$= 1.6 \times 10^{-19}$ coulomb

No. of electrons required = $\frac{20 \times 10^{-6}}{1.6 \times 10^{-19}}$

$= \frac{20}{1.6} \times 10^{13} = 1.25 \times 10^{14}$

12. (a) Let T be the tension in the string. Equation of motion of A is; $T = Ma$ .... (i). If a' is the acceleration of B, the equation of motion of B can be written as

$Ma' = F - T = F - Ma$

$a' = (F/M) - a$

13. (b) When sound travels from one medium to another its speed changes due to change in wavelength. Its frequency remains constant.
14. (b) Under isothermal conditions, there is no change in internal energy.
15. (d)

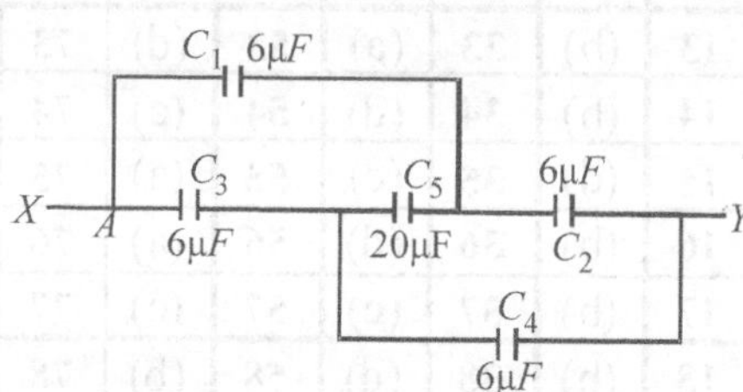

Equivalent circuit

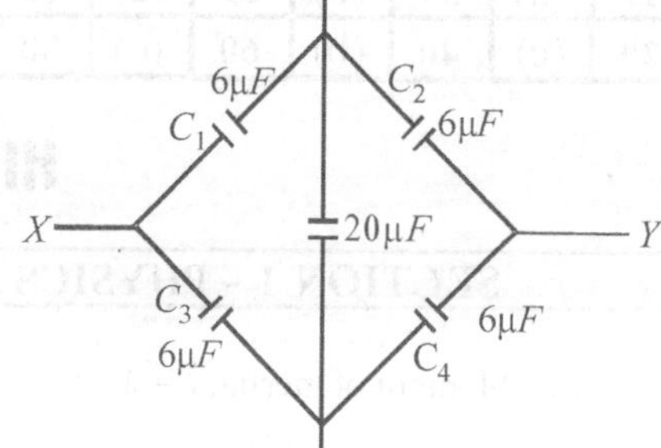

Here, $\frac{C_1}{C_3} = \frac{C_2}{C_4}$

Hence, no charge will flow through 20μF

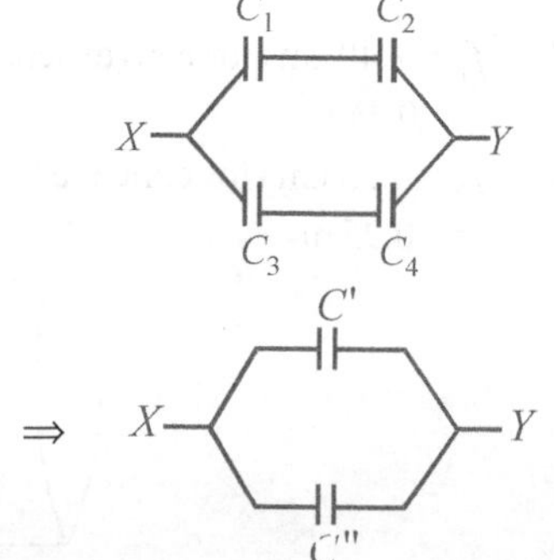

C_1 and C_2 are in series, also C_3 and C_4 are in series.
Hence, $C' = 3\ \mu F, C'' = 3\ \mu F$
C' and C'' are in parallel.
Hence net capacitance $= C' + C'' = 3 + 3$
$= 6\ \mu F$

16. (b) $E = \dfrac{p^2}{2m}$

or, $E_1 = \dfrac{p_1^2}{2m_1}, E_2 = \dfrac{p_2^2}{2m_2}$

or, $m_1 = \dfrac{p_1^2}{2E_1}, m_2 = \dfrac{p_2^2}{2E_2}$

$m_1 > m_2 \Rightarrow \dfrac{m_1}{m_2} > 1$

$\therefore \dfrac{p_1^2 E_2}{E_1 p_2^2} > 1 \Rightarrow \dfrac{E_2}{E_1} > 1 \quad [\because p_1 = p_2]$

or, $E_2 > E_1$

17. (b)

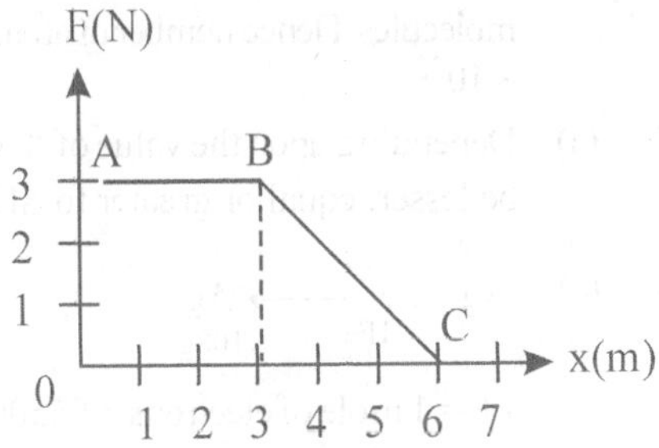

Work done = area under F-x graph
= area of trapezium OABC

$= \dfrac{1}{2}(3+6)(3) = 13.5\ \text{J}$

18. (b) Total kinetic energy of a rolling ball
K.E. = linear K.E. + rotational K.E.

$\text{K.E.} = \dfrac{1}{2}mv^2 + \dfrac{1}{2}I\omega^2$

$= \dfrac{1}{2}mv^2 + \dfrac{1}{2} \times \dfrac{2}{5}mr^2 \cdot \dfrac{v^2}{r^2}$

$= \left(\dfrac{5+2}{10}\right)mv^2 = \dfrac{7}{10}mv^2$

Fraction of total energy associated with

rotational kinetic energy $= \dfrac{\frac{1}{5}mv^2}{\frac{7}{10}mv^2}$

$= \dfrac{1}{5} \times \dfrac{10}{7} = \dfrac{2}{7}$

19. (d) Net charge inside the surface is zero. So, flux through the surface is zero.

20. (c) Microwave region wavelength $= 10^{-3}$ m to 1m

22. (c) $v_e = \sqrt{2gR}$
The escape velocity is independent of the angle at which the body is projected.

23. (c)

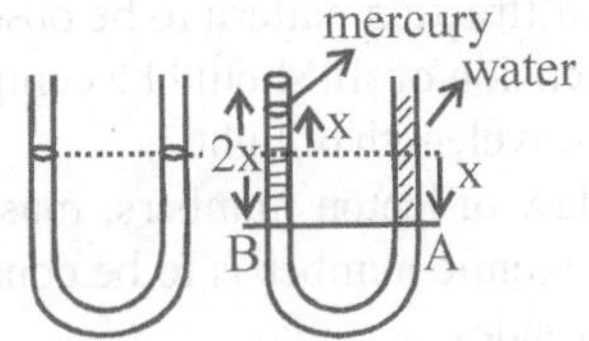

At point A and B pressure will be same. So
pressure at A = $P + h_1\rho_1 g$
pressure at B = $P + h_2\rho_2 g$
$P + h_1\rho_1 g = P + h_2\rho_2 g$
or, $11.2 \times 1 \times g = 2x \times 13.6 \times g$

or, $x = \dfrac{11.2}{2 \times 13.6} = 0.41$ cm

24. (d) $F = YA\alpha t = (2.0 \times 10^{11})(10^{-6})(1.1 \times 10^{-5})(20) = 44$ newton

25. (c) Radiation is the fastest mode of transfer of heat.

26. (c) The loop (i) is for soft iron and the loop (ii) is for steel in fig.

27. (d) $P = \dfrac{1}{3}\rho\bar{v^2} = \dfrac{2}{3} \times \left(\dfrac{1}{2}\rho\bar{v^2}\right) = \dfrac{2}{3}E$

28. (b) Rate of cooling $\propto$ temperature difference between system and surrounding.
As the temperature difference is halved, so the rate of cooling will also be halved

29. (a) A is positive as electric lines are coming out of it. B is negative as electric lines are entering into it.

30. (a) $y = 2\sin\pi\left\{200t - \dfrac{x}{15}\right\}$

$= 2\sin\left\{200\pi t - \frac{x\pi}{15}\right\}$

Comparing it with the equation of a wave,

$y = a\sin(\omega t - kx)$

$\omega = 200\pi; k = \frac{\pi}{15}$

$v = \frac{\omega}{k} = \frac{200\pi \times 15}{\pi} = 3000 \text{ cm/sec}$

31. (c) $r = E/I = 1.5/3 = 0.5$ ohm.

33. (a) The wheatstone bridge is balanced, when P/Q = R/S, In this case $\frac{3}{4} = \frac{4}{8}$, so bridge is balanced & 7Ω resistance is not effective)

34. (d) For diffraction pattern to be observed, the dimension of slit should be comparable to the wavelength of light.

35. (c) By law of lepton numbers, mass number and atomic number is to be conserved on both sides.

36. (d) Since the magnetic field is uniform the flux f through the square loop at any time t is constant, because

$f = B \times A = B \times L^2 = \text{constant}$

$\therefore\ \varepsilon = -\frac{d\phi}{dt} = \text{zero}$

37. (c) When the ray enters a glass slab from air, its frequency remains unchanged. Since glass slab in an optically denser medium, the velocity of light decreases and therefore we can conclude that the wavelength decreases.

$(\because\ v = \nu\lambda)$

39. (b) As R ∝ V^2/P or R ∝ 1/P, so resistance of heater is less than that of fan.

40. (b) $F = Bil = 2 \times 1.2 \times 0.5 = 1.2\ N$

41. (d) Since $p = nh\nu$

$\Rightarrow n = \frac{p}{h\nu} = \frac{2 \times 10^{-3}}{6.6 \times 10^{-34} \times 6 \times 10^{14}}$

$= 5 \times 10^{15}$

42. (a) The range of wavelengths of the lines of Lyman series varies from 912 A° to 1216 A°.

43. (c) P.E. = 2T.E.

$\Rightarrow$ P.E. = $2E_0$

In case of circular motion of satellite around the earth / planets revolving around sun/ electrons revolving in circular orbit P.E. = 2 T.E. and |K.E.| = |T.E.|].

44. (d) Positive terminal is at lower potential (0V) and negative terminal is at higher potential 5V.

45. (d) (A) is a NAND gate so output is $\overline{1 \times 1} = \bar{1} = 0$

(B) is a NOR gate so output is $\overline{0 + 1} = \bar{1} = 0$

(C) is a NAND gate so output is $\overline{0 \times 1} = \bar{0} = 1$

(D) is a XOR gate so output is $0 \oplus 0 = 0$

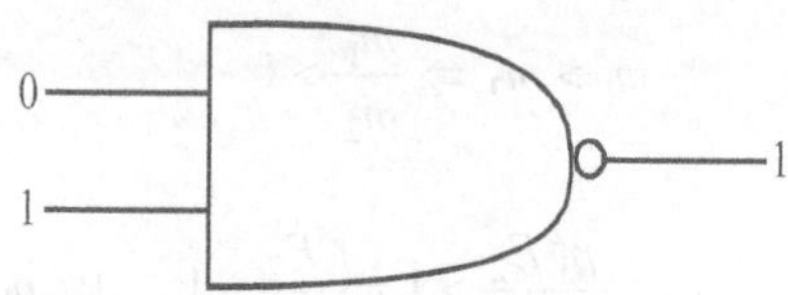

SECTION 2 - CHEMISTRY

46. (c) 22.4 L of CO_2 at STP = 1 mole = 6.023×10^{23} molecules. Hence number of atoms $3 \times 6.023 \times 10^{23}$.

47. (a) Depending upon the value of $T\Delta S$, ΔG can be lesser, equal or greater to ΔH

48. (a) $\underset{1F}{Ag^+ + e^-} \longrightarrow \underset{108g}{Ag}$

1 F = 1 mole of electrons = 96500 C

0.01F = 1.08 g Ag; Ag left = 1.08 – 1.08 = 0

49. (a)

	$KMnO_4$	MnO_2	Mn^{2+}	$Mn(OH)_3$	MnO_4^{2-}
O.S of Mn	+7	+4	+2	+3	+6
electrons needed	0	3	5	4	1

50. (a) All are examples of electrophilic addition reactions and involve the formation of stable carbocation leading to the formation of addition product according to Markovnikov's rule

52. (c) $N_2 + O_2 \rightarrow 2NO$

O.N. of N changes from 0 to +2 (oxidation) and O.N. of O changes from O to –2 (reduction).

53. (d) For a reaction to be at equilibrium $\Delta G = 0$.
Since
$\Delta G = \Delta H - T\Delta S$ so at equilibrium
$\Delta H - T\Delta S = 0$
or $\Delta H = T\Delta S$
For the reaction
$\frac{1}{2}X_2 + \frac{3}{2}Y_2 \longrightarrow XY_3$;
$\Delta H = -30 kJ$ (given)
Calculating ΔS for the above reaction, we get

$$\Delta S = 50 - \left[\frac{1}{2}\times 60 + \frac{3}{2}\times 40\right] JK^{-1}$$

$= 50 - (30 + 60)\ JK^{-1} = -40\ JK^{-1}$
At equilibrium,
$T\Delta S = \Delta H$
$\therefore \quad T \times (-40) = -30 \times 1000$ [$\because$ 1kJ = 1000J]

or $T = \frac{-30 \times 1000}{-40}$ or 750 K

54. (c) $k = \frac{1}{R} \times \text{Cell constant} = \frac{0.47}{31.6} = 0.01487$

55. (a) IE_1 is always less than IE_2.

56. (a) $Cl_3CCH(OH)_2 + NaOH \longrightarrow CHCl_3 + HCOONa + H_2O$

57. (c)

$$HC \equiv CH \xrightarrow[20\% H_2SO_4]{1\% HgSO_4} \underset{[A]}{CH_3CHO} \xrightarrow[H_2O]{CH_3MgX}$$

$$\underset{[B]}{CH_3CHOHCH_3} \xrightarrow{[O]} \underset{\text{Acetone [C]}}{CH_3COCH_3}$$

58. (b) Apply $\frac{P_1V_1}{T_1} = \frac{P_2V_2}{T_2}$

59. (c) The pH of 1×10^{-8} M HCl is always less than 7. Since the solution is of acid. It can be shown as
$[H^+] = 10^{-7}$ from H_2O and 10^{-8} from HCl
$[H^+] = 10^{-8}(10 + 1)$
$[H^+] = 10^{-8} \times 11$
$-\log[H^+] = 8 \log 10 - \log 11$
$pH = 8 - \log 11$
$pH = 6.93$.

60. (c) Let the rate law equation be
$r = k[A]^x[B]^y$ ....(1)
when A = 2A,
$4r = k[2A]^x[B]^y$ ....(2)
Divide(2) by (1) $\frac{4r}{r} = \frac{k[2A]^x[B]^y}{k[A]^x[B]^y}$
$\therefore 4 = (2)^x$, Hence $x = 2$
when B -= 2B,
$r = k[A]^x[2B]^y$ ....(3)
$\therefore$ 1 = Devide (3) by (1)

$$\frac{r}{r} = \frac{k[A]^x[2B]^y}{k[A]^x[B]^y}(2)^y \therefore y = 0$$

$\therefore$ Rate $= k[A]^2[B]^0$

61. (a) F and O belong to 2nd period whereas Cl and Br belong to 3rd and 4th periods respectively. Hence the sequence of the E.N. is
F > O > Cl > Br

62. (b) Since 1° is more reactive than 2° toward HI, latter can react only when both 1° alcoholic groups have reacted.

63. (c) The compounds containing active H-atoms (H atoms attached to N, O or S) react with CH_3COCl to form acetyl derivatives.

64. (b) Reverse reaction, Le Chatelier's principle
$\Delta n = 2 - 1 = 1$.

65. (b) $6 = -\log 10^{-5} + \log\frac{\text{Salt}}{\text{Acid}} = 5 + \log\frac{\text{Salt}}{\text{Acid}}$

$\log\frac{\text{Salt}}{\text{Acid}}$ must be 1.

$\therefore \frac{\text{Salt}}{\text{Acid}} = \frac{10}{1}$ or 10 : 1.

66. (b) For 10° rise of temperature the rate is almost doubled.

67. (a) In BCl_3 there is sp^2 hybridisation (planar) in NCl_3 hybridisation is sp^3 hence shape is pyramidal

68. (d) $\overset{6}{C}H_3\overset{5}{C}H_2\overset{4}{C}H = \overset{3}{C}H\overset{2}{\underset{\|}{C}}\overset{1}{C}H_3$ (with O double-bonded at C-2)
(hex 3-ene-5-one)

69. (d) Negative reaction with 2, 4- dinitrophenylhydrazine indicates absence of aldehydic (option *a*) and ketonic (option *c*) group, while negative reaction with metallic sodium indicates absence of an alcoholic group (option *b*), hence the organic compound, C_3H_6O is $CH_2 = CHOCH_3$ (option *d*).

70. (c) SiF_4 and CO_2 have zero dipole moment due to symmetrical tetrahedral and linear structures, NO_2 and O_3 both have V-shape and certain value for dipole moment.

71. (b) Number of unit cells

$$= \frac{\text{Total number of molecules}}{\text{Rank of Unit Cell (Z)}}$$

$$\therefore \text{Number of atoms} = 12.08 \times 10^{23} \times 2$$

$$= 24.16 \times 10^{23} \quad (\text{for BCC } Z = 2)$$

72. (c) Oxidation potential of M is more than Ni and less than Mn. Hence reducing power Mn > M > Ni

73. (d) SO_2 changes to H_2SO_4 (O.N. changes from +4 to +6 oxidation)

$2KI \rightarrow I_2$

(O.S. changes from –1 to 0 oxidation)

$PbS \rightarrow PbSO_4$

(O.S. changes from –2 to +6 oxidation)

$Ag_2O \rightarrow 2Ag$

(O.S. changes from +1 to 0 Reduction)

74. (d) Pyridine and thiourea form volatile cupric cyanide and also give bluish green flame like volatile cupric halides.

75. (a)

$$\begin{array}{l} CHO \\ | \\ CHOH \\ | \\ (CHOH)_3 \\ | \\ CH_2OH \end{array} \xrightarrow{3Ph\,NH\,NH_2} \begin{array}{l} CH{=}N.NHPh \\ | \\ CH{=}N.NHPh \\ | \\ (CHOH)_3 \\ | \\ CH_2OH \end{array} + PhNH_2 + NH_3$$

76. (a) $\mu_{rms} \propto \sqrt{\frac{1}{m}}$ at STP

78. (a) Relative lowering of V.P. = mole fraction of Solute

$$\frac{p^o - p_s}{p^o} = x_2$$

$$\frac{143 - p_s}{143} = \frac{0.5/65}{158/154}; \; p_s = 141.93 \text{mm}$$

79. (b) Cyanide process is for gold (I-D); floatation process - pine oil (II-B); Electrolytic reduction - Al (III-C); Zone refining - Ge (IV-A).

80. (b) Only structure (b) has a conjugated system, which is necessary for resonance.

81. (b) It corresponds to Balmer. Series $n_1 = 2$ and $n_2 = 5$.

82. (c) Liquid - liquid system is known as emulsion.

83. (c) $CaCO_3 \xrightarrow{\Delta} CaO + CO_2$;

$CaO + H_2O \rightarrow \underset{(B)}{Ca(OH)_2}$

$Ca(OH)_2 + CO_2 \rightarrow \underset{A}{CaCO_3} + H_2O$

84. (d) d_5 —— strong ligand field

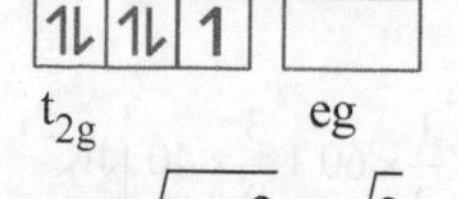

$\mu = n\sqrt{n+2} = \sqrt{3} = 1.73$BM

d^3 – in weak as well as in strong field

1	1	1		
t_{2g}			eg	

$\mu = \sqrt{3(5)} = \sqrt{15} = 3.87$ B.M.

d^4 – in weak ligend field

1	1	1	1
t_{2g}			eg

$\mu = \sqrt{4(6)} = \sqrt{24} = 4.89$

d^4 – in strong ligand field

⇅	1	1	
t_{2g}			eg

$\mu = \sqrt{2(4)} = \sqrt{8} = 2.82$

85. (c)

Species	N_2	O_2	N_2^-	O_2^-
Bond order	3	2	2.5	1.5

The O-O bond is O_2^- decreases

86. (d) Hybridisation in $Si(CH_3)_4$ (½ [4 + 4 + 0 – 0] = 4) is sp^3 which is tetrahedral.

87. (b)

	A + B	⇌	C + D
Initial	0.9 0.9		0 0
At eqm.	0.3 0.3		0.6 0.6

$$K_c = \frac{0.6 \times 0.6}{0.3 \times 0.3} = 4$$

88. (c) For *bcc* lattice, $4r$ = diagonal of the cube $= a\sqrt{3}$

$$\text{or}, \; r = \frac{350 \times 1.732}{4} = 151.5 \text{pm}$$

89. (b) In general,

$$\underset{\text{Aromatic aldehyde}}{\text{ArCHO}} \xrightarrow{CN^-} \underset{\text{Acyloins}}{\text{Ar}\overset{\text{OH}}{\overset{|}{\text{CH}}}\overset{\text{O}}{\overset{\|}{\text{C}}}\text{Ar}}$$

SECTION 3 - BOTANY

100. (a) Amphitropous : Both body of ovule and embryo sac are curved. The embryo sac assumes horse-shoe shape. *e.g. Papaveraceae.*

103. (b) DDT causes egg shell thinning in birds because it inhibits Calcium ATPase.

104. (d) Gene therapy is an experimental technique that uses genes to treat or prevent disease. The first clinical gene therapy was given for treating adenosine deaminase deficiency. A four-year old girl became the first gene therapy patient on September 14, 1990 at the NIH Clinical Center. Adenosine deaminase deficiency, also called ADA deficiency or ADA-SCID is an autosomal recessive metabolic disorder that causes immunodeficiency. ADA deficiency is due to a lack of the enzyme adenosine deaminase.

108. (a) Secondary wall situated near the plasma membrane after the formation of primary wall.

114. (d) The male gamete will be haploid (n). 2 polar nuclei will be diploid (2n). Endosperm formed by fusion of male gamete with two polar nuclei will be pentaploid.

$$\underset{(n)}{\text{male gamete}} + \underset{(2n)+(2n)}{\text{2 polar nuclei}} \rightarrow \underset{(5n)}{\text{Endosperm}}$$

116. (d) An area with particular types of trees dedicated to local deities or ancestral spirits that are protected by local communities through social traditions and taboos incorporating spiritual and ecological values are called as sacred groves.

Sacred groves act as an ideal centre for biodiversity conservation. Several plants and animals that are threatened in the forest are still well conserved in some of the sacred groves. It has been observed that several medicinal plants that are not to be found in the forest are abundant in the sacred groves. Further, rare, endangered, threatened and endemic species are often concentrated in sacred groves.

117. (c) In *Cocos nucifera* (coconut) milky endosperm is found in which many nuclei, vitamins and growth hormone e.g., cytokinins, auxin and induced cytokinin is found.

129. (b) With increasing humidity, the rate of transpiration decreases linearly, because the high saturation of water vapour in the atmosphere prevents the evaporation of more water from the leaf interior to the exterior.

130. (a) The gases that makes highest relative contribution to the green house gases are carbon dioxide (CO_2) and methane (CH_4).

132. (d) Biolistic is direct gene transferred method for constructing recombinant DNA. The gene gun was invented by John C. Sanford with Edward Wolf. A gene gun can be used to genetically infect cells or whole organisms with foreign DNA by aiming the barrel of the gun and firing. The microshot projectiles in the biolistic gene gun are made of microscopic (or nano) sized gold or platinum powders. These expensive powders are soaked in DNA or RNA (in raw or plasmid form) that are engineered for insertion into the genome of the cells or organisms under the gun.

133. (a) Statement (i) and (iv) are / is correct

Regulation of *lac* operon by repressor is referred as negative regulation. In negative regulation, a repressor molecule binds to the operator of an operon and terminates transcription. In positive regulation, an activator interacts with the RNA polymerase in the promoter region to initiate transcription.

Human genome contains some 20,000-25,000 billion genes bases.

SECTION 4 - ZOOLOGY

143. (a) ANF stimulates the loss of sodium in urine while aldosterone absorbs sodium from glomerular filtrate.

148. (d) Cortisol and testosterone are steroid hormones which are lipid soluble and easily pass through the cell membrane of a target

cell and bind to specific intracellular receptor.

149. (a) The caecum is a pouch-like portion of the large intestine which hosts some symbiotic micro-organism. The caecum absorbs water and salts from undigested foods before they continue on to the large intestine.

154. (c) Brown fat provides 20- times more energy than white fat.

156. (a) The Chylomicrons are formed inside enterocytes and are absorbed in lacteals

165. (a) Epinephrine and norepinephrine are secreted by adrenalmedulla in response to stress of any kind and during emergency situations and are called emergency hormones or hormones of flight, or fight.

166. (c) Troponin is a protein which is found on actin filament and myosin protein is found in myosin filament. Both actin and myosin are complex proteins in striated muscles. Thymosin is a hormone secreted by the thymus that stimulates development of T-cells. Prolactin is a hormone released by the pituitary gland that stimulates breast development and milk production in women. Rhodopsin, also known as visual purple, is not a hormone. It is a biological pigment in photoreceptor cells of the retina that is responsible for the first events in the perception of light.

NEET (NATIONAL ELIGIBILITY ENTRANCE TEST) Full Test - 9

ANSWER KEY

1	(a)	21	(c)	41	(a)	61	(d)	81	(d)	101	(a)	121	(a)	141	(d)	161	(a)
2	(c)	22	(d)	42	(a)	62	(b)	82	(b)	102	(d)	122	(c)	142	(c)	162	(b)
3	(c)	23	(c)	43	(a)	63	(c)	83	(d)	103	(a)	123	(c)	143	(c)	163	(d)
4	(b)	24	(d)	44	(b)	64	(d)	84	(d)	104	(a)	124	(b)	144	(b)	164	(d)
5	(c)	25	(b)	45	(a)	65	(c)	85	(a)	105	(d)	125	(a)	145	(d)	165	(d)
6	(a)	26	(a)	46	(b)	66	(a)	86	(c)	106	(d)	126	(d)	146	(b)	166	(c)
7	(c)	27	(b)	47	(c)	67	(c)	87	(c)	107	(b)	127	(a)	147	(a)	167	(d)
8	(a)	28	(c)	48	(b)	68	(d)	88	(a)	108	(d)	128	(a)	148	(a)	168	(b)
9	(d)	29	(c)	49	(d)	69	(a)	89	(d)	109	(d)	129	(b)	149	(d)	169	(b)
10	(b)	30	(d)	50	(a)	70	(c)	90	(a)	110	(a)	130	(d)	150	(b)	170	(d)
11	(d)	31	(c)	51	(a)	71	(a)	91	(c)	111	(d)	131	(d)	151	(b)	171	(c)
12	(a)	32	(c)	52	(c)	72	(c)	92	(d)	112	(d)	132	(d)	152	(d)	172	(c)
13	(c)	33	(c)	53	(d)	73	(a)	93	(d)	113	(b)	133	(b)	153	(c)	173	(d)
14	(d)	34	(a)	54	(a)	74	(d)	94	(a)	114	(c)	134	(b)	154	(b)	174	(a)
15	(b)	35	(c)	55	(d)	75	(b)	95	(b)	115	(d)	135	(a)	155	(d)	175	(a)
16	(b)	36	(c)	56	(a)	76	(b)	96	(d)	116	(c)	136	(a)	156	(c)	176	(a)
17	(d)	37	(b)	57	(c)	77	(b)	97	(d)	117	(c)	137	(a)	157	(d)	177	(b)
18	(d)	38	(c)	58	(b)	78	(b)	98	(a)	118	(d)	138	(b)	158	(c)	178	(c)
19	(a)	39	(a)	59	(a)	79	(d)	99	(a)	119	(d)	139	(b)	159	(a)	179	(a)
20	(b)	40	(d)	60	(c)	80	(a)	100	(c)	120	(a)	140	(c)	160	(b)	180	(b)

HINTS & SOLUTIONS

SECTION 1 -PHYSICS

1. (a) Number of significant figures in 23.023= 5
Number of significant figures in 0.0003 = 1
Number of significant figures in 2.1×10^{-3} =2

2. (c) New K.E., E' = 4E

$$p = \sqrt{2mE} \text{ and } p' = \sqrt{2mE'}$$

$$\frac{p}{p} = \sqrt{\frac{2m \times 4E}{2mE}} = 2$$

$$\frac{p'}{p} - 1 = 2 - 1$$ [on substrating 1 in both sides.]

$$\frac{p-p}{p} \times 100 = (2-1) \times 100 = 100\%$$

3. (c) If T is the tension in the string , Then
$T = mg$ (For outer masses)

$2T\cos\theta = \sqrt{2}\,mg$ (For inner mass)

Eliminating T, we get

$$2(mg)\cos\theta = \sqrt{2}\,mg$$

or $\cos\theta = 1/\sqrt{2} \Rightarrow \theta = 45°$

4. (b) $S_1 = \frac{1}{2} g \left(\frac{1}{2}\right)^2, S_1 + S_2 = g(1)^2$

$$\frac{1}{2} g \left(\frac{3}{2}\right)^2 = S + S_2 + S_3$$

5. (c) Bulk Modulus

$$= \frac{\text{Pressure}}{\text{Volume Strain}} = \frac{\text{Pressure}}{0}$$

Bulk Modulus = ∞

[As liquid is uncompressible, $\Delta V = 0$]

6. (a) The unit vector of any vector $\vec{A}$ is defined as $\hat{A} = \frac{\vec{A}}{|\vec{A}|}$

8. (a) $\xi = \frac{W}{Q} \Rightarrow V = \frac{W}{Q} \Rightarrow W = QV$

9. (d) Acceleration is the same everywhere equal to 'g'.

10. (b) $T = m\omega^2 r$

12. (a) Due to the charge inside a sphere of radius r only.

13. (c) Mass = 150 gm = $\frac{150}{1000}$ kg

Force = Mass × acceleration

$= \frac{150}{1000} \times 20\text{N} = 3\text{N}$

Impulsive force = $F.\Delta t = 3 \times 0.1 = 0.3\ N$

14. (d) Applying Snell's law at P,

$${}^1\mu_2 = \frac{\sin i}{\sin r_1} = \frac{\mu_2}{\mu_1} \quad ...(1)$$

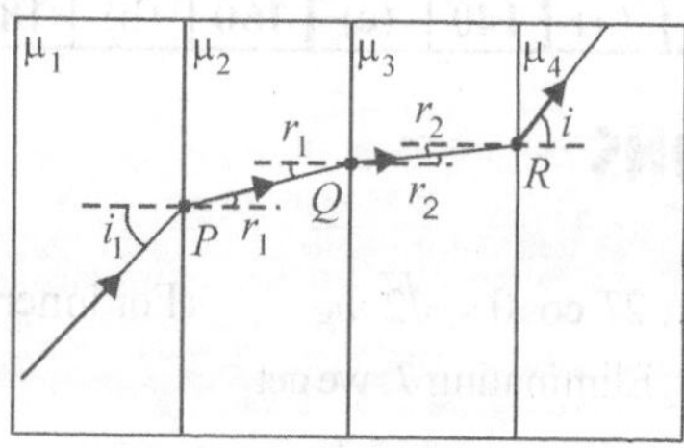

Applying Snell's law at Q,

$${}^2\mu_3 = \frac{\sin r_1}{\sin r_2} = \frac{\mu_3}{\mu_2} \quad ...(2)$$

Again applying Snell's law at R

$${}^3\mu_4 = \frac{\sin r_2}{\sin i} = \frac{\mu_4}{\mu_3} \quad ...(3)$$

Multiplying (i), (ii) and (iii), we get

$\mu_4 = \mu_1$

If the emergent ray is parallel to incident ray after travelling a number of parallel interfaces then the refractive index of the first and the last medium is always same.

15. (b) Paramagnetic liquid tends to flow from region of weaker magnetic fields to stronger magnetic fields.

16. (b) Acceleration $= \frac{d^2s}{dt^2} = \frac{2}{3}\text{m/s}^2$

Force acting on the body

$= 3 \times \frac{2}{3} = 2$ newton

Displacement in 2 sec $= \frac{1}{3} \times 2 \times 2 = \frac{4}{3}$m

Work done $= 2 \times \frac{4}{3} = \frac{8}{3}$ J

17. (d) Torque $= \frac{dL}{dt} = \frac{4A - A}{4} = \frac{3}{4}A$

[dL is change in angular momentum]

18. (d) All the given phenomena are explained by kinetic theory of gases.

19. (a) Resistance of bulb

$$= \frac{V^2}{P} = \frac{200 \times 200}{40} = 1000\Omega$$

New power $= \frac{V^2}{R} = \frac{100 \times 100}{1000} = 10\text{W}$.

20. (b) Let W newton be the load suspended. Then

$$Y = \frac{(W/A_1)}{(\ell_1/L)} = \frac{W L}{A_1 \ell_1} \quad ...(1)$$

and $Y = \frac{(W/A_2)}{(\ell_2/L)} = \frac{W L}{A_2 \ell_2}$ ....(2)

Dividing equation (1) by equation (2), we get

$$1 = \left(\frac{\ell_2}{\ell_1}\right)\left(\frac{A_2}{A_1}\right) = \left(\frac{\ell_2}{\ell_1}\right)\left(\frac{2}{1}\right)$$

$\therefore \frac{\ell_1}{\ell_2} = \frac{2}{1}$ or $\ell_1 : \ell_2 = 2 : 1$

21. (c)

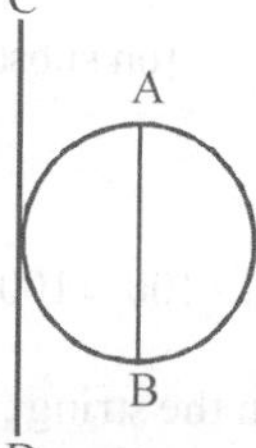

We know that moment of inertia of a disc about its diameter is $= \frac{1}{4}MR^2$.

In the figure AB is a diameter. Which passes through centre of mass of the disc. CD is another axis which is parallel to AB and about which we shall have to calculate moment of inertia.

Now, from the theory of parallel axis

$$I = I_g + Md^2 = \frac{1}{4}MR^2 + MR^2 = \frac{5MR^2}{4}$$

22. (d) We know that $\frac{g'}{g} = \frac{R^2}{(R+h)^2}$

$$\therefore \frac{g/9}{g} = \left[\frac{R}{R+h}\right]^2$$

$$\therefore \frac{R}{R+h} = \frac{1}{3}$$

$\therefore h = 2R$

23. (c) Work done = Area under curve *ACBDA*

24. (d) The apparent frquency of sound of both the trains will be changed.

The apparent frequency of sound of train coming in $= 240 \times \frac{320}{320-4}$

$$= \frac{240 \times 320}{316} = 243$$

For apparent frequency of train going away

$$= 240 \times \frac{320}{320+4} = 237$$

When these sounds are heard simultaneously beat is listened.

No. of beats per sec = 243 – 237 = 6.

25. (b) The Shunt of milliammeter will be of larger resistance, because a large part of the main current will pass through the coil of the galvanometer which will work as a smaller range ammeter.

26. (a) $y = 0.03 \sin \pi(2t - 0.01x)$ is equation of a progressive wave.

Comparing it with the standard form of equation

$y = a \sin(\omega t - kx)$

$\omega = 2\pi$; $k = 0.01\pi$

$$k = \frac{2\pi}{\lambda} = 0.01\pi$$

$$\Rightarrow \lambda = \frac{2}{0.01} = 200 \text{ m.}$$

27. (b) Three 5Ω resistors are in series. Their total resistance = 15Ω. Now it is in parallel with 5Ω resistor, so total resistance,

$$\frac{1}{R} = \frac{1}{5} + \frac{1}{15} = \frac{3+1}{15} = \frac{4}{15}$$

$$R = \frac{15}{4}$$

$$\therefore I = \frac{V}{R} = \frac{3}{15/4} = \frac{3 \times 4}{15} = 0.8A$$

28. (c) The direction of oscillations of *E* and *B* fields are perpendicular to each other as well as to the direction of propagation. So, electromagnetic waves are transverse in nature.

The electric and magnetic fields oscillate in same phase.

29. (c) Work done,

$$W = \Delta U = U_f - U_i = 0 - \left[\frac{-GMm}{R}\right]$$

$$W = \frac{6.67 \times 10^{-11} \times 100}{0.1} \times \frac{10}{1000}$$

$= 6.67 \times 10^{-10}$ J

30. (d) Since $T_n = \frac{T_i + T_c}{2}$ = Neutral temperature

$$T_n = \frac{T_i + 0}{2} = \frac{T_i}{2}$$

[T_c = 0°C = temperature of cold junction]

31. (c) Since efficiency of engine is $\eta = 1 - \frac{T_2}{T_1}$

According to problem,

$$\frac{1}{6} = 1 - \frac{T_2}{T_1} \quad ...(1)$$

When the temperature of the sink is reduced by 62°C, its efficiency is doubled

$$2\left(\frac{1}{6}\right) = 1 - \frac{T_2 - 62}{T_1} \quad ...(2)$$

Solving (1) and (2) $T_2 = 372$ K

$T_1 = 99°C$ = Temperature of source.

32. (c) The intermediate image in compound microscope is real, inverted and magnified.

33. (c) We know that for short dipole,

field at axial point, $E_a = \frac{2p}{d^3}$

and field at equatorial point, $E_q = \frac{p}{d^3}$

So, $E_a = 2E_q$

34. (a) 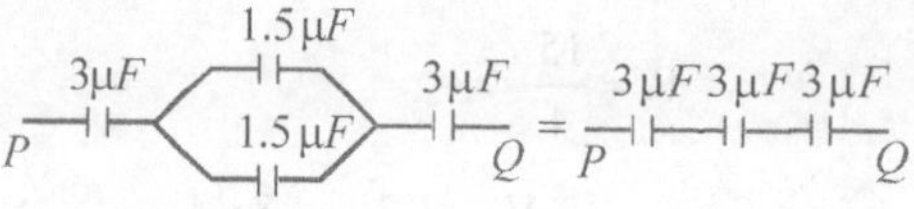 [Circuit: P —3μF— (1.5μF ∥ 1.5μF) —3μF— Q = P —3μF—3μF—3μF— Q]

$$\frac{1}{C_{effective}} = \frac{1}{3} + \frac{1}{3} + \frac{1}{3} = 1$$

$\Rightarrow C_{effective} = 1\,\mu F$

35. (c) Let $I_1 = I$ and $I_2 = 4I$

$$I_{max} = \left(\sqrt{I_1} + \sqrt{I_2}\right)^2 = \left(\sqrt{I} + \sqrt{4I}\right)^2$$

$$= \left(3\sqrt{I}\right)^2 = 9I$$

$$I_{min} = \left(\sqrt{I_1} - \sqrt{I_2}\right)^2 = \left(\sqrt{I} - \sqrt{4I}\right)^2 = I$$

36. (c) When the electron jumps from any orbit to second orbit, then wavelength of line obtained belongs to Balmer series.

37. (b) From Wien's displacement law,

$\lambda_{max} T$ = constant

$\therefore (\lambda_{max})_1 T_1 = (\lambda_{max})_2 T_2.$

$$\therefore (\lambda_{max})_2 = \frac{4800 \times 6000}{3000} Å = 9600 Å$$

38. (c) $r = \frac{mv}{qB}$ or $r \propto v$

As v is doubled, the radius also becomes double. Hence radius = 2 × 2 = 4 cm

39. (a) Power $\propto$ no. of electrons emitted (N)

$$P \propto \frac{1}{r^2} \Rightarrow N \propto \frac{1}{r^2}$$

40. (d) Since $\phi_{total} = \phi_A + \phi_B + \phi_C = \frac{q}{\varepsilon_0}$

where q is the total charge.

As shown in the figure, flux associated with the curved surface B is $\phi = \phi_B$

Let us assume flux linked with the plane surfaces A and C be

$\phi_A = \phi_C = \phi'$

Therefore,

$$\frac{q}{\varepsilon_0} = 2\phi' + \phi_B = 2\phi' + \phi$$

$$\Rightarrow \phi' = \frac{1}{2}\left(\frac{q}{\varepsilon_0} - \phi\right)$$

41. (a) $B = \frac{\mu_0 In}{2\pi a}$ Add vectorially.

42. (a) The work function has no effect on photoelectric current so long as $h\nu > W_0$. The photoelectric current is proportional to the intensity of incident light. Since there is no change in the intensity of light, hence $I_1 = I_2$.

43. (a) The side-wall of optical fibre provides total internal reflection of beam of light incident on it.

44. (b) When *p-n* junction is reverse biased, the flow of current is due to drifting of minority charge carriers across the junction.

45. (a) The current will flow through R_L when the diode is forward biased.

SECTION 2 - CHEMISTRY

46. (b) $E = h\upsilon = \frac{ch}{\lambda}$; and $\upsilon = \frac{c}{\lambda}$

$$8 \times 10^{15} = \frac{3.0 \times 10^8}{\lambda}$$

$$\therefore \lambda = \frac{3.0 \times 10^8}{8 \times 10^{15}} = 0.37 \times 10^{-7}$$

$= 37.5 \times 10^{-9}$ m $= 4 \times 10^1$ nm

48. (b) The right sequence of $I.E_1$ of Li < B < Be < C.

49. (d) Magetic moment $\mu = \sqrt{n(n+2)}$

where n = number of unpaired electrons

$\sqrt{15} = \sqrt{n(n+2)}$ $\therefore$ n = 3

51. (a) O_2^- has Bond order 1.5 (see text)

52. (c) $w = -P\Delta V = -10^{-5}(1 \times 10^{-2} - 1 \times 10^{-3})$

$= -900$J

53. (d) $\frac{4}{3}Al + O_2 \longrightarrow \frac{2}{3}Al_2O_3; \Delta G = -827 kJmol^{-1}$
Number of electrons involved is 4 $\left(\frac{2}{3} \times 6 = 4\right)$

$\Delta G^o = -nFE^o$ or $\frac{827000}{4 \times 96500} = E^o$

$\therefore E^o = 2.14V$

54. (a) $4FeS + 7O_2 \rightarrow 2Fe_2O_3 + 4SO_2$
$SO_2 + H_2O \rightarrow H_2SO_3 \rightleftharpoons 2H^+ + SO_3^{2-}$
It is dibasic

55. (d) The two components should be $(CH_3)_3CONa + (CH_3)_3CBr$. However, tert-alkyl halides tend to undergo elimination reaction rather than substitution leading to the formation of an alkene, $Me_2C = CH_2$

56. (a) See Le Chatelier's principle.

57. (c) According to Raoult's law $p_A = p^o_A \times x_B$ (partial pressure of solvent is equal to pure pressure of solvent multiplied by mole fraction of solvent.

58. (b) Follow ECS, more the reduction potential, the stronger the oxidising power and vice versa.

59. (a) Cu, Hg and Ag are attacked by conc acids but gold is not attacked. $AuCl_3$ forms $H[AuCl_4]$ a complex with HCl which is used for toning in photography.

60. (c) In presence of non-protic solvent such as $CHCl_3$ or CCl_4, concentration of electrophilc (Br^+) is less, hence reaction stops at the monobromo stage

61. (d) $K_{eqm} = \frac{K_f}{K_b} = \frac{2}{1} = 2$

62. (b) In AB AB packing spheres occupy 74%. 26% is empty.

63. (c) The unreacted AgBr is removed by soluble complex formation
$AgBr + 2Na_2S_2O_3 \rightarrow Na_3[Ag(S_2O_3)_2] + NaBr$

64. (d) The fluorine has low dissociation energy of F - F bond and reaction of atomic fluorine is exothermic in nature

65. (c) α-D-glucose and β-D-glucose differ in the arrangement of groups around one carbon atom (C_1) i.e. they differ in configuration at C_1, hence these are also known as anomers

66. (a) For CuS solubility is $(10^{-31})^{1/2}$;
For $Ag_2S = \left(\frac{k_{sp}}{4}\right)^{\frac{1}{3}} = \left(\frac{10^{-44}}{4}\right)^{\frac{1}{3}}$ and for

$HgS = \left(10^{-54}\right)^{\frac{1}{2}}$

67. (c) Sugar forms homogeneous solution hence no Tyndall effect is exhibited

68. (d) Most electropositive metals are obtained by electrolysis of their fused ionic salts.

69. (a) The complex chlorodiaquatriammine cobalt (III) chloride can have the structure $[CoCl(NH_3)_3(H_2O)_2]Cl_2$

71. (a) Ionic character follows the order
HF > HCl > HBr > HI

72. (c) pH = 3. ∴ $[H^+] = 10^{-3}$; pH = 6 ∴ $[H^+] = 10^{-6}$.
Hence $[H^+]$ reduced by 10^{-3} times.

75. (b) Recall that presence of electron-withdrawing group increases, while presence of electron-releasing group decreases the acidity of carboxylic acids.

$ClCH_2COOH > H-\overset{O}{\overset{||}{C}}-OH > CH_3-\overset{O}{\overset{||}{C}}-OH > C_2H_5-\overset{O}{\overset{||}{C}}-OH$

(electron-withdrawing gp.) (Electron-releasing character increasing from Left to Right)

76. (b) $Mg + 2HCl \rightarrow MgCl_2 + H_2 \uparrow$
1 mole — 1 mole
$\frac{1}{2}$ mole — $\frac{1}{2}$ mole (12g of Mg = $\frac{1}{2}$ mol)

77. (b) Desired eqn;
$C_2H_4 + 3O_2 \rightarrow 2CO_2 + 2H_2O$
$\Delta H = 2(-394) + 2(-286) - (52) = -1412 KJ$

78. (b) LHS : $H_2(P_1) \rightleftharpoons 2H^+ + 2e^-$ Oxidation
RHS : $2H^+ + 2e^- \rightleftharpoons H_2(P_2)$ Reduction
Net reaction :
$H_2(P_1) \rightleftharpoons H_2(P_2)$

$E = E^o - \frac{RT}{nF} \ln . \frac{P_2}{P_1} = 0 + \frac{RT}{2F} \ln \frac{P_1}{P_2}$

79. (d) Ozone absorbs U.V. radiations harmful to human life.

80. (a) Aldehydes, other than formaldehyde, when treated with RMgX give 2° alcohols

81. (d) O.N. of N in NO_2 and N_2O_4 is +4
∴ difference is zero.
O.N. of P in P_2O_5 and P_4O_{10} is +5
∴ difference is zero
O.N. of N in N_2O is +1 and in NO is +2. The difference is 1
O.N. of S in SO_2 is +4 and in SO_3 is +6. The difference is +2.

82. (b) $M = \frac{K_b \times w \times 1000}{\Delta T_b \times W}$

$= \frac{2.16 \times 0.11 \times 1000}{0.1 \times 15} = 158.4$

83. (d) In acid hydrolysis of methyl acetate all are present in one phase (liquid)

84. (d) $k = \frac{2.303}{t}\log\frac{a}{a-x}$

$= \frac{2.303}{24}\log\frac{1}{\frac{1}{8}} = \frac{2.303}{24}\log 8$

86. (c) Nitrogen can form NCl_3, N_2O_5 Ca_3N_2 and not NCl_5 since it has no d atomic orbitals in valence shell

87. (c) Copper is present in brass, bronze and german silver (see list - of alloys)

88. (a) The compounds of the type $M(AA)_2B_2$ exhibit both geometrical and optical isomerism

89. (d) More the stability of the carbocation, higher will be the reactivity of the parent chloride. Allyl chloride > Vinyl chloride > Chlorobenzene

SECTION 3 - BOTANY

98. (a) Natural auxin are naturally occurring auxin plants hormone which is called phytohormones. Eg. IAA and IBB. These are synthesized in shoot apices, leaf primordial and developing seeds from tryptophan. Some of the important synthetic auxins are 2, 4-D; 2, 4, 5-T (2, 4, 5- trichlorophenoxy acetic acid).

99. (a) Generative cell which is a cell of the male gametophyte or pollen grain in seed plants that divides to give rise directly or indirectly two sperms.

100. (c) Genotype is the genetic make up of an individual irrespective of the mendelian characters or genes impressing.

Parent : AABB × aabb

Gametes : (AB) (ab)

F_1 : AaBb

	AB	Ab	aB	ab
AB	AABB	AABb	AaBB	AaBb
Ab	AABb	AAbb	AaBb	Aabb
aB	AaBB	AaBb	aaBB	aaBb
ab	AaBb	Aabb	aaBb	aabb

101. (a) In negative (repressible) operon, the repressor co-repressor complex binds with the operator. The free repressor cannot bind to the operator.

102. (d) In submerged hydrophytes whole plant body remains under water. *e.g. Ceratophyllum* and *Utricularia*. In these plants, stomata is absent and gaseous exchange takes place through general body surface.

103. (a) *Azolla* and Blue green algae- *Anabaena* form biofertilizer in rice fields.

104. (a) Transgenic plants are the ones generated by introducing foreign DNA into a cell and regenerating a plant from that cell.

105. (d) The aim of Montreal protocol was to protect the ozone layer in the stratosphere by decreasing and eventually eliminating the use of of ozone depleting substances like CFCS.

109. (d) In collenchyma cell walls show localized thickenings due to presence of approximately 45% pectin, 35% hemicellulose and 20% cellulose.

114. (c) All others are diploid. Best method of haploid plant culture is pollen grain culture.

115. (d) In the DNA molecule, there are two strands which run anti-parallel one is 5' - 3' direction and other in 3' -5' direction, the two chains are held together by hydrogen bonds

between their bases. Adenine (A), a purine of one chain his exactly opposite thymine (T), a pyrimidine of the other chain. Similarly, cytosine (C), a pyrimidine lies opposite guanine (G), a purine. This allows a sort of lock & key arrangment between large sized purine & small sized pyrimidine. It is strengthened by the appearance of hydrogen bonds between the two.

116. (c) 2, 4-Dichlorophenoxy acetic acid is used for causing defoliation of forest trees.

2, 4-D or 2, 4 dichlorophenoxy acetic acid is an auxin hormone. It stimulates the growth activities of the cells of the root due to which roots get destroyed and thus plants finally destroys. 2, 4-D is used as a defoliant for broad leaved dicots.

117. (c) If the forest cover is reduced to half, large areas will become deserts.

118. (d) Fluorides of carbon is the major pollutant from jet plane emission.

119. (d) Exhaustible resources are those natural resources that are likely to be exhausted due to their continuous use. Forest is a renewable exhaustible resource that can be again and again produced. Coal, petroleum and minerals are non-renewable exhaustible natural resources that cannot be again produced.

123. (c) Being a xerophytic plant, *Nerium* bears multiple epidermis to cut the rate of transpiration.

126. (d) The sibling species are those species, which are born from a similar kind of parents having similar morphology. But they are reproductively isolated.

127. (a) DDT cannot be further metabolized into any stable by product.

128. (a) Nitrogenase is an enzyme involved in biological nitrogen fixation. Enzyme nitrate reductase is involved in conversion of nitrate to nitrite. Conversion of ammonia to nitrate is carried out by *Nitrosomonas* and *Nitrobacter.*

129. (b) Ex-situ conservation is the conservation of selected organism in places outside their natural homes. They include off site collection and gene banks.

SECTION 4 - ZOOLOGY

137. (a) All sponges are marine and have collared cells without any exception. Sponges are animals of the phylum Porifera. They are multicellular organisms which have bodies full of pores and channels allowing water to circulate through them, consisting of jelly-like substence sandwiched between two thin layers of cells.

144. (b) In human female, the periodic discharge of blood, mucus and cellular debris from uterine wall from non-pregnant women of sexual maturity is known as menstrual cycle. Few hours before the start of mensuration, the spiral arterioles constrict one by one resulting into the mucosa. The mucosa shrinks and the death of the blood deprived tissues takes place.

145. (d) Phenylketonuria is a human genetic disorder in which the body does not contain the enzyme phenylalaninc hydroxylase, necessary to metabolize phenylalanine to tyrosine, and converts phenylalanine instead to phenylpyruvic acid. As PKU is an autosomal recessive genetic disorder each parent must have at least one defective allele of the gene for PAH, and the child must inherit a defective allele from each parent. As such, it is possible for a parent with a PKU phenotype to have a child without PKU if the other parent possesses at least one functional allele of the gene for PAH. A child of two parents with the PKU phenotype will always receive two defective alleles so will always have PKU. The gene for PAH is located on chromosome 12, at location 12q22-q24.2.

146. (b) Mutation is any random sudden heritable change occurring in the genetic material. Sexual selection refers to selection of a mate by an organism.

147. (a) Geometric representation of age structure a characteristic of population.

148. (a) Monoclonal antibodies are identical molecules specific for one type of antigen. These are obtained by injecting the target antigen into a rat or mouse. Sometimes later, the spleen cells producing antibodies are isolated and fused with myeloma cells to produce monoclonal antibodies.

158. (c) The amount of yolk and its distribution determines the type and depth of cleavage. Cleavage is effected due to yolk resulting in different types of segmentation or cleavage.

159. (a) If the vasa efferentia in the human reproductive system get blocked the gametes will not be transformed from testes to epididymis.

160. (b) Medical termination of pregnancy is considered safe up to twelve weeks of pregnancy.

161. (a) Antibodies bind to specific antigens to produce large insoluble complexes which render them harmless and facilitate their destruction by other cells of the immune system this process is called agglutination.

162. (b) Ultrasound imaging has frequency range 1 - 15 MHz which has become part of our lives in the last decade. We are now all familiar with the blurry black and white sonograms (ultrasound pictures) that show the unborn baby inside the mother's uterus. Although it may take a little imagination on our part to understand that a round circle is the baby's head, that does not prevent a proud mother-to-be from showing off her first baby picture.

163. (d) Cry I endotoxine obtained from *Bacillus thuringiensis* are effective against bollworms. A bollworm is a common term for any larva of a moth that attacks the fruiting bodies of certain crops, especially cotton.

172. (c) The intensity of movement of basilar fibres regulates the loudness or amplitude of sound.

179. (a) Chromosomes occur in homologous pairs. Somatic cells have diploid number of chromosomes. Humans have 23 pairs *i.e.* 46 chromosomes. 22 pairs of autosomes and 1 pair of sex chromosomes i.e. XX in females and XY in males.

NEET (NATIONAL ELIGIBILITY ENTRANCE TEST) Full Test - 10

ANSWER KEY																	
1	(d)	21	(b)	41	(b)	61	(a)	81	(c)	101	(b)	121	(b)	141	(a)	161	(a)
2	(c)	22	(d)	42	(d)	62	(c)	82	(b)	102	(b)	122	(d)	142	(c)	162	(b)
3	(a)	23	(a)	43	(b)	63	(d)	83	(d)	103	(c)	123	(b)	143	(c)	163	(c)
4	(c)	24	(d)	44	(c)	64	(b)	84	(c)	104	(d)	124	(a)	144	(a)	164	(d)
5	(c)	25	(b)	45	(a)	65	(c)	85	(d)	105	(b)	125	(b)	145	(d)	165	(d)
6	(d)	26	(b)	46	(c)	66	(a)	86	(c)	106	(d)	126	(d)	146	(a)	166	(c)
7	(c)	27	(d)	47	(d)	67	(d)	87	(c)	107	(c)	127	(b)	147	(d)	167	(b)
8	(c)	28	(a)	48	(c)	68	(a)	88	(a)	108	(d)	128	(b)	148	(d)	168	(a)
9	(c)	29	(a)	49	(d)	69	(d)	89	(b)	109	(a)	129	(a)	149	(c)	169	(a)
10	(b)	30	(b)	50	(d)	70	(d)	90	(d)	110	(a)	130	(d)	150	(b)	170	(d)
11	(b)	31	(d)	51	(d)	71	(c)	91	(c)	111	(a)	131	(a)	151	(a)	171	(c)
12	(a)	32	(d)	52	(a)	72	(c)	92	(a)	112	(a)	132	(b)	152	(d)	172	(d)
13	(a)	33	(b)	53	(c)	73	(a)	93	(a)	113	(c)	133	(b)	153	(c)	173	(d)
14	(a)	34	(d)	54	(c)	74	(d)	94	(a)	114	(d)	134	(b)	154	(c)	174	(b)
15	(a)	35	(a)	55	(a)	75	(b)	95	(b)	115	(c)	135	(c)	155	(b)	175	(a)
16	(a)	36	(d)	56	(a)	76	(a)	96	(b)	116	(b)	136	(b)	156	(c)	176	(c)
17	(d)	37	(a)	57	(c)	77	(d)	97	(d)	117	(b)	137	(b)	157	(a)	177	(c)
18	(c)	38	(c)	58	(b)	78	(b)	98	(c)	118	(b)	138	(b)	158	(b)	178	(c)
19	(b)	39	(d)	59	(a)	79	(a)	99	(c)	119	(d)	139	(b)	159	(a)	179	(a)
20	(b)	40	(c)	60	(c)	80	(d)	100	(a)	120	(b)	140	(b)	160	(b)	180	(a)

HINTS & SOLUTIONS

SECTION 1 -PHYSICS

1. (d) $f = c\, m^x k^y$;

Spring constant k = force/length.

$[M^0L^0T^{-1}] = [M^x (MT^{-2})^y] = [M^{x+y} T^{-2y}]$

$\Rightarrow x + y = 0, -2y = -1$ or $y = \frac{1}{2}$

Therefore, $x = -\frac{1}{2}$

3. (a) Slope of x-t graph gives velocity. Between O to A slope is decreasing, for AB slope is zero, for BC slope is increasing, CD is straight line i.e. slope is unchanged.

5. (c) Weightlessness means that there is no reaction on a body from the floor. Since both the artificial satellite & the astronaut have same centripetal acceleration (as in a lift; which is falling freely, we does not feel any weight, because both lift & we fall with same acceleration). So the astronaut does not feel any weight inside the space craft.

6. (d) Let the temp of water $= 100 \times 1 \times (50 - \theta)$

Heat gained by ice $= 10 \times 80 + 10 \times 1 \times \theta$

Now heat lost = heat gained.

$5000 - 1000 = 800 + 100$

$4200 = 110\theta$

$\theta = \frac{4200}{110} = 38.2°C$

7. (c) $\int E.dx = \frac{Q_1}{\epsilon_0} \Rightarrow E = \frac{Q_1}{4\pi r^2 \epsilon_0}$

8. (c) As the aeroplane and released bomb has the same horizontal velocity.

9. (c) $\frac{K_1}{K_2} = \frac{p_1^2}{m_1} \times \frac{m_2}{p_2^2}$ $[\because p = mv \Rightarrow K = \frac{p^2}{2m}]$

Hence, $\frac{p_1}{p_2} = \sqrt{\frac{M_1}{M_2}} = \sqrt{\frac{1}{4}} = \frac{1}{2}$

10. (b) Since areas are equal, when a planet revolves around the sun, areal velocity is constant i.e.,

$\frac{d\vec{A}}{dt}$ = constant or $t_1 = t_2$

12. (a) For closed organ pipe

$\frac{\lambda}{4} = \ell \Rightarrow \lambda = 4\ell$; $n = \frac{V}{\lambda} = \frac{V}{4\ell}$

For open organ pipe,

$\frac{\lambda}{2} = \ell \Rightarrow \lambda = 2\ell$; $n' = \frac{V}{\lambda} = \frac{V}{2\ell}$

$n' = 2n$; $n : n' = 1 : 2$

13. (a) In an isothermal process, the temperature remains constant. So, the internal energy remains same.
If the gas is expanded by adding more molecules to it then the internal energy of the gas increases.

14. (a) $I = \frac{E}{R+r}$

$= \frac{E}{[(\sqrt{R})^2 + (\sqrt{r})^2 - 2\sqrt{2r}] + 2\sqrt{Rr}}$

$= \frac{E}{(\sqrt{R} - \sqrt{r})^2 + 2\sqrt{Rr}}$

I will be maximum if $(\sqrt{R} - \sqrt{r}) = 0$ or $R = r$

15. (a) Velocity at A, $\vec{v} = v\hat{j}$

Velocity at B, $\vec{v}_2 = -v\hat{i}$

Change in velocity,

$\Delta\vec{v} = \vec{v}_2 - \vec{v}_1 = -v\hat{i} - v\hat{j}$

$|\Delta\vec{v}| = \sqrt{(-v)^2 + (-v)^2} = v\sqrt{2}.$

16. (a) Moment of inertia of a body is given by the expression

$$I = \sum_{i=1}^{n} m_i r_i^2$$

This is the sum of product of mass of each particle and square of its distance from the axis. Naturally if distribution of mass changes, value of r will change & hence I will change.

17. (d) Moment of inertia of disc about its axis

$= \frac{1}{2} mr^2 = \frac{1}{2} \times 0.4 \times \left(\frac{100}{100}\right)^2$

$= \frac{1}{2} \times 0.4 = 0.2\ kgm^2$

18. (c) Field at Q is E. So, force on Q = QE
This force will be applied on –2Q. Also according to Coulomb's law. So, field at –2Q is $\frac{QE}{-2Q} = -\frac{E}{2}$.

19. (b) Solid lithium doesn't fuse with hydrogen gas. For fusion, both should be in gaseous state. So the statement is false.

20. (b) $\frac{1}{2}mv^2 = \frac{1}{2}kx^2$

$\Rightarrow mv^2 = kx^2$

or $0.5 \times (1.5)^2 = 50 \times x^2$

$\therefore x = 0.15$ m

21. (b) When radius is doubled, increase in surface area $= 2 \times 4\pi\ (4R^2 - R^2)$

$= 2 \times 4\pi \times 3R^2 = 24\pi R^2$

Work done = Increase in area × T

$= 24\pi R^2$ T.

22. (d) $E = 10^7$ V/m

$r = 0.10$ m

$E = \frac{V}{r}$

$V = Er = 10^7 \times \frac{10}{100}$

$V = 10^6\ V$

23. (a) We know that

Maximum velocity = ωA

Their velocity are same

$\omega_1 A_1 = \omega_2 A_2 = \omega_3 A_3$

24. (d) $R = \frac{\rho l}{\pi r^2}$. But $m = \pi r^2\ ld \therefore \pi r^2 = \frac{m}{ld}$

$\therefore R = \frac{\rho l^2 d}{m}$, $R_1 = \frac{\rho l_1^2 d}{m_1}$, $R_2 = \frac{\rho l_2^2 d}{m_2}$

$$R_3 = \frac{\rho l_3^2 d}{m_3}$$

$$R_1 : R_2 : R_3 = \frac{l_1^2}{m_1} : \frac{l_2^2}{m_2} : \frac{l_3^2}{m_3}$$

$$R_1 : R_2 : R_3 = \frac{25}{1} : \frac{9}{3} : \frac{1}{5} = 125 : 15 : 1$$

25. (b) Work done = Change in energy

$$= \frac{1}{2}\left(C + \frac{C}{2}\right)V^2 = \frac{1}{2}\left(\frac{3C}{2}\right)V^2 = \frac{3CV^2}{4}$$

26. (b) The relation between Y, k, η for some isotropic solid material is given by

$$Y = \frac{9K\eta}{3K + \eta} = \frac{9K}{\frac{3K}{\eta} + 1}$$

$$\Rightarrow \frac{3K}{\eta} + 1 = \frac{9K}{Y}$$

$$\Rightarrow \frac{3K}{\eta} = \frac{9K}{Y} - 1 = \frac{9K - Y}{Y}$$

$$\Rightarrow \eta = \frac{3KY}{9K - Y}.$$

27. (d) In elastic collision, the velocities get inter changed if the colliding objects have equal masses.

28. (a) $m_B g = \mu_s m_A g \quad \{\because m_A g = \mu_s m_A g\}$
$\Rightarrow m_B = \mu_s m_A$
or, $m_B = 0.2 \times 2 = 0.4$ kg

29. (a) **In adiabatic process**

Heat change $\Delta Q = 0$
By first law of thermodynamics,

$$\Delta Q = \Delta U + W$$

$$\Rightarrow 0 = \Delta U + W \Rightarrow \Delta U = -W$$

In isothermal process ;
Temperature change, $\Delta T = 0$
Change in internal energy,

$$\Delta U = nC_v \Delta T = 0$$

By $\Delta Q = \Delta U + W \Rightarrow \Delta Q = W$

30. (b)

[Circuit: current I enters, I' flows through galvanometer 25Ω; shunt 2.5Ω in parallel]

$$I' = \frac{I \times 2.5}{(25 + 2.5)} = I \times \frac{25}{275} = \frac{1}{11} \times I$$

$$\Rightarrow \frac{I}{I_0} = \frac{1}{11}.$$

31. (d) $P = 2\left[\frac{100}{10} + \frac{100}{-20} + 0\right]$

$P = 10$ dioptre

32. (d) Current is max at resonance condition

$$\text{So, } \omega = \frac{1}{\sqrt{LC}} = \frac{1}{\sqrt{0.5 \times 8 \times 10^{-6}}}$$

$$\omega = \frac{1}{\sqrt{4 \times 10^{-6}}}$$

$$\omega = \frac{1 \times 1000}{2}$$

$$\omega = 500 \text{ Hz}$$

33. (b) Angle between magnetic meridian and geographical meridian is called angle of declination.

34. (d) The fringe width in Young's double slit experiment is

$$\beta = \frac{\lambda D}{d}$$

where λ is the wavelength of light used. D is the distance between slit and screen. d is the distance between the slit.

$$\beta' = \frac{\lambda(2D)}{d/2} = 4\frac{\lambda D}{d} = 4\beta$$

35. (a) A current carrying wire in the neighbourhood produces magnetic field and electric field both.

36. (d) For maxima, $d \sin\theta = n\lambda$

$$\sin\theta = \frac{n\lambda}{d} = \frac{8\lambda}{10\lambda} \Rightarrow \sin\theta = \frac{4}{5}$$

$$\Rightarrow \tan\theta = \frac{4}{3}$$

$$\text{Also } \tan\theta = \frac{y}{D}$$

$$\therefore y = \frac{4D}{3}$$

37. (a) By Einstein's equation of photo-electric effect, the maximum kinetic energy of emitted photo-electrons is given by

$E_k = h\nu - W$ or $E_k = \frac{hc}{\lambda} - W$

Where, h = Planck's constant
ν = frequency of incident light
W = work function of metal
λ = wavelength of incident light

$$E_k = \frac{6.6\times10^{-34}\times3\times10^{8}}{2000\times10^{-10}\times1.6\times10^{-19}}\text{eV} - 4.2\text{eV}$$

So, $E_k = 2\text{eV}$

38. (c) Radio waves are low frequency electro-magnetic waves. So, their wavelength will be comparatively large.

39. (d) The value of red, violet and green are : 2, 7 and 10^5. And silver represents a tolerance 10%. Thus $R = 27\times10^5 \pm 10\%\ \Omega$.

40. (c) "O" act as focal point.

$$\frac{1}{f} = (\mu-1)\left(\frac{1}{R_1}-\frac{1}{R_2}\right)$$

$$\frac{1}{40} = (\mu-1)\left(\frac{1}{10}+\frac{1}{10}\right)$$

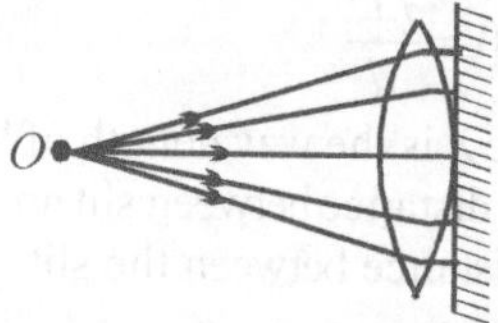

$$\frac{1}{40} = (\mu-1)\left(\frac{2}{10}\right) \Rightarrow \mu = \frac{9}{8}$$

41. (b) For first line of Balmer series

$$\frac{1}{\lambda} = R\left(\frac{1}{4}-\frac{1}{9}\right) \Rightarrow R = \frac{36}{5\lambda}$$

$\therefore$ Wavelength of the first line, λ_L of the Lyman series is given by

$$\frac{1}{\lambda_L} = R\left(1-\frac{1}{4}\right) = \frac{36}{5\lambda}\times\frac{3}{4} = \frac{27}{5\lambda}$$

$$\Rightarrow \lambda_L = \frac{5\lambda}{27}$$

42. (d) The required wavelength should be from 10^{-3}m to 0.1m.

43. (b) Impedance $= \sqrt{R^2+\omega^2L^2}$

$$= \sqrt{R^2+4\pi^2\nu^2L^2}$$

44. (c) Silicon is an indirect-band gap semi-conductor.

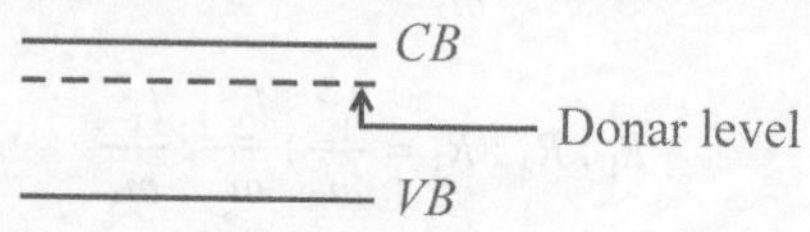

45. (a) This truth table is of the identity, $Y = \overline{A.B}$, hence it is NAND gate.
Here, the output is high even if all inputs are low or one input is low.

SECTION 2 - CHEMISTRY

46. (c) $\sigma^2\,\sigma^{*2}\,\sigma^2\,\sigma^{*2}\,\sigma^2\,\frac{\pi^2}{\pi^2}\,\frac{\pi^{*2}}{\pi^{*2}}\,\sigma^*$

49. (d) All statements are correct.

50. (d) In carboxylates (conjugate base of carboxylic acids), resonance is more significant because the two resonating structures are similar, while in phenoxide, the resonating structures are not equivalent, alkoxide ions do not show resonance.

51. (d) $N_2O_4 \rightleftharpoons 2NO_2$

$1-\alpha \qquad 2\alpha$

Total moles at eq $= 1-\alpha+2\alpha = 1+\alpha$

$$K_c = \frac{p^2_{NO_2}}{p_{N_2O_4}} = \frac{\left(\frac{2\alpha}{1+\alpha}.P\right)^2}{\frac{1-\alpha}{1+\alpha}.P} = \frac{4\alpha^2P}{1-\alpha^2}$$

52. (a) BHT is the important anti oxidant used in food.

54. (c) The condition for geometrical isomerism is

$$\begin{matrix}a\\b\end{matrix}\!>C=C<\!\begin{matrix}a\\b\end{matrix} \quad \text{or} \quad \begin{matrix}a\\b\end{matrix}\!>C=C<\!\begin{matrix}e\\d\end{matrix}$$

56. (a) It is not a buffer (see conditions for buffer), pH < 7. NaCl gives neutral solution. HCl will give acidic solution.

57. (c) Adsorption lowers the activation energy.

59. (a) $[Co(NH_3)_5CO_3]ClO_4$. Six monodentate ligands are allached to Co hence C. N. of Co = 6;
O. N. $= x + 5\times(0) + 1\times(-2) + 1\times(-1) = 0$
$\therefore x = +3$; electronic configuration of Co^{3+}[Ar] $3d^64s^0$ hence number of d electrons is 6 : All d electrons are paired due to strong ligand hence unpaired electrons zero.

61. (a) Solubility of AgI in KI soln. = $\frac{k_{sp} \text{ of AgI}}{[I^-]}$

$= \frac{1.0\times10^{-16}}{10^{-4}} = 1.0\times10^{-12}\ mol^{-1}$

62. (c) It is electrophoresis (see definition of electrophoresis)

63. (d) The correct formula of inorganic benzene is $B_3N_3H_6$ so (a) is incorrect statement

Boric acid (H_3BO_3 or $B(OH)_3$, i.e. B bonded to three –OH groups) is a lewis acid so (b) is incorrect statement.

The coordination number exhibited by beryllium is 4 and not 6 so statement (c) is incorrect.

Both $BeCl_2$ and $AlCl_3$ exhibit bridged structures in solid state so (d) is correct statement.

66. (a) $\because \Delta n = -1$

$\therefore K_p = 26 \times (0.0821 \times 523)^{-1} = 0.61\ atm^{-1}$

67. (d) Lesser the intermolecular forces, the more the volatile character.

68. (a) The stability of + 2 O.S. follows the order

$Pb^{2+} > Sn^{2+} > Ge^{2+}$

Hence reducing power $Ge > Sn > Pb$

69. (d) $K_4[Fe(CN)_6] + 6H_2SO_4 + 6H_2O \rightarrow$

$2K_2SO_4 + FeSO_4 + 3(NH_4)_2SO_4 + 6CO$

This is one of the methods for the preparation of carbon monoxide.

70. (d) Aniline is insoluble in water, because its –NH_2 group can't form H- bond with water due to bulky phenyl group.

71. (c) Reaction (c) is endothermic. Electrolysis or decomposition of H_2O endothermic in nature.

72. (c) For isotonic solutions $\pi_1 = \pi_2$ or $c_1 = c_2$;

$\frac{34.2}{342} = 0.1;\ \frac{18.0}{180} = 0.1.$

Hence 18 gL^{-1} glucose is 0.1 M.

73. (a) Galvanization

76. (a) $\lambda = \frac{h}{mv} = \frac{6.63\times10^{-34}}{0.060\times10} = 1.1\times10^{-33}$ m

77. (d) Let bond energy of A_2 be x then bond energy of AB is also x and bond energy of B_2 is x/2.

Enthalpy of formation of AB is - 100 kJ/mole:

$A_2 + B_2 \rightarrow 2AB;\quad \frac{1}{2}A_2 + \frac{1}{2}B_2 \rightarrow AB;\ \Delta_4 = -100kJ$

$\text{or} -100 = \left(\frac{x}{2} + \frac{x}{4}\right) - x \therefore -100 = \frac{2x + x - 4x}{4} \therefore x = 400kJ$

78. (b) The $Be(OH)_2$ and $Al(OH)_3$ are amphoteric in nature.

79. (a) The two structures involve only movement of electrons and not of atoms or groups, hence these are resonating structures.

80. (d) Formic acid has aldehydic grouping

$(H-\overset{\overset{\large O}{||}}{C}-OH)$.

81. (c) Not more than two electrons can be present in same atomic orbital. This is Paulis exclusion principle.

82. (b) For orthorhombic $\alpha = \beta = \gamma = 90°$.

85. (d) Since Sc^{3+} does not contain any unpaired electron it is colourless in water.

86. (c) Linear alkylbenzenesulphonates (LAB) carrying phenyl chains at secondary positions are biodegradable.

87. (c) KBr does not exhibit Frenkel defect.

88. (a) $t = \frac{2.303}{k}\log\frac{a}{a-x}$

$= \frac{2.303}{6}\log\frac{0.5}{0.05} = 0.384\,min.$

90. (d) In Ziegler Natta catalyst the active species the Ti^{III}.

SECTION 3 - BOTANY

92. (a) Statement (i) and (ii) are correct. *Riccia* is liverwort in which simplest sporophyte consists of capsule only while *Polytrichum* is moss in which sporophyte consists of foot, seta & capsule. *Volvox* is a fresh water green colonial alga. Reproduction is both sexual and asexual. Sexual reproduction is oogamous. Slime moulds are consumer decomposer protists. They possess characters of plants (cellulosic cell wall), animals (phagotrophic nutrition) and fungi (spores).

93. (a) The fluidity of membranes in a plant in cold weather may be maintained by increasing the number of phospholipids with unsaturated hydrocarbon tails.

94. (a) The pure water, at atmospheric pressure has

zero water potential. The addition of any solute particles reduces the free energy of water. Thus, the water potential will be negative.

99. (c) Movement of water (solvent) depends on diffusion pressure deficit, which is D.P.D = O.P – T.P
If O.P. = T.P. So D.P.D = 0
So no movement takes place. Water moves towards low D.P.D to high D.P.D.

108. (d) Four important functional aspects of the ecocystem are (i) Productivity (ii) Decomposition, (iii) Energy flow and (iv) Nutrient cycling.

119. (d)
(a) Gametophyte of bryophytes bears protonemal & leafy stage.
(b) In gymnosperm female gametophyte is not free living.
(c) They are present in *Marchantia* or which is a bryophyte.
(d) Origin of seed habit started in pteridophyte *Selaginaella*.

123. (b) The common bottle cork is the product of phellogen. Phellogen produces cork or phellem on the outer side. It consists of dead and compactly arranged rectangular cells that possess suberised cells walls. The cork cells contain tannins. Hence, they appear brown or dark brown in colour. The cork cells of some plants are filled with air *e.g.*, *Quercus suber* (Cork Oak or Bottle Cork).

126. (d) Hots are the geographical area where biodiversity is maximum. Two hotspots in India are Western ghats and North eastern himlayan region.

128. (b) Compensation point is existed in morning and evening time. At this time the rate of photosynthesis (intake of CO_2) and rate of respiration (outcome of CO_2) is equal. At this point no exchange of CO_2 is possible through stomata.

131. (a) Cobalt stimulates growth of legumes such as beans, clover and alfalfa. This stimulation of growth by cobalt is due to its use, not by the plant itself, but by nitrogen-fixing bacteria that live in roots of plants.

133. (b) The absorption of water still occurs when concentration of outer soil water is more than root hair cell sap.

134. (b) Cytokinin are usually amino purine so they resemble nucleic acids in structural aspects.

SECTION 4 - ZOOLOGY

139. (b) Long term intake of alcohol causes damage to liver which is known as cirrhosis of liver with continued alcohol intake, there is destruction of hepatocytes and fibroblasts (cell which form fibres) and stimulates of collagen protein formation.

154. (c) *Saccharomyces cervisiae* is a yeast used in making bread (Baker's yeast) and commercial production of ethanol.
Paramecium & *Plasmodium* are of animal kingdom while *pencillium* is a fungi. Lichen is composite organism formed from the symbiotic association of an algae and a fungus. *Nostoc* & *Anabaena* are examples of kingdom monera.

162. (b) Biowar or biological war or bioterrorism is the development of biological weapons against people, their crops and animals.

171. (c) Haematuria is the presence of blood cells in urine.

177. (c) The figure shows the tubectomy. This is a surgical method to prevent pregnancy in women. In tubectomy small part of the fallopian tube is removed or tied through a small cut in the abdomen or through vagina. It is very effective method but reversibility is very poor.

178. (c) Blastocyst is a thin-walled hollow structure in early embryonic development that contains a cluster of cells called the inner cell mass from which the embryo arises. The outer layer of cells gives rise to the placenta and other supporting tissues needed for fetal development within the uterus while the inner cell mass cells give rise to the tissues of the body. The blastocyst reaches the womb (uterus) around day 5, and implants into the uterine wall on about day 6.

179. (a) Preformed antibodies need to be injected to treat the bite by a viper snake. It is also a type of immunization which is called as passive immunization.
Antibodies, produced by B-cells, are typically made of basic structural units—each with two large heavy chains and two small light chains. B cells differentiate into plasma cells that secrete antibodies. Antibodies are proteins that bind to specific antigens and mark them for destruction by, for example, marking them more recognizable to phagocytic cells. Rejection of a kidney graft is not a function of B lymphocyte.